沉积盆地构造学进展

[美] Cathy Busby　Antonio Azor　主编

张功成　崔　敏　阳怀忠　董　伟　等译

石油工业出版社

内 容 提 要

本书从活动构造、构造年代学、物理与数值模拟以及不同大地构造背景下盆地构造的研究等方面综合论述了最新的沉积盆地构造分析技术和研究方法，代表了盆地构造学的研究现状。

本书主要作为高等院校石油、地质专业硕士和博士研究生的教学参考书，也可供相关专业的研究生和从事石油勘探的技术人员参考使用。

图书在版编目（CIP）数据

沉积盆地构造学进展/（美）巴斯比（Busby, C.），（美）阿佐尔（Azor, A.）主编；张功成等译. —北京：石油工业出版社，2016.6
书名原文：TECTONICS OF SEDIMENTARY BASINS（RECENT ADVANCES）
ISBN 978-7-5183-1228-3

Ⅰ.沉⋯
Ⅱ.①巴⋯②阿⋯③张⋯
Ⅲ.沉积盆地—沉积构造—研究
Ⅳ.P531

中国版本图书馆 CIP 数据核字（2016）第 074149 号

Tectonics of Sedimentary Basins：Recent Advances
Edited by Cathy Busby and Antonio Azor
ISBN 978-1-4051-9465-5
Copyright ⓒ 2012 by John Wiley & Sons Ltd
All Rights Reserved. Authorised translation from the English language edition published by John Wiley & Sons Limited. Responsibility for the accuracy of the translation rests solely with Petroleum Industry Press and is not the responsibility of John Wiley & Sons Limited. No part of this book may be reproduced in any form without the written permission of John Wiley & Sons Limited.

本书经 John Wiley & Sons Limited 授权翻译出版，简体中文版权归石油工业出版社有限公司所有，侵权必究。
北京市版权局著作权合同登记号：01-2013-1871
Copies of this book sold without a Wiley sticker on the cover are unauthorized and illegal

---

出版发行：石油工业出版社
　　　　　（北京安定门外安华里 2 区 1 号　100011）
　　网　　址：www.petropub.com
　　编辑部：（010）64523543
　　图书营销中心：（010）64523633
经　销：全国新华书店
印　刷：北京中石油彩色印刷有限责任公司

2016 年 6 月第 1 版　2016 年 6 月第 1 次印刷
787×1092 毫米　开本：1/16　印张：46.75
字数：1190 千字

定价：280.00 元
（如出现印装质量问题，我社图书营销中心负责调换）
版权所有，翻印必究

译者的话

近年来随着盆地构造学研究的进一步深入，出现了许多新的技术方法和研究成果，尤其是在活动构造、构造年代学、构造物理与数值模拟和不同大地构造背景下盆地构造的研究等方面取得了长足的进步和重大成果。国际上盆地构造相关的教科书相对较少，而且大多集中在单个盆地或局部构造的研究；此外，关于盆地构造研究的综述性论著也鲜有出版，缺乏盆地构造的最新研究方法和综合性研究论著。然而，高校和科研院所的学生和科研工作者需要掌握和学习这些最新的盆地构造分析技术和研究方法，了解全球不同构造背景下盆地构造的研究进展。在这一背景下，Cathy Busby 教授和 Antonio Azor 教授组织了盆地构造学领域一批杰出专家教授编写了《Tectonics of Sedimentary Basins Recent Advances》一书，总结了近15年来沉积盆地构造学的最新进展。

中海油研究总院勘探研究院组织翻译的《沉积盆地构造学进展》（《Tectonics of Sedimentary Basins Recent Advances》），恰好满足了目前教学和科研工作中盆地构造学的迫切需要，可以帮助高校和科研院所的相关教学以及科研工作。中海油研究总院勘探研究院的张功成教授级高级工程师、崔敏、阳怀忠、董伟、祁鹏、韩银学、纪沫、郭帅、赵钊、孙钰皓、蔡露露、陈莹、郭佳、王鹏、郭瑞、曾清波、廖计华、朱石磊，中国科学院地质与地球物理研究所胡圣标研究员、饶松，西北大学屈红军教授、蒲仁海教授、白升、牛宁、董治斌、王云、杨志文、章志明，同济大学邵磊教授、蔡国富、崔炳松、李阳、李政斯、吴哲、张浩、赵梦，吉林大学刘长利等分别翻译和校对了有关章节。张功成、崔敏、阳怀忠和董伟对本书作了统一校对整理。

最后，感谢本书的主编 Cathy Busby 和 Antonio Azor 给我们带来的盆地构造学的最新进展，展现了全球沉积盆地构造学研究的新技术方法和最新的研究成果。感谢 Wiley-Blackwell 允许出版中文译本。

张功成
2016.6

前　　言

　　20世纪60年代的板块构造革命首次为造山带和盆地提供了统一的框架模型。此后的20世纪70—80年代，出现了大量里程碑式的论文。当时（20世纪80年代晚期），Ingersoll 和 Busby 正在讲授"沉积盆地构造学"这门课程，他们发现关于盆地构造地质学的教科书或综述性论文非常缺乏。迫于无奈，教授们只得为学生们罗列出很长的文献清单供学生们参考阅读，并且尽力帮助学生们总结。鉴于此，Busby 和 Ingersoll 教授决定为 Blackwell 公司撰写一本主要针对研究生和地质学家的，关于盆地构造学的教科书，两位教授共用了5年的时间才撰写完成，这本书（Busby 和 Ingersoll，1995）在1995年出版。这本书主要是关于沉积盆地构造学的总结性简述，其网络版由出版商提供。我们建议本科水平的地质专业学生须先阅读1995年的版本，研究生以及专业人员在必要时也可参阅。

　　15年来，我们对于板块构造对沉积盆地形成及演化的控制作用的研究有了许多新进展，主要集中在活动构造方面，由于全球定位系统和地层（或地表）定年技术的进步，我们能够在数千年的时间尺度上将现今的板块运动与构造发育进行对比。基于数值—物理模拟和精细的现场监测，我们对构造事件的沉积响应有了进一步的认识。通过地壳到上地幔构造的地震勘测以及盆地充填的三维地震勘测，有些地区还有岩心作为证据，对地下岩石的研究有了重要进展。碎屑矿物的同位素（如：锆石的 U-Pb）研究现在广泛用于构造事件重建，包括：大尺度的盆地转换、盆地周缘的剥蚀样式、陆上沉积物的搬运过程重建。古地磁法现在被广泛应用于地层的精确定年与校正、大量火山物质来源和流经路径的确定，以及不同构造旋转程度的评估。另一种理解构造地质问题的新方法是 ArcGIS 数据库对地质年代学、地球化学、生物地层学和古地磁学数据的运用，以及人造卫星、航空照片和地质图的使用。同时，丰富的复杂火山—火山碎屑搬运—沉积体系的精确模型，使得我们能够对更多具有准确充填时间的盆地类型进行精确地构造重建。此外，地质动力过程的数值模拟和类比模拟要比以前更复杂了。

　　本书是关于沉积盆地构造学内容的更新和修正，因此命名为"沉积盆地构造学进展"。我们的任务是在盆地构造学领域组建全明星专家阵容，给他们一个场馆展示"前缘"成果。与1995年版本不同，本书没有将构造学这个主题的全部内容综合到一起。15年前出版的1995年版本，在当时是很有用的，但是随着网络版形式的出现，这种综合性已经不那么重要了。另外，我们认为新版本代表了盆地构造学的研究现状，撰写新版很快（约2年时间），因为本书包含许多较短的章节，这些较短的章节只被一、两位创作出版。笔者们简述了过去15年重要的成果。在本书中一般未阐述基本素材，但是基本素材的来源标注清楚，读者可以根据标注找到所有各主题重要的研究进展文章。我们认为这些阐述结果的章节展现了采用的创新方法和结论，使得我们对构造地质过程有了新的理解和认识，并且由此延伸到其他领域。下面对本书涉及的主要研究主题作了简单的小结，按照先后顺序阐述了各部分的基本内容。

　　（1）第一部分：绪论。

第一部分是沉积盆地构造学的综述，由 Ingersoll 撰写（第 1 章）。本章是 Ingersoll 教授对他在美国地质学会学报（Ingersoll，1988）上发表的文章和沉积盆地构造学（Ingersoll 和 Busby，1995）一书中关于沉积盆地构造学内容的更新和修正。本章为读者解释了本书后续章节中常见的地质作用和术语。

（2）第二部分：新技术与模拟。

由于新技术不断涌现，与 15 年前相比已经有了很大的进步，参与撰写的各位专家们一致认为《沉积盆地构造学进展》一书有必要包含论述这些技术的章节（1995 年版没有涉及技术的章节）。为使各章节尽可能简短，我们有意地减少了技术应用的讨论部分，仅提供应用实例的参考文献。本部分综述了沉积盆地构造分析技术的重要进展。碎屑锆石地质年代学方法（第 2 章）和陆地宇宙成因核素技术（第 3 章）现在已被广泛应用。同时，磁性地层学技术、地震解释技术和盆地—地层模拟技术也有了很大的进步（第 4 章至第 8 章）。本部分未涉及已有几十年发展历史的庞大而多样化的化学地层学技术，再版时将论述这部分内容。

根据构造背景，按离散和聚敛两大类边缘撰写了本书的第三和第四部分。

（3）第三部分：裂谷型、后裂谷型、张扭型和走滑型盆地。

第三部分，首先论述了典型的正在活动的裂谷——东非裂谷（第 9 章），然后论述了"成熟"大陆裂谷环境的张扭裂谷盆地（加利福尼亚湾，第 10 章），以及还在形成过程中的张扭裂谷（沃克裂谷，第 11 章）。本部分没有涉及转换边缘，如想全面了解走滑断裂体系，请参阅 Paul Mann 教授 2007 年发表的文章第 142 页（Geological Society of London Special Publication，卷 290）。为了与构造和沉积主题保持一致，本书第三部分也包括了被动边缘的内容，论述了海底扩张后的变形（第 12 章和第 13 章）。

（4）第四部分：聚敛边缘。

第四部分内容包括海沟至靠近大陆一侧的构造，以及海底俯冲背景至碰撞背景下的构造。论述内容包括俯冲开始过程（第 14 章）；现今研究的最复杂的俯冲体实例（第 15 章）；板块俯冲引起的弧前变形（第 16 章）；将火山弧拖入大陆俯冲带的盆地记录（第 17 章）；以安第斯弧前盆地为例的海底俯冲（第 18 章）；美国西南部的伸展和张扭弧内盆地（第 19 章）；与俯冲相关的碰撞前陆盆地的综述（第 20 章）。文中阐述了俯冲和碰撞环境中，位于造山带顶部的内陆盆地（安第斯山脉和青藏高原），时间尺度从几百万年（第 21 章）到几十万年及几千年（第 22 章）。文中还阐述了西班牙贝蒂克（Betic）山的山间盆地，该盆地被认为是受到地幔剥离和（或）板片后退或拆离作用的制约，最终形成于造山运动晚期的伸展盆地（第 23 章）。随后论述了位于大陆一侧构造背景下的前陆盆地。第 24 章论证了加拿大西部前陆盆地发生挠曲沉降的样式，并推测它主要受控于大洋板块俯冲，而在科迪勒拉山尺度上受控于地体增生事件的影响。第 25 章对比了典型连续前陆盆地（玻利维亚）和破碎前陆盆地（阿根廷）的特征。此后，阐述了逆冲楔/前陆系统的常用动力学模型（第 26 章）和挤压环境下，生长断层相关褶皱的模型（第 27 章）。第四部分没有涉及大洋—岛弧聚敛边缘的盆地构造研究，这些内容会在再版时加以论述。

（5）第五部分：板内盆地及其他类型盆地。

本书的最后部分没有将沉积盆地构造只局限于离散型和聚敛型板块构造背景。相对于盆地的规模和长期的构造及地层记录而言，板内多阶盆地非常重要（第 28 章）。在超大陆形成的背景下描述了格伦维尔构造事件的大量沉积记录（第 29 章）。第 30 章论述了克拉通盆地，克拉通盆地被认为是发育在较厚的、相对稳定的大陆岩石圈上、长期存在的圆形或椭圆形地壳坳陷，其主要形成于较低应变率的板块长期伸展的背景下。最后，第 31 章论述了不同构造环境下内流盆地中常见的特有地层和沉积相特征。

在本书编写及出版的过程中，还有一些重要的进展值得关注。Ingersoll 教授的术语修正章节（第 1 章）包括含有油气的盐体动力盆地。新技术得到了迅速发展，如运用古土壤、化石、硅酸盐和火山玻璃中稳定同位素对盆地进行研究，通过沉积盆地充填能够推断出古高程等，这些内容可参阅期刊（Review in Mineralogy 和 Geochemistry，2007）。全球地震层析成像技术正在飞速发展，运用该方法可以推断地幔和地表过程之间的关系。例如，由地幔—岩石圈下沉引起地表沉降的重要程度如何？是否会因为它们太小以及太短暂就不重要了？巨大的海底高原的俯冲作用如何控制大陆的隆起和沉降？如何从地表构造盆地形成过程获得俯冲过程的地形成像记录？本书讨论了板片后退作用，那么其他过程（如：停滞板片和破碎板片）的作用是什么样呢？

与所有的其他地质研究一样，本书是关于沉积盆地构造学的研究进展。我们认为最近几年来沉积盆地构造学研究取得了很大的进步，希望本书能让读者觉得非常有用。我们真诚地感谢为本书提供宝贵意见的各位专家。

我们尤其感谢西班牙教育部资助 Cathy Busby 和 Antonio Azor 于 2007—2008 年及 2010 年在格拉纳达大学地质动力学部门的工作。

主编：Cathy Busby 和 Antonio Azor
2011 年 2 月

章节评论人

匿名评审专家 3 人

Ramon Arrowsmith, Arizona State University
Peter Burgess, Royal Holloway University
Kevin Burke, University of Houston
Reynaldo Charrier, University of Chile
Peter Clift, University of Aberdeen
Christopher Connors, Washington and Lee University
Rob Crossley, Fugro Robertson Ltd
Peter DeCelles, University of Arizona
Alex Densmore, Durham University
Mark Deptuk, Canada Nova Scotia Offshore Petroleum Board
William R. Dickinson, University of Arizona
Christopher Fedo, University of Tennessee
Stanley C. Finney, California State University, Long Beach
William Galloway, University of Texas Institute for Geophysics
Miguel Garcés, University of Barcelona
Martin Gibling, Dalhousie University
Adrian Hartley, University of Aberdeen
Richard Heermance, California State University, Northridge
William Helland-Hansen, University of Bergen
Paul Heller, University of Wyoming
Matthew Horstwood, British Geological Survey
Brian Horton, University of Texas at Austin
Raymond V. Ingersoll, University of California, Los Angeles
Cari Jonson, University of Utah
Teresa Jordan, Cornell University
Paul Kapp, University of Arizona
Tim Lawton, New Mexico State University
Andrew Leier, University of Calgary
Nathaniel Lifton, Purdue University

Juan M. Lorenzo, Louisiana State University
Paul Mann, University of Texas Institute for Geophysics
Mariano Marzo, University of Barcelona
Margot McMechan, Geological Survey of Canada
Andrew Miall, University of Toronto
Ivan Marroquin, Paradigm Geophysical
Nick Mortimer, GNS Science
Lorean Moscardelli, University of Texas at Austin
Michael Murphy, University of Houston
Andrew Meigs, Oregon State University
Nadine MacQuarrie, Princeton University
Neil Opdyke, University of Florida

Michael Oskin, University of California, Davis
Chris Palola, University of Minnesota
Kevin Pickering, University College London
Marith Reheis, US Geological Survey
Ken Ridgway, Purdue University
Scott Samson, Syracuse University
David Scholl, US Geological Survey
John Shimeld, Geological Survey of Canada
Glen Stockmal, Geological Survey of Canada
Manfred Strecker, Universität Potsdam
Michael Taylor, University of Kansas
Reinoud Vissers, Utrecht University
Martha Withjack, Rutgers University

(张功成 译校)

目　录

第一部分　绪　论

第 1 章　沉积盆地的构造特征及其术语修正……………… RAYMOND V. INGERSOLL（3）

第二部分　新技术与模拟

第 2 章　碎屑锆石的 U-Pb 测年：现状和展望 ……………… GEORGE GEHRELS（61）
第 3 章　陆相岩层宇宙成因核素技术在评估构造活跃地区地表年龄和沉积物暴露史
　　　　中的应用 ……………………………………………………… JOHN C. GOSSE（80）
第 4 章　磁性地层学的原理与应用
　　　　………………………………… GUILLAUME DUPONT-NIVET, WOUT KRIJGSMAN（100）
第 5 章　三维地震解释技术在盆地分析中的应用
　　　　………………………………… CHRISTOPHER A-L. JACKSON, KARLA E. KANE（115）
第 6 章　沉积盆地内构造成因的冲积相砾石的搬运与保存
　　　　………………………………………… PHILIP A. ALLEN, PAUL L. HELLER（133）
第 7 章　构造地层模型中源—汇体系内沉积物的体积分配研究：方法与结论
　　　　——以拉腊米型大陆架—深水盆地为例… CRISTIAN CARVAJAL, RON STEEL（157）
第 8 章　前陆盆地中岩石圈与地表地质作用的模拟
　　　　………………………………… DANIEL GARCIA-CASTELLANOS, SIERD CLOETINGH（182）

第三部分　裂谷型、后裂谷型、张扭型和走滑型盆地

第 9 章　大陆裂谷盆地：来自东非裂谷的新认识
　　　　………………………………… CYNTHIA EBINGER, CHRISTOPHER A. SCHOLZ（221）
第 10 章　沉积物输入以及斜向板块运动对斜向离散活动板块边缘盆地发育的影响
　　　　——以加利福尼亚湾和索尔顿海槽为例
　　　　………………………………… REBECCA J. DORSEY, PAUL J. UMHOEFER（250）
第 11 章　活动张扭型陆内盆地——以美国大盆地西部的沃克通道为例
　　　　………………………………… ANGELA S. JAYKO, MARCUS BURSIK（270）
第 12 章　东北大西洋和南大西洋边缘的后裂谷变形："被动边缘"真的是被动吗？
　　　　………………………………………………………… DOUGLAS PATON（295）
第 13 章　早白垩世加拿大东部近海斯科舍被动边缘盆地构造变形对沉积的影响
　　　　………………………………… GEORGIA PE-PIPER, DAVID J. W. PIPER（318）

第四部分　聚敛边缘

第14章　板块边缘的转化及其沉积响应 ·················· KATHLEEN M. MARSAGLIA（339）
第15章　日本西南部俯冲带沉积环境的演化：来自南海（Nankai）海槽发震带试验中 Kumano 断面的最新结果
　　　　·················· MICHAEL B. UNDERWOOD, GREGORY F. MOORE（362）
第16章　板块俯冲作用对大陆弧前盆地的影响：以阿拉斯加南部为例
　　　　·········· KENNETH D. RIDGWAY, JEFFREY M TROP, EMILY S. FINZEL（382）
第17章　弧—陆碰撞背景下的盆地特征 ········· AMY E. DRAUT, PETER D. CLIFT（406）
第18章　智利北部塔潘帕·德尔·塔马鲁加尔（Pampa del Tamarugal）弧前盆地：构造与气候的相互作用 ············· PETER NESTER, TERESA JORDAN（433）
第19章　伸展型和张扭型大陆弧盆：以美国西南部为例 ············ CATHY J. BUSBY（449）
第20章　前陆盆地系统综述：不同构造背景下响应的多样性
　　　　·································· PETER G. DECELLES（475）
第21章　安第斯山和青藏高原腹地新生代盆地演化 ················ BRIAN K. HORTON（501）
第22章　盆地对青藏高原腹地活动伸展和走滑变形的响应
　　　　············ MICHALEL H. TAYLOR, PAUL A. KAPP, BRIAN K. HORTON（526）
第23章　西班牙东南部贝蒂克（Betic）山间盆地的地层、沉降及构造演化史
　　　　················ JOSE RODRIGUEZ-FERNANDEZ, ANTONIO AZOR,
　　　　　　　　　　　　　　　　　　　　　JOSE MIGUEL AZANON（545）
第24章　中白垩世加拿大西部前陆盆地内浅海相地层的构造沉降、沉积作用以及不整合之间的动力学关系：与科迪勒拉构造的关系
　　　　············ A. GUY PLINT, ADITYA TYAGI, PHIL J. A. MCCAUSLAND,
　　　　JESSICA R. KRAWETZ, HENG ZHANG, XAVIER ROCA, BOGDAN L.
　　　　VARBAN, Y. GREG HU, MICHAEL A. KREITNER, MICHAEL J. HAY（565）
第25章　连续型前陆盆地和破碎型前陆盆地的构造、地貌及沉积特征：以玻利维亚和阿根廷西北部安第斯山脉中部东侧为例
　　　　·················· MANFREN R. STRECKER, GEORGEE, HILLEY,
　　　　　　　　　BODO BOOKHAGEN, EDWARD R. SOBEL（596）
第26章　逆冲楔/前陆盆地系统················· HUGH SINGLAIR（613）
第27章　挤压背景下生长断层相关褶皱的二维运动学模型 ············ JOSEP POBLET（631）

第五部分　板内盆地及其他盆地类型

第28章　板内多阶盆地·············· CAPIL JOHNSON, BRADLEY D. RITTS（661）
第29章　大型格伦维尔沉积事件：罗迪尼亚超大陆形成的记录
　　　　············· ROBERT RAINBIRD, PETER CAWOOD, GEORGE GEHRELS（679）
第30章　克拉通盆地 ··············· PHILIP A. ALLEN, JOHN J. ARMITAGE（702）
第31章　内流盆地 ································· GARY NICHOLS（724）

第一部分 绪 论

第1章 沉积盆地的构造特征及其术语修正

RAYMOND V. INGERSOLL

(Department of Earth and Space Sciences, University of California, Los Angeles, California)

摘 要：现今的板块构造是深入了解沉积盆地构造特征的重要研究内容。在离散、聚敛、转换、板内、混合和混杂的大地构造背景下都可以形成沉积盆地。盆地的构造背景是可以发生转变的，这取决于其下伏地壳的类型、构造位置、沉积物供给以及所继承的原型盆地的性质。沉积盆地的沉降主要是由以下几方面因素引起的：（1）地壳减薄；（2）地幔岩石圈增厚；（3）沉积物和火山负载；（4）构造负载；（5）地壳下的负载；（6）软流圈流动；（7）地壳密度增大。不同沉积盆地的规模、演化周期和保存潜力差异很大。活动构造背景下形成的盆地，特别是在洋壳上形成的盆地，其演化周期短，保存潜力低；而形成于板内背景的盆地演化周期长，保存潜力高。

大陆裂谷可以演化成初始大洋盆地，进而形成大洋盆地。根据盆地的结构特征，可以将与大陆边缘相邻的大洋盆地划分为：陆架—陆坡—陆隆型、转换型和陆堤型三种。未演化为大洋的大陆裂谷可形成为夭折裂谷，进而发展为陆内盆地和坳拉槽。当大洋盆地内部或附近所有的板块边缘都不再活动时，就形成休眠大洋盆地，其下伏的洋壳被陆壳包围。

在会聚背景下，可以形成沉积盆地的位置有海沟、海沟斜坡、弧前、弧间、洋内弧后和陆内弧后。在复杂的岛弧—海沟体系动力学背景下，形成了多种与岛弧相关的盆地类型。值得一提的是，由于岛弧—海沟体系总体为挤压背景，所以沿着岩浆弧或在岩浆弧后形成了多种类型的构造响应。强烈伸展的岛弧内发育的裂谷通常演化为弧后洋盆，并最终扩张形成新的洋壳。在中性岛弧的弧后地区可能存在任何类型的地壳，这些地壳在俯冲带刚开始发育的时候就卷入了构造变形。强烈挤压的岛弧可以演化为弧后逆冲带和相关的后前陆盆地，或者后陆盆地；在极端的情况下，在早期克拉通阶段可能发育有破碎后前陆盆地。

非俯冲大陆地壳或者岛弧地壳被携带到俯冲带附近后，就会发生地壳的碰撞。碰撞一般从一个点开始，然后逐渐扩展，形成缝合带。在原始碰撞点的两侧形成残余大洋盆地，然后残余大洋盆地很快被来自缝合带的沉积物所充填。随着碰撞的继续，残余大洋盆地的沉积物供给量会持续增加，同时盆地发生收缩，直到最终出现增生体的缝合和仰冲为止。碰撞的同时，在大陆地壳俯冲板块上形成前陆盆地，而在仰冲板块上形成了碰撞后前陆盆地，也可能形成碰撞裂谷、破碎前陆和后陆盆地。

在转换背景以及与会聚有关的复合走滑断裂体系背景下，力学机制的改变与断裂走向的不规则变化、岩石类型和板块运动有关。力学机制的变化可形成走滑伸展、走滑挤压和走滑旋转环境，并形成与之相关的复合的、多样的、短期发育的沉积盆地。

最近，以前未命名的两种盆地类型越来越受到关注，即盐岩盆地（与盐体构造活动有关，主要沿着具有陆堤结构的板块边缘分布）和陨石撞击盆地（由地球之外的陨石撞击形成）。随着控盆作用的中止，无论是在离散、会聚、转换背景，还是混合背景下，沉积物都会在随后形成的盆地内聚集。

对所有沉积盆地进行分类和综述的最终目的是通过运用现在的盆地演化模型，提高对古构造和古地理重建的认识水平。利用多个学科的研究，对这些模型进行检验将提高我们对地球演化史的认识。

关键词： 盆地命名　板块构造背景　沉降机制　保存潜力　古构造重建

1.1　引言

十几年前，笔者在 Dickinson（1974b，1976a）根据构造对盆地进行分类的基础之上，对盆地进行了详细的分类，并修正了部分专业术语（Ingersoll，1998；Ingersoll 和 Busby，1995）。最近，盆地的分类又出现了一些新认识和模型；此外，也出现了一些新的专业术语。因此，我们现在有必要对沉积盆地的分类进行强调和修正，并讨论如何将其与盆地的构造特征联系起来。

之前有关盆地分类的文章中，笔者采用了 Dickinson（1974b，1976a）对盆地的命名和分类。这种盆地分类方案建立在现实存在的板块构造作用和构造特征之上。板块构造作用和构造特征最终控制了不同构造背景下，沉积盆地发育的位置以及盆地的形成和演化。板块的水平运动、地热随着时间的变化、地壳的伸展和收缩、重力均衡调整、地幔动力学和地表地质作用，甚至地球之外的事件都会影响沉积盆地。而对沉积盆地其他方面的研究，必然会使盆地的模型更加复杂。虽然我们应该利用统一的方案，根据盆地的形成过程和结果，建立盆地的模型；但是复杂的现实地质条件使得我们对于沉积盆地的认识不断提高，从而不得不建立复杂的模型来解释一些现象。因为一些现实存在的沉积盆地中，具有不符合现有盆地分类的特征，所以，许多新的沉积盆地类型被添加到 Ingersoll 和 Busby（1995）提供的分类中。Gould（1989，1998）提出："分类学并不仅仅是用于避免目录编辑中出现混乱的方法，而是关于自然规律的基础理论"。希望我的论述可以提供以下两个用途：（1）减少命名混乱；（2）为理解控制沉积盆地形成和演化的因素提供框架。

1.2　分类原则

沉积盆地的一级分类标准（Dickinson，1974b，1976a）是：（1）距离盆地最近的板块边缘的类型；（2）盆地与板块边缘的距离；（3）盆地基底的类型。因此，根据标准（1）和标准（2），盆地的一级分类包括离散型、板内型、聚敛型、转换型、混杂及混合型（表1-1）。以上的盆地分类，取决于基底的类型（大洋型、过渡型、大陆型和异常地壳）、构造位置、沉积物供给和继承性特征。

表 1-1　盆地分类（含现代和古代实例）

背景	盆地类型	定义	现今实例	古代实例	模拟模型
离散型	大陆裂谷	大陆地壳内部的裂谷，伴有双峰式火山活动	里奥格兰德裂谷	元古宙基维诺裂谷	3B
	新生大洋盆地和大陆边缘	发育在新洋壳之上，侧向为年轻大陆边缘的初生盆地	红海	侏罗纪东格陵兰	3C

续表

背景	盆地类型	定义	现今实例	古代实例	模拟模型
板内型	板内大陆边缘				
	陆架—陆坡—陆隆结构	成熟的板内裂谷的大陆架边缘，紧邻陆壳洋壳分界	美国东海岸	古生代美国和加拿大境内的科迪勒拉山脉	3D
	转换结构	源于转换板块边缘的板内大陆边缘	西非南海岸	前寒武纪—早古生代阿拉巴马—俄克拉何马转换地区	3E
	陆堤结构	进积于洋壳之上的板内大陆架边缘	密西西比河的海湾海岸区	早古生代加拿大的阿巴拉契亚地带	3F
	克拉通内盆地	下部发育古裂谷的宽阔克拉通盆地	乍得盆地	古生代密歇根盆地	3A
	大陆台地	被薄而广泛的沉积覆盖的稳定克拉通	巴伦特海	中古生代北美大陆中部	3A
	活动大洋盆地	在离散板块边缘形成的、以洋壳为基底的盆地，与岛弧—海沟体系无关	太平洋	多种类型的蛇绿岩组合	3G
	大洋岛屿、海底山、无震洋脊和海底高原	在大洋背景下形成的沉积平原和台地，与岛弧—海沟体系无关	皇帝—夏威夷海山	中生代 Snow Mountain 火山的复合体（加利福尼亚州北部的 Franciscan）	3G
	休眠大洋盆地	以洋壳为基底，既无扩张，也无俯冲	墨西哥湾	古生代塔里木盆地（中国）（?）	3H
聚敛型	海沟	由大洋岩石圈俯冲作用形成的深海槽	智利海沟	白垩纪 Shumagin 岛（阿拉斯加南部）	4A
	海沟—斜坡盆地	在俯冲杂岩上发育的局部构造坳陷	美州中部的海沟	白垩纪威尔士板片（加利福尼亚州中部）	4B
	弧前盆地	发育在岛弧—海沟之间的盆地	苏门答腊大陆架	白垩纪大峡谷（加利福尼亚州）	4B
	弧内盆地				
	大洋型弧内盆地	沿着大洋中的岛弧台地发育的盆地，包括被超覆和超覆的火山	Lzu Bonin 海沟	Copper 和 Gopher 山脉的组合（侏罗纪，加利福尼亚州）	4A
	大陆型弧内盆地	沿着大陆边缘岛弧台地发育的盆地，包括被超覆和超覆的火山	尼加拉瓜湖	早侏罗世内华达山脉（加利福尼亚州东部）	4C
	弧后盆地				
	大洋型弧后盆地	在洋内岩浆弧后发育的洋盆（包括活动弧和残留弧之间的弧间盆地）	马里亚纳弧后	侏罗纪 Josephina 的蛇绿岩（加利福尼亚州北部）	4A, B
	大陆型弧后盆地	大陆边缘岩浆弧后发育的大陆盆地，无前陆褶皱冲断带	桑达大陆架	晚三叠世—早侏罗世美国境内的科迪勒拉山脉	4C

续表

背景	盆地类型	定义	现今实例	古代实例	模拟模型
聚敛型	后前陆盆地				
	弧后前陆盆地	大陆边缘弧—沟体系内，大陆一侧发育的前陆盆地	安第斯山脉	白垩纪美国境内的科迪勒拉山的塞维尔前陆	4E
	碰撞后前陆盆地	板块碰撞时，在上覆板块之上发育的前陆盆地（可能含有后弧的前体部分）	塔里木盆地西部（中国）	三叠—侏罗纪鄂尔多斯盆地（中国）	4F
	破碎后前陆盆地	在弧后前陆背景下，以基底为核心的隆起间形成的盆地	Sierras Pampeanas 盆地（阿根廷）	晚白垩世—古近纪美国科迪勒拉山脉内的拉腊米盆地	4D
	残余大洋盆地	在碰撞陆缘，及弧—沟体系内形成的收缩洋盆，最终在缝合带内被俯冲和变形	孟加拉湾	宾夕法尼亚纪—二叠纪沃希托盆地	4E
	前前陆盆地	当大陆或者岛弧发生碰撞时，在俯冲板块内的，陆壳之上发育的前陆盆地	波斯湾	中新生代瑞士磨拉石盆地	4F
	楔形顶部盆地	形成于逆冲席之上，并且随着逆冲席移动的盆地	白沙瓦盆地（巴基斯坦）	新近纪亚平宁山脉（意大利）	4F
	后陆盆地	形成于前陆褶皱逆冲带后侧，盆地下伏的地壳增厚	阿尔蒂普拉诺高原（玻利维亚）	新近纪札达盆地（中国西藏）	4D
转换型	走滑伸展盆地	沿着走向滑动断层系拉张而形成的盆地	死海	石炭纪马达兰盆地（圣劳伦斯海湾）	5A
	走滑挤压盆地	沿着走向滑动断层系挤压而形成的盆地	圣巴巴拉盆地（前陆类型）（加利福尼亚州）	中新世山脊盆地（断层转向型）（加利福尼亚州）	5B
	走滑旋转盆地	走向滑动断层系内，地块绕垂直轴旋转而形成的盆地	西阿留申前弧前盆地（?）	中新世洛杉矶盆地（托潘加盆地）（加利福尼亚州）	5C
混杂及混合型	坳拉槽	与造山带成高角度相交的重新活动的古裂谷	密西西比河湾	古生代阿纳达科坳拉槽	6A
	碰撞裂谷	与造山带成高角度相交的裂谷，之前没有造山运动（与坳拉槽不同）	贝加尔裂谷（远源的）（西伯利亚）	莱茵地堑（近源的）（欧洲）	6B
	碰撞破碎前陆盆地	由于远程碰撞作用使得地壳变形而形成的各种类型的盆地	柴达木盆地（中国）	宾夕法尼亚纪—二叠纪，美国科迪勒拉山脉内古落基山盆地	6B
	盐岩盆地	由于盐体的变形而形成的盆地，通常发育于大陆堤和周缘前陆盆地内	墨西哥湾深部的小盆地	白垩纪—古近纪 La Popa 盆地（墨西哥）	3F
	陨石撞击盆地	由地球外的物质撞击地球表面而引起的沉陷	陨石坑（亚利桑那州）	白垩纪—古近纪希克苏鲁伯盆地（墨西哥）	3E
	继承性盆地	在局部造山或地裂活动停止后，在山间环境下形成的盆地	盆岭省南部（亚利桑那州）	古近纪的 Sustut 盆地（?）（不列颠哥伦比亚）	5C

笔者根据盆地各沉积时期的特征对盆地进行了分类和命名。在板块构造控制下形成的地层层序具有多地区和多时期分布的特点。单一的地层层序可代表几种不同的构造背景。"沉积盆地的演化可以看作是，在一系列不连续的板块构造背景下，各个板块之间连续相互作用的结果"（Dickinson，1974b）。

理解"盆地"的含义很重要，本文中的盆地是指沉积岩或火山岩地层的堆积；盆地的三维结构表现为碟形、楔形、席状以及不规则状。此外，基底沉降、沉积物搬运受到阻挡、先存空间的充填或源—汇系统的相对运动都可以形成盆地。

1.3 沉降机制和保存潜力

地表沉积物的沉降机制可以归纳为七个方面（Dickinson，1974b，1976a，1993；Ingersoll 和 Busby，1995）（表 1-2）：（1）伸展作用、侵蚀或者岩浆侵位导致的地壳减薄；（2）在冷却过程中地幔和岩石圈的增厚；（3）沉积岩或火山岩的负载（局部地壳均衡补偿或区域岩石圈挠曲）；（4）地壳和岩石圈的构造负载；（5）地壳和岩石圈的壳下负载；（6）软流圈流动的动力学作用；（7）地壳密度的增大。图 1-1 阐述了早期扩张阶段地壳减薄形成的盆地（例如裂谷和走滑拉分盆地）和海底扩张初始阶段地幔—岩石圈增厚形成的盆地（位于离散大陆边缘周围，在裂谷期向漂移期转换过渡阶段所形成的盆地，这些离散大陆边缘最终演化为陆内边缘）。大陆板块—大洋板块边缘的沉积作用主要由河流和三角洲提供物源（比如大陆堤和残余洋盆地）。构造运动控制的盆地类型与地壳消减带控制的盆地类型一致（比如海沟和前陆盆地）。通常，其他三类沉降机制是次要的。

表 1-2 下沉机制

地壳减薄	伸展作用、侵蚀或者岩浆侵位导致的地壳减薄
岩石圈变厚	在冷却过程中地幔和岩石圈的增厚
沉积物和火山的负载	在沉积作用和火山作用的影响下，局部地壳均衡补偿或区域岩石圈挠曲
构造负载	在仰冲作用和俯冲作用的影响下，局部地壳均衡补偿或区域岩石圈挠曲
地壳和岩石圈的壳下负载	板块底部岩石圈密度变大引起的岩石圈挠曲
软流圈的流动	软流圈流动的动力学作用，通常由地壳俯冲而引起地壳下冲或者会聚而引发软流圈流动
地壳密度增大	由温度或压力的变化，或者高密度的熔融物侵位到低密度的地壳中，而引起地壳密度增大

不同构造类型的沉积盆地，在规模、演化周期和保存潜力上存在很大的差异（图 1-2）（Ingersoll，1998；Ingersoll 和 Busby，1995；Woodcock，2004）。大多数沉积盆地在沉积后，在较短的时间内被破坏（比如大多数位于洋壳或者快速抬升的造山带中的盆地）。相对来说，在陆壳伸展过程中或者伸展之后（大陆裂谷演化为海底扩张或者不演化为海底扩张）形成的盆地具有较好的保存潜力，因为这些盆地随着裂谷的裂开，不断地发生沉降并被埋藏于沉积物之下。另一方面，一部分沿着大陆边缘分布的地层发生俯冲，被下拉到海沟之下，使得它们保存在中等至很深的地壳中，从而形成变形复杂的地体。这种变质沉积岩和变质火山岩地体，以及大量的残余洋盆沉积物，是陆壳的重要组成部分。这些地体的基底（洋壳）大多在俯冲过程中，被消减掉了（Graham，1975；Ingersoll，1995，2003）。

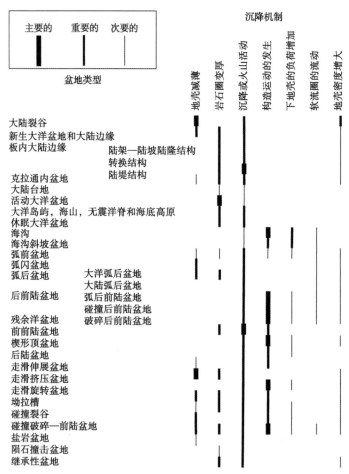

图 1-1 所有沉积盆地类型的沉降机制

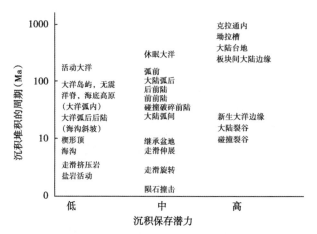

图 1-2 典型沉积盆地的发育周期与保存潜力的关系

盆地的保存潜力主要是指盆地在形成时和形成后，不发生隆升和侵蚀的平均时间。沉积物质或者火山物质在盆地被破坏期间或破坏之后，会以增生杂岩的形式得以保存（尤其是大洋盆地内沉积的地层）。

在基底稳定的情况下，边缘的盆地具有很高的保存潜力，但是在缝合带内和缝合带以下的板块边缘盆地，往往发生变形和变质作用，使其很难识别出来

1.4 离散背景

1.4.1 裂谷的发育过程和大陆分离

是主动裂陷作用（地幔对流驱动）还是被动裂陷作用（岩石圈驱动）控制了大陆裂谷的形成，一直饱受争议（Sengor 和 Burke，1978；Ingersoll 和 Busby，1995；Sengor，1995）。如果不考虑裂谷形成的机制，那么大陆分离可能形成两种结果："成功的"大陆裂谷演化成为海底扩张初期的大洋盆地（Ingersoll 和 Busby，1995；Leeder，1995），随后演化成为拥有对称特征的板块边缘主动大洋盆地（图 1-3）；"夭折的"裂谷不会演化为大洋盆地，而是演变为古裂谷，通常会被克拉通盆地所覆盖（Sengor，1995）。Ingersoll 和 Busby（1995）、Leeder（1995）和 Sengor（1995）阐述了大陆裂开期间及其之后，大陆拉张、盆地形成、构造演化以及大陆裂谷的多种后期演变过程。本文重点阐述在大陆裂谷向板块边缘演化（裂谷向漂移的转变）的过程中涉及的一些专业术语和模型。

1.4.2 大陆裂谷

大多数与大陆裂谷有关的盆地（图 1-3b）（Dickinson 的"地表裂谷论"，1974b；Ingersoll，1988）都表现为半地堑结构，并发育于正断层上盘（Leeder 和 Gawthorpe，1987；Leeder，1995；Gawthotpe 和 Leeder，2000）。Gawthotpe 和 Leeder（2000）总结了大陆裂谷盆地的构造—沉积演化的概念模型，该模型包括了盆地演化过程中，空间形态的变化。文中还讨论了构造、地貌、气候以及海—湖平面变化对盆地演化的影响。

包括高角度正断层的模型在内，所有模型均由 Gawthorpe 和 Leeder（2000）提供。在这些半地堑中，大多数的沉积物源来自于上盘；但盆地形成初期，来自于断层下盘的粗碎屑物质，则仅仅分布在盆缘断层周围的小而陡的冲积扇或者扇三角洲内。相反，克拉通内拆离盆地（形成于低角度正断层之上）接受了大量位于断层下盘的、以粗碎屑为主的沉积物沉积（Friendmann 和 Burbank，1995）。Gawthorpe 和 Leeder 的半地堑模型还考虑了以下因素：调节带、转换斜坡、背斜形地堑盆地以及向斜形地垒盆地（Rosendabl，1987；Faulds 和 Varge，1998；Igersoll，2001；Mack，2003）。

1.4.3 新生大洋盆地和大陆边缘

随着大陆岩石圈被拉张减薄，地幔软流圈明显上升并且接近地表（图 1-3c）。在大陆裂谷向海底扩张转换的过程中，形成了过渡型地壳，这种地壳要么是伸展型陆壳（类大陆的），要么是富沉积物的玄武质地壳（类大洋的）（Dickinson，1974b；Ingersoll，2008b）。大陆裂谷只有在缺乏有效沉积的情况下才会演化为海底扩张。在这种情况下洋壳就成为了软流圈在上拱熔融过程中，唯一与软流圈相互作用的固体物质（Einsele，1985；Nicolas，1985）。因此，在真正的海底扩张开始之前，于初始大洋盆地的边缘会形成具有一定宽度的过渡型地壳。

在这些过渡型地壳的形成和两个大陆边缘的分离过程中，新生大洋盆地开始发育["原始的大洋海湾"和"狭窄的大洋" Dickinson（1974b）；"原始的大洋裂谷槽"，Ingersoll（1988）]。红海是典型的新生大洋盆地，发育于主动海底扩张。此外，在红海的大陆边缘

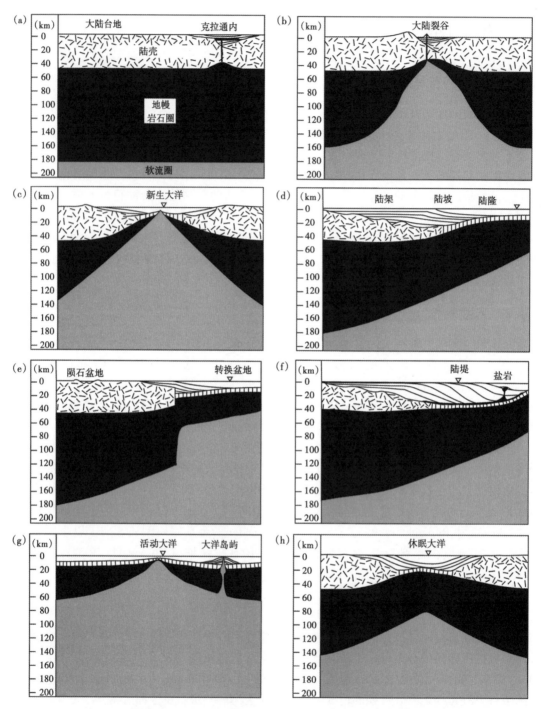

图1-3 现今离散背景、板内背景和其他混杂背景下,真实沉积盆地的模型

在板块分离引起的减压熔融过程中,地幔岩石圈持续减薄;而随着离散作用的终止,地幔岩石圈开始冷却加厚。同时也展示了两种特殊类型的盆地(陨石撞击盆地和盐岩盆地)。陨石撞击的位置是随机的;它们可能在地表的任意位置发生,但正常情况下,克拉通区域内的陨石盆地更容易保存(e)。盐岩盆地一般发育在下部埋藏有盐体的地区,陆堤(f)是形成该类型盆地的最佳区域。横线代表陆壳;垂线代表洋壳;黑色代表上地壳和次生的火成岩;橙色代表软流圈和下地壳;黑色代表盐体(仅限于活动的盐体)

发育有碎屑岩和碳酸盐岩沉积，以及隆起的裂谷肩（Cochran，1983；Bohannon，1986a，1986b；Coleman，1993；Leeder，1995；Purser 和 Bosence，1998；Bosworth，2005）。在裂谷盆地向新生大洋盆地转换期间会形成厚层的蒸发岩沉积。如果存在适度的干旱环境，且与其他大洋沟通受限，并缺乏碎屑物质的输入，那么在新生大洋盆地的演化过程中，也会形成蒸发岩沉积（Dickinson，1974b）。加利福尼亚海湾就是拉张型新生大洋盆地的典型实例（Atwater，1989；Lonsdate，1991；Atwater 和 Stock，1998；Axen 和 Fletcher，1998）。

1.5 板内背景

1.5.1 板内大陆边缘

随着两个大陆和洋中脊的不断分离，新生大洋盆地最终演化成为宽广的大洋（大西洋型）。在这样的演化过程中，裂谷大陆边缘不断地远离扩张洋中脊，期间新生的裂谷大陆边缘隆起的两翼不断地变冷和下沉。这个过程涉及了"裂谷向海底扩张"的过渡，即离散型大地构造背景向板内大地构造背景的演化（Dickisson，1974b，1976a；Ingersoll，1988；Bond，1995；Ingersoll 和 Busby，1995）。Withjack 等（1998）讨论了在转换过程中时间因素和地质过程的复杂性。

盆地的沉降机制包括：（1）陆壳在隆升和裂陷的过程中遭受拉张和侵蚀减薄；（2）抬升后，由于板内大陆边缘远离扩张中脊而导致的热沉降；（3）局部地壳和岩石圈在板内大陆边缘的演化过程中发生沉积负载（Bond 等，1995；Ingersoll 和 Busby，1995）。下地壳或者地壳深处的流动或密度增大也可以在局部形成沉降。

1.5.1.1 陆架—陆坡—陆隆结构

在向海减薄的陆壳之上，大多数成熟的板内大陆边缘，都发育向海增厚的楔形大陆架沉积（图1-3d）。过渡型地壳（类陆壳的和类洋壳的；Dickinson，1974b，1976a）位于厚层大陆架沉积到薄层大陆坡沉积的过渡带之下，这些沉积物依次与洋壳上的厚层浊流陆隆沉积和深海平原沉积相接（Bond 等，1995；Ingersoll 和 Busby，1995）。现代大西洋大陆边缘的大多数地区都存在这样的组合。在低纬度地区，因为碳酸岩盐沉积环境占支配地位，所以缺乏陆源碎屑的供给。

1.5.1.2 转换结构

板内大陆边缘起源于转换边缘，而不是具有狭窄柱状沉积物和过渡地壳的裂谷边缘（图1-3e）。从开始发生转换运动（同时在相邻的板块边缘开始发生"裂谷到漂移"的转换）到板内形成沉积（沿转换边界发生大洋中脊扩张之后），间隔数千万年（Bond 等，1995；Turner 等，2003；Wilson 等，2003）。西非的南海岸就是实例；新元古代到早古生代的亚拉巴马州—俄克拉何马州转换边界是地史时期的例子（Thomas，1991）。

1.5.1.3 陆堤结构

在板内大陆边缘内部，主要河流的分布，一般受到与大陆边缘呈高角度相交的古裂谷的控制（Burke 和 Dewey，1973；Dickinson，1974b；Audley-Charles 等，1977；Ingersoll 和 Busby，1995）。最好的例子就是现今的尼日尔三角洲（Burke，1972）和密西西比河三角洲（Worral 和 Snelson，1989；Salvador，1991；Galloway 等，2000）。在这些地区，大陆架上的沉积物已经进积到了洋壳之上。这是因为在大陆架内部，沉积物的厚度已经超过了负载均衡

(16~18km；Kinsman，1975）所允许的最大沉积物厚度（图1-3f）。在墨西哥湾，除了密西西比河以外，许多河流对大陆边缘的进积都起到了重要的作用；此外，这是一种陆堤结构，它不同于典型的陆架—陆坡—陆隆结构和转换结构。

1.5.2 克拉通内盆地

大多数的克拉通内盆地（例如密执安盆地）都发育在古裂谷之上（Derito等，1983；Quinlan，1987；Klein，1995；Sengor，1995；Howell和Van der Pluijim，1999）（图1-3a）。克拉通内盆地的重新沉降，与邻近造山带的造山运动所诱发的岩石圈应力变化有关（Derito等，1983；Howell和Van der Pluijim，1999）。当岩石圈的刚性降低时，盆地便发生沉降，使得上地壳（残余的古裂谷）内未补偿区域接受沉积。在造山运动期间，岩石圈的刚性增强，导致局部重力均衡又被打破。所以，克拉通内盆地经历几百个百万年的演化之后才能达到重力均衡（DeRito等，1983；Ingersoll和Busby，1995；Howell和Van der Pluijm，1999）。

1.5.3 大陆台地

克拉通的地层层序主要反映了全球的构造事件和海平面的升降变化（Sloss，1988；Bally，1989）。此外，地幔动力学、局部和区域的构造事件也影响了大陆台地的发育（Cloetingh，1988；Burgess和Gurnis，1995；Van der Pluijim等，1997；Burgess，2008）。与克拉通内盆地相比，大陆台地（图1-3a）上具有一定厚度的沉积物覆盖了整个大陆。台地上的地层可以过渡为大陆边缘、克拉通内盆地、前陆盆地以及与大陆边缘相邻的其他大地构造环境下的地层（Ingersoll和Busby，1995；Burgess，2008）。远端前陆和台地的地层差异较大，特别是当海平面很高的时候，台地形成大量的碳酸盐岩，而远端前陆则发育宽广的前陆挠曲。在台地上，由海平面升降所形成的韵律层很明显（Heckel，1984；Klein，1992；Klein和Kupperman，1992），此外，古纬度和古气候也影响了台地中地层的特征（Berry和Wilkinson，1994）。大陆台地在超大陆时期一般都遭受了暴露和剥蚀；随后在超大陆解体100Ma后，发育最大海泛面（Heller和Angevine，1985；Cogne等，2006）。

1.5.4 活动大洋盆地

随着大洋洋壳的扩张并远离洋中脊，大洋岩石圈的水深逐渐增加，热流值也呈现指数型衰减。此外，洋壳的年龄也逐渐变老（Sclater等，1971；Parsons和Sclater，1977；Stein和Stein，1992）（图1-3g）。随着洋壳从洋中脊不断地扩张，洋壳不断地远离洋中脊，其年龄也逐渐增大，并开始缓慢的下沉。此外，洋壳之上也不断地聚集远洋和半远洋的沉积物（Berger，1973；Heezen等，1973；Winterer，1973；Berger和Winterer，1974）。在碳酸盐岩补偿深度之上（CCD）碳酸盐岩发生沉淀聚集，在生物碳酸盐岩高产区，这个补偿深度会降低；在未一致认可的二氧化硅补偿深度（SCD）之上发生二氧化硅沉淀聚集；而在SCD之下，仅有深海黏土富集。这样就可以预测，从单纯洋壳到发育海相沉积时的年龄、深度以及古纬度。通过对岩浆岛弧和大陆边缘附近形成的火山碎屑和浊积岩的混杂堆积的研究，可以预测大洋板块的地层层序（Cook，1975；Ingersoll和Busby，1995）。

1.5.5 大洋岛屿、海底山、无震洋脊以及海底高原

随着大洋板块迁移远离扩张中脊，大洋岛屿、海底山、无震洋脊以及海底高原开始发生

热沉降。根据岩浆的发展情况，独立于扩张中脊（比如热点）的热异常点会产生新的岛屿、洋脊和海底高原。此时，该类型的盆地可能存在复杂的沉降史。Clague（1981）将海底火山的发育过程划分为三个连续的阶段：地表型火山、浅水型火山、深水或者半深海型火山（Ingersoll，1988；Ingersoll 和 Busby，1995）。随着不断地侵蚀和沉降，在岛屿边缘也许会形成生物礁和环礁，这主要取决于纬度、气候和相对海平面的高低（Jenkyn 和 Wilson，1999；Dickinson，2004）。会聚型边缘海的地貌差异会增大（例如北美滨海山脉的 Wrangellia；Ricketts，2008），其规模可以是小型的海山，也可以是大型的铁镁质火成岩省，例如翁通爪哇海底高原和相关的地形（Taylor，2006）。

1.5.6 休眠大洋盆地

休眠大洋盆地的基底是既不扩张也不沉降的洋壳；换句话说，它不发育活动的板块边缘（Ingersoll 和 Busby，1995）（图 1-3h）。与休眠大洋盆地相比，活动大洋盆地至少存在一个主动扩张洋脊（例如大西洋、太平洋和印度洋），而残余大洋盆地表现为至少存在一个俯冲带的小型收缩洋盆（例如孟加拉海湾和休恩湾）。"休眠"是指在大洋盆地中或者周缘没有造山活动和地震活动；"大洋"是指盆地的基底是大洋岩石圈，而克拉通内盆地的基底之下，一般为部分裂陷的大陆岩石圈（Ingersoll 和 Busby，1995）。

休眠大洋盆地存在两种对比鲜明的成因机制：（1）初始大洋盆地的扩张中脊停止活动（例如墨西哥湾；Pindell 和 Dewey，1982；Pindell，1985；Dickinson 和 Lawton，2001），或者（2）在大陆和岛弧（例如黑海；Okay 等，1994）或者南里海盆地（Brunet 等，2003；Vincent 等，2005）拼接缝合过程中，没有被消减的弧后盆地（伸展的或者中性的）。在海底扩张停止后，休眠大洋盆地的基底和原始地层一般被深藏在地下深处达数个百万年，所以大洋盆地的起源很难确定（例如中国西部的塔里木盆地和准噶尔盆地）（Sengor 等，1996）。随着板块活动停止，除了残余热的冷却导致地壳增厚以外，在盆地内或者周边的沉积物的负载成为了盆地沉降的主要动力（Ingersoll 和 Busby，1995）。休眠大洋盆地的演化周期可达数百个百万年，而它们的规模也有大有小。墨西哥湾是已知的最大的休眠大洋盆地，其北部边缘被沉积物快速充填（墨西哥沿岸的大陆架），而南部仍然出露被薄层沉积物覆盖的洋壳（Buffler 和 Thomas，1994；Galloway 等，2000；Dickinson 和 Lawton，2001）。南里海盆地规模比较小，仅部分地区有沉积物覆盖（局部厚度超过 20km；Brunet 等，2003），现今仍然是大洋盆地。相比较而言，塔里木盆地的沉积物厚度和南里海盆地的沉积物厚度相当，但是塔里木盆地却是整个盆地都被沉积物覆盖了。这三种盆地的基底都是大洋地壳，例如塔里木盆地的基底为海底高原（Sengor 等，1996）；这些盆地较长的冷却史表明其底部存在巨厚的刚性上地壳（Ingersoll 和 Busby，1995）。当休眠大洋盆地被充填到海平面时，表面上看起来类似于克拉通内盆地。在休眠大洋盆地内，发育在稳定大洋地壳之上的沉积地层厚度达 16~20km；而在克拉通盆地内，发育在陆壳之上的沉积地层却只有几千米厚，而且在克拉通内盆地中心下部发育有一个或多个古裂谷。因此，当板块内的应力变化影响到休眠大洋盆地及其周缘时，薄弱的盆地边缘一般最先发生变形；而克拉通内盆地的变形主要集中在其下伏的古裂谷边缘。在挤压变形时形成的前陆盆地往往分布于休眠大洋盆地的边缘之上（例如现代的塔里木盆地边缘）。克拉通内盆地可能发生沉降或者构造反转（例如现代的北海）（Cooper 和 Williams，1989；Cameron 等，1992）。

1.6 聚敛背景

1.6.1 岛弧—海沟体系

岛弧—海沟体系可以分为三种基本类型：（1）伸展型的岛弧—海沟体系；（2）中性的岛弧—海沟体系；（3）挤压型的岛弧—海沟体系（Dickinson 和 Seely，1979；Dewey，1980）（图1-4）。具有明显走滑特征的岛弧—海沟体系可以当作是第四种类型（Dorobek，2008）。

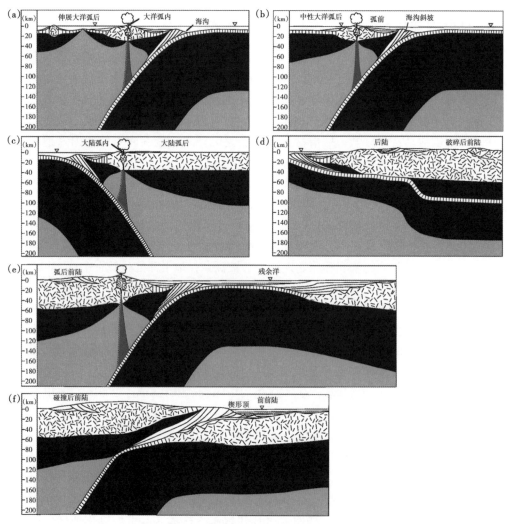

图 1-4 现今会聚背景下真实的沉积盆地模型

图a的左边存在一个残余的岛弧。海沟、海沟斜坡和弧前盆地只在图a和b中标记出来，但是它们与所有类型的岛弧—海沟体系都有关系。弧间盆地可以与任何类型的岩浆岛弧伴生，但是它们通常在伸展背景和中性背景中出现（a、b和c）。弧后盆地可以形成于挤压型岛弧—海沟体系（d、e），或者碰撞体系（f和图1-6）。残余大洋盆地可以在任意两个地壳发生碰撞的边缘处形成；此外图e表明，残余大洋盆地还可以在挤压型岛弧—海沟体系与板内边缘发生碰撞处形成。楔形盆地可以在任何受挤压背景下形成，图f中的前前陆盆地就是一个实例。如果中性的大陆型的岛弧—海沟体系（c）的构造背景转变为张性环境，那么它就会演变为伸展型大洋系统（a）。图例与图1-3相同；红色代表板块活动引起的岩石熔融虽然走滑断层可以在所有类型的岛弧—海沟体系中发育，但是在存在斜向挤压的背景下，走滑断层往往成对出现（Berk，1983）。根据对现代岛弧—海沟体系的分析（此外，Cruciani 等，2005 年的研究成果，也对现代岛弧—海沟体系的形成做出合理的解释），笔者认为，虽然很多因素控制了岛弧—海沟体系特征，但是最重要的因素是：（1）会聚的速率；（2）板块的年龄；（3）板块的倾角（Molnar 和 Atwater，1978；Uyeda 和 Kanamori，1979；Jarrard，1986；Kanamori，1986）。通过对现代地球一系列的分析研究，学者们提出了一个重要的问题：现代的扩张中脊和岛弧—海沟体系的格局是地球历史时期的稳定格局吗？或者只是不同时期构造特征的组合？几乎所有向东倾斜的现代岛弧（例如马里亚纳海沟）都发育在伸展背景下，并伴有以较高的角度向下俯冲的老岩石圈。而几乎所有向西倾斜的岛弧（例如安第斯山脉）都发育在挤压背景下，并伴有以较缓的角度向下俯冲的新岩石圈。大多数向南倾斜的岛弧（例如阿留申群岛）都发育在非挤压和非伸展背景下，并伴有以适中的角度向下俯冲的中度年龄的岩石圈。没有向北倾斜的岛弧。所以，在这种情况下，就很难区分板块的年龄、板块的倾角、倾向和不同岛弧—海沟体系之间的共变关系。目前学者们已经达成一致的是（相反的观点请参见 Schellart，2007，2008 的文献）：由于向西运动的潮汐滞后于地球向东的自转，所以岛弧—海沟体系的朝向可能是控制岛弧—海沟体系特征的根本因素（Bostrom，1971；Moore，1973；Dickinson，1978；Doglioni，1994；Doglioni 等，1999）。因为地球自始至终都是自西向东自转的，所以现今的特征就代表了整个地史时期内的特征。因此，古代岛弧—海沟体系的模式就能解释它们活动时的朝向。由于缺乏岛弧—海沟体系的基本特征的认识，使得前人提出了许多错误的古代造山带模型（Dickinson，2008）。

Dickinson（1974a，1976b）、Ingersoll（1988）、Ingersoll 和 Busby（1995）和 Dorobek（2008）等人总结了与岛弧—海沟体系有关的盆地所赋存的大地构造背景和沉降机制。因为岛弧火山为邻近的盆地提供了大量的沉积物，所以 Ingersoll 和 Busby（1995）以及 Smith 和 Landis（1995）研究了岛弧火山的形成和剥蚀特征。

弧前盆地、弧间盆地和弧后盆地之间的差别有时不是很明显。弧间盆地是指岛弧台地周围厚层的火山—火山碎屑以及其他沉积物堆积，而岛弧台地是由上超或被上超的火山形成的。地层中是否存在远源的火山岩以及相关的侵入岩是辨别弧间盆地的标志性特征，因为由岛弧衍生的火山碎屑物质可以搬运到弧前盆地、弧后盆地以及其他盆地。作为一个常见的专业词汇，"岛弧块体"是指由岛弧岩浆作用形成（Dickinson，1974a，1974b）的地壳，并且该地壳比岛弧台地的范围要大。Dickinson（1995）也讨论了弧前盆地和弧间盆地的区别。许多弧后盆地是由岛弧台地中的裂谷形成的（Marsaglia，1995），而且这些盆地早期属于弧间盆地。由于板块碰撞、板块重组以及由板块运动特征发生改变使得岛弧—海沟体系发生渐变或者突发的重新组合，造成弧前环境、弧间环境和弧后环境会暂时发生改变或者彼此相互叠置。

1.6.2 海沟

Karig 和 Sharman（1975）、Schweller 和 Kulm（1978）、Thornburg 和 Kulm（1987）以及 Underwood 和 Moore（1995）总结了活动海沟中，沉积作用和构造活动的动力学属性（图 1-4a）。当俯冲速率、俯冲角度、洋壳上的沉积物厚度、沉积速率以及沉积范围恒定时，海沟沉积就达到了动态均衡状态。Thornburg 和 Kulm（1987）研究了纵向物源（轴向海沟楔）和横向物源（海沟扇）相互作用的动力学特征。随着海沟中横向沉积物的供给越来越多，轴向海沟楔的槽道被迫沿着海和海沟楔方向加宽。海沟充填过程的动力学对比不仅有助于海沟的水深和沉积系统的研究，同时也有助于增生结构的研究（Thornburg 和 Kulm，1987；Underwood 和 Moore，1995）。这个动力学模型对重建海沟的沉积和构造过程也许会很有帮助，其意义在古代俯冲杂岩的研究中，已经体现出来了。

Scholl 等（1980）建立了一个概念模型，这个模型与俯冲带的增生过程和沉积参数有关，这些增生过程和沉积参数影响了弧前和海沟的特征。Cloos 和 Shreve（1988a，1988b）建立了俯冲带深部的定量模型，俯冲带的深部特征影响了变形作用和变质作用，此外还制约了弧前的整体形态。因为在深水背景下研究现代的沉积系统，对比现代和古代沉积系统具有一定难度，而且俯冲背景下构造变形很复杂，所以重建古海沟中的过路沉积系统很困难（Underwood 和 Moore，1995）。无论如何，科技进步以及学者们对现代和古代的沉积系统的持续研究，提升了我们对沉积系统和构造系统的认识水平（Maldonado 等，1994；Mountney 和 Westbrook，1996；Leverenz，2000；Kopp 和 Kukowski，2003）。

1.6.3　海沟—斜坡盆地

Moore 和 Karig（1976）提出了海沟坡上的小型封闭盆地的沉积模型（图 1-4b）。俯冲复合体上的变形形成了复杂的海底地形；使得浊积岩被拦截在海沟斜坡盆地内。由于刮擦下来的沉积物，在脱水和变形过程中，发生隆升同时断层空间增大，使得斜坡盆地的平均宽度、沉积物厚度和年龄沿着斜坡上倾方向，逐渐增大。位于古代俯冲复合体上的海沟斜坡盆地，正常情况下会被相对未变形的近源浊积岩所充填，而周围则是高度变形的不同来源的增生物质。海沟斜坡盆地与增生物质之间在沉积与构造上存在联系。苏门答腊岛附近的尼亚斯岛上沉积物厚度迅速增加，Moore 和 Karig 为其设计了一个模型。该模型不适用于弧前欠沉积地区。考虑到尼亚斯岛地区和安达曼群岛的特殊情况，Allen 等（2008）以及 Hall 和 Smyth（2008）对该模型提供了一些改进，并进行了解释。无论如何，Moore 和 Karig 的模型为低海沟斜坡上发育沉积盆地的研究提供了基本要素，并且是重建古俯冲复合体的基础。

Underwood 和 Moore（1995）、Aalto 和 Miller（1999）、Underwood 等（2003）以及 Allen 等（2008）论述了现代和古代海沟斜坡盆地的其他的实例，并讨论了它们对于重建古构造的意义。

1.6.4　弧前盆地

Dickinson 与 Seely（1979）和 Dickinson（1995）提出了岛弧—海沟体系的分类，他们的分类与 Dewey（1980）的分类相似，并且认为板块构造活动控制了沉降活动的开始和弧前的发育（图 1-4b）。控制弧前盆地几何形态的因素有：（1）初始大地构造背景；（2）俯冲板块上沉积物的厚度；（3）向海沟供给沉积物的速率；（4）向弧前区域供给沉积物的速率；（5）俯冲速率和方向；（6）俯冲的初始时间。由于当俯冲开始后，海沟的进积增生和岩浆弧的向后迁移，使得岛弧—海沟的宽度逐渐变大（Kickinson，1973）。厚层沉积物增生进积的速率非常快。岛弧—海沟变宽的结果就是弧前盆地也不断变大（例如大峡谷前陆盆地；Ingersoll，1979，1982；Dickinson，1995）。

弧前盆地包括以下几种类型（Dickinson 和 Seely，1979；Dickinson，1995）：（1）地块内弧前盆地（转换为弧内类型）；（2）增生型弧前盆地（海沟斜坡）；（3）残留型弧前盆地（位于大洋或者过渡型地壳之上，当俯冲开始后，被拖拽到海沟后侧）；（4）建造型弧前盆地（位于岛弧地块和俯冲复合体之间）；（5）复合型（以上几种类型的组合）。残留型和建造型弧前盆地倾向于演化为复合型盆地；一般来说，在这个演化过程中，伴随着弧前盆地的充填和水体变浅。

Stern 和 Bloomer（1992）论述了在马里亚纳海沟俯冲带开始活动时，推覆板块前缘的地壳扩展（渐新世）过程。这种类型的扩展仅仅会在洋内俯冲带开始作用后，10~20 个百万年内，在老岩石圈之上形成新的地壳。俯冲开始后，板片开始向后旋转，使得推覆板块中最脆弱的部分处于俯冲边缘处。随后，由于冷大洋岩石圈俯冲的影响，弧前地区也冷却下来；因此，成熟的洋内弧前区域的基底就变成了冷的硬地壳，从而阻止了地壳的继续扩展（Vink 等，1984；Steckler 和 tenBrink，1986；Dickinson，1995）。增生楔形体的浅层发育很多正断层（Platt，1986；Underwood 和 Moore，1995），但是在现今弧前地区，却没有发育任何的新生地壳裂谷，且古代弧前环境下也没有发育（Ingersoll，2000）。相比较而言，成熟洋内系统的岛弧轴线是推覆板块中最薄弱的部分，伸展作用通过弧间和弧后扩张来调节（Marsaglia，1995）。

最近的一些关于现代的和古代的前弧盆地的研究已经验证了 Dickinson（1995）所论述的综合模型的有效性（Einsele 等，1994；Van der Werff，1996；Mountney 和 Westbrook，1997；Constenous 等，2000；Trop，2008）。

1.6.5 弧内盆地

一般来说，综合火山地质、沉积学和盆地分析的研究很少，所以对岩浆岛弧（图 1-4a）内盆地的起源研究都很简单（Ingersoll，1988；Ingersoll 和 Busby，1995）。此外，由于岛弧具有高热流值、地温梯度变化剧烈和火山活动强烈等特征，而且岛弧一般在演化史中属于地壳收缩阶段；所以，沉积学家要想研究弧内盆地，就必须对热液蚀变、变质作用和变形作用等进行综合分析。Fisher 和 Schmincke（1984）、Cas 和 Wright（1987）、Cas 和 Busby-Spera（1991）、Fisher 和 Smith（1991）以及 Smith 和 Landis（1995）完美地总结了 1995 年以前的相关成果。

1.6.5.1 大洋型弧内盆地

岛弧内至少存在三种火山物质和碎屑物质的沉积中心（Ingersoll 和 Busby，1995；Smith 和 Landis，1995）。在火山和火山边缘之间地势较低的区域，可以形成沉积中心，但是只有位于海平面以下的沉积中心，具有很高的保存潜力（一般在大洋岛弧中）。岛弧火山活动中轴位置的迁移，使得大洋岛弧台地之上的活动岛链和废弃岛链之间也形成了地势较低的区域，这些区域也可以形成具有很高保存潜力的沉积中心。Smith 和 Landis（1995）将这三类弧间盆地描述为"以火山为界的盆地"。同样，也涉及到了"以断层为界的盆地"。这些盆地是由构造活动引起的迅速沉降而形成的，而不是由大陆火山活动引起的。这种模式成功地解释了盆地边缘的地形差异（Ingersoll 和 Busby，1995）。

对大洋弧内沉积类型做研究的地质学家还有 Robertson 和 Degnan（1994）、Fackler-Adams 和 Busby（1998）、Sowerbutts 和 Underhill（1998）、Sowerbutts（2000）、Clift 等（2005）以及 Busby 等（2006）。

1.6.5.2 大陆型弧内盆地

按照形成盆地的规模，将大陆型岛弧内，能够聚集并保存了厚层沉积物的机制进行了排序，它们依次是（Busby-Spera，1988b；Busby-Spera 等，1990）：（1）板块边缘的扩张或者走滑伸展；(2) 局部区域内发生的深成或者浅成岩体侵入而导致的扩张；(3) 大量凝灰岩的喷发导致局部破火山口发生沉降。板块边缘规模的扩张或者走滑伸展，产生了连续或者半连续延伸达几百至几千千米的大陆岛弧沉降带，这个沉降带记录了长达十多个百万年的高速率的沉降。虽然侵入体或者深成侵入体顶部扩张的影响（Tobisch等，1986）很难与大陆边缘大规模的扩张区别开来，但是前者形成的盆地应该是在更短的时间内形成（小于几个百万年），而且也不能接受岛弧外的沉积物。大陆火山形成的弧内盆地，规模小（10~60km宽），沉积物厚（1~4km），且主要充填物为凝灰岩。此外，当火山口塌陷后，许多火山和沉积物在塌陷的火山口周围聚集（Riggs和Busby-Spera，1991；Lipman，1992；Schermer和Busby，1994）。Funk等（2009）对尼加拉瓜和萨尔瓦多共和国境内，新生代弧内盆地的构造活动进行了详细的研究。

1.6.6　弧后盆地

存在两种弧后盆地：(1) 大洋岩浆弧之后的大洋型弧后盆地；(2) 在缺乏前陆褶皱冲断带的大陆边缘弧之后的大陆型弧后盆地（Ingersoll和Busby，1995）。许多弧后盆地是由裂谷或者海底扩张形成，初期是伸展性的盆地（图1-4a）（Marsaglia，1995）。这些盆地一般起源于岛弧裂谷，并且沿着岛弧中轴（弧间）分布，或者迅速移动到中轴的前部或者后部。"弧间盆地"（Karig，1970）这个词已经广泛地被"弧后盆地"这个词取代，但是仍然使用这一名词来描述沿着裂谷分布或者在岛弧中轴附近，弧后盆地之后发育的残余岛弧。这个残余岛弧的存在或者保存并不是区分弧后盆地的必要条件（Taylor和Karner，1983）。

许多弧后盆地不是伸展性的（Marsaglia，1995），而是形成于中性应力背景下（图1-4b）。最常见的非伸展性弧后盆地存在于古大洋盆地中，其形成于板块重组时期（例如白令海峡）。非伸展性弧后盆地也可以发育为大陆地壳（1-4c）（例如印度尼西亚的桑达大陆架）。在上浮地壳的早期碰撞过程中，大洋岛弧—海沟体系的卷入可以导致弧后地区的收缩（例如大安的列斯群岛和印度尼西亚东部）（TenBrink等，2009）；这种弧后收缩代表了两极倒转期间，俯冲作用处于初级阶段。

1.6.6.1　大洋型弧后盆地

现代大洋型弧后盆地可以通过大洋盆地岩石学特征，或者它们在活动还是不活动的岛弧—海沟体系的位置，来与其他的大洋盆地区分开（Taylor和Karner，1983；Marsaglia，1995）。这样的辨别特征通常不适用于古代的大洋型弧后盆地，因为这些盆地现今一般位于造山区域并经历了变质作用和构造变形作用，例如形成了蛇绿岩。蛇绿岩剖面顶部沉积物的特征和年代学特征，是鉴定板块初始构造背景的重要标志（Tanner和Rex，1979；Hopson等，1981，2008；Kimbrough，1984；Busby-Spera，1988a；Robertson，1989）。

在弧后火山碎屑沉积平原及其基底的研究中，以墨西哥的中侏罗世岩体最为详细（Busby-Spera，1987，1988a；Critelli等，2002）。研究结果与Karig和Moore（1975）关于大洋型弧后盆地与陆源沉积物的汇入相隔绝的观点一致，具有以下简单的、统一的沉积组合：(1) 横向上和垂向上岩相不同，这是因为厚层的火山碎屑沉积平原进积到宽广的弧后盆地；这样的沉积平原可以从火山岛弧向外延伸100多千米，并且在5Ma之内加积到5km厚（Lonsdale，1975）；(2) 沉积平原的上覆层是薄层的泥岩和砂岩，这些泥岩和砂岩主要是火山活动或者扩张活动停止后，岛弧被侵蚀而产生的物质。这种循环反映了海底扩张在大洋型弧后盆地具有短期阶段性活动的特点，这些沉积物是在10~15Ma或者更短的时间内形成的

(Taylor 和 Karner，1983)。这样所导致的结果就是，伸展性的大洋型弧后盆地的演化周期一般比弧内盆地短（图1-2）。演化周期短表明伸展性的大洋型弧后盆地具有短期阶段性的特点，这种盆地是弧后盆地中最常见的类型。相比较而言，尤其是在大陆型的岛弧中，每十多个百万年，岛弧就会经历一次阶段性的伸展。虽然弧后盆地以及盆地中的沉积物对造山带的形成有很大贡献，但是大多数的古代大洋型弧后盆地已经被俯冲消减；弧后盆地的弧前一侧会被部分地保存在地质记录中（Busby-Spera，1988a）。

Marsaglia（1995）论述了现代和古代的、大洋型和大陆型的以及伸展性和中性弧后盆地的特点，但是她的研究重点是西太平洋的伸展性弧后盆地（还可以参考Klein，1985）。最近更多的出版物总结了地中海西部复杂的伸展性弧后盆地的发育和演化（Maillard和Mauffert，1999；Pascucci等，1999；Mattei等，2002；Rollet等，2002）。Sibuet等（1998）总结了冲绳海槽的构造和岩浆演化过程，Critelli等（2002）分析了塞德罗斯岛的侏罗纪弧后盆地，Baja California利用砂岩岩石学特征，结合地层学和沉积岩石学对大洋型弧后盆地进行了研究。很少有人关注非伸展性弧后盆地（例如白令海峡的阿留申盆地）。这些非伸展性弧后盆地是洋内俯冲带开始活动后，大洋岛弧后侧的大洋地壳被捕获而形成的盆地（Ben-Avraham和Uyeda，1983；Tamaki和Honza，1991）。加勒比海的一部分、菲律宾盆地的西部、鄂霍次克海的一部分以及千岛群岛盆地的北部同样也存在大洋地壳演化为弧后盆地的过程（Uyeda和Ben-Avraham，1972；Scholl等，1975；Ben-Avraham和Uyeda，1983；Marsaglia，1995）。当它们演化为休眠大洋盆地时，这些弧后盆地一般比图1-2中指出的盆地具有更长的演化周期和更好的保存潜力。弧后背景下捕获的洋壳和其他洋壳一样复杂，含有海底高原、大陆碎块和转换断层（Marsaglia，1995）。

1.6.6.2 大陆型弧后盆地

印度尼西亚的桑达陆架是现代中性应力下典型的大陆型弧后盆地（Hamilton，1979；Ingersoll，1988；Ingersoll和Busby，1995）。DeCelles和Giles（1996）认为桑达陆架是弧后前陆盆地的最早期发展阶段，但是Moss和McCarthy（1997）否定了这种解释和意见，他们认为，在印度尼西亚的岩浆岛弧之后，没有发育弧后的缩短作用（DeCelles和Giles也是这样认为的，1997）。Moss和McCarthy（1997）认为，之前在桑达陆架的部分区域，存在着伸展作用。无论如何，伸展性弧后盆地可以演化为中性弧后盆地，而中性弧后盆地又可以演化为弧后前陆盆地。在后弧或者弧后构造背景下压力环境经常发生变化。

在美国西部有一系列相似的中生代后弧—弧后前陆盆地发育（Dickinson，1981a，1981b；Lawton，1994；Ingersoll，1997，2008a）。在二叠—三叠纪的索诺玛造山运动中，俯冲作用开始后，形成了一个大陆边缘岩浆弧（Hamilton，1969；Schweickert，1976，1978；Busby-Spera，1988b）。虽然在后弧的一些地区，有地壳伸展作用（Wyld，2000，2002），但是在中性后弧的动力背景下，在中三叠世到晚侏罗世期间，浅海相和非海相沉积占主导地位（Dickinson，1981a，1981b；Lawton，1994；Ingersoll，1997，2008a）。在侏罗纪，随着地壳缩短作用的开始，后弧地区逐渐演化为弧后前陆（Oldow，1984；Oldow等，1989；Lawton，1994；Wyld，2002）。因此，桑达陆架区域的伸展作用、中性作用、缩短作用的相对时间与美国西部中生代弧后—后前陆中，解释的层序事件相似。

1.6.7 后前陆盆地

由于部分大陆地壳俯冲于岛弧造山带之下，使得受到挤压的岛弧—海沟体系演化为弧后前陆盆地（Dickinson，1974b；Dewey，1980；Ingersoll，1988；Ingersoll 和 Busby，1995；DeCelles 和 Giles，1996）。"前陆盆地"是板块构造提出之前的一个术语，用来描述一个位于造山带和克拉通之间的盆地（Allen 等，1986）。通过对比形成于大陆碰撞时，俯冲板块之上的"周缘"前陆盆地，Dickinson（1974b）提出，使用"弧后的"（Retroarc）这个词用于描述形成于受挤压弧后前陆盆地。这样，虽然"后弧"（Backarc）与"弧后"（Retroarc）字面上的意思相同，但是前者用于伸展性和中性岛弧—海沟体系，后者用于挤压性岛弧—海沟体系。

Willett 等（1993）、Johnson 和 Beaumont（1995）、Beaumont 等（1996）以及 Naylor 和 Sinclair（2008）改进了 Dickinson（1974b）最初的命名，将前陆盆地中"弧后前陆"简化为"后前陆"，同时将"周缘前陆"修改为"前前陆"。本书采用了这一命名法，该方法很清楚地规定了后前陆盆地形成于会聚型板块边缘处的上覆板块，而前前陆盆地则形成于下伏板块之上。后前陆一般比前前陆有着更长的演化历史，因为前者一般在大洋地壳开始沉降时就已经开始发育（例如安第斯山脉的后前陆），而前前陆直到上覆的陆壳进入俯冲带时才开始发育（例如诱发的碰撞）（Dickinson，1974b；Ingersoll，1988；Cloos，1993；Ingersoll 和 Busby，1995）。为了更好地解释这些区别，笔者认为后前陆应该被细分为弧后前陆（形成于大陆边缘岛弧之后，例如安第斯山脉）和碰撞后前陆（在大陆碰撞时期形成于上覆的大陆板块上，例如南阿尔卑斯前陆盆地）。无论是否有更多限制性的词，用于说明陆壳或者洋壳在造山运动中是否发生俯冲，一般情况下"后前陆"这个名词可用于任何会聚板块边缘处，上覆板块上发育的前陆。大约三分之一的活动型岩浆岛弧与弧后前陆有关，而岛弧活动一般终止于前前陆和碰撞后前陆发育的地区。

1.6.7.1 弧后前陆盆地

Jordan（1981）分析了与爱达荷—怀俄明逆冲带伴生的不对称型的白垩纪弧后前陆盆地。他用一个二维弹性模型模拟了在逆冲断层负载和沉积物负载的作用下，宽缓的岩石圈是如何发生挠曲的（图 1-4e）。随着逆冲作用向东迁移，最大挠曲点也不断向东移动。由于逆冲断层负载所引起的侵蚀作用和沉积作用出现局部不均衡，导致沉降区域变宽，此外晚白垩世海平面的升降使得沉降区域变得更宽。通过对比盆地模型和盆地基底等厚线图的几何形态，可以检测出岩石圈挠曲刚度的数值。现今安第斯山脉内逆冲带和前陆盆地的特征，与白垩纪爱达荷—怀俄明盆地系统的特征相似（Jordan，1995）。地形受到逆冲断层几何学特征和重力均衡沉降的控制。

Jordan（1981）和 Beaumont（1981）提出的模型也适用于其他的弧后前陆盆地（Jordan，1995；DeCelles 和 Giles，1996；Catuneanu，2004）。这些模型或者衍生的模型证明了，前陆褶皱冲断带的构造活动是前陆盆地发生沉降的主要原因（Price，1973）。沉积作用的重建、沉积过程的循环、软流圈的动力学影响（Curnis，1993；Burgess 等，1997）以及海平面升降，是控制海退—海进的层序重要因素，但位于岛弧—海沟体系之后挤压构造活动才是真正的驱动力。白垩纪位于美国北部的海槽，就主要是由于挤压构造活动形成的（与高频海平面变化共同作用）（Dickinson，1976b，1981a）。有关爱达荷—怀俄明冲断带的逆冲时间和对逆冲作用的初始沉积响应细节已经被讨论过（Heller 等，1986），此外，挤压构造活动在形成弧后前陆盆地中的基本作用已经明确（Price，1973；DeCelles 和 Giles，1996）。

Jordan（1995）修改了他对白垩纪北美弧后前陆的认识，结合对美国南部新近纪到全新世弧后前陆的分析，论述了弧后前陆盆地的通用模型。DeCelles 和 Giles（1996）综合分析了前陆盆地系统，并将前陆盆地系统细分为四个不同的沉积带：逆冲楔顶部带、前渊带、前隆带以及隆外凹陷带。这四种沉积带在现今安第斯山脉东部中心的弧后前陆中都有发育（Horton和DeCelles，1997）。DeCelles和Horton（2003）运用这个模型解释了玻利维亚古近—新近纪前陆的地层特征，并推断大约前陆地壳已经向西俯冲了1000km，最终下插到安第斯造山带之下。Fildani等（2003）、Abascal（2005）、Gomez等（2005）、Hermoza等（2005）、Horton（2005）以及Uba等（2005）详细分析了安第斯山脉弧后前陆系统的各个不同部分。

Decelles和Giles（1996）的次级分类方法同样可以运用到所有类型的前陆盆地系统中。因为逆冲楔顶部带的性质与断层的动力学特征有直接的联系，所以，我们在后面要进行详细分析。其他三种构造带由逆断层带上部挠曲载荷形成。大多数前陆演化模型中未考虑逆冲楔顶部带和隆外凹陷带的演化（Decelles和Giles，1996）。含有逆冲楔顶部带的前陆模型必须在横截面上使用双倍的楔形而不能简化为单楔形（Decelles和Giles，1996）。Dorobek和Ross（1995）分析了许多现今模型和实例的特点，提高了我们对前陆盆地的理解。

1.6.7.2 碰撞后前陆盆地

南阿尔卑斯山脉中碰撞后前陆盆地的形成与欧洲板块俯冲于阿德里亚板块之下形成阿尔卑斯造山带是同步的（Bertotti等，1998；Carrapa，2009）（图1-4f）。晚白垩世，这个前陆在弧后背景下形成，最终演化为与阿尔卑斯造山带和北阿尔卑斯（磨拉石）前前陆盆地同期发育的碰撞后前陆盆地（Bertotti等，1998）。Bertotti等（1998）指出，随着时间的推移，阿德里亚板块的挠曲作用逐渐减弱。Carrapa和Carcia-Castellanos（2005）的研究表明波河平原西部，古近—新近纪的山前盆地是由渐新世—中新世，位于黏弹性板块上的阿尔卑斯后前陆挠曲形成的。Zattin等（2003）运用威尼斯盆地物源区的数据来研究南阿尔卑斯碰撞后前陆盆地的变形次序。亚平宁山脉的造山运动使得波河谷前前陆盆地上覆于老的后前陆盆地之上，这就使其成为了一个混合型前陆盆地（Ingersoll和Busby，1995；Miall，1995）。

中国中部鄂尔多斯盆地内，三叠—侏罗系的前陆盆地层序是华北板块和扬子板块拼合在一起的过程中，所形成的碰撞后前陆盆地沉积（Sitian等，1995；Ritts等，2009）。

1.6.7.3 破碎后前陆盆地

挤压的岛弧—海沟体系之下的低角度的俯冲，可能会导致弧后前陆盆地内的基底卷入构造变形（图1-4d）（Dickinson和Snyder，1978；Jordan，1995）。这种变形最著名的古代例子是美国西部的落基山脉区域；在现今的安第斯山脉前陆也找到了相似的变形特征（Jordan等，1983a，1983b；Jordan和Allmendinger，1986；Jordan，1995）。

Chapin和Cather（1981）、Dickinson等（1988，1990）、Cather和Chapin（1990）、Dickinson（1990）、Hansen（1990）以及Lawton（2008）综述了晚白垩世至始新世（拉腊米）的科罗拉多台地和落基山脉内，控制沉积作用和盆地形成的因素。他们都认为在这一时期，形成了各种类型的隆起和盆地，但是他们在古水系、科罗拉多台地东部边缘走向滑动变形的相对重要性，以及拉腊米造山运动是形成于两个不同的阶段还是同一应力背景下连续性的构造响应，存在争议。Yin和Ingersoll（1997）和Ingersoll（2001）建立了拉腊米地区地壳应力和新墨西哥州与南科罗拉多州盆地演化的模型，结果表明它们处于同一应力背景下。Hoy和Ridgway（1997）研究了怀俄明州内前陆隆起边缘的复合构造、地层学和沉积学之间的关系。Cardozo和Jordan（2001）、Davila和Astini（2003）、Sobel和Strecker（2003）以及Hilley和Strecker（2005）研究了阿根廷的破碎后前陆盆地及与之共生的隆起。

1.6.8 残余大洋盆地

当地壳俯冲到上覆（未沉降的）大陆或者岩浆岛弧地壳之下时就形成了强烈变形的缝合带（Cloos，1993）。缝合带可以包括裂开的大陆边缘和大陆边缘岩浆弧拼接（闭合的大洋盆地），或者岛弧和大陆边缘拼接等的多种类型（图1-4e、f）。大陆间的碰撞是不规律的，沿着缝合线的走向，碰撞时间、构造变形、沉积物扩散样式及保存特征都具有很大的差异（Dewey和Burke，1974）。

Graham等（1975）和Ingersoll等（1995，2003）用新生代的喜马拉雅—孟加拉系统的演化来类比晚古生代阿巴拉契亚—沃希托系统的演化，据此提出了与连续型缝合造山带相关的通用沉积扩散模型。"大多数的沉积物是从大陆碰撞形成的造山高地中搬运下来的，然后沿着三角洲复合体的轴向进入到残余洋盆地，形成浊积岩并发生后续变形，最后随着碰撞缝合带的伸长，卷入造山带内"（Graham等，1975，第273页）。虽然许多"复理石"和"磨拉石"沉积是在不同的构造背景下形成的（Ingersoll等，1995，2003；Miall，1995），但是在本模型中，复理石和磨拉石沉积，是在与构造缝合带共生的造山运动中同时形成的。

美国北部岛弧—大陆碰撞的实例中，存在各种规模的残余洋盆复理石，包括奥陶纪的阿巴拉契亚造山带（Rowley和Kidd，1981；Stanley和Ratcliffe，1985；Lash，1988；Bradley，1989；Bradley和Kidd，1991）和泥盆纪—密西西比纪在滨岸山脉的安特勒造山带（Speed和Sleep，1982；Dickinson等，1983）。在这两个实例中，很难根据"复理石"的沉积环境来明显地区分（Ingersoll和Busby，1995；Ingersoll等，1995；Miall，1995）残余大洋盆地和初期前前陆盆地。

Ingersoll等（1995，2003）回顾了古代的和现今存在的若干个残余洋盆，研究表明残余大洋盆地中的海底扇是地球上沉积厚度最大的地区。孟加拉海底扇是现今的最大的沉积体，印度河海底扇是第二大沉积体；两者都起源于目前地球上最大的抬升区域，西藏地台和喜马拉雅山脉。西藏北部三叠纪松潘—甘孜的复合体以及阿肯色州、俄克拉何马州和得克萨斯州的石炭—二叠纪的沃希托河—马拉松的复理石也沉积在残余洋盆地中，与它们相邻的都是被抬升了的大陆缝合带。虽然这些沉积体在大小上可与孟加拉海底扇和印度河海底扇相比，但是由于在缝合期间变形强烈，导致它们很难被重建（Ingersoll等，2003）。没有已知的沉积机制可以产生如此大的沉积体。此外，Ingersoll等也讨论了（2003）伴随大陆—大陆或者大陆—大洋—岛弧的碰撞的一些的残余洋盆。

1.6.9 前前陆盆地

当大陆碰撞发生在裂开的大陆边缘与岛弧—海沟体系内的俯冲带之间时，在裂谷边缘就会发生构造负载，然后下沉到海平面以下，而后又上升至地表（Dickinson，1974b；Ingersoll，1988；Maill，1995）。侵位动力载荷之下，弹性地壳的弯曲，可以形成前前陆盆地（图1-4f）。当前陆上发生动力载荷侵位时，发生在动力载荷和前缘隆起前部的正断层作用，是对挠曲的初始响应（Bradley和Kidd，1991；Maill，1995；Decelles和Giles，1996）。

虽然区分古代前前陆盆地和碰撞后前陆盆地（图1-4f）有一定难度，但是可以根据以下的特征区分（Ingersoll，1988；ingersoll和Busby，1995）：（1）岩浆岛弧的极性；（2）与早前前陆初始阶段共生的大洋俯冲杂岩的存在；（3）前前陆盆地的水深更大（前渊阶段）；（4）缝合带的不对称型（离前前陆盆地更近）；（5）后前陆盆地发育时间长（岛弧演化时间长），而前前陆盆地发育具有不连续性（没有前兆的大洋边缘闭合）；（6）火山碎屑可能进入后前陆，特别是在前陆发育早期，而前前陆盆地极少有火山碎屑搬运进来。

Stockmal 等（1986）设计了一个前前陆盆地形成的二维动力学模型，该模型将盆地的演化划分为几个有限的裂谷作用阶段。该模型修正了 Speed 和 Sleep（1982）的模型，论述了裂谷边缘的年龄和地形对地壳挠曲和盆地发育的影响。在古大陆边缘（120Ma）发生俯冲的早期阶段，年龄的影响主要表现在，时代越老的大陆边缘，其前隆挠曲越高以及海沟充填也较厚。随后的演化就不受大陆边缘年龄的影响了。前陆盆地的沉降对逆冲断层的负载很敏感，其负载深度可能超过 10km。在挤压阶段，地壳的厚度可以达到 70km（例如喜马拉雅山脉）。据推测，在外来岩体和近端的前陆盆地的变形过程中或变形后，发生了十多千米的抬升和侵蚀。大多数侵蚀碎屑没有沉积于前陆内的隆起之上；而是发生垂向搬运，进入残余大洋盆地（Graham 等，1975；Ingersoll 等，1995，2003；Miall，1995）。在宽广的、逐渐减弱的大陆边缘裂谷中，发育的具有低地形的厚层逆冲断层，被拉入俯冲带内（Stockmal 等，1986）。

Maill（1995）总结了所有"与碰撞有关的前陆盆地"，其包括了前前陆盆地和碰撞后前陆盆地。Dorobek 和 Ross（1995）以及 Mascle 等（1998）提供了一些关于前陆盆地的研究成果。关于前陆盆地研究的文献数量迅速增加，包括比利牛斯山（Arenas 等，2001；Jones 等，2004）、阿尔卑斯山（Sinclair，1997；Gupta 和 Allen，2000；Allen 等，2001；Pfiffner 等，2002；Kempf 和 Pfiffner，2004）、亚平宁山（Bertotti 等，2001；Lucente，2004）、喀尔巴阡山（Zoetemeijer 等，1999；Tarapoanca 等，2004；Leever 等，2006）、扎格罗斯山（Alavi，2004）、喜马拉雅山（Pivnik 和 Wells，1996；DeCelles 等，1998，2001；Najman 和 Garzanti，2000；Najman 等，2004）、龙门山（Yong 等，2003；Meng 等，2005）、中国台湾西部（Chen 等，2001；Lin 和 Watts，2002）、巴布亚岛（Galewsky 等，1996；Haddad 和 Watts，1999）、阿巴拉契亚山（Thomas，1995；Castle，2001）以及元古宙前陆的研究（Saylor，2003）。

1.6.10 楔形顶部盆地

Ori 和 Friend（1984）将"背驮型盆地"定义为逆冲岩席移动过程中形成和充填的盆地。Decelles 和 Giles（1996）认为"逆冲楔顶部"是一个简化的描述性词语，包括了"背驮型盆盆地"和"逆冲楔顶部盆地"两部分（图 1-4f）。楔形顶部盆地内沉积物的聚集发生在动态的大地构造背景下，其沉积物大多数来自于相关的褶皱冲断带，还有一部分来自于岛弧和基底隆起（Critelli 和 Le Pera，1994；Trop 和 Ridgway，1997）。褶皱冲断带可以在前前陆、后前陆或者走滑挤压背景下发育（Ingersoll 和 Busby，1995）。楔形顶部盆地与海沟斜坡盆地有着共同的特征。中国台湾南部的海底特征表明了吕宋岛弧西部的弧前—海沟斜坡—海沟向台湾碰撞带内的造山楔—楔形顶部—前渊过渡（Chiang 等，2004）。这种过渡发生在吕宋岛弧之下的台湾缝合带上。这个缝合带是洋壳试图向亚洲大陆地壳俯冲而形成的。由于逆冲断层带的不断发育，使得海沟斜坡盆地和楔形顶部盆地的保存潜力较低，因此，它们一般只存在于年轻的造山系统中（Burbank 和 Tahirkheli，1985）（图 1-2）。

Jordan（1995）、Miall（1995）、Nilsen 和 Sylvester（1995）以及 Talling 等（1995）对楔形顶部盆地及挤压过程的相互关系提出了另一种认识。正如前缘断层向前陆盆地不断推进一样，前陆沉积也可以向前推进，演变成楔形顶部沉积（DeCelles 和 Giles，1996；Pivnik 和

Khan，1996）；由于摩擦系数低，在逆冲前缘会发生快速的"雪橇运移"，这时，在各种楔形顶部盆地中都可能发育前陆沉积（Evans 和 Elliott，1999；Ford，2004）。Horton（1998）的研究表明，位于玻利维亚南部的科迪勒拉山脉东部地区，从渐新世到晚中新世晚期发育有同造山期的楔形顶部沉积。他认为这些沉积物的长期保存（大约 30Ma）可以反映这些地区处于半干旱气候；它们的存在增加了造山楔的质量，反过来促进了逆冲断层前缘的扩展（Horton，1998）。楔形顶部盆地可以形成于薄皮和厚皮（基底拆离）的挤压环境下（Casas-Sainz 等，2000）。

1.6.11 后陆盆地

Horton（本书第 21 章）描述了两类后陆盆地：非碰撞型的弧后造山带形成的盆地（例如安第斯山脉）和碰撞型造山带形成的盆地（例如喜马拉雅与西藏高原）。由于"后陆"是指造山带"内部"的部分，与褶皱和断层的倾向方向相反，后陆这个词是相对于褶皱逆冲断层带的方向而言的。在字面上说，位于褶皱逆冲断带之后的造山带的所有部分组成了后陆，这样就忽略了它最初的起源；例如，在双向造山带中，后前陆是前前陆盆地中后陆的一部分（阿尔卑斯山和比利牛斯山）。Horton（本书第 21 章）和笔者定义的"后陆盆地"，都是限定在造山带内的盆地，而不是其他文章中所说的后陆盆地。

通常在高海拔地区，后陆盆地内通常发育非海相沉积，并形成较厚的陆壳（Horton，本书第 21 章）（图 1-4d）。尽管一些盆地的形成跨越了数千万年（例如阿尔蒂普拉高原；Horton 等，2002），但是它们能保存下来的可能性很小，而且形成后，稳定存在时间也相当短（图 1-2）。伸展、挤压、走滑过程中，伴随断层作用而形成的地壳减薄，沉积作用和火山活动以及构造活动引起的沉降（图 1-1），都可以为后陆盆地创造可容空间。Horton（本书第 21 章）描述了两种后陆盆地的演化模式：（1）由于断层的重新活动而形成的盆地；（2）叠覆在早期前陆盆地之上的变形盆地。

Burchfiel 等（1992）、Garzione 等（2003）、Alcicek（2007）、Decelles 等（2007）、Giovanni 等（2010）以及 Saylor 等（2010）论述了内陆盆地的另外一些实例。

1.7 转换型

1.7.1 走滑体系

与走滑断层有关的沉积盆地的复杂性和多变性，不亚于其他类型的盆地（Ingersoll 和 Busby，1995）。洋壳上转换断层的表现形式，主要取决于板块构造的模型；而陆壳上的走滑断层则过于复杂，所以很难有合适的刚性板块模型来与之匹配。

由于大型地壳级断层滑动方向的调整（Crowell，1974a，1974b；Reading，1980），使得大陆地壳上的走滑断层很可能经历了扩张与挤压的转变。这样，走滑断层周围盆地的打开和闭合，可以类比于大洋盆地打开和闭合（威尔逊旋回），只是空间和时间尺度较小（Wilson，1966；Dewey 和 Burke，1974）。南加利福尼亚州，新近纪到全新世的演化完美地诠释了这一过程（Crowell，1974a，1974b；Schneider 等，1996；Ingersoll 和 Rumelhart，1999；Kellogg 和 Minor，2005；Ingersoll，2008b）。

尽管大多数盆地是混合型的，但是与走滑断层相关的盆地可以归结为几个端元类型。走滑伸展（包括张裂）型盆地形成于张应力背景下，而走滑挤压盆地形成于压应力的背景下（Crowell，1974b）。在旋转板块内，垂直于轴线并且与地壳旋转有关联的盆地（"转换旋转"Ingersoll，1988）经历了伸展、挤压、走滑等构造作用（Ingersoll 和 Busby，1995）。

Christie-Blick 和 Biddle（1985）以及 Nilsen 和 Sylyester（1995）认为走滑盆地的构造和地层研究，在很大程度上是在 Crowell（1974a，1974b）先前工作的基础之上发展起来的。他们阐述了走滑断层周围构造的复杂性和可能出现的盆地类型。控制构造样式的主要因素是：（1）相邻板块会聚和离散的程度；（2）位移大小；（3）变形岩石的物性；（4）先存构造（Christie-Blick 和 Biddle，1985）。挤压和沉积物负载导致伸展作用和挠曲负载，在这一过程中或之后发生的地壳减薄，热力学会造成沉积盆地发生沉降。沉降非常迅速，但是由于横向的热力学条件较差，所以在相对于狭窄的走滑伸展盆地内，其总沉降量明显小于长轴的垂直裂谷。各种与走滑断层相关的沉积盆地，在以下几个方面存在差异（Christie-blick 和 Biddle，1985）：（1）整体的不协调性；（2）纵向和横向盆地的不对称性；（3）阶段性的快速沉降；（4）横向上沉积相的突然改变和局部不整合；（5）在同一区域不同盆地中地层、地貌和不整合具有明显差异。

以上这些走滑体系的特点已经被古代的和现今的、国内的和国外的断层体系研究所证实（Barnes 等，2001，2005；Koukouveslas 和 Aydin，2002；Hsiao 等，2004；Okay 等，2004；Seeber 等，2004；Wakabayashi 等，2004）。

1.7.2 走滑伸展盆地

走滑伸展盆地（图 1-5a）形成于左行左旋式或右行右旋的断层结合处（Crowell，1974a，1974b；Reading，1980；Christie-Blick 和 Biddle，1985；Nislsen 和 Sylvester，1995）。在比较了走滑伸展盆地不同阶段的特征之后，Mann 等（1983）提出了一种模式来研究这些盆地。走滑伸展盆地的演化包括以下五个阶段：（1）在初始的主断层弯曲释放处发生伸展断层作用；（2）形成纺锤形盆地，通常被斜滑断层所分开；（3）进一步拉伸，产生缓 S 形或者缓 Z 形盆地；（4）发展成为菱形断陷，常伴有两个或更多的次圆状深渊；（5）继续伸展，在被转换断层错开的扩展中心处，形成狭长的洋壳。在 3 到 5 阶段玄武质火山作用和侵入作用变得重要（Crowell，1974b）。由于走滑伸展盆地形成的时间较短，而且走滑方向多变，所以大多数拉分盆地的长宽比很低（Mann 等，1983）。Mann（1997）论述了正常情况下，构造逃逸区内大型走滑伸展盆地的形成过程。长期活动的走滑伸展板块边缘可能演化为走滑伸展型的初始大洋盆地（例如加利福利亚海湾）或者板块内部的转化大陆边缘（例如非洲西部的南海岸）。

物理模拟模型为走滑拉分盆地的形成和发展研究提供了重要依据（Dooley 和 Mcclay，1997；Rahe 等，1998）。前人已经利用综合地球物理和地质方法，对一些年轻的走滑伸展盆地和断裂区，进行了地表和海底环境研究，例如新西兰的 Hope 断裂（Wood 等，1994），死海的转换断层（Katzman 等，1995；Hurwitz 等，2002；Lazar 等，2006）以及马尔马拉海的安那托利亚北部断裂（Okay 等，1999；Rangin 等，2004）。Dorsey 等（1995）讨论了加利福利亚海湾地区走滑伸展断层边缘，断裂控制的快速沉降对扇形三角洲沉积的影响。Waldron（2004）论述了新斯科舍地区走滑伸展盆地的中宾夕法尼亚纪斯泰勒顿地层有多期叠置的复杂演化史。这个古走滑伸展盆地的整体构造和地层发育特征与 Mann 等（1983）、Dooley 和 Mcclay（1977）以及 Rahe 等（1998）建立的模型一致。

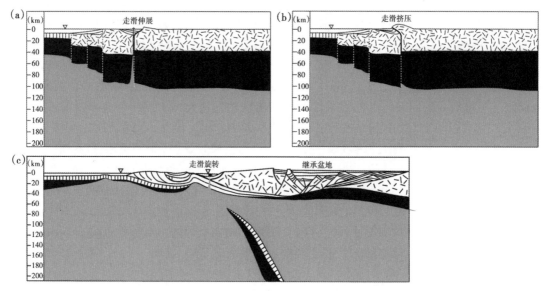

图 1-5 在变质和混杂背景下真实的沉积盆地模型

这些剖面是根据中新世到全新世的南加利福尼亚州演化过程绘制的（Ingersoll, 2008b）；在其他的构造背景下，走滑伸展盆地和走滑挤压盆地都比较普遍，但走滑旋转盆地比较少见；(c) 中描述了现今地球上，比较独特的三大板块的相互作用。构造活动终止后，可能会形成沉积盆地，(c) 中显示了美国盆岭省南部盆地，未变形的新近系至全新统上超到前期伸展背景下倾斜的断块和地层之上；图例与图 1-3 相同

1.7.3　走滑挤压盆地

走滑挤压盆地（图 1-5b）包括两种类型：(1) 由构造负载引起的弯曲沉降型盆地，这些盆地位于强烈挤压弯曲带周缘，变形强烈并发育在逆冲断层的边缘地区（例如北洛杉矶盆地，南加利福尼亚；Schneider 等，1996）；(2) 中等挤压弯曲形成的断层楔形盆地，即单个块体向挤压弯曲带运动，就会导致一个或两个边缘地区的迅速抬升，同时使盆地发生快速沉降（例如新近纪的南加利福尼亚中脊盆地）（Crowell, 1974b, 2003a, 2003b）。类型一的盆地模型中发育了与之前提到的前陆模型相似的挠曲负载，但其规模要小一些。

大洋中脊盆地是全世界出露最全面，研究最详尽的走滑挤压盆地之一（Crowell 和 Link, 1982; Crowell, 2003a）。Crowell（2003b）建立了一个大洋中脊盆地演化的动态模型（5~12Ma）。该模型中，位于盆地东北部西南走向的圣加百利断层和西北走向的一系列断裂之间，发育了一个狭窄的地壳板片。当旋转量完全转化为现代的圣安德烈亚斯断层后，洋脊盆地就停止活动了（Crowell 和 Link, 1982; Ingersoll 和 Rumelbart, 1999; Crowell, 2003）。圣加百利断层活动的最终结果是盆地西南部被抬升，同时在盆地边缘发育维奥林角砾岩。盆地内发生沉降的地层，移动到隆起边缘处，并开始接受东北部物源区的大量沉积。在接受了来自隆起和倾斜地层所形成的沉积之后，早期沉积中心开始越过弯曲中心向东南方向移动。虽然露头上出露的地层厚度超过 11km，但是盆地充填的垂向厚度大约只有它的三分之一。在类似的大地构造背景下，断层限定的古狭窄盆地中，会发育异常厚的粗碎屑岩（Ingersoll 和 Busby, 1995; 讨论了 May 等, 1993 年发表文章；May 在该文中否定了洋脊盆地演化的走滑伸展模型）。

McClay 和 Bonora（2001）提出了限制性叠置区的模型。目前学者们对一些海底和陆上较新的走滑挤压构造进行了详细的研究，比如新西兰的阿尔卑斯断层（Norris 和 Cooper，1995；Barnes 等，2005）；日本的神户和北大阪盆地（Itoh 等，2000）；西班牙的西瓦奥盆地（Erikson 等，1998）；委内瑞拉的马图林前陆盆地（Jacome 等，2003）和南弗兰克盆地（Bry 等，2004）。Trop 等（2004）论述了渐新世期间，沿着阿拉斯加德纳里断层的科罗拉多河谷盆地走滑挤压的成因。Meng 等（2005）论述了四川盆地西南部（华南板块）是如何从中生代的前前陆盆地开始演化。

1.7.4　走滑旋转盆地

来自加利福尼亚州南部的古地磁数据显示，从中新世至今，数个地块上广泛地发生过地磁的顺时针旋转现象（局部地方超过90°）（Luyendyk 等，1980；Hornafius 等，1986；Luyendyk，1991）。Luyendyk 和 Hornafius（1987）建立了该地区的几何模型，从而根据试验推断了旋转和非旋转地块边界断层的滑移量和方向、地块间裂开（盆地）及相互叠覆（逆掩断层）区域的范围。Dickinson（1996）沿着圣安德列斯转换断层系统，计算了由加利福尼亚州南部的走滑旋转构造作用所形成的总滑移量。这一结论有助于解释太平洋—北美板块运动的差异，同时为解释北美大陆边缘内部具有偏移的现象提供了前提（Dickinson 和 Wernicke，1997）。

Nicholson 等（1994）建立了一个微板块捕捉模型，用以解释北美板块、太平洋板块和法拉隆板块之间的复杂相互作用。三个板块开始作用后的30Ma 内（Atwater，1970，1989；Bohannon 和 Parsons，1995），加利福尼亚州南部海岸向太平洋板块之上的转换，具有明显的三段性的原因。第一阶段（12~18Ma）导致了垂直轴向的快速顺时针旋转，并在复杂拆离盆地内形成了托潘加组（Ingersoll 和 Rumelhart，1999；Ingersoll，2008b；图1~5c）。Crouch 和 Suppe（1993）提出横断山脉的重新旋转，形成了大规模的、核杂岩式的伸展作用。加利福尼亚州的南部边陲和洛杉矶盆地底部发育有卡特琳娜片岩，Crouch 和 Suppe 将这种现象解释为下盘的变质构造岩，在滑脱层之下发生构造剥蚀。

模型要想成功地解释洛杉矶地区异常复杂的盆地成因，就需要结合 Luyendyk 和 Hornafius（1987）以及 Dickinson（1996）的走滑旋转模型、Crouch 和 Suppe（1993）的滑脱模型、Nicholson 等（1994）的微板块捕捉模型，并对洛杉矶和相关盆地进行详细的地层学、沉积学和构造史研究（Wright，1991；Ingersoll 和 Rumelhart，1999；Ingersoll，2008b）。

1.8　混杂及混合型背景

1.8.1　坳拉槽

在大陆裂谷期，形成的三条裂谷间的夹角，通常近似的为120°，这可能是最省力的组合（Burke 和 Dewey，1973）。不管其初始过程是"主动"还是"被动"的（Sengor 和 Burke，1978；Morgan 和 Baker，1983），在大多数情况下，大陆的分离阶段都会形成两条裂谷，而第三条则没有发生海底扩张，最终演变为古裂谷（Sengor，1995），Hoffman 等（1974）通过对一个元古宙盆地的研究，认为古裂谷最终也可以演化为沉积盆地。把 Athapuscow 裂陷槽分为五个发展阶段，然后将其稍加修改，就可以成为大多数坳拉槽（在高角度造山带附近发育的线性沉积槽）演化的模型了（图1-6a）：（1）裂开阶段；（2）过渡阶段；（3）坳陷阶段；（4）激活阶段；（5）造山阶段。

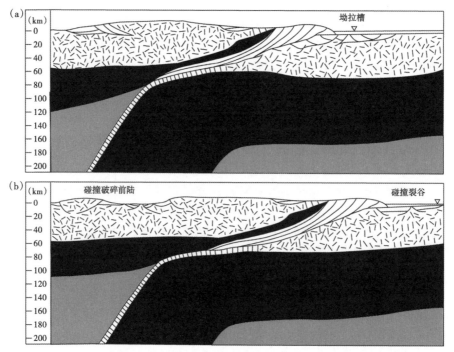

图1-6 陆—陆碰撞环境下真实的沉积盆地模型，在这种环境下可以形成混合盆地

横截面上所示的坳拉槽及碰撞裂谷并未表现出能体现其演化史的时间与空间关系（详见文章中的讨论）。位于板块内大陆边缘的古裂谷口处的三角洲，可以形成大陆堤沉积，而坳拉槽通常与这些沉积有关。因此，在图a中表现为巨厚的前碰撞地层。古断裂带的正断层通常在坳拉槽阶段重新活动（同碰撞期）。碰撞裂谷不存在前碰撞地层或者再活动构造。图b中右侧的碰撞裂谷类似于莱茵地堑前前陆地区的特征。左侧碰撞型破碎前陆可能包含一个远端的后—前陆碰撞裂谷，例如贝加尔断裂带。图例同图1-3

Sengor（1995）论述了形成古裂谷的各种机制，包括上拱作用、断裂作用、漂移作用（Hoffman等，1974）、薄膜应力、裂谷尖夭折和大陆旋转。所有以上的过程都可以形成"通过毗邻的大盆地的凹角，进入克拉通地块内的、狭长的，并且相当直的坳陷"（Shatsky，1964，引自Sengor，1995，P78）。

演化为洋盆的裂谷通常被新形成的大洋和大陆边缘上隆起的陆棚—斜坡所覆盖（图1-3c、d）。在这些地区，与大陆边缘毗邻的古裂谷演化为捕获了大陆内部大部分水系的凹陷；在这些凹陷中形成的主要三角洲，组成了大陆堤（例如尼日尔三角洲）（Dickinson，1974b；Ingersoll，1988；Ingersoll和Busby，1995）（图1-3f）。在大陆边缘发生活动或碰撞过程中，特别是在大陆海岬地方，裂开的大陆边缘沉积棱柱体发生强烈变形（Dewey和Burke，1974；Graham等，1975）。随着造山作用的进行，古裂谷演变为坳拉槽，期间可能会经历挤压、伸展或者平移变形。

Sengor等（1978）和Sengor（1995）提出了一个识别在大洋附近形成的，随后由于地壳碰撞（碰撞裂谷），然后关闭（坳拉槽）形成古裂谷的方法。这两种裂谷都与造山带呈高角度相交；坳拉槽的演化，与碰撞之前周围盆地的演化相一致；但是碰撞裂谷在碰撞前是不存在的。要辨别它们必须要有来自地层学的资料支持，因为初始裂谷的暂时性相互作用是检

验它们地球动力学起因的基本依据（Ingersoll 和 Busby，1995）。坳拉槽倾向于，在沿着大陆边缘开裂的凹角处形成（Dewey 和 Burke，1974），而碰撞裂谷则更容易在相反的海岬处形成，而且海岬是碰撞大陆变形最强烈的地区（Sengor，1976，1995）。但是由于前碰撞期的几何学很难进行重建（Thomas，1983，1985），所以使用这个辨认标准一定要谨慎。

1.8.2 碰撞裂谷

碰撞裂谷类似于坳拉槽（高角度造山带中的断裂），但是前造山阶段是不存在的（如图1-6b）。它通常形成于陆壳俯冲之前（在碰撞期与其他大陆或岛弧一同形成）。由于构造背景和样式的不同，目前有两个典型的实例，即中新生代的莱茵地堑以及晚新生代的贝加尔断裂。莱茵地堑内形成了一个类似于阿尔卑斯碰撞造山带的近端走滑伸展型碰撞裂谷（Sengor，1976）。这个碰撞裂谷形成于前前陆盆地背景下的板块俯冲带内（欧洲）。至今仍在活动的贝加尔断裂，同样也是一个走滑伸展型碰撞裂谷，但它处于喜马拉雅碰撞带的远端（Ingersoll 和 Busby，1995）。这是中亚碰撞破碎型前陆，上覆板块的一部分（图1-6b）。因此，它们是碰撞裂谷的端元形式：莱茵地堑形成于前前陆盆地的近端，而贝加尔断裂形成于后—前陆的远端。Sengor（1995）讨论了包括以上部分的许多实例。

1.8.3 碰撞破碎前陆盆地

不同形状和大小的大陆碰撞，通常会形成相当复杂的古造山带和相关沉积（Dewey 和 Burke，1974；Graham 等，1975，1993；Molnar 和 Tapponnier，1975；Sengor，1976，1995；Tapponnier 等，1982）。就如同 Tapponnier 等（1982）利用塑性黏土模型所证实的：印度板块和亚洲板块的碰撞产生了主要的陆内走滑断层及相应的前陆盆地、裂谷盆地、走滑伸展盆地、走滑挤压盆地及走滑旋转盆地（Graham 等，1993；Allen 等，1999；Yin 和 Harrison，2000；Howard 等，2003）。而所有这些盆地类型，都可能在前陆盆地（例如莱茵地堑）或者弧后盆地（贝加尔裂谷）的碰撞背景下形成。

关于碰撞破碎型前陆盆地及隆起的最佳古代实例是宾夕法尼亚纪—二叠纪的落基山造山带（Kluth 和 Coney，1981；Kluth，1986；Dickinson 和 Lawton，2003；Blakey，2008）。落基山脉的变形与劳亚古陆和冈瓦纳古陆的最终闭合是在晚石炭世至二叠纪同时发生的（Graham 等，1975；Kluth 和 Coney，1981；Kluth，1986；Ingersoll 等，1995，2003；Dickinson 和 Lawton，2003；Miall，2008）。通常，具有走滑挤压和走滑伸展性质的前陆和裂谷盆地，会分别被相邻的卷入基底的隆起所证实（Soreghan，1994；Geslin，1998；Hoy 和 Ridgway，2002；Barbeau，2003）。基底复活的特征决定了许多落基山脉内盆地和隆起的位置和特征。基底复活包括元古宙古裂谷的活化，最终形成坳拉槽（Sengor，1995；Marshak 等，2000；Dickinson 和 Lawton，2003；Blakey，2008；Miall，2008）。

其他的有关破碎型前陆盆地的研究包括晚二叠世北美大陆（Cannon，1995）的变形、晚古生代阿巴拉契亚的变形（McBride 和 Nelson，1999；Murphy 等，1999；Root 和 Onasch，1999）、晚古生代欧洲大陆的变形（Stollhofen 和 Stanistreet，1994；Mattern，2001；Vanbrabant 等，2002）以及中生代中亚大陆的变形（Sobel，1999；Vincent 和 Allen，1999；Kao 等，2001；Johnson，2004；Ritts 等，2009）。

1.8.4　盐岩盆地

随着对深海大陆边缘（尤其是大陆基，如北墨西哥湾）的不断勘探，已经证实盐岩的变形对产生封闭沉积盆地具有重要作用（图1-3f）（Worrall 和 Snelson，1989；Winker，1996；Prather 等，1998；Badalini 等，2000；Beaubouef 和 Friendmann，2000）。Hudec 等（2009）阐述了这类小型盆地的沉降机理，他们认为这类盆地可以被看作是小型的地壳盆地。他们认为这类小型盆地的沉降可以由以下几种因素形成：（1）密度的差异；（2）底辟的缩短；（3）伸展型底辟的下降；（4）盐岩地层的分解；（5）沉积地层的负荷；（6）盐下地层的变形。他们也讨论了区分这些机理的标准。

对古构造背景的研究证实，盆地的形成和盐体的构造演化具有独特性（Giles 和 Lawton，2002；Rowan 等，2003）。密度差异可以引起盐岩体内构造力和重力的变化，而盐岩在软弱沉积层对这些力的运动学响应方面起到了重要作用。所有那些与盐体活动过程（墨西哥湾盆地是典型的小型盐岩盆地，此外，还有多种盐岩相关盆地）有着直接的关系盆地，都被称为是"盐岩盆地"。

1.8.5　陨石撞击盆地

白垩系—古近系界线处，出现了大面积的铱元素异常（Alvarez 等，1980），使得学者们开始重视外星物质对地球演化史的影响。外星物质不仅对主要的演化变迁（例如，大规模灭绝事件）有着很大的影响，也可以诱发海啸、滑坡、尘雨以及其他陨石引发的事件，从而在大范围内形成沉积（Bourgeois 等，1988；Alvarez 等，1992；Smit 等，1996；Bralower 等，1998）。墨西哥北岸下的西克苏鲁伯陨石坑，是解释全球范围内，在白垩系—古近系界线处，由陨石产生沉积物的确凿证据（Hildebrand 等，1991；Pope 等，1991）。具有撞击坑式充填及相关特征明显的沉积盆地即被称为"陨石撞击盆地"（图1-3e）。

陨石撞击盆地如今在地球的很多地方被识别出来，如西克苏鲁伯、切萨匹克湾（Shah 等，2005；Gohn 等，2006；Hayden 等，2008）、北海（Stewart 和 Allen，2002，2005）和巴支伦海（Tsikalas 等，1998；Dypvik 等，2004）。在这类盆地中，学者们已经对一些盆地进行了详细的地层学研究和盆地分析（Marin 等，2001；Parnell，2005；Hayden，2008），其中有一些是富含油气的盆地（Grajalas-Nishimura 等，2000）。越来越多的文献阐述了陨石撞击盆地的重要性（Glickson 和 Haines，2005；Evans 等，2008）。Stewart 在2003年论述了辨别此类盆地的原则。

1.8.6　继承性盆地

早先关于继承性盆地的定义（King，1996）是：火山活动有限，并伴随着极小范围的抬升，覆盖在变形和侵位优地槽之上的深陷槽（Ingersoll，1988；Ingersoll 和 Busby，1995）。这个定义现在需要做出更改，即"深陷"和"优地槽"应该被分别替换为"山间"及"地体"。在板块构造背景下，继承性盆地首先在山间的环境下形成，一般位于不活跃的褶皱冲断带、缝合带、转换带及非克拉通化的古裂谷的顶部。继承性盆地的出现，表明了造山活动以及断裂活动的终止（Ingersoll 和 Busby，1995）。因此，有了它们的年代学限制，可以确定缝合带、变形以及裂陷作用发生的时间。它们在地体分析中有着特别的意义；而且它们代表了叠置的沉积组合，这些组合能够揭示地体增生的最短时间（Howell，1985；Ricketts，

2008)。

目前很少有关于现今存在的继承性盆地模型的著作发表。Eisbbacher 在 1974 年总结出了加拿大科迪勒拉古盆地的模型。该类著作的缺乏，可能反映了继承性盆地以及其构造背景的多样性。从某种意义来说，所有的盆地都可以看作是继承性盆地，因为它们在盆地的基底上会发育一些后续的造山活动或者断裂活动。事实上，在 1951 年，Kay 认为加利福尼亚的后—内达华盆地就是一个继承性盆地，但现在这个盆地被认为是在新生代被转换构造叠加而成的（Ingersoll，1982；Ingersoll 和 Schweickert，1986；Dickinson，1995）。Dickinson 讨论了有关"闭合的弧前盆地"的例子，这些盆地的残余物沿着闭合带出现；同时，沉积物也在继承盆地里沿着闭合带出现（Ricketts，2008）。现今继承性盆地这个词是指在后造山和断裂活动范围内，没有归结在任何板块框架内的一类盆地（Ingersoll 和 Busby，1995）。例如，大部分盆岭省南部的盆地在中新世之后就没有发生任何构造运动（Wernicke，1992；Dickinson，2006）。因此，本区内，现今的山间盆地就可以认为是继承性盆地（Ingersoll 和 Busby，1995）（图 1-5c）。

1.9 讨论

本文的读者也许会被控制沉积盆地构造演化过程的复杂性，以及复杂的盆地分类弄得不知所措。我们对于这些过程及结果了解得越多，那么所建立的模型就会越复杂，每个盆地都会显得独一无二（Dickinson，1993）。这种结果通常既令人兴奋又令人沮丧。兴奋是因为发现了一些新的事实和领悟；沮丧则是因为需要学习大量的新信息。新模型的出现往往都伴随着具有深刻意义的简化和归纳。观察、建模和实验的结合是一种反复的、自我调整的过程。

对所有沉积盆地进行分类和综述的最终目的是通过运用现在的盆地演化模型，提升对古构造和古地理重建的认识。对这些相关特征的识别有助于古构造的重建，包括缝合带（Burke 等，1977）、岩浆岛弧（Sengor，1991）、褶皱冲断带（McClay，1992）和变质带（Miyashiro，1973）。一个熟练的盆地分析人员必须会综合应用以下资料：地球化学、地球物理、石油地质、古生物以及其他一系列的学科的资料。此外，其他领域的工作者应该把注意力放在那些由沉积特征所限制的古构造重建上。笔者希望这篇文章可以为地球演化模型的跨学科发展和试验提供帮助。

致谢（略）

参 考 文 献

Aalto, K. R., and Miller, W., Ⅲ (1999) Sedimentology of the Pliocene Upper Onzole Formation, an inner-trench slope succession in northwestern Ecuador. Journal of South American Earth Sciences, 12, 69-85.

Abascal, L. del V. (2005) Combined thin-skinned and thick skinned deformation in the central Andean foreland of northwestern Argentina. Journal of South American Earth Sciences, 19, 75-81.

Alavi, M. (2004) Regional stratigraphy of the Zagros foldthrust belt of Iran and its proforeland evolution. American Journal of Science, 304, 1-20.

Alcicek, M. C. (2007) Tectonic development of an orogentop rift recorded by its terrestrial sedimentation pattern: the Neogene Esen basin of southwestern Anatolia, Turkey. Sedimentary Geology, 200, 117–140.

Allen, M. B., Vincent, S. J., and Wheeler, P. J. (1999) Late Cenozoic tectonics of the Kepingtage thrust zone: interactions of the Tien Shan and Tarim basin, northwest China. Tectonics, 18, 639–654.

Allen, P. A., Homewood, P., and Williams, G. D. (1986) Foreland basins: an introduction. International Association of Sedimentologists Special Publication, 8, 3–12.

Allen, P. A., Burgess, P. M., Galewsky, J., and Sinclair, H. D. (2001) Flexural–eustatic numerical model for drowning of the Eocene perialpine carbonate ramp and implications for Alpine geodynamics. Geological Society of America Bulletin, 113, 1052–1066.

Allen, R., Carter, A., Najman, Y., Bandopadhyay, P. C., Chapman, H. J., Bickle, M. J., Garzanti, E., Vezzoli, G., Ando, S., Foster, G. L., and Gerring, C. (2008) New constraints on the sedimentation and uplift history of the Andaman–Nicobar accretionary prism, South Andaman Island. Geological Society of America Special Paper, 436, 223–255.

Alvarez, L. W., Alvarez, W., Asaro, F., and Michel, H. V. (1980) Extraterrestrial cause for the Cretaceous–Tertiary extinction. Science, 208, 1095–1108.

Alvarez, W., Smit, J., Lowrie, W., Asaro, F., Margolis, S. V., Claeys, P., Kastner, M., and Hildebrand, A. R. (1992) Proximal impact deposits at the Cretaceous–Tertiary boundary in the Gulf of Mexico: a restudy of DSDP Leg 77 Sites 536 and 540. Geology, 20, 697–700.

Arenas, C., Millan, H., Pardo, G., and Pocovi, A. (2001) Ebro basin continental sedimentation associated with late compressional Pyrenean tectonics (north-eastern Iberia): controls on basin margin fans and fluvial systems. Basin Research, 13, 65–89.

Atwater, T. (1970) Implications of plate tectonics for the Cenozoic tectonic evolution of western North America. Geological Society of America Bulletin, 81, 3513–3535.

Atwater, T. (1989) Plate tectonic history of the northeast Pacific and western North America, in Winterer, E. L., Hussong, D. M., and Decker, R. W., eds., The eastern Pacific Ocean and Hawaii. The geology of North America, vol. N. Boulder, CO, Geological Society of America, 21–72.

Atwater, T., and Stock, J. (1998) Pacific–North America plate tectonics of the Neogene southwestern United States: an update. International Geology Review, 40, 375–402.

Audley-Charles, M. G., Curray, J. R., and Evans, G. (1977) Location of major deltas. Geology, 5, 341–344. Axen, G. J., and Fletcher, J. M. (1998) Late Miocene–Pleistocene extensional faulting, northern Gulf of California, Mexico and Salton Trough, California. International Geology Review, 40, 217–244.

Badalini, G., Kneller, B., and Winker, C. D. (2000) Architecture and process in the late Pleistocene Brazos–Trinity turbidite system, Gulf of Mexico continental slope, in Weimer P., Slatt R. M., Coleman J., Rossen N. C., Nelson H., Bouma A. H., Styzen M. J., Lawrence D. T., eds., Deepwater reservoirs of the world, 20th Annual Research Conference Proceedings. Society of Economic Paleontologist and Mineralogists, GulfCoast Section, Houston, TX, 16–34.

Bally, A. W. (1989) Phanerozoic basins of North America, in Bally, A. W., and Palmer, A. R., eds., The geology of North America: an overview. The Geology of North America, A. Boulder, CO, Geological Society of America, 397–447.

Barbeau, D. L. (2003) Aflexural model for the Paradox basin: implications for the tectonics of the Ancestral Rocky Mountains. Basin Research, 15, 97–115.

Barnes, P. M., Sutherland, R., Davy, B., and Delteil, J. (2001) Rapid creation and destruction of sedimentary basins on mature strike-slip faults: an example from the offshore Alpine fault, New Zealand. Journal of Structural Geology, 23, 1727–1739.

Barnes, P. M., Sutherland, R., and Delteil, J. (2005) Strikeslip structure and sedimentary basins of the southern Alpine fault, Fiordland, New Zealand. Geological Society of America Bulletin, 117, 411–435.

Beaubouef, R. T., and Friedmann, S. J. (2000) High resolution seismic/sequence stratigraphic framework for the evolution of Pleistocene intraslope basins, western Gulf of Mexico: Depositional models and reservoir analogs, in Weimer P., Slatt R. M., Coleman J., Rossen N. C., Nelson H., Bouma A. H., Styzen M. J., Lawrence D. T., eds., Deepwater reservoirs of the world, 20th Annual Research Conference Proceedings. Society of Economic Paleontologist and Mineralogists, GulfCoast Section, Houston, TX, 40–60.

Beaumont, C. (1981) Foreland basins. Geophysical Journal of the Royal Astronomical Society, 65, 291–329.

Beaumont, C., Ellis, S., Hamilton, J., and Fullsack, P. (1996) Mechanical model for subduction–collision tectonics of Alpine-type compressional orogens. Geology, 24, 675–678.

Beck, M. (1983) On the mechanism of tectonic transport in zones of oblique subduction. Tectonophysics, 93, 1–11.

Ben-Avraham, Z., and Uyeda, S. (1983) Entrapment origin of marginal seas. American Geophysical Union Geodynamics Series, 11, 91–104.

Berger, W. H. (1973) Cenozoic sedimentation in the eastern tropical Pacific. Geological Society of America Bulletin, 84, 1941–1954.

Berger, W. H., and Winterer, E. L. (1974) Plate stratigraphy and the fluctuating carbonate line. International Association of Sedimentologists Special Publication, 1, 11–48.

Berry, J. P., and Wilkinson, B. H. (1994) Paleoclimatic and tectonic control on the accumulation of North American cratonic sediment. Geological Society of America Bulletin, 106, 855–865.

Bertotti, G., Picotti, V., and Cloetingh, S. (1998) Lithospheric weakening during "retroforeland" basin formation: tectonic evolution of the central South Alpine foredeep. Tectonics, 17, 131–142.

Bertotti, G., Picotti, V., Chilovi, C., Fantoni, R., Merlini, S., and Mosconi, A. (2001) Neogene to Quaternary sedimentary basins in the south Adriatic (Central Mediterranean): foredeeps and lithospheric buckling. Tectonics, 20, 771–787.

Blakey, R. C. (2008) Pennsylvanian–Jurassic sedimentary basins of the Colorado Plateau and southern Rocky Mountains, in Miall, A. D., ed., The sedimentary basins of the United States and Canada. Sedimentary Basins of the World, vol. 5. Amsterdam, Elsevier, 245–296.

Bohannon, R. G. (1986a) Howmuchdivergence has occurred between Africa and Arabia as a result of the opening of the Red Sea? Geology, 14, 510-513.

Bohannon, R. G. (1986b) Tectonic configuration of the western Arabian continental margin, southern Red Sea. Tectonics, 5, 477-499.

Bohannon, R. G., and Parsons, T. (1995) Tectonic implications of post-30 Ma Pacific and North American relative plate motions. Geological Society of America Bulletin, 107, 937-959.

Bond, G. C., Kominz, M. A., and Sheridan, R. E. (1995) Continental terraces and rises, in Busby, C. J., and Ingersoll, R. V., eds., Tectonics of sedimentary basins. Oxford, Blackwell Science, 149-178.

Bostrom, R. C. (1971) Westward displacement of the lithosphere. Nature, 234, 536 - 538. Bosworth, W., Huchon, P., and McClay, K. (2005) The Red Sea and Gulf of Aden basins. Journal of African Earth Sciences, 43, 324-378.

Bourgeois, J., Hansen, T. A., Wiberg, P. L., andKauffman, E. G. (1988) A tsunami deposit at the Cretaceous-Tertiary boundary in Texas. Science, 241, 567-570.

Bradley, D. C. (1989) Taconic plate kinematics as revealed by foredeep stratigraphy, Appalachian orogen. Tectonics, 8, 1037-1049.

Bradley, D. C., and Kidd, W. S. F. (1991) Flexural extension of the upper continental crust in collisional foredeeps. Geological Society of America Bulletin, 103, 1416-1438.

Bralower, T. J., Paull, C. K., and Leckie, R. M. (1998) The Cretaceous-Tertiary boundary cocktail: Chicxulub impact triggers margin collapse and extensive sediment gravity flows. Geology, 26, 331-334.

Brunet, M-F., Korotaev, M. V., Ershov, A. V., and Nikishin, A. M. (2003) The South Caspian basin: a review of its evolution from subsidence modeling. Sedimentary Geology, 156, 119-148.

Bry, M., White, N., Singh, S., England, R., andTrowell, C. (2004) Anatomy and formation of oblique continental collision: South Falkland basin. Tectonics, 23, TC4011, 20p.

Buffler, R. T., and Thomas, W. A. (1994) Crustal structure and evolution of the southeastern margin of North America and the Gulf of Mexico basin, in Speed, R. C., ed., Phanerozoic evolution of North American continent - ocean transitions. Geological Society of America, Decade of North American Geology Summary Volume to accompany the DNAG Continent - Ocean Transects Series, 219-264.

Burbank, D. W., and Tahirkheli, R. A. K. (1985) The magnetostratigraphy, fission - track dating, and stratigraphic evolution of the Peshawar intermontane basin, northern Pakistan. Geological Society of America Bulletin, 96, 539-552.

Burchfiel, B. C., Chen, Z., Hodges, K. V., Liu, Y., Royden, L. H., Deng, C., and Xu, J. (1992) The south Tibetan detachment system, Himalayan orogen: Extension contemporaneous with and parallel to shortening in a collisional mountain belt. Geological Society of America Special Paper, 269, 41 p.

Burgess, P. M. (2008) Phanerozoic evolution of the sedimentary cover of the North American craton, in Miall, A. D., ed., The sedimentary basins of the United States and Canada.

Sedimentary basins of the world, vol. 5. Amsterdam, Elsevier, 31–63.

Burgess, P. M., and Gurnis, M. (1995) Mechanisms for the formation of cratonic stratigraphic sequences. Earth and Planetary Science Letters, 136, 647–663.

Burgess, P. M., Gurnis, M., and Moresi, L. (1997) Formation of sequences in the cratonic interior of North America by interaction between mantle, eustatic, and stratigraphic processes. Geological Society of America Bulletin, 109, 1515–1535.

Burke, K. (1972) Longshore drift, submarine canyons, and submarine fans in development of Niger delta. American Association of Petroleum Geologists Bulletin, 56, 1975–1983.

Burke, K., and Dewey, J. F. (1973) Plume-generated triple junctions: key indicators in applying plate tectonics to old rocks. Journal of Geology, 81, 406–433.

Burke, K., Dewey, J. F., and Kidd, W. S. F. (1977) World distribution of sutures: the sites of former oceans. Tectonophysics, 40, 69–99.

Busby, C. J., Adams, B. F., Mattinson, J., and De Oreo, S. (2006) View of an intact oceanic arc, from surficial to mesozonal levels: Cretaceous Alisitos arc, BajaCalifornia, Mexico. Journal of Volcanology and Geothermal Research, 149, 1–46.

Busby-Spera, C. J. (1987) Lithofacies of deep marine basalts emplaced on a Jurassic backarc apron, Baja California (Mexico). Journal of Geology, 95, 671–686.

Busby-Spera, C. J. (1988a) Evolution of a Middle Jurassic back-arc basin, Cedros Island, Baja California: evidence from a marine volcaniclastic apron. Geological Society of America Bulletin, 100, 218–233.

Busby-Spera, C. J. (1988b) Speculative tectonic model for the early Mesozoic arc of the southwest Cordilleran United States. Geology, 16, 1121–1125.

Busby-Spera, C. J., Mattinson, J. M., Riggs, N. R., and Schermer, E. R. (1990) The Triassic –Jurassic magmatic arc in the Mojave – SonoranDeserts and the Sierran – Klamath region: similarities and differences in paleogeographic evolution. Geological Society of America Special Paper, 255, 325–338.

Cameron, T. D. J., Crosby, A., Balson, P. S., Jeffrey, D. H., Lott, G. K., Bulat, J., and Harrison, D. J. (1992) United Kingdom offshore regional report: the geology of the southern North Sea. HMSO for the British Geological Survey, London, 152 p.

Cannon, W. F. (1995) Closing of the Midcontinent rift: a farfield effect of Grenvillian compression. Geology, 22, 155–158. Cardozo, N., and Jordan, T. (2001).

Causes of spatially variable tectonic subsidence in the Miocene Bermejo foreland basin, Argentina. Basin Research, 13, 335–357.

Carrapa, B. (2009) Tracing exhumation and orogenic wedge dynamics in the European Alps with detrital thermochronology. Geology, 37, 1127–1130.

Carrapa, B., and Garcia-Castellanos, D. (2005) Western Alpine back-thrusting as subsidence mechanism in the Tertiary Piedmont basin (western Po Plain, NW Italy). Tectonophysics, 406, 197–212.

Cas, R., and Busby-Spera, C., eds. (1991) Volcaniclastic sedimentation. Sedimentary Geology, 74, 362 p.

Cas, R. A. F., and Wright, J. V. (1987) Volcanic successions modern and ancient. Boston, Allen and Unwin, 528 p.

Casas-Sainz, A. M., Cortes-Garcia, A. L., and Maestro-Gonzalez, A. (2000) Intraplate deformation and basin formation during the Tertiary within the northern Iberian plate: origin and evolution of the Almazan basin. Tectonics, 19, 258-289.

Castle, J. W. (2001) Appalachian basin stratigraphic response to convergent-margin structural evolution. Basin Research, 13, 397-418.

Cather, S. M., and Chapin, C. E. (1990) Paleogeographic and paleotectonic setting of Laramide sedimentary basins in the central RockyMountain region: alternative interpretation. Geological Society of America Bulletin, 102, 256-260.

Catuneanu, O. (2004) Retroarc foreland systems: evolution through time. Journal of African Earth Sciences, 38, 225-242.

Chapin, C. E., and Cather, S. M. (1981) Eocene tectonics and sedimentation in the Colorado Plateau-Rocky Mountain area. Arizona Geological Society Digest, 14, 173-198.

Chen, W-S., Ridgway, K. D., Horng, C-S., Chen, Y-G., Shea, K-S., and Yeh, M-G. (2001) Stratigraphic architecture, magnetostratigraphy, and incised-valley systems of the Pliocene-Pleistocene collisional marine foreland basin of Taiwan. Geological Society of America Bulletin, 113, 1249-1271.

Chiang, C-S., Yu, H-S., and Chou, Y-W. (2004) Characteristics of the wedge-top depozone of the southern Taiwan foreland basin system. Basin Research, 16, 65-78.

Christie-Blick, N., and Biddle, K. T. (1985) Deformation and basin formation along strike-slip faults. Society of Economic Paleontologists and Mineralogists Special Publication, 37, 1-34.

Clague, D. A. (1981) Linear island and seamount chains, aseismic ridges and intraplate volcanism: results from DSDP. Society of Economic Paleontologists and Mineralogists Special Publication, 32, 7-22.

Clift, P. D., Draut, A. E., Kelemen, P. B., Blusztajn, J., and Greene, A. (2005) Stratigraphic and geochemical evolution of an oceanic arc upper crustal section: the Jurassic Talkeetna Volcanic Formation, south-central Alaska. Geological Society of America Bulletin, 117, 902-925.

Cloetingh, S. (1988) Intraplate stress: a new element in basin analysis, in Kleinspehn, K. L., and Paola, C., eds., New perspectives in basin analysis. New York, Springer-Verlag, 205-230.

Cloos, M. (1993) Lithospheric buoyancy and collisional orogenesis: subduction of oceanic plateaus, continental margins, island arcs, spreading ridges, and seamounts. Geological Society of America Bulletin, 105, 715-737.

Cloos, M., and Shreve, R. L. (1988a) Subduction-channel model of prism accretion, melange formation, sediment subduction, and subduction erosion at convergent plate margins: 1. background and description. Pure and Applied Geophysics, 128, 455-500.

Cloos, M., and Shreve, R. L. (1988b) Subduction-channel model of prism accretion, melange formation, sediment subduction, and subduction erosion at convergent plate margins: 2.

implications and discussion. Pure and Applied Geophysics, 128, 501–545.

Cochran, J. R. (1983) A model for development of Red Sea. American Association of Petroleum Geologists Bulletin, 67, 41–69.

Cogne, J. P., Humler, E., and Courtillot, V. (2006) Mean age of oceanic lithosphere drives eustatic sea-level change since Pangea breakup. Earth and Planetary Science Letters, 245, 115–122.

Coleman, R. G. (1993) Geologic evolution of the Red Sea. New York, OxfordUniversity Press, 186 p.

Constenius, K. N., Johnson, R. A., Dickinson, W. R., and Williams, T. A. (2000) Tectonic evolution of the Jurassic – CretaceousGreatValley forearc, California: implications for the Franciscan thrust-wedge hypothesis. Geological Society of America Bulletin, 112, 1703–1723.

Cook, H. E. (1975) North American stratigraphic principles as applied to deep-sea sediments. American Association of Petroleum Geologists Bulletin, 59, 817–837.

Cooper, M. A., and Williams, G. D., eds. (1989) Inversion tectonics. Geological Society of London Special Publication, 44, 375 p.

Critelli, S., and Le Pera, E. (1994) Detrital modes and provenance of Miocene sandstones and modern sands of the southern Apennines thrust-top basins (Italy). Journal of Sedimentary Research, A64, 824–835.

Critelli, S., Marsaglia, K. M., and Busby, C. J. (2002) Tectonic history of a Jurassic backarc-basin sequence (the Gran Canon Formation, Cedros Island, Mexico) based on compositional modes of tuffaceous deposits. Geological Society of America Bulletin, 114, 515–527.

Crouch, J. K., and Suppe, J. (1993) Late Cenozoic tectonic evolution of the Los Angeles basin and inner California borderland: a model for core complex-like crustal extension. Geological Society of America Bulletin, 105, 1415–1434.

Crowell, J. C. (1974a) Sedimentation along the San Andreas fault, California. Society of Economic Paleontologists and Mineralogists Special Publication, 19, 292–303.

Crowell, J. C. (1974b) Origin of late Cenozoic basins in southern California. Society of Economic Paleontologists and Mineralogists Special Publication, 22, 190–204.

Crowell, J. C., ed. (2003a) Evolution of Ridge basin, southern California: an interplay of sedimentation and tectonics. Geological Society of America Special Paper, 367, 247 p.

Crowell, J. C. (2003b) Tectonics of Ridge basin, southern California, in Crowell, J. C., ed., Evolution of Ridge basin, southern California: an interplay of sedimentation and tectonics. Geological Society of America Special Paper, 367, 157–203.

Crowell, J. C. and Link, M. H., eds. (1982) Geologic history of Ridge basin southern California. Book 22. Pacific Section, Society of Economic Paleontologists and Mineralogists, Los Angeles, 304 p.

Cruciani, C., Carminati, E., and Doglioni, C. (2005) Slab dip v. lithosphere age: no direct function. Earth and Planetary Science Letters, 238, 298–310.

Davila, F. M., and Astini, R. A. (2003) Early Middle Miocene broken foreland development in the southern Central Andes: evidence for extension prior to regional shortening. Basin

Research, 15, 379–396.

DeCelles, P. G., and Giles, K. A. (1996) Foreland basin systems. Basin Research, 8, 105–123.

DeCelles, P. G., and Giles, K. A. (1997) Foreland basin systems: Reply. Basin Research, 9, 171–176.

DeCelles, P. G., and Horton, B. K. (2003) Early to middle Tertiary foreland basin development and the history of Andean crustal shortening in Bolivia. Geological Society of America Bulletin, 115, 58–77.

DeCelles, P. G., Gehrels, G. E., Quade, J., and Ojha, T. P. (1998) Eocene–early Miocene foreland basin development and the history of Himalayan thrusting, western and central Nepal. Tectonics, 17, 741–765.

DeCelles, P. G., Robinson, D. M., Quade, J., Ojha, T. P., Garzione, C. N., Copeland, P., and Upreti, B. N. (2001) Stratigraphy, structure, and tectonic evolution of the Himalayan fold-thrust belt in western Nepal. Tectonics, 20, 487–509.

DeCelles, P. G., Quade, J., Kapp, P., Fan, M. J., Dettman, D. L., and Ding, L. (2007) High and dry in central Tibet during the Late Oligocene. Earth and Planetary Science Letters, 253, 389–401.

DeRito, R. F., Cozzarelli, F. A. and Hodge, D. S. (1983) Mechanism of subsidence of ancient cratonic rift basins. Tectonophysics, 94, 141–168.

Dewey, J. F. (1980) Episodicity, sequence and style at convergent plate boundaries. Geological Association of Canada Special Paper, 20, 553–573.

Dewey, J. F., and Burke, K. (1974) Hot spots and continental break–up: implications for collisional orogeny. Geology, 2, 57–60.

Dickinson, W. R. (1973) Widths of modern arc–trench gaps proportional to past duration of igneous activity in associated magmatic arcs. Journal of Geophysical Research, 78, 3376–3389.

Dickinson, W. R. (1974a) Sedimentation within and beside ancient and modern magmatic arcs. Society of Economic Paleontologists and Mineralogists Special Publication, 19, 230–239.

Dickinson, W. R. (1974b) Plate tectonics and sedimentation. Society of Economic Paleontologists and Mineralogists Special Publication, 22, 1–27.

Dickinson, W. R. (1976a) Plate tectonic evolution of sedimentary basins. American Association of Petroleum Geologists Continuing Education Course Notes Series 1, 62 p.

Dickinson, W. R. (1976b) Sedimentary basins developed during evolution of Mesozoic–Cenozoic arc–trench system in western North America. Canadian Journal of Earth Sciences, 13, 1268–1287.

Dickinson, W. R. (1978) Plate tectonic evolution of north Pacific rim. Journal of the Physics of the Earth, 26, Supplement, S1–S19.

Dickinson, W. R. (1981a) Plate tectonic evolution of the southern Cordillera. Arizona Geological Society Digest, 14, 113–135.

Dickinson, W. R. (1981b) Plate tectonics and the continental margin of California, in Ernst, W. G., ed., The geotectonic development of California. Rubey, vol. 1. Englewood Cliffs, NJ, Prentice Hall, 1–28.

Dickinson, W. R. (1990) Paleogeographic and paleotectonic setting of Laramide sedimentary basins in the central RockyMountain region: reply. Geological Society of America Bulletin, 102, 281-282.

Dickinson, W. R. (1993) Basin geodynamics. Basin Research, 5, 195-196. Dickinson, W. R. (1995) Forearc basins, in Busby, C. J., and Ingersoll, R. V., eds., Tectonics of sedimentary basins. Oxford, Blackwell Science, 221-261.

Dickinson, W. R. (1996) Kinematics of transrotational tectonism in the CaliforniaTransverseRanges and its contribution to cumulative slip along the San Andrea transform fault system. Geological Society of America Special Paper, 305, 46 p.

Dickinson, W. R. (2004) Impacts of eustasy and hydro-isostasy on the evolution and landforms of Pacific atolls. Palaeogeography, Palaeoclimatology, Palaeoecology, 213, 251-269.

Dickinson, W. R. (2006) Geotectonic evolution of the Great Basin. Geosphere, 7, 353-368.

Dickinson, W. R. (2008) Tectonic lessons from the configuration and internal anatomy of the Circum-Pacific orogenic belt, in Spencer, J. E., and Titley, S. R., eds., Ores and orogenesis: Circum-Pacific tectonics, geologic evolution, and ore deposits. Arizona Geological Society Digest 22, 5-18.

Dickinson, W. R., and Lawton, T. F. (2001) Carboniferous to Cretaceous assembly and fragmentation of Mexico. Geological Society of America Bulletin, 113, 1142-1160.

Dickinson, W. R., and Lawton, T. F. (2003) Sequential intercontinental suturing as the ultimate control for Pennsylvanian Ancestral Rocky Mountains deformation. Geology, 31, 609-612.

Dickinson, W. R., and Seely, D. R. (1979) Structure and stratigraphy of forearc regions. American Association of Petroleum Geologists Bulletin, 63, 2-31.

Dickinson, W. R., and Snyder, W. S. (1978) Plate tectonics of the Laramide orogeny. Geological Society of America Memoir, 151, 355-366.

Dickinson, W. R., and Wernicke, B. P. (1997) Reconciliation of San Andreas slip discrepancy by a combination of interior Basin and Range extension and transrotation near the coast. Geology, 25, 663-665.

Dickinson, W. R., Harbaugh, D. W., Saller, A. H., Heller, P. L., and Snyder, W. S. (1983) Detrital modes of upper Paleozoic sandstones derived from Antler orogen in Nevada: implications for nature of Antler orogeny. American Journal of Science, 283, 481-509.

Dickinson, W. R., Klute, M. A., Hayes, M. J., Janecke, S. U., Lundin, E. R., McKittrick, M. A., and Olivares, M. D. (1988) Paleogeographic and paleotectonic setting of Laramide sedimentary basins in the central Rocky Mountain region. Geological Society of America Bulletin, 100, 1023-1039.

Dickinson, W. R., Klute, M. A., Hayes, M. J., Janecke, S. U., Lundin, E. R., McKittrick, M. A., and Olivares, M. D. (1990) Paleogeographic and paleotectonic setting of Laramide sedimentary basins in the central RockyMountain region: reply. Geological Society of America Bulletin, 102, 259-260.

Doglioni, C. (1994) Foredeeps versus subduction zones. Geology, 22, 271-274.

Doglioni, C., Harabaglia, P., Merlini, S., Mongelli, F., Peccerillo, A., and Piromallo, C.

(1999) Orogens and slabs vs. their direction of subduction. Earth-Science Reviews, 45, 167–208.

Dooley, T., and McClay, K. (1997) Analog modeling of pullapart basins. American Association of Petroleum Geologists Bulletin, 81, 1804–1826.

Dorobek, S. L. (2008) Carbonate-platform facies in volcanicarc settings: characteristics and controls on deposition and stratigraphic development. Geological Society of America Special Paper, 436, 55–90.

Dorobek, S. L., and Ross, G. M., eds. (1995) Stratigraphic evolution of foreland basins. SEPM (Society for Sedimentary Geology) Special Publication, 52, 310 p.

Dorsey, R. J., Umhoefer, P. J., and Renne, P. R. (1995) Rapid subsidence and stacked Gilbert-type fan deltas, Pliocene Loreto basin, Baja California Sur, Mexico. Sedimentary Geology, 98, 181–204.

Dypvik, H., Sandbakken, P. T., Postma, G., and Mork, A. (2004) Early post-impact sedimentation around the central high of the Mjolnir impact crater (Barents Sea, Late Jurassic). Sedimentary Geology, 168, 227–247.

Einsele, G. (1985) Basaltic sill-sediment complexes in young spreading centers: genesis and significance. Geology, 13, 249–252.

Einsele, G., Liu, B., Durr, S., Frisch, W., Liu, G., Luterbacher, H. P., Ratschbacher, L., Ricken, W., Wendt, J., Wetzel, A., Yu, G., and Zheng, H. (1994) The Xigaze forearc basin: evolution and facies architecture (Cretaceous, Tibet). Sedimentary Geology, 90, 1–32.

Eisbacher, G. H. (1974) Evolution of successor basins in the Canadian Cordillera. Society of Economic Paleontologists and Mineralogists Special Publication, 19, 274–291.

Erikson, J. P., Pindall, J. L., Karner, G. D., Sonder, L. J., Fuller, E., and Dent, L. (1998) Neogene sedimentation and tectonics in the Cibao basin and northern Hispaniola: an example of basin evolution near a strike-slip-dominated plate boundary. Journal of Geology, 106, 473–494.

Evans, K. R., Horton, J. W., Jr., King, D. T., Jr., and Morrow, J. R., eds. (2008) The sedimentary record of meteorite impacts. Geological Society of America Special Paper, 437, 213 p.

Evans, M. J., and Elliott, T. (1999) Evolution of a thrustsheet-top basin: the Tertiary Barreme basin, Alpes-de-Haute-Provence, France. Geological Society of America Bulletin, 111, 1617–1643.

Fackler-Adams, B., and Busby, C. J. (1998) Structural and stratigraphic evolution of extensional oceanic arcs. Geology, 26, 735–738.

Faulds, J. E., and Varga, R. J. (1998) The role of accommodation zones and transfer zones in the regional segmentation of extended terranes. Geological Society of America Special Paper, 323, 1–45.

Fildani, A., Cope, T. D., Graham, S. A., and Wooden, J. L. (2003) Initiation of the Magallanes foreland basin: timing of the southernmost Patagonian Andes orogeny revised by detrital zircon provenance analysis. Geology, 31, 1081–1084.

Fisher, R. V., and Schmincke, H-U. (1984) Pyroclastic rocks. New York, Springer-Verlag, 472 p.

Fisher, R. V., and Smith, G. A., eds. (1991) Sedimentation in volcanic settings. SEPM (Society for Sedimentary Geology) Special Publication, 45, 257 p.

Ford, M. (2004) Depositional wedge tops: interaction between low basal friction external orogenic wedges andflexural forelandbasins. BasinResearch, 16, 361–375.

Friedmann, S. J., and Burbank, D. W. (1995) Rift basins and supradetachment basins: intracontinental extensional end-members. Basin Research, 7, 109–127.

Funk, J., Mann, P., McIntosh, K., and Stephens, J. (2009) Cenozoic tectonics of the Nicaraguan depression, Nicaragua, and Median Trough, El Salvador, based on seismicreflection profiling and remote-sensing data. Geological Society of America Bulletin, 121, 1491–1521.

Galewsky, J., Silver, E. A., Gallup, C. D., Edwards, R. L., and Potts, D. C. (1996) Foredeep tectonics and carbonate platform dynamics in the Huon Gulf, Papua New Guinea. Geology, 24, 819–822.

Galloway, W. E., Ganey-Curry, P. E., Li, X., and Buffler, R. T. (2000) Cenozoic depositional history of the Gulf of Mexico basin. American Association of Petroleum Geologists Bulletin, 84, 1743–1774.

Garzione, C. N., DeCelles, P. G., Hodkinson, D. G., Ojha, T. P., and Upreti, B. N. (2003) East-west extension and Miocene environmental change in the southern Tibetan plateau: Thakkhola graben, central Nepal. Geological Society of America Bulletin, 115, 3–20.

Gawthorpe, R. L., and Leeder, M. R. (2000) Tectono-sedimentary evolution of active extensional basins. Basin Research, 12, 195–218.

Geslin, J. K. (1998) Distal Ancestral Rocky Mountains tectonism: evolution of the Pennsylvanian-PermianOquirrh-WoodRiver basin, southern Idaho. Geological Society of America Bulletin, 110, 644–663.

Giles, K. A., and Lawton, T. F. (2002) Halokinetic sequence stratigraphy adjacent to the El Papalote diapir, northeastern Mexico. American Association of Petroleum Geologists Bulletin, 86, 823–840.

Giovanni, M. K., Horton, B. K., Garzione, C. N., McNulty, B., and Grove, M. (2010) Extensional basin evolution in the Cordillera Blanca, Peru: stratigraphic and isotopic 34 records of detachment faulting and orogenic collapse in the Andean hinterland. Tectonics, 29, TC6007, 21 p.

Glickson, A. Y., and Haines, P. W., eds. (2005) The maticissue: Shoemaker memorial issue on the Australian impact record: 1997–2005 update. Australian Journal of Earth Sciences, 52, 475–798.

Gohn, G. S., Koeberl, C., Miller, K. G., Reimold, W. U., and the Scientific Staff of the Chesapeake Bay Impact Structure Drilling Project (2006) Chesapeake Bay Impact Structure Deep Drilling Project completes coring. Scientific Drilling (3), 34–39.

Gomez, E., Jordan, T. E., Allmendinger, R. W., and Cardozo, N. (2005) Development of the Colombian foreland- basin system as a consequence of diachronous exhumation of the northern

Andes. Geological Society of America Bulletin, 117, 1272-1292.

Gould, S. J. (1989) Wonderful life: the Burgess Shale and the nature of history. New York, W. W. Norton, 347 p. Graham, S. A., Dickinson, W. R., and Ingersoll, R. V. (1975) Himalayan-Bengal model for flysch dispersal in Appalachian- Ouachita system. Geological Society of America Bulletin, 86, 273-286.

Graham, S. A., Hendrix, M. S., Wang, L. B., and Carroll, A. R. (1993) Collisional successor basins of western China: impact of tectonic inheritance on sand composition. Geological Society of America Bulletin, 105, 323-344.

Grajalas-Nishimura, J. M., Cedillo-Pardo, E., Rosales-Dominguez, C., Moran-Zenteno, D. J., Alvarez, W., Claeys, P., Ruis-Morales, J., Garcia-Hernandez, J., Padilla-Avila, P., and Sanchez-Rios, A. (2000) Chicxulub impact: the origin of reservoir and seal facies in the southeastern Mexico oil fields. Geology, 28, 307-310.

Gupta, S., and Allen, P. A. (2000) Implications of foreland paleotopography for stratigraphic development in the Eocene distal Alpine foreland basin. Geological Society of America Bulletin, 112, 515-530.

Gurnis, M. (1993) Depressed continental hypsometry behind oceanic trenches: a clue to subduction controls on sea-level change. Geology, 21, 29-32.

Haddad, D., and Watts, A. B. (1999) Subsidence history, gravity anomalies, and flexure of the northeast Australian margin in Papua New Guinea. Tectonics, 18, 827-842.

Hall, R., and Smyth, H. R. (2008) Cenozoic arc processes in Indonesia: identification of the key influences on the stratigraphic record in active volcanic arcs. Geological Society of America Special Paper, 436, 27-54.

Hamilton, W. (1969) Mesozoic California and the underflow of Pacific mantle. Geological Society of America Bulletin, 80, 2409-2430.

Hamilton, W. (1979) Tectonics of the Indonesian region. United States Geological Survey Professional Paper 1078, 345.

Hansen, W. R. (1990) Paleogeographic and paleotectonic setting of Laramide sedimentary basins in the central Rocky Mountain region: alternative interpretation. Geological Society of America Bulletin, 102, 280-282.

Hayden, T., Kominz, M., Powars, D. S., Edwards, L. E., Miller, K. G., Browning, J. V., and Kulpecz, A. A. (2008) Impact effects and regional tectonic insights: Backstripping the Chesapeake Bay impact structure. Geology, 36, 327-330.

Heckel, P. H. (1984) Changing concepts of Midcontinent Pennsylvanian cyclothems, North America, in Congres International de Stratigraphie et de Geologie du Carbonifere, 9th. Compte Rendu, 3. Carbondale, Southern IllinoisUniversity Press, 535-538.

Heezen, B. C., MacGregor, I. D., Foreman, H. P., Forristal, G., Hekel, H., Hesse, R., Hoskins, R. H., Jones, E. J. W, Kaneps, A., Krasheninnikov, V. A., Okada, H., and Ruef, M. H. (1973) Diachronous deposits: a kinematic interpretation of the post Jurassic sedimentary sequence on the Pacific Plate. Nature, 241, 25-32.

Heller, P. L., and Angevine, C. L. (1985) Sea-level cycles during the growth of Atlantic-type oceans. Earth and Planetary Science Letters, 75, 417–426.

Heller, P. L., Bowdler, S. S., Chambers, H. P., Coogan, J. C., Hagen, E. S., Shuster, M. W., Winslow, N. S., and Lawton, T. F. (1986) Time of initial thrusting in the Sevier orogenic belt, Idaho-Wyoming and Utah. Geology, 14, 388–391.

Hermoza, W., Brusset, S., Baby, P., Gil., W., Roddaz, M., Guerrero, N., and Bolanos, M. (2005) The Huallaga foreland basin evolution: thrust propagation in a deltaic environment, northern Peruvian Andes. Journal of South American Earth Sciences, 19, 21–34.

Hildebrand, A. R., Penfield, G. T., Kring, D. A., Pilkington, M., Camargo Z., A., Jacobsen, S. B., and Boynton, W. V. (1991) Chicxulub crater: a possible Cretaceous/Tertiary boundary impact crater on the Yucatan Peninsula, Mexico. Geology, 19, 867–871.

Hilley, G. E., and Strecker, M. R. (2005) Processes of oscillatory basin filling and excavation in a tectonically active orogen: Quebrada del Toro basin, NW Argentina. Geological Society of America Bulletin, 117, 887–901.

Hoffman, P. F., Dewey, J. F., and Burke, K. (1974) Aulacogens and their genetic relation to geosynclines with a Proterozoic example from Great Slave Lake, Canada. Society of Economic Paleontologists and Mineralogists Special Publication, 19, 38–55.

Hopson, C. A., Mattinson, J. M., and Pessagno, E. A., Jr. (1981) Coast Range ophiolite, western California, in Ernst, W. G., ed., The geotectonic development of California (Rubey, vol. 1). Englewood Cliffs, NJ, Prentice Hall, 418–510.

Hopson, C. A., Mattinson, J. M., Pessagno, E. A., and Luyendyk, B. P. (2008) CaliforniaCoastRange ophiolite: composite Middle and Late Jurassic oceanic lithosphere. Geological Society of America Special Paper, 438, 1–101.

Hornafius, J. S., Luyendyk, B. P., Terres, R. R., and Kamerling, M. J. (1986) Timing and extent of Neogene tectonic rotation in the western Transverse Ranges, California. Geological Society of America Bulletin, 97, 1476–1487.

Horton, B. K. (1998) Sediment accumulation on top of the Andean orogenic wedge: Oligocene to late Miocene basins of the Eastern Cordillera, southern Bolivia. Geological Society of America Bulletin, 110, 1174–1192.

Horton, B. K. (2005) Revised deformation history of the central Andes: inferences from Cenozoic foredeep and intermontane basins of the Eastern Cordillera, Bolivia. Tectonics, 24, TC3011, 18 p.

Horton, B. K., and DeCelles, P. G. (1997) The modern foreland basin system adjacent to the Central Andes. Geology, 25, 895–898.

Horton, B. K., Hampton, B. A., LaReau, B. N., and Baldellon, E. (2002) Tertiary provenance history of the northern and central Altiplano (central Andes, Bolivia): A detrital record of plateau-margin tectonics. Journal of Sedimentary Research, 72, 711–726.

Howard, J. P., Cunningham, W. D., Davies, S. J., Dijkstra, A. H., and Badarch, G. (2003) The stratigraphic and structural evolution of the Dzereg basin, western Mongolia: clastic sedimentation, transpressional faulting and basin destruction in an intraplate, intracontinental setting. Basin Research, 15, 45–72.

Howell, D. G., Jones, D. L., and Schermer, E. R. (1985) Tectonostratigraphic terranes of the circum–Pacific region. Circum–Pacific Council for Energy and Mineral Resources Earth Sciences Series, 1, 3–30.

Howell, P. D., and van der Pluijm, B. A. (1999) Structural sequences and styles of subsidence in the Michigan basin. Geological Society of America Bulletin, 111, 974–991.

Hoy, R. G., and Ridgway, K. D. (1997) Structural and sedimentological development of footwall growth synclines along an intraforeland uplift, east–central Bighorn Mountains, Wyoming. Geological Society of America Bulletin, 109, 915–935.

Hoy, R. G., and Ridgway, K. D. (2002) Syndepositional thrust–related deformation and sedimentation in an Ancestral Rocky Mountains basin, central Colorado trough, Colorado, USA. Geological Society of America Bulletin, 114, 804–828.

Hsiao, L-Y., Graham, S. A., and Tilander, N. (2004) Seismic reflection imaging of a major strike–slip fault zone in a rift system: Paleogene structure and evolution of the Tan–Lu fault system, LiaodongBay, Bohai, offshore China. American Association of Petroleum Geologists Bulletin, 88, 71–97.

Hudec, M. R., Jackson, M. P. A., and Schultz-Ela, D. D. (2009) The paradox of minibasin subsidence into salt: Clues to the evolution of crustal basins. Geological Society of America Bulletin, 121, 201–221.

Hurwitz, S., Garfunkel, Z., Ben–Gai, Y., Reznikov, M., Rotstein, Y., and Gvirtzman, H. (2002) The tectonic framework of a complex pull–apart basin: seismic reflection observations in the Sea of Galilee, Dead Sea transform. Tectonophysics, 359, 289–306.

Ingersoll, R. V. (1979) Evolution of the Late Cretaceous forearc basin, northern and central California. Geological Society of America Bulletin, 90, part 1, 813–826.

Ingersoll, R. V. (1982) Initiation and evolution of the Great Valley forearc basin of northern and central California, U. S. A. Geological Society of London Special Publication, 10, 459–467.

Ingersoll, R. V. (1988) Tectonics of sedimentary basins. Geological Society of America Bulletin, 100, 1704–1719.

Ingersoll, R. V. (1997) Phanerozoic tectonic evolution of central California and environs. International Geology Review, 39, 957–972.

Ingersoll, R. V. (2000) Models for origin and emplacement of Jurassic ophiolites of northern California, in Dilek, Y., Moores, E. M., Elthon, D., and Nicolas, A., eds., Ophiolites and oceanic crust: new insights from field studies and the Ocean Drilling Program. Geological Society of America Special Paper, 349, 395–402.

Ingersoll, R. V. (2001) Structural and stratigraphic evolution of the Rio Grande rift, northern New Mexico and southern Colorado. International Geology Review, 43, 867–891.

Ingersoll, R. V. (2008a) Subduction–related sedimentary basins of the U. S. A. Cordillera, in Miall, A. D., ed., The sedimentary basins of the United States and Canada. Sedimentary Basins of the World, vol. 5. Amsterdam, Elsevier, 395–428.

Ingersoll, R. V. (2008b) Reconstructing southern California, in Spencer, J. E., and Titley, S. R., eds., Ores and orogenesis: Circum–Pacific tectonics, geologic evolution, and ore deposits.

Arizona Geological Society Digest 22, 409-417.

Ingersoll, R. V., and Busby, C. J. (1995) Tectonics of sedimentary basins, in Busby, C. J., and Ingersoll, R. V., eds., Tectonics of sedimentary basins. Oxford, Blackwell Science, 1-51.

Ingersoll, R. V., and Rumelhart, P. E. (1999) Three-stage evolution of the Los Angeles basin, southern California. Geology, 27, 593-596 (Correction, 27, 864).

Ingersoll, R. V., and Schweickert, R. A. (1986) A platetectonic model for Late Jurassic ophiolite genesis, Nevadan orogeny and forearc initiation, northern California. Tectonics, 5, 901-912.

Ingersoll, R. V., Graham, S. A., and Dickinson, W. R. (1995) Remnant ocean basins, in Busby, C. J., and Ingersoll, R. V., eds., Tectonics of sedimentary basins. Oxford, Blackwell Science, 363-391.

Ingersoll, R. V., Dickinson, W. R., and Graham, S. A. (2003) Remnant-ocean submarine fans: largest sedimentary systems on Earth, in Chan, M. A., and Archer, A. W., eds., Extreme depositional environments: mega end members in geologic time. Geological Society of America Special Paper, 370, 191-208.

Itoh, Y., Takemura, K., Ishiyama, T., Tanaka, Y., and Iwaki, H. (2000) Basin formation at a contractional bend of a large transcurrent fault: Plio-Pleistocene subsidence of the Kobe and northern Osaka basins, Japan. Tectonophysics, 321, 327-341.

Jacome, M. I., Kusznir, N., Audemard, F., and Flint, S. (2003) Formation of the Maturin foreland basin, eastern Venezuela: thrust sheet loading or subduction dynamic topography. Tectonics, 22, TC1046, 17 p.

Jarrard, R. D. (1986) Relations among subduction parameters. Reviews of Geophysics, 24, 217-284.

Jenkyns, H. C., and Wilson, P. A. (1999) Stratigraphy, paleoceanography, and evolution of Cretaceous Pacific Guyots: relics from a greenhouse Earth. American Journal of Science, 299, 341-392.

Johnson, C. L. (2004) Polyphase evolution of the East Gobi basin: sedimentary and structural records of Mesozoic-Cenozoic intraplate deformation in Mongolia. Basin Research, 16, 79-99.

Johnson, D. D., and Beaumont, C. (1995) Preliminary results from a planform kinematic model of orogen evolution, surface processes and the development of clastic foreland basin stratigraphy, in Dorobek, S. L., and Ross, G. M., eds., Stratigraphic evolution of foreland basins. SEPM (Society for Sedimentary Geology) Special Publication, 52, 3-24.

Jones, M. A., Heller, P. L., Roca, E., Garces, M., and Cabrera, L. (2004) Time lag of syntectonic sedimentation across an alluvial basin: theory and example from the Ebro basin, Spain. Basin Research, 16, 467-488.

Jordan, T. E. (1981) Thrust loads and foreland basin evolution, Cretaceous, western United States. American Association of Petroleum Geologists Bulletin, 65, 2506-2520.

Jordan, T. E. (1995) Retroarc foreland and related basins, in Busby, C. J., and Ingersoll, R. V., eds., Tectonics of sedimentary basins. Oxford, Blackwell Science, 331-362.

Jordan, T. E., and Allmendinger, R. W. (1986) The Sierras Pampeanas of Argentina: a modern analogue of Rocky Mountain foreland deformation. American Journal of Science, 286, 737-764.

Jordan, T. E., Isacks, B. L., Ramos, V. A., and Allmendinger, R. W. (1983a) Mountain building in the central Andes. Episodes, 1983, 20-26.

Jordan, T. E., Isacks, B. L., Allmendinger, R. W., Brewer, J. A., Ramos, V. A., and Ando, C. J. (1983b) Andean tectonics related to geometry of subducted Nazca plate. Geological Society of America Bulletin, 94, 341-361.

Kanamori, H. (1986) Rupture process of subduction-zone earthquakes. Annual Reviews of Earth and Planetary Science, 14, 293-322.

Kao, H., Gao, R., Rau, R-J., Shi, D., Chen, R-Y., Guan, Y., and Wu, F. T. (2001) Seismic image of the Tarim basin and its collision with Tibet. Geology, 29, 575-578.

Karig, D. E. (1970) Kermadec arc-New Zealand tectonic confluence. New Zealand Journal of Geology and Geophysics, 13, 21-29.

Karig, D. E., and Moore, G. F. (1975) Tectonically controlled sedimentation in marginal basins. Earth and Planetary Science Letters, 26, 233-238.

Karig, D. E., and Sharman, G. F., III (1975) Subduction and accretion in trenches. Geological Society of America Bulletin, 86, 377-389.

Katzman, R., tenBrink, U. S., and Lin, J. (1995) Three-dimensional modeling of pull-apart basins: implications for the tectonics of the Dead Sea basin. Journal of Geophysical Research, 100, 6295-6312.

Kay, M. (1951) North American geosynclines. Geological Society of America Memoir 48, 143 p.

Kellogg, K. S., and Minor, S. A. (2005) Pliocene transpressional modification of depositional basins by convergent thrusting adjacent to the "Big Bend" of the San Andreas fault: an example from Lockwood Valley, southern California. Tectonics, 24, TC1004, 12 p.

Kempf, O., and Pfiffner, O. A. (2004) Early Tertiary evolution of the north Alpine foreland basin of the Swiss Alps and adjoining areas. Basin Research, 16, 549-567.

Kimbrough, D. L. (1984) Paleogeographic significance of the Middle Jurassic Gran Canon Formation, Cedros Island, Baja California, in Frizzell, V. A., Jr., ed., Geology of the Baja California Peninsula. Pacific Section, Society of Economic Paleontologists and Mineralogists, Los Angeles, 107-118.

King, P. B. (1966) The North American Cordillera. Canadian Institute of Mining and Metallurgy Special, 8, 1-25.

Kinsman, D. J. J. (1975) Rift valley basins and sedimentary history of trailing continental margins, in Fischer, A. G. and Judson, S., eds., Petroleum and global tectonics. Princeton, NJ, PrincetonUniversity Press, 83-126.

Klein, G. deV. (1985) The control of depositional depth, tectonic uplift, and volcanism on sedimentation processes in the back-arc basins of the western Pacific Ocean. Journal of Geology, 93, 1-25.

Klein, G. deV. (1992) Climatic and tectonic sea-level gauge for midcontinent Pennsylvanian cyclothems. Geology, 20, 363-366.

Klein, G. D. (1995) Intracratonic basins, in Busby, C. J., and Ingersoll, R. V., eds., Tectonics of sedimentary basins. Oxford, Blackwell Science, 459-478.

Klein, G. deV., and Kupperman, J. B. (1992) Pennsylvanian cyclothems: methods of distinguishing tectonicallyinduced changes in sea level from climatically-induced changes. Geological Society of America Bulletin, 104, 166-175.

Kluth, C. F. (1986) Plate tectonics of the Ancestral Rocky Mountains. American Association of Petroleum Geologists Memoir 42, 353-369.

Kluth, C. F., and Coney, P. J. (1981) Plate tectonics of the ancestral Rocky Mountains. Geology, 9, 10-15.

Kopp, H., and Kukowski, N. (2003) Backstop geometry and accretionary mechanics of the Sunda margin. Tectonics, 22, TC1072, 16.

Koukouvelas, I. K., and Aydin, A. (2002) Fault structure and related basins of the North Aegean Sea and its surroundings. Tectonics, 21, TC1046, 17.

Lash, G. G. (1988) Along-strike variations in foreland basin evolution: possible evidence for continental collision along an irregular margin. Basin Research, 1, 71-83.

Lawton, T. F. (1994) Tectonic setting of Mesozoic sedimentary basins, Rocky Mountain region, United States, in Caputo, M. V., Peterson, J. A., and Franczyk, K. J., eds., Mesozoic systems of the Rocky Mountain region, USA. Rocky Mountain Section, Society for Sedimentary Geology, Denver, 1-25.

Lawton, T. F. (2008) Laramide sedimentary basins, in Miall, A. D., ed., The sedimentary basins of the United States and Canada. Sedimentary Basins of the World, vol. 5. Amsterdam, Elsevier, 429-450.

Lazar, M., Ben-Avraham, Z., and Schattner, U. (2006) Formation of sequential basins along a strike-slip fault: geophysical observations from the Dead Sea basin. Tectonophysics, 421, 53-69.

Leeder, M. R. (1995) Continental rifts and proto-oceanic rift troughs, in Busby, C. J., and Ingersoll, R. V., eds., Tectonics of sedimentary basins. Oxford, Blackwell Science, 119-148.

Leeder, M. R., and Gawthorpe, R. L. (1987) Sedimentary models for extensional tilt-block/half-graben basins. Geological Society of London Special Publication, 28, 139-152.

Leever, K. A., Bertotti, G., Zoetemeijer, R., Matenco, L., and Cloetingh, S. A. P. L. (2006) The effects of lateral variation in lithospheric strength on foredeep evolution: implications for the East Carpathian foredeep. Tectonophysics, 421, 251-267.

Leverenz, A. (2000) Trench-sedimentation versus accreted submarine fan: an approach to regional-scale facies analysis in a Mesozoic accretionary complex: "Torlesse" terrane, northeastern North Island, New Zealand. Sedimentary Geology, 132, 125-160.

Lin, A. T., and Watts, A. B. (2002) Origin of the west Taiwan basin by orogenic loading and flexure of a rifted continental margin. Journal of Geophysical Research, 107, n. B9, 2185.

Lipman, P. W. (1992) Magmatism in the Cordilleran United States; progress and problems, in Burchfiel, B. C., Lipman, P. W., and Zoback, M. L., eds., The Cordilleran orogen: conterminous U. S. Geology of North America, G3. Boulder, CO, Geological Society of America, 481-514.

Lonsdale, P. (1975) Sedimentation and tectonic modification of the Samoan archipelagic apron. American Association of Petroleum Geologists Bulletin, 59, 780–798.

Lonsdale, P. (1991) Structural patterns of the Pacific floor offshore of peninsular California. American Association of Petroleum Geologists Memoir 47, 87–125.

Lucente, C. C. (2004) Topography and palaeogeographic evolution of a middle Miocene foredeep basin plain (northern Apennines, Italy). Sedimentary Geology, 170, 107–134.

Luyendyk, B. P. (1991) A model for Neogene crustal rotations, transtension, and transpression in southern California. Geological Society of America Bulletin, 103, 1528–1536.

Luyendyk, B. P., and Hornafius, J. S. (1987) Neogene crustal rotations, fault slip, and basin development in southern California, in Ingersoll, R. V., and Ernst, W. G., eds., Cenozoic basin development of coastal California (Rubey, vol. 6). Englewood Cliffs, NJ, Prentice Hall, 259–283.

Luyendyk, B. P., Kamerling, M. J., and Terres, R. (1980) Geometric model for Neogene crustal rotations in southern California. Geological Society of America Bulletin, 91, Part I, 211–217.

Mack, G. H., Seager, W. R., and Leeder, M. R. (2003) Synclinal-horst basins: examples from the southern Rio Grande rift and southern transition zone of southwestern New Mexico, USA. Basin Research, 15, 365–377.

Maillard, A., and Mauffret, A. (1999) Crustal structure and riftogenesis of the Valencia Trough (north-western Mediterranean Sea). Basin Research, 11, 357–379.

Maldonado, A., Larter, R. D., and Aldaya, F. (1994) Forearc tectonic evolution of the South Shetland margin, Antarctic Peninsula. Tectonics, 13, 1345–1370.

Mann, P. (1997) Model for the formation of large, transtensional basins in zones of tectonic escape. Geology, 25, 211–214.

Mann, P., Hempton, M. R., Bradley, D. C., and Burke, K. (1983) Development of pull-apart basins. Journal of Geology, 91, 529–554.

Marin, L. E., Sharpton, V. L., Fucugauchi, J. U., Smit, J., Sikora, P., Carney, C., Rebolledo-Vieyra, M. (2001) Stratigraphy at ground zero: a contemporary evaluation of well data in the Chicxulub impact basin. International Geology Review, 43, 1145–1149.

Marsaglia, K. M. (1995) Interarc and backarc basins, in Busby, C. J., and Ingersoll, R. V., eds., Tectonics of sedimentary basins. Oxford, Blackwell Science, 299–329.

Marshak, S., Karlstrom, K., and Timmons, J. M. (2000) Inversion of Proterozoic extensional faults: an explanation for the pattern of Laramide and Ancestral Rockies intracratonic deformation, United States. Geology, 28, 735–738.

Mascle, A., Puigdefabregas, C., Luterbacher, H. P., and Fernandez, M., eds. (1998) Cenozoic foreland basins of western Europe. Geological Society of London Special Publication, 134, 427 p.

Mattei, M., Cipollari, P., Cosentino, D., Argentieri, A., Rossetti, F., Speranza, F., and Di Bella, L. (2002) The Miocene tectono-sedimentary evolution of the southern Tyrrhenian Sea: stratigraphy, structural and palaeomagnetic data from the on-shore Amantea basin (Calabrian arc, Italy). Basin Research, 14, 147–168.

Mattern, F. (2001) Permo-Silesian movements between Baltica and western Europe: tectonics and "basin families". Terra Nova, 13, 368-375.

May, S. R., Ehman, K. D., Gray, G. G., and Crowell, J. C. (1993) Anew angle on the tectonic evolution of the Ridge basin, a "strike-slip" basin in southern California. Geological Society of America Bulletin, 105, 1357-1372.

McBride, J. H., and Nelson, W. J. (1999) Style and origin of mid-Carboniferous deformation in the Illinois basin, USA: Ancestral Rockies deformation? Tectonophysics, 305, 249-273.

McClay, K. R., ed. (1992) Thrust tectonics. London, Chapman and Hall, 447 p.

McClay, K., and Bonora, M. (2001) Analog models of restraining stepovers in strike-slip fault systems. American Association of Petroleum Geologists Bulletin, 85, 233-260.

Meng, Q-R., Wange, E., and Hu, J-M. (2005) Mesozoic sedimentary evolution of the northwest Sichuan basin: implication for continued clockwise rotation of the South China block. Geological Society of America Bulletin, 117, 396-410.

Miall, A. D. (1995) Collision-related foreland basins, in Busby, C. J., and Ingersoll, R. V., eds., Tectonics of sedimentary basins. Oxford, Blackwell Science, 393-424.

Miall, A. D. (2008) The southern midcontinent, Permian basin, and Ouachitas, in Miall, A. D., ed., The sedimentary basins of the United States and Canada Sedimentary Basins of the World, vol. 5. Amsterdam, Elsevier, 297-327.

Miyashiro, A. (1973) Metamorphism and metamorphic belts. New York, John Wiley, 492 p.

Molnar, and Atwater, T. (1978) Interarc spreading and Cordilleran tectonics as alternates related to age of subducted oceanic lithosphere. Earth and Planetary Science Letters, 41, 330-340.

Molnar, P., and Tapponnier, P. (1975) Cenozoic tectonics of Asia: effects of a continental collision. Science, 189, 419-426.

Moore, G. F., and Karig, D. E. (1976) Development of sedimentary basins on the lower trench slope. Geology, 4, 693-697.

Moore, G. W. (1973) Westward tidal lag as the driving force of plate tectonics. Geology, 1, 99-100.

Morgan, P., and Baker, B. H. (1983) Introduction: processes of continental rifting. Tectonophysics, 94, 1-10.

Moss, S. J., and McCarthy, A. J. (1997) Foreland basin systems: discussion. Basin Research, 9, 171-172.

Mountney, N. P., and Westbrook, G. K. (1996) Modelling sedimentation in ocean trenches: the Nankai Trough from 1 Ma to the present. Basin Research, 8, 85-101.

Mountney, N. P., and Westbrook, G. K. (1997) Quantitative analysis of Miocene to Recent forearc basin evolution along the Colombian convergent margin. Basin Research, 9, 177-196.

Murphy, J. B., Keppie, J. D., and Nance, R. D. (1999) Fault reactivation within Avalonia: plate margin to continental interior deformation. Tectonophysics, 305, 183-204.

Najman, Y., and Garzanti, E. (2000) Reconstructing early Himalayan tectonic evolution and paleogeography from Tertiary foreland basin sedimentary rocks, northern India. Geological Society of America Bulletin, 112, 435-449.

Najman, Y., Johnson, K., White, N., and Oliver, G. (2004) Evolution of the Himalayan foreland basin, NW India. Basin Research, 16, 1–24.

Naylor, M., and Sinclair, H. D. (2008) Pro- vs. retro-foreland basins. Basin Research, 20, 285–303. Nicolas, A. (1985) Novel type of crust produced during continental rifting. Nature, 315, 112–115.

Nicholson, C., Sorlien, C. C., Atwater, T., Crowell, J. C., and Luyendyk, B. P. (1994) Microplate capture, rotation of the western Transverse Ranges, and initiation of the San Andreas transform as a low-angle fault system. Geology, 22, 491–495.

Nilsen, T. H., and Sylvester, A. G. (1995) Strike-slip basins, in Busby, C. J., and Ingersoll, R. V., eds., Tectonics of sedimentary basins. Oxford, Blackwell Science, 425–457.

Norris, R. J., and Cooper, A. F. (1995) Origin of small-scale segmentation and transpressional thrusting along the Alpine fault, New Zealand. Geological Society of America Bulletin, 107, 231–240.

Okay, A. I., Sengor, A. M. C., and Gorur, N. (1994) Kinematic history of the opening of the Black Sea and its effect on the surrounding regions. Geology, 22, 267–270.

Okay, A. I., Demirbag, E., Kurt, H., Okay, N., and Kuscu, I. (1999) An active, deep marine strike-slip basin along the north Anatolian fault in Turkey. Tectonics, 18, 129–147.

Okay, A. I., Tuysuz, O., and Kaya, S. (2004) From transpression to transtension: changes in morphology and structure around a bend on the North Anatolian fault in the Marmara region. Tectonophysics, 391, 259–282.

Oldow, J. S. (1984) Evolution of a late Mesozoic back-arc fold and thrust belt in northwestern Great Basin, U. S. A. Tectonophysics, 102, 245–274.

Oldow, J. S., Bally, A. W., Ave Lallement, H. G., and Leeman, W. P. (1989) Phanerozoic evolution of the North American Cordillera; United States and Canada, in Bally, A. W., and Palmer, A. R., eds., The geology of North America; an overview (The Geology of North America Volume A). Boulder, CO, Geological Society of America, 139–232.

Ori, G. G., and Friend, P. F. (1984) Sedimentary basins formed and carried piggyback on active thrust sheets. Geology, 12, 475–478.

Parnell, J., Osinski, G. R., Lee, P., Green, P. F., and Baron, M. J. (2005) Thermal alteration of organic matter in an impact crater and the duration of postimpact heating. Geology, 33, 373–376.

Parsons, B., and Sclater, J. G. (1977) An analysis of the variation of ocean floor bathymetry and heat flow with age. Journal of Geophysical Research, 82, 803–827.

Pascucci, V., Merlin, S., and Martini, I. P. (1999) Seismic stratigraphy of the Miocene-Pleistocene sedimentary basins of the northern Tyrrhenian Sea and western Tuscany (Italy). Basin Research, 11, 337–356.

Pfiffner, O. A., Schlunegger, F., and Buiter, S. J. H. (2002) The Swiss Alps and their peripheral foreland basin: stratigraphic response to deep crustal processes. Tectonics, 21, TC1054, 16.

Pindell, J. L. (1985) Alleghenian reconstruction and subsequent evolution of the Gulf of Mexico,

Bahamas, and proto-Caribbean. Tectonics, 4, 1-39.

Pindell, J., and Dewey, J. F. (1982) Permo-Triassic reconstruction of western Pangea and the evolution of the Gulf of Mexico/Caribbean region. Tectonics, 1, 179-211.

Pivnik, D. A., and Khan, M. J. (1996) Transition from foreland- to piggyback-basin deposition, Plio-Pleistocene upper Siwalik Group, ShingharRange, NW Pakistan. Sedimentology, 43, 631-646.

Pivnik, D. A., and Wells, N. A. (1996) The transition from Tethys to the Himalaya as recorded in northwest Pakistan. Geological Society of America Bulletin, 108, 1295-1313.

Platt, J. P. (1986) Dynamics of orogenic wedges and the uplift of high-pressure metamorphic rocks. Geological Society of America Bulletin, 97, 1037-1053.

Pope, K. O., Ocampo, A. C., and Duller, C. E. (1991) Mexican site for K/T impact crater. Nature, 351, 105.

Prather, B. E., Booth, J. R., Steffens, G. S., and Craig, P. A. (1998) Classification, lithologic calibration, and stratigraphic succession of seismic facies of intraslope basins, deep-water Gulf of Mexico. American Association of Petroleum Geologists Bulletin, 82, 701-728.

Price, R. A. (1973) Large-scale gravitational flow of supracrustal rocks, southern Canadian Rockies, in DeJong, K. A., and Scholten, R., eds., Gravity and tectonics. New York, John Wiley and Sons, 491-502.

Purser, B. H., and Bosence, D. W. J., eds. (1998) Sedimentation and tectonics in rift basins: Red Sea: Gulf of Aden. New York, Chapman and Hall, 663.

Quinlan, G. M. (1987) Models of subsidence mechanisms in intracratonic basins, and their applicability to North American examples. Canadian Society of Petroleum Geologists Memoir 12, 463-481.

Rahe, B., Ferrill, D. A., and Morris, A. P. (1998) Physical analog modeling of pull-apart basin evolution. Tectonophysics, 285, 21-40.

Rangin, C., LePichon, X., Demirbag, E., and Imren, C. (2004) Strain localization in the Sea of Marmara: propagation of the north Anatolian fault in a now inactive pull-apart. Tectonics, 23, TC2014, 18.

Reading, H. G. (1980) Characteristics and recognition of strike-slip fault systems. International Association of Sedimentologists Special Publication, 4, 7-26.

Ricketts, B. D. (2008) Cordilleran sedimentary basins of western Canada record 180 million years of terrane accretion, in Miall, A. D., ed., The sedimentary basins of the United States and Canada Sedimentary Basins of the World, vol. 5. Amsterdam, Elsevier, 363-394.

Riggs, N. R., and Busby-Spera, C. J. (1991) Facies analysis of an ancient, dismembered, large caldera complex and implications for intra-arc subsidence: Middle Jurassic strata of Cobre Ridge, southern Arizona, USA. Sedimentary Geology, 74, 39-68.

Ritts, B. D., Weislogel, A., Graham, S. A., and Darby, B. J. (2009) Mesozoic tectonics and sedimentation of the giant polyphase nonmarine intraplate Ordos basin, western North China block. International Geology Review, 51, 95-115.

Robertson, A. H. F. (1989) Palaeoceanography and tectonic setting of the JurassicCoastRange ophiolite, central California: evidence from the extrusive rocks and the volcaniclastic sediment cover. Marine and Petroleum Geology, 6, 194-220.

Robertson, A., and Degnan, P. (1994) The Dras arc complex: lithofacies and reconstruction of a Late Cretaceous oceanic volcanic arc in the Indus suture zone, Ladakh Himalaya. Sedimentary Geology, 92, 117-145.

Rollet, N., Deverchere, J., Beslier, M-O., Guennoc, P., Rehault, J. P., Sosson, M., and Truffert, C. (2002) Back arc extension, tectonic inheritance, and volcanism in the Ligurian Sea, Western Mediterranean. Tectonics, 21 (3), 23.

Root, S., and Onasch, C. M. (1999) Structure and tectonic evolution of the transitional region between the central Appalachian foreland and interior cratonic basins. Tectonophysics, 305, 205-223.

Rosendahl, B. R. (1987) Architecture of continental rifts with special reference to East Africa. Annual Review of Earth and Planetary Sciences, 15, 445-503.

Rowan, M. G., Lawton, T. F., Giles, K. A., and Ratliff, R. A. (2003) Near-salt deformation in La Popa basin, Mexico, and the northern Gulf of Mexico: a general model for passive diapirism. American Association of Petroleum Geologists Bulletin, 87, 733-756.

Rowley, D. B., and Kidd, W. S. F. (1981) Stratigraphic relationships and detrital composition of the medial Ordovician flysch of western NewEngland: implications for the tectonic evolution of the Taconic orogeny. Journal of Geology, 89, 199-218.

Salvador, A., ed. (1991) The Gulf of Mexico basin [The Geology of North America, J]. Boulder, CO, Geological Society of America, 568.

Saylor, B. Z. (2003) Sequence stratigraphy and carbonatesiliciclastic mixing in a terminal Proterozoic foreland basin, Urusis Formation, Nama Group, Namibia. Journal of Sedimentary Research, 73, 264-279.

Saylor, J., DeCelles, P., Gehrels, G., Murphy, M., Zhang, R., and Kapp, P. (2010) Basin formation in the highHimalaya by arc-parallel extension and tectonic damming: Zhada basin, southwestern Tibet. Tectonics, 29, TC1004, 24.

Schellart, W. P. (2007) The potential influence of subduction zone polarity on overriding plate deformation, trench migration and slab dip angle. Tectonophysics, 445, 363-372.

Schellart, W. P. (2008) Overriding plate shortening and extension above subduction zones. a parametric study to explain formation of the AndesMountains. Geological Society of America Bulletin, 120, 1441-1454.

Schermer, E. R., and Busby, C. (1994) Jurassic magmatism in the central Mojave Desert: implications for arc paleogeography and preservation of continental volcanic sequences. Geological Society of America Bulletin, 106, 767-790.

Schneider, C. L., Hummon, C., Yeats, R. S., and Huftile, G. L. (1996) Structural evolution of the northern Los Angeles basin, California, based on growth strata. Tectonics, 15, 341-355.

Scholl, D. W., Buffington, E. C., and Marlow, M. S. (1975) Plate tectonics and the structural evolution of the Aleutian- Bering Sea region. Geological Society of America Special Paper, 151, 1-31.

Scholl, D. W., Von Huene, R., Vallier, T. L. and Howell, D. G. (1980) Sedimentary masses and concepts about tectonic processes at underthrust ocean margins. Geology, 8, 564-568.

Schweickert, R. A. (1976) Shallow-level plutonic complexes in the eastern Sierra Nevada, California and their tectonic implications. Geological Society of America Special Paper, 176, 58.

Schweickert, R. A. (1978) Triassic and Jurassic paleogeography of the Sierra Nevada and adjacent regions, California and western Nevada, in Howell, D. G., and McDougall, K. A., eds., Mesozoic paleogeography of the western United States. Pacific Section, Society of Economic Paleontologists and MineralogistsPacificCoast Paleogeography Symposium 2, 361-384.

Schweller, W. J. and Kulm, L. D. (1978) Depositional patterns and channelized sedimentation in active eastern Pacific trenches, in Stanley, D. J. and Kelling, G., eds., Sedimentation in submarine canyons, fans, and trenches. Stroudsburg, PA, Dowden, Hutchinson and Ross, Inc., 311-324.

Sclater, J. G., Anderson, R. N., and Bell, M. L. (1971) Elevation of ridges and evolution of the central eastern Pacific. Journal of Geophysical Research, 76, 7888-7915.

Seeber, L., Emre, O., Cormier, M-H., Sorlien, C. C., McHugh, C. M. G., Polonia, A., Ozer, N., and Cagatay, N. (2004) Uplift and subsidence from oblique slip: the Ganos-Marmara bend of the North Anatolian transform, western Turkey. Tectonophysics, 391, 239-258.

Sengor, A. M. C. (1976) Collision of irregular continental margins: implications for foreland deformation of Alpine-type orogens. Geology, 4, 779-782.

Sengor, A. M. C. (1995) Sedimentation and tectonics of fossil rifts, in Busby, C. J., and Ingersoll, R. V., eds., Tectonics of sedimentary basins. Oxford, Blackwell Science, 53-117.

Sengor, A. M. C. and Burke, K. (1978) Relative timing of rifting and volcanism on Earth and its tectonic implications. Geophysical Research Letters, 5, 419-421.

Sengor, A. M. C., Burke, K., and Dewey, J. F. (1978) Rifts at high angles to orogenic belts: tests for their origin and the upper Rhine graben as an example. American Journal of Science, 278, 24-40.

Sengor, A. M. C., Cin, A., Rowley, D. B., and Nie, S. (1991) Magmatic evolution of the Tethysides: a guide to reconstruction of collage history. Palaeogeography, Palaeoclimatology, Palaeoecology, 87, 411-440.

Sengor, A. M. C., Graham, S. A., and Biddle, K. T. (1996) Is the Tarim basin underlain by a Neoproterozoic oceanic plateau? Geological Society of America Abstracts with Programs, 28 (7), A-67.

Shah, A. K., Brozena, J., Vogt, P., Daniels, D., and Plescia, J. (2005) New surveys of the Chesapeake Bay impact structure suggest melt pockets and target-structure effect. Geology, 33, 417-420.

Shatsky, N. S. (1964) O progibach Donetskogo typa, in Akademik N. S. Shatsky, Izbrannie Trudi, 2. "Nauka," Moscow, 544-552.

Sibuet, J-C., Deffontaines, B., Hsu, S-K., Thareau, N., Le Formal, J-P., Liu, C-S., and

ACT party (1998) Okinawa trough backarc basin: early tectonic and magmatic evolution. Journal of Geophysical Research, 103, 30245-30267.

Sinclair, H. D. (1997) Tectonostratigraphic model for underfilled peripheral foreland basins: an Alpine perspective. Geological Society of America Bulletin, 109, 324-346.

Sitian, L., Shigong, Y., and Jerzykiewicz, T. (1995) Upper Triassic-Jurassic foreland sequences of the Ordos basin in China. SEPM (Society for Sedimentary Geology) Special Publication, 52, 233-241.

Sloss, L. L. (1988) Tectonic evolution of the craton in Phanerozoic time, in Sloss, L. L., ed., Sedimentary cover: North American craton: U. S. The Geology of North America, D-2. Boulder, CO, Geological Society of America, 25-51.

Smit, J., Roep, T. B., Alvarez, W., Montanari, A., Claeys, P., Grajales-Nishimura, J. M., and Bermu'dez-Santana, J. (1996) Coarse-grained, clastic sandstone complex at the K/T boundary around the Gulf of Mexico: Deposition by tsunami waves induced by the Chicxulub impact? Geological Society of America Special Paper, 307, 151-186.

Smith, G. A., and Landis, C. A. (1995) Intra-arc basins, in Busby, C. J., and Ingersoll, R. V., eds., Tectonics of sedimentary basins. Oxford, Blackwell Science, 263-298.

Sobel, E. R. (1999) Basin analysis of the Jurassic-Lower Cretaceous southwest Tarim basin, northwest China. Geological Society of America Bulletin, 111, 709-724.

Sobel, E. R., and Strecker, M. R. (2003) Uplift, exhumation and precipitation: tectonic and climatic control of late Cenozoic landscape evolution in the northern Sierras Pampeanas, Argentina. Basin Research, 15, 431-451.

Soreghan, G. S. (1994) Stratigraphic responses to geologic processes: Late Pennsylvanian eustasy and tectonics in the Pedregosa and Orogrande basins, Ancestral Rocky Mountains. Geological Society of America Bulletin, 106, 1195-1211.

Sowerbutts, A. (2000) Sedimentation and volcanism linked to multiphase rifting in an Oligo-Miocene intra-arc basin, Anglona, Sardinia. Geological Magazine, 137, 395-418.

Sowerbutts, A. A., and Underhill, J. R. (1998) Sedimentary response to intra-arc extension: controls on Oligo-Miocene deposition, Sarcidano sub-basin, Sardinia. Journal of the Geological Society of London, 155, 491-508.

Speed, R. C., and Sleep, N. H. (1982) Antler orogeny and foreland basin: a model. Geological Society of America Bulletin, 93, 815-828.

Stanley, R. S., and Ratcliffe, N. M. (1985) Tectonic synthesis of the Taconian orogeny in western New England. Geological Society of America Bulletin, 96, 1227-1250.

Steckler, M. S., and tenBrink, U. S. (1986) Lithospheric strength variations as a control on new plate boundaries: examples from the northern Red Sea region. Earth and Planetary Science Letters, 79, 120-132.

Stein, C. A. and Stein, S. (1992) A model for the global variation in oceanic depth and heat flow with lithospheric age. Nature, 359, 123-129.

Stern, R. J., and Bloomer, S. H. (1992) Subduction zone infancy: examples from the Eocene Izu-Bonin-Mariana and Jurassic California arcs. Geological Society of America Bulletin, 104,

1621-1636.

Stewart, S. A. (2003) How will we recognize buried impact craters in terrestrial sedimentary basins? Geology, 31, 929-932.

Stewart, S. A., and Allen, P. J. (2002) A 20-km-diameter multi-ringed impact structure in the North Sea. Nature, 418, 520-523. (Also see Discussion and Reply, 428.)

Stewart, S. A., and Allen, P. J. (2005) 3D seismic reflection mapping of the Silverpit multi-ringed crater, North Sea. Geological Society of America Bulletin, 117, 354-368.

Stockmal, G. S., Beaumont, C., and Boutilier, R. (1986) Geodynamic models of convergent margin tectonics: transition from rifted margin to overthrust belt and consequences for foreland-basin development. American Association of Petroleum Geologists Bulletin, 70, 181-190.

Stollhofen, H., and Stanistreet, I. G. (1994) Interaction between bimodal volcanism, fluvial sedimentation and basin development in the Permo-Carboniferous Saar-Nahe basin (south-west Germany). Basin Research, 6, 245-267.

Talling, P. J., Lawton, T. F., Burbank, D. W., and Hobbs, R. S. (1995) Evolution of latest Cretaceous-Eocene nonmarine deposystems in the Axhandle piggyback basin of central Utah. Geological Society of America Bulletin, 107, 297-315. Tamaki, K., and Honza, E. (1991) Global tectonics and formation of maginal basins: role of the western Pacific. Episodes, 14, 224-230.

Tanner, P. W. G., and Rex, D. C. (1979) Timing of events on a Lower Cretaceous island-arc marginal basin system on South Georgia. Geological Magazine, 116, 167-179.

Tapponnier, P., Peltzer, G., LeDain, A. Y., Armijo, R., and Cobbold, P. (1982) Propagating extrusion tectonics in Asia: new insights from simple experiments with plasticine. Geology, 10, 611-616.

Tarapoanca, M., Garcia-Castellanos, D., Bertotti, G., Matenco, L., Cloetingh, S. A. P. L., and Dinu, C. (2004) Role of the 3-D distributions of load and lithospheric strength in orogenic arcs: polystage subsidence in the Carpathians foredeep. Earth and Planetary Science Letters, 221, 163-180.

Taylor, B. (2006) The single largest oceanic plateau: Ontong Java-Manihiki-Hikurangi. Earth and Planetary Science Letters, 241, 372-380.

Taylor, B., and Karner, G. D. (1983) On the evolution of marginal basins. Reviews of Geophysics and Space Physics, 21, 1727-1741.

TenBrink, U. S., Marshak, S., and Bruna, J-L. G. (2009) Bivergent thrust wedges surrounding oceanic island arcs: insight from observations and sandbox models of northeastern Caribbean plate. Geological Society of America Bulletin, 121, 1522-1536.

Thomas, W. A. (1983) Continental margins, orogenic belts, and intracratonic structures. Geology, 11, 270-272.

Thomas, W. A. (1985) The Appalachian-Ouachita connection: Paleozoic orogenic belt at the southern margin of North America. Annual Review of Earth and Planetary Sciences, 13, 175-199.

Thomas, W. A. (1991) The Appalachian-Ouachita rifted margin of southeastern North America.

Geological Society of America Bulletin, 103, 415–431.

Thomas, W. A. (1995) Diachronous thrust loading and fault partitioning of the Black Warrior foreland basin within the Alabama recess of the late Paleozoic Appalachian: Ouachita thrust belt. SEPM (Society for Sedimentary Geology) Special Publication, 52, 111–126.

Thornburg, T. M., and Kulm, L. D. (1987) Sedimentation in the Chile Trench: depositional morphologies, lithofacies, and stratigraphy. Geological Society of America Bulletin, 98, 33–52. (Also see Discussion and Reply, 99, 598–600.)

Tobisch, O. T., Saleeby, J. B., and Fiske, R. S. (1986) Structural history of continental volcanic arc rocks, eastern Sierra Nevada, California: a case for extensional tectonics. Tectonics, 5, 65–94.

Trop, J. M. (2008) Latest Cretaceous forearc basin development along an accretionary convergent margin: southcentral Alaska. Geological Society of America Bulletin, 120, 207–224.

Trop, J. M., and Ridgway, K. D. (1997) Petrofacies and provenance of a Late Cretaceous suture zone thrust – top basin, Cantwell basin, central Alaska Range. Journal of Sedimentary Research, 67, 469–485.

Trop, J. M., Ridgway, K. D., and Sweet, A. R. (2004) Stratigraphy, palynology, and provenance of the Colorado Creek basin, Alaska, USA: Oligocene transpressional tectonics along the central Denali fault system. Canadian Journal of Earth Sciences, 41, 457–480.

Tsikalas, F., Gudlaugsson, S. T., and Faleide, J. I. (1998) Collapse, infilling, and postimpact deformation of the Mjolnir impact structure, Barents Sea. Geological Society of America Bulletin, 110, 537–552.

Turner, J. P., Rosendahl, B. R., and Wilson, P. G. (2003) Structure and evolution of an obliquely sheared continental margin: Rio Muni, West Africa. Tectonophysics, 374, 41–55.

Uba, C. E., Heubeck, C., and Hulka, C. (2005) Faceis analysis and basin architecture of the Neogene subandean synorogenic wedge, southern Bolivia. Sedimentary Geology, 180, 91–123.

Underwood, M. B., and Moore, G. F. (1995) Trenches and trench-slope basins, in Busby, C. J., and Ingersoll, R. V., eds., Tectonics of sedimentary basins. Oxford, Blackwell Science, 179–219.

Underwood, M. B., Moore, G. F., Taira, A., Klaus, A., Wilson, M. E. J., Fergusson, C. L., Hirano, S., Steurer, J., and Leg 190 Shipboard Scientific Party (2003) Sedimentary and tectonic evolution of a trench-slope basin in the Nankai subduction zone of southwest Japan. Journal of Sedimentary Research, 73, 589–602.

Uyeda, S., and Ben-Avraham, Z. (1972) Origin and development of the Philippine Sea. Nature, 240, 176–178.

Uyeda, S., and Kanamori, H. (1979) Back-arc opening and the mode of subduction. Journal of Geophysical Research, 84, 1049–1061.

Vanbrabant, Y., Braun, J. and Jongmans, D. (2002) Models of passive margin inversion: implications for the Rhenohercynian fold-and-thrust belt, Belgium and Germany. Earth and Planetary Science Letters, 202, 15–29.

Van der Pluijm, B. A., Craddock, J. P., Graham, B. R., and Harris, J. H. (1997) Paleostress in cratonic North America: implications for deformation of continental interiors. Science, 277, 794–796.

Van der Werff, W. (1996) Variation in forearc basin development along the Sunda Arc, Indonesia. Journal of Southeast Asian Earth Sciences, 14, 331–349.

Vincent, S. J., and Allen, M. B. (1999) Evolution of the Minle and Chaoshui basins, China: implications for Mesozoic strike–slip basin formation in central Asia. Geological Society of America Bulletin, 111, 725–742.

Vincent, S. J., Allen, M. B., Ismail–Zadeh, A. D., Flecker, R., Foland, K. A., and Simmons, M. D. (2005) Insights from the Talysh of Azerbaijan into the Paleogene evolution of the South Caspian region. Geological Society of America Bulletin, 117, 1513–1533.

Vink, G. E., Morgan, W. J., and Zhao, W-L. (1984) Preferential rifting of continents: a source of displaced terranes. Journal of Geophysical Research, 89, 10072–10076.

Wakabayashi, J., Hengesh, J. V., and Sawyer, T. L. (2004) Four-dimensional transform fault processes: progressive evolution of step-overs and bends. Tectonophysics, 392, 279–301.

Waldron, J. W. F. (2004) Anatomy and evolution of a pullapart basin, Stellarton, Nova Scotia. Geological Society of America Bulletin, 116, 109–127.

Wernicke, B. P. (1992) Cenozoic extensional tectonics of the U. S. Cordillera, in Burchfiel, B. C., Lipman, P. W., and Zoback, M. L., eds., The Cordilleran orogen: conterminus U. S. The Geology of North America, G-3. Boulder, CO, Geological Society of America, 553–581.

Willett, S., Beaumont, C., and Fullsack, P. (1993) Mechanical model for the tectonics of doubly vergent compressional orogens. Geology, 21, 371–374.

Wilson, J. T. (1966) Did the Atlantic close and then re-open? Nature, 211, 676–681. Wilson, P. G., Turner, J. P., and Westbrook, G. K. (2003) Structural architecture of the ocean–continent boundary at an oblique transform margin through deep-imaging seismic interpretation and gravity modeling: Equatorial Guinea, West Africa. Tectonophysics, 374, 19–40.

Winker, C. D. (1996) High-resolution seismic stratigraphy of a late Pleistocene submarine fan ponded by salt-withdrawal minibasins on the Gulf of Mexico continental slope. Proceedings of 1996 Offshore Technology Conference, Houston, 619–662.

Winterer, E. L. (1973) Sedimentary facies and plate tectonics of equatorial Pacific. American Association of Petroleum Geologists Bulletin, 57, 265–282.

Withjack, M. O., Schlische, R. W., and Olsen, P. E. (1998) Diachronous rifting, drifting, and inversion on the passive margin of central eastern North America: an analog for other passive margins. American Association of Petroleum Geologists Bulletin, 82, 817–835.

Wood, R. A., Pettinga, J. R., Bannister, S., Lamarche, G., and McMorran, T. J. (1994) Structure of the Hanmer strikeslip basin, Hope fault, New Zealand. Geological Society of America Bulletin, 106, 1459–1473.

Woodcock, N. H. (2004) Life span and fate of basins. Geology, 32, 685–688.

Worrall, D. M., and Snelson, S. (1989) Evolution of the northern Gulf of Mexico, with emphasis on Cenozoic growth faulting and the role of salt, in Bally, A. W., and Palmer, A. R., eds., The geology of North America: an overview. The Geology of North America, A. Boulder, CO,

Geological Society of America, 91–138.

Wright, T. (1991) Structural geology and tectonic evolution of the Los Angeles basin, California. American Association of Petroleum Geologists Memoir, 52, 35–134.

Wyld, S. J. (2000) Triassic evolution of the arc and backarc of northwestern Nevada, and evidence for extensional tectonism. Geological Society of America Special Paper, 347, 185–207.

Wyld, S. J. (2002) Structural evolution of a Mesozoic backarc fold-and-thrust belt in the U. S. Cordillera: new evidence from northern Nevada. Geological Society of America Bulletin, 114, 1452–1468.

Yin, A., and Harrison, T. M. (2000) Geologic evolution of the Himalayan-Tibetan orogen. Annual Reviews of Earth and Planetary Sciences, 28, 211–280.

Yin, A., and Ingersoll, R. V. (1997) A model for evolution of Laramide axial basins in the southern Rocky Mountains, U. S. A. International Geology Review, 39, 1113–1123.

Yong, L., Allen, P. A., Densmore, A. L., and Qiang, X. (2003) Evolution of the Longmen Shan foreland basin (western Sichuan, China) during the Late Triassic Indosinian orogeny. Basin Research, 15, 117–138.

Zattin, M., Stefani, C., and Martin, S. (2003) Detrital fission-track analysis and sedimentary petrofacies as keys of Alpine exhumation: the example of the Venetian foreland (European southern Alps, Italy). Journal of Sedimentary Research, 73, 1051–1061.

Zoetemeijer, R., Tomek, C., and Cloetingh, S. (1999) Flexural expression of European continental lithosphere under the western outer Carpathians. Tectonics, 18, 843–861.

（崔敏 董治斌 译，张功成 屈红军 校）

第二部分　新技术与模拟

第 2 章 碎屑锆石的 U-Pb 测年：现状和展望

GEORGE GEHRELS

(Department of Geoscience, University of Arizona, Tucson, USA)

摘　要：得益于离子探针和激光剥蚀电感耦合等离子体质谱仪的普遍使用，碎屑锆石的地质年代学正迅速发展成为研究碎屑岩地层的源区和最大沉积年龄的有利工具。在大多数应用中，它们具有足够的有效性和准确性。尽管碎屑锆石年龄是逐年产生的，但是在数据采集、筛选、描绘和对比等方面均存在很大的不确定性。这些方法的未来发展方向包括更高的定年精度、从年龄谱中提取重要信息的更好的工具、从定年颗粒中获取其他类型的信息（如 Hf、O、Li 同位素特征，稀土元素模式，冷却年龄，构造信息），以及能够实现全球沉积层序碎屑锆石年龄共享的数据库平台的有效性。

关键词：锆石　地质年代学　碎屑岩　源区　分析方法

2.1 引言

近 20 年来，碎屑锆石地质年代学得到了迅速的发展，从一项应用极其有限的技术发展成为一种研究沉积单元及其源区的近乎独立的方法。这项被誉为"锆石革命"的重大进展，得益于技术的进步，使得研究者能够有效地确定单颗锆石晶体的 U-Pb 年龄（Davis 等，2003；Kosler 和 Sylvester，2003）。在此之前，大部分碎屑锆石分析只能在多颗粒锆石碎片上进行，大多数的做法是对相同特征（如颜色、形态、磨圆等）的颗粒群进行分析，寄希望于每个碎片的所有颗粒具有相同的年龄（LeDent 等，1964；Hart 和 Davis，1969；Girty 和 Wardlaw；1984；Erdmer 和 Baadsgaard，1987；Gehrels 等，1990；Ross 和 Bowring，1990）。这样，尽管可以获得充分的年龄分布，但并不足以识别特定的物源年龄。

20 世纪 80—90 年代初，随着离子探针（Froude 等，1983；Dodson 等，1988；Ireland，1992）和热电离质谱仪（ID-TIMS）（Davis 等，1989）的应用，使得单颗粒锆石晶体的分析成为可能，其中热电离质谱分析技术进展包括 ^{205}Pb 靶的制备（Krogh 和 Davis，1975；Parrish 和 Krogh，1987）和溶解低品位颗粒的聚四氟乙烯微型胶囊的设计（Parrish，1987）。80 年代末至 90 年代初，离子探针和热电离质谱技术在碎屑锆石地质年代学研究中被广泛使用。20 世纪 90 年代末至 21 世纪初，激光剥蚀电感耦合等离子体质谱（LA-ICPMS）成为 U-Pb 分析中最普遍的工具（Fryer 等，1993；Machado 和 Gauthier，1996；Horn 等，2000；Kosler 等，2001；Li 等，2001；Machado 和 Simonetti，2001；Horstwood 等，2003），而碎屑锆石地质年代学的研究得到了迅速的发展。

目前，碎屑锆石地质年代学主要应用于以下四个方面：

（1）源区的研究，即通过比较碎屑矿物年龄和潜在源区岩层年龄，确定最终的沉积物源。

（2）源区确定之后，可以进一步确定源区地质体的年龄和特征。

（3）通过比较碎屑矿物的年龄，研究沉积单元的关系，可以研究不同沉积单元之间可能存在的联系。

（4）确定最老的沉积年龄，即碎屑岩单元中最年轻组分可能提供了最老的沉积年龄。

尽管我们正经历着碎屑锆石分析数量的激增，然而我们还不能回答一些关于采集、展示和解释 U-Pb 地质年代学数据等基本问题，最近 Horstwood 等（2009）也对其中一些问题进行了概括，例如：

（1）碎屑锆石研究中最优化的装置是什么？

（2）应该使用哪些数据？如何评估和过滤这些数据？

（3）每个研究实例中应该进行多少个点的分析？如何选择和分析颗粒？

（4）展示碎屑锆石数据最有效的方式是什么？U-Pb 谐和图解、T-W 图解、直方图、年龄分布图还是累计概率图？

（5）描述一组碎屑锆石年龄数据（如最年轻年龄）最好的方法是什么？

（6）对比几个样品年龄分布特征最好的方法是什么？

本章内容可以视为关于这些问题的一个进展报告，重点介绍了收集和解释 U-Pb 信息的方法现状，并对碎屑矿物研究未来发展机遇进行了简单的展望。Fedo 等（2003）介绍了碎屑锆石地质热年代学的发展历史和应用，Gehrels（2000）、Anderson（2005）以及 Nemchin 和 Cawood（2005）讨论了碎屑锆石的数据统计分析方法，读者可以参考。

2.2 碎屑锆石研究的最优装置

通常情况下，用于碎屑锆石定年的装置有三种，以下对这三种装置的分析方法进行了简要概述，并评价了其在碎屑锆石 U-Pb 分析中的优势和弊端。

2.2.1 同位素稀释—热电离质谱（TD-TIMS）

同位素稀释热电离质谱分析包括：溶解完整晶体或晶体碎片、添加同位素示踪剂（通常为 ^{205}Pb 和 ^{235}U）、U 和 Pb 的化学分离以及同位素热电离质谱分析（Bowring 和 Schmitz，2003；Parrish 和 Noble，2003；Mattinson，2005）。其中，化学溶解和分离耗时较长，且需要在超净环境下进行，以减小 Pb 和 U 的污染而获得纯净的分析物，最终得到高精度的同位素比值（约 0.1%，2σ）和 Pb/U 浓度。因此，同位素稀释热电离质谱方法获得的 U-Pb 年龄具有最高的精确度和准确度，这对于高时间分辨率的应用非常重要。然而，大多数情况下，碎屑锆石研究并不需要如此高的精度。

2.2.2 二次离子质谱（SIMS 或离子探针）

二次离子质谱一般在晶体的抛光面上进行，这些晶体与标样一起安装在环氧基树脂上，标样与被分析晶体是同种矿物且年龄和同位素组成已知（Ireland 和 Williams，2003）。因为溅射仅仅在低温高真空环境下发生，所以 Pb 和 U 的本底值很低，且分析所需的矿物用量很小（约 1ng）。这使得针对直径 10~30μm、厚 1μm 的晶体碎片直接进行 U-Pb 定年成为可能。因为不可能添加同位素示踪剂，所以碎屑锆石年龄通过标样进行校正（变换标样和实际样品,把标样校正到已知年龄,把相同的校正因子应用到实测样品)。这种方法获得的年龄精度为1%~2% (2σ)。同位素测试按照U-Th-Pb顺序进行,分析时间为15min。对于需要高的空间分辨率的研究(尤其在深度域),比如分析复杂的锆石晶体,这是一种理想方法。同时,离子探针可以分析其他同位素,所以我们也可以根据Ti和Zr浓度、稀土元素浓度和氧同位素的特征来约束岩石成因(Mojzsis等,2001;Valley,2003;Wooden等,2007)。

2.2.3 激光剥蚀电感耦合等离子体质谱(LA-ICPMS)

激光剥蚀电感耦合等离子体质谱方法与二次电离质谱非常类似,如碎屑锆石年龄通过标样进行校正、在晶体抛光面进行分析(Kosler和Sylvester,2003)、获得的年龄精度同样为1%~2%(2σ;Machado和Simonetti,2001;Horstwood等,2003;Kosler和Sylvester,2003;Chang等,2006;Gehrels等,2008;Horstwood,2008)。这种分析方法的优势之一是分析时间较短。此外,由于装备了足够多的分离器和收集器,所以能够同时测量U和Pb的含量。但是,这种方法仍然存在一些弊端,如等离子化作用中包含了高速的高温高压Ar、Pb和Hg的本底增加。为了获得高的信号,本底需要快速的电离,大多数情况下分析矿物的用量也更大(直径10~30μm、厚1μm的晶体碎片)。这种快速的电离作用加快了分析效率,一般分析时间仅为几分钟。激光剥蚀电感耦合等离子体质谱同样适合分析其他元素,如Hf-Lu-Yb(Hf同位素确定)和稀土元素(Machado和Simonetti,2001;Woodhead等,2004;Gerdes和Zeh,2006;Flowerdew等,2007;Mueller等,2007;Yuan等,2008,Kemp等,2009)。

2.3 碎屑锆石年龄的选择、评估和过滤

地质年代学中U-Pb体系非常普遍,这是因为(1)包含两套衰减体系(^{238}U—^{206}Pb,^{235}U—^{207}Pb);(2)两套衰减体系的半衰期与地球年龄相匹配;(3)几乎所有的地壳岩石具有相同的^{238}U/^{235}U比值(137.88;Steiger和Jäger,1977),从而将两套衰减体系相关联;(4)非放射性同位素^{204}Pb可以计算在晶体形成时期Pb的存在。因此,U-Pb年龄可以用^{206}Pb*/^{238}U和^{207}Pb*/^{235}U的交会图展示(*表示去除了原始Pb之后;Wetherill,1956;图2-1)。此外,谐和图也展示了^{206}Pb*/^{207}Pb*,即从原点到分析点直线的斜率(^{206}Pb*/^{207}Pb* = ^{206}Pb*/^{238}U/ [^{207}Pb*/^{235}U×137.88])(图2-1)。

实际操作中,对于ID-TIMS,^{206}Pb*/^{238}U是通过与包含了已知量Pb和U(如^{205}Pb和^{238}U)的示踪剂的溶液的对比进行确定的,SIMS和LA-ICPMS则是通过与标样对比。^{206}Pb*/^{207}Pb*通常只需要进行细微的校正,因为在仪器分馏过程中,^{206}Pb相对于^{207}Pb的量非常少。一般情况下,^{207}Pb*/^{235}U不能直接测量,而是通过实测的^{206}Pb*/^{238}U、^{206}Pb*/^{207}Pb*和已知的^{238}U/^{235}U比值计算得到。相对于^{208}U,^{235}U的含量很低(1/137.88),因为^{235}U的测定会大大增加^{207}Pb*/^{235}U定年中的不确定性,所以很多情况下谐和图解中根据^{206}Pb*/^{207}Pb*(许多实验室也用^{207}Pb*/^{206}Pb*表示)和^{206}Pb*/^{238}U投图。图2-1展示了^{206}Pb*/^{238}U、^{207}Pb*/^{235}U和^{206}Pb*/^{207}Pb*的交会图以及谐和线。

^{206}Pb*/^{238}U、^{207}Pb*/^{235}U和^{206}Pb*/^{207}Pb*的不确定性一般会在谐和图中产生一个锆石形状的多边形,通常用累计概率密度函数表示,其不确定性的相对值与年龄相关;对于年轻样品,^{206}Pb*/^{238}U年龄精度最高,^{206}Pb*/^{207}Pb*精度最低;反之,对于古老样品,^{206}Pb*/^{207}Pb*

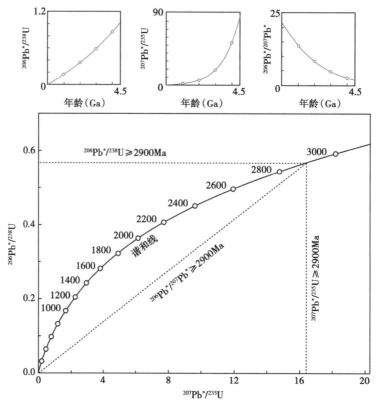

图 2-1 常用的三种地质年代计的 Pb*/U 谐和图解

谐和线表明三种地质年代计的年龄一致，注意 $^{206}Pb^*/^{238}U$ 和 $^{206}Pb^*/^{207}Pb^*$ 是两个独立的测量值，而 $^{207}Pb^*/^{235}U$ 是根据以上比值和现今 $^{238}U/^{235}U$ 比值（137.88）计算获得

年龄的精度要高于 $^{206}Pb^*/^{238}U$（见图 2-2）；$^{207}Pb^*/^{235}U$ 年龄的精度一般介于二者之间。理论和实际均表明，当年龄为 1.4Ga 时，三种年龄的不确定性一致（Gehrels，2000；Nemchin 和 Cawood，2005；Gehrels 等，2008）。由于这些变化，当年龄为 1.4Ga 时，U-Pb 体系的年龄分辨率最低；对于较年轻年龄，$^{206}Pb^*/^{238}U$ 精度较高；对于较古老年龄，$^{206}Pb^*/^{207}Pb^*$ 精度较高。

如果在误差范围内，$^{206}Pb^*/^{238}U$、$^{207}Pb^*/^{235}U$ 和 $^{206}Pb^*/^{207}Pb^*$ 年龄相似，那么在谐和图解中，分析点正好落在谐和线上，即所谓的"谐和"（Wetherill，1956）。然而，通常情况下，分析点往往落在谐和线下面，这种情况下成为"不谐和"，这导致了年龄值 $^{206}Pb^*/^{238}U$ < $^{207}Pb^*/^{235}U$ < $^{206}Pb^*/^{207}Pb^*$，这往往是由于年轻的热活动/水热活动中 Pb 的丢失或者更老矿物的继承造成的。谐和图解中很少会出现"反谐和"分析点，这种情况很可能是由于 $^{206}Pb^*/^{238}U$ 测试不精确造成的。不谐和度可以用 $^{206}Pb^*/^{238}U$ 年龄除以 $^{206}Pb^*/^{207}Pb^*$ 年龄来表述，在完全谐和的情况下，不谐和度等于"0"（不谐和度 = 100 - 100 × $^{206}Pb^*/^{238}U$ 年龄除以 $^{206}Pb^*/^{207}Pb^*$ 年龄）。相反，该比值也可以表示谐和度，在完全谐和的情况下，谐和度等于 100%（不谐和度 = 100 × $^{206}Pb^*/^{238}U$ 年龄除以 $^{206}Pb^*/^{207}Pb^*$ 年龄）。

因为随着年龄的增加，锆石对于 Pb 的丢失更加敏感（由于在放射性衰变过程中，晶格破坏随之增加），所以对于前寒武纪锆石，不谐和度仅为百分之几到百分之几十。图 2-3 中锆石结晶年龄为 1500Ma，在 500Ma 经历了 Pb 的丢失，$^{206}Pb^*/^{238}U$、$^{207}Pb^*/^{235}U$ 和 $^{206}Pb^*/^{207}Pb^*$ 年龄都小于锆石的结晶年龄，其中 $^{206}Pb^*/^{207}Pb^*$ 年龄与真实年龄最接近。因此，对于年龄小于 1.4Ga 的锆石，在合理的不谐和度（约 10%~30%）情况下，虽然 $^{206}Pb^*/^{238}U$ 年龄精确更高，但是 $^{206}Pb^*/^{207}Pb^*$ 年龄通常更加准确。$^{206}Pb^*/^{207}Pb^*$ 和 $^{206}Pb^*/^{238}U$ 年龄精度的分界点一般在 1.0~0.8Ga（Gehrels，2000；Gehrels 等，2008）。

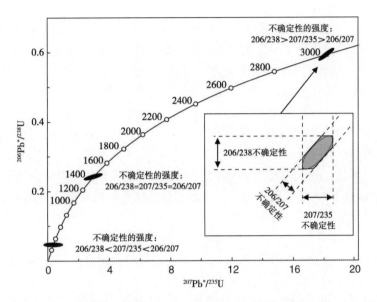

图 2-2 显生宙、元古宙和太古宙年龄不确定性椭圆的方向和大小
右面插图是三种地质年代计不确定性的图形描述

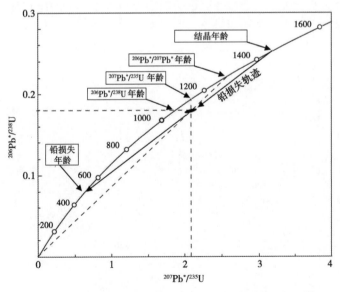

图 2-3 存在 Pb 丢失的中元古宙分析点碎屑锆石定年图解
对于 1400—800Ma 年龄范围的样品，尽管 $^{206}Pb^*/^{207}Pb^*$ 通常更加精确，但 $^{206}Pb^*/^{207}Pb^*$ 年龄较 $^{206}Pb^*/^{207}Pb^*$、$^{207}Pb^*/^{235}U$ 年龄更加准确

对于 Pb 丢失造成不谐和的分析点，$^{206}Pb^*/^{207}Pb^*$ 年龄甚至要比真实的结晶年龄年轻。即使分析的精度不变，随着不谐和度的增加和 Pb 丢失事件年龄的增加，$^{206}Pb^*/^{207}Pb^*$ 年龄的精度会随之降低。另外一种情况，如果不谐和是由于继承矿物引起的，那么$^{206}Pb^*/^{238}U$、$^{207}Pb^*/^{235}U$ 和$^{206}Pb^*/^{207}Pb^*$ 年龄都不准确，且不谐和度越大，不准确度随之增加。因此，通常情况下，碎屑锆石研究中需要根据不谐和度来对年龄数据进行过滤。但是，不谐和过滤合适的截点是多少呢？

不谐和过滤合适的等级需要基于具体的研究目的和复杂程度，根据具体的数据组进行确定。例如，研究中同时出现了显生宙和太古宙年龄，且其相对比例很重要，应该取较大的不谐和截点（30%），这样大部分前寒武纪年龄才能得以保留。如果研究的目的是为了检验某个特定的新太古代年龄，那么，为了获得最佳年龄分辨率，应该使用较小的不谐和截点（如 10%），甚至可以只使用谐和数据。最后，如果研究过程中只出现了年轻年龄（<100Ma），那么就很难获得可靠的$^{206}Pb^*/^{207}Pb^*$ 年龄，而且不和谐截点也很难确定，因此对于年轻年龄使用不谐和过滤，并不可靠。

此外，由于结晶之后不久发生 Pb 的丢失，和/或继承了老的组分，这些都将导致分析结果一致但获得的年龄并不准确，因此，不谐和过滤也不能保证获得准确无误的年龄。解决这一问题的方法是重点分析属于集群的点，因为 Pb 丢失和继承通常引起分析点以实际年龄为中心向外分散，这需要一个合理的方法对集群进行过滤（这里所说的群集是指在 2σ 不确定性下，三个或三个以上叠置的分析点）。不属于集群的分析点可能也是准确的，不应该被抛弃，但是它们的真实年龄的意义现在还不明确。

图 2-4 显示了应用集群过滤的一个实例，这组数据中包含了两组年龄：基于$^{206}Pb^*/^{207}Pb^*$ 叠置年龄获得的较老的年龄和基于$^{206}Pb^*/^{238}U$ 叠置年龄得出的较新的年龄。对于较古老的分析点，三组年龄群集被解释为记录了可靠的物源年龄，但分配给单个分析点和一对分析点时，这个解释并不合理。对于较年轻的分析点，可以解释为两组包含三个及其以上分析点的集群，分配给单个分析点和一对分析点时，解释也不合理。然而，随着不谐和度的增加

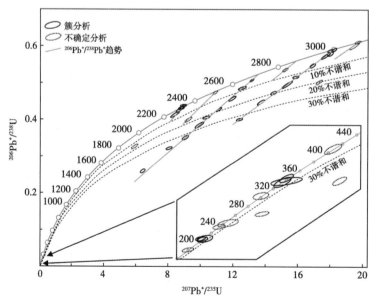

图 2-4　利用集群和不谐和过滤来评估碎屑锆石年龄的谐和图解

和 Pb 丢失会使事件的年龄增大（Pb 丢失轨迹偏离$^{206}Pb^*/^{207}Pb^*$ 谱线，图 2-4），根据群集过滤将产生更多的问题。在后面这种情况下，即使多个颗粒具有相同的结晶年龄，也不能根据$^{206}Pb^*/^{207}Pb^*$ 来定义群集。在明显的群集里面，群集分析仍然有可能掩盖真实的年龄变化。

Nemchin 和 Cawood（2005）建议使用另一种方法，即根据不谐和度进行单重分析，使得不谐和度很高的分析点对最终的年龄分布曲线贡献很小。这将解决必须确定一个特定不谐和过滤临界值的问题。但仍然需要确定适当的加权方式，以将对最终的年龄分布影响并不显著的不谐和分析点考虑在内。

由于这些复杂性，对每个数据组进行单独评价来确定每个分析点的稳定性就显得尤为重要。很显然，如果图 2-4 中显示的数据来自单个样品，不谐和过滤将使最终的年龄分布偏向年轻年龄，过滤临界值越小，偏差越大。同样地，当 Pb 丢失时代较新，群集过滤很适用，如果 Pb 丢失时代较老，这种方法便不能识别同期颗粒。一个合理的初始方法是使用中等的不谐和临界值（10%~30%），或适当的不谐和加权方式（Nemchin 和 Cawood, 2005）以消除或大大降低不谐和颗粒的影响，而不显著偏离数据组，然后依据群集评价能够被接受的数据，以突出最可靠的年龄群集。

目前，通过对单颗粒的多重分析来增加年龄值分配的可靠性正逐渐成为可能（Nemchin 和 Cawood, 2005；Simonetti 等, 2008；Johnston 等, 2009）。对于存在 Pb 丢失的颗粒，可以识别出经历了很少量 Pb 丢失的区域（如 U 浓度低的部分），利用 Pb 不同程度丢失量来生成 Pb 丢失轨迹，这条轨迹有着稳定的上截距。对于锆石多期生长的情况，单颗粒定年产生的结晶历史，不是单个年龄值，这种技术同样有用。图 2-5 是这种年龄影像方法的实例之一，其展示了单晶体上 140 个不同点位的分析结果。获得的年龄模式，与阴极发光成像进行叠加，将获得有利于重建源区的丰富的历史信息。

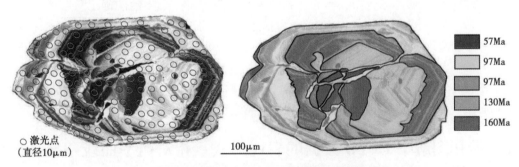

图 2-5　利用单颗粒多重少量分析来重建复杂的锆石生长历史

（据 Gehrels 等, 2009；阴极发光成像由加州大学洛杉矶分校离子探针设备 Axel Schmitt 获得）

2.4　分析过程中单个样品分析点数量和颗粒的选择

碎屑锆石研究中通常会关注两个基本问题：一是分析点的数量，二是如何选择颗粒（Gehrels, 2000；Fedo 等, 2003；Anderson, 2005）。这两个问题的回答很大程度上取决于研究的目的。

如果用碎屑锆石定年来约束最大沉积年龄，则应该集中分析颜色最浅、圆度最低的晶体颗粒，因为随着年龄的增大，颜色会逐渐加深（Silver 和 Deutsch，1963）。优先分析颜色最浅的晶体往往会产生更多的最年轻年龄。此外，由于圆度高，再循环的几率会增加，所以随着圆度的增高，年龄也会随之增大。分析过程中颗粒的数量应该取决于样品中最年轻颗粒的比例和最大沉积年龄所需要的可信度。

如果碎屑锆石分析是为了检验某预期组分的存在，那么同样应该按照上述的一般准则来选择锆石颗粒，即锆石年龄随着颜色的加深和圆度的增高而增大，因此，选择预期特征的锆石颗粒将最大可能地识别出预期组分的存在。分析颗粒的数量同样取决于预期组分的比例和识别预期组分所需要的可信度。

如果碎屑锆石分析是为了确定源区的基本特征，或者进行地层对比，那么自然需要选择反映原始样品锆石晶体真实的年龄分布和年龄比例的锆石颗粒。因此，这就必须谨慎，在加工过程中应避免如磁力分离过程中的对样品进行施压、按大小排序（无论是有意或在处理过程中），或者手工挑选（Gehrels，2000；Sircombe 和 Stern，2002；Fedo 等，2003；Anderson，2005；Nemchin 和 Cawood，2005）。分析过程中，应该从所有晶体中随机选择颗粒，不能考虑颗粒的大小、颜色、形状和圆度等。然而，必须考虑这种方法的例外：一是非常小的颗粒往往无法分析；二是破裂颗粒的年龄不可信（由于次生矿物、沿着破裂面 Pb 的丢失又或者是入射激光或一次离子束的异常行为）；三是避开包裹体；四是避开具有复杂年龄分带的颗粒，除非每个组分可以单独进行分析。因此，碎屑锆石晶体定位分析时最好使用 CL 成像。

源区研究中，所需分析点的数量取决于样品中存在的不同年龄组的数量和比例、晶体是否受到 Pb 丢失和/或继承的影响、所使用的分析方法的精度，以及得出某个特定的结论或检验某个特定假设所需要的置信水平。Anderson（2005）使用标准的二项式概率公式计算表明，对于占颗粒总数量 5% 的某个组分，进行 60 个点的分析，能够识别的概率为 95%。同样地，进行 117 个点的分析，能够识别占颗粒总数量 5% 的每一个组分的概率为 95%（Vermeesch，2004）。在实际研究中，因为计算阈值时，假定每一个年龄的确定只会产生一个确切年龄，所以这些阈值被低估。正如前面所描述的，由于 Pb 丢失和/或继承等的复杂性，通常情况并非如此。如果用年龄群集来标定，为了足够证明小的年龄总体被识别，就需要进行更多点的分析。大多数源区研究中，一个合理的初始方法是进行 100 个点的分析，使用不谐和过滤，这样较老分析点的数量降低的并不显著（如 10%~30%），而对属于集群的分析点影像很大（Gehrels 等，2008）。但是，这样的做法显然无法确切识别一组碎屑锆石的次要成分。

2.5 谐和图解中 U-Pb 数据的表现形式：相对概率曲线和累计概率曲线

如上所述，Pb^*/U 谐和图解是碎屑锆石数据最常见的表现和评价形式（图 2-6a）。然而，对于具有较宽年龄范围的样品，标准的 Pb^*/U 谐和图解往往掩盖了年轻颗粒。在这种情况下，有两种选择：一是坐标轴使用对数刻度（图 2-6b），二是使用 T-W 图解（Tera 和 Wasserburg，1972），即绘制 $^{207}Pb^*/^{206}Pb^*$ 与 $^{238}U/^{206}Pb^*$ 的交会图（图 2-6c）。

碎屑锆石年龄分布可以用简单的直方图表示，但年龄分布图（或相对年龄概率图）可以展示更加丰富的信息，因为这样关于各分析点的不确定性的信息也能够包含其中（Sircombe，2004；Sircombe 和 Hazelton，2004；Ludwig，2008）。年龄分布曲线由以下几部分组成：(1) 根据报告的年龄和不确定性，对各分析点设定正态（高斯）分布；(2) 对单一曲

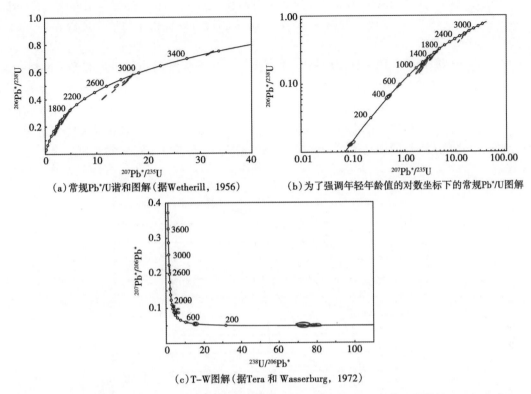

(a) 常规Pb*/U谐和图解（据Wetherill，1956）　　(b) 为了强调年轻年龄值的对数坐标下的常规Pb*/U图解

(c) T-W图解（据Tera 和 Wasserburg，1972）

图 2-6　一组碎屑锆石年龄的三种谐和图解（据 Ludwing，2008；三种图解均由软件 Isoplot 生成）

线内所有可以接受的分析点的概率分布求和；（3）如果归一化，用分析点的数量去除曲线下面的面积。图 2-7 展示了一系列归一化之后的年龄分布曲线，这样，某个样品年龄概率

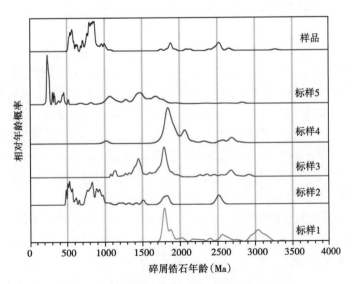

图 2-7　归一化年龄分布（相对年龄概率）图

该图展示了某样品与其他几个参考数据的年龄分布，图件由程序生成，
该程序可以在网站 www.geo.arizona.edu/alc 下载

分布曲线可以很方便地与其他几条参考曲线进行对比。年龄分布曲线可以用程序（Ludwig，2008；Sircombe，2004）生成，网站 www.geo.arizona.edu/alc 提供了将概率曲线整理并归一化的程序。

在一些具体应用中，用累计年龄概率曲线来展示碎屑年龄谱显得更加有利。图 2-8 展示的是一些累计概率曲线，它们与图 2-7 展示的数据一样。这两种平面图展示了完全相同的信息，即相对年龄曲线中的峰值对应着累计概率曲线中陡峭的一段。累计概率曲线同样可以用程序图形化，该程序可以在网站 www.geo.arizona.edu/alc 下载。如果需要将大量的曲线进行对比，或者感兴趣的样品年龄相似而百分比不同，这种曲线显得非常有用。

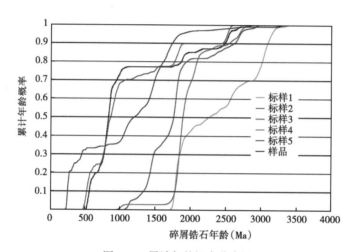

图 2-8　累计年龄概率分布图

该图展示的数据同图 2-7，图件由程序生成，该程序可以在网站 www.geo.arizona.edu/alc 下载

2.6　描述一组碎屑锆石年龄（如最小年龄）的最优方法

一般情况下，讨论碎屑锆石样品单个年龄并没有意义（或意义不大），所以描述分析点的年龄组或者年龄群集就显得很有必要，这可以通过年龄分布直方图很容易地展现，其缺点是引入了虚拟年龄边界且不能包含关于年龄不确定性的有用信息（如 Sircombe，2004）。在年龄分布图中使用年龄分组是另一个增加信息量的描述方法（图 2-7）。Agepick 程序首次尝试实现年龄系统描述，该软件可以在网站 www.geo.arizona.edu/alc 下载。这个程序通过三次或更多次叠置分析识别包含年龄概率的各年龄范围来强调主年龄组，然后给出各年龄范围分析点的个数，如 1682~1735Ma 范围共 8 个年龄值。同时，该程序还能给出各年龄范围的峰值年龄及其年龄值个数，如峰值年龄 1712Ma 共包含 4 个年龄值。

报告某个年龄范围最大和最小年龄，或者年龄概率中的峰值年龄的缺陷在于缺乏各年龄值相关的不确定性，即使清晰地知道了各分析点相关的不确定性之后依然如此。于是，通常需要计算一组年龄值的加权平均年龄和不确定性，Isoplot 可以实现这一功能（Ludwig，2008）。这一方法的建立基于假设所有包含的分析点都完全同源，年龄的分散仅仅是由于分析过程中的不确定性造成的。因为没有理由假定所有具有相似年龄的碎屑锆石都完全同源，且一些年龄值的不准确是由于 Pb 的丢失和/或继承等原因造成的，因此，在使用这种方法时需要特别谨慎。这一注意事项已经应用于所有的平均化过程，从计算简单的平均值到更加复杂的分析，如 Isoplot 中"TuffZirc"和"Unmixing"程序（Ludwig，2008）。

此外，一些补充信息可以用于评估年龄的复杂性，以及确定一组锆石颗粒是否同源。例如，U 的浓度和 U/Th 比值可以用于识别异常（非同源）晶体，确定异常年龄是否是由于 Pb 的丢失或者变质重结晶/增生作用造成的。识别这一信息的有效方法是绘制年龄与 U 的浓度和 U/Th 比值的交会图：年龄和 U 的浓度的相关性指示较年轻的年龄可能受到 Pb 丢失的影响，而年龄与 U/Th 比值的相关性则指示一些分析点受到变质重结晶/增生作用的影响。年龄和分析点位置（中心或边缘）的交会图可以反映继承性核部的存在或者颗粒边缘 Pb 的丢失。诸如此类的交会图均可用"AgePick"程序生成，该程序可以在网站 www.geo.arizona.edu/alc 下载。正如下文所述，这种多维分析将在今后得到进一步发展，以获得更多的定年晶体同位素、组成和结构的信息，并确定锆石颗粒组是否真的是同源的，深化对地质年代学复杂性的理解。

2.7 对比多样品的年龄分布的最优方法

获得碎屑锆石数据组之后，往往需要与该研究区内甚至是其他地区的样品进行比较。通常可以通过标准年龄分布（图 2-7）或者累计概率曲线（图 2-8）进行视觉对比。例如，从视觉比较来看，图 2-7 和图 2-8 中展示的年龄分布曲线与参考样品 2 类似，而与其他参考样品不同，该结论的得出基于这样的事实：被分析样品和参考样品 2 在大致相同的比例下包含相似的年龄。但是，叠置年龄和相似比例哪个更重要？又该如何定量描述呢？

Gehrels（2000）通过两个年龄分布曲线年龄叠合度的数值分析进行了早期定量比较的尝试。在叠合度分析中，如果在两个样品中没有相同年龄出现，叠合度取值为"0"；如果所有的年龄在两个样品中都有出现，叠合度取值为"1"。这种方式提供了特定年龄存在或者缺失的衡量方法。为了进一步考虑叠置年龄的比例，发展出第二个指标来即描述叠置年龄以相似比例存在的程度（"0"表示不同的比例或者非叠置年龄，"1"表示叠置年龄具有完全相同的比例）。以上标准都是描述性的，并不包含年龄分布中本身包含的关于叠合度和相似性的相关信息。叠合度和相似性分析软件可以从网站 www.geo.arizona.edu/alc 下载。

K-S 统计方法提供了多个年龄分布曲线的统计对比，它可以描述具有不同原始总体的年龄分布曲线的概率。基本标准是 P 值，它表征两个样品并非统计学上的不同的概率。例如，$P>0.05$、可信度>95%表明两个样品并非统计学不同。Berry 等（2001），DeGraaff-Surpless 等（2003）以及 Dickinson 和 Gehrels（2009）使用这种分析方法，对来自几组样品的碎屑锆石数据进行了有效对比。值得注意的是，K-S 统计对于年龄比例非常敏感。图 2-9 展示的两个样品的年龄分布曲线，它们包含的年龄完全相同，只是在两个年龄的比例上有 20%的差异，样品 1 包含 40 个 100Ma 的年龄值和 60 个 200Ma 的年龄值，而样品 2 包含 60 个 100Ma 的年龄值和 40 个 200Ma 的年龄值，由于这组数据的 P 值小于 0.05，K-S 统计指示这两个样品不具备相同的总体（95%可信度）。如果这种对比用于源区研究，一个合理的结论是这两个样品并非来自同一物源区。尽管统计学上准确，但是大部分研究者将会质疑，因为这两个样品包含完全相同的年龄组，只是在比例上存在细微差别而已。进行 K-S 统计分析的程序可以通过网站 www.geo.arizona.edu/alc 下载。

碎屑锆石数据组的对比还可以通过核心函数评估来完成（Sircombe 和 Hazelton，2004）。

这种方法能够产生一个数值来描述两条年龄分布曲线的差异，但是该值除了对年龄差异敏感以外，对年龄比例同样高度敏感。

很显然，目前急切需要更好的工具来对多个碎屑锆石数据组进行对比！

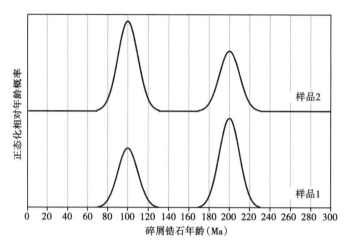

图 2-9　K-S 分析实例

假定两个样品各进行 100 次分析，样品 1 包含 40 个 100±10Ma 的年龄值和 60 个 200±10Ma 的年龄值，而样品 2 包含 60 个 100±10Ma 的年龄值和 40 个 200±10Ma 的年龄值。K-S 分析指示这两个样品不具备相同的总体（95%可信度）。如果这种对比用于源区研究或者地层对比，将会得出该样品不相关的结论。

这个例子说明 K-S 分析对于年龄比例高度敏感

2.8　未来发展机遇

近 15 年来，在碎屑锆石地质年代学迅速发展的同时，分析方法仍然存在一些挑战。其未来的发展机遇可以分为以下三个领域。

2.8.1　提高精度、准度、效率和空间分辨率

以上谈到的诸多挑战可以简单地概括为如何能够更加有效地获得 U-Pb 年龄、更加精细的时间分辨率以及更高的精度和准度。对于 SIMS 和 LA-ICPMS 而言，精度和准度提高到 1%~2%以上，可以更加有效地评价不谐和度和集中度，在大多数情况下这样可以提高数据的质量。这可以通过分析之前的锆石原位化学磨蚀实现（Mattinson，2005；Horstwood 等，2008）。提高这些分析技术的空间分辨率和效率以及可能的数据自动采集（Frei 和 Gerdes，2009；Holden 等，2009），将实现单颗粒的常规多次分析，这将提供一种强有力的工具来识别经历过 Pb 丢失或者具有多期年龄域的颗粒。从目前来看，存在的诸多复杂性尽管损害了数据组的质量，但对复杂性的详细描述将提供一个非常强大的工具来确定源区和物源历史。

Simonetti 等（2008），Cottle 等（2009，2008），Johnston 等（2009）和 Gehrels 等（2009）最近报道了采用 LA-ICPMS 方法通过提高空间分辨率来研究复杂晶体和小锆石（如页岩和粉砂岩中）的实例。

2.8.2　碎屑锆石数据库分析和归档工具

由于大量的碎屑锆石分析被逐年开展并报道，因此急切需要建立碎屑锆石数据库，以实

现全球数据共享以及与其他数据库的融合。截至目前，运行于某个国家、地区（省）的地质年代学数据库已经很多，如美国国家地质年代学数据库（Sloan 等，2003）、加拿大地质年代学知识库（http：//gdr.nrcan.gc.ca/geochron/index_e.php）、巴西国家地质年代学数据库（Silva 等，2003）和澳大利亚锆石地质年代学数据库（www.ga.gov.au/oracle/ozchron/）。然而，对于全球范围的源区研究，国际数据库系统的建立将发挥重要作用。理想的地质年代学系统可能是这样的，在该系统中，地质年代学数据可以通过完善的数据库查询方法进行分析，然后与所有可用的地质学、地球化学、地层学和古生物学数据结合在一起。作为 Strat-DB 计划和 Earth-Chem 计划（www.earthchem.org/earthchem-Web/index.jsp）的一部分，目前已经在建立地质年代学数据库方面取得了一定进展。

2.8.3 与其他同位素、物质组成、热年代学和构造信息的结合

碎屑锆石年代学的另外一个重要的发展方向体现在地质年代学数据与地球化学、同位素、热年代学和构造信息的结合。目前，感兴趣的地球化学数据包括 Ti 的浓度来确定结晶温度（Hanchar 和 Watson，2003；Watson 等，2006；Fu 等，2008）、稀土元素的浓度（Hoskin 和 Ireland，2000；Hoskin 和 Schaltegger，2003；Wooden 等，2007）。最常见的，与 Lu-Hf、Li、O 和 Sm-Nd 同位素的结合，可以为岩石成因提供重要的约束（Kinny 和 Maas，2003；Valley，2003；Hawkesworth 和 Kemp，2006；Scherer 等，2007；Ushikubo 等，2008）。U-Pb 年龄与热年代学信息（如（U-Th）/He 和裂变径迹）的比较，在物源和源区演化方面有重要的应用（Rahl 等，2003）。考虑到其他矿物（如磷灰石）中同样可以进行 U-Pb、He 和裂变径迹年龄的测定（Carrapa 等，2009），通过分析砂岩样品中不同封闭温度的年代学数据，可以获得关于物源和源区时间—温度历史的丰富信息。此外，光谱分析中蕴含着丰富的化学组成和构造信息（Corfu 等，2003；Nasdala 等，2003）。

2.8.4 未来发展机遇小结

以上论述了碎屑锆石年代学发展方向，总结一下包括以下几点：
（1）通过单颗粒的多重分析，确定精确的结晶年龄，进而约束源区和物源信息；
（2）U-Pb 数据自动采集（Frei 和 Gerdes，2009；Holden 等，2009）；
（3）从定年颗粒中获取额外信息（如 Hf、Li、O 同位素，稀土元素模式，Ti 浓度，He 和/或裂变径迹冷却年龄等），以助于确定物源和源区历史；
（4）开发能够解释和集成多维信息的图形化和定量化研究的工具；
（5）建立综合地层、古生物、地球化学和其他信息的碎屑锆石地质年代学的全球数据库。

2.9 结论

通过新的地质年代学方法可以确定源区、相关性和最大沉积年龄。这些方法的迅速发展为地质学家提供了一种强有力的新工具来研究沉积盆地和沉积岩及其来源。其中，最令人振奋的进展之一是碎屑锆石的应用，因为它是地壳岩石中的常见矿物，往往经历了多旋回的风化、搬运和成岩作用，因而通常可以提供强有力的结晶和冷却年龄（Fedo 等，2003）。尽管存在这些优势，但在碎屑锆石研究迅速发展的背后，我们不能否认在数据采集、解释、描述

和综合等基本方面，仍然存在不确定性（Horstwood 等，2008）。正因为这样，对于致力于碎屑锆石数据的研究者来说，参与样品处理、数据采集、数据解释（特别是复杂情况下）和评价年龄意义的整个过程就显得尤为重要。研究者只有对分析过程中的每一个步骤都有了完全的理解，才能对数据的优势和局限性做出充分的评价。同时，从事 U-Pb 同位素测试工作的实验人员，应该继续致力于开发更有力的工具来评价、展示、对比和集成数据，以及发展新的方法来收集和综合相关的地球化学和构造信息。通过这些努力，碎屑矿物地质年代学有望从一种大有前途的新技术发展成为能够破译出赋存在沉积岩中丰富的历史信息的基本工具。

致谢（略）

参 考 文 献

Anderson, T. (2005) Detrital zircons as tracers of sedimentary provenance: limiting conditions from statistics and numerical simulation. Chemical Geology, 216 (3-4), 249-270.

Berry, R. F., Jenner, G. A., Meffre, S., and Tubrett, M. N. (2001) ANorth American provenance for Neoproterozoic to Cambrian sandstones in Tasmania? Earth and Planetary Science Letters, 192, 207-222.

Bowring, S. A., and Schmitz, M. D. (2003) High-precision U-Pb zircon geochronology and the stratigraphic record, in Hanchar, J. M., and Hoskin, P. W. O., eds., Zircon: Reviews in Mineralogy and Geochemistry, 53, 305-326.

Carrapa, B., DeCelles, P., Reiners, P., Gehrels, G., and Sudo, M. (2009) Apatite triple dating and white mica $^{40}Ar/^{39}Ar$ thermochronology of syntectonic detritus in the central Andes: Amultiphase tectonothermal history. Geology, 37 (5), 407-410.

Chang, Z., Vervoort, J. D., McClelland, W. C., and Knaack, C. (2006) U-Pb dating of zircon by LA-ICP-MS. Geochemistry, Geophysics, Geosystems, 7, Q05009. doi: 10.1029/2005GC001100.

Corfu, F., Hanchar, J. M., Hoskin, P. W. O., and Kinny, P. (2003) Atlas of zircon textures, in Hanchar, J. M., and Hoskin, P. W. O., eds., Zircon: Reviews in Mineralogy and Geochemistry, 53, 469-500.

Cottle, J. M., Horstwood, M. S. A., and Parrish, R. R. (2009) Anew approach to single shot laser ablation analysis and its application to in-situ geochronology. Journal of Analytical Atomic Spectrometry. doi: 10.1039/b821899d.

Cottle, J., Parrish, R., and Horstwood, M. (2008) A new method for analyzing thin (>2 micron) zircon rims by LA-ICP-MS. Geochimica et Cosmochimia Acta, 72 (12S) A183.

Davis, D. W., Poulson, K. H., and Kamo, S. L. (1989) New insights into Archean crustal development from geochronology in the Rainy Lake area, Superior Province, Canada. Journal of Geology, 97, 379-398.

Davis, D. W., Williams, I. S., and Krogh, T. E. (2003) Historical development of U-Pb geochronology, in Hanchar, J. M., and Hoskin, P. W. O., eds., Zircon: Reviews in Mineralogy and Geochemistry, 53, 145-181.

DeGraaff-Surpless, K., Mahoney, J. B., Wooden, J. L., and McWilliams, M. O. (2003)

Lithofacies control in detrital zircon provenance studies: insights from the Cretaceous Methow basin, southern Canadian Cordillera. GSA Bulletin, 115, 899-915.

Dickinson, W. R., and Gehrels, G. E. (2009) U-Pb ages of detrital zircons in Jurassic eolian and associated sandstones of the Colorado Plateau: evidence for transcontinental dispersal and intraregional recycling of sediment. Geological Society of America Bulletin, 121, 408-433.

Dodson, M. H., Compston, W., Ireland, I. S., and Wilson. J. F. (1988) A search for ancient detrital zircons in Zimbabwean sediments. Journal of the Geological Society of London, 145, 977-983.

Eglington, B. M. (2004) DateView: a windows geochronology database. Computers and Geosciences, 30 (8), 847-858.

Erdmer, P., and Baadsgaard, H. (1987) 2. 2 Ga age of zircons in three occurrences of Upper Proterozoic clastic rocks of the northern Cassiar terrane, Yukon and British Columbia. Canadian Journal of Earth Sciences, 24, 1919-1924.

Fedo, C. M., Sircombe, K., and Rainbird, R. (2003) Detrital zircon analysis of the sedimentary record, in Hanchar, J. M., and Hoskin, P. W. O., eds., Zircon: Reviews in Mineralogy and Geochemistry, 53, 277-303.

Flowerdew, M., Millar, I., Curtis, M., Vaughan, A., Horstwood, M., Whitehouse, M., and Fanning, C. M. (2007) Combined U-Pb geochronology and Hf isotope geochemistry of detrital zircons from early Palaeozoic sedimentary rocks, Ellsworth-Whitmore Mountains block, Antarctica. Geological Society of America Bulletin, 119 (3-4), 275-288.

Frei, D., and Gerdes, A. (2009) Precise and accurate in situ U-Pb dating of zircon with high sample throughput by automated LA-SF-ICP-MS. Chemical Geology, 261 (3-4), 261-270.

Froude, D. O., Ireland, T. R., Kinny, P. D., Williams, I. S., Compston, W., Williams, I. R., and Meyers, J. S. (1983) Ion microprobe identification of 4, 100-4, 200 Myr-old terrestrial zircons. Nature, 304, 616-618.

Fryer, B. J., Jackson, S. E., and Longerich, H. P. (1993) The application of laser ablation microprobe-inductively coupled plasma mass spectrometry (LAM-ICPMS) to in situ (U) -Pb geochronology. Chemical Geology, 109, 1-8.

Fu, B., Page, F. Z., Cavosie, A. J., Fournelle, J., Kita, N. T., Lackey, J. S., Wilde, S. A., and Valley, J. W. (2008) Ti-in-zircon thermometry: applications and limitations. Contributions to Mineralogy and Petrology, 156 (2), 197-215.

Gehrels, G. E. (2000) Introduction to detrital zircon studies of Paleozoic and Triassic strata in western Nevada and northern California, in Soreghan, M. J. and Gehrels, G. E., eds., Paleozoic and Triassic paleogeography and tectonics of western Nevada and northern California. Geological Society of America Special Paper 347. Boulder, CO, Geological Society of America, 1-18.

Gehrels, G. E., McClelland, W. C., Samson, S. D., Patchett, P. J., and Jackson, J. L. (1990) Ancient continental margin assemblage in the northern Coast Mountains, southeast Alaska and northwest Canada. Geology, 18, 208-211.

Gehrels, G. E., Valencia, V., and Ruiz, J. (2008) Enhanced precision, accuracy, efficiency, and spatial resolution of U－Pb ages by laser ablation－multicollector－inductively coupled plasma－mass spectrometry. Geochemistry, Geophysics, Geosystems, 9, Q03017. doi: 10. 1029/2007 GC001805.

Gehrels, G., Rusmore, M., Woodsworth, G., Crawford, M., Andronicos, C., Hollister, L., Patchett, J., Ducea, M., Butler, R. Klepeis, K. Davidson, C., Mahoney, B., Friedman, R., Haggart, J., Crawford, W., Pearson, D., and Girardi, J. (2009) U－Th－Pb geochronology of the Coast Mountains Batholith in north－coastal British Columbia: constraints on age, petrogenesis, and tectonic evolution. Geological Society of America Bulletin, 121 (9–10), 1341–1361. doi: 10. 1130/B26404. 1.

Gerdes, A., and Zeh, A. (2006) Combined U–Pb and Hf isotope LA– (MC) –ICP–MS analyses of detrital zircons: comparison with SHRIMP and new constraints for the provenance and age of Armorican metasediment in central Germany. Earth and Planetary Science Letters, 249, 47–61.

Girty, G. H., and Wardlaw, M. S. (1984) Was the Alexander terrane a source of feldspathic sandstones in the Shoo Fly Complex, northern Sierra Nevada, California. Geological Society of America Bulletin, 96, 516–521.

Hanchar, J. M., and Watson, E. B. (2003) Zircon saturation thermometry, in Hanchar, J. M., and Hoskin, P. W. O., eds., Zircon. Reviews in Mineralogy and Geochemistry, 53, 89–112.

Hart, S. R., and Davis, G. L. (1969) Zircon U–Pb and wholerock Rb–Sr ages and early crustal development near Rainy Lake, Ontario. Geological Society of America Bulletin, 80, 595–616.

Hawkesworth, C. J., and Kemp, A. I. S. (2006) Using Hf and oxygen isotopes in zircons to unravel the record of crustal evolution. Chemical Geology, 226, 144–162.

Holden, P., Lanc, P., Ireland, T. R., Harrison, T. M., Foster, J. J., and Bruce, Z. (2009) Mass–spectrometric mining of Hadean zircons by automated SHRIMP multi－collector and single–collector U/Pb zircon age dating: the first 100, 000 grains. International Journal of Mass Spectrometry, 286 (2–3), 53–63.

Horn, I., Rudnick, R. L., and McDonough, W. F. (2000) Precise elemental and isotope ratio determinations by simultaneous solution nebulization and laser ablation–ICP–MS: application to U–Pb geochronology. Chemical Geology, 164, 281–301.

Horstwood, M. S. A. (2008) Data reduction strategies, uncertainty assessment, and resolution of LA– (MC) –ICP–MS isotope data, in Sylvester, P., ed., Laser ablation ICP–MS in the Earth sciences: current practices and outstanding issues. Short Course Series, 40. Quebec, QC, Mineralogical Association of Canada, 283–303.

Horstwood, M., Cottle, J., and Parrish, R. (2008) Improving the utility of detrital zircon studies through chemical abrasion. Geochimica et Cosmochimia Acta, 72 (12S) A392.

Horstwood, M. S. A., Foster, G. L., Parrish, R. R., Noble, S. R., and Nowell, G. M. (2003) Common–Pb corrected in situ U–Pb accessory mineral geochronology by LA–MC–ICPMS. Journal of Analytical Atomic Spectrometry, 18, 837–846.

Horstwood, M. S. A., Kosler, J., Jackson, S., Pearson, N., and Sylvester, P. (2009) Investigating age resolution in laser ablation geochronology. EOS, Transactions, 90 (6), 47.

Hoskin, P. W. O., and Ireland, T. R. (2000) Rare earth element chemistry of zircon and its use as a provenance indicator. Geology, 28 (7), 627-630.

Hoskin, P. W. O., and Schaltegger, U. (2003) The composition of zircon and igneous and metamorphic petrogenesis, in Hanchar, J. M., and Hoskin, P. W. O., eds., Zircon: Reviews in Mineralogy and Geochemistry, 53, 27-62.

Ireland, T. R. (1992) Crustal evolution of New Zealand: evidence from age distributions of detrital zircons in Western Province paragneisses and Torlesse greywacke. Geochimica et Cosmochimica Acta, 56, 911-920.

Ireland, T. R., and Williams, I. S. (2003) Considerations in zircon geochronology by SIMS, in Hanchar, J. M., and Hoskin, P. W. O., eds., Zircon: Reviews in Mineralogy and Geochemistry, 53, 215-241.

Johnston, S., Gehrels, G., Valencia, V., and Ruiz, J. (2009) Small - volume U - Pb geochronology by Laser Ablation-Multicollector-ICP Mass Spectrometry. Chemical Geology, 259, 218-229.

Kemp, A. I. S., Foster, G. L., Schersten, A., Whitehouse, M. J., Darling, J., and Storey, C. (2009) Concurrent Pb-Hf isotope analysis of zircon by laser ablation multi-collector ICP-MS, with implications for the crustal evolution of Greenland and the Himalayas. Chemical Geology, 261, 244-260.

Kinny, P. D., and Maas, R. (2003) Lu-Hf and Sm-Nd isotope systems in zircon, in Hanchar, J. M., and Hoskin, P. W. O., eds., Zircon: Reviews in Mineralogy and Geochemistry, 53, 327-341.

Kosler, J., and Sylvester, P. J. (2003) Present trends and the future of zircon in U - Pb geochronology: laser ablation ICPMS, in Hanchar, J. M., and Hoskin, P. W. O., eds., Zircon: Reviews in Mineralogy and Geochemistry, 53, 243-275.

Kosler, J., Tubrett, M., and Sylvester, P. (2001) Application of laser ablation ICPMS to U-Th-Pb dating of monazite. Geostandards Newsletter, 25, 375-386.

Krogh, T. E., and Davis, G. L. (1975) The production and preparation of ^{205}Pb for use as a tracer for isotope dilution analyses. Carnegie Institute of Washington Yearbook, 74, 416-417.

LeDent, D., Patterson, C., and Tilton, G. R. (1964) Ages of zircon and feldspar concentrates from North American beach and river sands. Journal of Geology, 72, 112-122.

Li, X., Liang, X., Sun, M., Guan, H., and Malpas, J. G. (2001) Precise ^{206}Pb/^{238}U age determination on zircons by laser ablation microprobe - inductively coupled plasma - mass spectrometry using continuous laser ablation. Chemical Geology, 175, 209-219.

Ludwig, K. R. (2008) Isoplot 3.6. Special Publication No. 4. Berkeley, CA, Berkeley Geochronology Center, 77 p.

Machado, N., and Gauthier, G. (1996) Determination of ^{207}Pb=^{206}Pb ages on zircon and monazite by laser ablation ICPMS and application to a study of sedimentary provenance and metamorphism in southeastern Brazil. Geochimica et Cosmochimica Acta, 60, 5063-5073.

Machado, N., and Simonetti, A. (2001) U-Pb dating and Hf isotopic composition of zircon by laser-ablation-MC-ICP-MS, in Sylvester, P., ed., Laser Ablation ICPMS in the Earth

Sciences: Mineralogical Association of Canada, Short Course Series, 29, 121–146.

Mattinson, J. M. (2005) Zircon U–Pb chemical abrasion ("CATIMS") method; combined annealing and multi–step partial dissolution analysis for improved precision and accuracy of zircon ages. Chemical Geology, 220, 47–66.

Mojzsis, S. J., Harrison, T. M., and Pidgeon, R. T. (2001) Oxygen–isotope evidence from ancient zircons for liquid water at the Earth's surface (4300) Myr ago. Nature, 409, 178–181.

Mueller, P. A., Foster, D. A., Mogk, D. W., Wooden, J. L., Kamenov, G. D., and Vogl, J. J. (2007) Detrital mineral chronology of the Uinta Mountain Group: implications for the Grenville flood in southwestern Laurentia. Geology, 35 (5), 431–434.

Nasdala, L., Zhang, M., Kempe, U., Panczer, G., Gaft, M., Andrut, M., and Plotze, M. (2003) in Hanchar, J. M., Hoskin, P. W. O., and Kinny, P. (2003) Atlas of zircon textures, in Hanchar, J. M., and Hoskin, P. W. O., eds., Zircon: Reviews in Mineralogy and Geochemistry, 53, 427–467.

Nemchin, A. A., and Cawood, P. A. (2005) Discordance of the U–Pb system in detrital zircons: Implication for provenance studies of sedimentary rocks. Sedimentary Geology, 182, 143–162.

Parrish, R. R. (1987) An improved microcapsule for zircon dissolution in U–Pb geochronology. Chemical Geology, 66, 99–102.

Parrish, R. R., and Krogh, T. E. (1987) Synthesis and purification of ^{205}Pb for U–Pb geochronology. Chemical Geology, 66, 103–110.

Parrish, R. R., and Noble, S. R. (2003) Zircon U–Th–Pb geochronology by isotope dilution–thermal ionization mass spectrometry (ID–TIMS) in Hanchar, J. M., and Hoskin, P. W. O., eds., Zircon: Reviews in Mineralogy and Geochemistry, 53, 183–213.

Rahl, J. M., Reiners, P. W., Campbell, I. H., Nicolescu, S., and Allen, C. M. (2003) Combined single–grain (U–Th)/He and U/Pb dating of detrial zircons from the Navajo Sandstone, Utah. Geology, 31 (9), 761–764.

Ross, G. M., and Bowring, S. A. (1990) Detrital zircon geochronology of the Windermere Supergroup and the tectonic assembly of the southern Canadian Cordillera. Journal of Geology, 98, 879–893.

Scherer, E. E., Whitehouse, M. J., and Munker, C. (2007) Zircon as a monitor of crustal growth. Elements, 3 (1), 19–24.

Silva, L. C., Rodrigues, J. B., Silveira, L. M. C., and Pimentel, M. M. (2003) The Brazilian National Geochronological Database: Chronobank, in da Silva Rosa, M., ed., IV South American Symposium on Isotope Geology: Salvador, 115–116.

Silver, L. T., and Deutsch, S. (1963) Uranium–lead isotopic variations in zircon: a case study. Journal of Geology, 71, 721–758.

Simonetti, A., Heaman, L., and Chacko, T. (2008) Use of discrete–dynode secondary electron multipliers with faradays: a "reduced volume" approach for in situ U–Pb dating of accessory minerals within petrographic thin sections by LA–MC–ICP–MS, in Sylvester, P., ed., Laser ablation ICP–MS in the earth sciences: current practices and outstanding issues. Short Course

Series, 40. Quebec, QC, Mineralogical Association of Canada, 241-264.

Sircombe, K. N. (2004) AGEDISPLAY: an EXCEL workbook to evaluate and display univariate geochronological data using binned frequency histograms and probability density distributions. Computers and Geosciences, 30 (1), 21-31.

Sircombe, K. N., and Hazelton, M. L. (2004) Comparison of detrital zircon age distributions by kernel functional estimation. Sedimentary Geology, 171, 91-111.

Sircombe, K. N., and Stern R. A. (2002) An investigation of artificial biasing in detrital zircon U-Pb geochronology due to magnetic separation in sample preparation. Geochimica et Cosmochimica Acta, 66 (13), 2379-2397.

Sloan, J., Henry, C. D., Hopkins, M., and Ludington, S. (2003) National Geochronological Database: US Geological Survey, Open-File Report 03-236.

Steiger, R. H., and Jager, E. (1977) Subcommission on geochronology: convention on the use of decay constants in geo- and cosmochronology. Earth and Planetary Science Letters, 36, 359-362.

Tera, F., and Wasserburg, G. J. (1972) U-Th-Pb systematic in three Apollo 14 basalts and the problem of initial Pb in lunar rocks. Earth and Planetary Science Letters, 14, 281-304.

Ushikubo, T., Kita, N., Cavosie, A., Wilde, S., Rudnick, R., and Valley, J. (2008) Lithium in JackHills zircons: evidence for extensive weathering of Earth's earliest crust. Earth and Planetary Science Letters, 272 (3-4), 666-676.

Valley, J. (2003) Oxygen isotopes in zircon, in Hanchar, J. M., and Hoskin, P. W. O., eds., Zircon: Reviews in Mineralogy and Geochemistry, 53, 343-385.

Vermeesch, P. (2004) How many grains are needed for a provenance study? Earth and Planetary Science Letters, 224, 441-451.

Watson, E. B., Wark, D. A., and Thomas, J. B. (2006) Crystallization thermometers for zircon and rutile. Contributions to Mineralogy and Petrology, 151, 413-433.

Wetherill, G. W. (1956) Discordant uranium-lead ages. International Transactions of the American Geophysical Union, 37, 320-326.

Wooden, J. L., Mazdab, F. K., and Barth, A. P. (2007) Using the temperature and compositional characteristics of zircon and sphene to better understand the petrogenesis of Mesozoic magmatism in the Transverse Ranges, California. Proceedings of the Ores and Orogenesis Conference, Tucson, AZ, 154.

Woodhead, J., Hergt, J., Shelley, M., Eggins, S., and Kemp, R. (2004) Zircon Hf-isotope analysis with an excimer laser, depth profiling, ablation geometries, and concomitant age estimation. Chemical Geology, 209, 121-135.

Yuan, H., Gao, S., Dai, M., Zong, C., Gunther, D., Fontaine, G., Liu, X., and Diwu, C. (2008) Simultaneous determinations of U-Pb age, Hf isotopes, and trace element compositions of zircon by excimer laser-ablation quadrupole and multicollector ICP-MS. Chemical Geology, 247, 100-118.

（饶松译，胡圣标 崔敏校）

第 3 章 陆相岩层宇宙成因核素技术在评估构造活跃地区地表年龄和沉积物暴露史中的应用

JOHN C. GOSSE

(Department of Earth Sciences, Dalhousie University, Halifax, Canada)

摘 要：在过去的半个世纪里，构造活跃地区构造运动标志的识别和应力大小的计算取得了长足的进步，但是很难获得新构造运动变形区岩层年龄，使得学者们对新构造运动速率和构造事件发生时间的研究明显滞后。陆相岩层宇宙成因核素技术（TCN）的发展与完善，为直接确定断层面年龄和计算断层破裂速率，确定构造变形的时间与幕次提供了有力手段，可以更清晰地展示出构造运动与沉积物通量之间的关系。陆相岩层宇宙成因核素技术（TCN）的有效测年范围从几十年（10）至几千万年（10^7），弥补了大地测量时间尺度较短与与造山带地壳动力学研究相关的时间较长的空白。本章是对目前 TCN 测年技术在评估构造活跃地区地表侵蚀和沉积物暴露史中应用的总结。

关键词：前陆盆地 挠曲 造山带 地球动力学 地层学

3.1 TCN 技术研究概况

20 世纪 80 年代中期以来，放射性碳测年校准工作的完成，代表着地质年代学研究的一次巨大进步，在时间尺度上促进了与高温热年代学相重叠的地壳变形研究。光学和红外波段的释光测年，尤其是单颗粒等分法，在测定几千年（10^3）至几十万年（10^5）时间尺度内的黄土和冲积物年龄方面取得了很好效果。铀系测年法的有效测年范围为几万年（10^4）至几十万年（10^5），目前对于晚更新世以来的钙华、碳酸盐岩土壤和蛋白石的年龄测定已经可以精确到几百年。磷灰石、锆石和（火山碎屑及其他火山沉积物中的）磁铁矿的（U-Th-Sm）/He 低温热年代学定年技术和包括火山玻璃的裂变径迹分析在内的其他测试方法，加强了我们对几万年（10^4）至几千万年（10^7）时间范围内火山活动和应变标志的年龄测定。陆相岩层宇宙成因核素技术（TCN）提供了：（1）一种测定地表和沉积物年龄的方法，其测年范围在几十年至几百万年之间，甚至更长的时间范围（$>10^6$），并且与其他地质年代学测试方法的测年范围相重叠；（2）一种独立的，评估局部地区或整个流域地表岩层侵蚀速率的方法。这种方法可以用来研究构造活跃地区，沉积物通量的相对变化趋势及其与构造活动的时空匹配关系（表 3-1）。

表 3-1 活动构造定年和侵蚀测量的宇宙成因同位素

核素	通常使用的元素和矿物	主要生成方式	半衰期	分析方法	注释
^{3}He	多种核素；橄榄石、角闪石、辉石	核素>3，裂变，放射性成因	稳定	Noble gas MS	一般使用基性火山岩；避免老矿物重结晶的年龄
^{10}Be	硅、氧、石英	裂变，μ介子作用	约1.39Ma	AMS	最年轻的年龄：几十年尺度，夏威夷熔岩流；不确定的半衰期；TCN广泛应用
^{14}C	硅、氧、石英	裂变，μ介子作用	5.73ka	AMS	饱和早，可以用于25ka以来的暴露；最大减少残留浓度的影响
^{21}Ne	硅、石英、长石	裂变，捕获的宇宙成因	稳定	Noble gas MS	最古老的年龄值：40Ma，阿塔卡马沙漠的冲积物；避免老矿物中的结晶年龄
^{26}Al	硅、石英	裂变，μ介子作用	0.72Ma	AMS	经常与^{10}Be一起在TCN使用
^{36}Cl	多种核素；全岩；长石、方解石	大于36的核素裂变，热中子捕获；μ介子作用	0.31Ma	AMS	可以用于几乎所有岩石
^{38}Ar	钾长石	大于38的核素裂变，μ介子作用	稳定	Noble gas MS	可以用于地表隆起速率测算；避免老矿物的结晶年龄

3.1.1 基本原理

目前，已发表的几篇TCN技术的综述文章，详细地阐述了宇宙辐射的相关概念、地质年龄与剥蚀速率的计算方法、地球化学样品的制备以及不确定性因素等方面的内容（Lal，1991；Gosse和Phillips，2001；Elias，2007；Dunai，2010）。从许多最近发表的TCN技术相关文献来看，许多人都使用微软Excel加载项（Vermeesch，2007）或在线计算器（Balco等，2008）等功能简单但操作界面友好的方式，来计算地质年龄、侵蚀速率和误差。读者可直接获取这些成果资料，包括核素生成速率、数据处理、计算公式、同位素化学与分析方面的具体信息。本文将着重介绍TCN技术的评估方法，以及构造运动速率对新构造运动所引起的晚新生代以来的构造变形。本文引用的大部分文献都是最新的综述，并且参考了近20年来，研究生学位论文中所采用的采样方法，吸取了其中失败的实验教训。TCN技术发展十分迅速（它是从基础物理学发展而来，包括核截面、核素半衰期、特定点核素生成速率等方面的内容。随着校准工作的完善，该方法的可靠性逐步提高，为我们更好地理解古地磁强度变化提供了技术支持）。本章的内容是现阶段TCN技术及其在构造领域应用的一个缩影。

核素（具有给定质子数和中子数的一类原子核所组成的元素）可以是放射成因（如放射性元素衰变为另一种元素）、核反应成因（如一个放射性粒子或一个原子核与靶核碰撞）或者宇宙成因的。宇宙成因核素是由入射的初级和次级宇宙射线粒子与靶核之间的任意核反应所形成的。

大多数初级宇宙射线（GCR）是由H、He和较重的粒子组成的，它们的加速度是从某些恒星的耀斑或超新星爆炸事件中获得的，而这些宇宙入射辐射的能谱范围可超过20个数量级。在数百万年及更短的时间尺度内，尽管太阳和地球的磁场偏转改变了大部分的辐射通量，但是太阳系宇宙辐射（GCR）的通量被认为是恒定且各向同性的。只有拥有足够能量的宇宙射线才能穿透地磁场，并且越靠近赤道，入射粒子穿透地磁场所需能量越高。其结果就是，赤道附近进入大气层的初级宇宙射线通量低而频谱高（由于赤道附近截止能量较高，很少有低能粒子可以穿透地磁场到达大气层）。这些进入大气层的初级宇宙射线与大气中的原子核相互作用生成次级宇宙射线，包括电磁辐射（如伽马射线）、强子辐射（如质子和中子）和介子辐射（k介子、π介子、μ介子）。一些二次粒子继续与大气分子发生作用，每次作用都形成一个二次粒子锥形喷流或"宇宙射线簇射"。随着粒子间的相互作用，二次粒子通量以一定速率衰减，称为粒子间相互作用的平均自由程（也称为衰减长度），它的大小取决于粒子的类型和能量（核截面）。从大气层的顶部至海平面平均会发生超过10次的粒子间相互作用。较高海拔地表的矿物就会接受较高的宇宙射线通量。在大气层的底部，多数二次粒子为μ介子——多是由π介子衰变而来——其质量为质子质量的1/8.9。二次μ介子和强子（核子）与大气层下的物质继续作用，在地表矿物和水体中形成岩层宇宙成因核素。

由于入射的初级和次级宇宙射线能谱范围宽（<0.1eV至10^{21}eV），其经不同作用机制生成的二次粒子也十分宽泛。这些作用机制包括：（1）散裂反应，为近地表数米的岩石中宇宙成因核素生成的主要作用机制，涉及的高能宇宙核子拥有足够的动能，其动能超过靶核的核子结合能；（2）特定的靶核捕获热（慢）中子［如^{35}Cl捕获一个热中子变为^{36}Cl（伴随μ介子散失）］；（3）各种μ介子相互作用，主要是捕获慢负μ介子或快μ介子相互作用，产生额外的快核子。主要受其大小的影响，μ介子与物质作用较弱，因此它比二次核子穿透的深度更大。由于μ介子流穿透深度大，因此它是计算长期剥蚀速率要考虑的重要因素，可用于同低温热年代学得出的历史推断作对比，也可用于计算埋藏年龄。对处于长期暴露环境的冲积层，其中形成的，伴随有大量μ介子产生的宇宙成因核素（TCN），甚至可以在地下30m处富集。

二次粒子与暴露矿物相互作用可以产生几乎所有的核素。近地表矿物中宇宙成因核素（TCNs）可以是稳定的（如惰性气体^{3}He、^{21}Ne和^{38}Ar），也可以是放射性的（^{10}Be、^{14}C、^{26}Al、^{36}Cl）。地质学家面临的挑战是如何准确识别以下宇宙成因核素（TCNs）：（1）稳定或有足够慢的衰变速率，可以在整个地质时间尺度内适用；（2）按一定的速率生成，并且其赋存的介质允许在所需的精度内对其含量进行测定；（3）其生成的量远高于所有非宇宙成因核素的量。用加速器质谱（AMS）测定放射性宇宙成因核素需要同时分析该元素的另一个同位素，并且两个同位素的比值必须在光谱仪的检测范围内。例如，测定铝硅酸盐中^{26}Al/^{27}Al就非常困难，因为原生^{27}Al的含量很高，^{26}Al/^{27}Al就不在光谱仪的检测范围内。

3.1.2 样品制备

要使用质谱法（MS）或加速质谱法（AMS）测定样品中的惰性气体，样品制备应尽量提高样品纯度以使仪器效率最优；收集足够质量的样品以获得所需的测量精度；降低等分子量干扰（如加速质谱法中^{10}B对^{10}Be的干扰、^{26}Mg对^{26}Al的干扰、^{36}S和^{36}Cl的相互干扰），这样就可以确保一个恒量和高束电流，从而在获得高精度数据的同时，减少测量时间。对于放射性核素的测量，小一点的样品可以缩短样品准备的时间和酸耗，降低样品污染的可能性，但样品的量必须满足测量精度的需求。样品中惰性气体质谱分析需要先作矿物提纯，并且在实验之前或实验过程中，先测定非宇宙成因的气体体积。现今的各种测定方法至少都可达到 2%（1σ）的测定精度。

3.1.3 数据处理与解释

暴露矿物中宇宙成因核素 i 的浓度为 N_i（原子数/质量），其大小与矿物中宇宙成因核素的形成速率 P_i 及矿物在地表面的暴露时间 T 成正比。例如在冲积扇表层荒漠卵石覆盖层中，处于暴露环境的橄榄石矿物中生成的宇宙成因核素 ^{3}He，其浓度

$$N_i = P_0 T + C_n + C_i \tag{3-1}$$

其中 P_0 为地表所有作用机制生成宇宙成因核素的总速率（原子数/质量×时间）；C_n 为非宇宙成因的 i 核素浓度；C_i 为在欲测定年龄的构造事件之前暴露地表形成的残留 i 核素浓度。地表面之下有作用机制生成宇宙成因核素的总速率为

$$P_z = \sum_1^x P_{x,o} e^{-\frac{z}{\Lambda_x}} \tag{3-2}$$

其中 z 为质量深度（长度与密度的函数，单位 g/cm^2）；x 为核素的生成机制（高能核子、热中子和 μ 介子作用；Granger 和 Muzikar，2001）；Λ 为每种作用机制 x 的平均衰减长度（单位 g/cm^2）。

已知在海平面和高纬度地区（选定的参考位置），大部分地表宇宙成因核素的生成速率 P_0 在 5%~10%（1σ）之间。然而核素生成速率及不同作用机制的相对贡献，在空间上的分布受地磁场和大气屏蔽的控制（Dunai 等，2000；Desilet 和 Zreda，2003；Pigati 和 Lifton，2004；Lifton 等，2005；Balco 等，2008）。在过去的 10 年中，CRONUS-Earth 和 CORNUS-EU 国际计划得到学者们的大量关注，大量的研究丰富和完善了我们对宇宙成因核素（TCN）生成速率参考值、空间尺度转换和生成速率随时间的变化关系方面的认识。对于某一指定的宇宙成因核素（TCN），其生成速率在全球约 20 个不同地区、不同地点和不同暴露时长下进行了校订，得到了暴露史的时限和可靠的独立年龄。通过空间尺度转换将不同地区测定的生成速率校正到参考位置处，为未知地区的研究提供参考。在大部分的情况下，简单的地心偶极子模型无法较好地定义地磁场的分布特征，并且我们对地史时期特别是在全新世之前的偶极子轴的位置变化、地磁场强度和包括准静态非偶极组分在内的高次谐波磁场效应仍知之甚少。在过去的 20 年内，宇宙成因核素（TCN）的生成速率依据地磁场在时间、空间上的变化所产生的影响做出了一些调整，这种调整在局部与时间整合的核素生成速率上可以造成超过 20% 的变化，这种效应在一定的暴露时长条件下随海拔的升高而增加（Lifton 等，2008）。然而，调整的幅度还没有达成共识，所以出版物中应指明 $P_{x,o}$、Λ_x、衰变常数、计算方法和至少所有的浓度和质谱标准数据（Dunai 和 Stuart，2009；Frankel 等，2010），以备在我们对地磁场变化的认知提高而重新调整与时间整合的核素生成速率后，可以重新使用这些数据计算年龄。地球大气层对宇宙射线的屏蔽作用不再由 Lal（1991）提出的海拔缩放对称参数模型来定义，而应该使用较复杂的现代（Stone，2000）和冰川（Staiger 等，2007）大气模型。与时间整合的核素生成速率也将随构造运动或均衡说在高地暴露期间而发生变化，间歇性局部或完整的覆盖地表［如被树木（Plug 等，2007）、火山灰、水、冰、沙丘或黄土覆盖］，以及由于成土作用或混合作用导致地下样品之上的岩层体积密度的变化等。

采样点之上岩层或沉积物的逐步侵蚀对样品中宇宙成因核素（TCN）的浓度影响是可以

预测的,因为宇宙成因核素的浓度是随埋藏深度的增加而降低的[由核素生成机制的衰减长度所控制(见公式3-2),并应排除低侵蚀率下捕获热中子成因的^{36}Cl核素的影响]。核素浓度随埋藏深度增加而减小,是指随着侵蚀作用的进行,埋藏的矿物会逐渐上升至地表,其接受的平均宇宙射线通量低于未受侵蚀地表矿物接受的平均通量。我们稍后再讨论^{36}Cl核素与侵蚀作用之间的特殊关系。宇宙成因核素的浓度 N 对侵蚀作用的敏感性,对确定最大和平均侵蚀速率十分有用,但是如果无法独立地确定侵蚀速率或者在研究的时间尺度内侵蚀速率不稳定,那么这种敏感性会变为获得准确暴露时间的障碍。地表的加积作用(土壤剖面的顶部颗粒的淀积作用)增加了样品的质量深度。最终,当质量深度 z 达到屏蔽厚度(如大于5倍衰减长度,但依靠特定的 TCN 生成机制和作用时长),样品中的宇宙射线通量就变得可忽略不计,若为稳定 TCN 则其浓度将保持不变,若为放射性 TCN 其浓度会逐渐降低,公式如下:

$$N_i = \sum_o^x \left(\frac{P_{i,x,o}}{\lambda_i + \frac{\varepsilon}{\Lambda_x}} e^{-\lambda_i t_b} + \frac{P_{i,x(z_b)}}{\lambda_i}[1 - e^{-\lambda_i t_b}] \right) \tag{3-3}$$

其中,λ_i 为放射性核素 i 的衰变常数;t_b 为埋藏总时长;z_b 为埋藏深度;ε 为埋藏前地表侵蚀速率[g/(cm$^2 \cdot$ a)](Balco 和 Rovey,2008)。未经过埋藏的较浅深度的样品中两种 TCN 浓度的比值为一近似的常数。在埋藏过程中,两种 TCN 中至少有一种为放射性核素;因两者衰变速率不同,而使二者的浓度比值随埋藏时间而改变。因此两种长寿命的放射性核素的浓度比值,如 ^{26}Al/^{10}Be 已应用于确定埋藏时长(如之前暴露地表的矿物被冰、沉积物、水或者岩浆覆盖的时间)。通常用较短寿命的同位素作为比值的分子。含有短一半衰变周期的同位素比值,如 ^{14}C,可能相对较早达到饱和,因此只在一些特殊情况下使用。需要注意上述公式中 TCN 的生成从沉积物沉积下来后开始计算,不包含之前暴露所产生的残留核素浓度 C_i。若沉积物残留同位素浓度比值与总浓度比值相同,那么这就无关紧要。然而,沉积物有可能在最终沉积之前经历了复杂的暴露历史(一次或多次暴露和埋藏的循环),这意味着沉积物中初始 TCN 浓度比值低于总浓度比值。对于不知道初始浓度(或初始比值)的难题,将在本章"等时线法确定沉积物埋藏年龄"一节中介绍。

这些导致 TCN 生成速率和浓度在时间和空间上变化的因素是多数 TCN 技术的不确定性的来源。由实地测量(样品的深度、厚度、几何形态)、样品准备与分析造成的误差普遍偏低(多为百分之几),放射性核素衰变常数的误差也仅为百分之几。对于测定在最终沉积或抬升暴露之前已经暴露过的沉积物或岩石表面的年龄时,通常需要知道残留 TCN 的浓度 C_i。而残留 TCN 浓度的确定是非常困难的,尽管已有了估算残留浓度的技术,但这些技术多需要额外的假设条件。残留浓度也许是研究构造活跃地区暴露时间的最大不确定性来源。像大多数地质年代计一样,在解释年龄和剥蚀速率之前必须验证一些假设,必须了解样品的地质背景。

最后通过介绍对该技术命名的一些看法来结束本节。一些不同的首字母缩写词已被用于描述宇宙核素技术,包括 CRN(Cosmogenic Radionuclide)定年、SED 定年(Surface Exposure Dating)、CSEA(Cosmogenic Surface Exposure Age)和 CSE(cosmogenic surface exposure)定年。第一种命名在广义上是不适用的,因为有三种在地质上应用的宇宙成因核素是稳定同位素(^{3}He、^{21}Ne 和 ^{38}Ar)。第二种命名容易造成混淆,因为采样不仅可从地表采集,也可从地下采集，而且该技术并不仅用于定年。第三种和第四种命名有语法错误。只有缩写TCN是最合适的并且在国际上通用，是CRONUS-Earth组织所推荐的命名方案。TCN强调了受地外和大气影响在地壳表层数十米内生成的原地核素，不限制该方法的采样地点（地表或地下）、应用（测定剥蚀速率或定年）和核素稳定性（惰性气体或放射性核素）。此外，地形表面和以沉积物或岩石构成的地表面有着显著的区别（Lal 和 Chen，2005），测定两者的年龄须用TCN技术的不同方法。TCN测定地表或沉积物暴露年龄的适用范围如图3-1所示。

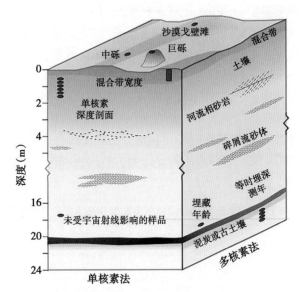

图 3-1　TCN 分析沉积物和表面的策略

应用单核素和多核素 TCN 进行晚新生代的地表和沉积物暴露测年的方法；
红色的是土壤和古土壤，所有的方法不太可能应用于单一地点

3.2　TCN 定年技术在活动构造研究中的应用

大地测量分析技术的进步、地形图分辨率的提高、复杂应变标志偏移测量精度的提高（Gold 等，2009）与 TCN 技术共同推动了断层运动学和地球动力学一些基础问题研究的进步和争议的解决。至今 TCN 技术已在十几个构造活跃地区进行了应用。冲积扇和冰碛物的 TCN 定年为我们了解西藏和南亚大陆动力学，研究晚新生代东欧与南欧、美国西南部、阿拉斯加、安第斯及新西兰等地区的构造运动做出了巨大的贡献。

TCN 技术在活动构造研究中一个最显著的应用例子就是验证了西藏地壳动力学中两个观点相悖的假说。其中一个较早提出的假说（Molnar 和 Tapponnier，1975）认为亚洲与印度板块之间 2000km 的会聚带中，大部分的应变集中于向东挤入的青藏高原边界的大断层中。另一个假说认为这种变形多分布于整个高原之上（England 和 McKenzie，1982）。两种观点都获得了数值模拟、类比模拟及大地测量等大量证据的支撑。根据两种假说中的边界走滑断层（如高原西北的阿尔金断裂和西南的喀喇昆仑断裂）滑动速率不同，提出了一个验证手段：挤出假说需要较高的断层滑动速率，并且沿整个断层方向的滑动速率的大小相近；而应

变内部分布式假说预测的边界走滑断层滑动速率是在长时间内平均的、相对较缓慢的，且断层滑动速率在空间上是变化的。在20世纪80年代初，最适用限定断层滑动速率的地质年代计是有机物放射性碳定年。在那个还未拥有加速质谱法的年代，仍需要采集大量的有机样品用于β计数，而实际情况是断裂冲积层中缺乏合适的取样对象；这些都限制了假说的验证过程。被认为是"末次盛冰期"或更晚的冰川地貌（未定年），其偏移也被用于研究断层滑移速度。随AMS和TCN技术发展，主要由法国和美国学者发起的TCN年代学和新构造运动研究热潮，测量了边界断层（和近来测量的高原内部断层）千年尺度内的滑移速率。在200ka至晚全新世之间的时间尺度内，由TCN技术获得了大量的通过测量河流阶地的偏移或期次、冲积扇沉积和冰碛物得到的滑动速率数据（表3-2）。在TCN技术的约束下，测得阿尔金断裂和喀喇昆仑断裂在晚新生代（沿不同时期，断层的不同段）滑动速率的范围分别是9~27mm/a和4~12mm/a。当该结果与晚全新世和现代运动速率（通过放射性碳定年和现代大地测量得到的运动速率）相对比时，发现并没有得到一个简单的趋势，仍无法解决两种假说的争议。虽然这些断层上滑动速率在时间和空间上的明显变化可能是真实存在的，但也不排除非理想采样方法和TCN数据的错误解释造成定年存在巨大误差的可能性（Brown等，2005）。下述关于TCN技术应用方法的评论会为解决活动构造年代学问题提供一些思路和指导，并提供一些替代方法以解决过去TCN技术应用失败地区的问题。

表3-2 由TCN定年技术得出的阿尔金断裂和喀喇昆仑断裂的走滑速率

作者	测量介质	走滑速率（mm/a）	时间范围（ka）	位置
阿尔金断裂				
Mériaux等（2004）	石英砾石：残留的河流地貌	26.9±6.9	113—6	图拉地区（约87°E）
Mériaux等（2005）	河道/阶地抬升的断距	17.8±3.6	6—5	阿克塞地区（约94°E）
Xu等（2005）	石英砾石阶地	19±4	68—7.1	东尔臣河谷
		11±3.5	14.7—13.3	芦草湾村附近
Gold等（2009）	河流阶地旁的隆起	9.0±1.3~15.5±1.7	6—0.5	图拉地区
喀拉昆仑断裂				
Brown等（2002）	冲积扇旁边的碎屑	4±1	14—11	中段
Chevalier等（2005）	冰碛砾石：地表鹅卵石	10.7±0.7	<150	南段
Brown等（2005）	Chevalier等的数据重解	4.7~4.9	325—45	南段
Chevalier等（2006）	横向展布的冰碛物和冲积扇/阶地	11.8±4.7~14.3±4.2	<200	从扎西岗狮泉河拐弯处到冈仁波齐山

目前，在活动构造研究中TCN定年技术多用于测定应变速率（表3-3）。最典型的方法就是测定应变标志［如阶地、扇体、岩浆、滨岸（湖）或冰碛物等］的暴露时间，以提供应变事件幕次的时间。它也可以直接测定出断层面上的暴露时间。最佳方法的选择主要取决于定年的对象是谁（地表年龄还是地表沉积物的年龄）以及影响原地核素生成速率的所有

因素。下述不同方法是按照复杂程度和所需费用的高低来排序的。

表 3-3 活动构造研究中 TCN 技术的应用

应用实例	地貌	常用核素	时间尺度(a)	参考文献
发生构造变形的标志物的暴露年龄	冲积扇	没有特殊要求	$10^3 \sim 10^7$	Brown 等（1998，2002，2005）；Daëron 等（2004）；Dunai 等（2005）；Dühnforth（2007）；Lee 等（2009）
	湖相或者滨海相		$10^3 \sim 10^6$	Owen 等（2007）
	河流阶地		$10^3 \sim 10^7$	Gold（2009）；Kozacl 等（2009）；Mériaux（2004，2005）；Xu 等（2005）
	冰碛物		$10^3 \sim 10^5$	Chevalier（2005，2006）；Mériaux（2009）
断面	基岩滑动面		$10^3 \sim 10^6$	Palumbo（2004）
构造形变标志物的埋藏年龄		^{26}Al/^{10}Be	$10^4 \sim 10^6$	Balco 和 Rovey（2008）；Granger 和 Smith（2000）；Granger 和 Muzikar（2001）；Granger（2006）
构造运动导致的河流下切	河床上的砾石	没有特殊要求	$10^3 \sim 10^6$	Adams（2009）；Anders（2005）；Pederson 等（2006）
岩石抬升	冰碛阶地		$10^3 \sim 10^6$	Perg（2001）；Brown 等（2002）；Kim 和 Sutherland（2004）
	废弃的岩石表面		$10^3 \sim 10^6$	Bennett 等（2005）
地表抬升		^{38}Ar/^{3}He	$10^5 \sim 10^7$	Renne 等（2001）
侵蚀作用	汇水区的侵蚀	没有特殊要求	$10^3 \sim 10^4$	Bierman 和 Nichols（2004）；Binnie（2009）；Brown 等（1998—2009）；Granger 等（1996）；Schaller 等（2002）；Wittmann 等（2007）
盆地充填的年代			$10^2 \sim 10^7$	Anderson（1996）；Balco 和 Rovey（2008）；Matmon（2009）；Dunai 等（2005）；Phillips 等（1997）

3.2.1 使用单一核素来确定岩石表面年龄

使用单一 TCN 核素法可以确定大部分地貌的岩石表面年龄，如熔岩断层崖、基岩断错脊、河流河滩以及海蚀地貌等（Kim 和 Sutherland，2004）。对目的层段附近的加密取样可以提高精度，并在许多情况下来减小总的误差。如果 $n>2$，并且所有样品都有相似的年龄（变异系数<测量精度），那么这些因素导致的误差（例如侵蚀、埋藏或者 TCN 从之前的暴露面的残留浓度）可以忽略。这些因素对这些样品造成的影响都相同。通过获得断层面的暴露时间来估计其走滑速率是一个重要并且可直接测量断层走滑速率的方法。Palumbo 等（2004）利用 ^{36}Cl 在意大利亚平宁山脉 Magnola 断层的下盘发育的 12m 宽的石灰岩断面中，发现了 5~7 次全新世的震滑事件。该方法的缺点和明显的不确定性在于，每次断层错动前下盘面的 TCN 核素产量难以确定，而且考虑到崩塌楔厚度后，也很难估算崩塌楔对于宇宙射线屏蔽的影响（会随着年代的不同而变化）。

3.2.2 使用单一核素来确定沉积物表面年龄

除了少数例外情况（如：极端干旱地区），冲积扇、阶地、抬升的三角洲、冰碛、岩崩以及其他包含碎屑的地貌表面都不是一成不变的。在成土作用以及与冰融和生物搅动（动物和植物）相关的过程中，其混合区带向上不过分米级别。沉积物表面可能会暂时被黄土、灰尘或者其他物质覆盖，也可能被卷入基底又或者发生化学或物理的侵蚀。沉积物表面自身也可以发生侵蚀。因此，地貌表面的测定年龄往往不是原生的地表年龄，因为很多沉积后的动力过程可以影响表面 TCN 法的测量值。在一般情况下，沉积物表层 TCN 暴露测年会产生一个可以忽略的最小年龄（见"利用深度剖面进行沉积物测年"一节）。

沉积物表层测年可以利用多种采样介质，包括混杂的卵石（来比较 OSL 法和铀系测年，Anders 等，2005；Pederson 等，2006），混杂或者单个的中砾石（Daëron 等，2004；Kozacı 等，2009），以及顶部的单个巨砾（Bennett 等，2005；Lee 等，2009；图 3-1）。但是沙漠中小砾石层中碎屑的暴露史存在一些不确定性，因为砾石层的来源和长时间的（至少几万年）稳定性还存在争议（Quade，2001；Matmon 等，2009）。巨砾的优势在于它们不会混杂或者发生古剥蚀，而且被沉积物、灰尘或者雪埋藏的程度要远小于卵石和更细的碎屑。了解整个冲积扇表面的沉积史（特别是全新世以来）相当重要，因为最上部的冲积扇单元可能不是同一时间的产物（看似一个年龄层内包含了多个不同年龄的碎屑流（Dühnforth 等，2007））。山前的地表过程研究也揭示了沉积速率与沉积机制的复杂性（Nichols 等，2007）。理想状态下，地表之下混合带中的某一个混合样品总的平均浓度值是唯一的（Perg 等，2001）。所以，尽管前人研究中很少使用，但在明确该地区继承假设的前提下，还是可以尝试在混合带采用混合的方法（一个单个样品包含一个或多个土层）进行取样（Brown 和 Bourles，2002）。

3.2.3 使用多种核素来确定沉积物表面年龄

在某些地表（岩石或者沉积物）可能已经被大面积掩埋或者侵蚀的情况下，可以使用双 TCN 核素法（其中一个必须是放射性核素）来确定沉积物表面的平均年龄。多核素测年的样品类型与单核素的方法一样，最常用的 TCN 值是石英中的 $^{26}Al/^{10}Be$，其他元素的使用要依具体的化学和同位素组成来具体分析。对于一个给定的暴露时间，通过计算方程（3-2）或者方程（3-3），就可以预测随着时间的变化，其数值的变化。在实际操作中，在已知某种核素浓度的情况下，时间是可以通过推测计算得到，所以我们将某一种同位素浓度与该比值的关系做成图版。在简单的暴露史模型中，埋深和侵蚀的浓度（忽略任何影响造成的衰减）与时间是不成比例的。在该方法中要对海拔和高纬度进行标准化，这样可以将不同年代和地点的样品（综合时间的生产率）绘制在同一比值的 TCN 浓度关系图版中；否则必须指明其假定的生产率。顶部厚的红色曲线（图 3-2）代表了简单模型中该比值随时间的变化。

在初始阶段 $^{26}Al/^{10}Be$ 是稳定的，直到其中半衰期最短的 ^{26}Al 开始衰变，其浓度的降低开始显著影响该比值；最终，当两种同位素达到平衡时，该比值就不再变化。而对于 ^{10}Be，这可能需要 8Ma 来完成这一过程。如果表面被侵蚀，那么达到平衡可能会更快。对于侵蚀面来说，该比值的曲线会比简单生产曲线略低（图 3-2 显示为 0.1~1mm/ka）。图中侵蚀区以下的样品（图 3-2 中的阴影部分）表明它们不管是否曾经被埋藏事件间断过，都至少经历过一次暴露事件，或者最近发生过一次侵蚀厚度明显（米级）的侵蚀事件。在简单生产曲线之上的样品一般都认为其受到了化学制剂或者质谱分析的影响。在早期的文献中（1989—2000），$^{26}Al/^{10}Be$ 与 ^{10}Be 关系的图版忽视了 μ 介子的影响。随着研究的发展，学界越来越意识到了 TCN 法中 μ 介子组分的不确定性，而且在最近几年，我们的认识和实际工作都有了逐步的改善。但是，当我们开始探索埋深大于几米的核素生产量情况时，发现埋藏史曲线的形态对于深度十分敏感，包括给定位置不同深度的样品在同一图版中都不一样。

由于获取足够高的精度以及期望极低的侵蚀速率的难度较大，所以 $^{26}Al/^{10}Be$ 很少同时用于确定剥蚀速度和年龄，但该方法在最早使用时就以此为目的。热中子产生的 ^{36}Cl 对于剥蚀的敏感度较好，所以成为了侵蚀面定年最有用的放射性核素。该核素的浓度不随深度呈指数降低，而是由于水和潮湿空气中质子的干扰以及近地面的热中子散失，在岩石和沉积物上部分米级的区域内，热中子产生的核素会随深度增加而增加。$^{36}Cl/^{10}Be$ 指数更适合于同时测定年龄和地表平均（稳定的、渐进的）侵蚀速率（Phillips 等，1997），因为相对于 $^{26}Al/^{10}Be$，该比值的侵蚀更广，可以勉强达到 2σ 误差椭圆的程度（图 3-2）。

在初次暴露后，表面的埋藏程度或多或少都可以通过 TCN 比值来反映，也可以通过 TCN 来计算最小埋藏时间，该项技术

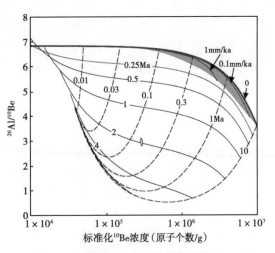

图 3-2　密度为 $1.9g/cm^3$ 的砾石层下 30m 处的 $^{26}Al/^{10}Be$ 与 ^{10}Be 图版

Balco 等（2008）为 Lal（1991）和 Stone（2000）每个时间恒定的模型提出的生产系统中，初始 ^{10}Be 生产率为 4.9 个原子/g，而表面的 $^{26}Al/^{10}Be$ 比值为 6.75，μ 介子则来自负介子和快介子（Heisinger 等，2002a，2002b；Hidy 等，2010）。粗红线上部区域代表了样品经历的侵蚀和埋藏可以忽略。图版上曲线的比值可以直接分析求得向右变大的暴露年龄（用 Ma 标示）；图版中阴影区的比值经历了侵蚀和轻微埋藏。代表零埋藏和稳定侵蚀（0.1~1.0mm/ka）曲线样品已标明。阴影下部的样品经历了复杂的暴露过程，其中至少被一期埋深中断。蓝细曲线代表的埋藏时间是从 0.25Ma 到 8Ma，代表了最小的埋藏时间。假设样品的 1σ 误差椭圆可以解释为 0.2Ma 的最小暴露时间和 2Ma 最小埋藏时间。但是，这也可能是埋藏后的多次暴露，这样总的时间跨度就大于 2.2Ma。如果合并的沉积物有一个较低的初始比值（因为复杂暴露早于最终的沉积），这样埋藏的时间所占比例就会减少

被称为"埋藏期测定"（见后两小节）。如果表面经历了间隙性埋藏（例如，表面因为剥蚀和熔化而不可见），可测得的暴露年龄应视为最小年龄，因为放射性核素在我们未知的埋藏时段会有衰减。用稳定 TCN 法获得的年龄也是最小值（埋藏期没有增加），但是会比放射性核素得到的年龄要老。在某些情况下，这个结果可能包括了暴露和埋藏的整个历史。

3.2.4　使用深度剖面来估计沉积物的年龄

残留浓度，即沉积物中矿物所具有的初始 TCN 浓度，该 TCN 浓度存在于任何沉积物中，并且其发生暴露的时间早于其他任何沉积。如果没有补偿的话，初始的 TCN 浓度将有一个忽略侵蚀和埋藏影响的最老年龄。这种残留浓度在高地形区可能最大（例如，若盆地内冲积扇表面的沉积物来源于高海拔区，明显具有较高的二级宇宙射线通量，而在被侵蚀的汇水区较低）。有三种方法可以用来评价残留量：（1）使用现代沉积物通量来测量沉积物刚

沉积时的浓度，使用该方法的前提是沉积物刚沉积时，具有同样的侵蚀速率和表面过程[例如，Brown 等（1998）对于喀喇昆仑山断层滑动速率的测量]；（2）测量已获得年龄层且位于屏蔽深度（沉积物中大于 10m）的沉积物浓度，其前提条件是汇水区的侵蚀早于其沉积时，且具有相似的速率。这两种方法的前提条件都难以实现，无法计算再沉积的沉积量，而且在一个广泛的流域内或者小区域内，特定沉积物的沉积明显依赖于气候。（3）该方法最优，也是最昂贵的方法。对深度剖面的沉积物进行合并采样（图 3-1；Anderson 等，1996）。单一核素深度剖面法的优势在于：它不需表面过程和侵蚀过程的速率长期恒定；混合区之下的深度要测量 3~7 个合并样本（建议至少测 5 个，当然越多越好）来确定浓度—深度曲线，如果获得了未离散的年龄值，则通过该值即可获得所需要的残留、侵蚀/加积（假设沉积物容重为单独测量）结果。浓度与深度图的渐近线指示的是平均残留浓度，而曲线形状则提供了侵蚀和加积信息。必须保证有足够的样品颗粒数量，只有这样才能获得有代表性的数据。在冰川环境下（高侵蚀速率），每个样品需要至少 40 个颗粒，来提供足够的平均值（例如，40 个卵石，或者满足矿物地球化学测量数量的多种粒径的颗粒，见样品制备部分）。但是，对于更多低侵蚀速率的干旱地区来说（较高的残留值），可能就会需要更多的颗粒（尽可能大的残留浓度范围），成千上万或更多分选良好的颗粒能达到更好的效果，特别是原生沉积构造保存完好的颗粒（Hidy 等，2010）。

在这三种方法中，最重要的是尽量保证残留浓度不要随深度的变化而变化。取样风险最小化的方法有以下两种：（1）在狭窄的深度范围取样，比如说小于 1m；（2）根据地质模式，避免在不同沉积环境中取样（例如，一些样品取自河流相，而另外的样品取自碎屑流沉积）（图 3-1）；在 21 世纪初发表的某地被断层错断的冲积扇的研究中，TCN 深度剖面法的年龄与其他测年吻合；在某些情况下，断距的不确定性甚至大于断面暴露年龄的不确定性（Gold 等，2009）。

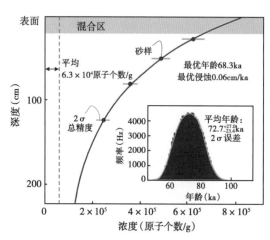

图 3-3 根据 Providence 山西南的冲积扇深度剖面中石英颗粒中 ^{10}Be 建立的浓度和年龄模型（AMS 分析在 LLNL，化学分析在 Dalhousie 地质年代学中心完成）样品采集于生物扰动区之下，每个合并取样的样品约 6cm 厚，至少 2kg 重；没有沙漠表层砾石的采样。生产率根据 Balco 等（2008），而年龄模型和不确定性源自 Hidy 和 Gosse（文章已录用）的十万次模拟；在这次模拟中密度恒定（1.9g/cm³）

理想状态下，深度—浓度曲线可以得到唯一的暴露时间、残留值、侵蚀或者加积的状态。混合区带之下的浓度随深度会有所变化：（1）残留浓度有所差别（特别是较新沉积物中残留浓度在总浓度中的所占比例极大时）；（2）密度随深度会有小的变化，特别是由成土作用造成的变化；（3）未确认的化学或含水量的变化，它可以影响热中子俘获 ^{36}Cl 的速率。为了获得剖面年龄有效、准确混合区下 1~1.5m 范围的沉积物必须立即取样（图 3-3）。如果可以判断出残留浓度在总量中占有重要比重，或者残留浓度，可以计算古盆地的平均侵蚀速率（见下文第 3.3 节），并且如果地质调查支持该深度联合残留浓度较稳定，那么深度剖面应该更深些，最好可以到混合区下 2m。有几种方法已经应用于深度剖面数据的解释，包括无侵蚀情况下仅有核裂变发生或者核裂变与 μ 介子交互作用下的简单的标准深度拟合曲线。被夹在最大侵蚀和最小侵蚀之间的数据具有一定的不确定性。使用迭代递减的 χ^2 分析法来获得最佳拟合曲线，以残留浓度、年龄值、侵蚀以及加积速率作为拟合变量。近期深度剖面研究以测量沉积物密度随深度和时间的变化为主。前人在研究当中尝试以更严格的蒙特卡洛模拟来构建深度剖面的概率分布函数（Phillips 等，1998）。最近研究的计算方法将随机分布、正态分布误差以及密度概率函数和 χ^2 图版中所有随机的中和系统的误差放大（Hidy 等，2010；图 3-3）。在某些侵蚀速率约束条件下（基于独立研究的土壤地貌学或者 ^{36}Cl 深度剖面控制下），成百上千条精度可控的曲线，可以给出最可能的年龄值范围，而不是简简单单由一条最佳拟合曲线（最小 χ^2）来确定（图 3-3）。

3.2.5 使用埋深测年来确定沉积物年龄

Klein 等（1986）首次借助 TCN 法，通过埋藏幕来推算复杂的暴露史。他们发现利比亚沙漠中的 12 个样品的 ^{26}Al/^{10}Be 值低于生产率，推测其至少发生了一期埋藏事件。之后研究中的 ^{26}Al/^{10}Be 被用于确定河床基岩被（弱侵蚀）冰盖覆盖的覆盖史，有助于分析南北极冰川发育史（Nishiizumi 等，1991）和确定沉积物年龄。^{26}Al/^{10}Be 值也用来确定喀斯特溶洞中废弃河道中沉积物的年龄进而研究北美中大陆水系演化，并且同时也用于确定洞穴中考古文物的年龄（Granger 和 Smith，2000；Granger，2006）。埋藏测年法要求 TCN 中至少要有一个放射性核素（见"用多个核素确定表面年龄"一节；图 3-2）。当沉积物从暴露状态突然被埋藏，则放射性元素和稳定元素的浓度都会减少（图 3-2）。根据测量值和初始值的差可以推测出唯一的埋藏时间，但在许多情况下沉积物沉积前初始的残留值是未知的；完全埋藏是十分重要的。完全埋藏需要其上覆盖物具有一定的堆密度（比基岩更多的沉积物）。μ 介子能穿透超过 10m 的岩石，虽然在更深处 μ 介子的通量还不是很清楚，但是大于 20m 岩石处可以满足埋藏小于 5Ma 的时间的通量。随着埋藏时间增加，初始的 μ 介子浓度会因衰减而减少。在某些情况下，埋藏瞬间发生时，其埋藏时间可以成功获得（如，沉积物进入了一个很深的洞穴或者被熔岩覆盖）。然而，如果埋藏期相对于沉积速率足够长，通过其他沉积物来测定沉积物快速埋藏的年龄还是有可能的。埋藏期和其沉积物初始浓度决定了放射性元素的选择。如果是不完全埋藏，而沉积速率和侵蚀史都是独立的，这样还是可以估计埋藏时间，其中最大的问题是确定初始值。如果河流或者风的沉积物在沉积前经历了长期的堆积（部分或者完全的埋深）或者搬运，两种同位素的比例将不同于初始比值。当测年层位被其他沉积物覆盖，如果不清楚现有值或部分埋藏是否早于最终沉积（残留率即生产率），则埋藏年龄应该取最大值。

3.2.6 沉积物等时法确定埋藏时间

Balco 近期发表的一系列论文中介绍了一种确定埋藏时间的新方法（Balco 和 Rovey，2008），该方法强调了确定埋藏前沉积物残留同位素比值的重要性（图 3-4）。这种方法可能会成为未来几年的常规方法，特别是已埋藏的沉积物中有古土壤发育的地方。长时期埋藏的沉积物对于现有值的作用不敏感，因此可以更多地使用传统方法（见前一节）。等时方法建立在土壤生产率随深度和埋藏的改变（TCN 值在较小深度范围是恒定的）而导致浓度变化关系的基础上。图 3-4 中蓝线表示的是从未被埋藏过的表面之上几米内沉积物的 TCN 浓度。无论哪种同位素的浓度，在取样深度方向上都会降低（曲线向下指向初始位置）。这条

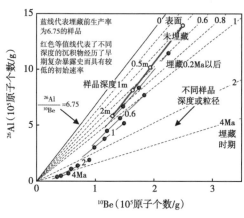

图 3-4 用 ^{26}Al—^{10}Be 图版来说明 TCN 等时线法埋藏定年的概念（据 Balco 和 Rovey，修改，2008）

对暴露的土壤（从未埋藏）的连续取样样品的比值得到的蓝色实线是深度与暴露时间的函数；蓝色虚线代表了古土壤中样品在某段时间内一次连续的埋藏；红色曲线代表了不同深度的沉积物经历了早期埋藏后在古土壤中被风化、再沉积的样品。红色样品的初始比值小于其对应同位素的生产比值；由等时线的位置及斜率同时可以求得初始值和埋藏时间

线从未发生过埋藏的深度剖面上进行多重取样来确定，值为 6.75（例如 $^{26}Al/^{10}Be$ 的产率）。如果沉积物首次埋藏并一直未被侵蚀，浓度和斜率（$^{26}Al/^{10}Be$）是唯一的，并可以表示不同的埋深时间（蓝色虚线）。然而，如果沉积物继承了初始浓度，并且如果该沉积物最早的埋深早于最终的沉积，则比值对应图中的不同曲线（红色）。图 3-4 空心圆代表沉积物继承了初始浓度并且埋藏早于沉积（在这种情况下，初始比值为 4.2），样品取自剖面上还未被埋藏的四个深度。一旦土壤被埋藏（成为古土壤），以红圈定义的红细线代表着随埋藏时间该比值下降的曲线。位于同一深度的三个样品在没有测得同位素比值变化的情况下，简单的长期埋藏或者由于最后埋藏事件以及早期其他事件的一些元素组合来区分 TCN 比值小于 6.75 是不可能的。尽管对于初始值的解释在推测埋藏时间上有显著控制，但等时法不需要独立的已知初始比值。随着埋藏时间的增加，需要注意的是埋藏测年可能对于低初始比值不敏感。在一些高残留（可能与颗粒粒径有关）和低侵蚀速率的地区，另一种方法是在不同深度不同粒径进行混合取样，使得每种粒径级别都有不同的残留 $^{26}Al/^{10}Be$ 浓度（用不同的粒径而不是不同深度）。等时线法的应用有三大挑战：第一（最好早于取样），必须确定初始埋藏面在埋藏前是否经过足够长的暴露（发育好的古土壤或者发育有厚层泥炭层）来获得足够的 ^{26}Al 和 ^{10}Be 浓度，从而可以在埋藏期来衰减；第二，取样间距不好把握，较密的取样间隔（深度小于 1m）将提高这些样品有相近的初始比值，而等时线法要求更宽的取样空间；第三，如果埋藏不完全（例如部分掩埋），模型需要更多的限制条件，而且沉积物堆密度和侵蚀/加积史也需要考虑。

3.3 在发生过侵蚀作用或者有过下切作用地区构造分析中的应用

TCN 浓度随深度的变化可以作为发生侵蚀层位的有效指示。对于侵蚀速率的估计可以从单一露头或者卵石层到汇水盆地级别（$>10^4 km^2$）。对于准确侵蚀速率的计算来说，时间跨度依赖于 TCN 方法、侵蚀类型以及侵蚀速率。在极高的侵蚀速率，或者有大断块参与的侵蚀的情况下，可以估算短期的侵蚀速率。下面描述的过程要有合适的长期稳定的侵蚀，速率至少要达到毫米/年的级别。在负载沉积物的流量发生短期变化的构造盆地中，或者是在探索物源区抬升与汇水区沉积间关系的盆地中，针对盆地汇水区的平均侵蚀速率的分析已经得到了广泛应用。例如，平均坡度和侵蚀速率的正相关关系，并不总是适用于地质构造活跃区陡峭的斜坡汇水区。再如，加利福尼亚州的 San Bernardino 山地区，侵蚀速率是由岩石构造抬升控制的，通过河道的下切以及块体整体的坡移来实现（Binnie 等，2007）；这种模式类似于其他学者提出的喜马拉雅山（Burbank 等，1996）和安第斯山（Antinao 和 Gosse，2009）模式。

河床基岩的侵蚀速率或者沉积层的侵蚀速率可以通过 TCN 法估算。在简单研究中（$C_i = 0$，侵蚀以外的环境影响可以忽略），TCN 浓度是生产率（已知）、年代（可能已知）和侵蚀速率（未知）的函数。如果暴露的时间可以独立测定（如表面经过 $^{40}Ar/^{39}Ar$ 法定年的熔岩），则该地层在暴露的时间内的平均侵蚀速率就可以精确求得（方程 3-2）。如果年龄难以获得，这里还有两种方法来确定侵蚀速率：（1）假设一个无限的年龄来解决最大侵蚀速率的问题；（2）利用半衰期较短的（如石英中的 ^{14}C）并且已达到侵蚀稳定浓度（浓度已经长期平衡并且受控于侵蚀作用）的放射性元素。在一些实际研究当中，如果存在疑似发生过部分或者偶然埋藏，又或者沉积物保留了一些先前暴露的地表残留浓度的情况，正如本章节中所描述的"利用深度剖面推测沉积物年龄"，这时可以通过一个深度剖面来解决平均侵蚀速率的问题。两种放射性元素很少被用来确定侵蚀速率，主要是因为敏感性的精度问题。但是，正如上文讨论的，对侵蚀速率敏感性更高的热中子产生的 ^{36}Cl 还未广泛采用。TCN 法可以为检验构造事件（如岩石抬升和剪切速率的趋势，断层扩展方向或者三联点的迁移以及陡峭浅板块的转换的趋势）来估算现今（Brown 等，1995；Granger 等，1996）以及古汇水盆地（Schaller 等，2002）全流域的平均侵蚀速率。与很多个体风化层表面采样不同，该方法的基本前提是在给定区域内，所有风化层通过滑移、越岸径流等方式影响地表径流颗粒并且每个颗粒都是表面侵蚀速率的同位素标记物，同时也是主要的浓度累计方式（Bierman 和 Nichols，2004）。风化层经历了高侵蚀速率（例如突兀或陡峭的斜坡），会产生低浓度的颗粒，因此，树木覆盖的低梯度表面或许侵蚀较慢，并且这样将会有助于颗粒获得高 TCN 浓度。利用测高的方法来获得整个汇水区的生产速率就是通过假设核素浓度恒定的状态下（主要由侵蚀控制）计算侵蚀速率。在沿流域的某些点采集的沉积物中的 TCN 浓度（储存的）可以因此解释为全流域的平均侵蚀速率或者估计地区上游的剥蚀速率（Wittmann 等，2007）。在解释储存浓度时注意事项包括：（1）缺乏矿物分布的均质性，所以需要额外建模；（2）由于大量块体坡移或者近期的构造活动（冰川活动或者风尘活动使得浓度没有达到稳定状态）而不存在假设的稳定状态；（3）总汇水区大部分地区没有沉积物生成（如盆地内或者沉积物长期沉积区）；（4）采样时距离天然或者人工坝体很近或者在不同侵蚀速率河流的交汇处；（5）石英相对其他矿物有较长的停留时间，特别是在高化学风化速率区（Riebe 等，2001）。具有合适的平均侵蚀速率的侵蚀时间会随着侵蚀速率的变大而下降，但是通常较慢侵蚀速率的千年或万年尺度至现代沉积物可以在晚第四纪产生一个平均侵蚀速率。古汇水区的平均侵蚀速率可以通过已知年龄的沉积物的残留浓度来计算；正如沉积获得的 TCN，其中的放射性元素和 μ 介子在没有完全屏蔽的情况下，会随着海拔和盆地地形的变化以及衰减而有变化。

3.4 展望

近年来随着地质年代学的快速发展，使其在活动构造领域的应用取得了显著效果；特别是以河流为研究对象的地质年代学，在构造形变史的研究中取得了很好的效果。20 世纪 90 年代和 21 世纪初的研究文章中，利用平底河谷（河床基岩阶地面侧向展布以及河流袭夺造成的废弃）下切速率作为源区抬升、盆地沉降以及造山消减的标志的研究文章不断增加；

放射性碳同位素、阴极发光、铀系元素以及 TCN 法等多种手段共同使用来确定河流冲积层沉积的年代。在其他一些研究案例中，宽谷的年代可以直接用 TCN 法获得（Adams 等，2009）。最近，TCN 法已经成为热年代学的补充试验方法，例如（U-Th-Sm）/He 法来检验造山带级的地形演化假设以及活动构造区的剥蚀和抬升速率（Adams 等，2009）。未来希望可以借助 TCN 速率分析精确度的提高，直接通过 TCN 生产率与大气质量灵敏度的关系来确定古海拔（Renne 等，2001）。

致谢（略）

参 考 文 献

Adams, B., Dietsch, C., Owen, L. A., Caffee, M. W., Spotila, J., and Haneberg, W. C. (2009) Exhumation and incision history of the Lahul Himalaya, northern India, based on (U-Th) /He thermochronometry and terrestrial cosmogenic nuclide methods. Geomorphology, 107, 285-299.

Anders, M. D., Pederson, J. L., Rittenour, T. M., Sharp, W. D., Gosse, J. C., Karlstrom, K. E., Crossey, L. J., Goble, R. J., Stockli, L., and Yang, G. (2005). Pleistocene geomorphology and geochronology of eastern Grand Canyon: linkages of landscape components during climate changes. Quaternary Science Reviews, 24, 2428-2448.

Anderson, R. S., Repka, J. L., and Dick, G. S. (1996) Explicit treatment of inheritance in dating depositional surfaces using in situ ^{10}Be and ^{26}Al. Geology, 24, 47-51. doi: 10. 1130/0091-761319960240047.

Antinao, J-L., and Gosse, J. (2009) Large rockslides in the Southern Central Andes of Chile (32-34. 5 S): tectonic control and significance for post-Miocene landscape evolution. Geomorphology, 144, 117-133.

Balco, G., Stone, J., Lifton, N., and Dunai, T. (2008) A complete and easily accessible means of calculating surface exposure ages or erosion rates from ^{10}Be and ^{26}Al measurements. Quaternary Geochronology, 3, 174-195. doi: 10. 1016/j. quageo. 2007. 12. 001.

Balco, G., and Rovey, C. W. (2008) An isochron method for cosmogenic-nuclide dating of buried soils and sediments. American Journal of Science, 308, 1083-1114, doi 10. 2475/10. 2008. 02.

Bennett, E. R., Youngson, J. H., Jackson, J. A., Norris, R. J., Raisbeck, G. M., Yiou, F., and Fielding, E. (2005) Growth of South Rough Ridge, Central Otago, New Zealand: using in situ cosmogenic isotopes and geomorphology to study an active, blind reverse fault. Journal of Geophysical Research, 110, B02404. doi: 10. 1029/ 2004JB003184.

Bierman, P., and Nichols, K. (2004) Rock to sediment - slope to sea with 10Be - rates of landscape change. Annual Review of Earth and Planetary Sciences, 32, 215-255.

Binnie, S. A., Phillips, W. M., Summerfield, M. A., and Fifield, L. K. (2007) Tectonic uplift, threshold hillslopes, and denudation rates in a developing mountain range. Geology, 35 (8), 742-746.

Binnie, S. A., Phillips, W. M., Summerfield, M. A., Fifield, L. K., and Spotila, J. A.

(2010) Tectonic and climatic controls of denudation rates in active orogens: the San Bernardino Mountains. California Geomorphology, 118 (3-4), 249-261.

Brown, A. G., Carey, C., Erkens, G., Fuchs, M., Hoffmann, T., Macaire, J-J., Moldenhaue, K-M., and Walling, D. E. (2009) From sedimentary records to sediment budgets: multiple approaches to catchment sediment flux Geomorphology, 108, 35-47.

Brown, E. T., Stallard, R. F., Larsen, M. C., Raisbeck, G. M., and Yiou, F. (1995) Denudation rates determined from the accumulation of in situ - produced ^{10}Be in the LuquilloExperimentalForest, Puerto Rico. Earth and Planetary Science Letters, 129, 193-202.

Brown, E. T., Bourles, D. L., Burchel, B. C., Oidong, D., Jun, L., Molnar, P., Raisbeck, G. M., and Yiou, F. (1998) Estimation of slip rates in the southern Tien Shan using cosmic ray exposure dates of abandoned alluvial fans. Geological Society of America Bulletin, 110 (3) 377-386.

Brown, E. T., Bendick, R., Bourles, D. L., Gaur, V., Molnar, P., Raisbeck, G. M., and Yiou, F. (2002) Slip rates on the Karakoram fault, Ladakh, India, determined using cosmic ray exposure dating of debris flows and moraines. Journal of Geophysical Research, 197, 7-13.

Brown, E. T., and Bourles, D. L. (2002) Use of a new ^{10}Be and ^{26}Al inventory method to date marine terraces, Santa Cruz, California, USA: comment. Geology, 30, 1147-1148.

Brown, E. T., Molnar, P., and Bourles, D. L. (2005) Technical comment on "Slip - rate measurements on the Karakorum Fault may imply secular variations in fault motion." Science, 309, 1326b.

Burbank, D. W., Leland, J., Fielding, E., Anderson, R. S., Brozovic, N., Reid, M. R., and Duncan, C. (1996) Bedrock incision, rock uplift and threshold hillslopes in the northwestern Himalayas. Nature, 379, 505-510.

Chevalier, M. L., Ryerson, F. J., Tapponnier, P., Finkel, R. C., Van Der Woerd, J., Li, H., and Qing, L. (2005) Slip - rate measurements on the Karakoram fault may imply secular variations in fault motion. Science, 307, 411-414. doi: 10. 1126/science. 1105466.

Chevalier, M., Tapponnier, P., van der Woerd, J., Finkel, R. C., Rerson, F. J., Li, H., and Liu, Q. (2006) Determination, by 10Be cosmogenic dating, of slip-rates on the Karakorum Fault (Tibet) and paleoclimatic evolution since 200 ka. AGU, Fall Mtg 2006, abstr #T21E-08.

Daeron, M, Benedetti, L, Tapponnier, P. Sursock, A., and Finkel, R. (2004) Constraints on the post 25-ka slip rate of the Yammouneh fault (Lebanon) using in situ cosmogenic ^{36}Cl dating of offset limestone-clast fans. Earth and Planetary Science Letters, 227, 105-119.

Desilets, D., and Zreda, M. (2003) Spatial and temporal distribution of secondary cosmic - ray nucleon intensities and applications to in situ cosmogenic dating. Earth and Planetary Science Letters, 206 (1-2), 21-42.

Dunai, T. J. (2000) Scaling factors for production rates of in-situ produced cosmogenic nuclides: a critical reevaluation. Earth and Planetary Science Letters, 176 (1), 157-169.

Dunai, T. J. (2010) Cosmogenic Nuclides: Principles, Concepts and Applications in the Earth Surface Sciences. CambridgeUniversity Press. 198 p.

Dunai, T. J., Gonzalez-Lopes G. A., and Juez-Larre J. (2005) Oligocene-Miocene age of aridity in the Atacama Desert revealed by exposure dating of erosion-sensitive landforms. Geology, 33 (4), 321-324.

Dunai, T. J., and Stuart, F. M. (2009) Reporting of cosmogenic nuclide data for exposure age and erosion rate determinations. Quaternary Geochronology, 4 (6), 437-440.

Duhnforth, M., Densmore, A. L., Ivy-Ochs, S., Allen, P. A., and Kubik, P. W. (2007) Timing and patterns of debris flow deposition on Shepherd and Symmes creek fans, Owens Valley, California, deduced from cosmogenic 10Be. Journal of Geophysical Research, 112, F03S15. doi: 10. 1029/2006JF000562.

England, P. C., and McKenzie, D. P. (1982) A thin viscous sheet model for continental deformation, Geophys. Journal of the Royal Astronomical Society, 70, 295-321.

Frankel, K. L., Finkel, R. C., and Owen, L. A. (2010) Terrestrial cosmogenic nuclide geochronology data reporting standards needed. EOS Transactions, 91 (4), 31. doi: 10. 1029/2010EO040003.

Gold, R. D., Cowgill, E., Arrowsmith, J. R., Gosse, J., Chen, X, and Wang, X-F. (2009) Riser diachroneity, lateral erosion, and uncertainty in rates of strike-slip faulting: a case study from Tuzidun along the Altyn Tagh Fault, NW China. Journal of Geophysical Research, 114, B04401. doi: 10. 1029/2008JB005913.

Gosse, J. C., and Phillips, F. M. (2001) Terrestrial in situ cosmogenic nuclides: theory and application. Quaternary Science Reviews, 20, 1475-1560.

Granger, D. E. (2006) A review of burial dating methods using ^{26}Al and ^{10}Be, in Siame, L. L., Bourle's, D. L., and Brown, E. T., editors, In-situ-produced cosmogenic nuclides and quantification of geological processes. Geological Society of America Special Paper 415, 1-16. doi: 10. 1130/2006. 2415 (01).

Granger, D. E., Kirchner, J. W., and Finkel, R. (1996) Spatially averaged longterm erosion rates measured from in-situ produced cosmogenic nuclides in alluvial sediment. Journal of Geology, 104, 249-257.

Granger, D. E., and Muzikar, P. F. (2001) Dating sediment burial with cosmogenic nuclides: theory, techniques, and limitations. Earth and Planetary Science Letters, 188 (1-2), 269-281.

Granger, D. E., and Smith, A. L. (2000) Dating buried sediments using radioactive decay and muogenic production of ^{26}Al and ^{10}Be. Nuclear Instruments and Methods in Physics Research B, 172, p. 822-826. doi: 10. 1016/ S0168-583X (00) 00087-2.

Heisinger, B., Lal, D., Jull, A. J. T., Kubik, P., Ivy-Ochs, S., Knie, K., and Nolte, E. (2002a) Production of selected cosmogenic radionuclides by muons: 2. Capture of negative muons. Earth and Planetary Science Letters, 200, 357-369.

Heisinger, B., Lal, D., Jull, A. J. T., Kubik, P., Ivy-Ochs, S., Neumaier, S., Knie, K., Lazarev, V., and Nolte, E. (2002b) Production of selected cosmogenic radionuclides by muons: 1. Fast muons. Earth and Planetary Science Letters, 200, 345-355.

Hidy, A., Gosse, J. C, Pederson, J., and Finkel, R. (2010) A constrained Monte Carlo approach to modeling exposure ages from profiles of cosmogenic nuclides: an example from Lees Ferry, AZ. Geochemistry, Geophysics, Geosystems, 11, 1-18. Q0AA10. doi: 10. 1029/2010GC003084.

Hidy, A., Gosse, J. C, Pederson, J., Mattern, P., Finkel, R. (2010) A constrained Monte Carlo approach to modeling exposure ages from profiles of cosmogenic nuclides: an example from Lees Ferry, AZ. Geochemistry, Geophysics, Geosystems, 11m Q0AA10. doi: 10. 1029/2010GC003084.

Kim, K. J., and Sutherland, R. (2004) Uplift rate and landscape development in southwest Fiordland, New Zealand, determined using ^{10}Be and ^{26}Al exposure dating of marine terraces. Geochimica et Cosmochimica Acta, 68, 10, 2313-2319.

Klein, J., Giegengack, R., Middleton, R., Sharma, P., Underwood, J. R. J., and Weeks, R. A. (1986) Revealing histories of exposure using in situ produced ^{26}Al and ^{10}Be in Libyan desert glass, Radiocarbon, 28, 547-555.

Kozacı, Ö., Dolan, J. F., and Finkel, R. C. (2009) A late Holocene slip rate for the central North Anatolian fault, at Tahtaköprü, Turkey, from cosmogenic ^{10}Be geochronology: Implications for fault loading and strain release rates. Journal of Geophysical Research, 114, B01405. doi: 10. 1029/2008JB005760.

Lal, D. (1991) Cosmic ray labeling of erosion surfaces: in situ nuclide production rates and erosion models. Earth and Planetary Science Letters, 104, 424-439.

Lal, D., and Chen, J. (2005) Cosmic ray labeling of erosion surfaces. II: Special cases of exposure histories of boulders, soils and beach terraces. Earth and Planetary Science Letters, 236 (3-4), 797-813.

Lee, J., Garwood, J, Stockli, D. F., and Gosse, J. (2009) Quaternary faulting in Queen Valley, California-Nevada: Implications for kinematics of fault-slip transfer in the eastern California shear zone-Walker Lane belt. GSA Bulletin, 121, 599-614.

Lifton, N. A., Bieber, J. W., Clem, J. M., Duldig, M. L., Evenson, P., Humble, J. E., and Pyle, R. (2005) Addressing solar modulation and long-term uncertainties in scaling secondary cosmic rays for in situ cosmogenic nuclide applications. Earth and Planetary Science Letters, 239, 140-161.

Lifton, N. A., Smart, D. F., and Shea, M. A. (2008) Scaling time-integrated in situ cosmogenic nuclide production rates using a continuous geomagnetic model. Earth and Planetary Science Letters, 268 (1-2), 190-201.

Matmon, A., Simhai, O., Amit, R, Haviv, I., Porat, N., Eric McDonald, E., Benedetti, L., and Finkel, R. (2009) Desert pavement – coated surfaces in extreme deserts present the longest-lived landforms on Earth. GSA Bulletin, 121 (5-6), 688-697. doi: 10. 1130/B26422. 1.

Meriaux, A-S., Ryerson, F. J., Tapponnier, P., Van der Woerd, J., Finkel, R. C., Xu, X., Xu, Z., and Caffee, M. W. (2004) Rapid slip along the central Altyn Tagh Fault: morphochronologic evidence from Cherchen He and Sulamu Tagh. Journal of Geophysical Research, 109, B06401. doi: 10. 1029/2003JB002558.

Meriaux, A-S., et al. (2005) The Aksay segment of the northern Altyn Tagh Fault: Tectonic geomorphology, landscape evolution, and Holocene slip rate. Journal of Geophysical Research, 110, B04404. doi: 10.1029/2004JB003210.

Meriaux, A-S., Sieh, K., Finkel, R. C., Rubin, C. M., Taylor, M. H., Meltzner, A. J., and Ryerson, F. J. (2009) Kinematic behavior of southern Alaska constrained by westward decreasing postglacial slip rates on the Denali Fault, Alaska. Journal of Geophysical Research, 114, B03404. doi: 10.1029/2007JB005053.

Molnar, P., and Tapponnier, P. (1975) Cenozoic tectonics of Asia: effects of a continental collision. Science, 189 (4201), 419–426.

Nichols, K. K., Bierman, P. R., Eppes, M. C., Caffee, M., Finkel, R., and Larsen, J. (2007) Timing of surficial process changes down a Mojave Desert piedmont. Quaternary Research, 68 (1), 151–161.

Nishiizumi, K., Kohl, C. P., Arnold, J. R., Klein, J., Fink, D., and Middleton, R. (1991) Cosmic ray produced ^{10}Be and ^{26}Al in Antarctic rocks: exposure and erosion history. Earth and Planetary Science Letters, 104 (2/4) 440–454.

Owen, L. A., Bright, J., Finkel, R. C., Jaiswal, M. K., Kaufman, D. S., Mahan, S., Radtke, U., Schneider, J. S., Sharp, W., Singhvi, A. K., and Warren, C. N. (2007) Numerical dating of a Late Quaternary spit-shoreline complex at the northern end of Silver Lake playa, Mojave Desert, California: a comparison of the applicability of radiocarbon, luminescence, terrestrial cosmogenic nuclide, electron spin resonance, U-series and amino acid racemization methods. Quaternary International, 166 (1), 87–110.

Palumbo, L., Benedetti, L., Bourles, D., Cinue, A., and Finkel, R. (2004) Slip history of the Magnola fault (Apennines, Central Italy, from ^{36}Cl surface exposure dating: evidence for strong earthquakes over the Holocene. Earth and Planetary Science Letters, 225, 163–176.

Pederson, J. L., Anders, M. D., Rittenhour, T. M., Sharp, W. D., Gosse, J. C., and Karlstrom, K. E. (2006) Using fill terraces to understand incision rates and evolution of the Colorado River in eastern Grand Canyon, Arizona. Journal of Geophysical Research, 111, F02003. doi: 10.1029/2004JF000201.

Perg, L. A., Anderson, R. S., and Finkel, R. C. (2001) Use of a new 10Be and ^{26}Al method to date marine terraces, Santa Cruz, California, USA. Geology, 29 (10), 879–882.

Phillips, W. M., McDonald, E. V., Reneau, S. L., and Poths, J. (1998) Dating soils and alluvium with cosmogenic ^{21}Ne depth profiles: case studies from the Pajarito Plateau, New Mexico, USA. Earth and Planetary Science Letters, 160, 209–223.

Phillips, F. M., Zreda, M. G., Gosse, J. C., Klein, J., Evenson, E. B., Hall, R. D., Chadwick, O. A., and Sharma, P. (1997) Cosmogenic ^{36}Cl and ^{10}Be ages of Quaternary glacial and fluvial deposits of the Wind River Range, Wyoming. Geological Society of America Bulletin, 109, 1453–1463.

Pigati, J. S., and Lifton, N. A. (2004) Geomagnetic effects on time-integrated cosmogenic nuclide production with emphasis on in-situ ^{14}C and ^{10}Be. Earth and Planetary Science Letters, 226, 193–205.

Plug, L., Gosse, J., West, J., and Bigley, R. (2007) Attenuation of cosmic ray flux in temperate forest. Journal of Geophysical Research, 112, F02022. doi: 10. 1029/ 2006JF000668.

Quade, J. (2001) Desert pavements and associated rock varnish in the Mojave Desert: how old can they be? Geology, 29 (9), 855-858. doi: 10. 1130/0091 7613 (2001) 029.

Renne, P. R., Farley, K. A., Becker, T. A., and Sharp, W. D. (2001) Terrestrial cosmogenic argon. Earth and Planetary Science Letters, 188 (3-4), 435-440.

Riebe, C. S., Kirchner, J. W., and Granger, D. E. (2001) Quantifying quartz enrichment and its consequences for cosmogenic measurements of erosion rates from alluvial sediment and regolith. Geomorphology, 40, 15-19.

Schaller, M., von Blanckenburg, F., Veldkamp, A., Tebbens, L. A., Hovius, N., and Kubik, P. W. (2002) A 30, 000 yr record of erosion rates from cosmogenic ^{10}Be in Middle European river terraces. Earth and Planetary Science Letters, 204, 307-320.

Staiger, J., Gosse, J., Toracinta, R., Oglesby, R., Fastook, J., and Johnson, J. (2007) Atmospheric scaling of cosmogenic nuclide production: the climate effect. Journal of Geophysical Research - Solid Earth, 112, B02205. doi: 10. 1029/2005JB003811.

Stone, J. (2000) Air pressure and cosmogenic isotope production. Journal of Geophysical Research, 105 (B10) 23753-23759.

Vermeesch, P. (2007) CosmoCalc: an Excel add - in for cosmogenic nuclide calculations, Geochemistry, Geophysics, Geosystems, 8, Q08003. doi: 10. 1029/2006GC001530.

Wittmann, H. F., von Blanckenburg, T., Kruesmann, K. P., Norton, and Kubik, P. W. (2007) Relation between rock uplift and denudation from cosmogenic nuclides in river sediment in the Central Alps of Switzerland. Journal of Geophysical Research, 112, F04010. doi: 10. 1029/ 2006JF000729.

Xu, X-W., Tapponnier, P., Van Der Woerd, J., Ryerson, F. J., Wang, F., Zheng, R., Chen, W., Ma, W. T., Yu, G. H., Chen, G. H., and Meriaux, A. S. (2005) Late Quaternary sinistral slip rate along the Altyn Tagh Fault and its structural transformational model. Science in China Series D: Earth Sciences, 48, 384-397.

(郭帅 译，董阳阳 董文武 校)

第 4 章 磁性地层学的原理与应用

GUILLAUME DUPONT-NIVET, WOUT KRIJGSMAN

(Paleomagnetic Laboratory, Fort Hoofddijk, Faculty of Geosciences-Utrecht University, Utrecht, the Neetherlands)

摘　要：磁性地层学是指利用岩石对地磁场极性倒转事件的记录来确定岩石地层序列的一种技术。磁性地层学适用于多种岩性（火山岩和沉积岩）和环境（大陆、湖泊和海洋），已经作为一种标准的研究手段，广泛应用于地球科学的众多研究领域。应用磁性地层学进行研究，有一个重要的前提条件：岩石必须准确记录地层沉积时的古地磁场变化情况，而且这个记录也必须可以通过各种研究古地磁和岩石磁学的技术测量获得。磁性地层学作为一种确定年代的工具，被广泛应用于构造地质学的研究中。例如，通过地磁变化的累计速率来推测构造变形幕的时限，同时可以应用于同构造期的构造运动测年等。磁性地层学也可以结合碎屑颗粒的热年代学，来确定地层从接受沉积到抬升暴露之间的折返时间；同时，磁性地层学所记载的精确的环境特征也是反映构造运动与气候变迁之间关系的最好资料，有助于我们从控制气候变化的角度来分析构造运动。我们将通过几个实例来展示以上所述磁性地层学的应用。

关键词：磁性地层学　古地磁　年龄　相关　时间尺度

4.1　引言

地球磁场最明显的特征就是不定间隔的极性倒转，这些倒转产生了类似于"条形码"一样的记录，正极性与反极性交替出现，而宽度则反映了每期倒转时间的长短（图4-1）。随着多种独立测年技术的应用与发展，测量精度不断提高，使精确获得倒转年代（极性倒转的时间间隔）成为可能。因此，学者们建立了全球标准的白垩纪至今的地磁极性年表（GPTS），并正在建立前白垩纪地磁极性年表。将岩石序列中所识别出的极性模式与地磁极性年表相对照，可以根据极性倒转点的绝对年代来推测岩石的年龄。但是，在建立这种地磁极性的对比时会出现很多的限制与缺陷（地磁累计速率变化、地层记录间断、地磁极性年表对比时特殊点缺失及采样分辨率较低等），然而这些问题一旦能够得到较好地解决，那么磁性地层学就可以在沉积盆地的构造研究方面取得广泛的应用，比如测定沉积物年龄、沉积速率变化、盆地展布、区域甚至全球级别的地层对比、断层断距对比、确定地层从开始沉积到抬升暴露之间的埋藏时间、构造与气候间的交互作用等。下面我们将通过几个最近的研究实例来介绍这些研究方法以及其应用。我们推荐给读者几部最近出版的比较优秀的关于磁性地层学方法的出版物，作为深入学习的资料（Butler，1992；Opdyke 和 Channell，1996；Tauxe，1998）。

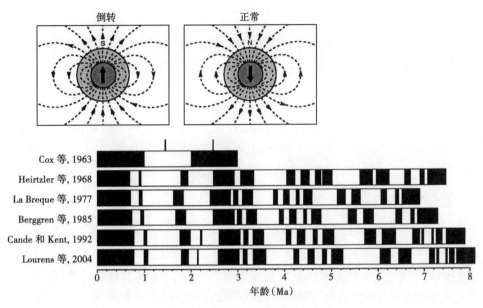

图 4-1 地球磁场极性倒转记录

上图：地轴偶极子场的正极性和反极性的状态；下图：地磁极性年表（GPTS）提供了正极性（黑色）和反极性（白色）相间隔的全球对比标准。地磁极性年表定年根据最新的古地磁资料和更精确的测年技术（放射性定年、生物地层学、天文调谐等）来校正

4.2 原理与方法

4.2.1 地球磁场的特征

地磁场是由外地核的液态金属对流所产生的。在地球表面，我们可以简单地将地磁场理解为一个偶极子场，即相当于在地球中心，存在一条磁棒（图 4-1）。这个偶极子场大概占了我们已发现磁场强度的 90%，而剩下的 10% 源自更高级的非偶极子场。在任意一个较小的时间尺度上，地磁场的强度与方向会有变化（即所谓的长期变化），与之相对应的地心偶极轴（地磁两极）明显与地球自转轴（地理两极）不重合。但是，如果将这种长期变化平均在几千年内来看的话，地磁与地理上的南北极是很难分辨的。所以，从地质时间尺度上讲，两极的概念包含了以上的两种含义。

一次典型的地磁场极性倒转一般需要几千年的时间，而且在全球范围可以认为是同步发生的。地史时期的地磁倒转事件可以被划分为统一的地磁年代单元，在过去的 35Ma 时间里，平均每期持续大概 300ka。但是，每期的时间跨度和本身就已不规律的地磁场间没有任何相关性，时间跨度从几千年到 20ka 不等。过去几百万年以来详细的古地磁记录显示，较小时间尺度（特别是小于 10ka）地磁场方向与强度的变化不会导致完全的地磁倒转事件。在这些所谓的地磁偏移事件发生后，地磁场会重新回到之前的稳定状态而不会发生倒转事件。全球已经发现了许多这种偏移事件，并且这种偏移事件也可以用来进行磁性地层学的年代测定。关于这种偏移事件的成因现在仍然存在争议，也是目前研究的重点，但是最近建模的相关成果表明核—幔边界处的热流体的侧向变化可能是引起这种变化重要的原因。

4.2.2 地磁极性年表

确定地质年代表是一项长期的、不断修订的过程。地磁极性年表最初是以大洋磁异常条带的时间为基础建立的，这些倒转事件的年龄是由放射性测年校准后的年龄值，进行有限线性内插估值而来。最新的方法则来自基于层序地层学中的"米兰科维奇旋回"建立的天文测年极性年表，并由此建立了新近纪的天文极性年表（APTS）（Lourens等，2004），前新近纪的天文极性年表的制定也正在进行中。天文极性年表的建立对于磁性地层学的优势在于，其磁异常条带相对宽度和倒转事件年龄可以被更精细地定义（大约5000a）。而最新的前新近纪地磁极性年表还是沿用Gradstein等2004年的修订版。

4.2.3 磁性地层学的取样

原则上讲，古地磁场可以通过地史时期岩石中的地磁记录进行重建。几乎所有岩石中都含有磁性矿物，例如，磁铁矿、赤铁矿、硫复铁矿和磁黄铁矿等。在这些岩石的形成过程中，这些磁性矿物（更准确地说是磁畴）被地磁场磁化，保留了最初的磁场方向，即原生天然剩余磁化强度（NRM）。岩石中保存的天然剩磁强度与背景地磁场相比很弱。当然，不同岩石间剩磁值也有不同，可以相差几个数量级。一般来说，含磁铁矿玄武岩和富火山质的沉积物中所含的古地磁信息要远多于含赤铁矿或者含铁硫化物的岩石。

进行磁性地层学取样的原则就是优选那些较好保存古地磁信息的岩石。同时，根据经验来说，颗粒越细的岩石可以提供更好的结果。某种程度上说，我们需要的是细粒、单畴磁性的碎屑颗粒，因为它们在大时间尺度上比粗颗粒、多畴磁性碎屑更稳定。所以，如有可能的话，细粒的火山岩及泥岩将是采样的首选。砂岩和中—粗粒的喷出岩可能也会包含有好的古地磁信息，特别是它们含有细粒的杂基物质时；但是这里存在的最大问题是获得的结果可能无法进行分析解释。除了颗粒大小，我们也可以根据沉积物的颜色来指导采样。一般来说，红色碎屑沉积物中含有的氧化铁，保存有较稳定的古地磁信息。白色沉积物，例如石膏和碳酸盐岩中的古地磁信息既弱又不易解释。但是，还是有很多的例外情况。很多极好的古地磁测试结果来自于石灰岩和粗砂岩。在任何情况下，避免取被风化的岩石作为样品，以此来规避古地磁信息被当今地磁场削去的风险。新鲜的露头最理想，但是严格来讲都会要求挖掘探槽来进行取样。

常规古地磁样品取样，使用水冷式金刚石钻头的手持钻机（图4-2），比较坚硬的岩石（玄武岩、致密胶结的砂岩、石灰岩等）一般要用重型汽油动力的钻机（通常配有定制的锯链机），但是，较软或是比较脆的岩石（泥岩、板岩等）用普通的电钻（要求有便携发电机和可充电蓄电池

图4-2　磁性地层学取样工具

a—电钻；b—适配器；c—钻头；d—空气压缩机（通常被水泵所替代）；e—发电机；f—定向器；g—罗盘；h—样品剥离工具；i—标记工具；j—钻取的样品；k—镐；l—铲；m—水壶

组）就可以了。在野外或是实验室中，有时水冷式钻机的冷却水会溶解、冲蚀那些较软的岩石，这时应使用压缩空气式钻机。标准古地磁取样的岩心柱直径约为 2.5cm，为了便于测量，至少要有 2cm 长；但是为了获得更多的样品当然是越长越好。古地磁样品在采样时需要标注样品的产状以便恢复磁信号的原始方向。在测定样品产状时，需要一个磁罗盘或者太阳罗盘以及一个磁斜计，用来测量磁偏角和样品岩心柱轴心的方位角，一般需要精确到 2°。测量完产状后，在岩心柱上标记贴签，并用金属箔或是纸包装好以便保存；在遇到特殊岩性不便钻孔取样时，则需要进行定向块状取样（需记录块状样品的走向与倾角），待回到实验室内再进行处理。对于未成岩沉积物，可以将样品从露头切下或是用 PVC 取样盒将其压成型作为定向取样的样品。

  一份成功的磁性地层学研究成果很大程度上取决于取样的方法。最理想的样品就是一块足够长、足够连续并且足够新鲜的样品。岩石颗粒越细，越能反映远源环境。找到辅助的年龄值约束条件（已测定年龄的凝灰岩地层、生物地层）很重要，特别是当这些条件都限定同一层段。我们要尽量避免记录中出现间断（不整合、断层等）的情况。当要对几条剖面拼合成的复合剖面进行采样时，每一个子剖面的对比都可能会出现错误（地层的缺失或者重复）；如果可能，我们可以对平行的两条或者多条含有相同层段的剖面同时进行取样，通过这样的对比来得到可信的磁性地层学结果；同时，通过这样的对比也可能发现一些对于构造研究有意义的穿时层段。地层取样的分辨率要以能控制目的层段的最小时间单元为目标（研究报告显示的最小时间单元是 20ka）；即使考虑密集取样，从大量的样品中也可能得不出可解释的信息。对倒转事件获得更多的细节信息，其分辨率也得到了提高。取样分辨率可以通过沉积物累计速率的正演来估计，比如确定年代界限、区域的地层对比以及大区域的构造地层格架等。标有精确的古地磁样品取样点位的准确而详细的岩石地层记录是磁性地层学研究的必要内容，全剖面的地层产状都要测量，以便应用精确的构造修正。如果剖面中包含有倾没褶皱，则地层产状和可能的褶皱轴心产状都要测量，以便作整体的构造校正。

4.2.4 实验室处理程序：获取原生磁信号

  我们一般使用两种磁力仪来测量岩石样品中的剩磁强度：旋转式磁力仪和低温超导式磁力仪。20 世纪 60 年代后发展了许多种类的旋转式磁力仪，但所有的这种磁力仪在测量时仍需要岩石样品沿一个轴来旋转，因为只有旋转时垂直于旋转方向的磁性物质才能产生电流从而被检测线圈记录。随着测量时间的增加，剩磁强度会不断降低（例如，低信号样品测量要求至少持续 30min）。在 70 年代早期，低温超导磁力仪被应用于较低磁强度样品的测量，相较旋转式磁力仪，低温超导磁力仪的测量可以做到更快、更准确。低温超导磁力仪采用了一种叫作超导量子干涉仪的磁场传感器，其工作须在液氦温度条件下（4°K）进行。低温超导磁力仪对于磁性地层学来说再合适不过，它可以在 1min 内精确测量低原生剩磁强度岩石样品。

  理想状态下，保存在岩石中的原生古地磁信号可以通过磁力仪进行一次测量就能获得；但是实际情况下，原始的剩磁通常会被后期的地质事件（化学风化、成岩作用、变质作用、火山活动以及区域构造活动等）所破坏甚至是完全覆盖。幸运的是，这些后期附加的改造甚至是部分覆盖的影响都可以通过多种退磁的方法予以去除。

  标准的退磁程序是通过逐步加温或者衰减交变磁场来逐步地将岩石样品中的剩磁去除。这两种方法分别是热退磁法和交变退磁法。热退磁法相对于交变退磁法，用时更长，程序也更复杂，劳动强度也更大，但是能得到更为丰富可靠的信息。每一步退磁程序后，用磁力仪测量剩余的剩磁其结果也都会有变化，我们就要分析这些方向和强度的变化。退磁程序必须要在背景场接近零的条件下完成，以保证在烘箱和交变线圈内的剩余磁场尽可能的低。

岩石样品中总的剩磁是一个三维矢量，大部分岩石中都有几个地磁分量，不同分量具有不同的退磁温度。如果这些分量形成于不同的地质年代，那么它们通常也就记录了不同时期的古地磁场方向。这就是说，根据一个样品中总的退磁结果可以得到反极性或者正极性的分量，在理想的状态下，每一个分量的方向与强度都可信。但是，可能存在两个甚至是多个分量的退磁温度重叠的情况，这就使得它们之间难以区分。如果样品的原生分量部分或者全部被一个未分辨的后期分量改造，可以通过低温对样品进行退磁分离。

通过逐步退磁得到的不同分量最佳的展示方式是使用 Zijderveld 图表。这个图表包括两个子表，是三维方向退磁路径在水平和垂直两个面上的投影（图4-3a）。这种图表的优势在于可以从多样的线性组合中分辨强度和角度的关系。不同分量的方向可以通过 Zijderveld 图表的退磁过程与主分量分析的最佳拟合直线来得求。最终的古地磁方向由磁偏角、磁倾角以及最大角偏差组成（MAD）。

4.3 磁性地层学结果的可靠性验证

退磁过程中获得的磁性特征可以确定样品中磁性矿物类型，而且这些磁性特征也可以通过其他不同的岩石磁性实验来提取。岩性不同，其磁性特征也将不同。这种特性应当引起注意，可以以此来确定这些分量是原生的还是次生的。例如，退磁过程中如果次生叠加退磁温度达到了450℃，那么所有退磁温度低于450℃的方向都不可靠（反向极性的样品在退磁温度达到450℃时会显示叠加的正向极性）。

可以通过一系列野外稳定性检验，进一步获取原生特征剩磁分量。这是基于赤平投影上的剩磁方向分布（图4-3b）分析得到的。图中远离古地磁长期变化散布区的数据（一般来说偏离平均方向45°的方向）都应该被舍去。

下面是几种最常用的磁性地层学野外检验方法：

褶皱检验（图4-3c）：如果从褶皱的某一岩层的不同产状部位进行采样获得的剩磁方向在倾斜校正后变得集中，则岩石的磁化方向是褶皱前获得的，而且是稳定的。严格来讲，褶皱检验没有直接证明分量的原始来源，只是证明了它的形成时间早于褶皱。但是，剩磁在褶皱的过程中得以保存，褶皱检验对原生剩磁具有极大的置信度，对褶皱形成具有丰富的年代约束信息（如生长褶皱和垮塌地层）。

倒转检验：实测的古地磁方向有不同的极性。特别是反极性的出现（即统计误差范围内的反向），可以看作是原生剩磁组分的指示。事实上，如果次生剩磁组分与原生剩磁组分叠加不能被完全抵消，平均正极性与平均负极性不完全相反，这类数据则不能通过倒转检验；因此，如果现今有大量叠加上的次生正极性，数据将不能通过倒转检验。有必要指出的是，即使剩磁极性分布很好，一些磁性地层的数据依然不能通过倒转检验。在这种情况下，磁性地层在原则上可以与地磁极性年表进行对比。

一旦有了地层古地磁方向的原始特征，就可以重建磁性地层记录（图4-4）。极性带通常至少由两个可识别的不同层位的同一极性而建立（一个标本不能建立极性带）。辅以已存在的年龄界限，需要找到最长的极性带作为可靠初始值（图4-4对比实例）与地磁极性年

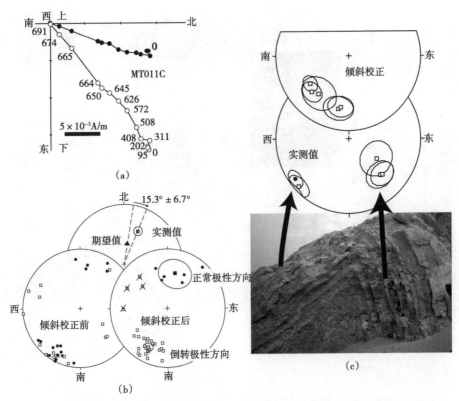

图 4-3 岩石退磁及其检验取样（据 Dupont-Nivet 等，2002，修改）

（a）一代表性样品的退磁矢量图（Zijderveld 图）。（b）特征剩磁方向，下面两个等面积赤平投影图分别显示了原位方向（倾斜校正前）（左）和倾斜校正方向（右），下半球以实心符号表示，上半球以空心符号示，黑色方块显示的是正向极性的倾斜校正，灰色方块则是反向极性的校正，椭圆形所圈是 95% 的置信区间，叉号代表了废弃样品方向；上图，实测平均方向用灰色方块表示（95% 置信区间），而期望方向（由已知的 20Ma 时欧亚大陆的古位置推算）用黑色三角形表示。（c）褶皱检验由该膝折褶皱两翼或者其他同斜层剖面进行取样，等面积投影显示了该点样品原位和倾斜校正后的平均方向（95% 置信区间）

表时代对比。对比中，如发现任何不匹配（极性带的丢失或者重复）都应彻底地对样品岩石磁性特征、采样分辨率、可能的缺失或者断层进行检查，仔细地综合评价所有可用的地层学信息；特别需要注意的是极性倒转应与沉积环境（岩相）的明显变化或者与推测的断层保持一致性。由极性带对比得出的沉积速率变化较大的层段，须根据实际地层情况进行讨论。为了补充这些重要的定性分析，Tauxe（1998）的实验给出某一给定的磁性地层是否可以正确地代表该段极性带的方法。也就是说，它检验的是足够的样品分辨率，而不是地层的完整程度（地层中不整合的出现），也不是特定磁性地层对比的好坏。

通过下列的基本指导可获得可靠的磁性地层学结果：

（1）磁化组分通过多级退磁进行隔离，这些组分在图上明显呈线性，且易于识别，主分量分析得到的剩磁方向结果至少经过四级退磁，排除原点及 MAD 低于 30°；

（2）剩磁特征（ChRM）方向通过一个或几个野外检验；

（3）实测的平均剩磁方向与现今磁场方向不同，与该采点期望方向和沉积时期相近；

（4）原生剩磁特征的磁载体组分可基于退磁特征和岩石磁性实验清晰地确立；

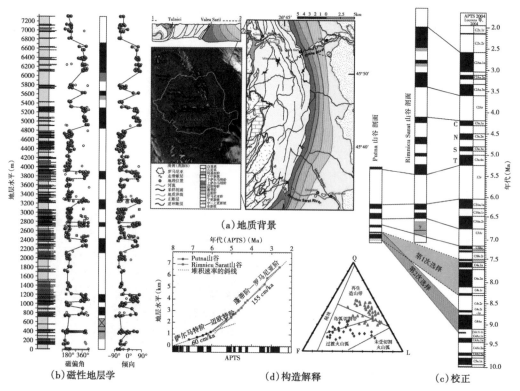

图 4-4 实例:喀尔巴阡山脉(Carpathian)前陆盆地的演化

(a) 研究区位于罗马尼亚喀尔巴阡山脉(Carpathian),其东部喀尔巴阡山脉前缘 Rimnicu Sarat 山谷和 Putna 山谷为采样剖面的位置所在,分别用绿线和红线标出;沿 Putna 山谷的横剖面显示研究区地层接触关系。(b) Rimnicu Sarat 山谷剖面实测的极性带与岩性剖面,其岩性从三角洲泥岩到沙坝及河道砂岩;在极性柱中,黑色及白色分别表示正向极性和反向极性间隔;实心点表示可靠的退磁方向图;白色正方形代表低温组分磁性比高温组分磁性具更明显的反向极性方向。(c) 2.3km 厚的 Putna 剖面及 7.3km 厚的 Rimnicu Sarat 剖面与地磁极性年表(Lourens 等,2004)对比图,剖面极性记录与地磁极性年表中同一时代界限用实线连接。柱中极性亚带的命名依据 APTS,C(Cochiti)、N(Nunivak)、S(Sidufjall)及 T(Thvera)由吉尔伯特时(Gilbert Chron)中常用的亚时而来;浅灰色显示古老 Putna 山谷较大相关,深灰色指示较小相关。(d) 堆积速率(未进行去压实校正)随时间(据地磁极性年表标定)变化图,显示倾角在 6Ma 明显发生变化,这一变化与 QFL 图解变化对应,指示构造事件

(5) 提供有其他独立评价的近似地层年代数据(如生物地层年代、放射性同位素年代);

(6) 磁性地层与极性时间范围相匹配,如果有一个极性带缺失(在对应地层内或者该时间范围内的缺失)或者实测的极性带长度出现大的偏差,须给出可靠的解释;

(7) 任何一个极性带都须基于高质量的多重剩磁特征(ChRM)方向,这些方向须为明显的正向极性或反向极性[选择的剩磁特征(ChRM)方向的平均正向或反向须在至多 45°偏角内];

(8) 极性带独立于岩性;

(9) 建立连续的剖面;

(10) 通过两个或两个以上的平行剖面确定极性类型。

4.4 应用与研究实例

磁性地层学测年有着广泛的应用，特别是在构造研究方面。我们将从以下几个研究实例来着重展示磁性地层学的主要用途，即通过沉积速率推测构造变形时间、构造测年，结合碎屑颗粒的热年代学的研究分析环境记录与构造活动或者是气候变化的相关性。其他类型的应用目前只是磁性地层学的"副产品"，待将来发展成熟后再作详细介绍。

4.4.1 沉积速率的变化与构造变形：来自喀尔巴阡山前陆盆地的研究

磁性地层学一项最常规的应用就是通过测定沉积速率的变化来推测与盆地相关的构造事件。喀尔巴阡山前陆盆地是很好的例子（Vasiliev等，2004）。以前，罗马尼亚喀尔巴阡山前陆盆地新近纪的沉积物测年没有形成可靠的年代模型，其构造演化也没有年代学的数据约束。通过两条剖面的取样和磁性地层学分析，得到的结果与地磁极性年表中 7.0~2.5Ma 段有极好的对应（图4-4）。同时，古地磁数据也可以来限制区域垂直轴旋转（Dupont-Nivet等，2005）。有了年龄的约束就可以得到每个极性带的沉积速率，在 5.8~6.0Ma 期间沉积速率明显增大（图4-4d）。需要注意的是，所求的沉积速率是针对未去压实沉积物的，但其假设前提却是去压实校正后所保留的沉积速率的变化。这一假设是与地层的顶、底位置及颗粒大小的不同所造成的压实作用有差异。理想状态下，厚度是去压实校正最先考虑的，但是这或许又会带来别的问题。此外，沉积速率的解释可能也有迷惑性和多解性，其构造意义或许也不明显。造成沉积速率变化的原因可以有很多：盆地沉降、物源的输入、盆地的开启或闭合以及构造和气候的变化等。在实际工作中，盆地的构造和沉积学的其他信息（物源、断裂时间、邻近断块或是盆内沉积物的抬升暴露的时间等）也都需要综合考虑。在该地区研究中参考了利用砂岩岩石学特征、泥岩地球化学资料并结合了地磁磁化率的研究得到的物源方面的研究成果，其结果清晰地显示物源区由活跃的火山弧向再生造山带的变化，期间伴有沉积速率的增加。构造和热年代学数据表明这次变化是由于东喀尔巴阡山推覆体的抬升和剥蚀造成的。

4.4.2 天山构造带构造活动的年代厘定

磁性地层学利用同构造期（比如生长地层和生长褶皱）沉积物中保存的构造特征来测年，已经被证明是获得构造活动时间的一种特别有效的手段。这些构造特征为剖面恢复提供了构造活动年代的时间限制，因此也使得推算剖面的缩短率成为可能。内陆造山带——天山构造带极好地厘定了构造年代，是磁性地层学应用的典范（图4-5）。近年来开展了大量针对天山南北两侧前陆盆地内沉积物的磁性地层学的研究工作。因为该地区没有火山岩层的控制，再加上可定年的古生物化石群落的稀少，所以同位素测年和古生物地层学的方法就不适用于该地区。但是研究区内连续性极好的红土剖面可以为磁性地层学的研

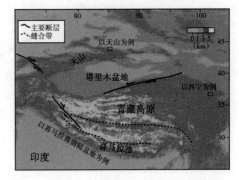

图4-5 印度—亚洲板块碰撞区数字高程模型
（标有天山、喜马拉雅山和西宁盆地位置）

究提供很好的古地磁记录。在喀尔巴阡山的例子中,他们研究测得的沉积速率的变化是由构造抬升造成的,构造运动的时间与后陆地区热年代学得到的抬升暴露的折返阶段有明显的相关性(Sobel 等,2006)。另外,该地区发育有分布广泛、向上变粗的西域砾岩;该地层底部年龄值范围较广,反映了应力机制的复杂性,包含了对全球气候影响、区域构造抬升以及局部构造变形影响(Heermance 等,2007;Charreau 等,2008;Li 等,2011)。磁性地层学对于精确建立新近纪天山地区构造变形的连续时间格架至关重要,通过不断获得褶皱以及生长地层的年代,来与地震剖面一起确定前陆盆地的演化序列(图 4-6)。通过生长地层测年值来重建构造期次比较有效(Poblet,第 27 章)。但是,如果地层在抬升、剥蚀的同时伴随褶皱发生,则残余地层中的最小年龄代表了最初形成褶皱时的年代(例如,较老的生长地层

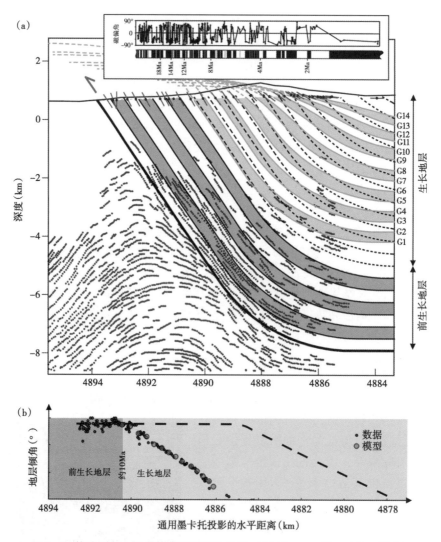

图 4-6 构造测年:天山构造带的同生地层(据 Charreau 等,2008,修改)
(a)上方是磁倾角的测量和推断的地质年代,并与地表测得的倾角建立的关系图;下方是贯穿天山前陆盆地北部的地震剖面图;前生长地层(黑色连续虚线)以及生长地层(连续灰线)由断层弯转褶皱推测。(b)断层弯转褶皱倾角的实测值(红点)和期望值(连续线或者灰点)(以弯曲褶皱和有限宽度的脊线为前提);虚线表示由前同生地层断层弯转褶皱推测的倾角;UTM—通用横轴墨卡托投影

如果还有保留，那么该地层就应该靠近褶皱的轴部，但是此时的褶皱还未形成）。理想情况下，地表的观测结果可以与地震剖面相对比，来识别和确定地下的生长地层（Charreau等，2008）。这种方法有助于校正褶皱的生长对于同构造期沉积速率的影响、剖面的恢复以及相关的缩短率的求取等（图4-6）。

4.4.3 磁性地层学及碎屑热年代学的研究：以喜马拉雅山前陆盆地为例

磁性地层学经常结合碎屑的热年代学（磷灰石、锆石的裂变径迹分析）来确定碎屑颗粒从物源区暴露剥蚀的冷却年龄到进入沉积盆地后的沉积年龄间的"时滞"（图4-7）。喜马拉雅山前陆盆地的Siwaliks组中新世沉积物中有相邻的喜马拉雅山隆升的证据（Beek等，2006；Najman，2006；Ojha等，2009），是确定"时滞"的理想研究区。沉积物从山侧面的剥蚀区到盆地内的搬运时间可以忽略，因此"时滞"可以反映沉积期内物源区的抬升暴露速率。反过来说，"时滞"更是与物源区的气候变化和构造运动关系密切（Bernet等，2006；Reiners和Brandon，2006）。"时滞"可以通过裂变径迹的年龄与磁性地层学的地层年龄共同推算得到。裂变径迹的标准误差大概在±15%，并且这个误差会带到后面的物源区分析和盆地抬升速率估算中。相比裂变径迹的误差，由磁性地层学得到的地层沉积年代的误差相对较小。但是，如果说裂变径迹的年龄（反映物源区开始剥蚀）要小于沉积时的年龄，这可能就是磁性地层学测年中出现了问题。总的来说，主要有以下三种可能（Bernet等，2006）：第一种情况，逐渐减小的"时滞"表明了物源区全程都是加速抬升的（提供物源）；

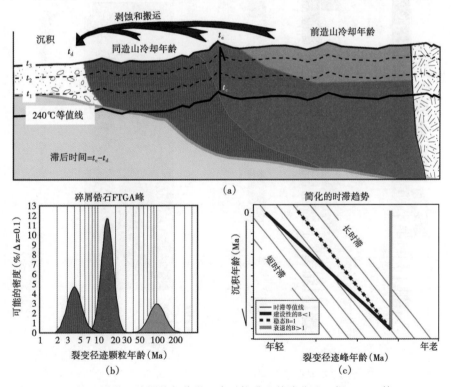

图4-7　磁性地层学和碎屑热年代学：喜马拉雅山前陆盆地（据Bernet等，2006）
(a) 在碎屑锆石的裂变径迹分析中通常应用"时滞"概念，它用简化的垂向折返路径表示；(b) 概率密度合成示例图，鉴于是二项式峰值最佳拟合，因此可能来自 (a) 图中的造山带；(c) 三种"时滞"趋势可由一系列标本的锆石裂变径迹峰值年龄观测得到

第二种情况，长期较稳定的"时滞"表明了连续性的抬升剥蚀处于一个稳定的速率（稳定状态）；第三种情况，"时滞"逆剖面向上持续增加（破坏状态）：锆石可能来自于母源区的一次短暂的快速冷却事件，也可能来自于沉积母岩区的再旋回，还可能来自于厚层的未再沉降的火山岩的风化。在尼泊尔中部的 Surai 剖面有关喜马拉雅山的结果（Beek 等，2006）显示，碎屑年龄与最初根据 GPTS（Cande 和 Kent，1995）建立的磁性地层学年龄无法建立关系；而随着磁性地层学年龄数据的增加（Ojha 等，2009），重新根据 APTS（Lourens 等，2004）对比的磁性地层年龄与碎屑颗粒年龄建立了关系。根据磁性地层学的沉积物年龄的约束而建立的时间格架和磷灰石裂变径迹的年龄，得到了 Surai 剖面稳定的"时滞"为 0.8 ± 0.5Ma，推测物源区从大约 7Ma 以 1.8km/Ma 的速度快速隆升。

4.4.4 气候变化还是构造运动？来自于西宁湖相沉积盆地的古环境记录

磁性地层学也可以通过研究沉积作用与沉积物堆积来识别全球气候变化的原因和对区域构造影响。下面这个来自亚洲的研究实例中，气候模型显示新生代印度板块和欧亚板块的碰撞伴随着青藏高原的隆升造成了季风的增强，大陆更加干燥，也导致了剥蚀的加快（图 4-8）。

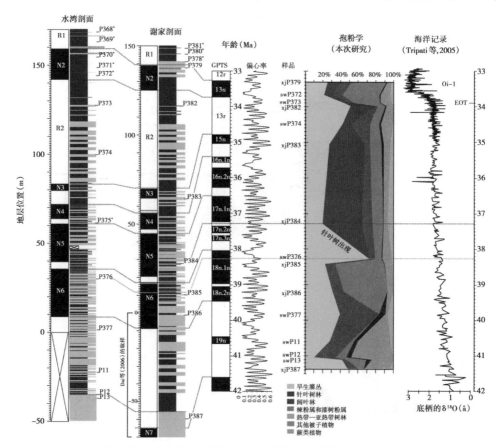

图 4-8 气候与构造活动的古环境记录：青藏高原东北部西宁盆地陆相湖泊沉积
（据 Dupont-Nivet 等，2008b，修改）

从两条剖面（谢家和水湾）取样与海相稳定同位素的对比（Tripati 等，2005）。在 34Ma 石膏沉积的出现与全球气候变迁相一致（Dupont-Nivet 等，2007）。孢粉记录显示高海拔针叶林的出现与青藏高原的隆升有关（Dupont-Nivet 等，2008b）。年龄模型是以古地磁层序 GPTS 对比（Gradstein 等，2004）以及偏心处理（Laskar 等，2004）为基础的从沉积记录中识别出的古环境变化可以是由区域构造作用导致的，但是同时也可以指示全球气候的变化情况。识别气候变化影响的关键在于要对有足够分辨率和精度的地质记录区域有相应的测年记录，使之能够与海域记录相对比。西宁盆地位于青藏高原东北部，有着罕见的长期连续地层记录（30Ma 至今），是通过沉积记录研究始新世到渐新世构造运动与气候过程关系的极佳研究目标（Dupont-Nivet 等，2007）。地层序列中厚层石膏层的消失揭示了区域沉积环境发生过显著的变化，可以将其看作区域气候干旱的标志（该证据在其他文章中也有记录）。这一变化的发生时间可以由磁性地层学精细测得，并与全球气候变化事件进行精确对比得出：发生在始新世、渐新世交替的 34Ma 的全球变冷事件，导致了南极大陆冰盖的形成。此外，同一剖面的花粉记录显示在 38Ma 出现了高纬度的花粉物种（Picea conifers）（Dupont-Nivet 等，2008b），但是该记录与已知的全球气候变化事件均不相符，这些高纬度花粉的出现更有可能是因为青藏高原隆升造成的。

4.4.5 垂直轴旋转以及古纬度的确定

古地磁学除了可以提供年龄上的约束，由它获得的古地磁方向还可以提供地层沉积时的垂直轴旋转和古纬度的信息。磁性地层学中的垂直轴旋转是构造测年的极佳选择，相对于沉积速率来讲，它不受剥蚀和气候的影响而更加独立。由不同盆地对比来看，该方法可以在区域构造时间校正上很好地进行定量分析。本次研究优选了 ChRM 方向（特别是 MAD<15°），并且根据长期变化来收集平均方向变化。西宁盆地内相当长的陆上地层沉积的磁性地层学标本提供了数百万年的古地磁方向记录（Dupont-Nivet 等，2008a），已经识别出了区域的顺时针旋转，但是未能建立精确的测年记录。由一系列最优 ChRM 方向计算得到滑动平均值，可为沉积旋回的时间偏差提供依据（图 4-9）。

原则上根据著名的偶极子方程，古纬度（lat）可直接由古地磁磁倾角（I）得到：

$$\tan(I) = 2\tan(\text{lat})$$

对于古环境研究（比如古气候、古生物学）来说，这是特别有价值的信息。但是应该注意到这一方法在应用上的重要缺陷：沉积记录中磁倾角趋于压扁，即普遍存在的磁倾角偏低的问题。沉积记录中的磁倾角比期望值偏低的可能原因是在沉积与随后的压实过程中颗粒的定向排列。近年来，磁倾角偏低的情况可以通过恢复古地磁方向压扁分布的方法得以成功校正（Tauxe，2005）。因为这种"拉长磁倾角"校正方法要求大量有效方向的记录（约100），而磁性地层学研究正好会产生大量独立的古地磁场测量数据，所以这种校正方法尤为适合（图 4-7b）。

4.4.6 磁性地层学的延伸应用及展望

磁性地层学已经被广泛应用于构造地质学研究的多个方面。即使不进行 GPTS 比对，它也可以成为盆内或盆间地层对比的基本手段之一。它可以用来估计穿时层的横向展布范围，也可以用来计算断层断距。在可预见的未来发展方向上，磁性地层学以研究沉积旋回尺度为基础来获取更高分辨率的测年记录，从而使其变得更独立、更可靠（Krijgsman 等，2004）。它可以用来作为沉积记录的额外独立的约束条件，可以在缺少放射性火山岩标定的情况下确定地层旋回；同时也有助于识别气候变化及对构造活动的影响（Dupont-Nivet 等，2007）。除了磁性地层学，只要能确定所有剖面样品的岩石磁性参数（NRM、磁化率及其各向异性等），就可以分析构造与气候变化的关系（Gilder 等，2001）。

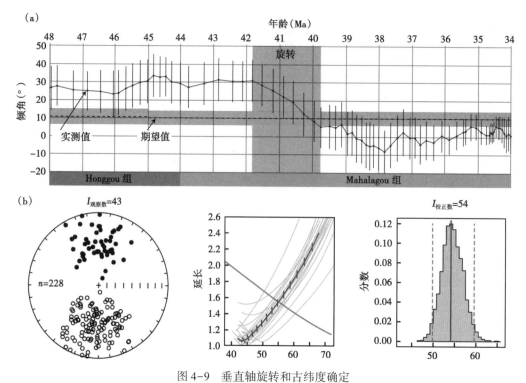

图 4-9　垂直轴旋转和古纬度确定

（a）垂直轴旋转分析。西宁盆地谢家剖面（48—34Ma）磁性地层学特征剩磁方向的滑动平均值（Dai 等，2006）。用滑动窗技术，由低至高依次求取 15 个连续的特征剩磁方向；将 95% 置信的平均磁偏角投到谢家剖面磁性地层学年龄关系图上，使其与 95% 置信的期望磁偏角（灰色阴影部分虚线所示）相对比，发现大部分顺时针旋转发生在图示垂直时间间隔的阴影部分。同一采点，期望的磁倾角及磁偏角由已知的欧亚大陆古位置推算得到。（b）用拉长磁倾角的方法来校正磁倾角偏低的情况。左边赤平投影图显示由原始数据得到的平均磁倾角（43±5°）要远小于期望磁倾角（60±2°）。中间的图反映了受压扁因子（0.35~1.00）影响的平均磁倾角的拉长数据（黑色曲线），同样，通过统计分析获得合成数据（浅灰色曲线）（Tauxe，1998）。校正的磁倾角由地磁场模型中的期望拉长数据（深灰色曲线）交叉获得。右边直方图显示校正后的磁倾角分布，平均值为 54°，在 50°~60° 范围内达到 95% 置信水平。校正后的平均值与期望值相符

致谢（略）

参 考 文 献

Beek, P. v. d., Robert, X., Mugnier, J-L., Bernet, M., Huyghe, P., and Labrin, E. (2006) Late Miocene: Recent exhumation of the central Himalaya and recycling in the foreland basin assessed by apatite fission-track thermochronology of Siwalik sediments, Nepal. Basin Research, 18, 413-434.

Bernet, M., Beek, P. v. d., Pik, R., Huyghe, P., Mugnier, J-L., Labrin, E., and Szulc, A. (2006) Miocene to recent exhumation of the central Himalaya determined from combined detrital zircon fission-track and U/Pb analysis of Siwalik sediments, western Nepal. Basin Research, 18, 393-412.

Butler, R. F. (1992) Paleomagnetism: Magnetic domains to geologic terranes. Boston, Blackwell Scientific Publications, 238 p.

Cande, S. C., and Kent, D. V. (1995) Revised calibration of the geomagnetic polarity timescale for the Late Cretaceous and Cenozoic. Journal of Geophysical Research B: Solid Earth, 100, 6093-6095.

Charreau, J., Avouac, J-P., Chen, Y., Dominguez, S., and Gilder, S. (2008) Miocene to present kinematics of faultbend folding across the Huerguosi anticline, northern Tianshan (China) derived from structural, seismic, and magnetostratigraphic data. Geology, 36, 871-874. doi: 10. 1130/G25073A. 1.

Dai, S., Fang, X., Dupont-Nivet, G., Song, C., Gao, J., Krijgsman, W., Langereis, C., and Zhang, W. (2006) Magnetostratigraphy of Cenozoic sediments from the XiningBasin: tectonic implications for the northeastern Tibetan Plateau. Journal of Geophysical Research, 111, B11102. doi: 10. 129/2005JB004187.

Dupont-Nivet, G., Dai, S., Fang, X., Krijgsman, W., Erens, V., Reitsma, M., and Langereis, C. (2008a) Timing and distribution of tectonic rotations in the northeastern Tibetan Plateau. Special Paper 444. Investigations into the Tectonics of the Tibetan Plateau, 73-87.

Dupont-Nivet, G., Guo, Z., Butler, R. F., and Jia, C. (2002) Discordant paleomagnetic direction in Miocene rocks from the central Tarim Basin: evidence for local deformation and inclination shallowing. Earth and Planetary Science Letters, 199, 473-482.

Dupont-Nivet, G., Hoorn, C., and Konert, M. (2008b) Tibetan uplift prior to the Eocene-Oligocene climate transition: evidence from pollen analysis of the XiningBasin. Geology, 36, 987-990. doi: 10. 1130/GS25063A. 1.

Dupont-Nivet, G., Krijgsman, W., Langereis, C. G., Abels, H. A., Dai, S., and Fang, X. (2007) Tibetan plateau aridification linked to global cooling at the Eocene - Oligocene transition. Nature, 445, 635-638.

Dupont-Nivet, G., Vasiliev, I., Langereis, C. G., Krijgsman, W., and Panaiotu, C. (2005) Neogene tectonic Magnetostratigraphic Methods and Applications 93 evolution of the southern and eastern Carpathians constrained by paleomagnetism. Earth and Planetary Science Letters, 236, 374-387.

Gilder, S., Chen, Y., and Sen, S. (2001) Oligo - Miocene magnetostratigraphy and rock magnetism of the Xishuigou section, Subei (Gansu Province, western China) and implications for shallow inclinations in central Asia. Journal of Geophysical Research, 106, 30505-30522.

Gradstein, F. M., Ogg, J. G., and Smith, A. G. (2004) The geomagnetic polarity time scale. Cambridge, CambridgeUniversity Press, 589 p.

Heermance, R. V., Chen, J., Burbank, D. W., and Wang, C. (2007) Chronology and tectonic controls of late tertiary deposition in the southwestern Tian Shan foreland, NW China. Basin Research, 19, 599-632.

Kirschvink, J. L. (1980) The least-square line and plane and the analysis of paleomagnetic data. Geophysical Journal of the Royal Astronomical Society, 62, 699-718.

Krijgsman, W., Gaboardi, S., Hilgen, F. J., Iaccarino, S., de Kaenel, E., and van der Laan, E.

(2004) Revised astrochronology for the Ain el Beida section (Atlantic Morocco): no glacio-eustatic control for the onset of the Messinian Salinity Crisis: Stratigraphy, 1, 87–101.

Laskar, J., Robutel, P., Joutel, F., Gastineau, M., Correia, A. C., and Levrard, B. (2004) A long-term numerical solution for the insolation quantities of the Earth. Astronomy and Astrophysics, 428, 261–285. doi: 10. 1051/0004- 6361: 20041335.

Li, C., Dupont-Nivet, G., and Guo, Z. (2011) Magnetostratigraphy of the Northern Tian Shan foreland, Taxi He section, China. Basin Research, 23, 101–117, doi: 10. 1111/j. 1365-2117. 2010. 00475. x.

Lourens, L. J., Hilgen, F. J., Laskar, J., Shackelton, N. J., and Wilson, D. S. (2004) The Neogene Period, in Gradstein, F. M., Ogg, J. G., and Smith, A. G., eds., Ageologic time scale. Cambridge, CambridgeUniversity Press, 409–440.

Najman, Y. (2006) The detrital record of orogenesis: a review of approaches and techniques used in the Himalayan sedimentary basins. Earth-Science Reviews, 74, 1–72.

Ojha, T. P., Butler, R. F., DeCelles, P. G., and Quade, J. (2009) Magnetic polarity stratigraphy of the Neogene foreland basin deposits of Nepal. Basin Research, 21, 61–90.

Opdyke, N. D., and Channell, J. E. T. (1996) Magnetic Stratigraphy. San Diego, Academic Press, 346 p. Panaiotu,

C. E., Vasiliev, I., Panaiotu, C. G., Krijgsman, W., and Langereis, C. (2007) Provenance analysis as a key to orogenic exhumation: a case study from the East Carpathians (Romania): Terra Nova, 19, 120–126.

Reiners, P. W., and Brandon, M. T. (2006) Using thermochronology to understand orogenic erosion. Annual Reviews of Earth and Planetary Sciences, 34, 419–466.

Sobel, E. R., Chen, J., and Heermance, R. V. (2006) Late Oligocene-Early Miocene initiation of shortening in the Southwestern Chinese Tian Shan: implications for Neogene shortening rate variations. Earth and Planetary Science Letters, 247, 70–81.

Tauxe, L. (1998) Paleomagnetic principles and practice. Dordrecht/Boston/London, Kluwer Academic Publisher, 299 p.

Tauxe, L. (2005) Inclination flattening and the geocentric axial dipole hypothesis. Earth and Planetary Science Letters, 233, 247–261.

Tripati, A., Backman, J., Elderfield, H., and Ferretti, P. (2005) Eocene bipolar glaciation associated with globlo carbon cycle changes. Nature, 436. doi: 10. 1038/ nature03874.

Vasiliev, I., Krijgsman, W., Langereis, C., Panaiotu, C. E., Matenco, L., and Bertotti, G. (2004) Towards an astrochronological framework for the Eastern Paratethys Mio-Pliocene sedimentary sequences of the Focsani basin (Romania). Earth and Planetary Science Letters, 227, 231–247.

Zijderveld, J. D. A. (1967) A. c. demagnetisation of rocks: analysis of results, in Collinson, D. W., and Al., E., eds., Methods in Palaeomagnetism. Amsterdam, Elsevier, 254–286.

(郭帅 译，王云 董治斌 校)

第5章 三维地震解释技术在盆地分析中的应用

CHRISTOPHER A-L. JACKSON[1], KARLA E. KANE[2]

(1. Department of Earth Science and Engineering, Imperial College, London, UK;
2. Staoil (UK) ltd, London, UK)

摘　要：虽然利用二维地震数据能够对沉积盆地的形成和充填进行区域性研究，但是，它们在空间上的低分辨率限制了对地质体的详细分析。随着三维地震采集技术和处理技术的提高，大面积的沉积盆地被高品质三维工区覆盖。随着复杂精密的软件和低价位、高性能工作站的发展，三维地震数据的解释达到前所未有的精细程度。本章一方面为地震解释员介绍一些关键的地震解释技术；另一方面，在多尺度上对如何应用这些技术分析沉积盆地的构造和充填进行实例说明。本章对构造图和地层厚度图也进行了分析，这是因为这些图件对确定盆地构造、同沉积可容空间的时空变化、大地构造和剥蚀等都非常重要。本章对特殊地质体的地震属性也进行了分析和讨论，这些属性包括基于网格和基于三维空间的几何属性和振幅属性，以及由网格和三维空间派生的基于振幅的属性。虽然地震属性分析是一个强大的盆地分析工具，但是，地下地质问题的答案并不存在于工作站和地震属性体中，工作站只是一个工具。要把地震数据成功地运用到盆地分析中，关键因素包括：(1) 具有所用解释软件的相关知识；(2) 了解所用地震属性的物理基础；(3) 能理解数据和解释质量对其生成的属性的影响；(4) 对地球物理数据进行慎重周到的地质解释。随着空间覆盖率和分辨率的不断提高，地震反射数据将有潜力像其他许多基于露头或分析实验的技术那样，在沉积盆地地质分析方面成为非常有用的工具。

关键词：沉积盆地　地震反射数据　地震解释　构造地质　地层分析

5.1 引言

20世纪70年代初，由于地震采集和数据处理方法的改进，二维地震数据的质量得到很大提高，这极大地推动了二维地震资料的地质解释（Sheriff 和 Geldart，1995；Yilmaz，2001；Bacon 等，2007）。特别重要的是，由于地震反射数据体信噪比的提高，能同时得到地下构造和地层的较为清晰的影像。随着地震反射成像可信度的增加，人们发现，根据反射终止关系，地震同相轴反射能分成多个有成因联系的组合，这预示一门新学科——地震地层学的诞生（Payton，1977；Vail，1987 相关章节）。结合地震层序，人们可以根据连续性、振幅、频率等内部反射特征来推断与这些地震层序相关的岩性和沉积环境。虽然二维地震数据提供了观察沉积盆地的形成和充填的区域性视角，但是，它们的空间分辨率（通常在几千米的级别）却限制了对空间展布小于测线间隔的地质体的详细分析（图5-1；Cartwright 和 Huuse，2005）。

从20世纪80年代至今，三维地震数据得以大量采集，大面积的沉积盆地（>1×10^4km^2）被高品质的地震工区覆盖（如垂向和横向分辨率都在几十米数量级）（Weimer 和

 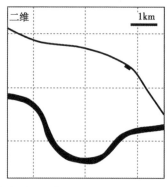

图 5-1 三维地震数据与二维地震数据空间分辨率对比（据 Cartwright 和 Huuse，2005，修改）
左图是用线距为 12.5~50.0m 的常规三维地震数据绘制的多条断层和一条水道。右图则是用线距为 2km 的常规二维地震数据绘制的相同构造。注意：运用三维地震数据能获得较高的空间精度和较复杂的地质体，而从二维地震数据所得到的空间图像却有一定的失真

Davis，1996；Dorn，1998；Davies 等，2004；Cartwright 和 Huuse，2005）。随着三维地震数据的采集和数据处理技术的提高，以及复杂精密的软件和低价高能工作站的发展，人们对三维地震数据的解释达到前所未有的精细。现阶段三维地震数据的垂向和横向分辨率都在几十米这个数量级，这在工业生产（Demyttanaere 等，1993；Weimer 和 Davis，1996；Dart 等，2004；Davies 等，2004；James 等，2004）和学术研究（见 Cartwright 和 Huuse 所作评论，2005）方面都产生了巨大影响。严格地说，现今三维地震数据的空间覆盖量已经能够达到研究沉积盆地几何特征和演化规律的要求，其尺度与本书中提到的其他方法相同。

本章主要介绍地震解释员所需的一些关键技术。本书的目的不是综述地震反射的基本原理（参见 Sheriff 和 Geldart 所作评论，1995；Yilmaz，2001；Bacon 等，2007），也不是详细回顾解释员所需的地震全部技术（Hart，2000；Brown，2004；Chopra 和 Marfurt，2005），而是列举一些关键技术，介绍他们如何在多尺度上理解沉积盆地的结构和充填（Hart，1999，2000；Cartwright 和 Huuse，2005；Chopra 和 Marfurt，2005；Davies 和 Posamentier，2005；Posamentier 等，2007）。首先，我们使用标准地震层位制作构造图（时间域或深度域）和厚度图（等时线或等厚线），并讨论这些图件的作用。这些图件对确定沉积盆地结构和地层厚度变化非常重要，进而分析同沉积可容纳空间的时空变化、大地构造和地层剥蚀（Hart，2000；Brown，2004；Bacon 等，2007）。其次，我们也对如何运用地震属性（从基本地震参数中衍生出的属性，如反射系数）分析地震反射数据中的地质体进行讨论。

5.2 构造图和厚度图

5.2.1 构造图

时间构造图或深度构造图是地震解释研究中最基础的图件。它们除了描述现今地质构造外，还是重建沉积盆地时空演化的基础。使用二维地震数据生成的构造图是一个空间连续分布的面（Hart，1999，2000），是通过一系列相互交叉的二维测线的地震层位数据解释拾取并网格化得到（运用插值算法）的。使用三维地震数据生成构造图，其使用的数据通常也

是各个层位的解释拾取，或者是三维地震体中所选测线上的"种子点"自动追踪，并在整个数据体中搜索和延伸（参见 Hart 提到的方法论，1999，2000；Brown，2004）。这样，"自动追踪"法能把三维地震数据内的每个共中心点炮集结合在一起。由于"时间"和"品质"是地震解释人员最关心的两个方面，所以自动追踪方法的好处是非常明显的。即（1）层位自动追踪比手动解释快很多；（2）由于自动追踪取决于地震数据的品质，因此，自动追踪的层位要比手动追踪的层位更精确。然而，解释人员也应该认识到，自动追踪在反射同相轴横向不连续的地区也会出现错误，最常见的原因是由于复杂的构造和地层接触关系（如削蚀、反射终止等）与数据品质的变差。

最初的地震解释和层位追踪的准确性对盆地结构和演化的研究至关重要，这是因为许多其他数据（等时图、振幅图、地震属性提取）的正确与否完全取决于解释人员所作构造图的精度。比如，如果构造图的解释较差，那么从构造图上提取的倾角图和振幅图（见本章的后面部分）将不具有或具有很少的地质意义。正如 Hart（1999）强调的那样，解释人员所做的构造图总是会被问及这样的问题："它有什么地球物理和地质意义吗？"

5.2.2 厚度图

厚度图是展现地震地层单元内部或之间空间厚度变化的图件，是地震解释流程中又一关键因素。有两种厚度图件可被使用：（1）等厚图，它反映深度域的垂直厚度，单位米或英尺，可通过在时深转换后的地震体中解释层位得到，也可以通过先在时间地震体中解释层位而后再进行时深转换得到；（2）等时图，它反映时间域的垂直厚度，单位毫秒（双程旅行时，毫秒；或 TWT，ms），通过在时间地震体中解释层位得到，不作时深转换。必须注意：真实的地层厚度，如从钻井数据中得到的厚度，与从地震等厚图或等时图中所得到的地层厚度之间存在误差。这与用在（1）地震数据体的深度偏移的，或（2）地震层位时深转换的地震速度的不确定性有关。

厚度图常常反映可容空间的时空变化，而可容空间的变化又可直接与大地构造引发的沉降和抬升类型相关联。因此，厚度图能够在多尺度上通过多种构造样式反映盆地内部构造的时空演化（Apotria 和 Wilkerson，2002；Jackson 等，2008）。例如，在北海盆地，人们使用时间构造图和等时图研究盆地的构造、与裂陷相关的断层和与裂后反转相关的褶皱的时空演化（图 5-2；Jackson 和 Larsen，2008，2009）。

5.3 地震属性分析

现今基于工作站的地震解释软件能够为使用者提供许多强大、高效的地震属性技术。在对属性分析结果进行解释的时候，我们必须要知道是对地震数据的哪一部分进行分析，结果是如何表现的，以及反映了怎样的地质意义（Brown，1996，2004，2005；Hart，1999，2002；Chopra 和 Marfurt，2005）。此外，解释人员还应该知道对于不同的地质环境使用解释工具包中的哪个功能最为适合，这将有利于提高工作效率和解释质量。虽然有许多的叠后地震属性可以使用，但在下面的介绍中把地震解释人员特别常用的属性细分为四种主要类型：（1）基于网格的几何属性；（2）基于空间的几何属性；（3）从网格和空间属性派生的基于振幅的属性；（4）其他属性。

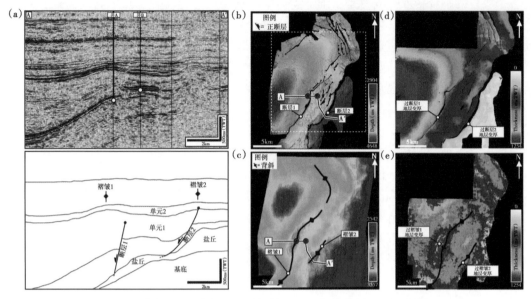

图 5-2 三维地震数据的地震剖面图、时间构造图和等时厚度图在盆地分析中的应用
(据 Tackson 和 Larsen, 2008, 2009)

资料来自挪威近海的北海盆地南维京地堑。(a) 地震剖面(上)和地质解释剖面(下),剖面中发育由重力驱动的成铲状的与断裂相关的正断层(断层 1 和 2)和与后期反转相关的背斜(褶皱 1 和 2)。单元 1 被认为形成于裂谷期,而单元 2 则被认为形成于反转期。该地震剖面位置参见时间构造图 5-1b 和图 5.1c。(b) 单元 1 底面(同裂谷底面)时间构造图,该图显示裂陷期的盆地几何形态和图 5-1a 中的那两条正断层(断层 1 和 2);注意:(b) 和 (c) 中的红点表示井所在位置 [就是在 (a) 中显示的那些井],这些井用于标定构造解释和相应的地震(等时)厚度。(c) 单元 2 底面(反转期底面)时间构造图,该图显示在裂陷期形成的断层上盘发育与反转相关的背斜构造(褶皱 1 和 2)。(d) 单元 1 的时间厚度图,该图反映同裂陷单元在断层 1 和断层 2 切过的地方同沉积增厚的特征。(e) 单元 2 的时间厚度图,该图反映同反转期单元体在图中两处主要的与反转相关的背斜处同沉积变薄的特征。参见 Jackson 和 Larsen (2008, 2009) 全面的描述和讨论

5.3.1 基于网格的几何属性

基于网格的几何属性在多种尺度上,在许多不同构造和地层中都适用。由于这些属性是从解释的地震层位中直接生成的(或通过它们衍生得到),因此,必须认识到所作图件的质量和图件的可解释性,与地震数据的品质和原始解释数据的合理性有直接的关系(Dalley 等,1989;Hesthammer 和 Fossen,1997;Hesthammer,1998;Brown,2004)。四种基于网格的几何属性对地震解释人员特别有用:(1)倾角;(2)方位角;(3)倾角—方位角;(4)曲率。

1)倾角

该属性计算解释层位上每一采样点的倾角,如果层位是深度域,那么倾角的单位是度,如果层位是时间域,那么倾角的单位是毫秒每米(ms/m)或者毫秒每英尺(ms/ft)(瞬时倾角;Dalley 等,1989;Mondt,1993)(图 5-3)。该属性生成的倾角图显示为层位倾角大小的空间变化率(图 5-4b)。然而,倾角图并不包含构造的倾角方向或构造的倾斜大小等信息(如一个断层的总偏移量或偏移方向)。该属性的使用有时会受到限制,如在地质体的倾角或倾斜方向与区域倾角相似的地区(通常被认为倾角饱和),或者在地质体的倾角和倾斜方向与区域倾角都相似但是方向相反,地质体的影像都会比较差(表 5-1)。

倾角属性图已经被广泛应用到沉积盆地构造轮廓的描绘中（如裂缝、断裂和褶皱）（图5-4b；jackson，2007）。Hoetz和Watters（1992）用该方法展示了荷兰境内正断层的几何形态和分布状况，其相对尺度较小（小于50m），形成原因与断陷和盐底辟塌陷相关。在Nun河油田（尼日利亚陆上），倾角被用于研究中等尺度（大于100m）的铲状生长断层的方位、分布和封闭性（Bouvier等，1989；Rijks和Jauffred，1999）。在北海盆地（欧洲西北部）地震数据质量变化显著的地区，倾角属性在半地堑边界的地壳尺度级别的（几千米断距）正断层，以及把碎屑岩储层分隔成几个独立部分的较小构造（几十米尺度）（Jones和Knip，1996；Hesthammer和Fossen，1997；Hesthammer，1998）的成像都非常有用。就在前不久，倾角属性还成功运用在海底陆坡沉积环境的地层分析中（Posamentier，2003；Posamentier和Kolla，2003；Posamentier等，2007）。比如，在研究侵蚀水道边缘的陡峭倾角、水道边缘天然堤顶部的倾角变化和在迎流面和背流面具有变倾角的沉积波动时，用该技术都能得到特别好的影像。

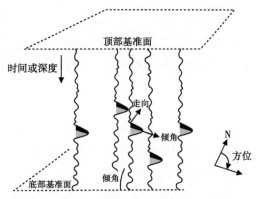

图5-3 计算地震层位倾角和方位的简易图
（据Mondt，1993，修改）

表5-1 不同的构造形态以及三种基于网格的几何属性（倾角、方位角、倾斜方位角和曲度）
在正断层成像上的相对优缺点（据Roerts，2001，修改）

构造形状	倾角	方位角	倾斜方位角	曲率
(a)	√	√	√	√
(b)	√	×	√	√
(c)	×	√	√	√
(d)	×	×	×	√

（a）平缓地层和反向陡峭断层；（b）平缓地层和顺向陡峭断层；（c）陡峭地层和反向陡峭断层；（d）平缓地层和顺向平缓断层。同时，还应注意到，只有在曲率图中能在所有情况中显示出断层。

2）方位角

该属性是计算每个采样点与正北方的夹角度数（瞬时方位角；Dalley等，1989）（图5-2），该属性反映在解释层位上各点倾角的空间变化情况（图5-4c；Jackson，2007）。然而，方位角图不包含任何构造的倾角大小信息（如断层总断距），并且，在目标地质体的倾角方向与层位的区域倾角相似时（表5-1），方位角图的特征不明显。

方位角属性已经广泛应用在描述沉积盆地的构造倾向上。前面已经列举过与断陷相关的具不同方位和倾角方向的不同规模的正断层（如从几百米到几千米的断距和走向长度；Bouvier 等，1989；Hoetz 和 Watters，1992；Jones 和 Knipe，1996；Hesthammer 和 Fossen，1997；Rijks 和 Jauffred，1999）。方位角属性在地层分析中虽很少被使用，但也有一定的意义（Mondt，1993；Posamentier 和 Kolla，2003；Posamentier 等，2007）。比如，由于河道砂体的差异压实，其上覆地层会发育微小的背斜褶皱，这可能导致与之关联的反射具有各种不同的倾角，这些倾角能通过方位角属性成像（Mondt，1993）。

3) 倾角—方位角

这是一个混合属性，结合了倾角和方位角。它比单独的倾角和方位角有更多的优点，因为，只在一张图上就能同时显示出倾角和方位角属性（图5-4d）（Dalley 等，1989；Brown，2004）。在研究构造走向（主要正断层；Hhetz 和 Watters，1992；Mondt，1993；Jones 和 Knipe，1996；Hesthammer 和 Fossen，1997a 和 1997b；Hesthammer，1998）、深水沉积环境的地层特征（Nissen 等，1999；Posamentier 和 Kolla，2003）和碳酸盐岩沉积环境（Skirius 等，1999）的地层特征方面，倾角—方位角图非常有用。

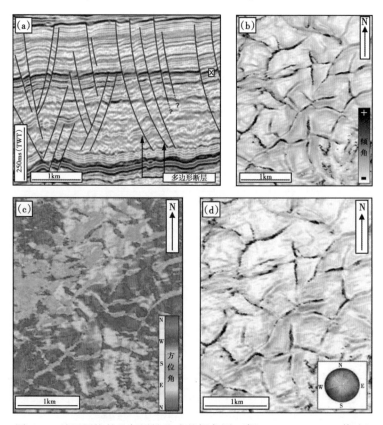

图 5-4 基于网格的几何属性生成的倾角图（据 Jackson，2007，修改）

(a) 来自北海盆地的垂向地震剖面，该剖面展示古近—新近系小断距正断层的分布情况。请注意图中用作绘制平面图的 X 反射同相轴。(b) X 反射同相轴的倾角图展示了断裂系统的多边形样式。(c) X 反射同相轴的方位角图不仅展示断裂系统的多边形样式，还展示每条断裂的倾角方向。(d) X 反射同相轴的倾角—方位角图既展示断裂系统的整体几何形态，又展现每条断裂的倾角方向。注意：不同的图使用了不同的色标

4）曲率

该属性是指一条曲线偏离直线（二维）的程度或者一个面（或地震层位）偏离平面（三维）的程度（Roberts，2001；Sigismondi 和 Soldo，2003）（图 5-5），它是一个很少被使用的并可能是被低估的属性。曲率的作用是减少区域层位倾角的影响以及对平面的偏离，这种偏离可能被解释成地质体。曲率与倾角的差别在于在高倾角地区层面上一些细小的倾角变化在曲率图上能显示而在倾角图上可能被掩盖（表 5-1）。以前，曲率被用于研究一系列构造和地层的几何特征和分布规律，包括复杂的断裂模式（Roberts，2001；Sigismondi 和 Soldo，2003；Marroquin 和 Hart，2004；Hart，2006；Sagan 和 Hart，2006）、深水水道复合体（图 5-6）、碳酸盐岩建隆（Hart 和 Sagan，2007），以及与碳酸盐喀斯特相关的塌陷岩脉。

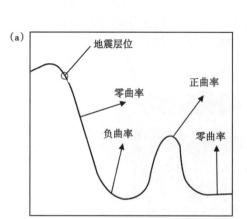
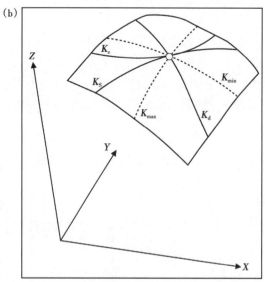

图 5-5　面曲率的二维和三维展示（据 Boberts，2001；Sigismondi 和 Soldo，2003，修改）
（a）面曲率的二维展示。注意曲率为"负值"、"正值""零值"时，曲线所在的位置。（b）面曲率的三维展示。能定义许多不同的曲率类型（如 K_s＝走向曲率；K_d＝倾角曲率；K_c＝等值线曲率；K_{max}＝最大曲率；K_{min}＝最小曲率；更多讨论参见 Roberts，2001）

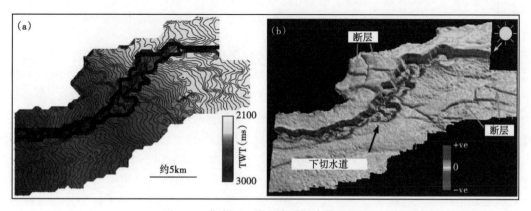

图 5-6　曲率用于深水水道复合体的研究
（a）古近—新近纪水道堤坝复合体的顶面时间构造图，注意该图在下切水道两侧的分辨率较差；
（b）叠置在时间构造图上的曲率图。注意该图对水道边缘发育的细小正断层有非常好的显示

5.3.2 基于体的几何属性

与上面介绍的基于网格的属性相比，体的属性在很大程度上不会受到解释员的主观偏差所影响，这是因为体的属性是直接从原始反射数据体衍生得到而不是从解释的地震层位得到。但是，由于各种属性都受到原始数据体内的噪声影响，因此，从衍生属性体中得到的地质特征应该小心使用（如 Hesthammer，1998）。

最常用的基于体的几何属性是用来显示反射数据体内部的不连续性的，是通过比较相邻地震道的波形得到的（图5-7；Bahorich 和 Farmer，1995；Marfurt 等，1998，1999）。Herein 把这种属性定义为相干性，人们常常也把它叫作连续性、相似性、协变性或者无序性。有时这些术语只是由用到不同解释软件品牌带来的纯粹的语义学差别，而在另一些情况下在如何计算地震道变化上也存在真正的不同。简单地说，在采样间隔内如果地震道之间的波形相似，那么计算的相关性就高（即数据之间相关）。相反，如果地震道之间波形有明显的不同，那么计算的相关性就低（即数据之间不相关）（图5-7）。在生成的数据体中（通常为相干体），解释人员的任务应该是在连续性中识别不连续点，这些不连续点可能与特殊的构造（如裂缝或断裂）或地层（如水道、峡谷等）有关（图5-8）。

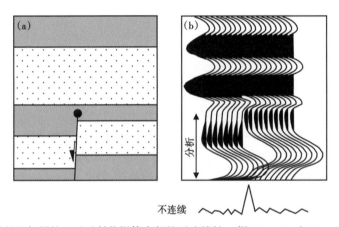

图5-7 基于体的几何属性显示反射数据体内部的不连续性（据 Bahorich 和 Famer，1995，模型）
(a) 同沉积生长断裂错动砂（点充填）泥（灰色）层的地质剖面；(b) 图a所示模型在地震上的简单响应。
注意在不连续处出现的尖峰值，这个尖峰是在断层发育处分析时窗内的低相关性区

大量的实例表明在盆地分析中运用相干属性能得到较好的结果。在构造发育区，Bohorich 等（Bahorich 和 Farmer，1995；Haskell 等，1999）运用相干属性能迅速获得一个数千米规模的盐体的几何特征及其与上覆正断层的空间搭配关系；Lonergan 等（1998）在一个更小的尺度上对古近—新近纪泥岩层序内的微断距断层的各种几何特征进行了研究。在地层研究中，相干属性被成功用来识别一系列深水碎屑岩沉积特征，包括块体搬运沉积、水下侵蚀水道和席状扇体（Nissen 等，1999；Posamentier 和 Kolla，2003；Saller 等，2004）。在碳酸盐岩沉积环境，Skirius 等（1999）和 Zampetti 等（2004，2004b）运用相关性成功地对碳酸盐岩斜坡和塔礁建隆的地貌特征进行了成图。在许多这样的实例中，相干体切片都被作为一种描述特殊地质体的重要方法。用相干体制作切片有三种方法：沿固定时间或深度的水平切割（时间或深度切片；Brown，2004；图5-8）、直接沿解释的地震层位切割（沿层切片；Brown，2004）、沿两个参考解释层位之间的"伪层位"切割（等比例切片；见 Zeng 等，

1998a, 1998b)。

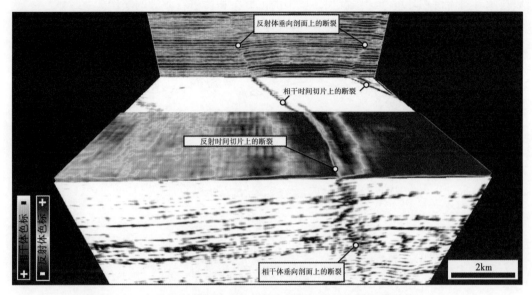

图 5-8 标准反射数据体和相干体在垂向和时间切片上的联合显示

这种显示展现相干性在伸展盆地内部构造方面成像的应用。注意在垂向切片和相干时间切片上正断层都非常清楚。左边的这条断层在反射时间切片和垂向相干切片上显示都不明显。对相干体而言，时间切片通常比垂向切片要好

5.3.3 从面和体中提取的基于振幅的属性

依靠其物性（通常与岩性有关），地质体可在地震数据中显示出振幅异常。不同的振幅提取算法其目的都是想精确地描绘振幅的平面分布，因为振幅的平面分布特征可能是地质体的一种反映（图 5-9 和图 5-10）。振幅提取的算法很多（如能量半衰期、最大正振幅或最大负振幅、反射强度、平均振幅等），更麻烦的是，这些属性在不同的软件中通常命名不同。均方根（RMS）（Brown，2005）是最常用的振幅属性之一。该属性通过扫描用户给定的层段，并以平方计算每个采样点的振幅能量，且不考虑振幅的极性。因此，能量相对强的正、负振幅就从相对于"背景"的弱振幅中脱颖而出。

振幅提取的方法主要有两种：(1) 沿层提取，在三维数据体内直接沿解释层提取，该层位是沿着（或接近）目标地质体的地层面；(2) 基于时窗的提取，与直接沿解释层提取不同，该方法从给定的间隔或时窗提取振幅（图 5-10）。在目标地层单元较厚（其厚度大于地震同相轴）或者在层位与目标地质体重合并且直接对层位成图比较困难的情况下，基于时窗的振幅提取是非常有效的技术。

制作基于时窗的振幅图有三种方法。第一种方法是在固定时间窗口内提取振幅（1000~1100ms）。该方法只在目的地层相对平坦的情况才有用，因为地层相对平坦才不会使所开的时窗切割由地震反射层确定的地质等时线。第二种方法是沿着地震解释层位上下开一包含该层位的时窗，并提取振幅（图 5-10）。第三种方法是沿着地震解释层位给一固定偏移作为时窗，并提取振幅（图 5-9、图 5-10cii）。地震解释员对时窗大小或"厚度"的选择对这三种方法的效果都至关重要。例如，随着时窗的加大，与目标地质体无关的振幅信息也会参加到成图中，从而显著增加振幅图受到"污染"的可能。本书向读者推荐 Zeng 等（1998a，

1998b)、Zeng（2007）和 Hadler-Jacobsen 等（2007）的文章，他们提供了很好的不同振幅提取技术对河流沉积环境和深水沉积环境中的富砂沉积体进行刻画的实例。

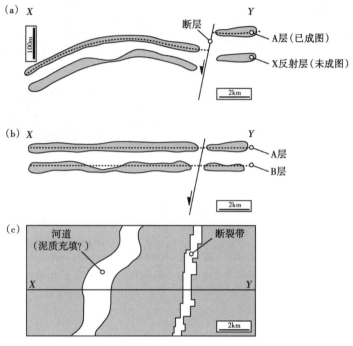

图 5-9　基于层位切片控制的振幅提取略图

(a) 三维数据体地震剖面图。地震层位 A（A层）在整个三维体内已经追踪。A层下伏一套同相轴（反射面X），由于该轴没有追踪解释，因此不能准确成图；同时，该轴还有明显的侵蚀特征。在图右侧发育一条正断层，该断层错断反射轴并产生褶皱。(b) 拉平层位 A 的地震剖面图。层位 B 是层位 A 下移 50m 后形成的层位。
(c) 对层位 B 提取的振幅图。在图上，水道沿着低振幅带成像，振幅图也被正断层撕破

5.3.4　其他的地震属性和解释方法

1）地震道波形聚类法

该类型假定地震道波形的细小变动与研究单元或层段的岩石属性变化有关。这是一个"混合"属性的例子，因为它既包含了振幅属性又包含了频率属性。地震数据是振幅和频率的混合体，因此波形属性在分析地震相方面具有强大的优势（Addy，1998；Brown，2004；Zeng，2004；Marroquin 等 2009）。该方法包括地震道波形的分类和分析，通常是针对解释层位在给定层段或时窗（即在解释层位的上部、下部或偏移一定时间）。根据波形细微的变动划分出许多不同的类型，每种类型对应一种颜色并平面成图，以展示地震相的变化，最终反映沉积环境的变化（图 5-11）。应当注意，地震道波形分类受到地震数据品质的限制，因为数据体内的噪声也会同目的地质体一起被分类，从而可能把噪声错误解释成目标地质体。而且，由于分析时窗往往是根据解释层位确定，因此，任何解释上的错误都将导致不正确的地震道波形的提取和分类，进而使地震相（以及推测的地质体）在成图受到"污染"。

2）"甜点"法

尽管波形分类属性在实际的研究中较少使用，但普遍认为该属性在刻画沉积盆地富含砂的沉积体时非常有潜力。"甜点"由频率和反射能量这两种属性共同描述（Radovich 和 Oliv-

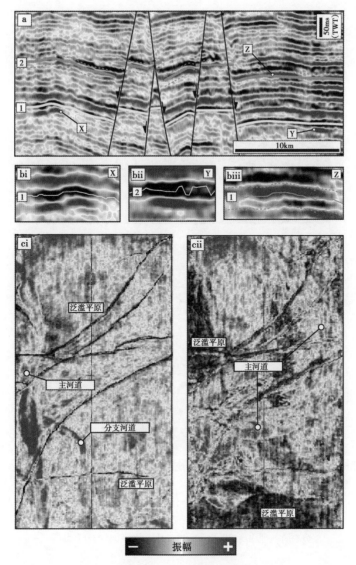

图 5-10 针对地层和构造的基于时窗的振幅成像技术实例（对比图 5-9）

(a) 来自三维数据体的地震剖面。层 1 在数据体内已全区追踪。从图上能见到 3 个侧向不连续的振幅异常体：两个直接位于层 1 上（X 和 Y），一个（Z）位于层 1 上方约 125ms 处（双程旅行时）。(b)（i）异常体 X 放大图；(ii) 异常体 Y 放大图；(iii) 异常体 Z 放大图；(c)（i）沿层 1 的振幅图，图上显示两组水道系统及其相关的越岸沉积；(ii) 沿层 2 的振幅图，图上水道的几何形态与下伏水道的有显著变化，层 2 是层 1 上移 125ms（双程旅行时）后生成的投影层或"伪"层，其振幅图是上下共开 24ms 时窗提取绘制的（参考图 5-10a 上的蓝色实线和虚线）

eros），它在数学上的识别依靠反射能量（一个常用的砂岩指标，总是正值且与相位无关）除以瞬时频率的平方根（通常瞬时频率不仅与地震频带宽度有关，且受层厚影响）。在碎屑岩沉积环境，泥岩为主的层段被认为不是好的"甜点"，由于其振幅低（声波阻抗差小）、频率高（反射轴相对较近）。相反，在砂岩为主并夹有泥岩的层段，由于振幅强（与包含的泥岩间的强声波阻抗差）、频率低（反射轴相对较宽），因此是好的"甜点"。在研究中发现，这种简化的关系可能会随着地层中存在的各种流体而变得复杂。这种复杂性可用来定量描述和理解含油气砂体的净毛比变化，这在深水沉积环境和海岸平原沉积环境都得到实证

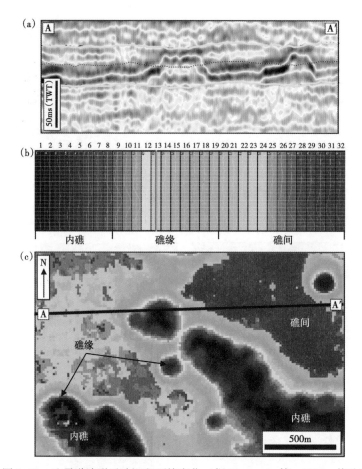

图 5-11 地震道波形反映沉积环境变化（据 Marroquin 等，2009，修改）
（a）加拿大陆上地区多个生物礁建隆的地震剖面图。注意在各个生物礁主体（内礁）和毗邻区（礁间）之间地震特征的横向变化。两条黄层之间是进行地震道波形分析的区域。（b）对（a）中两黄层间的数据进行地震道（即地震相）分类，共识别出 32 种类型，它们分别代表内礁、礁缘和礁间沉积环境。（c）地震道波形分类图（即地震相图）显示这三种主要沉积相的分布范围

（Coff，2004；Hart，2008a）。由 Hart（2008a）建立的简单地震正演模型对"甜点"法的物理基础进行说明，并指出其在地层研究方面非常重要（Hart，2008）。Hart（2008a）在刻画河道砂体时指出强噪音、复杂的地层接触关系和复杂构造特征会限制"甜点"法的运用。

3）频谱分解

这是一种新兴的地震解释方法，能对时间地层和不连续地质体进行二维和三维成图。该属性可认为是一种"时窗类"属性，因为它是针对时间段进行分析，该时间段以地震解释层位为参考。正常频段或全频段的地震数据在时间段内经过离散傅里叶变换（DFT）分解成一系列特殊频率的频谱（要想详细了解其方法参见 Partyka 等，1999）。频谱分解的目的是对不同厚度的层用离散频率成像。尽管该方法在碳酸盐沉积体系（Skirius 等，1999）成像研究方面成效显著，其更成功的是在分析深水沉积（Partyka 等，1999；Henderson 等，2007）和河流沉积体系方面。这些研究已经说明频谱分解技术在研究垂向分辨率低于全频分辨率的复杂地层时有良好效果。

5.4 结论

本章对运用不同地震解释技术在多种尺度上研究沉积盆地的形成和充填的优缺点进行了概述。但由于篇幅的限制，对其他许多技术没有进行讨论。它们包括透视技术和体素拾取技术（James 等，2004）以及各种基于频率和相位的属性技术（Radovich 和 Oliveros，1998；Hart 和 Chen，2004；Smythe 等，2004），这些技术在沉积盆地的地震解释过程中都被证明是有用的工具。因此，本书建议读者去了解这些不同属性的物理基础和在地震解释中的使用。

虽然地震解释和相关属性分析是非常强大的盆地分析工具，但是，地下地质问题的答案并不存在于工作站或者衍生的属性中。工作站只是工具，就像扫描电子显微镜、X 射线衍射或古生物地层分析一样，同时，我们也要避免成为"键盘工"，忘记思考而不停寻找解释的"致命银弹"（Brown，1996，2005；Hart，1999，2000）。成功运用地震数据对盆地的构造和地层进行研究的关键是：(1) 具有所用解释软件的相关知识；(2) 了解所用地震属性的物理基础；(3) 能理解数据和解释质量对其生成的属性的影响；(4) 对地球物理数据进行慎重周到的地质解释。此外，在地震解释期间人们常常忽视这样一个事实，就是在研究衍生属性图和剖面的地质意义时，结合时间构造图和厚度图，标准二维地震剖面和基础的地震地层原理也能起到关键的作用。而且，在地震解释过程中，地震数据和地震衍生属性与井数据的标定特别重要（Maynard，2006；图 5-2）。

与之前几位作者（Davies 等，2004；Cartwright 和 Huuse，2005）讨论过的一样，地震反射数据在与露头、井和其他工具相比时有一定局限。这些局限主要是受纵横向分辨率和不同岩石单元与地震响应之间复杂关系的影响。然而，地震数据（2D 和 3D）比大多数其他类型的数据在空间覆盖上都有优势，并且能够对大的（如地下几千平方千米面积）构造和地质目标进行三维显示，这些大的地质体即使在最大的野外露头上也几乎是观察不到的。随着空间覆盖率和分辨率的不断提高，地震反射数据将有潜力像其他许多基于露头的或分析实验的技术那样在沉积盆地地质分析方面成为非常有用的工具。

参 考 文 献

Addy, S. K. (1998) Neural network classification method helps seismic interval interpretation. Oil and Gas Journal, 96, 47–59.

Apotria, T., and Wilkerson, M. S. (2002) Seismic expression and kinematics of a fault–related fold termination: Rosario structure, Maracaibo Basin, Venezuela. Journal of Structural Geology, 24, 671–687.

Bacon, M., Simm, R., and Redshaw, T. (2007) 3–D seismic interpretation. Cambridge, CambridgeUniversity Press. Bahorich, M., and Farmer, S. (1995) 3–D seismic discontinuity for faults and stratigraphic features–the coherence cube. The Leading Edge, 14, 1053–1058.

Bouvier, J. D., Kaars–Sijpesteijn, C. H., Kluesner, D. F., Onyejekwe, C. C., and Van der Pal, R. C. (1989) Three–dimensional seismic interpretation and fault sealing investigation, Nun River Field, Nigeria. AAPG Bulletin, 73, 1397–1414.

Brown, A. R. (1996) Seismic attributes and their classification. The Leading Edge, 15, 1090.

Brown, A. R. (2004) Interpretation of three-dimensional seismic data, 6th ed. AAPG Memoir, 42, SEE Investigations in Geophysics, 9. Tulsa, OK, American Association of Petroleum Geologists. Brown, A. R. (2005) Pitfalls in 3D seismic interpretation. The Leading Edge, 24, 716-717.

Cartwright, J. A., and Huuse, M. (2005) Seismic technology: the geological Hubble. Basin Research, 17, 1-20.

Chopra, S., and Marfurt, K. J. (2005) Seismic attributes-a historical perspective. Geophysics, 70, 3-28.

Dalley, R. M., Gevers, E. C. A., Stampfli, G. M., Davies, D. J., Gastaldi, C. N., Ruijtenberg, P. A., and Vermeer, G. J. O. (1989) Dip and azimuth displays for 3D seismic interpretation. First Break, 7, 86-95.

Dart, C., Cloke, I. Herdlevær, A ., Gillard, D., Rivenæs, J., Otterlei, C., Johnsen, E., and Ekern, A. (2004). Use of 3D visualization techniques to unravel complex fault patterns for production planning: Njord Field, Halten Terrace, Norway, in Davies, R. J., Cartwright, J. A., Stewart, S. A., Lappin, M., and Underhill, J. R., eds., 3D seismic technology: application to the exploration of sedimentary basins. Memoir, 29. London, Geological Society, 49-261.

Davies, R. J., Stewart, S. A., Cartwright, J. A., Lappin, M., Johnston, R., Fraser, S. I., and Brown, A. R. (2004) 3D seismic technology: are we realising our full potential? in Davies, R. J., Cartwright, J. A., Stewart, S. A., Lappin, M., and Underhill, J. R., eds., 3D seismic technology: application to the exploration of sedimentary basins. Memoir, 29. London, Geological Society, 1-10.

Davies, R. J., and Posamentier, H. W. (2005) Geologic process in sedimentary basins inferred from three-dimensional seismic imaging. GSA Today, 15, 4-9.

Demyttanaere, R. R. A., Sluijk, A. H., and Bentley, M. R. (1993) A fundamental reappraisal of the structure of the Cormorant Field and its impact pm field development strategy, in Parker, J. R., ed., Petroleum geology of Northwest Europe: proceedings from the 4th conference. London, Geological Society of London, 1151-1157.

Dorn, A. G. (1998) Modern 3-D seismic interpretation. The Leading Edge, 17, 1262-1272.

Goff, D. (2004) Estimating net: gross from data histograms: examples from deepwater turbidites. Tulsa, OK, American Association of Petroleum Geologists. Search and Discovery Article 90037. http://www.searchanddiscovery.net/documents/abstracts/2004regional_west_africa/abstracts/goff.htm? q=%2Btext%3Arefine.

Hadler-Jacobsen, F., Gardner, M. H., and Borer, J. M. (2007) Seismic stratigraphic and geomorphic analysis of deepmarine deposition along the West African continental margins, in Davies, R. J., Posamentier, H. W., Wood, L. J., and Cartwright, J. A., eds., Seismic geomorphology: applications to hydrocarbon exploration and production. Special Publication, 277. London, Geological Society, 47-84.

Hart, B. S. (1999) Definition of subsurface stratigraphy, structure and rock properties from 3D seismic data. Earth-Science Reviews, 47, 189-218.

Hart, B. S. (2000) 3-D seismic interpretation: a primer for geologists. SEPM Short Course, 48, 123 p. Hart, B. S. (2002) Validating seismic attribute studies – beyond statistics. The Leading Edge, 21, 1016–1021.

Hart, B. S. (2006) Seismic expression of fracture-swarm sweet spots, Upper Cretaceous tight-gas reservoirs, San Juan Basin. AAPG Bulletin, 90, 1519–1534.

Hart, B. S. (2008a) Channel detection in 3-D seismic data using sweetness. AAPG Bulletin, 92, 733–742.

Hart, B. S. (2008b) Stratigraphically significant attributes. The Leading Edge, 27, 320–324.

Hart, B. S., and Chen, M-A. (2004) Understanding seismic attributes through forward modelling. The Leading Edge, 23, 834–841.

Hart, B. S., and Sagan, J. A. (2007) Curvature for visualisation of seismic geomorphology, in Davies, R. J., Posamentier, H. W., Wood, L. J., and Cartwright, J. A., eds., Seismic geomorphology: applications to hydrocarbon exploration and production. Special Publication, 277. London, Geological Society, 139–150.

Haskell, N., Nissen, S., Hughes, M., Grindhaug, J., Dhanani, S., Heath, R., Kantorowicz, J., Antrim, L., Cubanski, M., Nataraj, R., Schilly, M., and Wigger, S. (1999) Delineation of geological drilling hazards using 3-D seismic attributes. The Leading Edge, 18, 373–382.

Henderson, J., Purves, S., and Leppard, C. (2007) Automated delineation of geological elements from 3D seismic data through analysis of multichannel, volumetric spectral decomposition data. First Break, 25, 87–93.

Hesthammer, J. (1998) Evaluation of timedip, correlation and coherence maps for structural interpretation of seismic data. First Break, 16, 151–167.

Hesthammer, J., and Fossen, H. (1997a) The influence of seismic noise in structural interpretation of seismic attribute maps. First Break, 15, 209–219.

Hesthammer, J., and Fossen, H. (1997b) Seismic attribute analysis in structural interpretation of the Gullfaks Field, northern North Sea. Petroleum Geoscience, 3, 13–26.

Hoetz, H. L. J. G., and Watters, D. G. (1992) Seismic horizon attribute mapping for the Annerveen Gasfield, The Netherlands. First Break, 10, 41–51.

Jackson, C. A-L. (2007) Application of three-dimensional seismic data to documenting the scale, geometry and distribution of soft-sediment features in sedimentary basins: an example from the Lomre Terrace, offshore Norway, in Davies, R. J., Posamentier, H. W., Wood, L. J., and Cartwright, J. A., eds., Seismic geomorphology: applications to hydrocarbon exploration and production. Special Publication, 277. London, Geological Society, 253–267.

Jackson, C. A-L., and Larsen, E. (2008) Temporal constraints on basin inversion provided by 3D seismic and well data: a case study from the South Viking Graben, offshore Norway. Basin Research, 20, 397–417.

Jackson, C. A-L., and Larsen, E. (2009) Temporal and spatial evolution of a gravity-driven normal fault array; Middle– Upper Jurassic, South Viking Graben, northern North Sea. Journal of Structural Geology, 31, 388–402.

Jackson, M. P. A., Hudec, M. R., Jennette, D. C., and Kilby, R. E. (2008) Evolution of the

Cretaceous Astrid thrust belt in the ultra deep-water lower Congo Basin, Gabon. AAPG Bulletin, 92, 487-511.

James, H., Bond, R., and Eastwood, L. (2004) Direct visualisation and extraction of stratigraphic targets in complex structural settings, in Davies, R. J., Cartwright, J. A., Stewart, S. A., Lappin, M., and Underhill, J. R., eds., 3D seismic technology: application to the exploration of sedimentary basins. Memoir, 29. London, Geological Society, 227-234.

Jones, G., and Knipe, R. J. (1996) Seismic attribute maps: application to structural interpretation and fault seal analysis in the North Sea Basin. First Break, 124, 449-461.

Lonergan, L., Cartwright, J. A., and Jolly, R. (1998) The geometry of polygonal fault systems in Tertiary mudrocks of the North Sea. Journal of Structural Geology, 20, 529-548.

Marfurt, K. J., Kirlin, R. L., Farmer, S. L., and Bahorich, M. S. (1998) 3-D seismic attributes using a semblance-based coherency algorithm. Geophysics, 63, 1150-1165.

Marfurt, K. J., Sudhaker, V., Gersztenkorn, A., Crawford, K. D., and Nissen, S. E. (1999) Coherency calculations in the presence of structural dip. Geophysics, 64, 104-111.

Marroquin, I. D., Brault, J-J., and Hart, B. S. (2009) A visual data-mining methodology for seismic facies analysis: part 2-application to 3D seismic data. Geophysics, 74, 13-23.

Marroquin, I. D., and Hart, B. S. (2004) Seismic attributebased characterization of coalbed methane reservoirs: an example from the Fruitland Formation, San Juan basin, New Mexico. AAPG Bulletin, 88, 1603-1621.

Maynard, J. R. (2006) Fluvial response to active extension: evidence from 3D seismic data from the Frio Formation (Oligo-Miocene) of the Texas Gulf of Mexico Coast, USA. Sedimentology, 53, 515-536.

Mondt, J. C. (1993) Use of dip and azimuth horizon attributes in 3D seismic interpretation. SPE Paper 20943-PA, 253-257.

Nissen, S. E., Haskell, N. L., Steiner, C. T., and Coterill, K. L. (1999) Debris flow outrunner blocks, glide tracks, and pressure ridges identified on the Nigerian continental slope using 3-D seismic coherency. The Leading Edge, 18, 595-599.

Partyka, G., Grindley, J., and Lopez, J. (1999) Interpretational applications of spectral decomposition in reservoir characterisation. The Leading Edge, 18, 353-360.

Payton, C. E. (1977) Seismic-stratigraphy-applications to the exploration of sedimentary basins. AAPG Memoir, 26, Tulsa, OK, American Association of Petroleum Geologists, 516 p.

Peyton, L., Bottjer, R., and Partyka, G. (1998) Interpretation of incised valleys using new 3-D seismic techniques: a case history using spectral decomposition and coherency. The Leading Edge, 17, 1294-1298.

Posamentier, H. W. (2003) Depositional elements associated with a basin floor channel-levee system: case study from the Gulf of Mexico. Marine and Petroleum Geology, 20, 677-690.

Posamentier, H. W., and Kolla, V. (2003) Seismic geomorphology of depositional elements in deep-water settings. Journal of Sedimentary Research, 73, 367-388.

Posamentier, H. W., Davies, R. J., Cartwright, J. A., and Wood, L. (2007)Seismic geomorphology-an overview, in Davies, R. J., Posamentier, H. W., Wood, L. J., and Cartwright, J. A., eds., Seismic geomorphology: applications to hydrocarbon exploration and production. Special Publication, 277. London, Geological Society, 1-14.

Radovich, B. J., and Oliveros, R. B. (1998) 3 - D sequence interpretation of seismic instantaneous attributes from the Gorgon field. The Leading Edge, 17, 1286-1293.

Rijks, E. J. H., and Jauffred, J. C. E. M. (1999) Attribute extraction: an important application in any detailed 3-D interpretation study. The Leading Edge, 10, 11-19.

Roberts, A. (2001) Curvature attributes and their application to 3D interpreted horizons. First Break, 19, 85-100.

Sagan, J. A., and Hart, B. S. (2006) Three-dimensional seismic-based definition of fault-related porosity development: Trenton - Black Rover interval, Saybrook, Ohio. AAPG Bulletin, 90, 1763-1785.

Saller, A. H., Noah, J. T., Ruzuar, A. P., Schneider, R. (2004) Linked lowstand delta to basin-floor fan deposition, offshore Indonesia: an analog for deep-water reservoir systems. AAPG Bulletin, 88, 21-46.

Sheriff, R. E., and Geldart, I. P. (1995) Exploration seismology. Cambridge, CambridgeUniversity Press. Sigismondi, M. E., and Soldo, J. C. (2003) Curvature attributes and seismic interpretation: Case studies from Argentina basins. The Leading Edge, 22, 1122-1126.

Skirius, C., Nissen, S., Haskell, N., Marfurt, K., Hadley, S., Ternes, D., Michel, K., Reglar, I., D'Amico, D., Deliencourt, F., Romero, T., D' Angelo, R., and Brown, B. (1999) 3-D seismic attributes applied to carbonates. The Leading Edge, 18, 384-393.

Smythe, J., Gersztenkorn, A., Radovich, B., Li, C-F., and Liner, C. (2004) Gulf of Mexico shelf framework interpretation using a bed-form attribute from spectral imaging. The Leading Edge, 23, 921-926.

Sullivan, E. C., Marfurt, K. J., and Bhumentritt, C. (2007) Seismic geomorphology of Palaeozoic collapse features in the Fort Worth Basin (USA), in Davies, R. J., Posamentier, H. W., Wood, L. J., and Cartwright, J. A., eds., Seismic geomorphology: applications to hydrocarbon exploration and production. Special Publication, 277. London, Geological Society, 187-222.

Vail, P. R. (1987) Seismic stratigraphy using sequence stratigraphy; Part 1, Seismic stratigraphy interpretation procedure, in Bally, A. W., ed., Atlas of seismic stratigraphy. AAPG Studies in Geology, 27. Tulsa, OK, American Association of Petroleum Geologists, 1-10.

Weimer, P., and Davis, T. L. (1996) Applications of 3 - D seismic data to exploration and production. AAPG Studies in Geology, 42. Tulsa, OK, American Association of Petroleum Geologists, 270 p.

Yilmaz, O. (2001) Seismic data analysis and processing, inversion, and interpretation of seismic data. Tulsa, OK, Society of Exploration Geophysicists, 2027 p.

Zampetti, V., Schlager, W., van Konijnenburg, J-H., and Everts, A. J. (2004a) Architecture and growth history of a Miocene carbonate platform from 3D seismic reflection data; Luconia province, offshore Sarawak, Malaysia. Marine and Petroleum Geology, 21, 517-534.

Zampetti, V., Schlager, W., van Konijnenburg, J-H., and Everts, A. J. (2004b) 3-D seismic characterization of submarine landslides on a Miocene carbonate platform (Luconia Province, Malaysia). Journal of Sedimentary Research, 74, 817-830.

Zeng, H. (2004) Seismic geomorphology-based facies classification. The Leading Edge, 23, 644-688.

Zeng, H. (2007) Seismic imaging for seismic geomorphology beyond the seabed: potential and challenges, in Davies, R. J., Posamentier, H. W., Wood, L. J., and Cartwright, J. A., eds., Seismic geomorphology: applications to hydrocarbon exploration and production. Special Publication, 277. London, Geological Society, 15-28.

Zeng, H., Backus, M. M., Barrow, K. T., and Tyler, N. (1998a) Stratal slicing, part I: realistic 3-D seismic model. Geophysics, 63, 502-513.

Zeng, H., Henry, S. C., and Riola, J. P. (1998b) Stratal slicing, part II: real seismic data. Geophysics, 63, 514-522.

(刘长利 译,郭瑞 校)

第6章 沉积盆地内构造成因的冲积相砾石的搬运与保存

PHILIP A. ALLEN[1], PAUL L. HELLER[2]

(1. Department of Earth Science and Engineering, Imperial College, South Kensington Campus, London, UK; 2. Department of Geology and Geophsics, University of Wyoming, Laramie, USA)

摘 要：将粗砾石从物源区搬运到沉积盆地通常被视为构造活动的一个典型特征。同构造期砾岩的识别标志包括典型的结构样式、砾石前缘的进积作用以及向下游方向变细的速率。然而，对于根据此类沉积物来解释构造活动的发生时间，需要考虑将沉积物从物源区搬运到沉积中心输送体系的整体动力学格架。砾石形成和搬运相关过程的时间长短具有非常重要的意义，因为这些相关过程会削弱其作为构造活动记录仪的岩石记录的真实性。

粗粒沉积物所释放出的信息在其被搬运过程中发生了转换，而转换这些信息最重要的一种作用就是河流相沉积体系中局部保存的缓冲作用。搬运到沉积盆地内的砾石会对地层产生长期影响，这种影响受控于物源释放的信号频率与盆地接收信息响应时间之间的关系。目前关于砾石穿越盆地沉积产生的时间滞后方面的研究很少，一项关于西班牙前陆盆地的研究结果显示延迟时间大约为0.1~1Ma，并且这一时间随着砾石向下游方向搬运距离的增加而不断增加。

剥蚀和沉积地貌受构造活动的扰动，例如由于断层拼接造成的主边界断层滑动速率的改变，经历了短暂的响应期，其持续时间取决于气候、岩石侵蚀程度和沉积物输送体系的规模，但是通常为1Ma的级别。数值模拟和野外考察揭示地层序列差异性的一个重要部分是由新的稳定状态建立之前的瞬时活动造成的。

搬运到沉积盆地的砾石在向下游搬运过程中不断分散，但是仍然被选择性地分选出来形成地层。向下游变细的速率主要由沉积物供给量、粒径分布规律、构造沉降速率的空间分布等因素所决定。靠近物源区快速的构造沉降会导致向下游粒度迅速变细以及搬运距离变短，反之，较低的构造沉降速率会造成搬运距离变长并相应地降低粒度向下游变细的速率。

物源区剥蚀地貌和沉积盆地沉积地貌之间的耦合关系是目前研究的热点和难点领域之一。针对剥蚀作用和沉积作用对构造变形的定位以及再活化的影响也正在开展不同尺度的调查和研究。

关键词：同构造期砾岩　地层学　构造地质学　气候　地貌

6.1 引言

沉积盆地保存了构造活动影响地球表面的最连续的记录。通常假定构造运动塑造了地形，以地形为媒介，地层发生了风化和剥蚀等作用，由此形成了典型的沉积作用。因此，趋向于假设在构造运动与生成的沉积物之间存在一个简单的、线性的因果关系。然而，构造运动及气候因素的信息在沉积物从山区汇水源区搬运到沉积盆地最终沉积区的过程中会被转换、缓冲、扩大或者改变（图6-1和图6-2）。在沉积物从源到汇搬运过程中，构造运动形成的沉积信息可能被彻底转换，这一转换主要是通过剥蚀作用和沉积作用之间的滞后时间、非线性与双向耦合来实现的。这些转换作用包含了构造沉降导致近物源区粗粒碎屑物的沉积，整个搬运体系中沉积物的局部保存与释放，较老盆地中沉积物在年轻盆地中的再次沉积，稳定状态的剥蚀作用的时间滞后以及盆山耦合多种过程之间的相互作用，尤其是构造运动与地表地质过程的相互作用。这些相互作用发生在从单个断块到整个造山带及其附属盆地的一系列时间和空间尺度内。结果就是同构造期沉积作用的时间和类型可能明显不同于常规的解释盆地沉积充填的简单因果模型。

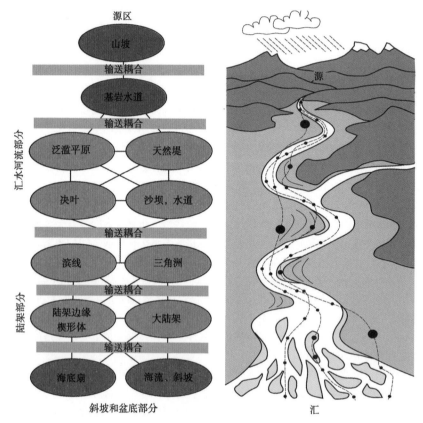

图 6-1 沉积物搬运示意图（据 Allen，2008；Malmon 等，2003，修改）

右图：沉积物沿着轨迹（虚线）从物源区搬运到汇集区。一些轨迹与短期搬运有关，在沉积物搬运路径上有短暂的保存阶段（小圈），而其他一些轨迹则与长时间搬运有关，在瞬变状态下有较长的保存阶段（大圈）。瞬变状态下沉积物长时间的保存表明沉积物供给信号有缓冲。左图：物源（橘色），沉积物临时保存点（绿色代表瞬时状态），固定保存点（蓝色代表吸收状态）。各部分宏观地貌间沉积物的搬运可能很复杂，并包含了动态信息反馈

本章选择性地回顾了构造成因的沉积信息在沉积物输送体系内被转换的一些作用过程，以及地表沉积物流量如何有选择性地"被取样"来建造下伏地层。本文针对陆相盆地的砾岩沉积物开展研究，强调了沉积物对构造运动响应的时间和空间尺度。通过模型和露头研究，总结了砂砾岩分布的各种控制因素，包括沉积物输送体系动力、构造沉降以及构造成因的岩石隆升和剥蚀。

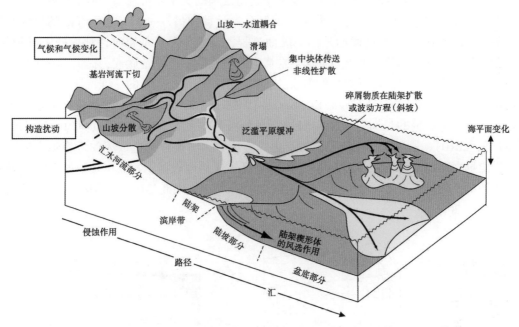

图 6-2 沉积物输送体系从源到汇的基本原理示意图（据 Somme 等，2009，修改）
沉积物输送体系受构造和气候驱动及扰动。它包含了一系列动态地貌、相互耦合的子体系或部分。
剥蚀作用将沉积物供给到一个主输送体系（轨道），然后再到长期汇集区

6.2 沉积输送体系与构造活动

大多数的盆地沉积特征是由创造长期可容纳空间的构造沉降速率 $\sigma(X)$ 与沉积物通量 Q_s 之间的总体平衡决定的。这种平衡控制了沉积物的空间分布、结构样式以及沉积物的沉积速率，并对沉积环境类型起主要控制作用。盆地地层记录不仅是未覆盖的物源区和盆地沉降区的历史，也是沉积物输送体系内沉积物释放、输运和暂时保存等动力学作用的历史（图 6-1 和图 6-2）。

被 Leeder（1999）称为"碎屑工厂"的沉积盆地腹地的地貌差异和生物地球化学作用体系是沉积物输送体系的剥蚀驱动力（图 6-2；Allen，2008）。这些腹地是颗粒与溶解状的河流成因的大规模沉积物通量的源区。根据地层来解释构造运动时间需要了解从腹地到沉积盆地的长期的大规模沉积物通量，以及腹地对于气候和构造运动等动力因素变化的响应时间。目前，大多数研究旨在认识适应于沉积盆地的中—大空间尺度的剥蚀作用与构造作用之间的耦合关系。

沉积物输送体系的概念是将沉积物的来源、输送和最终堆积过程作为一个单向的、整合的过程体系，所以沉积物可以从源到汇进行追踪。沉积物从物源区沿着输送通道被搬运到沉积区，也包括了在复杂路径上的间歇性输送，因此搬运时间的范围较宽。沉积颗粒的搬运时间包括沉积物"短暂的"就地保存时间到经过一定地质时期的堆积而处于一个"吸收状态"的时间（Malmon 等，2003；Allen，2008a）（图 6-1）。吸收状态存在于发生构造沉降或全球海平面上升的地方，这些可为沉积物的长期堆积提供可容纳空间，例如海岸三角洲或深海环境。

沉积物释放、搬运、间歇性保存以及地层建造之间的关系非常复杂，但是有两点非常突出。首先，沉积物颗粒沿着搬运通道从一个保存点运动到另外一个点是固有的随机性问题（Malmon 等，2003）。沉积物搬运是一个缓冲过程，就地保存只是暂时性地捕获了沉积物，之后会释放一些沉积颗粒。因此，依据颗粒搬运轨迹数量的统计学属性，提供了一个将沉积物输送体系作为一个整体进行测量的方法。其次，沉积物输送体系的三维特征与地面的沉积物通量转移到下伏地层有关（Toro-Escobar 等，1996；Allen，2008）。这种转移从根本上来说是由构造运动产生的长期可容纳空间的空间分布和幅度决定的（Fedele 和 Paola，2007；Duller 等，2010），所以下游颗粒粒径的变化趋势能够揭示这种构造成因的可容纳空间与沉积物供给量之间的平衡信息（图6-3）。下游选择性地从地表沉积物流中提取某些特定粒径的碎屑颗粒，同样也决定了不同沉积岩相粒级的有效性，因此是沉积结构的一个主控因素（Strong 等，2005）。

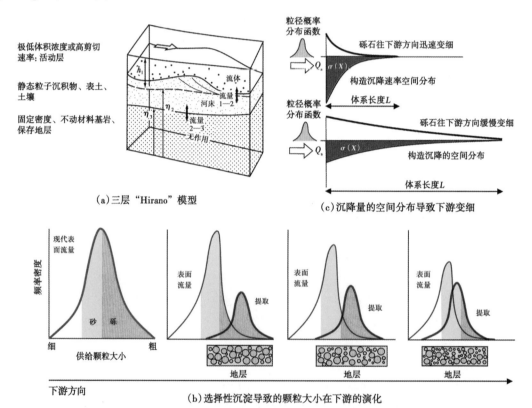

图 6-3 沉积物释放、搬运、间歇性保存以及地层建造之间的关系

(a) 一个三层的 "Hirano" 模型说明了以下三者的区别：具有低体积浓度及高剪切速率的表面活跃层，准静态的底层，具有惰性物质的不活跃层或固定密度的基底。颗粒沉积物可以从一个层输送到另一个层，从而引起这三个层的高度和厚度发生变化。在地层研究中，底层通常被假定厚度不变，有利目标的质量传递从流体流向基底。(b) 供给的沉积物从表层选择性抽提某些物质来建造地层，使得供给颗粒粒径的概率密度函数向下游方向发射转变。(c) 构造沉降量在空间分布的不同样式引起了地层中沉积物颗粒粒径向下游方向变小，其中沉积物供给的流量和粒径分布保持不变

沉积物输送体系由几个具有独特地貌作用过程的分段组成（图6-2），这些地貌过程控制着沉积碎屑物的剥蚀、搬运和沉积（Moore，1969；Paola，2000；Tinker 等，2008；Somme 等，2009）。沉积物阶梯式地穿过这些分段到达吸收状态，但是地貌子系统间的转换机制可能很复杂。本章仅限于考虑包括山体汇水区的上游部分和冲积盆地。

对于理解陆相沉积物输送体系动力学机制主要存在两个不确定的区域。第一个区域涉及产生沉积物供给的剥蚀机制。了解剥蚀地貌对于诸如构造与气候变化等干扰因素的响应及其时间尺度对于理解整个沉积物输送体系是至关重要的，因为剥蚀与沉积是一对耦合机制（Humphrey 和 Heller，1995；Babault 等，2005；Densmore 等，2007a）。由剥蚀机制生成的，并由体积、粒径和组分的概率密度函数（pdf）计量的沉积物（Paola，1990），通过沉积物输送体系影响了其分散过程中的许多方面（图 6-3）。第二个区域涉及到来自上游汇水区域的沉积物供给信息如何通过沉积物输送体系被转移，因为沉积物的暂时保存状态暗示了任何流出信号都被缓冲了（Métivier 和 Gaudemer，1999；Castelltort 和 Van den Driessche，2003；Jones 等，2004；Allen，2008b）。这种缓冲作用可能既包含了由于时间滞后而导致的供给信号转换，又包含了由于振幅衰减和波长增加导致的供给信号转换（Swenson，2005）。如果发生这种情况，沉积时间序列，例如海岸三角洲或深海相沉积物的长期堆积速率，将会是一个被高度转换的来自于上游汇水区的沉积物供给记录。这将会使为了确定驱动机制而简单地反演沉积区地层样式变得困难。

综合考虑沉积物输送体系的方法有几个明显优势：用质量平衡法模拟沉积物的沉积是有效的（Whipple 和 Trayler，1996；Allen 和 Hovius，1998；Allen 和 Densmore，2000）；便于理解由构造运动与气候变化产生的外部动力叠加到内部动力上的影响（Carretier 和 Lucazeau，2005；Densmore 等，2007a）；使得研究来自上游汇水区的瞬时沉积物供给信息是如何通过沉积物输送体系传播出去的变成可能；便于在生物地球化学周期内示踪与河流搬运流量相关的沉积物颗粒和溶质（包含碳质颗粒）（Galy 等，2007；Hilton 等，2008）。现在几乎所有的河流体系都受到了人类活动的严重影响（Meybeck 等，2003；Meybeck 和 Vorosmarty，2005；Syvitski 和 Milliman，2007），在人类活动成为主要影响因素之前，想要理解地貌及其沉积物输送体系是如何发挥作用的，地层及其他长期记录是必不可少的。

构造活动可以在不同尺度上对沉积物输送体系产生影响。从根本上讲，构造活动引起的岩石隆升造成了地形隆起和剥蚀通量，同时构造活动导致的沉降能产生可容纳空间和长期的沉积物保存。将地貌演化模型叠加到特定构造位移场上研究抬升正断层上盘的地貌演化引起了人们的极大关注（图 6-4；Allen 和 Densmore，2000；Hardy 和 Gawthorpe，2002；Densmore 等，2007a；Carretier 等，2009）。这种模型要求地貌规律要考虑山坡与河道的耦合。这些规律通常遵循水动力定律的各种公式（Whipple 和 Tucker，1999，2002；Snyder 等，2000；Tucker 和 Whipple，2002）。其他模型聚焦于一系列相互作用的张性断层的演化，生成了一个复杂的构造模板，其上的地形地貌与河流体系不断演化（Cowie 等，2006）。大量的研究强调在控制下盘汇水区域的位置和大小的过程中，以及在张性背景下沉积物通过河流搬运到上盘沉积中心的过程中，断裂体系生长变化是非常重要的（Roberts 和 Jackson，1990；Gawthorpe 和 Hurst，1993；Leeder 和 Jackson，1993；Gupta 等，1999）。类似的研究还有，旨在通过明确限制性盆地或"微型盆地"构造演化所控制的不规则深海地形，从而确定浊流的海底路径（Sinclair，1994；Grecula 等，2003；Hodgson 和 Haughton，2004；Violet 等，2005；Lamb 等，2006；Toniolo 等，2006）。同样地，还有少量的针对挤压构造背景下的类似研究也已经开展（Johnson 和 Beaumont，1995；Clevis 等，2004）。

对于剥蚀规律的应用表明，地貌达到剥蚀速率与岩石隆升速率平衡的稳定状态需要一定

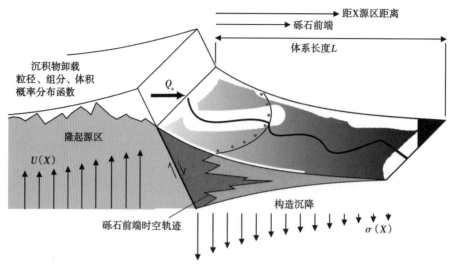

图 6-4 张性断层的构造位移与构造隆升下盘的剥蚀和邻区上盘盆地的沉积有关
沉积物在下盘被以一种特殊概率密度函数释放出来,该函数与沉积物粒径、体积和成分有关,并在上盘散布,
形成了一个被称作砾石前缘的移动边界。在上盘沉积地层中,砾岩前缘存在一个时—距轨道

的时间（Paola 等,2009）。超过这一剥蚀响应时间,沉积物输送到相邻盆地的速率迅速改变,而一旦接近,响应时间就会趋于稳定。同样地,沉积盆地也有响应时间,当大于这一时间时,盆地沉降速率与沉积物供给速率达到平衡,以致地表搬运沉积物的斜坡趋近于一个稳定的构型（Paola 等,1992）。剥蚀与沉积的耦合机制表明,需要考虑整个系统的响应时间尺度,才能在整个耦合体系中感受到源区构造运动的影响。本章是认识富砾石的沉积物输送体系中响应时间尺度的一次尝试。

6.3 地貌调节的时间尺度

地貌和邻近扇体在构造边界条件下对扰乱的响应时间可以通过数值模型进行估算,这些扰乱包括边界断层滑动速率的改变（Whipple 和 Tucker,1999; Allen 和 Densmore,2000; Densmore 等,2007a; Whittaker 等,2007a,2007b,2010; Allen,2008a),或者断层末端的侧向加积等（Densmore 等,2007b）。例如,主断层滑动速率的改变可能是由断层各分段之间生长、相互作用、联动过程造成的（Gawthorpe 和 Leeder,2000; Cowie 等,2006）。观测到的数据证实,响应的时间尺度应为 1Ma 左右。例如,横切意大利中部亚平宁山脉伸展断块的河流汇水区域,目前仍然对一个发生在约 0.8Ma 前的滑动速率的改变有持续响应。在河流长轴剖面中,可通过峡谷的上凸识别出某些短暂的地质活动（Whittaker 等,2007a,2007b,2010）。这种瞬时性的地质活动,不仅可以通过改变滑动速率来改造地貌形态,同时也可以将粗粒沉积物幕式搬运到上盘地层中。因此,瞬时性的地质活动是将粗粒沉积物搬运到上盘地层沉积中心的动力。

一个汇水扇体系对外部力量的响应部分取决于边界条件,例如前陆冲积体系的长度,沿着宽阔的山前地带响应时间会明显增加,并使得山体汇水动力受扇体形成过程所控制（Babaule 等,2005; Carretier 和 Lucazeay,2005）。区别这种外部动力的响应与内部动力的响应（Humphrey 和 Heller，1995；Coulthard 等，2002）充满了挑战。

在由一系列长度为千米级的背斜所组成的褶皱冲断带内，同样可以估算构造扰动与沉积响应之间的滞后时间，例如伊朗与巴基斯坦西部的扎格罗斯山脉（Mann 和 Vita-Finzi，1988）。在扎格罗斯山脉，褶皱的高度为2~3km，波长为10km。四个褶皱的形成时间超过2Ma（Tucker 和 Slingerland，1996）。研究人员采用一个数值地貌演化模型模拟出了逐渐暴露的一系列背斜，这些背斜由不同抗侵蚀能力的岩性组成。正如研究人员所定义的，要使得沉积物通量与某一给定的构造隆升速率达到平衡（95%），确实存在一个响应时间，这一时间长短取决于岩石强度。岩石强度较差的岩性响应时间约为0.1Ma年，而强度较好的岩性响应时间约为4Ma年。这一结果表明对于抗侵蚀的岩性，物源区逆断层位移速率的快速改变（<<1Ma时间段内），难以在邻近的楔顶和前陆盆地地层中被检测出来。相反地，物源区强度较差的岩性，在同样的逆断层位移过程中会留下明显的地层记录，尤其是在近源的盆地背景下。

6.4　砾石沉积的滞后时间

基于砂岩和砾岩组分变化的物源研究可以提供物源区的一些信息，因而被作为构造活动的指示器（Jordan 等，1988）。在粗粒沉积体系中，随着腹地汇水区母岩的暴露，通常假设砾石会在瞬间遍布整个沉积体系。通过对现代河流中砾石的分布特征研究发现，至少有少量新搬运来的碎屑成分是被快速输送进来的，因为每次洪水事件中，碎屑物通常从一个点坝到另一个点坝向下游方向搬运（Schick 等，1987a，1987b）。根据加利福尼亚海湾一个大型边界正断层上砾石层的蚀顶现象研究表明，一种新的碎屑类型可以在物源区暴露约0.01Ma之内出现在山前约10km范围内的某个位置（Mortimer 和 Carrapa，2007）。脉冲式构造隆升的开始与物源区新的岩石类型的暴露之间可能同样存在一个滞后时间，这主要是由岩石单元的厚度和分布范围、造山带的结构样式与气候的风化和剥蚀效率决定的（Clift 等，2004；DeCelles，1988；DeCelles 等，1995；Paola 和 Swenson，1998）。另外，汇水区域的河流袭夺可以迅速地分流排水，随之发生组分的改变（Willgoose 等，1991；Tucker 和 Slingerland，1996；Clift 和 Blusztajn，2005），这与局部的构造运动无关。由于砂质颗粒在平均水平的河流流量内很容易被搬运，因此只要识别出砂质颗粒中典型的矿物学特征，那么根据砂岩组分，尤其是重矿物组合就能够甚至敏感地指示出源区组分的变化。

热年代学滞后时间的概念非常有利于解释造山带及其沉积盆地演化（Braun 等，2006）。此处采用的滞后时间是沉积物的地层年代与沉积物颗粒的热年代或冷年代的差值，它记录了样品在物源区暴露期间经历了一个特殊等温期的时间。虽然滞后时间包含了沉积物搬运时间和可能的间歇性保存时间，但通常假设这一时期非常短（Brandon 和 Vance，1992），因此滞后时间仍然提供了物源区暴露速度的信息。所以，滞后时间的系统性变化可以通过由气候或构造驱动的剥蚀速度的变化来解释（Ruiz 等，2004）。根据母岩指示剂、特殊矿物颗粒同位素定年以及生物地层对沉积物年代的约束等进行的砂岩成分研究表明，在某些情况下矿物颗粒可以通过构造隆升穿越低温区域（位于地下几千米的深度）到达地表，然后经过输送体系分散并快速掩埋于深海扇中；在这种极端的情况下，滞后时间可能为0.5Ma或更短（Copeland 和 Harrison，1990；Heller 等，1992）。

滞后时间的系统性变化发生在造山带及其前陆盆地体系的大规模、长期演化期（Jamieson 和 Beaumont，1989；Beaumont 等，1999；Willett 等，2001；Willett 和 Brandon，2002）。在造山运动早期的建造阶段，构造岩体涌入的速度超过了地表沉积物的剥蚀速度，导致地壳增厚、地表隆升，以及地势的起伏。由于山坡和基岩上的河流加快了剥蚀速率，山区进入稳定状态，或者说动力平衡状态，此状态下构造通量与剥蚀通量大致平衡（Willett 和 Brandon，2002）。在此稳定状态阶段，造山带停止生长。当涌入造山带的构造通量减少以及剥蚀通量超过构造通量之后，造山带则进入破坏阶段。这种造山带演化模式应该伴随着碎屑颗粒的滞后时间记录先下降再升高。这种模式见于西欧的阿尔卑斯山脉附近的前陆盆地地层中（Bernet 等，2004；Spiegel 等，2004），但是还需要更多的数据来证明这是否是造山运动的一个普遍特征。值得一提的是，大规模的山体地形演化、剥蚀模式以及沉积物通量可能同样被全球或与地形耦合的气候变化强烈影响，正如西欧阿尔卑斯山脉的例子中所表现的那样（Cederbom 等 2004；Willett 等，2006；Whipple，2009）。

6.5　构造活动山前带砾石的搬运

砾石从造山带被搬运到盆地通常被视作为同构造沉积作用的典型特征，因此可以指示构造活动的发生时间。粗碎屑的搬运与沉积物供给和盆地沉降的相互作用直接相关，并被沉积物输送体系的内部动力所缓冲。这些控制因素间的相互作用影响了沉积盆地中沉积物搬运的速率、时间和样式（图6-5）。

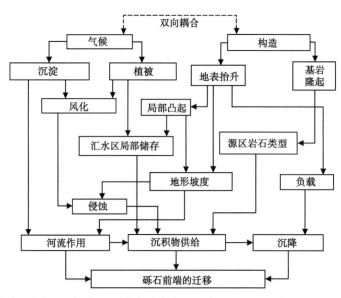

图6-5　关于构造和气候对砾岩搬运到冲积盆地中的影响之间的一些关联和反馈（箭头）的例子
虽然不全面，但是这些关联代表了直接驱动砾石前缘在盆地中移动的各种控制因素，
这些因素影响了沉积物供给、流体能量和盆地沉降

河流相砾石沉积的性质通常造成冲积盆地中的岩相在图上可分为两部分。盆地近源部分（最靠近山前带）的河流相砾石沉积物通常以粗砾岩为主。这些粗砾岩相对快速地向下游方向转变为以砂岩为主的河流相沉积，并含有少量分散的砾石。这两种岩相间的转换通常是突变的，不管是超过数千米的大型河流，还是只有几百米的小型河流（Sambrook-Smith 和 Ferguson，1995；Ferguson 等，1996）均是如此。这种迅速转换是砾石前缘的标志，其构成了一个移动的边界，并被用于勾绘盆地近源部分进积和加积作用的运动轨迹，类似于编制边缘海背景下的"滨线轨迹"（Helland-Hansen 和 Martinsen，1996；Helland-Hansen 和 Hampson，2009）（图 6-4）。

有限的露头研究表明，在沉积盆地中一个新组分的初次出现（可指示地质历史时期瞬时的构造运动）和与之伴随的砾石前缘的进积之间，存在一个明显的滞后时间。通过在图上标识砾石前缘随时间变化的位置（图 6-6），Jones 等（2004）在与西班牙东北部的加泰罗尼亚海岸山脉相邻的始新世—渐新世前陆盆地的砾石进积中发现了一个滞后时间。他们利用磁性地层学作为年代地层格架，在一种新的砾石组分初次出现时间（由逆冲运动造成蚀顶）与含有大量砾石沉积的砾石前缘的到达时间之间，确定了滞后时间。研究人员发现，滞后时间与距离同期逆断层前缘的远近密切相关。当逆冲带向盆地方向扩展时，前陆盆地中逆冲断层前缘距离指定点的距离在不断减小。因此，当剖面中位置最低的，以砂岩为主的部分沉积到达 Bot 镇时（图 6-6），逆冲断层前缘在 15~20km 处。由于逆冲断层向盆地逐步推进，逆冲相关的岩石隆起与砾石前缘到达之间的滞后时间，从逆冲断层前缘距离盆地 8km 时的大于 1Ma 年减小到距离盆地 5~6km 时的几十万年。出人意料的是，当逆冲断层前缘推进到 Bot 剖面 2~3km 以内的某点时，逆冲活动时间和砾石前缘到达时间之间的滞后时间再次变得很长，大于 1Ma。笔者将这一较长的滞后时间归因于沿着逆冲前缘带发生了阻挡而导致盆地强烈的三维分隔（Gupta，1997；Humphrey 和 Konrad，2000），这使补给河流偏离了 Bot 地区。因此，靠近构造带的三维效应可以对构造活动与沉积响应之间的滞后时间产生很大的影响。

图 6-6 照片和描出的线显示了西班牙 Bot（靠近甘德萨）的奥尔塔—德萨冲积体系底部砾岩前缘的轨迹（箭头所示）（据 Jones 等，2004）

图中的线标示了砾岩沉积序列的整体进积轨迹，它可以被细分为一系列向上变粗的旋回

来自构造活跃腹地的砾岩层是否表现为显著的砾石前缘进积还是加积阶段，取决于沉降和砾石供给量之间的相互关系（Heller 等，1988；Paola，1988；Flemings 和 Jordan，1989；图 6-7）。在物源区构造加速隆升阶段，砾石供给量（Q_s）十分充足，砾石前缘向盆地方向

进积，指示同构造沉积作用。反之，如果构造活动显示为盆地沉降速率增加，而沉积物供给量没有发生相应的变化，砾石前缘则会后退，并由于可容纳空间的增加而不断加积。显然，对于根据砾石进积或加积解释构造活动时间，依赖于沉积物供给或构造沉降是否起主控作用。Heller 和 Paola（1992）指出，确定主控因素的一个方法就是考虑砾石加积速率与进积速率之间的关系（图6-7）：两者之间为正相关关系，表明沉积物供给量发挥主要控制作用（即"同构造"），而负相关关系表明盆地沉降占主导。用术语来讲，如果沉积物供给量的增加是由岩石加速隆升和源区加速剥蚀造成的，那么砂砾岩进积符合"同构造"成因的经典定义。另一方面，如果砾石进积是由沉降速率的降低造成的，且沉降速率降低是由于构造活动强度降低引起的，那么砾石进积则是"反构造"成因的结果。大量采用这种方法或类似质量平衡方法的研究，在针对相邻造山带构造活动时间的问题上已经产生了违反直觉的解释（Sinclair 等，1991；Ballato 等，2008）。

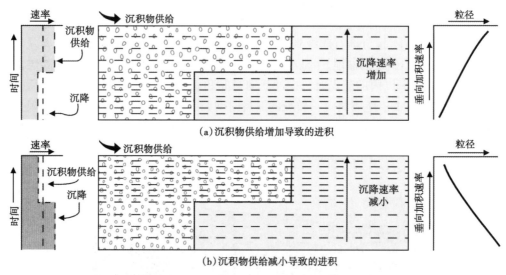

图 6-7　长周期的砾石进积（卵石模式）和沉积速率（虚线代表时间线）之间的关系（据 Heller 和 Paola，1992）
(a) 沉积物供给随着时间的推移而增加；(b) 沉降速率降低，但沉积物供给不变

6.6　选择性沉积作用形成的砾石向下游方向变细

远离构造活动高地的砾石搬运是指一定粒径范围内的碎屑颗粒释放到沉积物输送体系中。无论是现代以砂砾为主的河流，还是地层中保存的古代河流都表现为向下游方向颗粒粒径减小（Ferguson 等，1996；Robinson 和 Slingerland，1998；Duller 等，2010）。尽管磨蚀作用对于解释某些情况下底层沉积向下游方向变细很重要，但是通常认为向下游迅速变细的最好解释为，粗颗粒被表面流选择性筛选出来而搬运到地层的底部沉积（Parker，1991；Paola 等，1992；Hoey 和 Bluck，1999）（图6-3a）。因此，显然表面流的粒径分布与底层的粒径分布是不同的（图6-3b）。因为河流筛选不同粒径到不同岩相中的能力取决于粒径的分布，并且河流类型也取决于粒径的混合样式，选择性沉积作用不仅在决定砾岩前缘的位置中起了

重要的作用，而且在决定河流地层结构样式变化以及岩相分异中也扮演着重要的角色（Strong 等，2005）。

Fedele 和 Paola（2007）提出了一个砾岩和砂岩（各自）向下游变细的模型。Duller 等（2010）根据在西班牙北部 Poble 盆地始新世砾岩冲积层的现场试验扩展了这一定量模型（Mellere 和 Marzo，1992；Mellere，1993；Beamud 等，2003），现场试验应用了这一向下游变细的模型来研究这个小型楔顶盆地沉积物供给与构造沉降之间的平衡关系（图 6-8）。他们认为粒径向下游变小（图 6-8c）主要由以下因素控制：（a）沉积物供给的粒度概率密度函数，这一函数很难约束；（b）输送到盆地中的沉积物流量的量级；（c）生成可容纳空间的构造沉降的空间分布。值得注意的是，质量平衡方法中没有要求水力学和沉积物机制的细节。因此，如果不同地层单元的空间展布和体积可以通过盆地充填来估算，那么有可能通过沉积物供给速率（Q_s）和沉降速率（$\sigma_{(X)}$）的最优组合来解释观察到的粒径向下游变化的趋势。在 Poble 盆地充填中两个较低和较高的地层年代之间（二者的时间间隔约为 0.5Ma），

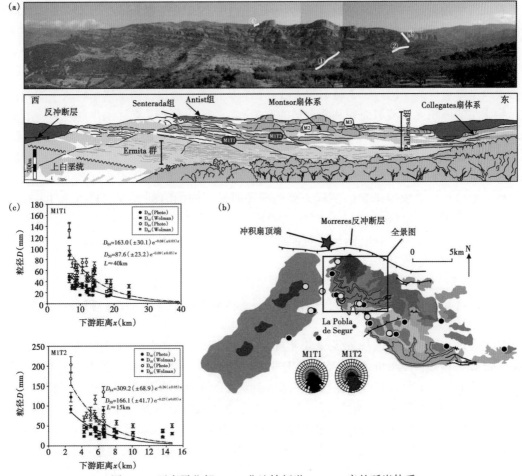

图 6-8　西班牙北部 Pobla 盆地始新世 Montsor 扇的砾岩体系

（a）盆地充填全景图，标出的线指示了地层术语，为了在 Montsor1 单元中作粒度分析而选取的上部和下部地层标记为 M1T1 和 M1T2（据 Beamud 等，2003；Mellere，1993）；（b）西班牙加泰罗尼亚莱里达省中部砾岩沉积平面图，显示了再造扇顶端的位置以及 Montsor1 单元两个地层的古水道流向玫瑰图（据 Duller 等，2010）；
（c）M1T1 和 M1T2 层中 D_{84} 和 D_{50} 的下游粒径趋势

Duller等（2010）发现随着输送到盆地的沉积物供给规模变大以及沉积体系长度的减小，砾石向下游方向变细的速率在升高。他们将这些变化归因于最大沉降速率的双倍增加和沉积物流量约20%的增加，这两个因素可能都与比利牛斯山脉造山带附近构造活动时间缩短有关。在这种情况下，如果发生选择性沉积作用，沉积物流量增加最初与沉积体系长度减小有关，表明在构造成因的可容纳空间中砾岩发生了近源堆积。

6.7 盆地响应时间

预测岩相迁移的简单几何模型是建立在沉积物供给与盆地可容纳空间之间的质量平衡的基础之上，它忽略了沉积物输送体系中的瞬时动力（Allen，2008b）；他们的结果假设了一个平衡或稳定状态的系统。然而，山坡—河谷体系进行自组织将沉积物从物源区穿过盆地进行疏散是需要时间的。只有当河流坡度和流量提供足够的流体强度时，河流才能搬运砾石。砾石可以沿着搬运路径充填阶地、扇体、泛滥平原以及河道点沙坝，并就地保存。同样地，将粗粒沉积物搬运到简单几何模型预测地点的过程中可能存在一个短的、但意义重大的滞后时间。当河流体系规模较小且效率低，同时盆地长度很大时，这些滞后时间会很长。

为了克服这些局限性，动态盆地模型建立了一个沉积物输送过程近似的模型，最初采用了扩散近似模型。只要沉积物搬运速率或流量与沉积物搬运体系的地形梯度相关，任何沉积物搬运模型都可以被简化为一个简单的扩散方程。从根本上来说，河流体系扩散率的控制因素是供水量的体积（与流域面积和降水量的有关）。它控制了河流流量，并直接影响了流体动力，最终控制了沉积物的搬运（Paola，2000）。如果大型盆地的主干河流的流量低且/或沉积物搬运效率低（例如辫状河），最终将花费更长的时间来达到一个代表沉积物供给与盆地沉降稳定响应的平衡剖面。

确定地层对构造成因砾岩的响应，存在两种明显不同的时间尺度（Kooi和Beaumont，1996；Davy和Crave，2000；Carretier等，2009）：一种时间尺度对应地形对高地的构造或气候驱动机制的响应，另一种时间尺度对应均衡盆地响应。盆地响应时间（T_{eq}）较为普及，由公式 $T_{eq}=L^2/k$ 得到（Paola等，1992），其中 L 为沉积体系的长度，k 为其扩散系数（km^2/a）。均衡盆地响应时间是指，用物理学术语来讲就是，如果一个盆地以恒定的速率演化，且沉降量和沉积物输入量的分配保持不变，地表形态（河床纵剖面）将会在 T_{eq} 时间内达到一个稳定状态。

如果盆地与汇水区域的规模及平均年降水量能够被估算的话，那么计算 T_{eq} 就很简单（Métivier等，1999；Castelltort和Van den Driessche，2003）。Paola等（1992）提出如果汇水区域和盆地的长度 L 具有相似的数量级，那么辫状河的平衡时间接近于 $10L/r$，而曲流河约为 $1.5L/r$，其中 r 为降水率。例如，一个发源于汇水区域的辫状河体系降水率为 $1m/a$，所属盆地的长度为 $10\sim100km$，那么它的均衡时间为 $0.1\sim1Ma$（Paola等，1992；Castelltort和Van Den Driessche，2003）。这意味着如果构造边界条件发生变化，例如造山带构造活动时间或断裂幕的缩短速率是周期性的，且发生过程经历的时间尺度比响应时间尺度短得多，该河流体系将会持续的瞬变且永远不会达到一个稳定状态的剖面。相应形成的砾石地层分布可能没有简单稳定的进积响应，这与大多数盆地充填的几何模型所预测的不一样（Paola等，1992）。

盆地响应时间（T_{eq}）概念的重要性在于它为评估盆地地层中观察到的砾岩分布的意义

提供了一个框架（Marr等，2000）。当一种支配力量变化时，例如沉积物供给量，在远小于T_{eq}的时间尺度内，以正弦曲线样式快速变化，河流体系输送砾石穿越整个盆地的动力和效率在砾岩分布中起了主导性的作用。另一方面，如果支配力量的变化发生在较长的时间尺度内，远长于T_{eq}，那么岩石记录中见到的砾岩分布将会与建立在沉积物供给量和盆地沉降量之间质量平衡基础上的简单几何模型所预测的结果十分吻合。因此，我们可以考虑到两种驱动力，当其控制变量（T）的变化频率远大于盆地均衡时间（即$T>>T_{eq}$）时为缓慢驱动力，当$T<<T_{eq}$时为快速驱动力（Marr等，2000）。对于一个长100km的沉积物输送体系来说，砾石扩散系数（K）为0.01km^2/a，砂岩扩散系数（K）为0.1km^2/a，输送砾石的T_{eq}为1Ma，而输送砂岩的T_{eq}为0.1Ma年。对于构造成因砾岩，缓慢驱动力为$T>>1Ma$，而快速驱动力为$T<<1Ma$。在这种情况下，时间尺度远小于1Ma的构造活动变化的地层记录与构造活动长期变化（>>1Ma）生成的地层记录区别很大。

在缓慢的驱动力条件下，砾岩前缘的运动与沉积物供给量的变化是相协调的。沉积物高流量时期远端堆积物和近端堆积物是相协调的（图6-7a）。当构造沉降发生变化时，快速沉降期发生近端沉积作用（图6-7b），引起了砾石前缘后退，这与Heller等（1988）两阶段地层模型的结果完全一致。反之，当沉降速率减小时，砾石前缘进积。对于沉积物流量恒定但砾石成分增加的情况下，砾石前缘向盆地方向移动，形成了一个向上变粗的地层特征。因此，在缓慢驱动力的情况下，地层充填特征直观明了。

在快速的驱动力条件下，沉积物供给量的快速变化导致了砾岩前缘不协调的移动。当供给量减少时，近源端坡度降低，且不整合面在近端被削截并伴随砾石前缘的加积。当供给量增加时，近源端坡度增加，并引起砾石前缘向物源区后退。沉降速率的快速变化对砾石前缘的位置影响很小。

很难解释是否任何在野外观察到的砾岩进积均反映了构造活动的增强或减弱，而沉积物供给量、沉降量、沉积物输送之间耦合的相互作用更增加了这种解释的难度。构造腹地的岩石加速隆升必然会导致剥蚀速率的加速，从而沉积物供给量也会相应增加。同时，如果岩石隆升速率超过剥蚀速率，那么会存在地表隆升。地形载荷的增加引起的挠曲作用导致了相关盆地沉降速率的增加（Allen和Allen，2005；第5章）。因此，当沉积物供给量增加，盆地沉降量也会增加来捕获释放的沉积物。砾石前缘是否进积、退积或保持不变，取决于岩石隆升、剥蚀速率以及造山带或盆地系统挠曲强度（硬度）（Burns等，1997）。另外，由于挠曲响应的幅度随着载荷距离的增加而迅速衰减，沉积物供给量与沉降量之间的相互作用与造山带的规模非线性相关。例如，横跨一个宽广的造山带的构造隆升（例如安第斯山脉和喜马拉雅山脉）会为前陆地区带来一个巨大的沉积物供给的增加，但是对于由挠曲沉降生成的可容纳空间起到相对较小的作用。这将会导致宽广的造山带对盆地造成过补偿，这与楔形动力和气候变化无关。

将这一想法延伸，砾石穿越整个盆地扩散的最大范围应该发生在造山带变形生长减弱最快时，但是剥蚀作用持续地将沉积物从构造高地搬运走（Heller等，1988；Burns等，1997）。造山带载荷的剥蚀搬运引起了盆地的挠曲回弹。盆地近源端隆升会导致剥蚀和再活化先前沉积的砾石。这些砾石将会以一个薄层舌状体的形式向盆地方向推进。在这种情况下，砾岩最大进积的时间标志着邻近造山带构造活动性最小的时间。相似的过程可以解释广泛分布的河流相砾岩的"后造山褶皱"，其沉积作用与比利牛斯造山带构造缩短结束时的相一致（Babaule等，2005）。

6.8 耦合响应

地形演化与盆地沉降模型为理解砾石搬运的控制因素和时间尺度提供了很有用的框架，更为稀有的是这些模型试图将这些过程进行耦合和整合。构造变形和剥蚀/沉积作用之间耦合的基础是，与地表地质过程相关的物质重新分配改变了岩石圈的应力状态，这可能引起内部变形和总体均衡调整（Beaumont 等，1999）。同样可能的是剥蚀和沉积作用导致下伏大陆岩石圈热状态的变化，尤其是放射性热成因元素生成的地方，这些元素高度集中在近地表地层中。这也很可能会影响大陆的应力状态与浮力以及构造的定位与反转（Sandiford 和 McLaran，2002）。

快速变形区域剥蚀时间的延长可以导致热的、力学性质薄弱的岩石向着地表方向形成水平对流，引起增强的构造定位和剥露作用，正如在喜马拉雅山南迦帕尔巴特峰与新西兰阿尔卑斯山南部等地区所提到的那样（Zeitler 等，2001；Koons 等，2003）。在大的空间尺度内，剥蚀和沉积作用已经被证实对于造山带的演化起非常重要的作用（Beaumont 等，1992；Avouac 和 Burov，1996；Willett，1999）。原因在于变形与地表地质过程之间的耦合在较小的长度规模内（<100km）是存在疑问的，例如岩石强度和抗弯刚度的影响。然而，在采用地壳弹性变形的数值模型中利用中等规模的长度进行计算，就可以表现出耦合效应，例如对逆冲相关的背斜和逆掩褶皱带的生长过程的模拟中（Simpson，2004a，2004b，2006）。举例来说，这些尺度的耦合作用可以解释为何大型横向河流可以横切双倾伏褶皱构造的轴向最高点，正如在扎格罗斯和亚平宁山脉等自然造山带中观察到的（Oberlander，1985；Alvarez，1999；Simpson，2004b）。由峡谷下切作用引起的剥蚀卸载通过一个浅的薄弱层中的补偿造成了生长背斜的上拱（例如地下盐丘）。

与构造位移场有关的耦合模型，如盆地沉降与高地汇水区沉积物的流出，提供了一种可以有效评估多种时空尺度地质过程的方法。例如，Densmore 等（2007a）展示了一个横跨张性断层的源—扇耦合体系的数值模型，该耦合体系受到滑动速率变化的扰动。他们采用搬运规律来处理断层下盘的剥蚀作用，该规律结合了分散过程（山坡）与集中过程（河道）的相关内容（Smith 和 Bretherton，1972；Simpson 和 Schlunegger，2003）。由于下盘剥蚀而释放出来的沉积物被搬运到扇体上，而扇端的位置则受控于来自汇水区的沉积物供给量、扇体坡度以及构造沉降样式。因此，根据体积平衡，扇体的搬运机制（碎屑流、层流）以及影响扇体几何形态的沉积物颗粒大小都可以被忽略（Paola 等，1992；Stock 等，2007）。

经历一个 5Ma 的起转时间之后，因为一个因子 2，滑动速率从 1mm/a 增加到 2mm/a，这使扇体坡度变得更陡，并使扇顶端的沉积物供给量增加，所有参数都在时间常量（$1-\exp(-t/\zeta)$）约为 0.6Ma 时达到一个新的平衡（图 6-9）。上盘沉降量的增加引起扇端的瞬时后退，但是当由于滑动速率的改变引起沉积物流量逐渐增长时，扇端进积并以一定角度上超到早期沉积体上，最终超出其前期扰动的位置。滑动速率瞬时变化之后的退积-进积事件的总时间为 2.5Ma。这一数值试验可以得到两个很有趣的结论：（1）构造边界条件的变化需要百万年数量级的时间才能完全传递到剥蚀机制上，而对于沉积响应则需要更长的时间，如果边界条件再次以比这个响应时间更快的频率发生改变，源—扇体系将会持续地处在一个瞬变状态，这种动力机制下的地层响应将很难被反演；（2）断层下盘隆升速率增加产生的滞后响应造成了扇体的进积，并因此造成沉积物流量增加，且不需要指示构造休眠。

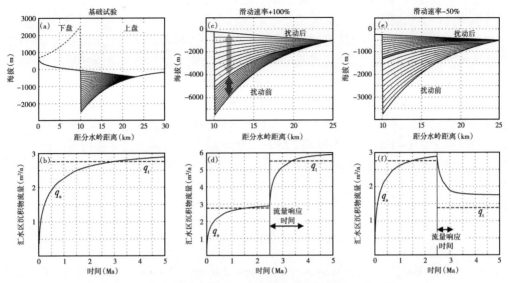

图6-9 通过对以垂直断层为边界的张性下盘作地形演化模拟，
总结构造位移、剥蚀和沉积间的相互作用（据Densmore等，2007a，修改）
注意滑动速率瞬时变化的滞后沉积响应（约1Ma）

在模型中观察到，快速加深和退积的地层样式之后紧接着的是变浅和进积的地层样式，这与意大利亚平宁山脉中部断层滑动速率增加引起的独立半地堑地层中观察到的现象一样（Whittaker等，2011）。同样地，在一个完整的裂谷盆地尺度上看也是如此，断裂的生长与连接引起了初始裂陷期，继之为主裂陷期和裂后期（Prosser，1993）。

根据Densmore等（2007a）的研究，当断层滑动速率从1mm/a减半到0.5mm/a，时间常量约为0.5Ma时，来自汇水区的沉积物供给量和扇体坡度都将衰减到一个更低的均衡值（图6-9e、f）。首先，由于可容纳空间生成速率降低，扇端快速进积，但随后由于沉积物供给量衰减，进积作用减弱而到达一个固定的位置。进积作用与一个低的扇斜坡共同生成了一个席状地层单元。

所有这些试验包含了滑动速率增大和减小导致的扇体进积，类似于扰动时间的滞后，但是沉积单元的几何形态不同。位于退积面之上的具有高沉积速率的楔状地层单元是构造压力增加引起的，而具有低沉积速率的席状地层单元是构造压力减少引起的。因此，砾石进积的几何形态、楔状及席状、连同下伏不整合面的分布，可以很好地指示近物源盆地构造活动时间（Paola等，1992）。然而，要注意在物理试验（Kim和Paola，2007）和数值模型中（Humphrey和Heller，1995）都观察到从高地汇水区域释放到扇体中的沉积物的自旋回周期性。因此，如何区别外部构造动力与内部自旋回是一个具有挑战性的问题。

6.9 结论

在大陆背景下，构造活动对于沉积物输送体系的运行起至关重要的作用。构造通量引起的源区岩石隆升是沉积物输送体系的剥蚀动力。构造位移与构造载荷同样会生成可容纳空间，使沉积物在地质过程中堆积充填盆地。但是对于根据保存的地层样式所解译的动力机制，却由于剥蚀作用与沉积地形对构造扰动响应的复杂性而存在疑问。另外，区分构造驱动还是气候和内部非强制动力驱动无关紧要。

　　构造扰动的响应与时间尺度有关。这些时间尺度一方面涉及高地汇水区域对空间样式或构造活动速率改变的响应，例如边界断层滑动速率的改变。数值方法联合地形演化模型揭示典型的响应时间尺度为百万年的级别。另一方面，时间尺度还与沉积物释放信息的转换有关，这种转换主要由河流相盆地中沉积物间歇性的搬运与保存引起的。因此，一个物源区信息通过沉积物输送体系时会被过滤和缓冲。构造事件的发生与稳定状态的沉积物响应之间的滞后时间尺度为 0.1~1Ma 年。不能确定的是，如果在小于上述时间尺度的时间内，地层记录是否还能被简单地转换为高频的构造变化。

致谢（略）

参 考 文 献

Allen, P. A. (2008a) From landscapes into geological history. Nature, 451 (7176), 274–276.

Allen, P. A. (2008b) Time scales of tectonic landscapes and their sediment routing systems, in Gallagher, K., Jones, S. J., and Wainwright, J., eds, Landscape evolution: denudation, climate and tectonics over different time and space scales. Special Publication, 296. London, Geological Society, 7–28.

Allen, P. A., and Allen, J. R. (2005) Basin analysis: principles and applications, 2nd ed. Oxford, Blackwell Science. Allen, P. A., and Densmore, A. L. (2000) Sediment flux from an uplifting fault block. Basin Research, 12, 367–380.

Allen, P. A., and Hovius, N. (1998) Sediment supply from landslide-dominated catchments: Implications for basin-margin fans. Basin Research, 10, 19–35. Alvarez, W. (1999) Drainage on evolving fold-thrust belts: A study of transverse canyons in the Apennines. Basin Research, 11, 267–284.

Avouac, J. P., and Burov, E. B. (1996) Erosion as a driving mechanism for intracontinental mountain growth. Journal Geophysical Research, 101, 17747–17769.

Babault, J., Van Den Driessche, J., Bonnet, S., Castelltort, S., and Crave, A. (2005) Origin of the highly elevated Pyrenean peneplain. Tectonics, 24, TC2010. doi: 10.1029/2004TC001697.

Ballato, P., Nowaczyk, N. R., Landgraf, A., Strecker, M. R., Friedrich, A. M., and Tabatabaei, S. H. (2008) Tectonic control on sedimentary facies pattern and sediment accumulation rates in the Miocene foreland basin of the southern Alborz mountains, northern Iran. Tectonics, 27 (TC6001) 1–20.

Beamud, E., Garces, M., Cabrera, L., Munoz, J. A., and Almar, Y. (2003) A new late Eocene continental chronostratigraphy from NE Spain. Earth and Planetary Science Letters, 216, 501–514.

Beaumont, C., Fullsack, P., and Hamilton, J. (1992) Erosional control on active compressional orogens, in McClay, K. R., ed., Thrust tectonics. NewYork, Chapman and Hall, 1–31.

Beaumont, C., Kooi, H., and Willett, S. D. (1999) Coupled tectonic-surface process models with applications to rifted margins and collisional orogens, in Summerfield, M. A., ed., Geomorphology and global tectonics. Chichester, John Wiley, 29–55.

Bernet, M., Brandon, M. Y., Garver, J. I., and Molitor, B. (2004) Fundamentals of detrital zircon fission track analysis for provenance and exhumation studies with examples from the European Alps, in Bernet, M., and Spiegel, C., eds., Detrital thermochronology: provenance analysis, exhumation and landscape evolution of mountain belts. Boulder, CO, Geological Society of America, 25–36.

Brandon, M. T., and Vance, J. A. (1992) New statistical methods for analysis of fission track grain–age distributions with applications to detrital grain ages from the Olympic subduction complex, western Washington State. American Journal of Science, 292, 565–636.

Braun, J., Van Der Beek, P., and Batt, G. (2006) Quantitative thermochronology: numerical methods for the interpretation of thermochronoloigcal data, in Bernet, M., and Spiegel, C., eds., Detrital thermochronology: provenance analysis, exhumation and landscape evolution of mountain belts. Boulder, CO, Geological Society of America, 131–150.

Burns, B. A., Heller, P. L., Marzo, M., and Paola, C. (1997) Fluvial response in a sequence stratigraphic framework: example from the Montserrat Fan Delta, Spain. Journal Sedimentary Research, 67, 311–321.

Carretier, S., and Lucazeau, F. (2005) How does alluvial sedimentation at range fronts modify the erosional dynamics of mountain catchments? Basin Research, 17, 361–381.

Carretier, S., Poisson, B., Vassallo, R., Pepin, E., and Farias, M. (2009) Tectonic interpretation of erosion rates at different spatial scales in an uplifting block. Journal Geophysical Research–Earth Surface, 114, F02003.

Castelltort, S., and Van Den Driessche, J. (2003) How plausible are high–frequency sediment supply driven–cycles in the stratigraphic record? Sedimentary Geology, 157, 3–13.

Cederbom, C. E., Sinclair H. D., Schlunegger, F., and Rahn, M. K. (2004) Climate-induced rebound and exhumation of the European Alps. Geology, 32, 709–712.

Clevis, Q, De Boer, P. L., and Nijman, W. (2004) Differentiating the effect of episodic tectonism and eustatic sealevel fluctuations in foreland basins filled by alluvial fans and axial deltaic systems: insights from a threedimensional stratigraphic forward model. Sedimentology, 51, 809–835.

Clift, P. D., and Blusztajn, J. (2005) Reorganization of the western Himalayan river system after five million years ago. Nature, 438, 1001–1003.

Clift, P. D., Layne, G. D., and Blusztajn, J. (2004) The erosional record of Tibetan uplift in the East Asian marginal seas, in Clift, P. D., Wang, P., Hayes, D., and Kuhnt, W., eds., Continent–ocean interactions in the East Asian Marginal Seas. Washington, DC, American Geophysical Union, 255–282.

Copeland, P., and Harrison, T. M. (1990) Episodic rapid uplift in the Himalaya revealed by $^{40}Ar = ^{39}Ar$ analysis of detrital K–feldspar and muscovite, Bengal fan. Geology, 18, 354–357.

Coulthard, T. J., Macklin, M. G., and Kirkby, M. J. (2002) Simulating upland river catchment and alluvial fan evolution. Earth Surface Processes and Landforms, 27, 269–288.

Cowie, P. A., Attal, M., Tucker, G. E., Whittaker, A. C., Naylor, M., Ganas, A., and Roberts, G. P. (2006) Investigating the surface process response to fault interaction and linkage using a numerical modelling approach. Basin Research, 18, 231-266.

Davy, P., and Crave, A. (2000) Upscaling local-scale transport processes in large-scale relief dynamics. Physics Chemistry Earth 25. Decelles, P. G. (1988) Lithologic provenance modeling applied to the Late Cretaceous synorogenic Echo Canyon Conglomerate, Utah: A case of multiple source areas. Geology, 16, 1039-1043.

DeCelles, P. G., Lawton, T. F., and Mitra, G. (1995) Thrust timing, growth of structural culminations, and synorogenic sedimentation in the type Sevier orogenic belt, western United States. Geology, 23, 699-702.

Densmore, A. L., Allen, P. A., and Simpson, G. (2007a) Development and response of a coupled catchment - fan system under changing tectonic and climatic forcing. Journal of Geophysical Research Earth Surface, 112, F01002. doi: 10.1029/2006/JF000474.

Densmore, A. L., Gupta, S., Allen, P. A., and Dawers, N. H. (2007b) Transient landscapes at fault tips. Journal Geophysical Research - Earth Surface, 112, F03S08. doi: 10.1029/2006JF000560.

Duller, R. A., Whittaker, A. C., Fedele, J. J., Whitchurch, A. L., Springett, J., Smithells, R., Fordyce, S., and Allen, P. A. (2010) From grain size to tectonics. Journal Geophysical Research-Earth Surface, 115, F03022. doi: 10.1029/ 2009JF000409.

Fedele, J. J., and Paola, C. (2007) Similarity solutions for fluvial sediment fining by selective deposition. Journal Geophysical Research-Earth Surface, 112, F02038.

Ferguson, R., Hoey, T., Wathen, S., and Werritty, A. (1996) Field evidence for rapid downstream fining of river gravels through selective transport. Geology, 24, 179-182.

Flemings, P. B., and Jordan, T. E. (1989) A synthetic stratigraphic model of foreland basin development. Journal of Geophysical Research, 94, B4, 3851-3866.

Galy, V., France-Lanord, C., Beyssac, O., Faure, P., Kudrass, H., and Palhol, F. (2007) Efficient organic carbon burial in the Bengal fan sustained by the Himalayan erosional system. Nature, 450. doi: 10.1038/nature06273.

Gawthorpe, R. L., and Hurst, J. M. (1993) Transfer zones in extensional basins - their structural style and influence on drainage development and stratigraphy. Journal Geological Society, 150, 1137-1152.

Gawthorpe, R. L., and Leeder, M. R. (2000) Tectono-sedimentary evolution of active extensional basins. Basin Research, 12, 195-218.

Grecula, M., Flint, S., Potts, G., Wickens, H. D., and Johnson, S. (2003) Partial ponding of turbidite systems in a basin with subtle growth-fold topography: Laingsburg-Karoo, South Africa. Journal of Sedimentary Research, 73, 603-620.

Gupta, S. (1997) Himalayan drainage patterns and the origin of fluvial megafans in the Ganges foreland basins. Geology, 25, 11-14.

Gupta, S., Underhill. J. R., Sharp, I. R., and Gawthorpe, R. L. (1999) Role of fault interaction in controlling synrift dispersal patterns: Miocene, Abu Alaqa Group, Suez Rift, Sinai, Egypt.

Basin Research, 11, 167–189.

Hardy, S., and Gawthorpe, R. (2002) Normal fault control on bedrock channel incision and sediment supply: insights from numerical modelling. Journal Geophysical Research, 107, 2246. doi: 10. 1029/2001JB000166.

Helland-Hansen, W., and Hampson, G. J. (2009) Trajectory analysis: concepts and applications. Basin Research, 21, 454–483.

Helland – Hansen, W., and Martinsen, O. J. (1996) Shoreline trajectories and sequences: description of variable depositional– dip scenarios. Journal of Sedimentary Research, B66 670–688.

Heller, P. L., Angevine, C. L., Winslow, N. S., and Paola, C. (1988) Two-phase stratigraphic model of foreland-basin sequences. Geology, 16, 501–504.

Heller, P. L., and Paola, C. (1992) The large-scale dynamics of grain-size variation in alluvial basins, 2: Application to syntectonic conglomerate. Basin Research, 4, 91–102.

Heller, P. L., Renne, P., and O'Neil, J. (1992) River mixing rate, residence time and subsidence rates from isotopic indicators: Eocene sandstones of the U. S. Pacific Northwest. Geology, 20, 1095–1098.

Hilton, R. G., Galy, A., and Hovius, N. (2008) Riverine particulate organic carbon from an active mountain belt: Importance of landsliding. Global Biogeochemical Cycles, 22, DOI: 10. 1029/2006GB002905.

Hodgson, D. M., and Haughton, P. D. W. (2004) Impact of syndepositional faulting on gravity current behaviour and deep-water stratigraphy: Tabernas-Sorbas Basin, SE Spain, in Lomas, S. A., and Joseph, P., eds., Confined turbidite systems. Special publication, 222. London, Geological Society, 222, 135–158.

Hoey, T. B., and Bluck, B. J. (1999) Identifying the controls over down-system fining of river gravels. J. Sedimentary Research, 69, 40–50.

Humphrey, N., and Heller, P. L. (1995) Natural oscillations in coupled geomorphic systems: Analternative origin for cyclic sedimentation. Geology 23, 499–502.

Humphrey, N. F., and Konrad, S. K. (2000) River incision or diversion in response to bedrock uplift. Geology, 28, 43–46.

Jamieson, R. A., and Beaumont, C. (1989) Deformation and metamorphism in convergent orogens: a model for uplift and exhumation of metamorphic terrains, in Daly, J. S., Cliff, R. A., and Yardley, B. W. D., eds., Evolution of metamorphic belts. London, Geological Society, 117–129.

Johnson, D. D., and Beaumont, C. (1995) Preliminary results from a planform kinematic model of orogen evolution, surface processes and the development of clastic foreland basin stratigraphy, in Dorobek, S. L., and Ross, G. M., eds., Stratigraphic evolution of foreland basins. Special Publication, 52. Tulsa, OK, Society of Economic Paleontologists and Mineralogists, 3–24.

Jones, M. A., Heller, P. L., Roca, E., Garces, M., and Cabrera, L. (2004) Time lag of syntectonic sedimentation across an alluvial bas in: theory and example from the Ebro Basin, Spain. Basin Research, 16, 489–506.

Jordan, T. E., Flemings, P. B., and Beer, J. A. (1988) Dating thrust-fault activity by use of foreland-basin strata, in Kleinspehn, K. L., and Paola, C., eds., New perspectives in basin analysis. New York, Springer-Verlag, 307-330.

Kim, W., and Paola, C. (2007) Long-period cyclic sedimentation with constant tectonic forcing in an experimental relay ramp. Geology, 35, 331-334.

Kooi, H., and Beaumont, C. (1996) Large-scale geomorphology: classical concepts reconciled and integrated with contemporary ideas via a surface processes model. Journal Geophysical Research, 101, 3361-3386.

Koons, P. O., Norris, R. J., Craw, D., and Cooper, A. F. (2003) Influence of exhumation on the structural evolution of transpressional plate boundaries: an example from the Southern Alps, New Zealand. Geology, 31, 3-6.

Lamb, M. P., Toniolo, H., and Parker, G. (2006) Trapping of sustained turbidity currents by intraslope minibasins. Sedimentology, 53, 147-160.

Leeder, M. R. (1999) Sedimentology and sedimentary basins: from turbulence to tectonics. London, Blackwell Science. Leeder, M. R., and Jackson, J. (1993) Interaction between normal faulting and drainage in active extensional basins, with examples from western United States and mainland Greece. Basin Research, 5, 79-102.

Malmon, D. V., Dunne, T., and Reneau, S. L. (2003) Stochastic theory of particle trajectories through alluvial valley floors. Journal Geology, 111, 525-542.

Mann, C. D., and Vita-Finzi, C. (1988) Holocene serial folding in the Zagros, in Audley-Charles, M., and Hallam, A., eds., Gondwana and Tethys. Special Publication, 37. London, Geological Society, 51-59.

Marr, J. G., Swenson, J. B., Paola, C., And Voller, V. R. (2000) A two-diffusion model of fluvial stratigraphy in closed depositional basins. Basin Research, 12, 381-398.

Mellere, D. (1993) Thrust-generated, back-fill stacking of alluvial fan sequences, south-central Pyrenees, Spain (La Pobla de Segur Conglomerates), in Frostick, L. E., and Steel, R. J., eds., Tectonic controls and signatures in sedimentary successions. Special Publication, 20. Ghent, Belgium, International Association of Sedimentologists, 259-276.

Mellere, D., and Marzo, M. (1992) Los depositon aluviales sintectonicos de la Pobla de Segur: alogruposy su significado tectonoextratigrafico. Acta Geologica Hispanica, 27, 145-159.

Metivier, F., and Gaudemer, Y. (1999) Stability of output fluxes of large rivers in South and East Asia during the last 2 million years: implications for floodplain processes. Basin Research, 11, 293-304.

Metivier, F., Gaudemer, Y., Tapponier, P., and Klein, M. (1999) Mass accumulation rates in Asiaduring the Cenozoic. Geophysical Journal International, 137, 280-318.

Meybeck, M., Laroche, L., Darr, H. H., and Syvitski, J. P. M. (2003) Global variability of total suspended solids and their fluxes in rivers. Global Planetary Change 39, 65-93.

Meybeck, M., and Vorosmarty, C. (2005) Fluvial filtering of land-to-ocean fluxes: from natural Holocene variations to Anthropocene. Comptes Rendus Geoscience, 337, 107-123.

Moore, G. T. (1969) Interactions of rivers and oceans: Pleistocene petroleum potential. American

Association Petroleum Geologists Bulletin, 53, 3421-2430.

Mortimer, E., and Carrapa, B. (2007) Footwall drainage evolution and scarp retreat in response to increasing fault displacement: Loreto fault, Baja California Sur, Mexico. Geology, 35, 651-654.

Oberlander, T. M. (1985) Origin of drainage transverse to structures in orogens, in Morisawa, M., and Hack, J. T., eds., Tectonic geomorphology. Binghampton Symposia in Geomorphology, International Series, 15. London, Allen and Unwin, 155-182.

Paola, C. (1988) Subsidence and gravel transport in alluvial basins, in Kleinspehn, K. L., and Paola, C., eds., New perspectives in basin analysis. New York, Springer Verlag, 231-243.

Paola, C. (1990) A simple basin-filling model for coarsegrained alluvial systems, in Cross, T. A. ed., Quantitative dynamic stratigraphy. Englewood Cliffs, NJ, Prentice-Hall, 363-374, 625p.

Paola, C. (2000) Quantitative models of sedimentary basin filling. Sedimentology, 47, 121-178.

Paola, C., Heller, P. L., and Angevine, C. L. (1992) The largescale dynamics of grain-size variation in alluvial basins, 1: theory. Basin Research, 4, 73-90.

Paola, C., Straub, K., Mohrig, D., andReinhardt, L. (2009) The "unreasonable effectiveness" of stratigraphic and geomorphic experiments. Earth Science Reviews, 97, 1-43.

Paola, C., and Swenson, J. B. (1998) Geometric constraints on composition of sediment derived from erosional landscapes. Basin Research, 10, 37-47.

Parker, G. (1991) Selective sorting and abrasion of river gravel. I: Theory. Journal Hydraulic Engineering, 117, 131-149.

Prosser, S. (1993) Rift-related linked depositional ssystems and their seismic expression, in Williams, G. D., and Dodds, A., eds., Tectonics and seismic sequence stratigraphy. Special Publication, 71. London, Geological Society, 35-66.

Roberts, S., and Jackson, J. (1990) Active normal faulting in mainland Greece: an overview, in Roberts, A. M., Yelding, G., and Freeman, B., eds., The geometry of normal faults. Special Publication, 56. London, Geological Society, 125-142.

Robinson, R. A. J., and Slingerland, R. L. (1998) Origin of fluvial grain size trends in a foreland basin: The Pocono Formation of the central Appalachian Basin. Journal Sedimentary Research, 68, 473-486.

Ruiz, G. M. H., Seward, D., and Winkler, W. (2004) Detrital thermochronology - a new perspective on hinterland tectonics, an example from the Andean Amazon Basin, Ecuador. Basin Research, 16, 413-430.

Sambrook-Smith, G. H., and Ferguson, R. I. (1995) The gravel-sand transition along river channels. Journal Sedimentary Research, 65, 423-430.

Sandiford. M., and McLaren, S. (2002) Tectonic feedback and the ordering of heat producing elements within the continental lithosphere. Earth and Planetary Science Letters, 204, 133-150.

Schick, A. P., Lekach, J., and Hassan, M. A. (1987a) Bed load transport in desert floods: observations in the Negev, in Thorne, C. R., Bathurst, J. C., and Hey, R. D., eds., Sediment transport in gravel-bed rivers. London, John Wiley and Sons Ltd., 617-642.

Schick, A. P., Lekach, J., and Hassan, M. A. (1987b) Vertical exchange of coarse bedload in desert streams. Special Publication, 35. London, Geological Society, 7-16.

Simpson, G. D. H. (2004a) Dynamic interactions between erosion, deposition, and three-dimensional deformation in compressional fold belt settings. Journal Geophysical Research, 109, F03007.

Simpson, G. D. H. (2004b) Role of river incision in enhancing deformation. Geology, 32, 341-344.

Simpson, G. D. H. (2006) Modelling interactions between fold-thrust belt deformation, foreland flexure and surface mass transport. Basin Research, 18, 1-19.

Simpson, G. D. H., and Schlunegger, F. (2003) Topographic evolution and morphology of surfaces evolving in response to coupled fluvial and hillslope sediment transport. Journal Geophysical Research, 108, B6, 2300. doi: 10. 1029/2002JB002162.

Sinclair, H. D. (1994) The influence of lateral basinal slopes on turbidite sedimentation in the Annot Sandstones of SE France. Journal of Sedimentary Research, 64, 42-54.

Sinclair, H. D., Coakley, B. J., Allen, P. A., and Watts, A. B. (1991) Simulation of foreland basin stratigraphy using a diffusion model of mountain belt uplift and erosion: An example form the Central Alps, Switzerland. Tectonics, 10, 599-620.

Smith, T. R., and Bretherton, F. P. (1972) Stability and the conservation of mass in drainage basin evolution. Water Resources Research, 8, 1506-1529.

Snyder, N. P., Whipple, K. X., Tucker, G. E., and Merritts, D. J. (2000) Landscape response to tectonic forcing: Digital elevation model analysis of stream profiles in the Mendocino triple junction region, northern California. Bulletin Geological Society America, 112, 1250-1263.

Somme, T. O., Helland-Hansen, W., Martinsen, O. J., and Thurmond, J. B. (2009) Relationships between morphological and sedimentological parameters in source-tosink systems: a basis for predicting semi-quantitative characteristics in subsurface systems. Basin Research, 21 (4), 361-387.

Spiegel, C., Siebel, W., Kuhlemann, J., and Frisch, W. (2004) Towards a comprehensive provenance analysis: a multimethod approach and its implications for the evolution of the central Alps, in Bernet, M., and Spiegel, C., eds., Detrital thermochronology: provenance analysis, exhumation and landscape evolution of mountain belts. Boulder, CO, Geological Society of America, 37-50.

Stock, J. D., Schmidt, K. M., and Miller, D. M. (2007) Controls on alluvial fan long-profiles. Geological Society America Bulletin, 120, 619-640. doi: 10. 1130/B26208. 1.

Strong, N., Sheets, B. A., Hickson, T. A., and Paola, C. (2005) A mass balance framework for quantifying down-system changes in fluvial architecture, in Blum, M., et al. eds., Fluvial aedimentology Ⅶ. Special Publication, 35. Ghent, Belgium, International Association of Sedimentologists, 243-253.

Swenson, J. B. (2005) Fluviodeltaic response to sea level: Amplitude and timing of shoreline translation and coastal onlap. Journal of Geophysical Research - Earth Surface, 110, F03007.

Syvitski, J. P. M., and Milliman, J. D. (2007) Geology, geography, and humans battle for dominance over the delivery of fluvial sediment to the coastal ocean. Journal of Geology 115, 1–19.

Tinker, J, De Wit, M., and Brown, R. (2008) Linking source and sink: Evaluating the balance between onshore erosion and offshore sediment accumulation since Gondwana break-up, South Africa. Tectonophysics, 455, 94–103.

Toniolo, H., Lamb, M. P., and Parker, G. (2006) Depositional turbidity currents in diapiric minibasins on the continental slope: formulation and theory. Journal of Sedimentary Research, 76, 783–797.

Toro-Escobar, C., Parker, G., and Paola, C. (1996) Transfer function for deposition of poorly sorted gravel in response to streambed aggradation. Journal Hydraulic Research, 34, 35–53.

Tucker, G. E., and Slingerland, R. (1996) Predicting sediment flux from fold and thrust belts. Basin Research, 8, 329–349.

Tucker, G. E., and Whipple, K. X. (2002) Topographic outcomes predicted by stream erosion models: sensitivity analysis and intermodel comparison. Journal GeophysicalResearch, 107 (B9), 2309. doi: 10. 1029/2002JB002125.

Violet, J., Sheets, B., Pratson, L., Paola, C., and Parker, G. (2005) Experiment on turbidity currents and their deposits in a model 3D subsiding minibasin. Journal of Sedimentary Research, v. 75, p. 820–843.

Whipple, K. X. (2009) The influence of climate on the tectonic evolution of mountain belts. Nature Geoscience, 2, 97–104.

Whipple, K. X., and Tucker, G. E. (1999) Dynamics of the stream power incision model: Implications for height limits of mountain ranges, landscape response time scales, and research needs. Journal Geophysical Research, 104, 17661–17674.

Whipple, K. X., and Tucker, G. E. (2002) Implications of sediment flux – dependent river incision models for landscape evolution. Journal Geophysical Research, 107 (B2), 2039. doi: 10. 1029/2000JB000044.

Whipple, K. X., and Trayler, C. R. (1996) Tectonic control on fan size: the importance of spatially variable subsidence rates. Basin Research 8, 351–366.

Whittaker, A. C., Cowie, P. A., Attal, M., Tucker, G. E., and Roberts, G. (2007a) Bedrock channel adjustment to tectonic forcing: Implications for predicting river incision rates. Geology, 35, 103–106.

Whittaker, A. C., Cowie, P. A., Attal, M., Tucker, G. E., and Roberts, G. (2007b) Characterizing the transient response of rivers crossing active normal faults: New field observations from Italy. Basin Research, 19, 529–556. doi: 10. 1111/j. 1365–2117. 2007. 00337.

Whittaker, A. C., Attal, M., and Allen, P. A. (2010) Characterising the origin, nature and fate of sediment exported from catchments perturbed by active tectonics. Basin Research, 22, 809–828. doi: 10. 1111/j. 1365– 2117. 2009. 00447. x.

Whittaker, A. C., Duller, R. A., Springett, J., Smithells, R., Whitchurch, A. L., and Allen, P. A. (2011) Decoding downstream trends in stratigraphic grain size as a function of tectonic subsidence and sediment supply. Bulletin Geological Society America. doi: 10. 1130/ B30351. 1.

Willgoose, G., Bras, R. L., and Iturbe, I. (1991) A coupled channel network growth and hillslope evolution model. Water Resources Research, 27, 1671–1684.

Willett, S. D. (1999) Orogeny and orography: the effects of erosion on the structure of mountain belts. Journal of Geophysical Research, 104, 28957–28981.

Willett, S. D., Slingerland, R. J., and Hovius, N. (2001) Uplift, shortening and steady-state topography in active mountain belts. American Journal of Science, 301, 455–480.

Willett, S. D., and Brandon, M. T. (2002) On steady states in mountain belts. Geology, 30, 175–178.

Willett, S. D., Schlunegger., F., and Picotti, V. (2006) Messinian climate change and erosional destruction of the central European Alps. Geology, 34, 613–616.

Zeitler, P. K., and others (2001) Crustal reworking at Nanga Parbat, Pakistan: metamorphic consequences of thermalmechanical coupling facilitated by erosion. Tectonics, 20, 712–728.

(陈莹 郭佳 译，陈莹 校)

第7章 构造地层模型中源—汇体系内沉积物的体积分配研究：方法与结论
——以拉腊米型大陆架—深水盆地为例

Cristian Carvajal, Ron Steel

（Jackson School of Geosciences, University of Texas, Austin, USA）

摘　要：本文以拉腊米型沃沙基—大分水岭盆地为例，阐述如何将沉积物体积分配原理融入到盆地分析中，以便更好地理解盆地的演化史，解释层序演化的驱动机制，恢复沉积物从源到汇的分配过程。从 Lewis—Fox Hills（马斯特里赫特阶）到拉腊米盆地的源—汇体系经历了两个阶段的演化。阶段一：持续的逆冲推覆抬升和地壳载荷沉降形成了陆架边缘，陆架边缘之上堆积了厚层的斜坡地层和海相顶积层，形成了不断向盆地深水区进积的陆架边缘楔。阶段二：随着斜坡地层和海岸平原顶积层体积的减少，陆架边缘的轨迹不断向前推进，且随斜坡高度的降低而逐渐趋于稳定。上述结果是由较低的可容空间和不断增加的沉积物供给速率造成的，同时陆架边缘楔向盆地远端的低沉降量区不断迁移也起到一定作用；此时逆冲抬升作用据信已经降低或者停止。从阶段一到阶段二沉积物供给量不断增加，平均速率为 $(4\sim16)\times10^6 t/a$。阶段一期间，沉积物供给量可高达 $200\sim2000 t/km^2 \cdot a$。盆地的最大地形起伏在洼地处介于 $100\sim500m$ 之间，在高地介于 $500\sim1000m$ 之间，阶段一的某个时期，山区地形高差可达 $1000\sim3000m$，这种情况一直持续到阶段二。拉腊米源—汇体系类似于现今东亚的一些河流体系，这些河流将巨量的沉积物搬运至洋盆。这一研究实例表明，将沉积物体积分配原理与动态地层分析相结合有利于改善构造地层模型，表征古源—汇体系。

关键词：从源到汇沉积物体积　拉腊米（Laramide）造山运动　沃沙基（Washakie）盆地　Lewis 陆架边缘

7.1　引言

沉积相（砂泥岩型）以及沉积物体积分配原理（Cross 和 Lessenger，1998；Siggerud 和 Steel，1999）是源—汇体系理论的核心组成之一，同时为其将来的研究奠定了坚实的理论基础。本文针对盆地内沉积物体积概算，沉积物在从源到汇运移通道上在不同区带间如何分布和储存，以及体砂岩与泥岩相如何分配等方面进行了大量估算。同时，计算中有必要考虑构造运动、海平面变化以及气候变化等因素的影响（Posamentier 和 Vail，1988；Hovius，1998；Syvitski 和 Milliman，2007；Steel 等，2008）。构造会对汇水盆地、地形、抬升速率以及沉降造成影响（Jordan，1981；Winker，1982；Heller 等，1988；DeCelles 和 Giles，1996；Rowan 等，2004），同样会对沉积物的形成和保存产生影响。气候对降水量和温度起主要的控制作用，进而会影响剥蚀量、沉积物生成量以及河流流域。相对海平面变化和沉积物通量本身会对大陆架、斜坡、盆地底部（Burgess 和 Hovius，1998；Porebski 和 Steel，2006；Carvajal 等，2009；Carvajal 和 Steel，2009）的沉积物储存（Curtis，1970；Schlager，1993）产生主要的控制作用。上述所有变量的相互作用产生了沉积相和体积分配。对其进行定量估算，结合盆地分析，将有助于建立一个改良的具有从源—汇体系概念的构造—地层模型。

本章节将阐明一个用来计算沉积相和体积分配的方法，并将该方法与构造—地层模型和从源到汇的特征描述相结合。本研究基于沃沙基—大分水岭盆地中Lewis—Fox Hills的陆架边缘楔的分析。沃沙基—大分水岭盆地位于美国怀俄明州南部，是一个拉腊米型盆地（Winn等，1987；McMillen和Winn，1991；Pyles和Slatt，2002，2007；Carvajal和Steel，2006）（图7-1）。

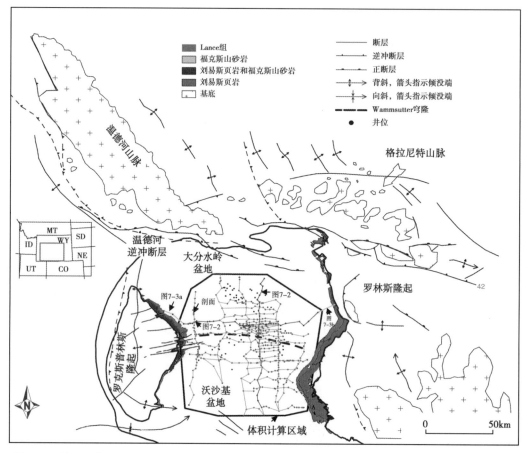

图7-1 该地理位置图显示了沃沙基—大分水岭盆地中Fox Hills的Lance组（薄层）以及Lewis组地层露头出露位置、周围的隆起，以及接近 *Baculites grandis* 时期的Lewis泥岩最大海泛面的范围（深蓝色线）（编图Love和Christiansen，1985；Blackstone，1991；Hettinger和Roberts，2005）
图上同时显示了数据库井位（点）、对比剖面（灰线），与菊石带相对比的对比剖面（浅蓝色线；图7-3a和图7-3b）以及图7-2的对比剖面（红线）

7.2 Lewis—Fox Hills从源到汇体系

位于美国怀俄明州中南部的Lewis—Fox Hills从源到汇体系发育于拉腊米造山运动的早期（马斯特里赫特阶）（图7-1）。厚层基底卷入型逆冲断层导致了温德河山脉、格拉尼特山、罗林斯隆起区的抬升（Dickinson 等，1988；Steidtmann 和 Middleton，1991；Dickinson，2004）以及相邻的沃沙基—大分水岭盆地沉积中心的沉降（图 7-1）。目前上述盆地由万苏特隆起分隔成两凹一隆的格局，但在马斯特里赫特阶早期为一个单一的深水盆地。欠压实厚度约 400m 的陆架斜坡地形（Asquith，1970）以较高的进积速率（大于 47km/Ma）向南充填到盆地中，从而揭示来自北部的物源有大量的沉积物供给（Carvajal 和 Steel，2006）（图 7-2）。陆架边缘的地层可以进一步细分为 Lance 组、Fox Hills 组砂岩以及 Lewis 组泥岩（图 7-2），分别代表了海岸平原、滨海—海岸线以及大陆架—深水环境（Gill 等，1970；Steidtmann，1993；Winn 等，1987）。因此，上述三个地层在马斯特里赫特阶盆地充填时期由北向南穿时抬升（图 7-2）。

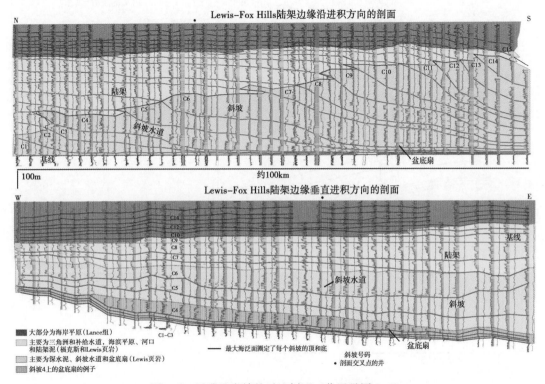

图 7-2 过盆地充填的对比剖面（位置见图 7-1）

GR/SP（左道）曲线中的黄色部分代表砂岩，灰色代表泥岩，黑色代表煤层；近似岩性实际颜色；右道曲线（如果有）代表电阻率

7.3 斜坡沉积地层的划分与沉积物体积计算方法

该方法包括三个基本步骤：（1）建立一个盆地尺度的高精度地层格架；（2）划分沉积—储存单元；（3）计算各单元的岩石和沉积物体积。

7.3.1 建立高精度地层格架：绘制斜坡沉积地层图

以 16 个泥岩层段作为标志层建立盆地充填序列，由于这些标志层形成于高频的盆地海泛时期，由此确定最大洪泛面，由深海一直对比到盆地陆架，建立起大陆斜坡沉积的地层格架（Carvajal 和 Steel，2006）。用盆地内 520 口井的 GR、电阻率、SP（当没有 GR 时）曲线来建立地层对比格架。对比结果可靠，基于以下证据：（1）生物地层证据表明（Kauffman 等，1993；Cobban 等，2006）最大洪泛面界面 8（mfs8）与盆地东西部边缘的 *B. clinolobatus* 露头发现可对比（图 7-3）（Weimer，1961；Gill 等，1970）；（2）连续选取大量平行物源方向（N—S）和垂直物源（W—E）的剖面（图 7-1）；（3）多条闭合对比剖面；（4）密集井网；（5）泥岩与凝灰岩处明显的低 GR 值和高电阻率值曲线等特征。尽管有多条交叉对比剖面和闭合剖面来尽量降低误差，然而在井较少的地区及海岸平原区域，地层对比的可靠性必然有一定程度的降低。

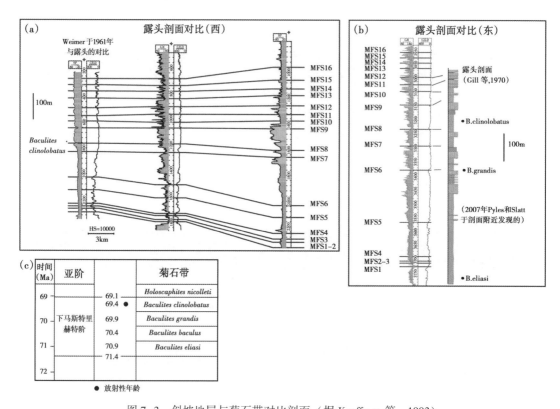

图 7-3 斜坡地层与菊石带对比剖面（据 Kauffman 等，1993）

(a) 位于西部；(b) 位于东部（位置见图 7-1），两个盆地陆架边缘内 MFS8 与 *B. clinolobatus* 可对比；
(c) 下马斯特里赫特阶菊石带的时间尺度

以砂为主的泥包砂岩相，笔者定义为斜坡沉积（据 Rich1951 修改，但仍然包含了同期的顶积层与盆底地层），这与 Galloway 的成因层序（1989）很相似。斜坡沉积主要是由相对海平面升降期间（或正常/强制性海退期间；Posamentier 和 Allen，1999）退积三角洲和海滨平原（以及毗邻的海岸平原）建造（Steel 等，2008）。陆架边缘三角洲到达斜坡外缘时会触发沉积物重力流，形成陆坡和盆底加积沉积（图 7-2 和图 7-4），进而形成大型的砂质深水扇。海侵带来的岸线后退及深水砂体向海搬运，导致砂岩沉积中心的向陆迁移，形成河口与障壁—潟湖体系。最大海侵时期在海洋和近海环境形成了一个泥岩凝缩段，之后在岸线体系重新发生海退后，形成了一个新的斜坡沉积。尽管这些斜坡沉积泥岩界面在盆地内普遍存在，但是其内部结构千差万别，这是由较长期的海侵和海退的内部响应造成的。

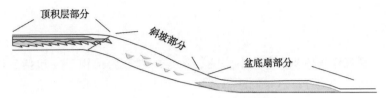

顶积层部分包括：三角洲、海滨平原和河口沉积以及它们等时的河流、海岸平原和陆架地层
斜坡部分包括：斜坡泥、主要分布于斜坡水道中的浊积砂岩
盆底扇部分包括：朵体中的水道和席状砂以及深水泥

图 7-4 陆架边缘不同分区的示意图

测井曲线显示斜坡沉积具有独特的岩性特征（Steel 等，2008；图 7-2）。海岸平原或斜坡顶积层的近源部分具有鼠形、箱形和钟形的测井曲线形态，这是由砂质河道切割泛滥平原（包含煤层）的典型特征。斜坡顶积层远端具有更多的漏斗形测井曲线特征，有时会有箱形—钟形的顶盖，这种特征反映了向上变粗、变厚的进积型三角洲或海滨平原顶部被河道削截的现象（图 7-2）。深水陆坡沉积物以泥岩为主，但是包含了大量的箱形—钟形或偶见尖峰状的曲线段，其侧向的不连续性是由斜坡倾向和走向方向上厚度突变造成的。这些测井响应是陆坡上的浊积水道砂岩的典型特征（图 7-2）。向陆坡底部和盆底方向，箱形—锯齿形测井曲线特征广泛且连续分布，沿着几十千米范围逐渐变细并形成宽阔的朵叶（图 7-2；Carvajal 和 Steel，2009），表明由浊流沉积逐渐转变为砂质深水扇沉积。

7.3.2 斜坡沉积与菊石带的相关性

将斜坡沉积与露头发现的菊石带相对比可以约束斜坡地层的年代范围（图 7-1 和图 7-3）。斜坡沉积地层 8 可以与盆地东西部边缘的 *Baculites clinolobatus* 露头相对比（Weimer，1961；Gill 等，1970）。除了在东邻盆地较年轻的沉积地层中有发现外（Perman，1990），在研究区目的层没有发现年代更晚的菊石带。因此，笔者将斜坡沉积地层 8—15 的地层年代定在 *B. clinolobatus* 化石带限定的年龄中，与之前学者定义的年代一致（Perman，1990；Pyles 和 Slatt，2007）。

斜坡沉积地层 5 和 6 可与 *B. grandis* 相对比（Gill 等，1970；Pyles 和 Slatt，2007）（图 7-3）。1976 年 Reynolds 在遥远的北部露头（Lost Soldier 地区；图 7-6）发现了 *B. baculus*（在菊石带之前），但是由于距离过远，露头/钻井控制较少，构造演化以及在马斯特里赫特阶期间深水斜坡的发育，造成 *B. baculus* 很难与研究区的目的层相对比。*B. baculus* 可能早于研究区斜坡沉积（图 7-3；Pyles 和 Slatt，2007）或者很有可能与 C1、C2 相当。发现于 Lewis 组泥岩底部附近的 *B. eliasi* 很可能早于研究区的斜坡沉积（图 7-3），Lewis 组泥岩位于下伏边缘海杏树组之上几十英尺，标志着海侵的开始，说明当时盆地没有发生明显的加深。因此，斜坡沉积地层 1—6 的年代在 *B. grandis* 和 *B. baculus*—*B. grandis* 所确定的年代之间。斜坡沉积地层 7 或属于上述年代区间亦或属于 *B. clinolobatus* 带。每一个 *B. baculus*、*B. grandis* 和 *B. clinolobatus* 菊石带都大约为 0.5Ma（Cobban 等，2006）。因此，斜坡沉积 8—15 的持续时间约为 0.5Ma，斜坡沉积层 1—6 约为 0.5~1.0Ma。

7.4 分区

为了便于计算沉积物体积,纵向上将陆架边缘楔划分为顶积层(包括海岸平原、大陆架以及从之前陆架—台地迁移过来的滨岸带)、斜坡和盆底扇(图 7-2 和图 7-4)等若干单元。大陆架和斜坡之间的界限近似于陆架边缘的轨迹(Steel 和 Olsen,2002)。斜坡砂体主要为斜坡水道中的砂岩,但也可能包含了由斜坡上部的陆架边缘三角洲前缘滑塌而来的砂岩。由于深水扇通常推进至斜坡底部形成坡底近端扇裙,等厚图上深水扇近源部分的边界就大致相当于斜坡与盆地底部的边界。笔者之所以倾向于采用这种界面而非严格选取斜坡到盆底的坡降梯度,是因为后者要求剖面具有准确的海底梯度变化数据,但遗憾的是研究人员并未掌握这个数据。更重要的是,深水扇的结构、沉积过程以及储集性能与斜坡水道差别巨大,所以笔者更倾向于将上述两种沉积体分开计算。

在对比剖面上追踪每个斜坡沉积地层内沉积单元的边界,并辅以砂岩、泥岩、地层等厚图等图件在全盆地延伸追踪。每个沉积单元在平面图上都可追踪出多边形形态,在剖面上具有一定的厚度,这样就在三维空间中确定了各单元的体积。

计算岩石和沉积物体积详述如下。

在每个斜坡沉积地层单元中都计算了砂岩、泥岩和煤层的体积。测井曲线标准化主要有以下几个步骤:(1)将数据库中异常的测井曲线剔除;(2)根据典型的 GR 曲线(砂泥岩间的界线为 75API,砂岩和煤层之间的界线为 25API)和典型的 SP 曲线(砂泥岩间的界线为 -10mV)将测井曲线标准化;(3)单独检查每条标准化后的曲线,从而保证每个沉积单元内的砂岩、泥岩、煤层分隔值准确无误,校正不准确的分隔值或者消除不准确的测井曲线。经过识别,只有煤层有很低的 GR 值(小于 25API),测井曲线呈典型的尖峰状。用 Petra 软件计算体积,每个斜坡沉积地层内的计算都包括:(1)每种岩性厚度的(例如砂岩和泥岩);(2)根据每口井的测井曲线段厚度,建立研究区厚度网格;(3)计算每个沉积单元多边形内网格的体积。为了节约计算和追踪多边形的时间,斜坡体积是由总体积减去陆架和盆地底部的体积得到的。

砂岩和泥岩体积被用来估算未经压实的沉积物原始体积,这有助于计算沉积物质量。颗粒体积(Liu 和 Galloway,1993,1997)是由岩石总体积减去孔隙和外部衍生的胶结物(例如来自岩石外部)体积得到的。乘以一个密度值 $2.65g/cm^3$,颗粒的体积可以转换成质量。本文采用每种岩性的胶结物和孔隙平均百分含量。Lewis 组泥岩中约有 20 个砂岩储层的孔隙百分比范围在 8%~25% 之间(大部分小于 18%),平均为 13.7%(Hettinger 和 Roberts,2005)。砂岩包含的胶结物百分比约为 23%(Van Horn 和 Shannon,1989)。没有数据表明这些胶结物中有多少来自外部,所以计算中采用一个保守的方法,即假设所有胶结物均来自外部。泥岩的孔隙度,据文献报道范围为 3%~18%,平均值为 12.5%(Almon 等,2001,2002)。平均泥质胶结物约为 6%(Bill Dawson,非正式出版物),并且大部分被认为是内源的,所以泥质胶结可以被忽略。因此,沉积物体积计算代表的是平均值及可能被低估的砂岩体积,因为有些砂岩胶结物很可能来自岩石内部(即来自骨架颗粒的转变)。

7.5 结果：各单元的沉积物体积以及随时间的变化趋势

在下文中，笔者将描述斜坡沉积地层中各单元相对的沉积物保存结果（表 7-1 和图 7-5），之后将讨论沉积斜坡区完整性的不确定性对计算结果的影响。

表 7-1 斜坡沉积（C）中砂岩体积（V_{ss}）、泥岩体积（V_{sh}）、煤层体积（V_c）

		A (km²)	V_{ss} (km³)	V_{ss} (%)	V_{sh} (km³)	V_{sh} (%)	V_c (km³)	TOT (km³)	TOT (%)	S_s/S_h	S_s (10⁹t)	S_h (10⁹t)	TOT (10⁹t)
1	TOP	0	0	0	0	0	0	0	0		0	0	0
	SL	71	1.3	23.7	4.8	5.0		6.1	6	0.28	2.2	11.0	13.3
	BF	8708	4.3	76.3	90.6	95.0		94.8	94	0.05	7.1	210.1	217.2
	TOT		5.6		95.4		0	100.9		0.06	9.4	221.1	230.5
2	TOP	159	2.2	8.9	4.4	3.8	0	6.7	5	0.50	3.7	10.3	14.0
	SL	403	12.1	48.4	27.5	23.5		39.7	28	0.44	20.4	63.9	84.2
	BF	8217	10.7	42.7	85.1	72.7		95.8	67	0.13	17.9	197.3	215.2
	TOT		25.0	100.0	117.1	100.0	0	142.1		0.21	42.1	271.4	313.5
3	TOP	234	5.6	33.2	5.1	3.4	0	10.8	6	1.10	9.4	11.9	21.3
	SL	1312	11.2	66.1	73.3	48.8		84.5	51	0.15	18.8	170.1	188.9
	BF	7308	0.1	0.7	71.8	47.8		71.9	43	0.00	0.2	166.4	166.6
	TOT		16.9	100.0	150.2	100.0	0	167.2		0.11	28.4	348.4	376.8
4	TOP	1688	24.8	24.7	33.2	9.6	0.4	58.4	13	0.75	41.6	77.0	118.6
	SL	1656	26.3	26.3	121.8	35.2		148.1	33	0.22	44.3	282.4	326.6
	BF	6888	49.1	49.0	191.2	55.2		240.3	54	0.26	82.5	443.4	525.9
	TOT		100.2	100.0	346.2	100.0	0.4	446.8		0.29	168.3	802.8	971.1
5	TOP	1916	39.4	54.9	75.4	20.5	0.1	114.8	26	0.52	66.2	174.7	240.9
	SL	1960	12.6	17.6	144.0	39.3		156.6	36	0.09	21.2	334.0	355.2
	BF	5131	19.7	27.5	147.4	40.2		167.2	38	0.13	33.2	341.8	375.0
	TOT		71.7	100.0	366.8	100.0	0.1	438.6		0.20	120.5	850.6	971.1
6	TOP	3008	68.0	43.7	111.7	23.4	0.2	180.0	28	0.61	114.3	259.0	373.3
	SL	2804	21.7	13.9	203.5	42.6		225.2	36	0.11	36.5	471.9	508.4
	BF	4059	65.9	42.3	163.0	34.1		228.9	36	0.40	110.7	378.0	488.7
	TOT		155.6	100.0	478.2	100.0	0.2	634.1		0.33	261.4	1108.9	1370.4
7	TOP	3394	55.6	65.6	98.4	37.9	0.3	154.3	45	0.57	93.5	228.2	321.7
	SL	2125	10.4	12.3	80.8	31.1		91.3	26	0.13	17.5	187.4	204.9
	BF	3645	18.8	22.2	80.5	31.0		99.3	29	0.23	31.6	186.8	218.3
	TOT		84.8	100.0	259.8	100.0	0.3	344.9		0.33	142.5	602.0	744.9

续表

		A (km^2)	V_{ss} (km^3)	V_{ss} (%)	V_{sh} (km^3)	V_{sh} (%)	V_c (km^3)	TOT (km^3)	TOT (%)	S_s/S_h	S_s (10^{9}t)	S_h (10^{9}t)	TOT (10^{9}t)
8	TOP	4170	116.0	64.6	164.9	36.8	2.3	283.3	45	0.70	195.0	382.5	577.4
	SL	2384	24.4	13.6	166.5	37.1		190.9	30	0.15	40.9	386.1	427.0
	BF	3000	39.3	21.9	117.0	26.1		156.2	25	0.34	66.0	271.2	337.2
	TOT		179.7	100.0	448.4	100.0	2.3	630.4		0.40	301.9	1039.8	1341.6
9	TOP	4538	61.4	46.6	92.4	23.2	1.5	155.4	29	0.66	103.2	214.3	317.6
	SL	2089	22.1	16.8	180.1	45.2		202.2	38	0.12	37.1	417.7	454.8
	BF	2519	48.2	36.6	126.1	31.6		174.3	33	0.38	81.0	292.4	373.3
	TOT		131.7	100.0	398.7	100.0	1.5	531.9		0.33	221.3	924.4	1145.7
10	TOP	5304	61.9	44.7	90.9	28.4	1.8	154.6	34	0.68	104.1	210.7	314.8
	SL	2246	24.0	17.3	134.1	41.9		158.0	34	0.18	40.3	310.9	351.2
	BF	1995	52.6	38.0	95.2	29.7		147.7	32	0.55	88.3	220.6	309.0
	TOT		138.5	100.0	320.1	100.0	1.8	460.4		0.43	232.7	742.2	974.9
11	TOP	5690	64.9	59.1	77.6	33.1	1.9	144.4	42	0.84	109.1	179.9	289.0
	SL	2143	29.0	26.4	119.8	51.1		148.8	43	0.24	48.8	277.7	326.5
	BF	1331	15.8	14.4	37.2	15.9		53.0	15	0.43	26.6	86.3	112.9
	TOT		109.8	100.0	234.6	100.0	1.9	346.3		0.47	184.5	543.9	728.4
12	TOP	6558	61.5	45.5	96.0	37.6	1.1	158.6	41	0.64	103.3	222.6	325.9
	SL	1959	28.4	21.0	96.6	37.9		125.0	32	0.29	47.7	223.9	271.6
	BF	1130	45.3	33.5	62.5	24.5		107.8	28	0.73	76.1	144.9	221.1
	TOT		135.2	100.0	255.1	100.0	1.1	391.4		0.53	227.2	591.4	818.6
13	TOP	7467	83.0	73.4	112.7	43.9	2.5	198.2	53	0.74	139.5	261.2	400.7
	SL	1821	23.0	20.3	120.6	47.0		143.6	39	0.19	38.7	279.6	318.3
	BF	399	7.1	6.3	23.5	9.2		30.7	8	0.30	12.0	54.5	66.5
	TOT		113.2	100.0	256.7	100.0	2.5	372.5		0.44	190.2	595.3	785.5
14	TOP	8177	100.9	91.3	131.3	58.1	2.9	235.1	69	0.77	169.5	304.5	474.0
	SL	1312	9.6	8.7	94.7	41.9		104.2	31	0.10	16.1	219.5	235.6
	BF	0	0	0	0	0		0	0		0	0	0
	TOT		110.4	100.0	226.0	100.0	2.9	339.3		0.49	185.5	524.0	709.6
15	TOP	8779	128.4	100.0	179.5	100.0	5.3	313.3	100	0.72	215.8	416.2	632.0
	SL	601	0	0	0	0		0	0	1.86	0	0	0
	BF	0	0	0	0	0		0	0		0	0	0
	TOT		128.4	100.0	179.5	100.0	5.3	313.3		0.72	215.8	416.2	632.0

TOP＝topset（顶积层），SL＝slope（斜坡），BF＝basin floor（盆地底部），TOT＝total（总体积）；分隔区域可能会上超，例如陆架边缘，如图7-4所示。

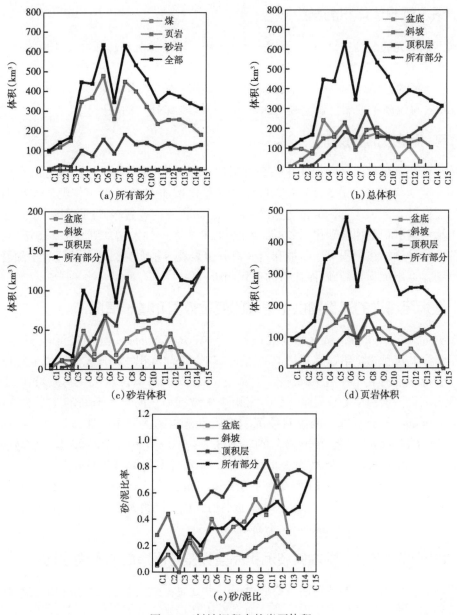

图 7-5 斜坡沉积中的岩石体积

7.5.1 斜坡沉积总体积随时间增加—减少的趋势

每个斜坡沉积的岩石总体积为 $101 \sim 634 km^3$（图 7-5a），从 C1 到 C8（$630km^3$）出现显著性的增加，之后从 C8 到 C15（$312km^3$）逐渐减少。

7.5.2 顶积层体积的不规则增长

计算结果显示，顶积层总体积一直增长到了 C8（$283km^3$），之后在 C9 后下降（$155m^3$），然后又连续增长到C15（$313km^3$）（图7-5b），这表明C9或C10时间段内盆地充填状况发生了明显变化，这将在下文中讨论。煤层的体积可忽略（小于1%）。

7.5.3 深水沉积体体积的减少

深水沉积单元表现为完全不同的体积变化趋势。斜坡总沉积体积介于100~200km³，随着时间变化有轻微的下降（C6—C15）。深水扇体积也显示出了从C4（240km³）到C12（108km³）下降的趋势。C4—C12深水扇在数据覆盖范围内几乎是完整的，然而C13突然下降到31km³，这反映了数据覆盖范围内深水扇（或至少部分）的缺失。

7.5.4 砂/泥岩比值的增加

总体积变化趋势与泥岩体积变化趋势（升降起伏）的相关性很好，但与砂岩体积变化趋势的相关性却很小（图7-5a）。砂岩体积持续稳定增加到C8，但不像泥岩体积那样急剧增长。然而，砂岩体积在C8—C15趋于保持在110km³不变或者轻微的下降。这意味着随着时间的推移斜坡沉积更加富砂化，砂/泥岩比值从0.1系统性地增加到0.7（图7-5e）。砂/泥岩比值的这种增加在顶积层、斜坡和盆底部分也表现得很明显，这可能代表了随着时间的推移供给到盆地中沉积物的砂/泥比值的变化趋势。

7.5.5 搬运至海洋的沉积物供给速率的变化范围和增长情况

通过 *B. baculus*—*B. grandis*（斜坡沉积1—6）和 *B. clinolobatus*（斜坡沉积8—15）来估算搬运至海洋的沉积物供给速率平均值的范围。对于海洋沉积物的质量，可以通过以下条件来建立一个范围（有利于控制不确定性）：（1）下限为只计算深水沉积物质量；（2）上限为计算深水沉积与顶积层的总质量。另外，由于斜坡沉积7的年代不确定（图7-3），研究人员将它的全部质量分别加到两组斜坡沉积物质量范围的上限中。通过计算，*B. baculus*—*B. grandis* 和 *B. clinolobatus* 的沉积物分别为（3.5~5.0）×10^{12}t 和（3.8~7.9）×10^{12}t。*B. baculus*—*B. grandis* 的时间跨度为0.5~1.0Ma，导致沉积物供给速率最小（1.0Ma）为4×10^{6}t/a，最大（0.5Ma）为10×10^{6}t/a。对于 *B. clinolobatus*（0.5Ma），沉积物供给速率为（8~16）×10^{6}t/a。这些计算表明搬运至海盆的平均沉积物供给速率为（4~6）×10^{6}t/a，且从 *B. baculus*—*B. grandis* 到 *B. clinolobatus* 不断增加。

7.5.6 由于数据覆盖范围内斜坡沉积完整性差异而造成的不确定性

研究认为，计算结果的敏感性主要取决于三个方面：主误差、斜坡沉积的完整性和总容积。为了如实地确定敏感性，认识清楚该地区地质环境的三个组成要素更显必要：（1）聚焦于由温德河、格拉尼特山脉、罗林斯隆起、大分水岭和沃沙基盆地组成的从源到汇体系；（2）陆架边缘向南进积、来自北部物源供给的深水扇（Ross，1995；Carvajal和Steel，2006，2009）以及图上的N—S向水道（Carvajal和Steel，2009）都可以证明陆架边缘的主要物源来自北部；（3）以东部水体更深、地层更厚为特征的盆地不对称沉降，导致沉积物体积从罗克斯普林斯隆起向东不断增加（图7-2）。向北、向西和向南是评价斜坡沉积完整性的主要方向；向东，大部分地层属于不同的源—汇体系。

南部，在研究区与图7-1南端（大约为科罗拉多—怀俄明州边界）之间约30km范围内，晚期的斜坡沉积（例如C12—C14）主要缺少深水沉积。再往南为桑德—沃什盆地，该盆地与沃沙基盆地在马斯特里赫特阶相连接，是另一个源—汇体系的一部分，该体系主要源自北部科罗拉多隆起，从而形成了西部的向北进积斜坡（Hettinger 和 Roberts，2005）。有学者建立的一条剖面显示（Hettinger 和 Roberts，2005），前文述及的缺失"边缘"地区并没有任何异常的水体加深或者扇体规模增加的情况（除非受剖面跨度所限），那么维持对图 7-2（N—S）剖面特征观察得到的认识。因此，给 C12—C14 体积分别增加一个普通扇的体积 150km³ 来补偿它们部分缺失的扇体，它们的体积变为小于 540km³，仍然小于 C6 和 C8 的峰值（约 630km³），尽管 C12—C14 部分缺失的扇体已由它们相对于 C6 和 C8 更加宽阔的顶积层作了补偿。

研究区北部和西部区域以顶积层沉积为主。深水地层向西部减薄、陆架边缘向南进积表明西部没有主要的物源供给体系（图 7-2；Carvajal 和 Steel，2009）。C1—C3 阶段，它们向西变密的特点说明西部不具有典型的顶积层。C1—C3 缺失的顶积层主要发育于北部（图 7-2），宽度为约小于 50km（C3 陆架边缘到隆起的距离）。另外，C1 部分缺失斜坡和扇体部分，C2—C3 缺失较小的东部斜坡部分（Carvajal 和 Steel，2009）。需要着重指出，C1—C3 的储集能力有限，因为它们的斜坡较小且较短（边缘此时处于雏形状态；图 7-2）。给 C1—C3 的每个斜坡沉积增加约 150km³（一个典型的宽阔顶积层体积），并给 C1 额外增加 200km³（一个小型斜坡沉积的典型深水沉积体积），C1—C3 的体积变为 300~450km³，小于 C6 和 C8 的体积（约 630km³），它们缺少同样的顶积层，因此它们的体积应该进一步增加以便支持 C1—C8 斜坡沉积体积的增加。

上述考虑同样有助于估算平均沉积物供给速率。对于 $B.\ baculus$—$B.\ grandis$ 沉积物质量范围，笔者分别给上限和下限增加大约 $0.4×10^{12}$t（C3 深水沉积质量）来补偿 C1 深水沉积质量，并且额外给上限增加约 $0.8×10^{12}$t（C4—C7 顶积层平均质量的三倍）来补偿 C1—C3 缺失的顶积层。它的质量范围变为 $(3.8~6.1)×10^{12}$t，并且平均沉积物供给速率增加到 $(4~12)×10^{6}$t/a。上限代表了极端的情况（0.5Ma 且包含全部的顶积层体积），然而，这个范围仍然表明 $B.\ clinolobatus$ 沉积物供给速率增加了 $(8~16)×10^{6}$t/a。此外，整个时间跨度内总的沉积物供给速率范围 $(4~16)×10^{6}$t/a 难以更高。例如，在 $(20~30)×10^{6}$t/a 范围内，海相沉积物质量可能接近两倍，这不切实际。上限（$16×10^{6}$t/a）可能会更高，但只增加很小的量，并且这很小的量不会对结果产生很大的影响。

总之，敏感性分析表明研究结果是相对可靠。分析结果显示，随着时间的变化，斜坡沉积总体积先增加后减少、顶积层体积持续增加及沉积物供给量在增加［大致范围为 $(4~16)×10^{6}$t/a］。Petter 等（2009，2011）应用斜坡沉积进积分析计算，得到沉积物通量范围与本次计算结果相似。

7.6　讨论：体积，构造—地层模型，从源到汇特征描述

本节尝试将沉积物体积和地层动力学综合分析，改进拉腊米型盆地构造—地层模型，更好地描述从源到汇特征。首先，笔者将讨论隆起的证据，然后是地层与隆起的关系。

7.6.1　拉腊米型盆地构造—地层模型

1）隆起和沉降

关于温德河山脉、格拉尼特山脉、罗林斯隆起区域内的马斯特里赫特阶隆起证据如下（图 7-6）：（1）研究区盆地内马斯特里赫特阶地层向北、东增厚（Johnson 等，2004），证

明该区沉降量在增加,而不断增加的沉降量是由基底卷入逆冲作用、地壳载荷、拉腊米隆起的抬升造成的,相对于西部的逆冲带,北部和东部当时是静止的(图 7-2;DeCelles,1994);(2)位于温德河山脉南部的太平洋溪流区域地下 Lance 组变薄,表明构造运动在持续进行(MacLeod,1981);(3) Lewis 组泥岩底部上超到一个侵蚀的/倾斜的不整合面上,对前马斯特里赫特阶地层造成削截,也意味着构造运动在持续进行(Reynolds,1976);(4)格拉尼特山脉东边的 Lance 组古河流的流向为 N—NE(Connor,1992),而在东沃沙基和大峡谷盆地内,它们的流向为 S—SW(Pyles 和 Slatt,2002;Carvajal 和 Steel,2009),这意味着在格拉尼特山脉和罗林斯区域内存在一个古隆起;(5)未成熟的 Lance 组河流相砂岩

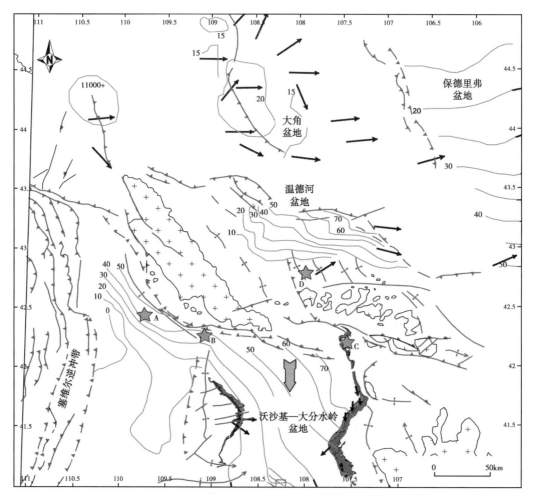

图 7-6 对马斯特里赫特阶隆起证据的总结(见图 7-1 图例)(数据来自 Reynolds,1976;
MacLeod,1981;Connor,1992;Hanson 等,2004;Johnson 等,2004;
Pyles 和 Slatt,2002;Carvajal 和 Steel,2006,2009)

蓝线为马斯特里赫特阶地层厚度等值线(单位:百英尺)。注意向着隆起地层明显变厚。星形:(A) Lance 组南—西向和东向倾斜前积层揭示了南东向的古水流(通过 FMI 成像测井资料得到);(B) 太平洋溪流区域:构造高地上部的 Lance 组(地下)变薄表明构造的生长;(C) Lost Soldier 区:Lewis 泥岩组上超到剥蚀的/角度不整合面的底部,削截了前马斯特里赫特阶岩石,表明构造的生长;(D) 北东向古水流和 Lance 组未成熟砂岩(含有斜长石和钾长石)揭示了物源来自格拉尼特山脉。灰色粗箭头代表陆架边缘进积的主方向,细箭头代表露头区的古水流数据

中钾长石的含量明显高于斜长石，表明基底侵入岩的暴露以及物源来自格拉尼特山脉（Connor，1992；先前提及的包含向北流的Lance组古河流砂岩）；（6）起源于温德河山脉与格拉尼特山脉南部的陆架边缘呈现典型的向南进积（McMillen和Winn，1991；Ross等，1995；Carvajal和Steel，2006，2009；Pyles和Slatt，2007）；（7）早古新世冲积扇中钾长石颗粒仅出现于温德河山脉的西部和南部，表明当时已经发生的基底暴露，揭示了马斯特里赫特阶的先前隆起（Steidtmann等，1986；Steidtmann和Middleton，1991）。隆起抬升包括温德河山脉、格拉尼特山脉、罗林斯隆起南部和西部沿着25°~30°断层发生的褶皱作用和逆冲作用（Berg，1962；Brown，1988；Blackstone，1991；Willis和Brown，1993）。隆起引起地壳载荷与岩石圈的挠曲（Hagen等，1985；Flemings等，1986；Shuster和Steidtmann，1988），导致盆地发生沉降的样式与前陆盆地很相似（Beaumont，1981；Jordan，1981；Heller等，1988），也与拉腊米型盆地有些相似。

2）陆架边缘地层

拉腊米型盆地的陆架边缘地层清楚地显示了该盆地充填的两个发展阶段（在这些学者的文中也可以看到相似的剖面：McMillen和Winn，1991；Ross等，1995；Pyles和Slatt，2007；Carvajal和Steel，2009）。盆地充填最关键的建造部分为斜坡沉积，其高度（约200~400m）表明盆地的水深大于400m（未经压实）。这样的水深意味着盆地初始沉降速率远远大于沉积物供给速率，从而在西部内陆地区可以堆积厚层的白垩纪浊流沉积：

（1）阶段一对应着斜坡沉积C1—C9的海退建造时期。这种斜坡沉积（图7-2 N—S剖面）显示出如下特点：①具有加积的结构样式（或进积的陆架边缘轨迹攀升较高）形成了厚层的顶积层；②斜坡高度持续稳定增加（例如从C4的约200m到C9的约360m）；③每个海退之后都会发育向陆侵入较远的海相泥岩舌状体；④福克斯山滨线体系发育良好、厚度大，横跨宽阔的早期陆架，该陆架由之前的大规模海侵向陆侵入沉积而成；⑤Lance组海岸平原相受限于顶积层向陆一侧的延伸范围（主要位于北部，图7-2中数据覆盖区之外的区域）。

（2）阶段二包括斜坡沉积C10—C15。特点如下（图7-2 N—S剖面）：①相对较薄的Lance—Fox Hills斜坡沉积顶积层，每一套顶积层都有明显的进积结构；②一个斜坡沉积的高度最初保持在360m或稍有升高（图7-2），后期就降到200m（例如C13—C14），后来有学者Hettinger和Roberts（2005）用剖面证明这一点，该剖面向南延伸更远，且具有相似的更短的斜坡；③每个顶积层段的底部的海相泥岩舌状体向陆侵入，但是比早期的舌状体海侵距离小；④变薄的福克斯山滨线体系穿越较窄的陆架（由较短距离的海侵造成）；⑤海岸平原延伸到每次海退后新增顶积层的外延（图7-2顶积层中的Lance组向盆地延伸很远）。

（3）阶段一和二，沉降、隆升、沉积物供给（图7-8）。在阶段一，加积结构样式、宽阔的海相顶积层段以及不断增长的斜坡沉积幅度都表明盆地的加深和相对海平面的急剧升高。水体加深与相对海平面的升高反过来又表明盆地的快速沉降（马斯特里赫特阶海平面升高难以加积形成厚层的地层；Miller等，1999，2004，2005），载荷驱动下形成的挠曲沉降模式，表明逆冲载荷和隆升的速率不断增加。相反地，阶段二的特征是由以下原因造成的：①正如上文提到的在 *B. clinolobatus* 时期输送到盆地的沉积物供给量的增加；②整体较低的且减小的相对海平面上升速率，造成盆地沉降，从C10中出现的下切谷看来，很可能是由海平面下降造成的（Carvajal和Steel，2009）。将沉积物供给量和沉降机制综合起来就可以得到进积结构样式、与宽阔海岸平原伴生的薄顶积层、随着时间推移变窄的陆架。到了阶段二的末期，斜坡沉积幅度的不断降低可能仅仅是由供给量增加和沉降量减小时期的盆地充填造成的。能够充填也是由于陆架边缘向着南部盆地区域进积，远离活动的抬升作用，那里发生载荷且沉降量较小。沉降量的减小表明随着时间的推移，虽然均衡补偿使山体和盆地回升很可能有助于进一步的隆升，在海平面下降时期增强下切谷的形成，但是逆冲作用驱动的隆升速率和载荷作用明显降低。

3）构造—地层模型与沉积物体积分配

（1）总体积增加—减少：阶段一中斜坡沉积总体积的增加反映了陆架边缘随着时间推移不断向陆推进。换言之，提供的斜坡顶积层厚度没有发生很大的变化，阶段一中顶积层的宽度、斜坡高度的增加、深水扇的发育揭示连续的斜坡沉积能够容纳顶积层沉积物不断增加的体积，但是难以支撑边缘的不断增长。正如下一章节分析结果所揭示的，很可能斜坡体积的增加受限于物源区同期隆升期间沉积物供给不断增加的产量和速率（Syvitski 等，2003）。阶段二期间，尽管具有更薄顶积层（更加平缓的陆架边缘轨迹）的陆架边缘持续增长，但是沉积物供给速率有可能增加，稳定下降的斜坡高度导致斜坡沉积可容纳的沉积物体积整体上趋于减小。

（2）顶积层体积不规则的增长：阶段一和阶段二期间顶积层体积不规则的增长反映了供给量增加和可容纳空间变化期间顶积层的加宽。阶段一期间，进积的不断发展使顶积层不断加宽，并且高的可容纳空间使其不断增厚，所以顶积层体积快速增长；然而在阶段二期间，可容纳空间的减小和沉积物供给的增加形成了较薄的顶积层，且其体积初始时很小但最后由于额外的进积作用而得到恢复性增长。

（3）深水沉积物体积的减小：深水沉积物体积随着时间推移不断减小或许反映了在陆架边缘生长早期，较窄的顶积层只有很小的储存能力，这有利于更多的沉积物过路不留而沉积到深水区。随着陆架边缘的进积和不断加深，它可以储存更多的顶积层沉积并且输送到深水区的沉积物更少。盆地逐渐地被充填，造成斜坡高度和长度减小，造导致充填末期深水区沉积物储存能力的降低。

（4）砂/泥比值的增加：沿着大峡谷盆地北部边缘的物源区地层厚度约为3000m（图7-7；Blackstone，1991）。形成一个大规模的向上变细且砂/泥比值降

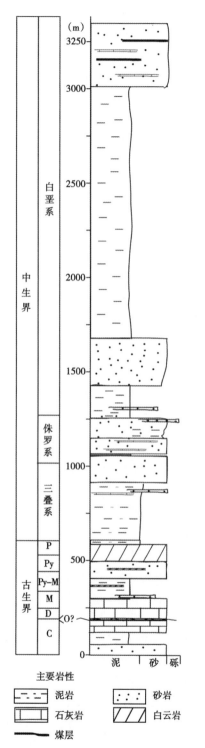

图 7-7　大峡谷盆地北部中生代和古生代地层剖面（据 Blackstone，1991 编制）

低的趋势，岩性从以砂岩和碳酸盐岩（古生代）为主向上变为以厚层泥岩为主（中生代）（Blackstone，1991）。很可能是富泥的地层起初没有上覆地层，而后来更多的砂岩下伏地层被暴露出来，笔者观察到，再沉积地层具有砂/泥比值向上不断增加的特征。Reynolds（1976）的文章指出在 Lost Soldier 背斜，Lewis 组泥岩底部的不整合面处有大约 1445m 的前 Lewis 组泥岩地层缺失。这一数字以及 3300m 总厚度的未覆盖地层表明，如果基底在早马斯特里赫特阶被暴露，盆地的平均隆升速率为 1~2mm/a。

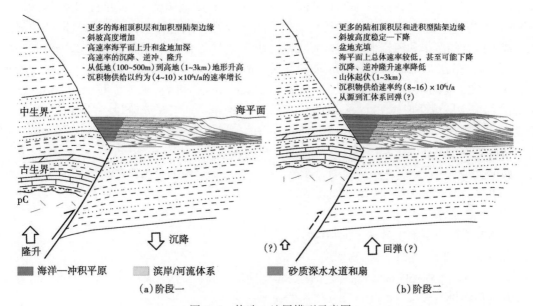

图 7-8　构造—地层模型示意图

研究中没有对顶积层和山体间的几何接触关系加以控制，并且上超很可能比地层旋转更加复杂，而且基岩在下马斯特里赫特阶期间就可能已经遭受暴露

（5）沉积物供给量的增加：本次研究的构造—地层模型揭示最大的隆升速率在阶段一的顶点和阶段二的起始点。整个阶段二持续的快速隆升导致了沉积物供给量的增加（Syvitsky 和 Milliman，2007）。同样地，汇水区的增加和未经压实沉积物的暴露带来的盆地回弹可能会进一步增加沉积物的供给量。

7.6.2　从源到汇拉腊米型盆地体系的特点

尝试应用现今源—汇体系的一些发现来推断沉积物供应量和地形，并将研究区体系与现今体系进行比较（Milliman 和 Meade，1983；Milliman 和 Syvitski，1992；Mulder 和 Syvitski，1996；Hovius，1998；Mulder 等，2003；Syvitski 等，2003，2004），但是首先要检查 Lewis Fox Hills 从源到汇体系是否已经关闭。

1）物源

根据 Gill 和 Cobban（1973）开创性的著作，一些研究人员逐渐认为研究区的物源是由怀俄明与蒙大拿州北部区域范围内一个向东流的三角洲的南向分支提供的，并延伸到达科塔的北部和南部（Winn 等，1987；Perman，1990；McMillen 和 Winn，1991）；该三角洲被称作雪利敦三角洲（图 7-6；Gill 和 Cobban，1973）。然而，在下马斯特里赫特阶，数据很大程度上反驳了上述解释，因为如前面章节所述，在温德河山脉、格拉尼特山和罗林斯隆起区域存在极为活跃的隆升活动，这些区域提供了大量的沉积物直接搬运到沃沙基—大峡谷盆地（图7-6；Connor，1992；Pyles 和 Slatt，2007；Carvajal 和 Steel，2009）。

2）汇水区域和沉积物通量

此次研究估算了汇水区域和沉积物产出的范围（河流卸载/区域）。现今基底隆升露头的面积约为15700km²（温德河山脉、格拉尼特山和罗林斯隆起），罗克斯林普斯隆起东部的沃沙基与大峡谷盆地的联合区域（主排水区位置）面积约为17700km²。这两个区域联合形成了汇水区域的上部边界面积为33400km²（图7-1）。在盆地初始充填期间，汇水区域适中（小面积的隆升区域及萌芽期的顶积层），但是由于隆升区域包括更大的面积且陆架边缘进积形成了更宽的海岸和河流平原，汇水区域面积变大（尽管汇水区只包含了部分陆相顶积层）。所以，在 *B. baculus*—*B. grandis* 期间，认为面积为 $(5\sim20)\times10^3 km^2$ 的汇水区域会形成 $200\sim2000 t/(km^2\cdot a)$ 的通量，并且在 *B. clinolobatus* 时期，认为面积为 $(20\sim35)\times10^3 km^2$ 的汇水区域会形成 $200\sim800 t/(km^2\cdot a)$ 的通量。这意味着阶段一的通量值要高于阶段二，反映了一个初始的隆升，隆升期间暴露的地层中包含压实很差的较易被侵蚀的沉积物，从而使一个小型的汇水区域产生大量的沉积物。

(1) 阶段一：

①更多的海相顶积层和加积型陆架边缘；

②斜坡高度增加；

③高速率的海平面上升和盆地加深；

④较高的沉降、逆冲、隆升速率；

⑤从低地（100~500m）到高地（1~3km）地形升高；

⑥沉积物供给以约 $(4\sim10)\times10^6 t/a$ 的速率增长。

(2) 阶段二：

①更多的陆相顶积层和进积型陆架边缘；

②斜坡高度稳定—下降；

③盆地充填；

④海平面上升总体速率较低，甚至可能下降；

⑤沉降、逆冲、隆升速率降低；

⑥山体地形（1~3km）；

⑦沉积物供给速率约 $(8\sim16)\times10^6 t/a$；

⑧从源到汇体系回弹（？）。

3）地形起伏

通过对载荷、汇水区域、沉积物产出量的估算，将研究的地质环境定义为"收缩型"（图7-9；Hovius，1998），与前述的拉腊米型地质环境高度一致，将 Lance 组汇水区定义为"多山型"（图7-9；最大地形起伏为1000~3000m；Milliman 和 Syvitsky，1992）。笔者建议可以使用 BQART 模型（Syvitski 等，2003，2004；Syvitski 和 Milliman，2007）更精确地估算平均地形起伏的最大值。BQART 模型对跨越了不同气候带、构造带和地理位置的488条现代河流进行了测试，它在河口区可以很好地预测沉积物载荷（Q_s，kg/s），内容如下：

$$Q_s = wBQ^{0.31}A^{0.5}RT, \quad T\geqslant 2℃$$

其中

(1) $w=0.02$；

(2) B（地质人为因素）$= IL(1-T_e)E_h$：

I 为冰川侵蚀因素：$I=1+0.09A_g$（A_g 为冰川覆盖面积的百分比），L 代表了盆地范围平均岩性，T_e 为湖相圈闭效率和人为储层，E_h 为土壤侵蚀的人为影响因素；

(3) Q = 河流产出量（km³/a）；

(4) A = 汇水区域面积（km²）；

(5) R = 汇水区最高地势（km）；

(6) T = 温度（℃）。

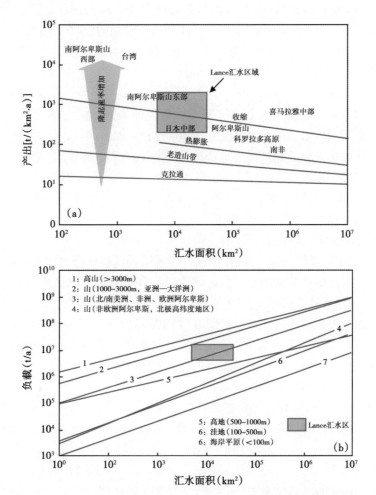

图 7-9 "收缩型"和"多山型"汇水区域构造背景估算模型

(a) 排水区域、沉积物产出量、构造背景（据 Hovius, 1998）揭示了 Lance 组汇水区域收缩的构造背景；(b) 排水区域、载荷、最大汇水区地势图（据 Milliman 和 Syvitski, 1992）揭示了 Lance 组汇水区域为山地。两个图都由现代河流测量而来

笔者已经估算面积和沉积物载荷，Q 可以从 Q（m³/s）$= 0.075A$（km²）中获得（Syvitski 和 Milliman，2007）。对于因素 B，基于研究区陆架边缘的地质情况和年代，认为 $I=1$（即"0"或者由于整体的温室效应形成的细微的冰川侵蚀），$L=2$（即基岩的岩性包括大量的沉积岩、河流相沉积及欠压实沉积物；更多的细节请看 Syvitski 和 Milliman 的文章，2007），$T_e=0$（湖相没有捕获沉积物或没有人为储层），$E_h=1$（土壤侵蚀中无人为影响）。

因此，对于研究区的陆架边缘，B=2。考虑到温度，古植物学的研究表明南怀俄明州马斯特里赫特阶的低地气候为温暖、稳定及湿润特征，大量的常绿阔叶植物和少量热带植物表明沉积量小于热带雨林（Upchurch 和 Wolfe，1993；Wolfe 和 Upchurch，1987）。热带植物和温带植物（年平均温度为20℃）的界线在古纬度 40~50℃之间，沃沙基和大峡谷盆地在古纬度 46~48℃之间。因此，研究中选择了一个平均的长期温度中值为20℃。

通过解答上述公式，研究人员分别得到了 B. baculus—B. grandis（采用（10~20）×10^3km^2 汇水面积）和 B. clinolobatus（采用（20~30）×10^3km^2 汇水面积）的平均地形起伏范围为 0.5~2.6km 和 0.9~2.5km。估算的地形与晚白垩世和古新世亚伯达、蒙大拿、怀俄明和科罗拉多盆地内同沉积环境组成的淡水软体动物所揭示的2500~3000m地形一致（Dettman 和 Lohmann，2000）。尽管假设只有一个主河流体系，如果存在两支河流（例如在 C9 存在两个主河流体系；详见 Carvajal 和 Steel，2009），那么两条河流的载荷与汇水区域会减小，但是地势估算仍然相同。有必要指出 B. baculus—B. grandis（阶段一）期间的最大地势很可能比 2.5km 小得多，因为该值是个极值，并不像假设的最大载荷（总斜坡沉积体积和最小时间为 0.5Ma）和较小的汇水面积（10×10^3km^2）这种情况。同样地，阶段二的最大地势明显大于1000m，这也是一种极端情况（最小载荷与较大的汇水面积）。地势估算的一个关键信息为从低地（100~500m，这个范围必定存在）到高地（500~1000m），并且在阶段一的某些点到山地（1000~3000m）的汇水面积，阶段二也仍然属于这个范畴。

4）Lewis—Fox Hills 与现代源—汇体系

对 Lance 组河流沉积物载荷的估算值（阶段一：（4~10）×10^6t/a；阶段二：（8~16）×10^6t/a）远小于密西西比河、布拉马普特拉河、奥里诺科河或亚马逊河这些载荷每年数亿吨的河流。显然这些河流的载荷能力非常大，因为它们有极大的汇水区域（几百万千米）。另一方面，Lance 河流域水量 [阶段一：200~2000t/（km^2·a）；阶段二：200~800t/（km^2·a）] 远大于密西西比河、奥里诺科河或亚马逊河 [分别为 120t/（km^2·a），150t/（km^2·a）和 190t/（km^2·a）]，并与恒河（520t/（km^2·a））和布拉马普特拉河（890t/（km^2·a））的流量相当。Lance 河流域高的产量很可能反映了它规模很小并且构造运动活跃，从而允许高比例的沉积物过路不留搬运到海相盆地沉积。

Lance 河流域与东亚河流有一些相似点。气候相似：东亚为热带，而 Lance 河流域为亚热带—热带（Upchurch 和 Wolfe，1993；Wolfe 和 Upchurch，1987）。Milliman 等（1999）报道称苏门答腊岛、爪哇岛、婆罗洲、西里伯斯岛、帝汶岛和新几内亚岛这些岛屿中每个岛屿一个组合的河流载荷为 (0.33~1.70)×10^9t/a，相当于一个约 (6~25)×10^6t/a 的河流平均载荷能力，与 Lance 河流载荷能力 (4~16)×10^6t/a 很相似。Lance 沉积物输出量 [200~2000t/（km^2·a）] 倾向于小于上述引用的岛屿 [1200~6200t/（km^2·a）]，或许反映了东亚河流的汇水区域包含了很多年轻的、易蚀的岩石，而它们中很多的最高地势都达到 1000~3000m，最大的河流汇水区域高于3000m。此外，东亚河流的数值代表了一个较短的时间段，而现在研究的数值估算超过了地质年代。东亚河流总体上提供了一些最大的输送到海洋中的沉积物体积（Milliman 和 Meade，1983；Milliman 等，1999），因此，它们与 Lewis—Fox Hills 从源到汇体系的相似性支持了 Carvajal 和 Steel（2006）的观点，即 Lewis—Fox Hills 陆架边缘是一个高供给量体系，至少对于陆架边缘来讲是这样。

7.7 结论

在怀俄明州的拉腊米型沃沙基—大峡谷盆地内,笔者举例说明了如何将沉积物体积分配法结合到盆地分析中,以便于总结一套古代从源到汇体系的特征,包含物源区隆升和盆地地层发育驱动机制。研究区从源到汇体系显示了两个阶段的演化。阶段一,不断增加的基底卷入、逆冲推覆隆起和地壳载荷导致了邻近盆地高的沉降速率,并且充填了一个具有向盆地深水区明显加积样式的陆架边缘沉积。陆架边缘的生长伴生了斜坡体积的增加和海相为主的顶积层。阶段二,陆架边缘轨迹变得进积作用更加明显,并且斜坡体积不断减小。该阶段伴生了宽阔的陆相顶积层和稳定—减小的斜坡高度,可能是由于可容纳空间减小和沉积物供给量的增加造成;推断逆冲推覆隆起在该时期不断降低。从阶段一到阶段二,平均沉积物供给速率从 $(4\sim10)\times10^6 t/a$ 增长到 $(8\sim16)\times10^6 t/a$(或稍多),且输出量在 $200\sim2000 t/(km^2\cdot a)$ 以内,盆地的最大地形起伏在洼地处介于 $100\sim500m$ 之间,在高地介于 $500\sim1000m$ 之间,阶段一的某个时期,山区地形高差可达 $1000\sim3000m$,这种情况一直持续到阶段二。拉腊米源—汇体系类似于现今东亚的一些河流体系,这些河流将巨量的沉积物搬运至洋盆。

致谢（略）

参 考 文 献

Almon, W. R., Dawson, W. C., Sutton, S. J., Ethridge, F. G., and Castelblanco, B. (2001) Sequence stratigraphy, petrophysical variation, and sealing capacity in deepwater shales, Upper Cretaceous Lewis Shale, south-central Wyoming, in Crockett, F., and Stilwell, D. P., eds., Wyoming gas resources and technology: guidebook. Casper, Wyoming Geological Association, 163–182.

Almon, W. R., Dawson, W. C., Sutton, S. J., Ethridge, F. G., and Castelblanco, B. (2002) Sequence stratigraphy, facies variation and petrophysical properties in deepwater shales, Upper Cretaceous Lewis Shale, south-central Wyoming, in 52nd annual convention of the Gulf Coast Association of Geological Societies, New Orleans, LA, 1041–1053.

Asquith, D. O. (1970) Depositional topography and major marine environments, Late Cretaceous, Wyoming. American Association of Petroleum Geologists Bulletin, 54 (7), 1184–1224.

Beaumont, C. (1981) Foreland basins. Geophysical Journal of the Royal Astronomical Society, 65, 291–329.

Berg, R. R. (1962) Mountain flank thrusting in Rocky Mountain foreland, Wyoming and Colorado. American Association of Petroleum Geologists Bulletin, 46, 2019–2032.

Blackstone, D. L., Jr. (1991) Tectonic relationships of the southeastern Wind River Range, southwestern Sweetwater Uplift, and Rawlins Uplift, Wyoming. Report of Investigations: United States, 24. Laramie, U. S. Geological Survey of Wyoming. Brown, W. G. (1988) Deformational style of Laramide uplifts in the Wyoming foreland, in Schmidt, C. J., and Perry, W. J., Jr., eds., Memoir 171 – Geological Society of America. Boulder, CO, Geological Society of America, 1–25.

Burgess, P. M., and Hovius, N. (1998) Rates of delta progradation during highstands; consequences for timing of deposition in deep-marine systems. Journal of the Geological Society of London, 155, Part 2, 217-222.

Carvajal, C., and Steel, R. J. (2009) Shelf-edge architecture and bypass of sand to deep water: influence of sediment supply, sea level, and shelf-edge processes. Journal of Sedimentary Research, 79, 652-672. doi: 10. 2110/jsr. 2009. 059.

Carvajal, C., Steel, R. J., and Petter, A. (2009) Sediment supply: the main driver of shelf-margin growth. Earth- Science Reviews, 96, 221-248. doi: 10. 1016/j. earscirev. 2009. 06. 008.

Carvajal, C. R., and Steel, R. J. (2006) Thick turbidite successions from supply-dominated shelves during sea-level highstand. Geology, 34, 665-668. doi: 10. 1130/ G22505. 1.

Cobban, W. A., Walaszczyk, I., Obradovich, J. D., and McKinney, K. (2006) A USGS zonal table for the Upper Cretaceous middle Cenomanian-Maastrichtian of the Western Interior of the United States based on ammonites, inoceramids, and radiometric ages. Open File Report 2006-1250. Washington, DC, U. S. Geological Survey, 46 p.

Connor, C. W. (1992) The Lance Formation; petrography and stratigraphy, Powder River basin and nearby basins, Wyoming and Montana. Washington, DC, U. S. Geological Survey Bulletin, 17 p.

Cross, T. A., and Lessenger, M. A. (1998) Sediment volume partitioning: rationale for stratigraphic model evaluation and high-resolution stratigraphic correlation, in Gradstein, F., Sandvik, K., and Milton, N., eds., Sequence stratigraphy: concepts and applications. Special Publication. Oslo, Norwegian Petroleum Society, 171-195.

Curtis, D. M. (1970) Miocene deltaic sedimentation, Louisiana Gulf Coast, in Morgan, J. P., ed., Deltaic sedimentation: modern and ancient. Special Publication, 15. Tulsa, OK, Society of Economic Paleontologists and Mineralogists, 293-308.

DeCelles, P. G. (1994) Late Cretaceous-Paleocene synorogenic sedimentation and kinematic history of the Sevier thrust belt, northeast Utah and southwest Wyoming. Geological Society of America Bulletin, 106, 32-56.

DeCelles, P., and Giles, K. A. (1996) Foreland basin systems. Basin Research, 8, 105-123.

Dettman, D. L., and Lohmann, K. C. (2000) Oxygen isotope evidence for high-altitude snow in the Laramide Rocky Mountains of North America during the Late Cretaceous and Paleogene. Geology, 28, 243-246.

Dickinson, W. R. (2004) Evolution of the North American Cordillera. Annual Review of Earth and Planetary Sciences, 32, 13-45.

Dickinson, W. R., Klute, M. A., Hayes, M. J., Janecke, S. U., Lundin, E. R., McKittrick, M. A., and Olivares, M. D. (1988) Paleogeographic and paleotectonic setting of Laramide sedimentary basins in the central Rocky Mountain region. Geological Society of America Bulletin (7), 1023-1039.

Flemings, P. B., Jordan, T. E., and Reynolds, S. (1986) Flexural analysis of two broken foreland basins; late Cenozoic Bermejo Basin and early Cenozoic Green River basin

(Abstract). AAPG Bulletin, 70, 591.

Galloway, W. E. (1989) Genetic stratigraphic sequences in basin analysis: I, architecture and genesis of floodingsurface bounded depositional units. American Association of Petroleum Geologists Bulletin, 73, 125–142.

Gill, J. R., and Cobban, W. A. (1973) Stratigraphy and geologic history of the Montana Group and equivalent rocks, Montana, Wyoming, and North and South Dakota. Professional Paper, 37. Washington, DC, U. S. Geological Survey.

Gill, J. R., Merewether, E. A., and Cobban, W. A. (1970) Stratigraphy and nomenclature of some Upper Cretaceous and lower Tertiary rocks in south – central Wyoming. Professional Paper, 53. Washington, DC, U. S. Geological Survey.

Hagen, E. S., Shuster, M. W., and Furlong, K. P. (1985) Tectonic loading and subsidence of intermontane basins; Wyoming foreland province. Geology, 13, 585–588.

Hanson, W. B., Vega, V., and Cox, D. (2004) Structural geology, seismic imaging and genesis of the giant Jonah Gas Field, Wyoming, USA, in Robinson, J. W., and Shanley, K. W., eds., Jonah Field: case study of a tight-gas fluvial reservoir: studies in geology, 52. American Association of PetroleumGeologists Bulletin, 21–35.

Heller, P. L., Angevine, C. L., Winslow, N. S., and Paola, C. (1988) Two-phase stratigraphic model of foreland-basin sequences. Geology, 16, 501–504.

Hettinger, R. D., and Roberts, L. N. R. (2005) Lewis total petroleum system of the Southwestern Wyoming Province, Wyoming, Colorado, and Utah. U. S. Geological Survey Digital Data Series: petroleum systems and geologic assessment of oil and gas in the Southwestern Wyoming Province, Wyoming, Colorado, and Utah. No. 39. Reston, VA, U. S. Geological Survey.

Hovius, N. (1998) Controls on sediment supply by large rivers, in Shanley, K. W., and McCabe, P. J., eds., Relative role of eustasy, climate and tectonism in continental rocks. Special Publication, 59. Tulsa, OK, Society of Economic Paleontologists and Mineralogists, 3–16.

Johnson, R. C., Finn, T. M., and Roberts, S. B. (2004) Regional stratigraphic setting of the Maastrichtian rocks in the Central Rocky Mountain region, in Robinson, J. W., and Shanley, K. W., eds., Jonah Field: case study of a tight – gas fluvial reservoir. AAPG Studies in Geology, 52. Tulsa, OK, American Association of Petroleum Geologists, 21.

Jordan, T. E. (1981) Thrust loads and foreland basin evolution, Cretaceous, western United States. AAPG Bulletin, 65, 2506–2520.

Kauffman, E. G., Sageman, B. B., Kirkland, J. I., Elder, W. P., Harries, P. J., and Villamil, T. (1993) Molluscan biostratigraphy of the Cretaceous Western Interior Basin, North America, in Caldwell, W. G. E., and Kauffman, E. G., eds., Evolution of the Western Interior Basin: Special Paper, 39. Toronto, ON, Geological Association of Canada, 397–424.

Liu, X., and Galloway, W. E. (1993) Sediment accumulation rate: problems and new approach, in Rates of geological processes-tectonics, sedimentation, eustasy and climate, implications of hydrocarbon exploration: Proceedings of the Fourteenth Annual Research Conference, Gulf Coast Section SEPM, 101–107.

Liu, X., and Galloway, W. E. (1997) Quantitative determination of tertiary sediment supply to the

North Sea Basin. American Association of Petroleum Geologists Bulletin, 81, 1482–1509.

Love, J. D., and Christiansen, A. C. (1985) Geologic map of Wyoming. Washington, DC, U. S. Geological Survey, scale 1: 500, 000.

MacLeod, M. K. (1981) The Pacific Creek Anticline: buckling above a basement thrust fault. University of Wyoming, Contributions to Geology, 19, 143–160.

McMillen, K. M., and Winn, R. D., Jr. (1991) Seismic facies of shelf, slope, and submarine fan environments of the Lewis Shale, Upper Cretaceous, Wyoming, in Weimer, P., and Link, M. H., eds., Seismic facies and sedimentary processes of submarine fans and turbidite systems. New York, Springer, 273–287.

Miller, K. G., Barrera, E., Olsson, R. K., Sugarman, P. J., and Savin, S. M. (1999) Does ice drive early Maastrichtian eustasy? Geology, 27, 783–786.

Miller, K. G., Sugarman, P. J., Browning, J. V., Kominz, M. A., Olsson, R. K., Feigenson, M. D., and Hernandez, J. C. (2004) Upper Cretaceous sequences and sea-level history, New Jersey coastal plain. Geological Society of America Bulletin, 116, 368–393.

Miller, K. G., Wright, J. D., and Browning, J. V. (2005) Visions of ice sheets in a greenhouse world. Marine Geology, 217, 215–231.

Milliman, J. D., Farnsworth, K. L., and Albertin, C. S. (1999) Flux and fate of fluvial sediments leaving large islands in the East Indies. Journal of Sea Research, 41, 97–107.

Milliman, J. D., and Meade, R. H. (1983) World-wide delivery of river sediment to the oceans. Journal of Geology, 91 (1), 1–21.

Milliman, J. D., and Syvitski, J. P. M. (1992) Geomorphic/ tectonic control of sediment discharge to the ocean: the importance of small mountainous rivers. The Journal of Geology, 100, 525–544.

Mulder, T., and Syvitski, J. P. M. (1996) Climatic and morphologic relationships of rivers; implications of sea-level fluctuations on river loads. Journal of Geology, 104 (5), 509–523.

Mulder, T., Syvitski, J. P. M., Migeon, S., Faugeres, J-C., and Savoye, B. (2003) Marine hyperpycnal flows: initiation, behavior and related deposits: a review. Marine and Petroleum Geology, 20 (6–8), 861–882.

Perman, R. C. (1990) Depositional history of the Maastrichtian Lewis Shale in south–central Wyoming: deltaic and interdeltaic, marginal marine through deep-water marine, environments. American Association of Petroleum Geologists Bulletin, 74 (11), 1695–1717.

Petter, A., Carvajal, C., Mohrig, D., and Steel, R. (2009) Sediment flux and load estimation for ancient shelfmargin successions. AAPG/SEPM Annual Meeting, Denver, CO, Search and Discovery Article no. 90090.

Petter, A., Kim, W., Muto, T., and Steel, R. (2011) Comment on "Clinoform quantification for assessing the effects of external forcing on continental margin development." Basin Research, 23, 118–121.

Porębski, S. J., and Steel, R. J. (2006) Deltas and sea-level change. Journal of Sedimentary Research, 76, 390–403.

Posamentier, H. W., and Allen, G. P. (1999) Siliciclastic sequence stratigraphy: concepts and applications, in Dalrymple, R. W., ed., Concepts in sedimentology and paleontology, vol. 7. Tulsa, OK, Society of Economic Paleontologists and Mineralogists, 204 p.

Posamentier, H. W., and Vail, P. R. (1988) Eustatic controls on clastic deposition: II, sequence and systems tract models, in Wilgus, C. K., Hastings, B. S., Ross, C. A., Posamentier, H. W., Van Wagoner, J., and Kendall, C. G. S. C., eds., Sea-level changes: an integrated approach. Special Publication, 42. Tulsa, OK, Society of Economic Paleontologists and Mineralogists, 125-154.

Pyles, D., and Slatt, R. (2002) Almond, Lewis, Fox Hills and Lance Systems in the Greater Green River Basin Wyoming, in Steel, R. J., Crabaugh, J., Carvajal, C., Slatt, R. M., Pyles, D., and Olson, M., eds., Introduction to the Lance, Fox Hills, and Lewis Formations on the Eastern Margin of the Great Divide and Washakie Basins, in American Association of Petroleum Geologists, Rocky Mountain Section, Annual Meeting (Laramie) Fieldtrip #1 Guidebook. Laramie, WY, 1-46.

Pyles, D., and Slatt, R. (2007) Stratigraphy of the Lewis Shale, Wyoming, USA: applications to understanding shelf edge to base-of-slope changes in stratigraphic architecture of prograding basin margins, in Nilsen, T. H., Shew, R. D., Steffens, G. S., and Studlick, J. R. J., eds., Atlas of deep-water outcrops: studies in geology, 56 [CD-ROM]. Tulsa, OK, American Association of Petroleum Geologists.

Reynolds, M. W. (1976) Influence of recurrent Laramide structural growth on sedimentation and petroleum accumulation, Lost Soldier area, Wyoming. American Association of Petroleum Geologists Bulletin, 60, 12-33.

Rich, J. L. (1951) Three critical environments of deposition and criteria for recognition of rocks deposited in each of them. Geological Society of America Bulletin, 62, 1-20.

Ross, W. C., Watts, D. E., and May, J. A. (1995) Insights from stratigraphic modeling; mud-limited versus sand-limited depositional systems. AAPG Bulletin, 79, 231-258.

Rowan, M. G., Peel, F. J., and Vendeville, B. C. (2004) Gravitydriven fold belts on passive margins, in McClay, K. R., ed., Thrust tectonics and hydrocarbon systems. Memoir, 82. Tulsa, OK, American Association of Petroleum Geologists, 157-182.

Schlager, W. (1993) Accommodation and supply: a dual control on stratigraphic sequences. Sedimentary Geology, 86, 111-136.

Shuster, M. W., and Steidtmann, J. R. (1988) Tectonic and sedimentary evolution of the northern Green River basin, western Wyoming, in J., C., and Perry, W. J., Jr., eds., Interaction of the Rocky Mountain Foreland and the Cordilleran thrust belt. Memoir, 171. Boulder, CO, United States, Geological Society of America, 515.

Siggerud, I. H., and Steel, R. J. (1999) Architecture and trace fossil characteristics of a 10.000-20.000 year, fluvialmarine sequence, SE Ebro Basin, Spain. Journal of Sedimentary Research, 69, 365-387.

Steel, R., Carvajal, C., Petter, A., and Uroza, C. (2008) Shelf and shelf-margin growth in scenarios of rising and falling sea level, in Hampson, G., Burgess, P. M., Steel, R., and others, eds., Recent advances in models of siliciclastic shallow-marine stratigraphy. Special

Publication, 90. Tulsa, OK, Society of Economic Paleontologists and Mineralogists, 47-71.

Steel, R. J., and Olsen, T. (2002) Clinforms, clinoform trajectories and deepwater sands, in Armentrout, J. M., and Rosen, N. C., eds., Sequence stratigraphic models for exploration and production: evolving methodology, emerging models and application histories [CD-ROM]. Proceedings of 22nd Research Conference, Gulf Coast Section Society of Economic Paleontologists and Mineralogists (CCSSEPM), 367-381.

Steidtmann, J. R. (1993) The Cretaceous foreland basin and its sedimentary record, in Snoke, A. W., Steidtmann, J. R., and Roberts, S. M., eds., Geology of Wyoming. Memoir, 5. Laramie, U. S. Geological Survey of Wyoming, 250-271.

Steidtmann, J. R., and Middleton, L. T. (1991) Fault chronology and uplift history of the southern Wind River Range, Wyoming: implications for Laramide and post-Laramide deformation in the Rocky Mountain foreland. Geological Society of America Bulletin, 103, 472-485.

Steidtmann, J. R., Middleton, L. T., Bottjer, R. J., Jackson, K. E., McGee, L. C., Southwell, E. H., and Lieblang, S. (1986) Geometry, distribution, and provenance of tectogenic conglomerates along the southern margin of the Wind River Range, Wyoming, in Peterson, J. A., ed., Paleotectonics and sedimentation in the Rocky Mountain region. Memoir, 41. Tulsa, OK, American Association of Petroleum Geologists, 321-332.

Syvitski, J. P. M., and Milliman, J. D. (2007) Geology, geography, and humans battle for dominance over the delivery of fluvial sediment to the coastal ocean. Journal of Geology, 115, 1-19.

Syvitski, J. P. M., Peckham, S. D., Hilberman, R., and Mulder, T. (2003) Predicting the terrestrial flux of sediment to the global ocean: a planetary perspective. Sedimentary Geology, 162 (1-2), 5-24.

Syvitski, J. P. M., Weaver, S. B., Nittrouer, C. A., Trincardi, F., and Canals, M. (2004) Strata formation on European Margins. Oceanography, 17 (4), 14-15.

Upchurch, G. R., Jr., and Wolfe, J. A. (1993) Cretaceous vegetation of the Western Interior and adjacent regions of North America, in Caldwell, W. G. E., and Kauffman, E. G., eds., Evolution of the Western Interior Basin. Special Paper, 39. Toronto, ON, Geological Association of Canada, 243-281.

Van Horn, M. D., and Shannon, L. T. (1989) Hay reservoir field: a submarine fan gas reservoir within the Lewis Shale, Sweetwater County, Wyoming, in Eisert, J. L., ed., Gas resources of Wyoming: Guidebook, 40. Casper, Wyoming Geological Association, 155-180.

Weimer, R. J. (1961) Spatial dimensions of Upper Cretaceous sandstones, Rocky Mountain Area, in American Association of Petroleum Geologists, eds., SP 22: geometry of sandstones bodies. American Association of Petroleum Geologists Bulletin, 82-97.

Willis, J. J., and Brown, W. G. (1993) Structural interpretations of the Rocky Mountain foreland; past, present, and future, in Stroock, B., and Andrew, S., eds., Wyoming Geological Association jubilee anniversary field conference: Guidebook: Casper, WY, United States, Wyoming Geological Association, 95-120.

Winker, C. D. (1982) Cenozoic shelf margins, northwestern Gulf of Mexico. American Association of Petroleum Geologists Bulletin, 66 (9), 1440.

Winn, R. D., Jr., Bishop, M. G., and Gardner, P. S. (1987) Shallow-water and sub-storm-base deposition of Lewis Shale in Cretaceous Western Interior seaway, southcentral Wyoming. American Association of Petroleum Geologists Bulletin, 71, 859–881.

Wolfe, J. A., and Upchurch, G. R., Jr. (1987) North American nonmarine climates and vegetation during the Late Cretaceous. Palaeogeography, Palaeoclimatology, Palaeoecology, 61, 33–77.

(陈莹 郭佳译，陈莹 校)

第8章 前陆盆地中岩石圈与地表地质作用的模拟

DANIEL GARCIA-CASTELLANOS[1], SIERD CLOETINGH[2]

(1. Instituto de Ciencias de la Tierra Jaume (ICTJA-CSIC), Barcelona;
2. Netherlands Research Center for Integrated Solid Earth Science (ISES), the Netherlands)

摘 要：本章综述了自20世纪90年代早期以来前陆盆地定量模拟的一些重要进展，重点阐述了岩石圈挠曲、剥蚀与河流搬运间的相互作用。挠曲可能由地形负载与板块拖曳所致，然而也可能由其他深部过程导致，比如板块拆沉、岩石圈拆沉以及弯折作用。需要以下几个方面的定量模拟特别强调：（1）挠曲模拟，弹性厚度估算以及岩石圈流变分层形式的解释；（2）详细的盆地充填几何样式/地层建模；（3）河流袭夺造成的盆地连通性的改变；（4）构造、河流水系、盆地连通性以及表层沉积物再分配间的相互作用；（5）岩石圈内部横向非均质性的影响，前陆盆地演化期间构造体制随时间的变化。流变性质的时空变化在以下几个方面发挥着关键的控制作用：（1）盆地的可容空间；（2）挠曲前隆演化；（3）河流袭夺对盆地连通性的改变。未来前陆盆地地球动力学研究主要的挑战在于，造山过程中地表作用与壳幔（解耦）耦合变形间相互作用的复杂动态三维研究。因此，将壳内解耦以及下地壳横向流动与岩石圈地幔拆沉区分出来很重要。数值模拟与物理模拟耦合结果显示，构造变形通过其对排水系统的影响从而影响着沉积物的空间再分配，这最终可能会导致构造样式沿走向改变。在前陆盆地演化的大陆阶段，气候对山间盆地尤其重要，在这些盆地中，内部排水系统可以大大增加沉积量。

关键词：河流水系 地壳动力学 气候 湖泊 内流盆地

8.1 引言

前陆盆地一词是 Dickinson（1974）提出来的，用来表述介于收缩造山带与未变形区域（前缘）间具有厚层沉积物堆积的区域。他将前陆盆地分成两个大的类型：陆—陆碰撞形成的周缘前陆盆地（例如，印度—恒河盆地与北阿尔卑斯盆地；图8-1）以及与大洋岩石圈俯冲有关的弧后前陆盆地（例如晚中生代—新生代落基山盆地）。前陆盆地的结构在横剖面上具有极不对称的特征，靠近造山带一侧沉积较厚，而沿被动（未变形）边缘呈楔形（Allen 等，1986）。受到早期海洋地质与地球物理研究的启发（Venign-Meinesz，1941），这样一个观点很快被广泛接受。即前陆盆地主要形成于前陆的挠曲均衡沉降（板块弯曲），这种沉降（弯曲）是由于造山带收缩，堆叠在造山带的构造单元的重量不断增长而形成的（Price，1973）。目前，前渊带内沉积物与水的重量、水平挤压、板块拖曳以及板块后撤时侧向软流圈推力都被认为是造成前陆盆地挠曲沉降的重要因素（图8-2）。

最近关于陆—陆碰撞初始俯冲的模拟研究（Johnson 和 Beaumont，1995）表明，周缘前陆盆地可以进一步划分为后前陆盆地（retro-foreland 和 basins）（形成于仰冲板块之上，例

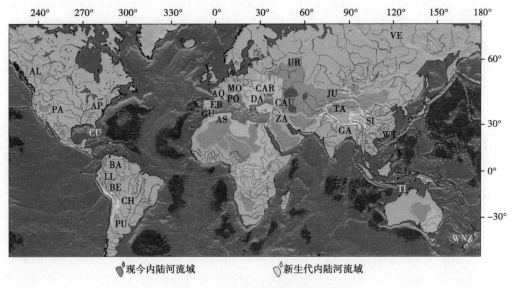

图 8-1 文中主要前陆盆地的位置

黄色与红色阴影部分分别代表当今内陆河流域以及新生代内陆河流域。内陆河流域独特的沉积物搬运方式影响着盆地的主体部分。从西到东的前陆盆地为：AL—艾伯塔盆地；PA—Paradox 盆地；AP—阿拉巴契亚；CU—古巴；LL—利亚诺斯（安第斯）；BA—巴里纳斯；BE—贝尼；CH—查科（安第斯）；PU—普纳（安第斯）；GU—瓜达尔基维尔（Betics）；EB—埃布罗（比利牛斯）；AQ—阿基坦；AS—阿尔及利亚—撒哈拉—阿特拉斯；MO—磨拉石盆地（北阿尔卑斯）；PO—宝盆与皮德蒙特盆地（南阿尔卑斯）；DA—Dacic 盆地（南喀尔巴阡）；CAR—次喀尔巴阡；CAU—高加索；ZA—扎格罗斯；UR—乌拉尔；JU—准噶尔；TA—塔里木；GA—恒河；SI—四川；VE—上扬斯克；WT—中国台湾西部；TI—帝汶；WNZ—新西兰西部

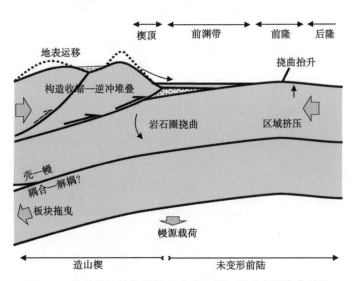

图 8-2 前陆盆地结构简图（未按比例）以及主要演化过程

图中的地壳/地幔部分主要适用于弧后盆地，而不适用于实际后障的结构

如，阿基坦盆地，注意不要和前面定义的弧后前陆盆地混淆）以及前前陆盆地（pro-foreland basins）（形成于俯冲板块之上，例如，埃布罗盆地（Ebro Basin）、北比利牛斯山脉（northern Pyrenees））。

还根据其与造山带中莫霍面位置的关系以及与后障（上覆板块）位置的关系划分出几种前陆盆地。真正的俯冲仅适用于地壳与地幔物质相互循环的大洋岩石圈。后前陆盆地，如加拿大落基山前陆盆地，发育于大洋俯冲带附近。在落基山盆地，大陆地壳被完整地保存下来，莫霍面从前陆到腹地连续，这表明，该盆地可能没有大陆俯冲，并且下地壳侧向流动不明显。在碰撞造山带，比如阿尔卑斯山或者比利牛斯山脉，莫霍面被地幔压头（mantle indenter）错断。俯冲板块的莫霍面深度向板块边界加深，并在俯冲板块大部分下地壳逐步堆叠的地方成为地壳根（Roure 等，1989，2010b）。对于板块后撤和弧后扩张的情况（例如，喀尔巴阡山脉（Tomek，1993）、亚平宁山脉（Patacca 等，2008），以及阿拉斯加的布鲁克斯山（Fuis 等，1997）），莫霍面没有垂向位移，如加拿大落基山脉。这种现象一个可能的解释是岩石圈地幔拆沉（Bocin，2010；Matenco 等，2010）。

由于挠曲的成因，前陆盆地与其他沉积盆地相比，一个普遍的特征是构造沉降加速（即一个凸起的沉降曲线），这本质上反映了逆冲载荷的幕式特征（Xie 和 Heller，2009），这种特征在弧后盆地与后前陆盆地中不那么明显（Naylor 和 Sicnclair，2008；图 8-3）。许多前陆盆地的演化经历了多期造山前的伸展，这种伸展可能引起古热流升高、沉积速率加快，从而影响前陆盆地演化中的垂直运动。例如阿基坦盆地（法国南部；Desegaulx 等，1991）与罗马尼亚喀尔巴阡山渊（Tarapoanca 等，2004）。另外，许多前陆盆地经历晚期垂向抬升与剥蚀，比如在北美科迪勒拉，从北部的加拿大的落基山山脉（Hardebol，2010 年）

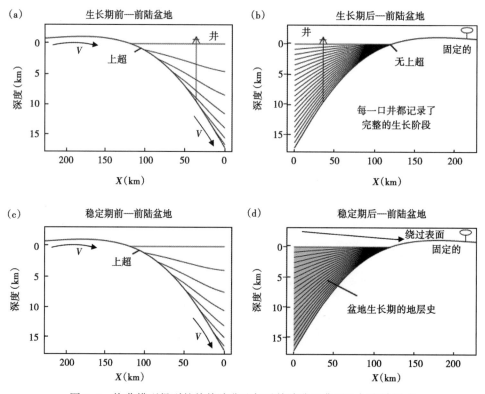

图 8-3 挠曲模型得到的前前陆盆地与后前陆盆地典型几何充填样式

发育阶段在 $X=0$ 处加一个垂向的载荷（a，b），稳态阶段在 $X=0$ 处加一个垂向的载荷（c，d）。水平层位的间隔是 3.55Ma，因此每个模拟的总时限是 71Ma。后前陆盆地综合钻井记录了完整的发育阶段，而在后前陆盆地中它只记录了盆地近期的水平对流到南部的墨西哥科尔多瓦地台（Roure，2008；Roure 等，2009），抬升可能与太平洋俯冲板块后方的地幔对流有关。在阿拉伯，阿尔及利亚北部以及亚平宁山脉也可以找到其他类似的例子，这些地方的抬升与板块拆沉以及弧后软流圈上涌有关（Van der Meulen 等，1999；Ali 和 Watts，2009；Roure 等，2010a；Tarapoanca 等，2010）。

非黏性流体上部的弹性薄板弯曲控制方程（Turcotte 和 Schubert，2001）通常被用来计算软流圈上部岩石圈的挠曲。该模拟表明，板块厚度越大，造山楔附近的挠曲沉降范围越宽，深度更浅，而且在前陆区存在一个几百米或者更小的前隆（Turcotte 和 Schubert，2001）。海相闭合海槽与海底火山岛链的板块挠曲地球物理研究模拟了不受地表剥蚀影响的前隆的高度（如 Watts，2001）。这些研究注意到隆起高度与邻区挠曲沉积中心深度的巨大反差。虽然前隆与前陆沉积中心间也存在这样的反差，但地表剥蚀常妨碍隆起高度的精确估算。结合模拟与实际观测结果可将前陆盆地划分为如图 8-2 中的四个带：楔顶带（变形楔之上的部位，包括背驮盆地）、前渊带（挠曲沉降最大的部位）、前隆（如果存在的话，就指前陆盆地的沉积覆盖层）、隆后（如果存在的话，就指前隆后方的沉积盖层）。

前陆盆地模拟首次提出了计算大陆岩石圈"有效弹性厚度（EET）"的数值方法（Beaumont，1981；Jordan，1981，1982；Flemings 和 Jordan，1989；Sinclair 等，1991），并用简单挠曲均衡模型建立了沉积充填与构造收缩间的一般关系。这些简单的方法只能解决地表搬运与构造变形相互作用中的一阶问题。比如，Beaumont（1981）提出进积楔下方的纯弹性岩石圈弯曲通常会产生超覆地层，而岩石圈内部黏性应力松弛可能会导致大规模退积以及沉积中心向造山带迁移。后面一种现象发现于亚伯达盆地并随后发现于许多前陆盆地系统（Carrapa 和 Garcia-Castellanos，2005）。然而，采用以上模型预测的盆地充填样式与实际的二维情况不相符（只模拟了一个垂直截面），在造山带盆地系统的尺度上也不能满足剥蚀与沉积间的质量守恒。

前陆盆地的多期演化特征解释了其几何形状的变化以及简单挠曲模型所预测的垂向运动特征（Cloetingh 和 Ziegler，2007；Roure 等，2010b）。对解释从前渊带几何形态或者重力场估计来的等效弹性厚度也十分重要。有效弹性厚度会随时空变化，也会随水平应力的时间变化而改变，但是岩石圈等温线的深度则不受这些因素影响。

本章回顾了前陆盆地弹性岩石圈挠曲的概念。首先讨论了岩石圈或地壳内部耦合或解耦机制的作用，盆地莫霍面几何形态（是否连续弯曲、是否垂向偏移）的重要性。然后讨论弹性挠曲模型在解释有效弹性厚度方面所存在的局限性。重要的是已经意识到，虽然这个概念很好地描述了弹性厚度，但它没有考虑到大陆岩石圈的脆—韧性分层结构。尤其是当弱下地壳存在时，它将对地表过程有很重要的影响。继承地形与流变性的侧向变化对盆地连通性与同造山期水道的改变起着主要的控制作用。笔者研究了岩石圈与地表过程间的响应，重点强调了沉积充填几何模型与前隆抬升。最后，综述了水道、气候与盆地演化间的关系。

8.2 岩石圈载荷、挠曲与流变性

8.2.1 弹性挠曲：结果与局限性

在过去 25 年里，前陆盆地的挠曲模拟研究致力于：（1）拟合沉积充填几何形态与造山带的构造历史来研究盆地沉降机制；（2）寻找薄板挠曲理论与野外实际间差异来进一步约束岩石圈力学性质或者深部过程的本质。

为了产生异常沉降，挠曲模拟绝大多数情况下需要增加额外的作用力，有时甚至是无约束的作用力。这些额外的作用力包括向上与向下的壳下负荷（通常称为"隐形负荷"）（图8-2）。这些负荷是由以下原因形成的：（1）俯冲板块的重量（Royden 和 Karner，1984；Royden，1993）；（2）地幔拆沉或板块破坏，比如次喀尔巴阡盆地（Tarapoanca 等，2004）、阿特拉斯山脉（Teson 等，2006），以及北阿尔卑斯前陆盆地（Sinclair，1997；Andeweg 和 Gloetingh，1998）；（3）岩石圈地幔厚度的侧向变化（Garcia-Castellanos 等，2002）；（4）地幔对流动力（安第斯；Davila 等，2005，2007）；（5）水平构造应力（宝盆（Po Basin）；Carrapa 和 Garcia Castellanos，2005）；（6）气候变化对楔平衡的扰动（北阿尔卑斯磨拉石盆地；Willett 等，2006）（位置见图8-1）。如果不考虑这些额外的作用力，挠曲均衡模型中用来解释盆地沉降的造山负荷就太小了，或者太大了，或者在不同阶段两者皆有可能。造山前的伸展作用会引起前陆盆地岩石圈有效弹性厚度的横向变化，从而可能会严重影响前渊带沉降的平面分布（图8-4；Tarapoanca 等，2004）。板内形变模型的结果表明水平构造应力造成的岩石圈变化不仅包括挠曲，还包括褶皱与弯曲，但薄板模型的结果中的岩石圈变化不包

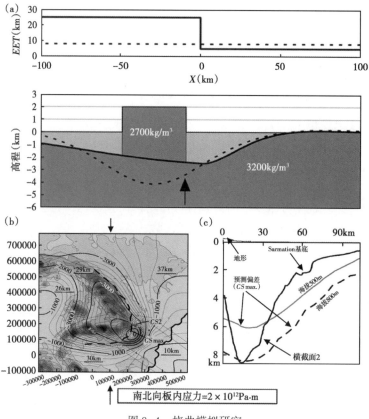

图 8-4 挠曲模拟研究

（a）一维挠曲计算结果图，如果 EET 突变的话，最大沉积则可超过区域负载。（b）次喀尔巴阡盆地 EET 横向变化（有下划线的数值）对沉降平面分布的影响，岩石圈内部强度分布造成了喀尔巴阡山脉远离地形负载处的集中沉降，该机制也可以与裂后热冷却以及板块拖曳沉积一起来解释为什么 Focsani 凹陷沉积厚度可以大于 12km；引自 Tarapoanca 等（2004）。（c）横截面（位置见图（b）中的CS2）展示了一个沉降剖面（Sarmatian 基底）以及两条计算的挠曲剖面

括后两种（Cloetingh 等，2002）。

至于岩石圈的力学性质，前陆盆地沉降正演模拟是估算大陆岩石圈有效弹性厚度的主要方法。这里提到的正演模型指的是通过改变作用于板块的构造应力与引力来重复计算挠曲剖面直到计算的基底深度与实际基底深度一致。有效弹性厚度被定义为与观察到的岩石圈弯曲拟合的最好的均匀弹性板块的厚度。其他估算有效弹性厚度的方法，比如基于地形与重力场间相关性的导纳法与相干法常常受到阻碍，因为它们计算前需要假设地形平坦并简化地壳几何形态。另外，这两种计算方法估算的有效弹性厚度通常与正演模型得到的结果不一致（Crosby，2007）。

表8-1与图8-5a展示了通过挠曲正演模拟获得的全球前陆盆地有效弹性厚度数据汇编。为了方便对比，每当不同的作者采用不同的弹性厚度数据时，杨氏模量都折合为 7×10^{10} Pa（详细信息见Turcotte和Schubert，2001）。将进行过挠曲研究的盆地归类并把它们的有效弹性厚度值平均，将平均值作为沿前陆盆地走向约300km范围内统一的有效弹性厚度。这样做是为了避免热点研究盆地有效弹性厚度值的重复（Watts，1992，2001）。这里的EET值不包括从重力谱分析（导纳法与相关法）得来的结果，因为重力谱分析得到的有效弹性厚度与正演模型计算的结果不一致（Crosby，2007）。经过这样的分组，从41个不同前陆盆地中总共得到了64个有效弹性厚度值（表8-1）。前陆盆地有效弹性厚度集中于10~25km之间，峰值位于10~15km之间（13个盆地）。那些位于喜马拉雅东缘与乌拉尔欧洲部分的前陆盆地具有最大的有效弹性厚度。这些高值反映了印度克拉通以及东欧地台岩石圈的刚度，它们都具有非常老的热年龄（Cloetingh和Burov，1996）。有效弹性厚度低于10~15km的盆地个数随有效弹性厚度值减小而减少，只有东瓜达尔基维尔盆地的平均有效弹性厚度小于5km。至于有效弹性厚度大于10~15km的盆地数量则逐渐减少为零。值得注意的是，前陆盆地有效弹性厚度的分布不存在如Watts（1992）描述的双峰分布，也不能用挠曲过程中地壳与岩石圈地幔变形间的耦合度变化来解释。图8-5的分布中存在两个缺口（有效弹性厚度为55~60km与90~95km），但由于其平均误差为11km（表8-1），所以这两个缺口数据的统计意义不大。

表8-1 前陆盆地挠曲正演模拟的有效弹性厚度值（EET）汇总

板块	载荷/造山带	盆地	EET (km)	误差± (km)	参考文献
南伊比利亚	Betics 西	瓜达尔基维尔	10	5	Van der Beek 和 Cloetingh（1992）；Garcia-Castellanos 等（2002）
南伊比利亚	Betics 西	瓜达尔基维尔	2.5	2.5	Van der Beek and Cloetingh（1992）；Garcia-Castellanos 等（2003）
中伊比利亚	伊比利亚中央系统	杜罗/塔霍/马德里	7	5	Andeweg（2002）
北伊比利亚	比利牛斯，伊比利亚山链，加泰罗尼亚海岸山脉	埃布罗盆地	22.5	10	Gaspar-Escribano 等（2001）；Zoetemeier 等（1990）
西欧洲地台	比利牛斯	阿基坦盆地	25.5	5	Brunet（1986）
西欧洲地台	西阿尔卑斯	北阿尔卑斯	25	15	Gutscher（1995）；Macario 等（1995）；Stewart and Watts（1997）；Ford 等（1999）；Karner and Watts（1983）

续表

板块	载荷/造山带	盆地	EET（km）	误差±（km）	参考文献
西欧洲地台	东阿尔卑斯	北阿尔卑斯	35	12	Gutscher（1995）；Steward and Watts（1997）；Karner and Watts（1983）
亚得里亚海北	阿尔卑斯/亚平宁山脉	宝盆西部	20	5	Carrapa and Garcia-Castellanos（2005）；Kroon（2002）
亚得里亚海北	阿尔卑斯/亚平宁山脉	宝盆西部	10	8	Tang 等（1992）；Kroon（2002）；Barbieri 等（2004）
亚得里亚海北	亚平宁山脉	亚得里亚海	28	3	Kroon（2002）
亚得里亚海中部	亚平宁山脉	亚得里亚海	6.5	3.5	Kroon（2002）
亚得里亚海	亚平宁山脉/迪纳拉造山带	亚平宁—迪纳拉盆地	14	10	Royden（1988）；Kruse and Royden（1994）
亚得里亚海南部	亚平宁山脉		28	3	Kroon（2002）
亚得里亚海南部	卡拉布里亚希腊构造带/西西里岛		11	8	Cogan 等（1989）；Moretti 和 Royden（1988）
默西亚地台	南喀尔巴阡山脉	Getic 凹陷	11	2	Matenco 等（1997）
东欧洲地台	南西喀尔巴阡山脉	东喀尔巴阡盆地	12	5	Matenco 等（1997）；Steward and Watts（1997）
东欧洲地台	喀尔巴阡山脉		35	10	Royden 和 Karner（1984）；Roure 等（1993）；Raileanu 等（1993）
东欧洲地台	北东喀尔巴阡山脉	东喀尔巴阡盆地	18	7	Matenco 等（1997）；Oszczypko 等（2005）
东欧洲地台	北西喀尔巴阡山脉	西喀尔巴阡盆地	12	4	Oszczypko 等（2005）
东欧洲地台	东高加索山脉	北高加索盆地	55	15	Ruppel 和 McNutt（1990）；Ershov 等（1999）
东欧洲地台	西高加索山脉	北高加索盆地	70	15	Ruppel 和 McNutt（1990）；Ershov 等（1999）
东欧洲地台	乌拉尔山脉		75	25	Kruse 和 McNutt（1988）
东欧洲地台	乌拉尔山脉		120	20	Piwowar 和 LeDrew（1996）
斯堪的纳维亚	苏格兰	苏格兰盆地	87	20	Piwowar 和 LeDrew（1996）
斯堪的纳维亚	苏格兰	苏格兰盆地	39	31	Poudjom-Djomani 等（1999）
蒙古中部			7	3	Bayasgalan 等（2005）
蒙古南部	阿尔泰—戈壁		15	5	Bayasgalan 等（2005）
蒙古西部			40	10	Bayasgalan 等（2005）

续表

板块	载荷/造山带	盆地	EET (km)	误差± (km)	参考文献
西伯利亚地台	维尔霍扬斯克	维尔霍扬斯克	50	10	McNutt 和 Kogan (1987); McNutt 等 (1988)
阿拉伯地盾	扎格罗斯山脉	扎格罗斯	50	25	Snyder 和 Barazangi (1986)
阿拉伯地盾	阿曼山	北阿曼	13	3	Ravaut 等 (1993)
土库曼斯坦—伊朗	科佩特山脉	里海	25	5	Artemjev 等 (1994)
亚洲	喀喇昆仑山		121	10	Caporali (1995)
塔里木南部	昆仑山脉	塔里木盆地	44	25	Lyon-Caen 和 Molnar (1984); Fan 和 Ma (1990); Teng (1991)
塔里木北部	天山	塔里木盆地	53	20	McNutt 和 Kogan (1987); McNutt 等 (1988); Liu 等 (2006)
东准噶尔		准噶尔盆地	12.5	12.5	Burov 等 (1990)
塔吉克凹陷	帕米尔		15	5	McNutt 等 (1988)
中国	印度尼西亚造山带	龙门山	48	5	Yong 等 (2003)
印度地盾	喀喇昆仑山		99	10	Caporali (1995)
印度地盾	喜马拉雅东部	恒河	90	15	Karner 和 Watts (1983); Lyon-Caen 和 Molnar (1985); Royden (1993)
印度地盾	喜马拉雅西部	恒河	34	6	Karner 和 Watts (1983); Lyon-Caen 和 Molnar (1985); Royden (1993)
印度地盾	喜马拉雅	恒河	25	5	Hetényi 等 (2006)
台湾西部	台湾	西台湾	15	6.5	Grotzinger and Royden (1990); Lin 和 Watts (2002)
巴布亚北部、几内亚东部			20	10	Haddad 和 Watts (1999)
巴布亚北部、几内亚东部			75	10	Haddad 和 Watts (1999)
新西兰		旺阿努伊盆地	17	8	Holt 和 Stern (1991); Stern 等 (1993)
东澳大利亚	新西兰南岛		22.5	12.5	Stern (1995)
北澳大利亚	班达弧—帝汶		79.1	10	Londoño 和 Lorenzo (2004); Tandon 等 (2000)
西北澳大利亚	班达弧—帝汶		55	25	Lorenzo 等 (1998)
东南极洲			62	37	Bott 和 Stern (1992)
美国地台	伯德伍德浅滩	南福克兰盆地	13	7	Bry 等 (2004)

续表

板块	载荷/造山带	盆地	EET (km)	误差± (km)	参考文献
美国地台南部（阿根廷）	东安第斯？火山？		14		Bahlburg 和 Furlong (1996)
美国地台南部（阿根廷）	安第斯	奥陶纪普纳盆地	14		Bahlburg 和 Furlong (1996)
美国地台南部（阿根廷）	安第斯	普纳盆地 22, 24 北	20		Prezzi (1999)
美国地台南部（玻利维亚）	安第斯	查科盆地	30.5	0.5	Coudert 等 (1995)
美国地台南部（秘鲁）	安第斯		40	15	Fan 等 (1996)
美国地台南部（厄瓜多尔）	安第斯		25	20	Stewart 和 Watts (1997)
美国地台南部（哥伦比亚）	安第斯	利亚诺斯盆地	70		Cardozo (1997)
南加利福尼亚	横向山脉北		50	5	Sheffels 和 McNutt (1986)
南加利福尼亚	横向山脉南		10	5	Sheffels 和 McNutt (1986)
北美地台	阿巴拉契亚山脉		105	25	Karner 和 Watts (1983); Hinze 和 Braile (1988); Rankin 等 (1991)
北美地台	阿巴拉契亚山脉		70	15	Steward 和 Watts (1997)
北美地台	落基山（古生代，原始的）	宾夕法尼亚州—二叠纪 Paradox	25		Barbeau (2003)
北美地台西部		爱达荷州怀俄明	22		Jordan (1981)
北美地台北西部	北极阿拉斯加	科尔维尔盆地	65	5	Nunn 等 (1987)

参考文献被分组了，沿每个指定前陆盆地走向约 300km 范围使用同一个 EET。误差估计基于 (a) EET 的横向变化；(b) 不同研究成果中的不同 EET；(c) 文献作者给出的误差；(d) 不符合以上三种情况的地区采用了最小误差，即 10%。杨氏模量被折合为 $7 \times 10^{10} Pa$。

8.2.2 弹性挠曲模型与更符合实际的大陆岩石圈流变模型间的联系

与大洋板块有效弹性厚度不同，大陆板块有效弹性厚度值与地震分布深度稍微相关，克拉通内部相关性更大（Watts, 2001）。这就限制了笔者进一步了解大陆板块有效弹性厚度的控制因素（Cloetingh 和 Burov, 1996）。大洋挠曲研究表明有效弹性厚度随构造板块年龄增长而增厚，并且由于黏性应力松弛，板块变弱而在负载停止后逐渐变薄（Nadai, 1963）。然而，这些规律通常不适用于大陆岩石圈（Watts, 2001）。由于大陆岩石圈具有石英—辉绿岩—橄榄石这样的组分结构，上地壳、下地壳以及岩石圈地幔的内部强度随深度增加，但同

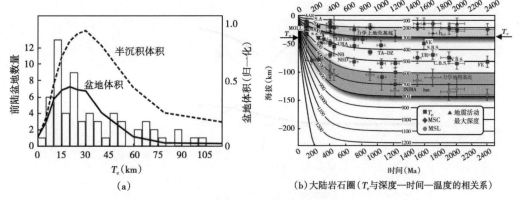

图 8-5 挠曲正演模拟

(a) 来自现有文献的前陆盆地弹性岩石圈厚度直方图（垂直条；见表8-1）。叠加的是来自地表过程与构造耦合模型的沿走向每单位长度的盆地体积（即盆地的横截面积）（黑线；Garcia-Castellanos，2002；图8-11）。
(b) 大陆岩石圈不同负荷时间、不同热模型条件下，观测的以及预测的 EET 数据、大机械强度地壳（MSC）的深度数据以及大岩石圈地幔强度的深度数据汇编。引自 Cloetingh and Burov（1996）

时强度又随地温梯度的增加而弱化。两种现象的同时存在使岩石圈内部出现两个薄弱带，造成上地壳与下地壳，以及下地壳与岩石圈地幔间的力学解耦。Burov 与 Diament（1992，1995）提出了一个公式来探索有效弹性厚度与板块间耦合度的关系，以及与板块弯曲过程中弹性（塑性）极限间的关系。由于岩石圈是由上地壳、下地壳、岩石圈地幔三个弹性板组成的，有效弹性厚度（T_e）是这三个厚度的总和：

$$T_e = h_{UC} + h_{LC} + h_{LM}$$

其中，h_{UC}，h_{LC}，h_{LM} 分别表示上地壳、下地壳与岩石圈地幔的力学厚度。如果这三层是完全独立的，那么：

$$T_e = \sqrt[3]{h_{UC}^3 + h_{LC}^3 + h_{LM}^3}$$

地壳与岩石圈地幔间的流变解耦可以影响有效弹性厚度。然而，在精确估计大陆岩石圈有效弹性厚度之前，应该在参考区域参数如板块年龄/负荷年龄的前提下先简单估算一个有效弹性厚度值（Watts，2001）。同时，综合考虑岩石圈结构、热体制以及组分的岩石圈强度正演模拟已经为当今岩石圈强度计算提供了一个大陆岩石圈尺度上的定量模型（例如，Tesauro 等，2009）。这种方法已经证明了欧洲岩石圈有效弹性厚度的横向不均一性，其新生代裂谷区的有效弹性厚度值小，欧洲缝合带向东的有效弹性厚度相对更大（Perez-Gussinye 和 Watts，2005）。

挠曲模型中脆—韧性强度与深度的相关性为阐明弹性板块弯曲、地震带厚度以及岩石圈挠曲应力间的关系提供了一个概念模型。绝大多数大陆岩石圈的地震活动集中在地壳中，这种观点说明了地壳与岩石圈地幔间的流变解耦，而不是岩石圈地幔强度弱。克拉通地区地幔岩石圈强度较大，因此有效弹性厚度更大（Watts 和 Burov，2003；图8-6）。Handy 与 Brun（2004）指出长期构造挠曲与地震活动间的相关性复杂，后者为岩石圈对应力的瞬态响应。

许多学者已运用上述多层流变分层模型研究了解耦以及随深度变化的应力对前陆盆地演化带来的影响（Lorenzo 等，1998；Garcia-Castellanos 等，2002；Zhou 等，2003）。例如，图

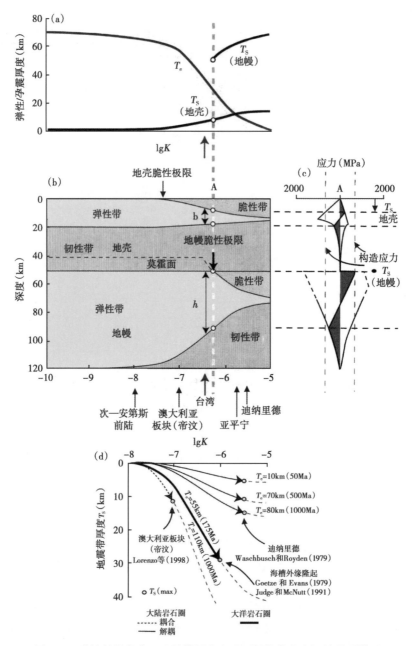

图 8-6 弹性板块弯曲、地震带厚度和岩石圈挠曲应力间的关系模型

(a) 弯曲板块地震活动带厚度（T_s）与弹性厚度（EET 或者 T_e）与曲度 K 的相关系（曲度是弯曲半径的倒数）。
(b) 板块脆韧性极限深度分布与曲度的关系式。(c) A 点以下的应力分布。(d) 地震带厚度与 EET、板块年龄以及板块曲度的关系。物理与数值模拟结果与实测情况一致。即 T_e 与 T_s 相关系不大（据 Watts 和 Burov，2003）。引用这些学者的原话，即"大陆内部缺少深的地幔地震更多的是因为地幔强度大，而不是通常人们认为的很软弱"

8-7 展示了俯冲于班达弧的澳大利亚地壳内部的应力分布，这种分布是由随深度变化的弹塑性流变模型推测而来的。

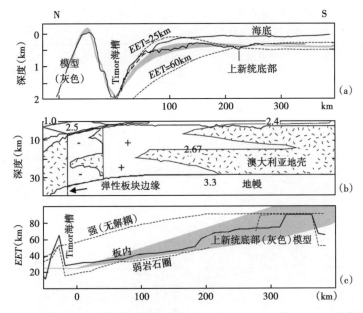

图 8-7 帝汶盆地上新统底部的挠曲模型（引自 Lorenzo 等，1998，有修改）
(a) 观测值与计算值偏差；(b) 张应力分布 (+) 与压应力分布 (-)；(c) 计算/采用 EET 值，来自双层弹—塑性深度相关挠曲模型

8.3 造山前伸展与后期挤压对前渊带演化的作用

许多前陆盆地都形成于先存的裂谷或裂陷大陆边缘（Stockmal 等，1986；Desegaulx 等，1991）。这种现象对岩石圈收缩模式和岩石圈沉降史以及热演化有一些重要的影响。图 8-8 展示了阿基坦盆地从裂谷到前渊的演化史（Desegaulx 等，1991），阿基坦盆地是比利牛斯山脉的后前陆盆地。在前陆盆地形成的初始阶段，比斯开湾地区造山前伸展产生的热效应还没有消退（图 8-8），因此该盆地经历了巨大的裂后热沉降（图 8-9）。随后于白垩纪—始新世前陆碰撞前经历了阿尔布期横向运动。造山前伸展也影响了罗马尼亚—喀尔巴阡前渊带，造成了黑海盆地西部的开启（Tarapoanca 等，2004）。

造山前伸展与前陆挠曲的叠加导致沉降模式与前陆盆地标准挠曲模型预测的结果有差异（Cloetingh 和 Ziegler，2007）。而且，下弯岩石圈流变性质的横向变化也会强烈影响沉降模式（图 8-4 和图 8-9）。

除了造山前伸展，晚期收敛速率发生变化所引起的挤压也可能增加前陆盆地的沉降量。在阿基坦盆地与喀尔巴盆地（Tarapoanca 等，2004）以及阿尔卑斯皮德蒙特盆地（Carrapa 和 Garcia-Castellanos，2005）能找到这样的例子。Cleotingh 等（2004）指出，大陆碰撞末期板块拆沉能强烈改变上覆板块的应力机制，造成高强度的压缩。在罗马尼亚—喀尔巴阡前渊带，末期碰撞使其前渊带的形状比单独的前陆挠曲模型预测的形状更对称。造山前伸展作用叠加上，前陆挠曲以及晚期挤压也可以解释超深前渊带的形成机理。例如罗马尼亚—喀尔巴阡地区，位于喀尔巴阡造山带侧翼的福克沙尼凹陷新近系沉积物超过 12km。热年代学研究表明，在过去的 12Ma 里，这里的剥蚀超过 4km（Sanders 等，1999；Merten 等，2010；Ne-

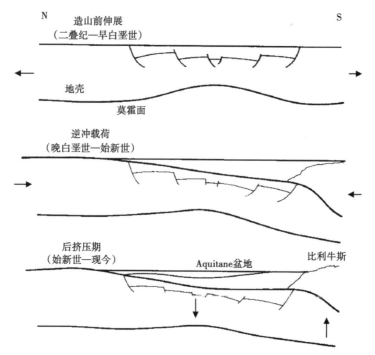

图 8-8 阿基坦（Aquitaine）前陆盆地演化简图（引自 Desegaulx 等，1991）

顶部图：（二叠纪—早白垩世）造山前伸展阶段；欧洲域裂陷。中部图：（晚白垩世—始新世）挤压；阿基坦（Aquitaine）前渊带的形成。底部图：（始新世—现今）后—挤压期，无弯曲

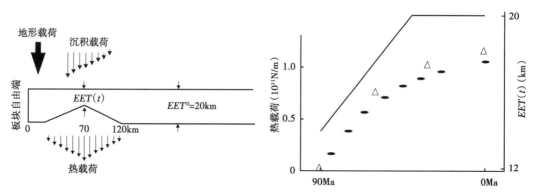

图 8-9 伸展造成的减弱与热沉降对前陆盆地沉降的影响（引自 Desegaulx 等，1991）

顶部图：热应力导致弹性板块减薄，使用了一个移动的地形（逆冲）载荷；底部图：三角形指示板块厚度随时间的变化、前造山带盆地中心的 $EET(t)$，点线表示热载荷演化

cea，2010）。

古近—新近纪皮德蒙特盆地是位于西阿尔卑斯的一个后前陆盆地，由于古利亚阿尔卑斯的叠加，它呈倾斜的非圆柱状（Bertotti 和 Mosca，2009）。因此，剖面模拟结果表明盆地形成过程中水平压力的存在显著增强了其顶超沉积（图 8-10；Carrapa 和 Garcia-Castellanos，2005）。新的沉积仅充填了位于靠近阿尔卑斯的盆地区域，离前陆越远，沉积物越老。

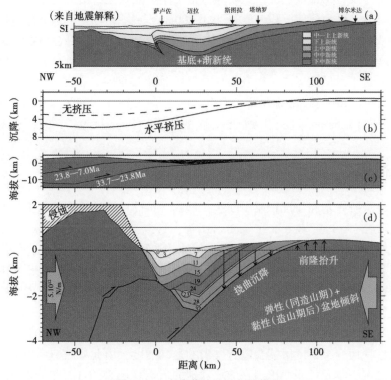

图 8-10 皮德蒙特盆地剖面模拟

(a) 地震研究约束下皮德蒙特盆地（Piedmont）的沉积充填几何样式（位置见图 8-11c）；(b) 计算垂向运动，虚线表示不考虑水平挤压的计算结果；(c) 黏弹性挠曲模型得到的最终几何充填样式（垂向没有放大）；(d) 盆地几何样式（垂向放大）（据 Carrapa 和 Garcia-Castellanos，2005，修改）

8.4 岩石圈与地表过程间的响应

前人进行了大量的模拟来研究前陆盆地的地貌演化与地表过程（Bonnet，2010；以及该文章里的参考文献）。这些研究的一个副产品是发展了一系列整合三维地表沉积搬运与盆地动态演化的模型（Johnson 和 Beaumont，1995）。耦合平面挠曲均衡与三维河流沉积物侵蚀/搬运的计算机模拟表明，如果没有挠曲沉降，河流搬运仅靠自己的力量不能在前陆产生大量的沉积物聚积（Garcia-Castellanos，2002）。这些结果与经典模式相似，即造山带前大面积岩石圈下弯引起的千米尺度的构造沉降是形成大容纳空间的前提。更具体地说，如果岩石圈的强度太低或太高，造山带附近形成的容纳空间则很小，那么河流将搬运走这些造山—盆地系统的沉积物。然而，早期的侵蚀扩散模型（Flemings 和 Jordan，1989；Sinclair 等，1991）表明，即使在岩石圈刚度很大的情况下也能产生大量的沉积物。更复杂的模型表明，在刚性非常高的情况下，如果没有差异垂向运动，河流将搬运这些沉积物。图 8-5 描绘了这些模型预测的盆地内可容空间与有效弹性厚度间的关系，该图与观察到的有效弹性厚度直方图极其相似。这说明了由于研究集中于大型的沉积盆地，图 8-5 中的有效弹性厚度值可能失之偏颇。问题是相对于大的和小的有效弹性厚度值（对应于小容积盆地），中等有效弹性厚度值（20~40km）（对应于"大"沉积盆地）的比例被高估了多少。例如，摩洛哥的阿特拉斯造山带，由于还没有用正演挠曲模型研究过这个小前陆盆地，所以它未包含于表8-1。没有相关的模拟可能是因为盆地沉积厚度（小于700m）与宽度太小（摩洛哥阿特拉斯，瓦尔扎扎特，小于35km）。初步挠曲模拟计算的阿尔扎扎特盆地的有效弹性厚度值小于4km（Teson等，2006），这样的低值可能与摩洛哥阿特拉斯山脉地幔上涌有关，这也解释了为什么该造山带收缩量小，但地形高，以及为什么其大地水准面高，而重力异常低（Teixell等，2005）。其他的例子如小克拉根福与施泰马尔盆地（澳大利亚，阿尔卑斯东部），它们的有效弹性厚度介于1~5km，这与上地壳强解耦有关（Nemes等，1997；Sachsenhofer等，1997）。

8.4.1 沉积充填几何样式模拟

一些学者结合研究基底与沉积层中断裂扩展的一些简单方法进行了造山带非瞬态负载的数值模拟（Toth等，1996；Garcia-Castellanos等，1997；Sassi与Rudkiewicz，2000；Albouy等，2002）。这些研究的目的是用沉积模式更好地约束构造演化，或用构造演化更好地约束沉积模式。为了达到这个目的，仅考虑负荷瞬间响应的纯弹性板块模型是不足的，还应该考虑岩石圈瞬间黏应力响应（Beaumont，1981）。与有效厚度随时间减薄的结果相似，盆地演化过程中的黏性应力松弛会导致盆地变窄变深。瓜达尔基维尔盆地（Garcia-Castellanos等，2002）与古近—新近纪皮埃蒙特盆地（Carrapa与Garcia-Castellanos，2005）黏弹性模型研究结果表明黏性盆地变窄速度可能会超过楔形的运动速度，由此导致造山带沉降区后撤。在这种情况下，老沉积单元可能会出现在前陆盆地的远端（图8-10）。在上述前陆盆地与艾伯塔盆地（Beaumont，1981）以及瑞士磨拉石盆地（Schlunegger等，1997）中发现了尖灭沉积单元向造山带转移现象。在艾伯塔盆地中，整个前陆与内陆的岩石圈被抬升（Hardebol，2010）。瑞士磨拉石盆地中，尖灭现象与汝拉山下部古生界地垒后米辛尼亚（Messinian）期反转有一定的联系。该反转解释了三叠纪滑脱褶皱的再褶皱，以及汝拉盆地与磨拉石盆地间的地形差异（Philippe，1994；Roure等，1994）。因此，考虑了黏应力松弛的模型，仍需要结合这些机制或者前面章节提到的额外的垂向或水平应力来拟合盆地结构。

三维野外与模拟的研究（Johnson与Beaumont，1995；Garcia-Castellanos，2002；Clevis等，2004）表明绝大多数前陆盆地系统的纵向（沿走向）沉积物搬运决定着非二维充填几何样式与沉积相分布。Clevis等（2004）开发了一个三维数值模拟来解释各种构造作用与海平面变化情况下，前陆盆地横向与轴向沉积体系控制的海相与冲积相沉积间的相互贯穿。根据它们的模拟结果，大尺度沉积堆叠的控制机制是沉积供应与挠曲沉降间的平衡。同时，各种构造活动与海平面波动叠加产生了轴向三角洲的分级模式以及小尺度冲积扇序列（许多前陆盆地的模式特征；Gupta，1997；Horton与DeCelles，2001）。该系列模型的端元包括一系列工具，这些工具可以进行详细的三维盆地地层建模，可以考虑沉积物搬运与沉降，还可以结合碎屑与碳酸盐岩沉积（Granjeon与Joseph，1999）。

8.4.2 前隆抬升

20世纪80年代采用挠曲模型研究盆地沉降的特殊目的是试图定义大陆前隆，即大洋弯曲沉降区附近的外部隆起。实际上，前陆盆地系统地质意义上的前隆是很难明确定义的，尤其是古系统中前隆的定义（Crampton和Allen，1995；DeCelles和Giles，1996；DeCelles，本书第20章）。Decelles与Giles（1996）指出，如果"额外的沉积负载干扰造山负荷的挠曲响应，前隆被埋藏或其形态被抑制"的话，周缘区的识别很困难。在任何情况下，由纯弹性挠曲模型预测的前隆的高度小于几百米，距离超过几百千米。因此，前隆侧面的倾斜不会超过1°。考虑弹性极限，以及基底顶部弯曲断层的控制，前隆抬升甚至更小（Tandon等，2000；Garcia-Castellanos等，2002）。

然而，由于区域应力场（Cloetingh等，1989）、逆冲作用影响下的应力累计和松弛（Peper等，1992）随时间的变化以及前陆盆地岩石圈流变性的空间变化（Waschbusch和Royden，1992），都会使得前隆的幅度随时间波动。数值模拟显示，即使当构造楔的位置稳定向前陆移动时，前隆的位置也不会稳定移动。相反，如果有效弹性厚度横向不均匀时，这种移动会减慢或加快（Waschbusch和Royden，1992），甚至会由于应力的黏性松弛而反向移向造山带（Beaumont，1981；Garcia-Castellanos等，2002）。因此，前隆抬升速率可能会随时空的改变而改变。实际上，目前活动的前隆可能标志着挠曲盆地几何形状当前的边界，然而盆地充填也可能记录了更老的、埋藏于前渊沉积中心的前隆。

同时，三维计算机实验结果表明（Garcia-Castellanos，2002；图8-11），内部固有（挠曲前）地形隆起比前隆造成的垂向抬升对水道系统的形成起着更重要的作用（前隆引起的垂向抬升通常小于200m）。因此，前隆不能简单地作为分水岭。下文将指出，作为前陆前部的变形区域，前隆位置的不连续将进一步使它对水道的影响复杂化。这些模型也考虑了前陆内部的地形梯度。最初下切侵蚀弱的河流系统可以发育成反映几百米前隆抬升（图8-11），甚至几十米后隆沉降的水系（Horton和DeCelles，1997；Roddaz等，2005）。这样的特殊地形也可能与前陆基底反转有关，例如美国的落基山、安第斯以及撒哈拉阿特拉斯，都是在曾经连续的前陆地区形成了高地形（Gries，1983；Colletta等，1997；Roure等，1997；Roure等，2010b）。

沿安第斯山脉东侧的一些前陆盆地中，沉积充填达到了前隆顶部，随后又被沉积盖层覆盖。计算机模拟表明（图8-11和图8-14），构造与侵蚀间平衡的改变（海平面不改变，没有前陆沉降）不能使活跃前隆沉积物的溢出（构造负载，前隆仍在抬升）。只要一个盆地是陆相的并且是外流的，水道系统就能适应构造与侵蚀的改变，克服造山带来的沉积物以及盆地容纳空间的减小，沿走向将多余的沉积运向开放海（Garcia-Castellanos，2002）。因此，盆地沉积物覆盖下的前隆是前期阶段的残留（构造负载离前陆更远时）或者是其他沉降机制造成的降于海平面以下的活跃前隆。后一种解释符合Davila等（2007）的说法，他指出幔源动态载荷造成了次安第斯周缘隆起的附加沉降，导致它被后来的沉积物覆盖。除了静态载荷以外，盆地下部动态载荷（俯冲板块上的黏性地幔角流；Catuneanu等，1997）可能会导致长波沉降以及前隆沉积溢出。前人的研究已经证实了地幔对流是形成这些动力的根本（Pysklywec和Mitrovica，1997）。静态载荷（均衡的）与动态载荷（如与地幔流的耦合）间的相互作用也可能控制了叠加模式的形成与保存。叠加模式不符合简单弹性挠曲模型的预测结果（Catuneanu等，1997）。未来前陆盆地分析中，区域上地幔流详细定量研究有助于更深入地理解这些系统。

8.4.3 水系、气候与盆地连通性

前陆盆地可以根据水系，区分单向构造负载（monovergent）与多向构造负载（multivergent）（图8-12）来分类。前一种是一般的外流盆地，根据沉积充填程度，这种外流盆地可细分为欠补偿、过补偿、溢出盆地。多数情况下，欠补偿盆地靠近洋—陆过渡带的前方，比如东地中海的东帝汶与爱奥尼亚盆地。它们位于海平面以下（波斯—阿拉伯湾与南阿尔卑

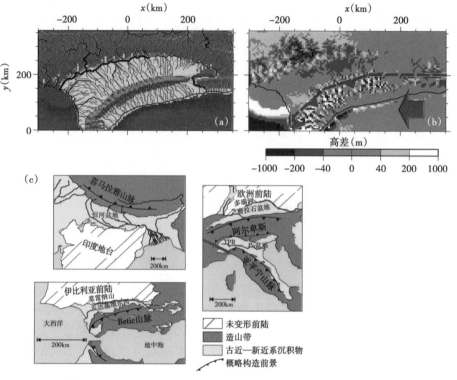

图 8-11 考虑了构造搬运、地表河流搬运以及挠曲均衡的模拟水系
(引自 Garcia-Castellanos, 2002, 修改)

受瓜达尔基维尔盆地(Guadalquivir)(南伊比利亚)的启发,该模拟在初始东—西走向的大陆边缘上叠加了左向移动板块的逆冲。(a) 最终的水系分布,考虑小于 60m 的地形干扰(黑线),不考虑小于 60m 的地形干扰(蓝线)。尽管改变了初始条件,但主要的河流还是系统地沿盆地外部界限流动。彩色阴影表示地表岩性:红层(棕色),同造山期沉积物(黄色),水(蓝色)。(b) 30Ma 后两种模拟地形的差异。虚线标出了最后一个容纳块体左向移动的活动逆断层的位置。初始任意小的干扰将改变短波域的最终地形以及河流的精确位置。虽然大规模的水系格局保存完好,当原始地形(图(a)中的黑线)落差较大时,前隆作为分水岭的作用就会大大降低。(c) 四个前陆盆地(分别为次喜马拉雅盆地,阿尔卑斯磨拉石盆地,Po 盆地—亚得里亚海前渊带(Po-Adriatic foredeep),以及瓜达基维尔盆地水系格局简图。除了有两方构造载荷的 Po 盆地(根据本文术语称为多方收敛盆地)外,其他三个盆地的外部被动边缘都有一条主河流。TPB—古近—新近纪皮德蒙特盆地。红线处为图 8-10 中的剖面位置

斯;Sircombe 和 Kamp,1998),但其前隆可能低于海平面(帝汶盆地与扎格罗斯盆地;DeCelles 和 Giles,1996;Londono 和 Lorenzo,2004),也可能高于海平面。溢出盆地主河流通常平行于盆地远端部分(靠近前隆)走向(阿尔卑斯—磨拉石、Betics—瓜达尔基维尔、喜马拉雅—恒河、南喀尔巴阡以及南次阿尔卑斯盆地)。计算机模拟显示,稳态条件下,这是均衡盆地向造山带倾斜以及河流沉积物向前陆搬运两种作用共同引起的地形变化的结果(Garcia-Castellanos,2002;图 8-11)。早期数值模拟表明,构造与地表过程间的相对速率对确定盆地是欠补偿还是过补偿、盆地沉积相分布以及主要河流水道的位置十分重要(Burbank 和 Anderson,2001)。在溢出盆地中,沉积充填与水系超过了前隆高度,比如艾伯塔盆地、玻利维亚亚马逊前陆盆地(Roddaz 等,2005)、玻利维亚中南部古近—新近纪地层以及安第斯东科迪勒拉(DeCelles 和 Horton,2003)。当前陆基底反转并将初始前陆盆地分为几

个次盆地时，溢出前陆盆地也可能有几个前陆区，比如巴里纳斯—马拉开波（委内瑞拉），马格达莱纳—利亚诺斯（哥伦比亚）、阿尔及利亚—撒哈拉以南、阿特拉斯前渊。

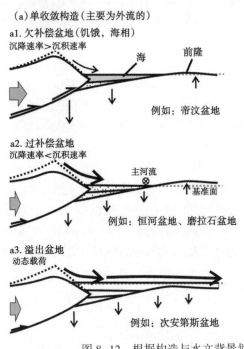

图 8-12 根据构造与水文背景划分的前陆盆地分类

(a) 只有一方构造载荷的盆地（单方收敛盆地）；(b) 多方构造载荷盆地（多方收敛盆地）。第一种盆地绝大多数与饥饿盆地（海相）及过补偿盆地的过渡带有水系流通。海平面变化以及构造载荷与剥蚀作用的相对重要性决定了盆地属于饥饿盆地还是过补偿盆地。附加沉降（比如幔源力造成的沉降）可能会导致沉积物溢出，即，地表沉积物搬运超过前隆高度，比如次安第斯盆地（例如，Roddaz 等，2005；Dávila 等，2007）。图 8-11 与图 8-14 里的数值模拟表明，盆地沉积物不可能超过前隆，除非有附加力（造山带载荷重力除外）的作用。多收敛盆地既可是外流的也可是内流的，但大多数多收敛盆地属于溢出盆地，它们的前隆被沉积物覆盖

在多向收敛前陆盆地中，孤立地形通常控制一个中央湖泊系统，这系统四周连接着径向的水系。盆地周边地形带来的湿空气降低了湖泊水位，抑制了湖泊水外流，这通常就形成内流河（如伊比利亚山脉—比利牛斯—埃布罗盆地、四川盆地、塔里木盆地）。早期水道封闭后，这些盆地的前隆被沉积物覆盖，盆地主要受气候与表面沉积物搬运控制。

埃布罗盆地是一个多向收敛盆地，沉积物来源于四周新生代山脉：比利牛斯山脉，伊比利亚山脉和加泰罗尼沿海山脉（Coney 等，1996；Arenas 等，2001）。该盆地是伊比利亚微板块内的一个沉积区，与海隔绝了 37Ma，至少 20Ma 内是一个内陆盆地。周围山脉的构造抬升造成了盆地的隔绝，干燥气候可能使盆地扩展。干燥气候使内陆湖泊水位降低，从而阻止其通向地中海。由于盆地的沉积充填高出海平面 600~800m，早期流入内部湖泊系统中的水最终于晚中新世通过加泰罗尼海岸山脉流向地中海，形成了现在的埃布罗河（Garcia-Castellanos 等，2003）。自晚中新世开始，埃布罗盆地支流的切割作用形成了其同构造期及后构造期独特的沉积充填样式。鉴于准确约束古抬升高度的现有局限性，这种建模很有意义。

基于磷灰石裂变径迹分析（AFT），Richardson 等（2008）得出了一个非常接近的四川盆地中生代—新生代演化过程（中国西南部，位置见图 8-1）。四川盆地，面积 22.5×

10^4km^2,沉积厚度厚达 4km,具有类似于埃布罗盆地的长期蒸发岩沉积。Richardson 等(2008)结合地层格架与钻井剖面报道了横穿该盆地大部分地区的 AFT 部分退火带,结果表明该盆地从 40Ma 后经历了加速冷却,大范围的沉积层遭受了 1~4km 的剥蚀。这种区域剥蚀被认为是内陆河流域通过长江三峡向外河流域转变的结果。即使盆地边界发生了构造收缩,水系的改变也造成了古近—新近系沉积物的减少。

20 世纪 90 年代与 21 世纪初期的模拟研究表明,剥蚀与沉积作用强烈影响了构造过程,调节着挤压带的变形(Beaumont 等,1992;Avouac 和 Burov,1996;Willett,1999;Jimenez-Munt 等,2005a)。图 8-13a 展示了抬升与剥蚀、前渊带沉积以及下地壳回流间响应的重要性。地表过程、均衡反应、地壳流体间的响应能加快地下最强抬升区地形的增长速率。比如中亚地区、科迪勒拉—次安第斯盆地,以及岩石圈拆沉地区,这些地区的地壳仍与上部板块附着。这种机制对于东阿尔卑斯也很重要,Laubscher(2010)提出,该地区只有地幔岩石圈俯冲。这时候下地壳可被认为是韧性的、流动于脆性上地壳与下伏脆性地幔间的物质。

逆冲动力特征、前陆盆地形成以及增生楔间的响应关系也受到大量数值模拟学者(Braun,2006;Simpson,2006;图 8-13b、c)与物理模拟学者(Persson 和 Sokoutis,2002;Smit 等,2003)的关注。古构造重建三维物理模拟技术与地表物质重新分配三维数值模拟技术有利于数值模拟与物理模拟的结合。模拟结果的验证需要高质量的数据,包括深地震反射剖面(Tomek,1993;Roure 等,2008),地震层序解释以及古温标(Roure 等,2010a)。

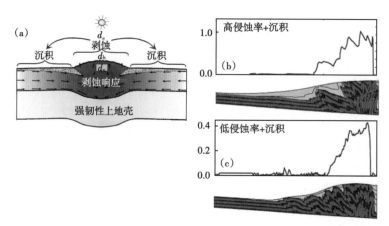

图 8-13 剥蚀/搬运与构造变形的两个例子
(a) Avouac 和 Burov(1996)解释了山链抬升、剥蚀,前渊带沉积以及下地壳回流间的响应(地壳尺度);
(b)、(c) 造山楔生长动态模型模拟的上地壳脆性变形,以及不同搬运速率的地表沉积物搬运
(Simpson,2006),黄色部分表示同造山期沉积

采用数字—物理综合模型研究了地表与构造过程间的相互作用,结果与物理变形模型(砂箱)以及河流搬运数值模型完全耦合(图 8-14;Persson 等,2004)。综合模拟中,用一个托架作为上边界层的压头,挤压最初平坦的沙层。收缩过程中,扫描仪将砂箱地形数字化,缩放后输入数字化地表过程模型。计算出的剥蚀与沉积按比例缩小,然后通过机械地向砂箱表面减少或增加沙量来呈现。动态地形顶部的水系演化导致表面沉积物再分配的横向差异,扰乱了初始的横向(沿走向)对称特征。图 8-14 中运用的模拟方法除了运用计算机计算挠曲均衡以外,其他都类似于 Persson 等(2004)的方法。最初的地形稍微(0.5°)向西(左)倾斜。实验过程中,盆地向的沉积物幕式搬运是由造山带水系重组造成的,而不是本

模型中未考虑的构造收缩率改变导致的。因此，除了构造与气候的幕式变化外，水系重组也是沉积物幕式搬运的机制（Sobel 等，2003）。前隆没有连续地向前陆移动，而是与活动冲断带前部以及弯曲域连续偏移相关，分别离散位于距模型下边界约 300km 以及 150km 处。Hippolyte 等（1994）与 Casero（2004）研究表明，亚平宁山脉米辛尼亚期（Messinian）至中上新世沉积中心经历了如上的连续横向跳跃。这些实验还表明剥蚀与沉积搬运的横向非均质性导致了横向构造穿时，扰乱了砂箱初始的东—西向对称，对变形分布产生了与造山作用早期二维数值模拟相似的影响（Willett，1999）。Persson 等（2004）的实验与图 8-14 中的一个实验，模拟了三维河流沉积物搬运，结果表明沉积物沿造山带轴向搬运以及沿前陆盆地搬运可能会导致盆—山系统构造演化的横向变化。但问题是怎样将沉积物搬运带来的影响与内部非均质性以及构造应力横向变化带来的影响区分开来。

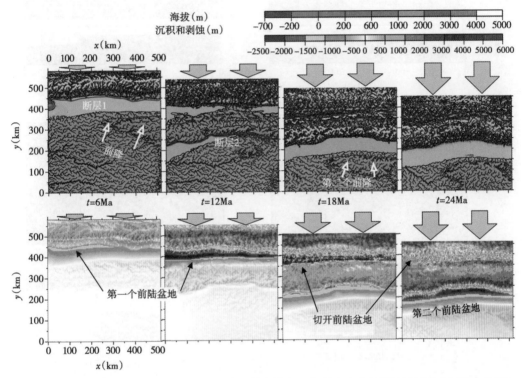

图 8-14 类似于 Persson 等（2004）的物理—数值模拟（但是本文在计算地形时考虑了挠曲均衡）用一个北部（上面）边界受挤压的砂箱来模拟地形与构造载荷。灰色箭头表示托架的累积运动。扫描沙体表面并将数据输入到一个计算河流下切与沉积物搬运的计算机软件（TISC）（Garcia-Castellanos 等，2002）。计算的剥蚀、沉积量按比例缩小，然后手动地用于实验模拟。图中显示了 6、12、18、24Ma 四个演化阶段。初始阶段（0Ma），表面平坦，稍微倾向于左边界，比右边界矮 100m。该造山带的水系南—北向发育，与构造走向垂直，一直到发育第二期更靠南的主要逆冲断层（$t=12$Ma）才有所改变。第二期断层产生的地形造成了沿造山带走向水系的大规模重组（$t=18$Ma，$t=24$Ma）。正如从各种地质背景中观察到的现象一样，由于构造缩短向更南方调节，前陆盆地产生的沉积物聚积被吞噬了（例如，DeCelles 和 Giles，1996）。前方两个主要变形造成了两个主要前隆的抬升。这两个前隆成为盆地与前陆的分水岭，初始东西倾向的水系基本上没被改造。背驮式盆地拼并与前隆的幕式运动与实际观测一致。比如，Hippolyte 等（1994）对南亚平宁山脉的研究结果

地表作用与构造过程的这些模拟中，一个特别有趣的结果是认识到了盆地连通性对约束盆地演化史的重要性。气候控制的盆地水量决定了它是开放水系（外流水系的）还是封闭

水系（内陆河），还决定了沉积物是沉积于盆地还是被搬运出盆地。虽然水系是否开放主要由构造变形分布决定，但图 8-15 表明，没有断层存在时，地表沉积物搬运导致的垂向运动对主要水系的改变也起着重要作用（Garcia-Castellanos 等，2003）。岩石圈刚度低的地区，

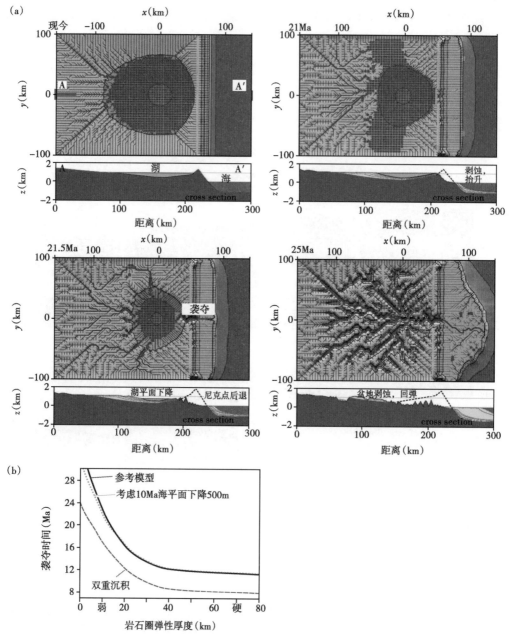

图 8-15　海拔约 1000m 的内陆湖水系被相邻以海平面为基准面的河流袭夺
（据新近纪埃布罗盆地特征）（引自 Garcia-Castellanos 等，2003，有修改）
(a) 袭夺过程的四个阶段；0—初始地形；21Ma 以后，虽然湖泊面积保持不变，但它的平面形态被边缘的沉积物移动而改造（蒸发量必须与沉积量相等）；短而急的入海溪流向源侵蚀以及沉积物堆积造成的海平面升高激发了河流袭夺；21.5Ma，袭夺过程只用了几十万年。(b) 袭夺时间与 EET 的关系式。EET 值小，促进了侵蚀地形屏障的局部均衡反弹，从而延缓了袭夺

内陆河流域盆地的袭夺与开放很慢。因为当岩石圈刚度很低时，地表作用与构造过程就会将两个流域间地形屏障剥蚀造成的垂直抬升最大化。Leever 等（2010）研究潘诺尼亚—次喀尔巴阡山脉系统（图 8-16）时详细探讨了等静压（非构造）控制下盆地连通性对盆地充填几何形态的影响。在这种特殊的构造背景中，喀尔巴阡造山带分隔了潘诺尼亚弱弧后岩石圈与刚度更高的达契奇前渊带（位置见图 8-16）。流变性差异明显影响了潘诺尼亚盆地到巴尔喀阡前渊带的沉积物搬运与分布。地中海许多前渊带都存在流变差异。在这些前渊带中，弧后扩张与邻区前陆逆冲褶皱带的发展密切相关，例如 Valencia Trough-Ebro 盆地系统。图 8-16 显示，刚度最低的岩石圈部分沉积物的容纳空间最大。也就是说，岩石圈刚度相对大的前渊带会减少从弱弧后来的沉积物搬运。如果岩石圈刚度大的前渊向相邻岩石圈相对弱的弧后排水，沉积搬运将会被加强。同时，如上文所述（图 8-15），分隔两个沉积中心的地形屏障内部的岩石圈刚度高意味着，侵蚀回弹更小盆地连通性反而更好。

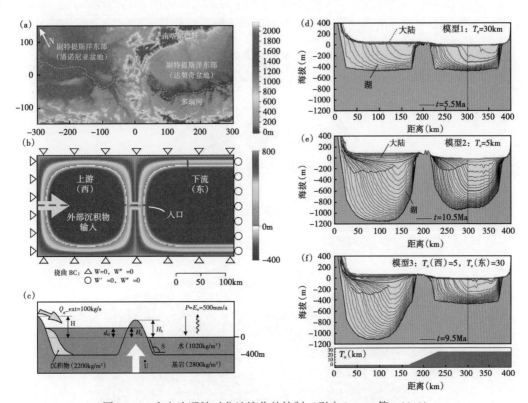

图 8-16 水文连通性对盆地演化的控制（引自 Leever 等，2010）

左图：(a) 喀尔巴阡—达契奇区域地形。(b) 地表过程模型的初始地形与设置。模型西部的沉积物输入（$1.3×10^3 km^3/Ma$）代表大潘诺尼亚水系域的通量。(c) 模型设置以及横截面参数。Q_s-ext—外部沉积物输入；P—沉积速率；e—蒸发率；S—斜度；H—周围地形的最高海拔；H_b—屏障高拔；H_g—入口海拔；D_h—袭夺时基准面间的高差；U—抬升速率。右图：沉积物搬运 15Ma 后的横截面。垂向线标注剖面方和从东西变为南—北。红实线表明上游盆地河流沉积（非湖相）的开始，虚线标注了陆架坡折。(d) — (f) 展示了三个不同 T_e 值条件下的横截面（位置见 (b)）：(d) T_e=30km，(e) 5km，(f) 横向变化，从潘诺尼亚盆地的 5km 变为达契奇盆地的 30km

8.5　讨论与结论

前陆盆地动态系统能为研究岩石圈与地表过程相互作用提供关键信息（图8-17）。多种因素都能约束岩石圈与地表过程的演化，包括构造、古地理、沉积学、古气候、势场（重力场与大地水准面），以及地震数据。从这些数据中可以提取下伏大陆岩石圈的热力学参数，这些参数对重建大陆古地形至关重要。挠曲均衡是一个简单的模型，也是研究地壳均衡的有效的一阶方法，比如基于模拟实验研究与其他过程的一阶相互作用（Garcia-Castellanos, 2002; Pelletier, 2007）。就其性质而言，前陆盆地一般会经历多期演化，并在很大程度上受控于其整体区域构造背景。尤其重要的是板块刚度与应力场随时间的变化和横向变化。这些横向变化解释了盆地垂向运动以及盆地几何形态与简单前陆挠曲模型预测结果间的偏差。岩石圈刚度的横向变化也严重影响着前陆盆地的水系演化以及其与相邻弧后区的连通性。造山前伸展引起的刚度随时间的变化严重影响着随后前陆盆地的沉降，因此后—前渊盆

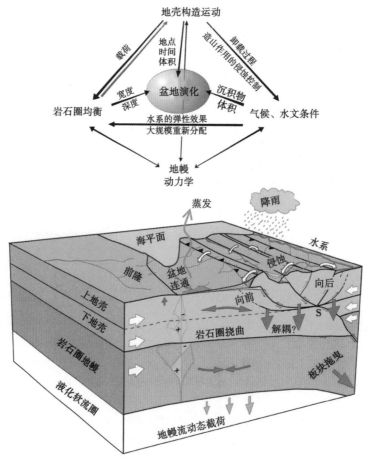

图 8-17　岩石圈与地表过程相互作用模拟图

（a）控制前陆盆地大规模演化的主要过程和一些主要的相互作用。（b）在构造缩短作用引起的区域动力（白色箭头）背景下所形成的主要地质过程和应力（灰色箭头）分布模式图。前隆之下的应力包络特征所显示的弯曲两解耦层（地壳和岩石圈地幔）在大陆岩石圈中伸展（+）和挤压（−）应力的分布。地壳构造和地幔构造、水系和水文连通性控制的沉降作用对盆地演化起到了重要的控制作用

地沉降史与挠曲前陆模型的模拟结果很不一致（Naylor 和 Sinclair，2008）。挠曲前隆对构造体制与岩石圈内部横向非均质性的改变很敏感，前隆强度也可能由于构造负载过程中黏性应力逐步松弛而增加。在某些情况下，前隆可以决定盆地的水系演化史。前陆盆地岩石圈与地表过程间存在很强的响应。数值与物理综合模拟是一种探索造山带—前陆盆地系统构造与气候耦合的有效方法。另外一个强烈影响盆地演化的内在因素是地形。挠曲前地形不仅可以修正挠曲计算中估算的构造负载，还可以叠加构造垂向运动对水系演化的影响。

陆内盆—山系统，比如瓦尔扎扎特盆地（Ouarzazate Basin）（阿特拉斯山脉，摩洛哥）以及塔霍与杜罗盆地（中伊比利亚）（Tajo 和 Duero basins（Central Iberia）），是早期中生代伸展盆地板内挤压与收缩的结果（Cloetingh 等，2002；Arboleya 等，2008）。据报道，在这些系统中，盆地沉降是对横向构造应力的响应。这些横向应力来自岩石圈或地壳褶皱，以及沿断层的少量收缩。前人也提出，山脉附近的小沉积中心可以解释为挠曲特征（Teixell 等，2005；Herrero 等，2010）。相反地，一些挠曲前陆盆地的例子反映了水平负载的影响（如古近—新近纪 Piedmont 盆地与 Guadalquivir 盆地）。与岩石圈褶皱类似，这些水平负载能加强盆地倾斜与前隆抬升。前陆盆地与板内盆地可以作为盆—山系统的两个端元。前陆盆地中垂向应力相对于水平应力更为重要，这可能是由大陆碰撞过程中地壳不断俯冲决定的。为了区分这些挤压盆地的端元盆地的变形，需要具备盆地充填以及深部地壳、岩石圈结构的高品质数据集。

前陆盆地水系演化的野外观察与数值模拟表明，盆地是欠补偿还是过补偿可能取决于地表与构造搬运哪一个作用更强烈以及海平面的变化。然而，河流沉积物搬运的数值模拟表明与造山带的重量无关动态载荷是盆地沉积物溢出的前提（沉积物厚度超过了前隆高度）。该模拟也指出了构造背景在决定其与相邻区域水系连通性方面的重要性。水系连通性控制了沉积物搬运的时间。然而，气候因素对于决定一个盆地是外流的还是内河流域以及大陆化盆地沉积时间也同样重要（Garcia-Castellanos，2006）。

为了进一步了解前陆盆地的动力机制，未来的盆地模拟应该结合岩石圈变形（岩石圈挠曲与褶皱、地壳冲断、沉积变形、岩石圈地幔沉降、下地壳与软流圈流体）与地表过程（结合一些气候过程）的整体动力机制。动态构造模拟显著促进了对造山作用的认识（Willett，1999；Jiménez-Munt 等，2005b；Willett 等，2006；Burov 和 Toussaint，2007；Garcia-Castellanos，2007；Whipple，2009），现在也开始促进对盆地的认识（Simpson，2010）。这些模型正在向可以考虑更多实际地表过程的三维模式发展（Braun，2006）。地表过程模拟显示，地貌中有大量的信息，这些信息有助于更好地定量理解盆—山系统。

致谢（略）

参 考 文 献

Albouy, E., Casero, P., Eschard, R., and Rudckiewicz, J. L. (2002) Tectonics and sedimentation in the Central Apennines. Società Geologica Italiana, 81st summer meeting, 10 – 12 September, 7-8.

Ali, M. Y., and Watts, A. B. (2009) Subsidence history, gravity anomalies and flexure of the United Arab Emirates (UAE) foreland basin. GeoArabia, 14, 17-44.

Allen, P. A., Homewood, P., and Williams, G. D., eds. (1986) Foreland basins: an introduction. International Association of Sedimentologists. Oxford, Blackwell Scientific.

Andeweg, B. (2002) Cenozoic tectonic evolution of the Iberian Peninsula. PhD thesis, VU University, the Netherlands.

Andeweg, B., and Cloetingh, S. (1998) Flexure and "unflexure" of the North Alpine German-Austrian Molasse Basin: constraints from forward tectonic modelling. Geological Society Special Publication, 134, 403–422.

Arboleya, M. L., Babault, J., Owen, L. A., Teixell, A., and Finkel, R. C. (2008) Timing and nature of Quaternary fluvial incision in the Ouarzazate foreland basin, Morocco. Journal of the Geological Society of London, 165, 1059–1073.

Arenas, C., Millán, H., Pardo, G., and Pocovi, A. (2001) EbroBasin continental sedimentation associated with late compressional Pyrenean tectonics (north–eastern Iberia): Controls on basin margin fans and fluvial system. Basin Research, 13, 65–89.

Artemjev, M. E., Kaban, M. K., Kucherinenko, V. A., Demyanov, G. V., and Taranov, V. A. (1994) Subcrustal density inhomogeneities of Northern Eurasia as derived from the gravity data and isostatic models of the lithosphere. Tectonophysics, 240 (1–4), 249–280.

Avouac, J. P., and Burov, E. B. (1996), Erosion as a driving mechanism of intracontinental mountain growth, Journal of Geophysical Research, 101, 17, 747–769. doi: 10. 1029/ 96JB01344.

Bahlburg, H., and Furlong, K. P. (1996) Lithospheric modeling of the Ordovician foreland basin in the Puna of northwestern Argentina: On the influence of arc loading on foreland basin formation. Tectonophysics, 259, 245–258.

Barbeau, D. L. (2003) A flexural model for the Paradox Basin: Implications for the tectonics of the Ancestral Rocky Mountains. Basin Research, 15, 97–115.

Barbieri, C., Bertotti, G., Di Giulio, A., Fantoni, R., and Zoetemeijer, R. (2004) Flexural response of the Venetian foreland to the South alpine tectonics along the TRANSALP profile. Terra Nova, 16, 273–280.

Bayasgalan, A., Jackson, J., and McKenzie, D. (2005) Lithosphere rheology and active tectonics in Mongolia: Relations between earthquake source parameters, gravity and GPS measurements. Geophysical Journal International, 163, 1151–1179.

Beaumont, C. (1981) Foreland Basins. Geophys. J. Roy. Astr. Soc., 65, 291–329.

Beaumont, C., Fullsack, P., and Hamilton, J. (1992), Erosion control of active compressional orogens, in: Thrust Tectonics, edited by K. R. McClay, 1–18, Chapman, and Hall, London.

Bertotti, G., and Mosca, P. (2009) Late-orogenic vertical movements within the arc of the SW Alps and Ligurian Alps. Tectonophysics, 475, 117–127. doi: 10. 1016/j. tecto. 2008. 08. 016.

Bocin, A. (2010) Crustal structure of the SE Carpathians and its foreland from densely spaced geophysical data. PhD thesis, VU University Amsterdam, 124 p.

Bonnet, S. (2010) Shrinking and splitting of drainage basins in orogenic landscapes from the migration of the main drainage divide. Nature Geoscience, 2, 766–770. DOI: 0. 1038/ NGEO666.

Bott, M. H. P., and Stern, T. A. (1992) Finite element analysis of TransantarcticMountain uplift and coeval subsidence in the Ross Embayment. Tectonophysics, 201, 41-356.

Braun, J. (2006) Recent advances and current problems in modelling surface processes and their interaction with crustal deformation. Geological Society Special Publication, 253, 307-325.

Brunet, M. F. (1986) The influence of the evolution of the Pyrenees on adjacent basins. Tectonophysics, 129, 343-354.

Bry, M., White, N., Singh, S., England, R., and Trowell, C. (2004) Anatomy and formation of oblique continental collision: South Falkland basin. Tectonics, 23, TC4011, 1-20.

Burbank, D. W., and Anderson, R. S. (2001) Tectonic Geomorphology. Blackwell Science. 274 pp.

Burov, E., and Toussaint, G. (2007) Surface processes and tectonics: Forcing of continental subduction and deep processes. Global Planet. Change, 58, 141-164.

Burov, E. B., and Diament, M. (1992) Flexure of the continental lithosphere with multilayered rheology. Geophysical Journal International, 109, 449-468.

Burov, E. B., and Diament, M. (1995) The effective elastic thickness (T_e) of continental lithosphere: What does it really mean? Journal of Geophysical Research, 100, 3905-3927.

Burov, E. B., Kogan, M. G., Lyon-Caen, H., and Molnar, P. (1990) Gravity anomalies, the deep structure, and dynamic processes beneath the Tien Shan. Earth and Planetary Science Letters, 96, 367-383.

Caporali, A. (1995) Gravity anomalies and the flexure of the lithosphere in the Karakoram, Pakistan. Journal of Geophysical Research, 100 (B8), 15075-15085.

Cardozo, N. (1997) Thermomechanical modeling of the Llanos Basin, Colombia, South America. American Geophysical Union. Spring meeting. Baltimore.

Carrapa, B., and Garcia-Castellanos, D. (2005) Western Alpine back-thrusting as subsidence mechanism in the Western Po Basin. Tectonophysics, 406, 197-212. doi: 10. 1016/j. tecto. 2005. 05. 021.

Catuneanu, O., Beaumont, C., and Waschbusch, P. (1997) Interplay of static loads and subduction dynamics in foreland basins: Reciprocal stratigraphies and the "missing" peripheral bulge. Geology, 25, 1087-1090.

Casero P. (2004) Structural setting of petroleum exploration plays in Italy. Italian Geological Society, Special Volume for the 32th International Geological Congress, Florence, 189-199+ Plates.

Clevis, Q., De Boer, P. L., and Nijman, W. (2004) Differentiating the effect of episodic tectonism and eustatic sea-level fluctuations in foreland basins filled by alluvial fans and axial deltaic systems: insights from a three-dimensional stratigraphic forward model. Sedimentology, 51, 809-835. doi: 10. 1111/j. 1365-3091. 2004. 00652. x.

Cloetingh, S., and Burov, E. B. (1996) Thermomechanical structure of European continental lithosphere: Constraints from rheological profiles and EET estimates. Geophysical Journal International, 124, 695-723.

Cloetingh, S., Burov, E., Andeweg, B., Beekmann, F., Andriessen, P. A. M., Garcia-

Castellanos, D., de Vicente, G., and Vegas, R. (2002) Lithospheric folding in Iberia. Tectonics, 21, 1041. doi: 10. 1029/2001TC901031.

Cloetingh, S. A. P. L., Burov, E., Matenco, L., Toussaint, G., Bertotti, G., Andriessen, P. A. M., Wortel, M. J. R., and Spakman, W. (2004) Thermo-mechanical controls on the mode of continental collision in the SE Carpathians (Romania). Earth and Planetary Science Letters, 218, 57–76.

Cloetingh, S., Kooi, H., and Groenewoud, W. (1989) Intraplate stresses and sedimentary basin evolution. In: Price, R. A. (ed.), Origin and evolution of sedimentary basins and their energy and Mineral resources. AGU Geophysical Monograph, 48, 1–16.

Cloetingh, S., and Ziegler, P. A. (2007) Tectonic Models for the Evolution of Sedimentary Basins. In: Treatise on Geophysics (ed. G. Schubert), vol. 6, pp. 485–611.

Cogan, J., Rigo, L., Grasso, M., and Lerche, I. (1989) Flexural tectonics of southeastern Sicily. J. Geodynamics, 11, 189–241.

Colletta, B., Roure, F., De Toni, B., Loureiro, D., Passalacqua, H., and Gou, Y. (1997) Tectonic inheritance, crustal architecture and contrasting structural styles along the northern and southern Andean flanks. Tectonics, 16, 777–794.

Coney, P. J., Munoz, J. A., McKlay, K. R., and Evenchick, C. A. (1996) Syntectonic burial and post-tectonic exhumation of the Southern Pyrenees foreland fold-thrust belt, Journal of the Geological Society of London, 153, 9–16.

Coudert, L., Frappa, M., Viguier, C., and Arias, R. (1995) Tectonic subsidence and crustal flexure in the Neogene Chaco basin of Bolivia. Tectonophysics, 243, 277–292.

Crampton, S. L., and Allen, P. A. (1995) Recognition of forebulge unconformities associated with early stage foreland basin development: example from the north Alpine foreland basin. Bulletin of the American Association of Petroleum, 79, 1495–1514.

Crosby, A. G. (2007) An assessment of the accuracy of admittance and coherence estimates using synthetic data. Geophysical Journal International, 171, 25–54.

Dávila, F. M., Astini, R. A., and Jordan, T. E. (2005) Subcrustal loads in the Andean foreland and pampean plain: stratigraphic, topographic and geophysical evidence. Revista de la Asociacion Geologica Argentina, 60, 775–786.

Dáavila, F. M., Astini, R. A., Jordan, T. E., Gehrels, G., and Ezpeleta, M. (2007) Miocene forebulge development previous to broken foreland partitioning in the southern Central Andes, west-central Argentina. Tectonics 26 (5), TC5016.

DeCelles, P. G., and Horton, B. K. (2003) Early to middle Tertiary foreland basin development and the history of Andean crustal shortening in Bolivia. Bulletin of the Geological Society of America, 115, 58–77. doi: 10. 1130/0016-760620031152. 0. CO; 2.

DeCelles, P. G., and Giles, K. A. (1996) Foreland basin systems. Basin Research, 8, 105–123.

Desegaulx, P., Kooi, H., and Cloetingh, S. (1991) Consequences of foreland basin development on thinned continental lithosphere: application to the Aquitaine basin (SW France). Earth and Planetary Science Letters, 106, 116–132.

Dickinson, W. R. (1974) Plate tectonics and sedimentation, in Dickinson, W. R., ed., Tectonics and sedimentation Special Publication, 22. Tulsa, Society for Sedimentary Geology, 1-27.

Ershov, A. V., Brunet, M-F., Korotaev, M. V., Nikishin, A. M., and Bolotov, S. (1999) Late Cenozoic burial history and dynamics of the Northern Caucasus molasse basin: implications for foreland basin modelling. Tectonophysics, 313. doi: 10. 1016/S0040-1951 (99) 00197-3.

Fan, G., Wallace, T. C., Beck, S. L., and Chase, C. G. (1996) Gravity anomaly and flexural model: constraints on the structure beneath the Peruvian Andes. Tectonophysics, 255, 99-109.

Fan, P., and Ma, B. L. (1990) Generan petroleum geology of the Tarim Basin, 1, 1-21. Beijing, Academia Sinica, Science Press.

Flemings, P. B., and Jordan, T. E. (1989) A synthetic stratigraphic model of foreland basins development. Journal of Geophysical Research, 94, 3851-3866.

Ford, M., Lickorish, W. H., and Kusznir, N. J. (1999) Tertiary foreland sedimentation in the Southern Subalpine Chains, SE France: A geodynamic appraisal. Basin Research, 11, 315-336.

Fuis, G. S., Murphy, J. M., Lutter, W. J., Moore, T. E., Bird, K. J., and Christensen, N. I. (1997) Deep seismic structure and tectonics of northern Alaska: crustal-scale duplexing with deformation extending into the upper mantle. Journal of Geophysical Research, 102, B9, 20, 873-896. doi: 10. 1029/96JB03959.

Garcia-Castellanos, D. (2002) Interplay between Lithospheric flexure and river transport in foreland basins. Basin Research, 14, 89-104. doi: 10. 1046/j. 1365-2117. 2002. 00174. x.

Garcia-Castellanos, D. (2006) Long-term evolution of tectonic lakes: Climatic controls on the development of internally drained basins, in S. D. Willett, N. Hovius, M. T. Brandon, and D. M. Fisher, eds., Tectonics, climate, and landscape evolution. Special Paper, 398. Boulder, CO, Geological Society of America, 283-294. doi: 10. 1130/2006. 2398 (17).

Garcia-Castellanos, D. (2007) The role of climate during high plateau formation: proposals from numerical experiments. Earth and Planetary Science Letters. doi: 10. 1016/j. epsl. 2007. 02. 039.

Garcia-Castellanos, D., Vergés, J., Gaspar-Escribano, J. M., and Cloetingh, S. (2003) Interplay between tectonics, climate and fluvial transport during the Cenozoic evolution of the Ebro Basin (NE Iberia). Journal of Geophysical Research, 108 (B7), 2347. doi: 10. 1029/2002JB002073.

Garcia-Castellanos, D., Fernàndez, M., and Torné, M. (1997) Numerical modeling of foreland basin formation: a program relating thrusting, flexure, sediment geometry and lithosphere rheology. Computers, and Geosciences, 23? (9), 993-1003. doi: 10. 1016/S0098-3004 (97) 00057-5.

Garcia-Castellanos, D., Fernàndez, M., and Torné, M. (2002) Modelling the evolution of the Guadalquivir foreland basin (South Spain). Tectonics, 21? (3). doi: 10. 1029/ 2001TC001339.

Gaspar-Escribano, J. M., van Wees, J. D., ter Voorde, M., Cloetingh, S., Roca, E., Cabrera, L., Munoz, J. A., Ziegler.

P. A., and Garcia-Castellanos, D. (2001) Three-dimensionalflexuralmodelling of the Ebro Basin

(NE Iberia). Geophysical Journal International, 145, 2, 349–368. doi: 10. 1029/ 2003TC001511.

Granjeon, D., and Joseph, P. (1999) Concepts and applications of a 3–D multiple lithology, diffusive model in stratigraphic modeling. Numerical Experiments in Stratigraphy: Recent Advances in Stratigraphic and Sedimentologic Computer Simulations, 62, 197–210.

Gries, R. (1983) Oil and gas prospectivity beneath Precambrian of foreland thrust plates in Rocky Montains. Bulletin of the American Association of Petroleum, 67, 1–28.

Grotzinger, J., and Royden, L. (1990) Elastic strength of the Slave craton at 1. 9 Gyr and implications for the thermal evolution of the continents. Nature, 347, 64–66.

Gupta, S. (1997) Himalayan drainage patterns and the origin of fluvial megafans in the Ganges foreland basin. Geology, 25, 11–14.

Gutscher, M. A. (1995) Crustal structure and dynamics in the Rhine Graben and the Alpine foreland. Geophysical Journal International, 122, 617–636.

Haddad, D., and Watts, A. B. (1999) Subsidence history, gravity anomalies, and flexure of the northeastAustralian margin in Papua New Guinea. Tectonics, 18, 827–842.

Handy, M. R., and Brun, J–P. (2004) Seismicity, structure and strength of the continental lithosphere. Earth and Planetary Science Letters, 223, 427–444.

Hardebol, N. (2010) The Foreland Belt of the SE Canadian Cordillera: from thrust–sheet to lithosphere controls on the burial and thermal evolution. PhD thesis, VU University, Amsterdam.

Herrero, A., Anlonso–Gavilan, G., and Colmenero, J. R. (2010) Depositional sequences in a foreland basin (north–western domain of the continental Duero basin, Spain). Sedimentary Geology, 220, 235–264. doi: 10. 1016/j. sedgeo. 2009. 11. 012.

Hetényi, G., Cattin, R., Vergne, J., and Nábělek, J. L. (2006) The effective elastic thickness of the India Plate from receiver function imaging, gravity anomalies and thermomechanical modelling. Geophysical Journal International, 167, 1106–1118.

Hinze, W. J., and Braile, L. W. (1988) The geology of North America, vol. D–2, Sedimentary cover–North American Craton. Boulder, CO, Geological Society of America, 5–24.

Hippolyte, J. C., Angelier, J., and Roure, F. (1994) A major geodynamic change revealed by quaternary stress patterns in the Southern Apennines (Italy). Tectonophysics, 230, 199–210.

Holt, W. E., and Stern, T. A. (1991) Sediment loading on the Western Platform of the New Zealand continent: implications for the strength of a continental margin. Earth and Planetary Science Letters, 107, 523–538.

Horton, B. K., and DeCelles, P. G. (1997) The modern foreland basin system adjacent to the Central Andes. Geology, 25, 895–898.

Horton, B. K., and DeCelles, P. G. (2001) Modern and ancient fluvial megafans in the foreland basin system of the central Andes, southern Bolivia: implications for drainage network evolution in fold–thrust belts. Basin Research, 13, 43–63.

Jiménez–Munt, I., Garcia–Castellanos, D., and Fernàndez, M. (2005a) Thin sheet modelling of lithospheric deformation and surface mass transport. Tectonophysics, 407, 239–255. doi: 10.

1016/j. tecto. 2005. 08. 015.

Jiménez-Munt, I., Garcia-Castellanos, D., Negredo, A., and Platt, J. (2005b) Gravitational and tectonic forces controlling the post-collisional deformation and the present day stress field of the Alps: constraints from numerical modelling. Tectonics, 24, TC5009. doi: 10. 1029/2004TC001754.

Johnson, D. D., and Beaumont, C. (1995) Preliminary results from a planform kinematic model of orogen evolution, surface processes and the developement of clastic foreland basin stratigraphy, in S. L. Dorobek, and G. M. Ross, eds., Stratigraphic evolution of foreland basins. Special Publication, 52. Tulsa, Society for Sedimentary Geology, 1–24.

Jordan, T. E. (1981) Thrust loads and foreland basin evolution, Cretaceous, western United States. Bulletin of the American Association of Petroleum, 65, 2506–2520. Jordan, T. E. (1982) Tectonic loading and foreland basin subsidence. International Congress on Sedimentology, 11, p. 131.

Karner, G. D., and Watts, A. B. (1983) Gravity anomalies and flexure of the lithosphere at mountain ranges. Journal of Geophysical Research B, 88, 10449–10477.

Kroon, I. (2002) Strength of the Adriatic lithosphere: inferences from tectonic modelling. PhD thesis, VU University, Amsterdam, 112 pp.

Kruse, S., and McNutt, M. (1988) Compensation of Paleozoic orogens: a comparison of the Urals to the Appalachians. Tectonophysics, 154, 1–17.

Kruse, S. E., and Royden, L. H. (1994) Bending and unbending of an elastic lithosphere: the Cenozoic history of the Apennine and Dinaride foredeep basins. Tectonics, 13, 278–302.

Laubscher, H. (2010) Jura, Alps and the boundary of the Adria subplate. Tectonophysics, 483, 223–239.

Leever, K., Matenco, L., Garcia-Castellanos, D., and Cloetingh, S. (2010) The evolution of the Danube gateway between Central and Eastern Paratethys (SE Europe): insight from numerical modelling of the causes and effects of connectivity between basins and its expression in the sedimentary record. Tectonophysics. doi: 10. 1016/j. tecto. 2010. 01. 003.

Lin, A. T., and Watts, A. B. (2002) Origin of the West Taiwan basin by orogenic loading and flexure of a rifted continental margin. Journal of Geophysical Research B, 107, 2–1.

Liu, S., Wang, L., Li, C., Zhang, P., and Li, H. (2006) Lithospheric thermo–rheological structure and Cenozoic thermal regime in the Tarim Basin, northwest China. Acta Geologica Sinica, 80, 344–350.

Londono, J., and Lorenzo, J. M. (2004) Geodynamics of continental plate collision during late tertiary foreland basin evolution in the Timor Sea: constraints from foreland sequences, elastic flexure and normal faulting. Tectonophysics, 392, 37–54.

Lorenzo, J., O' Brien, G., Stewart, J., and Tandon, K. (1998) Inelastic yielding and forebulge shape across a modern foreland basin: north west shelf of Australia, Timor Sea. Geophysical Research Letters, 25, 1455–1458.

Lyon-Caen, H., and Molnar, P. (1984) Gravity anomalies and structure of the western Tibet and southern Tarim Basin. Geophysical Research Letters, 11, 1251–1254.

Lyon-Caen, H., and Molnar, P. (1985) Gravity anomalies, flexure of the Indian plate, and the structure, support and evolution of the Himalaya and Ganga basin. Tectonics, 4, 513–538.

Macario, A., Malinverno, A., and Haxby, W. F. (1995) On the robustness of elastic thickness estimates obtained using the coherence method. Journal of Geophysical Research, 100, 15163–15172.

Matenco, L., Krézsek, C., Merten, S., Schmid, S., Cloetingh, S., and Andriessen, P. (2010) Characteristics of collisional orogens with low topographic build-up: an example from the Carpathians. Terra Nova, 22, 155–165.

Matenco, L., Zoetemeijer, R., Cloetingh, S., and Dinu, C. (1997) Lateral variations in mechanical properties of the Romanian external Carpathians: inferences of flexure and gravity modelling. Tectonophysics, 282, 147–166.

McNutt, M. K., Diament, M., and Kogan, M. G. (1988) Variations of elastic plate thickness at continental thrust belts. Journal of Geophysical Research B, 93, 8825–8838.

McNutt, M. K., and Kogan, M. G. (1987) Isostasy in the USSR II: interpretation of admittance data, in Fuchs, K. A. F., ed., Composition, structure and dynamics of the lithosphere-asthenosphere system. Washington, DC, American Geophysicists Union, 309–327.

Merten, S., Matenco, L., Foeken, J. P. T., Stuart, F. M., and Andriessen, P. A. M. (2010) From nappe-stacking to out-of-sequence post-collisional deformations: Cretaceous to Quaternary exhumation history of the SE Carpathians assessed by low-temperature thermochronology. Tectonics, 29, TC3013.

Moretti, I., and Royden, L. (1988) Deflection, gravity anomalies and tectonics of doubly subducted continental lithosphere: Adriatic and Ionian seas. Tectonics, 7, 875–893.

Nadai, A. (1963) Theory of flow and fracture of solids, vol 2. New York, McGraw-Hill, 705 p.

Naylor, M., and Sinclair, H. D. (2008) Pro- vs. retro-foreland basins. Basin Research, 20, 285–303.

Necea, D. (2010) High-resolution morpho-tectonic profiling across an orogen: tectonic-controled geomorphology and multiple dating approach in the SE Carpathians. PhD thesis, VU University Amsterdam, 147 p.

Nemes, F., Neubauer, F., Cloetingh, S., and Genser, J. (1997) The Klagenfurt Basin in the Eastern Alps: an intra-orogenic decoupled flexural basin? Tectonophysics, 282, 189–203.

Nunn, J. A., Czerniak, M., and Pilger, R. H. (1987) Constraints on the structure of Brooks Range and Colville Basin, Northern Alaska, from flexure and gravity analysis. Tectonics, 6, 603–617.

Oszczypko, N., Krzywiec, P., Popadyuk, I., and Peryt, T. (2005) Carpathian Foredeep Basin (Poland and Ukraine) – its sedimentary, structural and geodynamic evolution, in J. Golonka and F. J. Picha, eds., The Carpathians and their foreland: geology and hydrocarbon resources. AAPG, 84, 293–350.

Patacca, E., Scandone, P., Di Luzio, E., Cavinato, G. P., and Parotto, M. (2008) Structural architecture of the central Apennines: Interpretation of the CROP11 seismic profile from the Adriatic coast to the orographic divide. Tectonics, 27, TC3006. doi: 10.1029/

2005TC001917.

Pelletier, J. D. (2007) Erosion-rate determination from foreland basin geometry. Geology 35, 5-8.

Peper, T., Beekman, F., and Cloetingh, S (1992) Consequence of thrusting and intraplate stress fluctuation for vertical motions in foreland basins and peripheral areas. Geophysical Journal International, 111, 104-126.

Pérez-Gussinyé, M., and Watts, A. B. (2005) The long-term strength of Europe and its implications for plate-forming processes. Nature, 436, 381-384.

Persson, K. S., Garcia-Castellanos, D., and Sokoutis, D. (2004) River transport effects on compressional belts: first results from an integrated analogue-numerical model. Journal of Geophysical Research, 109? (B1), B01409. doi: 10. 1029/2002JB002274.

Persson, K. S., and Sokoutis, D. (2002) Analogue models of orogenic wedges controlled by erosion. Tectonophysics, 356, 323-336.

Philippe, Y. (1994) Transfer zones in the southern Jura belt (eastern France): Geometry, development and comparison with analogue modelling experiment. In Mascle A., ed., Hydrocarbon and petroleum geology of France. Special Publication, 4. European Association of Petroleum Geology, 269-280.

Piwowar, J. M., and LeDrew, E. F. (1996) Principal components analysis of arctic ice conditions between 1978 and 1987 as observed from the SMMR data record. Canadian Journal of Remote Sensing, 22, 390-403.

Poudjom-Djomani, Y., Fairhead, J. D., and Griffin, W. L. (1999) The flexural rigidity of Fennoscandia: reflection of the tectonothermal age of the lithospheric mantle. Earth and Planetary Science Letters, 174, 139-154. doi: 10. 1016/S0012-821X (99) 00260-5.

Prezzi, C. B. (1999) Diacronismo en la deformacion Mio-Pliocena de la Puna Argentina: un modelo flexural. Actas del Congreso Geologico Argentino, 14, 197-200.

Prezzi, C. B., Cornelius, E. U., and Gotze, H. J. (2009) Flexural isostasy in the Bolivian Andes: Chaco foreland basin development. Tectonophysics, 474, 526-543.

Price, R. A. (1973) Large scale gravitational flow of supra-crustal facies distribution of the Camp Rice and Palomas Formations, in De Jong, K. A., and Scholten, R. A., eds., Gravity and tectonics. New York, John Wiley and Sons, 491-502.

Pysklywec, R. N., and Mitrovica, J. X. (1997) Mantle avalanches and the dynamic topography of continents. Earth and Planetary Science Letters, 148, 347-455.

Raileanu, V., Talos, D., Varodin, V., and Stiopol, D. (1993) Crustal seismic reflection profiling in Romania on the Urziceni-Mizil line. Tectonophysics, 223, 401-409.

Rankin, D. W., Dillon, W. P., Black, D. F., Boyer, S. E., Daniels, D. L., Goldsmith, R., Grow, J. A., Horton, J. W., Jr., Hutchinson, D. R., Klitgord, K. D., McDowell, R. C., Milton, D. J., Owens, J. P., and Phillips, J. D. (1991) Continent-ocean transect E-4, Central Kentucky to Carolina Trough. Publication of Decade of North American Geology. Boulder, CO, Geological Society of America, 2 sheets.

Ravaut, P., Al Yahya'ey, A., Bayer, R., and Lesquer, A. (1993) Isostatic response of the Arabian platform to ophiolitic loading in Oman [Reponse isostatique de la plate-forme arabique au chargement ophiolitique en Oman]. Comptes Rendus – Academie des Sciences, ser. 2, 317, 463–470.

Richardson, N. J., Densmore, A. L., Seward, D., Fowler, A., Wipf, M., Ellis, M. A., Yong, L., and Zhang, Y. (2008) Extraordinary denudation in the SichuanBasin: insights from low-temperature thermochronology adjacent to the eastern margin of the Tibetan Plateau. Journal of Geophysical Research B, 113, B04409.

Roddaz, M., Viers, J., Brusset, S., Baby, P., and Herail, G. (2005) Sediment provenances and drainage evolution of the Neogene Amazonian foreland basin. Earth and Planetary Science Letters, 239, 57–78.

Roure, F. (2008) Foreland and hinterland basins: what controls their evolution? Davos Proceedings, Swiss Journal of Earth Sciences, Birkhauser Verlag, Basel. doi: 10.1007/s00015-008-1285-x.

Roure, F., Alzaga, H., Callot, J. P., Ferket, H., Granjeon, D., Gonzalez, G. E., Guilhaumou, N., Lopez, M., Mougin, P., Ortuno, S., and Séranne, M. (2009) Long lasting interactions between tectonic loading, unroofing, post-rift thermal subsidence and sedimentary transfers along the Western margin of the Gulf of Mexico: some insights from integrated quantitative studies. Tectonophysics, 475, 169–189.

Roure, F., Andriessen, P., Callot, J. P., Ferket, H., Gonzales, E., Guilhaumou, N., Hardebol, N., Lacombe, O., Malandain, J., Mougin, P., Muska, K., Ortuno, S., Sassi, W., Swennen, R., and Vilasi, N. (2010a) The use of paleo-thermo-barometers and coupled thermal, fluid flow and pore fluid pressure modelling for hydrocarbon and reservoir prediction in fold and thrust belts. Special Publication. London, Geological Society.

Roure, F., Cloetingh, S., Scheck-Wenderoth, M., and Ziegler, P. (2010b) Achievements and challenges in sedimentary basin analysis: a review, in S. Cloetingh, and G. Negendank, eds., New Frontiers in integrated solid earth sciences. International Year of Planet Earth. New York, Springer. doi: 10.1007/978-90-481-2737-5_5.

Roure, F., Choukroune, P., Berastegui, X., Munoz, J. A., Villien, A., Matheron, P., Bareyt, M., Séguret, M., Camara, P., and Déramond, P. (1989) ECORS deep seismic data and balanced cross-sections: geometric constraints on the evolution of the Pyrénées. Tectonics, 8? (1), 41–50.

Roure, F., Roca, E., and Sassi, W. (1993) The Neogene evolution of the outer Carpathian flysch units (Poland, Ukraine and Romania): kinematics of a foreland/fold-and-thrust belt system. Sedimentary Geology, 86, 177–201.

Roure, F., Brun, J. P., Colletta, B., and Vially, R (1994) Multiphase extensional structures, fault reactivation, and petroleum plays in the Alpine Foreland Basin of Southeastern France, in Mascle, A., ed., Hydrocarbon and petroleum geology of France. E. A. P. G. Special Publication no. 4. Paris, Springer-Verlag, 245–268.

Roure, F., Colletta, B., De Toni, B., Loureiro, D., Passalacqua, H., and Gou, Y. (1997) Within-plate deformations in the Maracaibo and East Zulia basins, Western Venezuela. Marine and Petroleum Geology, 14, 139–163.

Royden, L. H. (1988) Flexural behaviour of the continental lithosphere in Italy: constraints imposed by gravity and deflection data. Journal of Geophysical Research, 93, 7747–7766.

Royden, L. H. (1993) The tectonic expression slab pull at continental convergent boundaries. Tectonics, 12, 303–325.

Royden, L., and Karner, G. D. (1984) Flexure of the continental lithosphere beneath Apennine and Carpathian foredeep basins. Nature, 309, 142–144.

Ruppel, C., and McNutt, M. (1990) Regional compensation of the Greater Caucasus mountains based on an analysis of Bouguer gravity data. Earth and Planetary Science Letters, 98, 360–379.

Sachsenhofer, R. F., Lankreijer, A., Cloetingh, S., and Ebner, E (1997) Subsidence analysis and quantitative basin modelling in the Styrian Basin (Pannonian Basin system, Austria). Tectonophysics, 272, 175–176.

Samuelsson, J., and Middleton, M. F. (1999) The Caledonian foreland basin in Scandinavia: Constrained by the thermal maturation of the alum shale: a reply. Journal of the Geology Society of Sweden (GFF), 121, 157–159.

Sanders, C. A. E., Andriessen, P. A. M., and Cloetingh, S. A. P. L. (1999) Life cycle of the East Carpathian orogen: erosion history of a doubly vergent critical wedge assessed by fission track thermochronology. Journal of Geophysical Research, 104, 29095–29112.

Sassi, W., and Rudkiewicz, J. L. (2000) Computer modeling of petroleum systems along regional cross-sections in foreland fold and thrust belts, in Proceedings of geology and petroleum geology of the Mediterranean and Circum Mediterranean basins, Malta, 1–4 October, extended abstracts, C27, 1–4.

Schlunegger, F., Jordan, T. E., and Klaper, E. M. (1997) Controls of erosional denudation in the orogen on foreland basin evolution: The Oligocene central Swiss Molasse Basin as an example. Tectonics, 16, 823–840.

Sheffels, B., and McNutt, M. (1986) Role of subsurface loads and regional compensation in the isostatic balance of the Tranverse ranges, California: evidence for intracontinental subduction, Journal of Geophysical Research, 91, 6419–6431.

Simpson, G. H. D. (2006) Modelling interactions between fold–thrust belt deformation, foreland flexure and surface mass transport. Basin Research, 18, 125–143. doi: 10. 1111/j. 1365–2117. 2006. 00287. x.

Simpson, G. D. H (2010) Influence of the mechanical behaviour of brittle–ductile fold–thrust belts on the development of foreland basins. Basin Research, 22, 139–156. doi: 10. 1111/j. 1365–2117. 2009. 00406. x.

Sinclair, H. D. (1997) Flysch to molasse transition in peripheral foreland basins: The role of the passive margin versus slab breakoff. Geology, 25, 1123–1126.

Sinclair, H. D., Coakley, B., Allen, P. A., and Watts, A. B. (1991) Simulation of foreland basin stratigraphy using a diffusion model of mountain belt erosion: an example from the Alps of eastern Switzerland. Tectonics, 10, 599–620.

Sircombe, K. N., and Kamp, P. J. J. (1998) The South Westland Basin: Seismic stratigraphy, basin geometry and evolution of a foreland basin within the Southern Alps collision zone, New Zealand. Tectonophysics, 300, 359-387.

Smit, J. H. W., Brun, J. P., and Sokoutis, D. (2003) Deformation of brittle-ductile thrust wedges in experiments and nature. Journal of Geophysical Research, 108, B10. doi: 10.1029/2002JB002190.

Snyder, D. B., and Barazangi, M. (1986) Deep crustal structure and flexure of the Arabian plate beneath the Zagros collisional mountain belt as inferred from gravity observations.

Tectonics, 5, 361-373.

Sobel, E. R., Hilley, G. E., and Strecker, M. (2003) Formation of internally-drained contractional basins by aridity-limited bedrock incision. Journal of Geophysical Research, 108, 7, 25 pp.

Stern, T. A. (1995) Gravity anomalies and crustal loading at and adjacent to the Alpine Fault, New Zealand. New Zealand Journal of Geology and Geophysics, 38, 593-600.

Stern, T. A., Quinlan, G. M., and Holt, W. E. (1993) Crustal dynamics associated with the formation of the Wanganui Basin, New Zealand, in Ballance, P. F., ed., South Pacific sedimentary basins. Sedimentary Basins of the World.

Amsterdam, Elsevier, 213-223. Stewart, J., and Watts, A. B. (1997) Gravity anomalies and spatial variations of flexural rigidity at mountain ranges. Journal of Geophysical Research, 102, 5327-5352.

Stockmal, G. S., Beaumont, C., and Boutilier, R. (1986) Geodynamic models of convergent margin tectonics: the transition from rifted to overthrust belt and consequences for foreland basin development. American Association of Petroleum Geologists Bulletin, 70, 181-190.

Tandon, K., Lorenzo, J. M., and O'Brien, G. W. (2000) Effective elastic thickness of the northern Australian continental lithosphere subducting beneath the Banda orogen (Indonesia): inelastic failure at the start of continental subduction. Tectonophysics, 329, 39-60. doi: 10.1016/ S0040-1951 (00) 00187-6.

Tang, J., Lerche, I., and Cogan, J. C. (1992) Aninverse method for calculating basement geometries in foreland basins. Journal of Geodynamics, 15, 85-106.

Tarapoanca, C, M., Garcia-Castellanos, D., Bertotti, G., and Cloetingh, S. (2004) Subsidence mechanisms for the South-eastern Carpathians bend foreland. Earth and Planetary Science Letters, 221, 163-180. doi: 10. 1016/S0012-821X (04) 00068-8.

Tarapoanca, M., Andriessen, P., Broto, K., Cherel, L., Ellouz-Zimmermann, N., Faure, G. L., Jardin, A, Naville, C., and Roure, F. (2010) Forward kinematic modeling of a regional transect in the northern Emirates, using appatite fission track, age determination as constraints on paleo burial history. Arabian Journal of Geosciences, 3, 395-411.

Teng, J. W. (1991) Geophysical fields and hydrocarbon prospects of the Tarim Basin, vol. 2. Beijing, Academia Sinica, Science Press, 24-40.

Teixell, A., Ayarza, P., Zeyen, H., Fernandez, M., and Arboleya, M-L. (2005) Effects of mantle upwelling in a compressional setting: the Atlas Mountains of Morocco. Terra Nova, 17, 456-461.

Tesauro, M., Kaban, M. K., and Cloetingh, S. A. P. L. (2009) How rigid is Europe's lithosphere? Geophysical Research Letters, 36, L16303.

Teson, E., Teixell, A., Ayarza, P., Arboleya, M. L., Alvarez- Lobato, F., García-Castellanos, D., and Amrhar, M. (2006), Geometry and evolution of the Ouarzazate basin in the foreland of the High Atlas Mountains (Morocco), Geophysical Research Abstracts, 8, 01253 (EGU Meeting in Vienna, 2006).

Tomek, G. (1993) Deep crustal structure beneath the central and inner West Carpathians. Tectonophysics, 226, 417–431.

Toth, J., Kusznir, N. J., and Flint, S. S. (1996) A flexuralisostatic model of lithospheric shortening and forelandbasin formation: application to the Eastern Cordillera and Subandean belt of NW Argentina. Tectonics, 15, 213–223.

Turcotte, D. L., and Schubert, G. (2001) Geodynamics: applications of continuum physics to geological problems. New York, Wiley, 430 pp.

van der Beek, P. A., and Cloetingh, S. (1992) Lithospheric flexure and the tectonic evolution of the Betic cordilleras (SE Spain). Tectonophysics, 203, 325–344.

Van der Meulen, M. J., Kouwenhoven, J. J., van der Zwaan, G. J., Meulenkamp, J. E., and Wortel, M. J. R. (1999) Late Miocene uplift of the Romagnan Apennine and the detachment of subducted lithosphere. Tectonophysics, 315, 319–335.

Vening Meinesz, F. A. (1941) Gravity over Hawaiian archipielago and over Madeira area. Proc. Netherlands Acad. Wetensch., 44, 1–12.

Waschbusch, P. J., and Royden, L. H. (1992) Episodicity in foredeep basins. Geology (Boulder), 20, 915–918.

Watts, A. B. (1992) The effective elastic thickness of the lithosphere and the evolution of foreland basins. Basin Research, 4, 169–178.

Watts, A. B. (2001) Isostasy and flexure of the lithosphere. Cambridge, Cambridge University Press, 458 pp.

Watts, A. B., and Burov, E. B. (2003) Lithospheric strength and its relationship to the elastic and seismogenic layer thickness. Earth and Planetary Science Letters, 213, 113–131.

Whipple, K. X. (2009) The influence of climate on the tectonic evolution of mountain belts. Nature Geoscience, 2, 97–104. doi: 10. 1038/ngeo413.

Willett, S. D. (1999) Orogeny and orography: the effects of erosion on the structure of mountain belts, Journal of Geophysical Research, 104, 28957–28981.

Willett, S. D., Schlunegger, F., and Picotti, V. (2006) Messinian climate change and erosional destruction of the central European Alps. Geology, 34, 613–616.

Xie, X., and Heller, P. L. (2009) Plate tectonics and basin subsidence history. Geological Society of America Bulletin, 121, 55–64. doi: 10. 1130/B26398. 1.

Yong, L., Allen, P. A., Densmore, A. L., and Qiang, X. (2003) Evolution of the Longmen Shan Foreland Basin (Western Sichuan, China) during the Late Triassic Indosinian orogeny. Basin Research, 15, 117–138.

Zhou, Di, Yub, H-S., Xua, H-H., Shia, X-B., and Chou, Y-W. (2003) Modeling of thermo-rheological structure of lithosphere under the foreland basin and mountain belt of Taiwan. Tectonophysics, 374, 115-134.

Zoetemeijer, R., Desegaulx, P., Cloetingh, S., Roure, F., and Moretti, I. (1990) Lithospheric dynamics and tectonic-stratigraphic evolution of the Ebro Basin. Journal of Geophysical Research, 95 (B3), 2701-2711.

(胡圣标 译, 张功成 校)

第三部分
裂谷型、后裂谷型、
张扭型和走滑型盆地

第9章 大陆裂谷盆地：来自东非裂谷的新认识

CYNTHIA EBINGER[1], CHRISTOPHER A. SCHOLZ[2]

（1. Department of Earth and Environment Sciences, University of Rochester, Rochester, USA; 2. Department of Earth Sciences, Syracuse University, Syracuse, USA）

摘　要：东非裂谷系由"非洲之角"向南非延伸，贯穿了整个非洲大陆，这里火山地震活动活跃，发育有断层控制的沉积盆地和隆起的翼部。东非裂谷贯穿非洲超级地幔柱形成的宽阔穹隆，裂谷带内发育大量前裂谷期和同裂谷期的岩浆作用。本文回顾了形成宽阔高原、隆起的裂谷翼部和断裂控制的峡谷的裂谷过程，概述了裂谷带内的地壳垂向运动、气候、侵蚀和沉积作用间复杂的相互作用，厘定了裂谷带自博茨瓦纳至阿法尔裂谷带内裂谷构造的时空组合模式。总结了东非裂谷系中的晚新生代盆地的沉积地层格架的特征。这种相似的裂谷构造时空组合模式证实了断裂作用与沉积作用的相互关系，有助于建立裂谷盆地的结构构造、地层序列以及前裂谷期和同裂谷岩浆作用的演化模型。

关键词：裂谷盆地　裂谷结构　盆地分割　裂谷地层　同裂谷岩浆作用

9.1 引言

非洲中东部广阔的裂谷地带是重现人类早期进化史的天然剧场，同时它在过去的 30Ma 中，也为人类和动物的迁徙提供了通道（图 9-1）。地形的变化破坏了大气环流，加剧了亚洲的季风环流（Sepulchre 等，2006；Spiegel 等，2007）。伴随东非断裂系的形成、演化，并直到连通为一个裂谷系统，湖泊最初相互孤立，随后，随着河流不断地改道，河湖体系最终连通并改变了地貌。地震火山活动强烈的裂谷带为初步研究板块构造运动学和动力学机制以及建立裂谷盆地形成模型提供了理想的实验场。东非裂谷系延伸超过 3000km，包括晚于 1Ma 的断层和阿法尔坳陷内的初期洋壳扩张中心（图 9-1）。因此，东非大裂谷展示了同一地球动力学背景下，一个裂谷由初始发育到最终形成的全部历史。裂谷盆地中大型的露头和巨厚的沉积物记录了裂谷开始发育到大陆破裂过程中板块变形、岩浆活动、侵蚀作用还有大气环流间的相互作用，以及上述作用随气候变化发生的改变。在这些过程中产生的复杂反应明显不同于其他裂谷环境条件下形成的构造组合，特别是前裂谷期和同裂谷期岩浆不发育的地区，以及造山带中后裂谷作用发育的地区。

就大范围来看（>1000km），东非广阔的高原就是地球内部地幔动力学过程的反映（图 9-1）。岩浆作用是高原的重要组成部分之一，玄武质熔岩和长英质熔岩覆盖了直径超过 1000km 的区域，岩浆底侵作用分布更为广阔，厚度甚至超过 10km（White 和 McKenzie, 1989；Maguire 等，2006）。就伸展盆地系统范围（100~1000km）来看，岩石圈伸展和岩浆侵入导致了地壳垂向运动和水平压力梯度变化，从而引起了密度变化，也改变了远源场的板

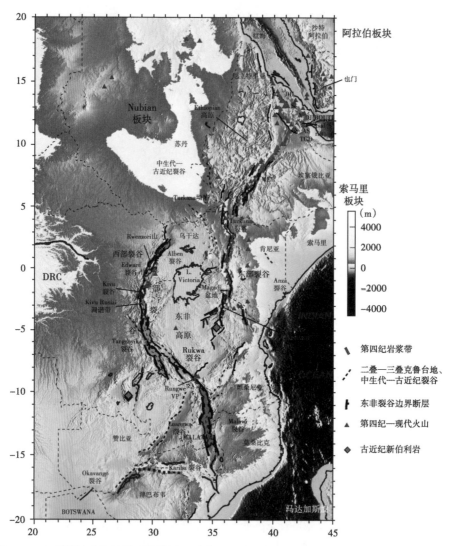

图 9-1 东非裂谷系航天雷达测绘图（引自 Ebinger, 1989a; Versfelt 和 Rosendahl, 1989; LeGall 等, 2004; Kinabo 等, 2007; Modisi 等, 2000)

图示范围包括史密森尼（美国博物馆）全球火山计划数据库 (www.volcano.si.edu) 中，渐新世以来的活火山和东非裂谷系边界断层，但不包括图中西南角的奥卡万戈裂谷区；非洲中部和东部的红海南部裂谷带、亚丁湾西部裂谷带、埃塞俄比亚裂谷带、西部裂谷带、东部裂谷带和发育于非洲南部高原上的初始裂谷带；NW 走向（南部）和 NE 走向（北部）的低隆起区即冈瓦纳裂解事件相关的二叠—三叠纪或侏罗—古近纪裂谷带发育的区域；在许多地区，渐新世以来的东非裂谷控制了这些古老裂谷的水系展布。白色粗线表示二叠—三叠纪和中生代—古近纪裂谷带的断层边界，这些断层被东非裂谷系的 N—S 走向断层截切

块应力场。就独立盆地及其翼部隆起范围（<100km）来看，正断层和走滑断层雁列展布，形成了半地堑型和地堑型盆地的地貌。长英质和玄武质火山作用和侵蚀作用改变了盆地和翼部的地貌，同时这些异常密度的构造也分割了裂谷带。

本章研究目的如下：(1) 总结大陆裂谷发育过程，包括广阔高原、裂谷带两翼和断层控制的裂谷；(2) 概括裂谷带内地壳垂向运动、气候、侵蚀和沉积作用间复杂的相互作用；(3) 确定东非裂谷系内部裂谷构造的时空组合特征；(4) 总结东非裂谷系内，晚新生代盆地的沉积格架。在概括断裂、气候和地表过程、岩浆作用以及地幔动力学机制在裂谷盆地构造演化和地层格架中所起作用的基础上，对比世界范围内活动裂谷系和古老裂谷系，建立东非裂谷系的构造地层格架。最后，本文探讨了边界断层对裂谷沉积物的控制，行星轨道周期对东非裂谷盆地内小尺度地层学的控制作用，及其对均时地层学模型的影响。结论部分指明了裂谷盆地研究工作未来的发展方向。

9.2 裂谷盆地形成与沉积充填

软流圈对流使原始150~250km厚的陆壳减薄至仅几十千米的薄弱陆壳，软流圈上涌、板块碰撞、岩墙上侵和/或岩石圈底部的牵引作用造成远源场板块活动、压力和应力梯度异常产生伸展应力，伸展作用使大陆开裂形成新的洋壳（Bott，1992；Bialas等，2010）。全球范围内衰亡裂谷的遗痕表明，只有部分大陆裂谷经历了伸展作用，局部应变达到破裂的临界点。破裂后不活动的裂谷带称为"被动边缘"，随着裂谷作用减弱和板块间发生热传递，"被动边缘"逐渐沉降到海平面以下。

McKenzie（1978）提出的经典岩石圈伸展模型影响了近20年来的被动大陆边缘构造演化研究。这一大陆裂谷构造的简单模式中，均一的岩石圈强烈拉伸形成盆地，晚期随热能量的消散而沉降。随后的理论和研究数据表明，在有限的裂谷发育时限内，裂谷作用过程中的小规模软流图对流和岩石圈应力场的改变，侵蚀和再沉积作用形成的沉积物分配响应也具有重要的意义（Braun和Beaumont，1989；Weissel和Karner，1989；Burov和Cloetingh，1997；Davis和Kusznir，2004）。根据地幔上涌在裂谷形成和演化过程中所起的作用，裂谷理论模型被定义为"主动型"和"被动型"，主动裂谷受控于地幔浮力形成区域性抬升作用，岩浆作用发育程度不均（Sengoer和Burke，1978）；被动裂谷则受控于板块边界的远源应力场。Huismans等（2001）指出被动裂谷也可能发育主动裂谷特征，表现出伸展作用与地幔动力学之间具有复杂的相互作用。

在以上研究基础上，随着计算机技术的迅速发展，科学家建立了裂谷盆地构造演化和地层组合的复杂三维模型（Frederiksen和Braun，2001；Huismans和Beaumont，2003；Lavier和Manatschal，2006）。全球被动大陆边缘和裂谷获得的大量高品质数据指出岩浆作用在裂谷带的形成和发育过程以及沉积盆地的抬升和沉降历史中所起的重要作用（Nielsen和Hopper，2002；Buck，2004；Van Avendonk等，2009）。岩浆侵入给具有流变学分层的大陆岩石圈增加了问题的复杂性，而这一研究在未来极具潜力（Bialas等，2010）。

大陆裂谷带由一系列平行于裂谷的延伸方向、不连续，且具有运动学成因关系的盆地组成（Bosworth，1983；Rosendahl，1987）。断陷沉积盆地受控于数百万年内继承性活动的断层作用，这一过程中伴生（或不伴生）有岩浆作用。本文首先总结了裂谷带内沉积盆地三维均时构造的一般模式，概括了典型裂谷变形作用在时间和空间上所起的约束作用，随后讨论了构造和气候的变化对地层组合和沉积相的改变和影响。

典型的不对称裂谷盆地通常受控于某一侧的大型同沉积断层系或者控盆断层，其断距由断层中心向两端逐渐减小（Densmore等，2007）。从大范围来看，上盘地层在断面根部发生向下的弯曲，使盆地另一侧地层形成宽缓的单斜褶皱（图9-2）。从小范围来看，断层上盘到控凹断层之间的30~70km宽的地层被一系列正断层切穿，塌陷形成一系列的凹陷（见本章"东非裂谷盆地演化"）。由于板块内部伸展应力大小有限，裂谷两翼的抬升主要受控于下盘沿控凹断层滑动的均衡效应（图9-3）。这种由于断层控制的盆地沉积充填造成的均衡补偿效应的幅度和空间分布主要受控于地壳减薄的几何学、裂谷充填物的密度和岩石圈的抗弯强度（Kusznir 和 Park，1987；Braun 和 Beaumont，1989；Weissel 和 Karner，1989）。强硬板块的回弹作用比软弱板块影响的范围更大（图9-2和图9-3）。与热隆升不同的是该类型裂谷两翼的力学隆升过程是不可逆的。

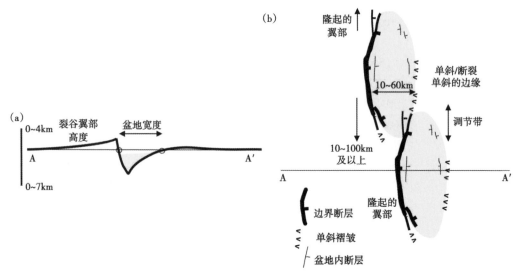

图9-2　构造活动大陆裂谷盆地概略图，边界断层剖面图（a）和平面图（b）（引自Ebinger等，1999）
(a) 大部分新生单相裂谷盆地属于半地堑盆地，大部分变形作用受控于沿边界断层发育的应力应变。均衡补偿弯曲导致裂谷翼部隆起，翼部的宽度和高度随板片强度的增加而增加。低凸起边缘表现为小型断层或单斜褶皱。
(b) 边界断层最大断距位于断层中部，断距向两侧逐渐减少。粗线显示A—A′剖面大概位置。坳陷以阴影显示

裂谷的两翼和盆地，在地貌上往往与大范围的高原穹隆相叠加；这种高山地貌的形成是由多个短期或长期的壳幔密度结构改变过程造成的。广泛分布的岩浆侵入体加厚了玄武岩省内的地壳厚度，这种地壳增厚作用在地表表现为形成了横跨数百千米的高约1000m的山脉（McKenzie，1984；Keir等，2009）。岩石圈减薄，地幔岩石圈被更热、密度更小的软流圈地幔所取代，形成比断陷盆地更大范围的穹隆构造。地壳和上地幔的密度随着温度的升高而降低；岩石圈地幔被更轻、更热的软流圈地幔所代替。软流圈地幔对板块的烘烤作用，在裂谷作用过程中引起了板块密度的降低。尽管其密度在某一深度上降低的可能性比较小，但就伸展岩石圈和未伸展岩石圈整体而言，累加的密度差异所引起的浮力差异，仍然能够决定地表是被抬升还是发生沉降（图9-3）。这些平面上密度的变化同样能够增强或减弱板块构造作用的驱动力（Buck，1991）。

软流圈和地幔深部的动力学过程能够控制空间尺度大于$10^2\sim10^3$km，时间尺度大于100Ma以内的抬升和沉降作用。随着裂谷作用、伸展区宽度和斜坡的进一步发展，裂谷带下方的软流圈开始出现小型对流，岩石圈物质的水平对流加强了烘烤作用并促进了穹隆两翼的抬升（Buck，1986；Van Wijk等，2008）。岩石圈—软流圈的突变型边界会持续影响早期的构造作用、局部岩浆作用、烘烤作用，以及晚期的多相共生体（King 和 Ritsema，2000）。最后，地幔对流作用引起的低密度热流体的上涌，造成大于1000km的大面积穹隆作用和玄武质岩浆的溢流（Lithgow-Bertelloni 和 Silver，1998；Gurnis等，2000）。在不同的动力学模

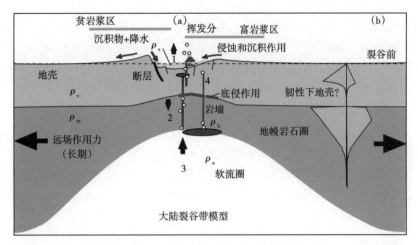

图 9-3 伸展作用过程中岩石圈构造模式图,粗箭头表示浮力方向（1-4）

符合 RHS 图示壳幔流变学曲线的屈服应力图解;大部分地区大陆下地壳的变形和流变学较弱,形成板状和铲状边界断层。a 代表了岩石圈减薄强烈的区域,b 代表裂谷前岩石圈,即"正常"岩石圈。ρ_c 为地壳密度,ρ_d 为岩浆密度,ρ_m 为地幔岩石圈密度,ρ_a 为软流圈密度。浮力 1 相当于断层滑移引起空气、水或沉积物替代地壳产生的浮力,反向的浮力 2 相当于地壳减薄带密度较大的地幔岩石圈代替地壳产生的浮力,浮力 3 相当于软流圈正常热流或热异常代替地幔岩石圈产生的浮力,浮力 4 是侵入的岩墙、岩席或岩浆房等岩浆作用产生的浮力。这四种力会叠加到构造应力中。侵蚀和沉积作用对物质质量仍起重新分配的改造作用。见 Buck（2004）和 Bialas 等（2010）对岩墙侵入过程中的裂谷推动力的详细分析

型中,隆起幅度和展布范围的差异、地幔对流和上覆板块耦合作用,很大程度上受控于岩石圈—软流圈边界的黏度结构（Gurnis 等,2000；Behn 等,2004）。裂谷活动或板块破裂的冷却作用,能够缓慢地将地幔岩石圈恢复到其裂开前的厚度和密度,同时伴随热烘烤区的沉降,这个沉降范围远大于断陷盆地的区域,而地壳厚度的变化则永远保存了下来。

世界范围内的大陆裂谷,在盆地的形成与连通的时空组合上具有相似性,而不取决于其地球动力学背景、先存地壳和岩石圈的不均一性。这一特征表明岩石圈的热力学性质在这一过程中起到了决定性的作用（Dunbar 和 Sawyer,1989；Buck,1991；Bassi,1995）。岩石圈的综合强度决定了裂谷盆地的范围、两翼隆起的高度和宽度,以及断层的延伸长度和影响深度（Jackson 和 Blenkinsop,1997；Ebinger 等,1999）（图 9-2）。板块强度越大,盆地及其隆起的两翼越宽阔,断层规模和长度越大,伴生的地震破裂作用也越频繁（Cowie 和 Scholz,1992；Jackson 和 Blenkinsop,1997）。例如,碰撞造山带的厚而热的地壳中常见短断层和狭窄的盆地,相对冷而厚的强硬岩石圈形成较长的控盆断层和宽阔的盆地。因此,盆地和断层的几何学特征是推断裂谷作用初始阶段构造背景的有力证据（Hutchinson 等,1992；Hayward 和 Ebinger,1996）。

生长断层系的连接方式决定了规则盆地的发育模式,而不取决于前裂谷期的基底性质（Hayward 和 Ebinger,1996；Contreras 等,2000）。如模型所示,其他影响因素,比如岩浆—断层的相互作用也可以影响断层的长度、空间展布和盆地的连通性（Behn 等,2006；Beutel 等,2010）。物理模拟和数值模型中有限的观测结果,表明在裂谷作用递进变形过程中,边界断层往往沿断层走向发育,并由数条短断层连接而成（Cowie 和 Shipton,1998；Densmore 等,2007）。苏伊士湾裂谷和东非裂谷的观测结果表明,断层的相互连接通常发生

在盆地演化开始的1~2Ma内（Morley，1988；Gawthorpe和Leeder，2000；Kinabo等，2007）。裂谷作用的初始阶段，边界断层和盆地边缘的复杂构造带往往平行于前裂谷构造，或为前裂谷构造的再活化的结果（Versfelt和Rosendahl，1989；Smith和Mosley，1993；Kinabo等，2007）。

由于反映新生活动裂谷带的地壳和上地幔减薄的二维图像数量非常少，所以裂谷作用过程中，中—下地壳中正断层的几何学特征和上地幔的应变分布问题目前存在争议。年轻的东非裂谷盆地发育于克拉通岩石圈背景之上，地壳强度大、地温梯度低，地表和下地壳的地震活动表明正断层发育仅影响到地壳的基底层次（Nyblade等，1996；Albaric等，2009）。下地壳强度小、温度高、变形离散，发育独立断层系，使得下地壳流动和岩浆侵入作用相互协调产生的应变，在中地壳形成边界断层（Kusznir等，1991；Buck，2004；Lavier和Manatschal，2006）。Jacksom和Blenkinsop（1997）认为马拉维南部长度超过100km，高约15m的断层形成于一次里氏8级地震。东非众多里氏7级以上的地震记录表明单次地震序列中大量的边界断层发生了滑动。

早期相互独立的边界断层在后期发生相互作用，并相互连接成为多样的调节带或转换带（Morley和Nelson，1990）。Walsh和Watterson（1991）提出"硬连接"和"软连接"的概念，硬连接是指断层面相连的情况，而软连接则包括两断层间岩石的塑性应变、转换斜坡和转换带（Larsen，1988；Peacock和Sanderson，1994）。转换斜坡和转换带往往与裂谷裂开的方向斜交，其几何学特征也随着断层的发育、连接、裂谷递进变形过程中裂谷中心的迁移而发生改变（Morley和Nelson，1990；Schlische，1995；Cartwright等，1996）（图9-2和图9-4）。因此，转换带的几何学和运动学特征是动态变化的（Densmore等，2007）。盆地间调节带中发育的转换断层和斜向走滑正断层，在不同的沉积盆地中也表现出不同的特征（Ebinger等，1989b）（图9-4b）。因此，调节带中边界断层的连通，可以打开或阻隔不同水系的流动，这主要取决于连通断层的倾角关系（Cowie等，2000；Commins等，2005；Densmore等，2007）。

图9-4 （a）阿莫洛边界断层照片，即典型的阶梯状断层，东非裂谷系轴向连通的边界断层（初始裂谷发育后的1~10Ma）；（b）Kivu-Rusizi调节带，位于吕西齐盆地基准面约780m之上，并向北延伸到基伍盆地基准面约1400m之上。照片上部可见裂谷面、断层三角面和斜向构造。这些转换断层调节了裂谷走向上不同方向的应力（据Ebinger，1989b）

拉张作用过程中是否伴随岩浆侵入，可以从侧面反映其发育规模和时空演化。Buck（2004）总结了克拉通岩石圈伸展过程中是否存在岩浆作用的差异，并指出在岩浆作用存在的情况下,裂谷作用所需的应力为无岩浆作用条件下的八分之一。岩浆侵入到韧性较强的岩石圈底部,诱发岩墙上涌,进入强度较大的岩石圈岩层。随伸展作用的增强,侵入体局部的减薄作用、热传递的平流和对流作用降低了板片的强度,岩浆进一步上侵到浅层(图9-3)。在裂谷递进变形过程中,应力也集中到这些热量高并且薄弱的岩浆侵入带内,这样就限制了岩浆作用的空间展布范围(Yamasaki和Gernigon,2009;Beutel等,2010)。玄武质岩浆侵入到地壳浅部层次增加了地壳的密度,导致局部发生沉降;岩浆底侵作用使地壳加厚,引起区域的浮力变化(图9-3)。高位岩浆房中岩浆的补给和挤出,导致地壳密度的时空演变,形成了随时间改变的隆升/沉降的更替模式(Pritchard和Simons,2002;Biggs等,2009)。

卫星大地测量监控和地震台站覆盖技术的改进,进一步揭示了裂谷过程的周期性。陆壳/洋壳裂谷作用的观测结果表明:板块运动所产生的远场应力场的累加过程,在相对短暂的裂谷事件开始前数十年已经开始形成(Doubre和Peltzer,2007;Nooner等,2009)。岩浆裂谷作用过程,不同于地震频发的断层滑动作用,它可以不通过地震释放应变(Wright等,2006;Calais等,2008)。盆岭省的大地应变测量学特征、地震活动特征和高精度热年代学填图对比工作表明,变形作用在空间上迁移的时间是0.1~1Ma(Niemi等,2004;Hammond和Thatcher,2007)。卫星大地监测系统和全球地震监测系统更直观地揭示了大陆裂谷作用在时间和空间上的特征。

裂谷系统内的侵蚀作用和沉积作用,促进并加剧了上述的伸展过程。隆起两翼的侵蚀作用对能干性和脆性较强的上地壳起到了减薄作用,减弱了局部岩石圈的强度(Burov和Cloetingh,1997)。这类可持续达数百万年以上的侵蚀作用,导致了边界断层的侵蚀回退和裂谷隆起两翼的水系逐渐远离裂谷(Van der Beek等,1998;Dansmore等,2004;Petit等,2009)。剥露过程中的均衡补偿作用使山体高度略为降低,但保留了裂谷两翼隆起的几何学形态。

裂谷盆地内部沉积地层的加厚,通常会降低弹性板块的厚度。原因可归纳为以下三点:裂谷结晶基底顶面深度大于相邻地壳,地温梯度高;沉积地层与地壳阻隔;沉积物对板块施加了额外的应力(Lavier和Steckler,1997;Bialas和Buck,2009)。沉积载荷以下的非弹性挠曲作用引起局部的弯曲(Burov和Cloetingh,1997)。

9.3 裂谷盆地的沉积充填:构造与气候的相互作用

大陆裂谷盆地内,沉积作用和充填作用的首要控制因素是伸展过程形成的构造分区和构造起伏。如前所述,沉降中心主要沿主边界断层的滑动方向发育。大多数东非地区的地震活动分布图表明这些沿构造线发育的脆性变形及其附属构造可由地表延伸至地下超过25km的深部(Nyblade等,1996;Albaric等,2009)。下地壳地震大部分发生在太古宙到古元古代的造山带内,该带内地温梯度较低,地壳厚度超过35km(Nyblade和Brazier,2002)。东非裂谷系东支和西支在地貌上易于识别,60~120km长的边界断层限定了一系列相互独立的半地堑,这些盆地现今首尾相连,形成了半连续的初始板块边界(Rosendahl,1978;Morley,1988)(图9-1和图9-2)。每个独立单元内,上盘的沉降中心位于断层中部,下盘的最大隆升也发育于各主要正断层的中部(Morley,1988;Gawthorpe等,1997)。这些边界断层限定了陡坡边缘、小型河道、块体流和重力流(Scholz等,1990;Tiercelin等,1992;Gawthorpe 和 Leeder，2000）。在上盘远离断层的一侧，水系的流域面积更广，沉积系统以前积为主，斜坡相对较缓，可容空间相对局限（Scholz，1995；Gawthorpe 等，1997）（图 9-5）。携带硅质碎屑沉积物的横向水系通常延伸到上盘边缘，甚至可以到达活动伸展区以外的地区（Well 等，1994；Johnson 等，1995；Reynolds，2008）（图 9-5）。

图 9-5　乌干达和刚果民主共和国内爱德华湖裂谷的美国地球资源卫星图像与 GTOPO30 数字高程叠加模型的图像

图示单一裂谷段的半地堑盆地，大部分的伸展作用和盆地沉降受控于湖泊西北缘边界断层的滑动；DRC（西部）盆地边缘下盘的隆起控制了西部水系的展布和地形；边界断层一侧的陡岸水系和塞姆利基河水系东部上盘弯曲的边缘处发育的卡津加运河（右上方）是部分被淹没的下切谷，也是更大的东部边缘流域盆地的一部分

相邻半地堑的连接构造（包括岩浆构造）很大程度上影响了区域河流系统，确定了进入盆地沉积中心的物源通道，进而控制了同裂谷沉积充填的地层几何形态。一旦边界断层发育达到其最大长度，盆地的范围就开始沿断层走向发育（Densmore 等，2004）（图 9-4b）。共用同一断层上盘，断层下盘叠置相对的两个盆地常常具有连续的地形地貌（图 9-6a）。两个沿走向叠置的半地堑，如果其共用一个断层下盘，那么就会形成地形较高的山岭或地垒，如坦噶尼喀湖的卡瓦拉岛山脊（Scholz 等，2003），或贝加尔湖的院士岭（Hutchinson 等，1992）。这些地形较高的区域将半地堑分隔为相对独立的沉积体系，起到分水岭的作用，形成完全独立的地层系统（图 9-6b）。相邻的半地堑和边界断层具有相同的极性，就可能会发育叠置区和转换断坡（Morley 和 Nelson，1990；Peacock 和 Sanderson，1994）（图 9-2）。这些地区将成为三角洲沉积的主要入口，大型流域体系由此扩展到盆地腹地，如艾伯特地堑（图 9-7）（Karp 等，出版中）。裂谷递进变形过程中，这些盆地间转换带的几何学和动力学往往是短暂的，随边界断层的演化、连接状态和区域应变的变化而改变。

内陆流域中大部分伸展系统的演化均与构造地貌密切相关，大型的河道通常注入裂谷盆地中地形最低的区域。最大型的河流系统和最大型的粗粒硅质碎屑沉积发育于各半地堑盆地的上盘、连续的半地堑盆地系统的末端、调节带、转换带、主要边缘断层倾没端或断层终点，以及下盘地形明显减弱的区域（Scholz，1995；Reynolds，2008）（图 9-5 和图 9-7）。

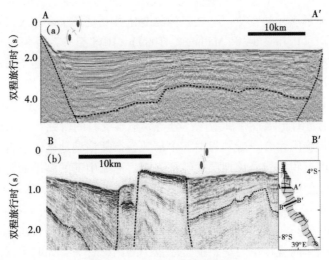

图 9-6 坦噶尼喀湖地震剖面

(a) 坦噶尼喀湖南部地震反射剖面,图示叠置裂谷的盆地结构;虚线表示边界断层面的位置,点线表示前裂谷基底,基底上发育宽缓背形和同裂谷沉积。(b) 坦噶尼喀湖中北部地震剖面,图示共用同一下盘相互叠置的边界断层,以及仅发育较薄同裂谷沉积的显著的基底下盘凸起

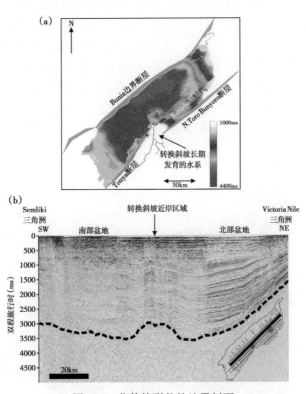

图 9-7 艾伯特裂谷的地震剖面

(a) 基底双程旅行时 (ms) 构造等时线图 (以湖面为基准面),图示艾伯特裂谷的基底凸起和盆地,数据来自海底二维地震资料解释成果;长期发育的水系注入南部边缘两条主要断层系间的盆地中。(b) 沿艾伯特裂谷轴向的海底二维地震反射剖面,图示主转换斜坡近岸的基底凸起和连接南部盆地和北部盆地的调谐带;右下角图显示二维地震测网和地震剖面的位置

229

沉积物的特征和沉积充填速率主要受控于大陆裂谷基准面（湖水面）的变化，而基准面的变化则主要受控于气候的变化。水文条件的变化主要受控于全球大气环流模式、区域地形和局部季风的影响（Perlmutter 和 Mathews，1989）。内陆裂谷系统与大气环流控制的水文特征关系密切，大型湖泊和冲积盆地基准面变化的幅度和频率往往高于海相盆地的变化，新生代海平面变化平均在+/-200m。裂谷系统沉积充填格架的时间跨度约 10^4~10^7a，东非大陆裂谷盆地内连续的沉积组合具有周期性的特征，遵循米兰科维奇定律（Olsen，1986）。Lezzar 等（1996）通过对坦噶尼喀湖地区高分辨率地震资料的研究，识别出了多个反映天体轨道韵律变化的沉积旋回组合。毗邻肯尼亚中部峡谷奈瓦沙湖地区，发育一套保存完好的中晚更新世湖泊和火山沉积地层。该地层的特征表明肯尼亚中部的湖平面和有效水系的空间展布在 0.02Ma 的时间范围内发生有规律的变化（Trauth 等，2003）。这一相对短期的变化被叠加到 5~8Ma 开始的东非大陆长期的干旱背景上（Vrba 等，1995；Behrensmeyer 等，1997；Behrensmeyer，2006）。行星轨道的韵律变化与高纬度气候作用相关（deMenocal，1995）。还有学者认为，气候变化的规律也与更新世全球范围的调整有关，如北半球冰期的开始，沃克环流的开始，中更新世的转换等（Tranth 等，2005）。

东非大部分长期存在的湖盆和潮湿环境的记录很少而且不连续。在非洲热带地区，只有马拉维和坦噶尼喀等较深的湖泊能够在干旱期发育有部分充填的特征。但是对这些湖泊中长期、连续和保存完好的沉积岩心的恢复研究需要大量的资金和先进的技术。2005 年，马拉维湖的科学钻探计划研究恢复了一系列马拉维湖中北部的岩心，其成果扩展了超过 0.5Ma 气候变化的连续记录（Scholz 等，2006）。这些岩心显示出明显的时间变化规律，每 0.02Ma 湖水平面变化 400~500m，与轨道偏心率最大值的时间相同（Scholz 等，2007）。偏心率降低时，干湿变化的时空变化也随之降低，如马拉维湖以潮湿环境为主。然而这常出现在裂谷西支具有较大聚水区的条件下，大陆干旱区其他湖盆是否存在这一现象仍有待验证。钻孔附近区域的高分辨率地震图像明确了沉积过程中明显的变化以及气候的影响（图9-8）。

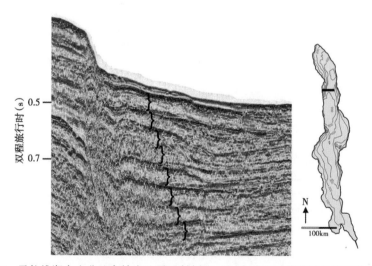

图9-8　马拉维湖中央盆地高精度地震反射剖面（图中展示了清晰的地震沉积旋回）
一个括号代表一个旋回；弱振幅、不连续反射代表湖水低位时期发育粗粒沉积物，连续强振幅反射为细粒盖层和高位期发育有富含有机物的半远洋沉积

9.4 东非裂谷盆地的演化

从地球动力学和大地构造背景方面论述。

埃塞俄比亚的阿法尔三联点，连接了东部裂谷系、西部裂谷系和广阔的埃塞俄比亚—也门以及东非高原（图9-1）。两高原间的坳陷是中生代发育不完全裂谷系的标志，使这一高原有可能成为南非延伸到红海统一隆起的有机组成部分，这个隆起称为非洲超级地幔柱省（Nyblade和Robinson，1994；Ritsema等，1998；Simmons等，2007）。观测模型和地球动力学模型提供了引人瞩目的证据，这些证据表明，在该区域内的一个或多个地幔柱之上发育裂谷作用。对于这一认识的争论主要集中在南非地下核幔边界广阔热软流圈物质低速带的平面展布、深度范围和连续性（Ebinger和Sleep，1998；Weeraratne等，2003；Simmons等，2007）。始新世至今所有火山喷发形成的火山物质的地球化学分析结果表明，埃塞俄比亚—也门地区的溢流玄武岩组合来源于地幔柱，但这一特征在埃塞俄比亚南部的东非高原变得不甚明显（Pik等，2006）。

东非裂谷系最早的火山活动发育于39~45Ma的埃塞俄比亚西南部和肯尼亚的最北端（Ebinger等，2000；Knight等，2003）（图9-1）。约45Ma前侵入东西裂谷间太古宙岩石圈的金伯利岩（Harrison等，2001）和裂谷西南翼的抬升（Batumike等，2007）指示交代作用早于高原隆起裂谷作用，这与地幔主导裂谷模型相一致。红海、亚丁湾最东端和埃塞俄比亚高原中部一线发育的大规模玄武质和长英质岩浆（Baker等，1996；Hofmann等，1997）（图9-1），与红海最南端的初始裂谷作用相一致，是西部裂谷带边界断层沿线线性火山口的标志（Wolfenden等，2005）。埃塞俄比亚广泛分布的玄武质岩浆持续到11Ma前，此时，埃塞俄比亚高原开始发育长英质岩浆（Kieffer等，2004）。埃塞俄比亚南部的东非高原地区，岩浆作用与裂谷断层系统初始形成均发生于16Ma，详述如下。

东非裂谷系最早的伸展作用起始于25Ma左右，位于现今图尔卡纳湖以西的不活跃的盆地，该地区在中生代时，曾发育岩石圈伸展引起的裂谷作用（Morley等，1992）（图9-1）。这一先存的岩石圈"薄弱带"为上升的地幔柱物质提供了空间，有利于以较小的伸展应力诱发小规模的减压熔融作用（Hendrie等，1994；Ebinger和Sleep，1998）。现今大部分肯尼亚的东部裂谷发育于15Ma间，沿其延伸方向喷发大量响岩（Hay等，1995）。东部裂谷系的镁铁质火山作用和断层作用逐渐向南传播，在1Ma内蔓延到坦桑尼亚中部（Foster等，1997）。西部裂谷因缺乏可供测年的矿物而无法确定断层和盆地形成的时限。维龙加省K-Ar年代学数据表明初始的火山作用开始于11Ma左右，附近基伍省的初始火山作用开始于10Ma（Kampunzu等，1998）。坦桑尼亚克拉通南部边界伦圭火山岩省的$^{40}Ar/^{39}Ar$年代学数据表明初始火山作用开始于8.6Ma（Ebinger等，1989a）。

高原隆起和裂谷短轴侧翼隆起的组合型式从根本上改变了东非的水系和气候。东部裂谷和西部裂谷的连接部位发育有宽而浅的维多利亚湖，裂谷翼部的隆起使向西和向东流的河流会聚起来，使水位上升至隆起而未变形的中部高原（Gani等，2007；Spiegel等，2007；Pik等，2008）。Pik等（2004，2008）剖析了广阔高原和裂谷翼部隆升的组合型式，认为埃塞俄比亚高原的隆升时限开始于25Ma。

东非裂谷系统发育的位置受控于先存的岩石圈构造（McConnell，1972）。东部裂谷和西部裂谷发育于未大规模变形（图9-1和图9-3）、能干性强（Petit和Ebinger，2000）、冷

（Nyblade 和 Pollack, 1993）而厚（Ritsema 等, 1998; Weeraratne 等, 2003）的太古宙坦桑尼亚克拉通东部和西部边缘。尽管东非裂谷构造横切了与冈瓦纳大陆裂解相关的二叠纪—三叠纪、侏罗纪—始新世的裂谷系统，但这些构造往往优先发育于上述机械破坏的薄弱地带（Hendrie 等, 1994; Morley, 1998）。埃塞俄比亚和肯尼亚裂谷带的层析模型和宽方位角地震数据反映了裂谷地堑下方的岩石圈地幔减薄和区域上热异常的软流圈（Green 等, 1991; Bastow 等, 2008; Huerta 等, 2009）。裂谷地带的地壳地震和大地电磁数据为"东非裂谷盆地演化"一节的每一实例均存在岩浆过程的影响提供了证据（图 9-9）。

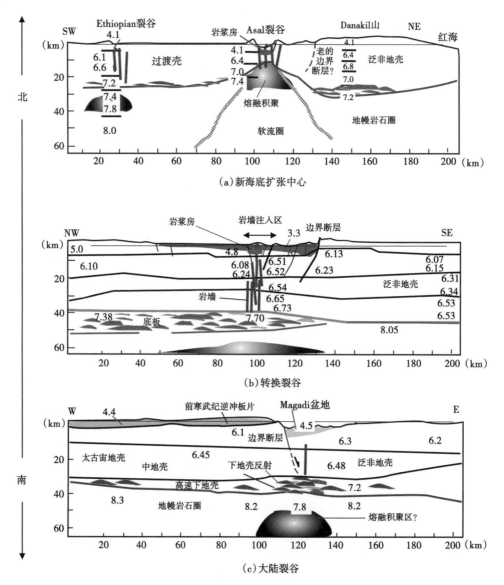

图 9-9　非洲—阿拉伯裂谷系大陆裂解过程剖面示意图

红色粗线为莫霍面，橘色线为 C 剖面底侵带的顶界面，紫色菱形为底侵区域或镁铁质下地壳岩席；简而言之，岩墙可认为起源于地幔和浅部岩浆房，但仍有可能存在复杂多阶段发育的岩浆库。(a) 阿萨尔裂谷新洋底扩张中心（据 Ruegg, 1975; Doubre 和 Peltzer, 2007）。(b) 埃塞俄比亚主裂谷转换裂谷（据 Mackenzie 等, 2005），大部分应变被新生火山带抵消（据 Keir 等, 2006）。(c) 肯尼亚马加迪盆地大陆型裂谷（据 Birt 等, 1997）

东非裂谷盆地的长度和宽度较大，盆地长度超过 80km，宽度为 40~70km（不包括多相裂谷盆地），说明这些年轻盆地下部岩石圈的能干性较强（Hayward 和 Ebinger，1996；Jackson 和 Blenkinsop，1997）（图 9-1、图 9-6 和图 9-7）。利用地震反射、接收函数和断层研究，在东非裂谷与白垩纪—古近纪裂谷叠加的图尔卡纳湖以南地区，对东部裂谷中段地壳伸展量进行估算，获得的地壳减薄量小于 10%（Birt 等，1997）（图 9-9a）。根据断层组合型式分析估算西部裂谷的伸展量小于 20%（Morley，1988）。大地测量估算埃塞俄比亚裂谷系现今板块的打开速度约 6mm/a（Bilham 等，1999），东非裂谷的打开速度约为 3mm/a（Fernandes 等，2004；Stamps 等，2008），与大洋磁条带异常显示的速度一致（Chu 和 Gordon，1999）。通过磷灰石裂变径迹分析获得裂谷两翼的隆升剥蚀量较小（<200m），说明观测到的起伏地形是岩石圈和构造过程的响应（Noble，1997；Van der Beek 等，1998）。

火山喷发中心的分布特征揭示了沿东非裂谷系走向发育的边界断层的时空演化关系（Hayward 和 Ebinger，1996；Keir 等，2009）。岩浆侵入带控制了由裂谷作用到洋底扩张作用过程中长轴方向的分割，详见下面的例子。最早的裂谷沉积序列表明，边界断层顶端是碱性玄武岩、响岩和粗面岩喷发集中的部位（Ebinger 等，1989a）。这些喷出岩沿调节带陡倾断层的出露，说明裂谷初始发育期岩墙作用促进了轴向断层的发育，但这种调节带内岩浆持续发育超过 10Ma 的模型仍有待证明（Beutel 等，2010）。

9.5 东非裂谷盆地演化

下面我们以东非裂谷系的典型实例来说明大陆裂谷向洋底扩张发展过程中，沉积盆地的演化历史。这些实例揭示了裂谷盆地从初始状态到大陆裂解的形成过程中断裂作用、岩浆作用和气候变化所起的作用。

9.5.1 第一阶段，初始断裂作用和沉降：以博茨瓦纳奥卡万戈裂谷为例

研究区内火山岩的年龄自北向南逐渐减小，这些火山岩和断层切穿了东部裂谷系地层，为识别非洲中南部隆起区的初始裂谷提供了证据。非洲中南部的地震台站记录了历史上活动地震和地表断层位移的证据。此外，多个地方也发现了大面积的陆内水系。非洲中部地区，如坦桑尼亚辛吉达省盆地、纳米比亚 Eiseb 地堑和奥卡万戈三角洲是东非及裂谷系轴向生长，发育初始裂谷带的标志（Scholz，1976；Modisi 等，2000；LeGall 等，2004；Wanke，2005）（图 9-1 和图 9-10a）。

奥卡万戈地区地表断层是从 9 万—4.1 万年前开始发育的，而奥卡万戈裂谷的盆地规模与东非裂谷 5~10Ma 形成的盆地相同（Ringrose 等，2005；Kinabo 等，2007）（图 9-10a）。各个独立的断层存在软连接或硬连接，长度可达 25~65km（Kinabo 等，2007），这恰好符合能干性强而且冷的克拉通岩石圈的伸展模式。通过分析航磁和重力数据，可以确定同伸展期沉积充填的厚度约 180~330m（Modisi 等，2000；Kinabo 等，2007），这表明奥卡万戈地堑的初始沉降仅局限在中—晚更新世。在小于 1Ma 的时间内，坦桑尼亚东北裂谷最南端也可见相同的组合型式（Foster 等，1997；LeGall 等，2004）。这些组合反映了断层快速发育并相互连通，最终形成较长的边界断层并控制了早期水系的分布（Morley，1988；Foster 等，1997；Kinabo 等，2007）。地表和浅部地壳不发育岩浆作用。此外，由于地球物理数据太少，无法估计下地壳和地幔岩石圈伸展过程中岩浆所起的作用（Kinabo 等，2007）。随着裂

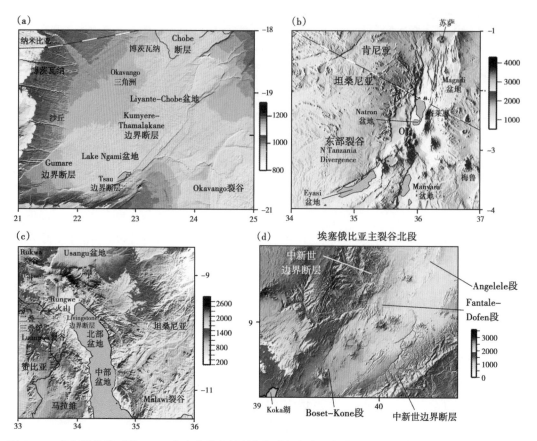

图 9-10 东非裂谷系（图 9-1）由南向北自新到老各裂谷段边界断层、盆地及其两翼几何学特征对比图
各裂谷段航天雷达地形测绘数字高程模型图示盆地及其两翼演化型式。a—c 比例尺相近，d 为局部放大图，可见中新世边界断层内的第四纪岩浆体。(a) 第一阶段，裂谷系内最年轻的伸展区——奥卡万戈裂谷带发育的初始边界断层和盆地；盆地深度<500m，边界断层长度约 100km，盆地宽度约 150km；盆地内部仅发育低凸起，不发育盆地内断层；断层引自 Kinabo 等（2007）。(b、c) 第二阶段，边界断层的轴向连通和斜向转换断层带形成后的大规模垂向构造作用，某些情况下以火山喷发中心的线状排列为标志。(b) 图示肯尼亚南部和坦桑尼亚北部小于 5Ma 的马加迪、纳特隆和马尼亚拉裂谷段。(c) 图示马拉维裂谷小于 9Ma 的卡龙加盆地；马拉维裂谷中央盆地不发育火山喷发中心，但马加迪盆地刚刚开始发育火山喷发中心；对比马拉维裂谷北部（c），东部裂谷段（b）发育更为狭窄的盆地以及大量的火山。第三阶段标志着由大陆型到洋底扩张的转换阶段，较之裂谷作用早期的边界断层，此时的岩浆侵入作用抵消了大部分的裂谷张力。(d) 以埃塞俄比亚中央裂谷北部为例，阶梯状断层切穿的第四纪火山和火山链，它是裂谷晚期发育的标志，阶梯状断层斜交于边界断层，发育时间相对较短，符合图 9-12 所示剖面的构造演化

谷作用和伸展作用的持续发育，盆地宽度也随长度的变化而不断增大，从而保持了盆地的纵横比。这些组合形式与发育于火山弧碰撞带之上的裂谷不同，岩石圈流变学和边界条件也大不相同（Dorsey 和 Umhoefer，本卷第 10 章）。似平行盆地组成的东非宽裂谷带是一系列多期伸展作用或裂谷迁移作用的结果，常见于肯尼亚和埃塞俄比亚复杂的图尔卡纳坳陷（Morley 等，1992；Hendrie 等，1994）。

沉积结构/沉积模型/沉积相主要受控于植被高度发育的曲流河或低弯度河流的陆上冲积三角洲。亚环境包括网状河，含泥炭弯曲河道和某些推进式弯曲河道（Stanistreet 和 McCarthy，1993）。季节性湖泊、永久湖沼和局部三角洲亚环境的发育位置受到十几米范围内地形

变化的影响，并控制了最早期盆地充填地层结构（图9-10a）。

9.5.2 第二阶段，分段发育的沉降、隆升和火山作用：以卡龙加盆地（西部裂谷）和纳特隆盆地（东部裂谷）为例

第一阶段正断层快速的发育连通，初步形成了盆地水系的组合形式。东非裂谷盆地进入了更长期的大规模垂向构造演化阶段。边界断层位移量的增加使沉积序列的厚度达到数千米甚至更多。边界断层控制了盆地大部分的伸展量，仅少量断层是在边界断层弯曲地块形成坳陷的过程中形成的（图9-7和图9-10）。沿断层面倾向的地壳物理剥蚀作用，使其宽缓的翼部抬升比周围地区高出数千米，在这一过程中，密度流充填了坳陷，而侵蚀作用则削弱了隆升作用的响应。在新生裂谷盆地的横剖面上，其几何学结构不对称，最高的翼部发育于倾向盆地最深的一侧。岩浆活动加剧了翼部隆起的抬升，长时间存在的盆地内部隆起，阻塞了水系的轴向流动。这些构造发育的组合形式导致狭长而独立湖盆的形成，其动物群落和沉积相也记录了断层组合的独立或连通特征（图9-11）。

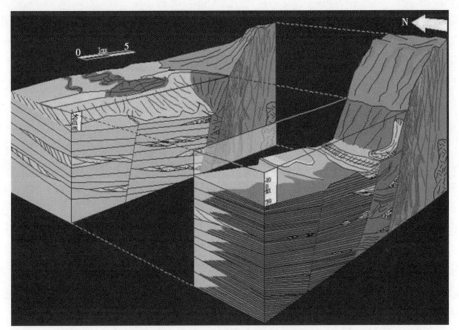

图9-11 马拉维湖北部主要构造、沉积环境和沉积相的概念模型
来自轴向河流体系的进积型三角洲沉积体系，断层控制的大型水下河道延伸至盆地内部的深水区域；
大部分较深的湖盆被细粒而富含有机质的半深水沉积覆盖

西部裂谷小于9Ma的卡龙加盆地和东部裂谷小于5Ma的纳特隆盆地，受到断裂和岩浆的共同作用。最早的岩浆喷发中心位于高角度断层系附近（Ebinger等，1989b；Foster等，1997）。大规模火山发育的地区往往是独立边界断层系统的交叉点。以伦盖碳酸盐岩火山（2962m高）为例，伦盖火山坐落于约100km长的马尼亚拉边界断层和约80km长的纳特隆边界断层之间的调节带处（图9-10b）。伦圭火山（2961m）、恩戈齐火山（2622m）和基埃焦火山（2175m高）是西部裂谷松圭边界断层和卡龙加边界断层间最大的活火山（图9-10c）。

盆地内的地球化学、地震、重力、大地测量、构造和地震数据，显示了构造变形和岩浆的作用范围仅限于裂谷的断层边界处，而其时间范围仅限于裂谷作用发育的最初几个百万年内，此时，裂谷两翼隆起的广阔高原上，不发育可识别的变形。应变集中在垂直于板片裂开方向的少量较长断层中，而早期的无定向性或离散断层没有继承性发育。裂谷下方往往发育少量熔体，这是由于软流圈上涌，充填了伸展作用所形成的空隙，热异常引起的减压熔融作用和板块基底岩石卷入引起了热流值变化（Vauchez 等，1997；Chesley 等，1999；Aulbach 等，2007）。发育于 5Ma 内的纳特隆裂谷中，存在 2007 个沿断层侵入的岩墙和碳酸盐岩火山，这些岩墙和火山爆发释放的主应变量，与地震活动不相符（Calais 等，2008），说明利用地震活动释放能量来估算裂谷打开速度所得的数值可能偏低。

马拉维湖的卡龙加盆地/北部盆地的地震反射剖面表明了地壳伸展作用的地表变形形式。盆地内发育次一级凹陷和正断层，在深水区伴生 10~20m 高的湖底高原（Mortimer 等，2007；Lyons 等，2010）（图 9-11）。在某些近岸区域，这些正断层延伸可达 30km，稳定发育了大型水道系统，形成了重要的斜坡沉积输送通道（Scholz 等，1990；Ng'ang'a，1993）。这些水道局部宽度 0.2~1.0km，在极为干旱的内陆逐渐消失。当湖平面变低时，马拉维湖变小，湖岸一直延伸到半地堑盆地最深的部位。在此期间，陆内河流和冲积扇受控于典型的断层控制的水道；此外，在湖盆的高位期，这些水道仍然控制了湖底的重力流和浊流，将沉积物搬运到数十千米以外的盆地沉降中心（Scholz 等，2007；Soreghan，1999）（图 9-11）。其他较深的裂谷湖泊也可以观察到相似的特征，如坦噶尼喀湖（Tiercelin 等，1992）和贝加尔湖（Nelson 等，1999）。

湖盆的主要差异源于气候的差别。在纳特隆半干旱流域，降雨量不稳定，一般小于 80cm/y（Prins，1987），而马拉维湖的卡龙加盆地流域年降雨量大于 150cm/a（马拉维地调局，1983）。两盆地间的另一主要差异在于火山作用的时空展布。火山及岩浆流动形成的地形阻隔改变了盆地的深度和盆地内水系的展布。伦圭层状火山的高部位坐落于 100km 长的盆地北端，发育在前寒武纪基底上的水系，袭夺了倾斜的中生代盆地中的水系。火山在其南翼局部地区形成了多雨的局部气候区，马拉维湖的局部水循环，并形成了局部的高沉积速率和广泛的前积扇体（Scholz 等，1990）（图 9-10c 和图 9-11）。尽管可能存在海底热泉，但大量淡水注入限制了现今的化学沉积作用。马拉维湖中的黏土物质以蒙皂石为主，北部盆地/卡龙加盆地则以高岭石和伊利石为主（Kalindekafe 等，1996），这种现象推测可能与伦圭火山剥蚀的火山沉积物不同有关。

相反，纳特隆裂谷和马加迪裂谷边缘以及内部的火山地盾和火山岩层在盆地流域内非常狭窄，古老的熔岩流沿边缘断层出露（图 9-9a 和图 9-10b）。纳特隆湖流域内几乎全部为小于 5Ma 的碱性玄武岩和碳酸盐岩的溶解物进入较浅的富含盐碱的湖泊系统而形成的（Yuretich，1982）。盆地内水系较小，玄武岩剥蚀输入的碎屑也较少，但热液沉积、熔结凝灰岩和滑坡沉积局部超出了盆地沉降的速率。区域上大量碱性火山岩的发育造成了纳特隆湖内大量的火山碎屑沉积，这使得富含 Na^+、HCO^- 和 Cl^- 的水体进入湖泊。尽管纳特隆湖区没有具体的记录，但纳特隆湖的重要组成成分是钠沸石、碳酸盐和蒸发岩，这与肯尼亚裂谷中部和南部的其他湖泊以及裂谷东支的其他地区具有相同的矿物组合。

9.5.3　第三阶段，边界断层到岩墙侵入的转换作用，以埃塞俄比亚主裂谷为例

随着板片的持续拉张，岩石圈—软流圈边界上升，减压熔融作用进一步发育。岩石圈最

大拉伸区顶部的中央裂谷的应变，受边界断层和岩浆侵入体共同调节的影响。岩浆侵入将热传导到地壳浅部，板块强度进一步降低。镁铁质岩浆和长英质地壳的横向密度差造成了应力集中，随后的岩墙侵入优先发育于固结的侵入体内部或周缘（Beutel等，2010）。随着时间的推移和热传导作用的加强，熔体上涌到较浅层次，弱化了岩石圈，加剧了应变集中（Buck，2004）。

埃塞俄比亚裂谷系揭示了大陆裂谷作用和海洋裂谷作用转换的一类陆上裂谷带（图9-1）。地幔层析成像模型表明，存在一个75km宽，扁平的低速带隆起到地表以下65km的地幔岩石圈底部（Bastow等，2008）。地震反射和宽方位角反射数据揭示了在隆起的北部高原以下的地壳底部存在一个约10km厚的高速强反射层，Mackenzie等（2005）将其解释为埃塞俄比亚主裂谷伸展作用前玄武质熔体冷却的结果（图9-9b和图9-10d）。因此，在27km厚的地壳中，超过三分之一的厚度可能由前裂谷期或同裂谷期的玄武岩浆溢流所形成的新生的岩浆物质组成（Maguire等，2006）。

裂谷的构造和地球物理横断面表明该盆地是一个深5km的非对称盆地，盆地向大规模边界断层倾斜，如第一、二阶段盆地所示（Mackenzie等，2005；Wolfenden等，2005）（图9-10d和图9-12）。埃塞俄比亚主裂谷的伸展作用大约开始于11Ma前，期间发育高角度初始边界断层，以一系列火山喷发中心为标志。自1.8Ma以来，岩浆作用和断裂作用仅局限于20km宽、60km长的"岩浆带"内，以线性展布的火山喷发中心、裂缝和中央裂谷的短断层为特征（Ebinger和Casey，2001；Casey等，2006）。地震反射的速度变化、三维地震层析成像和重力模型显示岩浆带位于地下约10km的高速、高密度长条状地质体以下，其应为冷却的熔融侵入体（Keranen等，2004；Tiberi等，2005）。构造和地震组合说明边界断层并不活跃，这是由于20km宽、50km长的岩浆带内，存在约10km厚的孕震地壳，孕震地壳内的岩墙作用抵消了应变（Keir等，2006）。此时的裂谷演化，埃塞俄比亚主裂谷发育大陆裂谷和海洋裂谷的变形特征，中中新世大规模边界断层和宽阔的翼部控制着长而宽的盆地内发育出的一个新的短而狭窄的岩浆轴向带（图9-10d）。

从45Ma至今，埃塞俄比亚裂谷的组合形式受到岩浆作用影响，但是其组合形式也不能完全反映东部裂谷盆地、西部裂谷盆地、非岩浆裂谷以及世界其他边缘裂谷的演化序列。这里的岩浆供给，有利于岩浆侵入所形成的应变释放和岩墙作用所引起的应力降低，不利于裂谷带内，大规模位移断层的发育。因此，边界断层并不活跃，岩浆侵入抵消了大于50%或更多的应变（Casey等，2006）。岩浆侵入体以上发育小规模断层和岩墙，分散了地壳顶部脆性层的应变。复合火山和裂开的盾形火山的多期次喷发，形成定向排列的火山口，与狭窄岩浆带上的裂缝相互叠加。随着时间的推移，在世界范围内，早期熔岩流向裂谷轴部弯曲，并在被动陆缘形成了向海倾斜的楔状基底。

火山口湖（如Shalla湖），以及由侵蚀的边界断层和盆地中部地震火山活动岩浆带形成的狭窄菱形地体是本裂谷阶段的沉降中心。在盆地边缘发育的早期地层，被沉降作用和断层的迁移作用改造，最终转变为盆地中部的狭窄区域。局部的隆升沉降和岩性组合受控于突发性的火山喷发及其喷发物的展布范围，以及相关的热液系统和气候变化的影响（LeTurdu等，1998）。因此，埃塞俄比亚主裂谷带代表了盆地演化的一个重要阶段，反映了这一时期盆地的形态、样式、侵入岩体和喷出岩体的体积变化，以及应变速率的增加（图9-12）。

这些组合形式在海底扩张开始的时候最为典型，如复杂的阿法尔裂谷带（Wolfenden等，2005；Quade和Wynn，2008）。国际上对阿法尔地区地球物理学、地质学和地球化学特

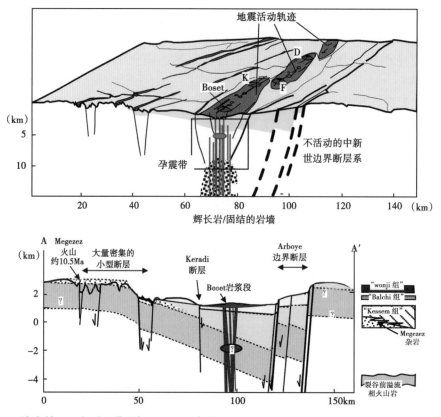

图 9-12 埃塞俄比亚主裂谷带北部地区上地壳构造三维几何模型（上图）和三维结构剖面（下图）
揭示了岩浆带（长度 50~60km）与现今不活动的边界断层系（长度 60~80km）之间的关系；三维图示范围与图 9-10d 大致相同；剖面图示意正面，垂向上以 10:1 比例尺放大，有利于显示地层样式（引自 Wolfenden 等，2005）；图 9-9b 显示了这一剖面更深层次的地壳构造特征；裂谷前以中生代—古近纪被动边缘序列与渐新世—中新世溢流玄武岩互层为主（不出露结晶基底）；盆地充填以火山碎屑岩和火山岩为主，含少量薄层富集火山碎屑岩的河流—湖泊相地层（克塞姆组），Megezez 是坐落于裂谷边缘的被侵蚀的中新世火山；Balchi 组和 Wonji 组是来自 Boset 火山、Kone 火山和间歇性喷发的火山熔岩和熔灰岩；Boset、Kone、Fantale 和 Dofan 火山是裂谷控制的成层性火山；见 Keir 等（2009）关于这一裂谷段地球物理资料的综合研究

征的研究成果为该地区地壳和上地幔构造提供了至关重要的约束，为我们在同一地质动力学背景下，研究克拉通裂谷从起始到消亡的演化历史提供了有利条件。

9.6 结论

东非裂谷系起源于元古宙的造山带，具有较深的太古宙克拉通基底，板块的能干性较强，控制了裂谷初始阶段岩石圈局部应变的非均一性。埃塞俄比亚—也门地区的玄武岩浆溢流作用开始于 45Ma，与非洲东部和中部克拉通区金伯利岩侵入时限相同，说明从红海南部到博茨瓦纳地区存在长期发育的、由深部地幔上涌驱动的伸展、隆升和岩浆作用。地幔的交代作用在这个厚而冷的克拉通岩石圈的初始裂谷作用中起到重要的作用。在初始裂谷阶段，裂谷内隆升作用和水系的控制作用较弱。

东非裂谷系内部裂谷构造的时空组合特征表明裂谷演化过程中断裂作用、岩浆作用、侵蚀作用和沉积作用具有密切的关系。裂谷作用开始的1Ma，厚而坚硬的克拉通岩石圈内发育边界断层向两侧延伸和/或相互贯通从而形成长的边界断层系统。这些断层系统开始控制盆地内的水系。沉积作用受控于地表的冲积扇和曲流河的发育。这些浅而广阔的盆地的地层组合受小规模的高地的影响。随着硬岩石圈裂谷作用的发育，边界断层系统抵消了大部分的伸展作用，从而形成了深而广阔的不对称盆地，其翼部宽而高，某些情况下可以形成较深的湖泊。盆地的水系开始广泛发育，较长的断层系统控制了水系的展布。气候在控制裂谷湖盆系统发育、火山的时空展布等方面起重要作用。地形高的部位被火山及其相关的熔岩叠加，改变了盆地深度和盆地内外的水系分布，使得湖泊系统发生盐碱化。

随后的板块加热变薄，减压熔融体积增加，岩浆侵入抵消了更多的应变。岩浆侵入体顶部的狭窄区域内，发育出小型的短断层。因为伸展作用只集中在裂谷中部的狭窄范围内，初始裂谷阶段的边界断层不再活跃，同时，新的短断层系和岩墙由于脆性地壳的拉伸作用而开始形成。沿裂谷走向排列的火山喷发出的熔岩相互堆叠，使软弱的板片向裂谷轴弯曲，形成世界范围内，在被动陆缘常见的，向海倾斜熔岩楔状体基底。沉积作用局限于破火山口的湖泊和主动裂谷翼部狭长楔状体处。盆地边缘较老的盆地地层，被沉降作用和断层活动破坏掉，并逐渐迁移到盆地中央的狭窄地带。阿法尔裂谷带内岩浆侵入作用长期发育并且多次活动。薄而重的地壳沉降于附近的海平面以下，促进了蒸发岩与玄武岩熔流互层盆地的形成。

东非裂谷系内，裂谷从初始发育到最终裂解的过程中，断裂、岩浆、沉积、气候和挥发分之间发生了复杂的相互作用，并受到裂谷初始期地幔交代作用、裂谷发育期不同程度的岩浆作用的强烈影响。裂谷系形成于不同的环境，如造山带或岩浆作用不发育的强亏损地幔之上，不同裂谷系中各阶段作用间的相互平衡明显不同。

致谢（略）

参 考 文 献

Albaric, J., Deverchere, J., Petit, C., Perrot, J., and Le Gall, B. (2009) Crustal rheology and depth distribution of earth-quakes: insights from the East African Rift, in Peron-Pinvidic, G., Shillington, D., van Wijk, J. and Gernigon, L., eds., Role of magmatism in continental lithosphere extension. Tectonophysics, 468, 28–41. doi: 10.1016/j.tecto.2008.05.021.

Aulbach, S., Rudnick, R. L., and McDonough, W. F. (2007) Li-Sr-Nd isotope signatures of the plume and cratonic lithospheric mantle beneath the margin of the riftedTanzanian craton (Labait). Contributions Mineralogy and Petrology, 155, 79–92.

Bassi, G. (1995) Relative importance of strain rate and rheology for the mode of continental extension. Geo-physical Journal International al, 122, 195–210.

Bastow, I. D., Nyblade, A. A., Stuart, G. W., Rooney, T., and M. H. Benoit (2008) Upper mantle seismic structure beneath the Ethiopian hotspot: rifting at the edge of the African low velocity anomaly. Geochem. Geophys. Geosyst., 9 (12). doi: 10.1029/2008GC002107.

Batumike, J. M., S. Y. O'Reilly, Griffin, W. L., and E. A. Belousova (2007) U-Pb and Hf-isotope analyses of zircon from the Kundelungu Kimberlites, D. R. Congo: Implica-tions for crustal evolution, Precambrian Research, 156, 195–225.

Behn, M. D., Conrad, C. P., and Silver, P. G. (2004) Detect ion of upper mantle flow associated with the African super-plume, Earth Planet. Sci. Lett., 224, 259-274.

Behn, M., Buck, W. R., and Sacks, S. (2006) Topographic controls on dike injection in volcanic rift zones. Earth Planet. Sci. Lett. 246, 188-196.

Behrensmeyer, A. K. (2006) Climate change and human evolution, Science, 311 (5760), 476-478.

Behrensme yer, A. K., Todd, N. E., Potts, R., and McBrinn, G. E. (2007) Late Pliocene faunal turnover in the Turkana Basin, Kenya and Ethiopia. Science, 278, 1589-1594. doi: 10. 1126/scienc e. 278. 5343. 1589.

Beutel, E., J. van Wijk, Ebinger, C., Keir, D., and Agostini, A. (2010) Formation and stability of magmatic segments in the Main Ethiopian and Afar rifts, Earth Planet. Sci. Letts. doi: 10. 1016/j. epsl. 2010. 02. 006.

Bialas, R. W., and Buck, W. R. (2009). How sediment promotes narrow rifting: Application to the Gulf of California, Tectonics, 28, TC4014. doi: 10. 1029/2008TC0 02394.

Bialas, R., Buck, W. R., and Qin, R. (2010) How much magma is required to rift a continent? Earth Planet. Sci. Lett., 292, 68-78.

Bilham, R., Bendick, R., Larson, K., Mohr, P., Braun, J., Tesfaye, S., and Asfaw, L. (1999) Secular and tidal strain across the Main Ethiopian Rift, Geophys. Res. Lett., 26 (18) 2789-2792.

Birt, C., Maguire, P. K. H., Khan, M. A., Thybo, H., Keller, G., and Patel, J. (1997) The influence of pre-existing struc-tures on the evolution of the southern Kenya rift valley: Evidence from seismic and gravit y studies: Tectonophy-sics, 278, 211-242.

Bosworth, W. (1985) Geometry of propagating continental rifts, Nature, 316, 625-627.

Bott, M. H. P. (1991) Sublithospheric loading and plate-boundary forces. Philosophical Transactions: Phys. Sci. and Eng., 337, 1645, 83-92.

Braun, J., and Beaumont, C. (1989) A physical explanation of the relation between flank uplifts and the break-up unconformity at rifted continental margins: Geology, 17, 760-764.

Buck, W. R. (1986) Small-scale convection induced by pas-sive rifting: The cause for uplift of rift shoulders, Earth Planet. Sci. Letts., 77, 362-372.

Buck, W. R. (1991) Modes of continental extension: Journal Geophysical Research, 96, 20161-20178.

Buck, W. R. (2004) Consequences of asthenospheric variability on continental rifting, in Karner, G., B. Taylor, N. Driscoll, and B. Kohlstedt (eds.) Rheology and deformation of the lithosphere at continental margins, Columbia Univ. Press, 92-137.

Burov, E. B., and Cloetingh, S. (1997) Erosion and rift dynamics: New thermomechanical aspects of post-rift evolution of extensional basins: Earth Planetary Science Letters, 150, 7-26.

Calais, E., N. d'Oreye, Albaric, J., Deschamps, A., Delvaux, D., J. Deverchere, Ebinger, C., Ferdinand, R., Kervyn, F., Macheyeki, A., Oyen, A., Perrot, J., Saria, E., Smets, B., Stamps, D., and Wauthier, C. (2008) Aseismic strain accommodation by dyking in a youthful continental rift, East Africa Nature, 456. doi: 10. 1038/nature07478.

Cartwrigh t, J., Mansfield, C., and Trudgill, J. (1996). The growth of normal faults by segment linkage. Special Publications, 99. Geological Society of London, 163-177.

Casey, M., Ebinger, C., Keir, D., Gloaguen, R., and Mohamed, F. (2006) Extension by dyke extension and faulting in an incipient oceanic rift: Ethiopian rift, Africa, in Yirgu, G., Ebinger, C., and Maguire, P. The Afar Volcanic Province within the East African Rift System. Special Publication, 259. Geological Society of London, 143-164.

Chesley, J., Rudnick, R., and C-T. Lee (1999) Re-Os system-atics of mantle xenoliths from the East African rift: age, structure, and history of the Tanzania craton, Geochim. Cosmochim. Acta, 63, 1203-1217.

Chu, C., and Gordon, R. (1999) Evidence for motion between Nubia and Somalia along the Southwest Indian Ridge: Nature, 398, 64-67.

Commins, D., Gupta, S., and Cartwright, J. (2005) Deformed streams reveal growth and linkage of a normal fault array in the Canyonlands graben, Utah, Geology, 33, 645-648. doi: 10.1130/G21433AR. 1.

Contreras, J., Anders, M. H., and Scholz, C. H. (2000) Growth of normal fault systems: observations from the Lake Malawi basin of the east African rift, Journal of Structural Geology, 22, 159-168.

Cowie, P. A., Gupta, S., and Dawers, N. H. (2000) Implica-tions of fault array evolution for synrift depocentre development: insig hts from a numerical fault growth model, Basin Research, 12, 241-261.

Cowie, P. A., and Scholz, C. H. (1992) Growth of faults by accumulation of seismic slip. J. Geophys. Res., 97 (B7) 11, 085-11, 095.

Cowie, P. A., and Shipton, Z. (1998) Fault tip displacement gradients and process zone dimensions Journal of Struc-tural Geology, 20, 983-997.

Davis, M., and Kusznir, N. (2004) Depth-dependent litho-spheric stretching, in Karner, G., B. Taylor, N. Driscoll, and B. Kohlstedt, eds., Rheology and deformation of the lithosphere at continental margins. New York, Columbia University Press, 92-137.

deMenocal, P. B. (1995) Plio-Pleistocene African climate. Science, 270, 53-59.

Densmore, A. L., Dawers, N. H., Gupta, S., Guidon, R., and Goldin, T (2004) Footwall topographic development during continental extension, J Geophys. Res., 109. doi: 10. 1029/2003JF000115.

Densmore, A. L., Gupta, S., Allen, P. A., and Gilpin, R. (2007) Transient landscapes at fault tips, J Geophys. Res., 112. doi: 10. 1029/2006JF000560.

Doubre, C., and Peltzer, G. (2007) Fluid-controlled faulting process in the Asal Rift, Djibouti, from 8 yr of radar interferometry observations, Geology, 35, 69-72.

Dunbar, J., and Sawyer, D. (1989) How pre-existing weaknesses control the style of continental breakup, J. Geophys. Res., 94, 7278-7292.

Ebinger, C. (1989a) Geometric and kinematic development of border faults and accommodation zones, Kivu-Rusizi rift, Africa, Tectonics, 8, 117-133.

Ebinger, C. (1989b) Tectonic development of the western branch of the East African rift system, Geol. Soc. Amer. Bull., 101, 885-903.

Ebinger, C., and Casey, M. (2001) Continental break-up in magmatic provinces: an Ethiopian example. Geology, 29, 527-530.

Ebinger, C. J., Jackson, J. A., Foster, A. N., and Hayward, N. J. (1999) Extensional basin geometry and the elastic lithosphere: Philosophical Transactions Royal Society London A., 357, 741-765.

Ebinger, C. J., and Sleep, N. H. (1998) Cenozoic magmatism throughout east Africa resulting from impact of a single plume: Nature, 395, 788-791.

Ebinger, C., Yemane, T., Harding, D., Tesfaye, S., Rex, D., and Kelley, S. (2000) Rift deflection, migration, and propagation: Linkage of the Ethiopian and Eastern rifts, Africa, Geol. Soc. Amer. Bull., 102, 163-176.

Fernandes, R. M. S., Ambrosius, R. A. C., Noomen, R., L. Bastus, Combrinck, L., Miranda, J. M., and Spakman, W. (2004) Angular velocity of Nubia and Somalia from continuous GPS data: implications on presentday relative kinem atics, EPSL, 222, 197-208.

Foster, A. N., Ebinger, C. J., Mbede, E., and Rex, D. (1997) Tectonic development of the northern Tanzanian sector of the East African rift system. Journal of the Geological Society of London, 154, 689-700.

Frederiksen, S., and Braun, J. (2001) Numerical modelling of strain localisation in the mantle lithosphere. Earth and Planetary Science Letters, 188, 241-251.

Gani, N. D., Gani, M. R., and M. Abdel-Selam (2007) Blue Nileincision on the Ethiopian Plateau: Pulse d plateau growth, Pliocene uplift, and hominin evolution. GSA Today, 17 (9). doi: 10. 1130/GSAT01709A. 1.

Gawthorpe, R. L., Sharp, I., Underhill, J. R. and Gupta, S. (1997). Linked sequence stratigraphic and structural evolution of propagating normal faults. Geology, 25, 795-798.

Gawthorpe, R. L., and Leeder, M. R. (2000) Tectono-sedimentary evolution of active extensional basins. Basin Research, 12, 195-218.

Green, V., Achauer, U., and Meyer, R. P. (1991) A three-dimensional seismic image of the crust and upper mantle beneath the Kenya rift. Nature, 354, 199-203. doi: 10. 1038/354199a0.

Gurnis, M., Mitrovica, J., Ritsema, J., and van Heijst, H-J. (2000) Constraining mantle density structure using geo-logical evidence of surface uplift rates: The case of the African superplume. G3, 1.

Hammond, W. C., and Thatcher, W. (2007) Crustal Deformation across the Sierra Nevada, Northern Walker Lane, Basin and Range transition, western United States measured with GPS, 2000-2004. J. Geophys. Res., 112, B05411. doi: 10. 1029/2006JB004625.

Harrison, T., Msuya, C. P., Murray, A. M., Fine-Jacobs, B., Baez, A. M., Mundil, R., and Ludwig, K. R. (2001) Paleontological investigations at the Eocene locality of Mahenge in north-central Tanzania, East Africa, in Gunnell, G. F. ed., Eocene biodiversity: unusual occur-rences and rarely sampled habitats. New York, Kluwer Academic/Plenum, 40-74.

Hay, D. E., Wendlandt, R. F., and Wendlandt, E. D. (1995) The origin of Kenya rift plateau-type flood phonolites: Evidence from geochemical studies for fusion of lower crust modified by alkali basaltic magmatism. J. Geophys. Res., 100 (B1), 411-422.

Hayward, N., and Ebinger, C. J. (1996) Variations in the along-axis segmentation of the Afar rift system: Tectonics, 15, 244-257.

Hendrie, D., Kusznir, N., Morley, C. K., and Ebinger, C. J. (1994) A quantitative model of rift basin development in the northern Kenya rift: evidence for the Turkana region as an "accommodation zone" during the Paleogene. Tectonophysics, 236, 409-438.

Hofmann, C., Courtillot, V., G. Feraud, Rochette, P., Yirgu, G., Ketefo, E., and Pik, R. (1997) Timing of the Ethiopian flood basalt event and implications for plume birth and global change. Nature, 389, 838-841. doi: 10. 1038/39853.

Huerta, A. D., Nyblade, A. A., and Reusch, A. M. (2009) Mantle transition zone structure beneath Kenya and Tanzania: more evidence for a deep-seated thermal upwelling in the mantle. Geophys. J. Int. doi: 10. 111/j. 1365-246X. 2009. 04092. x.

Huismans, R. S., Podladchikov, Y. Y., and Cloetingh, S. A. P. L (2001) Transition from passive to active rifting: Relative importance of asthenospheric doming and passive extension of the lithosphere. Journal of Geophysical Research, 106, 11271-11292.

Huismans, R. S., and Beaumont, C. (2003). Symmetric and asymmetric lithospheric extension: relative effects of frictional-plastic and viscous strain softening. Journal of Geophysical Research, 108, 2496. doi: 10. 1029/2002JB002026.

Hutchinson, D. H., Golmshtok, A. J., Zonenshain, L. P., Moore, T. C., Scholz, C. A., and Klitgord, K. D. (1992) Depositional and tectonic framework of the rift basins of Lake Baikal from multichannel seismic data. Geology, 20, 589-592.

Jackson, J. A., and Blenkin sop, T. (1997) The Bilila-Mtakataka fault in Malawi: an active, 100 km long normal fault segment in thick seismogenic crust. Tectonics, 16, 137-150.

Johnson, T. C., Wells, J. T., and Scholz, C. A. (1995) Deltaic sedimentation in a modern rift lake. GSA Bulletin, 107, 812-829.

Kalindekafe, L. S. N., Dolozi, M. B., and Yuretich, R. (1996) Distribution and origin of clay minerals in the sediments of Lake Malawi, in Johnson, T. C., and Odada, E. O. eds., The Limnology, climat ology and paleoclimatology of the East African lakes. Amsterdam, Gordon and Breach, 443-460.

Kampunzu, A. B., Kramers, J. D., and Makutu, M. N. (1998) Rb-Sr whole rock ages of the Lueshe, Kirumba and Numbi igneous complexes (Kivu, Democratic Republic of Congo) and the break-up of the Rodinia supercontinent. J. Afr. Earth Sci., 26, 29-36.

Karp, T., Scholz, C. A., and McGlue, M. M. (in press) Structure and stratigraphy of the Lake Albert Rift, East Africa: observations from seismic reflection and gravity data. AAPG Memoir - Lacustrine Sandstone Reservoirs.

Keir, D. B., Bastow, I. D., Daly, E., Cornwell, D., and Whaler, K. A. (2009) Lower-crustal earthq uakes near the Ethio-pian rift induced by magmatic processes. Geochem. Geophys. Geosyst. doi: 10. 1029/2009GC002382.

Keir, D., Ebinger, C., Stuart, G., Daly, E., and Ayele, A. (2006) Strain accommodation by magmatism and faulting at continental breakup: Seismicity of the northern Ethio-pian rift. J.

Geophys. Res., 111, B053 14. doi: 10. 1029/2005JB003748.

Keranen, K., Klemperer, S. L., Gloaguen, R., and EAGLE Working Group (2004) Three-dimensional seismic imaging of a protoridge axis in the Main Ethiopian rift. Geology, 32: 949-952.

Kieffer, B., Arndt, N., Lapierre, H., Bastien, F., Bosch, D., Pecher, A., Yirgu, G., Ayalew, D., Weis, D., Jerram, D. A., Keller, F., and Meugniot, C. (2004) Flood and shield basalts from Ethiopia: magmas from the African Super-swell. J. Petrol. 45, 793-834.

Kinabo, B. D., Atekwana, E. A., Hogan, J. P., Modisi, M. P., Wheaton, D. D., and Kampunzu, A. B. (2007) Early structural developm ent of the Okavango rift zone, NW Botswana. J. Afri. Earth Sci., 48, 125-136.

King, S. D., and Ritsema, J. (2000) African hot spot volca-nism: Small-scale convection in the upper mantle beneath cratons. Science, 290, 1137-1140.

Kusznir, N., and Park, R. (1987) The extensional strength of the continental lithosphere: its dependence on geothermal gradient, crustal composition, and thickness, in Coward, M., Dewey, J., and Hancock, P. eds., Continental extensional tectonics. Special Publication , 28. Geological Society of London, 35-42.

Kusznir, N., C. Vita-Finzi, Whitmarsh, R. B., England, P., M. H. P. Bott, Govers, R., Cartwright, J., and Murrell, S. (1991) The distribution of stress with depth in the lithosphere: thermo-rheological and geodynamic constraints (and discussion). Philosophical Transactions: Physical Sciences and Engineering, 337 (1645), 95-110.

Larsen, P. H. (1988) Relay structures in a Lower Permian basement-involved extension system, East Greenland. Journal of Structural Geology, 10, 3-8.

Lavier , L., and Steckler, M. (1997) The effect of sedimentary cover on the flexural strength of continental lithosphere. Nature, 389, 476-479.

Lavier , L. L., and Manatschal, G. (2006) A mechanism to thin the continental lithosphere at magma poor margins. Nature 440, 324-329.

LeGall, B., Gernigon, L., Rolet, J., Ebinger, C., Gloaguen, R., O. Nilsen, Dypvik, H., Deffontaines, B., and Mruma, A. (2004) Neogene - Recent rift propagation in Central Tanzania: morphostructural and aeromagnetic evidence from the Kilombero area. Geol. Soc. Amer. Bull., 116, 490-510.

LeTurdu, C., et al. (1998) The Ziway-Shala lake basin system, Main Ethiopian Rift: Influence of volcanism, tectonics, and climatic forcing on basin formation and sedimentation. Palaeogeography, Palaeoclimatology, Palaeoecology, 150, 135-177.

Lezzar, K. E., Tiercelin, J. J., Batist, M. D., and Cohen, A. S. (1996) New seismic stratigraphy and Late Tertiaryhistory of the North Tanganyika Basin, East African Rift system, deduced from multichannel and high resolution seismic reflection data and piston core evidence. Basin Research, 8, 1-28.

Lithgow-Bertelloni, C., and Silver, P. (1998) Dynamic topography, plate driving forces, and the African superswell. Nature, 395, 269-272.

Lyons, R. P., Scholz, C. A., Buoniconti, M. R., and Martin, M. R. (2010) Late Quaternary stratigraphic analysis of the Lake Malawi Rift, East Africa: an integration of drill-core and seismic-reflection data. Palaeogeography, Palaeoclimatology, Palaeoecology, in press.

Maguire, P. K. H., Keller, G. R., Klemperer, S. L., Mackenzie, G. D., Keranen, K., Harder, S., O'Reilly, B., Thybo, H., Asfaw, L., Khan, M. A., and Amha, M. (2006). Crustal structure of the Northern Main Ethiopian Rift from the EAGLE controlled source survey: a snap shot of incipient lithospheric break-up, in Yirgu, G. Ebinger, C. and Maguire, P. K. H. eds., The Afarvolcanic province within the East African Rift System. Special Publications, 259. Geological Society of London, 269-291.

Malawi Department of Surveys. (1983) National atlas of Malawi. Blantyre, Malawi, Malawi Department of Surveys, 79 p.

McConnell, R. B. (1972) The geological development of the Rift System of eastern Africa. Bull. Geol. Soc. Am, 83, 2549-2572.

McKenzie, D. (1978) Some remarks on the formation of sedimentary basins: Earth Planetary Science Letters, 40, 25-32.

McKenzie, D. P. (1984) A possible mechanism for epeirogenic uplift, Nature, 307, 616-618.

Modisi, M. P., Atekwana, E. A., Kampunzu, A. P., and Ngwisanyi, T. H. (2000) Rift kinematics during the incipient stages of continental extension: Evidence from the nascent Okavango rift basin, northwest Botswana, Geology, 28 (10) 939-942.

Morley, C. K. (1988) Variable extension in Lake Tanganyika: Tectonics, 7, 785-801.

Morley, C. K., and Nelson, R. A. (1990) Transfer zones in the East African Rift system and their relevance to hydro-carbon exploration in rifts, AAPG Bulletin, 74 doi: 10. 1306/0C9B2475-1710 -11D7-8645000102C1865D.

Morley, C. K., Wescott, W. A., Stone, D. M., Harper, R. M., Wigger, S. T., and Karanja, F. M. (1992) Tectonic evolution of the northern Kenyan Rift, Journal of the Geological Society, 149, 333-348. doi: 10. 1144/gsjgs. 149. 3. 0333.

Mortimer, E., Paton, D., Scholz, C. A., Strecker, M., and Blisniuk, P. (2007) Orthogonal to oblique rifting: effect of rift basin orientation in the evolution of the North Basin, Malawi Rift, East Africa, Basin Research, 19, 393-407.

Nelson, C. H., Karabanov, E. B., Colman, S. M., and Escutia, C. (1999) Tectonic and sediment supply control of deep rift lake turbidite systems: Lake Baikal, Russia, Geology, 27 (2), 163-166.

Ng'ang'a, P. (1993) Deltaic sedimentation in a lacustrine environment Lake Malawi, Africa, Journal of African Earth Sciences, 16, 253-264.

Nielsen, T. K., and Hopper, J. R. (2002) Formation of volcanic rifted margins: Are temperature anomalies required? Geophys. Res. Lett., 29 (21) 2022. doi: 10. 1029/2002GL015681.

Niemi, N. A., Wernicke, B. P., Friedrich, A. M., Simons, M., Bennett, R. A., and Davis, J. L. (2004) BARGEN continuous GPS data across the eastern Basin and Range province, and implications for fault system dynamics: Geophysical Journal International, 159, 842-862. doi: 10. 1111/j. 1365-246X. 2004. 02454 . x.

Noble, W., Foster, D. A., and Gleadow, A. J. W. (1997) The post Pan-African thermal and extensional history of crystalline basement rocks in eastern Tanzania, Tectonophysics, 275, 331 -350.

Nooner, S. L., Bennati, L., Calais, E., Buck, W. R, Hamling, I., Wright, T., and Lewi, E. (2009) Post-rifting relaxation in the Afar region. Ethiopia Geophysical Research Letters, 36.

Nyblade, A., and Robinson, S. (1994) The African Super-swell, Geophys. Res. Lett., 21 (9), 765-768.

Nyblade, A. A., Birt, C., Langston, C. A., Owen, T. J., and Last, R. J. (1996) Seismic experiment reveals rifting of craton in Tanzania, Eos. Transactions American Geophysic al Union, 77, 51, 517-517.

Nyblade, A. A., and Brazier, R. A. (2003) Precambrian lithospheric controls on the development of the east African Rift System. Geology, 30, 755-758.

Nyblade, A. A., and Pollack, H. N. (1993) A Global Analysis of Heat Flow From Precambrian Terrains: Implications 206 Part 3: Rift, Transtensional, Basin Settings for the Thermal Structure of Archean and Proterozoic Lithosphere, J. Geophys. Res., 98 (B7) 12207-12218.

Olsen, P. E. (1986) A 40-million-year lake record of Early Mesozoic orbital climatic forcing. Science, 234 (4778) 842-848.

Peacock, D. C. P., and Sanderson, D. J. (1994) Geometry and development of relay ramps in normal fault systems, AAPG Bulletin, 78, 147-165.

Perlmutter, M. A., and Matthews, M. D. (1989) Global cyclostratigraphy: a model, quantitative dynamic stratigraphy , ed., T. A. Cross. Englewood Cliffs, NJ, Prentice Hall.

Petit, C., and Ebinger, C. (2000) Flexure and mechanical behaviour of cratonic lithosphere: Gravity models of the East African and Baikal rifts, J. Geophys. Res., 105, 19151-19162.

Petit, C., Gunnell, Y., N. Gonga-Saholiariliva, Meyer, B., and J. S eguinot (2009) Faceted spurs at normal fault scarps: Insights from numerical modeling, J. Geophys. Res., 114, B05403. doi: 10. 1029/2008JB005955.

Pik, R., Marty, B., and Hilton, D. R. (2006) How many plumes in Africa? The geochemical point of view. Chemical Geology, 226, 100-114.

Pik, R., Marty, B., Carignan, J., and Lave, J. (2004) Stability of the Upper Nile drainage network (Ethiopia) deduced from (U-Th) /He thermochronometry: implications for uplift and erosion of the Afar plume dome. Earth Planet. Sci. Letts., 215, 73-88.

Pik, R., Marty, B., Carignan, J., Yirgu, G., and Ayalew, T. (2008) Timing of East African Rift developm ent in southern Ethiopia: implication for mantle plume activity and evolution of topography, Geology, 36, 167-170. doi: 10. 1130/G24233A. 1.

Prins, H. (1987). Nature conservation as an integral part of optimal land use in East Africa: The case of the Masai ecosystem of Northern Tanzania. Biological Conservation, 40, 141-161.

Pritchard, M., and Simons, M. (2002) A satellite geodetic survey of large-scale deformation of volcanic centres in the central Andes. Nature, 418, 167-171. doi: 10. 1038/nature00872.

Quade, J., and Wynn, J. (2008) The geology of early humans in the Horn of Africa, Geol. Soc. Amer. Spec. Pub. 446.

Reynolds, D. J. (2008) Structural and climatic controls on evolving drainage systems in extensional basins, Geological Society of London Conference: Rifts Renaissance: Stretching the Crust and Extending Exploration Frontiers, Houston, TX, August, 85.

Ringrose, S., Huntsman-Mapila, P., Kampunzu, H., Downey, W. D. Coetzee S., Vink, B., Matheson W., and Vanderpost, C. (2005) Geomorphological and geochemical evidence for palaeo feature formation in the northern Makgadikgadi sub-basin, Botswana, Palaeogeography, Palaeoclimatology and Palaeoecology, 217, 265–287.

Ritsema, J., Nyblade, A., Owens, T., Langston, C., and Van Decar, J. (1998) Upper mantle seismic velocity structure beneath Tanzania: implications for the stability of cratonic roots: Journal Geophysic al Research, 103, 21200–21214.

Rosendahl, B. R. (1987) Architecture of continental rifts with special reference to East Africa. Ann. Rev. Earth Planet. Sci., 15, 445–503.

Ruegg, J-C. (1975) Main results about the crustal and upper mantle structure of the Djibouti region TFAI, in Afar Between Continental and Oceanic Rifting, edited by A. Pilger and A. Rösler, Schweizerbart, Stuttgart. pp. 89–107.

Schlische, R. W. (1995) Geometry and origin of fault-related folds in extensional settings, AAPG Bulletin, 5.

Scholz, C. A. (1995) Deltas of the Lake Malawi Rift, East Africa: Seismic Expression and Exploration Implications, AAPG Bulletin, 79, 1679–1697.

Scholz, C. A., Cohen, A. S., Johnson, T. C., King, J. W., and Moran, K. (2006) The 2005 Lake Malawi Scientific Drilling Project, Scientific Drilling (2) 17–19.

Scholz, C. A., Johnson, T. C., Cohen, A. S., King, J. W., Peck, J., Overpeck, J. T., Talbot, M. R., Brown, E. T., Kalindekafe, L., Amoako, P. Y. O, Lyons, R. P, Shanahan, T. M., Castaneda, I. S., Heil, C. W., Forman, S. L., McHargue, L. R., Beuning, K. R., Gomez, J., and Pierson, J. (2007) East African megadroughts between 135–75 kyr ago and bearing on early-modern human origins, Proceedings of the National Academy of Sciences, 104, 16416–16421.

Scholz, C. A., King, J. W., Ellis, G. S., Swart, P. K., Stager, J. C., and Colman, S. M. (2003) Paleolimnology of Lake Tanganyika, East Africa, over the past 100 kYr, Journal of Paleolimnology, 30, 139–150.

Scholz, C. A., Rosendahl, B. R., and Scott, D. L. (1990) Development of coarse-grained facies in lacustrine rift systems: examples from East Africa. Geology, 18, 140–144.

Scholz, C. H. (1976) Evidence for incipient rifting in South Africa, Geophysical J. R. astr. Soc., 44, 135–144.

Sengor, A. M. C., and Burke, K. (1978) Relative timing of rifting and volcanism on Earth and its tectonic applications, Geophys. Res. Lett., 5, 419–421.

Sepulchre, P., Ramstein, G., Fluteau, F., Schuster, M., Tiercelin, J-J., and Brunet, M. (2006) Tectonic uplift and eastern Africa aridification, Science, 313, 1419–1423.

Simmons, N. A., Forte, A. M., and Grand, S. P. (2007) Thermochemical structure and dynamics of the African superplume, Geophysical Research Letters 34, L02301. doi: 10.1029/2006GL028009.

Smith, M., and Mosley, P. (1993) Crustal heterogeneity and basement influence on the development of the Kenya rift, East Africa. Tectonics, 12 (2), 591–606.

Soreghan, M. J., Scholz, C. A., and Wells, J. T. (1999) Coarse – grained deep – water sedimentation along a border fault margin of Lake Malawi, Africa: Seismic Strati – graphic Analysis, Journal of Sedimentary Research, 69, 832–846.

Spiegel, C., Kohn, B., Belton, D., and Gleadow, A. J. W. (2007) Morphotectonic evolution of the central Kenya rift flanks: Implications for late Cenozoic environmental change in East Africa. Geology, 427–430. doi: 10. 1130/G23108A. 1.

Stamps, S., Calais, E., Saria, E., Mbede, E., Hartnady, C., Nocquet, J–M., Ebinger, C., and Fernandes, R. (2008) A kinematic model for the East African rift system, Geophys. Res. Lett. L05304. doi: 10. 1029/2007GL032781.

Stanistreet, I. G., and McCarthy, T. S. (1993) The Okavango Fan and the classification of subaerial fan systems, Sedimentary Geology, 85, 115–133.

Tiberi, C., Ebinger, C., Ballu, V., Stuart, G., and Oluma, B. (2005) Inverse models of gravity data from the Red Sea–Gulf of Aden–Ethiopian rift triple junction zone, Geophys J. Int., 163, 775–787.

Tiercelin, J. J., Soreghan, M., Cohen, A. S., Lezzar, K. E., and Bouroullec, J. L. (1992) Sedimentation in large rift lakes: Example from the Middle Pleistocene–Modern deposits of the Tanganyika Trough, East African Rift system. Bull. Centres Rech. Explor. –Prod. Elf Aquitaine 16, 83–111.

Trauth, M. H., Deino, A., Bergner, A. G. N., and Strecker, M. R. (2003) East African climate change and orbital forcing during the last 175 kyr BP. Earth and Planetary Science Letters, 206, 297–313.

Trauth, M. H., Maslin, M. A., Deino, A., and Strecker, M. R. (2005) Late Cenozoic Moisture History of East Africa, Science, 309, 5743, 2051–2053.

Van Avendonk, H., Lavier, L., Shillington, D. J., and Manatschal, G. (2009) Extension of continental crust at the margin of the eastern Grand Banks, Newfoundland. Tectonophys. 468, 131–148.

van der Beek, P., Mbede, E., Andriessen, P., and Delvaux, D. (1998) Denudation history of the Malawi and Rukwa rift flanks from apatite fission track thermochronology, Journal African Earth Sciences, 26, 363–385.

Van Wijk, J., van Hunen, J., and Goes, S. (2008) Small–scale convection during continental rifting: Evidence from the Rio Grande rift. Geology, 36 (7), 575–578. doi: 10. 1130/G24691A. 1.

Vauchez, A., Barruol, G., and Tommasi, A. (1997) Why do continents break–up parallel to ancient orogenic belts? Terra Nova, 9, 62–66.

Versfelt, J., and Rosendahl, B. (1989) Relationships between pre – rift structure and rift architecture in Lakes Tanganyika and Malawi, East Africa, Nature, 337, 354–357.

Vrba, E. S., Denton, G. H., Partridge, T. C., and Burckle, L. H. eds. (1995) Paleoclimate and evolution with emphasis on human origins. New Haven, CT, Yale University Press.

Walsh, J., and Watterson, J. (1991) Geometric and kinematic coherence and scale effects in normal fault systems. Special Publication, 56. Geological Society of London, 193-203.

Wanke, H. (2005) The Namibian Eiseb Graben as an extension of the East African Rift: evidence from Landsat TM 5 imagery. South African Journal of Geology, 108 (4), 541-546.

Weeraratne, D. S., Forsyth, D. W., Fisher, K. M., and Nyblade, A. A., Evidence for an upper mantle plume beneath the Tanzanian Craton from Rayleigh wave tomography, Journal of Geophysical Research, 108 doi: 10. 1029/2002JB002273, 2003.

Weissel, J. K., and Karner, G. D. (1989) Flexural uplift of rift flanks due to mechanical unloading of the lithosphere under extension. Journal of Geophysical R esea rch, 94.

Wells, J. T., Scholz, C. A., and Johnson, T. C. (1994) High-stand deltas of Lake Malawi, East Africa: Environments of deposition and processes of sedimentation, in A. J. Lomando, B. C. Schreiber, and P. M. Harris, eds., Lacus-trine reservoirs and depositional systems. SEPM Core Workshop No. 19. Tulsa, OK, Society of Economic Paleontologists and Mineralogists, 1-35.

White, R., and McKenzie, D. (1989) Magmatism at rift zones: the generation of volcanic passive continental margins and flood basalts. Journal Geophysical Research, 94, 7685-7729.

Wolfenden, E., Ebinger, C., Yirgu, G., Renne, P., and Kelley, S. P. (2005) Evolution of the southern Red Sea rift: Birth of a magmatic margin. Geol. Soc. Amer. Bull., 117, 846-864.

Wright, T. J., Ebinger, C., Biggs, J., Ayele, A., Yirgu, G., Keir, D., and Stork, A. (2006) Magma-maintained rift segmentation at continental rupture in the 2005 Afar dyking episode. Nature, 442, 291-294.

Yamasaki, T., and Gernigon, L. (2009) Role of magmatism in continental lithosphere extension continental litho-sphere extension. Tectonophysics 468, 169-184.

Yuretich, R. F. (1982) Possible influences upon lake development in the East African rift valleys. Journal of Geology, 90, 329-337.

（纪沫 译，张功成 祁鹏 校）

第10章 沉积物输入以及斜向板块运动对斜向离散活动板块边缘盆地发育的影响
——以加利福尼亚湾和索尔顿海槽为例

REBECCA J. DORSEY[1], PAUL J. UMHOEFER[2]

(1. Department of Geological Sciences, University of Oregon, Eugene, USA;
2. Geology Program, School of Earth Sciences and Environmental Sustainability,
Northern Arizona University, Flagstaff, USA)

摘　要：新生代晚期，在加利福尼亚湾和索尔顿海槽区太平洋—北美板块边界发育张扭性盆地。轴部盆地沿板块主边界呈带状分布，宽约50~60km，南部为欠补偿的大洋扩张中心，垂直于北西走向大规模转换断层发育；北部为发育巨厚沉积的向北延伸的斜列拉分盆地，盆地中不发育正常洋壳。沿海湾—海槽侧翼发育边缘海盆地，主要包括拆离盆地（仅限北部）、张扭性盆地和典型的双断裂谷盆地。

前人研究表明，三个主要因素控制了研究区的构造样式、地层结构和沉积盆地总厚度：(1) 裂谷夹角（α），即板块边界的总体趋势和板块相对运动方向之间的锐角；(2) 邻区科罗拉多河和北部的其他小流域输入的大量沉积物；(3) 应变分解程度。有证据表明，北部海湾和索尔顿海槽，在裂谷夹角不小于30°处，以拆离断层和拆离盆地为主，而在加利福尼亚湾的中部和南部（裂谷夹角小于20°）不发育拆离断层。笔者认为快速扩张相关的高角度裂谷是北部地区拆离盆地形成的主要因素。

来自科罗拉多河的大量沉积物输入对地壳厚度、地壳组分、岩石圈力学和裂谷结构起到了主要控制作用。在沉积欠补偿的南加利福尼亚湾，板块边界从大陆裂谷逐渐转变为由正常洋壳和磁条带组成的海底扩张中心。中央海湾地区的瓜伊马斯扩张中心表现为年轻的洋壳和浅层侵入体。相反地，北部沉积物充填度较高的盆地则以厚层的新形成的过渡型地壳为特点，输入的和经岩浆改造的沉积物充填了以岩石圈破裂和斜向离散的方式形成的新空间。因此，沉积物输入速率决定了大陆裂解过程最终能否形成以正常的玄武质洋壳为特征的新洋盆。

关键词：加利福尼亚湾　斜向离散板块边界　裂谷结构　张扭盆地　科罗拉多河

10.1　引言

加利福尼亚湾和索尔顿海槽地区（图10-1）为研究斜向离散的板块边界盆地的发育过程提供了一个很好的实例。这一地区板块边界活动变形率高（51mm/a；Plattner等，2007），尽管不如陆上盆地易于观测，但通过近期的海洋地球物理研究也可以大体获得一些近海盆地特征的认识（在很大程度仍然缺乏近海盆地的详细研究）。构造洼地占据着这条板块边界，这里称其为"湾—槽通道"，包括一系列晚新生代张扭性盆地，是太平洋和北美板块之间的斜向右旋运动的响应（图10-1）。12-8Ma期间，地壳变形为一系列不同的伸展和张扭性构造，控制着盆地的几何形态、沉降速率和填充模式。输入的沉积物大多数来自于北部的科罗拉多河，少量来自于中央湾槽通道东部的较小水系，同时这些沉积物对盆地演化、地壳组分、地壳厚度和流变学特征起到了重要的控制作用。本文总结了沿湾槽通道的沉积盆地主要特点，并探讨了沉积物输入以及构造样式是如何变化影响这些盆地大小、形状、状态和演化的。

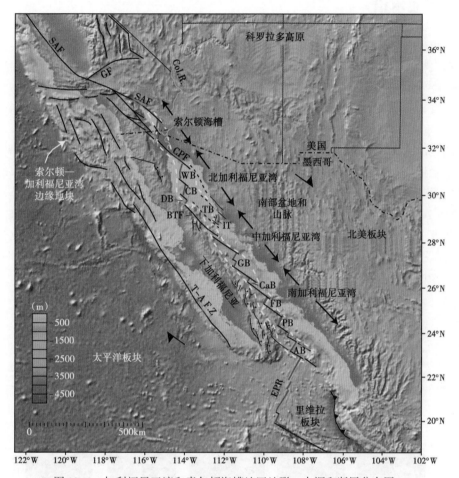

图 10-1　加利福尼亚湾和索尔顿海槽地区地形、水深和断层分布图

湾槽通道中的张扭盆地是响应沿太平洋—北美板块边界的斜向离散运动而形成的；水深沿着板块边界从南到北的有序减小主要是由于北部科罗拉多河大量的沉积物输入；黄色的虚线表示轴部盆地的大致位置。缩写：AB—阿拉孔盆地；BTF—巴伦娜转换断层；CaB—卡门盆地；CB—康萨格盆地；CPF—普列托断层；DB—德尔芬海盆；EPR—东太平洋隆起；FB—法伦盆地；GB—瓜伊马斯海盆；GF—加洛克断层；IT—蒂布龙岛；PB—佩斯卡德罗盆地；SAF—圣安德列斯大断层；T-A.F.Z—阿布雷奥约斯断裂带；TB—蒂布龙盆地；WB—瓦格纳盆地

研究表明，存在两个主要的参数控制了沿湾槽通道发育的盆地特征和演化：（1）板块边界整体走向与板块运动方向之间的锐角（α）；（2）来自科罗拉多河与北部其他水系的沉积物输入。假设板块运动平均角度为 310°，裂谷走向的一个微小变化导致了裂谷夹角从加利福尼亚湾中南部的 17°~18° 增加到北纬 30° 的北部地区的 33°~35°（图 10-1）。这两个参数（裂谷夹角和沉积物供应）的有序变化对构造样式、盆地形状、地壳厚度，以及板块边缘盆地从大陆裂谷到正常洋壳的海底扩张中心的转换程度起到了重要的控制作用。

另一个影响盆地构造样式的重要因素是沿板块边界变形范围的应变分解。本文中，笔者将应变分解定义为一种运动学类型，斜向应变分解为北西向转换断层的走向水平位移和北至北西走向正断层的倾向位移。非应变分解被定义为通过右旋、正性和左旋断层，以及不定向的斜向滑动断层控制的完整的张扭变形。

10.2 盆地术语

加利福尼亚湾和索尔顿海槽地区发育多种类型的走滑和伸展沉积盆地。图10-2列举了本文中用来描述盆地几何形态及类型的专业术语。大洋扩张中心以离散、消亡和在活动转换断层跃迁喷发的玄武质洋壳组成。拉分盆地，也被称为是菱形断陷或叠置盆地（Nilsen和Sylvester，1995），即在主要走滑断层系阶跃释放中形成的拉张区域（Burchfiel和Stewart，1966）。断层终端盆地形成于张扭变形与扩张的复合区域，主走滑断层逐渐并终止于分支断裂（Umhoefer等，2007）。尽管从简化的十字剖面上看类似于半地堑裂谷盆地（图10-2），断层终端盆地仍以复合的平面模式、多沉积物源、短暂的存在期（2~3Ma）、吉尔伯特式扇三角洲、不整合、横纵向上快速的岩相变化和控制盆地沉降的复合断层为主要特征（Umhoefer等，2007）。半地堑（正交型）裂谷盆地为倾向滑动的正断层扩张后形成的盆地，主要沿高角度主盆地边界正断层方向翘倾（Leeder和Gawthorpe，1987）。拆离盆地也是沿某一方向正交扩张后的产物，但盆地边界断层表现为低角度正断层或拆离断层（Friedmann和Burbank，1995）。

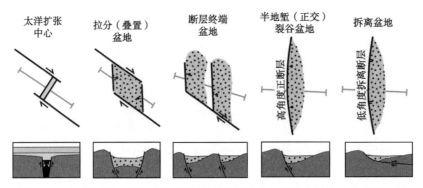

图10-2　加利福尼亚湾与索尔顿海槽地区主要的盆地类型以及文中使用术语

10.3 大地构造和区域构造综述

加利福尼亚湾和索尔顿海槽构成了太平洋和北美板块之间的斜向离散边界（图10-1）。这一纬度大部分的板块运动是通过转换断层和扩张中心，在湾槽通道中进行的，滑动速率为43~47mm/a（Plattner等，2007）。额外的4~6mm/a的滑动速率受控于近海加利福尼亚半岛西南方的阿布雷奥约斯断裂带（Plattner等，2007）（图10-1），该带北接南加利福尼亚大陆边缘复合的断裂网（Nicholson等，1994；Dixon等，2000）。在过去的12.5Ma中，区域的张扭作用使下加利福尼亚（墨）与墨西哥大陆斜向裂离（Atwater和Stock，1998；Oskin和Stock，2003）。近年来地震反射和折射研究为分析加利福尼亚湾的裂谷结构、地壳厚度和组分、盆地变形构造主控因素等提供了新的思路（Aragon-Arreola等，2005；Gonzalez等，

2005；Aragon-Arreola 和 Martin-Barajas，2007；Lizarralde 等，2007）。辅助的陆上研究确定了更精确的盆地形成时间和地壳变形的地层响应（Umhoefer 等，1994，2007；Axen 和 Fletcher，1998；Dorsey 和 Umhoefer，2000；Dorsey 等，2007）。

目前普遍认为太平洋—北美板块在 6Ma 时已局限于沿着现今湾槽通道的轴线位置运动（Oskin 等，2001），但 12.5~6Ma 板块边界变形的分布及运动学特征仍是不确定或者说存在争议的。举个实例，晚中新世板块运动被分解为下加利福尼亚西南阿布雷奥约斯断裂带走滑运动和现今加利福尼亚湾及周围地区的南西西至北东东向的拉张作用（Spencer 和 Normark，1979；Stock 和 Hodges，1989）。这个实例包括下加利福尼亚半岛和墨西哥大陆之间 300km 的北西向移动。另一实例为从 12.5Ma 开始在一个区域性的完整的右旋斜向剪切的单相内发生的变形，穿过了从下加利福尼亚半岛西南侧到墨西哥大陆的地带，包括横跨加利福尼亚湾的 450~500km 的分支（Gans，1997；Fletcher 等，2007）。以上实例仍未得到合理的解释，但不影响盆地特征的讨论。

湾槽通道内的沉积盆地可以被分为两种主要类型：轴部盆地和边缘盆地。轴部盆地沿主板块边界断层发育，宽度约 50~60km（图 10-1），并以水深的有序减小和盆地填充厚度从南到北的增加为特征。加利福尼亚湾南部的轴部盆地由较短的、沉积欠补偿且位于深水的（2000~3000m）大洋扩张中心组成，垂直于转换断层呈北东走向。在加利福尼亚湾中部，瓜伊马斯海盆的水深达 1600m，且有中等厚度的沉积。在加利福尼亚湾北部和索尔顿海槽，轴部盆地位于浅海相（<500m）及非海相环境中。轴部盆地在平面上表面为斜向分离的几何形态或菱形断陷，盆地中北西向转换断层与北—北北东向正断层相连接，其中叠置盆地包含有厚层沉积物而缺乏海底扩张或正常洋壳的证据。

边缘盆地位于轴部盆地带的外围，主要发育在加利福尼亚湾和索尔顿海槽的翼部（图 10-1）。一部分边缘盆地位于相对浅的水位和构造层，另一部分则因构造运动终止沉降而暴露于陆上，一些陆上盆地仍然活跃地补偿沉积。我们认为在边缘盆地存在三种主要的构造样式：（1）张扭性断层终端盆地，沿着或邻近走滑断层的末端发育（Umhoefer 等，2007）；（2）正交裂谷盆地，在高角度正断层的上盘形成；（3）拆离盆地，在低角度正断层的上盘形成（Friedmann 和 Burbank，1995）（图 10-2）。拆离盆地只沿着裂谷夹角为 33°~35° 的加利福尼亚湾北缘和索尔顿海槽发育，而在加利福尼亚湾的中南部不发育（裂谷对应的是 17°~18°）。

与北西向倾滑正断层（高或低角度）有关的盆地是应变分解的结果，对应的是板块运动中拉伸分量表现为北东向拉张作用，大致垂直于板块边界。这种类型的例子包括加利福尼亚湾中部的亚基盆地（Arreola 等，2005；图 10-4）、加拿大大卫拆离断层、索尔顿海槽内断层上盘的盆地（Axen 等，2000；图 10-5）、圣佩德罗断层以及下加利福尼亚北部的充填盆地（Stock 和 Hodges，1989；图 10-5）。不同的是，断层终端盆地形成于非分配应变环境中，区域的张扭作用会形成完整的斜向滑动断层和复合的断层网。

10.4 盆地研究总结

10.4.1 加利福尼亚湾南部区域

尽管只有阿拉孔隆起采用过现代地震资料研究（Sutherland，2006；Lizarralde 等，

2007），但分析认为加利福尼亚湾南部的阿拉孔、佩斯卡多、法伦轴部盆地是海底扩张的中心（图10-3）。轴部盆地是由长（60~140km）转换断层连接的短（10~45km）扩张中心（图10-3）。阿拉孔盆地的中心轴是一个真正的扩张构造脊，在扩张中心处有2.5km深；其他盆地有5~8km的宽槽，且达到了海平面以下深度约3~3.5km。阿拉孔隆起共发育6km洋壳，其中下地壳有5km，上地壳有1km且包含了一个薄沉积层（Sutherland，2006；Lizarralde等，2007）。磁异常分析及综合速度模型、断层成像、多波道地震数据的地壳结构研究表明在不对称扩张的早期阶段为3.0~2.4Ma，最后发生了扩张夭折。在2.4Ma一个小的洋脊跃迁激发了对称的海底扩张，并以46~47mm/a的速率持续到了现今（Sutherland，2006；Plattner等，2007）。

阿拉孔裂谷段包括了阿拉孔轴部盆地和共轭边界的翼部，以佩斯卡德罗和塔马约断裂带为界（图10-3）。这个裂谷段经历了约350km的大陆伸展（由平行于现今板块运动方向的横剖面计算得来），因此被定义为一个宽裂谷（Lizarralde等，2007）。沿着阿拉孔裂谷段发

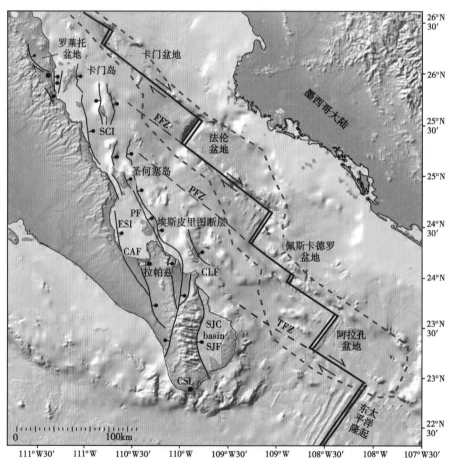

图 10-3 加利福尼亚湾南部区域断层及盆地分布图
轴部盆地是北东向的沉积欠补偿大洋扩张中心，垂直于北西向长转换断层；
蓝色虚线显示轴部盆地和边缘盆地间的大致边界
缩写：CAF—卡利萨尔断层；CIF—塞拉尔沃岛断层；CSL—圣卢卡斯；ESI—埃斯皮里图岛；FFZ—法伦断裂带；PF—帕蒂达断层；PFZ—佩斯卡德罗断裂带；SCI—圣卡塔利娜岛；SJC basin—圣何塞角盆地；SJF—圣何塞断层；TfZ—塔马约断裂带

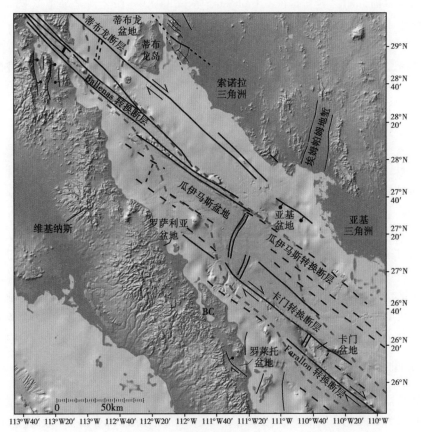

图 10-4 加利福尼亚湾中部地区断层及盆地分布图

该区域内的主要轴部盆地为瓜伊马斯盆地，一个以北西向长转换断层为大洋扩张中心边界的中度沉积海盆；
蓝色虚线显示轴部与边缘盆地间的大致边界；BC—巴依亚—康塞普西翁

育一系列不完全欠补偿盆地，沉积物厚度约 500~1500m。除两个结构复杂的张扭性盆地外（Sutherland，2006），大多数盆地表现为单一的半地堑结构。多波道地震数据显示了 500~700m 甚至更薄的同裂谷地层层序，其上覆更厚的裂后层序（Sutherland，2006；Brown，2007）。盆地及其相关断层研究表明，除了下面将讨论的少数在卡波斯到拉帕兹地区的活动更缓慢的盆地，断裂活动在加利福尼亚湾南部轴部盆地系统外部终止于几百万年前。部分薄大陆地壳（7km）的近北向断层和裂谷盆地意味着早期轴部盆地可能是与北西向的走滑断层相连接的拉分盆地（Lonsdale，1989；Sutherland，2006；Brown，2007）。

从圣何塞角盆地到拉帕兹地带，沿南加利福尼亚湾西南边缘的正交裂谷盆地群，发育于现今活动缓慢的正断层上盘（图 10-3）。这些盆地中的大多数持续活动并接受第四纪晚期沉积，揭示了盆地年轻的演化历史。圣何塞角盆地（SJC；图 10-3）发育中中新世到第四纪沉积，在 8Ma 间经历了从陆相到海相的变化（Carreno，1992；McTeague，2006），形成复合的半地堑，最终回归陆相沉积（Martinez-Guitierrez 和 Sethi，1997；McTeague，2006）。在盆地东部现有资料也证实了同裂谷正断层的活动和沉积（McTeague，2006）。

沿着加利福尼亚湾中西部边界的张扭断层终端盆地以多物源、短存在期（2~3Ma）、吉尔伯特式扇三角洲、不整合面、地层复合断块、快速侧向、垂向岩相变化以及对盆地沉降起控制作用的复合内部断层为其特征（Umhoefer 等，2007）。在圣何塞岛上（图 10-3），1km 厚的部分揭示了正交裂谷和拉分盆地之间的地层模式以及沿倾斜正断层的复杂沉积史（图 10-2）。在更北的地区，上新世罗莱托盆地形成了一个西翘的楔状不对称半地堑是西翘倾斜的罗莱托右旋正断层的响应（图 10-3；Umhoefer 等，1994；Dorsey 和 Umhoefer，2000）。大部分的盆地填充发生在断层快速滑动、沉降、海相吉尔伯特式扇三角洲沉积的短期事件中，时间在 2.46~2.36Ma 之间（Umhoefer 等，1994；Dorsey 和 Umhoefer，2000；Mortimer 等，2005）。

在拉帕兹—圣何塞角地区（典型的半地堑裂谷盆地）和圣何塞岛—罗莱托地区（张扭断层终端盆地）之间进行盆地类型的对比，可以解释边缘盆地和主板块边界断层之间的应变分解度的差别。在高分配度的拉帕兹—圣何塞角地区，盆地类型和古地震（Busch 等，2011；Maloney 等，2007）以及最近的地震资料都揭示了北—北西走向正断层占主导地位的东西向至北东—南西向拉张作用（Fletcher 和 Munguia，2000）。构造变形是独立的且由转换断层以及在阿拉孔盆地至少从 2.4Ma 开始的海底扩张中心的活动分配而成的（Lonsdale，1989；Sutherland，2006；Lizarralde 等，2007）。相对地，圣何塞岛—罗莱托边缘区域的断层作用和盆地结构的综合模式是一个更直接，复合联系到上新世的轴向加利福尼亚湾断层及盆地，伴随着从那时起的低断层活动速率（Mayer 和 Vincent，1999）。这意味着圣何塞岛—罗莱托区域板块边缘和相关轴向盆地与阿拉孔—佩斯卡多盆地西南的转换断层息息相关。

10.4.2　加利福尼亚海湾中部区域

瓜伊马斯盆地是南部沉积欠补偿盆地和北部填充盆地之间的轴部盆地，其最北端为以北西向转换断层为界的北东向海底扩张中心（图 10-1 和图 10-4）。瓜伊马斯盆地水深 1600m，沉积物厚 3km（Aragon-Arreola 等，2005），因此比南部轴部盆地的水深更浅，但沉积物充填更厚。盆地沉积物主要由深海及半深海相泥岩、硅藻软泥以及来自西索诺拉河的薄层富泥浊积岩组成（Curray 等，1982；Einsele 和 Niemitz，1982）。浊积砂岩主要为含细粒长石及含石英质火山碎屑岩，指示盆地填充为硅质碎屑组分来自于墨西哥大陆东部的河流而不是北部的科罗拉多河（Einsele 和 Niemitz，1982）。

Lizarralde 等（2007）认为除了地幔因素之外，瓜伊马斯盆地厚地壳的玄武岩浆高活动速率很可能是由于厚的盆地充填沉积物抑制了热液的循环。瓜伊马斯盆地为一狭长的裂谷，陆壳延伸的整体宽度小于 200km。瓜伊马斯盆地的沉积物上覆于 6~8km 的辉长岩地壳之上，比阿拉孔隆起 5km 的洋壳更厚。镁铁质地壳地层的形成时代为 3.6~2Ma（Lonsdale，1989）或 6Ma（Lizarralde 等，2007）。

中央海湾地区的边缘盆地呈现出应变分解和非分解构造活动的证据。索诺兰大陆架内的亚基边缘盆地多波道地震反射测线（图 10-4）显示了北东—南西向扩张形成了北西向延伸的，北东翘倾的半地堑裂谷盆地，其中从晚中新世到上新世积累了 4km 的沉积物（Aragon-Arreola 等，2005）。以该构造样式的应变分解为例，板块边界应变的拉张分量表现为正断层的倾角滑动，不发育轴向转换断层及转换断层控制的盆地。这些盆地和边界断层的活动终止于 2~3Ma 且西移至瓜伊马斯盆地（Aragon-Arreola 等，2005；Aragon-Arreola 和 Martin-Barajas，2007）。因此上新世断裂活动的向西变化和盆地形成可能与中央墨西哥湾的应变分解的结束一致。

伴随北北西向右旋正断层早期裂谷作用与斜向滑动（Wilson，1948；Ochoa 等，2000），加利福尼亚湾西边界（瓜伊马斯轴向盆地的西部）发育晚中新世圣罗萨莉亚盆地。小型复合断块、陆缘吉尔伯特三角洲和不整合面的存在暗示它可能为一张扭断层端元盆地（图10-2）。圣罗萨莉亚盆地最古老沉积的古地磁和 $^{40}Ar/^{39}Ar$ 测年定为距今 7.1Ma（Holt 等，2000）。因此在加利福尼亚湾北部和索尔顿海槽的海侵之前 1.0~0.6Ma，中央海湾发育一斜向裂陷作用相关的伸展海峡（McDougall 等，1999；Oskin 和 Stock，2003a；Dorsey 等，2007）。

10.4.3 加利福尼亚海湾北部区域

加利福尼亚海湾北部轴部盆地与中南部轴部盆地形成了鲜明的对比（图10-1和图10-5）。海底测量结果显示为浅水，在大多数地区水深为200m。沉积物充填比南部盆地要厚得多，经估算厚度为从 6.5~7.0km（Aragon-Arreola 和 Martin-Barajas，2007）到 10km（Gonzalez 等，2005）。科罗拉多河三角洲、重矿物组合、4.5km 深处的白垩纪有孔虫沉积和阿尔塔盆地地层层序的相互关系（Pacheco 等，2006）为源自科罗拉多河的沉积物为加利福尼亚湾北部盆地填充的主体提供了证据（Van Andel，1964）。加利福尼亚湾北部地壳基底位于海平面之下 15~20km（Gonzalez 等，2005）。地震 P 波速度及其斜度揭示这一地区的地壳由长英质绿片岩组成，这与索尔顿海槽的变质沉积岩类似（Fuis 等，1984）。然而，对比索尔顿海槽，还不清楚该区变质岩是由于这一地区下部地壳由科罗拉多河变质的玄武侵入岩组成（Fuis 等，1984），还是因为古老的花岗岩壳强烈减薄并从斜向裂谷带两侧剥落造成的（Gonzalez 等，2005）。

加利福尼亚湾北部轴部盆地是由四个南北走向的伸展次盆地（瓦格纳、康萨格、上德尔芬与下德尔芬盆地）组成的复合拉分盆地，连接了南部的巴莱纳斯转换断层和北部普列托断层（图10-5；Lonsdale，1989；Persaud 等，2003；Aragon-Arreola 和 Martin-Barajas，2007）。应变将巴莱纳斯转换断层转为一个复合的楔状分叉断层群，将宽阔的张扭变形区分成上德尔芬盆地和下德尔芬盆地。加利福尼亚湾东侧近海盆地的沉降在 3~2Ma 急剧减少，同时盆地边界断层活动转换为现今加利福尼亚湾西运动边界（Aragon-Arreola 等，2005；Aragon-Arreola 和 Martin-Barajas，2007）。

由于墨西哥国家石油公司探井中获取的微体化石岩屑丰度低、保存条件差，测算出的年龄不一致，北部轴部盆地海相沉积物的时期仍无法确定。部分学者通过现场解释古老微体化石，认为 1.5~2.0km 深的沉积为中—晚中新世的产物（Aragon-Arreola 和 Martin-Barajas，2007；Helenes 等，2009）。这种解释与 Oskin 和 Stock 对中—晚中新世火山岩的认识不一致，他们认为在 6.4Ma 时期索诺拉沿岸和蒂布龙岛与加利福尼亚半岛相毗邻，自 6.4Ma 起由于斜向扩张作用致加利福尼亚湾北部才被打开。该解释与资料记录的白垩系有孔虫改造事件不符，有孔虫证据研究认为在阿尔塔盆地 4.5km 的深度（Pacheco 等，2006）的沉积物是来自科罗拉多台地。这些研究结果支持蒂布龙盆地的海相沉积物大多数是上新统—更新统，中新统的微体化石是被改造的折中观点。

加利福尼亚湾北部的边缘盆地包括张扭盆地、裂谷和拆离盆地，形成年代从晚中新世直到现今（图10-5）。加利福尼亚东北部的北北西向的圣佩德罗断层是一个倾滑的铲式正断层，倾向位移是至少为 5km，位于加利福尼亚湾伸展区的西边界（Gastil 等，1973，1975；Stock 和 Hodges，1989）。圣佩德罗断层滑动开始于 11~6Ma 之间，并持续至今（Stock 和 Hodges，1989）。位于圣佩德罗断层东部的两个当前不活跃的拆离断层，形成了晚新生代拆离盆地的边界。圣罗萨盆地形成于 12.5~6Ma 之间，位于圣罗萨拆离断层上盘（图 10-5），沉积了非海相砾岩和砂岩，与周围地区形成晚中新世裂谷盆地一致（Bryant，1986；Oskin 和 Stock，2003c）。求维塔斯拆离断层走向北北东（15°~35°），断层上盘经历了东南向的斜向右旋运动（Black 和 Axen，2003；Black，2004）。求维塔斯拆离沉降开始于中新世末期，控制了 6Ma 时期的海侵作用，调节着海相硅藻岩沉积在断层近端角砾和长石质砂岩中的分布（Boehm，1984；Black 和 Axen，2003；Black，2004）。

　　断陷边缘盆地同样出现在蒂布龙岛和相邻的加利福尼亚湾东北部索诺拉海岸（图 10-5）。索诺拉大陆上的上中新统海相沉积包括北东倾向沉积和火山岩，反映了在 12~6Ma 之间的北

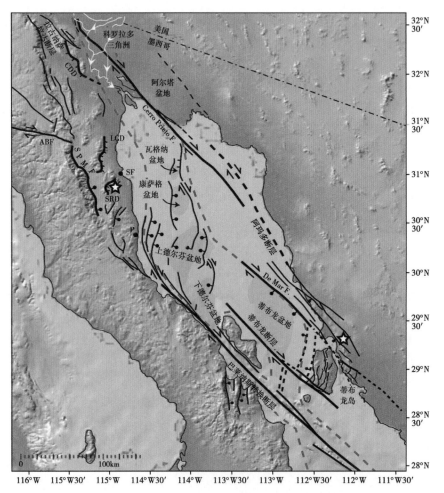

图 10-5　加利福尼亚湾北部断裂和盆地分布图

（据 Aragon-Arreola 和 Martin-Barajas，2007；Persaud 等，2003；Oskin 和 Stock，2003，2003b，2003c）加利福尼亚湾北部包括一些北—北北东向以转换层为界的斜向拉分盆地；德尔芬盆地活跃的扩散变形发生在紧密排列的斜向滑动断层之间，且没有证据表明该深度下存在洋壳（Persaud 等，2003）；地壳的大部分是沉积岩，这归因于科罗拉多河的高速率输入。缩写：ABF—阿瓜布兰卡断层；CDD—加拿大大卫拆离断层；LCD—求维塔斯拆离断层；P—普尔泰希托斯；SF—圣费利佩；SPMF—圣佩德罗断层；SRD—圣罗萨拆离断层。白色的星表示前中新世河流相砾岩具有明显的石灰岩碎屑，由于从 6.5Ma 开始的加利福尼亚湾北部打开事件，这些碎屑证明了在下加利福尼亚东北部和墨西哥大陆间 300km 的右旋错断的存在（Gastil 等，1973；Oskin 和 Stock，2003）；蓝色虚线表示轴向与边缘盆地间的大致边界

东—南西向扩张和宽裂谷盆地的形成（Gastil 和 Krumenacher，1977；Oskin 和 Stock，2003c）。中新世末期非海相张扭盆地索诺拉海岸的堆积物以北西向右旋断层和北北东向正断层为边界，包括了从 7.0~6.5Ma 开始的强烈右旋剪切活动中产生的一个低角度的拆离断层（Dorsey 等，2008；Bennett，2009）。上新统下部海相沉积出现在蒂布龙岛西南部中—上中新统火山岩附近，反映了响应 6.5~6.0Ma 沿着湾槽通道发生的太平洋—北美板块运动的海侵运动（Oskin 和 Stock，2003a，2003c）。

10.4.4 索尔顿海槽

索尔顿海槽是一个巨大的张扭性盆地，横跨加利福尼亚湾西北端活动板块边界（图 10-6）。来自科罗拉多河的沉积物控制了从早上新世起的盆地填充，形成了一个将索尔顿海从

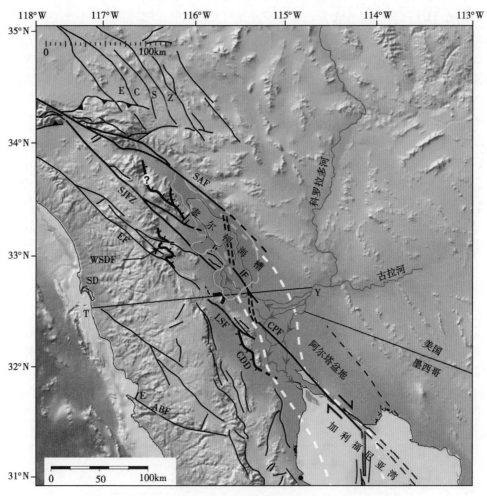

图 10-6 索尔顿海槽和加利福尼亚湾最北端断裂及盆地分布图
（据 Axen 和 Fletcher，1998；Dorsey 等，2011，修改）
白色虚线表示轴向盆地的位置。缩写：ABF—阿瓜布兰卡断层；CDD—加拿大大卫拆离断层；CPF—普莱托断层；
E—恩塞纳达；ECSZ—西加利福尼亚剪切带；EF—埃尔西诺断层；IF—帝国断层；LSF—拉古纳萨拉达断层；
SAF—圣安德列斯断层；SD—圣迭戈；SJFZ—圣哈辛托断裂带；SSPMF—圣佩德罗断层；T—提华纳；
WSDF—西索尔顿拆离断层；Y—尤马

加利福尼亚湾隔离开来的陆上三角洲（Merriam 和 Bandy，1965；Winker，1987）。科罗拉多河沉积物在深度较浅处因快速埋藏、岩浆活动、高热流而发生变质作用（Muffler 和 White，1969；Elders 和 Sass，1998），快速转换为索尔顿海之下的变质岩（Fuis 等，1984；Schmitt 和 Vazquez，2006；Dorsey，2010）。由地震和重力数据可知，索尔顿海槽轴部盆地之下的整个沉积基底由上部斜向裂谷带的沉积充填和下部镁铁质侵入共同形成，基底厚度为 5~12km，由 6Ma 之后的年轻地壳组成（Fuis 和 Kohler，1984；Fuis 等，1984；Fuis 和 Mooney，1991）。一些勘探井在阿尔塔盆地（图 10-5）4~5km 的深度钻遇前古近系花岗岩基底（Pacheco 等，2006），显示了沿着该地区板块边界不仅盆地深度变化快，而且构造复杂程度高。

  索尔顿海槽西南部为晚新生代边缘盆地，该盆地形成于两个低角度拆离断层的上盘：北部的东倾西索尔顿拆离断层（Axen 和 Fletcher，1998；Dorsey 等，2011）和南部的西倾加拿大大卫拆离断层（图 10-6；Axen，1995；Axen 等，2000；Martin-Barajas 等，2001）。在晚中新世 8—6.5Ma 之间扩张形成了北部地区裂谷相关冲积扇，随后在 6.3Ma 海侵进入索尔顿海槽地区（McDougall 等，1999；Dorsey 等，2007；McDougall，2008）。在早上新世科罗拉多河三角洲向南进积并更替了海相环境（Dibblee，1954，1984；Winker，1987；Winker 和 Kidwell，1996；Axen 和 Fletcher，1998；Dorsey 等，2011）。相似的实例如美国国境线以南的萨拉达潟湖周围的沉积岩露头等（图 10-6；Axen 等，2000；Martin-Barajas 等，2001）。

  索尔顿海槽西南部的北西向拆离断层的运动学特征变化大且难以识别。Axen 和 Fletcher（1998）发现西索尔顿拆离断层向北东向滑动，认为区域滑动应变分配中的板块边界应变发散分量被正交的北东向拉伸取代。但是，近来对西索尔顿拆离断层带的断层擦痕分析认为，上部板块的大部分离散运动转移到了东部或者南东东一侧（Steely 等，2009）。类似地，加拿大大卫拆离断层显示了伴随着断裂活动的断层擦痕的散布方向（Axen 和 Fletcher，1998）。因此，尽管索尔顿海槽西南部的拆离盆地类似于正交扩张和低角度正断层形成的盆地，但是它们的边界拆离断层经历了斜向右旋运动，未发生或只发生了很小的相对于板块边界的应变分解。

  索尔顿海槽地区在大约 1.4—1.1Ma 经历了一期较大的构造变革，受圣哈辛托、埃尔西诺和拉古纳萨拉达走滑断层带影响，上新世—更新世拆离断层和拆离盆地终止发育、分割并抬升（图 10-6；Morton 和 Matti，1993；Dorsey 和 Martin-Barajas，1999；Lutz 等，2006；Kirby 等，2007；Steely 等，2009）。这一转变与西索尔顿海槽从主控张扭作用到主控右旋扭动变形的转变同时发生，与早期的拆离盆地西南部的抬升和反转开始的时期一致。因此现今的索尔顿海槽的地形和盆地形成过程与上新世—更新世有明显差异。

10.5 讨论

10.5.1 裂谷夹角的影响

  黏土模拟实验表明，斜向裂谷夹角（α）小于 20°时受控于走滑断层变形，而裂谷角度大于 20°时受控于复合的连接走滑、正断和斜向滑动的断层网，断层走向和断裂带的复杂性取决于裂谷与板块运动方向的斜交程度（图 10-7；Withjack 和 Jamison，1986；Tron 和 Brun，1991；Clifton 等，2000）。数值模型也显示裂谷夹角小于 20°时为扭动主控型，裂谷夹角大于 20°时为拉张主控型（Tikoff 和 Teyssier，1994）。加利福尼亚湾和索尔顿海槽地区的断层几何学特征大体上支持了这一判断。洛莱托地区的张扭断层（加利福尼亚湾南部；裂谷夹角为17°~18°）系统类似于Withjack和Jamison（1986）提出的倾斜度为15°的理想模型，预测结果基于断裂带的演化和有限应变（Umhoefer和Stone，1996）。通过理论和模拟实验的预测认为，加利福尼亚湾北部地区远离裂谷轴部（边界）的部分（裂谷夹角为33°~35°）存在更大规模的正断层且比南部受到更多的张性应变。

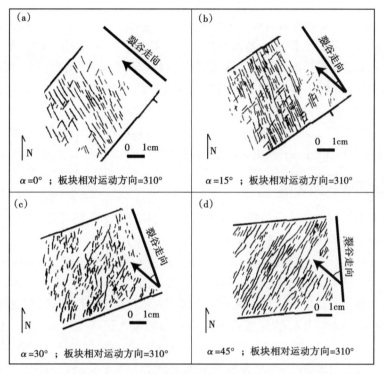

图10-7　基于不同裂谷夹角（α）的黏土模拟模型及由斜向裂谷作用产生的断裂模式图
（据Withjack和Jamison，1986，有修改）

箭头指示板块相对运动方向，310°保持不变。对于裂谷夹角为0°（纯走滑）和15°时（如a、b），应变主要表现为与裂谷走向近似平行的走滑断层；对于裂谷夹角为30°（c），张扭变形发生在由走滑断层、正断层和倾滑断层连成的复合网络中；对于裂谷夹角为大于45°（d），变形主要发生在垂直于板块运动方向的正断层上，裂谷走向变为裂谷夹角的一个函数（Withjack和Jamison，1986）。加利福尼亚湾南部裂谷走向为17°~20°（接近于图b），加利福尼亚湾北部到索尔顿海槽的裂谷走向整体约为30°（接近于图c）

需要补充的是，在强调的倾斜裂谷研究中的断层几何学特征方面，前人并未讨论过以下问题：是否缺乏低角度的正断层（拆离断层）。在过去6~12Ma的斜向离散板块边界的发育过程中，位于索尔顿海槽和东北加利福尼亚的5个大型拆离断层，再加上索诺拉海岸的1个较小的拆离断层，共同形成了北部地区（一个拉张主控型的裂谷倾角为33°~35°的地带）边缘盆地的断层边界。再者，圣佩德罗断层意味着浅层存在一个低角度正断层（Gastil等，1975），冈萨雷斯等（2005）认为北部加利福尼亚湾轴向盆地低于水平面拆离断层。相反，没有一个单一的拆离断层或拆离盆地在整个加利福尼亚湾地区的中南部被识别（裂谷倾角小于20°），后者中南部为一个扭动主控型的转换拉伸盆地。

我们认为拆离断层和拆离盆地的存在或缺失取决于裂谷夹角。一般认为，拆离断层会形成高热流和大陆的快速扩张区域（Buck，1991；Friedmann和Burbank，1995；Lavier和

Buck，2002）。高热流条件与板块边界是沿着早—中中新世岩浆弧的轴线形成环境相匹配（Stock 和 Hodges，1989）。我们认为湾槽地区拆离断层的形成需要大于北部扩张但小于南部扩张的临界速率，这归因于裂谷夹角的差异。沿着湾槽通道的走向，加利福尼亚和墨西哥大陆的运动速率（从南到北 43~47mm/a；Plattner 等，2007）相近，但从南到北走滑到拉张应变的转变可视为裂谷斜向拉张的函数变化。一个纯粹的走滑边界（夹角为 0°）只受走滑应变的影响而无拉张，一个纯粹的正交裂谷（夹角为 90°）则通过正断层的拉张调节所有板块运动。在裂谷夹角小于 20°的加利福尼亚湾中南部，走滑主控型张扭变形要么表现为斜向滑移断层与走滑断层（非分离样式；圣何塞岛到洛莱托地区）的连接，要么表现为北西向走滑断层的快速滑动及正断层的缓慢扩张的特征（拉帕兹到卡波斯地区；Fletcher 和 Munguia，2000）。在裂谷夹角大于 30°的湾槽北部地区，拉张作用在包括分解应变和非分解变形样式的总体应变中占到了很大的比重。我们认为北部地壳更高的扩张速率引起了更快的正断层滑动速率，很可能大于 1~2mm/a，因此达到了能在温度较高的厚地壳形成低角度正断层和拆离盆地的门限速度（Friedmann 和 Burbank，1995）。

10.5.2　沉积物输入的影响

索尔顿海槽和加利福尼亚湾北部的盆地填充和过填充了大量科罗拉多河的沉积物。陆相三角洲内微妙的地形结构将索尔顿海（海平面以下 70m）从加利福尼亚湾海域中隔离出来（Merriam 和 Bandy，1965；Winker，1987）。除了对地表特征和环境影响明显之外，快速的沉积物输入也直接影响到地壳的厚度和组成、岩石圈流变学特征、拉张机制以及裂谷结构。地球物理研究表明，前新生代加利福尼亚湾北部和索尔顿槽的大陆岩石圈在断陷盆地扩张之下完全破裂，新的地壳随年轻的同裂谷沉积物和幔源侵入而产生（Moore，1973；Fuis 等，1984；Nicolas，1985；Gonzalez 等，2005）。这些盆地中源自科罗拉多河的沉积物体积很大，估计在 $(2.2~3.4) \times 10^5 km^3$，在过去的 5~6Ma 以类似岛弧和海底扩张中心的活动速率形成了新过渡壳（Dorsey，2010）。沉积物的快速输入同样影响了上地幔的热构造和浮力的变化，扩大了应变的范围，促进了过渡地壳向窄裂谷模式的转换（Lizarralde 等，2007；Bialas 和 Buck，2009）。

加利福尼亚湾南部的欠补偿沉积区，板块边界完成了从大陆裂谷盆地到海底扩张中心的转换，并以镁铁质洋壳和磁条带为主要特征。相反，加利福尼亚湾北部和索尔顿海槽的补偿和过补偿盆地以由科罗拉多河沉积物输入形成的厚的新地壳为其填充方式，且以岩石圈破裂和板块离散为主要特征（Moore，1973；Fuis 等，1984；Nicolas，1985）。盆地完成从大陆裂谷到大洋扩张中心的转换程度，沿着板块边界从南到北的变化极具戏剧性，不仅盆地的发育都是从相同的时代开始的，而且从 6.4~6.1Ma 开始这些盆地离开主板块边界的距离是相同的（Oskin 等，2001；Oskin 和 Stock，2003）。北部盆地并非海相，因为沉积物的快速输入产生了新的地壳，阻止了正常的玄武岩洋壳在板块离散地带的形成。尽管之前存在的大陆岩石圈在北部完全破裂，但伸展作用并没有形成具有正常洋壳的海底扩张中心，取而代之的是新生成了一个较厚的过渡性地壳，该地壳由源自科罗拉多河的沉积物及幔源侵入体组成。

上述内容总结强调了沉积物输入是位于湾槽通道内的斜向离散盆地的地壳厚度和组分形成的首要控制因素。事实上，沉积物输入速率决定了裂谷作用是否可以完全取代过渡地壳并形成新的镁铁质洋壳。

10.6 结论

加利福尼亚湾和索尔顿海槽地区的沉积盆地揭示了与太平洋—北美板块斜向离散边缘之上的张扭变形密切相关的构造样式。盆地填充模式作为一个与科罗拉多河口距离相关的函数，存在系统性变化，在过去的 5~6Ma 中已经输送了大量的沉积物到这一地区。加利福尼亚湾南部轴部盆地由短的、沉积欠补偿的海洋扩张中心组成，北东走向并垂直于长转换断层；而北部轴向盆地主要是斜向拉分盆地，包含了厚的沉积充填但缺乏海底扩张中心或正常镁铁质洋壳的证据。

如此看来三个主要的参数控制着湾槽通道区的盆地的形态和演化：（1）板块边界整体走向和板块相对运动方向之间的锐角（裂谷夹角，α）；（2）来自北部科罗拉多河的沉积物输入；（3）应变分解度。拆离断层只在加利福尼亚湾北部和索尔顿海槽内裂谷夹角大于 30°的地区发育，而在加利福尼亚湾中南部裂谷角小于 20°的地区缺失。我们认为北部地区更大的裂谷夹角产生了更快的扩张速率，控制了低角度拆离断层和拆离盆地的形成，这一过程需要的扩张速率被认为大于 1~2mm/a。

湾槽通道内盆地的沉积物空间供给的变化对地壳厚度和组分、岩石圈流变学、地壳扩张机制和整个裂谷结构起到了绝对控制。在沉积欠补偿的加利福尼亚湾南部，板块边界完成了从大陆裂谷盆地到海底扩张中心的转换，并以正常的镁铁质洋壳和磁条带为主要特征。在加利福尼亚湾和索尔顿海槽的北部，沉积物供给足以保证盆地填充和过填充，由沉积物输入产生的新过渡性地壳阻止了正常洋壳和海底扩张中心的形成。因此沉积物输入的速率决定着大陆裂谷作用能否依预期形成新的以镁铁质地壳为底的洋盆。这个进程可能对于世界上其他的斜向离散边缘有着重要的意义，因为在那些板块边缘上同样也存在着由大型河流快速输入巨量沉积物的张扭型盆地。

致谢（略）

参考文献

Aragón-Arreola, M., Morandi, M., Mártin-Barajas, A., Deldago-Argote, L., and Gonzalez-Fernandez, A. (2005) Structure of the rift basins in the central Gulf of California: kinematic implications for oblique rifting. Tectonophysics, 409, 19–38.

Aragón-Arreola, M. and Martín-Barajas, A. (2007) Westward migration of extension in the northern Gulf of California, Mexico. Geology, 35, 571–574.

Atwater, T., and Stock, J. M. (1998) Pacific North America plate tectonics of the Neogene southwestern United States: An update: International Geology Review, 40, 375–402.

Axen, G. J. (1995) Extensional segmentation of the Main Gulf Escarpment, Mexico and the United States: Geology, 23, 515–518.

Axen, G. J., and Fletcher, J. M. (1998) Late Miocene–Pleistocene extensional faulting, northern Gulf of California, Mexico and Salton Trough, California. International Geology Review, 40, 217–244.

Axen, G. J., Grove, M., Stockli, D., Lovera, O. M., Rothstein, D. A., Fletcher, J. M.,

Farley, K., Abbott, P. L. (2000) Thermal evolution of Monte Blanco Dome; lowangle normal faulting during Gulf of California rifting and late Eocene denudation of the eastern Peninsular Ranges. Tectonics, April 2000 19, 197–212.

Bennett, S. E. K. (2009) Transtensional rifting in the late proto-Gulf of California near Bahía Kino, Sonora, Mexico. [M. S. thesis]: University of Norgth Carolina, 121 pp.

Bialas, R. W., and Buck, W. R. (2009) How sediment promotes narrow rifting: Application to the Gulf of California. Tectonics, 28 doi: 10. 1029/2008TC002394.

Black, N. (2004) Structure and hanging wall stratigraphy of the Las Cuevitas detachment, Central Sierra, San Felipe, Baja California, Mexico. [M. S. thesis]: UCLA, 102 pp.

Black, N., and Axen, G. (2003) Structure and hanging wall stratigraphy of the Las Cuevitas detachment, central Sierra San Felipe, Baja California: Geological Society of America Abstracts with Programs, 35 (4), 10.

Boehm, M. C. (1984) An overview of the lithostratigraphy, biostratigraphy, and paleoenvironments of the late Neogene San Felipe marine sequence, Baja California, Mexico. In: V. A. Frizzell, Jr. (Ed.) Geology of Baja California Peninsula. Field Trip Guidebook, Pac. Sect., Soc. Econ. Paleontol. Mineral., 39, 253–265.

Brown, H. E. (2007) Crustal rupture, creation, and subduction in the Gulf of California, Mexico and the role of gas hydrate in the submarine Storegga Slide, offshore Norway [Ph. D. thesis]: University of Wyoming, Laramie, 159 pp.

Bryant, B. A. (1986) Geology of the Sierra Santa Rosa Basin, Baja California, Mexico [M. S. thesis]: San Diego, San Diego State University, 75 p.

Buck, W. R. (1991) Modes of continental lithospheric extension. Jour. Geoph. Research, 96, 20161–20178.

Burchfiel, B. C., and Stewart, J. H. (1966) Pull-apart origin of the central segment of Death Valley, California. Geological Society of Americ Bulletin, 77, 439–442.

Busch, M. M., Arrowsmith, J. R., Umhoefer, P. J., Coyan, J. A, Maloney, S. J., and Guti_errez, G. M. (2011) Geometry and evolution of rift-margin, normal-fault-bounded basins from gravity and geology, La Paz – Los Cabos region, Baja California Sur, Mexico. Lithosphere, 3, 110–127.

Carreño, A. L. (1992) Neogene microfossils from the Santiago Diatomite, Baja California Sur, Mexico: Paleontología Mexicana, 59, 1–38.

Clifton, A. E., Schlische, R. W., Withjack, M. O., and Ackermann, R. V. (2000) Influence of rift obliquity on fault-population systematics: results of experimental clay models. Journal of Structural Geology, 22, 1491–1509.

Curray, J. R., Moore, D. G., et al. (1982) Guaymas Basin: sites 477, 478, and 481. Initial Reports of the DSDP, 64, 211–415.

Dibblee, T. W. (1954) Geology of the Imperial Valley region, California, Geology of Southern California, California Division of Mines Bulletin, 170, 21–28.

Dibblee, T. W. (1984) Stratigraphy and tectonics of the San Felipe Hills, Borrego Badlands, Superstition Hills, and vicinity. In Rigsby, C. A. (ed.) The Imperial Basin – tectonics, sedimentation, and thermal aspects. Los Angeles, Pacific Section S. E. P. M., 31-44.

Dixon, T., F. Farina, C. DeMets, F. Suarez Vidal, J. Fletcher, B. Marquez-Azua, M. Miller, O. Sanchez, and P. Umhoefer (2000) New kinematic models for Pacific- North America motion from 3 Myr to present, II: Tectonic implications for Baja and Alta California, Geophys. Res. Lett., 27, 3961-3964.

Dorsey, R. J. (2010) Sedimentation and crustal recycling along an active oblique-rift margin: Salton Trough and northern Gulf of California. Geology, 38, 443-446.

Dorsey, R. J. and Martín-Barajas, A. (1999) Sedimentation and deformation in a Pliocene-Pleistocene transtensional supradetachment basin, Laguna Salada, Northwest Mexico. Basin Research, 11, 205-222.

Dorsey, R. J. and Umhoefer, P. J. (2000) Tectonic and eustatic controls on sequence stratigraphy of the Pliocene Loreto basin, Baja California Sur, Mexico. Geolgical Society of America Bulletin, 112, 177-199.

Dorsey, R. J., Fluette, A., McDougall, K. A., Housen, B. A., Janecke, S. U., Axen, G. J., and Shirvell, C. R. (2007) Chronology of Miocene-Pliocene deposits at Split Mountain Gorge, southern California: A record of regional tectonics and Colorado River evolution. Geology, 35, 57-60.

Dorsey, R. J., Peryam, T. C., Bennett, S., Oskin, M. E., and Iriondo, A. (2008) Preliminary Basin Analysis of latest Miocene Nonmarine Deposits Near Bahia Kino, Coastal Sonora: A New Record of Crustal Deformation During Initial Opening of the Northern Gulf of California. Eos Trans. AGU, 89 (53) Fall Meeting Supplement, Abstract, T11A-1851.

Dorsey, R. J., Housen, B. A., Janecke, S. U., Fanning, M., and Spears, A. L. F. (2011) Stratigraphic record of basin development along a transtensional plate boundary: Late Cenozoic Fish Creek-Vallecito basin, southern California. Geological Society America Bulletin, 123, 771-793.

Einsele, G., and Niemitz, J. W. (1982) Budget of post-rifting sediments in the Gulf of California and calculation of the denudation rate in neighboring land areas. Initial Reports of the DSDP, 64 (2), 571-592.

Elders, W. A., and Sass, J. H. (1988) The Salton Sea scientific drilling project. Journal of Geophysical Research, 93 (B11), 12953-12968.

Fletcher, J. M., and Munguía, L. (2000) Active continental rifting in southern Baja California, Mexico: Implications for plate motion partitioning and the transition to seafloor spreading in the Gulf of California. Tectonics, 19, 1107-1123.

Fletcher, J. M., Grove, M., Kimbrough, D., Lovera, O., and Gehrels, G. E. (2007) Ridge-trench interactions and the Neogene tectonic evolution of the Magdalena shelf and southern Gulf of California: Insights from detrital zircon U-Pb ages from the Magdalena fan and adjacent areas. Geological Society of America Bulletin, 119, 1313-1336.

Friedmann, S. J., and Burbank, D. W. (1995) Rift basins and supradetachment basins: Intracontinental extensional end members. Basin Res., 7, 109-127.

Fuis, G. S., and Kohler, W. M. (1984) Crustal structure and tectonics of the Imperial Valley region, California, in Rigsby, C., ed., The Imperial Basin—tectonics, sedimentation, and thermal aspects, field trip guidebook. Los Angeles, Pacific Section SEPM, Society for Sedimentary Geology, 40, 1–13.

Fuis, G. S., and Mooney, W. D. (1991) Lithospheric structure and tectonics from seismic refraction and other data, in The San Andreas Fault System, California. U. S. Geol. Surv. Profess. 1515, 207–236.

Fuis, G. S., Mooney, W. D., Healy, J. H., McMechan, G. A., and Lutter, W. J. (1984) A seismic refraction survey of the Imperial Valley Region, California. Journal of Geophysical Research, 89, 1165–1189.

Gans, P. B. (1997) Large-magnitude Oligo-Miocene extension in southern Sonora: Implications for the tectonic evolution of northwest Mexico. Tectonics, 16 (3), 388–408.

Gastil, R. G., Lemone, D. V., and Stewart, W. J. (1973) Permian fusulinids from near San Felipe, Baja California: AAPG Bulletin, 57, 746–747.

Gastil, R. G., Phillips, R. P., and Allison, E. C. (1975) Reconnaissance geology of the State of Baja California. Boulder, CO, Geological Society of America Memoir 140, 170 p.

Gastil, R. G., and Krummenacher, D. (1977) Reconnaissance geology of coastal Sonora between Puerto Lobos and Bahia Kino. Geological Society of America Bulletin, 88, 189–198.

González-Fernandez, A., Danobeitia, J. J., Deldago-Argote, L., Michaud, F., Cordoba, D., and Bartolome, R. (2005) Mode of extension and rifting history of upper Tiburón and upper Delfin basins, northern Gulf of California. Journal of Geophysical Research, 110, 1–17.

Helenes, J., Carreño, A. L., and Carrillo, R. M. (2009) Middle to late Miocene chronostratigraphy and development of the northern Gulf of California. Marine Micropaleontology, 72, 10–25.

Holt, J. W., Holt, E. W., and Stock, J. M. (2000) An age constraint on Gulf of California rifting from the Santa Rosalía basin, Baja California Sur, Mexico. Geological Society of America Bulletin, 112 (4), 540–549.

Kirby, S. M., Janecke, S. U., Dorsey, R. J., Housen, B. A., Langenheim, V., McDougall, K., and Steely, A. N. (2007) Pleistocene Brawley and Ocotillo formations: Evidence for initial strike-slip deformation along the San Felipe and San Jacinto fault zones, southern California. Journal of Geology, 115, 43–62.

Lavier, L. L., and Buck, W. R. (2002) Half graben versus largeoffset low-angle normal fault: Importance of keeping cool during normal faulting. Journal of Geophysical Research, 107 doi: 10. 1029/2001JB000513.

Leeder, M. R., and Gawthorpe, R. L. (1987) Sedimentary models for extensional tilt-block/half-graben basins, in Coward, M. P. and Dewey J. F., eds., Continental extensional tectonics. Geological Society Special Publication 28, 139–152.

Lizarralde, D., Axen, G. J., Brown, H. E., Fletcher, J. M., Antonio González-Fernández, A., Harding, A. J., Holbrook, W. S., Kent, G. M., Paramo, P., Sutherland, F. and Umhoefer, P. J. (2007) Variation in styles of rifting in the Gulf of California. Nature, 448, 466–469.

Lonsdale, P. (1989) Geology and tectonic history of the Gulf of California, in Winterer, E. L., Hussong, D., and Decker, R. W., eds., The Eastern Pacific Ocean and Hawaii, vol. N

Boulder, CO, Geological Society of America, Geology of North America, 499-522.

Lutz, A. T., Dorsey, R. J., Housen, B. A., and Janecke, S. U. (2006) Stratigraphic record of Pleistocene faulting and basin evolution in the Borrego Badlands, San Jacinto fault zone, southern California. Geological Society of America Bulletin, 118, 1377-1397.

Maloney, S. J., Umhoefer, P. J., Arrowsmith, J. R., Gutiérrez, G. M., Santillanez, A. U., Rittenour, T. R. (2007) Late Pleistocene- Holocene faulting history along the Northern El Carrizal Fault, Baja California Sur, Mexico: earthquake recurrence at a persistently active rifted margin. Eos Trans. AGU88 (52) Fall Meet. Suppl., Abstract T41A-0358.

Martínez-Gutiérrez, G., and Sethi, P. S., 1997, Miocene- Pleistocene sediments within the San Jose del Cabo Basin, Baja California Sur, Mexico, in Johnson, M. E., and Ledesma- Vazquez, J., eds., Pliocene carbonates and related facies flanking the Gulf of California, Baja California, Mexico. Special Paper, 318. Boulder, CO, Geological Society of America, 141-166.

Mayer, L. and Vincent, K. R. (1999) Active tectonics of the Loreto area, Baja California Sur, Mexico. Geomorphology, 27, 243-255.

McDougall, K. A. (2008) Late Neogene marine incursions and the ancestral Gulf of California. In: Reheis, M., Herschler, R., and Miller, D. (eds.) Late Cenozoic Drainage History of the Southwestern Great Basin and Lower Colorado River Region: Geologic and Biotic Perspectives. Geological Society of Americ Special Paper 439, 355-373.

McDougall, K. A., Poore, R. Z., Matti, J. C. (1999) Age and paleoenvironment of the Imperial Formation near San Gorgonio Pass, Southern California. Journal of Foraminiferal Research, 29, 4-25.

McTeague, M. S. (2006) Marginal strata of the east central San Jose del Cabo basin, Baja California Sur, Mexico. Master's thesis, Northern Arizona University, Flagstaff, 152 pp.

Merriam, R. H. and Bandy, O. L. (1965) Source of upper Cenozoic sediments in the Colorado delta region. Journal of Sedimentary Petrology, 35, 911-916.

Moore, D. G. (1973) Plate-edge deformation and crustal growth, Gulf of California structural province. Geological Society of America Bulletin, 84, 1883-1905.

Mortimer, E., Gupta, S., and Cowie, P. A. (2005) Clinoform nucleation and growth in coarsegrained deltas, Loreto basin, Baja California Sur, Mexico: A response to episodic accelerations in fault displacement. Basin Research, 17, 337-359.

Morton, D. M. and Matti, J. C. (1993) Extension and contraction within an evolving divergent strike-slip fault complex: the San Andreas and San Jacinto fault zones at their convergence in southern California, in Powell, R. E., Weldon, R. J., II, and Matti, J. C., eds., The San Andreas fault system: displacement, palinspastic reconstruction, and geologic evolution: Memoir-Geological Society of America. Boulder, CO, Geological Society of America, 178, 217-230.

Muffler, J. L. P. and White, D. E. (1969) Active metamorphism of the upper Cenozoic sediments in the Salton Sea geothermal field and the Salton trough, southern California: Geological Society of America Bulletin, 80, 157-182.

Nicholson, C., Sorlien, C. C., Atwater, T., Crowell, J. C., Luyendyk, B. P. (1994) Microplate capture, rotation of the Western Transverse Ranges, and initiation of the San Andreas transform as a low-angle fault system, Geology, 22, 491-495.

Nicolas, A. (1985) Novel type of crust produced during continental rifting. Nature, 315, 112-115.

Nilsen, T. H., and Sylvester, A. G. (1995) Strike-slip basins, In: Busby, C. J. and Ingersoll, R. V. (Eds.) Tectonics of sedimentary basins. Oxford, Blackwell Science, 425-457.

Ochoa-Landín, L., Ruiz, J., Calmus, T., Pérez-Segura, E., and Escandón, F. (2000) Sedimentology and stratigraphy of the upper Miocene Boleo Formation, Santa Rosalía, Baja California, Mexico. Revista Mexicana de Ciencias Geológicas, 17 (2), 83-96.

Oskin, M., Stock, J., and Martín-Barajas, A. (2001) Rapid localization of Pacific-North America plate motion in the Gulf of California. Geology, 29, 459-462.

Oskin, M., and Stock, J. M. (2003a) Marine incursion synchronous with plate-boundary localization in the Gulf of California. Geology, 31, 23-26.

Oskin, M., and Stock, J. M. (2003b) Pacific-North America plate motion and opening of the Upper Delfin basin, northern Gulf of California. Geological Society of America Bulletin, 115, 1173-1190.

Oskin, M., and Stock, J. M. (2003c) Cenozoic volcanism and tectonics of the continental margins of the Upper Delfin basin, northeastern Baja California and western Sonora, in Kimbrough, D. L., Johnson, S. E., Paterson, S., Martin-Barajas, A., Fletcher, J. M., and Girty, G., eds., Tectonic evolution of northwestern Mexico and the southwestern USA: Geological Society of America Special Paper 374. Boulder, CO, Geological Society of America, 421-428.

Pacheco, M., Marin, A., Elders, W., Espinosa, J., Helenes, J., and Segura, A. (2006) Stratigraphy and structure of the Altar basin of NW Sonora: Implications for the history of the Colorado River delta and the Salton Trough, Revista Mexicana de Ceiencias Geologicás, 23, 1-22.

Persaud, P., Stock, J. M., Steckler, M. S., Martin-Barajas, A., Diebold, J. B., Gonzalez-Fernandez, A., andMountain, G. S. (2003) Active deformation and shallow structure of the Wagner, Consag, and Delfin basins, northern Gulf of California, Mexico: Journal of Geophysical Research, 108 doi: 10. 1029/2002JB001937.

Plattner, C., Malservisi, R., Dixon, T. H., LaFemina, P., Sella, G. F., Fletcher, J., and Suarez-Vidal, F. (2007) New constraints on relative motion between the Pacific Plate and Baja California microplate (Mexico) from GPS measurements. Geophys. J. Int., 170, 1373-1380. doi: 10. 1111/j. 1365-246X. 2007. 03494. x.

Schmitt, A. K., and Vazquez, J. A. (2006) Alteration and remelting of nascent oceanic crust during continental rupture: Evidence from zircon geochemistry of rhyolites and xenoliths from the Salton Trough, California. Earth and Planetary Science Letters, 252, 260-274.

Spencer, J. E., and Normark, W. R. (1979) Tosco-Abreojos fault zone: A Neogene transform plate boundary within the Pacific margin of south Baja California, Mexico: Geology, 7, 554-557.

Steely, A. N., Janecke, S. U., Dorsey, R. J. and Axen, G. J. (2009) Early Pleistocene initiation of the San Felipe fault zone, SW Salton Trough, during reorganization of the San Andreas fault system. Geological Society of America Bulletin, 121, 663-687.

Stock, J. M., and Hodges, K. V. (1989) Pre-Pliocene extension around the Gulf of California and the transfer of Baja California to the Pacific Plate: Tectonics, 8, 99-115.

Sutherland, F. H. (2006) Continental rifting across the southern Gulf of California. PhD thesis, University of California, San Diego, 173 p.

Tikoff, B., and Teyssler, C. (1994). Strain modeling of displacement: field partitioning in transpressional orogens. Journal of Structural Geology, 16, 1575-1588.

Tron, V., Brun, J. P. (1991). Experiments on oblique rifting in brittle – ductile systems. Tectonophysics 188, 71-84.

Umhoefer, P. J., Dorsey, R. J., and Renne, P. R. (1994) Tectonics of the Pliocene Loreto basin, Baja California Sur, Mexico, and evolution of the Gulf of California. Geology, 22, 649-652.

Umhoefer, P. J., and Stone, K. A. (1996) Description and kinematics of the SE Loreto basin fault array, Baja California Sur, Mexico: a positive field test of obliqueriftmodels. Journal of StructuralGeology, 18 (5), 595-614.

Umhoefer, P. J., Schwennicke, T., Del Margo, M. T., Ruiz- Geraldo, G., Ingle, J. C. Jr., and McIntosh, W. (2007) Transtensional fault – termination basins: an important basin type illustrated by the Pliocene San Jose Island basin and related basins in the southern Gulf of California, Mexico. Basin Research, 19, 297-322.

Van Andel, T. (1964) Recent marine sediments of Gulf of California, in Van Andel, T. and Shor, G. G., eds., Marine Geology of the Gulf of California: American Association of Petroleum Geologists, Memoir 3, 216-310.

Wilson, I. F. (1948) Buried topography, initial structures, and sedimentation in Santa Rosalia area, Baja California, Mexico: American Association of Petroleum Geologists Bulletin, 32, 1762-1807.

Winker, C. D. (1987) Neogene stratigraphy of the Fish Creek – Vallecito section, southern California: implications for early history of the northern Gulf of California and Colorado delta. Ph. D. dissertation, University of Arizona, Tucson, 494 p.

Winker, C. D., and Kidwell, S. M. (1996) Stratigraphy of a marine rift basin: Neogene of the western Salton Trough, California. In Abbott, P. L., and Cooper, J. D. (eds.) Field conference guidebook and volume for the annual convention, San Diego, CA, May, 1996 Bakersfield, CA, Pacific Section, American Association of Petroleum Geologist, 295-336.

Withjack. M. O. and Jamison, W. R. (1986) Deformation produced by oblique rifting. Tectonophysics, 126, 99-124.

(赵钊译，张功成 纪沫 祁鹏 校)

第 11 章 活动张扭型陆内盆地
——以美国大盆地西部的沃克通道为例

ANGELA S. JAYKO[1], MARCUS BURSIK[2]

(1. Regional Tectonics, U. S. Geological Survey, U. C. White Mountain Research Station, Bishop, USA; 2. Department of Geology, University at Buffalo, Buffalo, USA)

摘　要：沃克通道（Walker Lane）内沉积盆地的形态和面积是上新世—更新世张扭性变形和北美板块内华达山地块局部拆离共同作用的结果。多种不同的构造形态位于这个活动的转换拉张带内。沃克通道东北端部分被埋藏在喀斯喀特南部活火山下，附近的盆地填充并不完全。南侧盆地大小适中（25~45km 长，15~10km 宽），盆地翼部表现为南北向到北北东向为主 8~12km 宽的山区。这些盆地的走向为 300°左右，平行排列且呈顺时针旋转。阶跃区宽 85~100km，盆地沿东西向延长，面积较小（15~30km 长，5~15km 宽），局部被活火山中心占据。沃克通道最南端构造复杂，形成了构造高地。相邻盆地长 50~200km，宽 5~20km。沃克通道中张扭型盆地的变化很大程度上归因于应变分配。沃克通道内较大盆地有 2~6km 的位移相对于盆地边界断层，且有重力和钻井数据证实的碎屑沉积达 3km。盆地沉积可能包括火山沉积与湖泊沉积。其中火山沉积为双峰式玄武岩和流纹岩互层；而湖泊沉积反映了从冷淡水环境到含盐蒸发环境的较广范围。

关键词：张扭型盆地　沃克通道　欧文斯谷地堑　第四纪双峰大陆火山活动　新近纪压扭作用

11.1　引言

沃克通道（Walker Lane）位于美国大盆地的西缘，为陆内变形带提供了张扭盆地构造实例。本文包括以下几个主题：（1）沃克通道的区域构造背景；（2）沃克通道的特征描述；（3）应变调节的研究综述，该作用影响了张扭盆地性质，并为地层的活动过程提供了构造解释。本文还讨论了影响沃克通道内构造样式演化的因素。

11.1.1　构造背景

美国西南部的构造活动受控于太平洋和北美板块间的相互作用（图 11-1）。板块边界的特点是：南部表现为裂陷；西部海岸山脉因转换断层作用表现为走滑和压扭；东部沿沃克通道表现为张扭。太平洋—北美板块边界的北缘紧邻美国，位于门多西诺三联点上。残存的法伦板块地壳来自于古弧后拉张盆地的古喀斯喀特聚敛边界，离散板块边界的早期扩展影响着大盆地西部的构造地形变化。本文聚焦于主要构造特征，包括主动的和继承性的，这些特征影响着盆地变形、火山活动以及沃克通道上部地壳的动力学特征。

太平洋与北美板块间的分界面受圣安德列斯转换系统的控制。在南部，太平洋板块的东边界沿拉张的大洋扩张中心分布，其长度至少为 1.5×10^4 km（图 11-1）。加利福尼亚湾内的太平洋—北美板块边界是发散而狭长的，且受产生年轻洋壳的洋脊转换系统的控制

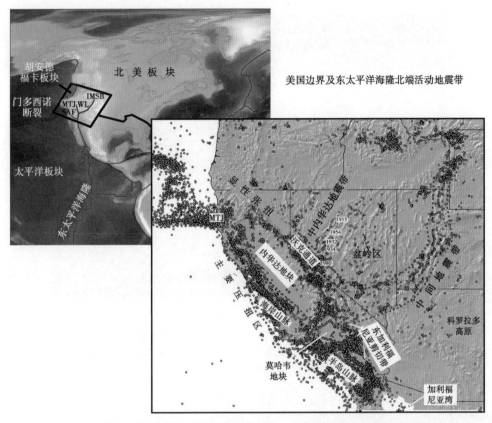

图 11-1　美国西部内华达地块、莫哈韦地块、科罗拉多高原、内华达和犹他
盆岭区地震活动性图（1960—2009 年）

插图为东太平洋海隆北端与北美板块碰撞的部分。MTJ—门多西诺三联点；
SAF—圣安德列斯断层；WL—Walker Lane；IMSB—山间地震带

（Vmhoefer 等，2002）。加利福尼亚湾北板块边界的克拉通内变形扩散和分流为三个有活动断层和较小火山活动的地震带（图 11-1）。太平洋—北美板块右旋位移穿过每一个地震带，其中圣安德列斯转换断层占位移量的 73%～75%，东加利福尼亚剪切带和沃克通道仅占 20%～22%，穿过盆岭省的尾部（山间地震带）占 1%～2%（Bennettet 等，1999；Dixon 等，2000）。这些地震带将各构造地形分离开（图 11-2；Wallace，1984）。向西倾斜的右旋内华达山地块，将圣安德列斯压扭区从沃克通道张扭系统中独立出来（图 11-2 和图 11-3；Frei，1986；Fay 和 Humphreys，2008）。通过这些地区构造样式和地形的对比可知，挤压应力并未通过陆壳转换系统进行长距离传导（图 11-3；Fay 和 Humphreys，2008；Heidbach 等，2008），地壳相较于岩石圈残余大洋板块更薄弱。内华达盆岭区的拉张作用正交于正断层，且斜交于沿加利福尼亚—内华达边界沃克通道内的断层（图 11-3）。

内华达地块是影响张扭作用向东延伸边界或离散边缘的主要陆壳地貌（Hammond 和 Thatcher，2007；Fay 和 Humphreys，2008；Kreemeret 等，2009）。地块与其东缘的隆起发育一致的相对连续的向西翘倾（Lindgren，1911；Christiansen，1966；Huber，1981，1990；Unruh，1991；Jayko，2009a，2009b）。全球定位系统和区域应力研究表明，内华达地块受太平洋板块运动的控制，主运动方向同为北西向但比太平洋板块运动速率要低得多（图 11-

图 11-2 美国边界地形图

显示西倾内华达山地块、Walker Lane 带、Walker Lane 构造带和盆岭省之间的变形区。
SNFF—内华达山前缘断层；GFZ—加洛克断裂带；MTJ—门多西诺三联点。本文讨论沃克通道内的
4 个张扭区：（1）金字塔湖区；（2）卡森区；（3）埃切斯尔区；（4）欧文斯河谷区

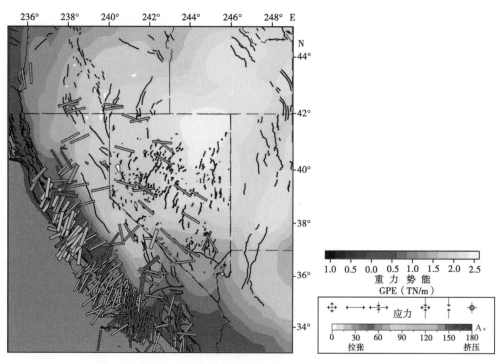

图 11-3 美国西部重力势能示意图（据 Fay 和 Humphreys，2008）

第四纪断层（黑线）来源于美国地质调查局断裂图 http://pubs.usgs.gov/fs/2004/3033/fs-2004-3033.html，
应力数据来自于应力图（Heidbach 等，2008；Fay 和 Humphreys，2008，修改）。应力栏从拉张应力指向挤压
应力方向。拉张作用在内华达盆岭区表现为正交于断裂，在沃克通道内表现为斜交于断裂

4；Hearn 和 Humphreys，1998；Dixon 等，2000；Kreemer 等，2009）。内华达地块拆离于稳定的北美板块，且与一些主要的岩石圈特征有关，这些特征表现为岩石学（Ducea 和 Saleeby，1996，1998）、地球物理学（Zandt，2003；Jones 等，2004；Zandt 等，2004）及地貌学特征（Christiansen，1966；Huber，1981，1990；Jayko，2009a，2009b）。内华达隆起部分是由于下部地壳或上地幔分离及沉陷（Ducea 和 Saleeby，1996）导致了大规模的垂向位移。沃克通道的动力机制可能来源于克拉通地块的张扭裂陷作用引起的地幔上涌（Crough 和 Thompson，1977；Mavko 和 Thompson，1983；Jayko，2009b）。这些过程与导致沃克通道张扭作用的板块运动是同期发生的。

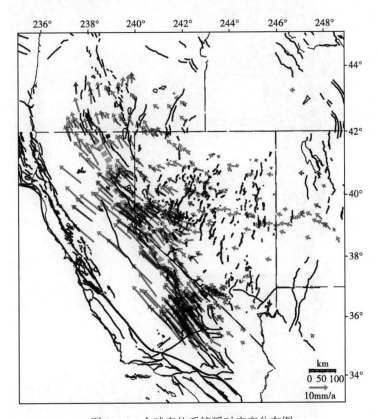

图 11-4 全球定位系统瞬时应变分布图

北美部分参考自 Kreemer 等（2009）；美国西部第四纪断层来自于美国地质调查局断裂图，http：//pubs.usgs.gov/fs/2004/3033/fs-2004-3033.html；彩色虚线表示 Walker Lane 带的西缘

11.1.2 Walker Lane 带

Walker Lane 带是一个构造地形学的定义，由盆岭地块和大型走滑断层系统组成（图 11-5；Stewart，1988）。由于板块的相互作用，该带在活动的右旋张扭区域内发生变形（Hearn 和 Humphreys，1998；Bennett 等，1999；Dixon 等，2000；Miller 等，2001；Unruh 等，2003；Faulds 等，2005；Fay 和 Humphreys，2008）。晚中新世的 Walker Lane 带内发生新生代张扭变形（Slemmons 等，1979；Stewart，1985，1988；Oldow，1992；Cashman 和 Fontaine，2000；Faulds 等，2005）。沃克通道北部边界（下文图 11-11；Stewart，1992）位于南喀斯喀特活动火山堆积物的东南部，转换断层从北部弧后扩张到南部张扭板块边界（Unruh 等，2003；Faulds 等，2005；Henry 等，2007；图 11-1 和图 11-2）。Walker Lane 大致平行于圣安德列斯断裂带，终止于门多西诺三联点以东 150km（图 11-2 和图 11-5）。Walker Lane 带宽度约为 50~80km，除去中新统和更古老的盆山地形，总长约 700km（Shaw，1965；

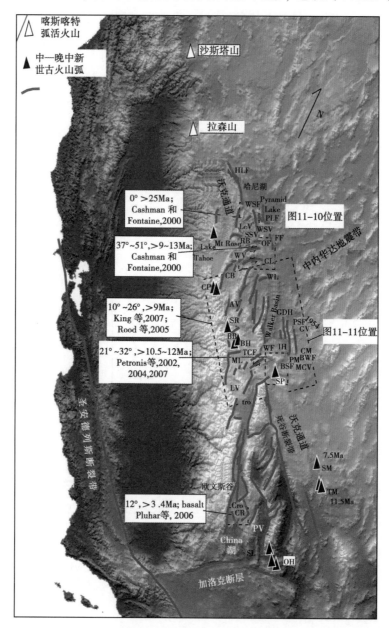

图 11-5 基于古地磁研究的晚中新世顺时针旋转和年轻地层的位置地形图

黑三角带——晚中新世火山弧中心；红线表示 Walker Lane 带内第四纪断层；旋转角度大体上为 20°~30°，但局部会达到 50°；晚中新世火山中心的研究来自于 Busby 等（2008a，2008b，2009）。LeV—莱蒙河谷；WSF—奥姆斯普林斯断层；WSV—奥姆斯普林斯河谷；SSV—西班牙斯普林斯谷；PLF—金字塔湖断层；FF—弗利断层；OF—奥林豪斯断层；CL—卡森构造线；WL—瓦布斯卡构造线；BSF—本顿斯普林斯断层；PS—石泉断层；WF—瓦苏克断层；BSF—Big Smokey 断层；CM—锡达山（Yount 等，1993）；CB—卡森盆地；CP—卡森山口；BH—博迭山；ML—莫诺湖；LV—长谷；SM—锡尔弗峰；TM—廷伯山；PV—帕那敏特谷；CR—科索山；SL—瑟尔斯湖；OH—欧塞德山

Hay,1976；Wright,1976；Bell 等,1999）。Walker Lane 带南部包括了美国本土海拔最高点与最低点，以及最大的落差。内华达前缘断层正位于沃克通道的西侧。东部边界由大量活动的不连续的右旋断层组成（图 11-5；Wallace,1984；Stewart,1988；Hearn 和 Humphreys,1998；Cashman 和 Fontaine,2000；Oldow 等,2001,2008；Bell 等,2004；Wesnousky,2005a,2005b；Slemmons 等,2008）。在南部，Walker Lane 转换带穿过东向的加洛克断层至东加利福尼亚剪切带，该带类似于右旋变形带且以莫哈韦地块东缘为边界（Stewart,1988；下文图 11-1 和图 11-2）。虽然两个变形带共享了瞬时应变位移（Miller 等,2001；Savage 等,2001），但产生的构造地形却截然不同，这说明了地块下伏岩石圈存在差异（Savage 等,2004；Zandt 等,2004）。

11.2 张扭盆地，沃克通道

11.2.1 概述

虽然 Walker Lane 带内构造连续性较差，但区域应力场的方向仍是相同的（图 11-3 和图 11-4；Fay 和 Humphreys,2008；Kreemer 等,2009）。导致 Walker Lane 带内的应力变化的因素可能包含以下几点：地壳厚度的差异、地壳岩石的非均质性（Stewart,1988）、上地幔和下地壳流变学变化（Tickoff 等,2004）、先存的新近纪弧内火山位置与构造特征（Busby 等,2008a,2008b）、内华达地块边界的自由边缘方向（Taylor 和 Dewey,2009），Walker Lane 带外的应变转移进入中内华达地震带或 Walker Lane 东部的其他盆山耦合带（Kreemer 等,2009），以及新近纪应变随时间变化的演化过程等（Faulds 等,2005；Jayko,2009b）。

张扭盆地形成于地壳脆性破坏强的断层边界地块。应变在断层系统中转换形成对转换斜坡、褶皱、拉分和右旋盆地。盆地形成因素也包括应变分解为倾滑和走滑的位移分量（图 11-6）。盆地由于断块翘倾导致形成非对称半地堑，这在张扭带的阶跃或继承系统中较为常见。脆性破坏常生成不规则的碎片型断层和断裂系统（Cladouhos 和 Marrett,1996；Scholz,1997；Clifton 等,2000；Ackermann 等,2001），导致形成不同大小的槽状盆地。盆地大小也可以作为张扭性系统演化阶段的函数。如小型雁列盆地形成于盆地变形的早期（Wu 等,2009），此时大型断层和地堑尚未完全成型。雁列盆地沿雷德尔断层方向形成并与运动方向倾斜，因此局部张扭系统中的早期盆地可能与后期形成的盆地之间存在着 20°~45°的夹角。单斜或披覆褶皱可能是隐伏正断层或者只是位移微小的断层在地表的表现形式（图 11-7）。

一级张扭盆地的规模和伸展率受震源带基底深度、地块边界断层产生时的地壳厚度（Ebinger 等,1999）、纯走滑位移的倾角（Wu 等,2009）、张扭断裂系统的演化等因素影响。先存地壳结构会影响早期的构造形态（Stewart,1985,1988,1992；Oldow,1992）。上地幔的属性差异（Wang 等,2002；Savage 等,2003）也通过断层滑动、断层倾向、断层长度、盆山方向及地块旋转等影响构造特征（Tickoff 等,2004）。张扭变形通常导致沿纵向、横向或任意方向的地块旋转（图 11-8a）。横向轴旋转引起相邻断块的多米诺式倾斜，且以半地堑的形式存在于张扭构造区内。

新近纪盆地碎屑沉积以 5°~20°的平缓翘倾广泛分布于 Walker Lane 带内。30°~65°或更大角度的陡倾局限于高应变和正断层与斜向走滑断层相交的区域。翘倾发生在多种构造环境中，包括拉分转换系统转换斜坡、铲式断层之上的滚动背斜、隐伏断层之上的披覆褶皱和半

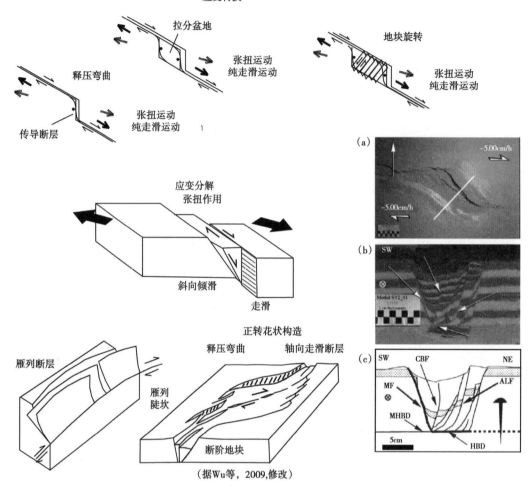

图 11-6 应变转换和应变分解模型示意图

包括释压弯曲、拉分盆地、地块旋转和反转花状构造；砂箱试验模型（a—c）显示张扭盆地（据 Rahe 等，1998）。MF—主断层；ALF—反向铲式断层；CBF—贯穿盆地的走滑断层；HBD—基底拆离；MHBD—水平拆离运动

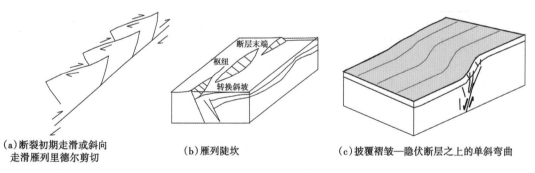

图 11-7 张扭盆地中的构造样式

（a）沿单一走滑断层排列的雁列断层；（b）雁列断层的陡坎和转换斜坡，披覆褶皱斜向正断层上部为单斜挠曲

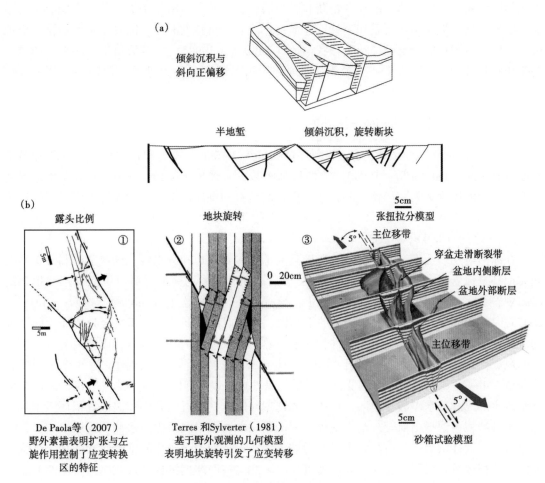

图 11-8 张扭变形的模型示意图

(a) 半地堑十字剖面，简图表示斜向正性分支。(b) 从左至右为：①右旋系统内的应变转移通过阶跃调节带实现，并发育正性断裂和雁列式裂缝系统；②地块旋转式应变转移（由沿断裂发育的小型裂缝研究得出的图表）；③张扭拉分盆地砂箱试验，彩色表示断层深度（据 Wu 等，2009）

地堑中的构造变形（图 11-7 和图 11-8）。垂向轴旋转型的盆地一般具有构造相对复杂的盆地边界。盆地边界可能在小而快速的沉降中心与褶皱—转换断层区之间变化。顺时针旋转区常位于右旋张扭构造带中（图 11-6 和图 11-8b）。旋转区部分受控于左旋或斜向走滑断层。因为右旋和左旋走滑断层可以形成于张扭构造系统，与断层倾向及应变转换有关的构造样式，例如强制褶皱和花状构造也常常出现。另外，释压弯曲可以产生反花状构造或负花状构造（图 11-6）。

小型碱性火山岩通常为橄榄玄武岩，沿着右旋与左旋转换系统的盲区挤出，贯穿了 Walker Lane 带的南部。从钙碱性火山作用到碱性玄武岩火山作用的变化，指示了压扭变形开始的时间，且叠加了更早的（晚中新世初期及更早）弧后拉张及火山作用（Busby 等，2009）。本文接下来将更多地讨论火山作用、张扭作用及释压弯曲。

11.2.2 张扭实验模型

张扭变形实验模型常用于解释断层、断裂系统及盆地的演化（Rahe 等，1998）。变形模型包括滑动块体、一个单一的断裂错断和黏土层，黏土层在走滑作用下，首先在低级走滑错动中生成雁列式里德尔断裂系和裂缝，并最终形成一个贯穿的断面。随着变形的加剧，模拟模型可以产生同根的分支断层及负花状构造（Rahe 等，1998）。一个低角度倾斜于运动方向的张扭拉张模型，可以产生一个位于盆地轴部，并且纵向贯穿盆地的走滑断层（图 11-6；Rahe 等，1998；Wu 等，2009）。

11.2.3 沉积格架，陆间张扭盆地

Walker Lane 带内的大型盆地边界断层位移量约 2~6km，重力和钻井资料显示其碎屑堆积厚达到 3km（Pakiser 等，1964）。震源带基底的平均深度约为 15~17km，但在盆地边界断层形成时期的厚度更大。尽管如此，一些更大的地堑的垂向位移大致为脆性地壳厚度的三分之一。另外，300~1500m 深的盆地沉积中心的沉积速率约小于 1m/ka，一般为 0.25~0.66m/ka 之间（Jayko，2005）。

在冰河时期，大型湖泊占据着盆岭区且局部连接着盆地和分水岭，这些盆地和分水岭在间冰期都是孤立存在的（Reheis 等，2002）。后冰期湖泊的均衡恢复可由更新世拉宏坦湖的岸线观测到，并为沃克通道内的张扭盆地提供一些证据（Adams 等，1999；Bills 等，2007）。Walker Lane 带位于内华达山的雨影区，因此以间冰期的旱区相集合为特征。间冰期的沉积格架包括：古土壤、风成、冲积、干荒盆地、淡和咸的湖泊沉积，以及河流沉积。大型的风成沙丘复合体出现在南方干旱盆地，位于低海拔、干旱逐渐向南推进并终止于死谷（Death Valley）——美国地形最低且最干旱的盆地之一（图 11-5）。图法（Tufa）是一个地下水流出形成的淡水碳酸盐地区，形成于冰河期与间冰期的近滨湖泊和湿地沉积环境中。

厚层冲积扇源于山前带，基本受控于活动断层。冲积扇沉积由冰水冲积互层组成，包括了径流、间歇性洪水、碎屑流和少量的滑坡沉积。而长期接受风化的富砂质碳酸盐古土壤一般位于冲积扇沉积层位，长期暴露而无沉积。因此这些间冰期沉积混杂在冰河时期的河流沉积、冰水沉积和湖泊沉积之中。

玄武岩以夹层形式出现在盆地地层组合中。大量的中新世、上新世玄武岩局部充填于张扭前负地形，常见于现今盆地的反转区域地层层序的底部（Jayko，2009a）。少量的玄武岩锥体和火山口都与拉分时期转换断层有关，释压弯曲同样广泛分布于 Walker Lane 带的中南部。因此，薄玄武岩溢流相是盆地局部层位的地层标志。酸性火山灰或凝灰岩在冲积扇碎屑沉积和沉积中心也较为常见。火山灰可能会从张扭带内的火山中心喷发出，也会从其他中心迁移至该区域。来自喀斯喀特弧和黄石热点的火山碎屑在沃克通道南部较常见（Sarna-Wojcicki 等，2005）。更多关于硅质长河谷和莫诺火山口的资料稍后将被提到。

11.2.4 沃克通道的盆地区

张扭盆地方向的变化可归因于受走滑控制的地壳区和斜向滑动正断层的应变分配，同时受地块垂向轴的顺时针旋转的影响（Stewart，1988；Cashman 和 Fontaine，2000；Petronis 等，2004，2007；Faulds 等，2005；Wesnousky，2005a，2005b）。四个主要的构造地形区——金字塔湖、卡森、埃切斯尔和欧文斯谷，代表了 Walker Lane 带内不同的盆地倾向、盆地大小和火山碎屑物质（下文图 11-2 和图 11-9；Stewart，1988；Cashman 和 Fontaine，2000；Petronis 等，2002，2007；Oldow，2003；Surpless，2008；Jayko，2009b；Oldow 等，2009）。演化时期边界断层上的主滑动作用反映了盆山系统的方向。盆地以右旋走滑和右旋

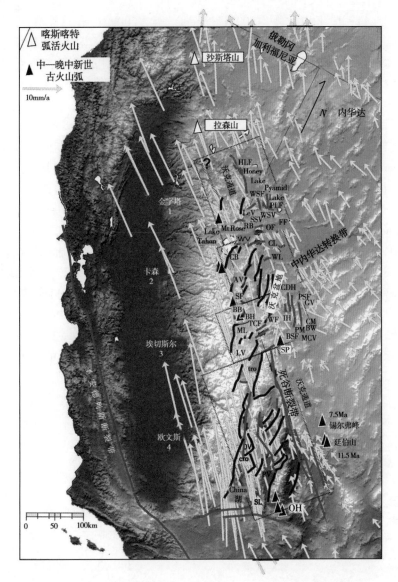

图 11-9 沃克通道内第四纪走滑断层（红色）及北美板块相对 GPS 位移示意图(据 Kreemer 等，2009)
走滑断层—红色；主要斜向滑动断层—黑色；虚线框表示研究区，箭头按比例长度显示位移速率；
10mm/a 的速率标记在侧边栏；标记如图 11-5

斜向滑动断层为边界，且以北西向为主，平行于运动学方向（Faulds 等，2005；Kreemer 等，2009）。由掀斜、顺时针地块转动和左旋滑动断层控制的盆地，通常发育一条位于盆地北缘的近东向边界，也可能同时出现在盆地的南北边界处（图 11-5）。在一些盆地尤其是沃克通道南部，右旋走滑断层发育在盆地的轴部。这些轴部走滑断层比边界断层要年轻，例如在欧文斯谷内。尽管更大的区域应力场相对于北美板块呈右旋状态，左旋断层在沿垂向轴顺时针旋转的区域内仍占主要地位。20°~30° 的顺时针断块旋转在南部的三个构造地形区被证实，下文描述的是北部区域内 3°~9° 的逆时针旋转（图 11-5 和图 11-6）。其中讨论的重点是沿沃克通道的构造地形区内的主要盆地及边界断层。

北部金字塔湖区（下文图11-2和图11-10）平行于迁移方向，以右旋走滑断层和盆山地形为特征（图11-9；Faulds等，2005；Kreemer等，2009）。在这个区域，盆地走向为N20°—45°W。主右旋走滑断层包括哈尼湖（Honey Lake）、金字塔湖、奥姆斯普林斯和弗利（Fernley）断裂带（Turner等，2008）。上中新统地层的逆时针旋转已被雁列式断层系所证实（Cashman和Fontaine，2000；Faulds等，2005）。在走滑系统的南缘分布着次一级的北向山脊、平原和小型盆地。北向盆地、平原山脊包括有皮特森山断层，该断层终止于东西向断裂带且起到了调节左旋滑动的作用。

卡森区（图11-2和图11-10）有以大量的东—北东向的右旋斜向断层及顺时针旋转的特点（Cashman和Fontaine，2000；Kreemer等，2009）。盆地区以北—北东向走向为主。盆地规模中等，长为25~45km，宽为10~15km，并由一个8~12km宽的狭窄山脊分隔。断块定向性和古地磁研究结果表明：35°~50°的顺时针旋转促进了盆地的打开（图11-10；Cashman和Fontaine，2000）。里诺—卡森盆地南北向延伸并终止于北东东—南西西向断层，调节了顺时针旋转（图11-10；Cashman和Fontaine，2000）。内华达山前缘断层陡崖起伏中等（约2000m），但热那亚断层的滑动率要高于南部（Ramelli等，1999）。活动的张扭变形在前缘断层的东部较明显（Slemmons等，1979；Oldow，1992，2003；Oldow等，2001）。顺时针旋转在这一区域南侧的斯维特沃特山区较常见，该区火山岩年龄约为10Ma（Busby等，2008a，2008b；图11-5表示斯维特沃特山区位置）且发生了10°~25°的偏转（King等，2007）。虽然该区域内有活动的热液系统，但并未出现第四纪或上新世的火山活动中心。

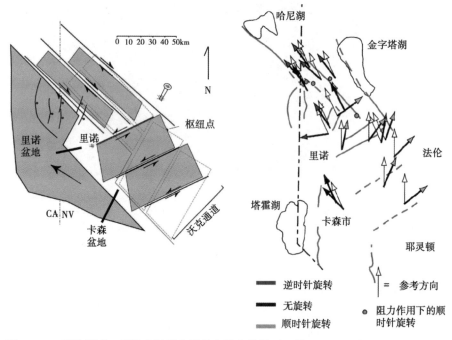

图11-10 断块模型，示沃克通道北部的走滑和旋转区（据Cashman和Fontaine，2000）

埃切斯尔区（Excelsior domain）（图11-2和图11-11）是一个释压弯曲区域，向东传递了大部分的走滑位移，且远离内华达山前缘断层方向（SNFF）。东西走向断层与双峰式火山活动控制了这一区域（Stewart，1988；Wesnousky，2005a，2005b；Hill，2006；Bursik，

2009；Oldow 等，2009）。埃切斯尔阶跃区（包括迈那挠曲）受控于东—北东向左旋倾斜主断层（图 11-11）。这一区域宽约 85~100km（Stewart，1985，1992；Wesnousky，2005a，2005b），并延伸出 15~30km 长、5~15km 宽的东西—北东东向小型走滑盆地，局部被活动的第四纪火山中心占据，包括长河谷火山洼地和莫诺—伊尼欧火山口（Hill，2006）。内华达陡崖被强烈分割且在断裂发育处表现为低—中等起伏的地层不连续。活动的第四纪东—北东向断层（Wesnousky，2005a，2005b）可能影响更古老的古生界和中生界的地壳结构（Stewart，1985，1988，1992）。在斯维特沃特山和银峰区顺时针旋转达 20°~30°（Petronis 等，2002，2007；Rood 等，2005；King 等，2007）。欧文斯谷及以南区域，广泛的晚中新世—上新世镁铁质熔岩的溢出标志着张扭盆地断裂系统的形成。

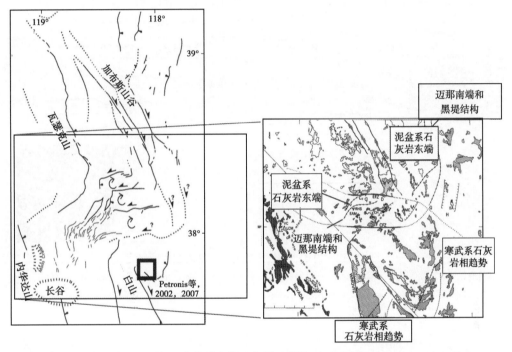

图 11-11　埃切斯尔阶跃及沃克通道中部示意图（据 Wesnousky，2005a；Stewart，1985）

阶跃区受东西向左倾断层的控制；$^{87}Sr/^{86}Sr$ 测试结果（Kistler，1968）证实沃克通道区内的古生界地层边界限定了地壳及地壳下部基底（Stewart，1985，插图）；第四纪断层和活跃的地震活动可以被用来分析更老时代的地层结构；插图显示了发生 20°~30°顺时针偏转的区域（据 Petronis 等，2007）

南端的欧文斯谷（下文图 11-13）的主要特征为：复合地堑、右旋走滑断层、碱性玄武岩转换断层以及包括科索火山区在内的上新世—晚第四纪双峰式火山活动（Bacon 等，1982；Reheis 和 Sawyer，1997）。在欧文斯谷的南部，内华达山前缘断层和附近的走滑断层相交织，产生了狭长的复合地堑（图 11-5 和下文图 11-13）。盆地与盆地边界断层保持着连续的地层起伏，并比卡森北部陡坡高 2~3 倍（图 11-12）。盆地整体呈扁圆状，长 50~200km、宽 15~20km。科索山区内 3.5Ma 的玄武岩流发生约 12°的顺时针旋转，并与右旋斜向走滑断层间的小型盆地同期形成（Pluhar 等，2006；图 11-5 和图 11-13）。以下部分将对于欧文斯地堑和显著的张扭特征进行更详细的描述。

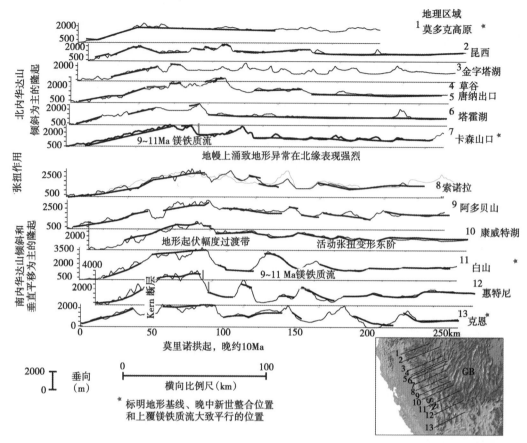

图 11-12 内华达山地形剖面显示基底起伏的变化

顶部是最北端的剖面，底部为南端剖面。剖面显示内华达山的中南部为宽广的背斜形态（据 Jayko, 2009b, 修改）；在过惠特尼和卡森地区的剖面上的橙色色棒显示陡坎高度与谷底的差异；加星标记的剖面中的晚中新世及上新世的分层基于已钻井的数据。GB—大盆地

11.2.5　张扭盆地——欧文斯地堑

　　欧文斯地堑具有大部分张扭盆地的典型特征（图 11-13）。欧文斯谷是 Walker Lane 带中最大的盆地以及最大的地堑结构。欧文斯地堑以两条活动的反向正断层、东侧的白山断层、西侧的内华达山前缘断层南翼为主要边界。地堑在局部范围内被更年轻的、穿过右旋走滑断层的近轴向盆地切割（Slemmons 等，2008；图 11-13）。美国边界最高的山峰达到了 4300m，最大地形起伏为 3100m，出现在沿着南内华达山和北白山的陡崖区。沉积中心与重力低值区有关，在地堑的南北端沉积物充填达到了 2000~3000m（Pakiser 等，1964；Bateman，1965；Jayko，2009a）。地堑的底部位于海拔 900~1300m 处。

　　白山断层的南端和内华达山前缘断层的北端在大松树火山区附近叠置形成了转换斜坡。基底岩层在阶跃构造区外露或只是被很浅的地层覆盖着。欧文斯谷断层在 1872 年活动引发了独立镇附近 7.5~7.9 级的地震（Beanland 和 Clark，1994；Slemmons 等，2008），并在阶跃带附近终止。欧文斯谷断层以反向逆牵引构造间的中部地堑为主。同时存在以第四纪活动断层为边界的几个地块，皆位于被晚新生代碎屑沉积物覆盖的地堑中（Slemmons 等，

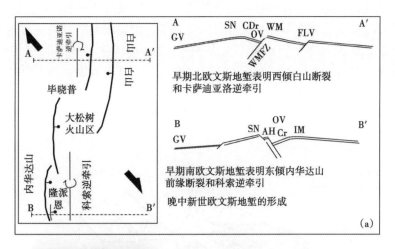

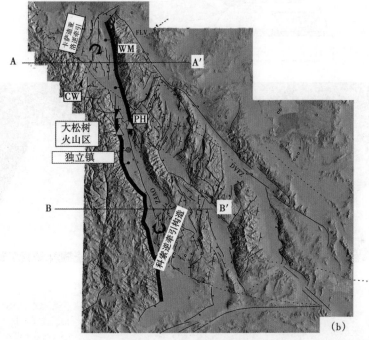

图 11-13 张扭盆地的特征

(a) 早期断层方向及晚中新世欧文斯地堑形成示意图（显著的特征是发育反向倾斜的正断层，该断层沿着白山及南内华达山前缘断层发育且在地块边缘发生逆牵引（据 Jayko，2009a，修改）。GV—大峡谷；SN—内华达山；CDr—卡萨迪亚洛逆牵引；WM—白山；OV—欧文斯谷；FLV—菲什谷；AH—亚拉巴马山；IM—伊尼欧山。(b) 图示内华达山前缘断层（阴影地形）和白山断层（黑色粗线标注）（彩色区为新生界岩层）；第四纪火山系统位于欧文斯地堑的反向端；随着张扭系统的演化，地堑两侧发育正断层，其后伴随着右旋走滑断层及沿着欧文斯谷轴向的断裂带的产生）

2008）。沿谷底断层的活动形成了局部的封闭次盆地、山地（图 11-13；Slemmons 等，2008）和火山。

卡萨迪亚洛逆牵引倾斜于北白山断裂带，科索逆牵引倾斜于地堑南缘的南内华达山前缘断层（图 11-13）。晚第四纪，大松树火山区的主要玄武岩火山活动发生在欧文斯谷断层的北缘，该处存在马尾构造和转换斜坡（Slemmons 等，2008）。

沿着较大的陡崖结构发育有转换斜坡及断块。凯奥特挠曲是一个转换斜坡，它沿着内华达山前缘断裂带连接了北欧文斯谷和长谷地区。内华达山前缘断层朝西穿过这个转换斜坡。部分晚中新世剥蚀面和上覆的10Ma的镁铁质流处于转换斜坡的顶部，且向北及向东倾斜于谷底（Dalrymple，1963；Jayko，2009a；图11-14）。晚中新世剥蚀面和上覆的10Ma的镁铁质流也向东倾斜于白山的东侧（图11-12）。沿着凯奥特挠曲转换斜坡西北侧陡崖发育的地貌类型，包括倾角变化的刻面脊酒杯状冲刷下切谷，以及响应气候变化形成的峡谷（图11-14）。

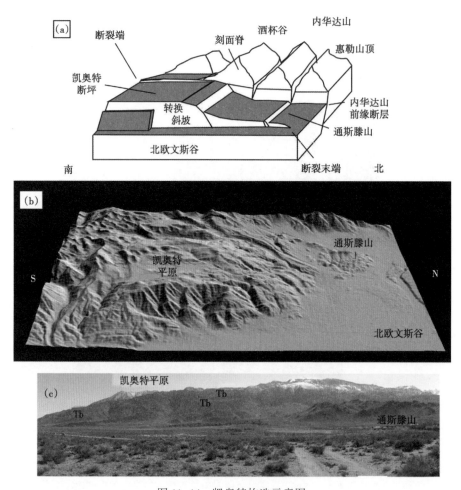

图 11-14 凯奥特构造示意图

（a）立体图示凯奥特转换斜坡、欧文斯谷西北端，蓝区表示被10Ma的玄武岩覆盖的晚中新世局部剥蚀残余，简图也显示与断层相关的不平衡地形；（b）凯奥特斜向晕渲地形图片示凯奥特平原古地表和沿通斯腾山的转换斜坡；（c）沿凯奥特平原的照片（镜头方向：南），Tb代表的是被玄武岩覆盖的翘倾的断阶平原

11.2.6 火山活动：长谷火山口和莫诺—伊尼欧火山口

沃克通道地区构造的主体形成于中新世盆岭区弧后扩张作用到沿门多西诺三联点的张扭作用（Lipman等，1972；Coney和Harms，1984）。前人对于沿东内华达山前缘的构造与岩浆的研究表明，在新近纪南部部分弧内岩浆扩散至北方（图11-5），这与莫诺背斜的发育可

能是同时发生的，因此推断地形变化与地幔上涌有关（图11-12；Jayko，2009b）。火山活动从中—晚中新世安山岩火山弧转变为晚中新世—更新世张扭双峰玄武—流纹岩（图11-15和图11-16）。张扭构造晚于中新世弧内岩浆活动，虽然少量的右旋走滑与10Ma的安山岩为同一时期（Busby等，2008b）；特定纬度的张扭作用发生在玄武—流纹岩的形成时期。

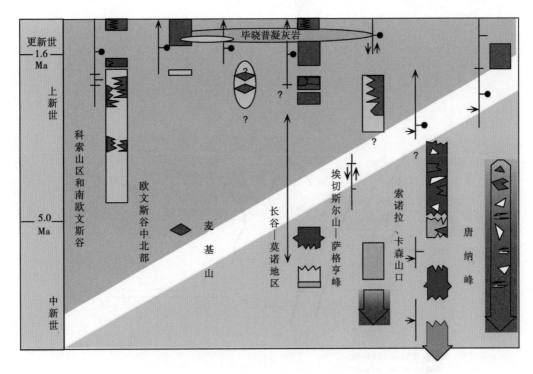

图11-15　东内华达山区构造与岩浆活动迁移图（据Bursik，2009）

示意图指向沿内华达山前缘从南（左）到北（右）的方向；粉色区主要为安山岩火山活动，绿色区为张扭与双峰火山活动。绿色—玄武岩；紫色—流纹岩—英安岩；红色—安山岩；蓝色—非特定火山岩。

这里未编入不同学者提出的与此有争论的描述

岩浆系统被认为是释压弯曲的产物（Bursik和Sieh，1989；Hill，2006；Putirka和Busby，2007；Bursik，2009），例如埃切斯尔阶跃区与莫诺背斜区。长谷第四纪熔岩和莫诺—伊尼欧火山（图11-10）代表着局部构造，为张扭构造活动区内上新世—更新世玄武岩演化的产物（Bailey等，1976；Metz和Mahood，1985；Oldow，2003）。长谷—莫诺—伊尼欧火山的构造环境不仅受控于埃切斯尔阶跃区东—北东向左旋断层（图11-16），也受到大型右旋北北西向山前断裂的影响（Bursik和Sieh，1989；Bursik，2009）。

长谷火山口南的内华达山主前缘断层为希尔顿溪断层（图11-16）。部分应变受控于火山口东侧的卡萨迪亚洛山断层。火山口北部断层系统更为复杂，但萨格亨峰、Hartley泉和银湖这三条断层调节了火山形成前的大部分地壳扩张活动。从前述的材料中可以估算火山垂向运动的平均偏移率（Rinehart和Ross，1964；Huber和Rinehart，1965；Gilbert等，1968；Krauskopf和Bateman，1977）。从这些资料估算出的垂直活动速率说明：希尔顿溪断层和萨格亨峰断层间释压褶皱区，充填了侵入的岩浆岩，表明区域伸展作用要早于长谷火山口的形成（图11-15和图11-16）。地质及地球物理的证据表明，释压弯曲帮助引导了岩浆的聚集并导致了毕晓普凝灰岩的喷发。毕晓普凝灰岩邻近希尔顿溪和卡萨迪亚洛山断层的交叉部位

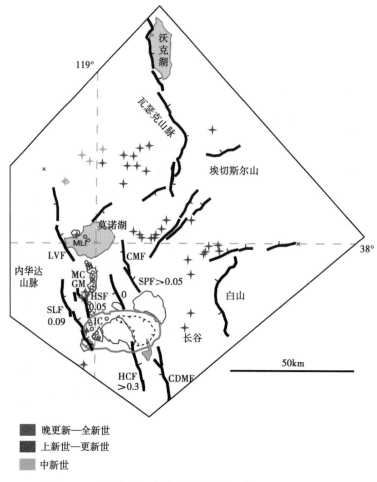

图 11-16　长谷火山洼地局部构造图（据 Bursik，2009）

与沃克通道有关的张扭构造区的火山洼地；黑线为主边界断层；运动的垂向分量如图示。NNW 向断层存在右旋运动的可变分量，NE 向断层存在左旋运动的分量（Bursik 和 Sieh，1989）。断层旁边的数字表示断层的垂直活动速率虚线表示最大火山陷落区（3km）（Carle，1988）；彩色表示不同的火山年龄。火山类型的标志：十字—玄武岩型；圆形—溢出型。实线轮廓表示更大的火山露头。LVF—利韦宁断层；MLI—莫诺湖群岛；MC—莫诺火山；GM—格拉斯山；IC—伊尼欧火山；CMF—考拉克山断层；SLF—锡尔弗湖断层；HSF—哈特利泉断层；SPF—萨格亨峰断层；HCF—希尔顿溪断层；CDMF—卡萨迪亚洛山断层。图中的比例及定位来自于未校正的航天飞机相片

（Hildreth 和 Mahood，1986）。喷出熔岩沿着边界断层向北，直到北格拉斯山与萨格亨峰前缘的交叉部位终止。重力数据显示整个主沉降地块的边界结构沿着断层的走向，勾勒出格拉斯山释压弯曲的范围（Carle，1988）。

莫诺火山岩浆系统被认为存在于释压弯曲的区域内（Bursik 和 Sieh，1989）（图 11-16）。哈特利泉断层在马尾构造南缘火山口附近终止。莫诺湖断层的雁列式拉张作用终止于北部地区。这些断裂是右旋走滑作用的证据。同样地，莫诺湖群岛在 Walker Lane 内的转换系统的定向扩张中发生火山喷发（迈娜挠曲；图 11-11）。这些系统的演化可能影响岩浆活动的迁移，贯穿于新近纪的长谷区并受控于转换系统的构造活动位移（Oldow 等，2008；Bursik，2009）。

莫诺火山岩浆房可能通过注入量的增加而扩大，沿着扩张盘和内华达前缘断层发生挤压构造运动。这种关系为火山活动的挤压构造地形提供了一些证据（Parsons 和 hompson，1991）。而更多的依据在于：在长谷和莫诺—伊尼欧地区，山岭的发育以累积的垂向位移为标志，在这一点上内华达山东部远不如北部和南部远离莫诺背斜的地方明显。

11.3 结论

对于岩石圈的应力变化以及 Walker Lane 带内不同构造地形显示的盆地特征，目前尚未有流变学研究或实验模型可以证明。Stewart（1988）指出埃切斯尔拉分盆地在古生代与中生代走向发生了大角度偏转，比现今已知的中晚新生代偏移还大。这说明继承性地壳结构可能会影响第四纪断裂活动的位置与方向。同样地，克拉通与大陆岩石区域图确定了沿着晚古生代和早中生代边界的削截，其发生在晚新生代时期欧文斯地堑形成的区域内。比起由同生构造体组成地壳的北部地区，南部盆地的面积大，这可能要归因于南部克拉通内更厚的早期大陆岩石圈（Ebinger 等，1999）。

北部地区的构造地形平行于运动迁移方向（Kreemer 等，2009）。卡森区顺时针构造旋转发生在邻近内华达山地块的自由边缘，并伴随着运动方向的小幅变动。自由边缘或运动边界控制着断裂的演化，进而影响着盆地几何形态（Taylor 和 Dewey，2009）。地块的顺时针旋转至少为 20°~30°且存在于整个张扭系统（图 11-5），在最近的 10Ma 发生了 11~13km/Ma 的右旋走滑（Hearn 和 Humphreys，1998；Hammond 和 Thatcher，2007）。Tickoff 等（2004）实验性的研究表明：在张扭系统中，比起断块旋转区域，走滑和斜向倾滑断裂活动区的地壳热度更低。因此，诸如 Walker Lane 带中南部的莫诺背斜区这样的地幔上涌地带（Jayko，2009）可能比北部拥有地热温度更低的地壳。

Busby 等（2008a，2008b）研究了沿着中北部沃克通道西缘的晚中新世弧中心。他们也提出了从钙质到碱性镁铁质火山活动的转变，这标志着从火山弧到弧内张扭三联点的变化。沃克通道内张扭系统也存在一些差异特征。张扭的发生与镁铁质、常规玄武岩或碱性玄武岩火山作用密切相关（Jayko，2009）。双峰、碱质火山作用发生在主要断层的交叉处，或者走滑断层和张扭系统的释压弯曲部位。地幔上涌引起了沃克通道南部的火山活动，一个细微的地形拱起即成为地幔上涌的地表特征，如莫诺背斜（Jayko，2009）。

致谢（略）

参 考 文 献

Ackermann, R. V., Schlische, R. W., and Withjack, M O. (2001) The geometric and statistical evolution of normal fault systems; an experimental study of the effects of mechanical layer thickness on scaling laws. Journal of Structural Geology, 23, 1803–1819.

Adams, K. D., Wesnousky, S. G., and Bills, B. G. (1999) Isostatic rebound, active faulting, and potential geomorphic effects in the Lake Lahontan basin, Nevada andCalifornia. Geological Society of America Bulletin, 111, 1739–1756.

Bacon, C. R., Giovannetti, D. M., Duffield, W. A., Dalrymple, G. B., and Drake, R. E.

(1982) Age of the Coso Formation, Inyo County, California. U. S. Geological Survey Bulletin 1527, 18 p.

Bailey, R. A., Dalrymple, G. B., and Lanphere, M. A. (1976) Volcanism, structure, and geochronology of Long Valley caldera, Mono County, California. Journal of Geophysical Research, 81, 725-744.

Bateman, P. C. (1965) Geology and tungsten mineralization of the Bishop District, California. U. S. Geological Survey Professional Paper 470, 208 p.

Beanland, S., and M. M. Clark (1994) The Owens Valley fault zone, eastern California, and surface rupture associated with the 1872 earthquake. U. S. Geological Survey Bulletin 1982, 29 p.

Bell, J. W., dePolo, C. M., Ramelli, A. R., Sarna-Wojcicki, A. M., and Meyer, C. E. (1999) Surface faulting and paleoseismic history of the 1932 Cedar Mountain earthquake area, west-central Nevada, and implications for modern tectonics of the Walker Lane. Geological Society of America Bulletin, 111, 791-807.

Bell, J. W., Caskey, J. S., Ramelli, A. R., and Guerrieri, L. (2004) Pattern and rates of faulting in the Central Nevada seismic belt, and paleoseismic evidence for prior belt like behavior. Bulletin of the Seismological Society of America, 94, 1229-1254.

Bennett, R. A., Davis, J. L., and Wernicke, B. P. (1999) Present-day pattern of Cordilleran deformation in the western United States. Geology, 31, 327-330.

Bills, B. G., Adams, K. D., and Wesnousky, S. G. (2007) Viscosity structure of the crust and upper mantle in western Nevada from isostatic rebound patterns of the late Pleistocene Lake Lahontan high shoreline. Journal of Geophysical Research, 112, B06405.

Bursik, M. (2009) A general model for tectonic control of magmatism: examples from Long Valley Caldera, USA and El Chichon, Mexico. Geofísica Internacional, 48, 171-183.

Bursik, M. I., and Sieh, K. E. (1989) Range front faulting and volcanism in the Mono Basin, eastern California. Journal of Geophysical Research, 94 (15), 587-609.

Busby, C., DeOreo, S., Skilling, I., Gans, P., and Hagan, J. C. (2008a) Carson Pass-Kirkwood paleocanyon system: paleogeography of the ancestral Cascades arc and implications for landscape evolution of the Sierra Nevada, California. Geological Society of America Bulletin, 120, 274-299.

Busby, C., Hagan, J. C., Putirka, Keith, Pluhar, C. J., Gans, P., Wagner, D. L., Rood, D., DeOreo, S., and Skilling, I. (2008b) The ancestral Cascades arc: Cenozoic evolution of the central Sierra Nevada, California and the birth of the new plate boundary. Geological Society of America Special Paper 438, 331-378.

Busby, C. J., and Putirka, K. (2009) Miocene evolution of the western edge of the Nevadaplano in the central and northern Sierra Nevada; palaeocanyons, magmatism, and structure. International Geology Review, 51, 670-701.

Carle, S. F. (1988) Three dimensional gravity modeling of the geologic structure of Long Valley caldera. Journal of Geophysical Research, 93, 13237-13250.

Cashman, P. H., and Fontaine, S. A. (2000) Strain partitioning in the northern Walker Lane, western Nevada and northeastern California. Tectonophysics, 326, 111-130.

Christiansen, M. N. (1966) Late Cenozoic crustal movements in the Sierra Nevada of California. Geological Society of America, 77, 163-182.

Cladouhos, T. T., and Marrett, R. (1996) Are fault growth and linkage models consistent with power-law distributions of fault lengths? Journal of Structural Geology, 18, 281-293.

Clifton, A. E., Schlische, R. W., Withjack, M. O., and Ackermann, R. V. (2000) Influence of rift obliquity on fault-population systematics; results of experimental clay models. Journal of Structural Geology, 22, 1491-1509.

Coney, P. J., and Harms, T. A. (1984) Cordilleran metamorphic core complexes; Cenozoic extensional relics of Mesozoic compression, Geology, 12, 550-554.

Crough, S. T., and Thompson, G. A. (1977) Upper mantle origin of Sierra Nevada uplift. Geology, 5, 396-399.

Dalrymple, G. B. (1963) Potassium-Argon dates of some Cenozoic volcanic rocks of the Sierra Nevada, California. Geological Society of America Bulletin, 74, 379-390.

Dixon, T. H., Miller, M., Farina, F., Wang, H., and Johnson, D. (2000) Present-day motion of the Sierra Nevada block and some tectonic implications for the Basin and Range province, North American Cordillera. Tectonics, 19, 1-24.

Ducea, M. N., and Saleeby, J. B. (1996) Buoyancy sources for a large unrooted mountain range, the Sierra Nevada, California: evidence from xenolith thermobarometry. Journal of Geophysical Research, 101, 8229-8241.

Ducea, M. N., and Saleeby, J. B. (1998) A case for delamination of the deep batholithic crust beneath the Sierra Nevada, California. International Geology Review, 40, 78-93.

Ebinger, C. J., Jackson, J. A., Foster, A. N., and Hayward, N. J. (1999) Extensional basin geometry and the elastic lithosphere. Philosophical Transactions, Royal Society. Mathematical, Physical and Engineering Sciences, 357, 741-765.

Faulds, J. E., Henry, C. D., and Hinz, N. H. (2005) Kinematics of the northern Walker Lane: an incipient transform fault along the Pacific-North American plate boundary. Geology, 33, 505-508.

Fay, N. P., and Humphreys, E. D. (2008) Forces acting on the Sierra Nevada block and implications for the strength of the San Andreas fault system and the dynamics of continental deformation in the western United States. Journal of Geophysical Research, 113, B12415, 18 p.

Frei, L. S. (1986) Additional paleomagnetic results from the Sierra Nevada: further constraints on Basin and Range extension and northward displacement in the western United States. Geological Society of America Bulletin, 97, 840-849.

Gilbert, C. M., Christiansen, M. N., Al-Rawi, Y., and Lajoie, K. R. (1968) Structural and volcanic history of Mono Basin, California-Nevada, in Coats, R. R., ed., Studies in volcanology: a memoir in honor of Howell Williams. Geological Society of America Memoir 116, 275-329.

Hammond, W. C., and Thatcher, W. (2007) Crustal deformation across the Sierra Nevada, northern Walker Lane, Basin and Range transition, Western United States measured with GPS, 2000—2004. Journal of Geophysical Research, vol. 112, no. B5, B05411.

Hay, E. A. (1976) Cenozoic uplifting of the Sierra Nevada in isostatic response to North American and Pacific plate interactions. Geology, 4, 763-766.

Hearn, E. H., and Humphreys, E. D. (1998) Kinematics of the southern Walker Lane Belt and motion of the Sierra Nevada block. Journal of Geophysical Research, 103, 27033-27049.

Heidbach, O., Tingay, M., Barth, A., Reinecker, J., Kurfe, D., and Muller, B. (2008) The 2008 release of the World Stress Map. www. world-stress-map. org.

Henry, C. D., Faulds, J. E., and dePolo, C. M. (2007) Geometry and timing of strike-slip and normal faults in the northern Walker Lane, northwestern Nevada and northeastern California: strain partitioning or sequential extensional and strike – slip deformation? In Till, A. B., Roeske, S., Sample, J., and Foster, D. A., eds., Exhumation associated with continental strike-slip fault systems. Geological Society of America Special Paper 434, 59-79.

Hildreth, E. W., and Mahood, G. A. (1986) Ring fracture eruption of the Bishop Tuff. Geological Society of America Bulletin, 97, 396-403.

Hill, D. P. (2006) Unrest in Long Valley caldera, California, 1978 – 2004, in Troise, C., DeNatale, G., Kilburn, C. R. J., eds., Mechanisms of activity and unrest at large calderas. Geological Society of London Special Publication 269, 1-24.

Huber, N. K. (1981) Amount and timing of late Cenozoic uplift and tilt of the central Sierra Nevada, California: evidence from the upper San Joaquin River. U. S. Geological Survey Professional Paper 1197, 28 p.

Huber, N. K. (1990) The late Cenozoic evolution of the Tuolumne River, central Sierra Nevada, California. Geological Society of America Bulletin, 102, 102-115.

Huber, N. King, and Rinehart, C. D. (1965) Geologic map of the Devils Postpile Quadrangle, Sierra Nevada, California. U. S. Geological Survey Map, scale1: 62500.

Jayko, A. S. (2005) Late Quaternary Denudation Rates, Death and Panamint Valleys, Eastern California. Earth Science Reviews, 73, 271-289.

Jayko, A. S. (2009a) Deformation of the Late Miocene to Pliocene Inyo Surface, eastern Sierra region, in Oldow, J., and Cashman, P., eds Late Cenozoic Structure and evolution of the Great Basin-Sierra Nevada Transition. Geological Society of America, Special Paper 447, 313-350.

Jayko, A. S. (2009b) The Mono Arch, eastern Sierra Region, California: dynamic topography associated with mantle upwelling? in The Sierra Nevada Plano, Ernst, G., ed., International Geology Review, 51, 702-722.

Jones, C. H., Farmer, G. L, and Unruh, J. (2004) Tectonics of Pliocene removal of lithosphere of the Sierra Nevada. Geological Society of America Bulletin, 116, 1408-1422.

King, N. M., Hillhouse, J. W., Gromme, S., Hausback, B. P., and Pluhar, C. J. (2007) Stratigraphy, paleomagnetism and anisotropy of magnetic susceptibility of the Miocene Stanislaus Group, central Sierra Nevada and Sweetwater Mountains, California and Nevada. Geosphere, 3, 646-666.

Krauskopf, K. B., and Bateman, P. C. (1977) Geologic map of the Glass Mountain Quadrangle,

Mono County, California, and Mineral County, Nevada. U. S. Geological Survey Map, scale 1:62, 500.

Kreemer, C., Blewitt, G., and Hammond, W. C. (2009) Geodetic constraints on contemporary deformation in the northern Walker lane: 2 Velocity and strain rate tensor analysis, in Oldow, J., and Cashman, P., eds., Late Cenozoic structure and evolution of the Great Basin-Sierra Nevada Transition. Geological Society of America, Special Paper 447, 313-350.

Lindgren, W. (1911) The Tertiary gravels of the Sierra Nevada of California. U. S. Geological Survey Professional Paper 73, 225 p.

Lipman, P. W., Prostka, H. J., and Christiansen, R. L. (1972) Cenozoic volcanism and plate-tectonic evolution of the Western United States. I. Early and Middle Cenozoic. Philosophical Transactions of the Royal Society of London, 271, 271-348.

Mavko, B. B., and Thompson, G. A. (1983) Crustal and Upper Mantle Structure of the northern and central Sierra Nevada. Journal of Geophysical Research, 88, 5874-5892.

Metz, J. M., and Mahood, G. A. (1985) Precursors to the Bishop Tuff eruption; Glass Mountain, Long Valley, California. Journal of Geophysical Research, 90, 11121-11126.

Miller, M. M., Johnson, D. J., Dixon, T. H., and Dokka, R. K. (2001) Refined kinematics of the eastern California shear zone from GPS observations, 1993-1998. Journal of Geophysical Research, 106, 2245-2263.

Oldow, J. S. (1992) Late Cenozoic displacement partitioning in the northwestern Great Basin: in Craig, S. D., ed., Structure, tectonics and mineralization of the Walker Lane. Walker Lane Symposium Proceedings Volume, Geological Society of Nevada, Reno, Nevada, 17-52.

Oldow, J. S. (2003) Active transtensional boundary zone between the western Great Basin and Sierra Nevada block, western U. S. Cordillera. Geology, 31, 1033-1036.

Oldow, J. S., Aiken, C. L. V., Hare, J. L., Ferguson, J. F., and Hardyman, R. F. (2001) Active displacement transfer and differential block motion within the central Walker Lane, western Great Basin. Geology, 29, 19-22.

Oldow, J. S., Geissman, J. W., and Stockli, D. F. (2008) Evolution and strain reorganization within late Neogene structural stepovers linking the central Walker Lane and northern eastern California shear zone, western Great Basin; lithospheric - scale transtension. International Geology Review, 50, 270-290.

Oldow, J. S., Elias, E. A., Ferranti, L., McClelland, W. C., and McIntosh, W. C. (2009) Late Miocene to Pliocene synextensional deposition in fault-bounded basins within the upper plate of the western Silver Peak-Lone Mountain extensional complex, west-central Nevada. Geological Society of America Special Paper 447, 275-312.

Pakiser, L. C., Kane, M. F. and Jackson, W. W. (1964) Structural Geology and volcanism of Owens Valley region, California a geophysical study; U. S. Geological Survey Professional Paper 438, 66.

Parsons, T., and Thompson, G. A. (1991) The role of magma overpressure in suppressing earthquakes and topography: worldwide examples. Science, 253, 1399-1402.

Petronis, M. S., Geismann, J. W., Oldow, J. S., and McIntosh, W. C. (2002) Paleomagnetic and $^{40}Ar/^{39}Ar$ geochronologic data bearing on the structural evolution of the Silver Peak extensional complex, west–central Nevada. Geological Society of America Bulletin, 114, 1108–1130.

Petronis, M. S., Geissman, J. W., and McIntosh, W. C. (2004) Transitional field clusters from uppermost Oligocene volcanic rocks in the central Walker Lane, western Nevada. Physics of the Earth and Planetary Interiors, 141, 207–238.

Petronis, M. S., Geissman, J. W., Oldow, J. S., and McIntosh, W. C. (2007) Tectonism of the southern silver peak range; paleomagnetic and geochronologic data bearing on the Neogene development of a regional extensional complex, central Walker Lane, Nevada; exhumation associated with continental strike–slip fault systems. Special Paper, Geological Society of America, 434, 81–106.

Pluhar, C. J., Coe, R. S., Lewis, J. C., Monastero, F. C., and Glen, J. M. G. (2006) Fault block kinematics at a releasing stepover of the eastern California shear zone; partitioning of rotation style in and around the Coso geothermal area and nascent metamorphic core complex. Earth and Planetary Science Letters, 250, 134–163.

Putirka, K., and Busby, C. J. (2007) The tectonic significance of high K_2O volcanism in the Sierra Nevada, California. Geology, 35, 923–926.

Rahe, B., Ferrill, D. A., and Morris, A. P. (1998) Physical analog modeling of pull-apart basin evolution. Tectonophysics, 285, 21–40.

Ramelli, A. R., Bell, J. W., dePolo, C. M., and Yount, J. C. (1999) Large-magnitude, Late Holocene earthquakes on the Genoa fault, west Central Nevada and eastern California. Bulletin of the Seismological Society of America, 89, 1458–1472.

Reheis, M. C., and Sawyer, T. L. (1997) Late Cenozoic history and slip rates of the Fish Lake Valley, Emigrant Peak, and Deep Springs fault zones, Nevada and California. Geological Society of America Bulletin, 109, 280–299.

Reheis, M. C., Sarna-Wojcicki, A. M., Reynolds, R. L., Repenning, C. A., and Mifflin, M. D. (2002) Pliocene to middle Pleistocene lakes in the western Great Basin; ages and connections. Smithsonian Contributions to the Earth Sciences, 33, 53–108.

Rinehart, C. D., and Ross, D. A. (1964) Geology and mineral deposits of the Mount Morrison Quadrangle, Sierra Nevada, California. U. S. Geological Survey Professional Paper 385, 106 pp.

Rood, D. H., Busby, C. J., Jayko, A. S., and Luyendyk, B. P. (2005) Neogene to Quaternary kinematics of the central Sierran frontal fault system in the Sonora Pass region; preliminary structural, paleomagnetic, and neotectonic results. Abstracts with Programs, Geological Society of America, 37, 65.

Sarna-Wojcicki, A. M., Reheis, M. C., Pringle, M. S., Fleck, R. J., Burbank, D., Meyer, C. E., Slate, J. L., Wan, E., Budahn, J. R., Troxel, B., and Walker, J. P. (2005) Tephra layers of Blind Spring Valley and related upper Pliocene and Pleistocene tephra layers, California, Nevada, and Utah; isotopic ages, correlation, and magnetostratigraphy. U. S. Geological Survey Professional Paper 1701, 63 p.

Savage, J. C., Gan, Weijun, and Svarc, J. L. (2001) Strain accumulation and rotation in the Eastern California shear zone. Journal of Geophysical Research, 106, 21995-22007.

Savage, B., Ji, Chen., and Helmberger, D. V. (2003) Velocity variations in the uppermost mantle beneath the southern Sierra Nevada and Walker Lane. Journal of Geophysical Research, 108 (B7), 16 p.

Savage, J. C., Svarc, J. L., and Prescott, W. H. (2004) Interseismic strain and rotation in the northeast Mojave Domain, eastern California. Journal of Geophysical Research, 109, 13 p.

Scholz, C. H. (1997) Scaling properties of faults and their populations. International Journal of Rock Mechanics and Mining Sciences & Geomechanics Abstracts, 34 (3-4), 9 p.

Shaw, D. R. (1965) Strike-slip control of Basin-Range structure indicated by historical faults in western Nevada. Geological Society of America Bulletin, 76, 1362-1378.

Slemmons, D. B., VanWormer, D., Bell, E. J., and Silberman, M. L. (1979) Recent crustal movements in the Sierra Nevada-Walker Lane region of California-Nevada: part 1, rate and style of deformation. Tectonophysics, 52, 561-570.

Slemmons, DB, Vittori, E., Jayko, A. S., Carver, G. A., and Bacon, S. N. (2008) Quaternary Fault and Lineament Map of Owens Valley, Inyo County, Eastern California. Geological Society of America Map and Chart Series 96, 25 pp. 2 sheets, scale: 1:100, 000.

Stewart, J. H. (1985) East-trending dextral faults in the western Great Basin: an explanation for anomalous trends of Pre-Cenozoic strata and Cenozoic faults. Tectonics, 4, 547-564.

Stewart, J. H. (1988) Tectonics of the Walker Lane Belt, western Great Basin; Mesozoic and Cenozoic deformation in a zone of shear. Rubey Volume, 7, 683-713.

Stewart, J. H. (1992) Walker Lane Belt, Nevada and California: an overview, in Craig, S. D., ed., Structure, tectonics and mineralization of the Walker Lane, Walker Lane Symposium Proceedings Volume. Geological Society of Nevada, Reno, Nevada, 1-16.

Surpless, B. (2008) Modern strain localization in the central Walker Lane, westerm United States: implications for the evolution of intraplate deformation in transtensional settings. Tectonophysics, 457, 239-253.

Taylor, T., and Dewey, D. (2009) Transtensional analyses of fault patterns and strain provinces of the Eastern California shear zone-Walker Lane on the eastern margin of the Sierra Nevada microplate, California and Nevada, International Geology Review, 51, 843-872.

Tickoff, B., Russo, R., Teyssier, C., and Andrea, T. (2004) Mantle driven deformation of orogenic zones and clutch tectonics, in Grocott et al., ed., Vertical Coupling an Decoupling in the Lithosphere. Geological Society Special Publication, 227, 41-64.

Turner, R., Keohler, R. D., Briggs, R. W., and Wesnousky, S. G. (2008) Paleoseismic and slip rate observations along the Honey Lake fault Zone, northeastern California, USA. Bulletin of the Seismological Society of America, 98, 1730-1736.

Umhoefer, P. J., Mayer, L., and Dorsey, R. J. (2002) Evolution of the margin of the Gulf of California near Loreto, Baja California Peninsula, Mexico. Geological Society of America Bulletin, 114, 849-868.

Unruh, J. R. (1991) The uplift of the Sierra Nevada and implications for Late Cenozoic epeirogeny in the western Cordillera. Geological Society of America, 103, 1395-1404.

Unruh, J. R., Humphrey, J., and Barron, A. (2003) Transtensional model for the Sierra Nevada frontal fault system, eastern California. Geology, 31, 327-330.

Wallace, R. E. (1984) Patterns and timing of late Quaternary faulting in the Great Basin Province and relation to some regional tectonic features. Journal of Geophysical Research, 89, 5763-5769.

Wang, K., Plank, T., Walker, J. D., and Smith, E. I. (2002) A mantle melting profile across the Basin and Range, SW USA. Journal of Geophysical Research, 107, no. B1, 21 p.

Wesnousky, S. G. (2005a) Active faulting in the Walker Lane. Tectonics, 24, 35 pp.

Wesnousky, S. G. (2005b) The San Andreas and Walker Lane fault systems, western North America; transpression, transtension, cumulative slip and the structural evolution of a major transform plate boundary. Journal of Structural Geology, 27, 1505-1512.

Wright, L. (1976) Late Cenozoic fault patterns and stress fields in the Great Basin and westward displacement of the Sierra Nevada block. Geology, 4, 489-494.

Wu, J. E., McClay, K., Whitehouse, P., and Dooley, T. (2009) 4D analogue modeling of transtensional pull-apart basins. Marine and Petroleum Geology, 26, 1608-1623.

Zandt, G. (2003) The southern Sierra Nevada drip and the mantle wind direction beneath the southwestern United States. International Geology Review, 45, 1-12.

Zandt, G., Hersh, G., Owens, T. J., Ducea, M., Saleeby, J., and Jones, C. H. (2004) Active foundering of a continental arc root beneath the southern Sierra Nevada in California. Nature, 431, 41-46.

(赵钊 译，纪沫 祁鹏 校)

第 12 章 东北大西洋和南大西洋边缘的后裂谷变形："被动边缘"真的是被动吗？

DOUGLAS PATON

(School of Earth and Environment, University of Leeds, USA)

摘　要：近来对于大陆岩石圈扩张机制与控制因素方面的研究取得了很多进展，但绝大多数的模型都假定裂后热沉降遵从指数衰减模式。这导致"被动边缘"这一术语被广泛用于描述该类成因的大陆边缘。本章将通过对东北大西洋和南大西洋边缘观察结果的讨论，指出这两种大陆边缘并不"被动"。东北大西洋发生过多次的快速沉降与隆升事件，这显然不能用简单的热沉降模式来解释，而是需要用地壳—地幔的交互作用来解释。此外，挤压构造的出现意味着洋脊推进中产生了水平缩短效应。在南大西洋，虽然资料相对匮乏，但仍有来自于陆上和近海的确凿证据表明南大西洋边缘经历了重要的后裂谷变形。虽然对于变形的精确时间目前仍无定论，但可以确定的是在晚白垩世和古近—新近纪都发生了变形。

关键词：被动边缘　后裂谷沉降　挤压变形　隆升

12.1　引言

大西洋南北两侧的大陆边缘普遍被认为是地壳减薄和火山沉积充填的产物。在过去的 30 年中，地壳减薄以及岩石圈拉伸的过程已被广泛地研究和记录，这些研究对我们理解陆缘形成过程及其影响起到了重要的推动作用。而这些仍基于岩石圈拉伸的两个阶段——裂谷作用及其后的热沉降。

Falvey（1974）从定性的角度提出岩石圈内的伸展作用可通过地壳的脆性变形以及地壳下岩石圈的塑性流动来进行调节，并用此观点解释许多大陆裂谷盆地的沉降史。McKenzie（1978）提出一维岩石圈拉伸的定量模型。模型假定地壳与岩石圈发生纯剪切和均匀拉伸作用，且软流圈被动上涌。这个模型的主要内容和原理是：岩石圈拉伸由两部分组成，其一是由岩石圈垂向减薄并以伸展系数 β 发生横向拉伸产生的受初始断层（与裂谷作用相关）控制的沉降；之后岩石圈冷却，等温线回落至裂谷前状态（图 12-1a），从而发生热沉降。与裂谷作用相关的沉降是瞬时性的，而裂后（后裂谷）热沉降随时间（约 50Ma）呈幂指数衰减，直至标准岩石圈热流值达到原值的 $1/e$，且沉降量降至最小。运用一维拉伸模型进行盆地分析的详细实例参见 Busby 和 Ingersoll（1995）、Allen 和 Allen（2005）等的研究。

这种均匀拉伸的理论此后一直被修正并应用于解释一系列大西洋型边缘背景下观察到的一维沉降模式，包括比斯开湾、加拿大东部大陆边缘、美国东北部大西洋海岸以及北海（Steckler 和 Watts，1978；Keen 和 Corsden，1981；Le Pichon 和 Sibuet，1981；Barton 和 Wood，1984；图 12-1b）。

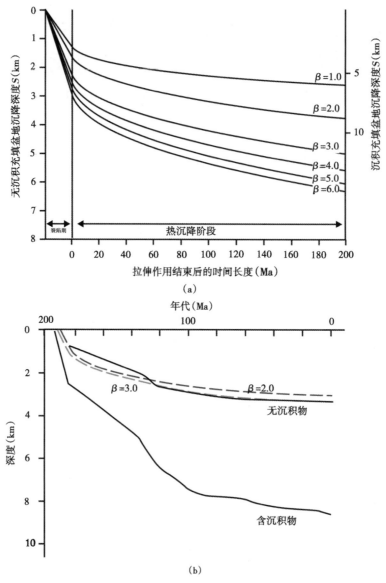

图 12-1 一维岩石圈拉伸的定量模型

(a) 一维沉降模型预测的不同伸展量的沉降曲线,瞬时裂谷作用之后为随时间幂指数衰减的热沉降(据 Sclater 等,1980);(b) 新斯科舍省边缘一口钻井的沉降曲线,反映了模型预测结果(虚线)和实测结果(实线)间的相似性(据 Keen 和 Cordsen,1981)

为了解释之前均匀拉伸模型所不能解释的地质现象,随后许多模型被提出以修正均匀拉伸理论。这些修正模型包括动态拉伸(包括岩石圈强度和黏度结构)、多期裂谷作用、与深度相关的拉伸作用、无放射性热流、纯剪切拉伸、瞬时拉伸和地幔柱诱发裂谷作用(Allen 和 Allen,2005)。这些模型中的绝大部分在改进中考虑到了对裂陷阶段的影响,并普遍假定热沉降阶段仍遵从随时间呈幂指数衰减。热沉降阶段产生的边缘通常被称为"大西洋型"或是"被动大陆"边缘,这源于它们的发育位置,或普遍假定其发育在减薄的陆壳之上。该区被认为地震活跃性较差且热流接近正常(Allen 和 Allen,2005)。在远离斜向扩张或转换断层区，除了"重力控制型变形"（盐构造、泥底辟、滑塌构造、滑坡和松软沉积物中的铲式生长断层），构造被认为相对平静（Miall，1990；Allen 和 Allen，2005）。

尽管拉伸模型成功应用于解释后裂谷一维沉降，但大量基于地震反射资料、井资料及回剥技术在欧洲西北部、北美东部和挪威大陆边缘的研究得出了与模型预测并不一致的结果。不一致性主要表现为与这些简单模型预测结果相比，发生了过多的沉降和隆升，抑或更少的沉降量。例如，据 Mutter（1984）估算，与模型预测结果相比，在挪威大陆边缘存在高达 1km 的沉降差。在过去的 10 年间，伴随着地震反射资料在质量、成像性和覆盖面积上的明显改进和探井的部署，发现被动大陆边缘不遵循简单拉伸模型的实例不断增加。欧洲西北部边界吸引了很多关注，但是其他被动陆缘特征具有相似性的证据越来越多。因此本文的核心问题就是：被动陆缘真的被动吗？讨论非均匀沉降的原因和影响时，需要结合回顾现有成果以及分析欧洲西北部和南大西洋边缘新的数据才能给出答案。

12.2 沉降与隆升的识别

在解决后裂谷变形问题之前，很重要的一点是考虑用来量化隆升与沉降的数据和技术。尽管近来许多研究是将陆上与近海资料进行综合分析，但对于边缘隆升沉降的幅度与时代的估算通常会聚焦于边界的近海部分。大多数技术方法是估算岩体与参照物的相对运动量，这个参照物可能是一个相邻的岩体（如地层几何学结构）或一个理想化的地温值（如镜质组反射率）。因此，这些技术要明确的是：岩体相对于参照物是否发生了沉降或隆升（England 和 Molnar，1990；Corcoran 和 Dore，2005）。

12.2.1 热学分析

热学分析是运用地层层序中的古温标（如镜质组反射率和凝灰石裂变径迹）来确定岩体的古地温峰值。之后与从常规地温梯度中预测的深度进行对比。镜质组反射率是富有机组分中镜质组的反射光强度与垂直入射光（通常波长 546nm）强度的百分比。由于有机质的反射特征与受热量相关，因此通过从典型样品中计算出平均镜质组反射率，并与实验数据相对比，即可估算出古地温（Sweeney 和 Burnham，1990）。尽管作为一门常规技术在许多盆地分析中得到应用，但古温标法仍存在一些局限性，最主要的是它只记录最大的热事件而非多个事件。

12.2.2 沉降分析

沉降分析基于将通过回剥和去压实技术计算出的垂直地层剖面的沉降量（Steckler 和 Watts，1978），与通过给定拉伸系数和裂谷作用持续时间得出的沉降曲线的值进行对比（McKenzie，1978；Jarvis 和 McKenzie，1980）。如上文所述，近年来对一维拉伸模型的改进，使其可以用来对多期裂谷事件、剥蚀作用进行分析，甚至可以延伸到二维空间模拟（Roberts 等，1998；Rowley 和 White，1998）。上述模型的主要局限之一为其结果过于依赖输入参数的精确性，包括岩石圈褶曲刚度、拉伸系数、古水深、压实系数以及生物地层定年数据等。

12.2.3 压实分析

在沉积盆地演化时期的沉积物埋藏导致了孔隙度的减少，因此将实测的孔隙度与预测的孔隙度曲线进行对比可以估算出埋藏深度（Sclater 和 Christie，1980）。另外，岩体的地震波速与孔隙度有关，可以将地震波速的变化与压实模拟的分析结果联系起来，以确定埋藏深度和剥蚀量（Corcoran 和 Dore，2005）。这个方法的局限性在于使用的是标准岩性的孔隙度及深度曲线等参数，并未考虑岩性的变化和局部的成岩压实作用。

12.2.4 构造—地层学

地震反射削截图和盆地级不整合界面被纳入到构造—地层分析中，以研究区域或局部范围内的海平面变化、构造沉降，以及热控边缘沉降之间的相互作用。研究主要受到高分辨率地震反射数据的成像精度，以及来自于钻井的生物地层资料品质的限制。一个常用的方法是，通过定量分析沉积时期的相对可容空间，或者通过辨别反射削截来识别隆升和剥蚀的时间间隔，以确定沉降量和沉降类型。这种方法只能得到一个相对的结果，而不是绝对值。在近岸地区，尽管水深难以确定，但如果来自地层学的证据可以被用于控制古水深，这个问题就迎刃而解了（Stoker 等，2005a；Paton 等，2008）。

12.2.5 陆上地貌学

对大西洋型边缘的研究一直局限于近海区域，直到近来，得益于一些地貌学资料，更多的关注开始聚焦于该类型边缘的陆上部分。垂直于边缘走向的最新地貌剖面资料显示，现今的边缘地貌往往与模型预测的并不相符（例如南大西洋边缘；Gallagher 和 Brown，1999）。在这些地形学研究技术中，常会使用凝灰石裂变径迹的数据。

12.3 东北大西洋非均匀裂后边缘沉降研究

12.3.1 区域背景

东北大西洋大陆边缘沉积盆地是自二叠纪（约 310Ma）至白垩纪晚期—古近纪早期一系列不连续裂谷事件的产物（Ziegler，1988）。其西北欧部分从博库派恩盆地、爱尔兰近海延伸至挪威中部，长约为 2500km，宽度在莫莱盆地小于 200km，最大约 500km（沃宁盆地）（图 12-2）。从构造角度看，该边缘发育大量被深水盆地所分割的高点，它们是不同时代岩石圈不均匀拉伸的产物（Dore 等，1999）。大量规模小、时代较早（二叠—三叠纪）的夭折裂谷的存在，表明在洋壳形成前存在长时间的裂谷作用。这些夭折裂谷盆地被分布更广泛的后裂谷盆地所覆盖。盆地的年龄显示裂谷作用具有向北西向迁移的特点，反映了从博库派恩—北海盆地（晚侏罗世）到罗科尔—法罗群岛—莫雷—沃宁盆地（早—中白垩世）的裂陷活动过程（Dore 等，1999）。

最终的裂陷事件伴随着古新世—始新世初期（65—55Ma）大量的火山喷发和侵入，形成了北大西洋火成岩区和现今大洋边缘。沿洋—陆边界存在的高速下地壳被证明为裂谷期岩浆底侵作用的产物（Saunders 等，1997）。陆上火山活动和海底倾斜反射体的存在为这种观点提供了支持。之后边界经历了裂后热沉降，导致了 4km 深的新生代沉积盆地的形成（如

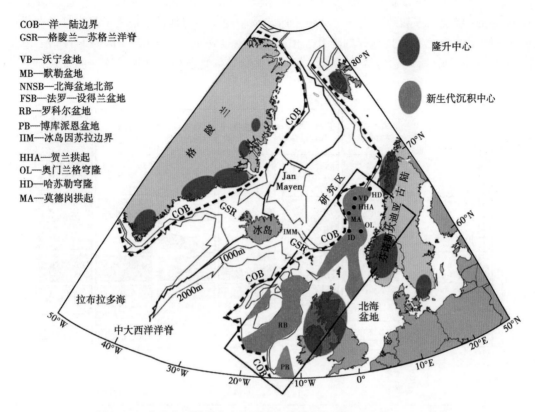

图 12-2 东北大西洋隆起与沉降中心位置图（据 Praeg 等，2005 修改）
其中多处不能被简单的一维拉伸模型所解释

罗科尔盆地；Ceramicola 等，2005）。多位学者（Praeg 等，2005；Dore 等，2008）认识到快速沉降事件、掀斜、穹隆构造无法被现有的被动边缘演化的模型所解释，而采用包括地壳—地幔交互作用和应力场的远程效应来研究上述地质现象。

12.3.2 沉降与隆升

从 19 世纪 80 年代开始，在北美东部和西北欧的大陆边缘，基于深海测量与沉降模型预测结果的对比研究，以及回剥技术的应用，促使研究者逐渐认识到简单的一维伸展模型的预测结果往往存在较大的偏差。Ceramicola 等（2005）将恢复后的古近纪不整合的古深度与井数据进行对比，发现在罗科尔盆地（至少 1.7km）、沃宁盆地（大于 1km）以及法罗群岛盆地（约 2.1km）存在比预测的更大的沉降。通过古近纪—现今的构造地层（65.5—0Ma；Stokcer 等，2005b）研究表明，该地区发育三幕区域构造事件：早新生代、晚始新世、中上新世（图 12-3）。这些不连续的事件归因于掀斜与沉降的多幕共轭活动。它们相对短暂的持续时间（小于 10Ma）表明其具有高达数百米/百万年的垂向运动速率，这明显与简单一维伸展模型预测的因岩石圈冷却造成的沉降速率不符（Praeg 等，2005；Stoker 等，2005b）。

在北海北部盆地与法罗群岛地区，通过回剥技术研究表明，早新生代裂后阶段发生了约 300~500m 的快速沉降，超过了简单拉伸模型的沉降预测（Turner 和 Scrutton，1993；Hall 和 White，1994）。法罗群岛地区的钻井资料表明：该快速沉降与边缘隆起（约 100m）发育时期一致，并形成了中古新统不整合（Turner 和 Scrutton，1993）（图 12-5）。边缘隆起事件在从芬诺斯坎迪亚古陆到苏格兰、英格兰西北部、爱尔兰的延长带最为明显，其中芬诺斯坎迪亚古陆据估算发生了 1.5km 的隆升（Riis，1996；Green 等，2002；Hendriks 和 Andriessen，2002）。盆地沉降与边缘隆升的耦合与该地区构造—地层研究结果一致（图 12-3）。这

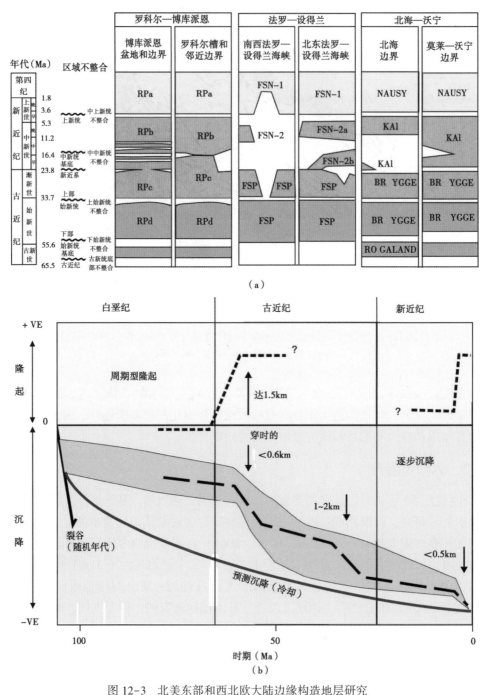

图 12-3 北美东部和西北欧大陆边缘构造地层研究

(a) 东北大西洋边缘的地层格架简图；(b) 预测沉降曲线与实测沉降/隆升量对比图
对比说明了多数快速沉降和隆升无法由简单模型预测；资料来源：Praeg 等，2005

些证实了在该地区发生了盆地的快速沉降、边缘的侵蚀削截,以及来自于内缘地区的大型楔状体的进积(Stoker 等,2001;Andersen 等,2002)。这一事件较为短暂(约5Ma),但在区域上并不是同步发生的;北部开始于晚古近纪(法罗群岛和北海北部),而在南部发生的更晚些(博库派恩和罗科尔盆地)。

罗科尔盆地上始新统掀斜事件就是很好的证据,上始新统不整合面显示了盆地区的快速加深(图12-3a、图12-4),中—晚始新世(约33Ma)的浅海相砾岩位于不整合面之下,而在不整合面之上则发育深水沉积。结合沉积学与几何学分析表明,在近海地区存在最大700m的沉降。掀斜作用使陆上古新统溢流玄武岩发生了倾角为4°的区域倾斜,从内缘区海拔500m降至盆底区海平面以下超过2km(Stoker 等,2005b)。

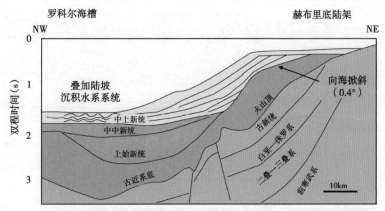

图12-4 过罗科尔海槽地震剖面图示5个区域不整合(据Stoker 等,2005a)
这些不整合已在东北大西洋被识别,位于不整合面上下的反射的交角结构展示了快速边界沉降与掀斜事件

北大西洋的晚新生代地层回剥也记录了一幕快速沉降事件,并通过上新统内部不整合面(约4Ma)之上的沉积充填反映出来(图12-5)。与晚始新世仅发育盆地沉降不同,古地热分析表明晚新生代快速沉降的同时还发生了边缘隆升。沿着芬诺斯坎迪亚古陆边缘,凝灰石裂变径迹数据揭示了构造隆升的变化,北部地区为1km,至挪威南部为1.5km(图12-2)。在芬诺斯坎迪亚古陆内,上新统不整合面及其下地层明显被削截,这证明了隆升作用的存在(Riis,1996)。过该区的剖面显示沿海岸边缘发育了一系列穹隆构造和高达1.5km的隆升,同时还发生了数百米的盆形沉降。这种千米规模的掀斜可与早新生代事件形成非常相似的地层结构,并导致近海高地和向海进积的陆架斜坡楔状沉积体的发育。

12.3.3 地壳—地幔交互作用

地壳—地幔交互作用造成了早新生代、晚始新世、晚新生代区域掀斜和隆升。最明显的交互作用是对流驱动变形,地幔柱的顶部侵位于裂陷大陆岩石圈之下,引起地壳抬升和同步的沉降。地壳—地幔交互作用的两个更进一步的力学机制可能发生,即在裂陷—漂移阶段的上地幔小规模的对流活动,以及在板块重组期间上地幔流的重构(Keen,1985;Stuevold 等,1992;King 等,2002)。

地幔柱在中下地壳层岩浆体的侵位中起主要作用。这引起了地壳岩石圈的均衡隆升。例如,5km厚的岩浆体的底侵作用可以引起600m的隆升(Brodie 和 White,1994)。虽然最初

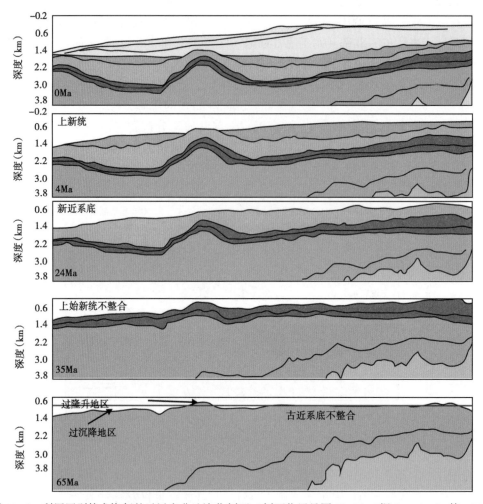

图 12-5 利用回剥技术恢复的过沃宁盆地演化剖面（剖面位置见图 12-2）（据 Ceramicola 等，2005）
将模拟结果与根据井数据得出的古水深进行对比，一期主要沉降事件并未在一维伸展模型中体现

对冰岛地幔柱的模拟认为发生了 2km 的大规模区域性持续隆升（White 和 McKenzie，1989），近来的研究结论却表明地幔柱可能只影响到一个窄带，且由于地幔柱流瞬时波动，隆升作用可能非常短暂（Jones 等，2002）。但是一些研究认为地幔柱模型无法解释个别区域现象（Praeg 等，2005）。如在东北大西洋边缘，虽然也经历了同期发生的隆升和快速盆地沉降，但边缘隆升和掀斜作用却发生于裂陷之后的 50Ma，并仅持续了 4Ma。

最近的研究通过热—力耦合数值模拟来分析地幔—地壳的交互作用（Keen 和 Boutilier，1995），预测不同规模的地幔对流。对流的规模取决于横向地温梯度，在地温梯度大的地方，如穿过裂谷或洋—陆边界的地区，模型预测的规模就较小（<100km；Keen，1985；Keen 和 Boutilier，1995；Korenaga 和 Jordan，2002）。这些小规模的地幔环流预计会引起千米级规模的垂向移动，尽管实际量与上地幔的黏度有关但影响有限。多个模型的预测结果也表明地幔流动不会均衡发生，因为其受岩石圈底部地形的重要影响，可能在横向上引起几百千米规模的上升流与下降流的变化。

Praeg 等（2005）把东北大西洋边缘变形的三个主要阶段归因于小规模的地幔环流（图

12-6）。古新世—早始新世主要的地幔—地壳交互作用发生在海底扩张中心。Praeg 等（2005）认为二次地幔流引起了大陆边缘/盆地地区向洋—陆边界东部掀斜。二次流可引起边缘隆升和盆地内的快速沉降。这些学者认为这一事件同时具有瞬时性与穿时性，因此可解释从法罗（晚古新世>55Ma）到博库派恩盆地（早始新世<55Ma）时序上的变化。后裂谷变形的第二阶段与板块重组事件有关，后者导致海底扩张终止于拉布拉多海内的格陵兰岛和劳伦古陆之间。因这与上始新世在该边缘只发育凹陷相一致（没有边缘隆升的证据），Praeg 等（2005）认为板块重组停止了早始新世小规模的环流，在边缘引起了区域规模的下沉（图12-6）。对于早上新世到现今的边缘隆起的解释仍须进一步探讨，它可能反映了由于东北大西洋变宽，或者其他全球板块重组事件，在大洋岩石圈和大陆岩石圈边界处发生了小规模的地幔再生。

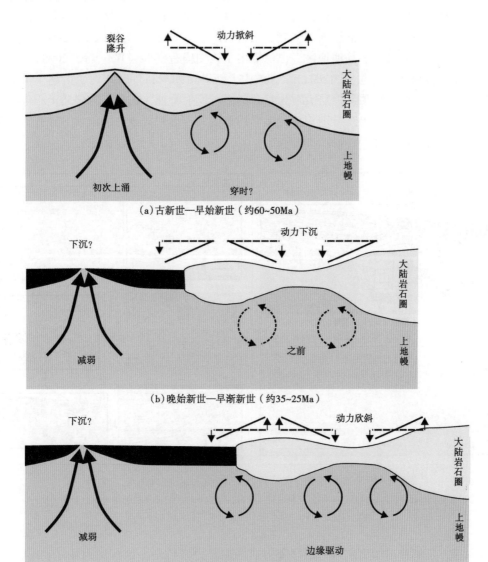

图 12-6 东北大西洋拉伸演化图（据 Praeg 等，2005）
图中的隆升与沉降事件是由陆壳与上地幔对流的交互作用引起的

12.3.4 穹隆构造

作为一个岩浆岩型边缘，在东北大西洋边缘发育大量的穹隆构造（如中挪威陆架上的加拉尔脊、维玛穹隆、伊萨克穹隆等），并被归因于大陆裂解过程中岩浆作用的产物，也就不足为奇（Dore 等，2008）。大量其他的穹隆构造，包括奥尔门兰格穹隆、哈福苏勒穹隆和莫德岗拱起，则是初始大洋形成后的产物（早始新世，距今50 Ma），与岩浆作用无关（图12-2）。这些构造以四面下倾闭合、盆地规模的构造反转以及下伏的先存正断层的反向活动为主要特征（Dore 等，2008）。

研究区最大的裂后穹隆是贺兰拱起，它长达200km，隆起幅度达500m（图12-7）。拱起受控于中中新世（约14Ma）重新活动的莱斯断层组（Brekke，2000）。结合钻井和高分辨率地震反射资料确定的地层结构，揭示了中中新世（约14Ma）不整合面的存在，以及渐新世—早中新世地层在褶皱枢纽处遭受剥蚀（图12-7b；Gomez 和 Verges，2005）。在奥门兰格穹隆和法罗—罗科尔等其他地区，也识别出了该期构造变形（Dore 等，2008）。构造变形作用一直持续到早上新世（约5Ma），导致中上新世（约4Ma）时期地层上超于下上新统不整合面之上。下上新统顶部地层表现为整合的平行反射特征，超覆于背斜之上，但其本身并未发生褶皱变形，显示出在早期构造变形作用形成的古地貌背景之下的填平补齐式的沉积特征（图12-7；Gomez 和 Verges，2005）。

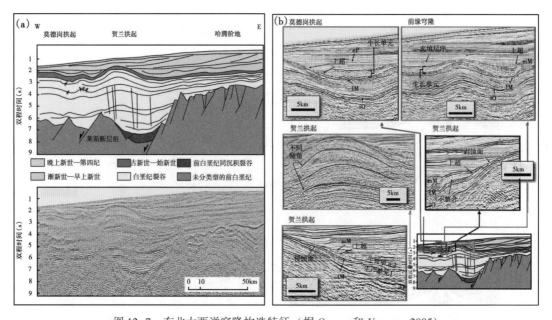

图12-7　东北大西洋穹隆构造特征（据 Gomez 和 Verges，2005）
（a）挪威边缘的裂后褶皱作用，以穿过贺兰阶地、汉森拱起和莫德岗拱起的地震剖面为例；
（b）褶皱的时代与形态通过反射波终止的几何特征表现

远源应力是解释诸如奥尔门兰格穹隆构造的主要力学机理。这些应力源于：（1）自北大西洋扩张中心的洋脊推进；（2）自阿尔卑斯—比利牛斯造山带的挤压造山作用；（3）两者的共同作用。

Dore 和 Lundin（1996）认为东北大西洋的多数穹隆构造是非火山成因的，包括贺兰和奥尔门兰格这两个最大的穹隆构造。他们认为这些构造与早渐新世洋脊推动导致的先存的基底断裂重新活动有关。但其后Dore等（2008）的研究与此产生了争论：不仅是中中新世的构造变形与早古近纪就开始的海底扩张不协调，洋脊推进力也过小而无法驱动已知的变形。Dore等（2008）认为这些穹隆构造发育的位置和时代与冰岛因苏拉边缘的形成有关，后者是一个位于冰岛和欧洲东北边缘间的500km宽的高原（图12-2）。这些学者认为从高原的海拔和规模上看，是可以产生比扩张洋脊更大的体积力，而且因苏拉边缘变形与穹隆构造同时发生。

之前提到，一些学者（Brekke，2000）认为由于远程效应，欧洲东北部的挤压作用与阿尔卑斯造山运动有关。而阿尔卑斯挤压作用的问题在于应力需要通过易于传递应力的中间区域传递，在这些区域内常常发生盆地的反转，而目前在阿尔卑斯挤压带北部并没有发现随时间的反转区（Dore等，2008）。

12.4 南大西洋非均匀裂后边缘沉降

前人有关后裂谷对南大西洋陆缘影响的记录要比北大西洋陆缘的少。这一方面是由于勘探资料的缺乏，也因为很多边缘盆地包含了盐组分，使得很难将盐构造与后裂谷边界构造变形区分开。非洲南部冈瓦纳超级大陆的中生代裂陷作用分为两个阶段：（1）约184Ma，从非洲德班到巴格；（2）约128±3Ma，从德班到几内亚，此时南大西洋大洋扩张运动开始。在大陆扩张时期，侧部隆起沿着大陆边缘发育，且前人的观点一致认为大陆边缘及其内部的地形遭受剥蚀，夷平将近海平面，导致在约100Ma的时候形成了一个相对平坦的"非洲表面"（Burke和Gunnell，2008；对"非洲表面"展开了细致的研究）。目前仍然存在争论的是：形成现今南非地形的隆升事件的发育时间尚不明确（图12-8）。

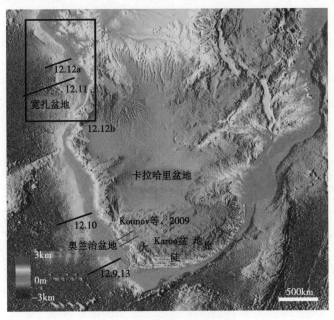

图12-8 非洲南部地形图

非洲南部受多处独特的高峻地形的控制；最典型的例子为大陡崖

12.4.1 区域隆起

非洲的等高线反映出一个没有挤压构造的大陆的最高海拔，比起其他大陆的0.2~0.5km，非洲大陆海拔可达0.4~0.6km（Burke和Gunnell，2008）。非洲南部大陆隆起最明显的表现为大陡崖，它平行于从西海岸纳米比亚延伸至东海岸林波波河的海岸线，总长3000km。这种独特的高峻地形不只局限于海岸地带，如图所示（图12-8），高海拔延续至陆块内部的许多地区。该图同样表明这不是一个单连续的面，而是围着卡拉哈里盆地形成5个不连续的隆起地形（海拔超过1.5km；Partridge和Maud，1987；Burke和Gunnell，2008）。地层学证据同样支持区域性隆起的观点，因为始新世和白垩纪海相沉积分布在非洲南部大部分地区海平面以上几百米处。区域范围内的关键问题为隆升的发生时间以及力学驱动机制。

大量的研究将凝灰石裂变径迹用于分析非洲南部隆升的幅度与时间。在早期的一项研究中，Gallagher和Brown（1999）认为南大西洋边缘大陆部分在中白垩世曾发生过脉冲式隆升。近来，Kounov等（2009）以两条垂直于南非西海岸的大陡崖的测线来确定隆起的时间。通过对凝灰石裂变径迹的样品分析，他们提出了研究区经历了"由热稳定期分隔的两期热冷却事件"（Kounov等，2009）。第一期热冷却作用（160—138Ma）对应于后卡罗超群的热松弛事件，之后发生了一幕局部隆升活动（138—115Ma）。第二期是发生于115—90Ma之间的快速冷却事件。虽然古地温梯度仍不确定，但他们还是通过模拟认为，在沿海岸地区存在一个剥蚀量达1.5~2.7km的剥蚀阶段，且在陡崖区该剥蚀量减至1km以下。在大西洋边缘和南部非洲其他地区的凝灰石裂变径迹研究，支持了这一结论。Tinker等（2008）通过来自于卡鲁盆地内已钻井样品的凝灰石裂变径迹分析，解释出了两期在时间上与Kounov（2009）研究结果一致的快速剥蚀事件。

相反地，包括Partridge和Maud（1987）、Burke（1996）、Burke和Gunnell（2008）等在内的多数研究认为隆升并未发生在中白垩世，现今地形是古近—新近纪的隆升作用的产物。Burke和Gunnell（2008）认同存在中白垩世的隆升，他们还认为继桑托事件之后的相关地形都在古近—新近纪隆起、盆地、地形起伏之前剥蚀殆尽。他们对于凝灰石裂变径迹的主要分歧在于1km的陡崖和据估计为20°的热传导梯度，在地表样品可以产生56℃的温度，这温度太低而无法使用凝灰石裂变径迹来确认。因此，"凝灰石裂变径迹数据无法将一阶和二阶模型有效地区分开。"Gallagher和Brown（1999）也强调了对裂变径迹能否识别年代更新的隆升事件的怀疑，并声称"基于针对低温（<50~60℃）和长持续时间（10~100Ma）情况下的裂变径迹退火模型的外推结果具有不确定性，大量年代更新（<20Ma）的测年数据非常值得怀疑"。Kounov等（2009）认为现今地形始于中白垩世构造变形。根据模拟结果，他们指出样品温度在中白垩世已经低于60℃，因此，如果剥蚀量小于2~3km，则其后发生的隆升活动都无法被识别出来。Burke和Gunnell（2008）利用非洲水系演化的分析，支持了其关于现今的地形源自古近—新近纪的假设。这一分析认为随海拔变化而发生的地形和分水岭的变化揭示了现今地形开始形成于30Ma（Burke和Wells，1989；Faure和Lange，1991）。

边缘隆升的时间目前仍然存在争议。鉴于陆上隆起剥蚀区是近海大陆边缘盆地的一个物源区，因此这些盆地是研究南大西洋边缘隆起形成时间的重要区域。

12.4.2 南非和纳米比亚边缘隆起

沿着南非的大西洋边缘，边缘构造变形的近海证据来自于构造地层学和热埋藏史的综合研究。证据显示该边缘与裂后阶段长期稳定的沉降有关。沉降导致大陆斜坡区沉积了较厚的沉积物（4km）。沉积物具有整合的地震反射结构，表明横跨120km宽的边缘发生了均匀沉降（图12-9）。边缘的地层结构是均匀热沉降与海平面变化共同作用的结果，可能引起最大洪泛面的加积与进积作用（Brown等，1995）。Paton等（2008）研究认为，与许多其他边缘相比，南非大西洋边缘白垩系并未发生大规模构造变形。他们将其归因于裂后早期因缺失合适的拆离层位而发生了均匀沉降。这也反映了晚白垩世之前构造变形的缺失。仅有的变形局限于小规模的断裂活动（<100 m），该活动由大陆架边缘破裂引起。

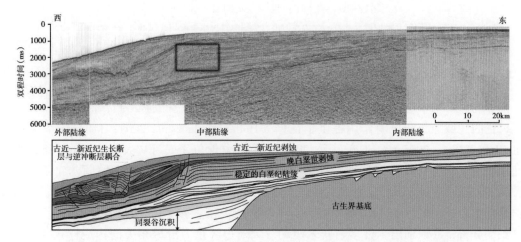

图12-9 过奥兰治盆地南部的地震及解释剖面（据Paton等，2008，有修改）
这项研究表明均匀的边缘沉降贯穿整个白垩纪，大部分沉积物聚集于内部陆缘；白垩纪末期，
内部陆缘发生了1km的隆升剥蚀，古近—新近纪沉积转移至外部陆缘

构造地层学研究（Paton等，2008）表明白垩纪末期（67Ma）沉积位置发生了重要变化。沉积作用从白垩纪坡折的中陆架区快速向西转移，导致了斜坡扇沉积体系的发育（图12-9）。沉积作用的迁移伴随着边缘中部与内部多处侵蚀不整合的发育。不整合的构造恢复揭示了隆升剥蚀的局部差别，坡折带为0，而近岸区达到了至少800m。除晚白垩世剥蚀之外，也存在白垩世末期的剥蚀和古近—新近纪隆升的证据。古近—新近纪隆起的大小难以量化，因为古近—新近纪沉积在边缘内部与中部非常薄。运用油井资料的热流分析和镜质组反射率的模拟可以识别达800m的隆起，但无法在晚白垩世和古近—新近纪隆起之间进行区分（图12-9）。尽管近岸区被薄的古近—新近系层序覆盖，但在其外侧仍存在一定的厚度。这意味着真正的古近—新近纪沉积物供给，可能来自于陆上隆起。与古近—新近纪隆起有关的不整合和削蚀作用可在更北的奥兰治盆地中被识别出来，而白垩纪隆起在平面上的分布更局限。

目前，对位于南非奥兰治盆地北部的纳米比亚边缘的后裂谷边缘变形的研究相对薄弱。尽管如此，从白垩纪和古近—新近纪大陆边缘的反射特征上看与南非非常相似（图12-10）。南纳米比亚盆地的地震剖面也显示了晚白垩世与古近—新近纪内部存在两个不整合。

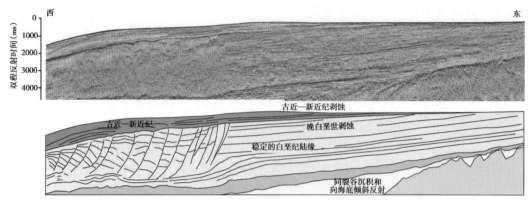

图 12-10 纳米比亚近海地震剖面与综合解释

此剖面位置位于图 12-9 剖面以北 500km，边界的形态很相似且反映了晚白垩世和古近—新近纪隆升事件

12.4.3 安哥拉和刚果边缘隆起

南非北部及纳米比亚盆地的大量研究为沿南大西洋边缘，特别是为安哥拉的宽扎盆地存在更广泛的区域隆起和掀斜作用提供了证据（Hudec 和 Jackson，2004；Jackson 等，2005；Walford 和 White，2005；Al-Hajri 等，2009）。宽扎盆地位于安哥拉大陆边缘的中部，形成于纽康姆阶（145—131Ma）时期裂谷作用。后裂谷早期盆地以阿普特阶—阿尔布阶蒸发岩沉积为主，厚度超过 1km。如此大的蒸发岩厚度与差异性热沉降和沉积负载相结合，即形成了典型的重力驱动型边缘。Hudec 和 Jackson（2004）对一条过盆地的 375km 长的剖面进行了详细的分析，并识别出了许多与盐岩驱动、大陆边缘重力滑塌相关的构造样式（图 12-11a），由陆到海依次包含了：前寒武系基底的露头、受到削蚀作用的楔状反射体、含有逆冲褶皱和褶曲的褶皱—褶断带、龟背状盐构造、拉伸驱动的筏移构造，由盐底辟、盐岩台地组成的盐控构造区、推覆体，以及相对无构造变形的深海平原。Hudec 和 Jackson（2004）通过对剖面进行构造恢复分析表明，热沉降驱动盐构造变形和后裂谷边缘隆起事件可以解释以上现象（图 12-11b、d）。由热沉降引起的第一期构造变形（阿普特阶—阿尔布阶；121—99Ma），导致盆地向海掀斜，同时上倾方向发育伸展构造系统。Hudec 和 Jackson（2004）认为，在经历短暂的构造平静期后，在宽扎盆地外缘可识别出一个发生在坎佩尼阶（75Ma）内相对短暂的以基底隆升形式出现的构造变形阶段，并认为在桑托期它们可能与宽扎盆地内部（约 84Ma）基底再次活动有关。最后阶段的构造变形发生于中新世（约 24Ma），虽然只引起了几百米的隆升，却导致了强烈的重力滑动作用。

在宽扎盆地内古近—新近纪隆起非常明显，地震反射的削截特征表明存在渐新世（35—30Ma）和上新世（3.5—1.8Ma）两个重要的不整合（图 12-12；Jackson 等，2005；Al-Hajri 等，2009）。通过深度剖面估算出与上新世不整合相关的剥蚀量近 1.6km。与渐新世（35—30Ma）不整合有关的剥蚀量难于确定。尽管陆架边缘斜坡上的沉积作用和剥蚀作用使得定量分析的难度加大，但 Cramez 和 Jackson（2000）仍通过地震反射的削截特征分析确定了该隆升作用，并估算出 150m 的剥蚀量。而 Walford 和 White（2005）通过对地震反射剖面的叠加速度分析，指出新近纪（上新世；3.5—1.8Ma）剥蚀量为 0.5~1.5km；而与渐新世不整合相关的剥蚀量达 2.5km。Al-Hajri 等（2009）将该方法推广到横跨整个南大西洋边缘的研究，并在定量分析上新世之后的剥蚀作用方面取得了与 Cramez 和 Jackson（2000）相一致的结果。Al-Hajri 等（2009）认为在刚果三角洲存在 500m 的剥蚀，在宽扎盆地存在 1km 的剥蚀，显示了隆起作用具有明显的横向变化（图 12-12）。

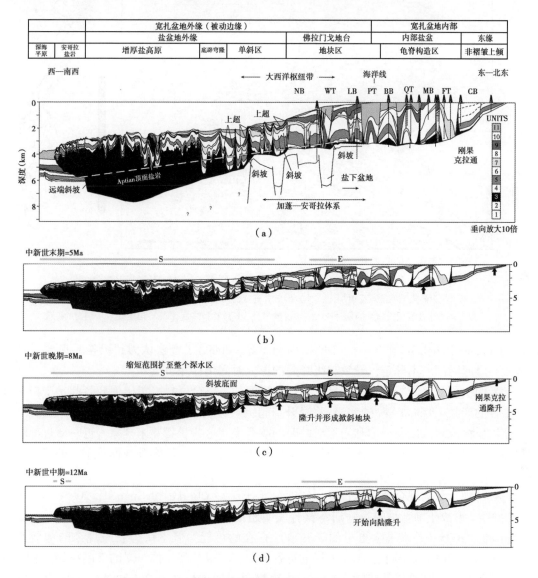

图 12-11　过宽扎盆地的剖面（据 Hudec 和 Jackson，1994）

（a）过宽扎盆地构造剖面表示与盐构造及不稳定边缘相关的构造特征；（b）是对图（a）的地震剖面的构造恢复；说明变形不单受盐构造影响，而是多种构造变形共同作用的结果

12.4.4　隆起的力学机制与时间

与西北欧大陆边缘相比，对南大西洋边缘的后裂谷构造变形研究和认识程度较低。尽管对于隆起的时间存在争议，但大部分学者都认为，从区域尺度来看，它是对岩石圈底部垂向应力的动态响应。地震层析成像技术表明在南部非洲岩石圈下部存在一个低速带，而自由重力异常也支持了该地区岩石圈之下存在超级隆起的观点（Nyblade 和 Robinson，1994；Burke，

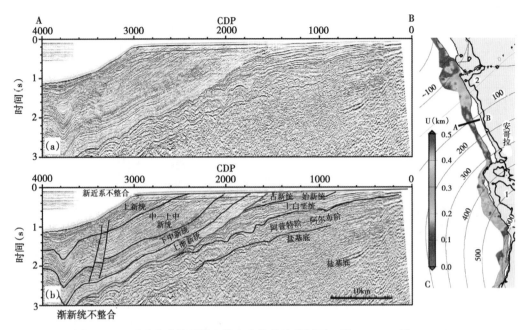

图 12-12　反映隆升作用具有横向变化的地震剖面（据 Al-Hajri 等，2009）
(a) 过安哥拉大陆边缘的地震反射剖面；(b) 经钻井标定的地震剖面解释，高亮部分为至少两期隆升；
(c) 根据均方根速度剖面估算的安哥拉大陆边缘的隆升（U）及剥蚀量（D）；黑实线表示剖面的位置

1996；Lithgow-Bertelloni 和 Silver，1998；Gurnis 等，2000）。普遍认为超级隆起形成了现今的地貌。隆起时间的不确定，导致超级隆起的形成时间仍存在争议。

学者们引用陆上的凝灰石裂变径迹数据和近海的不整合资料，指出现今隆起是中白垩世构造事件的产物。Tinker 等（2008）将隆升与超级隆起联系起来，并讨论了南部非洲隆起、大型镁铁质火成岩区以及金伯利岩形成的时间吻合度。这些学者并未排除古近—新近纪隆升的存在，但根据凝灰石裂变径迹的资料，他们认为古近—新近纪隆升规模较小，处于次要的地位。

另一种观点认为现今地貌与古近—新近纪隆升有关（Burke 和 Gunnell，2005；Al-Hajri 等，2009）。形成于渐新世的第一期不整合表明超地幔柱可能从 35~30Ma 开始活动。Burke 和 Gunnell（2005）并未排除更早的隆升事件存在的可能，但不认为它与现今的地貌有联系。而中白垩世事件归因于远源应力场的远程效应，它或与阿拉伯海岸的弧间碰撞（84Ma）有关，或是板块旋转重组的产物（Nurnberg 和 Muller，1991；Guiraud 和 Bosworth，1997）。

12.5　结论与意义

离散的大陆边缘在后裂谷热沉降阶段处于构造平静期这一假定，已不被近来的研究所认同。从东北和南大西洋边缘大量的实例可知该阶段发育多幕隆升、剥蚀和沉降事件的证据确凿，这明显与热指数衰减模型的预测并不一致。这些事件包括在挪威、南非、纳米比亚和安哥拉边缘发生的相对地形起伏小（100km）、持续时间短（<5Ma）的区域隆升和掀斜事件。在东北大西洋，有很多局部隆升事件引起构造穹隆的实例，诸如贺兰与奥尔门兰格穹隆。

早期简单的力学机制已无法解释所有的研究结果，对于其真正机制一直存在争论。很多力学机制被援引，诸如板块重组、地幔柱诱发隆升、来自于远源应力场的远程效应、洋脊推进形成的挤压力以及上地幔对流等。不考虑其形成机制，陆缘构造变形特征也具有重要意义，控制着陆缘演化的很多方面，包括沉积学、构造以及油气勘探前景。

东北大西洋的幕式构造变形事件对边缘水深的变化有重要的影响。古新世（Stoker 等，2005）的浅水碎屑沉积作用形成了一个向可容空间的进积体。边缘构造变形导致了从浅海相近源沉积到深水远端沉积的快速转换。在南大西洋的奥兰治盆地，边缘掀斜事件引起沉积作用从内缘到外缘的快速迁移，这个转变的一个直接结果是相对稳定的白垩纪边缘变为不稳定的古近—新近纪边缘（图12-13）。这种不稳定性导致深水褶皱冲断带的发育，从而形成了生长正断层和下倾方向逆冲断层相耦合的构造体系（图12-13；Butler 和 Paton，2010）。

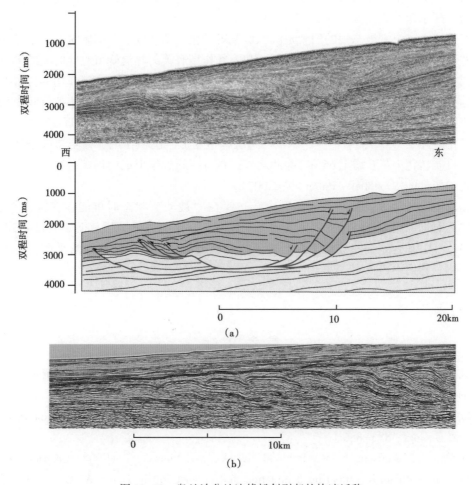

图 12-13 奥兰治盆地边缘掀斜引起的快速迁移

(a) 古近—新近纪奥兰治盆地快速沉积转换，导致了不稳定边缘及生长断层与逆冲断层耦合系统的形成；此沉积转换是裂后构造变形的产物（据 Paton 等，2008，修改）；(b) 纳米比亚近海有一个类似的逆冲断层系统，且这条地震剖面展示了逆冲系统中的叠瓦构造（据 Butler 和 Paton，2010；地震图像来自阿特拉斯公司，www.seismicatlas.org）

边缘构造变形会产生其他的后果，例如海洋环流的改变。以东北大西洋为例，Stoker 等（2005）描述了边缘水深的变化如何快速改变深水等深流沉积与搬运通道的位置。边缘隆升也能影响盆地的演化，例如隆升形成的高地可能成为局部物源区。而东北大西洋隆升形成的穹隆构造可能是末次冰期局部冰盖发育的部位（Eyles，1996）。

南部非洲内构造/隆起与气候的互相作用也被学者们广泛讨论（Burke 和 Gunnell，2008），讨论的焦点是隆升的时代。

从含油气潜力的角度来说，后裂谷构造变形对油气系统的各方面都有重要的意义，如烃源岩的沉积、储盖组合、圈闭的形成及成熟时期。在欧洲西北陆架，后裂谷变形区形成了一些非常有吸引力的勘探目标。在构造反转发生的地区，四周闭合、在内部存在轻微变形的穹隆叠加古新世浊流富砂沉积，易形成有利的含油气区带。如果邻区存在深埋的烃源岩，这些目标区将变得更为有利。而这其中主要的局限性在于油气运移与构造反转及圈闭形成时间是否匹配（Lundin 和 Dore，2002）。在南大西洋，这些隆升事件增加了边缘不稳定的重力势，这些重力助于深水褶皱冲断带的形成。随着对超深水环境与经济可行性的研究，其所处位置逐渐成为油气勘探的关注区（White 等，2003）。

油气系统中一个常被忽视的因素是沉积负载对下伏烃源岩成熟度的影响。一个已知的事实是这些盆地中的多数正处于后裂谷时期，和裂谷作用有关的热流峰值对此时期的影响无关紧要，而控制烃源岩热状态的主要因素为埋藏量。在奥兰治盆地沉积充填作用的位置变化十分迅速。整个白垩纪，边缘中部处于快速沉积区，而边缘外部则发生了较小的沉积作用。油气系统模拟表明，这种沉积的差异导致在边缘中部有机质向油气转化率接近100%，而在边缘外部转化率几乎为0。随着白垩纪末期的边缘构造变形，沉积作用迁移至边缘外部，导致该地区有机质转化率从0增至65%。除此之外，古近—新近纪形成的远端逆冲系统可为生成的油气提供局部的构造圈闭（Paton 等，2007）。

总结以上研究认为，虽然大西洋型大陆边缘在"被动"状态时可能存在构造平静期，但显然它们对远源应力场的远程效应或地幔—地壳交互作用也很敏感。正如 Hudec 和 Jackson（2004）认为的："一些被动边缘会长期在亚稳定的状态下保持精确的平衡，直到一个微小的变化令它们不再稳定。"

致谢（略）

参 考 文 献

Allen, P. A., and Allen, J. R. (2005) Basin analysis, principles and applications. Oxford, Blackwell, 552 p.

Al-Hajri, Y., White, N., and Fishwick, S. (2009) Scales of transient convective support beneath Africa. Geology, 883–886.

Andersen, M. S., Sørensen, A. B., Boldreel, L. O., and Nielsen, T. (2002) Cenozoic evolution of the Faroe Platform comparing denudation and deposition, in Doré, A. G., Cartwright, J. A., Stoker, M. S., Turner, J. P., and White, N., eds., Exhumation of the North Atlantic Margin: timing, mechanisms and implications for petroleum exploration. Special Publications, 196. London, Geological Society, 327–378.

Barton, P., and Wood, R (1984) Tectonic evolution of the North Sea basin: crustal stretching and subsidence. Geophysical Journal of the Royal Astronomical Society, 79, 987–1022.

Brekke, H. (2000) The tectonic evolution of the Norwegian Sea Continental Margin with emphasis on the Voring and Møre Basins, in Nottvedt, A., (ed.), Dynamics of the Norwegian Margin.

Special Publications, 167. London, Geological Society, 327-378.

Brodie, J., and White, N. (1994) Sedimentary basin inversion caused by igneous underplating: Northwest European continental shelf. Geology 22, 147-150.

Brown, L. F., Jr. Benson, J. M. Brink, G. J. et al. 1995. Sequence Stratigraphy in offshore South African Divergent Basins, An Atlas on Exploration for Cretaceous Lowstand Traps by SOEKOR (Pty) Ltd. American Association of Petroleum Geologists, Studies in Geology Series 41.

Burke, K. (1996) The African plate. South African Journal of Geology, 339-410.

Burke, K., and Gunnell, Y. (2008) The African erosion surface: a continental-scale synthesis of geomorphology, tectonics, and environmental change over the past 180 million years. Geological Society of America, 66 p.

Busby, C. J., and Ingersoll, R. V. (1995) Tectonics of sedimentary basins. Oxford, Blackwell Science.

Butler, R. H. B., and Paton, D. A. (2010) Evaluating lateral compaction in deepwater fold and thrust belts: How much are we missing from "nature's sandbox"? GSA Today, 20, 4-10.

Ceramicola, S., Stoker, M., Praeg, D., Shannon, P. M., De Santis, L., Hoult, R., Hjelstuen, B. O., Laberg, S., and Mathiesen, A. (2005) Anomalous Cenozoic subsidence along the "passive" continental margin from Ireland to mid-Norway. Marine and Petroleum Geology, 22, 1045-1067.

Corcoran, D. V., and Doré, A. G. (2005) A review of techniques for the estimation of magnitude and timing of exhumation in offshore basins. Earth-Science Reviews, 72, 129-168.

Cramez, C., and Jackson, M. P. A. (2000) Superposed deformation straddling the continental-oceanic transition in deep water. Angola, 17, 1095-1109.

Doré, A. G., Lundin, E. R., Kusznir, N. J., and Pascal, C. (2008) Potential mechanisms for the genesis of Cenozoic domal structures on the NE Atlantic margin: pros, cons and some new ideas. Special Publications, 306. London, Geological Society, 1-26.

Doré, A. G., and Lundin, E. R. (1996) Cenozoic compressional structures on the NE Atlantic margin: nature, origin, and potential significance for hydrocarbon exploration. Petroleum Geoscience 2, 299-311.

Doré, A. G., Lundin, E. R., Jensen, L. N., Birkeland, ϕ., Eliassen, P. E., and Fichler, C. (1999) Principal tectonic events in the evolution of the northwest European Atlantic margin, in Fleet, A. J., and Boldy, S. A. R., eds., Petroleum geology of Northwest Europe: Proceedings of the 5th Conference. London, Geological Society, 41-61.

England, P., and Molnar, P. (1990) Surface uplift, uplift of rocks, and exhumation of rocks. Geology, 1173-1177.

Eyles, N. (1996) Passive margin uplift around the North Atlantic region and its role in northern Hemisphere Late Cenozoic glaciation. Geology, 24, 103-106.

Falvey, D. A. (1974) The development of continental margins in plate tectonic theory. Aust. Petrol. Explor. Assoc. J., 14 (2), 95-106.

Gallagher, K., and Brown, R. (1999) Denudation and uplift at passive margins: the record on the Atlantic margin of southern Africa, Philosophical Transactions R. Soc. London, Ser. A, 367,

835–859.

Gómez, M., and Vergés, J. (2005) Quantifying the contribution of tectonics vs. differential compaction in the development of domes along the Mid-Norwegian Atlantic margin. Basin Research, 17, 289–310.

Green, P. F., Duddy, I. R., and Hegarty, K. A. (2002) Quantifying exhumation from apatite fission-track analysis and vitrinite reflectance data: precision, accuracy and latest results from the Atlantic margin of NW Europe, in Dore, A. G. D., Cartwright, J., Stoker, M. S., Turner, J. P., and White, N., eds., Exhumation of the North Atlantic Margin: timing, mechanisms and implications for petroleum exploration. Special Publications, 196. London, Geological Society, 331–354.

Guiraud, R., and Bosworth, W. (1997) Senonian basin inversion and rejuvenation of rifting in Africa and Arabia: Synthesis and application to plate-scale tectonics. Tectonophysics, 213, 131–134.

Gurnis, M., Mitrovica, J. X., Ritsema, J., and vanHeijst, H. -J. (2000) Constraining mantle density structure using geological evidence of surface uplift rates: the case of the African Superplume. Geochem. Geophys. Geosyst., 1.

Hall, B. D., and White, N. (1994) Origin of anomalous Tertiary subsidence adjacent to North Atlantic continental margins. Marine and Petroleum Geology 11 (6), 702–714.

Hendriks, B. W. H., and Andriessen, P. (2002) Pattern and timing of the post-Caledonian denudation of northern Scandinavia constrained by apatite fission-track thermochronology, in A. G. D. Dore, J. Cartwright, M. S. Stoker, J. P. Turner and N. White, eds., Exhumation of the North Atlantic Margin: timing, mechanisms and implications for petroleum exploration. Special Publications 196. London, Geological Society, 327–378.

Hudec, M. R., and Jackson, M. P. A. (2004) Regional restoration across the Kwanza Basin, Angola: Salt tectonics triggered by repeated uplift of a metastable passive margin. AAPG Bulletin, 971–990.

Jackson, M. P. A., Hudec, M. R., Hegarty, K. A. (2005) The great West-African Tertiary coastal uplift: Fact or fiction? A Perspective from the Angolan divergent margin. Tectonics, 24, TC6014.

Jarvis, G. T., and McKenzie, D. (1980) Sedimentary basin formation with finite extension rates. Earth and Planetary Science Letters, 48, 42–52.

Jones, S. M., White, N., Clarke, B. J., Rowley, E., and Gallagher, K. (2002) Present and past influence of the Iceland Plume on sedimentation, in Dore, A. G. D., Cartwright, J., Stoker, M. S., Turner, J. P., and White, N., eds., Exhumation of the North Atlantic Margin: timing, mechanisms and implications for petroleum exploration. Special Publications, 196. London, Geological Society, 13–25.

Keen, C. E. (1985) The dynamics of rifting: deformation of the lithosphere by active and passive driving forces. Geophysical Journal of the Royal Astronomical Society (1), 95–120.

Keen, C. E., and Boutilier, R. R. (1995) Lithosphere-asthenosphere interactions below rifts, in Banda, E., Torné, M., and Talwani, M., eds., Rifted ocean-continent boundaries.

Dordrecht, Kluwer Academic, 17-30.

Keen, C. E., and Cordsen, A. (1981) Crustal structure, seismic stratigraphy, and rift processes of the continental margin off eastern Canada: ocean bottom seismic refraction results off Nova Scotia. Can. J. Earth Sci., 18 (10) 1523-1538.

King, S. D. Lowman, J. P., and C. W. Gable (2002) Episodic tectonic plate reorganisations driven by mantle con vection. Earth and Planetary Science Letters 203, 83-91.

Korenaga, J., and Jordan, T. H. (2002) On the state of sublithospheric upper mantle beneath a supercontinent. Geophysical Journal International 149, 179-189.

Kounov, A., Viola, G., de Wit, M., and Andreoli, M. A. G. (2009) Denudation along the Atlantic passive margin: new insights from apatite fission-track analysis on the western coast of South Africa. Special Publications, 324. Geological Society of London, 287-306.

Le Pichon, X., and Sibuet J-C. (1981) Passive margins, a model of formation. Journal Geophysical Research, 86, 3708-3720.

Lithgow-Bertelloni, C., and Silver, P. G. (1998) Dynamic topography, plate driving forces and the African superswell. Nature, 395, 269-272.

Lundin, E., and Doré, A. G. (2002) Mid-Cenozioc postbreakup deformation in the "passive" margins, bordering the Norwegian-Greenland Sea. Marine and Petroleum Geology, 19, 79-93.

McKenzie, D. (1978) Some remarks on the development of sedimentary basins. Earth and Planetary Science Letters, 40, 25-32.

Miall, A. D. (1990) The principles of sedimentary basin analysis. New York, Springer, 668 p.

Mutter, J. C. (1984) Cenozoic and late Mesozoic stratigraphy and subsidence history of the Norwegian margin. Geological Society of America Bulletin, 95, 1135-1149.

Nurnberg, D., and Muller, R. D. (1991) The tectonic evolution of the South Atlantic from Late Jurassic to present. Tectonophysics, 191.

Nyblade, A. A., and Robinson, S. W. (1994) The African superswell. Geophysical Research Letters, 21, 765-768.

Partridge, T. C., and Maud, R. R. (1987) Geomorphic evolution of southern Africa since the Mesozoic. South African Journal of Geology, 90, 179-208.

Paton, D. A., di Primio, R., Kuhlmann, G., Van der Spuy, D., and Horsefield, B. (2007) Insights into the petroleum system evolution of the Southern Orange Basin, South Africa. South African Journal of Geology, 110, 261-274.

Paton, D., et al. (2008) Tectonically induced adjustment of passive-margin accommodation space; influence on the hydrocarbon potential of the Orange Basin, South Africa. American Association of Petroleum Geologists Bulletin, 92, 589-609.

Praeg, D., Stoker, M. S., Shannon, P. M., Ceramicola, S., Hjelstuen, B., Laberg, J. S., and Mathiesen, A. (2005) Episodic Cenozoic tectonism and the development of the NW European "passive" continental margin. Marine and Petroleum Geology 22, 1007-1030.

Riis, F. (1996) Quantification of Cenozoic vertical movements of Scandinavia by correlation of morphological surfaces with offshore data. Global and Planetary Change 12, 331-357.

Roberts, N. J., Kusznir, G., Yielding, and Styles, P. (1998) 2D flexural backstripping of extensional basins: the need for a sideways glance. Petroleum Geoscience, 4, 327–338.

Rowley, R., and White, N. (1998) Inverse modeling of extension and denudation in the East Irish Sea and surrounding areas. Earth and Planetary Science Letters, 161, 57–71.

Saunders, A. D., Fitton, J. G., Kerr, A. C., Norry, M. J., and Kent, R. W. (1997) The North Atlantic igneous province, in J. J. Mahoney and M. F. Coffin, eds., Large igneous provinces: continental, oceanic and planetary flood volcanism. American Geophysical Union 100, AGU Geophysical Monograph 100, 45–93.

Sclater, J. G., and Christie, P. A. B. (1980) Continental stretching: an explanation of the post-mid-Cretaceous subsidence of the Central North Sea basin. Journal of Geophysical Research 85, 3711–3739.

Sclater, J. G., Jaupart, C., and Galson, D. (1980) The heat flow through oceanic and continental crust and the heat loss of the Earth. Reviews Geophysics and Space Physics, 18, 269–311.

Steckler, M. S., and Watts, A. B. (1978) Subsidence of the Atlantic-type continental margin off New York. Earth and Planetary Science Letters, 41, 1–13.

Stoker, M. S., Hoult, R. J., Nielsen, T., Hjelstuen, B. O., Laberg, J. S., Shannon, P. M., Praeg, D., Mathiesen, A., van Weering, T. C. E., and McDonnell, A. (2005a) Sedimentary and oceanographic responses to early Neogene compression on the NW European margin. Marine and Petroleum Geology, 22, 1031–1044.

Stoker, M. S., Praeg, D., Hjelstuen, B. O., Laberg, J. S., Nielsen, T., and Shannon, P. M. (2005b). Neogene stratigraphy and the sedimentary and oceanographic development of the NW European Atlantic margin. Amsterdam, Elsevier. doi: 10.1016/j.marpetgeo.2004.11.007.

Stoker, M. S., van Weering, T. C. E., and Svaerdborg, T. (2001) A mid–late Cenozoic tectonostratigraphic framework for the Rockall Trough, in P. M. Shannon, P. D. W. Haughton, and D. Corcoran, eds., The petroleum exploration of Ireland's Offshore Basins. Special Publications, 188. London, Geological Society, 411–438.

Stuevold, L. M., Skogseid, J., and Eldholm, O. (1992) Post-Cretaceous uplift events on the Vøring continental margin. Geology, 20, 919–922.

Sweeney, J. J., and Burnham, A. K. (1990) Evaluation of a simple model of vitrinite reflectance based on chemical kinetics. American Association of Petroleum Geologists Bulletin, 1559–1570.

Tinker, J., de Wit, M., and Brown, R. (2008) Mesozoic exhumation of the southern Cape, South Africa, quantified using apatite fission track thermochronology. Tectonophysics, 455, 77–93.

Turner, J. D., and Scrutton, R. A. (1993) Subsidence patterns in western margin basins: evidence from the Faeroe–Shetland Basin, in Parker, J. R., ed., Petroleum geology of Northwest Europe: Proceedings of the 4th Conference. London, Geological Society, 975–983.

Walford, H. L., and White, N. J. (2005) Constraining uplift and denudation of west African continental margin by inversion of stacking velocity data. Journal of Geophysical Research B: Solid Earth, 110 (4), 1–16.

White, N., Thompson, M., and Barwise, T. (2003) Understanding the evolution of deep-water continental margins. Nature, 426, 334-343.

White, R. S., and McKenzie, D. (1989) Magmatism at rift zones: the generation of volcanic continental margins and flood basalts. Journal of Geophysical Research, 94 (6), 7685-7729.

Ziegler, P. A. (1988) Evolution of the Arctic-North Atlantic and the Western Tethys. AAPG Memoir 43. Tulsa, OK, American Association of Petroleum Geologists, 198 p. (+30 plates)

(赵钊 译,张功成 祁鹏 校)

第13章 早白垩世加拿大东部近海斯科舍被动边缘盆地构造变形对沉积的影响

GEORGIA PE-PIPER[1], DAVID J. W. PIPER[2]

(1. Department of Geology, Saint Mary's University, Halifax, Nova Scotica, Canada;
2. Geological Survey of Canada (Atlantic), Bedford Institute of Oceanography, Dartmouth, Nova Scotia, Candada)

摘 要：本文研究了被动陆缘基底同沉积构造活动对沉积充填的影响。斯科舍盆地在侏罗纪末期至早白垩世，即在海底开始扩张之后大约45Ma之内，盐构造活动造成外陆架斜坡迅速沉降，砂质三角洲在陆架上向前进积了数十千米。该时期砂质物源的供给速度非常快，比被动陆缘早期高出三至四倍，其主要原因是物源区的构造抬升，而并非被动陆缘大型河流入海口的摆动或气候变化。上述构造活动在陆地上存在直接证据，伴随着晚古生代断层的再次活动形成了一系列小型盆地，同时地垒抬升并在沉积物中出现粗碎屑沉积物夹层。对于三角洲砂岩的物源分析，本次研究采取了多种技术手段。砂泥岩全岩地球化学组分分析能够区分沉积物来源。白云母、独居石及锆石等矿物对于物理及化学风化作用的敏感性不一致，它们的地质年代学对比可分析沉积物是来自初期旋回物源还是多期旋回物源。白云母及独居石分析表明，大约一半的砂质沉积物为来自包括斯科舍内陆架在内的阿巴拉契亚山脉结晶基底的初期旋回物源。80%以上的锆石来自沉积岩或变质岩的多期次旋回物源。地震剖面揭示内陆架的构造变形为盆地提供了碎屑供给。沉积物堆积受盐构造活动的强烈影响，而盐构造活动是由三角洲沉积负载所驱动的。构造再次活动形成的陡峻地貌使沉积物以高密度流的形式搬运至前三角洲陆架区形成浊积岩。这些浊积岩是该盆地天然气富集的主要储层。研究表明，被动陆缘基底的再次活动与沉积物供给增加之间的复杂关系，需要展开多种技术手段进行研究。

关键词：构造 物源区 地质年代 沉积相 碎屑岩岩石学

13.1 引言

被动大陆边缘的沉积作用不是本文的研究重点。通常来说，成熟型被动陆缘的海平面变化掩饰了其构造活动对沉积的影响。在特定纬度范围内，生物及化学成因的沉积物受控于气候及海洋环境。然而，在某些被动陆缘的特定时期，构造事件能够大量增加碎屑沉积物供给，从而在沉积物结构的正常驱动因素中占主导地位。加拿大东部近海斯科舍盆地侏罗系顶部至白垩系下部的三角洲沉积就是一个典型例子（图13-1和图13-2）。

斯科舍盆地位于靠近中大西洋的东北部边界，三叠纪至早侏罗世在拉张作用下形成（Withjack等，2009）。斯科舍盆地裂谷作用大致开始于普林斯巴阶（Schettino和Turco，2009）（图13-3），隶属于大西洋早期裂谷事件。该期裂谷作用结束于纽芬兰岛至直布罗陀海峡的转换带，该转换带位于大浅滩转换边缘（图13-1）。

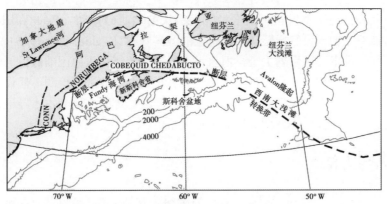

图 13-1 北大西洋西部加拿大被动陆缘斯科舍盆地区域背景与早白垩世主断层分布

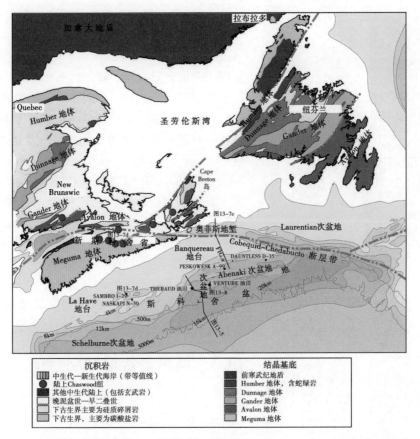

图 13-2 斯科舍盆地及阿巴拉契亚主要地体的气田、探井、基本地质特征与测线位置
（底图据 Williams 和 Grant，1998）

直至侏罗纪晚期，该陆架边缘为典型的亚热带被动陆缘，沉积了厚层的侏罗系陆架碳酸盐岩（Wade 和 Maclean，1990）。晚侏罗世至早白垩世发生显著变化，在陆架区沉积了数千米厚的砂质三角洲，意味着该时期陆源供给增加了 3 至 4 倍。前期文献明确了陆源供给增加的两个重要因素：其一是流经东加拿大地盾大部分地区的圣劳伦斯河的发育（Wade 和 Maclean，1990）；其二是纽芬兰岛大浅滩上阿瓦隆隆起及不整合的形成。该隆起及不整合的形

319

成与早白垩世裂谷作用相关，与纽芬兰岛大浅滩从伊比利亚裂离作用也相关，因为大西洋是向北裂开的。此外，晚侏罗世至早白垩世北美板块的北西向移动（Beck 和 Housen，2003）使得该地区由侏罗纪的干旱气候转变为白垩纪的湿润气候（Valdes 等，1996）。

本文中，我们设定了一系列关于早白垩世大量陆源碎屑注入原因的假设。假设物源注入与河流流经盆地的规模相关，我们可以利用注入沉积物的组分进行检测。假设与构造活动相关，沉积物组分同样可以揭示隆起的分布及成因。假设局部构造活动占主导地位，那么在沉积物分布及可容纳空间提供等方面又扮演了什么角色？

13.2 样品与方法

在斯科舍盆地，200多口钻井及密集的地震资料提供了可靠的地层及沉积相关信息（Wade 和 Maclean，1990）。该盆地的数据库可向公众开放相关数据资料（网址：http：//basin.gdr.nrcan.gc.ca/index_c.php）。大约40口钻井在下白垩统三角洲沉积中获取了常规岩心资料。我们对其中30口井的岩心进行了沉积学测井解释，并制作了碎屑岩岩石学样品。我们在陆上对应地层中的岩石也进行了取样及测井。这些岩石来自于两个砂质探槽、一个泥质探槽以及自然资源核心库（www.gov.ns.ca/natr/meb/one/dclhome.asp）250多口钻井中的40个代表性取心井。岩相分类及解释主要基于前人的研究成果（Piper 等，2004；Pe-Piper 等，2005a；Cummings 和 Arnott，2005；Gould 等，2010）。岩石学分析工作包括岩屑的显微观察、岩屑矿物"化学指纹特征"鉴别、独居石电子探针测年分析、白云母及锆石的单颗粒激光剥蚀测年、砂泥岩全岩地球化学分析。相关方法细节可见本文引用的相关文献。

13.3 斯科舍盆地地层特征

三叠纪至侏罗纪，大陆边缘属亚热带环境（Wade 和 Maclean，1990；Olsen 和 Et Touhami，2008）。三叠纪至早侏罗世的裂陷作用形成了一系列构造盆地，其中的 Laurentian、Abenaki、Sable 次盆地最终堆积了厚层沉积物（图13-2）。上述盆地充填了陆源碎屑红层沉积物，其上覆盖了以晚三叠世至早侏罗世的盐岩及赫唐阶拉斑玄武岩。中晚侏罗世，外陆架接受了碳酸盐岩沉积（易洛魁及阿贝内基组），向海一侧过渡为盆地相页岩，向陆一侧过渡为陆源碎屑岩（莫西干及米克马克组）（图13-3）。在斯科舍

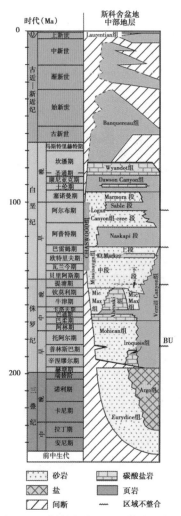

图13-3 斯科舍盆地地层综合柱状图
（Bμ=破裂不整合）
（据 Wade 和 MacLean，1990，
Cummings 和 Arnott，2005，修改；
时间尺度据 Gradstein 等，2004）

盆地晚侏罗世，上述陆源碎屑岩不断向海进积，在盆地东部其沉积厚度最大，形成了厚度达1km的提塘阶三角洲砂岩及米西索加组下部的页岩，是文彻及特博油田的重要储气层段。

早白垩世的米西索加和洛根坎宁组中，在三角洲背景下沉积了分布更为广泛的厚层陆源碎屑岩，其厚度可达数千米。在 Sable 及 Abenaki 次盆地，三角洲砂岩在晚侏罗世碳酸岩浅滩前向海进积了数十千米（图 13-4、图 13-5）。Missiauga 组中上部砂岩（贝里阿斯阶至巴雷姆阶；Williams 等，1990）向海一侧逐渐过渡为维里尔坎宁组页岩（图 13-3）。上覆阿普特至塞诺曼阶洛根坎宁组同样为三角洲沉积，主要由两期海侵页岩层（纳斯卡皮及赛布尔层）与两期含砂地层（克里及马莫拉层）交替组成。晚期为上白垩统道森坎宁组，该时期广泛海侵，以页岩沉积为主，其中发育两期广泛分布的白垩系沉积（彼得雷尔层及怀安多特组）。

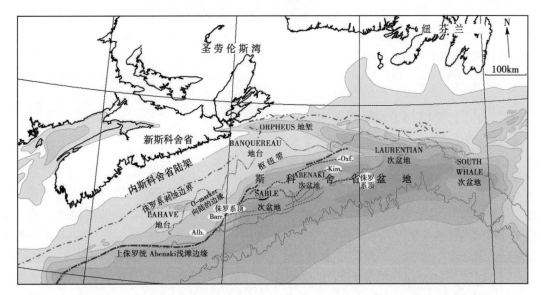

图 13-4　斯科舍盆地中晚侏罗世 Abenaki 碳酸盐岩浅滩、晚侏罗世—早白垩世进积三角洲向海一侧尖灭线及侏罗系、侏罗系剥蚀边界标志层向陆一侧剥蚀尖灭线（地层等厚线及水深线见图 13-2）
[据 Wade 和 MacLean（1990）及 MacLean 和 Wade（1992）编辑整理；白垩系三角洲据 Piper 等（2010）]

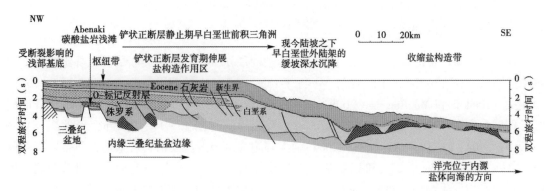

图 13-5　Sable 次盆地地西部地质剖面（原地震测线与解释层位可见于 www.cnsopb.ns.ca/call_for_bids_08_2/cnsopb/images/region/figure3.pdf）

与早白垩世三角洲沉积相对应的陆相沉积为 Chaswood 组的河流相沉积，发育于瓦兰今阶至阿普特阶，为一套 200m 厚的松散固结的砾岩、砂岩及泥岩（Stea and Pullan，2001；

Falcon-Lang 等,2007)。Chaswood 组保存于新斯科舍省及新布朗斯维克省几个小型盆地中,在邻近斯科舍盆地的新斯科舍省中部奥尔普斯地堑最为发育(图13-6)。

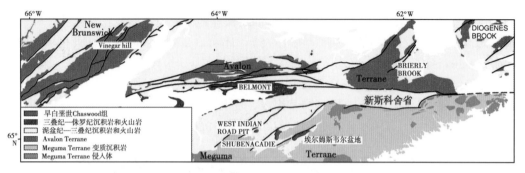

图 13-6 北东向同沉积断层控制下的 Chaswood 组地层分布

(同沉积断层参考于:Vinegar hill 断层,Piper 等(2007);Brierly Brook 断层,Pe-Piper 等(2005b);埃尔姆斯韦尔盆地,Hundert 等(2006);West Indian Road Pit,Gobeil 等(2006)。其他地区的同沉积断层尚未证实)

13.4 早白垩世区域构造变形

13.4.1 Chaswood 组中的证据

与斯科舍盆地 Missisauga 组及洛根坎宁组厚层三角洲沉积同期的 Chaswood 组(图13-6)河流相沉积中保存有同沉积构造变形的证据(图13-7d)。在同沉积断层两侧,地层厚度及沉积相都迅速变化(Gobeil 等,2006;Pe-Piper 等,2005b)。在一个 Chaswood 组的露头剖面,地层发生旋转,产状几乎直立,上覆为近水平状的砂岩及砾岩沉积(Gobeil 等,2006)。多口钻井揭示为大型滑坡沉积(Piper 等,2005)的长 5km、厚 2~5m 的旋转地块明确了该时期地层倾斜的斜率较大,使得先前沉积物多次发生崩塌。在 Chaswood 组沉积时期发生同沉积构造变形的所有盆地均是以北东向同沉积断层为边界,并平行于断层走向延伸(图13-6)。经钻井约束的地震解释表明,Chaswood 组地层下部的构造变形发生于该组上部地层沉积前,变形作用包括褶皱和断裂(Hundert 等,2007)。

13.4.2 斯科舍陆架的证据

斯科舍陆架内侧的基底及上覆下白垩统均被断层切割。许多该类断层为三叠纪裂陷期的继承性断层(图13-7b),走向北东,平行于 Chaswood 组中的断层。有些反转断层与盐岩活动相关(Rankin 和 MacRae 等,2008)。在奥尔普斯地堑,东西向断层占主导,不整合面削蚀了侏罗系顶部、Missisauga 组(图13-7c)及克里层(图13-3)的顶部变形地层(Weir Murphy,2004;Pe-Piper 和 Piper,2004)。大陆边缘沉降发生于斯科舍盆地枢纽带向海一侧(图13-5 和图13-7a),因此难以对基底构造进行识别分析。在早白垩世,枢纽带向陆一侧的大陆边缘均发生翘倾及间歇性剥蚀,如斯科舍盆地东部的地震测线所示,奥尔普斯地堑南部的下白垩统不断发生倾斜(图13-7a)。因此,下白垩统上超于上侏罗统的隐伏露头上(图13-7a),如图13-4 所示。早白垩世,沿走向往西,基岩在 Banquereau 台地周围暴露。此外,Missisauga 组中部与上部之间的 O—标记反射层剥蚀尖灭线位于上侏罗统剥蚀尖灭线的外侧。

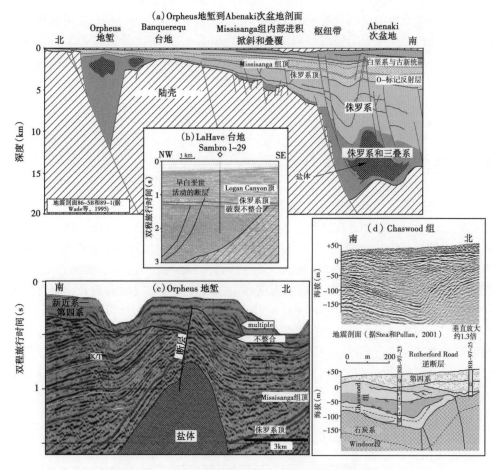

图 13-7 下白垩统同沉积基底构造变形的地震证据

(a) 东斯科舍陆架下白垩统的掀斜剥蚀（据 Wade 等，1995，重绘）；(b) LaHave 台地下白垩统中的正断层（据 MacLean 和 Wade，1993，修改）；(c) 在奥尔普斯地堑，受盐岩活动影响，Mississauga 组顶部不整合上超于侏罗系顶部不整合（据 Weir Murphy，2004，修改）；(d) 埃尔姆斯韦尔盆地 Chaswood 组中的同沉积褶皱（据 Piper，2005，修改）

13.4.3 早白垩世构造变形的区域证据

早白垩世构造变形的证据还广泛分布于加拿大东部其他地区及美国毗邻地区。在大浅滩，伊比利亚裂离形成的早白垩世拉张作用在贝里阿斯阶末期、欧特里夫阶及阿普特阶三个时期最为强烈（Tucholke 等，2007）。前两个拉张期与裂谷由南往北逐步打开有关；第三个时期的拉张作用与板块边缘次大陆地幔岩石圈的折返有关，在阿普特至阿尔布时期海底扩张开始时最为强烈。与奥尔普斯地堑同时期的区域不整合可见于大浅滩西南部（Pe-Piperand 和 Piper，2004），而同时期的构造变形可见于圣女贞德盆地的边界断层（Sinclair，1995）。在拉布拉多大陆边缘，裂陷作用发生于早白垩世，主转换构造走向北东（Balkwill 和 McMillan，1990）。

在新英格兰，存在早白垩世地壳块体差异升降及旋转的热年代学证据，例如康涅狄格州南部（Roden-Tice 和 Wintsch，2002）及缅因州的 Norumbega 断层（West 等，2008）（图 13-1）。早白垩世的反转断层在美国大西洋海岸平原广泛发育（Prowell，1988）。尽管没有经过明确的年代限定，通常认为芬迪湾地区早侏罗世之后的构造变形与 Chaswood 组揭示的早白垩世区域构造变形有关。芬迪湾的反转构造意味着这些构造变形与北北东向收缩有关，并诱发了东西向和北东—南西向断层的左旋走滑活动。(Baum 等,2008)。

关于区域构造变形的成因分析超出了本次研究的范围。这也许与早白垩世从格陵兰岛至波丘派恩浅滩的拉布拉多裂陷作用 (Balkwill 和 McMillan,1990) 或伊比利亚与大浅滩的裂离 (Tucholke 等,2007) 所形成的陆壳内构造体制有关。然而,有另一种观点认为,该构造变形在一定程度上可能与海洋的构造活动相关。原因在于地层变形的范围向南至少可延伸到康涅狄格州,而在与之对应的摩洛哥被动陆缘,存在与之同期的粗碎屑沉积物的输入 (Ghorbal 等,2008)。

13.5 斯科舍盆地盐构造

盐构造活动影响了斯科舍盆地枢纽带向海一侧的整个地区。原生的阿戈盐岩 (图13-3) 从盆地内侧地区大规模移动,在斯科舍斜坡 Sable 和 Abenaki 次盆地向海一侧形成了一系列盐底辟、盐墙及盐蓬构造 (图13-5)。中侏罗世开始,盆地东部的中上侏罗统三角洲沉积负载导致大部分盐岩在 Abenaki 次盆地移动 (图13-4) (Ings 和 Shimeld,2006)。盐岩变形作用持续至早白垩世,在 Sable 次盆地最为剧烈,在斜坡部位持续时间更长,一直到晚白垩世与古近—新近纪。晚侏罗世—早白垩世的盐构造活动使外陆架及上陆坡的上覆地层发生区域性伸展变形,并造成了在外陆架斜坡发育拆离铲式正断层和区域沉降 (Piper 等,2010)。由于正断层活动造成了可容纳空间增大,三角洲进积经常止步于大型断层上升盘,形成了一系列厚度大但范围窄的前三角洲砂岩序列沉积区,称之为"扩张区域"(图13-8)。与之类

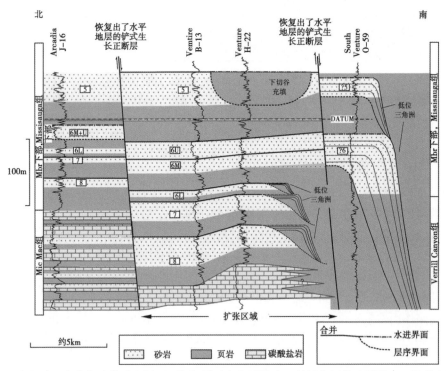

图13-8 文彻油田南北向连井对比剖面,恢复了两个主要铲式正断层之间的地层形态,注意文彻油田"扩张区域"巨厚的前三角洲砂岩,仅晚期砂岩(5与6层)越过南部生长断层向南进积

(据 Cummings 和 Arnott,2005,修改)

似的铲式断层控制三角洲进积的现代实例也可见,如爱琴海北部(Ferentions 等,1981;Piper 和 Perssoratis,1991)。

13.6 碎屑沉积物来源

13.6.1 简介

下白垩统三角洲中大量碎屑沉积物的出现缘于以下一个或几个因素:(1)携带沉积物的河流规模的增大;(2)物源区的构造抬升及去顶作用的增强;(3)气候变化带来的剥蚀作用的增速。碎屑沉积物与源岩匹配关系的分析能够揭示上述几种作用在提供物源方面的重要性。

下白垩统三角洲的潜在物源区为加拿大东南部的阿巴拉契亚造山带内以河流相砂岩充填为主的晚古生代继承性盆地,以及东加拿大地盾(Williams,1995)。早期关于 Missisauga 组碎屑矿物的地质测年工作从碎屑物的钾长石中通过 $^{40}Ar/^{39}Ar$ 测年获取了大约1000Ma 的年龄(Grist 等,1995),分析认为物源来自于格伦维尔省的加拿大地盾(图13-2)。该项工作也证实了 Wade 和 Maclean(1990)关于东加拿大地盾是重要物源区的推测。然而,有些白云母测年资料表明这些矿物来源于阿巴拉契亚造山带(Grist 等,1995)。

以下三种方法常被用来进行沉积物物源分析:碎屑矿物测年、特异性矿物分析及全岩地球化学分析。作为物源分析的指标,上述三种方法都有它的优势及不足,下面将逐一进行阐述。

13.6.2 碎屑矿物测年

矿物测年是碎屑沉积物物源分析最重要的方法。碎屑矿物的白云母、独居石、锆石都能进行测年,但是它们具有不同的物理及化学稳定性。许多稳定的锆石能够在老的沉积岩中重复旋回。独居石对化学风化作用较为敏感,而白云母对于物理磨蚀作用较敏感,因此它们主要出现在近源的初期旋回中。

单颗粒 $^{40}Ar/^{39}Ar$ 测年表明(Reynolds 等,2009),Sable 次盆地下白垩统砂岩中几乎所有的白云母年龄均位于420—240Ma 之间(图13-9和图13-10)。在 LaHave 台地,NaskapiN-30 井(图13-2)中白云母年龄位于372—350Ma 之间。在上述两个地区,石炭纪(360—300Ma)的测年资料取自于内陆架的基底变质沉积岩,在阿勒格尼可造山期该陆架经历了调整重置。泥盆纪(417—360Ma)测年资料取自于变质沉积岩及陆上阿卡迪亚造山带的花岗侵入岩。在斯科舍盆地,质量平衡计算需要发掘几十至几百米的早白垩世内陆架地层以提供白云母矿物(Reynolds 等,2009)。Chaswood 组的白云母似乎需要从石炭系砂岩中重新选取分析(Reynolds 等,2010)。白云母容易在搬运过程中被磨蚀,因此微量的年龄较大的矿物可能源于更内侧阿巴拉契亚造山带的搬运过程中的磨蚀作用。

前人已进行过碎屑独居石的电子探针测年研究(Pe-Piper 和 MacKay,2006;Triantaphllidis 等 2010)。新斯科舍省 Chaswood 组中的颗粒主要为奥陶—志留纪,反映了阿巴拉契亚山脉更内侧地区早古生代的深成作用及塔康造山运动。在新元古代末期的深成作用影响下,泥盆系及来自于阿瓦隆地区的颗粒数目较少,而前寒武纪的颗粒缺失。

在 Sable 次盆地,大部分独居石颗粒为泥盆纪,但塔康造山期、新元古代末期、中元古代及古元古代的颗粒各占独居石总量的10%左右(图13-9和图13-10)。比较而言,东部的 Abenaki 次盆地,中元古代至古元古代的独居石比例更高(约60%),揭示了加拿大地盾

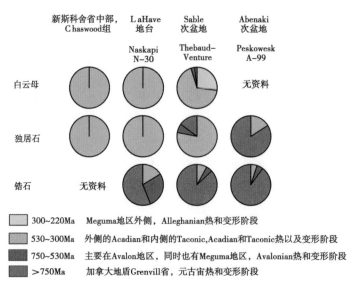

图 13-9 斯科舍盆地不同地区四个地质时期白云母、独居石及锆石的丰度对比

地质时期的划分根据热与变形阶段大致对应于阿勒格尼造山期（300—220Ma）、阿巴拉契亚阿卡迪亚与塔康造山期（530—300Ma）、阿瓦隆地区新元古代末期（750—530Ma）、加拿大地盾元古宙时期（>750Ma）

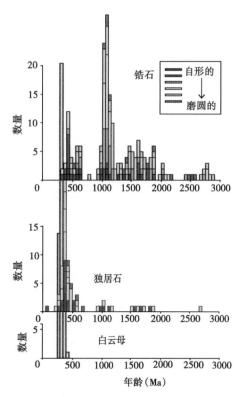

虽然独居石的耐物理磨蚀性相对较强，但其在酸性条件下容易发生化学分解。独居石可细分为全自形、半自形、圆形及不规则形。目前还没有关于形态变化与年代之间的系统研究，阿巴拉契亚山脉外侧中古生代短距离搬运的自形颗粒除外。可将这种变化解释为大部分独居石可能是初期旋回形成的。

目前已完成 10 个样品单颗粒锆石激光烧蚀测年工作，这些样品涵盖了全盆地地理与地层范围，这些资料还没有发表，但是 Sable 次盆地的数据如图 13-10 所示。在大多数样品中，前寒武纪的锆石占主导，其年龄峰值在 1700—1000Ma 之间，代表了阿巴拉契亚区源于劳伦系的岩石（Waldron 等，2008）。小部分样品揭示了 2000—600Ma 之间的一个小规模年龄峰值，代表了阿巴拉契亚外侧冈瓦那地层的岩石（Krogh 和 Keppie，1990）。这些峰值样品中锆石比例最高的唯一样品来自于 LaHave 台地的 Naskapi 钻井。所以在 500—300Ma 的锆石丰度相对较低的样品均代表了阿巴拉契亚的结晶基底，虽然这样丰度值之间也有差异。

同一样品或单元中，不同时期独居石及锆石

图 13-10 Sable 次盆地钻井中白云母、
独居石及锆石的单颗粒地质年代

对于独居石及锆石，颜色与矿物结构对应关系为：
红色=自形，蓝色=破碎或磨圆的

比例的比较能揭示多旋回再沉积物的重要性。例如，Sable 次盆地单颗粒的复合分析表明（图 13-10），78%的独居石为古生代，5%为新元古代，18%为地盾形成时期；与此形成对比的是，83%的锆石为地盾形成期，仅有 11%来自于阿巴拉契亚山脉的锆石属于古生代。这些变化有多个原因，不同源岩中独居石及锆石丰度的变化，以及颗粒的水动力分选作用。然而，在有些位置，锆石及独居石的年龄几乎完全匹配，例如 Peskowesk 井的洛根坎宁组。该井中阿巴拉契亚及地盾物源区的独居石、锆石丰度是相等的（图 13-9）。因此，在 Sable 次盆地，大部分独居石是初期旋回沉积，而只有小部分锆石是初期旋回沉积。大部分锆石破碎并磨圆（图 13-10 蓝色部分）并随早古生代或前寒武纪沉积物再次沉积。而早古生代或前寒武纪沉积物中包含高比例来自地盾源区的锆石。

LaHave 台地 Naskapi N-30 井 Mississauga 组上部地层中的一个样品揭示了单独利用锆石测年来进行物源分析的局限性。所有的独居石测年年龄在 431—305Ma，其丰度峰值在晚泥盆世至早石炭世。白云母也揭示了非常相似的年龄分布。然而，只有 16%的锆石揭示出了阿巴拉契亚物源区初旋回沉积的中古生代年龄特征（图 13-9）。剩余锆石年龄为前寒武纪，其丰度峰值为 2200—600Ma，与梅古马地区变质沉积岩的锆石相似（Krogh 和 Keppie，1990），因此解释为多旋回沉积。

综合看来，碎屑矿物的年代学分析表明，阿巴帕拉契亚造山带是斯科舍盆地砂岩的主要物源区。在 Sable 及 Abenaki 次盆地，全自形锆石及独居石（图 13-10 红色部分）表明沉积物少量直接来自于加拿大地盾。这点与西纽芬兰岛内侧暴露的锆石及独居石类似。即使存在大型的古圣劳伦斯河，它也并没有从加拿大地盾携带大量的砂岩。

13.6.3　碎屑矿物化学指纹图谱

碎屑矿物，包括许多重矿物，在沉积物物源分析中应用广泛。矿物丰度的变化是多种作用的结果，包括风化及搬运过程中的破坏及水动力分选（Morton 和 Hallsworth，1990）。上述各种作用能够通过比较相同矿物的不同化学成分的相对丰度变化（化学指纹图谱）进行归纳分类。化学指纹图谱是 Morton 和 Hallsworth（1994）创建的一种旨在分析重矿物变化的技术。在斯科舍盆地已经发现多种特征矿物，但是阿巴拉契亚基岩矿物成分数据库尚未建立，无法进行矿物成分对比。由于阿巴拉契亚地区亚平行于大陆边缘，难以从现代河流中建立类似于 Morton 等（2004）建立的用于确定北海周围石榴石来源的有用数据库。

在斯科舍盆地及 Chaswood 组中已经识别出石榴石、尖晶石、电气石、白云母、黑云母及长石等多种矿物组分（Pe-Piper 等，2009）。这些矿物组分可用于精细的物源分析。来自于 Abenaki 次盆地和 LaHave 台地的长石主要为钾长石，而斜长石更多来自于 Sable 次盆地。根据 Henry 和 Guidotti（1985）的标准，大量来自塞布尔次盆地的电气石主要来自变质沉积岩源区，而新斯科舍省 Chaswood 组的电气石中约 30%来自花岗岩母岩区。在 Abenaki 次盆地，电气石比较少见。

大部分特征矿物主要来自梅古马地区，见于 LaHave 台地 Naskapi N-30 井（图 13-2）。其中 35%的白云母富含钠，富含钠正是梅古马地区阿勒格尼剪切带白云母的特征（Reynolds 等，2009）。同时，石榴石与梅古马地区发现的石榴石也非常类似，Sable 次盆地两口井中的透闪石可能源自于梅古马地区（Pe-Piper 等，2009）。

碎屑矿物分析表明 LaHave 台地、Sable 次盆地及 Abenaki 次盆地接受了来自不同物源区的沉积物。这一点已被锆石及独居石测年资料证实。石榴石及电气石等特定矿物的化学分析能用于物源区解释。例如，尽管角闪石的出现意味着 Sable 次盆地西部的沉积物部分来自于梅古马地区，但大量不同类型石榴石的出现表明来自梅古马地区的沉积物占总沉积物含量的 20%以下。由于多旋回物源区矿物辨别困难及缺乏物源区数据库等原因，矿物化学指纹图谱的应用受到限制。

13.6.4 全岩地球化学

利用矿物测年及化学指纹图谱技术进行沉积物物源追踪的局限性在于这两种技术只能利用特征矿物进行分析。如 Pe-Piper 等（2008）针对斯科舍盆地所述，全岩地球化学能够分析碎屑沉积物的整体供应特征。该方法第一步首先要筛选出在成岩过程中经历了较强化学作用的岩石，Ca、Mg、Fe、P 的含量异常能够识别出含大量碳酸盐岩胶结物的岩石。与正常沉积物相比，斯科舍盆地下白垩统陆源沉积物 Ti 及 Fe 的含量异常偏高，一定程度反映出钛铁矿的富集（Pe-Piper 等，2005c）。通常这些沉积物的 Ca 含量较低，可能是因为阿巴拉契亚物源区缺少斜长石及镁铁质火成岩。

利用钛（常见于钛铁矿中）、锆及铬（常见于锆石与铬铁矿中）元素（图 13-11）的共变关系能够分析重矿物的水动力分选作用及初期旋回和多旋回矿物的比例关系。锆石、铬铁矿主要来源于花岗岩与蛇绿岩，因此其相关变量反映不了它们共同的来源，而反映了多旋回沉积物中超稳定矿物富集特征及沉积过程中的密度分异。相比之下，钛铁矿主要为初期旋回沉积的矿物（Pe-Piper 等，2005c），其与低富集程度锆石的相关变量可能是大小、密度等沉积分选作用的结果。但是锆石的富集并非与钛铁矿的富集一致。

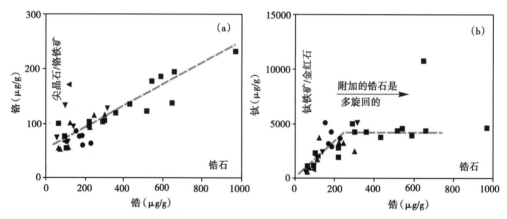

图 13-11　斯科舍盆地砂岩中钛、铬、锆元素的含量变化及它们在多旋回沉积分选中的指示意义（据 Pe-Piper 等，2008）

主成分的多变量统计分析表明，砂泥岩中主量元素的变化肯定了 Abenaki、Sable 次盆地及 LaHave 台地接受了不同成分的沉积物（Pe-Piper 等，2008）。仅 Rb、Sr、U、Th、Nb 及 Ti 等少量微量元素用于不同物源分析，其中 Nb 及 Ti 仅用于砂岩的物源分析。所有的微量元素在阿巴拉契亚地区的花岗岩中都很富集。

陆上的 Chaswood 组中，全岩地球化学的地层变化揭示了物源区间歇性构造活动带来的影响（Piper 等，2008）。由于 K、P、Sr、U 元素在 3 个局部不整合面的富集，成岩过程一定程度上掩盖了碎屑特征。在近海砂泥岩中，Ti（钛铁矿及相关产物中）、Zr（锆石中）、

Th 及 Y 元素通常受控于岩石中重矿物的富集程度。钛铁矿主要是初期旋回的重矿物，而大部分锆石源于多旋回沉积，因此，Ti/Zr 比能够用于分析来自结晶基底初期旋回沉积物的比例。泥岩中 Cr、Sr 元素的高度富集及 Ni/Co 的高比值似乎与铁镁质结晶基底的物源供给有关。地层中地球化学特征的变化意味着 Chaswood 组接受了三期沉积物供给，每期均沉积于一个区域不整合面之上（Piper 等，2008）。每一个循环都反映了以走滑断层为边界的地垒的隆起。这种隆起作用造成了石炭系砂岩剥蚀，随后是结晶基底的快速剥蚀，最终是活动断层附近的地表土层严重风化并大量供给。

在成岩作用过程中全岩地球化学的应用在主量及微量元素上受到限制，如富集、富含微量元素的重矿物的水动力分选等。然而，该分类能够分析初期旋回及多旋回的重矿物的相对富集特征（图13-11）。如果成岩作用对元素富集的影响较小，那么全岩地球化学就是分析沉积物供给过程中地层变化的有效方法。然而，Pe-Piper 等（2008）指出，大部分已发表的基于全岩地球化学方法的物源辨别图并不能区分斯科舍盆地的物源。

13.7　下白垩统沉积相分析

Cummings 和 Arnott（2005）及 Cummings 等（2006a，2006b）已对 Missisauga 组的沉积环境进行了分析。上述笔者指出大部分三角洲为受河流、河口沉积影响的潮汐三角洲沉积或以风暴作用为主的三角洲前缘沉积。低位期形成的下切谷（图13-8）（Drummond，1992；Cummings 和 Arnott，2005）主要充填了河口湾沉积（Cummings 等，2006b）。其预测了一个由大型古圣劳伦斯河构建的复杂低斜率陆架边缘三角洲，也就是一个典型的被动陆缘三角洲建造。

诸多证据链表明，上述沉积相解释需要进行修正。盆地东部的三角洲并未进积至大陆架边缘，而是停留在数百米深的外陆架斜坡（Piper 等，2010）。西部较远处的一些三角洲进积至古斜坡顶部，也就是陆架边缘的位置（Deputuck 等，2009）。岩石学及全岩地球化学证据表明，盆地长轴方向的物源供给存在变化，意味着阿巴拉契亚山脉的主物源区存在多条河流。白云母测年及 O-标记反射层的剥蚀边缘表明，在斯科舍盆地下白垩统沉积时期沉积物源供给快速，内陆架大部分地区不是广泛的海岸平原。新斯科舍省中部的 Chaswood 组中没有发现海侵证据，而是保留了辫状河砾质沉积（Gobeil 等，2006）。斯科舍盆地最靠近陆地部分的地震资料表明，在奥尔普斯地堑，多条浅水水道为辫状河沉积（Weir Murphy，2004）。

Missisauga 组及 Logan Canyon 组内厚层三角洲前缘砂岩的单层厚度在数分米至数米，明显沉积于单个沉积事件。Cummings 和 Arnott（2005）认为许多这种三角洲前缘砂岩受风暴作用的强烈影响，因此具有滨面沉积结构。Gould 等（2005）根据近期 MacEachern 等（2005）、Pattison 等（2007）、Myrow 等（2008）的工作，将这些砂岩重新解释为三角洲前缘浊积岩。这些浊积砂岩层形成于泛滥河流的高密度流，与高位期的河口湾潮汐沉积互层。从阿巴拉契亚山脉至斯科舍盆地的河流对于季风气候下的洪水泛滥很敏感（Herrle 等，2003）。这类高密度流沉积通常与风暴浪相关，因此造成了沉积物中既存在不定向悬浮沉积的证据，也存在受风暴浪影响的证据（Lamb 等，2008）。在前三角洲相带，沉积环境对于了解砂岩成岩作用非常重要。高沉积速率及孔隙水组分形成了独特的海底成岩环境，使埋藏成岩过程中孔隙更容易保存（Gould 等，2010）。

13.8 构造与沉积的相互影响

研究表明，在北大西洋中央早中生代裂谷和破裂不整合形成之后很长一段时间内，构造与沉积的相互作用非常明显。早白垩世，阿巴拉契亚山脉的构造变形可能就是斯科舍盆地物源供给激增3~4倍的主要原因。岩石学及测年资料表明，阿巴拉契亚地区是沉积物的主要来源，另外一个重要物源供给甚至来自于现今的斯科舍内陆架区。岩石学与地球化学资料同样揭示物源主要由多条河流携带供给，尽管位于布雷顿角岛及纽芬兰岛西侧之间的卡伯特海峡可能是曾经的物源主入口点（图13-12）。

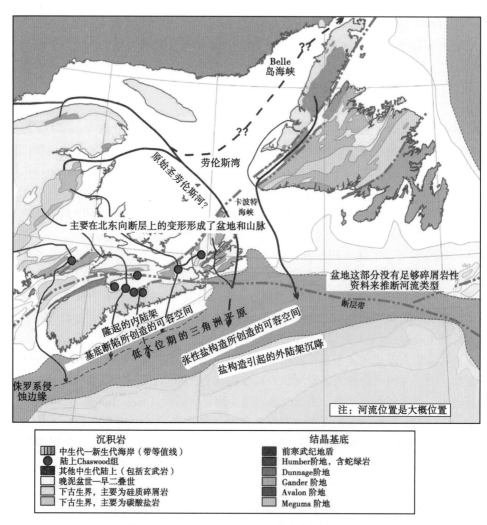

图13-12 斯科舍盆地早白垩世古地理特征（底图与图13-2相同）
揭示了主要构造影响及提供砂岩供给的河流的推测位置，成果基于碎屑岩石学及全岩地球化学资料

纽芬兰岛及大浅滩的抬升在早白垩世早期伊比利亚半岛的裂离之前，导致了晚侏罗世纽芬兰岛的大量物源供给到斯科舍盆地东部，造成了三角洲的进积作用（图13-4）。该期抬升作用的直接证据在陆上缺失，但可见于岩石学资料。

早白垩世的构造活动致使阿巴拉契亚往南直至康涅狄格州的广大地区形成了地块旋转。大部分构造变形是由于晚古生代走滑断层的复活作用引起的，在断层不同部位抬升与沉降程度也存在差异。同时期的变形主要发生于加拿大陆架边缘东部，往北直至拉布拉多裂陷盆地。Chaswood 组在瓦兰今阶开始沉积，盆地在早白垩世沿着北东向断裂发生构造变形。构造抬升及相关断裂作用延伸至内陆架，三期主要的区域不整合记录了在陆上 Chaswood 组及在斯科舍陆架内缘至少存在三期构造变形。

碎屑岩石学、测年资料及地球化学资料能够揭示早白垩世基底接受抬升剥蚀的一些细节。Sable 次盆地的大部分碎屑供给为多旋回沉积物源，可能大部分始于石炭纪。由于阿巴拉契亚物源的主导地位，年龄为格林维尔时期（约 1000Ma）的独居石及锆石可能并非来自于加拿大地盾本身，而是来源于纽芬兰岛西部地盾的内围层（图 13-12）。这些内围层作为沿着与 Chaswood 组盆地平行的北东向断层发育的垒堑系统的一部分发生抬升剥蚀。卡伯特海峡、贝勒岛海峡都可能曾经是古圣劳伦斯河的出口（图 13-12），但与纽芬兰西部北东向断层相关的构造活动降低了该河流进入圣劳伦斯湾的坡度，使得大部分负载得以沉积下来。

斯科舍盆地的沉积体系存在以 Chaswood 组辫状河粗砂岩及砾岩沉积为代表高坡度河流物源供给。季风气候引发的阵发性洪水泛滥形成了三角洲前缘的高密度流浊积岩。低坡度地形利于海岸平原及三角洲的进积，尤其是在海平面高位时期。在河口湾，高位期潮汐作用主导的沉积作用非常重要。三角洲前缘浊积岩是形成天然气厚砂岩储层的核心要素，由于海底成岩环境中绿泥石加大边的形成，该类砂岩的孔隙在埋藏成岩期得以保存。

早白垩世构造活动形成的新增可容空间较小。Chaswood 组厚度不到 200m，斯科舍盆地枢纽带内侧的构造变形主要导致了地层剥蚀及沉积物过路不沉积作用。下白垩统三角洲前积的可容空间源于盐构造活动。斯科舍盆地东部的侏罗系三角洲沉积负载引发了盐构造活动，而下白垩统三角洲的快速进积加速了盐岩构造变形。盐构造活动导致了外陆架斜坡的快速沉降，厚层三角洲在生长断层下降盘堆积，并向前进积至外陆架斜坡深水区（图 13-8）。主生长断层的沉降阻止了三角洲前积向前推进，形成一系列以断裂为边界范围狭窄的厚层砂岩连续沉积带，即"扩张区域"。

13.9　结论

本文分析了曾被认为是另一种被动陆缘三角洲的沉积学、地震反射结构及岩石学特征。它可能代表了其他大量砂岩快速输入的被动陆缘沉积（Poag 和 Sevon，1989；Ghorbal 等，2008）。地震剖面及地层分析表明，早白垩世陆上构造变形发生于盆地内砂岩物源供给 3~4 倍激增时期。一系列沉积岩石学研究表明，砂岩的主要物源区是阿巴拉契亚地区，其载体为 3 条以上的河流。由独居石、白云母及锆石的测年资料对比结果可知，大部分锆石尤其是 Sable 次盆地的锆石是多旋回沉积产物。山区河流系统易于形成阵发性洪水泛滥，沉积三角洲前缘高密度流浊积岩。大陆基底的构造活动形成的可容空间较小，大部分可容空间源于三角洲负载造成的盐岩活动。本文揭示了被动大陆边缘基底构造活动对沉积作用的复杂影响。

致谢（略）

参 考 文 献

Balkwill, H. R., and McMillan, N. J. (1990) Mesozoic-Cenozoic geology of the Labrador Shelf, in Keen, M. J., and Williams, G. L., eds., Geology of the continental margin off eastern Canada. Geological Survey of Canada, Geology of Canada, (2), 295-324.

Baum, M. S., Withjack, M. O., and Schlische, R. W. (2008) The ins and outs of buttress folds: Examples from the inverted Fundy Rift Basin, Nova Scotia and New Brunswick, Canada. Central Atlantic Conjugate Margins Conference, Halifax 2008, 53-61.

Beck, M. E., and Housen, B. A. (2003) Absolute velocity of North America during the Mesozoic from paleomagnetic data. Tectonophysics, 377, 33-54.

Cummings, D. I., and Arnott, R. W. C. (2005) Shelf margin deltas: a new (but old) play type offshore Nova Scotia. Bull. Can. Petrol. Geol., 53, 211-236.

Cummings, D. I., Hart, B. S., and Arnott, R. W. C. (2006a) Sedimentology and stratigraphy of a thick, areally extensive fluvial-marine transition, Missisauga Formation, offshore Nova Scotia, and its correlation with shelf margin and slope strata. Bull. Can. Petrol. Geol., 54, 152-174.

Cummings, D. I., Arnott, R. C. W., andHart, B. S. (2006b) Tidal signatures in a shelf-margin delta. Geology, 34, 249-252.

Deptuck, M. E., Kendall, K., and Smith, B. (2009) Complex deepwater fold-belts in the SW Sable Subbasin, offshore Nova Scotia. 2009 CSPG CSEG CWLS Convention, 4 p.

Drummond, K. J. (1992) Geology of Venture, a geopressured gas field, offshore Nova Scotia, in Halbouty, M. T., ed., Giant oil and gas fields of the decade 1978-1988. AAPG Memoir, 54, 55-71.

Falcon-Lang, H. J., Fensome, R. A., Gibling, M. R., Malcolm, J., Fletcher, K. R., and Holleman, M. (2007) Karst - related outliers of the Cretaceous Chaswood Formation of Maritime Canada. Can. J. Earth Sci., 44, 619-642.

Ferentinos, G., Brooks, M., and Collins, M. B. (1981) Gravityinduced deformation on the northern flank and floor of the Sporadhes Basin of the North Aegean Sea Trough. Mar. Geol., 44, 289-302.

Ghorbal, B., Bertotti, G., Foeken, J., and Andriessen, P. (2008) Unexpected Jurassic to Neogene vertical movements in "stable" parts of NW Africa revealed by low temperature geochronology. Terra Nova, 20, 355-363.

Gobeil, J. -P., Pe-Piper, G., and Piper, D. J. W. (2006) The Early Cretaceous ChaswoodFormation in theWest IndianRoad pit, central Nova Scotia. Can. J. Earth Sci., 43, 391-403.

Gould, K., Pe-Piper, G., and Piper, D. J. W. (2010) Relationship of diagenetic chlorite rims to depositional facies in Lower Cretaceous reservoir sandstones of the Scotian Basin. Sedimentology, 57, 587-610.

Gradstein, F. M., Ogg, J. G., Smith, A. G., Bleeker, W., and Lourens, L. J. (2004) A new Geologic Time Scale with special reference to the Precambrian and Neogene. Episodes, 27, 83-100.

Grist, A. M., Reynolds, P. H., Zentilli, M., and Beaumont, C. (1992) The Scotian Basin offshore Nova Scotia: thermal history and provenance of sandstones from apatite fission track and $^{40}Ar=^{39}Ar$ data. Can. J. Earth Sci., 29, 909-924.

Henry, D. J., and Guidotti, C. V. (1985) Tourmaline as a petrogenetic indicator mineral: an example from the staurolite-grade metapelites of NW Maine. Am. Mineral., 70, 1-15.

Herrle, J. O., Pross, J., Friedrich, O., Kössler, P., and Hemleben, C. (2003) Forcing mechanisms for mid-Cretaceous black shale formation: evidence from the Upper Aptian and Lower Albian of the Vocontian Basin (SE France). Palaeogeogr. Palaeoclimatol. Palaeoecol., 190, 399-426.

Hundert, T., Piper, D. J. W., and Pe-Piper, G. (2006) Genetic model and exploration guidelines for kaolin beneath unconformities in the Lower Cretaceous fluvial Chaswood Formation, Nova Scotia. Expl. Mining Geol., 15, 9-26.

Ings, S. J., and Shimeld, J. W. (2006) A new conceptual model for the structural evolution of a regional salt detachment on the northeast Scotian margin, offshore eastern Canada. AAPG Bulletin, 90, 1407-1423.

Krogh, T. E., and Keppie, J. D. (1990) Age of detrital zircon and titanite in the Meguma Group southern Nova Scotia, Canada: Clues to the origin of the Meguma Terrane. Tectonophysics, 177, 307-323.

Lamb, M. P., Myrow, P. M., Lukens, C., Houck, K., and Strauss, J. (2008) Deposits from wave-influenced turbidity currents: Pennsylvanian Minturn Formation, Colorado, U.S.A. J. Sedim. Res., 78, 480-498.

MacEachern, J. A., Bann, K. L., Bhattacharya, J. K., and Howell, C. D. (2005) Ichnology of deltas: organism responses to the dynamic interplay of rivers, waves, storms and tides, in Bhattacharya, J. K., and Giosan, L., eds., River deltas: concepts, models and examples. SEPM Special Publication 83. Tulsa, OK, Society of Economic Paleontologists and Mineralogists, 49-85.

MacLean, B. C., and Wade, J. A. (1992) Petroleum geology of the continental margin south of the islands of St Pierre and Miquelon, offshore eastern Canada. Bull. Can. Petrol. Geol., 40, 222-253.

MacLean, B. C., and Wade, J. A. (1993). Seismic markers and stratigraphic picks in Scotian Basin wells. Atlantic Geoscience Centre, Geological Survey of Canada.

Morton, A. C., and Hallsworth, C. R. (1994) Identifying provenance- specific features of detrital heavy mineral assemblages in sandstones. Sedim. Geol., 90, 241-256.

Morton, A. C., and Hallsworth, C. R. (1999) Processes controlling the composition of heavy mineral assemblages in sandstones. Sedim. Geol., 124, 3-29.

Morton, A., Hallsworth, C., and Chalton, B. (2004) Garnet compositions in Scottish and Norwegian basement terrains: a framework for interpretation of North Sea sandstone provenance. Mar. Petrol. Geol., 21, 393-410.

Myrow, P. M., Lukens, C., Lamb, M. P., Houck, K., and Strauss, J. (2008) Dynamics of a transgressive prodeltaic system: implications for geography and climate within a Pennsylvanian intracratonic basin, Colorado, U. S. A. J. Sedim. Res., 78, 512-528.

Olsen, P. E., and Et Touhami, M. (2008). Tropical to subtropical syntectonic sedimentation in the Permian to Jurassic Fundy Rift Basin, Atlantic Canada, in relation to the Moroccan conjugate margin. Fieldtrip guidebook, Atlantic Conjugate Margins conference, Halifax, NS, August, www. ldeo. columbia. edu/_polsen/nbcp/olsen et-touhami 08. pdf.

Pattison, S. A. J., Ainsworth, R. B., and Hoffman, T. A. (2007) Evidence of across-shelf transport of fine-grained sediments: turbidite filled shelf channels in the Campanian Aberdeen Member, Book Cliffs, Utah, U. S. A. Sedimentology, 54, 1033-1064.

Pe-Piper, G., and Mackay, R. M. (2006) Provenance of Lower Cretaceous sandstones onshore and offshore Nova Scotia from electron microprobe geochronology and chemical variation of detrital monazite. Bull. Can. Petrol. Geol., 54, 366-379.

Pe-Piper, G., and Piper, D. J. W. (2004) The effects of strikeslip motion along the Cobequid-Chedabucto-SW Grand Banks fault system on the Cretaceous-Tertiary evolution of Atlantic Canada. Can. J. Earth Sci., 41, 799-808.

Pe-Piper, G., Dolansky, L., and Piper, D. J. W. (2005a) Sedimentary environment and diagenesis of the mid-Cretaceous Chaswood Formation, Elmsvale Basin, Nova Scotia. Sedim. Geol., 178, 75-97.

Pe-Piper, G., Piper, D. J. W., Hundert, T., and Stea, R. R. (2005b) Outliers of Lower Cretaceous Chaswood Formation in northern Nova Scotia: results of scientific drilling and studies of sedimentology and sedimentary petrography. Geological Survey of Canada Open File 4845, 305 p.

Pe-Piper, G., Piper, D. J. W., and Dolansky, L. M. (2005c) Alteration of ilmenite in the Cretaceous sands of Nova Scotia, southeastern Canada. Clays Clay Mins., 53, 490-510.

Pe-Piper, G., Tsikouras, B., Piper, D. J. W., and Triantaphyllidis, S. (2009) Chemical fingerprinting of minerals in Lower Cretaceous sandstones, Scotian Basin. Geological Survey of Canada Open File 6288, 151 p.

Pe-Piper, G., Triantafyllidis, S., and Piper, D. J. W. (2008) Geochemical identification of clastic sediment provenance from known sources of similar geology: the Cretaceous Scotian Basin, Canada. J. Sedim. Res., 78, 595-607.

Piper, D. J. W., and Perissoratis, C. (1991). Late Quaternary sedimentation on the North Aegean Continental Margin, Greece. AAPG Bulletin, 75, 46-61.

Piper, D. J. W., Pe-Piper, G., and Ingram, S. I. (2004) Early Cretaceous sediment failure in the southwestern Sable sub-basin, offshore Nova Scotia. AAPG Bulletin, 88, 991-1006.

Piper, D. J. W., Pe-Piper, G., and Douglas, E. V. (2005). Tectonic deformation and its sedimentary consequences during deposition of the Lower Cretaceous Chaswood Formation, Elmsvale basin, Nova Scotia. Bull. Can. Petrol. Geol. 53, 189-199.

Piper, D. J. W., Pe-Piper, G., Hundert, T., and Venugopal, D. K. (2007) The Lower Cretaceous Chaswood Formation in southern New Brunswick: provenance and tectonics. Can. J. Earth Sci., 44, 665-677.

Piper, D. J. W., Pe-Piper, G., and Ledger-Piercey, S. (2008) Geochemistry of the Lower

Cretaceous Chaswood Formation, Nova Scotia, Canada: provenance and diagenesis. Can. J. Earth Sci., 45, 1083–1094.

Piper, D. J. W., Noftall, R., and Pe-Piper, G. (2010) Allochthonous prodeltaic sediment facies in the Lower Cretaceous at the Tantallon M-41 well: implications for the deepwater Scotian basin. AAPG Bull., 94, 87–104.

Poag, C. W., and Sevon, W. D. (1989) Arecord of Appalachian denudation in postrift Mesozoic and Cenozoic sedimentary deposits of the U. S. Middle Atlantic continental margin. Geomorphology, 2, 119–157.

Prowell, D. C. (1988) Cretaceous and Cenozoic tectonism on the Atlantic coastal margin, in: The Atlantic Continental Margin, ed. Sheridan, R. E., and Grow., J. A., Geological Society of America, The Geology of North America, I-2, 557–564.

Rankin, S., and MacRae, R. A. (2008). Salt-related growth fault history and structural inversion in the Penobscot area, western Abenaki Subbasin, offshore Nova Scotia (abstract). Atl. Geol., 44, 35.

Reynolds, P. H., Pe-Piper, G., Piper, D. J. W., and Grist, A. M. (2009) Single-grain detrital muscovite ages from Lower Cretaceous sandstones, Scotian basin, and their implications for provenance: Bull. Can. Petrol. Geol., 57, 25–42.

Reynolds, P. H., Pe-Piper, G., and Piper, D. J. W. (2010) Sediment sources and dispersion as revealed by single-grain $^{40}Ar=^{39}Ar$ ages of detrital muscovite from Carboniferous and Cretaceous rocks in mainland Nova Scotia. Can. J. Earth Sci., 47, 957–970.

Roden-Tice, M. K., and Wintsch, R. P. (2002) Early Cretaceous normal faulting in southern New England; evidence from apatite and zircon fission-track ages. J. Geol., 110, 159–178.

Schettino, M., and Turco, A. (2009) Breakup of Pangaea and plate kinematics of the central Atlantic and Atlas regions. Geophys. J. Intl., 178, 1078–1097.

Sinclair, I. K. (1995) Transpressional inversion due to episodic rotation of extensional stresses in Jeanne d'Arc basin, offshore Newfoundland. In: Basin inversion; Buchanan, J. G., and Buchanan, P. G., (eds.), Geol. Soc. London Spec. Publ., 88, 249–271.

Stea, R., and Pullan, S. (2001) Hidden Cretaceous basins in Nova Scotia. Can. J. Earth Sci., 38, 1335–1354.

Triantaphyllidis, S., Pe-Piper, G., MacKay, R., Piper, D. J. W., and Strathdee, G. (2010). Monazite geochronology from Lower Cretaceous sandstones of the Scotian Basin. Geological Survey of Canada Open File 6732, 450 p.

Tucholke, B. E., Sawyer, D. S., and Sibuet, J. -C. (2007) Breakup of the Newfoundland-Iberia rift. Geol. Soc. London Spec. Publ., 282, 9–46.

Valdes, P. J., Sellwood, B. W., and Price, G. D. (1996) Evaluating concepts of Cretaceous equability. Palaeoclimates, 2, 139–158.

Wade, J. A., and MacLean, B. C. (1990) Aspects of the geology of the Scotian Basin from recent seismic and well data. Chapter 5, in Geology of the continentalmargin off eastern Canada, ed. Keen, M. J., and Williams, G. L. Geological Survey of Canada, Geology of Canada, no. 2, 190–238.

Wade, J. A., MacLean, B. C., and Williams, G. L. (1995) Mesozoic and Cenozoic stratigraphy, eastern Scotian Shelf: new interpretations. Can. J. Earth Sci., 32, 1462-1473.

Waldron, J. W. F., Floyd, J. D., Simonetti, A., and Heaman, L. M. (2008) Ancient Laurentian detrital zircon in the closing Iapetus Ocean, Southern Uplands terrane, Scotland. Geology, 36, 527-530.

Weir Murphy, S. L. (2004) Cretaceous rocks of the Orpheus graben, offshore Nova Scotia. MSc thesis, Saint Mary's University.

West, D. P., Roden-Tice, M. K., Potter, J. K., and Barnard, N. Q. (2008) Assessing the role of orogen-parallel faulting in post-orogenic exhumation: low-temperature thermochronology across the Norumbega Fault System, Maine. Can. J. Earth Sci., 45, 287-301.

Williams, G. L., Ascoli, P., Barss, M. S., Bujak, J. P., Davies, E. H., Fensome, R. A., and Williamson, M. A. (1990) Biostratigraphy and related studies. Chapter 3, in Geology of the continental margin off eastern Canada, ed. Keen, M. J., and Williams, G. L. Geological Survey of Canada, Geology of Canada, no. 2, 87-137.

Williams, H., ed. (1995) Geology of the Appalachian-Caledonian Orogen in Canada and Greenland. Geological Survey of Canada, The Geology of Canada, 6, 1-944.

Williams, H., and Grant, A. C. (1998) Tectonic assemblages, Atlantic region, Canada. [1:3m map]. Geological Survey of Canada Open File 3657, 1 sheet.

Withjack, M. O., Schlische, R. W., and Baum, M. S. (2009) Extensional development of the Fundy rift basin, southeastern Canada. Geol. J., 44, 631-651.

(曾清波 译,张功成 祁鹏 校)

第四部分 聚敛边缘

第14章 板块边缘的转化及其沉积响应

KATHLEEN M. MARSAGLIA

(Department of Geological Sciences, California State
University Northridge, Northridge, USA)

摘　要：会聚型边界从开始形成到最终消亡，一直是地球动力学作用十分丰富的区域。人们对板块俯冲初始作用的过程知之甚少，学术界一直存在着自发机制和诱发机制的争论。地球动力学模型能够帮助预测在板块俯冲活动中的沉积—火山活动的地质记录，但是在世界范围内基本上很少有地质实例来验证这些地球动力学模型，尤其缺失前新生代的实例。在诱发型俯冲机制的地区，沉积作用应包括快速隆升的证据（不整合），以及之后的快速沉降的证据（向上变细变深的正韵律沉积序列），最后是岛弧体系的建造。在自发型俯冲机制的地区，古弧前地区表现为沉降和火山岛弧形成之前的早期岩浆活动。俯冲活动一旦开始，由于板块三联点的形成和迁移，其他构造活动就有可能会叠加到弧前建造的过程中。一些经过充分研究的现代—中生代的地质实例指示对不同类型俯冲的地质响应的差异主要取决于板块结构。尽管处于海沟—洋脊—海沟构成的三联点体系中的扩张洋脊与海沟之间的相互作用会导致异常的近海沟岩浆作用、弧前建造的热作用叠加和俯冲消减作用，但是在海沟—海沟—海沟构成的三联点体系控制的纯俯冲过程中，主要的地质响应与岛弧增生作用有关。在俯冲作用被板块转换运动取代的三联点体系（如海沟—断层—海沟）中，板块边缘的转化过程表现为逐步发展的隆起和弧前地区的构造变形，并伴随着岛弧岩浆活动的终止。会聚边界的消亡可以通过简单的废弃和沉降作用来完成，或者转化为转换板块边界，也可以在扩张洋脊的影响下表现为弧前地区的构造变形、隆升和热作用/火山活动的叠加。

关键词：俯冲　自发型俯冲　诱发型俯冲　三联点　沉积作用　蛇绿岩仰冲作用

14.1 引言

长期以来关于盆地及其构造演化的总结（Busby 和 Ingersoll，1995）一直都以静态研究为主，注重单一构造环境（Marsaglia 等，1995）。然而，板块边缘是地球动力作用十分丰富的区域，板块之间相互作用一旦开始，便随着时间不断演化，突然或者逐渐改变它们的构造属性，或者变得不再活跃，这些常常与板块三联点的迁移有关。板块边缘的构造属性的改变可以从被动（非板块边界）变为主动（转换或俯冲边界），从转换边界变成俯冲边界，反之亦然，又或者从扩张脊变为转换或俯冲带。这种板块边界属性的变化是由板块边界滑动速率和相对运动方向的改变所引起的（Casey 和 Dewey，1984）。识别岩石记录中此类事件的沉积学指纹对板块构造重建的解释十分重要，尤其是在白垩纪之前的板块运动实例中，只有极少甚至没有洋壳记录得到保存。

本章强调板块边界的转化及其在相关盆地中沉积学和地层学标志，每种情形包含了至少一个大陆岩石圈板块的实例。讨论的重点是会聚型边缘和俯冲带，从其位于最初板块边界的初始状态开始，到它们由于三联点迁移而改变，直至消亡。同时强调了文献中可以说明演变过程中地层记录（形成、演化、消亡）的实例。

14.2　初始会聚边缘

最初学者们认为俯冲带形成于古老而致密的洋壳，主动下沉于上地幔之下（Gurnis，1992），这也是威尔逊旋回（Wilson，1966）中的关键一步。在该旋回，大陆一侧形成大洋，大洋在之后的俯冲作用中逐步消失，直至最终发生陆—陆碰撞。正如 Sliver 和 Behn（2008）讨论的那样，与威尔逊旋回相关的大量的板块裂谷和俯冲作用具有波动性，显示了在相对重要的初始俯冲阶段也存在类似的波动性。现今的初始俯冲作用是相对少见的构造现象。

14.2.1　现代实例

与被动大陆边缘和消亡的洋中脊相比，转换带或断裂带更有可能作为早期俯冲带的发生地，常常是板块重组的结果（Mueller 和 Phillips，1991）。大多数现代新生的聚敛带是这种例子（图 14-1）：太平洋 Mussau 海沟（Seno 等，1994）、日本海（El-Fiky 和 Kato，2006）、新西兰南侧 Puysegur 洋脊（Gurnis 等，2004；Stern，2004）以及沿着 Macquarie 洋脊杂岩的南侧（Ruff 等，1989）。Mussau 海沟和 Macquarie 洋脊为洋内构造带，而 Puysegur 和日本海的例子则涉及了大洋和弧后地壳向陆壳之下的俯冲作用。

Mussau 海沟被认为沿着先存断裂带形成，是对加罗林群岛、印度—澳大利亚板块和太平洋板块之间三联点处板块—微板块重组事件的响应（Hegarty 等，1983）。已有地震和地球物理证据记录了沿着洋内 Mussau 海沟的挤压构造（图 14-1），但俯冲作用还没有发展到岩浆活动的阶段（Hegarty 等，1983）。另一个非岩浆活动的新生俯冲带例子在日本海东缘（图 14-1），该俯冲带早期表现为与日本弧后盆地形成有关的转换断层边界（Tamaki 和 Honza，1985；El-Fiky 和 Kato，2006）。该地区海底以发育一系列海脊和海沟为特征，并伴有与挤压相关的地震活动（6.9 级和 7.7 级）以及逆断层构造（Tamaki 和 Honza，1985）。

现存的俯冲初期例子中，新西兰西南部的 Puysegur 大陆边缘是最好的实例（Sutherland 等，2006）。该俯冲带沿着一条先前的转换板块边界发育，澳大利亚板块自中新世开始俯冲于太平洋板块之下，现已发展到了岩浆形成的阶段。这是个复杂的过程（图 14-2），包括了区域隆升、出露和剥蚀，以及之后的沉降和淹没。较新的块体（如菲奥德兰；House 等，2002）正在发生隆升，而南部老的块体已经经过抬升、波浪侵蚀，之后洋脊沉降至 1500m（Collot 等，1995）。

现代新生的俯冲带同样与大型陆壳块体的俯冲作用有关。地球物理证据认为翁通—爪哇高原与新几内亚的碰撞（图 14-1）导致了沿北所罗门海槽一个初始俯冲带的形成（Taira 等，2004）。此外，Stern（2004）认为在印度洋存在还未被认识到的，与碰撞作用有关的初始俯冲作用，该俯冲作用应该与喜马拉雅造山带有关。

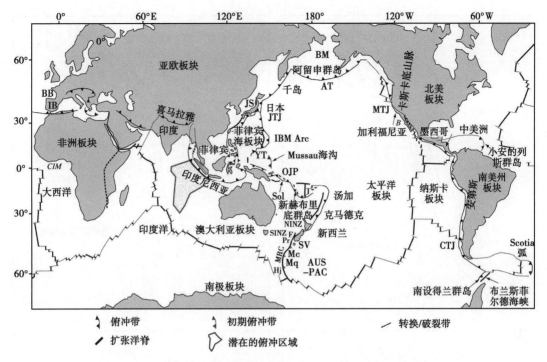

图 14-1　全球板块构造及其边界图，标记区域为文中讨论的俯冲起始和
现代三联点的位置（据 Stern，2004，改绘）

IBM—Izu-Bonin 弧；YT—Yap 海沟，Sol—索罗门弧；OJP—翁通—爪哇台地；MRC—麦考瑞脊杂岩体，F—Flordland，P—Puysegur，Mq—麦考瑞岛，Hj—Hjort，MTJ—Mendocino 三联点；JTJ—日本三联点；CTJ—智利三联点；BB—Biscay 湾；EPM—El Paso 山脉，Ca；JS＝日本海；IP—Iberian 半岛；NINZ—新西兰北岛，SINZ—新西兰南岛；BM—Berringinian 边缘；AT—阿留申海沟；CIM—科特迪瓦—加纳边缘；北美板块，南美板块，B—墨西哥下加利福尼亚半岛，印度洋中的密集点分布区为潜在消亡位置

14.2.2　俯冲初始阶段的模型

俯冲作用的初始阶段不是个简单的过程（McKenzie，1981；Kemp 和 Stevenson，1996；Gurnis 等，2004）。Stern（2004）认为俯冲作用的初始阶段主要有两种模型：（1）板块运动诱发的挤压和岩石圈破裂，或者（2）重力不稳定的（较老的、较低温的、较致密的）岩石圈自发坍塌。每一种模型都在下文进行更详细地描述。

如 Stern（2004）在图 14-3 中的描述，诱发型俯冲作用是由板块的传递或极性反转引起。该过程可以通过以一个地壳块体进入俯冲带来改变板块运动状态的方式，从而形成新的俯冲带。Cloos（1993）的"碰撞造山"模型认为该类型俯冲作用仅发生在厚度大于 15～20km 的陆壳或岛弧壳，或者是与夏威夷群岛规模类似的大型板内火山（高度大于 8km）或大洋高原（上文所示翁通—爪哇实例）相关的更厚的地壳（30km）。较小的海山和无地震活动的洋脊会影响仰冲板块（Marsaglia 等，1999；Collot 等，2001）、板块角度以及岛弧岩浆活动（Gutscher 等，1999；Saleeby，2003），但不会引起俯冲。板块极性倒转引起的俯冲作用与板块传递引起的俯冲作用很相似，但俯冲方向会发生突变。另一种诱发俯冲作用的情况出现在转换板块边界上，该处运动状态（欧拉极）的改变会导致两个板块发生会聚作用。

自发的俯冲作用可能起因于被动大陆边缘坍塌或转换带坍塌（图 14-3；Stern，2004）。

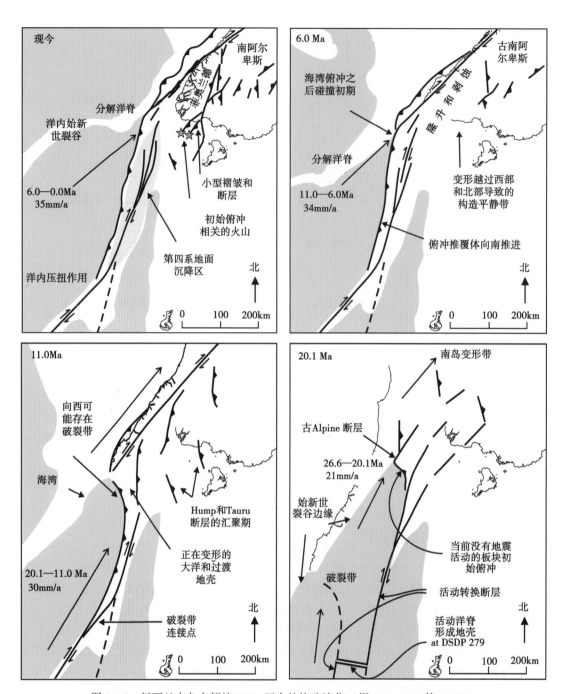

图 14-2 新西兰南岛南部约 20Ma 至今的构造演化（据 Sutherland 等，2006）

时间框架显示出非常复杂的俯冲带演化史，浅灰色表示海底洼地；浅蓝色是现今的浸没区域，曾经历上新世抬升剥露，到第四纪发生沉降（<1800m），Sutherland 等（2006）认为这与构造侵蚀或负板片浮力作用有联系；Sutherland 等（2006）推断相对年轻的火山（绿色星状）与俯冲相关

所以与诱发俯冲作用不同的是，板块运动的改变是在自发俯冲作用之后而不是之前（Hall 等，2003）。Stern（2004）讨论了自发型俯冲形成的条件，并强调一个发育较厚沉积负载（Cloetingh 等，1989）的薄弱带，必然会引起老的致密坚硬岩石圈发生俯冲作用。这种薄弱

带可能是一条先存断裂（Erickson，1993），或者由地壳拉张引起（Kemp 和 Stevenson，1996）。Hynes（2005）认为洋壳的软弱是由与软流圈沿着转换断层或断裂带上涌相关的年轻的玄武质岩浆活动造成的，Mueller 和 Phillips（1991）也支持这一说法。

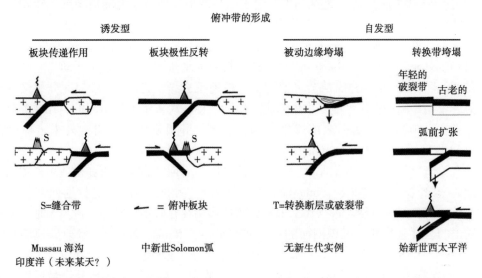

图 14-3　俯冲带发育模型（据 Stern，2004，修改）（红色三角为活跃岩浆岛弧）

Stern（2004）认为，Gurnis 等（2004）的模拟结果显示所有新生代俯冲作用的发育都是诱发型的，而不是自发型的。自发型俯冲作用最好的潜在例子可能被限制在新生代洋内系统，如 Izu-Bonin（图14-1）。在这一地区，沿着洋内的转换带，始新世出现了自发型俯冲作用（见 Gurnis 等（2004）的讨论；Stern，2004）。在这种情况下，俯冲作用由扩张洋脊与即将成为仰冲板块的块体之间的相互作用引起。Hall 等（2003）将这一类事件称为"灾变"，并指出仰冲板块上的地层记录了快速隆升（同构造期的始新世砾岩）以及随后达到1.5km 的沉降过程（Mizuno 等，1979；Mills，1980）。

14.2.3　初始俯冲作用的地层记录

由于发育方式有别，初始俯冲作用的地层记录在有关沉积盆地中会有不同特征。碰撞造山作用和板块传递作用的证据包括（Stern，2004）：（1）缝合线；（2）由漂浮地壳块体（大陆、岛弧或大型火成岩区）构成的相邻增生地块；以及（3）新生俯冲带形成的依次由岩浆弧、弧前和海沟组成的构造叠加带。在诱发型俯冲作用中，通常在仰冲板块的弧前地区存在早期挤压和隆升，之后演变为快速沉降和岩浆弧的发育（Gurnis，1992；Toth 和 Gurnis，1998；Hall 等，2003；Gurins 等，2004；Stern，2004）。因此，诱发型俯冲作用的沉积响应可能是角度不整合之上发育的浅海相—陆相沉积组合，以及之后的火山碎屑逐渐增多的深水沉积。与诱发型俯冲作用相比，上覆板块显著的伸展变形和强烈的岩浆活动证明了在自发俯冲过程中岩石圈发生了消减（Stern，2004）。在这种情况中，仅存在水体加深和伴生的岩浆活动的记录，而后者是其主要特征。不幸的是，俯冲作用发生后，初始俯冲的任何特征都会被破坏，使得较老的俯冲作用系统很难被识别（Gurnis 等，2004；Stern，2004）。

有些大陆边缘确实存在与现代新生俯冲带相似的，与板块重组相关的古老挤压—俯冲作用的构造证据。例如，在比斯开湾，伊比利亚半岛北缘（图14-1）漫长的被动边缘演化史曾被始新世短暂的俯冲作用中断（DSDP Leg 12 Shipboard Scientific Party，1972）。这次事件使得该大陆边缘异常狭窄陡峭，并在陆坡底部形成了一个被沉积物充填的较深海沟，大陆架也遭受了构造变形和隆升（Boillot 和 Capdevila，1977；Boillot 和 Malod，1979）。Malod 等（1982）利用潜水器揭示了这一地区陆坡遭受了强烈变形，使被动大陆边缘沉积物发生了叠瓦状逆冲推覆构造。在伊比利亚板块和欧洲板块沿着比利牛斯山脉会聚碰撞停止之前，据估计只发生了约 120~150km 甚至更少的洋壳消减，这不足以引起钙碱性岛弧型岩浆活动（Boillot 和 Capdevila，1977；Malod 等，1982）。新西兰北岛曾发生过岩浆活动，它经历了整个俯冲作用旋回，在大约 85Ma 从会聚型转变为被动型边缘，之后在约 25Ma 开始又从被动型转变为会聚型。俯冲作用可能是由中新世一次大规模板块重组事件引起的，在此期间，被动大陆边缘在岛弧型岩浆活动之前发生了构造变形，并被富含蛇绿岩的外来岩体所仰冲（Balance，1993；Kamp，1999；Whattam 等，2004，2005）。

其他有关新生代之前的例子非常少见，可能是因为后期的岩浆作用和构造事件覆盖了俯冲带初期的记录（Stern，2004；Sutherland 等，2006）。例如，在北美克拉通西部只有少量孤立的变质—沉积层序可能与古生代末期初始俯冲作用存在联系。其中一份关于加利福尼亚厄尔巴索山的二叠纪变质—沉积剖面的详细研究显示（Rains，2009），盆地的地层特征与 Gurnis（1992）和 Gurnis 等（2004）提出的诱发俯冲作用地球动力学模型相符合：在一个约 20Ma 的时间范围内，古弧前地区经历了快速抬升、沉降之后又再一次抬升，最后以岩浆岛弧的发育而结束。Rains（2009）研究表明，被动大陆边缘陆坡首先发生构造变形和隆升，并形成不整合面，随着水体逐渐加深，不整合面之上依次发育富含燧石—砾石的沉积、海相钙质到硅质泥岩沉积。之后伴随着火山碎屑的注入，盆地变浅，最终形成岛—弧体系并发育安山岩质熔岩流。

14.2.4 大西洋型初始俯冲作用

在大西洋型洋盆中，裂谷边界常被转换断层所错开。这些转换边界在海底扩张过程中逐渐变得静止。在非洲中部，沿科特迪瓦—加纳转换边缘的大洋钻探揭示了从转换边缘向被动边缘转化的沉积学记录。转化过程从一个发育局部隆起、高地温梯度和粗硅质碎屑注入的张扭性盆地开始，之后演变为被动边缘阶段的递进沉降（图 14-4；Mascle 等，1996；Basile 等，1998；Benkhelil 等，1998；Strand，1998）。伴随洋底扩张的开始和演化，边缘沉积作用受沿转换带发育的扩张脊的短暂影响变得复杂。

大洋的关闭要求俯冲作用沿着曾经的被动边缘开始，并能通过自发或诱发俯冲作用的方式，有效地激活古裂谷边缘或转换断层。一种替代方式是通过外来俯冲带的"传播"作用（Pindell 和 Kennan，2009），例如从太平洋逐渐侵入大西洋地区的小安的列斯岛弧俯冲带。由于转换被动边缘在地壳密度上突然和极端的变化，使其可能成为俯冲作用发生的有利位置。Stern（2004）指出，要使随着沿转换被动边缘的俯冲作用能够发生大洋关闭，关闭作用必须大体垂直于形成洋壳的洋中脊体系。在这种情况下，我们期望能在岩石记录中看到，早期的转换边缘到被动边缘的沉积序列（正如上文所描述的科特迪瓦—加纳转换边缘），以及晚期的构造变形和叠加的岛弧型岩浆活动。

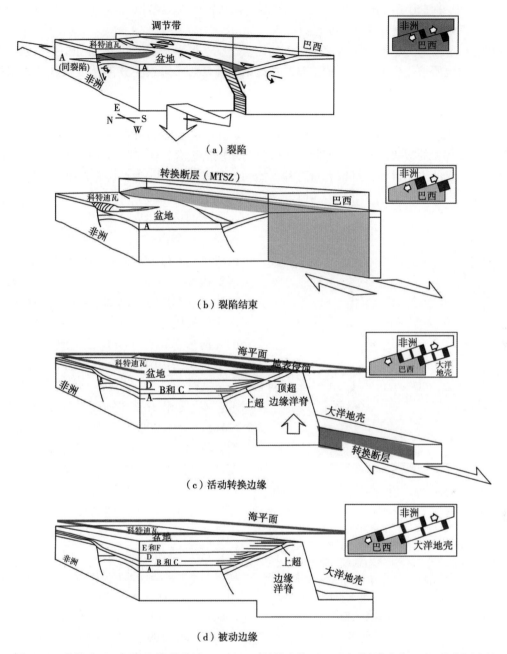

图 14-4 科特迪瓦—加纳边缘从裂谷 (a, b) 到转换边缘 (c) 再到被动边缘 (d) 演化历史的三维示意图（视角自西北；据 Mascle 等，1996 和 Basile 等，1993，修改）

14.3 三联点

一旦俯冲作用开始，俯冲带会逐渐或突然改变它的构造属性。转换—会聚复合型板块边界，是随着走向和时间而发生的渐进式板块边界转化的一种常见类型，它是斜向汇聚过程中角度变化的产物（Ryan 和 Coleman，1992；Tibaldi，1992）。在弧前地区或沿着岛弧轴部，平行于板块边缘的走滑断层作用，不管大小，均可转变为一个转换板块边界。Smith 和 Landis

（1995）总结并讨论了与斜向俯冲有关的盆地，如阿留申岛弧盆地（Geist 等，1988）。在板块的三联点处，这种转化会更加突然。

尽管在现今的全球板块构造体系中占的比重很小，但由于它们的不稳定和迁移性，三联点在过去的地质演化史中可能更加重要。在讨论俯冲带时不可避免要涉及三联点（扩张脊）的相互作用（DeLong 等，1979；Pavlis 等，1995；Brown，1998；Sisson 等，2003）。因此，确定三联点的地层—沉积特征具有重要地质意义。

根据 McKenzie 和 Morgan（1969）的描述以及 Cronin（1992）的进一步研究，共有 18 种与俯冲作用有关的板块三联点，即除了扩张脊与断层的组合，还包括海沟的组合。根据不同类型板块边界的不同排列，可以有近百种可能的组合方式，本章为了讨论方便，将它们简化为（图 14-5）：（1）包含 2 个断层和 1 个海沟的板块边界（FFT）；（2）包含 2 个海沟和 1 个断层的板块边界（TTF）；（3）包含 1 个洋脊、1 个海沟和 2 个断层的板块边界（RTF）；（4）包含 2 个海沟和 1 个洋脊的板块边界（TTR）；（5）包含 3 个海沟的板块边界（TTT）；（6）包含 2 个洋脊和 1 个海沟的板块边界（RRT）。尽管三联点可能的几何形态范围十分广泛，但在现今地球上的板块结构中仅存在其中的某一个子集。

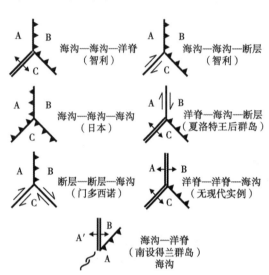

图 14-5 文中所讨论的板块三联点结构示意图
刺状线为海沟，刺点指示俯冲方向。箭头代表转换板块边缘，带箭头的双线为扩张脊。板块标记为 A，B，C。底部示例为特殊情况，迁移的洋脊—海沟相互作用导致 A 板块沿着之前海沟（曲线）逐渐生长（A' 的转化）。虽然没有现代洋脊—洋脊—海沟例子，已经利用这一结构来解释斐济附近的中新世 Ninerva 三联点（Sdrolias 等，2003）

文献中强调了板块三联点处发生的构造演化及相关的构造变形和岩浆活动（Madsen 等，2006）。三联点通常都不是一个独特的"点"，而是一个复杂的分布式应变区（Clark，1992；Allan 等，1993）。在 TTR 型（海沟—海沟—洋脊）三联点中，迁移在弧前地区的构造响应表现为与三联点同步迁移的洋脊交会或俯冲作用相关的异常隆升或岩浆活动（Marshak 和 Karig，1977），并受到热液活动的叠加和与深成作用有关的高温—低压变质作用的影响（DeLong 等，1979；Sisson 和 Pavlis，1993；Underwood，1993；Pavlis 等，1995）。

正如 Pavlis 等（1995）所总结的，三联点的迁移通过差异性隆升（物源）和沉降（沉积中心）影响了沉积物的分布以及岩浆活动中心的位置和迁移。盆地构型与沉积物搬运模式反映了三联点的类型和迁移方向。热隆升和挤压隆升是内侧板块（俯冲断层上盘）的主要特征。在由俯冲型向转换型转化的大陆边缘，或者在 TRT（海沟—洋脊—海沟）型或 TFT（海沟—断层—海沟）型三联点内俯冲板块浮力发生变化的区域，都有抬升的弧前盆地以及增生楔沉积的例子。隆升过程通常比较短暂，之后被沉降所取代。弧前地区强烈的隆升和构造作用导致了区域不整合的发育，沿三联点迁移方向这些不整合具有穿时性。隆起遭受剥蚀并向邻近深海环境（海沟或淹没块体）提供沉积供给，如海底扇复合体的发育。这些海相碎屑层序在成分上与"普通的"弧前沉积层序不同，因为它含有明显的来自增生楔的循环沉积单元（Marsaglia 和 Ingersoll，1992；Marsaglia 等，1992，1995）。最后，其火山碎屑含量达到最低值可能与洋脊俯冲作用沿岛弧轴产生的岩浆活动间隔期有关。这些地层特征在以下现代三联点地区得到了很好的研究和论述（图 14-1）：日本中部（Japan TTT）、北加利福尼亚（Mendocino FFT）以及智利沿海（TTR、TTF）。

日本三联点位于菲律宾、欧亚及太平洋板块交会处，其所处区域自中新世以来经历了复杂的构造演化史，包括三联点的迁移和弧后盆地的发育（Marsaglia 等，1992）。在日本中部的海沟—海沟—海沟（TTT）三联点体系中（图 14-1），板块三联点的动力学作用导致一个来自深部的年轻的花岗岩体（<2Ma）发生极快速的冷却隆升（Harayama，1992）。伊豆—小笠原岛弧（Izu-Bonin Arc）与本州岛弧（Honshu Arc）至今仍在伊豆（Izu）半岛发生碰撞作用。Ito 和 Masuda（1986）从相关邻近的沉积单元的研究中揭示了岛弧火山连续碰撞的历史。这些沉积单元由于后来的构造变形而隆升，现今沿着伊豆半岛出露（Ito 和 Masuda，1986；图 14-6 和图

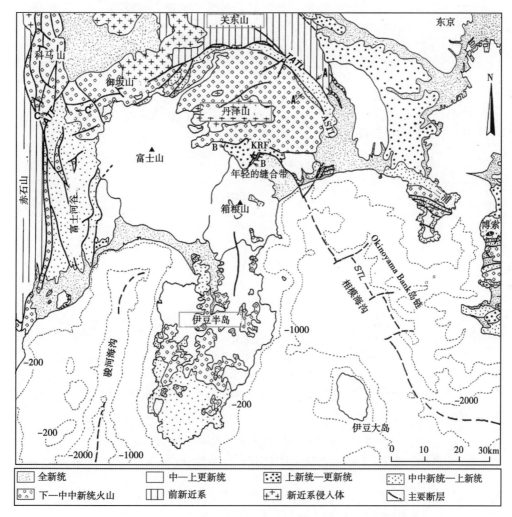

图 14-6　伊豆半岛三联点地质图板块（据 Ito 和 Masuda，1986）

在插图中表示出会聚型板块边界（刺状线）和会聚方向（开放箭头）。EUR—欧亚大陆，PHS—菲律宾海板块，PAC—太平洋板块，等深线单位为米。S. T. L—Sagami 构造线。图 14-7 中的演化剖面为近似地沿着 Izu 半岛的北—南向。彩色框标出与图 14-7 对应的标志物

14-7）。此外，这些沉积响应甚至在更远的来自伊豆碰撞带的砂质沉积物中也有所记录。该砂质沉积插入相邻的海沟并沿走向被搬运了数百千米（Marsaglia 等，1992）。

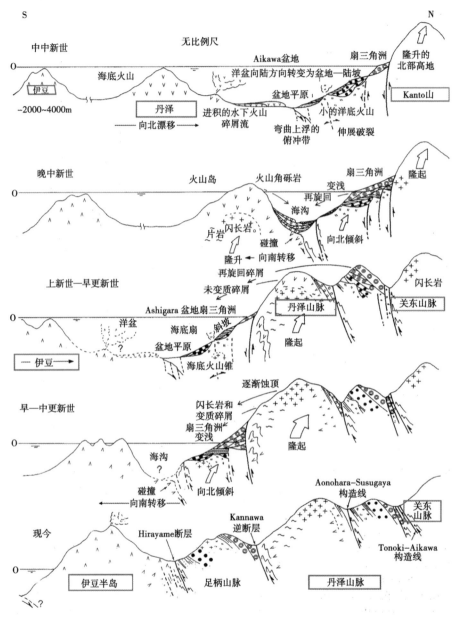

图 14-7　南北向横断面概略图（据 Masuda，1986）

显示出中中新世至今盆地的演化和三联点碰撞带的结构；彩色框标出与图 14-6 相对应的标志物

另一个已充分研究的例子是门多西诺 FFT 型（断层—断层—海沟）三联点，位于北美、太平洋和胡安德富卡板块会聚处（图 14-1、图 14-8、图 14-9）。在位于门多西诺三联点内或近北部地区的 Eel River 弧前盆地（Aalto 等，1995；Buiger 等，2002），沿着三联点迁移方向往北，该现代三联点以伴随着前陆盆地型构造变形的隆升作用为特征（Merritts 和 Bull，1989；Nilsen 和 Clarke，1989；Clarke，1992）。往南，三联点迁移的证据被部分记录在北

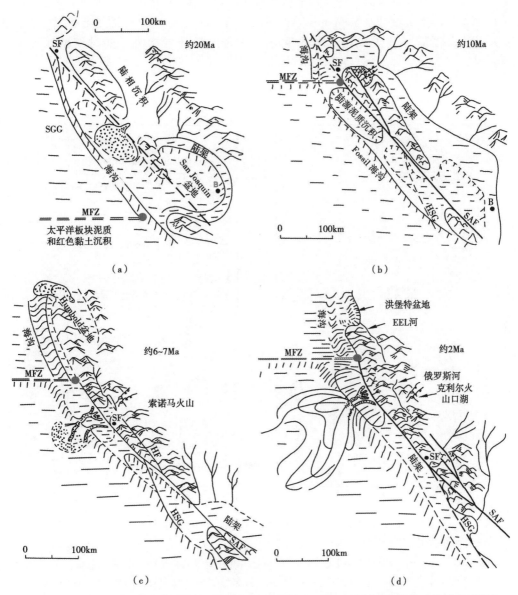

图 14-8 始新世—上新世加利福尼亚边缘的古地理重建，显示出在从俯冲向转换边缘转变过程中主要砂质（点）沉积中心和高地（据 Drake 等，1989，修改，详见其中引用）

B—贝克斯菲尔德；SF—圣弗朗西斯科；SAF—圣安德烈亚斯断层；PAB—Point Arena 盆地；MFZ—门多西诺断层带；SGG—Schooner 峡谷和 Gallaway 岩层。红色圆点标出了向北迁移并向内陆（向东）的门多西诺板块三联点

美板块上局部发育的沉积盆地的剥蚀残余地层中（Graham 等，1984；Nilsen 和 Clarke，1989）和太平洋板块上的大型海底扇内（Normark 和 Gutmacher，1985）。总的来说，三联点迁移的主要标志为迁移性隆升和较老弧前沉积的剥蚀，以及随后发生的沉降（Bachman 等，1984；Glazner 和 Loomis，1984；Nilsen 和 Clarke，1989；Loomis 和 Ingle，1994；Unruh 等，1995）和岛弧型岩浆活动的终止（Dickinson 和 Snyder，1979）。Lock 等（2006）的研究认为，与三联点相关的北美板块地壳增厚和迁移的双峰式沉降—隆升模型（动力地形）有关，后者控制了短暂的盆地的形成和沉积，以及水系和侵蚀作用的发育（图 14-10）。在

三联点南部，太平洋板块上的一个大型水下扇（Delgada fan）指示了在中新世—上新世时期边界碎屑输入的显著增加（McManus 等，1970），Drake 等（1989）将此与和三联点相关的北美海岸山脉隆升（北美板块）联系起来，认为该隆升可能由于同时期的海平面下降增强所致。

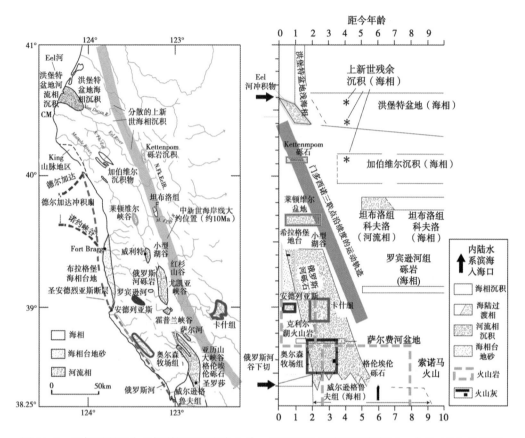

图 14-9　左侧图所示为北加利福尼亚海岸山脉，展现了圣安德烈亚斯断层，大约 10Ma 的海岸线位置俯冲后覆盖在增生楔上的沉积残余物，位于门多西诺海角（CM）外滨的门多西诺三联点。注入门多西诺三联点的大河包括北部的 Eel 河和南部的俄罗斯河。右侧是新近纪盖层残余沉积物随时间的纬度分布以及门多西诺三联点的迁移轨迹。河流相向海相转换的时间趋势记录了海岸山脉的逐渐出露的过程（据 Lock 等，2006，修改）

继续往南，三联点沿加利福尼亚边缘的迁移记录逐渐变少（图 14-8），这种变化某种程度上是因为早期的迁移构造仅局限于现今已被海水淹没的远离海岸的断层带中。对于这一地区的认识也受到海上钻井和地震资料的限制（Teng 和 Gorsline，1991）。在加利福尼亚中部和南部水下，古俯冲带和海沟的沉积充填可以在局部地震剖面上识别出来（Yeats 等，1981；McIntosh 等，1991；Meltzer 和 Levander，1991；Nicholson 等，1994）。该海沟可能已经因为微板块袭夺而废弃，或者作为转换运动的初始位置，由于三联点的不稳定性，伴随着板块边界（圣安德烈亚斯构造体系）横穿弧前区域向东迁移而被逐渐废弃（Ingersoll，1982）。年轻的俯冲洋壳和（或）洋脊相互作用的影响也改变了海沟—陆坡，该处的沉积剖面中记录了近海沟的岩浆活动和弧前隆升以及沿巴顿陡崖的侵蚀作用（Marsaglia 等，2006）。

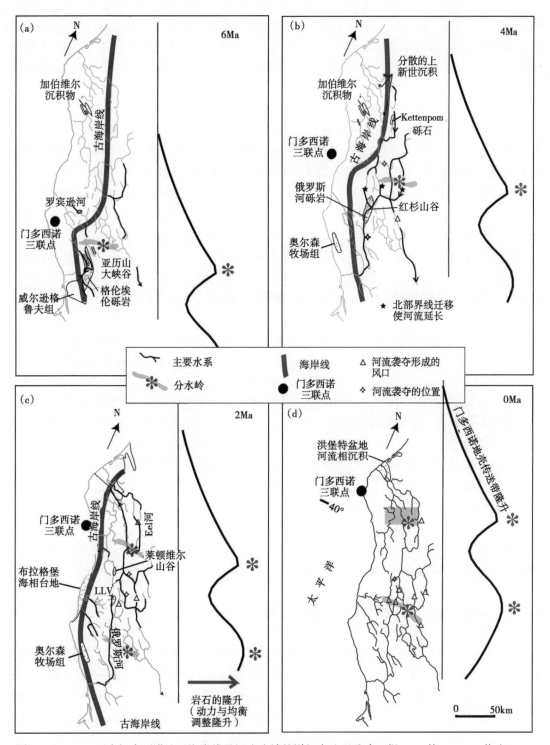

图14-10 6Ma至今门多西诺地区海岸线及河流流域的详细古地理重建（据Lock等，2006，修改）

每一阶段演化图的右侧的"岩石抬升"模式展示了一条双峰曲线，该曲线伴随三联点随时间向北迁移，用门多西诺地壳传送带（MCC）命名；该双峰现象是由于动力效应和均衡隆升效应（见2Ma图中的红色插入语）共同作用而形成；要注意与河流切割和海岸线移动相关的隆升的伴生迁移；LLV—小湖谷位于南美、南极和纳斯卡板块交会处的智利三联点，也可能存在一些反映三联点迁移的地质证据。沿着智利南部，与俯冲作用有关的侵蚀、狭窄的增生楔、靠近海沟的异常陆壳体以及沿着岛弧轴部的岩浆活动间歇，被认为与智利三联点向北移动有关（Forsythe 和 Prior，1992；Behrmann 等，1994）。在岩浆活动间歇期沉积记录表现为火山碎屑物输入的缺失或减少，以及外力火山碎屑的增加。在与板片窗（受软流圈上涌影响的洋脊俯冲作用形成的一个位于仰冲板块之下，但没有俯冲板块的区域）通道有关的智利弧后地区也存在三联点迁移和洋脊俯冲的记录。这些记录包括了岩浆活动、挤压构造和抬升作用等（Lagabrielle 等，2004）。最后，该三联点以发育非常年轻的（<2Ma）、被抬升的蛇绿岩套为特征。在南极和纳斯卡板块边界，延伸较短的洋脊和转换断层与智利海沟相互作用，形成了一系列复杂的三联点类型（TTF、TTR），该蛇绿岩很可能就是这些三联点的产物（Forsythe 和 Prior，1992）。由于板块的特殊几何形态，一些弧前地区经历了多期三联点迁移事件，产生了 Taitao 蛇绿岩套并将其抬升至地表（Forsythe 和 Prior，1992）。很多古老的蛇绿岩套可能与类似的三联点构造活动有关（Wakabayashi 和 Dilek，2003）。

14.4 蛇绿岩套的仰冲、俯冲作用和三联点之间的关系

蛇绿岩是古板块边缘上构造演化的常见标志，是非常值得关注的。蛇绿岩侵位常常与大洋关闭或俯冲增生有关，但也可以出现在一些其他构造背景中（Dewey，1976；Moores，1982；Casey 和 Dewey，1984；Wakabayashi 和 Dilek，2003）。Stern（2004）认为塞浦路斯和阿曼的大型蛇绿岩套发育于初生的岛弧型岩石圈，是自发俯冲作用的结果。Casey 和 Dewey（1984）认为在造山带鲜有从板内大西洋型边缘到安第斯型边缘转化的直接证据，并且提出很多仰冲蛇绿岩套与沿着早期板块边缘（特别是转换断层或三联点）的俯冲作用初始阶段有关。Wakabayashi 和 Dilek（2003）依据侵位机制对蛇绿岩套进行了分类，并重点强调了侵位于扩张脊与海沟（或转换断层）相互作用的复杂三联点处的蛇绿岩套。洋壳岩石的逆冲推覆和地表暴露是蛇绿岩侵位的关键阶段，最有可能作为侵位同期—后期的蛇绿岩（基性、超基性）碎屑保存在相关的沉积层序中（Marsaglia 等，2008）。另一种情况是，洋壳可能在俯冲作用开始阶段的板块抬升中就已发生暴露剥蚀，而没有发生逆冲推覆侵位（Macquarie 岛；Wakabayashi 和 Dilek，2003）。

14.5 主动边缘向被动边缘的转化

在活动边缘转变为被动边缘的区域，存在许多"废弃"俯冲带的记录。例如，研究认为白令边缘（图 14-1）逐步从会聚型边缘（60Ma 及以前）转化为转换型边缘（60—50Ma），并最终转化为被动边缘（50—40Ma）（Marlow 和 Cooper，1985）。在这一过程中板块边界向南跃迁至现在的阿留申板块边界的位置（Ben-Avraham 和 Uyeda，1983），并捕获了部分库拉板块的碎片。相关证据包括通过陆上滨岸地质和地震反射资料确定的白垩纪岛弧、古海沟，潜在的已转换或与转换作用有关的火山岩区，以及覆盖了整个现今被动边缘的较厚披覆沉积。

在南设得兰海沟（图 14-1）的实例中，有一种独特的迁移板块三联点类型（洋脊—海沟），逐渐停止了俯冲作用，并在其之后发育了一个被动边缘（Hole 和 Larter，1993）。当

菲尼克斯—南极扩张脊变为静止时，沿着该海沟的俯冲作用在约4Ma大幅减弱（Barker，1982）。从活动向非活动边缘转换的标志是被动边缘不发育海沟，因为边缘（早期的弧前区域）陆架到陆坡底部都经历了内部挤压作用。该环境以挤压性构造（褶皱和逆断层）为特征，伴随较小的正断层活动（Jabaloy等，2003），以及近海沟的岩浆活动（Hole和Larter，1993）。

另一南半球的例子来自于新西兰（图14-1）。在经历区域拉伸之后，随着冈瓦纳大陆的裂解和塔斯曼海的扩张，新西兰在古生代至中生代漫长的俯冲作用结束于中白垩世（Laird和Bradshaw，2004）。有人提出，与设得兰海沟体系类似，随着菲尼克斯—太平洋扩张脊被俯冲消减（Bradshaw，1989），或者由于靠近海沟导致俯冲板块被捕获而不再活动（Luyendyk，1995），使得俯冲作用在新西兰终止。白令海、南极和南设得兰实例中演化过程记录大多位于水下而难以获得，而新西兰由于后期构造事件的作用，板块边缘转化的地层记录遭受抬升和暴露剥蚀。这一构造响应表现为一个显著的不整合面，该不整合面是由于区域抬升和变形增生楔遭受剥蚀而形成的。在南岛利用锆石和生物化石确定其年龄为约100Ma（Laird和Bradshaw，2004）；而北部的磷灰石裂变径迹数据则指出其构造变形一直持续到85Ma（Mazengarb和Harris，1994；Kamp，1999）。不整合面之上依次被晚白垩世至渐新世陆坡—陆架背景之下的沉积所覆盖。上覆地层在区域上存在差异性，从南岛非海相的冲积相（Laird和Bradshaw，2004）过渡到北岛的海相沉积。在北岛的沉积序列整体向上变细，从底砾岩变为砂质浊积岩和页岩，随后是更多的灰质相和海绿石相沉积，反映了被动边缘的演化过程（Lewis和Laird，1986；Ballance，1993；Laird和Bradshaw，2004）。

14.6 总结和结论

目前对于俯冲作用起源的认识仍较薄弱，或许是因为在今天的地球上这一状态并不常见，仅在地球历史上某些时期很重要。初期的俯冲作用仅显示出构造变形和地震活动，只是随着时间的推移才逐渐发生响应板块俯冲的岩浆活动。地球动力学模型可以帮助我们预测这些事件在沉积—火山作用中的记录，但很少有地区对这些模型进行过验证，尤其是针对前新生代的例子。诱发型和自发型俯冲作用的启动过程仍存在争议。在诱发型俯冲作用的发生位置，沉积特征包括快速隆升（不整合）、快速沉降（向上变细变深的正韵律沉积序列）以及岛弧体系的形成。在自发型俯冲作用的区域，原始弧前地区经历了沉降作用和岩浆型岛弧形成之前的早期岩浆活动。未来需要做更多的工作去充分论证这一重要过程的沉积和地层响应。

一旦俯冲作用开始，由于板块三联点的形成和迁移，弧前地区将有可能发生其他构造叠加作用，三联点上的板块边缘的转化可以在一个点或者广阔地区发生，可能导致微板块上的盆地快速形成以及岩相发生快速变化。与俯冲作用相关的迁移三联点可能与海沟变形—反转、增生楔的抬升剥蚀、近海沟火山活动有关，也与仰冲板块上的盆地形成、变形和反转有联系。在活动边缘，三联点作为隆升、侵蚀和沉积物再分布的活动场所，扮演着大规模沉积作用的"循环装置"。主要的控制因素有：（1）三联点的类型（如TTT、TRT、TFT和FFT）；（2）三联点的几何形态（如板块交叉的角度）；（3）三联点的稳定性；（4）仰冲板块之上先存构造的活动；（5）仰冲板块的地壳性质（洋壳或者陆壳）；（6）碰撞后板块的运动学特征。在一个单纯俯冲作用（海沟—海沟—海沟）的三联点中，其主要特征与岛弧地区增生有关，而在海沟—洋脊—海沟型三联点中，海沟与扩张脊的相互作用可以引起异常的近海沟岩浆活动、弧前地区的热叠加作用，以及俯冲削减作用；在俯冲作用被转换板块运动取代的三联点中（如海沟—断层—断层），转化作用以弧前地区的阶段性抬升和构造变形为标志，并伴随着岛弧型岩浆活动的停止。还有很多问题尚未解答，包括沉积记录中保存三联点地质信息的可能性以及三联点地质信息的唯一性。

会聚型边缘可以若干方式消亡（变为被动边缘），比如可以通过简单的废弃和沉降，或者转化成转换板块边缘而停止会聚。在有扩张脊的地带，会聚边缘消亡前的转化作用可能引起弧前地区构造变形、抬升以及热—火山作用的叠加。

总的来说，从开始到终结，会聚型边缘都是动力学作用十分丰富的区域。依据岩石记录来揭示其演化过程是我们面临的重大挑战，这需要对现代发生的类似过程作更细致的研究。

致谢（略）

参 考 文 献

Aalto, K. R., McLaughlin, R. J., Carver, G. A., Barron, J. A., Sliter, W. V., and McDougal, E. (1995) Uplifted Neogene margin, outermost Cascadia – Mendocino triple junction region, California. Tectonics, 14, 1104–1116.

Allan, J. F., Chase, R. L., Cousens, B. L., Michael, P. J., Gorton, M. P., and Scott, S. D. (1993) The Tuzo Wilson volcanic field, NE Pacific, alkaline volcanism at a complex, diffuse, transform-trench-ridge triple junction. Journal of Geophysical Research, 98, 22367–22387.

Bachman, S. B., Underwood, M. B., and Menack, J. S. (1984) Cenozoic evolution of coastal northern California, in Tectonics and sedimentation along the California margin, Crouch, J. K., and Bachman, S. B., eds., Pacific Section SEPM, 38, 55–66.

Ballance, P. F. (1993) The paleo-Pacific, post-subduction, passive margin thermal relaxation sequence (Late Cretaceous- Paleogene) of the drifting New Zealand continent, in Ballance, P. F. (ed.) South Pacific Sedimentary Basins, Sedimentary Basins of the World, 2, 93–110.

Barker, P. F. (1982) The Cenozoic subduction history of the Pacific margin of the Antarctic Peninsula: ridge cresttrench interactions. Geological Society of London Journal, 139, 787–801.

Basile, C., Mascle, J., Popoff, M., Bouillin, J. P., Mascle, G. (1993) The Côte d'Ivoire- Ghana transform marg in: a marginal ridge structure deduced from seismic data. Tectonophysics, 222, 1–19.

Basile, C., Mascle, J., Benkhelil, J., and Bouillin, J. -P. (1998) Geodynamic evolution of the Côte D' Ivoire-Ghana transform margin: an overview of Leg 159 results. Proceedings of the Ocean Drilling Program, Scientific Results, 159, 101–110.

Behrmann, J. H., Lewis, S. D., Cande, S. C., and the ODP Leg 141 Scientific Party (1994) Tectonics and geology of spreading ridge subduction at the Chile Triple Junction; a synthesis of results from Leg 141 of the Ocean Drilling Program. Geologische Rundschau, 83, 832–852.

Ben-Avraham, Z., and Uyeda, S. (1983) Entrapment origin of marginal seas, in Hilde, T., and Uyeda, S. (eds.) Geodynamics of the western Pacific-Indonesia region, Geodynamics Series, 11, 91–104.

Benkhelil, J., Mascle, J., and Guiraud, M. (1998) Sedimentary and structural characteristics of the Cretaceous along the Côte D' Ivoire-Ghana transform margin and in the Benue Trough: a comparison. Proceedings of the Ocean Drilling Program, Scientific Results, 159, 93-99.

Boillot, G., and Capdevila, R. (1977) The Pyrenees: subduction and collision? Earth and Planetary Science Letters, 35, 151-160.

Boillot, G., and Malod, J. A. (1979) Subduction and tectonics on the continental margin off northern Spain. Marine Geology, 32, 53-70.

Bradshaw, J. D. (1989) Cretaceous geotectonic patterns in the New Zealand region. Tectonics, 8, 803-820.

Brown, M. (1998) Ridge-trench interactions and high-Tlow- P metamorphism, with particular reference to the Cretaceous evolution of the Japanese Islands. Geological Society of London, Special Publication 138, 137-169.

Burger, R. L., Fulthorpe, C. S., Austin, J. A., and Gulick, S. P. S. (2002) lower Pleistocene to present structural deformation and sequence stratigraphy of the continental shelf, offshore Eel River Basin, northern California. Marine Geology, 185, 249-281.

Busby, C. and Ingersoll R. V. (eds.) (1995) Tectonics of Sedimentary Basins, Blackwell.

Casey, J. F., and Dewey, J. F. (1984) Initiation of subduction zones along transform and accreting plate boundaries, triple-junction evolution, and forearc spreading centresimplications for ophiolite geology and obduction, in I. G Gass, S. J. Lippard, and A. W. Shelton, eds., Ophiolites and oceanic lithosphere, Geological Society Special Publication 13, 269-290.

Clarke, S. H., Jr., (1992) Geology of the Eel river basin and adjacent region: implications for late Cenozoic tectonics of the southern Cascadia subduction zone and Mendocino triple junction. AAPG Bulletin, 76, 199-224.

Cloetingh, S., Wortel, R., and Vlaar, N. J. (1989) On the initiation of subduction zones. Pure and Applied Geophysics, 129, 7-25.

Cloos, M. (1993) Lithospheric buoyancy and collisional orogenesis: Subduction of oceanic plateaus, continental margins, island arcs, spreading ridges, and seamounts. Geological Society of America Bulletin, 105, 715-737.

Collot, J-Y., Lamarche, G., Wood, R., Delteil, J., Sosson, M., Lebrun, J-F., and Coffin, M. (1995) Morphostructure of and incipient subduction zone along a transform plate boundary, Puysegur Ridge and Trench. Geology, 6, 519-522.

Collot, J-Y., Lewis, K., Lamarche, G., Lallemand, S. (2001) The giant Ruatoria debris avalanche on the northern Hikurangi margin, New Zealand; results of oblique seamount subduction. Journal of Geophysical Research, 106, B9, 19271-19297.

Cronin, V. S. (1992) Types and kinematic stability of triple junctions. Tectonophysics, 207, 287-301.

DeLong, S. E., Schwarz, W. M., and Anderson, R. N. (1979) Thermal effects of ridge subduction. Earth and Planetary Science Letters, 44, 239-246.

Dewey, J. F. (1976) Ophiolite obduction. Tectonophysics, 31, 93-120.

Dickinson, W. R., and Snyder, W. W. (1979) Plate tectonics of the Laramide orogeny.

Geological Society of America Memoir, 151, 355–366.

Drake, D. E., Cacchione, D. A., Gardner, J. V., and McCulloch, D. S. (1989) Morphology and growth history of Delgada Fan: implications for the Neogene evolution of the Point Arena Basin and the Mendocino triple junction. Journal of Geophysical Research, 94, 3139–3158.

DSDP Leg 112 Shipboard Scientific Party (1972) Site 119, in Laughton, A. S., Berggren, W. A., et al., eds., Initial Reports of the Deep Sea Drilling Project, 12, 753–901.

El-Fiky, G., and Kato, T. (2006) Secular crustal deformation and interplate coupling of the Japanese Islands as deduced from continuous GPS array, 1996—2001. Tectonophysics, 422, 1–22.

Erickson, S. G. (1993) Sedimentary loading, lithospheric flexure and subduction initiation at passive margins. Geology, 21, 125–128.

Forsythe, R., and Prior, D. (1992) Cenozoic continental geology of South America and its relations to the evolution of the Chile Triple Junction. Proceedings of the Ocean Drilling Program, Initial Reports, 141, 23–31.

Geist, E. L., Childs, J. R., and Scholl, D. W. (1988) The origin of summit basins of the Aleutian Ridge: implications for block rotation of an arc massif. Tectonics, 7, 327–341.

Glazner, A. F., and Loomis, D. P. (1984) Effect of subduction of the Mendocino fracture zone on Tertiary sedimentation in southern California. Sedimentary Geology, 38, 287–303.

Graham, S. A., McCloy, C., Hitzman, M., Ward, R., and Turner, R. (1984) Basin evolution during change from convergent to transform continental margin in central California. AAPG Bulletin, 68, 233–249.

Gurnis, M. (1992) Rapid continental subsidence following the initiation and evolution of subduction. Science, 255, 1556–1558.

Gurnis, M., Hall, C, and Lavier, L. (2004) Evolving force balance during incipient subduction: Geochemistry, Geophysics, Geosystems, 5. doi: 10. 1029/2003GC000681.

Gutscher, M. A., Olivet, J. L., Aslanian, D., Eissen, J. P., and Maury, R. (1999) The "lost Inca Plateau"; cause of flat subduction beneath Peru? Earth and Planetary Science Letters, 171, 335–341.

Hall, C. E., Gurnis, M., Sdrolias, M., Lavier, L. L., and Muller, R. D. (2003) Catastrophic initiation of subduction following forced convergence across fracture zones. Earth and Planetary Science Letters, 212, 15–30.

Harayama, S. (1992) Youngest exposed granitoid pluton on Earth: cooling and rapid uplift of the Pliocene-quaternary Takidani Granodiorite in the Japan Alps, central Japan. Geology, 20, 657–660.

Hegarty, K. A., Weissel, J. K., and Hayes, D. E. (1983) Convergence at the Caroline-Pacific Plate boundary: collision and subduction, in Hayes, D. E., ed., The tectonic and geologic evolution of Southeast Asian seas and islands, part 2. Geophysical Monograph, 27. American Geophysical Union, 326–348.

Hole and Larter (1993) trench-proximal volcanism following ridge-trench collision along the Antarctic Peninsula. Tectonics, 12, 897–910.

House, M. A., Gurnis, M., Kamp, P. J., and Sutherland, R. (2002) Uplift in Fiordland region, New Zealand implications for incipient subduction. Science, 297.

Hynes, A. (2005) Buoyancy of the oceanic lithosphere and subduction initiation. International Geology Review, 47, 938-948.

Ingersoll, R. V. (1982) Triple-junction instability as cause for late Cenozoic extension and fragmentation of the western United States. Geology, 10, 621-624.

Ito, M., and Masuda, F. (1986) Evolution of clastic piles in an arc-arc collision zone: late Cenozoic depositional history around the Tanzawa Mountains, Central Honshu, Japan. Sedimentary Geology, 49, 223-259.

Jabaloy, A., Balanyá, J-C., Barnolas, A., Galindo-Zaldívar, J., Hernández-Molina, F. J., Maldonado, A., Martínez-Martínez, J-M., Rodríguez-Fernández, J., Sanz de Galdeano, C., Somoza, L., Surinach, E., and Vázquez, J. T. (2003) The transition from an active to a passive margin (SW end of the South Shetland Trench, Antarctic Peninsula). Tectonophysics, 366, 55-81.

Kamp, P. J. J. (1999) Tracking crustal processes by FT thermochronology in a forearc high (Hikurangi margin, New Zealand) involving Cretaceous subduction termination and mid-Cenozoic subduction initiation. Tectonophysics, 307, 313-343.

Kemp, D. V., and Stevenson, D. J. (1996) A tensile flexural model for the initiation of subduction. Geophysical Journal, 125, 73-94.

Lagabrielle, Y., Suárez, M., Rossello, E. A., Hérail, G., Martinod, J., Régnier, M., and de la Cruz, R. (2004) Neogene to Quaternary tectonic evolution of the Patagonian Andes at the latitude of the Chile Triple Junction. Tectonophysics, 385, 211-241.

Laird, M. G., and Bradshaw, J. D. (2004) The break-up of a long-term relationship: the Cretaceous separation of New Zealand from Gondwana. Gondwana Research, 7, 273-286.

Lewis, D. W., and Laird, M. G. (1986) Gravity flow deposits, processes and tectonics subaerial to deep marine, northern South Island Cretaceous to recent, New Zealand: guide book for Field Excursion 31B. Twelfth International Sedimentological Congress, 24-30 August 1986, Canberra, Australia, 46 p.

Lock, J., Kelsey, H., Furlong, K., and Woolace, A. (2006) Late Neogene and Quaternary landscape evolution of the northern California Coast ranges: evidence for Mendocino triple junction tectonics. Geological Society of America Bulletin, 118, 1232-1246.

Loomis, K. B., and Ingle, J. C., Jr., (1994) Subsidence and uplift of the Late Cretaceous-Cenozoic margin of California: new evidence from the Gualala and Point Arena basins. Geological Society of America Bulletin, 106, 915-931.

Luyendyk, B. P. (1995) Hypothesis of Cretaceous rifting of east Gondwana caused by subducted slab capture. Geology, 23, 373-376.

Madsen, J. K., Thorkelson, D. J., Friedman, R. M., and Marshall, D. D. (2006) Cenozoic to Recent plate configurations in the Pacific Bas in: ridge subduction and slab window magmatism in western North America. Geosphere, 2, 11-34.

Malod, J-A., Boillot, G., Capdevila, R., Dupeuble, P-A., Lepvrier, C., Mascle, G., Muller,

C., Taugourdeau-Lantz, J. (1982) Subduction and tectonics on the continental margin off northern Spain: observations with the submersible Cyana, in Leggett, J. K., ed., Trench-forearc geology: Sedimentation and tectonics on modern and ancient active plate margins. Geological Society Special Publication 10, 309-315.

Marlow, M. S., and Cooper, A. K. (1985) Regional geology of the Beringian continental margin: in N. Nasu et al. (ed.) Formation of active ocean margins, 497-515.

Marsaglia, K. M. (1995) Chapter 8, Interarc and backarc basins, in Busby, C. and Ingersoll R. V. (eds.) Tectonics of Sedimentary Basins, Blackwell, 299-329.

Marsaglia, K. M., and Ingersoll, R. V. (1992) Compositional trends in arc-related, deep-marine sand and sandstone: a reassessment of magmatic - arc provenance. Geological Society of America Bulletin, 104, 1637-1649.

Marsaglia, K. M., Ingersoll, R. V., and Packer, B. M. (1992) Tectonic evolution of the Japanese islands as reflected in modal compositions of Cenozoic forearc and backarc sand and sandstone. Tectonics, 11, 1028-1044.

Marsaglia, K. M., Torrez, X., Padilla, I., and Rimkus, K. (1995) Provenance of Pleistocene and Pliocene sand and sandstone, ODP Leg 141, Chile Margin. Proceedings of the Ocean Drilling Program, Scientific Results, 141, 133-151.

Marsaglia, K. M., Mann, P., Hyatt, R., and Olson, H. (1999) Evaluating the influence of aseismic ridge subduction and accretion (?) on the detrital modes of forearc sandstone: an example from the Kronotsky Peninsula, Kamchatka forearc. Lithos, 46, 17-42.

Marsaglia, K. M. Davis, A. S., Rimkus, K., Clague, D. A. (2006) Evidence for interaction of a spreading ridge with the outer California borderland, Marine Geology, 229, 259-272.

Marsaglia, K. M., Bender, C., Dillon, J., Mazengarb, C., Marden, M., and Miranda, E. (2008) Syntectonic (?) mafic gravel associated with Miocene obduction and subduction initiation in New Zealand, Geological Society of America, Annual Meeting, Houston, TX, Abstract.

Marshak, R. S., and Karig, D. E. (1977) Triple junctions as a cause for anomalously near-trench igneous activity between the trench and volcanic arc. Geology, 5, 233-236.

Mascle, J., Lohman, G. P., Clift, P. D., et al. (1996) Proceedings of the Ocean Drilling Program, Initial Reports, 159, College Station, TX (Ocean Drilling Program).

Mazengarb, C., and Harris, D. H. M. (1994) Cretaceous stratigraphic and structural relations of Raukumera peninsula, New Zealand: stratigraphic patterns associated with the migration of a thrust system. Annales Tectonicae, 8, 100-118.

McIntosh, K. D., Reed, D. L., Silver, E. A., Meltzer, A. S. (1991) Deep structure and structural inversion along the central California continental margin from edge seismic profile RU-3. Journal of Geophysical Research, 96, B4, 6459-6473.

McKenzie, D. P. (1981) The initialization of trenches: a finite amplitude instability: in Talwani, M., and Pittman, W. C., III (eds.) Island arcs deep-sea trenches and back-arc basins. American Geophysical Union, Maurice Ewing series 1 57-61.

McKenzie, D. P., and Morgan, W. J. (1969) The evolution of triple junctions. Nature, 224, 125-

133.

McManus, D. A., Weser, O. E., von der Borch, C. C., Vallier, T., and Burns, R. E. (1970) Regional aspects of deep-sea drilling in the northeast Pacific. Initial Reports of the Deep Sea Drilling Project, 5, 621-636.

Meltzer, A. S., and Levander, A. R. (1991) Deep crustal reflection profiling offshore southern central California. Journal of Geophysical Research, 96, B4, 6475-6491.

Merritts, D., and Bull, W. B. (1989) Interpreting Quaternary uplift rates at the Mendocino triple junction, northern California, from uplifted marine terraces. Geology, 17, 1020-1024.

Mills, W. (1980) Analysis of conglomerates and associated sedimentary rocks of the Daito Ridge, Deep Sea Drilling Project Site 445, in Initial Reports of the Deep Sea Driling Project, 58, 643-657.

Mizuno, A., Okuda, Y., Niagumo, S., Kagami, H., and Nasu, N. (1979) Subsidence of the Daito Ridge and associated basins, North Philippine Sea, in Dickerson, P. W., ed., Geophysical and geological investigations of continental margins. AAPG Memoir 29. Tulsa, OK, American Association of Petroleum Geologists, 239-243.

Moores, E. M. (1982) Origin and emplacement of ophiolites. Reviews in Geophysics and Space Physics, 20, 735-760.

Mueller, S., and Phillips, R. J. (1991) On the initiation of subduction. Journal of Geophysical Research, 96, 651-665.

Nicholson, C., Sorlien, C. C., Atwater, T., Crowell, J. C., and Luyendyk, B. P. (1994) Microplate capture, rotation of the western Transverse Ranges, and initiation of the San Andreas transform as a low-angle fault system. Geology, 22, 491-495.

Nilsen, T. H., and Clarke, S. H., Jr. (1989) Late Cenozoic basins of northern California. Tectonics, 8, 1137-1158.

Normark, W. R., and Gutmacher, C. E. (1985) Delgada Fan, Pacific Ocean, in Bouma, A. H., Normark, W. R., and Barnes, N. E., eds., Submarine fans and related turbidite systems. New York, Springer-Verlag, 59-64.

Pavlis, T. L., Underwood, M., Sisson, V. B., Serpa, L. F., Prior, D., Marsaglia, K. M., Lewis, S. D., and Byrne, T. (1995) The effects of triple junction interactions at convergent plate margins: Report on the results of the joint JOI/USSAC and GSA Penrose conference, 44p.

Pindell, J., and Kennan, L. (2009) Tectonic evolution of the Gulf of Mexico, Caribbean and northern South America in the mantle reference frame: an update, in James, K., Lorente, M. A., and Pindell, J., eds., The origin and evolution of the Caribbean Plate. Special Publication. Geological Society of London.

Rains, J. (2009) Sedimentary record of middle Permian magmatic arc initiation, El Paso Mountains, Kern County, California, MS. Thesis, California State University Northridge, CA.

Ruff, L. J., Given, J. W., Sanders, C. O., and Sperber, C. M. (1989) Large earthquakes in the Macquarie Ridge complex: transitional tectonics and subduction initiation. Pure and Applied Geophysics, 129, 71-129.

Ryan, H. F., and Coleman, P. J. (1992) Composite transformconvergent plate boundaries:

description and discussion. Marine and Petroleum Geology, 9, 89-97.

Saleeby, J. (2003) Segmentation of the Laramide slab; evidence from the southern Sierra Nevada region. Geological Society of America Bulletin, 115, 655-668.

Sdrolias, M., Muller, R. D., and Gaina, C. (2003) Tectonic evolution of the southwest Pacific using constraints from backarc basins. Special Paper, 372. Geological Society of America, 343-359.

Seno, T., Stein, S., and Gripp, A. E. (1993) A model for the motion of the Philippine Sea Plate consistent with NUVEL-1 and geological data. Journal of Geophysical Research, 98, 17941-17948.

Silver, P. G., and Behn, M. D. (2008) Intermittent plate tectonics? Science, 319, 85-88.

Sisson, V. B., and Pavlis, T. L. (1993) Geologic consequences of plate reorganization: an example from the Eocene southern Alaskan fore arc. Geology, 21, 913-916.

Sisson, V. B., Pavlis, T. L., and Prior, D. J. (1994) Effects of triple junction interactions on convergent plate margins: results of a Penrose conference. Geological Society of America Today, 4, 248-249.

Sisson, V. B., Pavlis, T. L., Roeske, S. M., and Thorkelson, D. (2003) Introduction; an overview of ridge - trench interactions in modern and ancient setting, in Sisson, V. B., Roeske, S. M., and Pavlis, T. L. (eds.) Geology of a transpressional orogen developed during ridge-trench interaction along the North Pacific margin. Geological Society of America Special Paper, 371, 1-18.

Smith, G. A., and Landis, C. A. (1995) Intra-arc basins, in Busby, C. J., and Ingersoll, R. V., eds., Tectonics of sedimentary basins. Cambridge, MA, Blackwell, 263-298.

Stern, R. J. (2004) Subduction initiation: spontaneous and induced. Earth and Planetary Science Letters, 226, 275-292.

Strand, K. (1998) Sedimentary facies and sediment composition changes in response to tectonics of the Cote D'Ivoire - Ghana transform margin. Proceedings of the Ocean Drilling Program, Scientific Results, 159, 113-123.

Sutherland, R., Barnes, P., and Uruski, C. (2006) Miocene - Recent deformation, surface elevation, and volcanic intrusion of the overriding plate during subduction initiation, offshore southern Fiordland, Puysegur margin, southwest New Zealand. New Zealand Journal of Geology and Geophysics, 49, 131-149.

Taira, A., Mann, P., and Rahardiawan, R. (2004) Incipient subduction of the Ontong Java plateau along the North Solomon Trench. Tectonophysics, 389, 247-266.

Tamaki, K., and Honza, E. (1985) Incipient subduction and obduction along the eastern margin of the Japan Sea. Tectonophysics, 119, 381-406.

Teng, L. S., and Gorsline, D. S. (1991) Stratigraphic framework of the continental borderland basins, southern California. American Association of Petroleum Geologists Memoir, 47, 127-143.

Tibaldi, A. (1992) The role of transcurrent intra-arc tectonics in the configuration of a volcanic arc. Terra Nova, 4, 567-577.

Toth, J., and Gurnis, M. (1998) Dynamics of subduction initiation at preexisting fault zones.

Journal of Geophysical Research, 103, B8, 18053-18067.

Underwood, M. B., ed. (1993) Thermal evolution of the Tertiary Shimanto Belt, Southwest Japan: an example of ridge-trench interaction, Geological Society of America Special Paper, 172 p.

Unruh, J. R., Loewen, B. A., and Moores, E. M. (1995) Progressive arcward contraction of a Mesozoic-Tertiary fore-arc basin, southwestern Sacramento Valley, California. Geological Society of America Bulletin, 107, 38-53.

Wakabayashi, J., and Dilek, Y. (2003) What constitutes "emplacement" of an ophiolite? mechanisms and relationship to subduction initiation and formation of metamorphic soles, in Dilek, Y., and Robinson, P. T., eds., Ophiolites in Earth History. Special Publication, 218. Geological Society of London, 427-447.

Whattam, S. A., Malpas, J. G., Ali, J. R., Smith, I. E., and Lo, C-H. (2004) Origin of the Northland Ophiolite, northern new Zealand: discussion of new data and reassessment of the model. New Zealand Journal of Geology and Geophysics, 47, 383-389.

Whattam, S. A., Malpas, J., Ali, J. R., Lo., C-H., and Smith, I. A. (2005) Formation and emplacement of the Northland ophiolite, northern New Zealand: SW pacific tectonic implications. Journal of the Geological Society, 162, 225-241.

Wilson, J. T. (1966) Did the Atlantic close and then re-open? Nature, 211, 676-681.

Yeats, R. S., Haq, B. U., et al. (1981) Initial Reports of the Deep Sea Drilling Project, vol. 63. Washington, DC, U. S. Government Printing Office, 967 p.

(赵梦译，吴哲 崔敏 祁鹏 校)

第15章 日本西南部俯冲带沉积环境的演化：来自南海(Nankai)海槽发震带试验中 Kumano 断面的最新结果

MICHAEL B. UNDERWOOD[1], GREGORY F. MOORE[2]

(1. Department of Geological Sciences, University of Missouri-Columbia, Columbia, USA;
2. Department of Geology & Geophysics, University of Hawaii-Manoa, Honolulu, USA)

摘　要：南海（Nankai）海槽处于菲律宾板块向欧亚板块之下俯冲区域。沿大陆边缘的 Kii 半岛或 Kumano 盆地的狭长地带是 Nankai 海槽发震带试验（NanTroSEIZE）靶区。这一多学科、多期次、多航次的项目，主要目的是增进我们对沿着板块边界地震形成过程的了解。NanTroSEIZE 的 1 期钻探，包括综合大洋钻探计划的 314、315 和 316 航次。这三个航次的岩心和测井资料，提供了大量有关板块边缘地层和构造演化的新信息。本文通过三个部分来描述总结：Kumano（弧前）盆地；巨型分支断层带，即增生楔内重要的脱序冲断带；和增生楔内的前缘冲断带。钻探结果在很大程度上验证了三维地震数据的构造解释结果，并且也为研究很多地质事件的发生时间（如弧前盆地沉积的开始和巨型分支断层上的滑动历史）提供了宝贵的约束条件。我们观察到了多种类型的构造—沉积相互作用。前缘冲断带主要由大量的砂岩和砾岩组成，很可能是早更新世增生之前在海沟底部的轴向水道沉积而成的。增生楔向陆方向逐渐变老；穿时的不整合将增生楔和上覆的斜坡裙分开。Kumano 盆地的浊流沉积在经历以下一系列先期事件之后直到早更新世才开始：晚中新世和早更新世下伏增生楔的形成；增生楔之上不整合的剥蚀，存在大约 1.2Ma 的沉积间断；长期异常缓慢的半深海沉积，持续时间 1.67~3.8Ma；在 1.55Ma 沿着巨型分支断层的加速隆升增大了可容空间；通过内陆腹地的剥蚀以及峡谷和冲沟的下切进入海沟斜坡的上部，形成了稳固的浊流沉积物搬运体系。

关键词：综合大洋钻探计划　Nankai 海槽　弧前盆地　增生楔　岩性地层学

15.1 引言

南海（Nankai）海槽俯冲带，也许是世界上对海槽型沉积体系研究最深入的地区（图15-1）。该板块边界是菲律宾海板块俯冲到日本西南部的产物，其比较特殊的一个特征是缺乏相关的活动火山链（第四纪）。板块的会聚速率约 4.0~6.5cm/a，运动方向约 300°~315°（Seno 等，1993；Miyazaki 和 Heki，2001；Zang 等，2002）。Nankai 体系以构造变形前缘外侧厚的沉积物聚集（>1000m），以及增生楔下缘（toe）高的俯冲增生速率为特征（Underwood 和 Moore，1995；Clift 和 Vannucchi，2004），代表了增生—剥蚀统一体的一个端元。广义的类似条件在其他地方也存在，比如卡斯凯迪亚（Cascadia）盆地（Kulm 和 Fowler，1974；Carlson 和 Nelson，1987；MacKay 等，1992；Underwood 等，2005）、阿留申（Aleutian）海沟（Piper 等，1973；Scholl，1974；McCarthy 和 Scholl，1985；Lewis 等，1988；Gutscher 等，1998）、巽他（Sunda）海沟（Karig 等，1978；Moore 等，1980；

Beaudry 和 Moore，1985；Kopp 等，2001；Kopp 和 Kukowski，2003；Fischer 等，2007）和智利南部（Schweller 等，1981；Thornburg 和 Kulm，1987）。

四十多年来，南海（Nankai）一直是大洋科研钻探的焦点地区，先后经历了深海钻探计划（DSDP 钻孔 31 和 87）、大洋钻探计划（ODP 钻孔 131、190 和 196）（Taira 等，1991；Pickering 等，1993；Moore 等，2001；Mikada 等，2005）和综合大洋钻探计划（IODP 航次 314、315、316）（Tobin 等，2009a）。除了图 15-1 中的三个钻探横剖面（依次被命名为 Ashizuri、Muroto 和 Kumano 横剖面），科学家还获得了大量有关增生楔和 Shikoku 盆地俯冲沉积物的二维和三维地震反射图像（Aoki 等，1982；Moore 等，1990；Park 等，2002；Bangs 等，2004，2006；Gulick 等，2004；Moore 等，2007；Ike 等，2008a）。地球物理数据、界定沉积物年龄的技术手段以及岩心岩性特征的这一超级联合，为板块边缘地层和构造演化的研究，提供了广阔的前景。

Shikoku 盆地位于海沟的向海方向（图 15-1），晚渐新世由 Izu-Bonin 岛弧体系之后的弧后扩张和裂谷作用开始形成（Okino 等，1994；Kobayashi 等，1995；Sdrolias 等，2004）。扩张中心的后期重新定位和离轴海山的火山作用持续到中晚中新世。这一岩浆作用导致了火成岩基底结构形态的巨大变化，包括 Kinan 海山链的形成（Okino 等，1994）。Kashinosaki

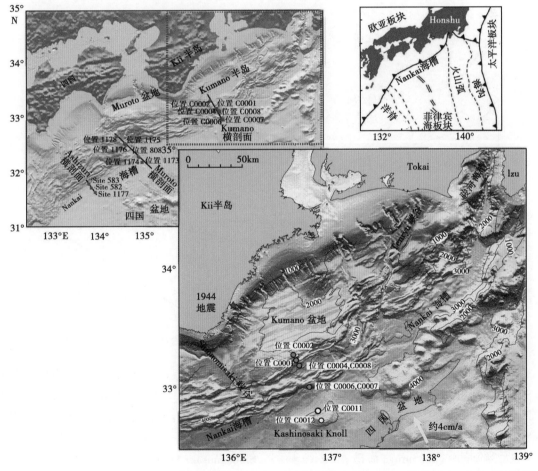

图 15-1 索引图显示了 Nankai 海槽发震带试验（Kii 半岛的滨外）的 Kumano 横剖面区，也显示了先前深海钻探计划和大洋钻探计划的 Muroto 和 Ashizuri 钻孔横剖面位置（据 Moore 等，2009，修改）

Knoll海山即代表了Kumano横剖面所在地区的一种基底特征（Ike等，2008a）。而盆地年轻基底的起伏形态，又反过来影响着盆地总的沉积厚度和每个沉积单元的沉积相特征（Ike等，2008b）。对基底地貌的典型沉积响应有：基底低的地区发育较厚的中新世浊流充填，基底高的地区则被中新世的半深海泥质密集段所覆盖（Underwood，2007；Ike等，2008b）。

伴随着俯冲作用的进行，深海盆地逐渐被沉积物所充填，导致海底的地貌起伏越来越不明显。因此，Shikoku盆地内的上新统—更新统，厚度比较均一，岩性也比较连续，主要由半深海泥组成，并含大量空落（air fall）火山灰夹层（Taira等，1991；Moore等，2001）。Shikoku盆地内的黏土矿物，随着时间也在不断变化。开始的时候，主要是中新世来源于岛弧的富蒙皂石集合体；新地层逐渐变为富含绿泥石和伊利石的集合体，主要来自隆升增生复合体的剥蚀（主要是沉积岩和变质沉积岩）。现今这些复合体，沿着日本的外带暴露地表（Steurer和Underwood，2003；Underwood和Steurer，2003；Underwood和Fergusson，2005）。随着俯冲作用将地层带到俯冲带前缘附近，Shikoku盆地相最终快速地淹没到向陆增厚的海沟浊积楔状体之下（图15-2）。该海沟浊积体总沉积厚度为0.5~1.3km（Taira和Ashi，1993；Mountney和Westbrook，1996）。

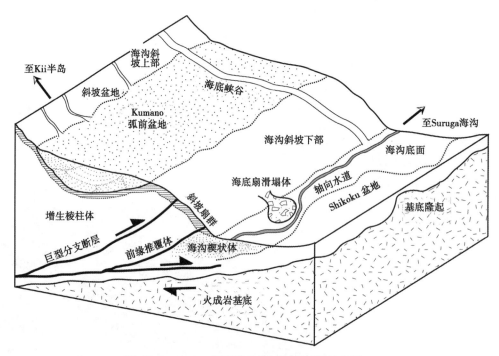

图15-2　Nankai海槽俯冲带的沉积环境示意图

南海（Nankai）海槽在其东北端和骏河（Suruga）海槽完全合并到一起。骏河（Suruga）海槽是一个巨大的海底沟谷体系。它沿着Izu-Bonin火山弧和Honshu之间的碰撞带西侧下切形成（图15-1）。碎屑颗粒的物源研究表明，第四纪海沟沉积楔主要来源于该碰撞带的快速隆升和碰撞（Marsaglia等，1992；Underwood和Fergusson，2005）。浊流通过一系列大型海底峡谷（Suruga海槽、Tenryu峡谷、Shionomisaki峡谷）向海洋提供了大量陆源碎屑沉积物，而这些陆源碎屑主要来自Honsu上的几个河流物源。这些深切水道的源头紧邻现代的海岸线，即使在海平面的高位期也一直活动着（Soh和Tokuyama，2002；Kawamura

等，2009）。总之，位于 Shikoku 盆地沉积之上的海沟—楔状浊流相粗粒沉积，构成了俯冲带体系的沉积物输入（图15-2）。沉积输入在三维空间的变化有利于塑造增生楔的结构构造、力学特征和水文地质特征（Underwood，2007）。而增生楔的构造演化，则反过来控制了上冲板块之上沉积中心的变迁，也控制了一系列重要的弧前盆地的演化（图15-2）。

Nankai 海槽发震带试验是目标宏伟的、多期次、多学科的研究项目。其目的是为了增进对活动俯冲型板块边界的地震形成过程的了解（Tobin 和 Kinoshita，2006）。Nankai 海槽发震带试验的横剖面区域，位于 Kii 半岛、Honshu、日本的向海方向（图15-1），并在最新的三维地震反射调查区之内（图15-3）。该研究区有几个突出的特征，包括一个大的弧前盆地（Kumano 盆地）和一条位于弧前盆地的向海方向潜在的发震巨型分支断层。Nankai 海槽发震带试验1期已于 2007—2008 年完成。综合大洋钻探计划（IODP）三个航次的钻孔共提供了 8 个位置的岩心和随钻测井数据，其钻探深度在海底之下的 329~1401m（图15-3）；试验的 2 期在海沟处朝大洋方向的 2 个参考点记录了俯冲带沉积物输入（航次 322；Underwood 等，2009）。最终的科技挑战目标，是利用超深直井钻穿海底之下 6~7km 的发震板块界面，然后在断层带内安装长期的钻孔观测装置（Tobin 和 Kinoshita，2006）。

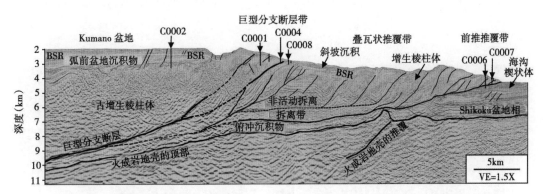

图 15-3 三维地震数据体中的联合地震测线（含基本地质解释），测线穿越（或邻近）所有南海海槽发震带试验1期的钻孔位置（图15-1）（据 Moore 等，2009；BSR—似海底反射层）

在1期钻探开始的时候，综合大洋钻探计划（IODP）的 314 航次，致力于获得弧前和增生楔的随钻测井数据（Tobin 等，2009b）。在随后的 315 航次中，对 6 个钻孔中的两个进行了取心，其中一个钻孔位于分支断层带之上（C0001），另一个钻孔（C0002）是在 Kumano 盆地向海一侧（Ashi 等，2009）。此后的 316 航次，有两个钻孔（C0006 和 C0007）位于增生楔的前缘冲断带；该航次的另外两个钻孔（位置 C0004 和 C0008）位于巨型分支断层之上或其浅部延伸区域（Screaton 等，2009a，2009b）。通过这三个航次的数据，可以让这一组科学家进一步明确边缘的沉积和构造演化特征，而这些特征以往仅靠地球物理数据难以限定。因此，这一章的研究目的有两个：强调这几次关键航程的发现；并将岩心数据标定到最新解释的三维地震上。

15.2　增生楔前缘

综合大洋钻探计划的 C0006 和 C0007 钻孔位于前缘冲断带向陆方向的第一个洋脊（图

15-4）。引入的变形前缘向海方向的沉积显示了典型的 Nankai 构造—沉积体系，总体上表现为火成岩基底之上依次被 Shikoku 盆地地层和第四系海沟增生楔沉积所覆盖，沉积厚度约 2.4km（图 15-3）。从地震数据上可以发现海沟楔状体的上部，含有大量成群的水道—天然堤复合体，也有明显的轴向水道从物源区沿着海底的高程梯度向下弯曲延伸，为 Suruga 海槽提供了物源供给（Shimamura，1989）。靠近斜坡的底部，存在一个发育完好的早期冲断带，它影响了上 Shikoku 盆地地层和海沟—楔状体下部的沉积（图 15-3）。

与前端的 Nankai 增生楔相比（Muroto 或 Ashizuri），Kumano 横剖面区内的前缘冲断带存在明显异常。沿着 Nankai 边缘的大部分地区，前缘断层的倾角很大（25°~35°），且在下倾合并到增生楔底部拆离面之前，向陆地仅延伸 1~2km。这一典型的"斜坡之上又有斜坡"的几何结构，形成了"海沟—楔状体之上又叠加了海沟—楔状体"的沉积样式（Moore 等，1990，2001）。相比而言，靠近 C0006 和 C0007 钻孔的前缘断层，表现为较缓的（大约 7°~8°）滑脱断层，向陆地延伸达近 6km。断层上盘为被抬升了的上 Shikoku 盆地相沉积和海沟—楔状体相沉积，下盘则为年轻的海沟沉积（图 15-4）。在向海的边缘，该断层与上盘和下盘地层平行（"断坪之上断坪"的几何学特征）。Kumano 的前缘冲断带，也含有大量向陆地陡倾的次级断层和局部的北冲断层。此外，在 C0006 钻孔向陆的方向上，明显存在一个小的楔形斜坡盆地（图 15-4），盆地内的沉积物被重新从邻近的向陆斜坡搬运到下伏的增生楔

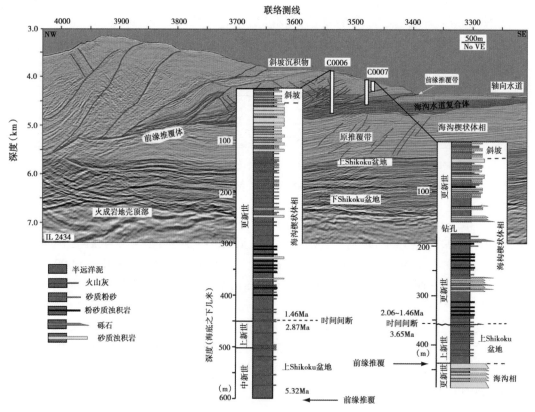

图 15-4　区域地震测线穿越了 IODP 的钻孔 C0006 和 C0007 钻孔，Nankai 增生楔的前缘推覆带
地震解释根据 Moore 等（2009）；图中地层的总结以 IODP 航次 314 和航次 316 的岩心和随钻测井资料为基础；年龄主要根据超微化石带

上。在前缘冲断带向陆方向小于5km的地方发育另一条显著的逆冲断层,其上盘在斜坡盆地充填之上至少平移了1.25km;该断层削掉了下伏背斜的顶部,成为一个脱序逆冲断层。脱序推覆体中的逆冲块体,和下盘块体相比,脱序逆冲断层的逆掩断块在地震上表现为弱振幅、弱连续的特征(图15-4)。前缘带这一奇特的结构特征,是由于洋脊或海山的俯冲形成的(正好位于三维地震调查区的西侧);在其附近的区域,斜坡的底部被可能是由于基底隆起俯冲所引起的大规模的海底滑塌所刮擦(Moore等,2009)。

C0006和C0007钻孔的岩心显示,最上部地层单元为一个向上变细的沉积序列,主要由半深海泥岩夹薄层的粉砂岩、砂质粉砂岩和细粒浊积岩组成,也含有微量的火山灰层——后者可能以空落火山灰的形式,进入大陆坡底沉积(航次316的科学家,2009b,2009c)。浊积岩具有底面突变、纹层面相互平行和正粒序的典型特征(图15-5)。砂质组分主要由石英、长石、岩屑颗粒和玻璃质碎屑组成。这些沉积物的最大年龄是晚更新世(0.436Ma,根据超微化石事件),推测其沉积环境从隆升的增生楔逐渐过渡为斜坡裙。

在该斜坡沉积壳层之下发育大约422—328m厚的增生海沟浊积体,含有若干富砂和富砾层(图15-4)。砂岩呈深灰色至黑色,由含有次生石英和长石的变质岩和火山岩岩屑组成。单个砂层的厚度为1~7m,内部无定形。薄层呈正粒序(图15-5)。砾岩为碎屑支撑,杂基较少,分选中等。碎屑是次圆—次棱角状(图15-5)。岩石碎屑(矿物颗粒的集合体)具有多种成分,包括燧石、砂岩、泥岩、糜棱状石英岩、富含石英的变质岩、火山岩岩屑、长英质的深成岩屑。砂岩和砾岩之间的富泥层段,被认为是水道间沉积和/或海沟—楔状体远端沉积。总的来讲,海沟—楔状体相沉积具有一种向上变粗、变厚的趋势,其最大年龄是早更新世(1.46~1.24Ma)。对于这种岩性变化的一种解释是,其沉积场所随着时间推移越来越靠近海沟的轴部。富砂和富砾层的重复,可能是由于增生之前、增生楔前缘内的推覆断裂作用导致;也可能是海沟轴向水道的周期性迁移或活化导致;还可能是两者共同作用的结果。虽然在这些旋回中气候肯定起了辅助作用,但来自超微化石和古地磁的年龄数据不够精确,难以将之与异周期动力的全球旋回进行对比。

在316航次中的一个比较重要的发现是,横跨增生海沟—楔状体的底部,存在一个沉积间断。该间断年龄存在显著变化,在C0006钻孔处为2.87—1.46Ma,而C0007钻孔处为1.46Ma或3.65—2.06Ma。界面之下地层主要由上Shikoku盆地相沉积组成,以生物扰动的半深海泥质为主,并含有火山灰夹层。那些泥/灰沉积的最大年龄值为晚中新世(5.32Ma)。在两口钻孔内Shikoku盆地相地层的顶部均被削掉,但让人难以理解的是,在地震剖面上界面上下地层呈整合接触关系(图15-4)。我们认为地层缺失的部分,在俯冲海山被埋藏到海沟楔状体下面之前,就被位于俯冲海山陡倾一侧的海底滑塌作用搬运带走了。地震剖面显示,位于Kumano横剖面区域内海沟向海的方向上的Kashinosaki Knoll海山的侧翼,存在类似的滑塌刮擦作用(Ike等,2008b)。在钻孔的深部,C0006钻孔处约为海底之下603m,而C0007钻孔处约为海底之下439m,前缘冲断带截断了上Shikoku盆地地层。随钻测井数据清晰地显示,前缘冲断带之下存在大量砂岩沉积(航次314科学家,2009b),但从下盘取出的未固结砂岩却很少。砂层比率较高,这与前缘冲断带地震资料解释相一致——逆冲推覆使得上Shikoku盆地相地层位于上盘,而下盘则为海沟—海底的轴向水道复合体(图15-4)。

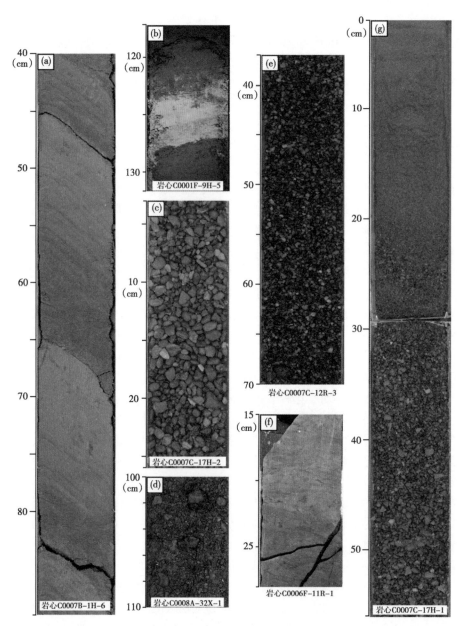

图 15-5 岩心图片显示了 Nankai 海沟俯冲带主要岩性的代表实例

(a) 斜坡扇相的薄层、细粒浊流沉积；(b) 斜坡扇相的空落火山灰沉积；(c) 增生海沟—楔状体浊积相内的多成分砾石；(d) 泥岩碎屑砾岩来自斜坡裙的块体搬运复合体（MTC）；(e) 增生海沟—楔状体浊积相内的粗砂到细砾；(f) 严重生物扰动的泥岩，具有螺旋潜迹的动物遗迹群，是该区大多数岩石地层内的典型半远洋夹层；(g) 增生的海沟—楔状体浊流相内的粒序层，细砾到细砂、粉砂

15.3 分支断层体系的向海边缘

综合大洋钻探计划（IODP）的 C0001、C0002 和 C0008 钻孔位于增生楔的斜坡上，正好处于一个主洋脊的东南部。该洋脊形成了 Kumano 盆地的向海边界（图 15-3）。这三个钻孔

在空间上横跨了巨型分支断层体系的向海边缘（图15-6）。巨型分支断层是一个区域上延伸较广的、脱序的断层体系。它切穿了增生楔较老的部分（Park等，2002），表现为复杂系列的交织断块，从海底向下可追踪到10km深处。在那里断层体系在板块界面（增生楔的底部和俯冲火成岩基底之间）的顶部发生分叉（图15-3）。增生楔内发育一系列较小的推覆体，具有高振幅、侧向连续的反射特征，很可能代表了增生的海沟—楔状体和Shikoku盆地地层。某些单个推覆体上的地层发生褶皱而卷入到上盘背斜中；斜倾角断层产生了一系列斜断坡（Oblique ramps），断坡的顶部被巨型分支断层所截断（Moore等，2007）。与在前缘增生楔看到的显著不同，浅层巨型分支断层上盘地震反射具有低振幅、侧向连续性差的特征（图15-6）。该地震反射特征，与经历过构造变形的、以泥岩为主的沉积序列一致，并非代表了广泛的浊积岩。靠近海底，增生地层内的这一断块，向上移动到年轻的斜坡盆地和斜坡裙沉积之上，后者向下和向陆方向至少分别可以追踪1250m（水平方向）和750m（垂直方向）。斜坡沉积物也可以静止于上盘断块之上，并在多个位置受滑塌作用影响。据Strasser等（2009）的研究，巨型分支断层之上的滑动始于约1.95Ma增生楔下缘附近的脱序冲断作用；在经历一段相对平静期后，约1.55Ma时断层再次活动，并演变为一个主要的分支断层。

岩心揭示了巨型分支断层之上的斜坡裙主要由富含碳酸盐岩、生物扰动的半深海泥岩组成，并含有火山灰的薄夹层（图15-6）。薄层和不规则斑点状的粉砂岩、砂质粉砂岩和细砂岩，在C0004和C0008钻孔较C0001钻孔常见。在C0001钻孔处，斜坡裙最底部10m厚地层主要由硅质碎屑砂岩构成，大部分是石英和长石碎屑颗粒，也含有一些沉积岩岩屑和低级变质岩岩屑（泥岩、泥板岩、燧石和石英—云母颗粒）。斜坡裙的最大沉积年龄从早更新世（1.60Ma）到晚上新世（2.06Ma），沉积厚度从C0004钻孔处的78m到C0001钻孔处的207m。斜坡裙沉积的泥质部分含有大量的钙质超微化石（方解石质量百分比高达25%），且向上方解石的含量逐渐增加。这一变化趋势可能是由于沉积场所被逐渐隆升，到达方解石补偿深度界面（CCD）之上导致。在Nankai俯冲带体系的其他地方，也能看见这一现象（Underwood等，2003）。

斜坡裙的底部存在一个复杂的角度不整合（图15-6）。在C0004钻孔处相应的沉积间断为2.06—1.60Ma（根据钙质超微化石测定），然而在C0001钻孔该间断年龄为3.79—2.06Ma。在C0001钻孔紧邻不整合面之下，发现了上新统—上中新统的增生楔地层，其时间范围是5.32—3.79Ma。地震振幅反射特征显示了断层上盘这一部分到分支断层系岩性较为单一；该处的岩心——主要由贫碳酸盐的半深海泥岩组成，也证实了这一观点（图15-6）。和富砂和富砾的增生海沟沉积（来自于前缘冲断带内的岩心）相比，富泥相带沉积于安静的沉积环境（图15-4）。其初始的沉积环境目前仍难以解释，但可尝试性归结于位于方解石补偿深度（CCD）之下的海沟底面异常富泥相沉积（航次315科学家，2009a）。不幸的是由于不稳定的钻井条件，该处的取心到海底之下456m就停止了。

在C0004钻孔，存在一个40m厚的块体搬运复合体（MTC）。它位于斜坡裙底部的不整合面之下，含有半深海泥质碎屑（图15-6）。块体搬运复合体的上部存在化学改造作用的证据（如黄铁矿）。同沉积角砾岩形成时间为2.87—2.06Ma，在某些地方是杂基支撑，而在其他地方则是颗粒支撑。这些泥质碎屑，可能来自斜坡裙的上坡暴露处，也可能是从增生楔暴露的顶部剥离下来的（航次316科学家，2009a；Strasser等，2009）。块体搬运复合体下部的地层，类似于C0001钻孔处的增生地层，主要由上新统的贫碳酸盐泥岩组成；只是年龄较后者小，为4.13—2.39Ma。

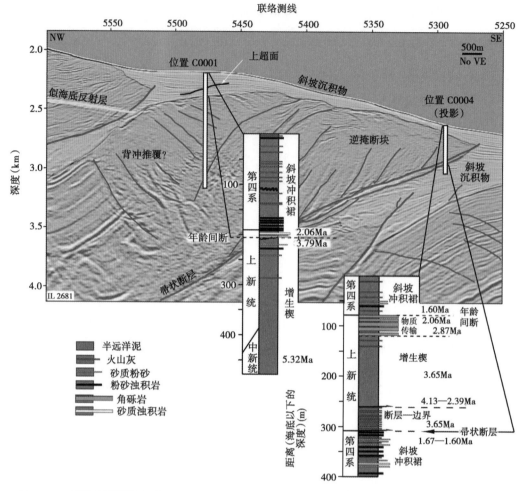

图 15-6 区域地震测线穿过 IODP 工区 C0001 和 C0002 以及 Nankai 增生楔的浅部巨型分支断层带

地震解释来自 Moore 等（2009）；地层总结是根据 IODP 航次 314、315 和 316 的
岩心和随钻测井数据；定年数据根据超微化石带

和 C0001 钻孔不同，C0004 钻孔在断层上倾没端附近钻穿了巨型分支断层（图 15-6），并钻遇了夹在两条近平行分支断层之间的约 50m 厚的上新统。这一被两条断层夹持的岩片最大年龄是 3.65Ma，发育贫碳酸盐的半深湖泥岩，并含有异常丰富的火山灰层。该岩片和增生楔内同类岩层的生物地层对比目前仍然没有定论；但我们发现横跨这两个近平行断层的底部，存在着地层时代倒转。下盘为早更新世（1.67—1.60Ma）地层；岩性和上部的斜坡裙类似，也由半深海泥质组成，其钙质超微化石含量中等。薄层的粉砂岩和粉砂质砂岩浊积体在该地层单元中也较常见。这一岩性组合的恢复支持了该区的构造解释，即巨型分支断层的上盘已经逆冲到了斜坡裙沉积之上。

C0008 钻孔在巨型分支断层末端向海方向进行了钻井取心（图 15-7）。该点的斜坡裙厚约 272m，含有两个次级地层单元。上部亚层主要由更新统（最大年龄是 1.67Ma）的半深海泥组成，钙质超微化石含量中等，含粉砂岩、砂质粉砂岩和火山灰的薄夹层（<10cm）。除了空落火山灰层（图 15-5），该层段还有一个 5m 厚的火山碎屑砂岩层。该火山碎屑砂岩层

含有大量的玄武岩碎屑（含褐色玻璃质）、浮岩和玻璃质碎片；火山灰主要集中在该层的顶部。大多数陆源粉砂岩和砂岩层主要由石英、长石和变质沉积岩岩屑组成，但也部分富含火山玻璃质碎片与浮岩碎块。由于含有大量自生黄铁矿，许多砂层的颜色接近黑色。总体而言，剖面向上砂和粉砂的比率降低，这一岩性组合是斜坡裙沉积的典型特征。

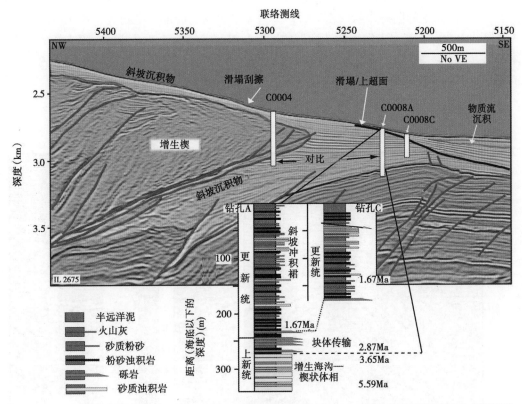

图 15-7　区域地震测线穿过 IODP 工区 C0004 和 C0008 以及 Nankai 增生楔的浅部巨型分支断层带
地震解释来自 Moore 等（2009）；C0008 处的地层总结是根据 IODP 航次 316 的岩心数据；年龄（Ma）根据超微化石带

　　下部亚层主要为泥质碎屑砾岩与半深海泥岩互层。砾石层的厚度为 2~80cm，一些坚硬泥质碎屑的最大粒径是 5cm。碎屑—杂基的组合方式从颗粒支撑到杂基支撑都有。碎屑颗粒的形态从圆状到次圆状（图 15-5）。该亚层被解释为块体搬运复合体（MTC），沉积于斜坡裙的底部（航次 316 科学家，2009d）。其最大年龄是 2.87Ma（类似于 C0004 钻孔的角砾岩沉积）。块体搬运复合体（MTC）与上部地层呈整合接触；其下部与下伏增生楔从地震反射的几何特征上看，也显现出平行整合接触。但紧邻接触界面之下的增生地层的顶部年龄是 3.65Ma（增生地层最底部的年龄为 5.59Ma），因此块体搬运复合体（MTC）与增生楔之间可能存在一定的沉积间断（约 0.78Ma）。

　　块体搬运复合体（MTC）之下的增生沉积物主要由细粒—粗粒的砂岩以及含砾砂岩组成，并含有少量的半深海泥质夹层。该砂岩具有多成分特征，含有大量的石英、长石和重矿物（主要是辉石），以及少量的褐色—黑色半透明颗粒（可能是一些基性到中性的火山玻璃质碎屑）和一些泥岩碎屑或海绿石。砾石包括磨圆的玄武岩、深成岩、片岩、脉石英、燧石、砂岩和泥岩的碎屑颗粒。该岩相在组分上与增生楔下缘附近发育的增生轴向水道沉积类似，因此可以推断，早上新世的轴向浊流与前缘刮擦具有类似成因。

15.4 Kumano 盆地

Kumano 盆地位于 Nankai 海槽上部边缘，是该区的几个著名弧前盆地之一。C0002 工作区位于 Kumano 盆地的向海边缘（图 15-1）。一条通过钻孔的地震主测线（图 15-8），揭示了一套厚约 1000m，基本未发生构造变形的地层。该套地层位于一个陡倾、不连续的地震反射体之上（Moore 等，2009）。这一构造变形样式（例如上盘背斜、向陆倾斜反射），与 Nankai 增生楔的其他部分，包括前缘增生楔的现今区带的构造样式基本一致。

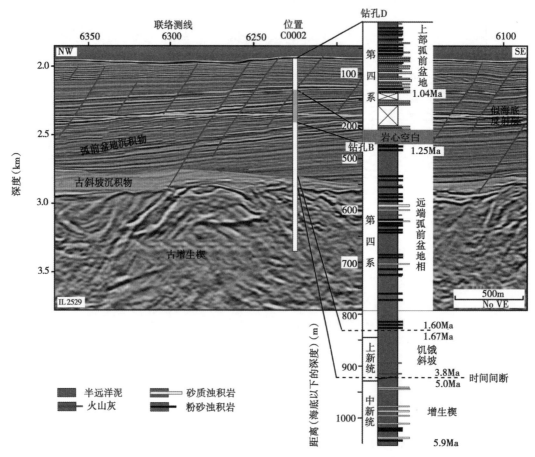

图 15-8 区域地震测线穿过 Kumano 弧前盆地 IODP 工区 C0002

地震解释来自 Moore 等（2009）；地层总结是根据 IODP 航次 314 和 315 的岩心和随钻测井数据；
年龄（Ma）根据超微化石带

增生楔内的取心率不高，但也显示增生沉积物主要由坚硬的、高度破碎的泥岩、粉砂岩和砂岩组成，并含有稀少的碳酸盐结核。泥岩中碳酸钙含量低，说明其沉积环境接近或位于方解石补偿深度界面（CCD）之下。增生楔地层为上中新统到下上新统，根据超微化石事件推断，其年龄范围是 5.9—5.0Ma。增生之前的原始沉积环境很难厘定，但很可能与古海沟楔状体的较细粒部分沉积有关联（航次 315 科学家，2009b）。

C0002 钻孔附近的增生楔顶面在地震剖面上表现出复杂的几何特征。界面之上为一套厚约 50~100m 的空白反射层段（图 15-8）。该层段局部向海倾斜；在其他位置则充填在构造低部位。其几何学特征，和不规则基底一致；其声学特征，显示其岩性较为单一。沉积作用可能发生在海沟斜坡的较低部位，且发生于后者被沿着巨型分支断层的滑动作用抬升之前（Moore 等，2009）。岩心资料证实，该层段主要为半深海泥岩组成的凝缩段，与下伏的增生楔被一角度不整合分开，相应的沉积间断跨度从上部约 3.8Ma 至下部 5.0Ma。界面之上为下更新统到中更新统（根据超微化石事件测定年龄为 3.79—1.67Ma）。在该时间段内沉积速率仅为 18~30m/Ma，比更新世斜坡裙沉积速率要低很多。泥岩中方解石的平均质量百分比约为 16%，反映其形成于方解石补偿深度（CCD）界面之上。该层段还存在一些其他值得注意的特征，包括 S 形的黏土充填脉构造（类似现象描述见 Ogawa 和 Miyata，1985；Pickering 等，1990；Brothers 等，1996）、广泛分布的 *Chrondrites* 和 *Zoophycos* 动物遗迹群、局部的海绿石富集和尖锐不规则的表面（被认为基底坚硬的反映）。

Kumano 盆地充填的大部分地层具有向陆地倾斜、高振幅、侧向连续的地震反射特征，被解释为浊积岩序列（Moore 等，2009）。浊积岩叠覆在老的凝缩段泥岩层之上（图 15-8），但岩心中没有证据表明它们之间存在不整合。浊积岩向陆地方向倾斜，可能由于沿着巨型分支断层的滑动促使盆地向海一侧发生隆升（Park 等，2002）。持续的隆升也导致了沉积作用向陆地迁移。盆地充填地层被非常年轻的正断层错动，甚至表层的沉积物也被许多正断层切穿和移动。岩心资料证实，浊流沉积作用开始于约 1.67—1.60Ma。从那之后，沉积速率明显加速，达到 400~800m/Ma。这些结果显示，弧前盆地比较年轻，直到中—晚更新世才开始出现高速的浊流沉积。

弧前盆地浊积岩的沉积特征（中等规模的旋回和独立的粒序层），在随钻测井记录中比较详细（航次 314 科学家，2009a）。尽管岩心资料精确可靠，但通过旋转取心系统获取砂岩岩心的回收率通常很低。最可靠的信息来自最上部 140m 的沉积物，这一段用液压系统进行了取心。该段主要由半深海泥与大量正粒序粉砂岩、砂质粉砂岩和砂岩薄夹层（1~15cm）组成，同时局部含有火山灰夹层（图 15-8）。获取的砂岩层最厚为 1.8m。这些岩层具有底面尖锐、弱的水平—平行纹层、正粒序以及顶部发散的特点。粉砂和砂的颗粒主要由石英、长石、沉积岩和变质沉积岩（页岩、泥板岩、燧石、石英—云母）岩屑以及各种各样的重矿物组成。粉砂质泥中的方解石含量一般比较低（平均质量百分数为 2%）；据推测这是由于生物成因的超微化石被高速输入的陆源碎屑所稀释造成的。在一个远端盆地—平原型沉积环境中，砂岩的沉积可能受控于沿着巨型分支断层发生的快速隆升作用形成的可容空间。该隆升作用约开始于 1.55Ma（Strasser 等，2009）。同时，来源于岸线的砂质和粉砂质浊流水道的搬运频率也取决于内陆的高剥蚀速率和有效输导体系对上陆坡的下切速率。该全新世输导体系以贯穿的海底峡谷和斜坡冲沟为代表（图 15-1），大约 1.6Ma 才开始高速活动，即下伏增生楔前缘增生之后约 3.4Ma，巨型分支断层上的初始滑动事件之后约 0.35Ma，输导体系才开始高速活动。

15.5 结论

综合三维地震调查中获得的地球物理数据以及综合大洋钻探计划（IODP）最近三次钻探航次中取得的岩心和测井数据，为 Nankai 俯冲带的岩石地层和构造演化提供了有价值的新信息。IODP 的船上数据经过提炼，有如下比较重要的发现。

在前缘断裂带岩心中获得的大多数信息与地震资料的构造解释和前缘增生体的基本模型一致。Kumano 横剖面区域内的增生楔,主要由整体向上变厚变粗的富砂和富砾沉积体组成。尽管由于其邻近 Suruga 海槽和 Tenryu 峡谷(物源供给通道),粗粒的水道—天然堤复合体沉积占优势,但其岩相特征和沿着 Nankai 海槽的 Muroto 和 Ashizuri 横剖面所取的岩心类似。逆冲断层罕见地将上 Shikoku 盆地相地层推覆到前缘增生楔的浅部。并且出人意料的是,在 Shikoku 盆地地层剖面顶部存在削截现象。缺失的地层部分,很可能在俯冲海山一侧海底滑塌作用导致的前缘增生之前就已被搬运走。该过程早于海山被逐渐埋于向陆增厚的海沟楔状体之下。

从巨型分支断层体系上倾没端附近的岩心中获取的大部分信息,和地震资料的构造解释一致。没有预料到的是,存在一个将斜坡裙和下伏增生楔分开的不整合。该不整合具有穿时性(向海方向变年轻),沉积间断的持续时间从 1.73Ma 到 0.46Ma。该不整合很可能是由增生楔抬升和削峭作用期间持续并广泛分布的滑坡和海底滑塌作用产生的。在两个钻孔的岩心中存在的块体搬运复合体(泥质碎屑角砾岩和砾岩)支持这一解释。浅部的分支断层上盘块体主要由贫碳酸盐的半深海泥岩组成,其细粒和岩性单一的特点与前缘冲断带富砂和富砾的增生海沟楔状沉积形成鲜明对比。

地震反射资料清晰地揭示了 Kumano 盆地充填的几何样式。岩心数据证实,盆地内几乎所有的沉积充填均发生在过去的 1.6Ma 之内。盆地下部的增生楔形成于晚中新世到早上新世(5.9Ma—5.0Ma)。弧前盆地首先经历了长期的缓慢的半深海沉积,该凝缩段沉积时间范围从约 3.79Ma 到约 1.67Ma,然后才是浊流沉积。一个角度不整合将饥饿的盆地相和下伏的增生楔分开;该不整合的沉积间断持续了约 1.2 Ma。向弧前盆地加速搬运沉积物,滞后于沿着分支断层隆升(开始于 1.5Ma 左右)产生的可容空间。大陆边缘广泛发育的海底峡谷和斜坡冲沟的下切作用,和更新世新生的日本群岛上物源区的快速隆升和剥蚀作用可能进一步增强了弧前盆地浊积砂的快速输入。

地震反射记录和岩心资料共同为研究 Nankai 边缘过去 6Ma 以来的演化提供了新视角。来自于 Nankai 海槽发震带试验 Kumano 横剖面的岩心资料支持俯冲带沉积体系怎样运行的一些基本论点。沿着俯冲带的走向,俯冲带物质输入的方式错综复杂,特别是对基底地貌和强烈影响增生楔演化的其他变量的响应,更加复杂。这种动态的相互作用在其他方向也存在,尤其是考虑到沿着巨型分支断裂发生的偏移、海底滑动造成的斜坡沉积物的重新搬运,以及浊积盆地可容空间的形成等。在解释岩石记录中类似古地质现象的时候,需要综合立体考虑构造与岩石地层的演化。

致谢(略)

参 考 文 献

Aoki, Y., Tamano, T., and Kato, S. (1982) Detailed structure of the Nankai Trough from migrated seismic sections. In Watkins, J. S., and Drake, C. L., eds., Studies in Continental Margin Geology. Am. Assoc. Petrol. Geol. Memoir, 34, 309-322.

Ashi, J., Lallemant, S., Masago, H., and the Expedition 315 Scientists (2009) Expedition 315 summary. In Kinoshita, M., Tobin, H., Ashi, J., Kimura, G., Lallemant, S., Screaton, E. J., Curewitz, D., Masago, H., Moe, K. T., and the Expedition 314/315/316 Scientists,

Proc. IODP, 314/ 315/316: Washington DC, Integrated Ocean Drilling Program Management International, Inc. doi: 10. 2204/iodp. proc. 314315316. 121. 2009.

Bangs, N. L., Shipley, T. H., Gulick, S. P. S., Moore, G. F., Kuromoto, S., and Nakamura, Y. (2004) Evolution of the Nankai Trough d_ecollement from the trench into the seismogenic zone: inferences from three-dimensional seismic reflection imaging. Geology, 32 (4), 273-276. doi: 10. 1130/G20211. 1.

Bangs, N. L. B., Gulick, S. P. SA., and Shipley, T. H. (2006) Seamount subduction erosion in the Nankai Trough and its potential impact on the seismogenic zone. Geology, 34 (8), 701-704. doi: 10. 1130/G22451. 1.

Beaudry, D., and Moore, G. F. (1985) Seismic stratigraphy and Cenozoic evolution of west Sumatra forearc basin. AAPG Bull., 69 (5), 742-759.

Brothers, R. J., Kemp, A. E. S., and Maltman, A. J. (1996) Mechanical development of vein structures due to the passage of earthquake waves through poorly consolidated sediments. Tectonophysics, 260 (4), 227-244. doi: 10. 1016/0040-1951 (96) 00088-1.

Carlson, P. R., and Nelson, C. H. (1987) Marine geology and resource potential of CascadiaBasin. In Geology and Resource Potential of the Continenta Margin of Western North America and Adjacent Ocean Basins-Beaufort Sea to Baja California, Scholl, D. W., Grantz, A., and Vedder, J., eds., Houston, TX, Circum-Pacific Counc. Energy Miner. Res., 523-536.

Clift, P., and Vannucchi, P. (2004) Controls on tectonic accretion versus erosion in subduction zones: implications for the origin and recycling of the continental crust. Rev. Geophys., 42: RG2001. doi: 10. 1029/2003RG000127.

Coulbourn, W. T. (1986), Sedimentologic summary, Nankai Trough Sites 582 and 583, and Japan Trench Site 584. In Kagami, H., Karig, D. E., and Coulbourn, eds., Init. Rpts. DSDP, 87: 909-926. Washington, DC, U. S. Goverment Printing Office.

Expedition 314 Scientists (2009a) Expedition 314 Site C0002. In Kinoshita, M., Tobin, H., Ashi, J., Kimura, G., Lallemant, S., Screaton, E. J., Curewitz, D., Masago, H., Moe, K. T., and the Expedition 314/315/316 Scientists, Proc. IODP, 314/315/316: Washington DC, Integrated Ocean Drilling Program Management International, Inc. doi: 10. 2204/iodp. proc. 314315316. 114. 2009.

Expedition 314 Scientists (2009b) Expedition 314 Site C0006. In Kinoshita, M., Tobin, H., Ashi, J., Kimura, G., Lallemant, S., Screaton, E. J., Curewitz, D., Masago, H., Moe, K. T., and the Expedition 314/315/316 Scientists, Proc. IODP, 314/315/316: Washington DC (Integrated Ocean Drilling Program Management International, Inc.) doi: 10. 2204/iodp. proc. 314315316. 118. 2009.

Expedition 315 Scientists (2009a) Expedition 315 Site C0001. In Kinoshita, M., Tobin, H., Ashi, J., Kimura, G., Lallemant, S., Screaton, E. J., Curewitz, D., Masago, H., Moe, K. T., and the Expedition 314/315/316 Scientists, Proc. IODP, 314/315/316. Washington, DC, Integrated Ocean Drilling Program Management International, Inc. doi: 10. 2204/iodp. proc. 314315316. 123. 2009.

Expedition 315 Scientists (2009b) Expedition 315 Site C0002. In Kinoshita, M., Tobin, H., Ashi, J., Kimura, G., Lallemant, S., Screaton, E. J., Curewitz, D., Masago, H., Moe, K. T., and the Expedition 314/315/316 Scientists, Proc. IODP, 314/315/316: Washington DC (Integrated Ocean Drilling Program Management International, Inc.) doi: 10. 2204/iodp. proc. 314315316. 124. 2009.

Expedition 316 Scientists (2009a) Expedition 316 Site C0004. In Kinoshita, M., Tobin, H., Ashi, J., Kimura, G., Lallemant, S., Screaton, E. J., Curewitz, D., Masago, H., Moe, K. T., and the Expedition 314/315/316 Scientists, Proc. IODP, 314/315/316: Washington, DC, Integrated Ocean Drilling Program Management International, Inc. doi: 10. 2204/iodp. proc. 314315316. 133. 2009.

Expedition 316 Scientists (2009b) Expedition 316 Site C0006. In Kinoshita, M., Tobin, H., Ashi, J., Kimura, G., Lallemant, S., Screaton, E. J., Curewitz, D., Masago, H., Moe, K. T., and the Expedition 314/315/316 Scientists, Proc. IODP, 314/315/316: Washington DC, Integrated Ocean Drilling Program Management International, Inc. doi: 10. 2204/iodp. proc. 314315316. 134. 2009.

Expedition 316 Scientists (2009c) Expedition 316 Site C0007. In Kinoshita, M., Tobin, H., Ashi, J., Kimura, G., Lallemant, S., Screaton, E. J., Curewitz, D., Masago, H., Moe, K. T., and the Expedition 314/315/316 Scientists, Proc. IODP, 314/315/316: Washington DC, Integrated Ocean Drilling Program Management International, Inc. doi: 10. 2204/iodp. proc. 314315316. 135. 2009.

Expedition 316 Scientists (2009d) Expedition 316 Site C0008. In Kinoshita, M., Tobin, H., Ashi, J., Kimura, G., Lallemant, S., Screaton, E. J., Curewitz, D., Masago, H., Moe, K. T., and the Expedition 314/315/316 Scientists, Proc. IODP, 314/315/316: Washington DC, Integrated Ocean Drilling Program Management International, Inc. doi: 10. 2204/iodp. proc. 314315316. 136. 2009.

Fisher, D., Mosher, D., Austin, J. A., Jr., Gulick, S. S. P., Masterlark, T., and Moran, K. (2007) Active deformation across the Sumatran forearc over the December 2004Mw 9. 2 rupture. Geology, 35 (2), 99-102. doi: 10. 1130/ G22993A. 1.

Gulick, S. P. S., Bangs, N. L. B., Shipley, T. H., Nakamura, Y., Moore, G., and Kuramoto, S. (2004) Three-dimensional architecture of the Nankai accretionary prism's imbricate thrust zone off Cape Muroto, Japan: prism reconstruction via en echelon thrust propagation. J. Geophys. Res., 109 (B2), B02105. doi: 10. 1029/2003JB002654.

Gutscher, M-A., Kukowski, N., Malavielle, J., and Lallemand, S. (1998) Episodic imbricate thrusting and underthrusting: Analog experiments and mechanical analysis applied to the Alaskan accretionary wedge. J. Geophys. Res., 103 (B5), 10161-10176.

Ike, T., Moore, G. F., Kuramoto, S., Park, J.-O., Kaneda, Y., and Taira, A. (2008a) Tectonics and sedimentation around Kashinosaki Knoll: a subducting basement high in the eastern Nankai Trough. Isl. Arc, 17 (3), 358-375. doi: 10. 1111/j. 1440-1738. 2008. 00625. x.

Ike, T., Moore, G. F., Kuramoto, S., Park, J.-O., Kaneda, Y., and Taira, A. (2008b)

Variations in sediment thickness and type along the northern Philippine Sea plate at the Nankai Trough. Isl. Arc, 17 (3), 342-357. doi: 10. 1111/ j. 1440-1738. 2008. 00624. x.

Kagami, H., Karig, D. E., Coulbourn, W., et al. (1986) Init. Repts. DSDP, 87. Washington, DC, U. S. Government Printing Office.

Karig, D. E., Ingle, J. C., Jr., et al. (1975) Init. Repts. DSDP, 31. Washington, DC, U. S. Government Printing Office. doi: 10. 2973/dsdp. proc. 31. 1975.

Karig, D. E., Suparka, S., Moore, G. F., and Hehanussa, P. E. (1978) Structure and Cenozoic evolution of the Sunda Arc in the central Sumatra region. In Watkins, J. S., Montadert, L., and Dickerson, P. W., eds., Geological and Geophysical Investigations of Continental Margins. Am. Assoc. Petrol. Geol. Memoir, 29, 223-237.

Kawamura, K., Ogawa, Y., Anma, R., Yokoyama, S., Kawakami, S., Dilek, Y., Moore, G. F., Hirano, S., Yamaguchi, A., Sasaki, T., and YK05-08 Leg 2 and YK06-02 Shipboard Scientific Parties (2009) Structural architecture and active deformation of the Nankai accretionary prism, Japan: Submersible survey results from the Tenryu submarine canyon. Geol. Soc. Am. Bull., 121, 1629-1646. doi: 10. 1130/B26219. 1.

Kobayashi, K., Kasuga, S., and Okino, K. (1995) ShikokuBasin and its margins. In Taylor, B., (Ed.), Back-arc basins. New York, Plenum, 381-405.

Kopp, H., Flueh, E. R., Klaeschen, D., Bialas, J., and Reichert, C. (2001) Crustal structure of the central Sunda margin at the onset of oblique subduction. Geophys. J. Int., 147, 449-474.

Kopp. H., and Kukowski, N. (2003) Backstop geometry and accretionary mechanics of the Sunda margin. Tectonics, 22, 1072. doi: 10. 1029/2002TC001420.

Kulm, L. D., and Fowler, G. A. (1974) Oregon continental margin structure and stratigraphy: Atest of the imbricate thrust model. In Burk, C. A., and Drake, C. L., eds., The geology of continental margins. New York, Springer- Verlag, 261-283.

Lewis, S. D., Ladd, J. W., and Bruns, T. R. (1988) Structural development of an accretionary prism by thrust and strike-slip faulting: Shumagin region, Aleutian Trench. Geol. Soc. Am. Bull., 100, 767-782.

MacKay, M. E., Moore, G. F., Cochrane, G. R., Moore, J. C., and Kulm, L. D. (1992) Landward vergence and oblique structural trends in the Oregon margin accretionary prism: Implications and effect on fluid flow. Earth Planet. Sci. Lett., 109, 477-491.

Marsaglia, K. M., Ingersoll, R. V., and Packer, B. M. (1992) Tectonic evolution of the JapaneseIslands as reflected in modal compositions of Cenozoic forearc and backarc sand and sandstone. Tectonics, 11, 1028-1044.

McCarthy, J., and Scholl, D. W. (1985) Mechanisms of subduction accretion along the central Aleutian Trench. Geol. Soc. Am. Bull., 96, 691-701.

Mikada, H., Moore, G. F., Taita, A., Becker, K., Moore, J. C., and Klaus, A., eds. (2005) Proc. ODP, Sci. Results, 190/196. wwwodp. tamu. edu/publications/190196SR/190196sr. htm.

Miyazaki, S., and Heki, K. (2001) Crustal velocity field of southwest Japan: subduction and arc-

arc collision. J. Geophys. Res., 106 (B3), 4305-4326. doi: 10. 1029/ 2000JB900312.

Moore, G. F., Curray, J. R., Moore, D. G., and Karig, D. E. (1980) Variations in geologic structure along the Sunda forearc, northeastern Indian Ocean. In Hayes, D. E., ed., The tectonic and geologic evolution of Southeast Asian seas and islands. Am. Geophys. Union Geophysical Monograph, 23, 145-160.

Moore, G. F., Shipley, T. H., Stoffa, P. L., Karig, D. E., Taira, A., Kuramoto, S., Tokuyama, H., and Suyehiro, K. (1990) Structure of the Nankai Trough accretionary zone from multichannel seismic reflection data. J. Geophys. Res., 95 (B6), 8753-8765. doi: 10. 1029/JB095iB06p08753.

Moore, G. F., Taira, A., Klaus, A., Becker, L., Boeckel, B., Cragg, B. A., Dean, A., Fergusson, C. L., Henry, P., Hirano, S., Hisamitsu, T., Hunze, S., Kastner, M., Maltman, A. J., Morgan, J. K., Murakami, Y., Saffer, D. M., Sánchez-Gomez, M., Screaton, E. J., Smith, D. C., Spivack, A. J., Steurer, J., Tobin, H. J., Ujiie, K., Underwood, M. B., and Wilson, M. (2001) New insights into deformation and fluid flow processes in the Nankai Trough accretionary prism: results of Ocean Drilling Program Leg 190. Geochem., Geophys., Geosyst., 2 (10), 1058. doi: 10. 1029/2001GC000166.

Moore, G. F., Bangs, N. L., Taira, A., Kuramoto, S., Pangborn, E., and Tobin, H. J. (2007) Three-dimensional splay fault geometry and implications for tsunami generation. Science, 318 (5853), 1128-1131. doi: 10. 1126/science. 1147195.

Moore, G. F., Park, J. O., Bangs, N. L., Gulick, S. P., Tobin, H. J., Nakamura, Y., Sato, S., Tsuji, T., Yoro, T., Tanaka, H., Uraki, S., Kido, Y., Sanada, Y., Kuramoto, S., and Taira, A. (2009) Structural and seismic stratigraphic framework of the NanTroSEIZE Stage 1 transect. In Kinoshita, M., Tobin, H., Ashi, J., Kimura, G., Lallemant, S., Screaton, E. J., Curewitz, D., Masago, H., Moe, K. T., and the Expedition 314/315/316 Scientists, eds., Proc. IODP, 314/315/ 316. Washington DC, Integrated Ocean Drilling Program Management International, Inc. doi: 10. 2204/iodp. proc. 314315316. 102. 2009.

Mountney, N. P., and Westbrook, G. K. (1996) Modeling sedimentation in ocean trenches: the Nankai Trough from 1 Myr to the present. Basin Research, 8, 85-101.

Ogawa, Y., and Miyata, Y. (1985) Vein structure and its deformational history in the sedimentary rocks of the Middle America Trench slope off Guatemala, Deep Sea Drilling Project Leg 84. In von Huene, R., Aubouin, J., Ed., et al., Init. Repts. DSDP, 84, Washington, DC, U. S. Government Printing Office, 811-829. doi: 10. 2973/ dsdp. proc. 84. 136. 1985.

Okino, K., Shimakawa, Y., and Nagaoka, S. (1994) Evolution of the ShikokuBasin. J. Geomagn. Geoelectr., 46, 463-479.

Park, J-O., Tsuru, T., Kodaira, S., Cummins, P. R., and Kaneda, Y. (2002) Splay fault branching along the Nankai subduction zone. Science, 297 (5584), 1157-1160. doi: 10. 1126/science. 1074111.

Pickering, K. T., Agar, S. M., and Prior, D. J. (1990) Vein structure and the role of pore fluids in early wet-sediment deformation, late Miocene volcaniclastic rocks, Miura Group, SE Japan. In Knipe, R. J., and Rutter, E. H., eds., Deformation Mechanisms, Rheology and

Tectonics. Geol. Soc. Spec. Publ., 54 (1), 417-430. doi: 10. 1144/GSL. SP. 1990. 054. 01. 38.

Pickering, K. T., Underwood, M. B., and Taira, A. (1993) Stratigraphic synthesis of the DSDP-ODP sites in the ShikokuBasin and Nankai Trough and accretionary prism. In Hill, I. A., Taira, A., Firth, J. V., et al. Proc. ODP, Sci. Results, 131, College Station, TX, Ocean Drilling Program, 313-330. doi: 10. 2973/odp. proc. sr. 131. 135. 1993.

Piper, D. J. W., von Huene, R., and Duncan, J. R. (1973) Late Quaternary sedimentation in the active eastern Aleutian Trench. Geology, 1 (1), 19-22.

Scholl, D. W. (1974) Sedimentary sequences in the north Pacific trenches. In Burk, C. A., and Drake, C. L., eds., The geology of continental margins. New York, Springer-Verlag, 493-504.

Screaton, E. J., Kimura, G., Curewitz, D., and the Expedition 316 Scientists (2009a) Expedition 316 summary. In Kinoshita, M., Tobin, H., Ashi, J., Kimura, G., Lallemant, S., Screaton, E. J., Curewitz, D., Masago, H., Moe, K. T., and the Expedition 314/315/316 Scientists, eds., Proc. IODP, 314/315/316. Washington, DC, Integrated Ocean Drilling Program Management International, Inc. doi: 10. 2204/iodp. proc. 314315316. 131. 2009.

Screaton, E. J., Kimura, G., Curowitz, D., and25others (2009b) Interactions between deformation and fluids in the frontal thrust region of the NanTroSEIZE transect offshore the Kii Peninsula, Japan: Results from IODP Expedition 316 Sites C0006 and C0007. Geochem. Geophys. Geosyst., 10, Q0AD01. doi: 10. 1029/2009GC002713.

Schweller, W. J., Kulm, L. D., and Prince, R. A. (1981) Tectonics, structure, and sedimentary framework of the Peru-Chile Trench. Geol. Soc. Am. Memoir 154, 323-349.

Seno, T., Stein, S., and Gripp, A. E. (1993) A model for the motion of the Philippine Sea plate consistent with NUVEL-1 and geological data. J. Geophys. Res., 98 (B10), 17941-17948. doi: 10. 1029/93JB00782.

Shimamura, K. (1989) Topography and sedimentary facies of the Nankai Deep Sea Channel. In Taira, A., and Masuda, F., eds., Sedimentary facies in the active plate margin. Tokyo, Terra Science, 529-556.

Soh, W., and Tokuyama, H. (2002) Rejuvenation of submarine canyon associated with ridge subduction, TenryuCanyon, off Tokai, central Japan. Marine Geol., 187, 203-220.

Steurer, J. F., and Underwood, M. B. (2003) Clay mineralogy of mudstones from the Nankai Trough reference sites and frontal accretionary prism. In Mikada, H., Moore, G. F., Taira, A., Becker, K., Moore, J. C., and Klaus, A., eds., Proc. ODP, Sci. Results, 190/196. College Station, TX, Ocean Drilling Program.

Strasser, M., Moore, G. F., Kimura, G., Kitamura, Y., Kopf, A. J., Lallemant, S., Park, J-O., Screaton, E. J., Su, X., Underwood, M. B., and Zhao, X. (2009) Origin and evolution of a tsunamigenic splay fault. Nature Geoscience, 2 (9), 648-652. doi: 10. 1038/ngeo609.

Taira, A., and Ashi, J. (1993) Sedimentary facies evolution of the Nankai forearc and its implications for the growth of the Shimanto accretionary prism. In Hill, I. A., Taira, A., Firth, J. V., et al. eds., Proc. ODP, Sci. Results, 131, College Station, TX, Ocean Drilling

Program, 331-341.

Taira, A., Hill, I., Firth, J. V., et al. (1991) Proc. ODP, Init. Repts., 131. College Station, TX, Ocean Drilling Program. doi: 10. 2973/odp. proc. ir. 131. 1991.

Thornburg, T. M., and Kulm, L. D. (1987) Sedimentation in the Chile Trench: depositional morphologies, lithofacies, and stratigraphy. Geol. Soc. Am. Bull., 98, 33-52.

Tobin, H. J., and Kinoshita, M. (2006) NanTroSEIZE: the IODP Nankai Trough Seismogenic Zone Experiment. Sci. Drill., 2, 23-27. doi: 10. 2204/iodp. sd. 2. 06. 2006.

Tobin, H., Kinoshita, M., Ashi, J., Lallemant, S., Kimura, G., Screaton, E., Moe, K. T., Masago, H., Curewitz, D., and the Expedition314/315/316Scientists (2009a) Nan TroSEIZE Stage 1 expeditions: introduction and synthesis of key results. In Kinoshita, M., Tobin, H., Ashi, J., Kimura, G., Lallemant, S., Screaton, E. J., Curewitz, D., Masago, H., Moe, K. T., and the Expedition 314/315/316 Scientists, eds., Proc. IODP, 314/315/316: Washington, DC, Integrated Ocean Drilling Program Management International, Inc. doi: 10. 2204/iodp. proc. 314315316. 101. 2009.

Tobin, H., Kinoshita, M., Moe, K. T., and the Expedition 314 Scientists (2009b) Expedition 314 summary. In Kinoshita, M., Tobin, H., Ashi, J., Kimura, G., Lallemant, S., Screaton, E. J., Curewitz, D., Masago, H., Moe, K. T., and the Expedition 314/315/316 Scientists, eds., Proc. IODP, 314/315/316: Washington, DC, Integrated Ocean Drilling Program Management International, Inc. doi: 10. 2204/iodp. proc. 314315316. 111. 2009.

Underwood, M. B. (2007) Sediment inputs to subduction zones: why lithostratigraphy and clay mineralogy matter. In Dixon, T. H., and Moore, J. C., eds., The seismogenic zone of subduction thrust faults. New York, ColumbiaUniversity Press, 42-85.

Underwood, M. B., and Fergusson, C. L. (2005) Late Cenozoic evolution of the Nankai trench-slope system: evidence from sand petrography and clay mineralogy. In Hodgson, D. M., and Flint, S. S., eds., Submarine slope systems: processes and products. Special Publication, 244. Geol. Soc. London, 113-129.

Underwood, M. B., and Moore, G. F. (1995) Trenches and trench-slope basins. In Busby, C. J., and Ingersoll, R. V., eds., Tectonics of sedimentary basins. Cambridge, MA, Blackwell Science, 179-220.

Underwood, M. B., and Steurer, J. (2003) Composition and sources of clay from the trench slope and shallow accretionary prism of Nankai Trough. In Mikada, H., Moore, G. F., Taira, A., Becker, K., Moore, J. C., and Klaus, A., eds., Proc. ODP, Sci. Results, 190/196 College Station, TX, Ocean Drilling Program.

Underwood, M. B., Moore, G. F., Taira, A., Klaus, A., Wilson, M. E. J., Fergusson, C. L., Hirano, S., Steurer, J., and the Leg 190 Shipboard Scientific Party (2003) Sedimentary and tectonic evolution of a trench-slope basin in the Nankai subduction zone of southwest Japan. J. Sediment. Res., 73 (4), 589-602.

Underwood, M. B., Hoke, K. D., Fisher, A. T., Davis, E. E., Giambalvo, E., Zuhlsdorff, L., and Spinelli, G. A. (2005) Provenance, stratigraphic architecture, and hydrogeologic influence of turbidites on the mid-ocean ridge flank of northwestern CascadiaBasin, Pacific

Ocean. J. Sediment. Res., 75, 149-164.

Underwood, M. B., Saito, S., Kubo, Y., and the Expedition 322 Scientists. (2010) Expedition 322 summary, in Saito, S., Underwood, M. B., Kubo, Y., and the Expedition 322 Scientists, eds., Proc. IODP, 322. Tokyo, Integrated Ocean Drilling Program Management International, Inc. doi: 10. 2204/iodp. proc. 322. 101. 2010.

Zang, S. X., Chen, Q. Y., Ning, J. Y., Shen, Z. K., and Liu, Y. G. (2002) Motion of the Philippine Sea plate consistent with the NUVEL-1A model. Geophys. J. Int., 150 (3), 809-819. doi: 10. 1046/j. 1365-246X. 2002. 01744. x.

(刘长利 译，阳怀忠 王鹏 校)

第16章 板块俯冲作用对大陆弧前盆地的影响：以阿拉斯加南部为例

KENNETH D. RIDGWAY[1], JEFFREY M TROP[2], EMILY S. FINZEL[3]

(1. Department of Earth & Atmospheric Sciences, Purdue University, West Lafayette, USA;
2. Department of Geology, Bucknell University, Lewisburg, USA;
3. Department of Earth & Atmospheric Sicengces, Purdue University, West Lafayette, USA)

摘 要：弧前盆地是发育于会聚边缘上部板块内的大型沉积物堆积场所，同时也是对俯冲作用的直接响应。这种盆地是岩浆弧—弧前盆地—增生楔组成的"三位一体"结构体系中的一部分，后者确定了多数与俯冲作用相关联的会聚板块边缘上部板块的构造格局。先前已有了许多关于弧前盆地的研究，其中探索了岩浆弧的结构、增生楔的出露与相邻弧前盆地内沉积充填之间的联系。这些研究为我们理解记录在弧前盆地中的以"正常"洋壳长期俯冲为特征的第一级构造过程提供了重要的框架。然而第二级俯冲过程，例如以海山、扩张且无地震活动的洋脊和大洋高原等形式存在的具有浮力的洋壳板片水平俯冲作用，却使得许多会聚大陆边缘变得十分复杂。这些二级构造过程能够在时空上完全改变上部板块的构造轮廓，并形成与"经典"的由岩浆弧—弧前盆地—增生楔组成的"三位一体"结构体系不相符的沉积盆地。

在这一章中，讨论了由古新世—始新世扩张洋脊俯冲作用以及随后的渐新世—全新世厚层洋壳俯冲作用而引起的阿拉斯加南部弧前盆地的变化。该厚洋壳目前正俯冲于阿拉斯加中南部下方，具有在地表可达30km、深部可达22km的最大厚度图像。来自阿拉斯加南部的研究表明，受板片水平俯冲作用改造的弧前盆地可能包含了与"经典"弧前盆地显著不同的沉积和火山地层记录。遭受改造的弧前盆地的演化过程和沉积特征包括：（1）具有浮力的、地形抬升且近似平行于板块边缘的扩张洋脊的水平俯冲作用，促使弧前盆地底部发生穿时性抬升，并使得较老的海相弧前盆地地层随洋脊俯冲而出露剥蚀。扩张洋脊通道导致了沉降作用、非海相地层和火山地层的沉积，局部沉积厚度超过了下伏的海相地层。（2）在扩张洋脊俯冲过程中，弧前盆地之下板片窗的插入，产生了盆内的局部高地和与之相邻的沉积中心，以及弧前盆地内或相邻区域的不连续火山中心。（3）厚层洋壳的水平俯冲作用同样能引起地表抬升和弧前盆地沉积地层的出露剥蚀。然而，板片水平俯冲区内厚洋壳的插入（即缺少板片窗）抑制了弧前盆地毗邻区与俯冲作用相关的岩浆活动。对阿拉斯加中南部之下宽度大于350km的厚洋壳俯冲而言，板片水平俯冲区之上的弧前盆地地层的出露，促进了沉积物向位于俯冲区周围的活跃盆地搬运、沉积。这些盆地记录了与发生于出露的、不活动的残余弧前盆地之下的板片水平俯冲作用同期的沉降和沉积速率的增长。

关键词：弧前盆地 扩张洋脊俯冲 板片水平俯冲 阿拉斯加 板片窗

16.1 引言

关于会聚边缘俯冲板块与上部板块内沉积盆地演化之间相互关系的认识，是理解板块边缘地球动力学的一个基础性突破。先前已有许多研究证实了弧前盆地"三位一体"结构体系的存在，该体系由岩浆弧、弧前盆地以及作为对会聚边缘俯冲作用第一级响应而发育的增生楔组成（Dickinson，1995）。第二级俯冲过程，比如扩张洋脊以及厚洋壳的板片水平俯冲作用（flat-slab subduction），在现代的会聚板块边缘也已有观测，应该在许多大陆边缘弧前盆地的地层中都有所记录。比如板片水平俯冲作用已经完全改变了阿拉斯加南部会聚板块边缘的动力学系统。为了证实这一观点，本文将讨论阿拉斯加南部新生代弧前盆地系统演化过程中浅部俯冲的两个特定时期：古新世—始新世扩张洋脊的俯冲消减作用以及渐新世—全新世亚库卡特微板块的俯冲消减作用，其中亚库卡特微板块为隶属于一块厚洋壳的活跃俯冲板片。

无论对于现代还是古代的会聚板块边缘，上部板块内板片水平俯冲作用过程的本质都是一个值得重点讨论的课题（Dickinson 和 Snyder，1979；Jordan 和 AJlmcndioger，1986；Hampel，2002；Espurt 等，2008）。基于地震活动及与之相关的构造变形的研究表明，板片水平俯冲作用会在会聚板块边缘形成宽广的扩散边界，并且增强挤压应力向内和向上传递进入上部板块（Lallemand 等，2005）。这一章的主旨在于展示扩张洋脊和厚洋壳的板片水平俯冲作用是如何导致在大陆边缘弧前盆地发展过程中盆地结构在时间和空间上的变化。

16.2 阿拉斯加南部弧前盆地系统的地质形态

阿拉斯加南部由复杂的构造带拼贴而成，包括中生代和新生代露头及活跃的沉积盆地、增生地体、岩浆活动带以及增生楔沉积（图 16-1 和图 16-2；Plafker 和 Berg，1994；Trap 和 Ridgway，2007）。在侏罗纪和白垩纪期间，一个侏罗纪的洋内岛弧以及三叠—侏罗纪的海底高原生长在阿拉斯加南部之前的大陆边缘（Wrangellia 在图 16-2a 上合成；Nokleberg 等，1985；Ridgway 等，2002；Trap 等，2005；Rioux 等，2007）。紧随该重要的地壳增长期之后，洋壳向北俯冲形成了一个大陆边缘岩浆弧、弧前盆地以及三叠—侏罗纪洋壳之上的增生楔（图 16-2b；Plafker 和 Berg，1994）。对该构造进行解释的地质学证据主要基于一条长度超过 2000km、距今 85—60Ma 的钙质碱性火成岩构造带，被视作与俯冲相关的岛弧岩浆作用的产物；另一产物为一条长度超过 2000km 的同时代变质沉积岩及变质火山岩构造带，被解释为增生楔地层（图 16-2；Plafker 等，1994；Sample 和 Reid，2003）。在岩浆弧与增生楔之间的海底斜坡上沉积了最大厚度超过 3km 的砂岩、泥岩及砾岩沉积（图 16-2、图 16-3、图 16-4；Trap 等，1999；Trap，2008）。这些地层现在在 Wrangell 山脉区域（图 16-2a 中的 WB；图 16-4a）、Copper River 盆地底面、Matanuska 山谷（图 16-2a 中的 MB；图 16-4b、c），以及 Cook Inlet 盆地底面沿着走向从东向西延绵超过 500km。扩张洋脊和厚洋壳在新生代发生的水平俯冲作用促使白垩纪末期的岩浆弧—弧前盆地—增生楔体系发生了实质性变化。这些改变包括海相弧前盆地地层的出露与侵蚀；以更集中的非海相沉积中心为特征的部分弧前盆地系统的沉积作用；板片窗火成岩向增生楔和弧前盆地地层侵入；以及主要断层系统的同沉积断层位移等（图 6-2b）。下面我们将描述有关弧前盆地变化过程及结果的新生代地层证据。

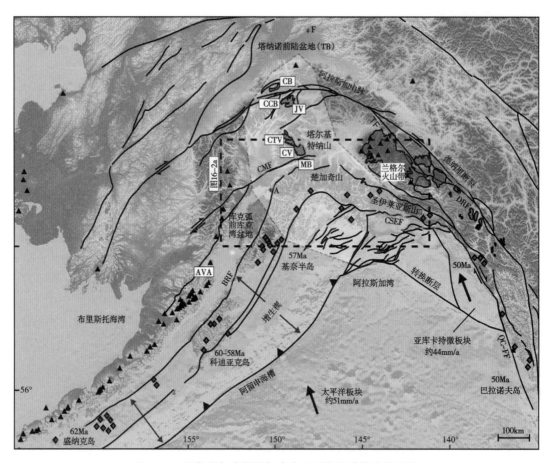

图 16-1　阿拉斯加南部及加拿大地区的地质构造格局图

包括文中讨论的主要断层、全新世的火山（黑色三角形）、现今的板块运动方向（源自 Leonard 等，2007）、亚库卡特微板块出露范围（淡黄色）以及解释的亚库卡特微板块俯冲作用范围（虚线内半透明白色区域）。亚库卡特微板块的俯冲带边界受到不严格的约束（改绘自 Eborhart-Phillips 等，2006；Fuis 等，2008）。近海沟的深成岩体（红色菱形）侵入增生楔岩体中，认为是伴随扩张洋脊俯冲的板片窗岩浆活动的产物，扩张洋脊近似平行于现今 Sanak 岛（图中左下角）在 62Ma 到现今 Baranof 岛（图中右下角）在 50Ma 的板块边缘位置（Baradley 等，2003）。认为向东展布的年轻的近海沟深成岩体代表了扩张洋脊俯冲带的向东迁移。活动的沉积盆地：TB—塔纳诺。出露的新生代沉积盆地：CCB—科罗拉多河；MB—马塔努斯卡。新生代主要的火山活动带：AVA—阿拉斯加半岛—阿留申群岛火山岛弧；CB—Cantwell 盆地火山区；CTV—中央 Talkeenta 山脉火山区；CV—Caribou Creek 火山区；JV—Jack 河火山区；WVB—兰格尔火山带。主要断层：CMF—Castle 山断层；CSEF—Chugach-St. Elias 断层；BRF—边界山脉断层；DRF—Duke 河断层；QC-FF—Queen Charlotto-Fairweather 断层；TF—Totschunda 断层；A—Anchorage 镇；F—Fairbanks 镇

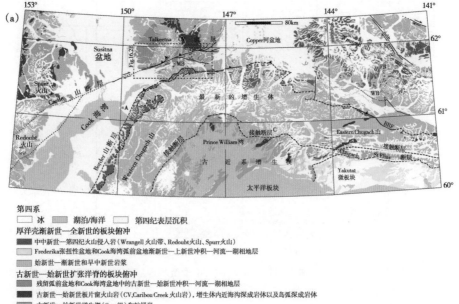

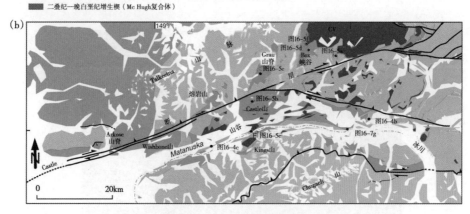

图 16-2 阿拉斯加弧前盆地地质图

(a) 阿拉斯加中南部综合地质图，表现出晚白垩世"正常"俯冲作用、古新世及始新世扩张洋脊的板片水平俯冲作用，以及渐新世—全新世亚库卡特微板块的俯冲作用的特征；图 16-1 中黑色虚线表示了该地质图所表示的位置。缩写表示：A—Anchorage；BRF—边界山脉断层；CV—Caribou Creek 火山区；CTV—中央 Talkeenta 火山区；FB—Frederika 盆地；MB—Matanuska 盆地；WB—Wrangell 山脉盆地。黑色细线定义的是一个 1:250000 的四边形区域（据 Trop 和 Ridgway，2007）。(b) Matanuska 山谷、Talkeetna 山脉南部及 Chugach 山脉南部的放大地质图。在这张地质图上，Talkeetna 山脉代表弧前区域出露的弧控部分，Matanuska 峡谷代表剥露的弧前盆地，Chugach 山脉代表剥露的增生楔。在古新世—始新世洋脊俯冲之前，最新的白垩系阿拉斯加南部弧前系统是由一个岩浆弧（Castle 山脉断层北侧的紫色和灰色区域）、海相弧前盆地（翠绿色）以及一个增生楔（浅绿色）组成。古新世—始新世的洋脊俯冲作用导致了白垩纪岛弧岩浆活动的停止以及板片窗岩浆岩喷发和侵入的开始。我们注意到板片窗岩浆活动的记录遍及穿弧前区域，包括了增生楔、弧前盆地及岛弧区。古新世洋脊俯冲通常也以沉积体系的改变为特征，弧前盆地的沉积从海相海底扇沉积系统（翠绿色）转变为非海相河流沉积体系（橘黄色）。地质图来自 Wilson 等（1998）。更多详细内容在文中予以介绍

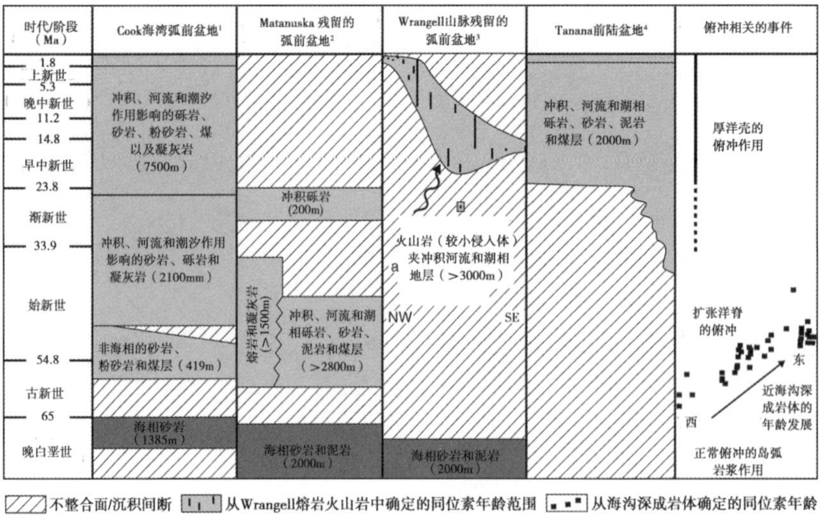

图 16-3 阿拉斯加中南部沉积盆地中的晚白垩世—新生代沉积地层，以及沿阿拉斯加南部边缘的俯冲参数变化（1：Flores 等，2004 年；2，3：Trop 和 Ridgway，2007 年；4：Ridgway 等，2007 年）

值得注意的是在晚白垩世，阿拉斯加南部的弧前盆地是以正常俯冲作用以及广泛沉积的海相砂岩、泥岩为特征。在古新世—始新世洋脊俯冲作用期间，沉积特征发生了变化，表现为在 Cook 湾及 Matanuska 弧前盆地中出现了非海相沉积。渐新世—全新世亚库卡特微板块的板片水平俯冲作用使得弧前盆地体系高度地分区。Matanuska 残余弧前盆地位于板片水平俯冲带上部区域，以出露和遭受侵蚀为特征。Cook 湾弧前盆地位于板片水平俯冲带西缘（图 16-1），以非海相沉积为特征。Wrangell 山脉残余弧前盆地位于板片水平俯冲带东缘（图 16-1），主要受到火山作用的控制。Tanana 弧前盆地位于板片水平俯冲带北缘（图 16-1），以非海相沉积为特征。更多的讨论将在文中详述。数据来源：近海沟深成岩体年龄依据 Brandley 等（2003）。Wrangell 熔岩年龄来自 Rithter 等（1990）、Ridgway 等（1992）、Skulski 等（1992）及 J. M. Trop（n. d.）

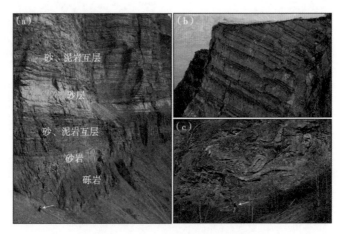

图 16-4 海底斜坡的沉积特征

（a）Wrangell 山脉上白垩 MacColl 山脊的近似深海扇沉积地层。Gcm—砾岩；Sm—砂岩；Sm/Fmm—砂泥岩互层。以一个人（白色箭头所指）为参照物。这些地层是在正常俯冲作用时期沉积于 Wrangell 山脉弧前盆地中（WB 如图 16-2a；Trop 等，1999 年）。（b）Talkeetna 山脉南部 Matanuska 造山带晚白垩世地层以中厚至厚层状泥岩（棕色层）为特征。单个的砂岩层被较薄的无定形泥岩层（黑色层）分隔。认为这些地层是形成于 Matanuska 弧前盆地的深海斜坡相沉积（MB 见图 16-2a；Trop，2008 年）。出露部分高达 18m。（c）Matanuska 弧前盆地上白垩统 Matanuska 组的内部变形（MB 见图 16-2a）。这一柔软沉积物变形构造是由于发育在不稳定深海斜坡上的泥砂岩互层的块体滑塌造成（据 Trop，2008）。出露面高约 20m，以人（白色箭头所指）为参考系

16.3 古新世—始新世扩张洋脊俯冲作用

16.3.1 简介

扩张洋脊俯冲是沿着现今大陆会聚边缘一种常见的地质构造运动，新不列颠岛南部所罗门海沟和安第斯—哥斯达黎加大陆边缘（例如，研究比较详细的 Fisher 等，2004；Taylor 等，2005；Pedoja 等，2006；Sak 等，2009）。新生代的扩张洋脊俯冲已经成为影响大部分北美科迪勒拉大陆边缘的重要地质作用（Wilson 等，2005；Madsen 等，2006；Pallares 等，2007）。前人关于扩张洋脊俯冲的鉴别证据包括在弧前盆地和增生楔内出露的火山岩和深成岩，例如这些火成岩所具有的来自亏损地幔的地球化学特征（Lytwyn 等，2000；D'Orazio 等，2001；Gorring 和 Kay，2001；Colo 和 Stewart，2008），以及增生楔内高温低压变质岩特征的深成岩等（Sisson 等，1989；Underwood 等，1999；Iwamori，2000）。扩张洋脊俯冲的相似地质证据广泛分布在阿拉斯加、英属哥伦比亚、华盛顿和俄勒冈州，这些地区也是现今研究关注的焦点（Haeussler 等，2003；Sisson 等，2003；Madse 等，2006）。然而，针对北美科迪勒拉山脉西北部沉积盆地内洋脊俯冲过程中沉积记录却比较少（Bradley 等，2003；While 和 Bradley，2006；Trap 和 Ridgway，2007）。

为了评估扩张洋脊俯冲对沉积盆地发育的影响，我们将概述最近以及目前正在进行的关于 Matanuska 弧前盆地的研究工作（图 16-1 和图 16-2a 的 MB）。该盆地古新统—始新统露头非常好，这些地层沉积与沿着阿拉斯加南部边缘的扩张洋脊俯冲作用同时代（图 16-3；Trap 等，2003；Trap 和 Plawman，2006；Kassab 等，2009；Kortya 等，2009）。增生楔区域内近海沟火山岩体详细的年代学研究，揭示了从 62Ma 到 50Ma 岩浆活动，向东扩展横跨阿拉斯加南部（图 16-1 和图 16-3；Bradley 等，2003）。在 Matanuska 盆地内和周围区域，古近系火山岩和侵入岩的地球化学和岩石学基础研究表明，它们来自于与扩张洋脊俯冲相关的亏损玄武岩岩浆源区（图 16-2a 的 CV 和 CTV；Cole 等，2006；Flanagan 等，2006；Cole 和 StPwarl，2008）。在洋脊俯冲过程中，两个俯冲板块之间的空隙允许板块下的软流圈地幔进入上部板块（Thorkolson，1996）；该过程中形成的侵入和喷发岩浆后来被称为板片窗火成岩（Slab window igneous）。Matanuska 盆地近海沟深成岩体和板片窗火成岩之间的古新世—始新世沉积地层，最大厚度大于 2800m（图 16-3）。这些地层出露在残留的弧前盆地位置，后者位于盆地以北的白垩纪—早古新世岛弧侵入岩和盆地以南的增生楔地层之间（图 16-2b）。古新世—始新世的沉积和火山岩地层沿着一个盆地范围内的角度不整合覆盖在白垩纪末期海相弧前盆地地层之上（图 16-3 和图 16-5a）。

16.3.2 Matanuska 盆地的沉积格架

Matanuska 盆地南部和北部出露的古新世—始新世粗粒沉积地层主要由河控冲积扇和辫状—网状河流系统沉积物形成（图 16-3 和图 16-5b；Trap 和 Ridgway，2000；Trap 等，2003）。在盆地北部，沉积地层中发育厚层的火山熔岩、凝灰岩和火山碎屑砂岩夹层，记录了近喷口溢流和火山碎屑喷发（图 16-5c 和图 16-5d）。与之相反，盆地南部沉积地层中缺少火山岩夹层，碳质泥岩、煤以及细粒至中粒砂岩是盆地轴部的沉积特征（图 16-5e）。这些细粒地层被解释为代表了发育良好漫滩环境的蜿蜒河流体系沉积。

图 16-5 扩张洋脊俯冲的实例

(a) 侏罗—白垩纪海相弧前盆地地层（强烈向右倾斜）与始新世板片窗火山岩厚层地层之间的角度不整合界面（白色箭头标示其大概位置）。同一不整合界面还将 Matanuska 盆地中的古新世—始新世非海相沉积以及火山沉积与中生代海相地层分隔开来。认为这一不整合界面是弧前盆地之下扩张洋脊浅部俯冲作用的结果。出露高度约 250m。(b) 始新世粗粒冲积扇和辫状河地层（Tw，Wishbone 组）以及上覆板片窗火山岩（Tv），这些保存在 Matanuska 盆地中的沉积物是与山脉俯冲作用同时期形成的。出露部分厚度可达 200m 左右。照片源自图 16-2b 的 Castle 山脉区域。(c) 在 Gray 山脊区域 Arkose 山脊地层内古新世—始新世河流相和湖相地层（棕色层）与凝灰岩（白色层）互层（图 16-2b）。根据上覆年龄及岩相展布趋势，认为大部分的凝灰岩来自板片窗 CaribouCreek 火山活动中心（CV，见图 16-2b）。以一个穿红色 T 恤的人（白色箭头指示）为参考系。(d) 在古新世—始新世 Arkose 山脊地层中用红色条带标记一系列层状凝灰岩和火山砂岩。上覆地层为厚层状、块状并具有交错层理的河流相砂岩及砾岩（棕色层）。红色箭头指示一系列岩墙（深棕色岩体）。研究区内普遍存在相似的岩墙及相关岩体侵入古新世—始新世残余弧前盆地地层中（图 16-2b）。照片源自图 16-2b 中的 Box Canyon 地区。以画圈的人作为参考系。(e) Chickaloon 地层中河流相砂岩沉积（S_m）和河漫滩沉积（F_{sm}）在古新世—始新世期间沿着 Matanuska 盆地的轴向得到沉积保存。以人作为参考系。(f) 始新世板片窗火山岩（微红棕色岩体）上覆于 Arkose 山脊地层中古新世—始新世冲积扇相及辫状河相沉积之上。照片源自图 16-2b 中 Box Canyon 区域。黑色箭头指示照片中央作为参考系的人（穿红色 T 恤）

古新世—始新世的非海相沉积地层以整体向上变粗的巨层序为特征，最大厚度可大于 2800m。多数剖面下部由泥岩、煤和砂岩夹层构成，它们依次被粗砾岩和砂岩覆盖（Trap 等，2003；Kassab 等，2009；Kortyna 等，2009）。我们认为这一巨厚层序是冲积扇和河流沉积体系向盆地进积的产物，作为对Castle山脉和Border Ranges（邦德山脉）断层系统的同沉积位移以及Caribou Greek火山岩区建造的响应（图16-2b）。生长向斜在北倾的Castle山脉断层体系下盘中得到了很好的保存，并被解释为是在始新世冲积扇和辫状河地层沉积过程中北端向上位移的结果。这里定义的下盘生长向斜是指，逆断层下盘地层发生了褶皱变形，在向斜构造形成时发生同沉积作用（Hoy和Ridgway，1997年；Ridgwayet等，1997）。沿着同时代Caribou Greek火山区（位于Matanuska盆地北部边缘的东部，CV见图16-2b）的南部边界出露的沉积地层提供了古水流和地层成分资料，这些资料揭示了古近纪火山中心曾提供了沉积物源，并影响了沉积体系的发育（图16-5d；Colo等，2006）。相反，沿着走向向西更远处出露的地层主要由来自侏罗—白垩纪残余岩浆弧侵蚀的碎屑组成（Arkose洋脊区见图16-2b；Trap和Ridgway，2000年；Trap等，2009）。与盆地南缘Border Ranges（邦德山脉）断层体系有关的沉积作用是增生楔抬升与沿北倾斜滑断层同沉积位移的共同产物（Little和Naeser，1989）。总之，古新世—渐新世的Matanuska盆地充填着非海相沉积体系，它们与大断层体系的位移、残余中生代火山弧和与扩张洋脊俯冲有关的古近纪活跃火山中心的剥蚀密切相关。

16.3.3 扩张洋脊俯冲的沉积学记录

从Matanuska弧前盆地古新世至始新世地层中获得的信息与地表抬升有着较好的一致性，后者与扩张洋脊俯冲消减有关。洋脊俯冲之后发生与洋脊通道相关的盆地沉降。在这部分讨论中，出露于横穿Border Ranges（邦德山脉）断裂的弧前盆地南部近海沟深成岩体年龄记录了扩张洋脊在Matanuska盆地外侧俯冲的大致时间（在图16-1中150°—145°的经度之间）。这些近海沟深成岩体的同位素年龄结合地质条件约束，指示出Matanuska盆地区域洋脊俯冲作用发生在大约57—56Ma（晚古新世）（图16-1和图16-3；Bradley等，2000；Cole等，2006）。通过年轻的、有浮力的俯冲地壳和上冲板块之间增强的耦合作用，毗邻扩张洋脊的年轻地壳的俯冲作用促使了挤压应力的形成和地表的抬升（Cloos，1993；Taylor等，2005）。更具浮力的、地形上更高的洋脊俯冲会引发弧前区域额外的表层抬升（Thorkelson，1996；Sak等，2009）。在我们的重建中，与洋脊俯冲作用相关的地表抬升的最古老地层证据是将古新世—始新世非海相地层及板片窗火山岩与下伏白垩纪末马斯特里赫特期（Maastrichtian）海底陆坡浊积岩顶部分隔开来的不整合面（图16-3和图16-5a；Trop，2008）。这个不整合面在阿拉斯加南部弧前盆地、西边的Cook海湾、直到东边的Wrangell山脉中的各露头及钻井中均有记录（图16-2a和图16-3）。我们将弧前盆地地层中发育的白垩纪末期—古新世的不整合面解释为与洋脊俯冲过程相关的地表抬升开始的标志。在阿拉斯加南部，俯冲的扩张洋脊大致平行于大陆边缘。扩张洋脊沿着弧前盆地走向逐渐向东迁移，促进了跨时代的地表抬升以及不整合面发育。在海相地层遭受抬升之后，Matanuska盆地在晚古新世至早始新世发育了厚层粗粒碎屑楔状沉积，后者分别从弧前盆地南部和北部向盆地方向进积（图16-5b）。

弧前盆地之下浮力洋脊的通道被认为能够激活该地区主要的断层系统（也就是Castle山脉和Border Ranges山脉断层）。沿着Castle山脉断层出露的下盘生长向斜，是同沉积位移的产物。总的来说，阿拉斯加南部弧前盆地系统在白垩纪末期—始新世这段时期内沉降历史，与通过新生大洋岩石圈和扩张洋脊的俯冲作用引起的构造抬升和不整合面发育有较好的一致性。该俯冲作用之后发生了与板片窗通道和俯冲洋脊拖曳后缘相关的沉降和沉积作用。我们对于 Matanuska 弧前盆地古新世至始新世洋脊俯冲沉积学记录的解释，与在现代会聚型大陆边缘活跃洋脊消减作用区的发现相一致。比如，全新世大型扩张洋脊的俯冲作用有助于许多现代板块边缘的净抬升速率超过末次盛冰期之后的海平面上升。因此，抬升的海相地层一般都出露在现代俯冲洋脊和海山的内侧（Gulscher 等，1999；Taylor 等，2005；Pedoja 等，2006；Saket 等，2009）。除此之外，弧前盆地的沉积充填也受到更新世—全新世扩张洋脊和海山俯冲作用的影响，这反映了在海底高地前缘俯冲期间的快速垂向隆升作用，之后沿着俯冲洋脊后缘发生沉降（Sak 等，2004）。

在 Matanuska 盆地扩张洋脊的俯冲作用常常与弧前区域广泛的火山活动同时发生。在盆地南部边缘，57—52 Ma 的岩脉群沿着 Border Ranges 山脉断裂系统侵入到中生代增生楔地层中（图 16—2b；Little 和 Naoser，1989）。相互交插、横切、叠置的古新世至始新世的板片窗火成岩地层和与之相关的浅成侵入岩体，以及同时期的非海相沉积地层在 Matanuska 盆地都很常见（图 16-2b、图 16-5b、d、f；Cole 等，2006；Flanagan 等，2006；Cole 和 Stewart，2008）。例如，在盆地的中心区域（图 16-2b 中的 Kings 山脉区域），地表有一个宽度约 4~5km 的侵入体。前期的地球化学研究将 Matanuska 盆地内的岩浆活动归因于与残余弧前盆地下洋脊俯冲作用相关的板片自由窗（"slab-free" window）的发育，这与加州西部新生代洋脊俯冲作用的研究成果一致。该研究表明板片自由窗为源自地幔的岩浆上涌进入上覆大陆岩石圈提供了更为直接的通道（Dickinson 和 Snyder，1979；Cole 和 Basu，1992，1995；Cole 和 Stewart，2008）。在阿拉斯加南部的实例中，一条板片窗火山岩带侵入到一条与白垩纪末期岩浆弧—弧前盆地—增生楔系统轴向相垂直的北西走向的露头带中（图 16-2a 中的 CV 和 CTV）。在盆地更内侧，出露于 Talkeetna 山脉北部（图 16-1 的Ⅳ）和阿拉斯加山脉中（图 16-1 的 CB）的同时期火成岩的地球化学组成与更富集幔源的衍生物一致，很可能是来自于白垩纪末期至古新世大陆边缘岛弧型岩浆作用的残留地幔楔（Cole 等，2007；Cole 和 Stewart，2008）。

16.3.4 小结

与许多俯冲相关的会聚大陆边缘类似，白垩纪末期阿拉斯加南部边缘发育了由岩浆弧—弧前盆地—增生楔组成的典型的"三位一体"结构系统（图 16-6a）。然而，沿着阿拉斯加南部新生代早期会聚边缘的扩张洋脊俯冲，明显地改造了白垩纪末海相弧前盆地系统。这些重要的改造包括：时间上以一个区域不整合面为界，由海相沉积体系转变为非海相沉积体系；邻近板片窗火山中心的厚层非海相碎屑楔状体的局部沉积（如图 16-2b 的箱形峡谷）；断裂系统重新活动并导致下盘厚层砾岩的沉积（图 16-2b 的 WishBone Hill 和 Castle Mountain）。这些地质特点与那些关于弧前盆地的经典模型具有很大差别。在弧前盆地的经典模型中，大多数硅质碎屑岩来自岛弧岩浆活动产生的火山岩，并沉积于海底斜坡和海底扇环境中（Ingersoll，1982，1983；Dickinson，1995）。受洋脊俯冲改造的弧前盆地广泛发育火山活动，导致该类盆地具有比经典的"冷"弧前盆地更高的热流（DeLong 等，1979）。高热流对该类弧前盆地中的石油和煤炭的成熟十分重要。比如，在阿拉斯加南部的弧前盆地系统中就蕴含着丰富的石油和煤炭资源（Magoon，1994；Wahrhaftig 等，1994；Flores 等，2004）。虽然要清楚说明扩张洋脊俯冲过程对弧前盆地发育演化的影响，还需要对其他经历过该过程的弧前盆地的地层记录进行更多的研究，但是对阿拉斯加南部弧前盆地系统的分析至少表明，洋脊俯冲过程改变了弧前盆地的地球动力学机制、沉积体系、区域地层和沉积物组分等。

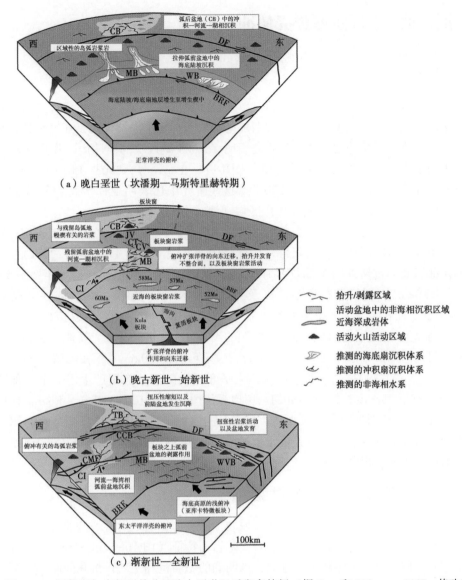

图 16-6 阿拉斯加南部弧前盆地晚白垩世以后发育特征（据 Trop 和 Ridgway，2007，修改）

(a) 阿拉斯加南部会聚大陆边缘白垩纪末期（坎潘期至马斯特里赫特期）弧前盆地以发育良好且侧向伸展的岩浆弧—弧前盆地（MB，WB）—上部板块中的增生楔为特征。弧前盆地沉积体系主要由海相、深海相海底扇系统组成，沉积物主要来源于火山弧。同时代发育的 Cantwell 盆地（CB）为一个弧后前陆盆地。BRF—边缘山脉断层；CB—Cantwell 盆地；DF—德纳里峰断层；MB—Matanuska 盆地；WB—Wrangell 山脉盆地。(b) 古新世—始新世扩张洋脊俯冲、板片岩浆窗岩浆活动，以及残留弧前盆地中河流—湖泊相的沉积（MB，CI）。注意在板式窗之上的岩浆活动，包括在残余弧前盆地（近海沟深成岩体）之内（CV，CTV）或之外。在盆地内部的岩浆活动（CB，JV）反映的是伴随早期岩浆活动的残余地幔楔。需要注意向东的年轻近海沟深成岩体，它们反映近似平行于板块边缘的洋脊的穿时俯冲。CB—Cantwell 火山岩；CI—Cook 湾盆地；CTV—Talkeetna 中央山脉火山岩；CV—Caribou Creek 火山区；JV—Jack 河火山岩；MB—Matanuska 地；DF—Denali 断层；BRF—边缘山脉断层；CMF—Castle 山脉断层；#A—Anchorage 镇。(c) 渐新世—全新世亚库卡特微板块板式俯冲，平板区域之上盆地内（MB，CCB）较老地层出露，盆地内沉积物沿着平板俯冲区边缘（CI，TB）分布，残余弧前盆地中的火山作用和走滑断层变形沿着平板区东缘（WVB）分布。CCB—科罗拉多河盆地；DF—Denali-Totschunda-Duke 河断层系统；TB—Tanana 盆地；WVB—Wrangell 火山带；其他缩写可见图 16-6a

16.4 渐新世—全新世厚洋壳俯冲作用

16.4.1 简介

在这部分内容中将分析渐新世—全新世，阿拉斯加南部会聚边缘之下亚库卡特微板块水平俯冲作用的沉积记录（图16-1）。阿拉斯加南部的板片水平俯冲作用是亚库卡特微板块俯冲消减作用的产物。近期的地球物理学研究表明Yakutat微板块的俯冲消减作用区域是由厚的具有浮力的地壳组成，类似于大洋高原（图16-1；Ferris等，2003；Eberhart Phillips等，2006；Christeson等，2008；Lowe等，2008）。这些研究中还表明亚库卡特微板块在无俯冲作用区域的厚度为25~30km，在俯冲作用区域则消减为11~22km。

在现今的阿拉斯加南部边缘存在自东向西的构造体制转化——由东部的转换构造转变为中部的水平俯冲，再向西转化为"正常"的洋壳俯冲（图16-1；Finzel等，2011）。在西北加拿大的东部以及阿拉斯加的东部发育有Wrangell火山带的火山活动（WVB见图16-1）、浅源地震（<50km）、右旋转换断层，以及Queen Charlotte–FairweatherDenali–Totschunda–Duke河流系统。而在包含了阿拉斯加中南部大部分地区的中部区域，则以相对较高的地形、缺少火山活动以及被亚库特微板块确定的缓倾的Wadati-Benioff带来与其他地区相区分。这一微板块在阿拉斯加下方水平延伸约250km，在深度达到150km时，在阿留申海沟内延伸至600km（Ferris等，2003）。在西部区域，太平洋板块的俯冲作用产生了一个更加陡倾的Wadati-Benioff带，其深度在约400km的阿留申海沟内可达到100~150km（Eberhart-Phillips等，2006）。沿着阿拉斯加南部会聚大陆边缘走向的这种变化，能够估算板片水平俯冲作用区域之上弧前盆地的剥露情况。有关板片水平俯冲作用的确切沉积记录存在于沿着水平俯冲作用区域周边分布的沉积盆地中。

16.4.2 板片水平俯冲区之上的盆地反转

在阿拉斯加南部，现今板片水平俯冲区域之上的上部板块以发育一些北美地区最高的地形地貌、大型地震活动以及渐新世甚至更老的基岩为特征（图16-1和图16-3）。现今位于板片水平俯冲区域之上的不活跃沉积盆地，以明确的中新世和上新世沉积间断为特征。例如在马塔努斯卡残余弧前盆地，第四纪地表沉积不整合覆盖于渐新世和更老的盆地地层之上（MB见图16-1及图16-3；Trop等，2003）。我们使用"残余"来描述新近纪马塔努斯卡盆地，是因为盆地现今不再发生沉降，且在早古新世岛弧型岩浆活动就已经停止。在板片水平俯冲区域的更内侧，第四纪地表沉积沿着阿拉斯加山脉南侧不整合覆盖于渐新世和更古老沉积地层之上（CCB见图16-1；Trop等，2001）。我们将这种区域性沉积间断解释为板片水平俯冲作用在该区所引发的渐新世后盆地反转和剥蚀作用的结果。这些地层资料与来自于整个板片水平俯冲区出露的岩石的始新世到上新世热年代数据相一致（例如Little和Naeser，1989；Fitzgerald等，1995；O'Sullivan和Currie，1996；Hoffman和Armstrong，2006；Berger等，2008；Enkelmann等，2008，2009；Arkle等，2009；Sendziak等，2009；Benowitz等，2011）。因此我们推断，地层剥露是亚库卡特微板块中具有浮力的厚洋壳向阿拉斯加南部下方插入的产物（Finzel等，2011）。

16.4.3 板片水平俯冲区周边的盆地演化及火山活动

库克湾弧前盆地沿着板片水平俯冲区西部边缘分布（图 16-1），西北和东南分别以阿留申岛弧和楚加奇增生楔为界，沉积了约 10km 厚的新生代非海相地层（图 16-3；Floreset 等，2004），其中古近系厚约 2500m，新近系厚约 7500m，并含有厚达 15m 的煤层（图 16-3 和图 16-7a）。这些地层记录显示了库克湾弧前盆地早中新世开始的沉降沉积速率的增加。该过程形成了一个远离现今板片水平俯冲区并向西南进积进入到相邻的库克湾盆地的大型碎屑楔状沉积体（图 16-6c；Finzel 等，2007，2009，2011）。

塔纳诺（Tanana）前陆盆地紧邻板片水平俯冲区的北部（图 16-1 和图 16-6c 的 TB）。沿着盆地的南部边缘发育了 2000~3000m 厚的新生代地层。沿着阿拉斯加山脉北侧分布的逆冲断层使得该套地层遭受剥蚀，并形成了系列小山丘（Ridgway 等；2002，2007）。近期研究表明，这些小山丘是活动的、向北推进的塔纳诺（Tanana）前陆盆地新近纪冲断构造带的一部分（Bemis 和 Wallance，2007；Lesh 和 Ridgway，2007）。该前陆盆地下部为 800m 厚的中中新统和上中新统，沉积于河流、湖泊以及泥炭沼泽环境（图 16-3 和图 16-7b）。厚层湖相泥岩以及超过 20m 厚煤层（图 16-7c、d）的发育表明沉积期间盆地发生了区域性沉降。盆地上部为 1200m 厚的上新世地层，沉积于向北进积的冲积扇及辫状河三角洲平原沉积环境（图 16-7e），反映了阿拉斯加山脉高地的侵蚀过程。我们将 Tanana 前陆盆地中的新近系解释为一个远离现今水平俯冲区北部边缘的大型碎屑楔状沉积体向北进积的产物（图 16-6c）。

板片水平俯冲区的东部边界为兰格尔火山带（WVB，图 16-1 和图 16-6c）和活动的走滑断层带。兰格尔火山带由熔岩流、熔岩丘、火山碎屑地层及最大厚度超过 3000m 的非海相沉积层序组成（图 16-3 和图 16-7f；Richter 等，1990）。兰格尔火山带东南部由始新—上中新统冲积—湖相沉积地层以及具有碱性—过渡性地球化学组成的火山喷发中心组成（Skulski 等，1991；Cole 和 Ridgway，1993；Ridgway 和 DeCelles，1993）；其西北部则以中中新—全新统冲积—湖相沉积地层及具有过渡性至钙—碱性地球化学组成的火山侵入岩为特征，该侵入岩为典型的与俯冲作用有关的火山岩套（Richter 等，1990；Preece 和 Hart，2004；Tidmore，2004；Delaney，2006；Trop 等，2007）。在兰格尔火山带西北部的 Frederika 盆地（图 16-2 中的 FB），冲积—湖相碎屑地层中包含厚层熔岩、凝灰岩以及火山碎屑泥岩夹层（图 16-7f），地层沿着北西向的转换断层以及南北向的正断层出露（Tidmore，2004；Delaney，2006；Trop 等，2007）。我们将兰格尔火山带内的地层学记录解释为沿着亚库卡特板块东部边缘发生的张扭构造以及部分熔融作用的结果。具有浮力的亚库卡特板块逐渐向西北俯冲至阿拉斯加之下，促使岩浆活动向北西方向迁移，同时沿着转换断层和正断层发育张扭盆地。由于在兰格尔火山带之下深度超过 50km 处缺乏现今俯冲板片存在的地球物理证据（Eberhart-Phillips 等，2006；Qi 等，2007），再加上兰格尔火山带西北部存在埃达克岩（Preece 和 Hart，2004），因此可以认为在亚库卡特板块缓倾的东部边缘发生了部分熔融作用。

图 16-7 板片水平俯冲区周边的盆地演化及火山活动实例

(a) 库克湾盆地中典型的中中新世 Beluga 组，包括非海相、厚层状、含交错层理砂岩及煤层，以人（白色箭头所指）为参考系，这些地层沿着平板俯冲带西边缘分布，沉积物的一个来源为位于平板俯冲带正上方的邻近的出露弧前盆地（MB 图 16-1）；(b) 塔纳诺盆地晚中新世河流相砂岩、粉砂岩以及煤层（黑色层）沉积（TB 可见图 16-1）。照片中层理清晰可见，这些地层沿着板片水平俯冲带北边缘分布（图 16-1），以照片中央的人为参考系；(c) 塔纳诺盆地中中新世 Suntrana 组为较厚煤层夹杂着大量具有交错层理的砂岩（上覆白色层），以人作为参考系；(d) 晚中新世 Grubstake 组（棕色层）中层状湖相泥岩与上覆塔纳诺盆地中的上新世 Nenana 砾石层相接连，图中站在右边中心的人临近接触带；(e) 厚达 1200m 的上新世 Nenana 砾岩层的下部，该地层是冲积扇和辫状河沉积，其物源来自盆地北部的阿拉斯加山脉，以人（位于图片右下方）为参考系；(f) Frederika 组（Tf）的中新世沉积地层以及兰格尔熔岩的熔岩流（Tw），出露高度约为 700m，这些地层是兰格尔火山带的一部分，确定了板片水平俯冲的东缘（在图 16-1 上标注为 WVB）；(g) 现今 Matanuska 河，它对位于板片水平俯冲带之上出露的弧前盆地沉积物进行侵蚀和搬运，它搬运的沉积物在位于板片水平俯冲带区域西缘的库克湾弧前盆地沉积下来（图 16-2），沿河出露的古新世—始新世 Chickaloon 组是在马塔努斯卡盆地洋脊俯冲早期沉积形成，视图为南西向

16.4.4 小结

阿拉斯加南部会聚型大陆边缘之下亚库卡特微板块的板片水平俯冲作用对于中新世—全新世的盆地演化具有重大影响，并且引起弧前盆地动力学机制沿着走向突然变化。作为水平俯冲作用区现今的西部边界（图16-6c），库克湾弧前盆地发育了厚度超过7500m的新近系（图16-3）。而在东部，位于板片水平俯冲区之上的马塔努斯卡弧前盆地则遭受剥露，表现为一个侵蚀区和沉积物路过区。我们认为，在中新世—全新世期间从水平俯冲区之上的剥露地区剥蚀的沉积物被向西搬运至库克湾弧前盆地，或向北搬运至纳塔诺前陆盆地（图16-6c 和图16-7g；Finzel 等，2009）。沿着水平俯冲区的东部边缘，板块边缘的火山活动及张扭性盆地的发育，导致了中新世—全新世走滑断层、火山岩区、冲积相—湖相火山碎屑地层叠置在白垩纪的海相弧前盆地地层之上（图16-1和图16-3）。弧前地区盆地动力学沿走向的这种变化表明，一个350km宽的厚洋壳碎片的侵入显著改造了先前超过500km长的海相弧前盆地系统。

16.5 结论

位于会聚型大陆边缘上部板块内的弧前盆地是巨大的沉积物堆积区，它反映了与俯冲作用有关的第一级地质构造特征（Dickinson，1995）。弧前盆地内的地层记录了上部板块的俯冲驱动过程，比如岩浆弧的生长及增生楔建造的发育。前人具有里程碑意义的研究明确地展示了岩浆弧—弧前盆地—增生楔是如何形成"三位一体"结构体系的——大部分会聚大陆边缘的上盘都以此为特征（Dickinson，1974，1976；Dickinson 和 Seeley，1979；Ingersoll，1982，1983）。许多关于弧前盆地的研究都聚焦于岛弧型岩浆活动、增生楔出露与弧前盆地沉积作用之间的相互作用（Dickinson，1995；Busby 等，1998；Clift 等，2000；McNeill 等，2000；Kimbrough 等，2001；DeGraaff-Surpless 等，2002；Unruh 等，2007；Fildani 等，2008；Trop，2008）。这些研究为分析以"正常"洋壳长期俯冲为特征的弧前地区的演化和地层提供了重要框架。在本章中，我们展示了扩张中心和厚层洋壳的不连续幕式水平俯冲作用如何影响了阿拉斯加南部会聚型大陆边缘沉积盆地的结构与沉积地层。板片水平俯冲过程显著地改变了上部板块的构造格局，并形成了与经典的岩浆弧—弧前盆地—增生楔模式所不同的沉积盆地。以下过程和特征可以用来区分"改造过"的弧前盆地与"经典"的弧前盆地。

（1）扩张洋脊和厚洋壳的板片水平俯冲都能够导致弧前盆地底部的构造抬升、较老海相弧前盆地地层的剥露，以及局部厚度可超过下伏海相地层的非海相地层沉积。与之相反，"经典"的弧前盆地虽也可发育浅滩相、并向上变为浅海和/或非海相地层，但仍以深海相沉积地层为主（Dickinson，1995；Busby 等，1998；Takashima 等，2004；Fildani 等，2008）。

（2）被板片水平俯冲作用改造过的弧前盆地以地层沿走向发生突变为特征。扩张洋脊俯冲期间，在弧前盆地之下发育板片窗的位置会产生局部的盆内高地和相邻的沉积中心，以及弧前盆地内部或相邻区域的不连续火山中心。厚洋壳区域伸展部分的浅部俯冲作用可能会使弧前盆地上覆于浅部板块的部分发生剥露；与之相反，弧前盆地系统沿着水平板片区域走向分布的部分，则成为了主要沉降区和沉积区。与经典的弧前盆地相比，扩张洋脊或厚洋壳的板片水平俯冲作用将原先长条形的海相弧前盆地系统分割为多个互相分隔的具有独特地层充填的非海相沉积中心。

（3）受板片水平俯冲作用改造的弧前盆地的沉积记录中存在区域和局部的不整合。就阿拉斯加南部会聚大陆边缘而言，俯冲的扩张洋脊与大陆边缘近乎平行，在这种情况下扩张洋脊沿着弧前盆地走向的逐渐迁移促使了不同时代的地表抬升和不整合面形成。与之类似，厚层洋壳区域性伸展部分的水平俯冲作用也导致水平俯冲区之上的区域性地表抬升和不整合面形成。在阿拉斯加南部下方的厚层洋壳水平俯冲的例子中，从水平俯冲区之上的暴露区域剥蚀下来的沉积物被搬运至沿水平板片周边分布的盆地中。

（4）与"经典"的弧前盆地有关的火成岩往往是岛弧岩浆活动的产物，且主要发育在盆地边缘内侧。相比之下，板片水平俯冲作用会导致在弧前区域火成岩的分布显著不同，甚至在某些情况下缺少岩浆活动。例如，就扩张洋脊俯冲来说，残余弧前盆地之下的板片自由窗（slab-free window）通道能导致火山岩侵位——火成岩侵入、叠置于先前海相弧前盆地之上，并为同期非海相沉积系统提供重要的盆内物源。而在厚洋壳俯冲的情况下，火山活动在板片水平俯冲区域上方停止，但可能沿着板片边缘广泛发育。以阿拉斯加南部为例，兰格尔火山带沿着邻近板片水平俯冲区东缘的渗漏转换断层（leaky transform fault）发育，并成为地球上体积最大的火山活动带之一。

（5）弧前盆地并不是特别有利的油气勘探区，这是由于普遍缺乏粗粒的非海相储层、利于富含有机质烃源岩富集的海底陆坡相非常有限，且较低的地温梯度抑制了烃源岩的成熟，以及存在孔隙保存不稳定的，硅质碎屑沉积物等造成的成分成熟度低（通常富火山质）（Dickinson，1995）。然而，受板片水平俯冲作用改造的弧前盆地，具有克服上述主要障碍的潜力，并能发育油气系统。对扩张洋脊俯冲而言，沿着洋脊通道在盆地内部发育了厚层粗粒非海相地层的沉积，同时盆地之下的板片自由窗通道在洋脊俯冲过程中提供热源从而加速了烃源岩的成熟。而在厚洋壳俯冲的情况下，板片水平俯冲区上部板块的区域剥露和侵蚀作用产生大量沉积物，这些沉积物质被搬运至板片水平俯冲区周围的活跃的沉降盆地中。区域剥露作用使得残留岛弧被深切、深成岩出露以及成分成熟度高的沉积物在弧前盆地系统沿走向的部分沉积。研究表明，在战略评估弧前盆地的勘探潜力时，应该充分考虑扩张洋脊和具有浮力的厚层板块的水平俯冲作用可能扮演的角色。

致谢（略）

参 考 文 献

Arkle, J. C., Armstrong, P. A., and Haeussler, P. J. (2009) The western Chugach Mountains and northern Prince William Sound (Alaska): Locus of subduction-related exhumation? Geological Society of America, Abstracts with Programs, 41, 290.

Bemis, S. P., and Wallace, W. K. (2007) Neotectonic framework of the north-central Alaska Range foothills, in Ridgway, K. D., Trop, J. M, Glen, J. M. G., and O'Neill, J. M., eds., Tectonic growth of a collisional continental margin: crustal evolution of southern Alaska. Geological Society of America Special Paper 431, 549–572.

Benowitz, J. A., Layer, P., Armstrong, P., Perry, S., Haeussler, P., Fitzgerald, P., (2011) Spatial variations in focused exhumation along a continental-scale strike-slip fault: The Denali fault of the eastern Alaska Range. Geosphere, 7 (2), doi: 10. 1130/GES00589. 1.

Berger, A. L., Spotila, J. A., Chapman, J. B., Pavlis, T. L., Enkelmann, E., Ruppert, N. A., and Buscher, J. T. (2008) Architecture, kinematics, and exhumation of a convergent orogenic wedge: A thermochronological investigation of tectonic-climatic interactions within the central St. Elias orogen, Alaska. Earth and Planetary Science Letters, 270, 13-24.

Bradley, D. C., Parrish, R., Clendenen, W., Lux, D., Layer, P. W., Heizler, M., and Donley, D. T. (2000) New geochronological evidence for the timing of early Tertiary ridge subduction in southern Alaska. U. S. Geological Survey Professional Paper 1615, 5-21.

Bradley, D. C., Kusky, T. M., Haeussler, P. J., Goldfarb, R. J., Miller, M. L., Dumoulin, J. A., Nelson, S. W., and Karl, S. M. (2003) Geologic signature of early Tertiary ridge subductionin Alaska, in Sisson, V. B., Roeske, S. M., and Pavlis, T. L., Geology of a transpressional orogen developed during ridge-trench interaction along the north Pacific margin. Geological Society of America Special Paper 371, 19-49.

Busby, C. J., Smith, D., Morris, W., and Fackler-Adams, B. N. (1998) Evolutionary model for convergent margins facing large ocean basins; Mesozoic BajaCalifornia, Mexico. Geology, 26, 227-230.

Christeson, G. L., van Avendonk, H., Gulick, S. P., Worthington, L., and Pavlis, T. (2008) Crustal structure of the Yakutat microplate: constraints from STEEP wideangle seismic data. Eos Transactions, American Geophysical Union, 89 (53), Fall Meeting Supplement, Abstract T53B-1942.

Clift, P. D., Degnan, P. J., Hannigan, R., and Blusztajn, J. (2000) Sedimentary and geochemical evolution of the Dras forearc basin, Indus suture, Ladakh Himalaya, India. Geological Society of America Bulletin, 112, 450-466.

Cloos, M. (1993) Lithospheric buoyancy and collisional orogenesis: subduction of oceanic plateaus, continental margins, island arcs, spreading ridges, and seamounts. Geological Society of America Bulletin, 105, 715-737.

Cole, R. B., and Basu (1992) Middle Tertiary volcanism during ridge-trench interactions in western California. Science, 258, 793-796.

Cole, R. B., and Basu, A. R. (1995) Nd-Sr isotopic geochemistry and tectonics of ridge subduction and middle Cenozoic volcanism in western California. Geological Society of America Bulletin, 107, 167-179.

Cole, R. B., and Ridgway, K. D. (1993) The influence of volcanism on fluvial depositional systems in a Cenozoic strike-slip basin, Denali fault system, Yukon Territory, Canada. Journal of Sedimentary Petrology, 63, 152-166.

Cole, R. B., and Stewart, B. S. (2008) Continental margin volcanism at sites of spreading ridge subduction: Examples from southern Alaska and western California. Tectonophysics, 464 (1-4), 118-136.

Cole, R. B., Nelson, S. W., Layer, P. W., and Oswald, P. J. (2006) Eocene volcanism above a depleted mantle slab window in southern Alaska. Geological Society of America Bulletin, 118 (1-2), 140-158.

Cole, R. B., Layer, P. W., Hooks, B., Cyr, A., and Turner, J. (2007) Magmatism and deformation in a terrane suture zone south of the Denali fault system, northern Talkeetna Mountains, Alaska, in Ridgway, K. D., Trop, J. M., Glen, J. M. G., and O'Neill, J. M., eds., Tectonic growth of a collisional continental margin: crustal evolution of southern Alaska. Geological Society of America Special Publication 431, 447–506.

DeGraaff-Surpless, K., Graham, S. A., Wooden, J. L., and McWilliams, M. O. (2002) Detrital zircon provenance analysis of the Great Valley Group, California: Evolution of an arc-forearc system. Geological Society of America Bulletin, 114 (12), 1564–1580.

Delaney, M. R. (2006) Sedimentology and stratigraphy of the Frederika Formation, Wrangell–St. EliasMountains, south central Alaska. B. S. thesis, BucknellUniversity, Lewisburg, 62 p.

DeLong, S. E., Schwarz, W. M., and Anderson, R. N. (1979) Thermal effects of ridge subduction. Earth and Planetary Science Letters, 44 (2), 239–246.

Dickinson, W. R. (1974) Plate tectonics and sedimentation, in Dickinson, W. R., Tectonics and sedimentation, Society of Economic Paleontologists and Mineralogists Special Publication 22, 1–27.

Dickinson, W. R. (1976) Sedimentary basins developed during evolution of Mesozoic–Cenozoic arc-trench system in western North America. Canadian Journal of Earth Sciences, 13, 1268–1287.

Dickinson, W. R. (1995) Forearc basins, in Busby, C. J., and Ingersoll, R. V., eds., Tectonics of sedimentary basins. Cambridge, Blackwell Science, 221–262.

Dickinson, W. R., and Seely, D. R. (1979) Structure and stratigraphy of forearc regions. American Association of Petroleum Geologists Bulletin, 63, 2–31.

Dickinson, W. R., and Snyder, W. S. (1979) Geometry of subducted slabs related to San Andreas Transform. Journal of Geology, 87, 609–627.

D'Orazio, M., Agostini, S., Innocenti, F., Haller, J. J., Manetti, P., and Mazzarini, F. (2001) Slab window–related magmatism from southernmost South America: the Late Miocene mafic volcanics from the Estancia Glencross Area (52°S, Argentina–Chile). Lithos, 57, 67–89.

Eberhart-Phillips, D., Christensen, D. H., Borcher, T. M., Hansen, R., Ruppert, N. A., Haeussler, P. J., and Abers, G. A. (2006) Imaging the transition from Aleutian subduction to Yakutat collision in central Alaska, with local earthquakes and active source data. Journal of Geophysical Research, 111, B11303. doi: 10. 1029/2005JB004240.

Enkelmann, E., Garver, J. I, and Pavlis, T. L. (2008) Rapid exhumation of ice-covered rocks of the Chugach–St. Elias orogen, southeast Alaska. Geology, 36, 915–918.

Enkelmann, E., Zeitler, P. K., Pavlis, T. L., Garver, J. I., and Ridgway, K. D. (2009) Intense localized rock uplift and erosion in the St. Elias orogen of Alaska. Nature Geoscience, 2, 360–363.

Espurt, N., Funiciello, F., Martinod, J., Guillaume, B., Regard, V., Faccenna, C., and Brusset, S. (2008) Flat subduction dynamics and deformation of the South American plate: Insights from analog modeling. Tectonics, 27, TC3011. doi: 10. 1029/2007TC002175.

Ferris, A., Abers, G. A., Christensen, D. H., and Veenstra, E. (2003) High resolution image of the subducted Pacific plate beneath central Alaska, 50–150 km depth. Earth and Planetary

Science Letters, 214, 575-588.

Fildani, A., Hessler, A. M., and Graham, S. A. (2008) Trenchforearc interactions reflected in the sedimentary fill of Talara basin, northwest Peru. Basin Research, 30, 305-331.

Finzel, E. S., Ridgway, K. D., Brennan, P., and Landis, P. (2007) Miocene and Pliocene sedimentary footprint of flat - slab subduction of the Yakutat terrane, southern Alaska. Geological Society of America, Abstracts with Programs, 39 (6), 491.

Finzel, E. S., Ridgway, K. D., LePain, D. L., and Valencia, V. (2009) Insights on provenance of forearc basin strata: UPb geochronology from Cenozoic strata in the Cook Inlet basin, Alaska. Geological Society of America, Abstracts with Programs, 41 (7), 303.

Finzel, E. S., Trop, J. M., Ridgway, K. D., and Enkelmann, E. (2011) Upper plate proxies for flat-slab subduction processes in southern Alaska. Earth and Planetary Science Letters, 303, doi: 10. 1016/j. epsl. 2011. 01. 014.

Fisher, D. M., Gardner, T. W., Sak, P. B., Sanchez, J. D., Murphy, K., and Vannucchi, P. (2004) Active thrusting in the inner forearc of an erosive convergent margin, Pacific coast, Costa Rica. Tectonics, 23. doi: 10. 1029/ 2002TC001464.

Fitzgerald, P. G., Sorkhabi, R. B., Redfield, T. F., and Stump, E. (1995) Uplift and denudation of the central Alaska Range: a case study in the use of apatite fission - track thermochronology to determine absolute uplift parameters. Journal of Geophysical Research, 100 (20), 175-191.

Flanagan, D., Fischietto, N. E., Rothfuss, J. L., Cole, R. B., and Chung, S. L. (2006) Eocene magmatism in a remnant forearc basin, MatanuskaValley, southern Alaska. Geological Society of America, Abstracts with Programs, 38 (7), 559.

Flores, R. M., Stricker, G. D., and Kinney, S. A. (2004) Alaska Coal Geology, Resources, and Coalbed Methane Potential. U. S. Geological Survey DDS-77, 125 p.

Fuis, G. S., Moore, T. E., Plafker, G., Brocher, T. M., Fisher, M. A., Mooney, W. D., Nokleberg, W. J., Page, R. A., Beaudoin, B. C., Christensen, N. I., Levander, A. R., Lutter, W. J., Saltus, R. W., and Ruppert, N. A. (2008) Trans-Alaska crustal transect and continental evolution involving subduction underplating and synchronous foreland thrusting. Geology, 36, 267-270.

Gorring, M. L., and Kay, S. M. (2001) Mantle processes and sources of Neogene slab window magmas from southern Patagonia, Argentina. Journal of Petrology, 42, 1067-1094.

Gutscher, M. A., Malavieille, J., Lallemand, S., and Collot, J. Y. (1999) Tectonic segmentation of the north Andean margin: impact of the Carnegie Ridge collision. Earth and Planetary Science Letters, 168, 255-270.

Haeussler, P. J., Bradley, D. C., Wells, R. E., and Miller, M. L. (2003) Life and death of the Resurrection plate: Evidence for it existence and subduction in the northeastern Pacific in Paleocene-Eocene time. Geological Society of America Bulletin, 115, 867-880.

Hampel, A. (2002) The migration history of the Nazca Ridge along the Peruvian active margin: a re-evaluation. Earth and Planetary Science Letters, 203, 665-679.

Hoffman, M. D., and Armstrong, P. A. (2006) Miocene exhumation of the southern

TalkeetnaMountains, south central Alaska, based on apatite (U-Th)/He thermochronology. Geological Society of America, Abstracts with Programs, 38 (5), 9.

Hoy, R. G., and Ridgway, K. D. (1997) Structural and sedimentological development of footwall growth synclines along an intraforeland uplift, east-central Bighorn Mountains, Wyoming. Geological Society of America Bulletin, 109, 915-935.

Ingersoll, R. V. (1982) Initiation and evolution of the GreatValley forearc basin of northern and central California, U. S. A., in Leggett, J. K., ed., Trench-forearc geology: Sedimentation and tectonics on modern and ancient active plate margins. Geological Society [London] Special Publication 10, 459-467.

Ingersoll, R. V. (1983) Petrofacies and provenance of late Mesozoic forearc basin, northern and central California. American Association of Petroleum Geologists Bulletin, 67, 1125-1142.

Iwamori, H. (2000) Thermal effects of ridge subduction and its implications for the origin of granitic batholith and paired metamorphic belts. Earth and Planetary Science Letters, 181, 131-144.

Jordan, T. E., and Allmendinger, R. W. (1986) The Sierras Pampeanas of Argentina-A modern analog of RockyMountain foreland deformation. American Journal of Science, 286, 737-764.

Kassab, C. M., Kortyna, C. D., Ridgway, K. D., and Trop, J. M. (2009) Sedimentology, structural framework, and basin analysis of the eastern Arkose Ridge Formation, Talkeetna Mountains, Alaska. Geological Society of America, Abstracts with Programs, 41 (7), 304.

Kimbrough, D. L., Smith, D. P., Mahoney, J. B., Moore, T. E., Grove, M., Gastil, R. G., Ortega-Rivera, A., and Fanning, C. M. (2001) Forearc-basin sedimentary response to rapid Late Cretaceous batholith emplacement in the Peninsular Ranges of southern and Baja California. Geology, 29, 491-494.

Kortyna, C. D., Trop, J. M., LeComte, A. A., Bauer, E. M., Kassab, C. M., Ridgway, K. D., and Sunderlin, D. (2009) Sedimentology, paleontology, and structural framework of the central Arkose Ridge Formation, Talkeetna Mountains, Alaska. Geological Society of America, Abstract with Programs, 41 (7), 304.

Lallemand, S., Heuret, A., and Boutelier, D. (2005) On the relationships between slab dip, back-arc stress, upper plate absolute motions, and crustal nature in subduction zones. Geochemistry, Geophysics, Geosystems, 6 (9). doi: 10. 1029/2005GC000917.

Leonard, L. J., Hyndman, R. D., Mazzotti, S., Nykolaishen, L., Schmidt, M., and Hippchen, S. (2007) Current deformation in the northern Canadian Cordillera inferred from GPS measurements. Journal of Geophysical Research, 112, B11401. doi: 10. 1029/2007JB005061.

Lesh, M. E., and Ridgway, K. D. (2007) Geomorphic evidence of active transpressional deformation in the Tanana foreland basin, south-central Alaska, in Ridgway, K. D., Trop, J. M., Glen, J. M. G., and O'Neill, J. M., eds., Tectonic growth of a collisional continental margin: Crustal evolution of southern Alaska. Geological Society of America Special Paper 431, 573-592.

Little, T. A., and Naeser, C. D. (1989) Tertiary tectonics of the BorderRanges fault system, Chugach Mountains, Alaska: Deformation and uplift in a forearc setting. Journal of Geophysical Research, 94, 4333-4360.

Lowe, L. A., Gulick, S. P., Christeson, G. L., van Avendonk, H., Reece, R., Elmore, R., and Pavlis, T. (2008) Crustal structure and deformation of the Yakutat microplate: New insights from STEEP marine seismic reflection data. Eos Trans. AGU, 89 (53), Fall Meeting Supplement, Abstract T53B-1941.

Lytwyn, J., Lockhart, S., Casey, J., and Kusky, T. (2000) Geochemistry of near trench intrusives associated with ridge subduction, Seldovia quadrangle, southern Alaska. Journal of Geophysical Research, 105, 27957-27978.

Madsen, J. K., Thorkelson, D. J., Friedman, R. M., and Marshall, D. D. (2006) Cenozoic to Recent plate configurations in the PacificBasin: ridge subduction and slab window magmatism in western North America. Geosphere, 1, 11-34.

Magoon, L. B. (1994) Petroleum resources in Alaska, in Plafker, G., and Berg, H. G., eds., The Geology of Alaska. Boulder, CO, Geological Society of America, The Geology of North America, G-1 905-936.

McNeill, L. C., Goldfinger, C., Kulm, L. D., and Yeats, R. S. (2000) Tectonics of the Neogene Cascadia forearc basin; investigations of a deformed late Miocene unconformity. Geological Society of America Bulletin, 112 (8), 1209-1224.

Nokleberg, W. J., Jones, D. L., and Silberling, N. J. (1985) Origin and tectonic evolution of the Maclaren and Wrangellia terranes, eastern Alaska Range, Alaska. Geological Society of America Bulletin, 96, 1251-1270.

O' Sullivan, P. B., and Currie, L. D. (1996) Thermotectonic history of Mt. Logan, Yukon Territory, Canada: implications of multiple episodes of middle to late Cenozoic denudation. Earth and Planetary Science Letters, 144, 251-261.

Pallares, C., Maury, R. C., Bellon, H., Royer, J., Calmus, T., Aguill_on-Robles, A., Cotten, J., Benoit, M., Michaud, F., and Bourgois, J. (2007) Slab-tearing following ridge-trench collision: evidence from Miocene volcanism in Baja California, M _ exico. Journal of Volcanology and Geothermal Research, 161, 95-117.

Pedoja, K., Ortlieb, L., Dumont, J. F., Lamothe, M., Ghaleb, B., Auclair, M., and Laborusse, B. (2006) Quaternary coastal uplift along the Talara Arc (Ecuador, Northern Peru) from new marine terrace data. Marine Geology, 228, 73-91.

Plafker, George, and Berg, H. C. (1994) Overview of the geology and tectonic evolution of Alaska: in Plafker, George, and Berg, H. C., eds., The Geology of Alaska. Boulder, CO, Geological Society of America, The Geology of North America, G-1, 989-1021.

Plafker, George, Moore, J. C., and Winkler G. R. (1994) Geology of the southern Alaska margin, in Plafker, G., and Berg, H. C., eds., The Geology of Alaska. Boulder, CO, Geological Survey of America, The Geology of North America, G-1, 389-450.

Preece, S. J., and Hart, W. K. (2004) Geochemical variations in the <5 Ma Wrangell volcanic field, Alaska: implications for the magmatic and tectonic development of a complex continental arc system. Tectonophysics, 392, 165-191.

Qi, C., Zhao, D., Chen, Y., and Ruppert, N. A. (2007) New insight into the crust and upper mantle structure under Alaska. Polar Science, 1, 85-100.

Richter, D. H., Smith, J. G., Lanphere, M. A., Dalrymple, G. B., Reed, B. L., and Shew, N. (1990) Age and progression of volcanism, Wrangell volcanic field, Alaska. Bulletin of Volcanology, 53, 29–44.

Ridgway, K. D., and DeCelles, P. G. (1993) Stream–dominated alluvial–fan and lacustrine depositional systems in Cenozoic strike–slip basins, Denali fault system, Yukon Territory. Sedimentology, 40, 645–666.

Ridgway, K. D., Skulski, T., and Sweet, A. R. (1992) Cenozoic displacement along the Denali fault system, Yukon Territory. EOS Trans. AGU, 73, 534.

Ridgway, K. D., Trop, J. M., and Sweet, A. R. (1997) Thrusttop basin formation along a suture zone, Cantwell basin, Alaska Range: Implications for development of the Denali fault system. Geological Society of America Bulletin, 109, 505–523.

Ridgway, K. D., Trop, J. M., Nokleberg, W. J., Davidson, C. M., and Eastham, K. D. (2002) Mesozoic and Cenozoic tectonics of the eastern and central Alaska Range: progressive basin development and deformation within a suture zone. Geological Society of America Bulletin, 114, 1480–1504.

Ridgway, K. D., Thoms, E. E., Lesh, M. E., Layer, P. W., White, J. M., and Smith, S. V. (2007) Neogene transpressional foreland basin development on the north side of the central Alaska Range, Usibelli Group and Nenana Gravel, Tanana basin, in Ridgway, K. D., Trop, J. M., Glen, J. M. G., and O'Neill, J. M., eds., Tectonic growth of a collisional continental margin: crustal evolution of southern Alaska. Geological Society of America Special Paper 431, 507–547.

Rioux, M., Hacker, B., Mattinson, J., Kelemen, P., Blusztajn, J., and Gehrels, G. (2007) The magmatic development of an intra-oceanic arc: high-precision U–Pb zircon and whole-rock isotopic analyses from the accreted Talkeetna arc, south-central Alaska. Geological Society of America Bulletin, 119, 1168–1184.

Sak, P. B., Fisher, D. M., and Gardner, T. W. (2004) Effects of subducting seafloor roughness on upper plate vertical tectonism: Osa Peninsula, Costa Rica. Tectonics, 23. doi: 10.1029/2002TC001474.

Sak, P. B., Fisher, D. M., Gardner, T. W., Marshall, J. S., and LaFemina, P. C. (2009) Rough crust subduction, forearc kinematics, and Quaternary uplift rates, Costa Rican segment of the Middle American Trench. Geological Society of America Bulletin, 121, 992–1012.

Sample, J. C., and Reid, M. R. (2003) Large-scale, latest Cretaceous uplift along the Northeast Pacific Rim; evidence from sediment volume, sandstone petrography, and Nd isotope signatures of the Kodiak Formation, Kodiak Islands, Alaska, in Sisson, V. B., Roeske, S. M., and Pavlis, T. L. eds: Geology of a transpressional orogen developed during ridge-trench interaction along the north Pacific margin. Geological Society of America Special Paper 371, 51–70.

Sendziak, K., Armstrong, P. A., and Haeussler, P. J. (2009) Constraints on exhumation of the western Chugach Mountains (Alaska) based on zircon fission-track analysis of modern glacial outwash. Geological Society of America, Abstracts with Programs, 41, 290.

Sisson, V. B., Hollister, L. S., and Onstott, T. C. (1989) Petrologic and age constraints on the origin of a lowpressure/ high-temperature metamorphic complex, southern Alaska. Journal of Geophysical Research, 94, 4392-4410.

Sisson, V. B., Pavlis, T. L., Roeske, S. M., and Thorkelson, D. J. (2003) Introduction: an overview of ridge-trench interactions in modern and ancient settings, in Sisson, V. B., Roeske, S. M., and Pavlis, T. L., eds., Geology of a transpressional orogen developed during ridge-trench interaction along the north Pacific margin. Geological Society of America Special Paper 371, 1-18.

Skulski, T., Francis, D., and Ludden, J. (1991) Arc-transform magmatism in the Wrangell volcanic belt. Geology, 19, 11-14.

Skulski, T., Francis, D., and Ludden, J. (1992) Volcanism in an arc-transform transition zone: the stratigraphy of the St. ClareCreek volcanic field, Wrangell volcanic belt, Yukon, Canada. Canadian Journal of Earth Science, 29, 446-461.

Takashima, R., Kawabe, F., Nishi, H., Moriya, K., Wani, R., and Ando, H. (2004) Geology and stratigraphy of forearc basin sediments in Hokkaido, Japan; Cretaceous environmental events on the north-west Pacific margin. Cretaceous Research, 25 (3), 365-390.

Taylor, F. W., Mann, P., Bevis, M. G., Edwards, R. L., Cheng, H., Cutler, K. B., Gray, S. C., Burr, G. S., Beck, J. W., Phillips, D. A., Cabioch, G., and Recy, J. (2005) Rapid fore-arc uplift and subsidence caused by impinging bathymetric features; examples from the New Hebrides and Solomon arcs. Tectonics, 24 (6), TC6005. doi: 10.1029/2004TC001650.

Thorkelson, D. J. (1996) Subduction of diverging plates and the principles of slab window formation. Tectonophysics, 255, 47-63.

Tidmore, R. S. (2004) Sedimentologic and petrologic investigation of the Miocene Frederika Formation: a record of the initial construction of the Wrangell volcanic field, Alaska. BS thesis, BucknellUniversity, Lewisburg, PA, 103 p.

Trop, J. M. (2008) Latest Cretaceous forearc basin development along an accretionary convergent margin: southcentral Alaska. Geological Society of America Bulletin, 120, 207-224.

Trop, J. M., n. d., Unpublished data, BucknellUniversity, Lewisburg, PA.

Trop, J. M., and Ridgway, K. D. (2000) Sedimentology, stratigraphy, and tectonic importance of the Paleocene-Eocene Arkose Ridge Formation, Cook Inlet-Matanuska Valley forearc basin, Alaska, in Pinney, D. S. and Davis, P. K., eds., Short notes on Alaskan Geology (1999) Fairbanks, Alaska, Division of Geological and Geophysical Surveys Professional Report, 119, 129-144.

Trop, J. M., and Plawman, T. (2006) Bedrock geology of the Glenn Highway from Anchorage to Sheep Mountain, Alaska: Mesozoic-Cenozoic forearc basin development along an accretionary convergent margin: field trip guidebook. 2006 Joint Meeting of the Geological Society of America (Cordillera Section) American Association of Petroleum Geologists (Pacific Section) and Society of Petroleum Engineers (Western Region), May, 20p.

Trop, J. M., and Ridgway, K. D. (2007) Mesozoic and Cenozoic tectonic growth of southern Alaska: a sedimentary basin perspective, in Ridgway, K. D., Trop, J. M., Glen, J. M. G., and O'Neill, J. M., eds., Tectonic growth of a collisional continental margin: crustal evolution of southern Alaska. Geological Society of America Special Paper 431, 55-94.

Trop, J. M., Ridgway, K. D., Sweet, A. R., and Layer, P. W. (1999) Submarine fan deposystems and tectonics of a Late Cretaceous forearc basin along an accretionary convergent plate boundary, MacColl Ridge Formation, Wrangell Mountains, Alaska. Canadian Journal of Earth Sciences, 36, 433-458.

Trop, J. M., Ridgway, K. D., and Spell, T. L. (2003) Synorogenic sedimentation and forearc basin development along a transpressional plate boundary, MatanuskaValley - TalkeetnaMountains, southern Alaska, in Sisson, V. B., Roeske, S., and Pavlis, T. L., eds: Geology of a transpressional orogen developed during ridge-trench interaction along the north Pacific margin. Geological Society of America Special Paper, 371, 89-118.

Trop, J. M., Ridgway, K. D., and Sweet, A. R. (2004) Stratigraphy, palynology, and provenance of the Colorado Creek basin, Alaska, U. S. A.: Oligocene transpressional tectonics along the central Denali fault system. Canadian Journal of Earth Sciences, 41, 457-480.

Trop, J. M., Szuch, D. A., Rioux, M., and Blodgett, R. B. (2005) Sedimentology and provenance of the Upper Jurassic Naknek Formation, Talkeetna Mountains, Alaska: bearings on the accretionary tectonic history of the Wrangellia composite terrane. Geological Society of America Bulletin, 117. doi: 10. 1130/B25575. 1.

Trop, J. M., Snyder, D. C., Hart, W. K., Idleman, B., and Delaney, M. R. (2007) Miocene intra-arc basin development within the Wrangell volcanic field, Frederika Formation and Lower Wrangell Lava, eastern Wrangell Mountains, Alaska. Geological Society of America, Abstracts with Programs, 38 (7), 491.

Trop, J. M., Kortyna, C. D., Valencia, V. A., Kassab, C. M., Bradley, D. C., and Wooden, J. L. (2009) Detrital zircon provenance analysis of Cretaceous-Oligocene sedimentary strata from the Matanuska Valley-Talkeetna Mountains forearc basin, Southern Alaska; Geological Society of America, Abstracts with Programs, 41 (7), 304.

Underwood, M. B., Shelton, K. L., McLaughlin, R. J., Laughland, M. M., and Solomon, R. M. (1999) Middle Miocene paleotemperature anomalies within the Franciscan Complex of Northern California: thermo-tectonic responses near the Mendocino triple junction. Geological Society of America Bulletin, 111, 1448-1467.

Unruh, J. R., Dumitru, T. A., and Sawyer, T. L. (2007) Coupling of early Tertiary extension in the Great Valley forearc basin with blueschist exhumation in the underlying Fransiscan accretionarywedge atMount Diablo, California. Geological Society of America Bulletin, 119, 1347-1367.

Wahrhaftig, Clyde, Bartsch-Winkler, S., and Stricker, G. D. (1994) Coal in Alaska, in Plafker, George, and Berg, H. G., eds., The geology of Alaska. Boulder, CO, Geological Society of America, The Geology of North America, G-1, 937-978.

White, T. S., and Bradley, D. C. (2006) Unconformity and subsequent basalt extrusion event associated with ridge subduction, Paleogene Alaska. Geological Society of America Abstracts with Programs, Speciality Meeting no. 2, 92.

Wilson, D. S., McCrory, P. A., and Stanley, R. G. (2005) Implications of volcanism in coastal California for the Neogene deformation history of western North America. Tectonics, 24, TC3008. doi: 10.1029/ 2003TC001621.

Wilson, F. H., Dover, J. H., Bradley, D. C., Weber, F. R., Bundtzen, T. K., and Haeussler, P. J. (1998) Geologic map of central (interior) Alaska. U. S. Geological Survey Open-File Report 98-133-A, 63 p.

(张浩 译, 阳怀忠 赵梦 吴哲 校)

第 17 章　弧—陆碰撞背景下的盆地特征

AMY E. DRAUT[1], PETER D. CLIFT[2]

(1. U. S. Geological Survey, Santa Cruz, California, USA;
2. School of Geosciences, University of Aberdeen, Andrdeen, UK)

摘　要：弧—陆碰撞通常发生在板块构造旋回中，并且导致造山带迅速形成和迅速崩塌，通常时间跨度为 5~15Ma。普遍认为弧—陆碰撞导致的大陆块体增生是控制地质历史时期陆壳构造和地球化学演化的一个重要过程。洋内岛弧与被动大陆边缘的碰撞（岛弧位于上部板块且面向大陆的情况）比洋内岛弧与活动大陆边缘的碰撞（需要多个会聚带，且会聚带内位于下部板块的岛弧退回到主动大陆边缘）涉及到更多形态上的变化，并且每种情况下盆地的保存潜力也不同。不管是在碰撞前还是碰撞后构造侵蚀边缘和构造增生边缘的海沟和弧前演化也存在很大的差异。

我们详细研究了弧—陆碰撞过程中海沟、沟—坡盆地、弧前盆地、弧内盆地和弧后盆地的演化过程。由于在碰撞过程中被迅速抬升和剥蚀，且在侵蚀大陆边缘遭受俯冲侵蚀而逐渐被破坏，沟—坡盆地被保存下来的可能性较低。弧—陆碰撞之后的海沟沉积物和沟—坡盆地偏向于在碰撞之前相当长时间内处于构造增生的大陆边缘得以保存。构造侵蚀边缘弧前盆地底部通常为坚硬的岩石圈，因此很可能在与被动大陆边缘碰撞时得以保存，甚至有时在整个碰撞造山过程中持续沉积。由于具有较低的挠曲刚度，弧内盆地经常发育较深，并且如果得以保存可记录长时期的岛弧和碰撞构造活动。弧后盆地则通常遭受俯冲作用，盆内沉积充填不是消失就是仅作为碎片保存在混杂岩序列中。来自碰撞造山带的大部分沉积物最终都进入到碰撞作用形成的前陆盆地中，并且可能在不遭受构造变形的情况下被大量保存下来。与陆—陆碰撞形成的前陆盆地相比，弧—陆碰撞形成的前陆盆地一般存在的时间较短，并可能在碰撞之后，由于在碰撞带外侧形成新的活动大陆边缘以及造山带的伸展坍塌，而发生局部反转。

关键词：弧陆碰撞　弧前盆地　弧后盆地　海沟—陆坡盆地　构造侵蚀

17.1　引言

火山岛弧周边的沉积记录是解释主动大陆边缘构造和岩浆活动历史的基础，既可作为自身演化的终止，又可作为理解地球陆壳演化的一种方式，大部分陆壳都是在这种环境下形成（Rudnic 和 Fountain，1995；Hawkesworth 和 Kemp，2006）。虽然在所有地质单元中，岛弧盆地包含了最完整的与板块会聚和陆壳生长有关的沉积和地球化学历史，但是由于其保存条件仍不及相对远离破坏性板块边界的盆地，因此在解释古老地质记录时存在较大的困难。尽管如此，岛弧盆地在某些缝合线地带仍得以保存（图 17-1）。从活动岛弧盆地的沉积记录中可以获得很多信息（图 17-1），从而得到一个更完整的关于岛弧盆地演化的认识。这反过来又使得这些地质单元在增生地体中更容易被识别出来，并用来进行古地理和构造的重建。

在 Busby 和 Ingersoll（1995）贡献性研究之后的 15 年里，由于围绕活动岛弧的海洋测绘技术的改进和广泛应用，对盆地的形成和演化的认识也取得了新的进展。这些方法生成了海底特征、构造以及基底地层的高分辨率图像（Gaedicke 等，2000；Wright 等，2000；Laursen 等，2002；Kopp 等，2006）。地球动力学模拟能力在近年来也同样得到增强，从而可以更好地认识地壳负载和盆地发育（Tang 和 Chemenda，2000；Londono 和 Lorenzo，2004；Waltham 等，2008）。对活动的和古造山带的地球物理成像极大地提高了人们对弧—陆碰撞带中构造过程的认识（Dadson 等，2003；Johnson 等，2008；Kimura 等，2008）。对现代和古代岛弧碰撞带的野外研究也扩大了我们对构造与气候耦合作用的理解（Dadson 等，2003；Johnson 等，2008；Kimura 等，2008）。

在这里我们简要回顾一下形成于洋内岛弧，和那些在弧—弧和弧—陆碰撞过程中演化的主要盆地类型，并通过研究增生岛弧地体来评估岛弧与大陆边缘碰撞过程中的盆地的归宿。我们的讨论借鉴了具体的例子，但并不详细综述所有的增生岛弧。弧—陆碰撞不仅常常破坏先前存在的盆地，也可产生出新的盆地，正如我们采用的吕宋岛弧与欧亚陆缘在中国台湾持续碰撞的例子。目前活跃的弧—陆碰撞的其他例子（图 17-1）发生在巴布亚新几内亚（美拉尼西亚岛弧；Cloos 等，2005）和日本（伊豆—小笠原岛弧等；Taira 等，2001）。了解可容空间的演变、结合对岛弧活动和碰撞过程所有阶段的岩石圈强度的约束，可以最大限度地从分析岛弧盆地的沉积和地球化学记录中得到认识。

最后，了解弧—陆碰撞的过程可以为陆壳形成和演化提供有价值的见解。伴随着拆沉过程中硅质岩浆作用和致密铁镁质下地壳的亏损（Bird，1978）或对流不稳定作用（Jull 和 Kelemen，2001），岛弧地体增生所致的大陆生长被认为是大陆形成其结构和地球化学组成的关键过程（Pearcy 等，1990；Rudnick 和 Fountain，1995；Suyehiro 等，1996；Holbrook 等，1999；Busby 等，2006；Brown，2009；Draut 等，2009）。另外，大多数陆壳物质是通过俯冲带返回上地幔的，其动力学机制在岛弧和海沟—陆坡盆地中有所记录。

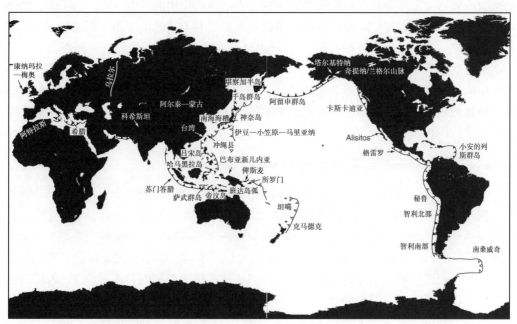

图 17-1　文中讨论的活动岛弧和增生岛弧地体全球分布图
俯冲带用空心三角形表示，指示侵蚀边缘；实心三角形代表增生边缘

17.2 洋内弧盆及其演化

Busby 和 Ingersoll（1995）已经对洋内岛弧系统作了详细讨论。我们简要地回顾一下沿着洋内岛弧形成的主要盆地类型，并讨论从增生岛弧地区了解到的在弧—陆碰撞过程中每种盆地类型的可能演变。我们的讨论仅限于洋内岛弧周围形成的盆地，这些盆地最终与大陆碰撞；大陆岛弧及其盆地超出了本研究的范围，其他文章已对此有很好的综述（Wilson，1991；Jordan，1995；Fildani 和 Hessler，2005；Centeno-Garcia 等，2008；Lamarche 等，2008）。在这里我们认为的"洋内"岛弧是在大洋岩石圈中由于俯冲作用开始而形成的，不包括基底内在碰撞事件之前就存在的陆壳。

大洋俯冲带的几何学形态是板块会聚速率、板片年龄和倾角，以及上覆板块流变性和浮力的函数（Turcotte 和 Schubert，1982；Jarrard，1986；Billen 和 Gurnis，2003）。图 17-2 展示了侵蚀—增生洋内俯冲系统的剖面图，图 17-3 则是各种弧盆类型所发育的具有代表性的沉积剖面。约四分之三的现代俯冲带是侵蚀类型（图 17-2a），而四分之一是增生类型（图 17-2b；Von Huene 和 Scholl，1991）。主要弧盆形成的地球动力学机制十分不同。海沟和海沟—陆坡盆地沿着两个板块之间的边界形成，是对俯冲板块挠曲作用和由底部构造侵蚀驱动的上盘伸展作用的响应（图 17-2；Von Huene 和 Scholl，1991；Underwood 和 Moore，1995；Underwood 等，2003）。弧前盆地形成于上覆板块内的岛弧和外弧高之间（海沟—陆坡坡折；Dickinson，1995）。弧内盆地作为岛弧的一部分形成于岛弧台地上的火山建造之间（Smith 和 Landis，1995）。弧后盆地可以形成于与新地壳形成时的高热流有关的岛弧裂谷作用，或者具有由捕获到的古洋壳（较岛弧发育早）形成的深水盆地特征（Karig，1971；Taylor 和 Karner，1983；Tamaki 和 Honza，1991；Clift，1995；Marsaglia，1995）。深盆可以在临近岛弧的位置形成，作为对火山岛弧建造负载导致的挠曲作用的响应（Waltham 等，2008），也可以由于岛弧的伸展作用形成。每种类型的盆地都具有不同的沉积相、构造控制作用以及下伏基底组成，是对弧—陆碰撞过程的不同响应。不同盆地类型在盆地保存潜力、变质作用，或者盆地所处岛弧与大陆边缘碰撞过程中盆地遭受的破坏作用等方面也存在差异。弧—陆碰撞中的盆地沉积是快速而短暂的，碰撞、造山运动和造山带坍塌通常仅持续 5~15Ma。然而，岛弧的爆发性火山活动，再加上造山带的快速隆升、侵蚀作用，为碰撞带周围的盆地提供了大量的沉积供给。

17.2.1 海沟和沟—坡盆地（trench-slope basins）

在岛弧和弧前区域的外侧，沉积物聚集在海沟中，并且，如果厚度超过 1km，通常会形成增生楔（Clift 和 Vannucchi，2004；图 17-2b）。与被动（大陆）边缘碰撞的洋内岛弧，即使岛弧活动的开阔洋盆阶段以俯冲侵蚀作用为特征，当碰撞作用临近时通常也会形成一个增生楔（图 17-2a）。然而，在其中任何一种情况下，海沟沉积序列的保存潜力都是受限制的。不管是在侵蚀的还是增生的俯冲带上，海沟陆坡上构造控制的盆地（图 17-2a）往往形成较小的沉积中心。位于汤加的沟—坡盆地，大小约 10km×50km，具有复杂的地层，反映了它们的动态构造环境（图 17-3；Clift 等，1998）。它们在较冷的弧前地带由脆性的变形地壳的伸展作用而形成。强烈的变形作用使得海沟—陆坡的挠曲强度较低，因此这些盆地具有非常深且较狭窄的面貌。在某些岛弧中，海沟和沟—坡盆地接受来自宽阔、抬升的弧前台地

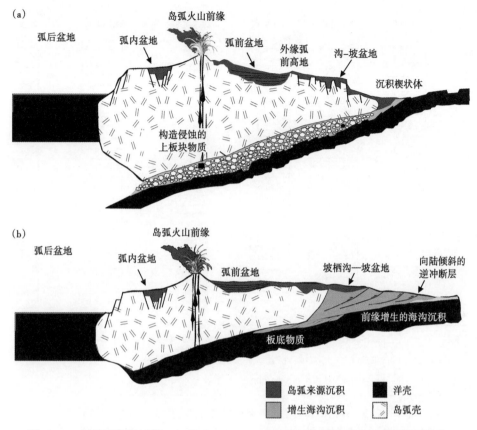

图 17-2 剖面示意图（据 Clift 和 Vannucchi，2004；经美国地球物理联盟许可转载）
(a) 经历俯冲侵蚀作用的典型西太平洋型洋内岛弧，展示了岛弧和岛弧外侧高地之间的盆地、弧后盆地的演化，以及伴随俯冲侵蚀构造作用的伸展作用。(b) 增生洋内岛弧系统（例如吕宋或小安的列斯），沉积物的俯冲形成增生楔前缘并位于弧前区之下。图中所示弧间盆地位于火山中心的内侧，通常沿火山轴分布

的沉积物的沉积（通过海底峡谷搬运），而并不一定是来自岛弧本身，因为台地和弧前盆地将岛弧前端与海沟分开（汤加、希腊和小安的列斯弧；Scholl 和 Vallier，1985；Marsaglia 和 Ingersoll，1992；Draut 和 Clift，2006）。例如，高分辨率测深地形图像显示出弧前汤加台地和切割台地的峡谷（Wright 等，2000）。沟—坡盆地的保存潜力非常低。在碰撞过程中，盆地被快速抬升并剥蚀，并在侵蚀陆缘甚至在洋内俯冲带由于俯冲侵蚀作用而逐步遭受破坏。

在构造侵蚀和构造增生的陆缘中，不论在碰撞之前还是在之后，海沟和弧前的演化（以及与之相关的沉积作用）都存在显著的差别（图 17-2）。沿着诸如吕宋岛北部、日本南海、苏门答腊岛或小安的列斯群岛等增生陆缘，可容空间可能受到限制，沉积物在增生楔上的坡栖盆地得以保存（Beaudry 和 Moore，1985；Underwood 等，2003）。当来自增生楔的沉积物开始出现和旋回时，隆起结束，这些沉积物可以与来自岛弧的火山碎屑物质混合，形成浅水相沉积。该过程通常出现在受大规模底侵作用影响的弧前楔体位置，或者发生在逆断层和反向逆冲断层将增生楔变短的位置。相反，作为对弧前地壳底部侵蚀和减薄作用的响应，沿构造侵蚀陆缘的海沟—陆坡快速下沉（西太平洋；图 17-2a），并在海沟附形成了大量的可容空间；同时海沟—陆坡也是缓慢的深水沉积区域。随着俯冲带侵蚀引起岛弧向着海沟迁

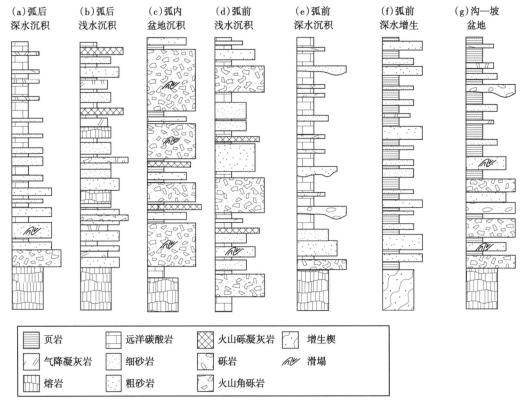

图17-3 洋内岛弧背景下各类沉积盆地具有代表性的垂直剖面

(a) 远端、深水弧后盆地：火山岩基底被角砾岩及之后的远洋碳酸盐岩、火山砂岩和凝灰岩上覆。随着盆地的扩大，碳酸盐岩的比例增加。(b) 近端、浅水弧后盆地：沉积物与远缘弧后盆地大致相同，但更粗。包括硅质近火山产物，有熔岩、凝灰岩和火山砾凝灰岩。(c) 弧间盆地：火山基底，上覆浅海相碳酸盐岩，指示沉降；以近端的火山产物、角砾岩、凝灰岩（浮石、火山砾凝灰岩），偶尔出现的浊积岩和碳酸盐岩为主。(d) 近端、浅水弧前盆地，与近端弧后盆地类似，基本上是火山周围的山麓冲积平原沉积，含角砾岩（块体流沉积）与浊积岩互层；(e) 侵蚀边缘的远端、深水弧前盆地：火山岩基底，上覆角砾岩、深水碳酸盐岩，以及来自岛弧的河道火山碎屑沉积，包括可能由于盆地极大的横向连续性而形成的海底扇；(f) 增生边缘的远端弧前盆地：基底为变形、向上倾斜的增生楔物质，上覆深水页岩，复理石浊积序列（砂岩和页岩），偶见凝灰岩；沉积背景以页岩为主，而不是碳酸岩，因为增生楔最厚部分形成的地方存在地面径流，这意味着泥沙的搬运将超过远洋碳酸盐岩沉积。(g) 沟—坡盆地：与火山岩基底一同展示，就像在汤加侵蚀边缘一样，尽管在诸如日本南海的增生环境中基底可以是增生楔的变形沉积物；以页岩为主的盆地还充填了砂岩和浊流沉积，包括水道蚀积砾岩；陡峭陆坡和活跃断层引起了滑塌，使沉积物被改造

移，海沟和它们的沉积记录逐渐被破坏（Collot等，2008）。因此，弧—陆碰撞之后的海沟沉积物和沟—坡盆地偏向于在碰撞之前相当长时间内处于构造增生的大陆边缘得以保存。如果增生楔物质保存在造山带中，尽管其遭受的构造变形可能更强，但仍很难将其与体积含量极少的沟—坡盆地沉积识别开来 Dickinson，1982）。

在许多古岛弧地体中均发现了来自构造增生陆缘的海沟沉积（图17-2b）；增生楔变质沉积岩往往比增生岛弧其他部分更容易发生改变和变形，甚至在碰撞之前就发生变质作用。这是由于它们被掩埋于弧前区域之下，增生的底板物质受到变形和加热（Sample 和 Moore，1987；Moore 等，1991）。与之相反，增生楔前缘往往只发生构造变形而不发生变质作用。例

如韦斯特波特（Westport）杂岩和爱尔兰西部南康尼马拉（South Connemara）群的变质沉积岩。它们被解释为与奥陶纪岛弧有关的增生楔前缘混杂堆积物质，该岛弧与劳伦大陆边缘在约470Ma发生碰撞（Dewey 和 Ryan，1990；Ryan 和 Dewey，2004）。从海洋环境的构造侵蚀到岛弧和被动陆缘的碰撞的迅速转变形成了俯冲混杂岩带，组分常常为蛇绿岩，位于增生岛弧和大陆之间且只有数百米厚，例如科希斯坦和欧亚大陆之间的 Shyok 缝合带（Robertson 和 Collins，2002），以及中国台湾的 Lichi 混杂堆积（Chang 等，2000）。该混杂堆积代表了碰撞时期的俯冲通道（subduction channel）充填，固定在板块运动有效停止的位置。

在乌拉尔山脉的马格尼托哥尔斯克增生岛弧地区，泥盆纪至早石炭世弧—陆碰撞的穿时产物（由 Puchkov 总结，2009）中就包含了乌拉尔南部增生杂岩中的变质沉积岩。该泥盆纪增生杂岩为确定弧—陆碰撞的时间提供了重要约束（Alvarez-Marron 等，2000；Brown 等，2006b）。在该处 5~6km 厚的杂砂岩浊积岩已经发生变形、角砾化，并被韧性剪切成为一个大型向斜构造，变质程度从近变质带到绿片岩相。Bowen 等（2006b）认为该增生楔中海沟沉积作用的终止（弗拉阶；385—375Ma）指示了由于波罗的海陆壳到达俯冲带而造成的隆升和广泛的盆地不稳定。

乌拉尔山脉南部增生杂岩的变形过程类似于现今在西太平洋和东南亚的新几内亚发生的弧—陆碰撞作用，在那里澳大利亚板块洋壳部分的向北俯冲始于 30 Ma（Quarles van Ufford 和 Cloos，2005）。在帝汶，伴随着基底卷入的逆冲作用使得增生楔呈叠瓦状，帝汶海槽内（通向班达岛弧的海沟）6km 厚的增生楔向南侵位在澳大利亚陆壳边缘之上（Snyder 等，1996）。往东在俾斯麦岛弧和澳大利亚陆缘相互碰撞的相关区域，增生楔前缘中的浊流和深海沉积作为 Erap 杂岩暴露在巴布亚新几内亚的菲尼斯特雷山脉（Abbott 等，1994），该杂岩被认为是从下部板块（澳大利亚板块）刮落下来的沉积物质。Erap 杂岩被切割成叠瓦状的逆冲断层片，地壳厚度也增加了一倍；俾斯麦岛弧沿着脱序逆断层仰冲在增生楔上，两者都侵位于部分澳大利亚大陆边缘之上（Abbott 等，1994）。先前底侵于相同岛弧下的沉积物质出露，形成巴布亚新几内亚北部的 Derewo 变质岩带（Warren 和 Cloos，2007）。在新几内亚地区保存了几个与弧—陆碰撞相关的同碰撞期盆地序列：一个是被解释为渐新世巴布亚俯冲带海沟充填（艾于勒海沟）的 7km 厚硅质碎屑的艾于勒群，它是由于弧—陆碰撞早期造山经历削截和侵蚀作用而形成；其次是海沟填充物质在中新世中期遭受抬升和构造变形（Quarles van Ufford 和 Cloos，2005）。

需要注意的是，增生杂岩通常形成于岛弧与被动大陆边缘碰撞的最终阶段，而开放大洋俯冲（open-ocean subduction）较普遍与俯冲侵蚀有关。这种差异很大程度上反映了每种环境中输入海沟的沉积物的相对流量。作为面向被动大陆边缘并与陆缘碰撞带接近的岛弧（中国台湾；图 17-4），增生过程起主导作用，因为进入海沟的沉积单元厚度远远超过 1km。由于弧—陆碰撞通常是倾斜的，并伴随着活动碰撞区域沿着走向逐渐迁移，因此在缝合作用之前增生的多数海沟沉积物通常沿走向发生（来自于活跃造山带的）再沉积（即堆积在大陆边缘的物质被大范围侵蚀并且沿着海沟走向再次沉积在海沟内）。在岛弧通过后退至主动大陆边缘来实现碰撞的这种情况下（与面向被动大陆边缘相反），沉积物不会在碰撞岛弧的海沟中大幅增加，因此不太可能形成大型增生杂岩（如将在下文讨论的侏罗纪塔尔基特纳岛弧；图 17-5）。

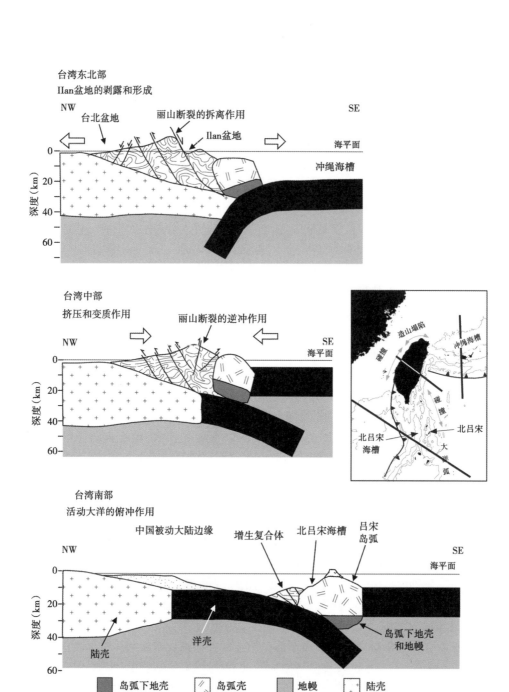

图 17-4 台湾弧—陆碰撞示意图（据 Clift 等，2008；美国地球物理联盟许可转载）
其中，吕宋岛弧与欧亚大陆被动大陆边缘发生碰撞，插图显示三条剖面的位置。
Ilan 盆地的形成是由于吕宋岛弧和中国大陆 SW 向迁移碰撞时台湾中央山脉的重力塌陷作用所致

17.2.2 弧前盆地

弧前盆地的可容空间受控于岛弧岩体的几何形态、海沟—陆坡坡折以及它们相对海平面位置（Dickinson，1995）。弧外高地（outer-arc high）的隆升和断层作用可以随时间改变盆

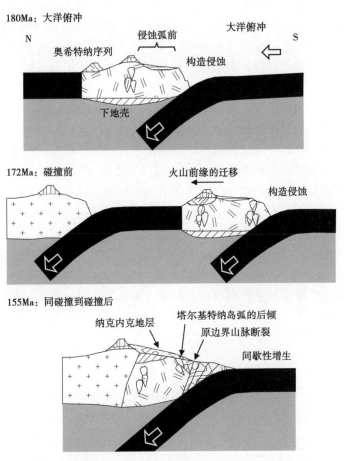

图 17-5　碰撞过程示意图（据 Clift 等，2005b；美国地球物理联盟许可转载）
展示出岛弧退后进入活动大陆边缘（大陆弧），以中侏罗世阿拉斯加中南部的塔尔基特纳岛弧为例。
变形集中在碰撞岛弧之间，而南部的洋内海沟保持开放和活跃

地形态（Ryan 和 Scholl，1989）。有人认为弧前盆地，尤其是大陆岛弧弧前盆地的构造松弛和沉降，是底部俯冲侵蚀引起的地壳减薄的作用结果（Oncken，1998；Clift 等，2003；Wells 等，2003）。虽然近期的数值模拟表明弧前盆地可以是俯冲带的板块内耦合作用的自然结果，不需要强调俯冲侵蚀的作用（Fuller 等，2006），但是弧前下方的俯冲侵蚀和底侵作用仍可能有利于垂直构造运动，并且对弧前盆地和弧外高地地形产生影响（Sample 和 Moore，1987；Moore 等，1991）。弧前盆地的形成不需要主动沉降，但也能反映构造高地的相对隆升，后者使得沉积物在隆起区和岛弧之间堆积。弧前盆地的几何形态根据盆地岩石圈的力学特征而大幅变化。上覆于构造增生杂岩之上的盆地具有较低挠曲强度（相对软弱的岩石圈），而位于成熟的、火成大洋弧前地壳上的盆地则不太可能发生变形，因为它们具有较冷的刚性岩石圈根（图 17-6）；即使与较软弱的增生杂岩并置，后者也有可能在碰撞造山作用下得以保存。

　　碰撞之前的弧前盆地沉积物（图 17-3d 至 f）包括来自火山中心陡坡的临近块体流沉积、远端火山碎屑物浊流沉积、来自盆地内靠海沟方向（外弧地势高处）再沉积的蛇绿岩碎屑物，以及只在弧前外部区域出现的深海沉积物；弧前外部可能是唯一记录未被物质坡移

（mass wasting）影响的远端火山灰沉降的地区（Dickinson，1974；Larue 等，1991；Marsaglia 和 Ingersoll，1992；Underwood 等，1995；Ballance 等，2004；Draut 和 Clift，2006）。海底峡谷能够将沉积物搬运至弧前盆地中（Kopp 等，2006），正如其他地区的峡谷补给海沟。弧前盆地可能包含了比岛弧前端任何位置活动持续时间更长的沉积记录，如马里亚纳岛弧，在长达 45Ma 时间里它持续为同一弧前盆地提供沉积供给。正如以上所讨论的海沟情况，构造侵蚀边缘与增生边缘（分别为图 17-2a 和图 17-2b）在可容空间上存在差别；由于持续的基底沉降作用，侵蚀边缘的弧前盆地具有更深的水体（图 17-3e）。

弧前盆地不仅能够在弧—陆碰撞过程中得以保存，而且在碰撞过程中持续沉降并接受沉积。这在一定程度上反映了大洋弧前岩石圈较强的力学特征（图 17-6）。Ryan（2008）以爱尔兰西部的奥陶纪格兰扁造山带为例，揭示了如果俯冲大陆边缘外侧的榴辉岩大幅度减小下伏板块的上浮作用并引起同期的弧前沉降，弧前盆地在碰撞过程中持续沉降并接受沉积将变为可能。在格兰扁造山带，South Mayo 海槽由一套厚达 9km 的弧前盆地沉积序列组成，包含了碰撞前、同碰撞期和碰撞后沉积地层以及火山岩沉积，且并没有发育大型不整合界面（Dewey 和 Ryan，1990；Draut 和 Clift，2001）。Ryan（2008）通过模拟分析提出，位于俯冲带上盘的弧前盆地不可能在碰撞过程中得到保存，除非在地势上受到均衡掩蔽的作用，比如在下盘形成大规模的榴辉岩地层建造（Cloos，1993；Dewey 等，1993）；否则即使弧前盆地在造山过程中没有被破坏，但由于碰撞过程盆地的抬升以及可容空间的减小，盆地内的沉积作用也极有可能停止。

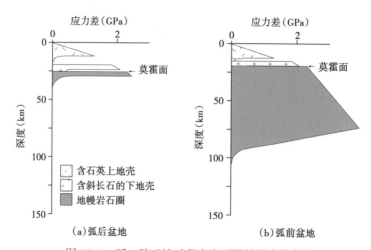

图 17-6　弧—陆碰撞过程中岩石圈的强度分布图

靠近岛弧火山前缘或位于弧后裂谷内的盆地（a）往往很弱（Watts 等，1982；Watts，2001），而弧前盆地（b）有较强的下伏地幔，因为热成熟度高，即更加冷却（Lin 和 Watts，2002；Crawford 等，2003）。因此，弧前盆地不容易在造山过程中遭受变形和破坏

抬升、倾斜的印度尼西亚中新世南萨武盆地就是这种类型的盆地之一。澳大利亚板块边缘陆壳的上浮作用导致的刚性弧前基底发生俯冲断层作用，使得盆地发生快速抬升（5mm/a）并出露地表（Van der Werff，1995）。目前该弧前盆地的部分区域仍为地震活动区（北萨武盆地），但是随着弧—陆碰撞的进行，可容空间逐渐萎缩。

或者，弧前区域及其盆地会在弧—陆碰撞中被破坏。Shemenda（1994）和 Boutelier 等（2003）的物理实验显示，如果碰撞作用力使岛弧前缘的上覆板块破裂，那么弧前区域将完全被岛弧岩体逆冲和俯冲，可能仅剩下弧前地壳最靠近海沟的部分以及即将被仰冲且合并进入造山带内的弧前盆地。虽然该模式在研究堪察加半岛古新世至始新世与陆壳倾斜碰撞的 Achaivayam-Valaginskaya 岛弧块体时就已经提出（Kon stantinovskaia，2001），但目前还没有发现能明确支持该模式的野外地质证据。Cloos（1993）对地壳厚度及密度的分析表明，部分西太平洋岛弧有足够薄的地壳（玄武岩地壳厚度应小于 17km，花岗岩岛弧型地壳厚度应小于 15km）能使岛弧作为一个整体被俯冲。

岛弧俯冲在几个碰撞带都有出现，例如伊豆—小笠原岛弧北部俯冲于本州岛之下（Otsuki，1990）以及阿留申岛弧俯冲于堪察加半岛之下（Scholl，2007）。然而，目前岛弧地壳是否真的被俯冲至深部还不清楚，因为它很可能被上覆岛弧的构造地壳之下的底侵所取代。而关于陆壳的质量均衡论也认为，大型岛弧俯冲作用并不常见（Clift 等，2009）。以伊亚岛弧为例，关于该岛弧大部分地壳是增生的且底侵至碰撞的日本边缘的认识还存在争议。在伊豆—本州碰撞地区，伊豆岛弧的下地壳在神奈山脉出露（Tani 等，2007），而弧前盆地在三浦半岛区域出露（Ogawa 等，1985），表明至少部分岛弧正经历增生过程。马鲁古海域的岛弧逆冲作用也已被认识到，此处的桑义赫（Sangihe）岛弧逆冲在哈马黑拉岛岛弧之上（Hall 和 Smyth，2008）。同样的过程也出现在加拿大东部纽芬兰地区奥陶纪的增生块体内，那里的一个岛弧在另一岛弧之下的部分俯冲作用被认为会引发快速沉降、充填，并在俯冲岛弧和弧后区域之上新形成的盆地内沉积黑色页岩（Zagorevski 等，2008）。岛弧地壳越年轻、越薄，它被俯冲消减的可能性越大；因此，相对于较老的厚层岛弧内的盆地，年轻的薄层岛弧上的盆地较难于在造山过程中保存下来（Cloos，1993）。无论它们最终是被俯冲消减还是在造山带中保存下来，弧前陆壳及盆内物质会完全变形并被碰撞区内发育走滑断层所错断，正如同晚中新世以来的千岛—北海道岛弧（Kusunoki 和 Kimura，1998）以及在堪察加地区 Mys 半岛处与陆壳发生斜碰撞的阿留申岛弧（Gaedicke 等，2000）。

关于碰撞前的弧前盆地另外一种结局可由中国台湾为例生动地表现出来（图 17-4 和图 17-7）。在该区域，增生杂岩体随着吕宋岛弧与中国被动大陆边缘逐步发生碰撞而增长（Suppe，1981；Huang 等，2000）。被动大陆边缘沉积物叠瓦作用与来自中国台湾的沉积物共同进入马尼拉海沟，导致巨大增生脊的形成，并最终演化为海岸山系（图 17-7）。尽管主要的逆冲方向朝向海沟，反冲断层仍然将增生沉积物推向古老的北吕宋海槽弧前盆地之上，导致该盆地被最终填埋并且沉积终止（Lundberg 等，1997；Hirtzel 等，2009）。盆地残余物质在近岸出露，但其整体宽度已大幅减小，并且其中保存的沉积物已普遍遭受强烈变形（图 17-7）。虽然岛弧局部地仰冲于盆地之上，但大部分岛弧壳体都位于山脉深部，并有效地增生（仰冲）于亚洲边缘之中（Clift 等，2009）。与由同碰撞期沉积物沿走向搬运而充填的弧前盆地及碰撞后塌陷盆地相比，中国台湾南部弧前盆地沉积物的保存能力适中。

现存的弧前盆地为研究弧—陆碰撞的时间和动力学机制提供了宝贵数据。上述 Mayo 南部海槽提供了在弧—陆碰撞过程中火山岩地球化学变化的详细记录，指示出陆壳俯冲的程度和时间（Draut 和 Clift，2001）。在乌拉尔山泥盆系 Aktau 组揭示了由燧石质和火山碎屑沉积组成的 5km 厚弧前盆地沉积（Brown 等，2006）。随着被动大陆边缘抵达该海沟位置，弧前盆地沉积作用增加，岛弧来源的火山碎屑浊流在俯冲板片上沉积。快速沉积、弱固结的弧前沉积物的软沉积变形以及大型滑塌沉积（约 10km²）都是由于长达 3~5Ma 的陆壳向岛弧之下的俯冲作用引发的地震活动而形成的（Brown 和 Spadea，1999）。

自侏罗纪以来在阿拉斯加南部，不同岛弧和微陆壳地体增生在北美主动大陆边缘

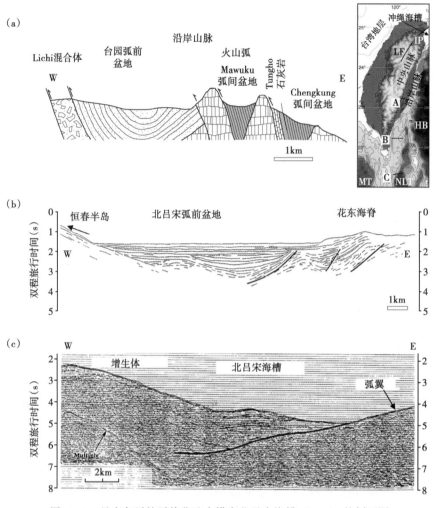

图17-7 吕宋岛弧的弧前盆地中横穿北吕宋海槽（NLT）的剖面图

插图显示了剖面位置。HB—Huatung 海盆；MT—Manila 海沟；IP—Ilan 平原；LF—Lishan 断层。
(a) 台湾中部弧前盆地缩略图（据 Huang 等，2000）；(b) 台湾南侧盆地，显示出更深的充填和更宽广的伸展，以及 (c) 位于远离碰撞带的深水区域，但是它的西部边缘已经受到 Manila 增生楔的影响

(Plafker 和 Berg，1994）。记录了该碰撞历史的沉积盆地出露地表，使得详细的相分析和沉积物搬运路径重建成为可能（Trop 等，2002；图17-5）。同碰撞期的弧前盆地沉积包含了上侏罗统的 Naknek 组—条沿着大陆边缘出露的长达1200km、厚度大于900m 的地层带，该地层带记录了在兰格尔联合地块末次增生至北美板块前的合并过程中塔尔基特纳—奇特诺岛弧与北美大陆或较小地体间的碰撞的地壳尺度压缩和深源物质剥露（Trop 等，2005）。Naknek 组相指示出陡峭盆底的沉积作用，伴随着重力流砾岩沉积和近源扇—三角洲单元过渡为深水浊流和前三角洲泥质沉积（图17-8）。上部不整合面反映出岛弧增生后的地面隆起和侵蚀作用。

通常情况下，被卷入到弧—陆碰撞作用的盆地的物源很大程度上取决于碰撞的构造地质学特征。在弧前区域面向碰撞带的岛弧/被动陆缘碰撞中，叠瓦状、经历构造变形和变质的被动大陆边缘沉积岩塑造了主要地形。这意味着碰撞同期或后期的弧前盆地沉积物，例如台

图 17-8 岛弧碰撞背景下的野外沉积照片

(a) 水下扇砾岩（碎屑是变质沉积岩）和 (b) 来自中国台湾的早同碰撞期 Paliwan 组砂质浊积岩。这些沉积物含有上新世晚期侵蚀于成熟造山带的物质，现在出露于海岸山脉。(c) 砾岩（大多数碎屑为火成岩）和 (d) 爱尔兰 Mayo 南部后碰撞早期奥陶系 Rosroe 组水道砂岩。Rosroe 组碎屑的物源主要来自变质的被动大陆边缘，但有高达 20% 的输入来自增生岛弧（Clift 等，2009b）。(e) 和 (f) 展示了阿拉斯加东南部同碰撞期侏罗纪 Naknek 组的盆地沉积（e 和 f 的图片得到 J. M. Trop 许可）：(e) 近缘扇三角洲弧前地层（左侧人物为比例尺）以差—中等分选、逆粒序砾岩（其花岗岩碎屑同位素年龄为侏罗世，与附近塔尔基特纳岛弧侵入体一致）为主，为海洋重力流沉积；(f) 远端、前三角洲弧前地层（人作为比例尺），以薄层、层状、底部突变、正粒序浊积砂岩（稳定单元）和悬浮物沉降形成的块状、黑色泥岩为特征。砂岩以长英质碎屑和花岗质岩石碎屑为主。泥岩含有侏罗纪菊石、放射虫和有孔虫化石湾的 Paliwan 组（Dorsey，1992）或爱尔兰 Mayo 南部的 Rosroe 组，大部分来源于变形大陆边缘的剥蚀（图 17-8）。Mayo 南部的沉积物源显示，岛弧本身只形成不大的地形，像现今的中国台湾一样。相反地，当岛弧碰撞后退进入主动大陆边缘时，比如本州岛—伊豆、科希斯坦或塔尔基特纳地区（图 17-5），沉积过程主要受控于碰撞并出露的岛弧岩体的侵蚀。这些地区的证据显示出，只要大洋岛弧的弧前盆地继续面朝开放洋盆，就能够保持相对不变形（Clift 等，2000）。纵使随着岛弧火山作用的成分变化沉积物源会有所不同，但这种弧前盆地中的沉积活动会继续，而弧后区域（面对碰撞带的区域）则强烈变形并被俯冲（在下文中讨论）。

其他保存在岛弧碰撞区的弧前盆地的沉积序列包括摩洛哥的 Sarhro 群（其褶皱杂砂岩浊积体和几千米厚的火山岩沉积记录了小阿特拉斯山造山带内约 660Ma 时期的弧—陆碰撞；Thomas 等，2002）、巴布亚新几内亚菲尼斯摩尔山脉始新统—中新统 Sarawaget 组，以及可能存在的反映了内蒙古东南部阿不泰造山带与华北微大陆之间的关闭洋盆的上二叠统源自岛弧的远端浊积体（Johnson 等，2008）。

17.2.3 弧内盆地

岛弧内部也可以形成盆地。与弧前盆地和弧后盆地相比，这种类型的盆地通常规模较小且水深较浅，并且可以"停"在海底高地之间（Smith 和 Landis，1995）。被火山中心或者断层所限的弧内盆地（Busby，2004），其构造成因在所有盆地中也是最复杂和最多样化的。这些盆地可以由于裂谷作用形成，后者在大洋岛弧的活动过程中可以出现多次（比如下加利福尼亚的 Alisitos 增生岛弧区；Busby 等，2006），并且当大陆岛弧内的弧内盆地打开时可能形成新的洋壳（如墨西哥 Guerrero 复合地体杂岩的演化过程；Cenleno-Gartia 等，2008）。其他弧内盆地受走滑断层作用（Sarewitz 和 Lewis，1991）或者由于板块运动随时间变化而发生岛弧大规模块体旋转的影响，发育为压扭或张扭盆地（Geist 等，1988）。影响弧内盆地保存的是，靠近岛弧火山前缘由于缺少较大的岩石圈根而挠曲刚度较低，导致盆地很容易发生构造变形、反转乃至被破坏。

弧内盆地沉积包括源自岛弧区的碎屑冲积平原内火山物质和火山碎屑物质，沉积物从火山喷发中心向浊积相和深水漂移相逐渐变细（如图 17-3c；Draut 和 Clift，2006）。下加利福尼亚出露的白垩系近源弧内盆地相包括硅质水下火山碎屑流沉积、凝灰岩以及凝灰浊积岩、玻璃质角砾岩，以及玄武质和硅质熔岩与海洋沉积物相混合的证据；这些盆地沉积物中的火山物质具有厚壳蛤礁沉积物和海相泥岩夹层，表明在盆地沉积过程中相对海平面发生了变化（Busby 等，2006）。弧—陆碰撞过程中，弧内盆地充填来自于造山带的碎屑物质，并且，如果这些物质保留下来，则可以记录岛弧碰撞过程中主动大陆边缘的构造以及地球化学变化。

Huang 等（1995）认为，台湾海岸山脉出露的上新世和更新世弧内盆地，是在主动弧—陆碰撞过程中的走滑转换断层作用下形成的。这些同碰撞期的弧内盆地和现代没有卷入到碰撞的弧内盆地规模相当（宽 1.5~10.0km，长 40km），并且预测这些弧后盆地会在吕宋岛弧和欧亚陆缘的碰撞过程中很快消亡，导致盆地形成和充填后马上快速反转。盆地相包括了深水复理岩，叠覆在火山岩基底顶部的浅海生物礁碳酸盐岩之上，表明盆地由快速沉降作用形成。这些弧后盆地伴随着仅持续了 0.8~3.1Ma 的沉积过程而发育、充填并反转。

17.2.4 弧后盆地

弧后盆地常出现于洋内岛弧之后（见图 17-2），既可形成于岛弧发育之后的裂谷、伸展作用（如在 Mariana 岛弧和 Tonga 岛弧就经历过两次裂谷作用；Hawkins，1974；Taylor，1992），也可作为早于岛弧和俯冲带形成的的古老洋底（Karig，1971；Taylor 和 Karner，1983）。弧后盆地以火山岩和岛弧的火山碎屑产物沉积为主，其沉积相指示了水深随着远离岛弧而增加。在岛弧裂谷作用形成的弧后盆地内，弧后地层中富含硅质火山活动产物（比如伊豆—小笠原岛弧；Nishimura 等，1992；Jizasa 等，1999；Fiske 等，2001）。

弧后盆地很难在经历弧—陆碰撞后被保存下来。岛弧—被动大陆边缘碰撞（如图17-4）之后常在增生岛弧外侧形成一个新的俯冲带，即原来的弧后盆地位置成为新的海沟并开始遭受俯冲。这种演化实例已多次在地质记录中被识别出来（Dewey 和 Ryan，1990；Teng，1990；Konstantinovskaia，2001；Van Staal 等，2007；Dickinson，2008）。在大洋岛弧通过弧后盆地关闭实现与主动大陆边缘碰撞的地区，例如下加利福尼亚白垩系—侏罗系 Alisitos 岛弧地块（Busby，2004；Busb 等，2006）和阿拉斯加南部侏罗系塔尔基特纳—奇特诺岛弧地块（如图 17-5；Plafker 和 Berg，1994）在增生时发生的情况，弧后地区必然在碰撞开始之前就经历了俯冲。虽然大部分弧后盆地及其沉积充填会因此遭到破坏，但是其残余部分可能会在新的缝合带内以变形的、刮擦块体的形式而保存下来。在遭受变形的弧后地区，大洋岛弧的深层岩体可能会出露并遭受剥蚀，并且这些剥蚀下来的碎屑物质可能会在局部以厚层浊积岩和砾岩的形式沉积在附近的斜坡扇杂岩内（日本的 Ashigara 盆地；Sob 等，1998）。在这种通过弧后盆地的关闭使岛弧增生到主动大陆边缘的情形下，通常发育多个俯冲带（除了主俯冲带，在大陆和弧后地区至少存在大规模的逆冲和会聚），而且很可能发生部分岛弧块体的大规模俯冲和再循环（Suyehiro 等，1996；Busby，2004）。

尽管弧后盆地被保存的潜力普遍很低，但在一些增生的岛弧地块中确实包含有几千米厚的弧后地层（如图 17-3a，b）。例如在下加利福尼亚的 Alisitos 增生岛弧上，侏罗系 Gran Canyon 组由直接覆盖在超俯冲带蛇绿岩套上的火山质和火山碎屑质的弧后物质组成（Kimbrough，1984；Busby，2004）。基底的热液蚀变反映了在热裂谷岛弧地壳上的沉积；而弧后沉积序列由以火山砾凝灰岩和凝灰角砾岩为主的深海火山碎屑沉积物组成；近源的英安火山碎屑流形成了几十米厚的粒序递变层（Busby，2004）。从这些弧后沉积层序中可识别出多幕岛弧裂谷作用。而多期次的裂谷作用将弧后地区破裂为一系列以断层为界的地块和盆地，并将弧后地区与活动的岛弧前缘隔离开来。随着来自岛弧的近源沉积输入由于裂陷作用而间断，弧后地区被细粒的外生火山碎屑、岩屑砂岩和粉砂岩所覆盖（Busby，2004）。

在阿拉斯加南部侏罗系岛弧系统后方发育大型增生盆地，其沉积充填跨越了从碰撞前到碰撞后数十个百万年（Trap 和 Ridgway，2007）。位于塔尔基特纳—奇特诺洋内岛弧后方的纳措廷和兰格尔山盆地（Trap 和 Ridgway，2002；Manuszak 等，2007）沉积了来源于中—晚侏罗世岛弧的海相火山碎屑砂岩、泥岩、燧石和凝灰岩。尽管从洋内活动向碰撞过渡的记录没有被保存下来，霍恩山附近出露的下侏罗统 Talkeetna 组仍然能反映出类似的沉积环境（弧后的远端、深水沉积）。在塔尔基特纳—奇特诺岛弧合并至兰格尔复合块体（WCT）内并停止活动（晚侏罗世）之后，兰格尔复合块体内的兰格尔山盆地作为对应于新生的奇萨纳活动岛弧（位于奇萨纳和已消亡的古岛弧之间）的一个弧前盆地，仍然持续活动（Trop 等，2002，2005；Trap 和 Ridgway，2007）。兰格尔山盆地的沉积地层记录了塔尔基特纳—奇特诺岛弧和兰格尔复合块体的抬升以及剥蚀过程。与此同时，作为奇萨纳岛弧后方的前陆盆地，纳措廷盆地也持续活动，并接受了约 3km 厚的向上变粗的沉积，代表了海底扇及上覆陆架沉积（Manuszak 等，2007）。（需要注意的是，该盆地早于奇萨纳岛弧形成；位于岛弧后方的盆地并不需要像 Mariana 和汤加岛弧的伸展弧后盆地，或下加利福尼亚地区 Alisitos 岛弧的弧后盆地那样，是岛弧裂谷和海底扩张的产物；Clift，1995；Busby，2004）。位于奇萨纳岛弧前部和后方的沉积盆地的沉积演化持续了整个白垩纪；Nutzolin 和兰格尔山盆地在晚白垩世—新近纪遭受抬升，并伴有同期大型右旋走滑断裂活动引起的褶皱作用（Trop 和 Ridgway，2007）。

17.3 弧—陆碰撞过程中形成的盆地

在活跃的弧—陆碰撞带除了发育"短命"的弧内盆地之外（Huang 等，1995），也可以像陆—陆碰撞那样，因地壳负载而发育前陆盆地。这些盆地形成所涉及的岩石圈通常具有较高的热成熟度（即更冷），能干性最强，因此具有高挠曲强度，从而形成宽阔的盆地。弧—陆碰撞带前陆地区的主要特征是，因为造山运动和随后的坍塌作用十分迅速（5~15Ma；Abbott 等，1997；Friedrich 等，1999），导致前陆盆地的形成和充填也很迅速；并且随着主动大陆边缘开始形成于造山带外侧，挠曲负载作用减弱，可能会导致盆地发生构造反转。这是因为俯冲作用的重建往往伴随着碰撞造山带的坍塌和快速剥蚀，这两者的结合减少了负载并降低了挠曲程度。澳大利亚与巴布亚新几内亚之间的碰撞导致了前陆盆地的形成，在菲尼斯泰尔岭（Finisterre Range）底部的 Leron 组发育了扇三角洲杂岩（Abbott 等，1994）。与此类似，帝汶—阿鲁（Timor-Aru）海槽代表了由于澳大利亚大陆边缘与班达岛弧碰撞产生的挠曲负载而形成的年轻的前陆盆地（Snyder 等，1996；Londono 和 Lorenzo，2004）。由于澳大利亚大陆边缘形成于侏罗纪，与卷入了年轻地壳的岛弧碰撞带相比更成熟和刚硬，因此形成了较为宽广的前陆盆地。

Brown 等（2006b）将乌拉尔南部的增生杂岩描述为一个发育在前陆盆地内的逆冲堆叠体。该前陆盆地位于中亚地区泥盆纪弧—陆碰撞带的大陆一侧，在其接受沉积的同时，早期形成的弧前盆地也发生沉积充填，二者物源一致。以中国台湾为例，在大多数（即便不是全部）先存岛弧盆地发生反转和消亡之后，前陆盆地挠曲作用的范围和深度也在碰撞作用达到顶点时也达到了最大值。它们被来自于造山带沉积物所充填，可能发育向上变浅的陆相和浅海相沉积，之下为造山早期的深海相复理石沉积（图17-7 和图17-8）。

随着倾斜弧—陆碰撞活动带沿着边缘移动（如中国台湾），当碰撞作用的挤压构造应力消失时，随着俯冲带极性倒转，碰撞后的造山带逐渐坍塌。造山带坍塌包含了软弱的、较热的造山带地壳的构造伸展，并在恢复沉积作用的位置形成盆地（Teng，1996）。这些盆地往往会在缝合带附近形成，并不会影响到更宽阔的前陆地区。然而，远离中部建造的物质再分配可减小俯冲的被动大陆边缘上的负载，从而引起板块的部分回弹和前陆盆地的反转。现代中国台湾北部宜兰平原/冲绳海槽（Ilan Plain/ Okinawa Trough）就是碰撞后期造山带坍塌盆地一个最明显的例子（图17-4）。在乌拉尔弧—陆碰撞缝合带，浅水碳酸盐岩不整合覆盖于岛弧地层单元之上，记录了碰撞后期造山带的坍塌（Brown 和 Spadea，1999）。同样，摩洛哥的小阿特拉斯山造山带在经历弧—陆碰撞之后，推断的造山后坍塌（580—550Ma）导致的伸展作用形成了断控磨拉石盆地，并接受酸性火山岩和火山碎屑沉积（Ouarzazate 群；Thomas 等，2002）。

17.4 实例分析：台湾弧—陆碰撞过程中的盆地演化

吕宋大洋岛弧与中国被动陆缘渐进迁移式碰撞过程中的清晰的构造关系（Suppe，1984；Teng，1990），揭示了在不同碰撞阶段发育的不同沉积盆地类型。值得注意的是碰撞之前、同碰撞期和碰撞之后的盆地沉积充填都主要来自造山带核部变质程度更高的山脉的侵蚀作用。由于岩体隆升速度快、岩性较脆，且处于有利于风化侵蚀的热带气候，中国台湾是现今地球上最大的沉积物源区之一（Milliman 和 Syvitski，1992）。沉积物沿走向通过马尼拉海沟和弧前区长轴（北吕宋海槽）向南搬运，将来自于造山带的沉积物从山脉搬运到碰撞之前的岛弧区域（图 17-7）。与此类似，沉积物通过受骊山断层（Lishan Fault）控制的兰阳溪（Lanyang River）沿着走向被搬运至冲绳海槽，后者被视为碰撞后的伸展坍塌盆地（图 17-4 和图 17-7；Clift 等，2008）。因此，由于大量沉积物沿走向搬运，中国台湾周围盆地（也可能是其他的弧—陆碰撞造山带）记录的侵蚀历史一般与盆地自身形成的构造演化阶段不同期。

如上所述，虽然北吕宋海槽的一些残余部分在中央山脉和海岸山脉之间得以保存，但碰撞前期形成的盆地被保存下来的可能性仍然最低。事实上，不论两侧山脉哪一边发生快速抬升，两侧山脉间的纵向峡谷内河流和冲积扇沉积都会持续发育。由于北吕宋海槽的深海沉积遭受构造逆冲而掩埋，所以可能在碰撞过程中会得以保存并在之后出露。出露于邻近造山带核部的海岸山脉的 Paliwan 组为同碰撞期早期沉积，主要由晚上新世近源变质岩组成（图 17-8b）（Dorsey，1992）。Paliwan 组碎屑岩的变质特征表明它们是在晚上新世剥蚀于一个成熟的造山带。由于物质通过渐进式碰撞系统有效地向北迁移，我们推断这些沉积岩当时沉积在最高山脉以南的一个盆地中，但是，与现今系统相似，沉积岩中也包含了从在当时位于北部的原始中央山脉循环而来的物质。从沉积作用开始，Paliwan 组砾岩便遭受变形和抬升。

台湾碰撞带的最大沉积体位于台湾海峡之下。在该处，前陆盆地（局部沉积厚度大于 5km）将渐新世中国南海裂谷沉积埋藏至约 2km 之下（Lin 等，2003）。边缘岩石圈的挠曲强度并不是特别高（$T_e = 8 \sim 13km$；Lin 和 Watts，2002），导致盆地相对狭窄（约70km），但具有很大的保存潜力。相反，在吕宋岛弧的弧后盆地（即菲律宾海的 Huatung 盆地），来自于海岸山脉的沉积物可能会随着俯冲极性反转而遭受俯冲并消失。Huatung 盆地（图 17-7）接受了来自造山带的大量沉积输入，但该地壳之后被俯冲至新形成的琉球岛弧大陆边缘之下。虽然部分物质会被刮擦到新生的大陆边缘增生楔中，但大多数被俯冲到岛弧之下而消失（Clift 和 Vannucchi，2004）。

17.5 总结

弧—陆碰撞通常发生在板块构造旋回过程中并导致造山带的迅速形成和快速坍塌，时间跨度往往只有 5~15Ma。通过弧—陆碰撞（伴随硅质岩浆岩活动和致密硅镁质下地壳的亏损）实现大陆增生，被认为是控制陆壳随时间演化的主要过程。弧—陆碰撞带的快速抬升产生了大量沉积物，这些沉积物被沿碰撞边缘向在碰撞前、碰撞后形成的盆地和前陆地区搬运。洋内岛弧与被动大陆边缘的碰撞（岛弧位于上部板块且面向大陆的情况）比洋内弧与活动大陆边缘的碰撞（需要多个会聚带，且会聚带内下部板块的岛弧退回到主动大陆边缘）涉及到更多形态上的变化，并且每种情况下盆地的保存潜力也不同。在后一种情况下，具有发育大规模俯冲和部分岛弧块体再循环的巨大的潜力。

不管是在碰撞前还是碰撞后构造侵蚀边缘和构造增生边缘的海沟和弧前演化（及其相关的沉积作用）存在很大的差异。在碰撞前的洋内岛弧活跃期，构造侵蚀边缘更大的水深导致了其深海相沉积的发育要比增生边缘中的慢。沟—坡盆地被保存的潜力很小。在碰撞过程中它们被迅速抬升和侵蚀，并且在构造侵蚀边缘，尤其是在洋内俯冲带，它们被俯冲侵蚀作用逐步破坏。在弧—陆碰撞之后，海沟沉积物和沟—坡盆地（变形，并且若位于底板还会发生变质）更偏向于在碰撞之前很长时间处于构造增生状态的边缘得以保存。

构造侵蚀大洋岛弧的弧前盆地底部通常为坚硬的岩石圈，因此很可能在与被动大陆边缘碰撞时得以保存，甚至由于在下伏板片中发育榴辉岩而在整个碰撞和造山过程中持续沉积。被保存下来的弧前盆地可能具有很高的沉积速率，沉积物主要来自于遭受构造变形、变质的被动大陆边缘和遭受暴露剥蚀的增生岛弧，并以浊流、块体流和局部发育的扇为特征。位于软弱的、变形的增生杂岩体之上的弧前盆地更容易在碰撞中受到破坏，因为它们更容易发生构造变形、抬升和侵蚀。由于具有较低的挠曲刚度，弧内盆地经常发育较深，并且，如果得以保存可记录长期的岛弧和碰撞构造活动。而碰撞前期的弧前和弧内盆地可能被充填并在碰撞中得到保存。弧后盆地则通常遭受俯冲作用，其沉积盖层不是消失就是仅作为碎片保存在混杂岩序列中。

来自碰撞造山带的大部分沉积物最终都进入到碰撞作用形成的前陆盆地中，并且可能在不遭受构造变形的情况下被大量保存下来。与陆—陆碰撞形成的前陆盆地（如扎格罗斯和喜马拉雅）相比，弧—陆碰撞盆地存在时间短，并可能在碰撞之后，在碰撞带外侧形成新的活动大陆边缘时由于造山带的坍塌而发生局部反转。

致谢（略）

参 考 文 献

Abbott, L. D., Silver, E. A., Anderson, R. S., Smith, R. B., Ingle, J. C., Kling, S. A., Haig, D., Small, E., Galewsky, J., and Sliter, W. (1997) Measurement of the tectonic uplift rate in a young collisional mountain belt, Finisterre Range, Papua New Guinea. Nature, 385, 501–507.

Abbott, L. D., Silver, E. A., and Galewsky, J. (1994) Structural evolution of a modern arc-continent collision in Papua New Guinea. Tectonics, 13, 1007–1034.

Alvarez-Marron, J., Brown, D., Perez-Estaun, A., Puchkov, V., and Gorozhanina, Y. (2000) Accretionary complex structure and kinematics during Paleozoic arc-continent collision in the southern Urals. Tectonophysics, 325, 175–191.

Ballance, F., Tappin, D. R., and Wilkinson, I. P. (2004) Volcaniclastic gravity flow sedimentation on a frontal arc platform: the Miocene of Tonga. New Zealand Journal of Geology and Geophysics, 47, 567–587.

Beaudry, D., and Moore, G. F. (1985) Seismic stratigraphy and Cenozoic evolution of West Sumatra forearc basin. AAPG Bulletin, 69 (5), 742–759.

Billen, M. I., and Gurnis, M. (2003) Comparison of dynamic flow models for the Central Aleutian and Tonga-Kermadec subduction zones. Geochemistry Geophysics Geosystems, 4 (1035). doi:

10. 1029/2001GC000295.

Bird, P. (1978) Initiation of intracontinental subduction in the Himalaya. Journal of Geophysical Research, 83, 4975-4987.

Boutelier, D., Chemenda, A., and Burg, J-P. (2003) Subduction versus accretion of intra-oceanic volcanic arcs: insight from thermo-mechanical analogue experiments. Earth and Planetary Science Letters, 212, 31-45.

Brown, D. (2009) The growth and destruction of continental crust during arc-continent collision in the southern Urals. Tectonophysics, 479, 185-196.

Brown, D., Juhlin, C., Tryggvason, A., Friberg, M., Rybalka, A., Puchkov, V., and Petrov, G. (2006a) Structural architecture of the southern and middle Urals foreland from reflection seismic profiles. Tectonics, 25 (TC1002). doi: 10. 1029/2005TC001834.

Brown, D., and Spadea, P. (1999) Processes of forearc and accretionary complex formation during arc-continent collision in the southern Ural Mountains. Geology, 27, 649-652.

Brown, D., Spadea, P., Puchkov, V., Alvarez-Marron, J., Herrington, R., Willner, A. P., Hetzel, R., Gorozhanina, Y., and Juhlin, C. (2006b) Arc-continent collision in the southern Urals. Earth-Science Reviews, 79, 261-287.

Busby, C. J. (2004) Continental growth at convergent margins facing large ocean basins—a case study from Mesozoic convergent-margin basins of BajaCalifornia, Mexico. Tectonophysics, 392, 241-277.

Busby, C. J., and Ingersoll, R. V. (1995) Tectonics of sedimentary basins. Oxford, Blackwell Science, 579 p. Busby, C., Adams, B. F., Mattinson, J., and Deoreo, S. (2006) View of an intact oceanic arc, fromsurficial level to mesozonal levels: Cretaceous Alisitos arc, Baja California. Journal of Volcanology and Geothermal Research, 149, 1-46.

Centeno-Garcia, E., Guerrero-Suastegui, M., and Talavera-Mendoza, O. (2008) The Guerrero Composite Terrane of western Mexico: collision and subsequent rifting in a supra-subduction zone, in Draut, A. E., Clift, P. D., and Scholl, D. W., eds., Formation and applications of the sedimentary record in arc collision zones. Geological Society of America Special Paper 436, 279-308.

Chang, C. P., Angelier, J., and Huang, C-Y. (2000) Origin and evolution of a melange; the active plate boundary and suture zone of the Longitudinal Valley, Taiwan. Tectonophysics, 325 (1-2), 43-62.

Clift, P. D. (1995) Volcaniclastic sedimentation and volcanism during the rifting of western Pacific backarc basins, in Taylor B., and Natland, J., eds., Active Margins and Marginal Basins of the Western Pacific. American Geophysical Union, Monograph vol. 88, 67-96.

Clift, P., and Vannucchi, P. (2004) Controls on tectonic accretion versus erosion in subduction zones; implications for the origin and recycling of the continental crust. Reviews of Geophysics, 42 (RG2001). doi: 10. 1029/ 2003RG000127.

Clift, D., MacLeod, C. J., Tappin, D. R., Wright, D., and Bloomer, S. H. (1998) Tectonic controls on sedimentation in the Tonga Trench and Forearc, SW Pacific. Geological Society of America Bulletin, 110, 483-496.

Clift, D., Degnan, J., Hannigan, R., and Blusztajn, J. (2000) Sedimentary and geochemical evolution of the Dras forearc basin, Indus suture, Ladakh Himalaya, India. Geological Society of America Bulletin, 112 (3), 450–466.

Clift, P. D., Pecher, I., Kukowski, N., and Hampel, A. (2003) Tectonic erosion of the Peruvian forearc, Lima Basin, by subduction erosion and Nazca Ridge collision. Tectonics, 22, 1023. doi: 10.1029/2002TC001386.

Clift, D., Draut, A. E., Kelemen, B., Blusztajn, J., and Greene, A. (2005a) Stratigraphic and geochemical evolution of an oceanic arc upper crustal section; the Jurassic Talkeetna Volcanic Formation, south-central Alaska. Geological Society of America Bulletin, 117 (7-8), 902–925.

Clift, D., Pavlis, T., DeBari, S. M., Draut, A. E., Rioux, M., and Kelemen, B. (2005b) Subduction erosion of the Jurassic Talkeetna – Bonanza Arc and the Mesozoic accretionary tectonics of western North America. Geology, 33 (11), 881–884.

Clift, D., Lin, A. T. S., Carter, A., Wu, F., Draut, A. E., Lai, T-H., Fei, L-Y., Schouten, H., and Teng, L. S. (2008) Orogenic collapse in the Ilan Plain Basin of northern Taiwan, in Draut, A. E., Clift, D., and Scholl, D. W., eds., Sedimentation in arc collisional settings. Special paper, 436. Boulder, CO, Geological Society of America, 257–278.

Clift, D., Schouten, H., and Vannucchi, P. (2009a) Arccontinent collisions, subduction mass recycling and the maintenance of the continental crust, in Cawood, P., and Kroener, A., eds., Accretionary Orogens in Space and Time. Special Publication, 318. Geological Society of London, 75–103.

Clift, P. D., Carter, A., Draut, A. E., Long, H. V., Chew, D. M., and Schouten, H. A. (2009b) Detrital U–Pb zircon dating of lower Ordovician syn–arc–continent collision conglomerates in the Irish Caledonides. Tectonophysics, 479, 165–174.

Cloos, M. (1993) Lithospheric buoyancy and collisional orogenesis: subduction of oceanic plateaus, continental margins, island arcs, spreading ridges, and seamounts. Geological Society of America Bulletin, 105, 715–737.

Cloos, M., Sapiie, B., Quarles van Ufford, A., Weiland, R. J., Warren, P. Q., and McMahon, T. P. (2005) Collisional delamination in New Guinea: the geotectonics of slab breakoff. Geological Society of America Special Paper 400, 51 p.

Collot, J-Y., Agudelo, W., and Ribodetti, A. (2008) Origin of a crustal splay fault and its relation to the seismogenic zone and underplating at the erosional north Ecuador and southwest Colombia margin. Journal of Geophysical Research, 113 (B12102). doi: 10.1029/2008JB005691.

Crawford, W. C., Hildebrand, J. A., Dorman, L. M., Webb, S. C., and Wiens, D. A. (2003) Tonga Ridge and LauBasin crustal structure from seismic refraction data. Journal of Geophysical Research, 108 (B4 2195). doi: 10.1029/2001JB001435.

Dadson, S., Hovius, N., Chen, H., Dade, W. B., Hsieh, M. L., Willett, S., Hu, J. C., Horng, M. J., Chen, M. C., Stark, C. P., Lague, D., and Lin, J. C. (2003) Links between erosion, runoff variability and seismicity in the Taiwan orogen. Nature, 426, 648–

651.

Dewey, J. F., and Ryan, D. (1990) The Ordovician Evolution of the South Mayo Trough, western Ireland. Tectonics, 9, 887-901.

Dewey, J. F., Ryan, D., and Andersen, T. B. (1993) Orogenic uplift and collapse, crustal thickness, fabrics and metamorphic phase changes: the role of eclogites, in Prichard, H. M., Alabaster, T., Harris, N. B. W., and Neary, C. R., eds., Magmatic Processes and Plate Tectonics, 76. London, Geological Society, 25-34.

Dickinson, W. R. (1974) Sedimentation within and beside ancient and modern magmatic arcs, in Dott, R. H., ed., Modern and ancient geosynclinal sedimentation; deposits in magmatic arc and trench systems. Special Publication, 19, SEPM, 230-239.

Dickinson, W. R. (1982) Compositions of sandstones in circum- Pacific subduction complexes and fore-arc basins. American Association of Petroleum Geologists Bulletin, 66, 121-137.

Dickinson, W. R. (1995) Forearc basins, in Busby, C. J., and Ingersoll, R. V., eds., Tectonics of sedimentary basins. Oxford, Blackwell Science, 221-261.

Dickinson, W. R. (2008) Accretionary Mesozoic-Cenozoic expansion of the Cordilleran continental margin in California and adjacent Oregon. Geosphere, 4, 329-353. doi: 10. 1130/GES00105. 1.

Dorsey, R. J. (1992) Collapse of the Luzon volcanic arc during onset of arc-continent collision; evidence from a Miocene- Pliocene unconformity, eastern Taiwan. Tectonics, 11, 177-191.

Draut, A. E., and Clift, D. (2001) Geochemical evolution of arc magmatism during arc-continent collision, South Mayo, Ireland. Geology, 29 (6), 543-546.

Draut, A. E., and Clift, D. (2006) Sedimentary processes in modern and ancient oceanic arc settings; evidence from the Jurassic Talkeetna Formation of Alaska and the Mariana and Tonga arcs, Western Pacific. Journal of Sedimentary Research, 76 (3-4), 493-514.

Draut, A. E., Clift, P. D., Amato, J. M., Blusztajn, J., and Schouten, H. (2009) Arc - continent collision and the formation of continental crust: a new geochemical and isotopic record from the Ordovician Tyrone Igneous Complex, Ireland. Journal of the Geological Society, London, 166, 485-500. doi: 10. 1144/0016-76492008-102.

Fildani, A., and Hessler, A. M. (2005) Stratigraphic record across a retroarc basin inversion: RocasVerdes - MagallanesBasin, Patagonian Andes, Chile. Geological Society of America Bulletin, 117, 1596-1614. doi: 10. 1130/ B25708. 1.

Fiske, R. S., Naka, J., Iizasa, K., Yuasa, M., and Klaus, A. (2001) Submarine silicic caldera at the front of the Izu- Bonin arc, Japan: voluminous seafloor eruptions of rhyolite pumice. Geological Society of America Bulletin, 113, 813-824.

Friedrich, A. M., Bowring, S. A., Martin, M. W., and Hodges, K. V. (1999) Short-lived continental magmatic arc at Connemara, western Irish Caledonides; implications for the age of the Grampian Orogeny. Geology, 27 (1), 27-30.

Fuller, C. W., Willett, S. D., and Brandon, M. T. (2006) Formation of forearc basins and their influence on subduction zone earthquakes. Geology, 34, 65-68.

Gaedicke, C., Baranov, B., Seliverstov, N., Alexeiev, D., Tsukanov, N., and Freitag, R.

(2000) Structure of an active arc-continent collision area: the Aleutian-Kamchatka junction. Tectonophysics, 325, 63-85.

Hall, R., and Smyth, H. R. (2008) Cenozoic arc processes in Indonesia: identification of the key influences on the stratigraphic record in active volcanic arcs, in Draut, A. E., Clift, D., and Scholl, D. W., eds., Formation and applications of the sedimentary record in arc collision zones. Special Paper, 436. Boulder, CO, Geological Society of America, 27-54.

Hawkesworth, C. J., and Kemp, A. I. S. (2006) Evolution of the continental crust. Nature, 443, 811-817.

Hawkins, J. W. (1974) Geology of the LauBasin, a marginal sea behind the Tonga arc, in Burk, C. A., and Drake, C. L., eds., The geology of continental margins. Springer- Verlag, New York, 505-520.

Hirtzel, J., Chi, W-C., Reed, D., Chen, L., Liu, C-S., and Lundberg, N. (2009) Destruction of Luzon forearc basin from subduction to Taiwan arc-continent collision. Tectonophysics. doi: 10. 1016/j. tecto. 2009. 01. 032.

Holbrook, W. S., Lizarralde, D., McGeary, S., Bangs, N., and Diebold, J. (1999) Structure and composition of the Aleutian island arc and implications for continental crust growth. Geology, 27, 31-34.

Huang, C-Y., Yuan, B., Song, S-R., Lin, C-W., Wang, C., Chen, M-T., Shyu, C-T., and Karp, B. (1995) Tectonics of short-lived intra-arc basins in the arc-continent collision terrane of the Coastal Range, eastern Taiwan. Tectonics, 14, 19-38.

Huang, C. Y., Yuan, B., Lin, C. W., and Wang, T. K. (2000) Geodynamic processes of Taiwan arc-continent collision and comparison with analogs in Timor, Papua New Guinea, Urals and Corsica. Tectonophysics, 325, 1-21. doi: 10. 1016/S0040-1951 (00) 00128-1.

Iizasa, K., Fiske, R. S., Ishizuka, O., Yuasa, M., Hashimoto, J., Naka, Y., Horii, Y., Fujiwara, Y., Imai, A., Koyama, S. (1999) A Kuroko-type polymetallic sulfide deposit in a submarine silicic caldera. Science, 283, 975-977.

Jarrard, R. D. (1986) Relations among subduction zone parameters. Reviews of Geophysics, 24, 217-284.

Johnson, C. L., Amory, J. A., Zinniker, D., Lamb, M. A., Graham, S. A., Affolter, M., and Badarch, G. (2008) Sedimentary response to arc-continent collision, Permian, southern Mongolia, in Draut, A. E., Clift, D., and Scholl, D. W., eds., Formation and applications of the sedimentary record in arc collision zones, 436. Boulder, CO, Geological Society of America Special Paper, 363-390.

Jordan, T. E. (1995) Retroarc foreland and related basins, in Busby, C. J., and Ingersoll, R. V., eds., Tectonics of sedimentary basins. Oxford, Blackwell Science, 331-362.

Jull, M., and Kelemen, P. B. (2001) On the conditions for lower crustal convective instability. Journal of Geophysical Research, 106 (B4) 6423-6446.

Karig, D. E. (1971) Origin and development of marginal basins in the western Pacific. Journal of Geophysical Research, 76, 2542-2561.

Kimbrough, D. L. (1984) Paleogeographic significance of the Middle Jurassic Gran Canon

Formation, Cedros Island, Baja California, in Frizzell, V. A. ed., Geology of the Baja California peninsula. Pacific Section, Society of Economic Paleontologists and Mineralogists, Book 39, Los Angeles, California, 107-118.

Kimura, G., Kitamura, Y., Yamguchi, A., and Raimbourg, H. (2008) Links amoung mountain building, surface erosion and growth of an accretionary prism in a subduction zone: an example from southwest Japan, in Draut, A. E., Clift, D., and Scholl, D. W., eds., Formation and applications of the sedimentary record in arc collision zones, 436. Boulder, CO, Geological Society of America, 391-403.

Konstantinovskaia, E. A. (2001) Arc-continent collision and subduction polarity reversal in the Cenozoic evolution of the Northwest Pacific: an example from Kamchatka. Tectonophysics, 333, 75-94.

Kopp, H., Flueh, E. R., Petersen, C. J., Weinrebe, W., Wittwer, A., and Meramex Scientists, (2006) The Java margin revisited: evidence for subduction erosion off Java. Earth and Planetary Science Letters, 242, 130-142.

Kusunoki, K., and Kimura, G. (1998) Collision and extrusion at the Kuril-Japan arc junction. Tectonics, 17, 843-858.

Lamarche, G., Joanne, C., and Collot, J-Y. (2008) Successive, large mass-transport deposits in the south Kermadec fore - arc basin, New Zealand—the Matakaoa submarine instability complex. Geochemistry Geophysics Geosystems, 9 (Q04001). doi: 10. 1029/2007GC001843.

Larue, D. K., Smith, A. L., and Schellekens, J. H. (1991) Oceanic island arc stratigraphy in the Caribbean region: don't take it for granite. Sedimentary Geology, 74, 289-308.

Laursen, J., Scholl, D. W., and von Huene, R. (2002) Neotectonic deformation of the central Chile margin: Deepwater forearc basin formation in response to hot spot ridge and seamount subduction. Tectonics, 21 (5), 1038. doi: 10. 1029/2001TC901023.

Lin, A. T., and Watts, A. B. (2002) Origin of the West Taiwan basin by orogenic loading and flexure of a rifted continental margin. Journal of Geophysical Research, 107, 2185. doi: 10. 1029/2001JB000669.

Lin, A. T., Watts, A. B., and Hesselbo, S. P. (2003) Cenozoic stratigraphy and subsidence history of the South China Sea margin in the Taiwan region. Basin Research, 15, 453-478.

Londono, J., and Lorenzo, J. M. (2004) Geodynamics of continental plate collision during late Tertiary foreland basin evolution in the Timor Sea: constraints from foreland sequences, elastic flexure and normal faulting. Tectonophysics, 392, 37-54.

Lundberg, N., Reed, D. L., Liu, C-S., and Lieske, J. (1997) Forearc-basin closure and arc accretion in the submarine suture zone south of Taiwan. Tectonophysics, 274, 5-23.

Manuszak, J. D., Ridgway, K. D., Trop, J. M., and Gehrels, G. E. (2007) Sedimentary record of the tectonic growth of a collisional continental margin: Upper Jurassic: Lower Cretaceous Nutzotin Mountain sequence, eastern Alaska Range, Alaska, in Ridgway, K. D., Trop, J. M., Glen, J. M. G., and O'Neill, J. M., eds., Tectonic growth of a collisional continental margin—crustal evolution of southern Alaska. Special Paper, 431. Boulder, CO, Geological Society of America, 345-377.

Marsaglia, K. M. (1995) Interarc and backarc basins, in Busby, C. J., and Ingersoll, R. V., eds., Tectonics of sedimentary basins. Oxford, Blackwell Science, 299-329.

Marsaglia, K. M., and Ingersoll, R. V. (1992) Compositional trends in arc-related, deep-marine sand and sandstone: a reassessment of magmatic – arc provenance. Geological Society of America Bulletin, 104, 1637-1649.

Milliman, J. D., and Syvitski, J. P. M. (1992) Geomorphic/ tectonic control of sediment discharge to the ocean; the importance of small mountainous rivers. Journal of Geology, 100, 525-544.

Moore, J. C., Diebold, J., Fisher, M. A., Sample, J., Brocher, T., Talwani, M., Ewing, J., von Huene, R., Rowe, C., Stone, D., Stevens, C., and Sawyer, D. (1991) EDGE deep seismic reflection transect of the eastern Aleutian arc – trench layered lower crust reveals underplating and continental growth. Geology, 19, 420-424.

Nishimura, A., Rodolfo, K., Koizumi, A., Gill, J., and Fujioka, K. (1992) Episodic deposition of Pliocene-Pleistocene pumice deposits of Izu-Bonin arc, Leg 126. Proceedings of the Ocean Drilling Program, Scientific Results, College Station, Texas, 126, 3-21.

Ogawa, Y., Horiuchi, K., Taniguchi, H., and Naka, J. (1985) Collision of the Izu arc with Honshu and the effects of oblique subduction in the Miura-Boso Peninsulas. Tectonophysics, 119, 349-379.

Oncken, O. (1998) Evidence fore precollisional subduction erosion in ancient collision belts: the case of the mid- European Variscides. Geology, 26, 1075-1078.

Otsuki, K. (1990) Westward migration of the Izu – Bonin Trench, northward motion of the Philippine Sea plate, and their relationships to the Cenozoic tectonics of Japanese island arcs. Tectonophysics, 180, 351-367.

Pearcy, L. G., DeBari, S. M., and Sleep, N. H. (1990) Mass balance calculations for two sections of island arc and implications for the formation of continents. Earth and Planetary Science Letters, 96, 427-442.

Plafker, G. B., and Berg, H. C. (1994) Overview of the geology and tectonic evolution of Alaska, in Plafker, G., and Berg, H. C., eds., The Geology of Alaska: the geology of North America, G-1. Boulder, CO, Geological Society of America, 989-1021.

Puchkov, V. N. (2009) The diachronous (step – wise) arc – continent collision in the Urals. Tectonophysics, 479, 175-184.

Quarles van Ufford, A., and Cloos, M. (2005) Cenozoic tectonics of NewGuinea. American Association of Petroleum Geologists Bulletin, 89, 119-140.

Robertson, A. H. F., and Collins, A. S. (2002) Shyok suture zone, N Pakistan; late Mesozoic- Tertiary evolution of a critical suture separating the oceanic Ladakh Arc from the Asian continental margin. Journal of Asian Earth Sciences, 20 (3), 309-351.

Rudnick, R. L., and Fountain, D. M. (1995) Nature and composition of the continental crust; a lower crustal perspective. Reviews of Geophysics, 33, 267-309.

Ryan, H. F., and Scholl, D. W. (1989) The evolution of forearc structures along an oblique convergent margin, central Aleutian arc. Tectonics, 8, 497-516.

Ryan, D. (2008) Preservation of forearc basins during island arc-continent collision: some insights from the Ordovician of western Ireland, in Draut, A. E., Clift, D., and Scholl, D. W., eds., Formation and applications of the sedimentary record in arc collision zones. Special Paper, 436. Boulder, CO, Geological Society ofAmerica, 1-9.

Ryan, P. D., and Dewey, J. F. (2004) The South Connemara Group reinterpreted: a subduction-accretion complex in the Caledonides of Galway Bay, western Ireland. Journal of Geodynamics, 37, 513-529.

Sample, J. C., and Moore, J. C. (1987) Structural style and kinematics of an underplated slate belt, Kodiak and adjacent islands, Alaska. Bulletin of the Geological Society of America, 99, 7-20.

Sarewitz, D. R., and Lewis, S. D. (1991) The Marinduque intra-arc basin, Philippines: basin genesis and in situ ophiolite development in a strike-slip setting. Geological Society of America Bulletin, 103, 597-614.

Scholl, D. W. (2007) Viewing the tectonic evolution of the Kamchatka - Aleutian (KAT) connection with an Alaska crustal extrusion perspective, in Eichelberger, J., Gordeev, E., Izbekov, P., and Lees, J., eds., Volcanism and Subduction: the Kamchatka Region. Geophysical Monograph Series, 172. WashingtonDC, American Geophysical Union, 3-35.

Scholl, D. W., and Vallier, T. L. (1985) Geology and offshore resources of Pacific island arcs—Tonga region, Earth Science Series. Houston, TX, Circum Pacific Council for Energy and Resources, 488.

Shemenda, A. I. (1994) Subduction: insights from physical modeling. Dordrecht, Kluwer, 215 p.

Smith, G. A., and Landis, C. A. (1995) Intra-arc basins, in Busby, C. J., and Ingersoll, R. V., eds., Tectonics of sedimentary basins. Oxford, Blackwell Science, 263-298.

Snyder, D. B., Prasetyo, H., Blundell, D. J., Pigram, C. J., Barber, A. J., Richardson, A., and Tjokosaproetro, S. (1996) A dual doubly vergent orogen in the Banda Arc continentarc collision zone as observed on deep seismic reflection profiles. Tectonics, 15, 34-53.

Soh, W., Nakayama, K., and Kimura, T. (1998) Arc-arc collision in the Izu collision zone, central Jaan, deduced from the Ashigara Basin and the adjcent Tanzawa Mountains. The Island Arc, 7, 330-341.

Suppe, J. (1981) Mechanics of mountain building and metamorphism in Taiwan. Memoir, 4, Geological Society of China, 67-89.

Suppe, J. (1984) Kinematics of arc-continent collision, flipping of subduction, and backarc spreading near Taiwan, in Tsan, S. F., ed., A special volume dedicated to Chun-Sun Ho on the occasion of his retirement. Geological Society of China Memoir, 6, 21-33.

Suyehiro, K., Takahashi, N., Ariie, Y., Yokoi, Y., Hino, R., Shinohara, M., Kanazawa, T., Hirata, N., Tokuyama, H., and Taira, A. (1996) Continental crust, crustal underplating, and low-Q upper mantle beneath an oceanic island arc. Science, 272, 390-392.

Taira, A. (2001) Tectonic evolution of the Japanese island arc system. Annual Reviews of Earth and Planetary Science, 29, 109-134.

Tamaki, K., and Honza, E. (1991) Global tectonics and formation of marginal basins—role of the western Pacific. Episodes, 14, 224-230.

Tang, J-C., and Chemenda, A. I. (2000) Numerical modeling of arc-continent collision: application to Taiwan. Tectonophysics, 325, 23-42.

Tani, K., Dunkley, D. J., Wysoczanski, R., and Tatsumi, Y. (2007) Does Tanzawa plutonic complex represent the IBM middle crust? New age constraint from SHRIMP zircon U-Pb geochronology. Geochimica et Cosmochimica Acta, 71 (15S) A1002.

Taylor, B. (1992) Rifting and the volcanic-tectonic evolution of the Izu-Bonin-Mariana arc, in Taylor, B., Fujioka, K. et al., Proceedings of the Ocean Drilling Program, Scientific Results, 126. Ocean Drilling Program, College Station, TX, 627-651.

Taylor, B., and Karner, G. D. (1983) On the evolution of marginal basins. Reviews of Geophysics and Space Physics, 21, 1727-1741.

Teng, L. S. (1990) Geotectonic evolution of late Cenozoic arccontinent collision in Taiwan. Tectonophysics, 183, 57-76.

Teng, L. S. (1996) Extensional collapse of the northern Taiwan mountain belt. Geology, 24, 949-952.

Thomas, R. J., Chevallier, L. P., Gresse, G., Harmer, R. E., Eglington, B. M., Armstrong, R. A., de Beer, C. H., Martini, J. E. J., de Kock, G. S., Macey, H., and Ingram, B. A. (2002) Precambrian evolution of the Sirwa Window, Anti-Atlas orogen, Morocco. Precambrian Research, 118, 1-57.

Trop, J. M., and Ridgway, K. D. (2007) Mesozoic and Cenozoic tectonic growth of southern Alaska—a sedimentary basin perspective, in Ridgway, K. D., Trop, J. M., Glen, J. M. G., and O'Neill, J. M., eds., Tectonic growth of a collisional continental margin—crustal evolution of southern Alaska. Special Paper, 431. Boulder, CO, Geological Society of America, 55-94.

Trop, J. M., Ridgway, K. D., Manuszak, J. D., and Layer, P. (2002) Mesozoic sedimentary-basin development on the allochthonous Wrangellia composite terrane, Wrangell Mountains basin, Alaska: A long-term record of terrane migration and arc construction. Geological Society of America Bulletin, 114, 693-717. doi: 10. 1130/0016-7606(2002)114.

Trop, J. M., Szuch, D. A., Rioux, M., and Blodgett, R. B. (2005) Sedimentology and provenance of the Upper Jurassic Naknek Formation, Talkeetna Mountains, Alaska: Bearings on the accretionary tectonic history of the Wrangellia composite terrane. Geological Society of America Bulletin, 117, 570-588. doi: 10. 1130/B25575. 1.

Turcotte, D. L., and Schubert, G. (1982) Geodynamics: applications of continuum mechanics to geological problems. New York, John Wiley, 450 p.

Underwood, M., Ballance, P., Clift, P., Hiscott, R., Marsaglia, K., Pickering, K., and Reid, P. (1995) Sedimentation in forearc basins, trenches, and collision zones of the western Pacific: a summary of results from the Ocean Drilling Program, in Taylor, B., and Natland, J., eds., Active margins and marginal basins of the Western Pacific. Geophysical Monograph, 88. Washington, DC, American Geophysical Union, 315-354.

Underwood, M. B., and Moore, G. F. (1995) Trenches and trench-slope basins, in Busby, C.

J., and Ingersoll, R. V., eds., Tectonics of sedimentary basins. Oxford, Blackwell Science, 179–219.

Underwood, M. B., Moore, G. F., Taira, A., Klaus, A., Wilson, M. E. J., Fergusson, C. L., Hirano, S., Steurer, J., and Leg 190 Shipboard Scientific Party (2003) Sedimentary and tectonic evolution of a trench–slope basin in the Nankai subduction zone of southwest Japan. Journal of Sedimentary Research, 73, 589–602.

van der Werff, W. (1995) Cenozoic evolution of the Savu Basin, Indonesia: forearc basin response to arccontinent collision. Marine and Petroleum Geology, 12, 247–262.

van Staal, C. R., Whalen, J. B., McNicoll, J., Pehrsson, S., Lissenberg, C. J., Zagorevski, A., van Breemen, O., and Jenner, G. A. (2007) The Notre Dame arc and the Taconic orogeny in Newfoundland, in Hatcher, R. D., Jr., Carlson, M. P., McBride, J. H., and Martinez Catalan, J. R., eds., 4-D Framework of continental crust. Memoir, 200. Boulder, CO, Geological Society of America, 511–552.

von Huene, R., and Scholl, D. W. (1991) Observations at convergent margins concerning sediment subduction, subduction erosion, and the growth of continental crust. Reviews of Geophysics, 29 (3), 279–316.

Waltham, D., Hall, R., Smyth, H. R., and Ebinger, C. J. (2008) Basin formation by volcanic arc loading, in Draut, A. E., Clift, D., and Scholl, D. W., eds., Formation and applications of the sedimentary record in arc collision zones. Special Paper, 436. Boulder, CO, Geological Society of America, 11–26.

Warren, Q., and Cloos, M. (2007) Petrology and tectonics of the Derewo metamorphic belt, west New Guinea. International Geology Review, 49, 520–553.

Watts, A. B. (2001) Isostacy and flexure of the lithosphere. Cambridge, Cambridge University Press, 458 p.

Watts, A. B., Karner, G. D., and Steckler, M. S. (1982) Lithospheric flexure and the evolution of sedimentary basins. Philosophical Transactions of the Royal Society of London Series a– Mathematical Physical and Engineering Sciences, 305 (1489), 249–281.

Wells, R. E., Blakely, R. J., Sugiyama, Y., Scholl, D. W., and Dinterman, P. A. (2003) Basin – centered asperities in great subduction zone earthquakes: a link between slip, subsidence, and subduction erosion? Journal of Geophysical Research, 108 (B10), 2507. doi: 10. 1029/2002JBOO2072.

Wilson, T. J. (1991) Transition from back–arc to foreland basin development in the southernmost Andes—stratigraphic record from the Ultima Esperanza district, Chile. Geological Society of America Bulletin, 103, 98–111. doi: 10. 1130/0016-7606 (1991) 103.

Wright, D. J., Bloomer, S. H., MacLeod, C. J., Taylor, B., and Goodliffe, A. (2000) Bathymetry of the Tonga Trench and forearc: a map series. Marine Geophysical Researches, 21, 489–511.

Wu, Y-M., Chang, C-H., Zhao, L., Shyu, J. B. H., Chen, Y-G., Sieh, K., and Avouac, J-P. (2007) Seismic tomography of Taiwan; improved constraints from a dense network of strong motion stations. Journal of Geophysical Research, 112 (B08312). doi: 10. 1029/

2007JB004983.

Zagorevski, A., Staal, C. v., McNicoll, J., and Rogers, N. (2008) Arc-arc collision and the assembly of the Annieopsquotch Accretionary Tract: Ordovician tectonics along the Laurentian margin in Newfoundland Appalachians, in Draut, A. E., Clift, D., and Scholl, D. W., eds., Formation and applications of the sedimentary record in arc collision zones. Special Paper, 436. Boulder, CO, Geological Society of America.

(李阳 译，阳怀忠 吴哲 赵梦 校)

第 18 章 智利北部塔潘帕·德尔·塔马鲁加尔 (Pampa del Tamarugal) 弧前盆地：构造与气候的相互作用

PETER NESTER, TERESA JORDAN

(Department of Earth and Atmospheric Sciences, Cornell University, Ithaca, USA)

摘　要：智利北部安第斯山脉西侧发育了晚渐新世以来的非海相潘帕·德尔·塔马鲁加尔大草原弧前沉积盆地（PdT 盆地）。该延伸长、平行海沟分布的地块内盆地位于陆壳之上，并在始新世岩浆弧作用下叠置在白垩系山根之上。盆地充填呈透镜状，在中央轴部厚度可达 1000～1800m，向东尖灭至阿尔蒂普拉诺高原和火山弧的西缘，向西抵达由中生代岛弧体系侵蚀残余形成的低山脉。大多数填充的沉积物源自安第斯山脉的侵蚀和西科迪勒拉火山弧的喷发作用。在盆地的前半段演化历史中（约 25—11Ma），整个盆地为一个统一沉积区；但在 11Ma 之后盆地北部沉积作用基本停止，南部沉积作用也减弱。盆地对从半干旱到极端干旱的气候演变和东部边缘约 3000m 构造与地表隆升表现出了显著的沉积响应，后者也为盆地同时期的沉积充填提供了物源。在早—中中新世，一系列小规模阶梯状排列的西倾单斜褶皱累计形成了约 2000m 的抬升；而在晚中新世到第四纪，盆地东部 30～45km 区域表现为一个长波长的西倾单斜褶皱的翼部，并产生了约 1000m 抬升。

关键词：弧前盆地　极端干旱气候　同构造沉积　非海相沉积环境　单斜褶皱

18.1　简介、主要特征和地质背景

约 30Ma（渐新世）以来，在智利北部安第斯山脉西侧发育了一个非海相弧前沉积盆地。潘帕·德尔·塔马鲁加尔（Pampa del Tamarugal）盆地（以下简称 PdT 盆地）是智利北部中央坳陷内最大的弧前盆地（图 18-1）。虽然现今已很少接受沉积，但仍保留着弧前盆地的形态。它记录了安第斯山脉西侧一系列环境和构造演化的主要地质表现，如响应于弧前构造的地貌演化、中安第斯山火山弧的喷发史和区域古气候、古水文演化历史等。

PdT 盆地平行于海沟和火山岛弧展布，长度大于 500km（约南纬 18°～22°），东西宽达 50～75km（图 18-2 和图 18-3）。虽然本次研究集中在南纬 19°的区域，但盆地从智利向北可延伸到秘鲁。盆地海拔低于 3000m 的区域整体被阿塔卡马沙漠（Atacama Desert）覆盖，气候极其干燥（图 18-1）；盆地西侧为海岸山脉，东部是安第斯山脉，为构造高地——前科迪勒拉高原（Precordillera）和阿尔蒂普拉诺高原（Altiplano），或火山体（西科迪勒拉山，Western Cordillera）（图 18-2）。

晚渐新世以来，纳斯卡板块（Nazca plate）沿着与智利北部大陆边缘北—北北西向海沟近于垂直的方向，向南美板块西缘之下会聚俯冲（图 18-1）。在约 26—5Ma，纳斯卡板块以正向至小于 10°的会聚角度和高会聚速率（约 120km/Ma）为特征俯冲，之后在上新世—第四纪会聚倾斜角度增加了约 5°，且俯冲速率降低（约 80km/Ma）（Somoza 和 Ghidella，

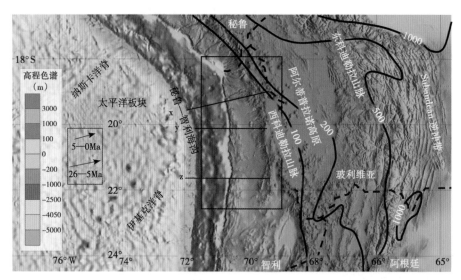

图 18-1　以色彩变化表示海拔的阴影地貌图反映了纳斯卡—南美板块边缘系统主要构造特征

大框表示图 18-2 所示的区域；小框表示了纳斯卡板块相对于南美洲板块的会聚向量，计算位置为南纬 22°。（据 Somoza 和 Ghidella，2005）；等值线以 mm/a 为单位显示平均年降水量；x、y 和 z 线是图 18-4 中的深海地形剖面

2005）。中央安第斯山脉在自 27Ma 以来的近 10Ma 的火山活动间歇的末期（James 和 Sacks，1999；Trumbull 等，2006；Kay 和 Coira，2009）海拔上升到了现今高度的四分之一（Charrier 等，1994；Gregory-Wodzicki，2000；Ghosh 等，2006；Garzione 等，2006，2008），宽度大于 600km（Eiger 等，2005；Barnes 和 Ehlers，2009）。同期，整个盆地开始接受沉积（Victor 等，2004；Charrier 等，2007；Jordan 等，2010）。之后中央安第斯山脉在宽度上变化较轻微，而海拔高度快速增加，成为地球上第二高的高原系统，且同时广泛发育岛弧和弧后岩浆活动（Allmendinger 等，1997；Eiger 等，2005；Barnes 和 Ehlers，2009）。在 PdT 盆地形成时，约 250km 的水平收缩作用形成了与海沟平行的上地壳冲断带，如阿尔蒂普拉诺、东科迪勒拉、Interandean 和 Subandean 构造带等（图 18-1）（Kley 和 Monaldi，1998；McQuarrie，2002；Eiger 等，2005）。

晚渐新世—第四纪，PdT 是一个位于陆壳之上的"地块内"盆地（intra-massif basin）（Dickinson 和 Seely，1979；Dickinson，1995；Smith 和 Landis，1995），该陆壳包括了贯穿始新世岩浆弧的白垩纪侵蚀山根（即盆地之下的地壳是一个弧块体）。弧块体内的弧前盆地通常为非海相充填。Dickinson 和 Seely（1979）将智利北部的弧前区域定义为脊状弧前区内的陆上高地盆地（terrestrial upland basin）。最近，越来越多的研究指出大多数俯冲边缘以长期的构造侵蚀作用而不是沉积增生作用为特征（von Huene 和 Scholl，1991；Clift 和 Vannucci，2004）。经典的弧前盆地模型并没有充分考虑这种在智利北部弧前系统普遍存在的情况（图18-4a）。

前弧区由沿着西科迪勒拉山脉分布的成层火山岩层构成（图 18-2）。在整个新生代，火山弧的位置整体发生了迁移（James 和 Sacks，1999；Kay 等，1999；Worner 等，2000；Trumbull 等，2006；Charrier 等，2007）。从约 65Ma 至 35Ma，火山弧位于现今的弧前盆地内，并在接下来的 10Ma 中没有发生火山活动（Trumbull 等，2006）。在约 25Ma 到约 15Ma 之间，火山弧的西缘处在现今 PdT 盆地北部的东缘，之后在晚中新世到第四纪向东迁移约

20km。与此相反，火山弧前缘位于该盆地南部以东几十千米处，甚至不在盆地的汇水区（图18-2）。新近纪的两种类型弧火山活动对弧前区域产生了很重要的影响。盆地北部东缘的安山质成层火山产生的熔岩流，与邻近的盆地充填形成互层，物源来自于遭受侵蚀的外来碎屑沉积（Pinto等，2004；Farias等，2005；Garcia和Herail，2005）；而火山碎屑喷发形成的熔结凝灰岩则在盆地内广泛分布（de Silva，1989；Worner等，2000）。在横截面上盆地充填呈透镜状（图18-3），地层厚度普遍超过1000m，在沉积中心厚度甚至达到1800m（图18-2）（Victor等，2004；Garcia和Herail，2005；Charrier等，2007；Nester，2008；Jordan等，2010）。盆地内主要为硅质碎屑地层，由大量外来碎屑物（Parraguez，1998；Tomlinson等，2001；Pinto等，2004）和火成碎屑沉积组成，后者主要为熔结凝灰岩（Dingman 和 Galli，1965；Pinto等，2004；Victor等，2004；Farias等，2005；Garcia和Herail，2005）。盆地西部以冲积扇和河流相为主，并伴随局部的湖泊相沉积（Parraguez，1998；Saez等，1999；Pinto等，2004；Farias等，2005；Garcia和 Herail，2005；Nester，2008；Jordan等，2010）。PdT盆地最老的碎屑地层单元形成于早渐新世（图18-3a）（Garcia和Herail，2005；Charrier等，2007）。在盆地北部，富含火山碎屑的渐新统和下中新统地层上超于构造抬升的盆地东部边缘，而中中新统则退覆于抬升的东缘（Pinto等，

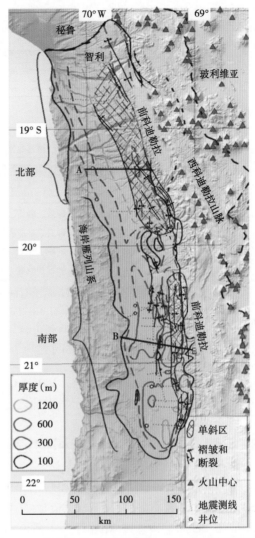

图18-2 部分智利、秘鲁最南端，以及玻利维亚西部的数字高程模型展示弧前区域、火山岛弧以及高安第斯高原
等厚线显示渐新世—新近纪的弧前盆地充填厚度，在盆地北部可能包括几百米的始新世地层；100m的黑色等高线是近似的PdT盆地空间范围。红色三角形表明晚中新世—第四纪火山；早中中新世火山岩地层的西部边界线近似位置为南纬20°东部的100m厚线；长波长单斜褶皱是安第斯西部斜坡位置（图18-4b），其中一个狭窄区域（交叉阴影）包含短波单斜褶皱，"r. fault"表示逆断层；现有的地下数据源于地震测线和石油勘探钻孔

2004；Farias等，2005；Garcia和 Herail，2005），且沉积速率在晚中新世到上新世急剧下降（图18-3a）。在盆地南半部分，最老沉积填充形成于晚渐新世（Dingman 和 Galli，1965；Tomlinson等，2001；Victor等，2004；Nester，2008；Jordan等，2009）（图18-3b）；在局部区域，上中新统上超于盆地东侧的高海拔区域（图18-3b）（Jordan等，2010）。

现今的PdT盆地地表海拔将近1000m（图18-1a和图18-4b）。在演化过程中盆地的古

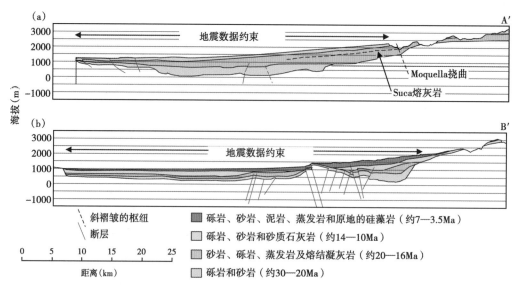

图 18-3 盆地地层单元剖面图，基于石油工业二维地震反射剖面和地表地质特征

垂直放大比例约 4∶1，剖面位置见图 18-2；(a) PdT 盆地北部，地面地质情况基于 Pinto 等（2004）、Parraguez（1998）、von Rotz 等（2005）和 Allmendinger 等（2005）的研究，地下地质情况来自 99-01 地震测线的地质解释；(b) 横穿盆地南部的剖面，根据 Nester（2008）和 Jordan 等（2010）对 99-06 地震测线的解释

海拔一直存在很大的争议（Farias 等，2005；Garcia 和 Herail，2006；Jordan 等，2010），但完全的非海相沉积表明盆地始终处于海平面之上。

18.2 PdT 盆地的弧前形态和构造研究

纳斯卡板块和南美板块在智利、秘鲁沿岸的接触带以构造侵蚀（或俯冲侵蚀）为典型特征。纳斯卡大洋岩石圈的俯冲过程造成了南美板块的净亏损（Mortimer 和 Saric，1975；von Huene 和 Scholl，1991）。在邻近 PdT 弧前盆地的位置，陆壳延伸到了海沟（图 18-1 和图 18-4）。构造侵蚀作用在俯冲的主安第斯构造旋回的整个 200Ma 时期内长期存在（Kukowski 和 Oncken，2006）。弧前地区的主要海底地貌特征显示出了伸展构造变形特征（Huene 和 Ranero，2003；Ranero 等，2006）（图 18-4）。随着大陆地幔和地壳因为俯冲侵蚀作用在陆坡之下发生滑离，上部的地壳发生拉伸和沉降。

陆上的情况则完全相反。在伸展变形带和 PdT 弧前盆地之间是阶地形态的科迪勒拉沿岸（Coastal Cordillera）。该山脉由古生代变质沉积岩和中生代弧后盆地火山岩组成，并且沉积充填受到中生代侵入体的切割。邻近 PdT 盆地的科迪勒拉沿岸（Coastal Cordillera）新近纪伸展变形（Mortimer 和 Saric，1975）很弱（Allmendinger 和 Gonzalez，2009）。在 PdT 盆地区域海岸山脉的海拔与盆地的海拔相似，或仅仅高出数百米（图 18-4b）。而在科迪勒拉沿岸和太平洋海岸线之间存在一个高差达 700~1500m 的非断崖型海岸陡崖（图 18-4b）（Mortimer 和 Saric，1975；Allmendinger 等，2005）。该地区还存在一个至今仍未被很好解释的大型地壳构造变形转换现象：在短短 15~50km 的横向范围，从一个地区的上陆坡之上发生超过 1800m 的沉降（Ranero 等，2006），到另一个地区地形突然抬升超过海平面 1000m 以上。

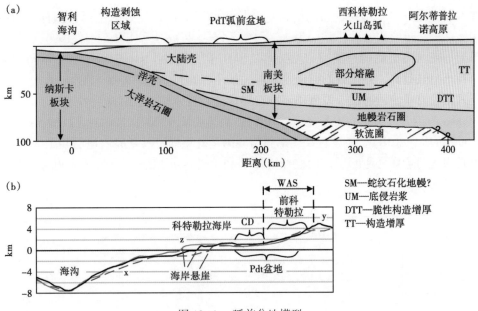

图 18-4　弧前盆地模型

(a) 在纳斯卡和南美洲岩石圈尺度下的弧前系统横断面 (据 Giese 等, 1999; Sick 等, 2006), 无垂直放大, 在新生代大陆地壳通过各种过程增厚; 大陆弧前区域之下的地壳存在一个古莫霍面 (虚线), Giese 等 (1999) 描述其地球物理特征为, 水合作用产生的蛇纹石化和下部火山弧的镁铁质岩浆底侵作用; Beck 和 Zandt (2002) 认为台地高原之下的地壳通过构造叠加达到目前的 70~80km 厚。(b) 垂直放大比例为 5:1 的水深和地形, 图 18-1 中显示了 x、y、z 测线的位置; 值得注意的是海洋弧前区域有一个较大的水深梯度, 该区域伸展作用远远大于构造侵蚀作用, 但海岸峭壁和中央坳陷 (CD) 之间有很小的海拔变化, 相当于一个很小伸展或缩短的区域; WAS 代表安第斯西部斜坡

18.3　PdT 盆地的基底和构造特征

在始新世至渐新世, 位于现今前科迪勒拉高原内的始新世构造山脊和沿 PdT 盆地轴向分布的早期火山中心残余被剥蚀为一个横跨 PdT 盆地和科迪勒拉沿岸 (Coastal Cordillera), 并向太平洋倾斜的低起伏的宽阔古近纪平原 (Mortimer 和 Saric, 1975)。平原的表面即构成了 PdT 盆地充填与下伏基底的接触面, 并且在接受约 1000m 的弧前盆地沉积之后, 该接触面在盆地之下 (海拔 100~400m) 发生挠曲变形, 而不再是向海洋倾斜 (图 18-3)。PdT 盆地基底以在中生代弧后盆地系统形成演化和始新世白垩系岩体侵入过程中形成的岩浆岩和沉积岩为主 (Serangeomin, 2002)。俯冲的纳斯卡板块位于弧前盆地之下 57~80km 深度位置。地壳厚度约 37~42km (图 18-4a)。Gieser 等 (1999) 和 Sickr 等 (2006) 通过地震折射、重力、电阻率、热流和海洋地震反射数据等资料的分析, 提出在地壳之下存在一个由蛇纹岩化地幔组成的相对薄弱带, 并缺失软流圈。

中央坳陷 (图 18-4b) 东部为安第斯山脉的西部斜坡, 它是一个整体向西 4°~5° 倾斜的平缓区 (Farias 等, 2005; Hoke 等, 2007; Jordan 等, 2010), 调节了从盆地到前科迪勒拉高原主峰海拔约 3000m 或阿尔蒂普拉诺高原 (取决于纬度) 约海拔 4000m 的地形起伏。火山弧的前锋位于阿尔蒂普拉诺高原的西缘, 其顶峰则上升到了高原之上, 并形成了被称为西科迪勒拉的高原地貌。

在 PdT 盆地东部被地层覆盖的前科迪勒拉山脉斜坡区，发育的主要构造样式为西倾的单斜褶皱。虽然单斜褶皱并非是广为认知的构造样式（Davis 和 Reynolds，1996），但却是智利北部上地壳内典型的构造类型（Mortimer 和 Saric，1975；Isacks，1988）。单斜褶皱仅具有一个西倾褶皱翼部，并发育与弧前盆地相关的生长地层（图 18-5、图 18-6）。尽管这种变形样式比较简单，但通过一系列单斜构造形成的大规模构造变形却导致了阿尔蒂普拉诺高原近 3000m 的抬升（Mortimer 和 Saric，1975；Victor 等，2004；Farias 等，2005；Hoke 等，2007；Jordan 等，2010）。通过对某些单斜褶皱的基底顶面以及基底内地层断距的研究表明，形成单斜构造的主要是逆断层，而这些逆断层在上部沉积地层中不明显（Victor 等，2004；Farias 等，2005）。

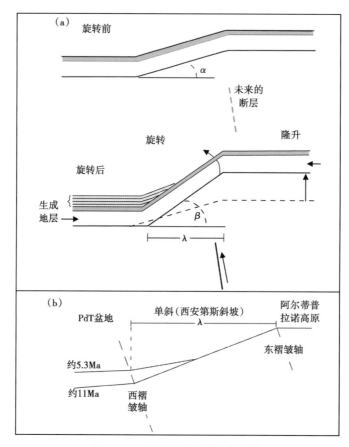

图 18-5 翘起引起的长波长变形

(a) 单斜褶皱形成的一般几何形态和表面变形；（据 Davis 和 Reynolds，1996；Patton，2004）；
(b) PdT 盆地南部的东面一侧的长波长变形，由地震反射和露头地层资料获得（据 Nester，2008；Jordan 等，2010）

两种规模的单斜构造代表了两个独立的演化阶段。短波长的单斜褶皱（翼部宽 1~4km，局部幅度 400~800m）（Pinto 等，2004；Victor 等，2004；Farias 等，2005）主要形成于早—中中新世。在约 11Ma 之后叠加了长波长的单斜褶皱构造活动（翼部宽 30~50km，幅度约 1000m）（Isacks，1988；Farias 等，2005；Hoke 等，2007；Jordan 等，2010）。在盆地轴部与前科迪勒拉高地之间 20~40km 宽的南北向地带内（图 18-2），短波长单斜构造引起的地

图 18-6 PdT 盆地东部的短波长单斜挠曲作用在盆地充填的同时发育地形起伏，特别是在中新世早期和中期（据 Nester，2008；Jordan 等，2010）

(a) 盆地北部的 Moquella 挠曲（据 Pinto 等，2004）：北北西方向，在峡谷壁上 440m 位置；上部出露较差地层（中中新统）倾斜幅度小于下伏悬崖形成单元（下中新统）。(b) 出露于盆地南部 Tambillo 峡谷的挠曲：虚线表示层理面，实线表示中新世早期不整合之下强烈倾斜地层与上覆适度倾斜地层之间的不整合面。注意右下角作为比例尺的人

形位移总计约 1700m（Victor 等，2004；Farias 等，2005）。中新世盆地充填之上的西倾山麓侵蚀平原显示，长波长的单斜构造变形导致了额外约 1100m 的抬升（Farias 等，2005；Hoke 等，2007；Jordan 等，2010）。

Victor 等（2004）推断这些小规模的单斜褶皱有一条共同主控断层，该断层为一条东倾的高角度逆冲断层，并一直延伸到中地壳。虽然没有直接观察到控制西缘长波长单斜褶皱形成的断层，但 Isacks（1988）和 Farias 等（2005）认为发生阿尔蒂普拉诺高原之下的下地壳流动可能是其形成的控制机制，后者在韧性变形区的西部边界产生了一个地壳尺度的断弯褶皱（即单斜构造）。

在 PdT 盆地的低部位存在小幅度中新世和甚至更年轻的构造变形作用。东西向的逆断层和褶皱在科迪勒拉沿岸导致晚新生代地层发生了数十至数百米的位移（Allmendinger 等，2005；Allmendinger 和 Gonzalez，2009），并在局部向盆地轴部延伸，但由于未影响到上覆上中新统而变得不明显（Nester，2008）。在盆地的南部区域，向东或向西倾的北向逆断层，

断距较小（大部分小于几百米），但错断了上新生界（Victor 等，2004；Nester，2008）。在盆地西南部，向东或向西倾的小规模北向正断层，在局部形成了深度小于100m的地堑（Nester，2008）。

目前已知有多种原因导致弧前盆地堆积了巨厚的地层（Dickinson，1995）：有的涉及了构造沉降作用，而其他的仅仅只是考虑了因地形阻挡而阻止了远距离搬运。因此，并没有一个通用的合理机制来解释PdT盆地的形成。由于缺少中生代之后的海相地层，因此在估算PdT盆地沉积充填过程中发生的绝对沉降量具有很大的不确定性。然而，巨厚的地层沉积需要数百米高的地形屏障，或是由在盆地沉积充填过程中盆地边缘构造变形和盆地中心相对沉降形成的地形起伏，或者是两者综合作用的结果。

Hartley等（2000）认为在PdT盆地没有发生过真正的构造沉降。这与缺少调节科迪勒拉沿岸相对于PdT盆地构造位移的渐新世—中新世断层相一致（Mortimer和Saric，1975），并且Victor等（2004）也发现在前科迪勒拉中新世的断层和褶皱作用引起的地壳加厚不足以形成明显的挠曲沉降。科迪勒拉沿岸和PdT盆地之间的初始地表起伏与PdT盆地沉积物沿着科迪勒拉沿岸东侧基底上超（Parraguez，1998）和PdT盆地内部水系指示的沉积相有关（Mortimer和Saric，1975；Naranjo和Paskoff，1985）。

18.4 PdT盆地地表演化和地貌特征

当潜在蒸发量大大超过平均年降水量时就形成"极端干旱"气候。一个典型的实例是智利北部地区。该区域海拔在3000m以下（Houston和Hartley，2003），平均年降水量低于20~40mm（Ewing等，2006）。极端干旱气候导致化学风化停止，同时在盐碱地沉积了来源于大气和地下水的蒸发盐矿物（Rech等，2003）。在这种气候之下，邻近山脉的沉积盆地会出现一种非常独特的现象，即由于陆上径流和河流搬运作用的相对缺乏，大大抑制了侵蚀作用以及削弱了盆地的碎屑物供给。

从地貌上看，现今的盆地大致以19°35′S为界分为南北两个区域（图18-2）。北部地区以被深切谷（低于邻近平原近1000m的深度）所分隔的广阔平原为特征，水系从高海拔的安第斯山脉向西流入太平洋（Garcia和Herail，2005）。这些平原具有复合成因。部分是覆盖于最新沉积填充之上的沉积表面残余（图18-3a），其他大部分平原则是在几次大的降水间隔期遭受了数米的剥蚀（Evenstar等，2009；Lehmann等，2009）。南部地区主要发育内部水系，并伴随有广泛的冲积扇，在罕见的（十年或千年一遇）潮湿间隔期仍然形成网状的碎屑沉积（Houston，2002；Nester等，2007）。在邻近海岸山脉的一条狭窄、不连续地带，来自于地下水的蒸发盐矿物沉积形成盐田。在盆地的西南端发育一条流经深切谷的单道河流（Loa），但它源自与PdT盆地没有地表水系相连的遥远源头（图18-2）。

深切谷切割平原的时代仍然是一个主要研究问题。一般认为下切作用始于约11—6.4Ma之间（Hoke等，2007），而大多数峡谷下切作用发生在5.5Ma之后（Farias等，2005）。一旦深切谷形成，相邻的PdT盆地沉积表面处于饥饿沉积状态，沉积物绕行进入太平洋海岸线（Garcia和Herail，2006）。利用宇宙射线产生的核素测年资料对北部地区地貌特征多样性研究表明，长期缺乏地表水累积的后果是导致了极其缓慢侵蚀作用，侵蚀率小于0.7m/Ma（Schlunegger等，2006；Kober等，2007）。在PdT盆地西南部和邻近科迪勒拉沿岸地区，地形自中中新世早期以来几乎没有发生变化（Cortes等，2006）。

盆地北部和南部之间的地表形态差异是由于北部具有更高的海拔和更广的河流流域造成的（图18-2）（Nester等，2007）。后者带来了更高的河流径流量，从而在北部形成了深切谷。与此相反，盆地南部由于缺乏河流径流量和缺乏侵蚀作用，以发育盆内水系为主。

18.5 古气候和沉积相

PdT盆地内早中新世古土壤揭示了半干旱环境（即平均年降水量250~500mm）（Rech等，2006，2009）（图18-7）。在从约15—12Ma时期里，盆地经历了由干旱气候转变为极端干旱气候的过程，导致潜水面下降和盐渍土的初始沉积（Alpers和Brimhall，1988；Rech等，2006，2009）。晚中新世至今，PdT盆地以极端干旱气候为主，由超干旱气候主导，间断出现短暂的微弱至中等潮湿环境（Kiefer等，1997；Latorre等，2005；Nester等，2007；Evenstar等，2009）。

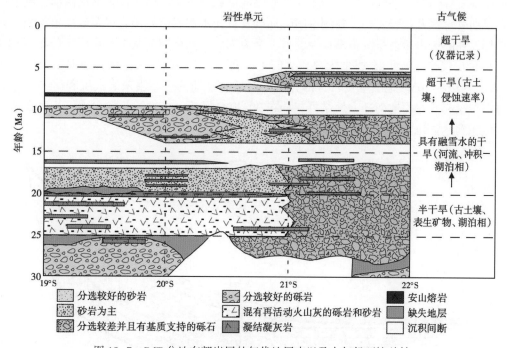

图18-7　PdT盆地东部岩层的年代地层表以及古气候环境总结

图表综合了来自海拔2000m南北走向的横断面的一系列空间重叠研究信息。来源：南纬20°20′北部的年龄约束来自Vergara等（1986）、Munoz和Sepulveda（1992）、Munoz和Charrier（1996）、Bouzari和Clark（2002）、Farias等（2005）、Victor等（2004）以及Munoz（2007）；南纬20°20′南部的年龄约束来自Blanco和Tomlinson（2006）、Hoke等（2007）以及Jordan等（2010）。古气候历史由前人研究推导出来（Alpers和Brimhall，1988；Parraguez，1998；Saez等，1999；Rech等，2003，2006；Dunai等，2005；Farias等，2005；Munoz，2007；Nester，2008；Evenstar等，2009；Lehmann等，2009）

盆地东部地层的时间分布揭示了极端气候演化的显著特征，可由熔结凝灰岩、火山熔岩和火山灰等的年代学资料来约束（Garcia和Herail，2006；Parraguez，1998；Pinto等，2004；Farias等，2005；Nester，2008；Jordan等，2010）（图18-3和图18-7）。大约25—16Ma有相对完整的冲积扇环境沉积记录，显示了半干旱气候条件（Rech等，2006，2009）。

然而，一系列的侵蚀面和地层缺失揭示了大约始于 15Ma 的沉积间断。在盆地东部大部分区域，新一轮的沉积作用始于约 14—12Ma 之间（Farias 等，2005；Nester，2008）。到 10Ma 冲积扇体系的沉积作用基本停止（图 18-7），只发育来自流域盆地极端干旱部分的少量碎屑沉积（Evenstar 等，2009；Rech 等，2009）。一个短暂而更加潮湿的气候出现在晚中新世末期（约 5—7Ma 之间），导致了盆地东部地区的下切侵蚀以及盆地中、西部地区冲积扇体系的重新堆积（Kiefer 等，1997；Evenstar 等，2009；Jordan 等，2010）。极端干旱气候对盆地西部影响较小，在那里科迪勒拉沿岸形成的地形屏障使其获得了少量的地表水、碎屑物、地下水和溶解物。在盆地北端，上中新统顶部地层以湿地硅藻土和上新统薄层盐田沉积为特征（Allmendinger 等，2005）。而在盆地南端，上中新统—上新统的浅湖相和盐田沉积最晚也始于 5Ma（图 18-3b 和图 18-8）（Mortimer 和 Saric，1975；Chong Diaz 等，1999；Saez 等，1999；Pisera 和 Saez，2003；Nester，2008）。

总的来说，PdT 盆地在晚渐新世—上新世发育与现今沉积环境类似的不对称沉积相。在盆地东部发育向西变细的冲积扇相粗粒硅质碎屑沉积（图 18-8）；而在盆地西部，湖泊环境在干旱期发育蒸发盐岩沉积，在潮湿间隔期则发育淡水泥岩、石灰岩以及硅藻土沉积。

爆发性的火山活动是沉积多样性的另一个原因。虽然熔结凝灰岩夹层广泛分布（图 18-7），但也仅仅在盆地北部（图 18-2）它们才占据了沉积物的很大比例。在那里中新统（图 18-3a；约 20—16Ma 地层单元）大部分由横跨整个盆地的熔结凝灰岩夹层和熔结凝灰岩碎屑再沉积组成。在盆地南部，熔结凝灰岩在 20°30′S 附近（东部）近源盆地沉积厚度达数十米，但在到达盆地轴部之前就已减薄尖灭。

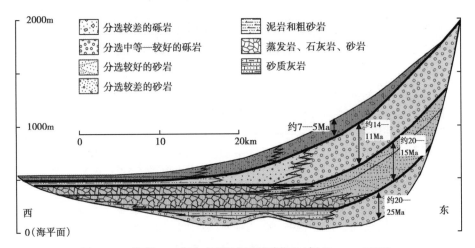

图 18-8　横穿 PdT 盆地南部的沉积相剖面（据 Nester，2008）

粗线指示层序边界，利用相同的四个时间间隔将沉积地层分隔开来，在图 18-3 中采用相同的颜色说明。也说明了大致的沉积年龄，与图 18-3 的颜色方案相匹配

18.6　结论

位于以构造侵蚀为主的俯冲系统和收缩机制成因的安第斯构造高原之间的非海相 PdT 盆地揭示了智利弧前地区的构造演化过程（图 18-9）。板块边界的构造侵蚀作用并没有影响到 PdT 盆地；PdT 盆地的形成是造成安第斯山脉隆升的地壳作用的次级效应，PdT 弧前盆地

的东侧地形受由一系列单斜褶皱形成的约3000m的构造抬升控制，其中大约1700m的抬升产生于27—11Ma之间，大约1100m的抬升产生于11Ma以来。岛弧火山活动和岛弧岩石侵蚀作用控制了盆地沉积填充。鉴于极端的区域气候环境，古气候在确定该弧前盆地的演化以及沉积相中非常重要。气候条件控制了进入盆地的沉积物通量：在半干旱—干旱气候时，大量的硅质碎屑物质进入盆地；而在极端干旱气候时，盆地沉积物匮乏（图18-9）。

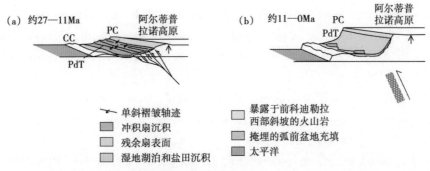

图18-9 PdT弧前沉积盆地晚渐新世—新近纪演化的简单示意图

PC—前科特迪勒；CC—科特迪勒沿岸。（a）渐新世—中中新世，由于隐伏逆断裂之上短波长单斜褶皱的生长，弧前盆地和阿尔蒂普拉诺高原之间的表层高差增加；这些隐伏逆断裂在深部收敛在一起，并延伸到中地壳（Victor等，2004）；半干旱—干旱的气候导致冲积相沉积广泛发育，使沉积上超和超覆于生长褶皱之上。（b）晚中新世、上新世和全新世，阿尔蒂普拉诺高原和前科特迪勒相对于弧前的抬升是对阿尔蒂普拉诺韧性中—下地壳增厚的响应，这里以广泛的韧性剪切带来作为示意图；这产生了一个地壳尺度的断弯褶皱（Isacks，1988；Farıas等，2005），以一个长波长的单斜来呈现，西翼为前科特迪勒斜坡；在超干旱的气候环境中，很少的沉积物供应到沉积盆地中；在图示中不仅缺少下切至太平洋的峡谷，同样也缺少PdT南部和北部的历史和特征

致谢（略）

参 考 文 献

Allmendinger, R. W., and Gonz_alez, G. (2009) Neogene to Quaternary tectonics of the Coastal Cordillera, northern Chile. Tectonophysics. doi: 10. 1016/j. tecto. 2009. 04. 019.

Allmendinger, R. W., Jordan, T. E., Kay, S. M., and Isacks, B. L. (1997) The evolution of the Altiplano–Puna plateau of the Central Andes. Annual Reviews of Earth and Planetary Science, 25, 139–174.

Allmendinger, R. W., Gonz_alez, G., Yu, J., Hoke, G., and Isacks, B. (2005) Trench–parallel shortening in the Northern Chilean Forearc: tectonic and climatic implications. GSAB, 117, 89–104.

Alpers, C. N., and Brimhall, G. H. (1988) Middle Miocene climatic change in the Atacama Desert, northern Chile: evidence from supergene mineralization at La Escondida. GSAB, 100, 1640–1656.

Barnes, J. B., and Ehlers, T. A. (2009) End member models for Andean plateau uplift. Earth–Science Rev., 97, 117–144.

Beck, S. L., and Zandt, G. (2002) The nature of orogenic crust in the central Andes. Journal of Geophysical Research–Solid Earth, 107 (B10). doi: 10. 1029/2000JB000124.

Blanco, N., and Tomlinson, A. (2006) Fm. Sichal: Sedimentación aluvial (Eoceno-Oligoceno) sintectónica al evento orogénico Incaico, Región de Antofagasta, Chile. Congreso Geológico Chileno, 11, 29-32.

Bouzari, F., and Clark, A. H. (2002) Anatomy, evolution, and metallogenic significance of the supergene orebody of the Cerro Colorado porphyry copper deposit, I Región, northern Chile. Econ. Geol., 97, 1701-1740.

Charrier, R., Muñoz, N., and Palma-Held, S. (1994) Edady contenido paleoflor_ ıstico de la Formación Chucaly condiciones paleoclimáticas para el Oligoceno tardío: Mioceno Inferior en el Altiplano de Arica, Chile. Proceedings 7th Congreso Geológico Chileno, Concepción, vol. 1, 434-437.

Charrier, R., Pinto, L., and Rodríguez, M. P. (2007) Tectonostratigraphic evolution of the Andean orogen in Chile, in Moreno, T., and Gibbons, W., eds., The geology of Chile. The Geological Society, London, 21-114.

Chong Díaz, G., Mendoza, M., Garcia-Veigas, J., Pueyo, J., and Turner, P. (1999) Evolution and geochemical signatures in a Neogene forearc evaporitic basin: the Salar Grande (Central Andes of Chile). Palaeogeography, Palaeoclimatology. Palaeoecology, 151, 39-54.

Clift, P., and Vannucci, P. (2004) Controls on tectonic accretion versus erosion in subduction zones: implications for the origin and recycling of the continental crust. Rev. Geophysics, 42, RG2001. doi: 10. 1029/ 2003RG000127.

Cortés, J., González, G., Dunai, T., and Carrizo S. D. (2006) Datación de superficies mediante 21Ne cosmogénico en la Cordillera de la Costa del norte de Chile. 11th Congreso Geologico Chileno, 2, 185-188.

Davis, G. H., and Reynolds, S. J. (1996) Structural geology of rocks and regions, 2nd ed. New York, John Wiley & Sons, 776 p.

de Silva, S. L. (1989) Altiplano-Puna volcanic complex of the central Andes. Geology, 17, 1102-1106.

Dickinson, W. R. (1995) Forearc basins, in Busby, C. J., and Ingersoll, R. V., eds., Tectonics of sedimentary basins. Oxford, Blackwell Science, 221-261.

Dickinson, W. R., and Seely, D. R. (1979) Structure and stratigraphy of forearc regions. AAPG Bull. 63, 2-31.

Dingman, R. J., and Galli, O. C. (1965) Geology and groundwater resources of the Pica area, Tarapaca Province, Chile. Reston, VA, USGS Bulletin, 113 p.

Dunai, T. J., González, G. A., and Juez-Larré, J. (2005) Oligocene-Miocene age of aridity in the Atacama Desert revealed by exposure dating of erosion-sensitive landforms. Geology, 33, 321-324.

Elger, K., Oncken, O., and Glodny, J., (2005) Plateau-style accumulation of deformation: Southern Altiplano. Tectonics, 24, TC4020. doi: 10/1029/2004TC001675.

Evenstar, L. A., Hartley, A. J., Stuart, F. M., Mather, A. E., Rice, C. M., and Chong, G. (2009) Multiphase development of the Atacama planation surface recorded by 3He exposure ages: implications for uplift and Cenozoic climate change in western South America. Geology,

37, 27–30.

Ewing, S. A., Sutter, B., Owen, J., Nishiizumi, K., Sharp, W., Cliff, S. S., Perry, K., Dietrich, W., McKay, C. P., and Amundson, R. (2006) A threshold in soil formation at Earth's arid-hyperarid transition. Geochimica et Cosmochimica Acta, 70, 5293–5322.

Farías, M., Charrier, R., Comte, D., Martinod, J., and Hérail, G. (2005) Late Cenozoic deformation and uplift of the western flank of the Altiplano: Evidence from the depositional, tectonic, and geomorphologic evolution and shallow seismic activity (northern Chile at 19 degrees 300S). Tectonics, 24, TC1005. doi: 10. 1029/ 2006TC002046.

García, M., and Hérail, G. (2005) Fault-related folding, drainage network evolution and valley incision during the Neogene in the Andean Precordillera of northern Chile. Geomorphology, 65, 279– 300.

Garcia, M., and Hérail, G. (2006) Fluvial incision and Neogene uplift of the western Central Andes, northern Chile. 11th Congreso Geológico Chileno, 2, 221–224.

Garzione, C. N., Molnar, P., Libarkin, J. C., and MacFadden, B. J. (2006) Rapid late Miocene rise of the Bolivian Altiplano: Evidence for removal of mantle lithosphere. EPSL, 241, 543–556.

Garzione, C. N., Hoke, G. D., Libarkin, J. C., Withers, S., MacFadden, B. J., Eiler, J. M., Gosh, P., and Mulch, A. (2008) Rise of the Andes. Science, 320 (5881), 1304–1307. doi: 10. 1126/science. 1148615.

Ghosh, P., Garzione, C. N., and Eiler, J. M. (2006) Rapid uplift of the Altiplano revealed through C-13-O-18 bonds in paleosol carbonates. Science, 311 (5760), 511–515.

Giese, P., Scheuber, E., Schilling, F., Schmitz, M., and Wigger, P. (1999) Crustal thickening processes in the Central Andes and the different natures of the Mohodiscontinuity. J. S. Am. Earth Sci., 12, 201–220.

Gregory-Wodzicki, K. M. (2000) Uplift history of the Central and Northern Andes: a review. GSAB, 112, 1091–1105.

Hartley, A. J., Chong, G., Turner, P., May, G., Kape, S. J., and Jolley, E. J. (2000) Development of a continental forearc: a Neogene example from the Central Andes, northern Chile. Geology, 28, 331–334.

Haselton, K., Hilley, G., and Strecker, M. R. (2002) Average Pleistocene climatic patterns in the southern Central Andes: controls on mountain glaciation and paleoclimate implications. J. Geol., 110, 211–226.

Hoke, G. D., Isacks, B. L., Jordan, T. E., Blanco, N., Tomlinson, A. J., and Ramezani, J. (2007) Geomorphic evidence for post-10 Ma uplift of the western flank of the central Andes 18 degrees 300–22 degrees S. Tectonics, 26, TC5021. doi: 10. 1029/2006TC002082.

Houston, J. (2002) Groundwater recharge through an alluvial fan in the Atacama Desert, northern Chile: mechanisms, magnitudes and causes. Hydrological Processes, 16, 3019–3035.

Houston, J., and Hartley, A. J. (2003) The Central Andean west–slope rainshadow and its potential contribution to the origin of hyper-aridity in the Atacama Desert. International Journal of Climatology, 23, 1453–1464.

Isacks, B. L. (1988) Uplift of the central Andean plateau and bending of the Bolivian orocline. JGR, 93, 3211-3231.

James, D. E., and Sacks, I. S. (1999) Cenozoic formation of the Central Andes: a geophysical perspective, in Skinner, B. J., ed., Geology and ore deposits of the Central Andes. Soc. Econ. Geol. Spec. Pub., 7, 1-26.

Jordan, T. E., Nester, P., Blanco, N., Hoke, G. D., Dávila, F., and Tomlinson, A. J. (2010) Uplift of the Altiplano-Puna Plateau: a view from the west. Tectonics, 29, TC5007. doi: 10.1029/2010TC002661.

Kay, S. M., and Coira, B. L. (2009) Shallowing and steepening subduction zones, continental lithospheric loss, magmatism, and crustal flow under the Central Andean Altiplano- Puna plateau, in Kay, S. M., Ramos, V. A., and Dickinson, W. R., eds., Backbone of the Americas: shallow subduction, plateau uplift, and ridge and terrane collision. Geol. Soc. Amer. Memoir 204, 229-259. doi: 10. 1130/2009. 1204 (11).

Kay, S. M., Mpodozis, C., and Coira, B. (1999) Neogene Mamatism, tectonism, and Mineral Deposits of the Central Andes 22 to 33 Slatitude. In Skinner, B. J., ed., Geology and ore deposits of the Central Andes. Soc. Econ. Geol. Spec. Pub., 7, 27-59.

Kiefer, E., Dorr, M. J., Ibbeken, H., and Gotze, H. J. (1997) Gravity-based mass balance of an alluvial fan giant: the Arcas Fan, Pampa del Tamarugal, Northern Chile. Revista Geologica Chile, 24, 165-185.

Kley, J., and Monaldi, C. R. (1998) Tectonic shortening and crustal thickness in the Central Andes; how good is the correlation? Geology, 26, 723-726.

Kober, F., Ivy-Ochs, S., Schlunegger, F., Baur, H., Kubik, P. W., and Wieler, R. (2007) Denudation rates and topography - driven rainfall threshold in northern Chile: Multiple cosmogenic nuclide data and sediment yield budgets. Geomorphology, 83, 97-120.

Latorre, C. L., Betancourt, J. L., Rech, J. A., et al. (2005) Late Quaternary history of the Atacama Desert, in Smith, M., and Hesse, P., eds., 23 degrees south: archaeology and environmental history of the Southern Deserts. Canberra, NationalMuseum of Australia, 73-90.

Lehmann, S., Rech, J. A., Currie, B. S., Jordan, T. E., and Riquelme, R. (2009) Redefining the Tarapacá pediplain; analysis of relict soils in the northern Atacama Desert, Chile (abstract). GSA Conv., 244.

McQuarrie, N. (2002) The kinematic history of the central Andean fold-thrust belt, Bolivia: implications for building a high plateau. GSAB, 114, 950-963.

Mortimer, C., and Saric, N. (1975) Cenozoic studies in northernmost Chile. Geol. Rundsch., 64, 395-400.

Munoz, N., and Charrier, R. (1996) Uplift of the western border of the Altiplano on a west vergent thrust system, Northern Chile. J. S. Am. Earth Sci., 9, 171-181.

Munoz, N., and Sepu'lveda, P. (1992) Estructuras compresivas convergencia al oeste en el borde oriental de la Depresion Central, norte de Chile (19 degrees 150S). Revista Geologica Chile, 19, 241-247.

Munoz, V. A. (2007) Evolución morfoestructural del piedemonte Altiplánico Chileno deurante el

Cenozoico Superior entre La Quebrada de Tarapacay La Quebrada de Sagasca (1945-2115S). Geologist thesis, Universidad de Chile.

Naranjo, J., and Paskoff, R. (1985) Evolución Cenozoica del piedemonte Andino en la Pampa del Tamarugal, norte de Chile (18-21S). 4th Congreso Geologico Chileno, 4 pp. 5/ 149-5/165.

Nester, P. (2008) Basin and paleoclimate evolution of the Pampa del Tamarugal forearc valley, Atacama Desert, northern Chile (PhD dissertation). CornellUniversity, Ithaca, NY.

Nester, P. L., Gayo, E., Latorre, C., Jordan, T. E., and Blanco, N. (2007) Perennial stream discharge in the hyperarid Atacama Desert of northern Chile during the latest Pleistocene. Proceedings of the National Academy of Sciences of the United States of America, 104, 19724-19729.

Parraguez, G. (1998) Sedimentología y geomorfología producto de la tectónica cenozoica, en la Depresión Central, I Región de Tarapacá, Chile. Thesis, Universidad de Chile.

Patton, T. L. (2004) Numerical models of growth - sediment development above an active monocline. Basin Research, 16, 25-30. doi: 10. 1046/j. 1365-2117. 2003. 00220. x.

Pinto, L., Hérail, G., and Charrier, R. (2004) Sedimentación sintectónica asociada a las estructuras neógenas en la Precordillera de la zona de Moquella, Tarapaca (1915S, norte de Chile). Revista Geologica Chile, 31, 19-44.

Pisera, A., and Saez, A. (2003) Paleoenvironmental significance of a new species of freshwater sponge from the Late Miocene Quillagua Formation (N Chile). J. South Am. Earth Sci., 15, 847-852.

Ranero, C. R., von Huene, R., Weinrebe, W., et al. (2006) Tectonic processes along the Chile convergent margin, in Oncken, O., Chong, G., Franz, G., Giese, P., Goetze, H-J., Ramos, V. A., Strecker, M. R. & Wigger, P., eds., The Andes: active subduction orogeny. Frontiers in Earth Sciences. Berlin, Springer, 91-121.

Rech, J. A., Quade, J., and Hart, W. S. (2003) Isotopic evidence for the source of Ca and S in soil gypsum, anhydrite and calcite in the Atacama Desert, Chile. Geochim Cosmochim Acta, 67, 575-586.

Rech, J. A., Currie, B. S., Jordan, T. E., and Riquelme, R. (2009) The antiquity of the Atacama Desert and its implication for the paleoelevation of the Central Andes (abstract). GSA Conv. 203-209.

Rech, J. A., Currie, B. S., Michalski, G., and Cowan, A. M. (2006) Neogene climate change and uplift in the Atacama Desert, Chile. Geology, 34, 761-764.

Saez, A., Cabrera, L., Jensen, A., and Chong, G. (1999) Late Neogene lacustrine record and palaeogeography in the Quillagua-Llamara basin, Central Andean forearc (northern Chile). Palaeo Palaeo Palaeo, 151, 5-37.

Schlunegger, F., Zeilinger, G., Kounov, A., Kober, F., and Husser, B. (2006) Scale of relief growth in the forearc of the Andes of Northern Chile (Arica latitude, 18°S). Terra Nova, 18, 217-223.

SERNAGEOMIN. (2002) Mapa Geologico de Chile. Servicio Nacional de Geología y Minería, Chile. Carta geológica de Chile, Serie Geología Básica, N 75, scale 1:1000000.

Santiago, SERNAGEOMIN. Sick, C., Yoon, M-K., Rauch, K., et al. (2006) Seismic images of accretive and erosive subduction zones from the Chilean margin, in Oncken, O., Chong, G., Franz, G., Giese, P., Goetze, H-J., Ramos, V. A., Strecker, M. R. & Wigger, P., eds., The Andes: active subduction orogeny. Frontiers in Earth Sciences. Berlin, Springer, 147-169.

Smith, G. A., and Landis, C. A. (1995) Intra-arc basins, in Busby, C. J., Ingersoll, R. V., eds., Tectonics of sedimentary basins. Oxford, Blackwell Science, 263-298.

Somoza, R., and Ghidella, M. E. (2005) Convergencia en el margen occidental de Am_erica del Sur durante el Cenozoico: subduccion de las placas de Nazca, Farallon y Aluk. Revista Asoc. Geol. Argentina, 60, 797-809.

Tomlinson, A. J., Blanco, N., Maksaev, V., Dilles, J. H., Grunder, A. L., and Ladino, M. (2001) Geología de la Precordillera Andina de Quebrada Blanca-Chuquicamata, Regiones I y II (20degs300-22degs300S). Servicio Nacional de Geología y Minería, IR-01-20.

Trumbull, R. B., Riller, U., Oncken, O., Scheuber, E., Munier, K., and Hongn, F. (2006) The time-space distribution of Cenozoic arc volcanism in the south-central Andes: a new data compilation and some tectonic implications, in Oncken, O., Chong, G., Franz, G., Giese, P., Goetze, H-J., Ramos, V. A., Strecker, M. R. & Wigger, P., eds., The Andes: active subduction orogeny. Frontiers in Earth Sciences. Berlin, Springer, 29-43. doi: 10. 1007/ 978-3-540-48684-8.

Vergara, M., C. Marangunic, H. Bellon, and R. Brousse (1986) Edades K-Ar de las ignimbritas de las Quebradas Juan de Morales y Sagasca, norte de Chile. Serie Comunicaciones Dep. Geología, Universidad Chile, 36, 1-7.

Victor, P., Oncken, O., and Glodny, J. (2004) Uplift of the western Altiplano plateau: Evidence from the Precordillera between 20 degrees and 21 degrees S (northern Chile). Tectonics, 23, TC4004. doi: 10. 1029/2003TC001519.

von Huene, R., and Ranero, C. R. (2003) Subduction erosion and basal friction along the sediment-starved convergent margin off Antofagasta, Chile. Journal of Geophysical Research-Solid Earth, 108. doi: 10. 1029/2001JB001569.

von Huene, R., and Scholl, D. W. (1991) Observations at convergent margins concerning sediment subduction, subduction erosion, and the growth of continental crust. Rev. Geophysics, 29279-29316.

von Rotz, R., Schlunegger, F., Heller, F., and Villa, I. (2005) Assessing the age of relief growth in the Andes of northern Chile: magnetopolarity chronologies from Neogene continental sections. Terra Nova, 17, 462-471.

Worner, G., Hammerschmidt, K., Jenjes-Kunst, F., et al. (2000) Geochronology ($^{40}Ar/^{39}Ar$, K-Ar and He-exposure ages) of Cenozoic magmatic rocks from Northern Chile (18°-22°S): implications for magmatism and tectonic evolution of the central Andes. Revista geológica Chile, 27, 205-240.

(李阳 译，阳怀忠 吴哲 校)

第 19 章　伸展型和张扭型大陆弧盆：以美国西南部为例

CATHY J. BUSBY

（Department of Earth Science, University of Califonia, Santa Barbara, USA）

摘　要：伸展型和张扭型大陆弧盆保存了很厚且连续的沉积层序，并对大陆的生长具有重要的贡献。因此，了解这些盆地的演化过程显得十分重要。在这一章中笔者以美国西南部中生代到新生代为例，描述了四种类型的大陆弧盆：（1）早期低位伸展大陆弧盆；（2）早期低位张扭大陆弧盆；（3）晚期高位伸展大陆弧盆；（4）晚期高位张扭大陆弧盆。

在早中生代泛古陆解体时期，古太平洋盆地很可能是由非常大的、古老且较冷的板块组成。这些板块在俯冲过程中的后撤作用，导致上部板块，特别是在沿着北美和南美西部边缘发育的热减薄岛弧区，发生伸展和沉降。晚三叠世至中侏罗世，美国西南部的早期低位伸展大陆弧盆以表壳岩为底。这表明岩浆作用之前未发生隆升作用，且盆地经历了快速沉降，局部位于海平面之下。因此这些盆地为克拉通来源的沉积物形成了一个沉积区而非阻挡区。这些盆地以发育大量的、广泛分布的大规模硅质破火山口为特征。破火山口的喷发产物则快速掩埋了断层崖和地垒块体，导致盆地内发育很少的外生碎屑沉积。晚侏罗世，由于墨西哥湾的打开并导致了斜向俯冲，沿着早期低位伸展大陆弧盆的轴部形成了早期低位张扭大陆弧盆。盆地在释压弯曲或释压叠阶部位发生沉降，而在邻近受限弯曲或叠阶部位发生隆升［"海豚效应（porpoising）"］，并伴随有同期活动的正断层和逆断层。隆升事件形成了大量的以古深切谷和大型滑坡断崖形式出现的不整合，同时导致巨型滑坡块体在沉降区堆积。冲起构造的侵蚀作用产生了大量的粗粒外生碎屑沉积。在该背景下硅质巨型大陆破火山口局限在释压叠阶部位的对称性盆地内持续发育。

从白垩纪到古新世，在逐渐年轻的板片的浅层俯冲形成的收缩应力体制作用下，岛弧向东迁移，这样就产生了一个广阔的高原，由于它与现代安第斯岛弧的阿尔蒂普拉诺高原相似，因此被称之为"内华达高原"（Nevadaplano）。之后在始新世—中新世期间，由于板片后撤、变陡，火山作用向西迁移，形成了晚期高位伸展大陆弧盆。与早期伸展弧盆相似，晚期伸展大陆弧盆也发育"超级火山（supervolcano）"硅质破火山口区，但它们主要局限在地壳增厚部位（内华达高原），而早期伸展弧盆阶段破火山口则可发育在岛弧的任何地方。与早期盆地明显不同的是，晚期盆地形成于遭受强烈侵蚀的基底之上，且喷发产物沿着在地壳缩短时期形成的下切谷流动。另外，晚期伸展大陆弧盆普遍缺乏海相沉积；并且，由于盆地海拔较其他地区高太多，也不发育源自克拉通供给的沉积。在约 12Ma，东西向伸展作用被北西—南东向张扭作用所取代，相对应地，太平洋板块相对于科罗拉多高原的运动方向也由向西变为更趋于向北运动，并导致了微板块的捕获。内华达微板块（Sierra Nevada microplate）开始形成，其后缘位于古喀斯喀特（Cascades）岛弧的轴部。晚期张扭大陆弧阶段以在张扭断层叠阶部分发育大型火山中心为特征。每次张扭作用均形成一个不整合，并带来一次岩浆活动。岩石圈规模的拉分构造引发深部熔体，导致"喷溢安山岩（flood andesite）"的喷发。由于盆地紧靠隆起，导致巨型滑坡沉积得以保存。源自内华达高原的东西向古水系被沿着新的板块边界发育的南北向水系所扰乱、取代。

相对于地球化学和地球物理特征,大陆岛弧的地层学和构造学方面的研究一直被忽视;然而,这些地质特征能为研究该区域构造演化提供至关重要的参考依据和约束条件。

关键词:伸展作用 岛弧 喀斯喀特山系 侏罗纪 科迪勒拉山系

19.1 引言

为了明确伸展型和张扭型大陆弧盆的主要特征,本文综合分析了针对美国西南部该类盆地构造和地层学研究的成果。首先概述了对大陆弧构造过程认识的演变,以及这种演变对理解美国西部大陆弧盆的影响。之后在本文主体部分描述了两种伸展和张扭盆地类型:(1)早期低位(low-lying)大陆弧盆(图19-1)和(2)晚期高位(high-standing)大陆弧盆(图19-2)。文中展示了过去25年来有关盆地结构和充填的图件,并为读者详细地提供了来自于公开出版物的主要资料(地质图和横剖面图、实测剖面、地质年代学、地球化学以及古生物学资料等)。在描述至今仍未得到很好研究的晚期盆地之前,首先通过岩性地层柱归纳早期盆地的鉴别特征。为了强调伸展型和张扭型大陆弧盆在俯冲早期与俯冲晚期存在的差异(表19-1),在本文的最后归纳了笔者新近发表的和未发表的关于晚期盆地研究的成果,并附有综合岩性地层柱状图。

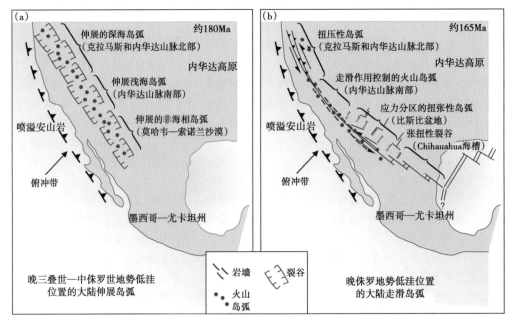

图19-1 美国西南部晚三叠世—侏罗纪低位伸展—张扭大陆弧

(a)晚三叠世至中侏罗世大陆弧以伸展为主,可能是在冷的古岩石圈俯冲过程中板片快速后撤形成的(Busby-Spera,1988;Busby等,2005);(b)晚侏罗世大陆弧记录了沿着与墨西哥湾裂谷相关的左旋走滑断裂的左旋斜向会聚(Silver和Anderson,1974);大陆边缘相对于这些断裂的曲度导致了北部的压扭作用和南部的张扭作用

(据Saleeby和Busby-Spera,1992)

许多现代大陆岛弧明显包括了收缩构造及其相关的盆地(Tibaldi等,2009),但通常难以保存伸展和张扭盆地,而后者却能在地质记录中得到很好的体现。伸展型大陆弧盆保存有

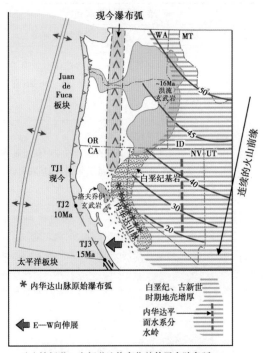 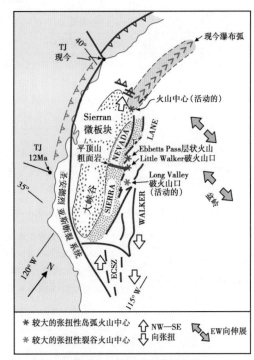

(a) 始新世—中新世地势高位的伸展大陆岛弧　　　　(b) 12Ma—现今张扭性大陆岛弧

图19-2　美国西部新生代高位伸展—张扭大陆弧

(a) Great 盆地内始新世至中新世迁移的岛弧火山活动；岛弧向西南方向的迁移，伴随着俯冲过程中板片后撤/变陡引起的伸展作用（Dickinson，2006；Cousens 等，2008；Busby 和 Putirka，2009）。该伸展作用叠加到高且广阔的内华达高原（Nevadaplano）上；后者是在晚中生代—古新世低角度俯冲引起的收缩应力机制下形成的（DeCelles，2004）。图中还显示了白垩系内华达山脉岩基的位置及其向北延伸进入内华达西北部，以及晚白垩世—新生代板片后撤之前低角度俯冲期间岩基去顶作用时活动的盆地的残余部分。板片后撤在16Ma完成，此时岛弧前缘到达内华达山脉东部位置；16Ma到11Ma之间的岛弧伸展作用可能是受圣安德烈亚斯断层系统的发育（三联点南部），以及北美板块和太平洋板块之间的相互作用控制的（Dickinson，2006）。除了加利福尼亚的洛夫乔伊玄武岩在岛弧前缘喷发，16Ma左右的喷溢玄武岩喷发于弧后位置（Garrison 等，2008；Busby 和 Putirka，2009）。海底重建显示了15Ma 和 10Ma 三联点位置（Dickinson，1997；Atwater 和 Stock，1998）；TJ1 位置（显示现今三联点位置，位于卡斯卡底俯冲带的圣安德烈亚斯断层带和门多西诺破裂带之间）以供参考。（b）内华达山脉微板块（Sierra Nevada microplate）后缘12Ma至现今的张扭岛弧火山活动。下述 Unruh 等（2003）广泛使用的图19-1，透视显示的是有关内华达山脉—北美最佳 Euler 极点的科迪勒拉西部的投影。Sierran 微板块位于圣安德烈亚斯断层和沃克通道之间，目前可容纳北美板块和太平洋板块之间20%～25%的板块运动，可能代表了未来的板块边界（参考 Busby 和 Putirka，2009）。板块构造重建展示了太平洋板块和内华达山脉微板块相对于科罗拉多高原，在10—12Ma从向西运动到向北运动的改变（参考 McQuarrie 和 Wernicke，2005）。这表明内华达山脉微板块在这个时候生成（见正文）。最大的火山中心位于张扭盆地中，包括一个活动的裂谷中心（Long Valley 破火山口）和一个活动的岛弧中心（Lassen），以及一个中新世岛弧破火山口（Little Walker）和中新世岛弧层状火山（Ebbetts Pass）。图19-8 显示最大的中新世岛弧火山与现代 Long Valley 破火山口的构造环境比较

很厚且连续的沉积序列（Smith 和 Landis，1995），并具有较高的岩浆产率（Taupo 火山带；White 等，2006）。因此，它们对大陆的生长具有重要的贡献。出于这个原因，了解这些盆地的演化过程显得十分重要。本文的目的就是给读者一个关于伸展和张扭大陆弧盆的地球动力学背景、构造特征、盆地结构、沉积学和火山学的整体认识。

表 19-1 大陆弧盆在俯冲早期与晚期的差异

	早期低位大陆弧盆			晚期高位大陆弧盆	
	冷的古大洋岩石圈俯冲			板片后撤	微板块的捕获
动力学背景和时代（美国西南部）	晚三叠世—中侏罗世伸展岛弧（图19-1a、图19-3、图19-4、图19-6a、b）	晚侏罗世张扭岛弧（位于伸展岛弧之上）（图19-1b、图19-5、图19-6c）		始新世—中新世伸展岛弧（迁移的）（图19-2a和图19-7）	12Ma—现今的张扭岛弧（微板块后缘）（图19-2b和图19-7）
盆地基底	强烈沉降的表壳岩（没有前期的抬升）	下沉的伸展岛弧地层和抬升的基底（相互毗邻）		在内华达高原白垩纪深成岩体遭受去顶作用，并被古河道/切谷下切	下沉的始新世—中新世伸展岛弧地层+抬升的基底块体
构造和不整合面	沿着同沉积正断裂发生均匀快速且持续的沉降；不整合面不发育	沿着同期的正断裂和逆断裂的不同规模的"海豚效应"；大量的深部不整合面	低角度俯冲作用造成的白垩纪地壳缩短和增厚	伴随板片后撤及软流圈上涌发生热抬升和东西向伸展；不整合面发育	NW—SE向张扭作用产生NNW向和NE向正断裂，并分别带有右旋和左旋的运动分量；脉冲式张扭活动形成多期不整合和脉冲式岩浆活动
古地理	岛弧断陷，局部下沉于海平面之下，并且作为克拉通来源沉积物的会聚区（不是障壁区）	复杂地形扰乱的大型沉积体系；局部沉积物源		古河道/切谷下切内华达高原西翼，并将沉积物向西搬运至加利福尼亚大峡谷	沃克通道西部向西流动的古河道/峡谷遭受紊乱；南北向水系形成
火山活动	广泛分布的硅质大型破火山口喷发	局限于释压叠阶部位的硅质破火山口；沿着释压弯曲部位较小的火山中心		内华达高原顶部硅质火山口；岛弧向西穿过较薄地壳时，中间组火山岩的喷发；16Ma沿着断控裂缝的Lovejoy喷溢玄武岩火山口	断层叠阶部位的大型火山中心（如约10Ma时的Little Walder破火山口；约2Ma的Ebbetts Pass层积火山；和现今的Lassen火山中心）；沿着岩石圈尺度的拉分构造引起厚地壳之下的深部熔融，导致包含"喷溢安山岩"的高钾火山活动
沉积	断崖和地垒块体被火山碎屑掩覆；导致较少的外生碎屑沉积	来自于盆内和盆外高地的大量外生碎屑和滑坡沉积		原生火山沉积；经古河道/切谷进入火山碎屑流和河流再沉积（向西）	来自盆外和盆内高地的滑坡沉积；南北向河流

19.2 区域构造背景及认识演变

在20世纪80年代末之前，"安第斯弧"这个术语被广泛用于在陆壳上形成的所有火山深成岛弧（volcanoplutonic arcs），导致普遍认为大陆岛弧是形成于收缩应力体制之下，并以位于隆起的高位区为特征的（Hamilton，1969；Burchfiel和Davis，1972）。当时只有少数学者提出，大陆弧也可能存在于伸展或张扭成因的受断裂边界控制的深坳陷内。从这个角度出发，现代的伸展和张扭岛弧仅仅能在中美洲火山弧（Burkart 和 Self，1985）、堪察加岛弧（Erlich，1979）和苏门答腊岛弧（Fitch，1972）中被识别出来。虽然在 Taupo 火山带也识别出了伸展活动，但一直被解释为其代表了一个弧后背景（Cole，1984），直到后来地质学和地质年代学研究证明它是一个伸展大陆弧盆（Houghton 等，1991）。由于伸展或张扭大陆弧的现代例子很少，所以对它们的成因知之甚少，并且针对从地质记录识别出来的伸展岛弧的构造模型也是主观推测的（Busby-Spera，1988）。尽管如此，要想很好地解释在地质记录中常见的、连续且很厚的大陆弧沉积序列，一个伸展构造背景是必需的（Busby-Spera，1988）。现有的模型无法很好地解释为大陆生长做出重大贡献的伸展大陆弧盆。

20 世纪 80 年代初，我们认识到伸展大洋岛弧是在"冷"的古大洋岩石圈俯冲过程中由于板片后撤（slab rollback）产生的。但这种机制最初并不是为大陆弧提出的（Molnar 和 Atwater，1978；Dewey，1980）。同时越来越明显的是，沿北美和南美的西部边缘，在大陆弧的早期阶段至少在部分地区具有快速沉降的特征（Busby-Spera，1983；Maze，1984；Burke，1988）。这表明现今世界缺少对某些地质过程的研究。

Jarrard（1986）对现今弧—沟系统构造特征的动力学控制机制分析表明，俯冲带的寿命和仰冲板块的挤压应力之间呈强烈的正相关。Jarrard（1986）的研究同样表明世界上大多数俯冲带已经持续了很长一段时间。由此笔者推测，在中生代早期泛古大陆裂解过程中，由巨大的古岩石圈板块组成的古太平洋盆地的俯冲在古太平洋盆地的西缘形成了伸展大陆弧（Busby 等，1998；图 19-1a）。由于现今大部分俯冲带都经历了长期活动，因此我们必须关注相关地质记录，以更多地了解在俯冲早期阶段伸展岛弧内的大陆生长（表 19-1；Busby，2004）。

从 20 世纪 90 年代到 21 世纪，现代和古代岛弧区的走滑断层系统得到越来越多的关注。这是因为：（1）斜向会聚比正交会聚更常见，并且在大多数大陆弧地区，仅仅与正交方向偏离 10°的斜向会聚就能导致在上部板块内发育走滑断层；（2）断层集中在岛弧的热减薄地壳上，特别是在比大洋岛弧地壳更软弱，且与俯冲板块耦合更好的陆壳之上（Dewey，1980；Jarrard，1986；McCaffrey，1992；Ryan 和 Coleman，1992；Smith 和 Landis，1995）。在现今的许多大陆弧区，如跨墨西哥火山带（Van Bemmelen，1949）、安第斯岛弧（Cembrano 等，1996；Thomson，2002）、苏门答腊岛弧（Bellier 和 Sebrier，1994）、Aeolian 岛弧（Gioncada 等，2003）、Calabrian 岛弧（Van Dijk，1994）和中菲律宾岛弧（Sarewitz 和 Lewis，1991）等，已经对走滑断层作了大量的研究和描述。然而，相对于转换边缘的走滑盆地，对于弧内走滑盆地的研究仍然非常少（Busby 和 Bassett，2007）。在本文中，笔者描述了沿着晚三叠世早期—中侏罗世伸展大陆弧盆地轴部发育的晚侏罗世走滑弧内盆地的主要特征。现今的加利福尼亚州和亚利桑那州的 Mojave-Sonoran 沙漠就位于这些盆内（图 19-1b）。这些弧内走滑盆地从其沉积填充和构造特征上可以与下伏的伸展岛弧盆地相区分（表 19-1）。

之后本文描述了俯冲晚期在增厚陆壳之下板片后撤/变陡背景下形成的伸展大陆弧盆地（图 19-2a、表 19-1）。在白垩纪—古新世期间，由于俯冲板片逐渐变浅，岛弧在收缩应力背景之下向东迁移，地壳处于增厚状态（Coney 和 Reynolds，1977；Humphreys，2008，2009；Dickinson，2006；DeCelles 等，2009）。这样就产生了高原，由于它与现代安第斯岛弧的阿尔蒂普拉诺高原相似，DeCelles（2004）称之为"内华达高原（Nevadaplano）"（图

19-2a）。之后，从始新世到中新世，美国西南部和墨西哥北部的火山活动向西迁移（图19-2a），这被解释为记录了在板片变陡时的岛弧岩浆作用（Coney和Reynolds，1977；Dickinson，2006；Cousens等，2008）。还有其他一些模型被提出来解释该火山作用（Humphries，2009），但笔者认为形成于这些火山带内的伸展盆地是弧内盆地（参见Busby和Putirka，2009摘要）。McQuarrie和Wernicke（2005）通过板块构造重建也揭示出了东西向的伸展作用（图19-2a）。

最后，本文描述了开始于12Ma，由于太平洋板块捕获微板块而形成的，或正在形成的张扭岛弧盆地（图19-2b、表19-1）。到16Ma，岛弧前缘（以及相关的伸展构造）向海沟方向延伸直达内华达山脉巨大的白垩系岩基（图19-2a）。从16Ma至12Ma，中新世弧内的热软化作用使得在坚硬大陆块体内部的大陆变的软弱（现今仍然无断层发育）。McQuarrie和Wernicke（2005）对板块构造重建表明，在10—12Ma期间，太平洋板块和内华达山脉微板块（Sierra Nevada microplate）相对于科罗拉多高原，由向西运动变为更多的向北运动（图19-2b）。大地测量研究表明，内华达微板块的张扭性东部边界（沃克通道；图19-2b）目前容纳了北美板块和太平洋板块之间约25%的板块运动量（Unruh等，2003）。东部边界更接近于"经典"的板块边界，这是因为与北部和西部边界相比，东部边界是离散的，而后两者由于分别受到收缩和张扭构造作用的叠加，具有弥散而又复杂的特点。因此，东部微板块边界是确定微板块形成时间的理想场所，同时也可以确定那些标志岛弧轴部内张扭微板块边界发育特征的形成时间。虽然裂谷作用未能沿着Walker Lane带形成新的海底，但笔者还是认为它与加利福尼亚海湾的裂谷有许多共同的特点。包括：（1）裂谷作用均开始于约12Ma；（2）裂谷作用均位于与俯冲作用相关的岛弧的热减薄的轴部，该岛弧因板片后撤而发育伸展作用；（3）由于岛弧西部边界巨厚白垩系岩基地壳的阻挡，岛弧停止向海沟方向迁移，导致热减薄作用增强。在加利福尼亚的墨西哥湾，可能由于扩张中心位于向墨西哥之下俯冲的板块和巨大的太平洋板块之间，裂谷作用导致了非常快速的海底扩张（至6Ma）。随后太平洋板块拖曳已消亡的上部板片（下加利福尼亚州）向西北方向运动（Busby和Putirka，2010a，2010b）。与之相反，大陆解体仍然在加利福尼亚地区进行；大陆表现为沿着现今Cascades岛弧的轴部被向北拉开（图19-2b）。

19.3 早期低位大陆弧盆

19.3.1 晚三叠世—中侏罗世伸展岛弧

图19-3a和表19-1展示了早期低位伸展大陆弧的古地理单元（图19-1a）。Busby-Spera（1988）将其称之为"断陷（graben depression）"以强调其低位特征（而不是隆起区顶部的地堑）。岛弧型火山分布在一个长度大于1200km的地堑内，其南部（黄色区域）位于海平面之上的古生代以来的克拉通环境，而北部（蓝色区域）则位于海平面之下的古生代冒地斜（减薄大陆）到增生地壳之上（图19-1a）。早—中侏罗世位于现今科罗拉多高原的沙漠区的超成熟的石英砂被吹进"岛弧断陷（arc graben depression）"内，随后由于构造沉降作用，使得该地区保存的石英砂岩甚至比现今高原的弧后地区更厚（图19-3b）。约300~1000m/Ma的快速构造沉降导致了厚达10km的非海相和浅海相沉积地层得到保存，并且不发育侵蚀不整合面（Busby等，1990，2005）。在亚利桑那州南部的大陆弧段，虽然没

有基岩出露，但中生代和新生代火山岩的同位素特征表明它们位于"断陷"之下，并限制了"断陷"的范围（Tosdal 等，1989），也就是说基岩被深埋于岛弧断陷之下。Haxel 等

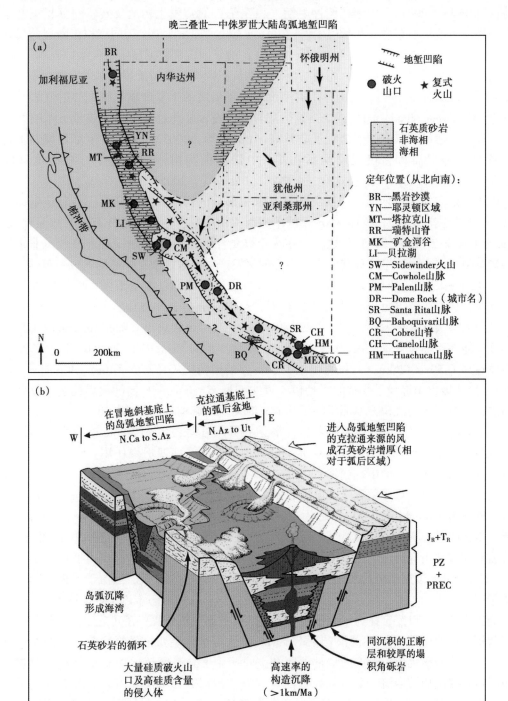

图 19-3　美国西南部晚三叠世至中侏罗世伸展大陆弧"地堑洼地"（据 Busby-Spera，1988）
（a）岛弧古地理图；（b）总结图 19-3a 所示区域的伸展大陆弧成因的证据（Busby-Spera，1988；Busby 等，1990）。将盆地从浅海相到非海相环境的充填变化总结在图 19-6a 和图 19-6b 中。N. Ca to S. AZ—加利福尼亚州北部到亚利桑那州南部；N. Az to UT—亚利桑那州北部到犹他州

（2005）最近的一篇文章也强调了亚利桑那州南部侏罗纪岩浆弧内的"深盆地"的重要性。不管是在大陆还是大洋背景下，硅质破火山口和硅质浅部侵入体的大量发育是伸展岛弧的典型特征（Busby，2004）。由于在俯冲晚期（白垩纪）通常发生构造反转，同沉积正断层已难以识别，但地层展布样式和断层岩屑楔（fault talus wedge）仍然为同沉积正断层的发育提供了有利证据。另外，Solomon 和 Taylor（1991）的氧同位素数据显示大气水侵入到了地壳，也为岛弧裂谷作用提供了证据，而铁氧化矿物的沉积则表明了岛弧裂谷作用发育在干旱环境（Barton 和 Johnson，1996）。火山作用类型也反映了当时的干旱环境：缺失如射气岩浆喷发空落浮岩（phreatoplinian fall）或未熔结熔灰岩等代表岩浆通过火山口湖或地下水喷发的沉积；相反地，熔结凝灰岩的熔结程度非常高（Busby 等，2005）。当然，干旱环境并不是伸展岛弧的固有特征，它仅仅反映了在早—中侏罗世美国西南部位于"马纬度（horse latitudes，即副热带高压带）"这样一个事实（Busby 等，2005）。

图 19-4a 显示了早期低位伸展大陆弧非海相部分的构造和沉积样式。沿着地堑边界断层带的大量分支断裂发育了许多小规模的单成因火山中心，并且常常伴有岩浆渗漏到地表，Riggs 和 Busby-Spera（1990）称之为"多火山口复合体（multi-vent complex）"。非常快速的构造沉降导致火山口迅速被其他火山中心的喷发产物或来自于克拉通的风成石英砂岩所掩埋。与 Taupo 火山带的伸展大陆弧类似，断崖被快速掩埋，以至于未发育明显的冲积扇沉积；同时，快速沉降也阻止了侵蚀不整合的发育（Busby 等，2005）。

硅质大型大陆破火山口（图 19-4b、19-4d），或"超级火山（supervolcano）"常见于早期低位伸展大陆弧背景（表 19-1）。例如，中侏罗世的 CobreRidge 破火山口（图 19-4d）规模就达到了 50km×25km，且发育了至少 3km 厚的熔结凝灰岩充填（Riggs 和 Busby-Spera，1991）。这些破火山口异乎寻常的大规模、直线型形态、NW—SE 向的边界构造、复杂的坍塌历史，以及巨厚的沉积充填都反映了其形成演化受区域构造控制的特点。

在 Brick Mine 凝灰岩（熔结凝灰岩，来源不明）喷发之前，Pajarito 凝灰岩（熔结凝灰岩）喷发间歇期的沉积作用记录了西北端破火山口经历了两期坍塌作用。Cobre Ridge 破火山口与 Altiplano-Puna 火山复合体（Volcanic Complex）类似，具有规模大、直线型形态和坍塌作用复杂的特点，只是后者受区域构造体系影响喷发中心更大。继 Van Bemmelen（1949）之后，Silva 和 Gosnold（2007）称之为"火山—构造断陷（Volcano-tectonic depression）"。因此，本文同意 Dickinson 和 Lawton（2001）的观点，即对 Cobre Ridge 破火山口而言，"区域构造对局部火山口坍塌的影响不可忽视"。美国西南部之下在俯冲早期可能形成了很多类似于 Cobre Ridge 破火山口的"地堑式火山口（graben calderas）"，但由于同火山期伸展构造作用或后期收缩构造作用（白垩纪）的切割改造，其几何形态难以很好地恢复。

蜂巢状破火山口复合体为早期低位伸展大陆弧的厚层沉积提供了可容空间。例如位于现今加利福尼亚西南部 Mojave 沙漠的下 Sidewinder 火山带就主要由呈蜂窝状分布的四个破火山口组成。这些破火山口形成的局部可容纳空间接受了总厚度超过 4km 的熔结凝灰岩、巨型角砾岩、安山质熔岩流、源自克拉通的风成石英砂屑岩，以及火山碎屑沉积（包括河流和泥石流沉积；图 19-4b）。同岩浆活动正断层使这些破火山口向其由于伸展作用而去顶的侵入岩根下沉（图 19-4c）。这些侵入岩体随后又被在晚侏罗世张扭构造体制下喷发的岛弧型火山岩不整合覆盖（图 19-5a；Schermer 和 Busby，1994）。

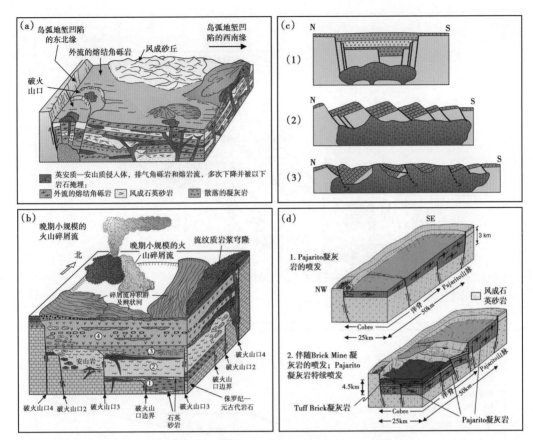

图 19-4 早期阶段低位伸展大陆弧盆的构造和地层类型
(亚利桑那州和加利福尼亚州三叠纪至中侏罗世)

(a) 亚利桑那州南部 Santa Rita 山脉区域伸展弧"地堑洼地"的北缘 (示意图,"SR"位置位于图 19-3a 中) (据 Riggs 和 Busby-Spera, 1990)。(b) Mojave 沙漠中部 Sidewinder 火山序列下部,巢状破火山口复合体的可容空间 (位于图 19-3a 中的西南方向;据 Schermer 和 Busby, 1994)。(c) 同岩浆时期的伸展和岩基剥露 (解释剖面) 展示了 Sidewinder 火山岩系破火山口下部和伴随的岩基岩石 (1) 岩基侵位过程中正断裂引起断错现象,(2) 导致岩基的部分剥露,(3) 张扭环境中 Sidewinder 火山岩系上部喷发之前 (图 19-5a) (位于图 19-3a 中的西南部;据 Schermer 和 Busby, 1994)。(d) 亚利桑那州南部中侏罗世 Cobre Ridge 破火山口,区域性正断层控制的直线型"地堑破火山口"(位于图 19-3a 中的 CR 位置;据 Riggs 和 Busby-Spera, 1991)

19.3.2 晚侏罗世张扭岛弧

晚侏罗世左旋张扭大陆弧盆沿着亚利桑那州和加利福尼亚州南部侏罗纪早期伸展大陆弧盆的轴部发育 (图 19-1b, 表 19-1; Saleeby 和 Busby-Spera, 1992; Schermer 和 Busby, 1994; Busby 等, 2005)。500km 长的板块边缘级别的独立岩墙群为研究岩墙群的喷发以及与之相当的浅成侵入构造提供了独特的视角,该岩墙群约在 148Ma 侵位于加利福尼亚岛弧段 (图 19-1b),其出露的构造线位于 Mojave 沙漠内 (图 19-5a;Schermer 和 Busby, 1994)。上 Sidewinder 火山带 (图 19-5a) 是一个宽双峰的,由岩墙、熔岩和火山口角砾岩组成的碱性岩套,主要由岛弧裂谷成因的粗面玄武岩、玄武岩至玄武质安山岩和钠闪碱流岩、钠闪碱流质粗面岩以及流纹岩组成 (Schermer 和 Busby, 1994)。与该地区中侏罗世深

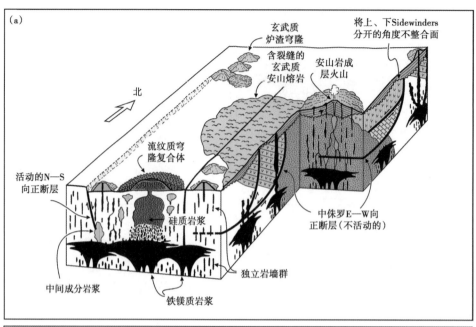

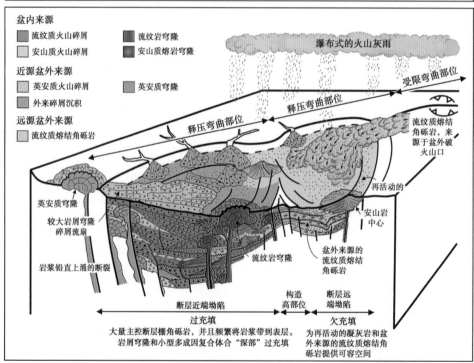

图 19-5 早期阶段低位张扭性大陆弧盆的构造和地层类型
(晚侏罗世加利福尼亚州和亚利桑那州的 Mojave 和 Sonora 沙漠)

(a) Mojave 沙漠中部的晚侏罗世 Sidewinder 火山岩系上部(据 Schermer 和 Busby, 1994),提供了一个有关板块边缘独立岩墙群(见图 19-1b 中的岩墙),以及与它同期侵入和喷发过程的观点。(b) 晚侏罗世亚利桑那州南部 Santa Rita Glance 砾岩为弧内走滑盆地的研究提供了一种观点(据 Busby 和 Bassett, 2007)。与较老、纯粹伸展性盆地(图 19-4)相比,弧内走滑盆地主要有与正断层走滑同时代的逆断层走滑、极其深的不整合、高比例的外来碎屑沉积(见正文)

成岩类似（Solomon 和 Taylor，1991），晚侏罗世深成岩的氧同位素也提供了大气水向深部渗透的证据，显示了该时期岛弧裂谷作用的发育。

巨型硅质大陆破火山口在这样的环境中（类似伸展岛弧）持续发育，但并不是广泛分布（表19-1），而是被限制在释压叠阶（releasing stepovers）部位的对称性盆地内。例如，Busby等（2005）就认为位于亚利桑那州南部的三个侏罗纪破火山口（Lipman 和 Hagstrum，1992）是沿着Sawmill Canyon左旋走滑断裂带发育的。之后这些破火山口沿着断裂带被切割改造，以至于无法识别其初始的形态（圆形或矩形）。

与释压叠阶部位形成的盆地和破火山口相比，释压弯曲（releasing bends）部位发育不对称盆地，并伴随有来自于复合断控火山口的频繁但规模小的火山喷发（图19-5b；表19-1；Busby 和 Bassett，2007）。目前对于沿着转换板块边界发育的释压弯曲盆地已有数十年的研究，其中包括Crowell（1974，1982，2003）对Ridge盆地的经典研究，该盆地形成于San Andreas断裂带活动早期阶段。相比之下，对岩浆岛弧区释压弯曲盆地特征的研究非常少。因此，本文总结了晚侏罗世张扭弧盆的构造、地层和火山活动特征，并将之与早—中侏罗世伸展岛弧进行了对比。而在进行对比之前，我们首先必须考虑由于北美板块大规模漂移导致的重大气候变化对沉积充填的影响。

晚侏罗世岛弧构造样式从伸展到左旋张扭的演变（至少部分上），是由于墨西哥湾的打开而引起的，后者导致了上部板块内发育左旋走滑断层活动，以及加利福尼亚州和亚利桑那州之下的斜向俯冲作用（图19-1b）。该过程伴随着北美板块快速从"马纬度（即副热带高压带）"向北漂移进入温带（Busby等，2005；图19-1b）。古地磁数据显示，北美西南部在大约185—160Ma之间从赤道纬度向北移动了约15°；之后甚至更快，大约从152—150Ma又向北漂移了13°（Busby等，2005）。弧后（位于现今美国西部的科罗拉多高原）和岛弧地区的沉积序列显示，到侏罗纪末，风成沉积已被河流沉积所取代。岛弧区火山作用类型也发生了根本的变化，从由熔结到超熔结的熔灰岩所反映的"干旱型喷发"转变为以岩浆通过火山口湖喷发为代表的"潮湿型喷发"，并发育射气岩浆喷发空落浮岩（phreatoplinian fall）和未熔结熔灰岩等（类似于现今的Taupo火山带；Busby等，2005）。因此，在晚侏罗世弧盆充填缺乏风成沉积（图19-5b），而发育"砂岩"沉积（Drewes，1971），后者实际上是由玻璃质碎片和自形晶体组成的被河流重塑沉积的凝灰岩。射气岩浆喷发凝灰岩分布十分广泛，这种极具特色的沉积由细纹层状或包卷纹理状白色至粉红色瓷状岩（porcellanite）组成。这些是由重大构造事件导致的气候变化对盆地充填影响的结果。在晚侏罗世张扭弧盆和早—中侏罗世伸展弧盆之间，构造控制作用存在很多差异（表19-1；Busby等，2005）。如图19-5b所示，张扭弧盆表现出不对称。盆地一侧是控盆边界走滑断层，另一侧则主要发育非区域性的小型断裂。详细的地层学分析表明，盆内断层性质随着时间在正滑断层和逆滑断层之间转换；部分倾滑断层与盆地其他地区的逆滑断层同时活动；这是走滑盆地的典型特征（完整的讨论见Bassett 和 Busby，2005）。由于盆地沉降主要发生在紧邻主断裂附近，因此盆地沉积充填向远离主走滑断裂带方向快速减薄。此外，因为被火山活动利用的断裂主要分布在走滑断层附近，所以与火山口相关的沉积（流纹熔岩丘和安山质熔岩锥）及其物源供给也沿着远离走滑断层方向减少。岩屑锥和冲积扇沉积（"外生碎屑沉积"；图19-5）局限发育在盆地靠近走滑断层的最深的一端，在离走滑断层长达2km的范围内包含有最大可达4m的块体或漂砾。位于走滑断裂之上的英安岩岩丘反复将块体—火山灰流输入盆地最深的一端。来自于走滑断层的、高的外生碎屑沉积供给，再加上大量的盆内火山活动，使得盆地毗邻走滑断层一侧处于"过补偿"状态；而盆地另一端（远离走滑断层一侧）的"欠补偿"沉积则为河流重塑的凝灰岩沉积提供了可容空间，同时也发育来自于盆地外部释压叠阶区破火山口喷发的火山碎屑流的沉积。

从张扭弧盆沉积充填横断面（垂直于主断裂方向）可识别出两种类型的侵蚀不整合（图19-5b所示的锯齿状黑线）：（1）远离走滑断层一侧的对称型侵蚀不整合，埋深200~600m，古斜坡坡度20°~25°。这些不整合代表了在受限弯曲（restraining bends）部位发生隆升期间下切盆地的深部河谷。（2）毗邻走滑断层一侧的非对称型不整合，以极高的垂直落差（460~910m）和非常高的古斜坡坡度为特征（邻近走滑断层的一侧达到40°~71°；图19-5b）。这些不整合代表了盆地沿着走滑断层受限弯曲部位运动时盆地隆升（反转）形成的断层崖和古滑坡断崖（paleo-landslide scars）（Busby和Bassett，2007）。以上特征可以与新西兰近海的活动走滑盆地相类比，在该盆地的一侧遭受上冲作用并发育滑坡断崖（landslide scars），而在盆地的另一侧则发生下沉并含有滑坡沉积（Barnes等，2001）。相一致地，在沿着Sawmill Canyon断裂带分布的盆地的其他地区，滑坡块体也常见于盆地的晚侏罗世沉积充填体系内（Busby，等2005）。由于盆地充填在受限弯曲部位会遭受破坏，因此，其仅仅可能在张扭系统中得到部分保存（Bassett和Busby，2005）。

考虑到亚利桑那州南部和墨西哥北部的岛弧火山岩（至少部分的）被形成于海平面高位期的早白垩世Mural灰岩所覆盖，本文将晚侏罗世张扭弧盆归入到了"早期低位弧盆"范畴（表19-1）。在早白垩世Mural灰岩沉积时的海平面高位期，Chihuahua海槽与墨西哥湾可能是相连通的（图19-1b；Dickinson等，1986，1987，1989）。

19.4 早期低位大陆弧盆小结

早—中侏罗世的早期低位伸展大陆弧和晚侏罗世的张扭大陆弧的主要特征见图19-6和表19-1。

沿着同沉积正断层，伸展岛弧发生均匀快速而又连续的沉降作用（Busby-Spera，1988；Busby-Spera等，1990；Riggs和Busby-Spera，1990；Riggs等，1993；Haxel等，2005）。在北部，岛弧发育在减薄的陆壳和过渡壳之上，以浅海相和深海相沉积为主（图19-1a）；而在南部岛弧位于克拉通之上，以非海相沉积为主。在海域部分（图19-6a），熔结凝灰岩沉积于深水破火山口，同时发育由硅质火山碎屑再沉积形成的浊积扇。蒸汽喷发导致喷发产物发生破碎作用（fragmentation），产生了玻璃质碎屑熔岩（hyaloclastite lavas）和玻璃质凝灰岩（hyalotuffs）。在非海相区域，火山沉积以大量熔结凝灰岩为主（图19-6b），并伴随有沿断裂分支分布的小规模安山岩喷发中心（图19-4b）。源自克拉通的超成熟沉积物被搬运至岛弧断陷内，并以风成岩的形式沉积于非海相区域，以大陆架上的浊积砂岩沉积于海相区域。不管在海相还是非海相区域，硅质破火山口数量多且规模大，因此，尽管有来自于正断层的滑坡巨角砾岩沉积，但是由于断层崖被大量的火山碎屑所掩埋，外生碎屑沉积所占比例依然很小。

相比之下，早期张扭岛弧的释压弯曲盆地（图19-6c）经历了快速的抬升和沉降交替，或"海豚效应（porpoising）"。这是沿着走滑断裂带沉降的释压弯曲和隆升的受限弯曲交替出现的结果。与伸展盆地不同，该类型盆地内断层性质随着时间在正断层和逆断层之间转换；并且在盆地一个部位的逆断层和另一个部位的正断层同时活动。抬升事件产生了许多大

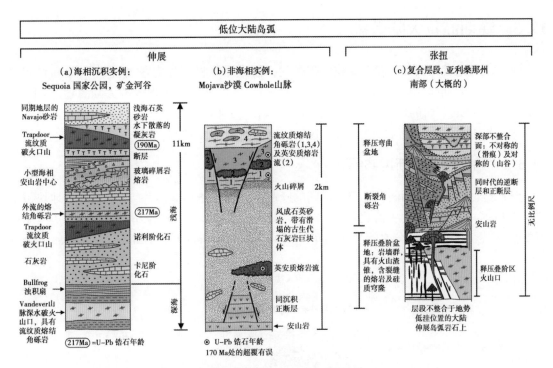

图 19-6 早期阶段地势低位伸展大陆弧的典型实测地层剖面（a 和 b），以及早期阶段地势低位张扭大陆弧的综合地层柱状图（c）

(a) 加利福尼亚 Sequoia 国家公园 Mineral King 位置的海相沉积（位于图 19-3a 的 MK 位置）。11km 厚的深海到浅海层段包括巨型大陆破火山口和溢流的熔结凝灰岩、较小的安山质层状火山堆积、浊积扇，以及很厚的浅海层序，表明快速的构造沉降，包含了石灰岩和风暴岩。除了层段顶部伴随石灰岩透镜体的石英粗砂碎屑岩没有受到影响，所有沉积物质都受到火山来源物质的再改造。这些也许是 Navajo 砂岩同时期的浅海相沉积，并表现为受洋流改造的风成砂岩（据 Busby-Spera，1983，1984a，1984b，1985，1986；Kokelaar 和 Busby，1992）。(b) Mojave 沙漠东部 Cowhole 山脉的非海相沉积（位于图 19-3a 中的 CM 位置；据 Busby 等，2002）。硅质熔岩流和熔结凝灰岩与大于 800m 厚的风成石英砂岩互层，记录了岛弧地堑洼地内克拉通来源的超成熟砂岩的堆集（图 19-3）。像柱状图 a 一样，可能由于一致的高沉降速率而不存在侵蚀不整合，虽然滑塌巨型块体从断层崖位置搬出，可能由于火山岩碎屑沉积物将断层崖掩埋，而不存在外来碎屑沉积物（据 Busby 等，2002）。(c) 低位大陆张扭弧盆地包括两种类型：（1）释压弯曲盆地，以"海豚效应（porpoising）"盆地为特征，具有同时代/交替的正断层走滑和逆断层走滑、众多小型单成因火山中心、深部不整合，以及大量的外来碎屑沉积（据 Busby 和 Bassett，2007）；（2）释压叠阶盆地，以大型大陆破火山口和岩墙群/广泛双峰式碱性成分的喷发复合体为主（据 Schermer 和 Busby，1994）

规模的以深古峡谷或大型滑坡断崖的形式出现的不整合，并造成巨大滑坡块体在沉降区沉积（图 19-6c；Bassett 和 Busby，2005；Busby 和 Bassett，2007）。沿受限弯曲发育的冲起构造之上的剥蚀作用，产生了大量的外生碎屑沉积物供给。释压弯曲盆地不发育硅质破火山口，而是沿着断裂发育复合的单成火山中心。

因为邻近没有冲起构造提供沉积物供给，所以早期张扭岛弧的释压叠阶盆地不发育逆断层构造和外生碎屑沉积（图 19-6b），取而代之的是，在部分地区以硅质破火山口沉积和熔结凝灰流溢流沉积为主，而在其他地区则以宽双峰岩墙群为主，并发育与之相关的熔岩流、熔岩丘和小规模的层状火山或火山渣锥（cinder cones）。

461

19.5 晚期高位大陆弧盆

本节将讨论美国西部的晚期高位大陆弧盆，按成因可将其分为两类：（1）始新世到中新世在板片后撤期间"内华达高原（Nevadaplano）"的伸展作用形成的大陆弧盆（图19-2a）；（2）12Ma 开始的内华达微板块从内华达高原西部边缘分离产生的张扭作用形成的大陆弧盆（Walker Lane 带，图19-2b）。这些盆地的特征见表19-1。

19.5.1 始新世至中新世伸展弧

晚白垩世至古新世期间，在整个广阔的隆起区（即内华达高原），白垩纪岛弧地块被削顶至岩基面，并伴随有"古河道"或"古切谷"达数百米的下切作用（如图19-7所示的不整合）。近南北向的古分水岭位于内华达高原（Nevadaplano）的顶部（图19-2a）；分水

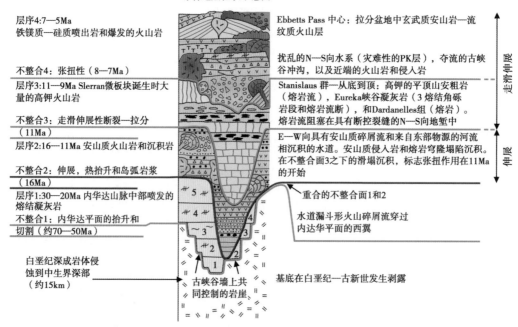

图19-7 晚期阶段地势高位伸展性和张扭性大陆弧盆的综合地层柱状图（示意图）
保存在内华达山脉中部、WalkerLane 带西部，以及加利福尼亚州的古峡谷和地堑中不整合围限的 Ancestral Cascades 岛弧层序（归纳自 Busby 等，2008a，2008b；Busby 和 Putirka，2009；Busby 等，2009a，2009b）。晚白垩世—古新世（不整合），古峡谷刻入一个类似阿尔蒂普拉诺的高原"内华达高原"的西翼，古峡谷的形成是对低角度俯冲的响应。这些古峡谷将熔结凝灰岩从熔结凝灰岩中部岛弧破火山口源区，搬运到加利福尼亚州中部（层序1）。由于在始新世至中新世，俯冲板片后退至更陡深度，伴随岛弧岩浆向西南方向波及的伸展和热抬升作用，形成地堑（图19-2a）。不整合面2记录了在16Ma岛弧前缘到加利福尼亚州东部—内华达州西部的信息，并且层序2以安山质火山岩为主。不整合面3和层序3记录了内华达山脉微板块从内华达高原西缘张扭性分离的开始（见正文）。不整合面4与岛弧抬升和伸展的第三阶段相对应，随后大型火山中心沿着断层发育（层序4，见正文）。进一步的讨论在正文中

岭西侧的古水道向西延伸直达加利福尼亚大峡谷（Great Valley），并被内华达高原顶部附近喷发的熔结凝灰岩所充填（Henry，2008；图19-7所示层序1）。在东部的盆地和山脉内，古河道遭受断层破坏作用并被埋藏于盆地之下；与之相比，古河道在内华达山脉（Sierra Nevada）则得到了较好的保存和出露。因此，内华达山脉内的古河道是研究内华达高原西侧的古地理及其新生代演化的最佳场所（Busby等，2008a，2008b；Busby和Putirka，2009；Hagan等，2009）。

如图19-7所示，长期持续沉积（30—20Ma）的熔结凝灰岩被形成于第一次岛弧火山活动之前（16—13Ma）的不整合面2所强烈切割（图19-7）。最新的稳定同位素研究表明，在岛弧横跨爱达荷州和内华达州向西南蔓延的同时，伴随了同时代的伸展作用和地表抬升，这被解释为是费拉隆（Farallon）板片后撤引起的热效应（Horton和Chamberlain，2006；Kent-Corson等，2006；Mulch等，2007）。Busby和Putirka（2009）综述了内华达山脉在岛弧火山活动开始时发育伸展作用的证据（图19-7）。因此，本文推断不整合面2记录了岛弧前锋向西推进至现今Walker lane构造带西部和内华达山脉东部时的隆升和伸展（图19-2）。

层序2由安山质岛弧火山岩组成，包括浅层侵入岩、块体—火山灰流凝灰岩、火山碎屑流沉积、河流相沉积和较少的熔岩流（图19-7）。火山中心通常较小并沿断层分布（Busby和Putirka，2009）。在内华达山脉的北部，层序2内含有Lovejoy玄武岩（图19-2a）。Lovejoy玄武岩是加利福尼亚州最大的溢流喷发（effusive eruptive）岩体（150km^3），它是沿着Walker Lane构造带最大的断裂带之一——Honey Lake断裂带的裂缝（Garrison等，2008）喷发的。Lovejoy玄武岩通过水道向内华达高原最西侧流动，并蔓延进入加利福尼亚的中央峡谷（Central Valley）（图19-2a）。虽然该裂缝喷发可能记录了内华达山脉前缘伸展作用的开始（Busby和Putirka，2009），但正如图19-7中所示，层序2中发育了大量的16-11Ma来自东部的河流相砂岩（层序2；图19-7），表明伸展作用强度还不足以破坏东西向的古河道/古切谷系统。与此相反，下面将要讨论的张扭作用则破坏了东西向的古水系（层序3和层序4，图19-7）。

19.5.2 12Ma以来的张扭弧

内华达山脉中部的不整合面3和层序3（图19-7）记录了内华达山脉微板块与内华达高原的西部边缘在古Cascades弧热减薄的轴部发生张扭式分离的开始（图19-2b；Busby和Putirka，2009）。在约12-11Ma，张扭作用导致了断层附近的地层发生同火山期掀斜，形成角度不整合；并抬升形成强烈的侵蚀不整合面（Busby等，2008a），同时发育大型滑坡，单个滑坡块体长度甚至可达1.6km（Busby等，2008a），紧随其后的是大规模的高钾火山作用（层序3；图19-7）。这些高钾火山岩喷发于古Cascades弧内，包括作为标准产地的"安粗岩"熔岩流（加利福尼亚Table Mountain粗安岩；Ransome，1898），以及独特且广泛分布的粗面英安岩熔结凝灰岩（Eureka峡谷凝灰岩）。张扭作用使得深部富含K$_2$O的低熔岩浆被向上输送并喷发，记录了内华达微板块的形成（Putirka和Busby，2007）。

以下是标志内华达微板块形成的构造作用或产物（Busby和Putirka，2009，2010a，b；Busby等，已录用）：

（1）沿断控裂缝发生强烈的溢流喷发（effusive eruptions），包括中间组分的"喷溢岩流（flood lava）"的裂缝喷发。

这些"喷溢安山岩（flood andesites）"是从8~12km长的裂缝中喷发的，后者位于沿着现今内华达山脉顶部和 Sonora Pass 山脉前缘分布的火山构造地堑内（图 19-8）。"喷溢安山岩"裂缝火山口由 100~200m 厚的块状（未成层状）环形火山渣沉积组成，厚达 5m 的棱角状致密块体分散在由未分选的火山渣砾、火山弹和火山灰组成的红色基质中。Table Mountain 安粗岩（图 19-7）大量进入地堑内沉积，厚度达 300~400m；但也有部分溢出地堑进入古河道内，沉积厚度一般小于 50m（图 19-8）。古河道漏斗流（funneled flows）向西穿过未发育断裂变形的内华达块体。一个规模达 20km³（笔者所见过的体积最大的中间组分熔岩流）的熔岩个体向西至少流动了 130km（Gorny 等，2009；Pluhar 等，2009），这对安山岩来说是一个极其罕见的远距离流动。规模大于 200km³ 的 Table Mountain 安粗岩熔岩流是约在 10.4Ma 和 10.2Ma 之间短短 0.028~0.23Ma 爆发喷发的（Busby 等，2008a；Hagan，2010；Busby 和 Putirka，2010a，2010b）。虽然自 Table Mountain 安粗岩喷发以来，图 19-8 所示的大部分断层重新活动，但至少有一半的位移量是在 Table Mountain 安粗岩喷发前或喷发中形成的（Hagan，2010；Koerne，2010；Busby 等，已录用）。

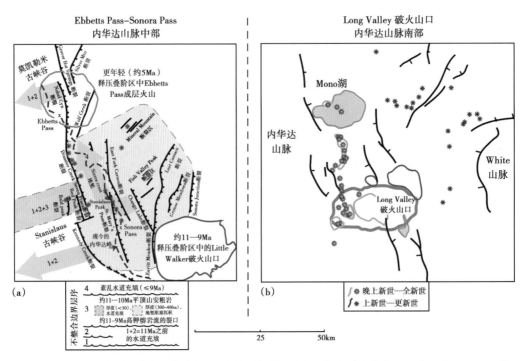

图 19-8　沿着内华达山脉微板块东部边界张扭性张扭叠阶区的大型火山中心，表现为 11—9Ma 的 LittleWalker 破火山口（图 a 基于 Busby 等的制图，出版中）与上新世—现今的 Long Valley 破火山口（图 b 据 Bursik，2009）相类似；位置见图 19-2b

在这两个区域，NNW 向正断层具有右旋运动的不同分量，并且 NE 向断裂具有左旋运动分量，熔岩流裂口沿着这些断层分布，并且破火山口由于伸展作用在东—西方向上伸长。在中新世弧实例中（a），极端溢流喷发沿着断层控制的裂缝发生，包括了"喷溢岩流"中间组分的裂缝喷发。这些大量聚集在同火山地堑内的裂口附近。内华达高原原始、东西向长水系受南北向地堑的影响逐渐被扰乱，这些都记录在地层中（层序 1—4，图 19-7 所描述）。现今 Sierran 顶部位置的东边现代水系为南北向（图 a 中显示）。EbbettsPass 层状火山最新发现更年轻的岛弧火山中心在张扭叠阶区显示（层序 4；图 19-7；据 Busby 和 Putirka，2009；Hagan，2010；见正文）

（2）在释压张扭叠阶断层（releasing transtensional stepover faults）最大位移处发育大规模的火山中心。

Little Walker 破火山口发育在 Table Mountain 安粗岩火山构造熔岩复合体（volcanotectonic lava complex）的最大伸展部位（图 19-8）（Busby 等，2008a；Hagan，2010）。破火山口在约 9.5-9.4Ma 喷发形成了 3 个大体积的熔结凝灰岩岩床（即 Eureka 峡谷凝灰岩；图 19-7）。在 Eureka 峡谷凝灰岩从破火山口喷发期间，粗安质到玄武质熔岩（trachyandesitic to basaltic lavas）持续从沿着断层分布的火山渣锥（cinder cones）中喷发，并向西北方向流出破火山口；与此同时，粗面英安质熔岩（trachydacitic lavas）也通过断层喷发，且在组分上难以与 Eureka 峡谷凝灰岩相区分（Koerner 和 Busby，2009；Hagan，2010）。沿着 Mineral Mountain 断裂带发育了一个由火山碎屑涌浪堆积（surge deposit）的凝灰岩环状建造组成的粗面英安岩（trachydacite）火山口（图 19-8）；而在其附近的一个火山口则由一个向上侵入到了熔岩丘的侵入体组成。粗面英安岩甚至可从 Little Walker 破火山口西侧达 15km 处的 Bald Peak-Red Peak 断裂带喷发（图 19-8）。构造和火山活动的这种关系与附近的第四纪 Long Valley 破火山口十分相似。在后者，Inyo-Mono 火山链沿着断裂从破火山口向北延伸（图 19-8）；与 11—9Ma 的 Little Walker 破火山口一样，尽管这些地区已不处于俯冲板片之上（由于三联点向北迁移；图 19-2a），大陆岩石圈沿着释压弯曲部位发生破裂（Bursik，2009）。

（3）南北向地堑的发育扰乱了东西向古水系，并成为熔岩和河道沉积的通道。

向西流过内华达山脉中部的古水道很好地反映了渐新世和 12Ma 之前的中新世的沉积（层序 1 和 2；图 19-8）。然而，由于南北向地堑的发育，层序 3 内只有很少的熔岩流流入这些古河道（图 19-8）。虽然与熔岩流相比，火山碎屑流具有更好的流动性，在东西向古河道沉积充填中也具有比熔岩流更好的体现，但整体上讲，相对于东西向古河道充填，二者在南北向地堑内的沉积更厚、分布更广泛。而层序 4 则完全局限分布在随南北向地堑而形成的南北向古河道内（图 19-8）。例如，与 Sonora Pass 和 Ebbetts Pass 地区包含层序 1 和层序 2 的古河道相比，Sierra Crest 地堑的 Disaster Creek 断裂部分缺失比 Stanislaus 群（层序 3）更老的地层单元（图 19-8）。这是因为该部分地堑位于约 11Ma 形成的两个古切谷之间。Sierra Crest 地堑 Disaster Creek 段内层序 3 的高钾火山岩遭受深下切（发育不整合面 4；图 19-7），并被层序 4 内河流相沉积（Disaster Peak 组，图 19-7）所覆盖。后者沉积于宽大于 3km、深 400m、古水流方向为由南向北的古河道内。与之相反，由于夺流作用，Sierra Crest 地堑西部的 Stanislaus 古峡谷明显缺失层序 4 地层（图 19-8）。现今 Little Walker 河流水系沿着图 19-8 所示区域的东部边界由南向北流动；东 Fork Carson 河流的源头位于沿着东 Fork Carson 断裂带发育的 SierraCrest 地堑内（图 19-8）。

从新生板块边界的 Sonora Pass 地区喷发而来的火山岩覆盖区域，通常比受其他任何沿着板块边界的岛弧火山中心影响的区域要大。然而，我们却在张扭叠阶部位（transtensional stepovers）识别出了另外两个大型岛弧火山中心：约 5Ma 的 Ebbetts Pass 层积型火山中心（Stratovolcano）和现今的 Lassen 火山中心（图 19-2b）。

Ebbetts Pass 火山中心发育在释压正断层叠阶（releasing normal fault stepover）部位，是一个锥形/层状火山复合体（Busby 和 Putirka，2009；Hagan，2010）。径向下倾的底部玄武质安山岩—安山岩熔岩流和火山渣空落沉积（scoria fall deposits）形成了一个直径 10km 的早期火山中心；之后火山中心核部遭受硅质岩栓（plugs）侵入，并随着硅质块状—火山灰流凝灰岩的喷发，生长为直径至少达 18km 的火山建造。同时火山中心还发育流纹质熔结凝灰岩、熔岩流、侵入体和两类辉石英安岩。Ebbetts Pass 火山中心明显占据了一个深的拉分盆地，因为其下部缺失厚约 500m 的层序 3 地层（Stanislaus 群）（Busby 等，2009b）。

Lassen 火山中心（图 19-2b）则代表了古 Cacscades 岛弧内 Ebbetts Pass 火山中心的一个现代实例；后者是一个位于沿着 Walker Lane 构造带东北边缘发育的正断层和张扭断层之上的岛弧火山（Muffle 等，2008；Janik 和 McLaren，2010）。在 Lassen 火山中心，一个"显著的不整合面"将 3.5Ma 以来的火山岩与下伏前新生代地层分开，这可能表明，"自从 3.5Ma 以来，Walker Lane 构造带北部与 Cascade 俯冲带的相互作用逐渐增强，从而形成了一个利于发育大型火山中心的张扭构造背景"（Muffler 等，2008）。

19.6 晚期高位和早期低位大陆弧盆对比

前文已经详细描述了四种类型大陆弧盆的基本特征，分别是：（1）早期低位伸展型，（2）早期低位张扭型；（3）晚期高位伸展型；（4）晚期高位张扭型。下面笔者将对比分析这几种类型盆地的异同。

晚期高位大陆弧盆和早期低位大陆弧盆存在三个方面的显著差异（表 19-1）：首先，晚期高位大陆弧盆不发育海相沉积。这是因为晚期盆地位于增厚地壳之上，常需要通过数百万年甚至数十个百万年的伸展作用才得以减薄。其次，晚期盆地不发育来自克拉通的沉积。这是因为其遭受抬升较大陆其他地区高。第三，晚期盆地形成于遭受强烈剥蚀的基底之上。与早期盆地明显不同，晚期盆地位于遭受剥露的基底地层之上，且喷发产物沿着在地壳缩短时期形成的下切谷流动（表 19-1）。

早期低位伸展大陆弧盆与其他三种类型大陆弧盆的差异也体现在三个方面：（1）与其他三种类型盆地相比，早期伸展大陆弧盆具有更高的沉降速率，从而保存了更厚的沉积充填；（2）在早期低位伸展大陆弧盆，由于地堑发生均一持续且快速的沉降作用，因此其沉积充填内缺乏侵蚀不整合；（3）由于断层崖和地垒被火山碎屑所掩埋，早期低位伸展大陆弧盆内外生碎屑沉积非常稀少（表 19-1）。这表明早期低位伸展大陆弧阶段岩石圈快速拉伸引起的沉降超过了热作用引起的表层抬升，因此下盘块体发生下沉并被火山碎屑掩埋，而不是经历抬升侵蚀。另外，尽管在基底中可能记录了在更早的造山事件中基岩遭受抬升，但由于火山活动并不是在抬升作用之后发生的，因此早期低位伸展大陆弧盆通常以表壳岩（supracrustal rock）为底。

与早期伸展大陆弧盆相似，晚期伸展大陆弧盆也发育"超级火山"硅质破火山口区（见表 19-1）。然而，在晚期伸展大陆弧阶段，破火山口很大程度上局限在地壳增厚区（内华达高原的顶部；图 19-2a）；而在早期伸展大陆弧阶段，破火山可以发育在岛弧的任何地区（图 19-3a）。

早期和晚期张扭盆地存在三个方面的相似性：（1）均在释压叠阶区发育破火山口；（2）在盆地毗邻挤出构造（pop-ups）部位发育大量的滑坡沉积；以及（3）沉积填充中包含大量的侵蚀不整合（见表 19-1）。与晚期张扭盆地相比，早期张扭盆地明显更靠近释压弯曲部位发育，因为这些盆地发育了更多的外生碎屑沉积。然而这并不是它们之间的本质差别。早期张扭盆地和晚期张扭盆地最大的不同在于，晚期盆地缺乏海相地层，并且它们是形成于被河道/切谷强烈切割地貌之上的，而这些古河道/切谷又为岛弧地层的沉积提供了额外的可容纳空间（作为对断层作用的补充）。

19.7 结论

本文首次尝试确定了在俯冲早期阶段形成的伸展和张扭大陆弧盆的主要特征，并将它们与在俯冲晚期形成的盆地进行对比；同时还列举了美国西部一些不同类型大陆弧盆的盆地结构和沉积充填特征。然而，与其他类型盆地相比，由于其研究的困难性，我们对大陆弧盆的认识还很有限。对大陆弧盆开展研究，必须精通沉积学、火山地质学，以及构造地质学和火成岩岩石学。火山岩地层以复杂出名，原生地层横向变化迅速，且通常叠加了后期改造和变质作用，并被侵入体所破坏。由于非海相盆地占主导地位，所以可用来确定地层年龄的化石很少，后期的改造和变质作用也妨碍了同位素年代学研究。尽管如此，对该类型盆地的研究仍然为分析区域构造演化提供了重要的参考依据和约束条件。虽然第四纪岛弧火山岩经历了较少的改造和变形作用，但它们并不能像本章所描述的中生代到新生代岛弧一样，为区域构造演化提供了完整时间序列和深部构造的视角。

与地球化学和地球物理相比，对大陆弧的地质方面的研究一直被忽视（Hildreth，2007）。虽然对第四纪 Cascades 岛弧（目前对大陆岛弧研究最深入的地区）的研究表明，其发育了超过 2300 个火山（其中层积型火山少于 30 个）（Hildreth，2007），但目前关于一个岩浆岛弧是由"一到两条均匀分布的层积型火山形成的火山链"构成的认识依然存在。保存在显生宙地质记录中的大陆岛弧岩石的多样性也反映了第四纪火山链的复杂性。

致谢（略）

参 考 文 献

Atwater, T., and Stock, J. (1998) Pacific-North America plate tectonics of the Neogene Southwestern United States: an update. International Geological Review, 40, 375-402.

Barnes, P. M., Sutherland, R., Davy, B., and Delteil, J. (2001) Rapid creation and destruction of sedimentary basins on mature strike-slip faults: an example from the offshore Alpine Fault. New Zealand. Journal of Structural Geology, 23, 1727-1739. doi: 10.1016/S0191-8141(01)00044-X.

Barton, M. D., and Johnson, D. A. (1996) An evaporitic source model for igneous-related Fe-oxide (-REE-Cu-Au-U) mineralization. Geology, 24, 259-262.

Bellier, O., and Sebrier, M. (1994) Relationship between tectonism and volcanism along the Great Sumatran fault zone deduced by SPOT image analyses. Tectonophysics, 233, 215-231.

Burchfiel, B. C., and Davis, G. A. (1972) Structural framework and evolution of the southern part of the Cordilleran orogen, western United States. American Journal of Science, 272, 97-118.

Burkart, B., and Self, S. (1985) Extension and rotation of crustal blocks in northern Central America and effect on the volcanic arc. Geology, 13, 22-26.

Burke, K. (1988) Tectonic evolution of the Caribbean. Annual Reviews of Earth and Planetary Sciences, 16, 201-230.

Bursik, M. (2009) A general model for tectonic control of magmatism: examples from Long Valley

Caldera (USA) and El Chichon (Mexico). Geofisica Internacional, 48 (1), 171-183.

Busby-Spera, C. J. (1983) Paleogeographic reconstruction of a submarine volcanic center: geochronology, volcanology and sedimentology of the Mineral King Roof Pendant, Sierra Nevada, California. PhD dissertation, PrincetonUniversity, Princeton, NJ, 290 p.

Busby-Spera, C. J. (1984a) Large-volume rhyolite ash-flow eruptions and submarine caldera collapse in the lower Mesozoic Sierra Nevada, California. Jour. Geophys. Res., 89 (B10) 8417-8427.

Busby-Spera, C. J. (1984b) The lower Mesozoic continental margin and marine intra-arc sedimentation at Mineral King, California, in Bachman, S., and Crouch, J., eds., Tectonics and sedimentation along the California margin. Los Angeles, Pacific Section, Society of Economic Paleontologists and Mineralogists, 135-156.

Busby-Spera, C. J. (1985) A sand-rich submarine fan in the lower Mesozoic Mineral King caldera complex, Sierra Nevada, California. Jour. Sed. Petrology, 55, 376-391.

Busby-Spera, C. J. (1986) Depositional features of rhyolitic and andesitic volcaniclastic rocks of the Mineral King submarine caldera complex, Sierra Nevada, California. Jour. Volcanology and Geothermal Res., 27, 43-76.

Busby-Spera, C. J. (1988) Speculative tectonic model for the Early Mesozoic arc of the southwest Cordilleran United States. Geology, 16, 1121-1125.

Busby-Spera, C. J., Mattinson, J. M., Riggs, N. R., and Schermer, E. R. (1990) The Triassic-Jurassic magmatic arc in the Mojave-Sonoran deserts and the Sierran-Klamath region; similarities and differences in paleogeographic evolution, in Harwood, D., and Miller, M. M., eds., Paleozoic and Early Mesozoic paleogeographic relations: Sierra Nevada, Klamath Mountains, and related terranes. Special Paper, 255, Boulder, CO, Geological Society of America, 325-338.

Busby, C. J., Schermer, E. R., and Mattinson, J. M. (2002) Extensional arc setting and ages of Middle Jurassic eolianites, Cowhole Mountains (Mojave Desert, CA), in Glazner, A., Walker, J. D., and Bartley, J. M., eds., Geologic Evolution of the Mojave desert and Southwestern Basin and Range. GSA Memoir, 195. Boulder, CO, Geological Society of America, 79-92.

Busby, C. J. (2004) Continental growth at convergent margins facing large ocean basins: a case study from Mesozoic Baja California, Mexico. Tectonophysics, 392, 241-277.

Busby, C. J., Bassett, K, Steiner, M. B., and Riggs, N. R. (2005) Climatic and tectonic controls on Jurassic intra-arc basins related to northward drift of North America, in Anderson, T. H., Nourse, J. A., McKee, J. W., and Steiner, M. B., eds., The Mojave-Sonora megashear hypothesis: development, assessment, and alternatives. Geological Society of America Special Paper, 393. Boulder, CO, Geological Society of America, 359-376.

Busby, C. J., and Bassett, K. (2007) Volcanic facies architecture of an intra-arc strike-slip basin, SantaRitaMountains, Arizona. Bulletin of Volcanology, 70 (1), 85-103.

Busby, C. J., Hagan, J., Putirka, K., Pluhar, C., Gans, P., Rood, D., DeOreo, S., Skilling, I., and Wagner, D. (2008a) The ancestral Cascades arc: implications for the development of the Sierran microplate and tectonic significance of high K_2O volcanism, in J. Wright and J. Shervais, eds., Ophiolites, arcs and batholiths. Special Paper, 438. Boulder, CO, Geological Society of America, 331–378.

Busby, C., DeOreo, S., Skilling, I., Gans, P., and Hagan, J. (2008b) Carson Pass–Kirkwood paleocanyon system: implications for the Tertiary evolution of the Sierra Nevada, California. Geological Society of America Bulletin, 120 (3–4), 274–299.

Busby, C. J., and Putirka, K. (2009) Miocene evolution of the western edge of the Nevadaplano in the central and northern Sierra Nevada: Paleocanyons, magmatism and structure, in, Ernst, G., ed., The rise and fall of the Nevada Plano. International Geology Reviews, 51 (7–8), 670–701.

Busby, C., Koerner, A., Hagan, J., Putirka, K., and Pluhar, C. (2009a) Volcanism due to transtension at the birth of the Sierra Nevada microplate: similarities to ongoing transtension at nearby LongValley. Boulder, CO, Geological Society of America, abstr.

Busby, C., Putirka, K., Hagan, J., Koerner, A., and Melosh, B. (2009b) Controls of extension on Miocene arc magmatism in the central Sierra Nevada (CA). American Geophysical Union fall meeting, abstr.

Busby, C., and Putirka, K. (2010a) Birth of a plate boundary: transtensional tectonics and magmatism, Sierra Nevada microplate and Gulf of California rift. Tectonic Crossroads: Geological Society of America Global Geoscience Meeting, Ankara, Turkey, October, abstr.

Busby, C., and Putirka, K. (2010b) Geologic signals of the initiation of continental rifting within the ancestral Cascades arc. Geological Society of America Annual Meeting, Boulder, CO, October, abstr. Busby, C., Koerner, A., and Hagan, J. (accepted) The Miocene Ancestral Cascades arc at SonoraPass, Sierra Nevada (California): lithostratigraphy, paleogeography and structure. Geosphere, special issue.

Cembrano, J., Herve, F., and Lavenu, A. (1996) The Liquine Ofqui fault zone: a long-lived intra-arc fault system in southern Chile. Tectonophysics, 259, 55–66.

Cole, J. W. (1984) Taupo–Rotorua depression: an ensialic marginal basin of the North Island, New Zealand, in Kokellar, B., and Howells, M., eds., Marginal basin geology. Special Publication, 16. Geological Society of London, 109–120.

Coney, P. J., and Reynolds, S. J. (1977) Cordilleran Benioff zones. Nature, 270, 403–406.

Cousens, B., Prytulak, J., Henry, C., Alcazar, A., and Brownrigg, T. (2008) Geology, geochronology, and geochemistry of the Mio–Pliocene Ancestral Cascades arc, northern Sierra Nevada, California and Nevada: the roles of the upper mantle subducting slab, and the Sierra Nevada lithosphere. Geosphere, 4, 829–853.

Crowell, J. C. (1974) Origin of late Cenozoic basins in southern California, in Tectonics and sedimentation. SEPM Special Publication, 826. Tulsa, OK, Society of Economic Paleontologists and Mineralogists, 190–204.

Crowell, J. C. (1982) The tectonics of Ridge basin, southern California, in Crowell, J. C., and Link, M. H., eds., Geologic history of Ridge basin, southern California. SEPM Series book 22 Pacific Section, 25–41.

Crowell, J. C. (2003) Introduction to geology of RidgeBasin, Southern California. Special Paper, 367. Boulder, CO, Geological Society of America, 1-15.

DeCelles, G. (2004) Late Jurassic to Eocene evolution of the Cordilleran thrust belt and foreland basin system, western USA. American Journal of Science, 304, 105-168.

DeCelles, P. G., Ducea, M. N., Kapp, P. A., and Zandt, G. (2009) Cyclicity in Cordilleran orogenic systems. Nature Geoscience, Advance Online Publication, 1 - 7. doi: 10. 1038/NGEO469.

de Silva, S. L., and Gosnold, W. A. (2007) Episodic construction of batholiths: insights from the spatiotemporal development of an ignimbrite flareup. Journal of Volcanology and Geothermal Research, 167, 320-355. doi: 10. 1016/j. volgeores. 2007. 07. 015.

Dewey, J. F. (1980) Episodicity, sequence and style at convergent plate boundaries, in Strangway, D. W., ed., The continental crust and its mineral deposits. Geological Association of Canada Special Paper 20, 553-573.

Dickinson, W. R. (1997) Tectonic Implications of Cenozoic volcanism in coastal California. Geological Society of America Bulletin, 109, 936-934.

Dickinson, W. R. (2006) Geotectonic evolution of the Great Basin. Geosphere, 2, 353-368. doi: 10. 1130/GES00054. 1.

Dickinson, W. R., and Lawton, T. F. (2001) Tectonic setting and sandstone petrofacies of the Bisbee basin (USAMexico). Journal of South American Earth Sciences, 14, 475-504. doi: 10. 1016/S0895-9811 (01) 00046-3.

Dickinson, W. R., Klute, M. A., and Swift, P. N. (1986) The Bisbee Basin and its bearing on Late Mesozoic paleogeographic and paleotectonic relations between the Cordilleran and Caribbean regions, in Abbott, P. L., ed., Cretaceous Stratigraphy of Western North America. Society of Economic Paleontologists and Mineralogists, Field-Trip Guidebook Pacific Section, 51-62.

Dickinson, W. R., Klute, M. A., and Bilodeau, W. L. (1987) Tectonic setting and sedimentological features of upper Mesozoic strata in southeastern Arizona, in Davis, G. H., and NandenDolder, E. M., eds., Geologic Diversity of Arizona and Its Margins: Excursions to Choice Areas. Arizona Geological Survey Special Paper 5, 266-279.

Dickinson, W. R., Fiorillo, A. R., Hall, D. L., Monreal, R., Potochnik, A. R., and Swift, P. N. (1989) Cretaceous strata of southern Arizona, in Jenny, J. P., and Reynolds, S. J., eds., Geologic Evolution of Arizona. Arizona Geological Society Digest, 17, 397-434.

Drewes, H. (1971) Geologic map of the Mount Wrightson Quadrangle, southeast of Tucson, Santa Cruz and Pima Counties, Arizona. US Geol Survey Map I-614. Erlich, E. N. (1979) Recent structure of Kamchatka and position of Quaternary volcanoes. Bulletin Volcaologique, 42, 13-42.

Fitch, T. J. (1972) Plate convergence, transcurrent faults, and internal deformation adjacent to Southeast Asias and the western Pacific. Journal of Geophysical Research, 77, 4432-4460.

Garrison, N., Busby, C. J., Gans, P. B., Putirka, K. and Wagner, D. L. (2008) A mantle plume beneath California? The mid-Miocene Lovejoy flood basalt, northern California, in

Wright, J., and Shervais, J., eds., Ophiolites, Arcs and Batholiths. Geological Society of America Special Paper 438, 551–572.

Gioncada, A., Mazzuoli, R., Bisson, M., Pareschi, M. T. (2003) Petrology of volcanic products younger than 42 ka on the Lipari–Vulcano complex (Aeolian Islands, Italy): an example of volcanism controlled by tectonics. J. Volcanol Geotherm Res, 122, 191–220.

Gorny, C., Busby, C., Pluhar, C., Hagan, J., and Putirka, K. (2009) An in–depth look at distal Sierran paleochannel fill: drill cores through the Table Mountain Latite near Knights Ferry, in Ernst, G., ed., The rise and fall of the Nevadaplano, International Geology Reviews, 51 (9–11), 824–842.

Hagan, J. C. (2010) Volcanology, structure, and stratigraphy of the central Sierra Nevada range fron (California) CarsonPass to SonoraPass. PhD dissertation, University of California, Santa Barbara, 257 p.

Hagan, J., Busby, C., Renne, P., and Putirka, K. (2009) Cenozoic evolution of paleochannels, Ancestral Cascades magmas, and range–front faults in the Kirkwood–Carson Pass–Markleeville region (Central Sierra Nevada), in Ernst, G., ed., The rise and fall of the Nevadaplano. International Geology Reviews (9–11), 777–823.

Hamilton, W. (1969) Mesozoic California and the underflow of Pacific mantle. Geological Society of America Bulletin, 80, 2409–2430.

Haxel, G. B., Wright, J. E., Riggs, N. R. Tosday, R. M., and May, D. J. (2005) Middle Jurassic Topawa Group, Baboquivari Mountains, south–central Arizona: volcanic and sedimentary record of deep basins within the Jurassic magmatic arc, in Anderson, T. H., Nourse, J. A., McKee, J. W., and Steiner, M. B., eds., The Mojave–Sonora megashear hypothesis: development, assessment, and alternatives. Geological Society of America Special Paper 393, 329–358.

Henry, C. D. (2008) Ash–flow tuffs and paleovalleys in northeastern Nevada: implications for Eocene paleogeography and extension in the Sevier hinterland, northern Great Basin. Geosphere, 4, 1–35.

Hildreth, W. (2007) Quaternary magmatism in the Cascades: geologic perspectives. USGS Professional Paper 1744, 125 p.

Horton, T. W., and Chamberlain, C. P. (2006) Stable isotopic evidence for Neogene surface downdrop in the central Basin and RangeProvince. GSA Bulletin, 118, 475–490.

Houghton, B. F., Hayward, B. W., Cole, J. W., Hobden, B., and Johnston D. M. (1991) Inventory of Quaternary volcanic centres and features of the Taupo Volcanic Zone (with additional entries for MayorIsland and the KermadecIslands). Geological Society of New Zealand Miscellaneous Publication 55, 156 p.

Humphreys, E. D. (2008) Cenozoic slab windows beneath the western United States. Arizona Geological Society Digest 22, 389–396.

Humphreys, E. (2009) Relation of flat slab subduction to magmatism and deformation in the western United States, in Kay, S. M., Ramos, V. A., and Dickinson, W. R., eds., Backbone of the Americas: shallow subduction, plateau uplift, and ridge and terrane collision.

Geological Society of America Memoir 204. Boulder, CO, Geological Society of America. doi: 10. 1130/2009. 1204 (04).

Janik, C. J., and McLaren, M. K. (2010) Seismicity and fluid geochemistry at Lassen Volcanic Naitonal Park, California: evidence for two circulation cells in the hydrothermal system. Journal of Volcanology and Geothermal Research, 189, 257–277.

Jarrard, R. D. (1986) Relations among subduction parameters. Reviews of Geophysics, 24, 184–217.

Kent-Corson, M. L., Sherman, L. S., Mulch, A., and Chamberlain, C. P. (2006) Cenozoic topographic and climatic response to changing tectonic boundary conditions in Western North America. Earth and Planetary Science Letters, 252, 453–466.

Koerner, A., and Busby, C. (2009) New evidence for alternating effusive and explosive eruptions from the type section of the Stanislaus Group in the "cataract" paloecanyon, central Sierra Nevada (CA), in G. Ernst, ed., The rise and fall of the Nevada Plano, International Geology Reviews (9–11), 962–985.

Koerner, Alice A. (2010) Cenozoic evolution of the SonoraPass to Dardanelles region, Sierra Nevada, California: Paleochannels, volcanism and faulting: unpublishedMS Thesis, University of California at Santa Barbara, 145 pp.

Kokelaar, B. P., and Busby, C. J. (1992) Subaqueous explosive eruption and welding of pyroclastic deposits. Science, 257, 196–201.

Lipman, P. W., and Hagstrum, J. T. (1992) Jurassic ash flow sheets, calderas, and related intrusions of the Cordilleran volcanic arc in southeastern Arizona: implications for regional tectonics and ore deposits. Geological Society of America Bulletin, 104, 32–39.

Maze, W. B. (1984) Jurassic La Quinta Formation in the Sierra de Perija, northwest Venezuela: geology and tectonic movement of red beds and volcanic rocks, in Bonini, W. E., Hargraves, R. B., and Shagam, R., eds., The Caribbean South American Plate Boundary and Regional Tectonics. Geological Society of America Memoir 162, 263–282.

McCaffrey, R. (1992) Oblique plate convergence, slip vectors, and forearc deformation. Journal of Geophysical Research, 97 (B6) 8905–8915.

McQuarrie, N., and Wernicke, B. P. (2005) An animated tectonic reconstruction of southwestern North America since 36 Ma. Geosphere, 1 (3), 147–172. doi: 10. 1130/ GES00016. 1.

Molnar, Peter and T. Atwater (1978) Interarc spreading and Cordilleran tectonics as alternates related to the age of subducted oceanic lithosphere. Earth and Planet. Sci. Lett., 41, 330–340.

Muffler, L., Blakely, R., and Clynne, M. (2008) Interaction of the Walker Lane and the Cascade volcanic arc, northern California. Eos Trans. American Geophysical Union, 89, no. 53, Abstr.

Mulch, A., Teyssier, C., Cosca, M. A., and Chamberlain, C. P. (2007) Stable isotope paleoaltimetry of Eocene core complexes in the North American Cordillera. Tectonics, 26. doi: 10. 1029/2006TC001995.

Pluhar, C., Deino, A., King, N., Busby, C., Hausback, B., Wright, T., and Fischer, C. (2009) Lithostratigraphy, magnetostratigraphy, geochronology and radiometric dating of the Stanislaus Group, CA, and age of the Little Walker Caldera, in G. Ernst, ed., The rise and fall of the Nevadaplano, International Geology Reviews (9-11), 873-899.

Riggs, N. R., and Busby-Spera, C. J. (1990) Evolution of a multi-vent volcanic complex within a subsiding arc graben depression: Mount Wrightson Formation, southern Arizona. Geological Society of America Bulletin, 102 (8), 1114-1135.

Riggs, N. R., and Busby-Spera, C. J. (1991) Facies analysis of an ancient large caldera complex and implications for intra-arc subsidence: middle Jurassic Cobre Ridge Group, Southern Arizona, USA. Sedimentary Geology, Special Issue on Volcanogenic Sedimentation, ed. R. Cas and C. Busby-Spera, 74 (1-4), 39-70.

Riggs, N. R., Mattinson, J. M., and Busby, C. J. (1993) Correlation of Mesozoic eolian strata between the magmatic and the Colorado Plateau: new U-Pb geochronologic data from southern Arizona. Geological Society of America Bulletin, 105 (9), 1231-1246.

Ryan, H., Coleman, P. J. (1992). Composite transform-convergent plate boundaries: description and discussion. Mar Petrol Geol 9, 89-97.

Saleeby, J. B., and Busby-Spera, C. J. (1992) Early Mesozoic evolution of the western US Cordillera, In Burchfiel, B. C., Lipman, W., and Zoback, M. L., eds., The Cordilleran Orogen: coterminus United States. Decade of North American Geology, Geological Society of America, G-3, 107-168.

Sarewitz, D. R., and Lewis, S. D. (1991) The Marinduque intra-arc basin, Philippines: basin genesis and in situ ophiolite development in a strike-slip setting. Geol Soc Am Bull, 103, 597-614.

Schermer, E. R. and Busby, C. J. (1994) Jurassic magmatism in the central Mojave Desert: implications for arc paleogeography and preservation of continental volcanic sequences. Geological Society of America Bulletin, 106, 767-790.

Silver, L. T., and Anderson, T. H. (1974) Possible left-lateral early to middle Mesozoic disruption of the southwestern North American craton margin. Geological Society of America Abstracts with Programs, 6 (7), 955-956.

Smith, G. A. and Landis, C. A. (1995) Intra-arc basins, in Busby, C. J., and Ingersoll, R. V., eds., Tectonics of sedimentary basins. Oxford, Blackwell Science, 263-298.

Solomon, G. C., and Taylor, H. P. (1991) Oxygen isotope studies of Jurassic fossil hydrothermal systems, Mojave Desert, southeastern California, in Taylor, H. P., O' Neil, J. R., Jr., and Kaplan, I. R., eds., Stabl isotope geochemistry: a tribute to Sam Epstein. The Geochemical Society Special Publication, (3), 449-462.

Thomson, S. N. (2002) Late Cenozoic geomorphic and tectonic evolution of the Patagonian Andes between latitudes 42S and 46S: an appraisal based on fission-track results from the transpressional intra-arc Liquine-Ofqui fault zone. Geological Society of America Bulletin, 114, 1159-1173.

Tibaldi, A., Pasquaré, FG., and Tormey, D. (2009) Volcanism in reverse fault settings, in New frontiers in integrated solid earth sciences. Berlin, Springer Science Business Media B. V.

Tosdal, R. M., Haxel, G. B., and Wright, J. E. (1989) Jurassic geology of the Sonoran Desert region, southern Arizona, southeastern California, and northernmost Sonora: construction of a continental-margin magmatic arc, in Jenny, J. P., and Reynolds, S. J., eds., Geologic evolution of Arizona. Arizona Geological Society Digest, 17, 397-434.

Unruh, J. R., Humphrey, J., and Barron, A. (2003) Transtensional model for the Sierra Nevada frontal fault system, eastern California. Geology, 31, 327-330.

van Bemmelen, R. W. (1949) The geology of Indonesia, 2 vols. The Hague, Government Printing Office.

Van Dijk, J. P. (1994) Late Neogene kinematics of intra-arc oblique shear zones: the Petilia-Rizzuto fault Zone (Calabrian arc, central Mediterranean). Tectonics, 13, 1201-1230.

White, S. M., Crisp, J. A., and Spera, F. J. (2006) Long-term eruption rates and magma budgets. Geochemistry, Geophysics and Geosystems, 7 (1), 20 p. doi: 10.1029/2005GC001002.

(李阳 译，阳怀忠 吴哲 校)

第 20 章 前陆盆地系统综述：不同构造背景下响应的多样性

PETER G. DECELLES

(Department of Geosciences, University of Arizona, USA)

摘　要：前陆盆地的"四分"方案（楔顶沉积带、前渊带、前隆带和隆外凹陷带）在全球范围内许多前陆盆地系统中得到广泛应用，但在地层记录上存在显著差异。这些差异是由构造背景和与褶皱—冲断带相关的属性特征决定的。水平缩短作用使得褶皱—冲断带持续增长，导致要求前陆岩石圈向褶皱—冲断带迁移。地形负载形成的挠曲波可在前陆岩石圈中侧向迁移达约 1000km，与挠曲波的波长相当。强烈的侧向迁移活动导致在地层记录中前陆盆地沉积带呈垂向叠置关系。标准的地层层序包含多个千米厚的向上变粗层序，在其底部以一个密集的地层凝缩带或一个大规模的不整合（由于前隆带的迁移）为标志，而在其顶部则是以伴随生长构造的粗颗粒近源相沉积（楔顶沉积带）为标志。前渊沉积常位于前隆不整合/凝缩带和楔顶沉积之间；隆外沉积带则可能在层序地层的最底部出现。楔顶带沉积因处于构造高部位而易受剥蚀，隆外带和前隆带沉积的保存在某种程度上依赖于构造背景。

褶皱—冲断带可分成三种主要类型：弧后型、碰撞（或周缘）型以及那些与后退式碰撞俯冲带相关的类型。弧后前陆盆地系统（如现代安第斯山系统）易受由俯冲洋壳板片与地幔楔之间的黏滞性耦合产生的、传播至前陆岩石圈的远源动力负载影响。这种长波长的沉降，叠加到地形挠曲波产生的沉降中，使发育良好的前隆带和隆外带沉积地层的保存成为可能。碰撞（周缘）前陆盆地系统（如现代喜马拉雅系统）因缺少动力沉降，使前隆带和隆外沉积带易受剥蚀而不易保存。后退式碰撞前陆盆地系统（如地中海区域）常常伴随大型俯冲板片负载，在前渊和楔顶沉积带沉积窄而巨厚的沉积物。这些前陆盆地系统以巨厚的前渊和楔顶沉积为特征，远远超过仅依据地形负载而获得的沉积厚度。在碰撞环境下岩石圈的刚性变化可能影响隆外和前隆带沉积地层的保存。若这些远端前陆盆地沉积没有得到保存，则造山运动事件几乎一半的历史（在地层记录中获得的）将可能遗失。

许多前陆地层层序为估算挠曲波在前陆地区迁移的速率提供了充足的信息，而迁移速率的估算又可分解成冲断带内的扩展速率和缩短速率。前陆盆地沉降曲线可反演生成理想的挠曲剖面，从而得到岩石圈的挠曲属性。然而，为了理解长期的地球动力学过程、挠曲刚度的空间变化、造山带负载规模的变化，以及冲断带的扩展速率和缩短速率的变化都需要对冲断—前陆盆地系统进行恢复重建。

关键词：前陆盆地　挠曲　造山带　地球动力学　地层学

20.1　引言

前陆盆地系统由在主褶皱—冲断带前缘陆壳上发育的沉降区构成。前陆盆地系统是地球上最大的沉积物堆积体之一，它们跨越整个大陆地块，并延伸到相邻的残留洋盆和被动大陆边缘的陆棚区域。因在成因上与冲断带和会聚边缘密切相关，前陆盆地系统成为地球动力学研究的关注点（Price，1973；Dickinson，1974）。理想的前陆盆地系统由四个在不同的局部运动学和沉降条件下形成的独立沉降带组成（图20-1）。楔顶沉积带（wedge-top）包括覆盖冲断带活动前缘部分的沉积，其结构和成分成熟度低，生长构造是其界定特征。前渊沉积带（foredeep）由冲断带负载而形成的挠曲"槽"（"moat"）（Price，1973）内的沉积构成。来源于冲断带的沉积物有可能越过前渊带，进入前隆带（forebulge）为代表的挠曲隆起区域，甚至超出前隆带进入宽而浅的次级挠曲沉降带，即隆外凹陷带（backbulge）（图20-1）。尽管前陆盆地系统的"四分"方案在全球范围内很多现代的例子中得到了证实（扎格罗斯、喜马拉雅、塔拉纳基、安第斯山、亚平宁、中国台湾、北澳大利亚等），但在不同的前陆构造背景下也存在显著的差异性，且这些差异在前陆地层记录中留下了明显的证据（Sinclair，1997；Catuneanu，2004）。也就是说，不存在一个适用于所有前陆盆地的简单通用模型。本章以大尺度对前陆盆地系统作了一个最新的综述，重点强调了构造作用，特别是对特殊构造背景的响应过程中盆地演化的差异性，以及冲断带与相邻前陆盆地系统之间的地球动力耦合机制。

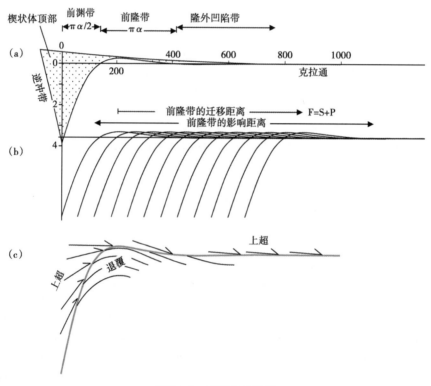

图 20-1　前陆盆地系统

（a）垂直邻近冲断带走向的理想前陆盆地系统横切面示意图（改自 Flemings 和 Jordan，1989；据 DeCelles 和 Giles，1996），注：垂直方向作了极端放大。（b）侧向迁移达 700km 的前隆带系统的速度和距离影响；挠曲波迁移距离（F）是冲断带缩短量（S）与冲断带增生距离（P）之和。（c）与侧向的前隆迁移有关的特征地层的上超和退覆模式

20.2　前陆盆地的构造背景

前陆盆地系统与冲断带关系密切。冲断带可分为三种主要类型：弧后型（retroarc）、碰

撞型（collisional）[或周缘型（peripheral）]和与后退式碰撞（retreating collisional）俯冲带相关的冲断带类型（图20-2）。而与转换断层系统相关的局部地壳缩短区则不在考虑范围

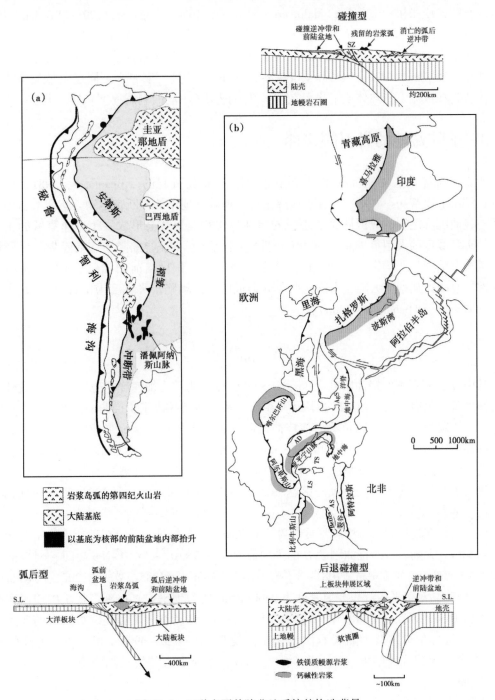

图20-2 三种主要前陆盆地系统的构造背景

（a）安第斯山脉弧后冲断带及前陆盆地系统。（b）碰撞型（橙色）和后退碰撞型（蓝色）背景下的前陆盆地系统，横切面示意图分别在其上面和下面。扎格罗斯山脉和喜马拉雅山脉是典型的碰撞型背景下的前陆盆地系统，后退碰撞型背景下前陆盆地系统的典型例子是地中海地区

477

内。弧后型和碰撞型冲断带形成于真正的板块会聚构造背景中，每个板块上的固定参考点均发生相对快速的会聚运动（如 Heuret 和 Lallemand，2005；Schellart，2008）。后退碰撞型冲断带则形成于俯冲速率大于会聚速率的板块构造背景之下，从而导致上覆板块在冲断带后方区域发育拉张或张扭构造作用（Malinverno 和 Ryan，1986；Doglioni，1991；Royden，1993；Jolivet 和 Faccenna，2000）。虽然在这些不同构造背景下形成的前陆盆地最终都归因于由于造山作用地壳增厚形成的负载而引起的挠曲沉降，但在每种构造背景下仍然存在显著的额外负载作用（Royden，1993；DeCelles 和 Giles，1996），并且，来源于不同造山带类型的同造山期沉积物在成分上和热体质上也存在差异性，记录了不同的剥露程度和母岩类型。

20.3 前陆盆地的岩石圈挠曲

所有前陆盆地系统的沉降都是由相邻冲断带负载引起的（Beaumont，1981；Jordan，1981；Karner 和 Watts，1983）。许多研究论述了前陆盆地中的冲断带运动学特征、造山负载作用与挠曲沉降之间的成因联系。理想的弹性板块挠曲负载会形成一个迅速衰减的正弦剖面：邻近负载区的大型负挠曲（前渊沉积带）、中等规模的正挠曲隆起（前隆带）以及最远端次级负凹陷隆外凹陷带（图 20-1）。挠曲（正或负）幅度自前渊沉积带向隆后沉积带约呈三次方衰减。典型的大陆岩石圈挠曲负载形成的前渊坳陷水平宽度为 $\pi\alpha$，其中 α 为挠曲参数（Turcotte 和 Schubert，2006）：

$$\alpha = \left[\frac{4D}{\Delta\rho g}\right]^{1/4}$$

其中 D 为岩石圈的挠曲刚度（flexural rigidity）；g 为重力加速度；$\Delta\rho$ 为地幔和盆地充填物的密度差。负载是由有效连续板块还是由破裂板块所支撑决定了挠曲剖面的波长：连续板块可承担比破裂板块更强的负载，使挠曲沉降延伸范围更广，从而形成更浅、更宽的坳陷（Flemings 和 Jordan，1989；Turcotte 和 Schubert，2006）。然而，关于前陆岩石圈是否表现出纯弹性形变，是否经历了黏弹性（visco-elastic）松弛并伴随有（或没有）与深度相关流变性变化，仍然是地质界持续争议的话题（例如 Quinlan 和 Beaumont，1984；Garcia-Castellanos 等，1997）。

线性负载在连续岩石圈中将形成宽度为 $0.75\pi\alpha$ 的挠曲前渊，但在破裂板块中前渊宽度只有 $0.5\pi\alpha$。大陆岩石圈挠曲刚度 D 的平均值介于约 $5\times10^{22} \sim 4\times10^{24}$ Nm 之间（对应的弹性厚度为 20~90km；如 Jordan，1981；Lyon-Caen 和 Molnar，1985；Watts，2001；Roddaz 等，2005）。因此，对于平均沉积充填密度为 2500kg/m³ 的盆地来说，在破裂板块中前渊宽度约为 110~350km，而在连续板块中则达约 170~515km。在破裂板块和连续板块中的前隆宽度均为 $\pi\alpha$，这与在对典型大陆岩石圈来说前隆宽度为约 220~690km 相一致。前隆的幅度（正）是最大前渊挠曲幅度（负）的约 4%~7%，在典型大陆岩石圈中换算为高度大约是 200~400m。

实际上，负载的形状和规模、前陆盆地系统的沉积充填量、盆地内的沉积物搬运类型（Garda-Castellanos，2002）、沿着板块向上挠曲部分的正断层活动（Bradley 和 Kidd，1991；Londono 和 Lorenzo，2004）、大陆岩石圈的构造不均一性（Flemings 和 Jordan，1989；Sinclair 等，1991；Waschbuscb 和 Royden，1992；Blisniuk 等，1998；Cardozo 和 Jordan，2001；

Cloetingh 等，2004）以及负载的大尺度三维形状（Chase 等，2009）均强烈影响着挠曲形态。尤其是随着挠曲波动在克拉通岩石圈内传播而重新活动的古基底构造（Blisniuk 等，1998；Cardozo 和 Jordan，2001），以及卷入到挠曲剖面内的古地形特征（Gupta 和 Allen，2000；Bilham 等，2003），对挠曲变形的影响，一直是个很大的地质难题。尽管影响因素如此复杂，近 30 年的研究表明影响大部分前陆盆地系统特征的第一要素仍然是弹性挠曲变形（综述见 Allen 和 Allen，2005）。

20.4 冲断带—前陆系统的侧向迁移

经过数十个百万年的生长，地球上的主要褶皱—冲断带往往卷入了达数百千米的地壳水平缩短作用。例如，喜马拉雅山脉冲断带发育了至少 500km、最高甚至高达 900km 的缩短（Coward 和 Butler，1984；Srivastava 和 Mitra，1994；DeCelles 等，2001；Robinson 等，2006；Murphy，2007）；安第斯山脉中央冲断带经历了近 400km 的缩短作用（Kley 和 Monaldi，1998；McQuarrie，2002；Arriagada 等，2008）；北美科迪勒拉山脉冲断带则涉及至少 350km 的缩短作用（DeCelles 和 Coogan，2006；Evenchick 等，2007）。如此大规模的水平缩短，至少需要挠曲波在岩石圈内发生距离与之相当的水平迁移。这段距离还包括挠曲波为了调节造山楔宽度增加（或传播距离）而发生的侧向位移。因此，从一阶方程上来考虑，挠曲波侧向迁移的总距离相当于冲断带传播距离和地壳总缩短量之和（图 20-1；DeCelles 和 De-Celles，2001）。对于地球大陆规模的冲断带—前陆系统来说，此距离大约介于 500~1000km 之间。前一节已讨论到，该距离的上限与前陆盆地系统内挠曲剖面的平均波长相当。由此，在主冲断带的整个演化时期内，相应的前陆盆地系统很可能发生水平距离等于、甚至超过自身半个波长的侧向迁移。反过来，这将会导致在地层记录中沉积带的垂向叠置（Flemings 和 Jordan，1989；Coakley 和 Watts，1991；Sinclair 等，1991；Verges 等，1998；Burkhard 和 Sommaruga，1998），相当于前陆盆地地层的瓦尔特定律。

在许多前陆盆地中均发育沉积带的这种"瓦尔特式（Waltherian）"序列（如图 20-3），包括北美科迪勒拉山（Plint 等，1993；DeCelles 和 Currie，1996；Fuentes 等，2009）、安第斯山（Jordan 等，1993；DeCelles 和 Horton，2003；Uba 等，2006）、北阿尔卑斯山（Sinclair，1997；Burkhard 和 Sommaruga，1998；Gupta 和 Allen，2000）、比利牛斯山（Verges 等，1998）、台湾岛（Yu 和 Chou，2001；Tensi 等，2006）、卡鲁（Catuneanu，2004）、冈底斯山（Leier 等，2007）和喜马拉雅山等（DeCelles 等，1998a；1998b）前陆盆地系统。其典型的地层模式是由总体向上变粗的数千米地层层序组成，下部发育因前隆带在前陆地区迁移形成的大型剥蚀不整合或地层凝缩段。这些凝缩段和剥蚀是由于前隆地区沉积可容空间的不足所致。非海相前隆带以强烈的成土作用、喀斯特风化作用（若存在碳酸盐岩基底）、河流侵蚀和超稳定砾岩滞留沉积（ultra-stable conglomeratelags）为特征（Herb，1988；Plint 等，1993；Demko 等，2004；Del Papa 等，2010）。当挠曲前隆带迁移至一个地区时，老的断裂将会被复活，形成的局部地形将会非常有利于在远离造山前锋区域发育粗颗粒沉积物（Blisniuk 等，1998；Burkhard 和 Sommaruga，1998）。海相前隆带则以浅海碳酸盐岩建隆、铁矿石以及海底侵蚀与饥饿沉积为标志（Tankard，1986；Dorobek，1995；Gupta 和 Allen，2000；Allen 等，2001）。假定挠曲波的平均迁移速率为 10~25mm/a，前隆带宽约 200~700km，则前隆不整合/凝缩带（DCZ）所代表的时间为数个到数十个百万年（图 20-3）。

随着冲断带向克拉通内部迁移，前隆不整合/凝缩带（DCZ）的年龄也向克拉通方向变年轻（Coakley 和 Walls，1991；Crampton 和 Allen，1995；Sinclair，1997；Burkhard 和 Sommaruga，1998）。在许多前陆盆地系统中，前隆不整合/凝缩带（DCZ）位于远源细粒沉积的隆外凹陷带之上，且普遍被前渊沉积带所覆盖，层序顶部为含有生长构造和典型粗粒近源相楔顶沉积（图 20-3；DeCelles，1994；Ford 等，1997；Williams 等，1998；Lawton 等，1999；

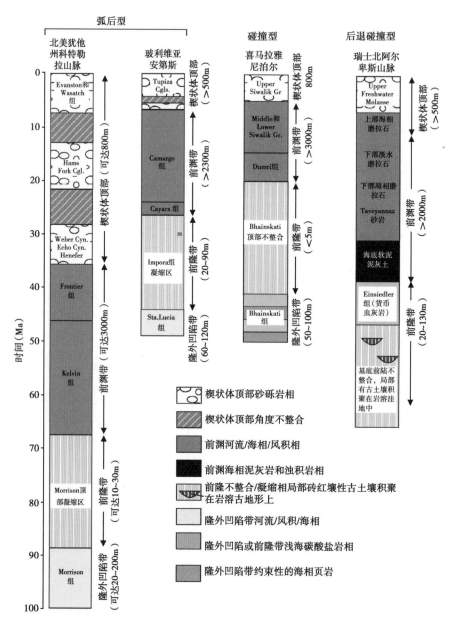

图 20-3　挠曲波在前陆岩石圈迁移时产生的前陆盆地沉降带垂向叠置而形成的"瓦尔特"层序例子
（据 Nepal, DeCelles 等，1980a；Bolivia, DeCelles 和 Horton，2003；Utah, DeCelles，1994；DeCelles 和 Currie，1996；以及阿尔派恩；Sinclair，1997；Beck 等，1998；Burkhard 和 Sommaruga，1998）
各实例均是在单个地点或区域的简化垂向地层层序，所有的例子中均有明显的穿时性存在，厚度是沉降带的区域平均值，纵轴代表时间而不是年龄。尼泊尔、北安第斯山和玻利维亚例子中年龄为新生代，犹他州的例子属于侏罗纪—古新世年龄

Chiang 等，2004）。受附近与褶皱—冲断带前缘传播相关的生长地貌的影响，楔顶沉积具有强烈的横向各向异性、大量的局部不整合和复杂的古地理等特征。

尽管该模型准确预测了在全球前陆盆地系统地层记录中发现的许多特征，但是在不同的构造背景之下，有些特征会发生变化，例如，在缺失来自冲断带的大量沉积供给和/或区域动力学沉降的情况下（Gurnis，1992；Liu 等，2005），地层记录中就很可能没有保存前隆和隆后沉积（Flemings 和 Jordan，1989；Sinclair，1997；Catuneanu，2004）。

20.5 弧后前陆盆地系统

弧后前陆盆地系统沿着科迪勒拉造山带（安第斯山型）内侧发育，位于迅速会聚的洋—陆俯冲带上部的大陆板块之上（图 20-2a；Dickinson，1974；Jordan，1995）。然而不是所有的洋—陆会聚板块边界都是以科迪勒拉型造山带和前陆盆地为特征的，某些情况下以上部板块的弧后扩张为主。最近对控制俯冲带之上大陆板块缩短的众多参数的分析表明，上部板块向海沟方向的快速运动，以及距俯冲板片边缘较长的横向距离对发育强烈的弧后收缩体制具有重要的作用（Heuret 和 Lallemand，2005；Schellart，2008）。形成的前陆盆地具有大陆级规模，与紧邻的造山带平行方向延伸近数千千米，垂直造山带方向延伸近几百千米。其现代原型盆地是安第斯山前陆盆地，后者沿着安第斯山脉冲断带东翼延伸超过 7000km，在其东侧是南美克拉通（图 20-2a；Jordan，1995）。

现代安第斯山脉前陆盆地系统发育一个 50~75km 宽的楔顶沉积带，一个 250~300km 宽的前渊沉积带，一个因大部分被埋藏而没有地貌特征显示的前隆带（5~10m 的大地水准面正异常）表明岩石圈发生了向上挠曲变形（Chase 等，2009）以及一个宽度大于 400km、以沼泽泛滥平原和河流沉积环境为主的隆外凹陷沉积带（图 20-4；Horton 和 Decelles，1997）。贝尼河盆地的弧形河流阶地（Aalto 等，2003），以及亚马逊流域盆地西部详细的沉积记录和挠曲模拟（Roddaz 等，2005）均表明安第斯前隆带的发育。在北美科迪勒拉山脉冲断带东部也发育类似的晚侏罗世—早新生代前陆盆地系统（DeCelles，2004；Miall 等，2009）。与太平洋扩张中心的碰撞，导致沿着北美大陆西部边缘的板块运动以转换断层为主，从而使北美前陆盆地系统开始处于休眠状态（Dickinson，2004）。安第斯山脉和科迪勒拉山脉前陆盆地系统共计充填了大于 $15×10^6 km^3$ 碎屑沉积物，同时蕴含丰富的油气。

在弧后背景中，上部板块与俯冲的大洋板块之间通过地幔楔形成黏滞性耦合，从而形成的附加负载会影响前陆盆地岩石圈（图 20-2；Mitrovica 等，1989；Gurnis，1992；Lithgow-Bertelloni 和 Gurnis，1997）。这类"动力学沉降（dynamic subsidence）"可以向盆地内侧蔓延长达 1000km，同时作为对地形负载朝海沟方向向下作用的响应，整个挠曲剖面发生倾斜（DeCelles 和 Giles，1996；Pang 和 Nummedal，1995；Catuneanu，2004；Liu 等，2005）。在这种背景下，前陆盆地将发育并保存完好前隆带和隆外凹陷沉积带（Catuneanu，2004）。若前陆地区位于海平面之上，前隆带将沉积一套叠置的过成熟古土壤层序，反映其长时间的低速率沉积，类似沉积在中安第斯山古近系前陆盆地层序（DeCelles 和 Horton，2003）、北美西部科迪勒拉山前陆盆地侏罗系—白垩系分界面（Demko 等，2004；Fucntes 等，2009）以及藏南冈底斯山前陆盆地下白垩统（Leier 等，2007）中均有发现。在海相弧后前陆盆地系统中，如北美科迪勒拉的中白垩统，前隆带迁移的轨迹以浅海相浅滩化（和可能的碳酸盐岩建造）、剥蚀，以及显著的退覆和超覆模式为标志（图 20-1；Tankard，1986；Plint 等，

1993；Yang 和 Miall，2009）。

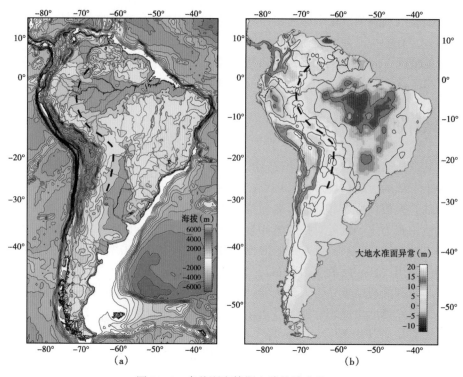

图 20-4　南美洲安第斯山脉前陆盆地

（a）南美洲地形，粗虚线代表 Chase 等（2009）显示的安第斯山脉以东约 400~500km 的地形异常高值；注意，因盆地的过补偿充填，前隆带在地形上并没有显示。（b）大地水准面异常图，单位为米，据 GRACE 数据（Chase 等，2009）。前陆地区的前隆带顶峰位于沿着粗虚线 5~10m 异常的范围内

20.6　碰撞前陆盆地系统

碰撞前陆盆地系统形成于陆—陆碰撞背景下的俯冲板块之上（图 20-2b；Dickinson，1974；Miall，1995；Jolivet 和 Faccenna，2000）。现代的例子有喜马拉雅山、扎格罗斯山和新几内亚前陆盆地系统（Dewey 等，1989）。现代陆—陆碰撞的缝合带均没有超过约 3000km，因此，地球上现代碰撞前陆盆地系统规模均没有美国弧后前陆盆地系统那么大。然而，从地中海西部一直延伸到缅甸断续分布的一系列大陆板块碰撞，组成延绵长约 8000km 的链状碰撞前陆盆地系统。

典型的活动碰撞前陆盆地系统是喜马拉雅山脉前陆盆地（Burbank 等，1996；Najman，2006）。该盆地系统实际上是由至少三个前陆盆地系统组成，即中喜马拉雅前陆盆地、孟加拉盆地和印度河盆地（图 20-5）。但只有喜马拉雅段是真正意义上的碰撞前陆盆地系统——孟加拉盆地和下印度河盆地发育在俯冲速率远大于当地板块会聚速率的岩石圈之上，实际上形成了后退碰撞式冲断带和前陆盆地（将在下一节讨论）。

喜马拉雅前陆盆地系统宽约 400~450km，长约 2000km（图 20-5）。该现代盆地具有活动的楔顶沉积带和前渊沉积带。Bilham 等（2003）认为北印度高地是喜马拉雅山脉冲断带

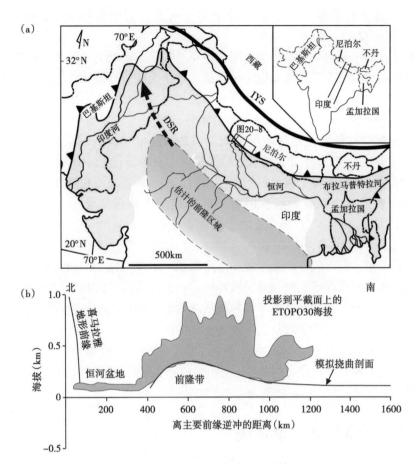

图 20-5 碰撞前陆盆地系统

(a) 巴基斯坦、印度和孟加拉国地区与印度—亚洲陆—陆碰撞有关的碰撞前陆盆地（黄色区域）。其中没有将前陆盆地系统中部的喜马拉雅部分和侧翼与后退式碰撞有关的印度和孟加拉盆地区分开来。粗实线代表印度—雅鲁缝合线（IYS）带；锯齿实线代表冲断带的前缘。标有 DSR 的粗虚线代表德里—萨戈达山脊，Duroy 等（1989）认为是前隆带。西尼泊尔上的四边形方框代表图 20-8 所在区域。印度北部的灰色阴影区代表依据 GRACE 大地水准面数据（见 www.csr.utexas.edu/grace/gravity）得出的喜马拉雅挠曲前隆的大致区域。(b) 沿着 25° 和 30° 纵剖面（见插图部分的剖面）的平均高程以及在尼泊尔与喜马拉雅山脉前锋的距离，据 Bilham 等（2003）。实线代表模拟的与时间有关的剥蚀面（有效弹性厚度为 105km），与拉成一个平面的高程数据（灰色区域）的包络线底部拟合较好。Bilham 等（2003）解释该上凸的剥蚀基底是对现代印度高原北部古地形下切的地貌响应，随着印度板块仰冲于喜马拉雅山脉冲断带前缘的挠曲前隆带之上而形成

前缘的前隆带；Duray 等（1989）的分析表明萨戈达山脊是该前隆带的延伸，并向北西方向倾伏在印度河前陆地区之下（图20-5）。北印度前隆带继承了前新生代古地貌发育向上挠曲的高低起伏不平的地形，并被新生流域切割至活动挠曲的水平面（图 20-5b；Bilham 等，2003）。挠曲的幅度约 400m，前隆带宽大于 600km。显著增加的地形起伏是从古生代—中生代印度北部高原继承而来（Bilham 等，2003）。印度地盾岩石圈的高挠曲刚度和喜马拉雅的高地形负载，形成巨大规模的北部印度前隆带。因此，现今的印度—恒河前陆盆地处于严重的（deeply?）欠补偿沉积状态，轴向发育的河流系统经过侧翼的下印度河盆地和孟加拉盆地，一个遭受剥蚀的前隆带以及一个无沉积的隆后区域流出盆地（图20-5a）。然而该类情况并非经常发生，冲断带前缘区域的始新统表明，沉积物来自于处于限制性浅海沉积环境的

早期喜马拉雅山脉，后者位于一个主体处于水下的前隆带的南部（DeCelles等，1998a，2004；Najman等，2005）。

与弧后前陆盆地不同的是，碰撞型前陆盆地不受控于远源动力学负载的影响。故此，控制挠曲剖面形状和规模的主要因素是造山带负载。如果前陆岩石圈刚性大，如现代喜马拉雅前陆盆地之下和其南部岩石圈，前隆带和隆外凹陷带将很难被保存下来，挠曲波的经过将会形成强烈的剥蚀不整合（Crampton和Allen，1995）。另一方面，如果前陆岩石圈仅仅是中等刚性的，如扎格罗斯前陆盆地之下的岩石圈（Snyder和Barazangi，1986；Watts，2001；Watts和Burov，2003），则隆外带甚至前隆带沉积将会在地层记录中保存下来。例如，在喜马拉雅前陆盆地系统，沿着整个喜马拉雅从西向东延伸，始新统海相隆外沉积被一个剥蚀不整合面，或者代表了整个渐新世沉积的极其成熟但非常薄的氧化土（Oxisol）层所覆盖（DeCelles等，1998a，2004）。在氧化土之上为向上变粗的中新世—上新世前渊带沉积以及局部的楔顶带沉积（图20-3；Burbank等，1996；DeCelles等，1998b）。显然，喜马拉雅前陆盆地经历了从一个可保存隆外凹陷带和前隆带沉积，向现今这些沉积带不发育明显沉积的转变。造成这种转变的原因可能是喜马拉雅冲断带负载规模的扩大和印度地盾岩石圈向南挠曲刚度的增加。能与始新世—渐新世喜马拉雅前陆盆地系统类比的一个现代实例是扎格罗斯前陆盆地，后者地形负载规模和阿拉伯半岛岩石圈的刚度均不如现今的喜马拉雅山—印度地区。然而，甚至扎格罗斯盆地系统也不是一个能与早期喜马拉雅盆地系统相类比的完美实例，因为在新近纪红海打开且靠近Afar热点期间，与阿拉伯地区区域掀斜相关的远端沉降影响了扎格罗斯的前陆地区（Ali和Watts，2009）。

20.7　后退式碰撞前陆盆地系统

在板块俯冲速率大于板块会聚速率的碰撞构造背景下，要维持持续的板块俯冲，需要俯冲板块的枢纽线向后退方向迁移（即与俯冲方向相反方向）。为了补偿板块之间形成的间隙，上部板块将经历拉张减薄作用（图20-2b；Malinverno和Ryan，1986；Doglioni，1991；Royden，1993）。该类冲断带—前陆盆地系统最著名的实例位于地中海地区。非洲板块北缘的岬角地带（如阿德里亚、阿拉伯）从白垩纪晚期开始与欧亚板块碰撞（Cavazza等，2004；Schmid等，2008），在岬角之间的区域形成碰撞造山带和新特提斯洋壳岩石滞留受限板片（stranding aerially restricted slabs）（图20-6；Jolivet和Faccenna，2000；Faccenna等，2004；Spakman和Wortel，2004）。尽管岬角之间的碰撞减弱了板块会聚速率，这些洋壳板片的密度驱动俯冲作用仍持续进行，且为了调节持续进行的俯冲作用，俯冲带枢纽线被迫向板块俯冲相反方向后退。上部板块上的褶皱—冲断带继续向前陆方向迁移，但其尾部经历了区域拉张和地壳减薄作用（Doglioni，1991；Cavinato和DeCelles，1999）。紧随这些迁移的俯冲带之后可能形成一些小的洋盆，并且在某种情况下，一些微地块从欧亚板块中分离出来并横跨中地中海快速迁移（如科西嘉—撒迪尼亚和卡拉布里亚微板块及巴利阿里群岛；Malinverno和Ryan，1986；Dewey等，1989；Bonardi等，2001；Gutscher等，2002；BoothRea等，2007；图20-6）。该过程与发生在西太平洋洋壳俯冲系统内上部板块的弧后扩张类似。

与碰撞型和弧后型前陆系统相反，后退式碰撞系统以形成弧长较短（约1000km）、弧度高（曲率高达180°）和海拔低（<2km）的褶皱—冲断带为特征，并以未变质沉积岩为主（Royden，1993）。年轻的例子有亚平宁山脉、喀尔巴阡山脉、贝蒂克—里夫造山带和可能

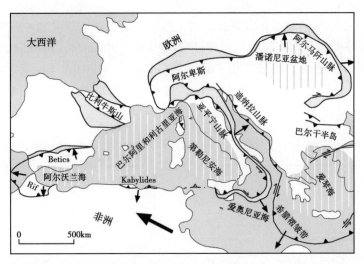

图 20-6　地中海地区中部和西部构造图（据 Jolivet 和 Faccenna，2000；Schmid 等，2008）
显示在高角度弧形的贝蒂克—里夫造山带、亚平宁—伊朗—卡比里德（Kabylidean）、喀尔巴阡山脉和希腊褶皱—冲断带前面（锯齿线，小箭头代表俯冲带后退/冲断带前进方向）的后退俯冲系统。前陆盆地用黄色标注，上地壳的地壳拉张区域用垂向直纹图样表示。大箭头代表现代非洲板块相对欧亚板块的运动方向

的北阿尔卑斯山褶皱—冲断带以及与它们相应的前陆盆地（图20-6）；较古老的例子包括安特勒造山带（Antler）（泥盆纪—密西西比期；Giles 和 Dickinson，1995）、塔科尼克山（Taconic）（奥陶纪；Jacobi，1981；Hiscott 等，1986）、沃希托山（Ouachita）（石炭纪；Houseknecht，1986）前陆盆地。与后退式俯冲带相关的前陆盆地系统，一般发育比地形负载推测厚度的两倍还厚的前渊带沉积，这也许需要俯冲板片负载的作用（Karner 和 Watts，1983；Royden，1993）。例如，位于亚平宁山脉冲断带前缘的前陆盆地从墨西拿阶（Messinian）开始沉积了一套局部厚度超过7km的楔顶和邻近前渊带沉积（Ori 等，1986；Bigi 等，1992）。Royden 和 Karner（1984）认为该极端沉降不能仅仅解释为由相对温和的亚平宁负载形成的挠曲沉降导致，而是俯冲的亚得里亚板片负载促成前陆岩石圈弯曲运动并在沉降中起着主控作用。在喀尔巴阡（Carpathian）前陆盆地也存在类似的情形（Roydenand 和 Karner，1984）。

在后退式碰撞背景下的前陆盆地系统以波长相对较短、狭窄的前渊带（约100km）和前隆带以及相对不发育的隆外凹陷带为特征，而其楔顶沉积却可能非常厚（如 Bigi 等，1992）。在前隆沉积带中通常发育碳酸盐岩建造；而在前渊沉积带中则普遍发育浊流沉积和其他海相沉积（Hiscott 等，1986；Houseknecht，1986；Wuellner 等，1986；Giles 和 Dickinson，1995；Sinclair，1997）。在后退式碰撞系统中，沿着碰撞带俯冲（后退）板片的挠曲刚度和侧向撕裂（lateral tears）也存在差异，而这些复杂的差异性强烈地分割着前陆盆地系统（Matenco 和 Bertotti，2000；Spakman 和 Wortel，2004；Ostaszewski 等，2008）。

20.8　前陆盆地的地层记录：反演构造过程

原则上，如果可以估算一个给定的造山带负载—前陆盆地系统的迁移速率，那么就可以反演其挠曲剖面并确定剖面上每一个点的沉降史。用时间来代替水平距离，挠曲剖面的一阶

导数即可反映沉降史（图20-7）。将模拟的沉降史与真实沉降史进行比较，可用来估算岩石圈的挠曲刚度。图20-8展示了该方法在尼泊尔地区喜马拉雅前陆盆地中的应用。根据GPS数据的估计，喜马拉雅山脉前陆盆地向南迁移的速率为20mm/a（Bettinelli等，2006），这与对新近纪长期的迁移历史的估算相一致（DeCelles等，1998b；Lave和Avouac，2000）。通过拟合得到的最佳挠曲刚度为1.0×10^{24}Nm，也与重力数据及前人用挠曲分析得出的结论相吻合（Lyon-Caen和Molnar，1985）。

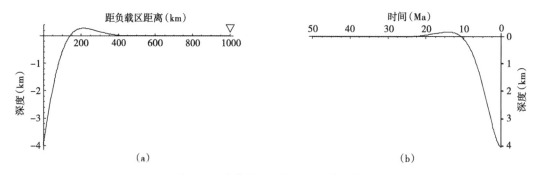

图20-7 挠曲剖面上某点的沉降史模拟

(a) 对破裂板块上线性负载响应的挠曲剖面演化，在x轴的"0"点位置，挠曲刚度为10^{23}Nm，盆地充填物与地幔的密度差为600kg/m³，盆地最大深度为4km。x轴上1000km处的三角形代表前陆地区挠曲剖面上的一点，挠曲剖面迁移至该点，将产生 (b) 所示的沉降史曲线。(b) (a) 中显示挠曲剖面的一阶导数，时间代替了空间，是为了模拟以20mm/a的固定速率经过1000km处倒三角形位置时的挠曲剖面迁移活动。该曲线是三角形处反演的沉降历史

这种简单方法存在的主要问题是，影响挠曲波长的造山带负载规模和前陆岩石圈挠曲刚度在一定规模地质历史时期内（≥50Ma）并不一定是个常量。在许多情况下，随着逐渐增长的造山带向更老、更硬的克拉通岩石圈迁移，造山带负载和挠曲刚度随时间会逐渐增大。因此，必须首先重新恢复造山系统，包括前陆盆地系统，从而来估计造山带负载规模随时间的变化。例如Jordan（1981）、Beaumont（1981）、Quinlan和Beaumont（1984）以及Liu等（2005）利用阿巴拉契亚和科迪勒拉造山带负载随时间的演化，来模拟前渊沉降带的迁移历史。该方法另外一个潜在的问题是在冲断带的活动历史中地壳缩短速率也会变化，从而造成挠曲波迁移速率可能极度不稳定。

将前陆盆地系统的四个部分结合起来进行恢复，能使挠曲波的迁移估计更具约束性，因为在前陆盆地地层的复原过程中，前渊带和前隆带均可用来定位随时间变化的挠曲波。其中一个特殊的值是前隆沉积带/不整合的持续时间。结合前陆系统的迁移速率，前隆沉积带的宽度可用来限定挠曲刚度；反过来，如果能够估算出挠曲刚度，那么可以利用一个已知垂向地层序列里的前隆沉积带持续时间来计算挠曲波的迁移速率。在一个仍然活动或没有被后期拉张作用完全破坏的造山系统中，挠曲波的迁移距离（F）大约是冲断带宽度（P）和缩短量（S）之和（图20-1b），即通过合理估算这三个变量（F, P, S）中的任意两个，就能计算出另外一个变量值（DeCelles和DeCelles，2001）。

图20-9是对尼泊尔地区喜马拉雅冲断带和前陆盆地系统恢复的挠曲模型。该模型是基于平衡区域横断面的增量重建而得（DeCelles等，2001；Robinson等，2006）。由于古地形难于确定，冲断负载被简化成一个矩形块体；不同的颜色代表了喜马拉雅冲断带中不同的构造地层亚带（特提斯喜马拉雅带、大喜马拉雅带、小喜马拉雅带和亚喜马拉雅带）。模型描

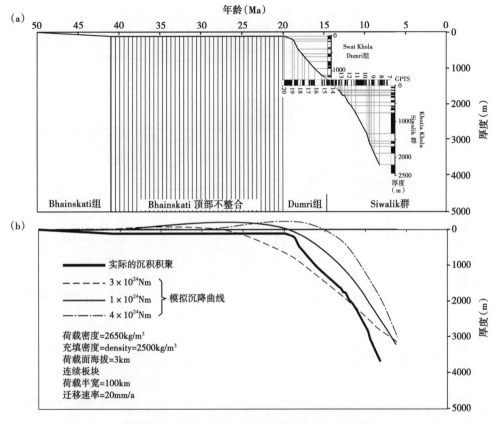

图 20-8 尼泊尔地区喜马拉雅前陆盆地沉降史模块

(a) 尼泊尔西部的喜马拉雅新生代前陆盆地时间和沉积厚度（未压实）曲线约 20-8Ma 数据根据尼泊尔西部 Swat Khola 和 KhutiaKhola（Ojha 等，2008）与 Lourens 等（2004）的全球古地磁磁性时间进行对比得出；50—20Ma 部分数据来自于 DeCelles 等（1998a）。(b) 粗线代表 (a) 里的曲线，细线代表应用图例里的挠曲刚度反演挠曲地形剖面得到的理论沉降曲线，挠曲波的迁移速率设置为估计的 20mm/a。将所有的挠曲模拟参数（挠曲刚度除外）设置成常数并在图例中标出。通过细微地调整负载的大小，改变最大的垂直预测沉降量，从而使之与真实的曲线进行更好的匹配。检验模拟的核心是是否显示出了合理的波长。结果显示，只有基于 1Nm×1024Nm 挠曲刚性得到的曲线能与实际的累积历史相匹配，伴随在约 20Ma 时有一个从前隆带向前渊沉降带迅速转化的过程

述了在前陆盆地挠曲波演化背景下的恢复重建参考点和喜马拉雅冲断带负载的增长。尽管该模型的结果与保存在小喜马拉雅带（Lesser Himalayan zones）和亚喜马拉雅带（Subhimalayan zones）内的前陆盆地的地层记录一致，且与在冲断带中估算的缩短量也一致，但大部分地层记录未得到证实。尤其是冲断带在中新世之前的缩短历史并不清楚。现在已知的有：(1) 前陆地区处在较浅的局限海沉积位置，部分物源来自始新世开始形成的冲断带（图 20-9c；DeCelles 等，2004；Najman 等，2005）；(2) 该地区在渐新世的大部分时期内被一个主要不整合斜削（图 20-9d、e），且在早中新世晚期处在前渊带（图 20-9f）。横切关系和前陆盆地内中新世至现今的地层记录很好地约束了剖面 g 和 h（图 20-9g，h）（Lyon-Caen 和 Molnar，1985；DeCelles 等，1998b；Szulc 等，2006；Ojha 等，2008）。虽然随时间变化的挠曲刚度没有被约束，但可以用于恢复重建的前陆沉积带位置匹配最好的挠曲剖面反演得出。

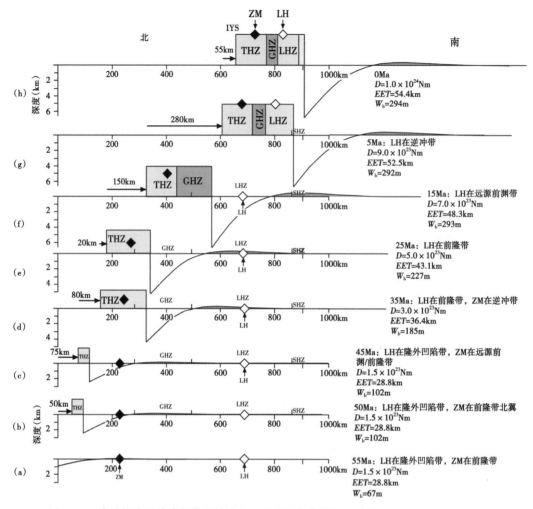

图 20-9 喜马拉雅山脉冲断带及前陆盆地系统的挠曲模型，采用了长方形负载并基于从
55Ma 开始的冲断带恢复重建模型（据 DeCelles 等，2001）

以作用在破碎板块的线性负载作为初始条件；所有的后续模型采用的是连续板块和长方形负载。负载的密度是 2650kg/m³；盆地充填物的密度为 2500kg/m³。各个阶段的地壳缩短量用最北边块体左侧的箭头标出。菱形标注 ZM 和 LH 分别代表古近—新近系中的遮普惹山和小喜马拉雅山的复原地点。THZ—特提斯喜马拉雅带；GHZ—大喜马拉雅带；LHZ—小喜马拉雅带。模型中也标注出了各个阶段变化的挠曲刚度（D），有效弹性厚度（EET）及前隆高度（W_b）

20.9 构造背景对前陆盆地系统的影响

前陆盆地系统的构造背景应该在地层记录中有所反映。碰撞系统缺乏一个动态沉降机制，因此常见前隆带剥蚀现象，尤其是在前陆岩石圈高挠曲刚度的情况下，如喜马拉雅前陆盆地系统。在可能遭受强烈动力沉降影响的弧后系统（Catuncanu，2004；Liu 等，2005），前隆沉积带保存完好，也发育相对厚的隆外沉积带（Currie，2002；Leier 等，2007）。后退式碰撞系统以发育窄而超厚的深海前渊沉积带和厚的楔顶沉积带为特征（Orict 等，1986）。在这种构造背景下前隆带主要以不整合为特征，隆外沉积稀少甚至缺失（Crampton 和 Allen，

1995；Sinclair，1997；Burkhard 和 Sommaruga，1998）。依据前陆盆地充填的碎屑岩成分的不同，可以进一步区分这些系统的基本地层特征。Garzanti 等（2007）详细论述了本章提及的弧后型、碰撞型和后退式碰撞型前陆盆地系统［即 Garzanli 等（2007）依次提到的亚平宁型、安第斯山型和阿尔卑斯型］之间存在的显著差异。

20.10 存在的问题

关于前陆盆地系统地层记录仍然存在的一些核心问题，譬如前隆带和隆外凹陷带沉积地层的保存、挠曲波长距离迁移的意义、前陆岩石圈内的复杂构造对前陆盆地系统的影响，以及前陆盆地中尚很少被研究到的构造与气候相互作用的关系等。

20.10.1 前隆和隆外沉积带的保存

由于隆外沉积带和前隆沉积带可能代表了附近造山带一半的时间演化，因此其地层记录的保存是一个很重要的课题。实际上，这些沉积带由于其厚度薄、结构和成分成熟度高以及"造山期前"的形态，在前陆盆地的地层分析中常常被忽略掉。

控制前陆盆地系统隆外带沉积保存潜力的两个主要因素是相对于前隆带顶端的沉积物充填程度和沉积物供应的方向（图 20-10；如 Giles 和 Dickinson，1995）。沉积物可以来自前陆盆地系统的一侧或者两侧；并且如果前隆带暴露并遭受剥蚀，也可以向隆外沉积带和前渊

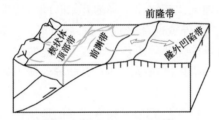
(a) 前渊带和前隆带欠充填：前隆带侵蚀、不整合发育，轴向前渊水系

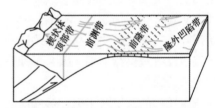
(b) 前渊带和前隆带过充填：前隆带发生埋藏并削平，沿着前隆带顶部发育地层凝缩，横载向前渊水系

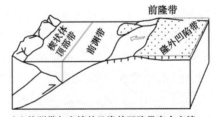
(c) 前渊带欠充填并且隆外凹陷带完全充填：前隆带克拉通一侧发生埋藏和地层凝缩；前隆带前渊一侧遭受剥蚀

图 20-10　非海相前隆盆地中多种加积/退积情形的示意图

（a）前渊带未充填至前隆带的顶端（"未充满"阶段）；前隆带暴露并形成一个分割前渊带和隆外凹陷带的地形水系。在前隆带地区，前造山期岩石遭受深度风化作用（垂直—阴影线），并发育不整合。取决于基准线的变化，隆外地区也可能遭受剥蚀并发育不整合。(b) 前隆带被来自褶皱—冲断带和克拉通区的沉积物埋藏（"充满"阶段）。尽管前隆带顶端发生加积作用，但沉积速率非常低且沉积物经历剧烈风化和凝缩（垂直阴影线）。（c）前渊带欠充填但隆外凹陷带被来自克拉通区或经平面外路径而来的褶皱—冲断带的沉积物充填至前隆顶端。局部发育沉积物从隆外凹陷带溢出至前渊带沉积

沉积带提供沉积供给（Blisniuk 等，1998）。如果前渊带和隆外沉积带相对于前隆带顶端没有被填充满，前隆带将会遭受暴露、剥蚀，尤其是在非海相背景下（图20-10a；Jacobi，1981；Coakley 和 Watts，1991；Crampton 和 Allen，1995）。在前隆带未被掩埋的前陆盆地系统中，前渊沉积带远端大部分沉积速率可能会非常低，使得在地层记录中常有古土壤层叠加在上面（Tandon 等，2008）。该情况下，大部分隆外带沉积将不可能在前陆盆地系统中的地层层序中保存下来。即使前隆带位于海平面以下，由于海底侵蚀在其顶端也会发生斜削作用（Crampton 和 Allen，1995）。如果前渊带和隆外沉积带沉积填充高于前隆顶端，前隆带将成为一个由多个凝缩层构成的沉积带，并且隆外带地层将可能被保存下来（图20-10b；Flemings 和 Jordan，1989；DeCelles 和 Horton，2003）。类似地，如果前渊带未被沉积充填满，但隆外沉积带被充填满，那么在经过前隆带顶端的剥蚀作用之后，隆外沉积带的部分沉积物将可被保存下来（图20-10c）。来自全球范围的前陆盆地资料表明，前隆带和隆外沉积带沉积在弧后背景中保存最完好，可能是因为长波长的动力沉降覆盖了整个挠曲剖面。此外，在后退式碰撞型前陆盆地系统中，通常易受剥蚀的冲断带前缘厚层楔顶沉积，可能因板片负载驱动的沉降作用而得以保存。

20.10.2 侧向迁移的意义

正如 Flemings 和 Jordan（1989）所指出的那样，理解冲断带—前陆盆地耦合系统的侧向迁移对正确解释前陆盆地的地层记录至关重要。在造山带中，常将快速沉降的起点看作是逆冲推覆和挠曲负载开始的时间。然而，在整个前陆地区快速挠曲沉降的开始时间具有很强的穿时性（Verges 等，1998），例如在逆冲作用早期，隆外或前隆带地区直到前渊带迁移至此才开始沉降。按照挠曲波迁移的平均速率来算，这可能需要经历数十个百万年的时间。如果隆外带沉积没有被保存下来，因为早期楔顶带和前渊带地层遭受剥蚀，可能导致造山运动早期地层记录的缺失。

对侧向迁移的认识对于中—远端前陆盆地沉积的粒度趋势和不整合特征的地层模型的建立也非常重要。基于前陆盆地向上和向下挠曲是对负载规模变化的响应的认识（Catuneanu 等，1997；Willis，2000），建立了目前流行的前陆盆地地层的两相模型，该模型认为细颗粒沉积是同构造成因而粗颗粒相是"反构造"成因的（Heller 等，1988）。对于一个给定挠曲刚度，增长的负载在前渊带将形成更大的挠曲沉降并接受来自冲断带附近的粗颗粒沉积物；因剥蚀而消减的负载作用将促使冲断带和邻近的前渊带发生均衡抬升，从而造成邻近区域可容空间的减小，并导致粗颗粒的沉积物向系统的远端部分进积。该模型概念上吸引人的地方是它是基于冲断带静态的位置和幕式稳定期而建立的。然而一些冲断带可能经历了多次构造不活动时期，持续的板块会聚将维持一个活动的地壳缩短冲断带，这又反过来驱使前陆盆地挠曲波在整个冲断带的演化过程中在岩石圈中侧向传播。即使在脱序逆冲断层（out-of-sequence thrusting）活动时期，前陆岩石圈也将继续向缩短的冲断带卷入岩石。前陆盆地系统的恢复重建（Homewood 等，1986；Pfiffner，1986；Sinclair，1997；Verges 等，1998；Currie，2002），包括冲断带自身的重建，对合理地解释盆地充填中长期的结构趋势是非常必要的。

20.10.3 构造的继承性和复活

位于挠曲大陆岩石圈之上的前陆盆地系统，对继承性基底构造的复活以及与被前隆带向上挠曲相关的小型断裂分割的微小区块非常敏感。与古克拉通构造和先存裂谷边缘相关的构造组构尤其容易复活（Schwartz 和 DeCelles，1988；Bradley 和 Kidd，1991；Meyers 等，1992；Crampton 和 Allen，1995；Blisniuk 等，1998；Gupta 和 Allen，2000；Zaleha，2001；Londono 和 Lorenzo，2004）。在前隆带地区微小的前陆内构造的影响最为强烈，因为挠曲抬升形成了网状拉张剪应力，使上地壳沿着微小但数量众多的正断层破碎。尽管这些构造幅度小（平均只有数十米），但它们可以控制前渊带远端和前隆带的水系和局部沉积物源。这些构造在油气勘探中也具有潜在的重要性，因为它们可以影响储层构型和分布。

20.10.4 待研究的领域

尽管对前陆盆地系统的研究较多，但一直忽视了对造山作用和气候过程之间相互作用的研究。大规模的冲断带，如安第斯山脉和喜马拉雅山脉，控制着地球上最大的地形气候梯度（orographic climate gradients）的发育（Grujic 等，2006；Strecker 等，2007；Bookhagen 和 Strecker，2008），并且在它们侧翼的前陆盆地中保存的内源碎屑造山物质、原地化学沉积组分（如胶结物和古土壤组分），以及矿物碎屑等，记录了气候变化和剥露作用。最近在地质年代学和热年代学的突破性成果，包括多重测年法（multi-dating methods），显示出在揭示复杂的气候变冷和起源历史上的巨大潜力（Bernet 等，2006；Carrapaet 等，2009；Najman 等，2009）。将这些新的资料与通过对前陆盆地沉积带地层分析建立起的地球动力学框架综合起来研究，将对造山作用和气候系统演化有一个整体的认识。

20.11　总结

（1）对现代和古代（地层学分析）前陆盆地系统四个部分的详细大量研究，为揭示造山运动的沉积演化史提供了有力的约束。前陆挠曲波的长距离（长约1000km）侧向迁移，形成了在"瓦尔特"层序上沉积带的垂向叠置，并可用冲断带的扩展和缩短速度加以解释，从而提供了充足的地层年代信息。其中特别有价值的是前隆带上发育的不整合/凝缩带，因为其持续时间可用来估计挠曲波的迁移速率。缺乏对前陆盆地层序中前隆带和隆外带沉积的认识，将阻碍对一个造山运动早期历史的地层辨识。

（2）弧后背景下的动力沉降将有助于加强前隆带和隆外沉积带地层的保存，而在碰撞型和后退式碰撞系统中，由于缺乏区域远源沉降机制以及它们发育高且/或窄的前隆带，前隆带和隆外带沉积地层常不易保存下来。然而，喜马拉雅前陆盆地系统早期（始新世—渐新世）发育薄的隆外沉积和一个大规模的不整合/古土壤凝缩带，后者是前隆带迁移的标志。在向现代盆地形态过渡的演化过程中，前隆带和隆外带均没有沉积，表明卷入印度—亚洲碰撞带中的印度板块岩石圈的刚度的增加，这也是中新世—现今前隆带和隆外带地层记录缺失的原因。

（3）作为穿时挠曲剖面的一个替代指标，前陆盆地沉降曲线的反演能够提供前陆岩石圈挠曲属性的一阶估算。但是，俯冲前陆岩石圈挠曲刚度的显著变化（通常增大）、俯冲岩石圈的迁移速率以及造山带负载的规模，是需要对冲断带—前陆盆地系统进行恢复重建，去完整刻画长期的地球动力学演变。

（4）微小的前陆内部构造的活动和先存构造的复活在前陆盆地中非常常见，尤其是在前隆带，尽管它们的规模通常较小，但这些构造通常能够控制着局部的水系、物源以及岩相的分布。

致谢（略）

参 考 文 献

Aalto, R. Maurice-Bourgoin, L., Dunne, T., Montgomery, D. R., Nittrouer, C. A., and Guyot, J-L. (2003) Episodic sediment accumulation on Amazonian flood plains influenced by El Nino/Souther Oscillation. Nature, 425, 493-497.

Ali, M. Y., and Watts, A. B. (2009) Subsidence history, gravity anomalies and flexure of the United Arab Emirates (UAE) foreland basin. GeoArabia, 14, 17-44.

Allen, P. A. and Allen, J. R. (2005) Basin analysis. Oxford, Blackwell, 549 pp.

Allen, P. A., Burgess, P. M., Galewsky, J., and Sinclair, H. D. (2001) Flexural-eustatic numericalmodel for drowning of the Eocene perialpine carbonate ramp and implications for Alpine geodynamics. GSA Bulletin, 113, 1052-1066.

Arriagada, C., Roperch, P., Mpodozis, C. and Cobbold, P. R. (2008) Paleogene building of the Bolivian orocline: tectonic restoration of the central Andes in 2-D map view. Tectonics, 27, TC6014, doi: 10. 1029/2008TC002269.

Beaumont, C. (1981) Foreland basins. Geophysical Journal of the Royal Astronomical Society, 65, 291-329.

Beck, C., Deville, E., Blanc, E., Phillippe, Y, and Tardy, M. (1998) Horizontal shortening control of Middle Miocene marine siliciclastic accumulation (Upper Marine Molasse) in the southern termination of the Savoy Molasse Basin (northwestern Alps. Southern Jura) in Mascle, A., et al. (eds.), Cenozoic foreland basins of western Europe. Geological Society Special Publications, 134, 263-278.

Bernet, M., Van der Beek, P., Pik, R., Huyghe, P., Mugnier, JL., Labrin, E., and Szulc, A. (2006) Miocene to Recent exhumation of the central Himalaya determined from combined detrital zircon fission-track and U-Pb analysis of Siwalik sediments, western Nepal. Basin Research, 18, 393-412.

Bettinelli, P., et al. (2006) Plate motion of India and interseismic strain in the Nepal Himalaya from GPS and DORIS measurements. J. Geod. doi: 10. 1007/s00190- 006-0030-3.

Bigi, G., Cosentino, D., Parotto, M., Sartori, R., and Scandone, (1992) Structural model of Italy and gravity map. Consiglio Nazionale delle Ricerche, Progetto Finalizzato Geodinamica, 1:500, 000, Florence, Italy.

Bilham, R., Bendick, R., and Wallace, K. (2003) Flexure of the Indian plate and intraplate earthquakes. Proceedings of IndianAcademy of Science (Earth and Planetary Sciences), 112 (3), 1-14.

Blisniuk, P. M., Sonder, L. J., and Lillie, R. J. (1998) Foreland normal fault control on NW-Himalayan thrust front development. Tectonics, 17, 766-779.

Bonardi, G. Cavazza, W., Perrone, V., and Rossi, S. (2001) Calabria-Peloritani terrane and northern Ionian Sea, in Vai G. B., Martini, I. P., eds., Anatomy of an orogen: the Apennines and adjacent Mediterranean basins. Dordrecht, Kluwer Academic Publishers, 287-306.

Bookhagen, B. and Strecker, M. R. (2008) Orographic barriers, high-resolution TRMM rainfall, hillslopes angles, and relief variations along the eastern Andes. Geophysical Research Letters, 35, L06403, doi: 10. 1029/ 2007GL032011.

Booth-Rea, G., Ranero, C. R., Martinez-Martinez, J. M., and Grevemeyer, I. (2007) Crustal types and Tertiary tectonic evolution of the Alborán sea, western Mediterranean. Geochemistry, Geophysics, Geosystems 8. doi: 10. 1029/ 2007/GC001639.

Bradley, D. C., and Kidd, W. S. F. (1991) Flexural extension of the upper continental crust in collisional foredeeps. Bull. Geol. Soc. Am., 103, 1416-1438.

Burbank, D. W., Beck, R. A., and Mulder, T. (1996) The Himalayan foreland basin, in Yin, A., and Harrison, T. M., eds., The tectonics of Asia. New York, CambridgeUniversity Press, 205-226.

Burkhard, M. and Sommaruga, A. (1998) Evolution of the western Swiss Molasse basin: structural relations with the Alps and the Jura belt, in Mascle, A., et al. (eds.), Cenozoic foreland basins of western Europe. Geological Society Special Publications, 134, 279-298.

Cardozo, N., and Jordan, T. (2001) Causes of spatially variable tectonic subsidence in the Miocene Bermejo foreland basin, Argentina. Basin Research, 13, 335-357.

Carrapa, B., DeCelles, P. G., Reiners, P. W., Gehrels, G. E., and Sudo, M. (2009) Apatite triple dating and white mica $^{40}Ar=^{39}Ar$ thermochronology of syntectonic detritus in the Central Andes: a multi-phase tectonothermal history. Geology, 37, 407-410.

Catuneanu, O. (2004) Retroarc foreland systems—evolution through time. Journal of African Earth Sciences, 38, 225-242.

Catuneanu, O., Beaumont, C., and Waschbusch, (1997) Interplay of static loads and subduction dynamics in foreland basins: reciprocal stratigraphies and the "missing" peripheral bulge. Geology, 25, 1087-1090.

Cavazza, W., Roure, F. M., and Ziegler, P. A. (2004) The Mediterranean area and the sutrounding regions: active processes, remnants of former Tethyan oceans and related thrust belts, in Cavazza, W., Roure, F. M., Spakman, W., Stampfli, F. M., and Ziegler, P. A., eds., The transmed atlas. Berlin, Springer, 1-29.

Cavinato, G. P., and DeCelles, G. (1999) Extensional basins in the tectonically bimodal central Apennines foldthrust belt, Italy: response to corner flow above a subducting slab in retrograde motion. Geology, 27, 955-958.

Chase, C. G., Sussman, A. J., and Coblentz, D. D. (2009) Curved Andes: geoid, forebulge, and flexure. Lithosphere, 1, 358-363. doi: 10. 1130/L67. 1.

Chiang, C-S., Yu, H-S., Chou, Y-W. (2004) Characteristics of the wedge-top depozone of the southern Taiwan foreland basin system. Basin Res., 16, 65-78.

Cloetingh, S. A. P. L., et al. (2004) Thermo-mechanical controls on the mode of continental collision in the SE Carpathians (Romania). Earth and Planetary Science Letters, 218, 57-76.

Coakley, B. J., and Watts, A. B. (1991) Tectonic controls on the development of unconformities: the North Slope, Alaska. Tectonics, 10, 101-130.

Coward, M. P. and Butler, R. W. H. (1984) Thrust tectonics and the deep structure of the Pakistan Himalaya. Geology, 13, 417-420.

Crampton, S. L., and Allen, P. A. (1995) Recognition of flexural forebulge unconformities in the geologic record. American Association of Petroleum Geologists Bulletin, 79, 1495–1514.

Currie, B. S. (2002) Structural configuration of the Early Cretaceous Cordilleran foreland–basin system and Sevier thrust belt, Utah and Colorado. J. Geology, 110, 697–718.

DeCelles, P. G. (1994) Late Cretaceous – Paleocene synorogenic sedimentation and kinematic history of the Sevier thrust belt, northeast Utah and southwest Wyoming. Geological Society of America Bulletin, 106, 32–56.

DeCelles, P. G., and Coogan, J. C. (2006) Regional structure and kinematic history of the Sevier fold–thrust belt, central Utah: implications for the Cordilleran magmatic arc and foreland basin system. Geological Society of America Bulletin, 118, 841–864. doi: 10. 1130/B25759. 1.

DeCelles, P. G., and Currie, B. S. (1996) Long–term sediment accumulation in the Middle Jurassic–early Eocene cordilleran retroarc foreland basin system. Geology, 24, 591–594.

DeCelles, G., and DeCelles, C. (2001) Rates of shortening, propagation, under thrusting, and flexural wave migration in continental orogenic systems. Geology, 29, 135–138.

DeCelles, P. G., Gehrels, G. E., Najman, Y., Martin, A. J., Carter, A., and Garzanti, E. (2004) Detrital geochronology and geochemistry of Cretaceous–Early Miocene strata of Nepal: implications for timing and diachroneity of initial Himalayan orogenesis. Earth and Planetary Science Letters, 227, 313–330; doi: 10. 1016/j. epsl. 2004. 08. 019.

DeCelles, G., and Giles, K. N. (1996) Foreland basin systems. Basin Research, 8, 105–123.

DeCelles, G., and Horton, B. K. (2003) Implications of earlymiddle Tertiary foreland basin development for the history of Andean crustal shortening in Bolivia. Geological Society of America Bulletin, 115, 58–77.

DeCelles, G., Gehrels, G. E., Quade, J., and Ojha, T. P. (1998a) Eocene–early Miocene foreland basin development and the history of Himalayan thrusting, western and central Nepal. Tectonics, 17, 741–765.

DeCelles, G., Gehrels, G. E., Quade, J., Ojha, T. P., Kapp, A., and Upreti, B. N. (1998b) Neogene foreland basin deposits, erosional unroofing, and the kinematic history of the Himalayan fold–thrust belt, western Nepal. Geological Society of America Bulletin, 110, 2–21.

DeCelles, G., Robinson, D. M., Quade, J., Ojha, T. P., Garzione, C. N., Copeland, P., and Upreti, B. N. (2001) Stratigraphy, structure, and tectonic evolution of the Himalayan fold–thrust belt in western Nepal. Tectonics, 20, 487–509.

DeCelles, P. G. (2004) Late Jurassic to Eocene evolution of the Cordilleran thrust belt and foreland basin system, western USA. American Journal of Science, 304, 105–168.

del Papa, C., Kirschbaum, A., Powell, J., Brod, A., Hongn, F., and Pimentel, M. (2010) Sedimentological geochemical and paleontological insights applied to continental omission surfaces: a new approach for reconstructing an Eocene foreland basin in NW Argentina. J. South Am. Earth Sci. 29, 327–345.

Demko, T. M., Currie, B. S., and Nicoll, K. A. (2004) Regional paleoclimatic and stratigraphic implications of paleosols and fluvial/overbank architecture in the Morrison Formation (Upper

Jurassic) western interior, USA. Sedimentary Geology 167, 115–135.

Dewey, J. F., Cande, S., and Pitman, W. C. (1989) Tectonic evolution of the India/Eurasia collision zone. Eclogae Geologicae Helvetiae, 82, 717–734.

Dickinson, W. R. (1974) Plate tectonics and sedimentation, in Dickinson, W. R. ed., Tectonics and sedimentation. SEPM Special Publication, 22. Tulsa, OK, Society of Economic Paleontologists and Mineralogists, 1–27.

Dickinson, W. R. (2004) Evolution of the North American Cordillera. Annual Reviews of Earth and Planetary Science, 32, 13–45.

Doglioni, C. (1991) Aproposal for the kinematic modeling of W–dipping subductions; possible applications to the Tyrrhenian–Apennines system. Terra Nova, 3, 423–434.

Dorobek, S. L. (1995) Synorogenic carbonate platforms and reefs in foreland basins: controls on stratigraphic evolution and platform/reef morphology, in Dorobek, S. L. and Ross, G. M., eds., Stratigraphic evolution of foreland basins. SEPM Special Publication, 52. Tulsa, OK, Society of Economic Paleontologists and Mineralogists, 127–147.

Duroy, Y., Farah, A., and Lillie, R. J. (1989) Reinterpretation of the gravity field in the Himalayan foreland of Pakistan, in Malinconico, L. L., and Lillie, R. J., eds., Tectonics of the Western Himalayas. Special Paper, 132. Boulder, CO, Geological Society of America 217–236.

Evenchick, C. A., McMechan, M. E., McNicoll, V. J., and Carr, S. D. (2007) A synthesis of the Jurassic–Cretaceous tectonic evolution of the central and southeastern Canadian Cordillera: exploring links across the orogen, in Sears, J. L., et al. eds., Whence the mountains? Special Paper, 433. Geological Society of America, 117–145.

Faccenna, C., Piromallo, C., Crespo–Blanc, A., Jolivet, L., and Rosetti, F. (2004) Lateral slab deformation and the origin of the western Mediterranean arcs. Tectonics, 23, TC1012, doi: 10.1029/2002TC001488.

Flemings, B., and Jordan, T. E. (1989) A synthetic stratigraphic model of foreland basin development. Journal of Geophysical Research, 94, 3851–3866.

Ford, M., Williams, E. A., Artoni, A., Verges, J., and Hardy, S. (1997) Progressive evolution of a fault–related fold pair from growth strata geometries, Sant Llorenç de Morunys, SE Pyrenees. J. Struct. Geol. 19, 413–441.

Fuentes, F., DeCelles, P. G., and Gehrels, G. E. (2009) Jurassic onset of foreland basin deposition in northwestern Montana, USA: implications for along–strike synchroneity of Cordilleran orogenic activity. Geology 37, 379–382.

Garcia–Castellanos, D. (2002) Interplay between lithospheric flexure and river transport in foreland basins. Basin Research 14, 89–104.

Garcia–Castellanos, D., Fernandez, M., and Torne, M. (1997) Numerical modeling of foreland basin formation: a program relating thrusting, flexure, sediment geometry and lithosphere rheology. Computers & Geoscience 23, 993–1003.

Garzanti, E., Doglioni, C., Vezzoli, G., and Andó, S. (2007) Orogenic belts and orogenic sediment provenance. Journal of Geology 115, 315–334.

Giles, K. A., and Dickinson, W. R. (1995) The interplay of eustasy and lithospheric flexure in forming stratigraphic sequences in foreland settings: an example from the Antler foreland, Nevada and Utah, in Dorobek, S. L., and Ross, G. M., eds., Stratigraphic evolution of foreland basins. SEPM Special Publication, 52. Tulsa, OK, Society of Economic Paleontologists and Mineralogists, 187–211.

Grujic, D., Coutand, I., Bookhagen, B., Bonnet, S., Blythe, A., Duncan, C. (2006) Climatic forcing of erosion, landscape, and tectonics in the Bhutan Himalayas. Geology 34, 801–804.

Gupta, S., and Allen, P. A. (2000) Implications of foreland paleotopography for stratigraphic development in the Eocene distal Alpine foreland basin, Geological Society of America Bulletin 112, 515–530.

Gurnis, M. (1992) Rapid continental subsidence following the initiation and evolution of subduction. Science, 255, 1556–1558.

Gutscher, M-A., Malod, J., Rehault, J. P., Contrucci, I., Klingelhoefer, F., Mendes-Victor, L., Spakman, W. (2002) Evidence for active subduction beneath Gibraltar. Geology 30, 1071–1074.

Heller, P. L., Angevine, C. L., Winslow, N. S., and Paolo, C. (1988) Two-phase stratigraphic model of foreland-basin sequences. Geology, 16, 501–504.

Herb, R. (1988) Eocaene paläogeographie and paläotektonik des Helvetikums. Eclogae Geologicae Helvetiae, 81, 611–657.

Heuret, A. and Lallemand, S. (2005) Plate motions, slab dynamics and backarc deformation. Physics of Earth and Planetary Interiors, 149, 31–51.

Homewood, P., Allen, P. A., and Williams, G. D. (1986) Dynamics of the Mosasse Basin of western Switzerland, in Allen, P. A. and Homewood, P., eds., Foreland Basins, Spec. Pub. Intern. Assoc. Sediment. 8, 199–218.

Horton, B. K, and DeCelles, P. G. (1997) The modern foreland basin system adjacent to the Central Andes. Geology, 25, 895–898.

Houseknecht, D. W. (1986) Evolution from passive margin to foreland basin: the Atoka Formation of the Arkoma basin, south-central U. S. A., in Allen, P. A. and Homewood, P., eds., ForelandBasins, Spec. Pub. Intern. Assoc. Sediment., (8), 327–345.

Jacobi, R. D. (1981) Peripheral bulge: a causal mechanism for the Lower Ordovician unconformity along the western margin of the northern Appalachians. Earth and Planetary Science Letters, 56, 245–251.

Jolivet, L., and Faccenna, C. (2000) Mediterranean extension and the Africa-Eurasia collision. Tectonics 19, 1095–1106.

Jordan, T. E. (1981) Thrust loads and foreland basin evolution, Cretaceous, western United States. American Association of Petroleum Geologists Bulletin, 65, 2506–2520.

Jordan, T. E. (1995) Retroarc foreland and related basins, in Busby, C. J., and Ingersoll, R. V., eds., Tectonics of Sedimentary Basins. Oxford, Blackwell Science, 331–362.

Jordan, T. E., Allmendinger, R. W., Damanti, J. F., and Drake, R. E. (1993) Chronology of motion in a complete thrust belt: the Precordillera, 30–31S, AndesMountains. Journal of

Geology, 101, 135-156.

Karner, G. D., and Watts, A. B. (1983) Gravity anomalies and flexure of the lithosphere at mountain ranges. Journal of Geophysical Research, 88, 10449-10477.

Kley, J., and Monaldi, C. R. (1998) Tectonic shortening and crustal thickness in the Central Andes: How good is the correlation? Geology, 26, 723-726.

Lave, J., and Avouac, J-P. (2000) Active folding of fluvial terraces across the Siwaliks Hills, Himalayas of central Nepal. Journal of Geophysical Research 106, 26561-26592.

Lawton, T. F., Roca, E., and Guimerá, J. (1999) Kinematicstratigraphic evolution of a growth syncline and its implicatioins for the devolopment of the proximal foreland basin, southestern Ebro basin, Catalunay, Spain. Geol. Soc. Am. Bull., 111, 412-431.

Leier, A., DeCelles, P. G., and Kapp, (2007) The Takena Formation of the Lhasa terrane, southern Tibet: the record of a Late Cretaceous retroarc foreland basin. Geological Society of America Bulletin, 119, 31-48.

Lithgow-Bertelloni, C. and Gurnis, M. (1997) Cenozoic subsidence and uplift of continents from time-varying dynamic topography. Geology 25, 735-738.

Liu, S-F., Nummedal, D., Yin, P-G., and Luo, H-J. (2005) Linkage of Sevier thrusting episodes and Late Cretaceous foreland basin megasequences across southern Wyoming (USA). Basin Research 17, 487-506.

Londono, J. and Lorenzo, J. M. (2004) Geodynamics of continental plate collision during late tertiary foreland basin evolution in the Timor Sea: constraints from foreland sequences, elastic flexure and normal faulting. Tectonophysics 392, 37-54.

Lourens, L. J., Hilgen, F. J., Laskar, J., Shackleton, N. J. and Wilson, D. (2004) The Neogene period, in Gradstein, F. M., Ogg, J. G., and Smith, A. G., eds., Ageologic time scale 2004. Cambridge, Cambridge University Press, chap. 20.

Lyon-Caen, H., and Molnar, (1985) Gravity anomalies, flexure of the Indian plate, and the structure, support and evolution of the Himalaya and Ganga basin. Tectonics 4, 513-538.

Malinverno, A., and Ryan, W. B. F. (1986) Extension in the Tyrrhenian Sea and shortening in the Apennines as result of arc migration driven by sinking of the lithosphere. Tectonics 5, 227-245.

Matenco, L. and Bertotti, G. (2000) Tertiary tectonic evolution of the external East Carpathians (Romania). Tectonophysics, 316, 255-286.

McQuarrie, N. (2002) The kinematic history of the central Andean fold-thrust belt, Bolivia: implications for building a high plateau. Geological Society of America Bulletin, 114, 950-963.

Meyers, J. H., Suttner, L. J., Furer, L. C., May, M. T., and Soreghan, M. J. (1992) Intrabasinal tectonic control on fluvial sandstone bodies in the Cloverly Formation (Early Cretaceous) west-central Wyoming, USA. Basin Research, 4, 315-334.

Miall, A. D. (1995) Collision-related foreland basins, in Busby, C. J., and Ingersoll, R. V., eds., Tectonics of sedimentary basins. Oxford, Blackwell Science, 393-424.

Mitrovica, J. X., Beaumont, C., and Jarvis, G. T. (1989) Tilting of continental interiors by the dynamical effects of subduction. Tectonics, 8, 1079-1094.

Murphy, M. A. (2007) Isotopic characteristics of the Gurla Mandhata metamorphic core complex: implications for the architecture of the Himalayan orogen. Geology, 35, 983-986.

Najman, Y. (2006) The detrital record of orogenesis: a review of approaches and techniques used in the Himalayan sedimentary basins. Earth Science Reviews 74, 1-72.

Najman, Y., Carter, A., Oliver, G., and Garzanti, E. (2005) Provenance of Eocene foreland basin sediments, Nepal: constraints to the timing and diachroneity of early Himalayan orogenesis. Geology 33, 309-312.

Najman, Y., Bickle, M., Garzanti, E., Pringle, M., Barfod, D., Brozovic, N., Burbank, D., and Ando, S. (2009) Reconstructing the exhumation history of the Lesser Himalaya, NW India, from a multitechnique provenance study of the foreland basin Siwalik Group. Tectonics, 28, TC5018. doi: 10. 1029/2009TC002506.

Ojha, T. P., Butler, R. F., DeCelles, P. G., and Quade, J. (2008) Magnetic polarity stratigraphy of the Neogene foreland basin deposits of Nepal. Basin Research. doi: 10. 1111/ j. 1365-2117. 2008. 00374. x.

Ori, G. G., Roveri, M., & Vannoni, F. (1986) Plio-Pleistocene sedimentation in the Apenninic-Adriatic foredeep (central Adriatic Sea, Italy) in Allen, A., and Homewood, P., eds., Foreland basins, Spec. Pub. Int. Ass. Sed. 8, 183-198.

Pang, M., and Nummedal, D. (1995) Flexural subsidence and basement tectonics of the Cretaceous western interior basin, United States. Geology, 23, 173-176.

Pfiffner, O. A. (1986) Evolution of the north Alpine foreland basin in the Central Alps, in Allen, P. A., and Homewood, P., eds., Foreland basins. Spec. Pub. Int. Ass. Sed. 8, 219-228.

Plint, A. G., Hart, B. S., Donaldson, W. S. (1993) Lithospheric flexure as a control on stratal geometry and facies distribution in Upper Cretaceous rocks of the Alberta foreland basin. Basin Research, 5, 69-77.

Price, R. A. (1973) Large scale gravitational flow of supracrustal rocks, southern Canadian Rockies, in DeJong, K. A., and Scholten, R. A., eds., Gravity and tectonics. New York, Wiley, 491-502.

Quinlan, G. M., and Beaumont, C. (1984) Appalachian thrusting, lithospheric flexure, and the Paleozoic stratigraphy of the Eastern Interior of North America. Canadian Journal of Earth Science, 21, 973-996.

Robinson, D. M., DeCelles, P. G., and Copeland, (2006) Tectonic evolution of the Himalayan thrust belt in western Nepal: implications for channel flow models. Geological Society of America Bulletin, 118, 865-885. doi: 10. 1130/B25911. 1.

Roddaz, M. Baby, P., Brusset, S., Hermoza, W., and Darrozes, J. M. (2005) Forebulge dynamics and environmental control in Western Amazonia: the case study of the Arch of Iquitos (Peru). Tectonophysics, 399, 87-108.

Royden, L. and Karner, G. D. (1984) Flexure of lithosphere beneath Apennine and Carpathian foredeep basdins: evidence for an insufficient topographic load. American Association of Petroleum Geologists, 68, 704-712.

Royden, L. H. (1993) The tectonic expression of slab pull at continental convergent boundaries. Tectonics, 12, 303-325.

Schellart, W. P. (2008) Overriding plate shortening and extension above suduction zones: a parametric study to explain formation of the Andes Mountains. Geological Society of America Bulletin, 120, 1441-1454.

Schmid, S. M., Bernoulli, D., Fugenschuh, B., Matenco, L., Schefer, S., Schuster, R., Tischler, M., and Usutaszewski, K. (2008) The Alpine – Carpathian – Dinaridic orogenic system: correlation and evolution of tectonic units. Swiss Journal of Geoscience, 101, 139-183.

Schwartz, R. K., and DeCelles, P. G. (1988) Foreland basin evolution and synorogenic sedimentation in response to interactive Cretaceous thrusting and reactivated foreland partitioning. Geological Society of America Memoir, 171, 489-513.

Sinclair, H. D., Coakley, B. J., Allen, P. A., and Watts, A. B. (1991) Simulation of foreland basin stratigraphy using a diffusion model of mountain belt uplift and erosion: an example from the central Alps, Switzerland. Tectonics, 10, 599-620.

Sinclair, H. D. (1997) Tectonostratigraphic model for underfilled peripheral foreland basins: an Alpine perspective. Geological Society of America Bulletin 109, 324-346.

Snyder, D. B., and Barazangi, M. (1986) Deep crustal structure and flexure of the Arabian plate beneath the Zagros collisional mountain belt as inferred from gravity observations. Tectonics 6, 361-373.

Spakman, W., and Wortel, R. (2004) A tomographic view on western Mediterranean geodynamics, in Cavazza, W., Roure, F. M., Spakman, W., Stampfli, F. M., and Ziegler, P. A. (eds.), The Transmed Atlas, Springer, Berlin, 31-52.

Srivastava, P., and Mitra, G. (1994) Thrust geometries and deep structure of the outer and lesser Himalaya, Kumaon and Garhwal (India): implications for evolution of the Himalayan fold-and-thrust belt. Tectonics, 13, 89-109.

Strecker, M. R., Alonso, R. N., Bookhagen, B., Carrapa, B., Hilley, G. E., Sobel, E. R., and Trauth, M. H. (2007) Tectonics and climate of the southern central Andes. Annu. Rev. Earth Planet. Sci., 35, 747-787.

Szulc, A. G., Najman, Y., Sinclair, H. D., Pringle, M., Bickle, M., Chapman, H., Garzanti, E., Ando, S., Huyghe, P., Mugnier, J-L., Ojha, T., and DeCelles, (2006) Tectonic evolution of the Himalaya constrained by detrital $^{40}Ar-^{39}Ar$, Sm-Nd and petrographic data from the Siwalik foreland basin succession, SW Nepal. Basin Research, 18, 375-392.

Tandon, S. K., Sinha, R., Gibling, M. R., Dasgupta, A. S., and Ghazanfare, (2008) Late Quaternary evolution of the Ganga Plains: myths and misconceptions, recent developments and future directions. Golden Jubilee Memoir of the Geological Society of India, 66, 259-299.

Tankard, A. J. (1986) On the depositional response to thrusting and lithospheric flexure: examples from the Appalachian and Rocky Mountain basins. International Association of Sedimentologists Special Publication, 8, 369-392.

Tensi, J., Mouthereau, F., and Lacombe (2006) Lithospheric bulge in the West Taiwan basin.

Basin Research, 18, 277–299.

Turcotte, D. L., and Schubert, G. (2006) Geodynamics: applications of continuum physics to geological problems. New York, John Wiley & Sons, 456 p.

Uba, C. E., Heubeck, C., and Hulka, C. (2006) Evolution of the late Cenozoic Chaco foreland basin, southern Bolivia. Basin Research, 18, 145–170.

Ustaszewski, K., Schmid, S. M., Fugenschuh, B., Tischler, M., Kissling, E., and Spakman, W. (2008) A map-view restoration of the Alpine-Carpathian-Dinaridic system for the Early Miocene. Swiss J. Geosci., 101, 273–294.

Vergés, J., Marzo, M., Santaeulária, T., Serra-Kiel, J., Burbank, D. W., Muñoz, J. A., Giménez-Montsant, J. (1998) Quantified vertical motions and tectonic evolution of the SE Pyrenean foreland basin, in Mascle, A., et al. (eds.) Cenozoic foreland basins of western Europe. Geological Society Special Publications, 134, 107–134.

Waschbusch, J., and Royden, L. H. (1992) Spatial and temporal evolution of foredeep basins: lateral strength variations and inelastic yielding in continental lithosphere. Basin Research, 4, 179–196.

Watts, A. B. (2001) Isostasy and flexure of the lithosphere. Cambridge, Cambridge UniversityPress, 478 p.

Watts, A. B. and Burov, E. B. (2003) Lithospheric strength and its relationship to the elastic and seismogenic layer thickness. Earth and Planetary Science Letters. 213, 113–131.

Williams, E. A., Ford, M., Verges, J., and Artoni, A. (1998) Alluvial gravel sedimentation in a contractional growth fold setting, Sant Llorenç de Morunys, southeastern Pyrenees, in Mascle, A., et al. (eds.) Cenozoic foreland basins of western Europe. Geological Society Special Publications, 134, 69–106.

Willis, A. (2000) Tectonic control of nested sequence architecture in the Sego Sandstone, Neslen Formation and upper Castlegate Sandstone (Upper Cretaceous) Sevier foreland basin, Utah, USA. Sedimentary Geology, 136, 277–317.

Wuellner, D. E., Lehtonen, L. R., & James, W. C. (1986) Sedimentary tectonic development of the Marathon and Val Verde basins, west Texas, U. S. A.: a Permo-Carboniferous migrating foredeep, in Allen, P. A., and Homewood, P., eds., Foreland basins. Spec. Pub. Int. Ass. Sed. 8, 347–368.

Yang, Y., and Miall, A. D. (2009) Evolution of the northern Cordilleran foreland basin during the middle Cretaceous. Geological Society of America Bulletin, 121, 483–501.

Yu, H. S., and Chou, Y. W. (2001) Characteristics and development of the flexural forebulge and basal unconformity of western Taiwan foreland basin. Tectonophysics, 333, 277–291.

Zaleha, M. J., Way, J. N., and Suttner, L. J. (2001) Effects of syndepositional faulting and folding on Early Cretaceous rivers and alluvial architecture (Lakota and Cloverly Formations, Wyoming, U. S. A.). J. Sed. Res., 71, 880–894.

(蔡国富 译，阳怀忠 吴哲 校)

第21章 安第斯山和青藏高原腹地新生代盆地演化

BRIAN K. HORTON

(Department of Geological Sciences and Institute for Geophysics,
Jackson School of Geosciences, University of Texas at Austin, Austin, USA)

摘　要：在新生代板块会聚时期安第斯山脉及喜马拉雅—西藏造山带的腹地区域形成了许多沉积盆地。在长期的陆壳构造变形及地表隆升过程中，这些腹地盆地充填了非海相的沉积物（通常形成于高海拔、低地势、内流水系，以及干旱和半干旱环境）。安第斯内陆盆地形成于西部的岩浆弧和东部趋向克拉通的褶皱—冲断带之间。在印度—欧亚碰撞带，腹地盆地作为生长中的青藏高原的组成部分，形成于喜马拉雅冲断带和北部的亚洲板块内陆之间。安第斯和西藏腹地沉积盆地在构造部位、海拔及地层演化方面，与它们所对应的褶皱—冲断带邻近的前陆盆地有明显不同。岩石圈的挠曲、断裂导致的沉降、地形阻塞促成了这些腹地盆地的沉积可容空间的发育，残余的前陆盆地也继承性地形成了多个年轻的腹地盆地。

关键词：阿尔蒂普拉诺高原　安第斯　青藏　褶皱—冲断带　前陆盆地　腹地　高原

21.1　引言

会聚造山带的"腹地"包括远离低海拔前陆盆地的内部隆起区域和褶皱—冲断带的外部山麓。位于区域基准面之上的腹地盆地容易遭受剥蚀且较难在地质记录中得到保存。然而，在大型现代造山带的腹地地区，尤其是在安第斯和喜马拉雅—西藏系统中，保留了横跨几十个百万年的长期沉积记录（Jordan 和 Alonso，1987；Allmendinger 等，1997；Yin 和 Harrison，2000；Tapponnier 等，2001）。这些地层中包含了可能形成于海平面 3~5km 之上的高海拔地区的非海相沉积（Garzione 等，2000，2006；Rowley 和 Currie，2006；DeCelles 等，2007a；Saylor 等，2009）。

本章概述了腹地盆地的构造演化史。安第斯山腹地和青藏高原分别是非碰撞弧后造山带和碰撞造山带的潜在端元。这两个系统均记录了早白垩世或新生代以来的地壳缩短、增厚和低地势腹地的地表隆升历史。安第斯山脉（图 21-1）是一个典型的洋—陆会聚板块边缘，为理解俯冲带、岩浆弧、弧后冲断带和弧后前陆盆地提供了有利条件（Jordan 等，1983；Isacks，1988；Horton 和 DeCelles，1997；James 和 Sacks，1999；Kay 等，1999；Beck 和 Zandt，2002；Kay 等，2005；McQuarrie 等，2005；Uba 等，2006）。喜马拉雅—西藏造山带（图 21-2）是一个陆—陆碰撞的典型实例，并且验证了收缩构造和周缘前陆盆地的概念（Molnar 和 Tapponnier，1975；Tapponnier 等，1982；Molnar 等，1993；Royden 等，1997；DeCelles 等，1998；Chemenda 等，2000；Hodges，2000；Lave 和 Avoune，2000；Beaumont 等，2001）。

本章主要关注安第斯和西藏地区典型腹地盆地的构造背景和综合的新生代地层，对于沉

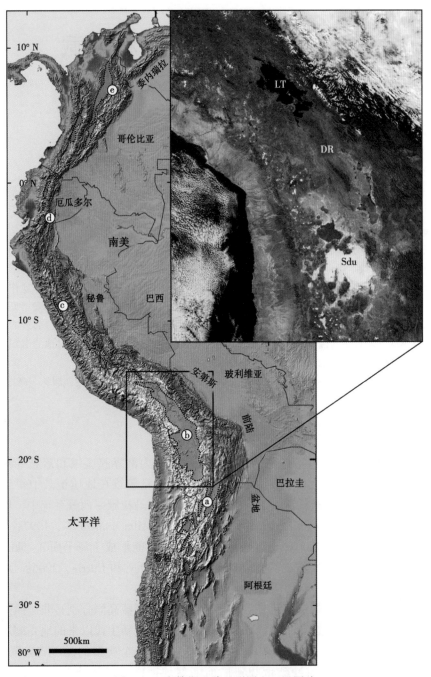

图 21-1 安第斯山脉地形图和卫星图片

数字高程模型（源自 NASA）刻画的安第斯腹地主要盆地（轮廓和阴影部分）集中在安第斯山脉中部和北部。(a) 克鲁塞斯盆地（阿根廷普纳高原）；(b) 高原盆地（玻利维亚阿尔蒂普拉诺高原）；(c) Callojon de Huaylas 盆地（秘鲁科迪勒拉布兰卡）；(d) 昆卡（Cuenca）盆地（厄瓜多尔安第斯山脉内部山谷（Interandean Valley））；(e) 马格达莱纳河谷（Magdalena Valley）盆地（哥伦比亚）。彩色卫片（源自 NASA Terra/MODIS）显示出安第斯腹地中部的盐湖 [SdU—乌尤尼盐湖（Salar de Uyuni）]、淡水湖 [LT—Tilicaca 湖（Lake Tilicaca）] 和河流 [DR—德萨瓜德罗河（Desaguadero river）]

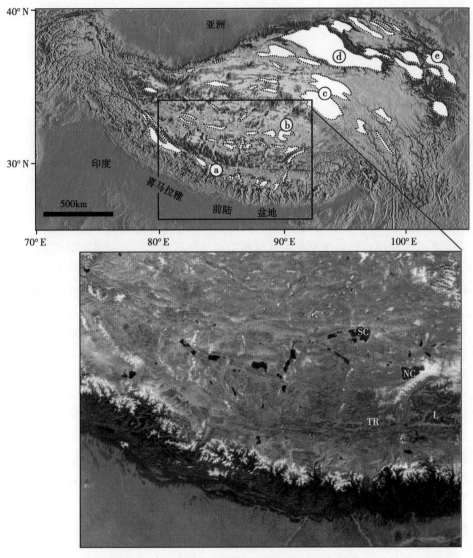

图 21-2 喜马拉雅—西藏造山带地形图和卫星图片

数字高程模型（据 Michael H.Taylor；改自 Yin，2000）显示出主要的青藏高原腹地盆地（轮廓和阴影部分），暖色代表高海拔。(a) Thakkhola 盆地（西藏南部）；(b) 伦坡拉盆地（西藏中部）；(c) 可可西里盆地（西藏中北部）；(d) 羌塘盆地（西藏北部）和（e）西宁盆地（西藏东北部）。彩色卫片（源自 NASA Terra/MODIS）显示出西藏中南部的内流水系湖盆（SC—西林措湖；NC—纳木错湖），位于西藏东南部拉萨（L）附近的外流水系和西藏南端的纵向水系（TR—藏布江河）

积学、沉积物源和沉积分布模式更详细的论述请见引用的参考文献。通过对两种会聚系统的归纳，很容易发现在腹地地区中发育着多种盆地演化模式，其中包括冲断负载导致的地壳挠曲、沿着走滑或伸展断裂的断裂沉降，以及在内流水系地区形成的阻塞地形。尽管一些腹地盆地是在变形的造山带基底之上新形成的，但是由于构造变形前缘远离造山带内部并向克拉通方向推挤，前陆盆地融入生长的造山带内，安第斯弧后腹地和西藏碰撞腹地具有一个共同的演化模式，即都经历了前期的前陆盆地系统向抬升的腹地盆地转变。

21.2 安第斯腹地盆地

21.2.1 构造背景

安第斯山脉（图21-1）是纳兹卡洋壳板片向东俯冲到南美洲板块之下形成的非碰撞构造变形的产物。Wadati 贝尼奥夫带沿着约 7000km 长的会聚板块边缘，倾向向东发生中度倾斜（约30°）至近水平的变化（James，1971；Barazangi 和 Isacks，1976）。板块俯冲和岛弧岩浆作用普遍开始于中生代。尽管受新生代地壳缩短的叠合作用影响，构造重建仍显示中生代经历了偏拉张的应力作用（Dalziel，1981；Mpodozis 和 Ramos，1989；Salfity 和 Marquillas，1994；Cooper 等，1995）。从早白垩世晚期非洲裂陷开始，南美洲板块明显向西运动（Coney 和 Evonchick，1994），但真正的挤压和造山运动是从新生代开始的。

安第斯地区地壳缩短归因于早新生代或晚白垩世的板块会聚，但其开始时间存在空间差异性（Steinmann，1929；Megard，1984；Dalziel，1986；Wilson，1991；Dengo 和 Covey，1993；Ramos 和 Aleman，2000；Horton 等，2001；McQuarrie 等，2005，2008；Mpodozis 等，2005）。一些学者认为构造缩短限于 23~30Ma，且有证据显示在晚渐新世加速缩短（Isacks，1988；Sempere 等，1990；Gubbels 等，1993；Hindle 等，2002）。沿着碰撞带东西向的最小缩短量具有差异性，变化范围为约 20~350km，在安第斯山脉中部（10°~30°S）观测到的缩短量和地表隆升最大（Kley，1996；Baby 等，1997；Kley 和 Monaldi，1998；Kley 等，1999；McQuarrie，2002a，2002b）。除了地壳缩短和增厚外，岩石圈的变薄、拆沉滑脱及其他一些地质过程在安第斯高原中部也非常重要（Kay 和 Kay，1993；Whitman 等，1996；Beck 和 Zandt，2002；Schurr 等，2006）。地壳缩短主要与南美洲板块向海沟方向（西向）的加速有关，其他影响因素包括剥蚀强度（Masek 等，1994；Horton，1999；Lamb 和 Davis，2003）、岩浆活动（Babeyko 等，2002；DeCelles 等，2009）、板片动力过程（Gutscher 等，2000；Ramos 等，2002）和下地壳流动（Lamb 和 Hoke，1997；Kley 和 Monaldi，1998）。

21.2.2 盆地概述

安第斯腹地盆地位于安第斯岩浆弧与克拉通方向（东向会聚）的冲断带尾翼（西部）之间的弧后腹地（图21-1）。这些腹地盆地的新生代填充物主要暴露在平行于区域构造走向的出露带上。出露区宽 10~100km，长 10~500km，以多种断裂、褶皱和地形地貌特征为界（Jordan 和 Alonso，1987；Kennan 等，1995；Marocco 等，1995；Horton，1998）。

盆地以充填中和充填后的构造发育为特征，其中包括褶皱—逆冲推覆构造、走滑断层和少量正断层。安第斯腹地盆地的面积从 100km² 到 50000km² 不等，沉积了大于 2~12km 厚的非海相层序，沉积速率通常大于 200m/Ma（Horton 等，2001；Garzione 等，2008）。该厚度和沉积速率与已知的安第斯前陆、楔顶及背驮盆地相吻合（Reynolds 等，1990；Jordan，1995；Echavarria 等，2003；Horton，2005）。

安第斯腹地地区的现代沉积包括湖泊相、河流相和风成沉积物，一般形成于干旱和半干旱环境。在内流水系地区，如 Altiplano-Punn 高原（安第斯中部）（图21-1），淡水湖（如塔塔湖）和咸水湖（如乌尤尼盐湖）与轴向河流系统（Desaguadero 河）共生。活动沉积也出现在腹地地区的内流水系，伴随平行于构造走向的轴向河流，如哥伦比亚的马格达莱纳河及秘鲁的乌卡亚利河。

21.2.3 代表性盆地

按沉积历史安第斯腹地盆地可分为五个带（图21-3）：普纳高原（阿根廷）、阿尔蒂普拉诺高原（玻利维亚）、科迪勒拉布兰卡山（秘鲁）、Interandean Valley（厄瓜多尔）和马格达莱纳河谷（哥伦比亚）。相关的变形历史综合图（图21-3）显示，安第斯山各部分地壳缩短的起始时间的估算具有很大的不确定性。

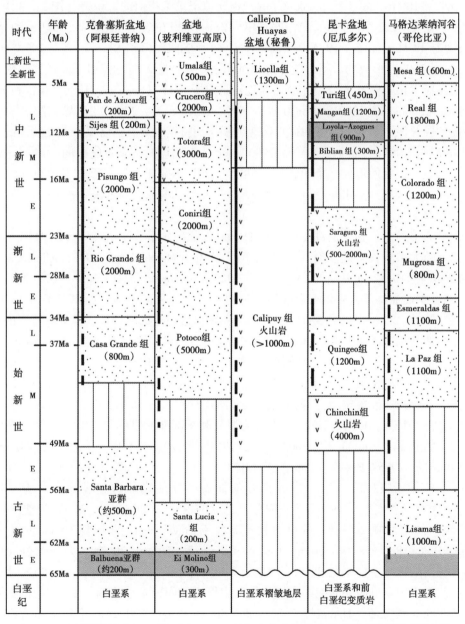

图21-3 安第斯腹地盆地（位于图21-1）的新生代非海相地层记录和推断的变形历史
地层：碎屑沉积（点状）；局限海相沉积（阴影部分）；同沉积火山作用（V形）；无保存记录（竖线）。
构造变形：挤压（窄黑条）；伸展（白条）；走滑变形（灰条）；时间约束不确定（虚线）

安第斯腹地盆地所在的构造部位及地层演化历史与位于亚马逊、巴拉那和奥里诺科河流域系统的低洼地带上的低海拔前陆盆地系统存在差异（Jordan，1995；Horton和DeCelles，1997；De Berc等，2005）。基于它们的构造边界、海拔和持续历史，这些腹地盆地可以与位于冲断带上的楔顶或背驮盆地相区别，楔顶或背驮盆地常处于逆冲带内由断裂形成的地形边界之间，且发育时间短暂（Boer等，1990；Marocco等，1995；Horton，1998，2005；Mosolf等，2011）。

（1）在阿根廷北部（21°—26°S）4~5km高的普纳高原上（图21-1），被分割开的盆地面积一般小于$10×10^4km^2$，且1~6km厚的盆地充填记录了相似的沉积史（图21-3a）。在许多盆地中，萨尔塔裂谷内的白垩纪非海相到滨海沉积在裂后热沉降（Jordan和Alonso，1987；Salfity和Marquillas，1994）或前陆盆地初始挠曲沉降（Kraemer等，1999；Carrapa和DeCelles，2008）形成的早新生代非海相沉积之上。冲断带向克拉通方向迁移，使早期的前陆盆地在新生代中期发生破裂，腹地盆地由此开始形成（Schwab，1985；Boll和Hernandez，1986；Coutand等，2001；Hongn等，2007；Carrapa等，2008）。这导致阻隔地形和内流水系的形成并保存至今，众多的孤立盆地沉积作用持续进行，这常见于蒸发盐湖或干盐湖内（Alonso等，1991；Vandervoort等，1995）。上地壳缩短时期新近纪山间障碍（orographic barriers）的发育，促使形成一个包含周期性阻塞地形和掘开山间双重过程的独特的地貌演化史（Sobel等，2003；Hilley和Strecker，2005；Coutand等，2006；Strecker等，2007；Hain等，2011）。

（2）在玻利维亚，位于南纬16°—21°的4km高的阿尔蒂普拉诺高原（图21-1）有一个腹地背景下的大型现代盆地，时代至少始于约25Ma（Marshall等，1992，Allmendinger等，1997；Lamb和Hoke，1997；Horton等，2002a；Hampton和Horton，2007；Murray等，2010）。与普纳高原相反，阿尔蒂普拉诺高原只含一个完整的盆地，面积大于$500×10^4km^2$，新生代层序厚达12km（图21-3b）。尽管局部存在白垩纪边缘海环境，但只有有限的证据支持白垩纪的伸展构造影响了普纳地区（Wolsink等，1995）。在该地区虽然走滑断层和正断层起到了非常重要的作用，但阿尔蒂普拉诺的新生代发育历史却包含一个由初始前陆盆地条件下形成的厚层非海相层序（Lamb和Hoke，1997；Elger等，2005）。阿尔蒂普拉诺盆地的地形隔离与主要逆冲推覆构造前缘向克拉通方向迁移有关（McQuarrie，2002a；DeCelles和Horton，2003；McQuarrie等，2005）。西部岩浆弧和东部冲断带的抬升和侵蚀活动在阿尔蒂普拉诺盆地的物源（Horton等，2002a）及高原边缘的低温热年代学记录上（Barnes等，2006；Gillis等，2006；Ege等，2007）均有体现。

（3）在安第斯腹地唯一一个发育良好的伸展型盆地毗邻科迪勒拉布兰卡（图21-1）——一个在秘鲁安第斯山脉上的海拔大于5km的山岭。Callejon de Huaylas表层拆离盆地（图21-3c）位于西倾20°—45°的科迪勒拉布兰卡活动正断层的上盘（Bonnot等，1988；Schwartz，1988；Pelford和Atherton，1992；McNulty和Farber，2002）。大于1.3km厚的层序覆盖在挠曲的侏罗—白垩纪和早新生代岛弧火山岩之上（Giovanni等，2010）。盆地开始发育的时间与晚中新世俯冲的纳斯卡板片的压扁作用时间以及位于南纬2°—14°的岛弧岩浆作用停止时间大致吻合（Hampel，2002；McNulty和Farber，2002）。正断层和盆地的演化是重力坍陷（Dalmayrac和Molnar，1981；Sebrier等，1988）和与倾斜板片动力作用有关的应变分解的结果。秘鲁的其他盆地（阿亚库乔、卡哈班巴、圣马科斯、纳莫拉盆地）与科迪勒拉布兰卡不同，这些盆地与当地的压扭和张扭构造有关（Megard等，1984；Marocco

等，1995；Wise 等，2008）。

（4）厄瓜多尔安第斯山脉内部山谷（Interandean Valley）的小型新近纪盆地（图 21-1）位于科迪勒拉弧西部和科迪勒拉冲断带东部之间的海拔约 3km 的高地上。单个的盆地（昆卡、Giron-Santa Isabel、Nabon、洛哈和 Malacatos-Vilcabamba 盆地）面积小于 1000km^2 且充填小于 2~5km 的中中新世至第四纪沉积（图 21-3）（Lavenu 等，1992，1995）。厄瓜多尔腹地地区缺失古近纪沉积，表明不连通的盆地的前新近纪构造活动之前经历了剥蚀或无沉积过程。尽管这些关系与前新近纪造山运动一致，但海相化石表明一些腹地地区在约 15—10Ma 之前仍然位于低海拔环境（Hungerbühler 等，1995，2002）。许多盆地是以潜在的前新生代缝合线附近的走滑断层和反转断层为边界的，表明存在与可能的断裂复活有关的压扭背景的沉降（Hungerbühler 等，2002；Winkler 等，2005）。

（5）哥伦比亚马格达莱纳河谷盆地（图 21-1 和图 21-3e）介于一个残余中生代—早新生代火山弧（科迪勒拉中部）和一个新生代褶皱—冲断带（科迪勒拉东部）之间。尽管经历哥伦比亚最西端复杂的地块增生，白垩纪和新生代盆地的演化还是归因于区域伸展向缩短过程的短暂过渡作用（Dengo 和 Covey，1993；Cooper 等，1995）。对于哥伦比亚安第斯山脉的中东部区域，在白垩纪同裂陷期和裂后沉降的海相沉积之后，伴随着一套早新生代挠曲沉降时期沉积的 2~5km 厚的非海相沉积。科迪勒拉中部的初始构造缩短和地壳负载使马格达莱纳河谷上的盆地向前陆盆地环境转变（Gómez 等，2003，2005a，2005b；Horton 等，2010）。碎屑物源记录（Nie 等，2010；Moreno 等，2011）表明，东科迪勒拉缩短所致的隆升，造成这个早新生代前陆盆地随后的分带，从而形成了一个孤立的腹地盆地并额外充填了 2~5km 厚的新近系沉积（Van Houten 和 Travis，1968；Van Houten，1976）。

21.3 青藏高原腹地盆地

21.3.1 构造背景

喜马拉雅山脉和青藏高原是地球上最高的山脉和最大的造山高原。喜马拉雅—西藏造山带的新生代建造是由印度板块相对亚洲板块的碰撞和持续的向北推进驱动的。对印度—亚洲板块碰撞时间的估计集中在约 55Ma（Searle 等，1987；Dewey 等，1989；Rowley，1996），也有人认为碰撞发生在约 70—60Ma（Jeager 等，1989；Yin 和 Harrison，2000）或约 34Ma（Aitchison 等，2007）。紧随初始碰撞之后，印度岩石圈向北俯冲 700km 至西藏南部和中部之下（Le Pichon 等，1992；Owens 和 Zandt，1997；DeCelles 等，2002）。

碰撞初期的南北向缩短集中在现今印度—亚洲（印度—雅鲁—藏布）缝合带周围的隆起腹地地区及北部的青藏高原。估计缩短作用的开启受前碰撞变形和抬升的影响而复杂化。在碰撞之前，北倾的印度洋壳板片俯冲说明了横贯喜马拉雅（冈底斯—拉达克—科希斯坦）的岩浆弧和相关的弧后缩短（Burg 等，1983；England 和 Searle，1986；Kapp 等，2007；Leier 等，2007a）。基于构造重建，碰撞期间喜马拉雅地区的总缩短量估计在 600~750km 之间（DeCelles 等，1998，2002；Robinson 等，2006），而青藏高原地区大约为 500—750km（Murphy 等，1997；Yin 和 Harrison，2000；Kapp 等，2005）。尽管古地磁数据显示西藏和稳定的亚洲板块内陆之间有远大于上述值的南北会聚距离（达 2500km）（Achache 等，1984），但构造重建不能计算出如此大的值。白垩系—新生界中磁倾角变浅而不是极端缩短，这可能用来解释该古地磁数据（Tan等，2003）。

21.3.2 盆地概述

喜马拉雅—西藏造山带内部的沉积盆地（图21-2）形成于印度—亚洲板块的持续会聚时期。青藏高原的腹地盆地分布在介于南部的印度—亚洲缝合带和中亚的克拉通内陆之间的区域内，该区域南北宽约1000km、东西长约2000km。许多盆地以沿着逆冲推覆构造和走滑断层形成的正地形地貌为边界。大多数西藏腹地盆地的出露发生在受限制的、以断层为边界的区域内（Métivier等，1998；Liu等，2001；Horton等，2002b；DeCelles等，2007b；Saylor等，2009）。

与盆地沉积充填有关的断层表明存在一个宽阔的向中亚方向的北向推进变形（Tapponnier等，2001；Royden等，2008）。然而，青藏高原最北端的一些构造在早新生代的活化（Yin等，2002，2007，2008a，2008b；Horton等，2004；Dupont-Nivet等，2004）表明，许多新生代的沉积发生在腹地环境而不是在前陆盆地中。西藏的腹地盆地显示它们之间的总厚度和沉积速率差别较大。一些盆地的厚度小于1km且沉积速率小于50~100m/Ma（Horton等，2002b），而另一些盆地含厚度达6~10km、沉积速率达1000m/Ma的层序（Métivier等，1998；Liu等，2001；Fang等，2003），与喜马拉雅前陆盆地（Burbank等，1996；DeCelles等，1998）相当。

现今高原中活动的沉积作用主要发生在干旱/半干旱条件下被分隔的、内部发育或不发育水系的含淡水和咸水湖的盆地中。现今河流作用在西藏东部占据着更为主要的作用，大型外流水系（如黄河、长江、湄公河、怒江、红河和雅鲁藏布江）形成的深部切割，塑造了晚新生代至今的地貌特征。然而，特定的区域也发生主动的沉积物充填，如西藏东部的现代若尔盖盆地中，黄河流域的部分地区有湿地河湖相沉积（Chen等，1999）。

21.3.3 代表性盆地

青藏高原的碰撞腹地盆地表现出多种多样的沉积和构造变形历史，在此总结五个地区（图21-4）：印度—亚洲缝合带（西藏南端）、拉萨—羌塘地块边界（西藏中部）、风火山、柴达木盆地和西藏东北部区域。这些地区的缩短开始时间与印度—亚洲板块初始碰撞大致对应。西藏高海拔腹地盆地（图21-2）的发育独立于其南部的喜马拉雅褶皱—冲断带，其构造背景和地层记录也不同于高原边缘的低海拔前陆盆地和碰撞后继盆地，如喜马拉雅前陆盆地、四川盆地、河西盆地、塔里木盆地、准噶尔盆地、吐鲁番盆地和许多蒙古盆地（Hendrix等，1992；Graham等，1993，2001；Burbank等，1996；DeCelles等，1998；Johnson，2004；Johnson和Ritts，本书）。

（1）在西藏的南端（图21-2），沿着北倾正断层形成的沉降，产生了一系列中中新世—第四纪的腹地盆地。单个的裂谷盆地，如Thakkhola地堑（图21-4a）在面积为约20000km^{2}的出露地区含厚达5km的非海相沉积（Garziono等，2003）。这些北倾的盆地不同于与印度—亚洲缝合带南部的西藏南部滑脱系统有关的东倾的伸展型盆地（Burchfiel等，1992；Hodges，2000）。西藏东西向的伸展构造由重力坍塌（Molnar等，1993）、轴向弧形拉张（McCaffrey和Nabalek，1998）、印度碰撞（Kapp和Guynn，2004）和沿着太平洋板块边缘的俯冲后退作用形成（Northrup等，1995；Yin，2000）。

（2）在印度—亚洲碰撞前，西藏南部和中部的拉萨地块上（图21-2），于白垩纪在冈

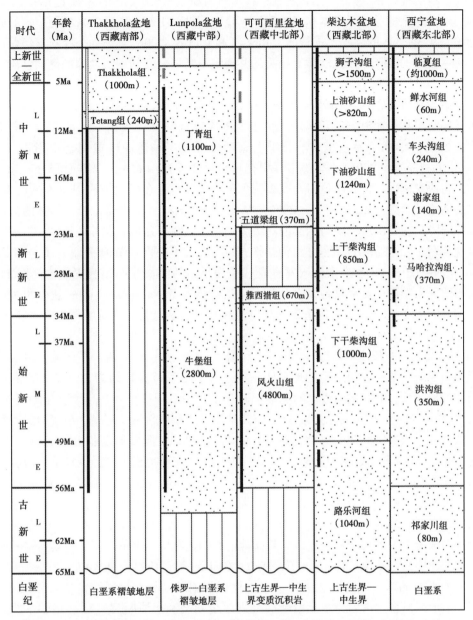

图 21-4 青藏高原腹地盆地（位于图 21-1）的新生代非海相地层记录和推断的变形历史
地层：碎屑沉积（点状）；无保存记录（竖线）。构造变形：挤压（窄黑条）；伸展（白条）；
走滑变形（灰条）；时间约束不确定（虚线）

底斯弧北部形成了一个弧后前陆盆地（Kapp 等，2007；Leier 等，2007a）。一些学者认为它是弧后扩张盆地的中生代生长（Zhang，2000；Chen 等，2003）和/或拉萨和羌塘块体碰撞时发育的一个周缘前陆盆地（Kapp 等，2005；Leier 等，2007b）。在羌塘地块中，约 3~4km 厚的沉积物从缩短有关的构造向南以及可能向北滑脱至这些海相和非海相沉积盆地中。紧随印度—亚洲板块的初始碰撞，前期的前陆盆地和伸展型盆地被不连续的盆地（包括杜巴和伦坡拉盆地；图 21-4b）和它们的边界逆冲断层所叠置（Xu，1984；Leeder 等，1988）。这

509

些新生代盆地由挠曲和/或下盘倾斜所致,在出露面积达 $2\times10^4 km^2$ 的区域上沉积了厚达 4~5km 的层序,并充填了近源扇体和湖相沉积。晚始新世至中新世碳酸盐岩的同位素研究表明盆地充填是在 3~5km 的高程上进行的(Rowley 和 Currie,2006)。另外一些新近纪盆地是由于伸展构造而致的沉降和在运动学上相关的共轭走滑断层之间的楔形块体的侧向挤出而形成(Taylor 等,2003,本书)。

(3)西藏中北部的风火山地区(图 21-2)由可可西里盆地复合体组成(图 21-4c),部分被逆冲断层相关地形分隔开来。三个主要的次级坳陷(风火山—汉台山、五道梁和卓乃湖次级坳陷)形成于始新世—渐新世会聚时期的褶皱—逆冲推覆构造带附近(Liu 和 Wang,2001;Liu 等,2001)。非海相沉积充填最大厚度达约 6km,出露面积分布超过 $10\times 10^4 km^2$。与西藏南部对比,同位素研究表明可可西里盆地是在高程为约 2km 左右时发育的,并经过后期抬升至约 5km(Cyr 等,2005;Wang 等,2008)。早中新世时,持续的盆地充填溢出地形屏障,使该次级坳陷与相邻的次级坳陷连通成统一盆地。在其他地区,沉积作用在渐新世晚期停止,并经历广泛的剥蚀,形成了成熟的夷平(剥蚀平原)面(Wang 等,2002)。在西藏的中—东和东南区域发育有一系列小型的细长非海相沉积盆地(图 21-2),与风火山地区大型可可西里盆地复合体部分对应。狭窄的 Nangqian-Yushu 盆地和贡觉盆地系统沉积厚度为 2~4km,平行构造走向且以逆冲断层和限制性走滑断层为边界(Horton 等,2002b;Spurlin 等,2005;Studnicki-Gizbert 等,2008)。

(4)西藏北部的柴达木盆地(图 21-2 和图 21-4d)是一个内流水系区域,以沿着主要逆冲断层和走滑断层的运动产生的地形阻塞为边界。尽管该面积约为 $10\times 10^4 km^2$ 的盆地含有活动的沉积区域,沿着盆地边缘的变形仍减小了原始沉积区域(Bally 等,1986;Yin 等,2002,2007,2008a)。柴达木盆地具有一个几乎完整的新生代沉积记录,尽管沿着西藏北部边缘地区发现了新生代早期的海相化石(Ritts 等,2008),但盆地以非海相沉积充填为主,厚度达 10km(Song 和 Wang,1993;Yin 等,2008b)。在新生代早期,沉降是由沿着南部边缘的地壳增厚而导致的挠曲下沉引起的(Bally 等,1986;Yin 等,2007,2008a,2008b)。西藏北部边界上的阿尔金走滑断层的错动促进了盆地的闭合,形成的地形阻隔使得后期的沉积充填得以发育(Métivier 等,1998;Tapponnier 等,2001;Yin 等,2002)。依据在青藏高原中所处的构造部位、更高的海拔以及与盆地边界构造在成因上的联系,柴达木盆地与亚洲中部的大型碰撞后继盆地(Hendrix 等,1992;Graham 等,1993)有区别。

(5)在西藏的东北角(图 21-2),新生代沉积记录以早期伸展性盆地的分割(分区)为特征。该地区发育一系列与主走滑断层(阿尔金断裂和海原断裂)运动相联系的逆冲推覆系统(祁连山、南山和六盘山)(Tapponnier 等,1990;Burchfiol 等,1991)。尽管这些构造一般是由新生代晚期的构造变形所致,但古地磁和地层记录显示是由前新近纪的构造变形造成的(Dupont-Nivet 等,2004;Horton 等,2004)。在印度—亚洲碰撞之前,侏罗—白垩纪 1~3km 厚的沉积层序部分归因于伸展和张扭作用(Vincent 和 Allen,1999)。这些大型盆地在新生代构造隆升时被肢解,促进了由多种逆冲推覆和走滑构造控制的局部盆地的沉积充填,如西宁盆地(图 21-4e)(Fang 等,2003;Horton 等,2004)。

21.4 讨论与结论

尽管非碰撞弧后造山带与碰撞系统在构造背景上存在根本的区别,安第斯山脉与喜马拉雅—西藏造山带的腹地地区仍然存在相似之处。两个系统均含高海拔的平原、内流水系区域、活动的腹地沉积作用以及新生代造山运动时老盆地充填物的出露。尽管许多安第斯山脉和青藏的腹地盆地具有各自的沉积历史，仍有一些过程控制了弧后和碰撞系统造山带内部的盆地演化。安第斯和西藏地区揭示了腹地盆地演化的两个特殊模式。

第一类包含在先前经遭受剥蚀和/或构造变形的区域上作为新激活特征发育起来的腹地盆地。在该类情况中，与造山带内部的构造缩短、伸展和走滑变形构造有关的特殊构造的初始滑动，促成了盆地的演化。陡峭的基底卷入型断层的反转滑动，在普纳高原（Schwab，1985；Hongn 等，2007）和西藏中部的杜巴—伦坡拉地区（Leeder 等，1988）形成了以单个挠曲盆地为边界的孤立山岭。腹地伸展作用使得在诸如秘鲁安第斯（McNulty 和 Farber，2002；Giovanni 等，2010）和西藏南端（Yin，2000；Garzione 等，2003）区域形成裂谷和拆离盆地。另外一些断裂成因和挠曲成因的盆地沉降有利于解释在张扭和压扭腹地环境下的盆地演化，尤其是在西藏中部（Taylor 等，2003，本书）和厄瓜多尔地区（Hungerbühler 等，2002；Winkler 等，2005）。

第二类常见的盆地演化模式涉及早期的前陆盆地演化，包括由于构造变形前缘向克拉通方向推进和前陆盆地联合进入生长的造山带中，楔顶沉降带进入腹地盆地中（图21-5 和图21-6）。腹地盆地在前陆盆地之上叠置或继承的主要例子包括阿尔蒂普拉诺—普纳高原（安第斯山脉中部）和西藏的南部和北部部分地区。然而，各个系统都有其独特性。首先，在安第斯高原中部（图21-1），造山前缘不需要作为一个变形的稳定前进波迁移。普纳高原经历了一个构造变形的非系统性模式演变，并在新生代构造缩短时期的不同时间形成了孤立的盆地（Coutand 等，2001，2006；Carrapa 和 DeCelles，2008）。阿尔蒂普拉诺腹地盆地形成于构造变形带前缘突然向克拉通方向（东向）跳跃了约200km 并导致沉积分布模式显著改变的时期（图21-5）（Horton 等，2002a；DeCelles 和 Horton，2003；McQuarrie 等，2005）。在安第斯山脉北部存在一个造山前缘的类似跳跃，促进了 Magdalena 山谷盆地的腹地生长阶段（Gómez 等，2003，2005a，2005b；Nie 等，2010；Moreno 等，2011）。其次，在青藏高原的南部和中部（图21-2），一个白垩纪的前陆盆地是该盆地两翼上的地壳负载的产物。北部与羌塘—拉萨块体碰撞相关的周缘负载，以及南部在印度—亚洲碰撞前与洋—陆俯冲边界

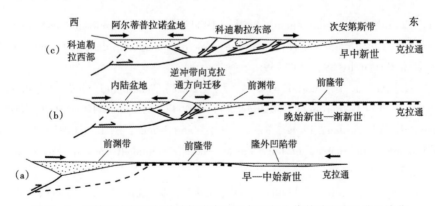

图 21-5 安第斯中部阿尔蒂普拉诺高原及邻区新生代前陆和腹地盆地演化
重建示意图（据 Decelles 和 Horton，2003，修改）
图中显示出沉积充填（点状）、剥蚀或严格限制性沉积充填的前隆带（粗虚线）、沉积扩散方式
（大箭头）和简化的深度逆冲构造

（和冈底斯弧）有关的弧后构造变形形成了岩石圈的挠曲（Leier 等，2007a）。腹地盆地环境只形成于新生代早期促使青藏高原大部分地区处于构造缩短状态的印度—亚洲碰撞之后。再次，西藏北部的可可西里盆地（图 21-2）代表着一个潜在的情况，即在高原向北生长的过程中造山带定向的反向逆冲作用促使前陆地区向腹地沉积转变（图 21-6）（Liu 等，2001；Yin 等，2007）。在更北的地区，沿着高原现今边界的走滑构造变形和缩短作用，形成了柴达木盆地（Metivier 等，1998；Yin 等，2002，2008a，2008b）。尽管这些来自安第斯和西藏的例子在各方面存在差别，但具有一个统一的要素，即最近活化的地形屏障抬升作用促使先存的前陆盆地和后继的腹地盆地没有经历大规模的剥蚀过程。这些新的屏障产生了地形降雨阴影作用（Sobel 等，2003；Strecker 等，2007），从而进一步促进了干旱/半干旱环境、内流水系和持续的隆升腹地盆地的充填，使腹地盆地在沉积厚度和延续时间上能与低海拔的前陆盆地相媲美。

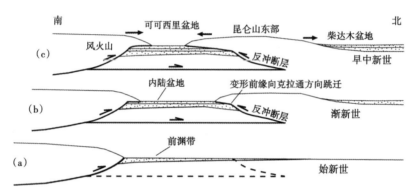

图 21-6　西藏北部可可西里盆地及邻区在新生代可能从前陆盆地向腹地
盆地转化的重建示意图（据 Yin 等，2007，修改）

图中显示出沉积充填（点状），沉积扩散方式（大箭头）和在深度上普遍的逆冲构造

致谢（略）

参 考 文 献

Achache, J., Courtillot, V., and Zhou, Y. X. (1984) Paleogeographic and tectonic evolution of southern Tibet since middle Cretaceous time: new paleomagnetic data and synthesis. Journal of Geophysical Research, 89, 311-339.

Aitchison, J. C., Ali, J. R., and Davis, A. M. (2007) When and where did India and Asia collide? Journal of Geophysical Research, 112, B05423. doi: 10.1029/2006JB004706.

Allmendinger, R. W., Jordan, T. E., Kay, S. M., and Isacks, B. L. (1997) The evolution of the Altiplano-Puna plateau of the Central Andes. Annual Review of Earth and Planetary Sciences, 25, 139-174.

Alonso, R. N., Jordan, T. E., Tabbutt, K. T., and Vandervoort, D. S. (1991) Giant evaporite belts of the Neogene central Andes. Geology, 19, 401-404.

Babeyko, A. Y., Sobolev, S. V., Trumbull, R. B., Oncken, O., and Lavier, L. L. (2002) Numerical models of crustal scale convection and partial melting beneath the Altiplano-Puna plateau. Earth and Planetary Science Lettes, 199, 373-388.

Baby, P., Rochat, P., Mascle, G., and Hérail, G. (1997) Neogene shortening contribution to crustal thickening in the back arc of the Central Andes. Geology, 25, 883-886.

Bally, A. W., Chou, I. M., Clayton, R., Eugster, H. P., Kidwell, S., Meckel, L. D., Ryder, R. T., Watts, A. B., and Wilson, A. A. (1986) Notes on sedimentary basins in China: report of the American Sedimentary Basins Delegation to the People's Republic of China. U. S. Geological Survey Open-File Report 86-237, 108 p.

Barazangi, M., and Isacks, B. L. (1976) Spatial distribution of earthquakes and subduction of the Nazca Plate beneath South America. Geology, 4, 686-692.

Barnes, J. B., Ehlers, T. A., McQuarrie, N., O'Sullivan, P. B., and Pelletier, J. D. (2006) Eocene to recent variations in erosion across the central Andean fold-thrust belt, northern Bolivia: implications for plateau evolution. Earth and Planetary Science Letters, 248, 118-133.

Beaumont, C., Jamieson, R. A., Nguyen, M. H., and Lee, B. (2001) Himalayan tectonics explained by extrusion of a low-viscosity crustal channel coupled to focused surface denudation. Nature, 414 (6865), 738-742.

Beck. S. L., and Zandt, G. (2002) The nature of orogenic crust in the Central Andes. Journal of Geophysical Research, 107, 2230. doi: 10. 1029/2000JB000124.

Beer, J. A., Allmendinger, R. W., Figueroa, D. E., and Jordan, T. E. (1990) Seismic stratigraphy of a Neogene piggyback basin, Argentina. American Association of Petroleum Geologists, 74, 1183-1202.

Blisniuk, P. M., Hacker, B. R., Glodny, J., Ratschbacher, L., Bi, S. W., Wu, Z. H., McWilliams, M. O., and Calvert, A. (2001) Normal faulting in central Tibet since at least 13. 5 Myr ago. Nature, 412 (6847), 628-632.

Boll, A., and Hernández, R. M. (1986) Interpretación estructural delárea Tres Cruces. Boletín de Informaciones Petroleras (Yacimientos Petrolíferos Fiscales) 7, 2-14.

Bonnot, D., Sébrier, M., and Mercier, J. (1988) Evolution geodynamique plio-quaternaire du bassin intracordillerain du Callejon de Huaylas et de la Cordillere Blanche, Perou. Geodynamique, 3, 57-83.

Burbank, D. W., Beck, R. A., and Mulder, T. (1996) The Himalayan foreland basin, in Yin, A., and Harrison, T. M., eds., The tectonics of Asia. New York, Cambridge University Press, 149-189.

Burchfiel, B. C., Chen, Z., Hodges, K. V., Liu, Y., Royden, L. H., Deng, C., and Xu, J. (1992) The south Tibetan detachment system, Himalayan orogen: Extension contemporaneous with and parallel to shortening in a collisional mountain belt. Geological Society of America Special Paper 269, 41 p.

Burchfiel, B. C., Zhang, P., Wang, Y., Zhang, W., Song, F., Deng, Q., Molnar, P., and Royden, L. (1991) Geology of the Haiyuan fault zone, Ningxia-Hui autonomous region, China, and its relation to the evolution of the northeastern margin of the Tibetan plateau. Tectonics, 10, 1091-1110.

Burg, J. P., Proust, F., Tapponnier, P., and Chen, G. M. (1983) Deformation phases and tectonic evolution of the Lhasa block (southern Tibet, China). Eclogae Geologicae Helvetiae, 76, 643-665.

Carrapa, B., and DeCelles, P. G. (2008) Eocene exhumation and basin development in the Puna of northwestern Argentina. Tectonics, 27, TC1015. doi: 10. 1029/2007TC002127.

Carrapa, B., Hauer, J., Schoenbohm, L., Strecker, M. R., Schmitt, A. K., Villanueva, A., and Gomez, J. S. (2008) Dynamics of deformation and sedimentation in the northern Sierras Pampeanas: an integrated study of the Neogene Fiambala basin, NW Argentina. Geological Society of America Bulletin, 120, 1518-1543.

Chemenda, A. I., Burg, J. P., and Mattauer, M. (2000) Evolutionary model of the Himalaya-Tibet system: geopoem based on new modelling, geological and geophysical data. Earth and Planetary Science Letters, 174, 397-409.

Chen, F. H., Bloemendal, J., Zhang, P. Z., and Liu, G. X. (1999) An 800 ky proxy record of climate from lake sediments of the Zoige Basin, eastern Tibetan Plateau. Palaeogeography, Palaeoclimatology, Palaeoecology, 151, 307-320.

Chen, X., Yin, A., Gehrels, G. E., Cowgill, E. S., Grove, M., Harrison, T. M., and Wang, X. F. (2003) Two phases of Mesozoic north-south extension in the eastern Altyn Tagh range, northern Tibetan plateau. Tectonics, 22, 1053. doi: 10. 1029/2001TC001336.

Coney, P. J., and Evenchick, C. A. (1994) Consolidation of the American Cordilleras. Journal of South American Earth Sciences, 7, 241-262.

Cooper, M. A., Addison, F. T., Alvarez, R., Coral, M., Graham, R. H., Hayward, A. B., Howe, S., Martinez, J., Naar, J., Penas, R., Pulham, A. J., and Taborda, A. (1995) Basin development and tectonic history of the Llanos basin, Eastern Cordillera, and Middle Magdalena Valley, Colombia. American Association of PetroleumGeologists Bulletin, 79, 1421-1443.

Coutand, I., Carrapa, B., Deeken, A., Schmitt, A. K., Sobel, E. R., and Strecker, M. R. (2006) Propagation of orographic barriers along an active range front: insights fromsandstone petrography and detrital apatite fissiontrack thermochronology in the intramontane Angastaco basin, NW Argentina. Basin Research, 18, 1-26.

Coutand, I., Cobbold, P. R., de Urreiztieta, M., Gautier, P., Chauvin, A., Gapais, D., Rossello, E. A., and López- Gamundí, O. (2001) Style and history of Andean deformation, Puna plateau, northwestern Argentina. Tectonics, 20, 210-234.

Cyr, A. J., Currie, B. S., and Rowley, D. B. (2005) Geochemical evaluation of Fenghuoshan Group lacustrine carbonates, north-central Tibet: implications for the paleoaltimetry of the Eocene Tibetan plateau. Journal of Geology, 113, 517-533.

Dalmayrac, B., and Molnar, P. (1981) Parallel thrust and normal faulting in Peru and constraints on the state of stress. Earth and Planetary Science Letters, 55, 473-481.

Dalziel, I. W. D. (1981) Back-arc extension in the southern Andes: a review and critical reappraisal. Philosophical Transactions of the Royal Society of London, Series A, Mathematical, Physical, and Engineering Sciences, 300 (1454), 319-355.

Dalziel, I. W. D. (1986) Collision and cordillera orogenesis: an Andean perspective, in Coward,

M. P., and Ries, A. C., eds., Collision tectonics. Geological Society of London Special Publication 19, 389-404.

de Berc, S. B., Soula, J. C., Baby, P., Souris, M., Christophoul, F., and Rosero, J. (2005) Geomorphic evidence of active deformation and uplift in a modern continental wedgetop – foredeep transition: example of the eastern Ecuadorian Andes. Tectonophysics, 399, 351-380.

DeCelles, P. G., Ducea, M. N., Kapp, P., and Zandt, G. (2009) Cyclicity in Cordilleran orogenic systems. Nature Geoscience, 2, 251-257.

DeCelles, P. G., Gehrels, G. E., Quade, J., Ojha, T. P., Kapp, P. A., and Upreti, B. N. (1998) Neogene foreland basin deposits, erosional unroofing, and the kinematic history of the Himalayan fold-thrust belt, western Nepal. Geological Society of America Bulletin, 110, 2-21.

DeCelles, P. G., and Giles, K. N. (1996) Foreland basin systems. Basin Research, 8, 105-123.

DeCelles, P. G., and Horton, B. K. (2003) Early to middle Tertiary foreland basin development and the history of Andean crustal shortening in Bolivia. Geological Society of America Bulletin, 115, 58-77.

DeCelles, P. G., Quade, J., Kapp, P., Fan, M. J., Dettman, D. L., and Ding, L. (2007a) High and dry in central Tibet during the Late Oligocene. Earth and Planetary Science Letters, 253, 389-401.

DeCelles, P. G., Kapp, P., Ding, L., and Gehrels, G. E. (2007b) Late Cretaceous to middle Tertiary basin evolution in the central Tibetan Plateau: changing environments in response to tectonic partitioning, aridification, and regional elevation gain. Geological Society of America Bulletin, 119, 654-680.

DeCelles, P. G., Robinson, D. M., and Zandt, G. (2002) Implications of shortening in the Himalayan fold-thrust belt for uplift of the Tibetan Plateau. Tectonics, 21, TC1062. doi: 10.1029/2001TC001322.

Dengo, C. A., and Covey, M. C. (1993) Structure of the Eastern Cordillera of Colombia: implications for trap styles and regional tectonics. American Association of Petroleum Geologists Bulletin, 77, 1315-1337.

Dewey, J. F., Cande, S., and Pitman, W. C. (1989) Tectonic evolution of the India-Eurasia collision zone. Eclogae Geologicae Helvetiae, 82, 717-734.

Dupont-Nivet, G., Horton, B. K., Butler, R. F., Wang, J., Zhou, J., and Waanders, G. L. (2004) Paleogene clockwise tectonic rotation of the Xining – Lanzhou region, northeastern Tibetan Plateau. Journal of Geophysical Research, 109, B04401. doi: 10.1029/2003JB002620.

Echavarria, L., Hern_andez, R., Allmendinger, R., and Reynolds, J. (2003) Subandean thrust and fold belt of northwestern Argentina: geometry and timing of the Andean evolution. American Association of Petroleum Geologists Bulletin, 87, 965-985.

Ege, H., Sobel, E. R., Scheuber, E., and Jacobshagen, V. (2007) Exhumation history of the southern Altiplano plateau (southern Bolivia) constrained by apatite fission track thermochronology. Tectonics, 26, TC1004. doi: 10.1029/ 2005TC001869.

Elger, K., Oncken, O., and Glodny, J. (2005) Plateaustyle accumulation of deformation: Southern Altiplano. Tectonics, 24, TC4020. doi: 10. 1029/2004TC001675.

England, P., and Searle, M. (1986) The Cretaceous-Tertiary deformation of the Lhasa block and its implications for crustal thickening in Tibet. Tectonics, 5, 1-14.

Fang, X., Garzione, C., Van der Voo, R., Li, J. J., and Fan, M. J. (2003) Flexural subsidence by 29 Ma on the NE edge of Tibet from the magnetostratigraphy of Linxia Basin, China. Earth and Planetary Science Letters, 210, 545-560.

Garzione, C. N., DeCelles, P. G., Hodkinson, D. G., Ojha, T. P., and Upreti, B. N. (2003) East-west extension and Miocene environmental change in the southern Tibetan plateau: Thakkhola graben, central Nepal. Geological Society of America Bulletin, 115, 3-20.

Garzione, C. N., Dettman, D. L., Quade, J., DeCelles, P. G., and Butler, R. F. (2000) High times on the Tibetan Plateau: paleoelevation of the Thakkhola graben, Nepal. Geology, 28, 339-342.

Garzione, C. N., Hoke, G. D., Libarkin, J. C., Withers, S., MacFadden, B., Eiler, J., Ghosh, P., and Mulch, A. (2008) Rise of the Andes. Science, 320 (5881) 1304-1307.

Garzione, C. N., Molnar, P., Libarkin, J. C., and MacFadden, B. J. (2006) Rapid late Miocene rise of the Bolivian Altiplano: evidence for removal of mantle lithosphere. Earth and Planetary Science Letters, 241, 543-556.

Gillis, R. J., Horton, B. K., and Grove, M. (2006) Thermochronology, geochronology, and upper crustal structure of the Cordillera Real: implications for Cenozoic exhumation of the central Andean plateau. Tectonics, 25, TC6007. doi: 10. 1029/2005TC001887.

Giovanni, M. K., Horton, B. K., Garzione, C. N., McNulty, B., and Grove, M. (2010) Extensional basin evolution in the Cordillera Blanca, Peru: stratigraphic and isotopic records of detachment faulting and orogenic collapse in the Andean hinterland. Tectonics, 29, TC6007. doi: TC2010TC002666.

Gomez, E., Jordan, T. E., Allmendinger, R. W., and Cardozo, N. (2005a) Development of the Colombian foreland-basin system as a consequence of diachronous exhumation of the northern Andes. Geological Society of America Bulletin, 117, 1272-1292.

Gomez, E., Jordan, T. E., Allmendinger, R. W., Hegarty, K. and Kelley, S. (2005b) Syntectonic Cenozoic sedimentation in the northern middle Magdalena Valley Basin of Colombia and implications for exhumation of the Northern Andes. Geological Society of America Bulletin, 117, 547-569.

Gomez, E., Jordan, T. E., Allmendinger, R. W., Hegarty, K., Kelley, S., and Heizler, M. (2003) Controls on architecture of the Late Cretaceous to Cenozoic southern Middle Magdalena Valley Basin, Colombia. Geological Society of America Bulletin, 115, 131-147.

Graham, S. A., Hendrix, M. S., Wang, L. B., and Carroll, A. R. (1993) Collisional successor basins of western China: impact of tectonic inheritance on sand composition. Geological Society of America Bulletin, 105, 323-344.

Graham, S. A., Hendrix, M. S., Johnson, C. L., Badamgarav, D., Badarch, G., Amory, J., Porter, M., Barsbold, R., Webb, L. E., and Hacker, B. R. (2001) Sedimentary record and tectonic implications of Mesozoic rifting in southeast Mongolia. Geological Society of America Bulletin, 113, 1560-1579.

Gubbels, T. L., Isacks, B. L., and Farrar, E. (1993) High-level surfaces, plateau uplift, and foreland development, Bolivian central Andes. Geology, 21, 695-698.

Gutscher, M. A., Spakman, W., Bijwaard, H., and Engdahl, E. R. (2000) Geodynamics of flat subduction: Seismicity and tomographic constraints from the Andean margin. Tectonics, 19, 814-833.

Hain, M. P., Strecker, M. R., Bookhagen, B., Alonso, R. N., Pingel, H., and Schmitt, A. K. (2011) Neogene to Quaternary broken foreland formation and sedimentation dynamics in the Andes of NW Argentina (25S). Tectonics, 30, TC2006. doi: 10. 1029/2010TC002703.

Hampel, A. (2002) The migration history of the Nazca Ridge along the Peruvian active margin: a re-evaluation. Earth and Planetary Science Letters, 203, 665-679.

Hampton, B. A., and Horton, B. K. (2007) Sheetflow fluvial processes in a rapidly subsiding basin, Altiplano plateau, Bolivia. Sedimentology, 54, 1121-1147.

Hendrix, M. S., Graham, S. A., Carroll, A. R., Sobel, E. R., McKnight, C. L., Schulein, B. J., and Wang, Z. (1992) Sedimentary record and climatic implications of recurrent deformation in the Tian Shan: evidence from Mesozoic strata of the north Tarim, south Junggar, and Turpan basins, northwest China. Geological Society of America Bulletin, 104, 53-79.

Hilley, G. E., and Strecker, M. R. (2005) Processes of oscillatory basin filling and excavation in a tectonically active orogen: Quebrada del Toro Basin, NW Argentina. Geological Society of America Bulletin, 117, 887-901.

Hindle, D., Kley, J., Klosko, E., Stein, S., Dixon, T., and Norabuena, E. (2002) Consistency of geologic and geodetic displacements during Andean orogenesis. Geophysical Research Letters, 10. 1029/2001GL013757.

Hodges, K. V. (2000) Tectonics of the Himalaya and southern Tibet from two perspectives. Geological Society of America Bulletin, 112, 324-350.

Hongn, F., del Papa, C., Powell, J., Petrinovic, I., Mon, R., and Deraco, V. (2007) Middle Eocene deformation and sedimentation in the Puna-Eastern Cordillera transition (23 degrees-26 degrees S): control by preexisting heterogeneities on the pattern of initial Andean shortening. Geology, 35, 271-274.

Horton, B. K. (1998) Sediment accumulation on top of the Andean orogenic wedge: Oligocene to late Miocene basins of the Eastern Cordillera, southern Bolivia. Geological Society of America Bulletin, 110, 1174-1192.

Horton, B. K. (1999) Erosional control on the geometry and kinematics of thrust belt development in the central Andes. Tectonics, 18, 1292-1304.

Horton, B. K. (2005) Revised deformation history of the central Andes: inferences from Cenozoic foredeep and intermontane basins of the Eastern Cordillera, Bolivia. Tectonics, TC3011, 24. doi: 10. 1029/2003TC001619.

Horton, B. K, and DeCelles, P. G. (1997) The modern foreland basin system adjacent to the Central Andes. Geology, 25, 895-898.

Horton, B. K., Dupont-Nivet, G., Zhou, J., Waanders, G. L., Butler, R. F., and Wang, J. (2004) Mesozoic-Cenozoic evolution of the Xining-Minhe and Dangchang basins, northeastern Tibetan plateau: magnetostratigraphic and biostratigraphic results. Journal of Geophysical Research, 109, B04402. doi: 10.1029/2003JB002913.

Horton, B. K., Hampton, B. A., LaReau, B. N., and Baldellón, E. (2002a) Tertiary provenance history of the northern and central Altiplano (central Andes, Bolivia): a detrital record of plateau-margin tectonics. Journal of Sedimentary Research, 72, 711-726.

Horton, B. K., Hampton, B. A., and Waanders, G. L. (2001) Paleogene synorogenic sedimentation in the Altiplano plateau and implications for initial mountain building in the central Andes. Geological Society of America Bulletin, 113, 1387-1400.

Horton, B. K., Saylor, J. E., Nie, J., Mora, A., Parra, M., Reyes-Harker, A., and Stockli, D. F. (2010) Linking sedimentation in the northern Andes to basement configuration, Mesozoic extension, and Cenozoic shortening: evidence from detrital zircon U-Pb ages, Eastern Cordillera, Colombia. Geological Society of America Bulletin, 122, 1423-1442.

Horton, B. K., Yin, A., Spurlin, M. S., Zhou, J., and Wang, J. (2002b) Paleocene-Eocene syncontractional sedimentation in narrow, lacustrine-dominated basins of eastcentral Tibet. Geological Society of America Bulletin, 114, 771-786.

Hungerbühler, D., Steinmann, M., Winkler, W., Seward, D., Eguez, E., Heller, F., and Ford, M. (1995) An integrated study of fill and deformation in the Andean intermontane basin of Nabon (late Miocene) southern Ecuador. Sedimentary Geology, 96, 257-279.

Hungerbühler, D., Steinmann, M., Winkler, W., Seward, D., Eguez, A., Peterson, D. E., Helg, U., and Hammer, C. (2002) Neogene stratigraphy and Andean geodynamics of southern Ecuador. Earth-Science Reviews, 57, 75-124.

Isacks, B. L. (1988) Uplift of the central Andean plateau and bending of the Bolivian orocline. Journal of Geophysical Research, 93, 3211-3231.

Jaeger, J. J., Courtillot, V., and Tapponnier, P. (1989) Paleontological view of the ages of the Deccan traps, the Cretaceous-Tertiary boundary, and the India-Asia collision. Geology, 17, 316-319.

James, D. E. (1971) Plate-tectonic model for the evolution of the central Andes. Geological Society of America Bulletin, 82, 3325-3346.

James, D. E., and Sacks, S. (1999) Cenozoic formation of the central Andes: a geophysical perspective, in Skinner, B. J., ed., Geology and ore deposits of the central Andes. Society of Economic Geologists Special Publication 7, 1-25.

Johnson, C. L. (2004) Polyphase evolution of the East Gobi basin: sedimentary and structural records of Mesozoic-Cenozoic intraplate deformation in Mongolia. Basin Research, 16, 79-99.

Jordan, T. E. (1995) Retroarc foreland and related basins, in Busby, C. J., and Ingersoll, R. V., eds., Tectonics of sedimentary basins. Oxford, Blackwell Science, 331-362.

Jordan, T. E., and Alonso, R. N. (1987) Cenozoic stratigraphy and basin tectonics of the Andes Mountains, 20-28 south latitude. American Association of Petroleum Geologists Bulletin, 71, 49-64.

Jordan, T. E., Isacks, B. L., Allmendinger, R. W., Brewer, J. A., Ramos, V. A., and Ando, C. J. (1983) Andean tectonics related to geometry of subducted Nazca plate. Geological Society of America Bulletin, 94, 341–361.

Kapp, P., DeCelles, P. G., Leier, A. L., Fabijanic, J. M., He, S., Pullen, A., and Gehrels, G. E. (2007) The Gangdese retroarc thrust belt revealed. GSA Today, 17, 4–9.

Kapp, P., and Guynn, J. H. (2004) Indian punch rifts Tibet. Geology, 32, 993–996.

Kapp, P., Taylor, M., Stockli, D., and Ding, L. (2008) Development of active low-angle normal fault systems during orogenic collapse: insight from Tibet. Geology, 36, 7–10.

Kapp, P., Yin, A., Harrison, T. M., and Ding, L. (2005) Cretaceous–Tertiary shortening, basin development, and volcanism in central Tibet. Geological Society of America Bulletin, 117, 865–878.

Kay, S. M., Godoy, E., and Kurtz, A. (2005) Episodic arc migration, crustal thickening, subduction erosion, and magmatism in the south-central Andes. Geological Society of America Bulletin, 117, 67–88.

Kay, R. W., and Kay, S. M. (1993) Delamination and delamination magmatism. Tectonophysics, 201, 177–189.

Kennan, L., Lamb, S., and Rundle, C. (1995) K–Ar dates from the Altiplano and Cordillera Oriental of Bolivia: implications for Cenozoic stratigraphy and tectonics. Journal of South American Earth Sciences, 8, 163–186.

Kley, J. (1996) Transition from basement–involved to thinskinned thrusting in the Cordillera Oriental of southern Bolivia. Tectonics, 15, 763–775.

Kley, J., and Monaldi, C. R. (1998) Tectonic shortening and crustal thickness in the Central Andes: how good is the correlation? Geology, 26, 723–726.

Kley, J., Monaldi, C. R., and Salfity, J. A. (1999) Along-strike segmentation of the Andean foreland: causes and consequences. Tectonophysics, 301, 75–94.

Kraemer, B., Adelmann, D., Alten, M., Schnurr, W., Erpenstein, K., Kiefer, E., Van den Bogaard, P., and Görler, K. (1999) Incorporation of the Paleogene foreland into the Neogene Puna plateau: the Salar de Antofalla area, NW Argentina. Journal of South American Earth Sciences, 12, 157–182.

Lamb, S., and Davis, P. (2003) Cenozoic climate change as a possible cause for the rise of the Andes. Nature, 425, 792–797.

Lamb, S., and Hoke, L. (1997) Origin of the high plateau in the Central Andes, Bolivia, South America. Tectonics, 16, 623–649.

Lave, J., and Avouac, J. P. (2000) Active folding of fluvial terraces across the Siwaliks Hills, Himalayas of central Nepal. Journal of Geophysical Research, 105, 5735–5770.

Lavenu, A., Noblet, C., Bonhomme, G., Eguez, A., Dugas, F., Vivier, G. (1992) New K–Ar age dates of Neogene to Quaternary volcanic rocks from the Ecuadorian Andes: implications for the relationship between sedimentation, volcanism and tectonics. Journal of South American Earth Sciences, 5, 309–320.

Lavenu, A., Noblet, C., and Winter, T. (1995) Neogene ongoing tectonics in the southern

Ecuadorian Andes: analysis of the evolution of the stress field. Journal of Structural Geology, 17, 47–58.

Le Pichon, X., Fournier, M., and Jolivet, L. (1992) Kinematics, topography, shortening, and extrusion in the India- Eurasia collision. Tectonics, 11, 1085–1098.

Leeder, M. R., Smith, A. B., and Yin, J. (1988) Sedimentology, palaeoecology and palaeoenvironmental evolution of the 1985 Lhasa to Golmud Geotraverse. Philosophical Transactions of the Royal Society of London, A327, 107–143.

Leier, A. L., DeCelles, P. G., Kapp, P., and Ding, L. (2007a) The Takena Formation of the Lhasa terrane, southern Tibet: the record of a Late Cretaceous retroarc foreland basin. Geological Society of America Bulletin, 119, 31–48.

Leier, A. L., DeCelles, P. G., Kapp, P., and Gehrels, G. E. (2007b) Lower Cretaceous strata in the Lhasa terrane, Tibet, with implications for understanding the early tectonic history of the Tibetan plateau. Journal of Sedimentary Research, 77, 809–825.

Liu, Z., and Wang, C. (2001) Facies analysis and depositional systems of Cenozoic sediments in the Hoh Xil basin, northern Tibet. Sedimentary Geology, 140, 251–270.

Liu, Z, Wang, C., and Yi, H. (2001) Evolution and mass accumulation of the Cenozoic Hoh Xil basin, northern Tibet. Journal of Sedimentary Research, 71, 971–984.

Marocco, R., Lavenu, A., and Baudino, R. (1995) Intermontane late Paleogene–Neogene basins of the Andes of Ecuador and Peru: sedimentologic and tectonic characteristics, in Tankard, A. J., Su_arez, R., and Welsink, H. J., eds., Petroleum basins of South America. American Association of Petroleum Geologists Memoir 62, 597–613.

Marshall, L. G., Swisher, C. C., III, Lavenu, A., Hoffstetter, R., and Curtis, G. H. (1992) Geochronology of the mammalbearing late Cenozoic on the northern Altiplano, Bolivia. Journal of South American Earth Sciences, 5, 1–19.

Masek, J. G., Isacks, B. L., Gubbels, T. L., and Fielding, E. J. (1994) Erosion and tectonics at the margins of continental plateaus. Journal of Geophysical Research, 99, 13941–13956.

McCaffrey, R., and Nabelek, J. (1998) Role of oblique convergence in the active deformation of the Himalayas and southern Tibet plateau. Geology, 26, 691–694.

McNulty, B., and Farber, D. (2002) Active detachment faulting above the Peruvian flat slab. Geology, 30, 567–570.

McNulty, B. A., Farber, D. L., Wallace, G. S., Lopez, R., and Palacios, O. (1998) Role of plate kinematics and plateslip- vector partitioning in continental magmatic arcs: evidence from the Cordillera Blanca, Peru. Geology, 26, 827–830.

McQuarrie, N. (2002a) The kinematic history of the central Andean fold–thrust belt, Bolivia: implications for building a high plateau. Geological Society of America Bulletin, 114, 950–963.

McQuarrie, N. (2002b) Initial plate geometry, shortening variations, and evolution of the Bolivian orocline. Geology, 30, 867–870.

McQuarrie, N., Horton, B. K., Zandt, G., Beck, S., and DeCelles, P. G. (2005) Lithospheric evolution of the Andean fold-thrust belt, Bolivia, and the origin of the central Andean plateau.

Tectonophysics, 399, 15-37.

McQuarrie, N., Barnes, J. B., and Ehlers, T. A. (2008) Geometric, kinematic, and erosional history of the central Andean Plateau, Bolivia (15-17 degrees S). Tectonics, 27, TC3007. doi: 10. 1029/2006TC002054.

Megard, F. (1984) The Andean orogenic period and its major structures in central and northern Peru. Journal of the Geological Society of London, 141, 893-900.

Megard, F., Noble, D. C., McKee, E. H., and Bellon, H. (1984) Multiple pulses of Neogene compressive deformation in the Ayacucho intermontane basin, Andes of central Peru. Geological Society of America Bulletin, 95, 1108-1117.

Metivier, F., Gaudemer, Y., Tapponnier, P., and Meyer, B. (1998) Northeastward growth of the Tibet plateau deduced from balanced reconstruction of two depositional areas: the Qaidam and Hexi Corridor basins, China. Tectonics, 17, 823-842.

Molnar, P., England, P., and Martinod, J. (1993) Mantle dynamics, the uplift of the Tibetan Plateau, and the Indian monsoon. Reviews of Geophysics, 31, 357-396.

Molnar, P. and Tapponnier, P. (1975) Cenozoic tectonics of Asia: effects of a continental collision. Science, 189, 419-426.

Moreno, C. J., Horton, B. K., Caballero, V., Mora, A., Parra, M., and Sierra, J. (2011) Depositional and provenance record of the Paleogene transition from foreland to hinterland basin evolution during Andean orogenesis, northern Middle Magdalena Valley Basin, Colombia. Journal of South American Earth Sciences, in press.

Mosolf, J. G., Horton, B. K., Heizler, M. T., and Matos, R. (2011) Unroofing the core of the central Andean foldthrust belt during focused late Miocene exhumation: evidence from the Tipuani-Mapiri wedge-top basin, Bolivia. Basin Research, 23, 346-360.

Mpodozis, C., Arriagada, C., Basso, M., Roperch, P., Cobbold, P., and Reich, M. (2005) Late Mesozoic to Paleogene stratigraphy of the Salar de Atacama Basin, Antofagasta, northern Chile: implications for the tectonic evolution of the central Andes. Tectonophysics, 399, 125-154.

Mpodozis, C., and Ramos, V. A. (1989) The Andes of Chile and Argentina, in Ericksen, G. E., Canas Pinochet, M. T., and Reinemund, J. A., eds., Geology of the Andes and its relation to hydrocarbon and mineral resources. Circum - Pacific Council for Energy and Mineral Resources, Earth Science Series, 11, 59-90.

Murphy, M. A., Yin, A., Harrison, T. M., Durr, S. B., Chen, Z., Ryerson, F. J., Kidd, W. S. F., Wang, X., and Zhou, X. (1997) Did the Indo-Asian collision alone create the Tibetan plateau? Geology, 25, 719-722.

Murray, B. P., Horton, B. K., Matos, R., and Heizler, M. T. (2010) Oligocene-Miocene basin evolution in the northern Altiplano, Bolivia: implications for evolution of the central Andean backthrust belt and high plateau. Geological Society of America Bulletin, 122, 1443-1462.

Nie, J., Horton, B. K., Mora, A., Saylor, J. E., Housh, T. B., Rubiano, J., and Naranjo, J. (2010) Tracking exhumation of Andean ranges bounding the Middle Magdalena Valley basin, Colombia. Geology, 451-454.

Northrup, C. J., Royden, L. H., and Burchfiel, B. C. (1995) Motion of the Pacific plate relative to Eurasia and its potential relation to Cenozoic extension along the eastern margin of Eurasia. Geology, 23, 719–722.

Owens, T. J., and Zandt, G. (1997) Implications of crustal property variations for models of Tibetan plateau evolution. Nature, 387, 37–43.

Petford, N., and Atherton, M. P. (1992) Granitoid emplacement and deformation along a major crustal lineament: the Cordillera Blanca, Peru. Tectonophysics, 205, 171–185.

Ramos, V. A., and Aleman, A. (2000) Tectonic evolution of the Andes, in Cordani, U. G., Milani, E. J., Thomaz Filho, A., and Campos Neto, M. C., eds., Tectonic evolution of South America. 31st International Geological Congress, Rio de Janeiro, Brazil, 635–685.

Ramos, V. A., Cristallini, E. O., and Perez, D. J. (2002) The Pampean flat-slab of the Central Andes. Journal of South American Earth Sciences, 15, 59–78.

Reynolds, J. H., Jordan, T. E., Johnson, N. M., Damanti, J. F., and Tabbutt, K. D. (1990) Neogene deformation of the flatsubduction segment of the Argentine – Chilean Andes: magnetostratigraphic constraints from Las Juntas, La Rioja province, Argentina. Geological Society of America Bulletin, 102, 1607–1622.

Ritts, B. D., Yue, Y. J., Graham, S. A., Sobel, E. R., Abbink, O. A., and Stockli, D. (2008) From sea level to high elevation in 15 million years: uplift history of the northern Tibetan Plateau margin in the Altun Shan. American Journal of Science, 308, 657–678.

Robinson, D. M., DeCelles, P. G., and Copeland, P. (2006) Tectonic evolution of the Himalayan thrust belt in western Nepal: implications for channel flow models. Geological Society of America Bulletin, 118, 865–885.

Rowley, D. B. (1996) Age of initiation of collision between India and Asia: a review of stratigraphic data. Earth and Planetary Science Letters, 145, 1–13.

Rowley, D. B., and Currie, B. S. (2006) Palaeo-altimetry of the late Eocene to Miocene Lunpola basin, central Tibet. Nature, 439 (7077), 677–681.

Royden, L. H., Burchfiel, B. C., King, R. W., Wang, E., Chen, Z., Shen, F., and Liu, Y. (1997) Surfacedeformationand lower crustal flow in eastern Tibet. Science, 276, 788–790.

Royden, L. H., Burchfiel, B. C., and Van der Hilst, R. D. (2008) The geological evolution of the Tibetan plateau. Science, 321 (5892), 1054–1058.

Salfity, J. A., and Marquillas, R. A. (1994) Tectonic and sedimentary evolution of the Cretaceous–Eocene Salta Group basin, Argentina, in Salfity, J. A., ed., Cretaceous tectonics of the Andes. Wiesbaden, Germany, Vieweg Publishing, 266–315.

Saylor, J. E., Quade, J., Dettman, D. L., DeCelles, P. G., Kapp, P. A., and Ding, L. (2009) The late Miocene through present paleoelevation history of southwestern Tibet. American Journal of Science, 309, 1–42.

Schellart, W. P. (2008) Overriding plate shortening and extension above subduction zones: a parametric study to explain formation of the Andes Mountains. Geological Society of America Bulletin, 120, 1441–1454.

Schurr, B., Rietbrock, A., Asch, G., Kind, R., and Oncken, O. (2006) Evidence for lithospheric detachment in the central Andes from local earthquake tomography. Tectonophysics, 415, 203-223.

Schwab, K. (1985) Basin formation in a thickening crust— the intermontane basins in the Puna and the Eastern Cordillera ofNWArgentina (central Andes). IV Congreso Geologico Chileno, Actas, 2, 138-158.

Schwartz, D. P. (1988) Paleoseismicity and neotectonics of the Cordillera Blanca fault zone, northern Peruvian Andes. Journal of Geophysical Research, 93, 4712-4730.

Searle, M. P., Windley, B. F., Coward, M. P., Cooper, D. J. W., Rex, A. J., Rex, D., Li, T. D., Xiao, X. C., Jan, M. Q., Thakur, V. C., and Kumar, S. (1987) The closing of Tethys and the tectonics of the Himalaya. Geological Society of America Bulletin, 98, 678-701.

Sebrier, M., Mercier, J. L., Machare, J., Bonnot, D., Cabrera, J., and Blanc, J. L. (1988) The state of stress in an overriding plate situated above a flat slab: the Andes of central Peru. Tectonics, 7, 895-928.

Sempere, T., Hérail, G., Oller, J., and Bonhomme, M. G. (1990) Late Oligocene – early Miocene major tectonic crisis and related basins in Bolivia. Geology, 18, 946-949.

Sobel, E. R., Hilley, G. E., and Strecker, M. R. (2003) Formation of internally drained contractional basins by araditylimited bedrock incision. Journal of Geophysical Research, 108, 2344. doi: 10. 1029/2002JB001883.

Sobolev, S. V., and Babeyko, A. Y. (2005) What drives orogeny in the Andes? Geology, 33, 617-620.

Song, T., and Wang, X. (1993) Structural styles and stratigraphic patterns of syndepositional faults ina contractional setting: examples from Quaidam basin, northwestern China. American Association of Petroleum Geologists Bulletin, 77, 102-117.

Spurlin, M. S., Yin, A., Horton, B. K., Zhou, J., and Wang, J. (2005) Structural evolution of the Yushu-Nangqian region and its relationship to syn-collisional igneous activity, east-central Tibet. Geological Society of America Bulletin, 117, 1293-1317.

Steinmann, G. (1929) Geologie von Peru. Karl Winter, Heidelberg, Germany, 448 p. Strecker, M. R., Alonso, R. N., Bookhagen, B., Carrapa, B., Hilley, G. E., Sobel, E. R., and Trauth, M. H. (2007) Tectonics and climate of the southern central Andes, Annual Review of Earth and Planetary Sciences, 35, 747-787.

Studnicki-Gizbert, C., Burchfiel, B. C., Li, Z., and Chen, Z. (2008) Early Tertiary Gonjo basin, eastern Tibet: sedimentary and structural record of the early history of India – Asia collision. Geosphere, 4, 713-735.

Tan, X. D., Kodama, K. P., Chen, H. L., Fang, D. J., Sun, D. J., and Li, Y. A. (2003) Paleomagnetism and magnetic anisotropy of Cretaceous red beds from the Tarim basin, northwest China: evidence for a rock magnetic cause of anomalously shallow paleomagnetic inclinations from central Asia. Journal of Geophysical Research, 108, 2107. doi: 10. 1029/2001JB001608.

Tapponnier, P., Xu, Z. Q., Roger, F., Meyer, B., Arnaud, N., Wittlinger, G., and Yang, J.

S. (2001) Oblique stepwise rise and growth of the Tibet plateau. Science, 294 (5547). 1671–1677.

Tapponnier, P., Meyer, B., Avouac, J. P., Peltzer, G., Guademer, Y., Guo, S., Xiang, H., Yin, K., Chen, Z., Cai, S., and Dai, H. (1990) Active thrusting and folding in the Qilian Shan, and decoupling between upper crust and mantle in northeastern Tibet. Earth and Planetary Science Letters, 97, 382–403.

Tapponnier, P., Peltzer, G., Le Dain, A. Y., Armijo, R., and Cobbold, P. (1982). Propagating extrusion tectonics in Asia: new insights from simple experiments with plasticine. Geology, 10, 611–616.

Taylor, M., Yin, A., Ryerson, F. J., Kapp, P., and Ding, L. (2003) Conjugate strike-slip faulting along the Bangong-Nujiang suture zone accommodates coeval east-west extension and north-south shortening in the interior of the Tibetan Plateau. Tectonics, 22, 1044. doi: 10.1029/2002TC001361.

Uba, C. E., Heubeck, C., and Hulka, C. (2006) Evolution of the late Cenozoic Chaco foreland basin, southern Bolivia. Basin Research, 18, 145–170.

Van Houten, F. B. (1976) Late Cenozoic volcaniclastic deposits, Andean foredeep, Colombia. Geological Society of America Bulletin, 87, 481–495.

Van Houten, F. B., and Travis, R. B. (1968) Cenozoic deposits, Upper Magdalena Valley, Colombia. American Association of Petroleum Geologists Bulletin, 52, 675–702.

Vandervoort, D. S., Jordan, T. E., Zeitler, P. K., and Alonso, R. N. (1995) Chronology of internal drainage development and uplift, southern Puna plateau, Argentine central Andes. Geology, 23, 145–148.

Vincent, S. J., and Allen, M. B. (1999) Evolution of the Minle and Chaoshui Basins, China: implications for Mesozoic strike-slip basin formation in Central Asia. Geological Society of America Bulletin, 111, 725–742.

Wang, C. S., Liu, Z. F., Yi, H. S., Liu, S., and Zhao, X. X. (2002) Tertiary crustal shortenings and peneplanation in the Hoh Xil region: implications for the tectonic history of the northern Tibetan Plateau. Journal of Asian Earth Sciences, 20, 211–223.

Wang, C. S., Zhao, X. X., Liu, Z. F., Lippert, P. C., Graham, S. A., Coe, R. S., Yi, H. S., Zhu, L. D., Liu, S., and Li, Y. L. (2008) Constraints on the early uplift history of the Tibetan Plateau. Proceedings of the National Academy of Science of the United States of America, vol. 105, 4987–4992.

Welsink, H. J., Martinez, E., Aranibar, O., and Jarandilla, J. (1995) Structural inversion of a Cretaceous rift basin, southern Altiplano, Bolivia, in Tankard, A. J., Suárez, R., and Welsink, H. J., eds., Petroleum basins of South America. American Association of Petroleum Geologists Memoir 62, 305–324.

Whitman, D., Isacks, B. L., and Kay, S. M. (1996) Lithospheric structure and along-strike segmentation of the central Andean plateau: seismic Q, magmatism, flexure, topography and tectonics. Tectonophysics, 259, 29–40.

Wilson, T. J. (1991) Transition from back-arc to foreland basin development in the southernmost

Andes: stratigraphic record from the Ultima Esperanza district, Chile. Geological Society of America Bulletin, 103, 98-111.

Winkler, W., Villagomez, D., Spikings, R., Abegglen, P., Tobler, S., and Eguez, A. (2005) The Chota basin and its significance for the inception and tectonic setting of the inter-Andean depression in Ecuador. Journal of South American Earth Science, 19, 5-19.

Wise, J. M., Noble, D. C., Zanetti, K. A., and Spell, T. L. (2008) Quechua II contraction in the Ayacucho intermontane basin: evidence for rapid and episodic Neogene deformation in the Andes of central Peru. Journal of South American Earth Sciences, 26, 383-396.

Xu, Z. (1984) Tertiary system and its petroleum potential in the Lunpola Basin, Xizang (Tibet). U. S. Geological Survey Open-File Report 84-420, 5 p.

Yin, A. (2000) Mode of Cenozoic east-west extension in Tibet suggesting a common origin of rifts in Asia during the Indo-Asian collision. Journal of Geophysical Research, 105, 21745-21759.

Yin, A., Dang, Y., Zhang, M., McRivette, M. W., Burgess, W. P., and Chen, X. (2007) Cenozoic tectonic evolution of Qaidam basin and its surrounding regions (part 2): wedge tectonics in southern Qaidam basin and the Eastern Kunlun Range, in Sears, J. W., Harms, T. A., and Evenchick, C. A., eds., Whence the mountains? Inquiries into the evolution of orogenic systems: a volume in honor of Raymond Price. Geological Society of America Special Paper 433, 369-390.

Yin, A., Dang, Y. Q., Wang, L. C., Jiang, W. M., Zhou, S. P., Chen, X. H., Gehrels, G. E., and McRivette, M. W. (2008a) Cenozoic tectonic evolution of Qaidam basin and its surrounding regions (part 1): the southern Qilian Shan- Nan Shan thrust belt and northern Qaidam basin. Geological Society of America Bulletin, 120, 813-846.

Yin, A., Dang, Y. Q., Zhang, M., Chen, X. H., and McRivette, M. W. (2008b) Cenozoic tectonic evolution of Qaidam Basin and its surrounding regions (part 3): structural geology, sedimentation, and regional tectonic reconstruction. Geological Society of America Bulletin, 120, 847-876.

Yin, A., and Harrison, T. M. (2000) Geologic evolution of the Himalayan-Tibetan orogen. Annual Review of Earth and Planetary Sciences, 28, 211-280.

Yin, A., Rumelhart, P. E., Butler, R., Cowgill, E., Harrison, T. M., Foster, D. A., Ingersoll, R. V., Zhang, Q., Zhou, X. Q., Wang, X. F., Hanson, A., and Raza, A. (2002) Tectonic history of the Altyn Tagh fault system in northern Tibet inferred from Cenozoic sedimentation. Geological Society of America Bullletin, 114, 1257-1295.

Zhang, K. J. (2000) Cretaceous paleogeography of Tibet and adjacent areas (China): tectonic implications. Cretaceous Research, 21, 23-33.

(蔡国富 译，吴哲 赵钊 崔敏 校)

第 22 章　盆地对青藏高原腹地活动伸展和走滑变形的响应

MICHALEL H. TAYLOR[1], PAUL A. KAPP[2], BRIAN K. HORTON[3]

(1. Department of Geology, University of Kansas, Lawrence, USA;
2. Department of Geosciences, University of Arizona, Tucson, USA;
3. Department of Geological Sciences and Institute for Geophysics, Jackson School of Geosciences, University of Texasat Austin, Austin, USA)

摘　要：喜马拉雅—西藏造山体系的内流腹地中发育有现代沉积盆地，其中包括沿西藏中部共轭走滑构造带分布的南北向裂谷盆地和走滑盆地。本章通过具体的实例来研究这两种盆地的几何形态、与断层活动的关系以及可能的演化史。特别值得注意的是这些活动盆地是沿着断裂系统发育的，该系统与碰撞系统中的腹地有动力学成因联系，同时遭受了同时代南北向缩短（部分与共轭走滑断裂相协调）和东西向伸展变形的构造作用。对西藏地区活动盆地形成后的构造演化过程的认识，可能会影响到对发育在增厚的热地壳之上的裂谷盆地和走滑盆地的保存潜力及成因的理解。

关键词：褶皱—冲断带　喜马拉雅山脉　腹地　高原　西藏　伸展构造

22.1　引言

腹地地区位于收缩造山带中的相对高海拔区域，可为非海相的沉积物充填提供可容空间。在造山生长早期腹地地区通常以逆冲推覆作用为特征。地壳增厚和（或者）岩石圈底部迁移导致的重力势能的不断增加，使得逆冲构造向走滑和正断层转化（Molnar 和 Tapponnier，1978；Dewey，1988；Hodges 和 Walker，1992；Allmendinger 等，1997）。本章对位于喜马拉雅—西藏造山带腹地，以断裂为边界的活动沉积盆地体系以及这些盆地在不断的伸展和走滑变形中的发育情况进行详细的回顾。

青藏高原是地球上最大的活动造山带，平均海拔超过 5km，地壳平均厚度达 85km。沿着南部、东北部以及西北部边缘的逆冲型地震是印度—欧亚大陆碰撞有关的活动逆冲断层的最好例证（Taylor 和 Yin，2009）。与此相反，青藏高原内部由于走滑断层和正断层相结合，遭受了同时代的南北向缩短和东西向伸展作用（图 22-1 和图 22-2；Taylor 和 Yin，2009）。西藏中部的活动构造包括近南北向的裂谷边界正断层，以及北西走向左旋走滑断层或北西走向的右旋走滑断层（图 22-1 和图 22-2）。这些构造对发育在青藏高原内流水系区域内并以断层为界的活动沉积盆地起到了一级控制作用。

近期研究表明，西藏内部的一些南北向裂谷中有变质核杂岩出露，这些裂谷以低角度活动（拆离）正断层为边界并且伴生拆离盆地（Kapp 等，2008；Murphy 等，2002；Pan 和 Kidd，1992）。值得注意的是，很长时间以来，地球动力学和地质学观点都认为伸展构造的

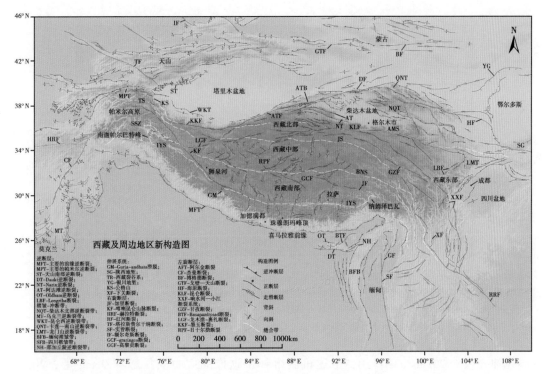

图 22-1　与印度—亚洲碰撞带及邻区相关的活动断裂的彩色阴影地形图（据 Taylor 和 Yin，2009）

图中主要信息来源来自以下资料（Allegre 等，1984；Armijo 等，1986，1989；Arrowsmith 和 Strecker，1999；Avouac 和 Peltzer，1993；Avouac 等，1993；Burchfiel 等，1991，1995，1999；Cowgill，2000，2004；Darby 和 Ritts，2002；Darby 等，2005；Gold 等，2006；Jackson，1992；Jackson 等，1995；Jackson 和 McKenzio，1984；Kapp 和 Guynn，2004；Kirby 等，2000；Lave 和 Avouac，2000；Meriaux 等，2004，2005；Murphy 等，2000；Peltzer 等，1989；Robinson 等，2004；Tapponnior 等，1981，2001；Tapponnier 和 Molnar，1979；Taylor 和 Peltzer，2006；Taylor 等，2003；tenBrink 和 Taylor，2002；Thatcher，2007；Thompson 等，2002；Wang 和 Burchfiel，2000；Wang 等，1998；Xu 等，2008；Yeats 和 Lillie，1991；Yin 等，2008；Yin 和 Harrison，2000），增加了本文的动力学解释。逆冲断裂在上覆板块上具倒钩图案，正断层在上盘具条状和球形图案，箭头显示走滑断裂的水平运动方向。白色虚线代表中生代缝合带：IYS—印度—雅鲁缝合带；BNS—班公—怒江缝合带；JS—金沙缝合带；SSZ—什约克缝合带；TS—Tanymas 缝合带；AMS—阿尼玛卿—南昆仑—慕士塔格缝合带

"核杂岩" 模式更容易发育在地热梯度大、地壳厚度大的高海拔地区（Buck，1991，1993；Coney 和 Harms，1984；Larhonbruch 和 Morgan，1990；Regenauer-Lieb 等，2006）。虽然此观点被很多人接受，但与厚地壳区伸展构造核杂岩模式有关的直接观察资料非常有限，因为研究的大量拆离系统大多暴露在低海拔地区，地壳厚度相对较小。如东巴布亚新几内亚、巴哈加利福尼亚、索尔顿海槽、死谷和爱琴海（Abors 和 Roecker，1991；Axen 等，1999；Baldwin 等，1993；Abers 等，2002；Cowan 等，2003；Hayman，2003；Lister，1984；Numelin 等，2007；Rietbrock 等.1996；sorel，2000）。

V 形共轭走滑断裂系是与西藏裂谷盆地有关的运动学特征（图 22-1）（Armijo 等，1986，1989；Taylor 等，2003）。人们认为，这些 V 形共轭走滑断裂系有助于三角形楔状块体向东移动（Taylor 等，2003）。三角形的盆地形成于向东挤出的楔状体尾部。伴随着向东挤出过程，印度与欧亚大陆的碰撞引起南北向挤压，边界断层以及周围的岩体发生了垂直轴向旋转。据 GPS 研究显示，目前西藏中部正经历地壳的南北向缩短（大约 10mm/a）及东

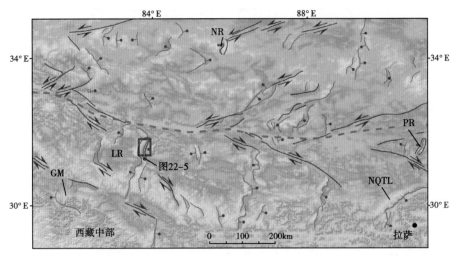

图 22-2 西藏中部色调地形图，显示北倾的裂谷体系和 V 形共轭走滑盆地

方框为图 22-3 及图 22-6 的位置，高程数据来自 Shuttle Radar Topography Mission 90m 分辨率数据（SRTM）。箭头代表走滑运动方向，正断层在上盘以条状和球形标记，虚线代表印度—雅鲁缝合带和班公—怒江缝合带的大体位置。NQTL—图 22-6 念青唐古拉的位置；GM-Curia Mandhata；LR-隆格尔山脉，位置见图 22-1

西向拉伸（大约 20mm/a）作用，上述机制解释了这样的应力场状态（Zhang 等，2004）。接下来，本文对南北向的西藏裂谷作了一个综述，介绍了盆地发育阶段中的不同状态，并且揭示了低角度正断层在塑造地貌中的潜在作用。然后，又总结了目前对于与青藏高原沉积盆地有关的走滑构造作用的认识，并描述了它的运动学特征，以说明它们如何与南北向的裂谷互相影响。此外观测结果表明，在这个地球上最大的挤压造山带中，逆冲推覆作用处于不活动状态，这一发现对于解释古代腹地盆地的区域构造背景有很大作用。

22.2 西藏裂谷

本文提到的西藏南部区域坐落于两个缝合带之间，南为印度—雅鲁藏布江缝合带（IYS），北为班公—怒江缝合带（BNS）。西藏中部是指南部 BNS 与北部阿尼玛卿—南昆仑缝合带之间的地区（图 22-1）。西藏南部裂谷地形特征显著，有些地方向南延伸越过 IYS 进入喜马拉雅（如 Thakkhola graben）（图 22-1 和图 22-2）。自从 30 年前的这一发现（Molnar 和 Tapponnier，1978），经过大量的地球物理研究，对西藏裂谷之下的岩石圈构造认识有了很大程度的提高（Klemperer，2006）。通过对拉萨北部的念青唐古拉边界的拆离断层下盘中暴露的倾斜地壳剖面进行调查，沿裂谷的地质研究提供了对西藏地壳中下层的地热和地球化学演化的进一步认识。下盘花岗岩的年龄表明，岩浆作用在印度与亚洲的碰撞过程中（古新世—晚中新世）呈半连续状态（Kapp 等，2005）。因为裂谷与活动的走滑断裂具有运动学上的联系，西藏首批新构造应力场详细调查揭示了它们与挤出构造具有明显的相关性（Armijo 等，1986）。另外一些研究表明在运动学上西藏裂谷的发育与喜马拉雅弧的演化有关（Seeber 和 Armbruster，1984）。随着喜马拉雅弧向南掩冲生长，喜马拉雅山脉弧使青藏原地壳东西向延展（Seeber 和 Armbruster，1984）。

裂谷发育的起始位置标志着最大挤压应力的方向由水平方向向垂直方向转变（Mercier

等，1987），同时西藏达到现今海拔高度，使亚洲季风加强（Harrison 等，1995；Molnar 等 1993）。正断层下盘岩石的低温热年代学研究表明，裂谷形成的起始时间为 4—10Ma（Kapp 等，2008；Stockli 等，2002）。有证据表明在约 14Ma 时，存在东西向拉张作用。这些证据包括沿 IYS 的南北向岩墙的年龄（Williams 等，2001；Yin 等，1994）、沿西藏中部双湖正断层生长的同造山期的白云母（Blisniuk 等，2001），以及 Thakkhola 地堑东部拉张断裂中热液成因的白云母年龄（Coleman 和 Hodges，1995）。对西藏南部和喜马拉雅北部盆地中新世以来古海拔研究表明，裂谷开始发育时，西藏海拔高程接近或等于现在的高程（Garzione 等，2000，2003；Spicer 等，2003；Currie 等，2005；Saylor 等，2009）。

目前对西藏南北向裂谷的研究认为，裂谷受控于统一的伸展变形过程（Kapp 等，2008）。也就是说，有些裂谷盆地断裂规模小、倾向滑动位移小（小于几千米），这样的盆地普遍被认为是未成熟的裂谷。未成熟裂谷以高角度正断层为界，内部盆地中心是干涸的湖相盆地。而成熟裂谷以低角度正断层为界，滑距大于 10km，上盘盆地正在经受隆升和切割。成熟裂谷的实例包括拉萨块体西部的隆格尔裂谷（Lunggar Rift）和拉萨北部的念青唐古拉裂谷（图 22-2）。这些裂谷的中心部位出露变质核杂岩，这些核杂岩以拆离断层和拆离盆地为界。地质条件和几何形态沿着裂谷走向的变化，展示了变质核杂岩、裂谷盆地以及地貌在伸展过程中的演化情况。

22.3 隆格尔裂谷

隆格尔裂谷（Kapp 等，2008）南北长约 70km，东西宽约 40km，最大海拔超过 6500m，发育众多的冰山切割隆格尔山脉。东北向裂谷中发育外流水系，裂谷中央有一条分水岭；活动沉积中心位于裂谷的北部和南部（图 22-3）。隆格尔山脉西部边缘是倾角小于 40°，东倾的正断层（隆格尔拆离断层），该断层使得下盘的糜棱片麻岩和各种经过变形的浅色花岗岩与上盘的古生代—新近纪岩层并列接触（图 22-3）。下盘糜棱片麻岩的片理普遍向东倾斜 30°—40°，表现出向东倾伏的拉伸线理。S—C 面理组构与顶部向东的位移相吻合。根据出露于拆离断层下盘东西向延伸的糜棱岩估算，最小的滑距约为 15km。隆格尔拆离断层滑距约 30km，其分支向南或向北转变成东倾的高角度正断层（图 22-3）。

根据断层下盘糜棱片麻岩和各种经过变形的淡色花岗岩中的锆石，计算得到的 U-Pb 锆石结晶年龄分别为 9Ma 和 15Ma（图 22-3）。该地区中部的磷灰石 U-Th/He 冷却年龄为 0.4~0.7Ma。这些数据表明，隆格尔拆离断层的下盘经历了中新世岩浆作用，之后于上新世—更新世迅速冷却（Kapp 等，2008）。

拆离断层中部接近上盘的位置为西倾的砾岩，沿剖面向上倾角逐渐减小（从 >50° 到约 10°），并且地层厚度向滑脱面逐渐增厚，这表明沉积与拆离作用同时进行。最古老的盆地沉积物包括湖相的泥岩与细—中粒互层砂岩。沿剖面向上粒级逐渐变大，成为含砾砂岩和砾岩，基质支撑，与邻近上盘的冲积扇和扇三角洲沉积相一致。现代盆地具有相似的沉积过程，同时沿现代山岭的前缘可见碎屑流和滑塌体。代表河流沉积过程的中—粗粒含砾砂岩和中—粗砾互层砾岩呈不整合覆盖在这些沉积组合之上。碎屑组成主要是花岗岩和不同变形程度的片麻岩，与下盘的来源一致。该套地层可能代表了沉积条件由湖相迅速转为碎屑流为主的冲积扇过渡，最后转向轴向河流体系。这一变化过程普遍存在于伸展期和滑脱拆离盆地（如 Leeder 和 Gawthorpe，1987；Friedmann 和 Burbank，1995）。

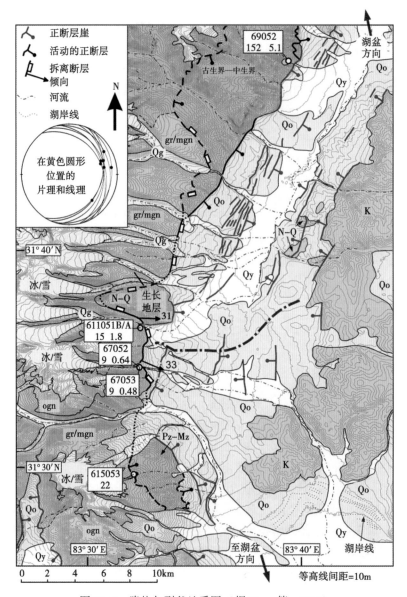

图 22-3 隆格尔裂谷地质图（据 Kapp 等，2008）

蓝色数字是 U-Pb 锆石年龄，单位为 Ma。红色数字代表（U-Th）He 磷灰石年龄，单位为 Ma。Pz-Mz—古生界—中生界，未区分；K—白垩系；Kgn—白垩系（?）火成片麻岩；gr/mgn—多样的变形花岗岩、糜棱质花岗岩和片麻岩；N-Q—新近系—第四系沉积；Qo—较老第四系沉积；Qy—较新第四系沉积；Qg—第四系冰川沉积

未被扰乱的冰碛石和冲击扇不整合覆盖于断层之上，表明这一现象表明，以山脉为界的隆格尔拆离断层目前处于不活动的状态（图 22-3 和图 22-4）。第四纪（?）松散沉积物和反向正断层普遍存在于断层上盘的盆地中，而裂谷盆地沉积被走向 NNE、倾角为 40°~60°的正断层切割。正如距离不活动山脉前缘 6km 处的陡崖式中的断层所展示，活动断层向盆地方向迁移（图 22-3 和图 22-4）。断层产生的陡崖与山脉前缘平行，这些陡崖系统中的净隆起量及断距，与断裂从底部进入低角度活动正断层的深度相一致（图 22-4）。在其他区域发现的这种几何关系被解释为正在发生活动拆离断层作用（Axen 等，1999）。

图 22-4 活动的陡崖型断层和隆格尔拆离断层下盘的西向视图

本地高约3m，沿着最新发育的大约2m高的陡崖形成。陡崖位于以山脉为边界的隆格尔拆离断层东部的4~6km范围内。注：山脉前缘断裂缺失，与断裂向盆地方向迁移以及盆地内自源自生沉积相一致。断层陡崖位置见图22-3

22.4 念青唐古拉裂谷

亚东—古鲁裂谷系统由三个部分组成：南部和北部裂谷为南北向展布，而中部裂谷为北东—南西向展布，与念青唐古拉山脉平行（图22-2）。念青唐古拉山脉长150km，宽35~40km，局部地形起伏达2km以上，而裂谷最宽处仅有10~15km（图22-6）。在念青唐古拉山脉中，白垩纪—晚中新世的花岗岩和正片麻岩出露于念青唐古拉拆离断层的下盘，该下盘以山脉为边界，南东倾向，倾角为22°~37°（Pan 和 Kidd，1992；Harrison 等，1995；Kapp 等，2005）。剖面上念青唐古拉拆离面的最大滑距为25~30km（Harrison 等，1995；Kapp 等，2005）。念青唐古拉拆离断裂带含有绿泥石角砾岩和碎裂岩，下伏小于2km厚的低角度糜棱剪切带；南东向倾伏的拉伸线理及S—C面理组构表明了其顶部向南东方向的剪切运动（Pan 和 Kidd，1992；Kapp 等，2005）。取自与伸展方向平行的几个穿过山脉的横剖面的热年代测量样品，约束了山脉的中—低温剥露热史。出露的下盘岩石中的磷灰石、锆石、金红石和独居石的（U-Th）/He 数据限定的岩石年龄为3~6Ma（沿倾向年龄变小），并且沿山脉前缘年龄小于1Ma。这些结果表明在上新世—更新世经历了剥蚀作用（Stockli 等，2002）。山脉南部，磷灰石（U-Th）/He 年龄从邻近山脊的10Ma 到山脉前缘糜棱带的不到5Ma（Stockli 等，2002）。这套数据与 $^{40}Ar/^{39}Ar$ 的结果（Harrison 等，1995）一致，这表明该区自中新世以后经历了大规模的剥蚀作用。

该区构造观测研究得出的热年代学结果表明，念青唐古拉拆离滑脱构造在过去大约7Ma中的滑距为25~30km，滑动速率约为4mm/a（Harrison 等，1995）。在其上盘新近纪—第四纪（？）冲积扇和河流砾石沉积中记录了念青唐古拉拆离滑脱构造更多的历史，上盘倾向NW，与新生代末期SE倾向拆离断层中的长期滑动相一致。穿过念青唐古拉裂谷盆地的地

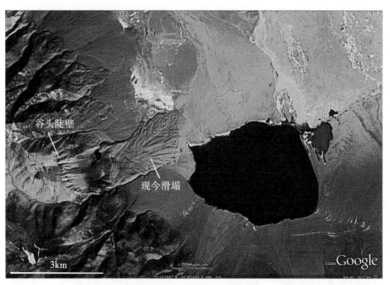

图 22-5 形成于活动的隆格尔拆离系统上盘超拆离盆地的全新世滑塌沉积
(与 Freidmann 和 Burbank, 1995 预测的相一致)

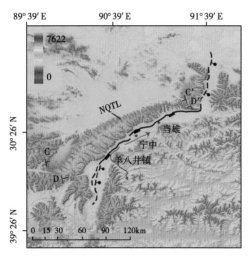

图 22-6 念青唐古拉山脉彩色阴影地形图
注意水系在中央裂谷的分开位置。C—C′和 D—D′显示图 22-7 中条带剖面的端点位置。位置见图 22-2

震反射剖面显示了一个大致倾向 SE 的反射层，解释为念青唐古拉拆离滑脱面（Cogan 等，1998）。当投影向上倾时，低角度反射体在宁中西北方向与山前断层崖相交叉（图 22-6），断崖总落差约 28m。在当雄附近，以山脉为边界的滑脱作用并不活跃，板块之上与山脉平行的活动正断层剥失掉约 4~5km 厚的第四纪（?）沉积物，并搬运到盆地中（图 22-6）。裂谷盆地的东南部，晚新生代的伸展构造作用对上盘岩石的影响并不明显（Kapp 等，2005）。

值得注意的是，在中部念青唐古拉拆离滑脱构造的东北走向段，上盘盆地被下切的地方其断距也最大。这种关系与半地堑演化的一般模式明显不同，一般模式中断层位移最大位置在盆地的沉积中心，即盆地沉降量最大的地区（Leeder 和 Gawthorpe, 1987）。上盘盆地的这些特征（在 30°20′N 附近）代表盆地内存在明显的分水岭，将相反的两条水系隔开，河流分别向南注入羊八井或向北注入当雄地区（图 22-6）。

22.4.1 隆格尔与念青唐古拉裂谷的共同特征

虽然对于西藏裂谷的认识有限，但成熟的裂谷具有很多的共同点：（1）伸展构造主要集中在山脉边界正断层的上盘或沿其上盘分布（地图显示小于 6km）。这种情况与科迪勒拉变质核杂岩相矛盾，后者拆离滑脱构造的上盘由于发育正断层而被极大地减薄（Coney,

1980；Wernicke，1981；Lister 和 Davis，1989）。（2）裂谷包括以山脉边界的低角度正断层，目前其表面部分并不活动。然而，地质特征和新构造关系表明，裂谷盆地中的高角度正断层正在活动，其底部进入浅层构造部位并产生滑脱面沿倾斜向下的伸展作用。（3）比较成熟的西藏裂谷盆地显示出分水岭特征，最大下切地区与最大伸展拉张地区大体一致，以糜棱片麻岩与冲积物并列接触为佐证。与西藏地区的未成熟裂谷形成对比，后者发育以山脉为边界的高角度正断层，位于裂谷中部的干涸沉积中心仍在活动，是典型的半地堑系统（Leeder 和 Gawthorpe，1987）。

22.5 数字高程模型分析

最近，数字高程模型（DEMs）的使用使得我们对于构造与地貌演化关系的理解更进了一步。为了了解西藏裂谷上盘流域分水岭与下盘的空间关系，我们分析了 90m 分辨率的 SRTM DEM 数据。基于 SRTM DEM，通过对盆地中第四纪沉积物的大致分布范围，以及对一条约 4km 宽的轴向跨越山脉顶部的条带进行数字化得到了下盘顶部和上盘盆地与裂谷平行的地形条带剖面。在以 North 和 Pung Co 裂谷为代表的未成熟裂谷中，下盘和上盘的高程数据的标绘与推测的最大拉张区有关（图 22-2 和图 22-7a）。在未成熟裂谷中，盆地高程最低处位于裂谷中部，与内流水系中以高角度正断层为边界的半地堑系统一致，活动的沉积中心与最大伸展拉张区重合。这种格局与较成熟的隆格尔和念青唐古拉裂谷明显不同。那些成熟的裂谷（图 22-7b）显示出的分水岭（而不是沉积中心）与推测的沿着山脉边界低角度正断层的最大伸展拉张区域一致。在以上实例中，现代河流以现代分水岭为界分别向南或向

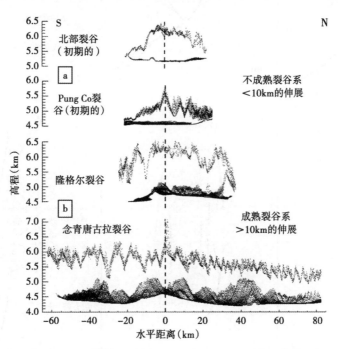

图 22-7 来自数字高程模型（90m 分辨率）的裂谷型山脉和盆地地形高程条带（据 Kapp 等，2008，修改）
竖直虚线代表推测的最大拉张区域。(a) 未成熟裂谷的盆地剖面在中部是平坦的，反映了区域的基准面。(b) 以拆离断层为边界的裂谷盆地剖面，在裂谷中部出现地形高点，反映了经历活动抬升和切割。剖面位置见图 22-2 及图 22-6

北流，表明由于局部性的空间改变，该区域水系发生了一次由内流向外流的改变。在隆格尔裂谷中，这个过程的地质证据是湖相裂谷盆地充填，并在分水岭附近被现代河流下切。这表明在裂谷发育的早期，裂谷盆地中部曾经是湖泊的沉积中心。总之，以上的观测结果表明，裂谷发育过程中发生了一次河流由内流向外流环境的转变，从而产生了下文中西藏裂谷演化的假说。

22.6 裂谷盆地的发育

我们关于西藏裂谷盆地发育的假说（图22-8）建立在一个前提之上，那就是不同伸展幅度的裂谷代表了区域伸展的过程，并且滑动是沿着拆离断层发生在深构造层上的（Kapp等，2008）。受之前描述的低角度正断层背景上，发育的裂谷和拆离盆地模型的启发（Schlische，1991；Freidmann和Burbank，1995），本文提出了下面的概念模型。假说的主要贡献是，在地图平面上提出了断层演化、地形以及盆地发育情况。首先，裂谷起始于一些半地堑盆地，以铲式高角度正断层为界，底部进入近水平的糜棱剪切带（图22-8a）。随着伸展构造作用的持续增大，构造卸载作用导致下盘均衡抬升并向后倾斜，最终废弃山脉边界高角度正断层的上倾部分，同时也导致在中上层地壳中的正断层隆起（图22-8b）。持续滑动

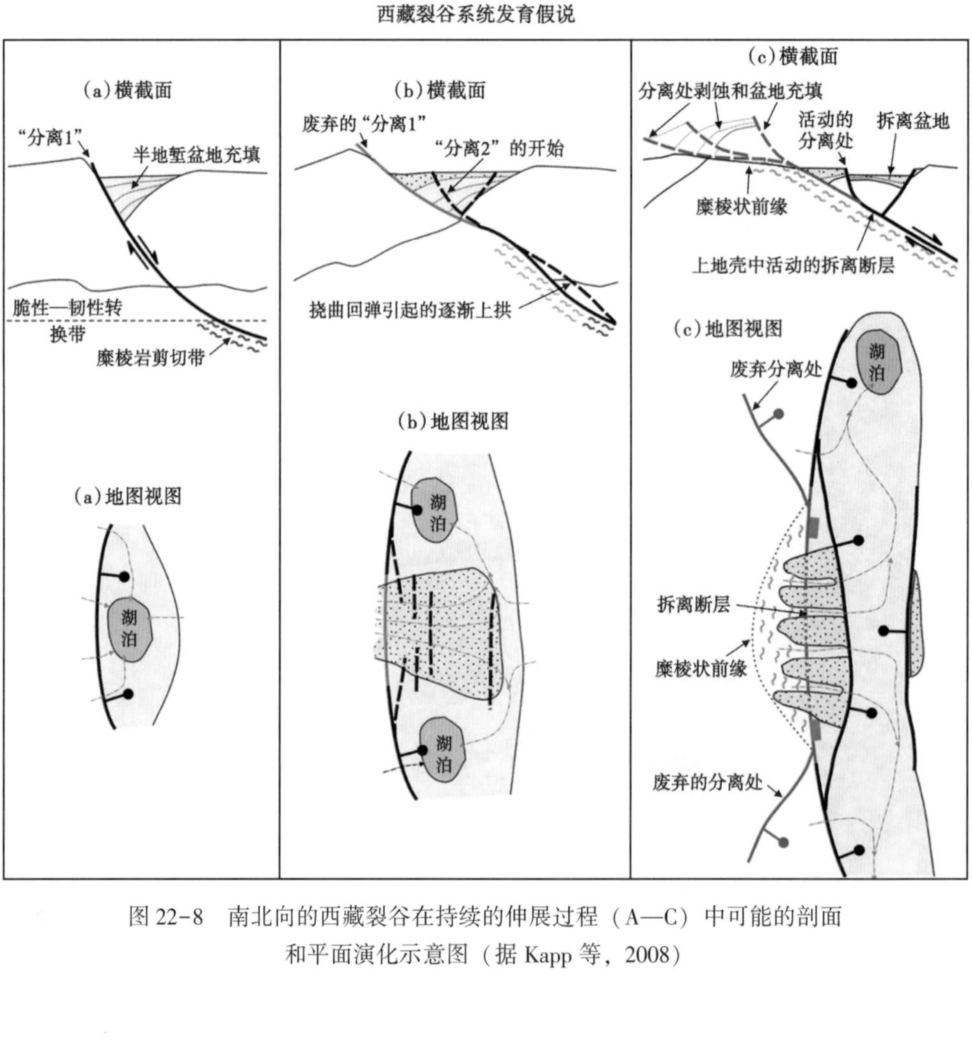

图 22-8 南北向的西藏裂谷在持续的伸展过程（A—C）中可能的剖面和平面演化示意图（据 Kapp 等，2008）

时期，在深部主要正断层的上盘和隆起附近，滑动拆离正断层渐进发育，形成了一个在脆性上地壳滑动的拆离断层。拆离作用发育过程中，裂谷盆地的局部区域进入下盘，经历抬升侵蚀，形成盆地循环或"同型装配"（Horton 和 Schmitt，1998），也导致裂谷变窄（图 22-8b）。此外，下盘最大滑动区的均衡反弹导致上覆裂谷盆地的抬升和侵蚀，以及盆地内流分水岭的形成（图 22-7b、c）。根据盆地的演化，该模型预测了断层有关的快速沉降和随后的均衡抬升，以及河流由内流向外流的迅速变化。该模式要求在断层伸长和盆地扩展过程中沿走向的横向迁移。因此，该模型对裂谷发育过程中的沉积分布和聚集提出大量但可预测的时空变化。西藏裂谷是包括走滑断层在内的，运动学上相联系的断裂系统的一部分。下面本文勾勒出了西藏中部的区域走滑构造系统。对西藏中部走滑断裂系统认识的重要性在于：(1) 它们形成了几个西藏最大的构造活动盆地；(2) 它们经历了西藏中部有间接记录的几次最大的地震；(3) 它们在一定程度上为理解岩石圈以片状或连续方式变形提供了帮助。

22.7 西藏中部走滑断层和相关盆地

在本节中，我们论述了一个位于西藏中部的与走滑构造有关的盆地系统，该盆地沿 BNS 分布，与南北向裂谷系统密切相关。沿 BNS 分布一系列右旋断裂，称为喀喇昆仑—嘉黎断裂带（KJFZ）。该系统呈阶梯状，由 NW 走向的右旋断裂组成，是西藏中部的南缘，其西起喀喇昆仑断裂，向东至少延伸至嘉黎断裂带，即喜马拉雅山脉东部附近（图 22-1；Armijo 等，1989）。人们认为该断裂系统有助于与拉萨地块有关的羌塘地块向东挤出（Armijo 等，1986，1989）。这表明单个右旋断裂的滑距至少约 65km，这是喀喇昆仑断裂位移的最小值（Murphy 等，2000；Robinson，2009）。最近，断裂系统以北的一系列 NE 走向左旋断裂也被纳入了这个断裂系统。总之，这些断裂形成了一个 V 形共轭断裂系统，称为西藏中部共轭走滑带（Taylor 等，2003）。西藏中部共轭走滑带东西长约 1200km，南北宽约 300km。然而，西藏中部单个走滑构造的滑距不足 20km，与西藏中部（上）地壳刚性块体的挤出一致，在平面图上东西长不足 150~200km（Taylor 等，2003；图 22-9）。

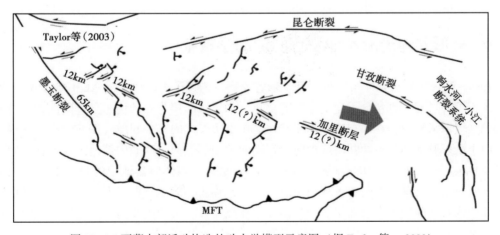

图 22-9 西藏中部活动构造的动力学模型示意图（据 Taylor 等，2003）
在西藏内部，共轭的走滑断层调节着同期的 E—W 向拉张和 N—S 向收缩。紧密分布的（<150km）共轭走滑断层促进分散的三角形楔形体的向东挤出

西藏中部的 V 形共轭走滑系统具有以下几个共同特征：（1）断层方向为 60°—75°，而不是最大压应力的 30°方向；（2）所有的 V 形共轭走滑受断层交点的限制，开口方向与当地地形斜坡一致（即向东，为艾里地壳均衡中的地壳减薄方向）；（3）所有走滑断裂终止于它们的结合点（即 V 形的顶点）；（4）V 形末端位于班公—怒江缝合带，形成了不连续的东西向盆地体系（图 22-1）。

人们认为穿过西藏中部的走滑断层是区域应力状态（补偿高地形所示）的表现形式（Mercier 等，1987）。高原上，由于地形很高，最大应力方向（$\sigma1$）是垂直的，与南北向裂谷的发育相一致。局部地区，BNS 缝合带形成了东西向的槽状洼地，使得应力方向（$\sigma1$）由垂直变为南北向，与走滑断裂的发育一致（Mercier 等，1987）。

近来利用 GPS 对西藏中部进行的大地测量学研究表明，在区域规模上断裂系统目前正在活动，东西扩张速率约为 15～20mm/a，南北缩短速率约为 10mm/a（Gan 等，2007；Wang 等，2001；Zhang 等，2004）。Taylor 和 Peltzer（2006）利用合成孔径雷达干涉测量技术对单个断裂进行了研究，估计右旋的南部格仁错断层滑动速率相对较高，达 7～10mm/a。利用同样的技术对北部左旋日干措断裂体系进行研究得出其保守滑动速率为 3.4mm/a（Taylor 和 Peltzer，2006）。总之，格仁错与日干措断裂在西藏组成了一个占主导地位并有地震活动的共轭走滑断裂带（图 22-1）（Taylor 等，2003；Taylor 和 Peltzer，2006；Sun 等，2008）。

Taylor 等（2003）及 Taylor 和 Peltzer（2006）认为，共轭走滑断裂是西藏中部沿班公—怒江缝合带分布的重要构造，在很大程度上是因为断裂的滑动距离超过 10km，滑动速率超过 3mm/a。长达 1200km 的西藏中部共轭走滑断裂带引发了一个问题，那就是它在为容纳印度—亚洲会聚起到了怎样的作用，更具体地说，它与南北向展布的西藏裂谷在运动学上具有怎样的关系。人们很早就认识到，BNS 南部的右旋断裂与南北向裂谷体系具有运动学上的联系（Armijo 等，1986，1989）（图 22-1）。由于东西向伸展变形构造分布于西藏中部，一系列以共轭走滑断裂为边界的小型楔状体的东向挤出也为东西向的伸展构造提供了空间。Taylor 等（2003）认为，东西伸展的整体模式也许表现为青藏高原在整体挤压应力状态下的向东扩展，以南北向缩短、东西向伸展和垂向减薄为特征。

22.8 V 形共轭走滑盆地发育的运动学特征

Taylor 等（2003）建立了一种运动学示意模型，描述了楔状体穿过西藏中部共轭走滑带的向东挤出的过程，也包含了盆地发育的信息。如果以断层为界的块体是半刚性的，沿 BNS 分布的共轭走滑断裂会在向东移动的楔状体的尾部边缘之间产生间隙（图 22-9）。这些间隔可以由沿东倾活动的逆冲断层或褶皱的运动而关闭。然而，西藏中部地区并没有逆冲地震聚集（Lnngin 等，2003；Molnar 和 Lyon-Caen，1989），东西向盆地中的晚新生代沉积中也缺少东西向褶皱，这与上面的解释相矛盾。然而，Taylor 等（2003）认为，围岩的竖直轴向旋转和楔状岩体的边界走滑断层，为与印度—亚洲碰撞有关的南北向缩短提供了空间。该运动学模型的几何形态（图 22-10）有助于解释众多的东西向内陆盆地，诸如沿 BNS 分布的西陵措（Siling Co）、伦坡拉及东措盆地（图 22-1；同样见本文中的 Horton）。Taylor 等（2003）认为，沿 BNS 带的局部低海拔是由以 V 形共轭走滑断层为界的小楔状块体的挤出引起的。盆地边界断层发生斜向滑动，盆地东西两侧的近南北向正断层发生倾向滑动（图 22-

10)。2008年1月9日发生于东措盆地南北向正断层的地震印证了这个解释。里氏6.4、5.9、5.4级的三次地震连续发生在与东北向左旋Riganpei Co断裂系统有运动学联系的北向构造之上,而Riganpei Co断裂与东措盆地有关。断裂系统的西南端,断裂走向向北偏移,其分支进入滑脱伸展型叠瓦状冲积扇的扇端(Taylor等,2003)。初步InSAR数据的弹性半空间模型表明,一条N25°—30°E走向、倾角65°的断层发生破裂(Sun等,2008)。第二次地震事件发生于西侧9km处的一条与该断层平行的倾角70°、向西倾的正断层之上。这一系列的地震表明,西藏中部的内陆盆地是活动的,并与相应的断裂系统相协调。

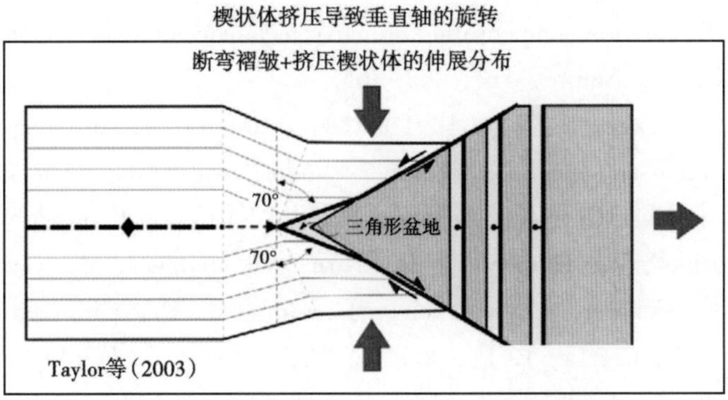

图22-10 Taylor等(2003)认为共轭走滑断层与相应盆地潜在的相互作用是在交会位置发育由于楔状体的挤出产生缺口,并且由于楔状体围岩块体沿着竖轴旋转而关闭,伴随走滑断裂通过断—弯褶皱机制的最小旋转。这种关系预测会在楔形体的尾部形成一个三角形盆地。挤出楔形体分散的E—W向拉伸作用将会形成更多的小型伸展性盆地

这些盆地演化开始于小型三角盆地的局部断层引起的沉降。随着断层的进一步活动,这个以断层为界的三角形盆地越来越大,并发育了穿时的地层沉积(Crowell,1974,1982)。在整个盆地发育的过程中,沿盆地边缘的断裂地形有助于促进内部引流。因此,虽然沿盆地边缘断层可以发育持续的河流系统,但湖相沉积更可能占主导地位。

22.9 讨论和结论

地球上最大的碰撞造山带内流腹地的现代盆地发育主要受控于运动学上相协调的正断层和走滑断层。活动中的裂谷盆地正在经历可容空间活动带向逐步抬升剥蚀过渡的阶段(图22-8)。次级的可容空间由位于挤出楔状体尾端的,以V形共轭走滑构造系为界的三角形盆地提供(图22-10)。该结论有助于理解西藏中部和南部的活动腹地盆地,也可能为解释古代腹地地区沉积物展布和充填的复杂模式提供线索。

西藏内流区的活动沉积盆地提供了盆地演化的现代状态,以及构造活动引起的地貌变化。模型预测显示,西藏裂谷中的伸展型盆地开始于断层引起的快速沉降,随后下盘均衡反弹,并伴随上盘盆地中的水流经历了一次由内流向外流的转变。预计相模式会在断裂延伸和盆地扩张期间沿走向发生侧向迁移,表明在沉积物分布和充填方面具有大量但可预测的时空多样性。裂谷沉积的中部发现了不整合,该不整合也许标志着拆离断层的开始,并且加强均衡反弹作用,此外还指示了核杂岩扩张模式的有利条件。走滑盆地发育开始于局部由断裂引起的三角形小盆地的沉降。随着走滑断裂的进一步活动,三角形断裂边界盆地逐渐变大,并

产生了穿时的沉积充填。在整个盆地发育的过程中，沿盆地边缘的断裂相关地形有助于促进内流水系的发育。因此，虽然沿盆地断层边缘可以持续发育河流系统，但根据对现代 BNS 带的观测，三角形盆地更可能以湖相沉积为主。

致谢（略）

参 考 文 献

Abers, G. A., Ferris, A., Craig, M., Davies, H., Lerner-Lam, A. L., Mutter, J. C., and Taylor, B. (2002) Mantle compensation of active metamorphic core complexes at Woodlark rift in Papua New Guinea. Nature, 418, 862–865.

Abers, G. A., and Roecker, S. W. (1991) Deep-structure of an arc-continent collision: earthquake relocation and inversion for upper mantle P and S-wave velocities beneath Papua-New-Guinea. Journal of Geophysical Research-Solid Earth and Planets, 96, 6379–6401.

Allegre, C. J., Courtillot, V., Tapponnier, P., Hirn, A., Mattauer, M., Coulon, C., Jaeger, J. J., Achache, J., Scharer, U., Marcoux, J., Burg, J. P., Girardeau, J., Armijo, R., Gariepy, C., Gopel, C., Li, T. D., Xiao, X. C., Chang, C. F., Li, G. Q., Lin, B. Y., Teng, J. W., Wang, N. W., Chen, G. M., Han, T. L., Wang, X. B., Den, W. M., Sheng, H. B., Cao, Y. G., Zhou, J., Qiu, H. R., Bao, P. S., Wang, S. C., Wang, B. X., Zhou, Y. X., and Ronghua, X. (1984) Structure and Evolution of the Himalaya-Tibet Orogenic Belt. Nature, 307, 17–22.

Allmendinger, R. W., Jordan, T. E., Kay, S. M., and Isacks, B. L. (1997) The evolution of the Altiplano-Puna plateau of the Central Andes. Annual Review of Earth and Planetary Sciences, 25, 139–174.

Armijo, R., Tapponnier, P., Mercier, J. L., and Han, T. L. (1986) Quaternary extension in Southern Tibet: field observations and tectonic implications. Journal of Geophysical Research-Solid Earth and Planets, 91, 13803–13872.

Armijo, R., Tapponnier, P., and Tonglin, H. (1989) Late Cenozoic right-lateral strike-slip faulting in southern Tibet. Journal of Geophysical Research, 94, 2787–2838.

Arrowsmith, R., and Strecker, M. R. (1999) Seismotectonic rangefront segementation and mountain growth in the Pamir-Alai region, Kyrgyzstan (Indo-Asian collision zone). Geological Society of America Bulletin, 111, 1665–1683.

Avouac, J. P., and Peltzer, G. (1993) Active tectonics of the southern Xinjiang, China: Analysis of terrace risers and normal fault scarp degradation along the Hotan-Qira fault system. Journal of Geophysical Research, 98, 21773–21807.

Avouac, J. P., Tapponnier, P., Bai, M., You, H., and Wang, G. (1993) Active thrusting and folding along the Northern Tien-Shan and Late Cenozoic rotation of the tarim relative to Dzungaria and Kazakhstan. Journal of Geophysical Research-Solid Earth, 98, 6755–6804.

Axen, G. J., Fletcher, J. M., Cowgill, E., Murphy, M., Kapp, P., MacMillan, I., and Ramos-Velazquez, E. (1999) Rangefront fault scarps of the Sierra El Mayor, Baja California: Formed above an active low-angle normal fault? Geology, 27, 247–250.

Baldwin, S. L., Lister, G. S., Hill, E. J., Foster, D. A., and McDougall, I. (1993) Thermochronological Constraints on the Tectonic Evolution of Active Metamorphic Core Complexes, Dentrecasteaux Islands, Papua-New-Guinea. Tectonics, 12, 611-628.

Blisniuk, P., Hacker, B., Glodny, J., Ratschbacher, L., B., S., Z., W., McWilliams, M., and Calvert, A. (2001) Normal faulting in central Tibet since at least 13.5 Myr ago. Nature, 412, 628-632.

Buck, W. R. (1991) Modes of continental lithospheric extension. Journal of Geophysical Research-Solid Earth, 96, 20161-20178.

Buck, W. R. (1993) Effect of lithospheric thickness on the formation of high-angle and low-angle normal faults. Geology, 21, 933-936.

Burchfiel, B. C., Brown, E. T., Qidong, D., Xianyue, F., Jun, L., Molnar, P., Jianbang, S., Zhangming, W., and Huichuan, Y. (1999) Crustal shortening on the margins of the Tien Shan, Xinjiang, China. International Geology Review, 41, 665-700.

Burchfiel, B. C., Peizhen, Z., Yipeng, W., Weiqi, Z., Fangmin, S., Qidong, D., Molnar, P., and Royden, L. (1991) Geology of the Haiyuan fault zone, Ningxia-Hui autonomous region, China, and its relation to the evolution of the northeastern margin of the Tibetan Plateau. Tectonics, 10, 1091-1110.

Burchfiel, C., Chen, Z., Liu, Y., and Royden, L. (1995) Tectonics of the Longmen Shan and adjacent regions, central China. International Geology Review, 37, 661-735.

Cogan, M. J., Nelson, K. D., Kidd, W. S. F., Wu, C., Wenjin, Z., Yongjun, Y., Jixiang, L., Brown, L. D., Hauck, M. L., Alsdorf, M. C., Edwards, M. A., and Kuo, J. T. (1998) Shallow structures of the Yadong-Gulu Rift, southern Tibet, from refraction analysis of Project INDEPTH common midpoint data. Tectonics, 17, 46-61.

Coleman, M. E., and Hodges, K. V. (1995) Evidence for Tibetan plateau uplift before 14 m. y. ago from a new minimum age for east-west extension. Nature, 374, 49-41.

Coney, P. J. (1980) Cordilleran metamorphic core complexes, in Crittenden, M. D., Coney, P. J., and Davis, G. H., eds., Cordilleran metamorphic core complexes. Geological Society of America Memoir, 153, 7-34.

Coney, P. J., and Harms, T. A. (1984) Cordilleran metamorphic core complexes: Cenozoic extensional relics of Mesozoic compression. Geology, 12, 550-554.

Cowan, D. S., Cladouhos, T. T., and Morgan, J. K. (2003) Structural geology and kinematic history of rocks formed along low-angle normal faults, Death Valley, California. Geological Society of America Bulletin, 115, 1230-1248.

Cowgill, E., Yin, A., Arrowsmith, J. R., Feng, W. X., and Zhang, S. H. (2004) The Akato Tagh bend along the Altyn Tagh fault, northwest Tibet 1: smoothing by vertical- axis rotation and the effect of topographic stresses on bend-flanking faults. Geological Society of America Bulletin, 116, 1423-1442.

Cowgill, E., Yin, A., Feng, W. X., and Qing, Z. (2000) Is the North Altyn fault part of a strike-slip duplex along the Altyn Tagh fault system? Geology, 28, 255-258.

Currie, B. S., Rowley, D. B., and Tabor, N. J. (2005) Middle Miocene paleoaltimetry of southern Tibet: implications for the role of mantle thickening and delamination in the Himalayan orogen. Geology, 33, 181-184.

Darby, B. J., and Ritts, B. D. (2002) Mesozoic contractional deformation in the middle of the Asian tectonic collage: the intraplate Western Ordos fold-thrust belt, China. Earth and Planetary Science Letters, 205, 13-24.

Darby, B. J., Ritts, B. D., Yue, Y. J., and Meng, Q. R. (2005) Did the Altyn Tagh fault extend beyond the Tibetan Plateau? Earth and Planetary Science Letters, 240, 425-435.

Dewey, J. F. (1988) Extensional Collapse of orogens. Tectonics, 7, 1123-1139.

Friedmann, S. J., and Burbank, D. W. (1995) Rift basins and supradetachment basins: intracontinental extensional end-members. Basin Research, 7, 109-127.

Gan, W. J., Zhang, P. Z., Shen, Z. K., Niu, Z. J., Wang, M., Wan, Y. G., Zhou, D. M., and Cheng, J. (2007) Present-day crustal motion within the Tibetan Plateau inferred from GPS measurements. Journal of Geophysical Research-Solid Earth, 112.

Garzione, C. N., DeCelles, P. G., Hodkinson, D. G., Ojha, T. P., and Upreti, B. N. (2003) East-west extension and Miocene environmental change in the southern Tibetan plateau: Thakkhola graben, central Nepal. Geological Society of America Bulletin, 115, 3-20.

Garzione, C. N., Dettman, D. L., Quade, J., DeCelles, P. G., and Butler, R. F. (2000) High times on the Tibetan Plateau: paleoelevation of the Thakkhola graben, Nepal. Geology, 28, 339-342.

Gold, R. D., Cowgill, E., Wang, X. F., and Chen, X. H. (2006) Application of trishear fault-propagation folding to active reverse faults: examples from the Dalong Fault, Gansu Province, NW China. Journal of Structural Geology, 28, 200-219.

Harrison, T. M., Copeland, P., Kidd, W. S. F., and Lovera, O. M. (1995) Activation of the Nyainqentanghla Shear Zone: implications for uplift of the Southern Tibetan Plateau. Tectonics, 14, 658-676.

Hayman, N. W., Knott, J. R., Cowan, D. S., Nemser, E., and Sarna-Wojcicki, A. M. (2003) Quaternary low-angle slip on detachment faults in Death Valley, California. Geology, 31, 343-346.

Hodges, K., and J. D. Walker (1992) Extension in the Cretaceous Sevier orogen, North American Cordillera. Geological Society of America Bulletin, 104, 560-569.

Horton, B. K, and Schmitt, J. G., (1998) Development and exhumation of a Neogene sedimentary basin during extension, Nevada. Geological Society of America Bulletin, 110, 163-172.

Jackson, J. (1992) Partitioning of strike-slip and convergent motion between Eurasia and Arabia in Eastern Turkey and the Caucasus. Journal of Geophysical Research-Solid Earth, 97, 12471-12479.

Jackson, J., Haines, J., and Holt, W. (1995) The Accommodation of Arabia-Eurasia Plate Convergence in Iran. Journal of Geophysical Research-Solid Earth, 100, 15205-15219.

Jackson, J., and McKenzie, D. (1984) Active tectonics of the Alpine Himalayan Belt between Western Turkey and Pakistan. Geophysical Journal of the Royal Astronomical Society, 77,

185-198.

Kapp, J., Harrison, T. M., Kapp, P., Grove, M., Lovera, O. M., and Ding, L. (2005) Nyainqentanglha Shan: a window into the tectonic, thermal, and geochemical evolution of the Lhasa block, southern Tibet. Journal of Geophysical Research, 110. doi: 10. 1029/2004JB003330.

Kapp, P., and Guynn, J. H. (2004) Indian punch rifts Tibet. Geology, 32, 993-996. Kapp, P., Taylor, M., Stockli, D., and Ding, L. (2008) Development of active low-angle normal fault systems during orogenic collapse: insight from Tibet. Geology, 36, 7-10. doi: 10. 1130/G24054A. 1.

Kirby, E., Whipple, K. X., Burchfiel, B. C., Tang, W. Q., Berger, G., Sun, Z. M., and Chen, Z. L. (2000) Neotectonics of the Min Shan, China: implications for mechanisms driving Quaternary deformation along the eastern margin of the Tibetan Plateau. Geological Society of America Bulletin, 112, 375-393.

Klemperer, S. L. (2006) Crustal flow in Tibet: geophysical evidence for the physical state of Tibetan lithosphere, and inferred patterns of active flow, in Law, R. D., Searle, M. P., and Godin, L., eds., Channel Flow, Ductile Extrusion and Exhumation in Continental Collision Zones. Special Publication, 268. London, Geological Society, 39-70.

Lachenbruch, A. H., and Morgan, P. (1990) Continental extension, magmatism and elevation; formal relations and rules of thumb. Tectonophysics, 174, 39-62.

Langin, W. R., Brown, L. D., and Sandvol E. A. (2003) Seismicity of central Tibet from Project INDEPTH III seismic recordings. Bulletin of the Seismological Society of America, 93, 2146-2159.

Lave, J., and Avouac, J. P. (2000) Active folding of fluvial terraces across the Siwaliks Hills, Himalayas of central Nepal. Journal of Geophysical Research, 105, 5735-5770.

Leeder, M. R., and Gawthorpe, R. L. (1987) Sedimentary models for extensional tilt-block/half-graben basins, in Coward, M. P., Dewey, J. F., and Hancock, P. L., eds., Continental extensional tectonics. Geological Society of London Special Publication, 28, 139-152.

Lister, G. S. (1984) Metamorphic Core Complexes of Cordilleran Type in the Cyclades, Aegean Sea, Greece. Geology, 12, 221-225.

Lister, G. S., and Davis, G. A. (1989) The origin of metamorphic core complexes and detachment faults formed during Tertiary continental extension in the northern Colorado River region, U. S. A. Journal of Structural Geology, 11, 65-94.

Mercier, J. L., Armijo, R., Tapponnier, P., Careygailhardis, E., and Lin, H. T. (1987) Change from Late Tertiary Compression to Quaternary Extension in Southern Tibet During the India-Asia Collision. Tectonics, 6, 275-304.

Meriaux, A. S., Ryerson, F. J., Tapponnier, P., Van der Woerd, J., Finkel, R. C., Xu, X. W., Xu, Z. Q., and Caffee, M. W. (2004) Rapid slip along the central Altyn Tagh Fault: morphochronologic evidence from Cherchen He and Sulamu Tagh. Journal of Geophysical Research- Solid Earth, 109, B06401. doi: 10. 1029/2003JB002558.

Meriaux, A. S., Tapponnier, P., Ryerson, F. J., Xu, X. W., King, G., Van der Woerd, J.,

Finkel, R. C., Li, H. B., Caffee, M. W., Xu, Z. Q., and Chen, W. B. (2005) The Aksay segment of the northern Altyn Tagh fault: tectonic geomorphology, landscape evolution, and Holocene slip rate. Journal of Geophysical Research-Solid Earth, 110, B04404. doi: 10.1029/2004JB003210.

Molnar, P., England, P., and Martinod, J. (1993) Mantle dynamics, uplift of the Tibetan Plateau, and the Indian Monsoon. Reviews of Geophysics, 31, 357-396.

Molnar, P., and Lyon-Caen, H. (1989) Fault plane solutions of earthquakes and active tectonics of the Tibetan Plateau and its margins. Geophysical Journal International, 99, 123-153.

Molnar, P., and Tapponnier, P. (1978) Active tectonics of Tibet. Journal of Geophysical Research, 83, 5361-5375.

Murphy, M., Yin, A., Kapp, P., Harrison, T. M., Manning, C., and Ryerson, F. (2002) Structural evolution of the Gurla Mandhata detachment system, southwest Tibet: implications for the eastward extent of the Karakoram fault system. Geological Society of America Bulletin, 114, 428-447.

Murphy, M. A., Yin, A., Kapp, P., Harrison, T. M., Lin, D., and Guo, J. H. (2000) Southward propagation of the Karakoram fault system, southwest Tibet: timing and magnitude of slip. Geology, 28, 451-454.

Numelin, T., Kirby, E., Walker, J. D., and Didericksen, B. (2007) Late pleistocene slip on a low-angle normal fault, Searles Valley, California. Geosphere, 3, 163-176.

Pan, Y., and Kidd, W. S. F. (1992) Nyainqentanglha shear zone: a late Miocene extensional detachment in the southern Tibetan plateau. Geology, 20, 775-778.

Peltzer, G., Tapponnier, P., and Armijo, R. (1989) Magnitude of Late Quaternary left-lateral displacements along the north edge of Tibet. Science, 246, 1285-1289.

Regenauer-Lieb, K., Weinburg, R., and Rosenbaum, G. (2006) The effect of energy feedbacks on continental strength. Nature, 442, 67-70.

Rietbrock, A., Tiberi, C., Scherbaum, F., and Lyon-Caen, H. (1996) Seismic slip on a low angle normal fault in the Gulf of Corinth: evidence from high-resolution cluster analysis of microearthquakes. Geophysical Research Letters, 23, 1817-1820.

Robinson, A., Yin, A., Manning, C., Harrison, T. M., Zhang, S. H., and Wang, X. F. (2004) Tectonic evolution of the northeastern Pamir: constraints from the northern portion of the Cenozoic Kongur Shan extensional system, western China. Geological Society of America Bulletin, 116, 953-973.

Robinson, A. C. (2009) Geologic offsets across the northern Karakorum fault: Implications for its role and terrane correlations in the western Himalayan-Tibetan orogen. Earth and Planetary Science Letters, 279, 123-130.

Saylor, J. E., Quade, J., Dellman, D. L., DeCelles, P. G., Kapp, P. A., and Ding, L. (2009) The late Miocene through present paleoelevation history of southwestern Tibet. American Journal of Science, 309, 1-42.

Schlische, R. W. (1991) Half-graben basin filling models: new constraints on continental extensional basin development. Basin Research, 3, 123-141.

Seeber, L., and Armbruster, J. G. (1984) Some elements of continental subduction along the Himalayan front. Tectonophysics, 105, 263-278.

Sorel, D. (2000) A Pleistocene and still-active detachment fault and the origin of the Corinth-Patras rift, Greece. Geology, 28, 83-86.

Spicer, R. A., Harris, N. B. W., Widdowson, M., Herman, A. B., Guo, S. X., Valdes, P. J., Wolfe, J. A., and Kelley, S. P. (2003) Constant elevation of southern Tibet over the past 15 million years. Nature, 421, 622-624.

Stockli, D., Taylor, M., Yin, A., Harrison, T. M., D'Andrea, J., Kapp, P., and Ding, L. (2002) Late Miocene-Pliocene inception of E-W extension in Tibet as evidenced by apatite (U-Th)/He data, Geological Society of America Abstracts with Programs, vol. 34. Denver, 411.

Sun, J., Shen, Z., Xu, X., and Burgmann, R. (2008) Synthetic normal faulting of the 9 January 2008 Nima (Tibet) earthquake from conventional and along-track SAR interferometry. Geophysical Research Letters, 35, L22308. doi: 10.1029/2008GL035691.

Tapponnier, P., Mercier, J. L., Armijo, R., Tonglin, H., and Ji, Z. (1981) Field evidence for active normal faulting in Tibet. Nature, 294, 410-415.

Tapponnier, P., and Molnar, P. (1979) Active faulting and Cenozoic tectonics of the Tien Shan, Mongolia, and Baykal regions. Journal of Geophysical Research, 84, 3425-3459.

Tapponnier, P., Xu, Z. Q., Roger, F., Meyer, B., Arnaud, N., Wittlinger, G., and Yang, J. S. (2001) Oblique stepwise rise and growth of the Tibet plateau. Science, 294, 1671-1677.

Taylor, M. (n. d.) Unpublished data, University of Kansas, Lawrence. Taylor, M., and Peltzer, G. (2006) Current slip rates on conjugate strike slip faults in central tibet using synthetic aperture radar interferometry. Journal of Geophysical Research, 111. doi: 10.1029/2005JB004014.

Taylor, M., and Yin, A. (2009) Active structures on the Tibetan Plateau and surrounding regions: relationships with earthquakes, contemporary strain, and Late Cenozoic volcanism. Geosphere, 5, 199-214.

Taylor, M., Yin, A., Ryerson, F., Kapp, P., and Ding, L. (2003) Conjugate strike slip fault accommodate coeval northsouth shortening and east-west extension along the Bangong-Nujiang suture zone in central Tibet. Tectonics, 22. doi: 10.1029/2002TC001361.

ten Brink, U. S., and Taylor, M. (2002) Crustal structure of central Lake Baikal: Insights into intracontinental rifting. Journal of Geophysical Research, 107. doi: 10.1029/2001JB000300.

Thatcher, W. (2007) Micoplate model for the present-day deformation of Tibet. Journal of Geophysical Research, 112. doi: 10.1029/2005JB004244.

Thompson, S., Weldon, R., Rubin, C., Abdrakhmatov, K., Molnar, P., and Berger, G. (2002) Late Quaternary sliprates across the central Tien Shan, Kyrgyzstan, central Asia. Journal of Geophysical Research, 107. doi: 10.1029/2001JB000596.

Wang, E., and Burchfiel, B. C. (2000) Late Cenozoic to Holocene deformation in southwestern Sichuan and adjacent Yunnan, China, and its role in formation of the southeastern part of the Tibetan plateau. Geological Society of America Bulletin, 112, 413-423.

Wang, E., Burchfiel, C., Royden, L., Liangzhong, L., Wenxin, L., and Zhiliong, C. (1998) Late Cenozoic Xianshuihe- Xiaojiang Red River and Dali fault systems of southern Sichuan and central Yunnan, China. Geological Society of America Special Paper, 327, 108.

Wang, Q., Zhang, P., Freymueller, J., Bilham, R., Larson, K., Lai, X., You, X., Niu, Z., Wu, J., Li, Y., Liu, J., Yang, Z., and Chen, Q. (2001) Present day crustal deformation in China constrained by global positionaing system measurements. Nature, 294, 574-577.

Wernicke, B. (1981) Low-angle normal faults in the Basin and Range Province: Nappe tectonics in an extending orogen:. Nature, 291, 645-648.

Wernicke, B. P., and Axen, G. J. (1988) On the role of isotasy in the evolution of normal fault systems. Geology, 16, 848-851.

Williams, H., Turner, S., Kelley, S., and Harris, N. (2001) Age and composition of dikes in southern Tibet: new constraints on the timing of east-west extension and its relationship to postcollisional volcanism. Geology, 29, 339-342.

Xu, X., Wen, X., Chen, G. H., and Yu, G. H. (2008) Discovery of the Longriba Fault Zone in Eastern Bayan Har Block, China and its tectonic implication. Science in China Series D-Earth Sciences, 51, 1209-1223.

Yeats, R. S., and Lillie, R. J. (1991) Contemporary tectonics of the Himalayan Frontal Fault System: folds, blind thrusts and the 1905 Kangra Earthquake. Journal of Structural Geology, 13, 215-225.

Yin, A., Dang, Y. Q., Wang, L. C., Jiang, W. M., Zhou, S. P., Chen, X. H., Gehrels, G. E., and McRivette, M. W. (2008) Cenozoic tectonic evolution of Qaidam basin and its surrounding regions (part 1): the southern Qilian Shan-Nan Shan thrust belt and northern Qaidam basin. Geological Society of America Bulletin, 120, 813-846.

Yin, A., and Harrison, T. M. (2000) Geologic evolution of the Himalayan-Tibetan orogen. Annual Review of Earth and Planetary Science, 28, 211-280.

Yin, A., Harrison, T. M., Ryerson, F. J., Chen, W. J., Kidd, W. S. F., and Copeland, P. F. (1994) Tertiary structural evolution of the Gangdese thrust system in southeastern Tibet. Journal of Geophysical Research, 99, 18175-18201.

Zhang, P., Shen, Z., Wang, M., Gan, W., Burgmann, R., and Molnar, P. (2004) Continuous deformation of the Tibetan Plateau from global positioning system data. Geology, 32, 809-812.

(蔡国富 崔炳松 译，吴哲 赵钊 崔敏 校)

第 23 章　西班牙东南部贝蒂克（Betic）山间盆地的地层、沉降及构造演化史

JOSE RODRIGUEZ-FERNANDEZ[1], ANTONIO AZOR[2],
JOSE MIGUEL AZANON[3]

(1. Instituto Andaluz de Ciencias de la Tierra (CSIC-UGR), Avenida de las Palmeras, Armilla, Granada, Spain; 2. Departamento de Geodinamica, Universidad de Granada, Campus de Fuentenueva, Granada, Spain; 3. Instituto Andaluz de Ciencias de la Tierra (CSIC-UGR), Campus de Fuentenueva, Granada, Spain and Departamento de Geodina'mica, Universidad de Granada, Campus de Fuentenueva, Granada, Spain)

摘　要：西班牙南部贝蒂克（Betic）山脉的最显著特征之一就是晚中新世的伸展构造，它形成于一个微弱挤压期之后。低角度正断层对于内部带的构成或 Alboran 带变质岩的剥蚀起到了重要作用，同时也影响到了沉积盆地的形成，这些盆地通常被称为贝蒂克山间盆地。

贝蒂克山间盆地位于阿尔沃兰带或者阿尔沃兰带与外带的连接区域。这些盆地具有共同的沉积演化特征，以区域性不整合面所划分的三个主要层序为特征。底部（晚塞拉瓦莱期—早托尔托纳期）以及顶部层序（晚托尔托纳期—更新世）由陆源碎屑沉积岩构成。中部层序（晚托尔托纳—晚墨西拿期）由海相沉积岩组成，代表了一次非常短暂的穿时海侵事件，这次海侵在 Betics 的东部及南部延续时间较长。盆地沉降速率的最大值出现在晚塞拉瓦莱期—早托尔托纳期（13—7Ma），其后至更新世沉积速率逐渐降低。盆地围区山脊的抬升与盆地的沉降大致同步。

贝蒂克山间盆地的沉降与阿尔沃兰带深部构造变质单元伸展作用开始的时间相同，即晚塞拉瓦莱期—早托尔托纳期。从托尔托纳晚期开始，非洲—欧洲板块会聚方向发生改变，使该阶段主要形成 E—W 向褶皱、NW—SE 向高角度正断层以及 E—W 向和 NW—SE 向的走滑断层。这些构造伴随同期的挤压、剥蚀以及伸展作用使盆地及其基底发生了复杂变形。随后，走滑断层对部分盆地边界进行了强烈的改造，破坏了此前的一些伸展构造。以地幔分层和/或俯冲板块后撤/板片拆离为特征的岩石圈规模的假说为解释这些贝蒂克山间盆地提供了依据。

关键词：伸展构造　走滑断层　Betic-Rif 造山带　西地中海　阿尔沃兰盆地

23.1 引言

山间盆地一词通常指周围被山脉环绕的盆地，在盆地的构造环境、沉降原因、基底背景以及沉积填充等方面没有其他特别的含义。故山间盆地可发育于各种各样的构造环境之中：如（1）走滑断裂环境（Sylvester, 1984, 1988; Christie-Blick 和 Biddle, 1985; Crowell, 1987）；（2）由于变质核杂岩体的剥蚀而形成的伸展环境（Cloetingh 等, 1996; Dunkl 等, 1998; Fritz 和 Messner, 1999）；（3）弧内环境（Busby 和 Ingersoll, 1995 及其参考文献）；或者是（4）褶皱—冲断带顶部（中亚地区；Cloetingh, 1988）。

对于贝蒂克（Betic）地区来说，山间盆地发育于新近纪—第四纪造山晚期上地壳减薄

以及内带发生剥蚀的大背景下。因此贝蒂克山间盆地无论在过去或是现在所代表的都是为群山环绕的凹陷地带。这些盆地主要位于山系的中部及东部（图 23-1 和图 23-2）。最大的阿尔沃兰盆地位于研究区南部，地中海最西端。其西边界为 Gibraltar 岛弧（一个斜向弯曲弧），向东朝 Algero-Balearic 盆地张开。

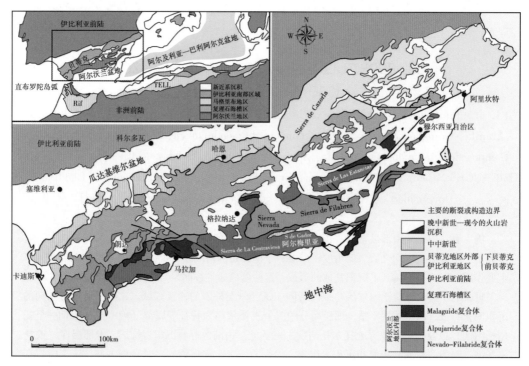

图 23-1　贝蒂克盆地的简化地质图（小插图所示为贝蒂克地区在西地中海的位置）

在阿尔沃兰盆地之下，陆壳向东减薄。从 Gibraltar 岛弧附近的 15km 减薄至 Algero-Balearic 盆地过渡地区的 6km（Booth-Rea 等，2007）。新近纪—第四纪（12.0—0.65Ma）Alboran 盆地的岛弧火山岩（Duggen 等，2003）与上述地壳减薄有关（Booth-Rea 等，2004）。阿尔沃兰盆地的底部是高压变质岩，属于贝蒂克内带的构造变质单元（Platt 等，1998；Comas 等，1999）。新近纪晚期 Alboran 盆地的沉积记录与贝蒂克南部山间盆地的记录相似（Rodriguez-Fernandez 等，1999）。

相对于了解较少的碰撞早期阶段（始新世；Lonergan，1993；Platt 等，2005），贝蒂克山间盆地为造山晚期伸展构造提供了较好的记录。不仅记录了盆地开始形成的时代，也记录了构造运动对盆地沉积体系及盆地建造的控制作用。依据生物地层定年和与主要断层体系的关系，可以推测出这些盆地最初发育于主要伸展滑脱构造的上盘，这些伸展滑脱构造在晚塞拉瓦莱期—早托尔托纳期（13—7Ma）处于活动状态。周围基底的剥蚀历史也可以从沉积记录及物源研究中获得，并且还可以与基底的热史数据相结合（Johnson，1997；Johnson 等，1997；Clark，2004；Reinhardt 等，2007；Clark 及 Dempster，2009）。近岸沉积岩的现今海拔数据记录完整，可以为推测盆地周围山脊的抬升速率提供约束（Braga 等，2003；Sanz de Galdeano 和 Alfaro，2004）。

本文旨在从沉积记录和构造样式方面对贝蒂克山间盆地的主要特征进行总结，以推测晚

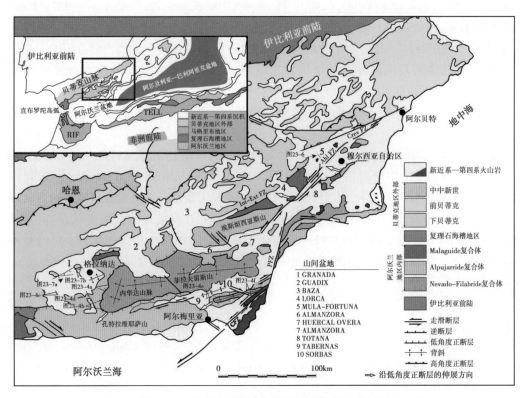

图23-2 贝蒂克地区中部以及东部地区的地质图

小插图所示为贝蒂克在西地中海的位置。文中主要山间盆地由数字1-10代表。Int-Ext FZ—内部—外部地区的连接；Crev FZ—Crevillente 断层带；Alh FZ—Alhama de Murcia 断层带；PFZ—Palomares 断层带；CFZ—Carboneras 断层带。地震剖面见图23-6及图23-7，位置见图23-4

期造山的主要构造事件时间以及盆地基底剥蚀的时间。这些研究成果将与以Betic为整体的现今构造演化模型相结合。

23.2 地质背景

西班牙东南部的贝蒂克山与位于摩洛哥北部的Rif山脉一起，形成了位于地中海最西端环绕阿尔沃兰海的弧形阿尔卑斯山（图23-1）。贝蒂克相当于这个造山带的北分支，并被划分为典型的外带（北部）与内带（南部）。其外带（图23-1、图23-2）与中生代—古近纪伊比利亚张裂型古边界一样，可以被细分为前贝蒂克单元和后贝蒂克单元，后者代表离伊比利亚古边界最远的区域。

内带也被称为阿尔沃兰带，包含三个构造叠加的复合体，主要由上地壳以及中地壳岩石组成，年代在古生代到中生代之间，在阿尔卑斯造山运动时期受到变质作用影响。三个复合体由下到上分别为Nevado-Filabride、Alpujarride以及Malaguide复合体。Nevado-Filabride复合体最近被认为是华力西期（古生代）的基底，早—中中新世在高压条件下发生变质，而后出露于一个大规模的背形核部（Platt等，2004，2006；Booth-Rea等，2005，2007）。Alpujarride复合体也受到高压变质作用影响，但是Malaguide复合体只受到了低级变质作用影响。堆叠形成的内带在上阿基坦阶—下波尔多阶整体向西移动，并在下波尔多阶（18Ma）

与外带发生碰撞（Durand-Delga，1980；Hermes，1985；Sanz de Galdeano，1987；Balanya 等，1997；Geel 和 Roep，1998，1999）。

在西部，内带与外带之间为复理石槽带（图 23-1），主要由中生代到新近纪的深水相沉积岩组成，在减薄陆壳或洋壳为基底的深窄盆地中沉积形成，其变形作用以及现今位置（主要位于贝蒂克最西部）主要是由于阿尔沃兰带向西移动而形成的（Wildi，1983）。这个西向运动持续至今，并被认为是形成直布罗陀岛弧造山楔的主要因素（Gutscher 等，2002，2009）。

Malaguide 单元与较高的 Alpujarride 单元之间的韧性伸展分离于渐新世—阿基坦阶（Aldaya 等，1991）或早—中中新世（Mayoral 等，1994；Lonergan 和 Platt，1995），是阿尔沃兰带开始发生构造剥蚀的证据。此次分离运动主要表现为在东贝蒂克地区从顶部向西北方向以及西南方向的运动。在西贝蒂克地区，Alpujarride 单元在晚波尔多期发生从顶部向 NNW 方向的伸展（Garcia-Dueiias 等，1992；Crespo-Blanc 等，1994）。根据磁异常条带的基本趋势（Rehault 等，1985），初次伸展幕与北阿尔及利亚盆地的张开是同时发生的（Martinez-Martinez 和 Azafion，1997）。

第二次更加明显的伸展作用发生于中—晚中新世（13—7Ma），其顶部朝 SWW 方向运动。该幕伸展作用是 Nevado-Filabride 复合体，即阿尔沃兰带更深的构造变质复合体发生剥蚀作用的结果（Balanya 和 Garcfa-Dueiias，1987；Monie 等，1991；Garda-Duenas 等，1992；Watts 等，1993；Platt 和 Whitehouse，1999；Martinez-Martinez 等，2002；Booth-Rea 等，2007）。

现今贝蒂克地区的应力场（Reicherter 和 Peters，2005；De Vicente 等，2008）主要为 NNW-SSE 至 N—S 向缩短，这是由于非洲板块与欧洲板块的会聚作用从晚托尔托纳阶之后较为活跃。期间形成的 E—W 向的褶皱，构成了内部区域的盆地与山脉的地理特征。与这次挤压运动同时发生的是一次方向上大致垂直的伸展运动（Sanz de Galdeano，1988；Galindo-Zaldivar 等，1993；Rodrfguez-Fernandez 和 Martm-Penela，1993；Buforn 等，1995；Sanz de Galdeano 和 Alfaro，2004；Stokes，2008）。这使得 Nevado-Filabride 复合体在 12-8Ma 期间发生了进一步的伸展剥蚀（Johnson 等，1997；Reinhart 等，2007；Clark 和 Dempster，2009），同时形成了主要的山间盆地。

23.3 贝蒂克山间盆地的沉积记录

贝蒂克山间盆地或位于内带，或沿着内带与外带的连接区域分布。其基底由内带的古生代以及三叠纪变质岩和/或外带的中生代和古近—新近纪沉积岩组成。此外，在一些盆地中，其基底是由代表先前残留沉积盆地的沉积岩石组成，而之前的沉积盆地主要年代为阿基塔阶—下塞拉瓦莱阶，与贝蒂克山间盆地无关。这些沉积基底岩石不包含来自阿尔沃兰带最底部的变质岩复合体（Nevado-Filabride）的碎屑，这表明该复合体在当时并没有发生剥蚀。

最重要的几个山间盆地如下（图 23-1 和图 23-2）：Granada、Guadix、Baza、Lorca、Mula、Fortuna、Almanzora、Huercal-Overa、Totana、Tabernas 以及 Sorbas 盆地。这些盆地在上塞拉瓦莱阶—下托尔托纳阶时期（13—7Ma）同时发育，并具有相同的沉积演化历史（图 23-3），具有由区域不整合面分隔的三个主要层序。这三个层序由下至上描述如下。

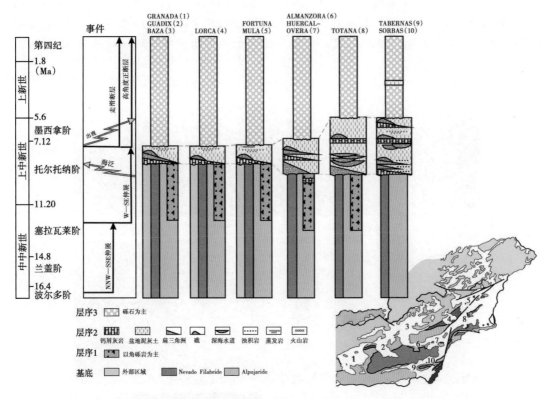

图23-3 主要贝蒂克山间盆地的综合地层剖面,包含文中所述所有重要事件的时间。
小图所示为带有数字的不同盆地位置

23.3.1 层序1

山间盆地中填充的最老沉积物为上塞拉瓦莱阶的粗粒碎屑岩(图23-3),沉积于陆相或不稳定的海陆过渡的环境中。这些沉积物组成了层序1的最下部,包含红色角砾岩、砾岩、生物碎屑钙质灰岩、淡水灰岩、砂岩以及含石膏泥岩(Volk 和 Rondeol,1964;Rodriguez 和 Fernandez,1982)。此层序底部沉积全部来源于阿尔沃兰带的两个上部复合体,这表明这些复合体在当时已经出现。

层序1的上部由冲积扇体系组成,沉积物来自盆地的周围高地。具体由含砂、粉砂、泥质夹层的砾岩组成,是一个向上变细的层序。层序1的上部由于其岩石的碎屑性质难以测年,在一些盆地中该部分的时代依然存在争议。忽略这些限制的话,层序1的上部可能具有等时性,其年龄可能为托尔托纳阶早期(11Ma)。层序1的上部记录了Nevado-Filabride复合体碎屑的初次出现,揭示出阿尔沃兰带最底部的变质复合体在此时开始出露。

从构造角度来说,层序1记录了晚塞拉瓦莱—早托尔托纳期(13—7Ma)伸展拆离作用在下盘发育的显著抬升区域中剥落的粗粒碎屑沉积。

23.3.2 层序2

层序2不整合于层序1之上,或直接覆盖于基底之上。这一层序主要由发育在陆缘斜坡台地温水中的生物碎屑钙质灰岩沉积组成(Puga-Bernabeu,2007),向远端变化为盆地沉积泥灰岩。层序之中的同期构造不整合面，成为这些盆地的边界（图23-4c）。在一些盆地中（如Sorbas, Tabernas），层序2含有由浊积砾岩以及砂岩填充的深水水道（Kleverlaan, 1987a, 1987b; Haughton, 2000, 2001）。层序2向上出现粗粒碎屑扇三角洲沉积楔，其顶

图23-4 贝蒂克山间盆地的发育示意图

(a) 分割位于Nevada山脉西翼Nevado-Filabride复合体和Alpujarride复合体的低角度正断层。(b) Nevada山脉西南端的倾斜航空照片；两个主要的构造边界分别对应分割Nevado-Filabride/Alpujarride复合体的低角度正断层，以及将Alpujarride基底与新近纪沉积填充分隔的高角度正断层；后者在多数地区都表现为不整合面接触。(c) 在Granada盆地西南边界，下托尔托纳阶沉积不整合覆盖于Alpujarride基底之上。(d) Granada盆地边缘的高角度正断层，断层面对Alpujarride复合体中的白云岩造成了影响。(e) Tabernas盆地边缘的走滑断层，这个断层段在上新世沉积作用下发生石化作用。(f) Carboneras断层带（CFZ）的墨西拿期沉积受到走滑断层的影响，上新世的褶皱沉积使这些走滑断层发生石化作用，具体位置见图23-2

部生长有珊瑚礁（Braga 等，1990）。这些扇三角洲与周围隆起上的古水道体系相连，其中滑塌、角砾岩及重力流的出现证明同时期发生过构造运动。

在托尔托纳末期，由于阿尔沃兰带抬升，一些盆地与海的过渡部位开始变得不稳定。其结果就是，由于反复的海侵，在盆地较深处沉积了蒸发岩（主要为盐及石膏）。而在最北部的盆地，即那些位于内部地区边缘的盆地（Granada，Lorca 和 Fortuna 盆地），可见托尔托纳晚期的蒸发岩，其年代早于"地中海盐度危机"事件（Rodriguez-Fornandez 等，1984；Garces 等，1998；Dinares 等，1999；Krijgsman 等，2000；Rodriguez-Fernandez 和 Sanz de Galdeano，2006）。Krijgsman 等（2000）用"托尔顿盐度危机"一词将此事件（7.8—7.6Ma）与著名的地中海盐度危机事件相区分，后一事件在距离地中海较近的盆地中也有记录，例如 Sorbas 盆地，其蒸发岩沉积年代为晚墨西拿期（6Ma）。

部分贝蒂克山间盆地的晚墨西拿期沉积岩是由冲积扇碎屑楔组成，上覆河流相和湖泊相沉积，包括淡水灰岩及蒸发岩。认为这些非海相沉积岩是由于相对快速的抬升运动造成的陆地形成或者出露而形成的（Rodrfguez-Fcrnandoz 等，1984；Montcnat 等，1990c）。这一过程在距离地中海较远的盆地中发生得较早。因此，层序 2 顶部具有穿时性，并且向东和向南变新（图 23-3）。对于东部的贝蒂克盆地，Meijninger（2006）也描述其具有相似的向东穿时性，并认为其与上地幔的拆离过程有关。

总而言之，层序 2 代表了阿尔沃兰带出露之后在晚托尔托纳期发生的短暂（2.0—1.8Ma）海泛（Meijningcr，2006；Rodriguez-Fernandez 和 Sanz de Galdeano，2006）。此次海泛的主要原因可能是层序 2 沉积之前及沉积过程中经历了构造伸展作用而产生了沉降（Rodriguez-Fermindez 和 Sanz de Galdeano，2006）。在中—晚中新世期间海平面升降非常平缓（<150m；Hardenbol 等，1998），因此可以不将其作为对这次海泛作用影响的主要因素。沉降的构造原因也与计算出的此层序高沉积速率相符合（Granada 盆地 217m/Ma；Rodriguez-Fernandez 和 Sanz de Galdeano，2006）。这次晚中新世的海泛在其他地中海盆地也有记录，例如克里特岛的 Iraklion 盆地，其许多构造、年代及地层特征都与贝蒂克山间盆地相似（Ten Veen 和 Postma，1999a，1999b）。

23.3.3 层序 3

此层序主要由上新世—更新世的红色角砾岩、砂岩以及黏土岩组成，在基底和层序 1、层序 2 之上呈不整合接触。这些陆相沉积包括辫状河、曲流河以及冲积扇相。沉积相的分布表明冲积扇主要分布于盆地边缘，为周围山脉提供排水系统，沿盆地方向与辫状河或曲流河相连接（Viseras，1991）。在盆地的低洼位置还发育有湖相细粒碎屑沉积或石灰岩沉积。更新世末期盆地内最后沉积了干旱气候环境下发育的成土钙质砾岩（Azanon 等，2006）。

在沉积作用之后，由于构造抬升及外围水系的发育，这些盆地都遭到了剥蚀作用，而其中一些盆地还发生了构造反转。盆地与盆地之间从内部水系向外部水系的转换具有穿时性，时间跨越从早上新世到晚更新世（Viseras 和 Fernandez，1992；Azann 等，2006；Stokes，2008；Perez Pena 等，2009）。

23.4 贝蒂克山间盆地的沉降史

对于贝蒂克山间盆地最初的普遍看法是这些盆地存在的时间很短，其活跃期短于 13~

12Ma。此外这些盆地都显示出在初始阶段沉降速度最快（晚塞拉瓦莱期—早托尔托纳期，13—8.5Ma）的特征。目前已有研究利用去压实和构造沉降曲线（经过或未经过古海平面及古水深校正）来定量研究这些垂向变化（Cloetingh 等，1992；Soria 等，1998，2001；Rodríguez-Fernández 等，1999；Garcés 等，2001；Hanne 等，2003；Braga 等，2003；Sanz de Galdeano 和 Alfaro，2004；Meijninger，2006；Rodríguez Fernández 和 Sanz de Galdeano，2006）。这些研究都采用 Airy 地壳均衡说，得出的沉降量结果大致相同，并认为其存在的微小差距是由于地层年代、沉积厚度以及沉积间断所导致。

研究认为，盆地最初、也是最快的发育阶段（晚塞拉瓦莱期—早托尔托纳期）记录下了构造的快速伸展过程（图 23-5）。在第二阶段记录下了盆地抬升及盆地"大陆化"，这一阶段沉降速率较低。第二沉降阶段在盆地中穿时发育，在西部及北部盆地（晚托尔托纳期）中的发育年代要早于东部及南部盆地（墨西拿期）。这一沉降阶段还与走滑断层开始活动的时间一致（见下一节）。墨西拿期及上新世—更新世的沉积岩由于沿盆地边缘发育的走滑断层作用而发生强烈变形，但是这一时期的沉积厚度并不大，表明走滑断层对于盆地沉降没有多少促进作用。此外并没有发现幕式沉降或者交替出现抬升与沉降的证据，因此，认为这是典型的走滑特征（Xie 和 Holler，2009；图 23-5）。

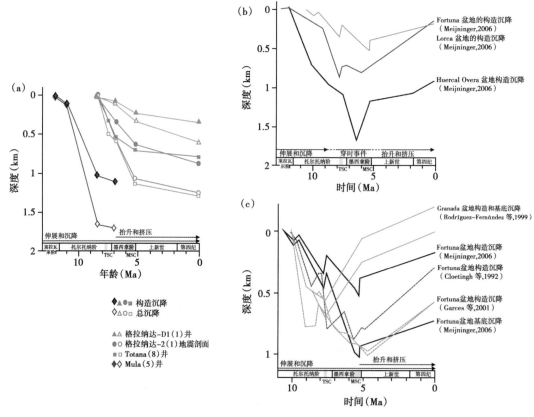

图 23-5 部分贝蒂克山间盆地的沉降曲线

(a) Granada、Totana 以及 Mula 盆地，分别标示了构造沉降及总沉降量，未进行古海平面及古水深校正。

(b) Fortuna、Lorca 及 Huercal Overa 盆地，已进行了古水深校正，据 Meijninger（2006）修改。

(c) Granada 及 Fortuna 盆地，已进行了古水深校正，据几位不同作者修改推测认为沉降的第三阶段存在于西部部分盆地中（Granada，Baza 盆地；图 23-5）。这一阶段以上新世—更新世较快的沉降速率为特征，其沉积中心位于 NW—SE 向高角度正断层的上盘中（Sanzdo Galdcano 等，1985；Rodriguez Fernandez 和 Sanz de Galdeano，2006；Perez-Pena，2009）。但是此阶段的沉降速率依然低于第一沉降阶段（晚塞拉瓦莱期—早托尔托纳期）。

总之，贝蒂克山间盆地的沉降模式具有一个普遍特征，即在晚塞拉瓦莱期—早托尔托纳期伸展构造运动最活跃的时期达到最大值，其后逐渐递减直至更新世。

23.5 贝蒂克山间盆地的构造背景及成因

对于贝蒂克山间盆地的成因必须在特定的背景之下进行研究，即在阿尔沃兰带和贝蒂克及 Rifean 古边界碰撞而导致造山带增厚之后，发生的地壳减薄和伸展作用背景。在这一背景之下存在两种关于盆地成因的主要假设。第一种认为与欧洲—非洲大陆会聚而导致的收缩作用相关，并由此形成了地壳规模的走滑断层（Montenat 等，1987，1990；Sanz de Galdeano 和 Vera，1992）。而近年来，针对盆地边缘断裂动力学的详细构造研究表明，这些盆地发育于由于地壳减薄而形成的主要 W—SW 向伸展拆离体系的上盘中（Martinez-Martinez 和 Azanon，1997；Meijninger 和 Vissers，2006；Rodriguez-Fernandez 和 Sanz de Galdeano，2006）。

贝蒂克山间盆地发育在减薄的构造变质单元之上，这些单元记录了复杂的演化史。伸展构造使得阿尔沃兰带单元出露，并与下贝蒂克单元一起构成了山间盆地的基底。阿尔沃兰带中可见塑性和脆性的伸展构造，既有不同时期分别形成的（Martinez-Martinez 及 Azaiion，1997），也有单个过程中由塑性向脆性变形的逐渐转变（Platt 和 Vissers，1989；Jabaloy 等，1992）。

脆性伸展变形以高角度和低角度的正断层为特征，这些断层控制着山间盆地沉积发育、同沉积作用变形及盆地的不稳定性。低角度正断层同时影响着盆地的基底和地层单元。在这方面，一些盆地（Granada，Lorca，Mula，Fortuna，Totana）的地震反射剖面可以有效地证明低角度正断层的重要性（图 23-6 和图 23-7）。这些低角度正断层呈现断坡—断坪的几何特征，在几组主要的 NNW 及 SWW 向拆离体系（图 23-2、图 23-4、图 23-6、图 23-7）中

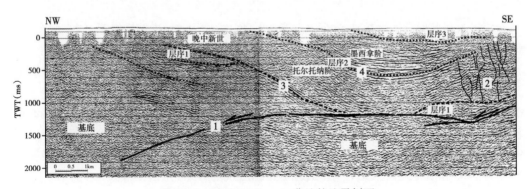

图 23-6 过 Mula-Fortuna 盆地的地震剖面

最重要的反射特征编号如下：（1）分隔基底与盆地充填的低角度正断层；（2）影响盆地东南边缘的一系列弯曲走滑断层；（3）盆地的底部接触：基底对应 Malaguide 复合体，而盆地充填底部对应托尔托纳阶沉积；
（4）托尔托纳阶—墨西拿阶边界，详细说明见正文，图 23-2

有规律地分布（Martinez-Martinez 等，2002；Booth-Rea 等，2004，2007）。对于 Nevado-Filabride 复合体来说，它的出露与晚塞拉瓦莱期—早托尔托纳期（13—7Ma）的伸展阶段有关，伴随着水平方向上至少 107~109km 的位移，其估计的 β 值为 3.5~3.9（Martinez-Martinez 等，2002）。这一伸展阶段主要集中于 Alpujarride 与 Nevado-Filabride 连接处（图 23-4a，b），产生了碎裂岩、断层泥及角砾岩，这些也可以在盆地的沉积充填中识别出来。通过对 Fortuna、Lorca 及 Huercal Overa 盆地沉积充填与基底的露头构造进行研究可估算伸展量，其 β 值为 1.1~2.0（Meijninger，2006）。

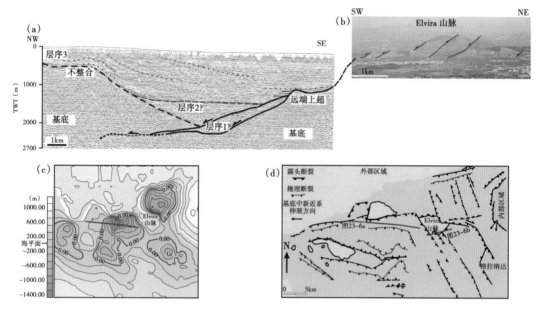

图 23-7　Granada 盆地的地震剖面和地质综合图

(a) 过 Granada 盆地北边缘的地震剖面显示了低角度正断层和充填了早新近纪沉积物的沉积中心，注意基底与断层上盘新近纪沉积之间的不整合接触及断层下盘的远端上超，位置见图 23-2；(b) 靠近图 23-7a 中地震剖面位置的 Elvira 山脉全景图，注意正断层露头以及它们与地震剖面中所显示的正断层的连续性；(c) 位于图 23-7a 中的 Granada 盆地北缘基底的等高线图（单位为米），红线代表海平面；(d) Granada 盆地北部的地质综合图显示了主要的野外露头和掩埋断层，红线位置见图 23-7a、b

在晚托尔托纳期，由于伊比利亚半岛与非洲板块的位置变化导致贝蒂克的主挤压轴由 NW—SE 向变为 NNW—SSE 向（Dewey 等，1989）。这一压力轴旋转活动导致一系列由挤压作用形成的近 E—W 走向的亚平行的千米级规模的直立褶皱（Weijermars 等，1985；Sanz de Galdeano，1987；Rodriguez-Fernandez 和 Sanz de Galdeano，1992；Martinez-Martinez 等，2002；Mejninger，2006；Booth-Rea 等，2007），不过依然有高角度正断层（图 23-4d）与走滑断层（图 23-4e、f）形成。对基底岩石的热年代学（Johnson，1997；Johnson 等，1997；Clark，2004；Reinhardt 等，2007；Clark 和 Dempster，2009）、古生物定年及沉积物源研究（Volk 和 Rondeal，1964；Rodriguez-Fernandez，1982；Montenat 等，1987）表明这些褶皱的年代为晚托尔托纳期。例如，Granada 盆地层序 2 下部褶皱变形，且与上覆上部地层呈现不整合接触关系，这样可以较准确地确定褶皱年龄。这些晚托尔托纳期的褶皱可能与同时期最北端的贝蒂克山间盆地的形成有关（Estevez 等，1982；图 23-2 及图 23-3）。

高角度正断层主要为 NW—SE 向，具 SW 向的下降断块（图 23-4d）。在西 Nevada 山脉，正断层可能与一系列 N—S 向褶皱（图 23-2）有关，这些褶皱可能是由于主要正断层体系的下盘发生差异性卸荷而形成（Martinez-Martinez 等，2002）。

走滑断层体系主要沿 E—W 向和 NE—SW 向分布，其中最重要的一系列断层（Crevillente、Alhama de Murcia、Palomares 和 Carboneras 断层）出露于东贝蒂克。这些 NE—SW 向走滑断层是左旋的，并被一些学者认为是地壳规模的边界（Jerez，1973；Leblanc 和 Olivier，1984；Montenat 等，1987；Weijermars，1987a；De Larouziere 等，1988；Keller 等，1995；Huibregtse，1998；Sanz de Galdeano，2008）。在贝蒂克内部地区，一系列 E—W 向右旋走滑断层将南 Nevada 山脉前部与南部的新近纪—第四纪盆地相分隔（图 23-1 与图 23-2）。这些伴随同沉积变形作用的右旋断层（Sanz de Galdeano，1987；Montenat 等，1990b；Haughton，2000，2001）与千米级规模的限制性弯曲作用共同出现于盆地边缘（Sanz de Galdeano 等，1985；Rodrfguez-Fernandez 等，1990）。

在评价走滑断层对于贝蒂克山间盆地的重要程度上存在着两个主要问题。其一为走滑断层的初始发育时期。对于这一点，一些学者（Wcijermars，1987b；Montenat，1990a；Boorsma，1993；Keller 等，1995）通过研究受断层区影响的沉积物年代，认为初始年代应为早中新世，而其他学者（Meijninger，2006；Meijninger 和 Vissers，2006）认为由于早期形成的伸展断层在走滑边界发生再活化（Crevillente 和 Alhama de Murcia 断层；图 23-2）并伴随同沉积变形作用，因而推测其年代为晚中新世—早上新世。值得注意的是，跨越 Carboneras 断层带的海上地震剖面（图 23-2）没有显示前墨西拿期有走滑断层的清晰证据。第二个问题集中于走滑断层的滑动距离，由于这些断层与主构造呈亚平行分布，故在断层两侧缺乏好的参照物。但是在 Carboneras 断层的例子中，利用石膏层作为参照，测算出侧向走滑断距大致为14km（Baena 等，1982）。考虑到该断层的重要性这一距离远远小于预期值（Montenat 等，1990a；Keller 等，1995）。最后，在走滑断层主控区并未识别出常见的垂向轴旋转块体这一特征（krijgsman 和 Garces，2004）。因此，走滑断层在盆地形成中扮演次级角色，主要组成了早期盆地边缘的伸展构造。而很多盆地中的确可见走滑断层位于盆地边界，使得盆地初期的伸展构造环境不明显。在这类盆地中，其伸展方向不能被准确确定，也使得对于这类盆地的成因存在较多争议。

总而言之，褶皱、正断层以及走滑断层自晚托尔托纳期开始，就在一个复杂的包含挤压、剥蚀及拉张的背景下影响着盆地及其基底。在这种背景下，山脉抬升与盆地沉降受较老低角度正断层影响较大，同时也受到由于非洲/伊比利亚半岛会聚作用产生的新的高角度断层的影响。从这方面来说，盆地的形成受 NW—SE 向高角度正断层的影响较大，而这些正断层切割了早期的低角度拉张断层。这些高角度正断层对西部盆地上新世—更新世主要沉降中心的形成具有重要作用（Rodriguez-Fernandez 和 Sanz de Galdeano，2006；Perez-Peiia，2009）。

对于贝蒂克山间盆地需要提出一个岩石圈规模的假说，该假说可能会带来关于伸展构造本身的争议。大多数模型都认为俯冲或（和）分层是西地中海地区地壳伸展构造发育的两种深部过程原因。而在这类模型中，直布罗陀岛弧下上地幔中的一个 E—SE 方向的倾斜高速体被认为是俯冲洋壳或者是分层的大陆岩石圈（Calvert 等，2000；Gutscher 等，2002；Spakrnan 和 Wortel，2004）。新近纪火山岩的地球化学数据也支持这一模型，指示出将阿尔沃兰盆地内部区域下部的分层洋壳岩石圈与 Betic-Rif 边缘之下陆壳岩石圈地幔相结合（Duggen 等，2003，2005）。其他模型认为造山带的伸展作用与俯冲作用的后撤和板片拆离，以及之后直布罗陀岛弧造山楔的发育有关（Morley，1993；Royden，1993；Lonergan 和 White，1997；Gutscher 等，2002；2009；Duggen 等，2003，2005；Spakman 和 Wortel，2004；Booth-Rea 等，2007）。

23.6 总结

贝蒂克山间盆地存在时间短，其活跃时期大约在12—13Ma（晚塞拉瓦莱期—更新世）。这些盆地普遍具有三个沉积层序，其间由区域性不整合面分开。层序1（晚塞拉瓦莱期—早托尔托纳期）与层序3（晚托尔托纳期—更新世）主要由陆相碎屑沉积岩组成。层序2（晚托尔托纳期—晚墨西拿期）主要由海相沉积岩组成，代表存在一次穿时的海侵事件，并且在向东和向南的方向海侵时间持续较长（图23-3）。贝蒂克山间盆地普遍在晚塞拉瓦莱期—早托尔托纳期的（13—7Ma）沉降速率达到最大值，而向更新世逐渐减小。盆地周围山脉的抬升速率与盆地沉降速率大致相同。

贝蒂克山间盆地最先发育于较大伸展拆离层的上盘，在晚塞拉瓦莱期—早托尔托纳期活动，这一活动也使得阿尔沃兰带的深部构造变质单元剥露出来。在晚托尔托纳期—晚墨西拿期，非洲—欧洲板块的会聚方向的变化使贝蒂克地区的应力方向和主要构造受到影响，其分布也发生变化。主要压应力转为NNW—SSE方向，产生了E—W向挤压褶皱、NW—SE向高角度正断层，以及E—W向和NE—SW向走滑断层。这些构造在挤压、剥蚀、拉张同时存在的复杂背景下对盆地及其基底产生了影响。走滑断层使一些盆地边界发生强烈变形，但是没有产生重要的沉积中心。因此走滑断层在盆地形成过程中扮演次级角色，起到了掩盖早期拉张构造的作用。

贝蒂克山间盆地证实了晚期造山运动的演化，认为其发育遵循了以地幔分层和（或）俯冲后撤/板片拆离为特征的模式。

致谢（略）

参 考 文 献

Aldaya, F., Alvarez, F., Galindo-Zaldívar, J., González Lodeiro, F., Jabaloy, A., and Navarro Vilá, F. (1991) The Malaguide-Alpujarride Contact (Betic Cordilleras, Spain): a brittle extensional detachment. Comtes Rendus de la Académie des Sciences Série II, 313, 1447–1453.

Azanon, J. M., and Goffé, B. (1997) Ferro- and magnesiocarpholite assemblages as record of high-P, low-T metamorphism in the Central Alpujarrides, Betic Cordillera (Se Spain). European Journal of Mineralogy, 9, 1035–1051.

Azanon, J. M., Tuccimei, P., Azor, A., Sánchez-Almazo, I. M., Alonso-Zarza, A. M., Soligo, M., and Perez-Pena, J. V. (2006) Calcrete features and age estimates from U/Th dating: implications for the analysis of Quaternary erosion rates in the Northern Limb of the Sierra Nevada Range (Betic Cordillera, Southeast Spain). Geological Society of America, Special Paper 416, 223–239.

Baena, J., García Rodríguez, J., Maldonado, A., Uchupi, E., Udías, A., and Wandossell, J.

(1982) Mapa Geológico De La Plataforma Continental Espanola Y Zonas Adyacentes. 1:200. 000. Almería-Garrucha-Chella-Los Genoveses, IGME. Madrid.

Balanyá, J. C., and García Duenas, V. (1987) Les Directions Structurales Dans Le Domaine D' Alborán De Parts Et D' outre Du Detroit De Gibralter. Comptes Rendus Acad. Sci. Paris, 304, 929-933.

Balanyá, J. C., García - Duenas, V., Azanón, J. M., and Sánchez - Gómez, M. (1997) Alternating Contrctional and Extensional Events in the Alpujárride Nappes of the Alborán Domain (Betics, Gibraltar Arc). Tectonics, 16, 226-238.

Boorsma, L. J. (1993) Syn-Tectonic Sedimentation in a Neogene Strike-Slip Basin (Serrata Area, Se Spain). Academic Proefschrift Thesis, Vrije.

Booth-Rea, G., Azanon, J. M., and Garca-Duenas, V. (2004) Extensinal tectonics in the Northeastern Betics (Se Spain): case study of extension in a multilayered upper crust with contrasting rheologies. Journal of Structural Geology, 26, 2039-2058.

Booth-Rea, G., Azanon, J. M., Martínez-Martinez, J. M., Vidal, O., and García-Duenas, V. (2005) Contrasting structural and P-T evolution of tectonic units in the Southeastern Betics: key for understanding the exhumation of the Alborán Domain Hp/Lt Crustal Rocks (Western Mediterranean). Tectonics, 24, 1-23.

Booth-Rea, G., Ranero, C. R., Martínez-Martínez, J. M., and Grevemeyer, I. (2007) Crustal types and Tertiary tectonic evolution of the Alborán Sea, Western Mediterranean. Geochemistry Geophysics Geosystems, 8, 1-25.

Braga, J. C., Martín, J. M., and Alcalá, B. (1990) Coral reefs in coarse-terrigenous sedimentary environments (Upper Tortonian, Granada Basin, Southern Spain). Sedimentary Geology, 66, 135-150.

Braga, J., Martín, J., and Quesada, C. (2003) Patterns and average rates of Late Neogene - Recent uplift of the Betic Cordillera, SE Spain. Geomorphology, 50, 3-26.

Buforn, E., Sanz de Galdeano, C., and Udías, A. (1995) Seismotectonics of the Ibero - Maghrebian Region. Tectonophysics, 248, 247-261.

Busby, C. J., and Ingersoll, R. V. (1995) Tectonics of sedimentary basins. Oxford, Blackwell Science.

Calvert, A., Sandvol, E., Seber, D., Barazangi, M., Roecker, S., Mourabit, T., Vidal, F., Alguacil, G., and Jabour, N. (2000) Geodynamic evolution of the Lithosphere and Upper Mantle beneath the Alborán Region of the Western Mediterranean: constraints from travel time tomography. Journal of Geophysical Research, 105, 10871-10898.

Christie-Blick, N., and Biddle, K. T. (1985) Deformation and basin formation along strike-slip faults, in Biddle, K. T., and Christie - Blick, N., eds., strike - slip deformation, basin formation and sedimentation, Spec. Publ. 37. Tulsa, OK, Society of Economic Paleontologists and Mineralogists, 1-34.

Clark, S. J. P. (2004) A Sedimentary Record of Orogenic Uplift, Granada Basin, Se Spain. PhD thesis, University of Glasgow.

Clark, S. J. P., and Dempster, T. J. (2009) The record of tectonic denudation and erosion in an emerging orogen: an apatite fission-track study of the Sierra Nevada, Southern Spain. Journal of the Geological Society, 166, 87-100.

Cloetingh, S. (1988) Intraplate stress: a new elements in basin analysis, in Kleinspehn, C., New perspective in basin analysis. New York, Springer-Verlag, 205-230.

Cloetingh, S., Vanderbeek, P. A., Vanrees, D., Roep, T. B., Biermann, C., and Stephenson, R. A. (1992) Flexural interaction and the dynamics of Neogene extensional basin formation in the Alborán-Betic Region. Geo- Marine Letters, 12, 66-75.

Cloetingh, S., Benavraham, Z., Sassi, W., Horvath, F., and Al, E. (1996) Dynamics of basin formation and strike-slip tectonics. Tectonophysics, 266, 1-10.

Comas, M. C., Platt, J. P., Soto, J. I., and Watts, A. B. (1999) The origin and tectonic history of the Alborán Basin: insights from Leg 161 results, in Zahn, M. C. R., and Klaus, A., eds., Proceedings of the Ocean Drilling Results. Scientific Results, 161, 555-580.

Crespo-Blanc, A., Orozco, M., and Garcia-Duenas, V. (1994). Extension versus compression during the Miocene tectonic evolution of the Betic chain: late folding of normal fault systems. Tectonics, 13, 78-88.

Crowell, J. (1987) Late Cenozoic Basins of Onshore Southern California: Complexity in the Hallmarks of Their Tectonis History, in Ingersoll, R., ed., Cenozoic basins development of Coastal California. Englewood Cliffs, NJ, Prentice Hall, 204-241.

De Larouziere, F. D., Bolze, J., Bordet, P., Hernández, J., Montenat, C., and Ott D' Estevou, P. (1988) The Betic segment of the Lithospheric Trans-Alborán shear zone during the Late Miocene. Tectonophysics, 152, 41-52.

De Vicente, G., Cloetingh, S., Munoz-Martín, A., Olaiz, A., Stich, D., Vegas, R., Galindo-Zaldlvar, J., and Fernandez- Lozano, J. (2008) Inversion of moment tensor focal mechanisms from active stresses around Microcontinent Iberia: Tectonic implications. Tectonics, TC1009.

Dewey, J. F., Helman, M. L., Turco, E., Hutton, D. W. H., and Knott, S. D. (1989) Alpine Tectonics (Ed. by M. P. Coward, and D. y. P. Dietrich, R. G), 45, Geological Society of London Special Publication. 265-283.

Dinarés, J., Ortí, F., Playá, I., and Rosell, L. (1999) Paleomagnetic chronology of the evaporitic sedimentation in the Neogene Fortuna Basin (Se Spain): Early restriction preceding the "Messinian Salinity Crisis." Palaeogeography, Palaeoclimatology, Palaeoecology, 154, 161-178.

Duggen, S., Hoernle, K., Van Den Bogaard, P., Rupke, L., Phipps, and Morgan, J. (2003) Deep roots of the Messinian salinity crisis. Nature, 422, 602-606.

Duggen, S., Hoernle, K., Van Den Bogaard, P., and Garbe-Scho¨Nberg, D. (2005) Post-collisional transition from subduction to intraplate - type magmatism in the Westernmost Mediterranean: Evidence for continentaledge delamination of subcontinental lithosphere. Journal of Petrology, 46, 1155-1201.

Dunkl, I., Grasemann, B., and Frisch, W. (1998) Thermal effects of exhumation of a metamorphic core complex on hanging wall syn-rift sediments: an example from the Rechnitz Window, Eastern Alps. Tectonophysics, 297, 31-50.

Durand-Delga, M. (1980) Lamediterranee Occidentale: Etape De Sa Genése Et Problémes Structuraux Liés A Celle-Ci, in Livre Jubilaire De La Soc. Geol. De France 1830—1980. Mem. H. Ser. 10, 203-224.

S. G. F. Estévez, A., Rodríguez-Fernández, J., Sanz de Galdeano, C., and Vera, J. A. (1982) Evidencia de una fase compresiva de edad Tortoniense en el Sector Central de las Cordilleras Béticas. Estudios Geologicos, 38, 55-60.

Fritz, H., and Messner, M. (1999) Intramontane basin formation during oblique convergence in the Eastern Desert of Egypt: magmatically versus tectonically induced subsidence. Tectonophysics, 315, 145-162.

Galindo-Zaldivar, J., González-Lodeiro, F., and Jabaloy, A. (1993) Stress and paleostress in the Betic-Rif Cordilleras (Miocene to the present). Tectonophysics, 227, 105-126.

Galindo-Zaldívar, J., Jabaloy, A., Serrano, I., Morales, J., González-Lodeiro, F., and Torcal, F. (1999) Recent and present-day stresses in the Granada Basin (Betic Cordilleras): example of a late Miocene-present-day extensional basin in a convergent plate boundary. Tectonics, 18, 686-702.

Garcés, M., Krijgsman, W., and Agustí, J. (1998) Chronology of the Late Turolian deposits of the Fortuna Basin (Se Spain): implications for the Messinian evolution of the Eastern Betics. Earth and Planetary Science Letters, 163, 69-81.

Garcés, M., Krijgsman, W., and Agustí, J. (2001) Chronostratigraphic Framework and Evolution of the Fortuna Basin (Eastern Betics) since the Late Miocene. Basin Research, 199-216.

Garcia-Duenas, V., Balanyá, J. C., and Martınez-Martinez, J. M. (1992) Miocene extensional detachments in the outcropping basement of the Northern Alborán Basin (Betics) and their tectonic implications. Geo-Marine Letters, 12, 88-95.

Geel, T., and Roep, T. B. (1998) Oligocene to Miocene Basin development in the Eastern Betic Cordilleras, SE Spain (Velez Rubio Corridor-Espuna): reflections of Western Mediterranean plate tectonic reorganization. Basin Research, 10, 325-343.

Geel, T., and Roep, T. B. (1999) Oligocene to Middle Miocene Basin development in the Velez Rubio Corridor: Espuna (Internal-External Zone Boundary; Eastern Betic Cordilleras, SE Spain). Geol. Mijnbouw, 77, 39-61.

Gutscher, M. A., Malod, J., Rehault, J. P., Contrucci, I., Klingelhoefer, F., Mendes-Victor, L., and Spakman, W. (2002) Evidence for active subduction beneath Gibraltar. Geology, 30, 1071-1074.

Gutscher, M. A., S, D., Westbrook, G. K., Gente, P., Babonneau, N., Mulder, T., Gonthier, E., Bartolome, R., Luis, J., Rosas, F., and Terrinha, P. (2009) Tectonic shortening and gravitational spreading in the Gulf of Cadiz Accretionary Wedge: observations from multi-beam bathymetry and seismic profiling. Marine and Petroleum Geology, 26, 647-659.

Hanne, D., White, N., and Lonergan, L. (2003) Subsidence analyses from the Betic Cordillera, Southeast Spain. Basin Research, 15, 1-21.

Hardenbol, J., Thierry, J., Farley, M., Jacquin, T., Graciansky, P-C., and Vail, P. (1998) Mesozoic and Cenozoic sequence chronostratigraphic chart, in P. C. Graciansky, J. Hardenbol,

T. Jacquin, and P. R. Vail, eds., Mesozoic and Cenozoic sequence chronostratigraphic framework of European basins, 60. Tulsa, OK, Society of Economic Paleontologists and Mineralogists.

Haughton, P. D. W. (2000) Evolving turbidite systems on a deforming basin floor, Tabernas, Se Spain. Sedimentology, 47, 497-518.

Haughton, P. D. W. (2001) Contained turbidites used to track sea bed deformation and basin migration, Sorbas Basin, South-East Spain. Basin Research, 13, 117-139.

Hermes, J. (1985) Algunos Aspectos De La Estructura DeLa Zona Subbética (Cordilleras Béticas, Espana Meridional). Estudios Geológicos, 41, 157-176.

Huibregtse, P., Vanalebeek, H., Zaal, M., and Biermann, C. (1998) Paleostress analysis of the Northern Nijar and Southern Vera Basins: constraints for the Neogene displacement history of major strike-slip faults in the Betic Cordilleras, Se Spain. Tectonophysics, 300, 79-101.

Jabaloy, A., Galindo-Zaldívar, J., and González Lodeiro, F. (1992) The Mecina extensional system: its relations with the Post - Aquitanian piggy - back basins and the paleostresses evolution (Betic Cordilleras). Geo-Marine Letters, 12, 96-103.

Jerez, L. (1973) Geología de la Zona Prebética en la Transversal de Elche de la Sierra y Sectores Adyacentes (Provincias de Albacete y Murcia). Tesis Doctoral, Universidad de Granada, Granada.

Johnson, C. (1997) Resolving Denudational histories in orogenic belts with apatite fission-track thermochronology and structural data: an example from Southern Spain. Geology, 25, 623-626.

Johnson, C., Harbury, N., and Hurford, A. J. (1997) The role of extension in the Miocene denudation of the Nevado- Filábride, Betic Cordillera (Se Spain). Tectonics, 16, 189-204.

Keller, J. V. A., Hall, S. H., Dart, C. J., and Mcclay, K. R. (1995) The geometry and evolution of a transpressional strikeslip system: the Carboneras Fault, Se Spain. Journal of the Geological Society, 152, 339-351.

Kleverlaan, K. (1989a) Neogene history of the Tabernas Basin, (Se Spain) and its Tortonian submarine fan development. Geol. Mjnb, 68, 421-432.

Kleverlaan, K. (1989b) Three distinctive feeder-lobe systems within one time slice of the Tortonian Tabernas Fan, Se Spain. Sedimentology, 36, 25-45.

Krijgsman, W., Raffi, I., Taberner, C., and Zachariasse, W. J. (2000) The Tortonian Salinity Crisis' of the Eastern Betics (Spain). Earth and Planetary Science Letters, 181, 497-511.

Krijgsman, W., and Garcés, M. (2004) Palaeomagnetic constraints on the geodynamic evolution of the Gibraltar Arc. Terra Nova, 16, 281-287.

Leblanc, D., and Olivier, P. (1984) Role of strike-slip faults in the Betic-Rifian Orogeny. Tectonophysics, 101, 345-355.

Lonergan, L. (1993) Timing and kinematics of deformation in the Malaguide Complex, Internal Zone of the Betic Cordillera, Southeast Spain. Tectonics, 12, 460-476.

Lonergan, L., and Platt, J. P. (1995) The Malaguide-Alpujarride boundary: a major extensional contact in the internal zone of the Eastern Betic Cordillera, Se Spain. Journal of Structural

Geology, 17, 1655-1671.

Lonergan, L., and White, N. (1997) Origin of the Betic-Rif mountain belt. Tectonics, 16, 504-522.

Martinez-Martínez, J. M., Soto, J. I., and Balanyá, J. C. (2002) Orthogonal folding of extensional detachments: structure and origin of the Sierra Nevada Elongated Dome (Betics, Se Spain). Tectonics, 21, 20.

Martınez-Martınez, J. M., and Azanon, J. M. (1997) Mode of extensional tectonics in the Southeastern Betics (Se Spain): implications for the tectonic evolution of the Peri-Alborán Orogenic System. Tectonics, 16, 205-225.

Mayoral, E., Crespo-Blanc, A., Díaz, M. G., Benot, C., and Orozco, M. (1994) Rifting miocéne du domaine d' Alborán datations de sédiments discordants sur les unités alpujarrides en extension (sud de la Sierra Nevada, Chaîne Bétique). Comptes Rendus de l' Academie des Sciences, Paris, Série II, 319, 581-588.

Meijninger, B. M. L. (2006) Late-Orogenic extension and strike-slip deformation in the Neogene of Southeastern Spain, Utrecht University, Utrecht.

Meijninger, B. M. L., and Vissers, R. L. M. (2006) Miocene extensional basin development in the Betic Cordillera, Se Spain revealed through analysis of the Alhama De Murcia and Crevillente Faults. Basin Research, 18, 547-571.

Monié, P., Galindo-Zaldivar, J., González Lodeiro, F., Goffe, B., and Jabaloy, A. (1991) ^{40}Ar/^{39}Ar Geochronology of Alpine Tectonism in the Betic Cordilleras (Southern Spain). Journal of the Geological Society, 148, 289-297.

Montenat, C., Ott D' Estevou, P., and Masse, P. (1987) Tectonic-sedimentary character of the Betic Neogene Basins evolving in a Crustal Transcurrent Shear Zone (Se Spain). Bull. Centres Rech. Explor. Elf-Aquitaine, 11, 1-22.

Montenat, C., Rodríguez-Fernández, J., Ott D' Estevou, P., and Sanz de Galdeano, C. (1990c) Geodynamic Evolution of the Betic Neogene Intramontane Basins (S. And Se. Spain). In: Iberian Neogene Basins (Ed. by J. Agusti, R. Doménech, and R. Juliá), Mem. Spe. 2, 6 -59.

Institut de Paleontología Miguel Crusafonf, Barcelona. Montenat, C., Ott D' Estevou, P., and Delort, T. (1990b) Le Bassin De Lorca. In: Les Bassins Néogénes Du Domaine Bétique Orientale (Espagne) (Ed. byMontenat, C.), 12-13, 261-280.

Institut Geologique Albert de Lapparent, Paris. Montenat, C., Ott D' Estevou, P., and de La Chapelle, G. (1990a) Le Bassin De Nijar-Carboneras Et Le Couloir Du Bas-Andarax, in C. Montenat, ed., Les Bassins Néogénes Du Domaine Bétique Orientale (Espagne). Doc. Et Trav. Igal, 12-13, 129-164.

Institut Geologique Albert de Lapparent, Paris. Morley, C. K. (1993) Discussion of origins of hinterland basins to the Rif-Betic Cordillera and Carpathians. Tectonophysics, 226, 359-376.

Perez-Pena, J. V. (2009) GIS-based tools and methods for landscape analysis and evaluation of active tectonics, University of Granada.

Perez-Pena, J. V., Azanon, J. M., Azor, A., Tuccimei, P., Della Seta, M., and Soligo, M.

(2009) Quaternary Landscape Evolution and Erosion Rates for an Intramontane Neogene Basin (Guadix-Baza Basin, Se Spain). Geomorphology, 106, 206-218.

Platt, J. P., and Vissers, R. L. M. (1989) Extensional collapse of a thickened continental lithosphere: a working hypothesis for the Alborán Sea and Gibraltar Arc. Geology, 17, 540-543.

Platt, J., Soto, J. I., Whitehouse, M. J., Hurford, A. J., and Kelley, S. P. (1998) Thermal Evolution, Rate of Exhumation, and Tectonic Significance of Metamorphic Rocks from the Floor of the Alborán Extensional Basin, Western Mediterranean. Tectonics, 17, 671-689.

Platt, J., and Whiterhouse, M. J. (1999) Early Miocene high-temperature metamorphism exhumation in the Betic Cordillera (Spain): evidence from U-Pb Zircon Age. Earth and Planetary Science Letters, 171, 591-605.

Platt, J., Anczkiewicz, R., Soto, J. I., Kelley, S. P., and Thirlwall, M. (2004). Two Phases of Continental subduction in the Betic Orogen, Western Mediterranean. Geol. Soc. Am. Abstr. Programs., 36, 409.

Platt, J. P., Kelley, S. P., Carter, A., and Orozco, M. (2005) Timing of tectonic events in the Alpujarride Complex, Betic Cordillera, Southern Spain. J. Geol. Soc. London, 162, 1-12.

Platt, J., Anczkiewicz, R., Soto, J. I., Kelley, S. P., and Thirlwall, M. (2006) Early Miocene continental subduction and rapid exhumation in the Western Mediterranean. Geology, 34, 981-984.

Puga-Bernabéu, A. (2007) Depositional Models of Temperate Carbonates: Insihts into in Situ and Redeposited Sediments from Southern Spain, South Australia and North New Zealand. Tesis Doctoral, Universidad de Granada, Granada.

Rehault, J. P., Boillot, G., and Mauffret, A. (1985). The Western Mediterranean basin. In: Geological evolution of the Mediterranean basin, Stanley, D. J. et al. eds., Springer-Verlag, New York, 101-129.

Reicherter, K. R., and Peters, G. (2005) Neotectonic evolution of the Central Betic Cordilleras (Southern Spain). Tectonophysics, 405, 191-212.

Reinhardt, L. J., Bishop, P., Hoey, T. B., Dempster, T. J., and Sanderson, D. C. W. (2007) Quantification of the transient response to base-level fall in asmallmountain catchment: Sierra Nevada, Southern Spain. J. Geophys. Res. 112.

Rodríguez-Fernández, J. (1982) El Mioceno Del Sector Central De Las Cordilleras Beticas. Tesis Doctoral, Universidad de Granada, Granada.

Rodriıguez-Fernández, J., Fernandez, J., Lopez-Garrido, A. C., and Sanz de Galdeano, C. (1984) The central sector of the Betic Cordilleras, a realm situated between the Atlantic and Mediterranean domains during the Upper Miocene. Ann. Geol. Pays. Hell., 32, 97-103.

Rodriguez-Fernández, J., Sanz de Galdeano, C., and Serrano, F. (1990) Le Coluoir Des Alpujarras. In: Les Bassins Neogenes Du Domaine Betique Oriental (Ed. by C. Montenat), Doc. Et Travaux De L' I. G. A. L, 12-13, 87-100. I. G. A. L, Paris.

Rodríguez Fernández, J., and Sanz de Galdeano, C. (1992) Onshore Neogene Stratigraphy in the North of the Alborán Sea (Betic Internal Zones: Paleogeographic Implications. Geo-Marine

Letters, 12, 123-128.

Rodríguez-Fernández, J., and Martín-Penela, A. J. (1993) Neogene evolution of the Campo-De-Dalías and the surrounding offshore areas (Northeastern Alborán Sea). Geodinamica Acta, 6, 255-270.

Rodriguez-Fernández, J., Comas, M. C., Soria, J., Martin-Perez, J. A., and Soto, J. I. (1999) The sedimentary record of the Alboran Basin: an attempt at sedimentary sequence correlation and subsidence analysis, in Proceeding of the Ocean Drilling Program. Sci. Results (Ed. by R. Zhan, M. C. Comas, and A. Klaus), 161, 69-75, College Station, TX.

Rodriguez-Fernandez, J., and Sanz de Galdeano, C. (2006) Late Orogenic Intramontane Basin Development: The Granada Basin, Betics (Southern Spain). Basin Research, 18, 85-102.

Royden, L. H. (1993) Evolution of retreating subduction boundaries formed during continental collision. Tectonics, 12, 629-638.

Sanz de Galdeano, C. (1987) Strike-slip faults in the Southern Border of the Vera Basin (Alméria, Betic Cordilleras). Estudios Geológicos., 43, 435-443.

Sanz de Galdeano, C. (1988) The fault system and the neotectonic features of the Betic Cordilleras: the Iberia Peninsula. Fifth E. G. T. Workshop. E. B. a. V. Mendes, Eur. Sci. Found. Estoril, 99-109.

Sanz de Galdeano, C. (1990) Estructura y Estratigrafía de La Sierra de Los Guajares y Sectores Próximos (Conjunto Alpujárride, Cordilleras Béticas). Estudios Geológicos, 46, 123-134.

Sanz de Galdeano, C. (2008) The Cádiz-Alicante Fault: an important discontinuity in the Betic Cordillera. Revista de la Sociedad Geológica de Espana, 21, 49-58.

Sanz de Galdeano, C. S., Rodríguez-Fernández, J., and López-Garrido, A. C. (1985) A strike-slip-fault corridor within the Alpujarra Mountains (Betic-Cordilleras, Spain). Geologische Rundschau, 74, 641-655.

Sanz de Galdeano, C., and Vera, J. A. (1992) Stratigraphic record and paleogeographical context of the Neogene Basins in the Betic Cordillera, Spain. Basin Research, 4, 21-36.

Sanz de Galdeano, C., and Alfaro, P. (2004) Tectonic significance of the present relief of the Betic Cordillera. Geomorphology, 63, 178-190.

Sanz deGaldeano, C., Delgado, J., Galindo-Zaldívar, J., Marín-Lechado, C., Alfaro, P., Garcia Tortosa, F. J., Lopez-Garrido, A. C., and. Gil, A. (2007) Anomalías Gravimétricas de la Cuenca de Guadix-Baza (Cordillera Betica, Espana). Boletín Geológico y Minero, 118, 763-774.

Smith, G. A., and Landis, C. A. (1995) Intraárc basins, in Busby, C., and Ingersoll, R., eds., Tectonics of sedimentary basins. Oxford, Blackwell Science, 263-298.

Soria, J., Viseras, C., and Fernández, J. (1998) Late Miocene-Pleistocene Tectono-sedimentary evolution and subsidence history of the Central Betic Cordillera (Spain): a case study in the Guadix Intramontane Basin. Geol. Mag, 135, 565-574.

Soria, J. M., Alfaro, P., Fernández, J., and Viseras, C. (2001) Quantitative Subsidence-Uplift Analysis of the Bajo Segura Basin (Eastern Betic Cordillera, Spain): tectonic control on the stratigraphic architecture. Sediment. Geol, 140, 271-289.

Spakman, W., and Wortel, M. J. R. (2004) A tomographic view on Western Mediterranean Geodynamics, in Cavazza, W., Roure, F., Spakman, W., Stamplei, G. M., and Ziegler, P. A., eds., The TRANSMED atlas: the Mediterranean region from crust to mantle. Berlin, Springer-Verlag, 31-52.

Stokes, M. (2008) Plio-Pleistocene drainage development in an inverted sedimentary basin: Vera Basin, Betic Cordillera, Se Spain. Geomorphology, 193-211.

Sylvester, A. G., ed. (1984) Wrench faults tectonics. AAPG. Reprint Series, 28, 374. Sylvester, A. G. (1988) Strike-slip faults. Geol. Soc. Ame. Bull, 100, 1666-1703.

Ten Veen, J. H., and Postma, G. (1999a) Neogene tectonic and basin fill patterns in the Hellenic Outer-Arc (Crete, Greece). Basin Research, 11, 223-241.

Ten Veen, J. H., and Postma, G. (1999b) Roll-back controlled vertical movement of outer-arc basins of the Hellenic Subduction Zone (Crete, Greece). Basin Research, 11, 243-266.

Viseras, C. (1991) Estratigrafía Y Sedimentología Del Relleno Aluvial De La Cuenca De Guadix (Cordilleras Béticas). PhD thesis, Universidad de Granada, Granada. Viseras, C., and Fernández, J. (1992) Sedimentary basin destruction inferred from the evolution of drainage systems in the Betic Cordillera, Southern Spain. Journal of the Geological Society, 149, 1021-1029.

Volk, H. R., and Rondeel, H. E. (1964) Zur Gliederung Des Jungtertiars Im Becken Von Vera, Sudspanien. Geologie en Mijnbouw, 43, 310-315. Watts, A. B., Platt, J. P., and Buhl, P. (1993) Tectonic evolution of the Alborán Sea Basin. Basin Research, 5, 153-177.

Weijermars, R. (1987a) A revision of the Eurasian-African Plate Boundary in the Western Mediterranean. Geologische Rundschau, 76, 667-676.

Weijermars, R. (1987b) The Palomares Brittle-Ductile Shear Zone of Southern Spain. Journal of Structural Geology, 9, 139-157.

Weijermars, R., Roep, T., Van Den Eeckhout, B., Postma, G., and Kleverlaan, K. (1985) Uplift history of a Betic fold nappe inferred from Neogene-Quaternary sedimentation and tectonics (in the Sierra Alhamilla and Almería, Sorbas and Tabernas Basins of the Betic Cordilleras, Se Spain). Geol. Mjnb, 64, 397-411.

Wildi, W. (1983) La Chaîne Tello-Rifaine (Algerie, Marroc, Tunisie): structure, Stratigraphie Et Evolusion Du Trias Au Miocene. Rev. Geol. Dyn. Geogr. Phys. 24.

Xie, X., and Heller, P. L. (2009) Plate tectonics and basin subsidence history. Geological Society of America Bulletin, 121, 55-64.

(崔敏 李政斯译，吴哲 赵钊校)

第24章　中白垩世加拿大西部前陆盆地内浅海相地层的构造沉降、沉积作用以及不整合之间的动力学关系：与科迪勒拉构造的关系

A. GUY PLINT[1], ADITYA TYAGI[1], PHIL J. A. MCCAUSLAND[1],
JESSICA R. KRAWETZ[2], HENG ZHANG[3], XAVIER ROCA[4],
BOGDAN L. VARBAN[4], Y. GREG HU[5], MICHAEL A. KREITNER[6],
MICHAEL J. HAY[7]

（1. Department of Earth Sciences, University of Western Ontario, London, Canada；
2. Canadian Natural Resources Limited, Calgary, Canada；
3. 3rd Avenue SW, Calgary, Canada；4. Imperial Oil Resources, Calgary, Canada；
5. Loring Tarcore Laboratories Ltd., NE, Calgary, Canada；
6. Suncor Energy Inc., Calgary, Canada；7. Talisman Energy Inc., Calgary, Canada）

摘　要：中白垩世的科罗拉多群以及同期地层中（中晚阿尔布期—早坎潘期，约104—83Ma）的岩石，主要由距离海岸几十米到几百千米背景下所形成的海相泥岩及粉砂岩组成。在北部与西部地区，发育有少量的近滨及海岸平原相沉积。根据海侵以及海泛面发育情况可以将岩石划分为不同的层段。段是地层的基本组成部分，其正常的年龄跨度为5万~20万年。有成因联系的段可以形成较大的非正式"单元"（时间跨度可达约40万~80万年），以及地层组（时间跨度为几百万年）。除了塞诺曼阶的Dunvegan组外，科罗拉多群的岩石均不发育斜层理，这表明沉积物供应速率超过了可容空间的增加速率，沉积物迅速填充满了由有效浪基面所控制的水下"可容空间"。这些地层组可以形成走向与倾向上延伸数百千米，厚度达100~900m的棱镜状楔形体。由这些楔形体组成的"单元"弧度更大，通常在走向上可延伸100~300km，这表明该地区受到了更大的局部荷载。沉积中心的横向迁移及地层单元之间等厚线的弯曲指示了沉降和负载的变化轨迹（例如1Ma内横向的移动距离可以达到100~300km），根据这些特征就可以推测活动逆冲断层的抬升以及形成过程。同一地层组的厚度在沿着走向100~200km的范围内，可以发育有多个局部的沉降中心，并且在一次海泛面发育的时间内（即大约1万年）可以沿走向移动多达300km，这些现象表明荷载在空间上与时间上都是不均匀分布的。

通常，楔形地层段会向上变为平面席状。楔状体的发育表明研究区具有较高的可容空间与加积速率，这使得近岸砂岩与砾岩在靠近造山带边缘发育。相反，席状体内部通常发育有大量的前积近岸砂岩，表明研究区内水体较浅，发育快速海退，可容空间增加速率较小。

科罗拉多组中已识别出两个大型沉积中心。N—S向的沉积中心位于艾伯塔省及不列颠哥伦比亚省的北部，区内发育了阿尔布阶到中塞诺曼阶，其中南部表现为不整合。南部的NW—SE向的沉积中心从晚塞诺曼期开始形成（约95Ma），沉积了晚塞诺曼期到早坎潘期（和之后的地层）的地层；在该沉积中心的北部，晚塞诺曼期—晚康尼亚克期不整合发育。95Ma左右沉积中心迅速地由北向南迁移。本文推测北部N—S向沉积中心的形成与Stikine地块和Yukon-Tanana地块向大陆

会聚时发生的碰撞和顺时针旋转有关。南部 NW—SE 向沉积中心的形成可能受到了增生地块向北移动过程中右旋挤压作用的影响。北美西缘的挠曲特征和前陆盆地沉积中心的迁移主要受控于太平洋板块向北美板块俯冲的影响。在科迪勒拉山的这么大的规模上，主要受到地块增生的影响，而在小于 300km 的范围内，逆冲推覆体负载、基底薄弱面和其他局部因素也控制了构造变形机制。

关键词：前陆盆地　白垩纪　加拿大　挠曲沉降　碎屑岩层序

24.1 引言

"沉降前渊中的沉积特征记录了整个东科迪勒拉褶皱带内造山作用的变化。"——Ray Price (1973)

目前有大量关于加拿大西部前陆盆地的钻井和岩心数据公开发表，使得该地区成为世界上研究地层结构与沉降史的天然实验室。本章总结了最近 25 年来有关白垩纪中期（晚阿尔布期—早坎潘期）科罗拉多群（由 Roca 等定义，2008），以及艾伯塔和不列颠哥伦比亚地区同期地层相关的研究成果（图 24-1）。在研究过程中，尽可能地细化地层层序，但是资料的时间分辨率仅为 2 万~10 万年。沉积于海平面附近的岩石的空间展布形态可以反映该地区的沉降演化模式。这种几何学方法可以将挠曲成因的相对海平面变化与全球海平面变化区分开（Ryer，1993）。

由于加拿大西部的陆内前陆盆地的沉降模式大致符合简单挠曲模型，使其成为了研究的热点地区。与之相反，大部分美国境内的前陆盆地由于受到多期深部断裂垂直运动的影响，使其地层变形十分复杂，盆地内发育了多个隆起和坳陷（Schwartz 和 DeCelles，1988）。

虽然科罗拉多群包含了大量浅海相砂岩，并形成了重要的储层（Viking 和 Cardium 组），但是大部分地层以海相泥岩及薄层泥岩—粉砂岩互层（传统上称之为页岩）为主，故将其作为"非储层"来看待。这种富泥岩的组合现今被认为是形成于水平搬运发育的浅水环境中，而不是垂向搬运的环境（Scheiber 等，2007；Var-

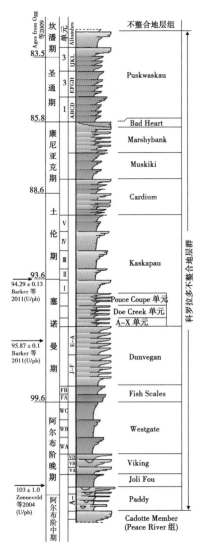

图 24-1　科罗拉多群综合地层图
（包含 Paddy 组；据 Roca 等，2008）

表明了白垩系各地层组、地层单元以及地层段之间的关系（Plint 等，1986；Plint，1990，2000；Donaldson 等，1998；Varban 和 Plint，2005；Kreitner 和 Plint，2000；Roca 等，2008；Hu 和 Plint，2009）。图中沉积相标记为：黄色—近滨砂岩；灰色和橄榄色—远滨泥质异粒岩相；绿色—潟湖相；橙色—鲕粒状铁矿。需要注意的是这是一张综合柱状图，并非在所有地区都可见图中所有地层。箭头上的年龄为艾伯塔地区最新的 U/Pb 年龄测试结果

ban 和 Plint，2008a；Plint 等，2009；Bhattacharya 和 MacEachern，2009）。

中侏罗世早期，大西洋的打开使得北美板块开始向西快速移动，而法拉隆板块的大洋岩石圈向东俯冲（Engebretson 等，1985；Monger，1993，1997；Monger 和 Price，2002）。落基山脉的科迪勒拉山系是许多外来地体向北美板块拼合增生的产物。地壳的重叠和增厚导致科迪勒拉东—中部抬升（欧明尼卡构造带及 Intermontane 构造带）。在中侏罗世晚期，研究区被抬升至海平面之上，开始为东侧新形成的前陆盆地提供物源（Monger，1993，1997；Evenchick 等，2007）。由于地壳持续缩短并发生挤压逆冲作用，侏罗纪至早白垩世沉积于前陆盆地西部的沉积物开始陆续仰冲并迁移，成为晚白垩世及古近纪碎屑沉积的物源。

科迪勒拉褶皱冲断带与相邻的中生代前陆盆地之间具有成因联系（Bally 等，1966；Price，1973）。许多学者都讨论过该盆地的形成过程，如 Kauffman 和 Caldwell（1993）、Monger（1993）、McMechan 和 Thompson（1993）、Lawton（1994）、DeCelles（2004）以及 Miall 等（2008）。早期对于前陆盆地的研究认为挠曲沉降主要归因于造山运动及盆地充填沉积物的静荷载（Beaumont，1981；Jordan，1981）。后来的学者们认识到前陆盆地地层的厚度与宽度太大，不能用简单的静荷载模型来解释，因为该模型可以用于解释挠曲前渊的形成，只能影响几百千米范围的构造变形。于是，学者们将一部分沉降归因于相对冷的洋壳俯冲所引起的地幔流作用在岩石圈基底上形成的"动力学地形"（Mitrovica 等，1989；Liu 及 Nummedal，2004）。其原理是板块俯冲角度逐渐减小的同时，上覆板块的动力学的挠曲波长也相应变长，从而使得构造变形范围变大。因此，白垩纪浅海沉积物在距离前缘隆起约 500km 的 Saskatchewan 及 Manitoba 地区超覆于加拿大地盾之上，这为法拉隆板块低角度俯冲引起的北美板块向西倾斜的论断提供了证据。

McMechan 及 Thompson（1993）提出前陆盆地内的地层可以为相邻褶皱逆冲带的演化研究提供丰富数据。这些数据可以研究以下地质事件：（1）变形的起止时间；（2）山体快速隆升与盆地沉降的时间间隔；（3）构造稳定期的持续时间；（4）个别构造运动的活动时间；（5）变质岩及深成岩的顶部剥蚀时间；（6）构造变形前缘的推进速度。McMechan 及 Thompson（1993）认为，加拿大西部前渊地区的沉积信息不能精确地解决这些问题。因此，充分细致的地层划分工作至少能够解决其中一些问题。

24.2 新的地层划分

我们研究的白垩纪浅海及边缘海相的碎屑地层包括科罗拉多群的大部分地层以及中晚阿尔布期—早坎潘期的同期地层（图 24-1）。前人的研究区域都分布于北纬 51°30′到 56°20′（约 700km）的前渊地区，沿褶皱逆冲带垂直方向延伸约 400km（图 24-2）。本文展示的等厚线图数据主要依据前人已发表的相关图件（Donaldson 等，1998，1999；Varban 和 Plint，2005，2008a，2008b；Kreitner 和 Plint，2006；Plint 和 Wadsworth，2006；Plint 和 Kreitner，2007；Tyagi 等，2007；Roca 等，2008；Hay 和 Flint，2009；Hu 和 Plint，2009；Plint 等，2009）。另外，文中也用到了未发表的新数据（Hu，1997；Varban，2004；Rylaarsdam，2006；Roca，2007；Zhang，2007；Tyagi，2009），详细情况见图表说明。本文地层学划分建立在详细的闭合校正网格之上，包括了数百米的测井曲线和分布在落基山脉山麓丘陵至皮斯河谷的两百多个露头剖面。下面是根据地层划分的结果，来阐述沉积与构造之间的关系。

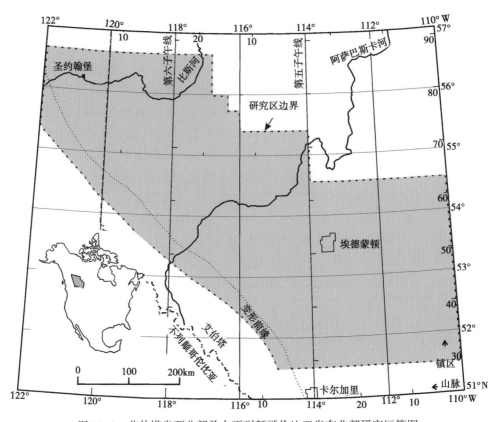

图 24-2 艾伯塔省西北部及大不列颠哥伦比亚省东北部研究区简图
图中可见主要河流和聚居地。图中给出了经纬度坐标以及领土勘测镇区/山脉系统坐标。阴影区为研究区域，范围约 $20×10^4 km^2$；目前没有人对整个区域进行过地层学研究

24.2.1 界面、地层段和地层单元

在靠近造山带的一侧前陆盆地的沉降速率最快，因此，海平面变化对该区域的影响有限，使得古河谷及河流间等陆上不整合面不发育。这些古河谷体系发育的地区，其地层厚度沿造山带方向没有明显变大，例如晚阿尔布期的 Viking 组（Pattison 和 Walker，1994；Roca，2007；Roca 等，2008）和塞诺曼阶的 Dunvegan 组（Plint，2000，2002，2003；Plint 和 Wadsworth，2003，2006；McCarthy 和 Plint，1998）。由此，可以推断该时期沉降速率很低，或者几乎没有沉降。

由于陆上不整合面不发育且识别困难，因此，识别海侵面是区域成图中最有效的方法。在下面的讨论中，由海侵面或海泛面分界的岩石组合称为地层段。通常几个地层段组合起来形成更大的地层组合，这些组合可能体现了海退的特征。这种"复合体"简称为"地层单元"（Varban 和 Plint，2005；Roca 等，2008；Hu 和 Plint，2009；图 24-1）。地层单元可能与构造沉降的暂停有关，因为在这期间沉积中心在时间与空间上相对稳定，同时经历了一个"由快变慢"的沉降变化。下面我们来对这一过程进行详细论述。

24.2.2 科罗拉多群的沉积环境

科罗拉多群（图 24-1）几乎全部由陆源碎屑岩组成。细粒—中粒砂岩揭示了波浪控制

的临滨及内陆架沉积环境。砾岩集中的地区是河口发育的位置（Hart 和 Plint，2003；Varban 和 Plint，2008a）。含丘状交错层理的分米级层状风暴砂岩可朝外滨方向延伸 30km（Hart 和 Plint，2003；Varban 和 Plint，2008a）。科罗拉多群的外滨沉积主要由薄互层的泥岩、粉砂岩以及极细砂岩组成。毫米—厘米级砂岩层内的波痕特征表明该地区广泛地受到了风潮及地球自转流的影响。泥质异粒岩相的发育表明当时的沉积环境为浅水缓坡陆架。成果图件表明中—极细砂岩沉积于离海岸 100km 的范围内，而粉砂岩则可以搬运至距离海岸大约 200km 的位置，再远的地区仅发育有黏土和极细粉砂级别的沉积物，并混有数量不等的有机质和颗石藻碳酸盐岩（Tyagi 等，2007；Varban 和 Plint，2008a）。

海岸平原相沉积主要集中在盆地的西侧，由洪泛平原、湖泊、决口扇和决口三角洲环境下的网状河砂岩组成（McCarthy 等，1999；LumsdonWest 和 Plint，2005；Rylaarsdam，2006；Roca，2007）。古土壤发育且成分未成熟表明研究区处于加积环境并且地下水位较高。古土壤仅发育于河间的地表表明其成土时间较长且水系发育（McCarthy 和 Plint，1998，2003）。网状河沉积发育表明冲积平原的坡度极低，约为几厘米/千米（Lumsdon-West 和 Plint，2005）。只有西侧最远处的局部河流沉积物粒度变粗，并含有砾岩，这一特征表明变形前缘的坡度较陡。

非碎屑沉积很少见。研究区内的富颗石藻碳酸质泥岩在下土伦阶及中圣通阶增多，主要沉积于离岸 200km 以外的区域，大约形成于 Greenhorn 和 Niobrara 旋回中的海侵高峰期（Bloch 等，1993；Kauffman 和 Caldwell，1993；Nielsen 等，2003；Tyagi 等，2007）。上康尼亚克阶 Bad Heart 组内发育有约几米厚的磷酸盐及鲕粒状铁矿，并覆盖于早白垩世的不整合面之上，其下为 Kaskapau 组的 Pouce Coupe 单元（图 24-1）。这种沉积相主要发育于极浅的、缺乏碎屑供应且会间歇性出现的前缘隆起上（Donaldson 等，1998，1999；PlinL，2000）。

24.2.3 地层模式与古水深

斜层理在科罗拉多群中很少出现，而在塞诺曼阶的 Dunvegan 组中则大量出现（Bhattacharya 和 Walker，1991；Plint，2000；Plint 等，2009），此外在康尼亚克阶的 Marshybank 组中也有小范围发育（Plint，1990；Plint 和 Norris，1991）。与此相反，本文研究表明 Paddy 组、Joli Fou 组、Viking 组、Westgate 组、Fish Scales 组、Dunvegan 组顶部、Kaskapau 组、Cardium 组、Muskiki 组以及 Puskwaskau 组内部发育平行或近平行层理，且没有证据表明这是一个"陆架—陆坡—盆地"体系。临滨及三角洲前缘砂岩在 Viking 组、Cardium 组等地层中表现为典型的幅度为 10~20m 左右的斜坡增生楔特征（Plint 等，1986；Hart 和 Plint，1993；Posamentier 和 Chamberlain，1993；Roca，2007），但在临滨的外侧，地层为平行接触。几何学特征表明除了极少数地区以外，研究区的海底地形平缓，并且缓坡坡度小，这使得离岸几百千米外的地区，水深不超过 50m（甚至更浅）（Varban 和 Plint，2008a；Hu 和 Plint，2009）。持续的风暴会使海底成为由浪基面所限制的"泥质可容空间边界"，从而产生与海平面近似平行的海底（Varban 和 Plint，2008a）。除了 Dunvegan 组下部，科罗拉多群其他地层的沉积物供应速率都基本上等于或大于可容空间增加速率（Plint 等，2009）。

24.2.4 等厚图的解释分析

科罗拉多群的大部分岩石都沉积在靠近海平面的低角度海岸平原或是浅海斜坡上。没有证据表明粗粒沉积物会像河口周围的沉积一样沉积于沉积中心处。也没有地层学或沉积学证据表明早期存在盆底的坳陷可以产生厚层沉积组合。与之相反的是，有证据表明陆架的水深浅，且受到波浪与风暴流的持续影响。斜层理的普遍缺乏表明沉积中心处于高速沉积体系中，其沉降的部分会被很快填满（Plint 等，2009）。因此，沉积中心往往不是水深最大的位置。据此笔者认为，巨厚的沉积物堆积表明该处可能是地层形成时的构造沉降中心，而不是该处发育早期构造"丘状体"或水槽而形成了巨厚沉积。

24.2.5 岩体形态：席状及楔状

地层单元的成因可以用前陆盆地自身的不对称沉降来简要解释。Ryer（1993）指出广泛分布的薄层席状地层单元是由海平面变化控制的，而楔状单元则是由不对称沉降作用产生。尽管我们绘制的等厚图的层序分辨率高于前人的研究成果（Ryer，1993；Pang 和 Nummedal，1995），但依然可以利用类似理论来分析等厚图。几乎所有在走向和倾向上延伸数百千米，且时间跨度大于 10 万年的席状岩体都很难用构造机制进行解释，因此有人提出了海平面变化的成因机制（Plint 和 Kreitner，2007）。但是我们的厚度图表明典型的楔形单元可以在小于 20 万年的时间内发育。构造和海平面变化机制共同控制了可容空间/沉积物供应量比率的变化，以及"高频率"的不整合地层段的形成和地层的展布范围和形态。

24.3 挠曲的沉积中心

24.3.1 时间尺度和沉降模式

本文对于科罗拉多群的分析结果表明沉降模式明显地取决于观测的时间尺度。为明确这一点，本文讨论了不同时间尺度的岩石样品：（1）几个百万年；（2）大约 40 万~80 万年；（3）约 5 万~20 万年。我们在研究加拿大西部地区时仅有很少的放射性测年数据可用，因此段和带的地质年代边界主要是借助于美国南部境内生物地层的校正以及（体现在最新的地质年代表中；Ogg 等，2009）放射性测年的插值数据（主要为 Ar-Ar）。艾伯塔省膨土岩中锆石 U-Pb 测年数据表明，膨土岩中的 Ar-Ar 年龄在估算白垩系年龄时系统地少了大约 1Ma（Barker 等，2011；图 24-1）。

跨越了几个百万年的地层组合。Kaskapau 组的 I—V 单元（以及南部 Blackstone 组的等时地层）时间跨度约 3.2Ma，Puskwaskau 组的 1—3 单元时间跨度约 2.5Ma，这两地层组成了相对简单的棱柱状楔形岩体（图 24-1、图 24-3a、图 24-4a）。需要注意的是，这些楔形体主要由薄互层泥岩和异粒陆架相岩石组成，沉积在离岸约 400km 处，水深不超过数十米。这种地层等厚图样式表明该地区的挠曲是受到一个至少 600km 长的线性荷载而形成的。

地层组合跨越了 40 万~80 万年（"单元"）。与地层组相比，部分"地层单元"反映了更复杂的沉降模式。例如 Doe Creek 地层单元和 Pouce Coupe 地层单元，以及上覆 Kaskapau 组 I—Ⅲ 地层单元的等厚图表明，连续地层单元的展布范围与展布方向都具有较大变化（图 24-1 及图 24-3b—f）。同样，当对 Puskwaskau 组的 1、2、3 单元（每个单元跨越约 75 万年）分别成图时，发现其沉降并非像图 24-4a 中所推测的呈现棱镜状模式，而是在连续单元中间呈现出弯曲的沉降轨迹，轨迹移动了约 200~350km。沉降轨迹在地层单元 1 沉积之后向北移动，然后在地层单元 2 沉积之后向南移动。

时间跨越小于 5 万~20 万年的地层组合（地层段）。在地层段的范围内也可以发生挠曲沉降中心的侧向移动。Puskwaskau 组中的 D—I 层段（平均每个跨越小于 20 万年）包含了

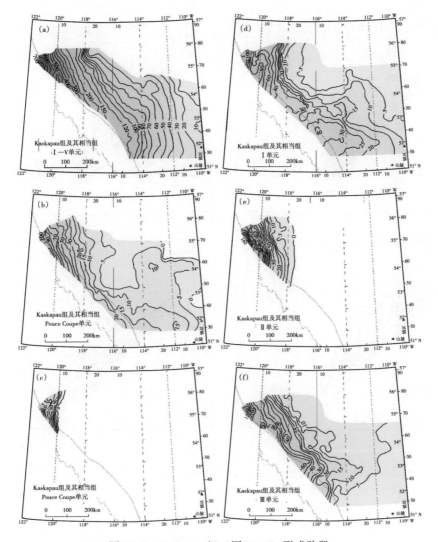

图 24-3 Kaskapau 组（图 24-1）形成阶段

(a) Kaskapau 组Ⅰ—Ⅴ单元的等厚图，可见明显的海相泥质地层形成了简单的棱状楔形体，时间跨度约为3Ma。
(b) Doe Creek 单元可见强烈弯曲的前渊带，前缘隆起区可能位于大约116°W附近，并最终在110°W附近尖灭。
(c) Pouce Coupe 地层单元主要集中于 120°W 以西的前渊带中，地层单元的顶部为早白垩世不整合面，代表了东部的一次沉积间断。(d) Ⅰ单元在形态上与 Doe Creek 单元相似。需要注意两者等厚线的区别，在西北端的等厚线为近 N—S 向展布，而南端则表现为 NW—SE 向。(e) Ⅱ单元局限于一个狭窄的沉积范围内，在 118°W 附近发生尖灭，等厚线表现为 NNW—SSE 走向。(f) Ⅲ单元的等厚线突然逆时针旋转至 NW—SE 走向，形成了宽约 300km 的前缘隆起区，其沉积厚度仅 5~10m。重点是（1）每个单元都具有明显的等厚线走向，(c) 和 (e) 主要为 N—S 向，而 (b) 和 (d) 在不同的区域分别呈现为 N—S 向和 NW—SE 向，这意味着存在两种相互独立的荷载，此外在地壳薄弱区域可能存在挠曲作用。(f) 表明西北向的荷载逐渐成为主导因素（并在科罗拉多群之后的地层沉积过程中持续起作用）。(2) 向盆地方向的地层单元尖灭的特点表明地层组的远端含有较多的、隐藏的、泥岩覆盖泥岩的平行不整合，可能跨越了几十万年，特别是 Pouce Coupe 地层单元及Ⅱ地层单元。Pouce Coupe 地层单元中的近滨砂岩较为异常，它从沉积中心东缘向西进积（Kreitner 和 lint，2006）。Pouce Coupe 地层单元顶部的不整合可能与海平面下降有关。除了邻近的前渊带，Pouce Coupe 地层单元的其余部分都暴露于地表。资料来源：南至镇区 65°带的图表数据都来源于 Tyagi（2009）；图（a）来自于 Varban 和 Plint（2005）；(d) — (f) 来自于 Varban（2004）；(b)、(c) 来自于 Kreitner 和 Plint（2006）。所有等厚图的单位都为米

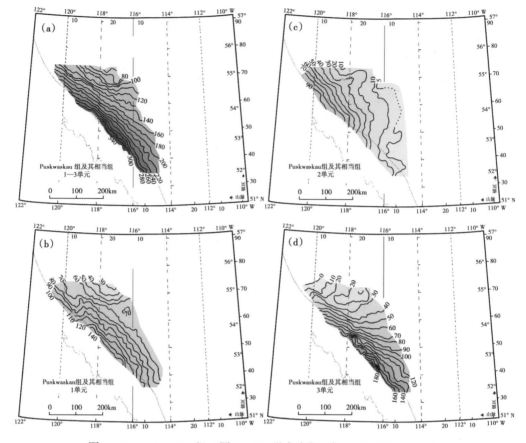

图 24-4 Puskwaskau 组（图 24-1）形成阶段（据 Hu 和 Plint，2009）

(a) 完整的 Puskwaskau 组等厚图（单元 1—3），呈现出明显的海相泥岩的棱形楔状体特征，时间跨度约 2.5Ma；
(b) 1 单元；(c) 2 单元；(d) 3 单元；在连续的地层段中，活动的沉积中心沿走向变化迅速。图 (b) — (d) 表明活动荷载的位置可以沿走向发生快速的改变

地层单元 1 的末期、整个地层单元 2 以及地层单元 3 的初期（图 24-5）就发生了挠曲沉降中心的侧向移动。这表明新的挠曲沉降中心从地质角度来说可能是突然出现的，因为它跨越了单个海泛面，并且沉积中心沿走向移动了几十到几百千米。在 Puskwaskau 组 H 段和 I 段（图 24-5e、f）沉积轨迹穿过一个海泛面，并发生侧向变化。其相对复杂程度可在 Kaskapau 组各个地层段的等厚图中看出（Varban 和 Plint，2008b）。

24.3.2 沉积中心的侧向移动和方向变化

挠曲沉积中心以及荷载的位置在科罗拉多群中发生了快速的变化。由于皮斯河湾构造（PRA）的间歇沉降使得科迪勒拉山荷载所形成的挠曲变形更加复杂化。皮斯河湾构造是一个存在时间很长，基本与大陆边缘垂直的一个基底构造，其一部分延伸到研究区的北部（见以下讨论；O'Connell，1994）。

沉降的动态模式可以由六张等厚图来说明（图 24-6a—f），它们代表了上阿尔布阶 Westgate 组到下塞诺曼阶 Fish Scales 组、Dunvegan 组的下部及上部，以及 Kaskapau 组的下部的沉积特征，总共跨越了约 6Ma（图 24-1）。这一系列图件表明，挠曲轴面由上阿尔布阶

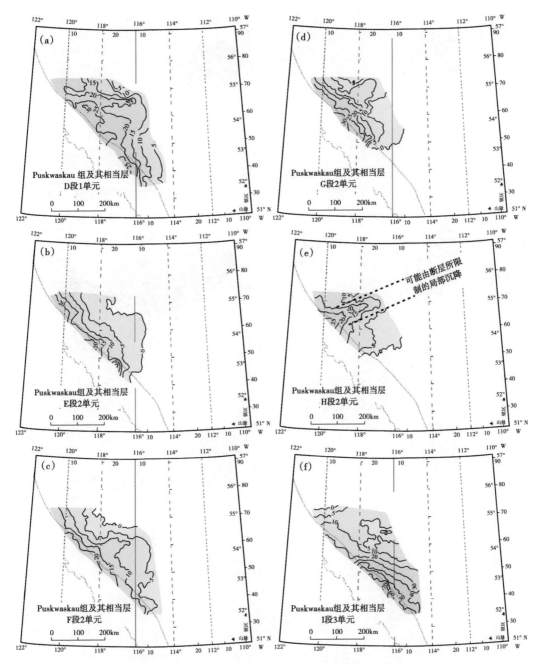

图 24-5　Puskwaskau 组中六个连续地层段（每个时间跨度小于 20 万年；图 24-1）
的等厚图呈现出沉积中心的快速变化的特征（据 Hu，1997）

(a) D 层段；(b) E 层段；(c) F 层段；(d) G 层段；(e) H 层段；(f) I 层段。需要注意的是从 D 层段到 H 层段，沉降中心向西北方向逐渐迁移，同时 E、G 和 H 层段向西南方向完全尖灭。I 层段记录了一次地质尺度上沉积中心向西南方向的瞬间改变的特征，以及在西北向上地层尖灭的特征。G 层段和 H 层段中 SW—NE 向展布的次级沉积中心的成因尚不清楚，但是可能与荷载引起的断层再活动有关，而这些断层曾切割了下伏的古生代地层，如图 (e) 所示

的 WSW—ENE 向（图 24-6a）逐步变为上塞诺曼阶的 NW—SE 向（图 24-6f）。南部沉积中心在上塞诺曼阶到下土伦阶的演化如图 24-3 所示。

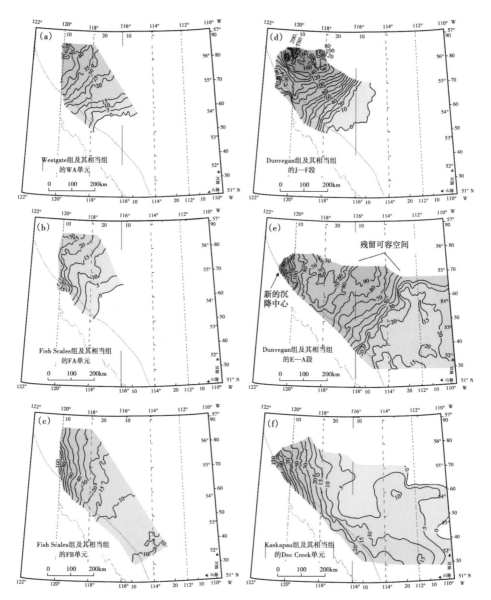

图 24-6 等厚线的走向逐渐发生旋转表明北部（阿尔布阶—中塞诺曼阶）
及南部沉积中心（上塞诺曼阶—坎潘阶）的演化过程

从 Westgate 组到 Kaskapau 组（时间跨度约 6Ma；图 24-1）的等厚图可以看出这一变化。(a) Westgate 组 WA 地层单元；(b) Fish Scales 组 FA 地层单元；(c) Fish Scales 组 FB 地层单元；(d) Dunvegan 组 J—F 层段；(e) Dunvegan 组 E—A 层段；(f) Kaskapau 组 Doe Creek 地层单元。Westgate 组 WA 地层单元向东南尖灭于 Viking 组顶部的陆上不整合面之上，表明西北部存在荷载。Fish Scale 组 FA 地层单元向东南尖灭（尖灭于一个水下沉积间断之上？），等厚线走向表明西北部与东南部都存在荷载。Fish Scale 组 FB 地层单元的等厚线曾经转为 N—S 向。Dunvegan 组 J—F 层段表明这一时期研究区回到了西北部存在荷载的状态，广泛发育向东南方向进积的三角洲层序。在 Dunvegan 组上部（E—A 层段），东南向的沉积中心可以在一定程度上反映前三角洲沉积时区域剩余可容空间的填充过程。但是，一个新的 N—S 向沉积中心出现在研究区的 NW 部，该区域的沉降主要发生在 Dunvegan 组 A 层段（Hay 和 Plint，2009）。南部沉积中心的持续演化如图 24-3 所示。资料来源：(a) — (c) 来源于 Zhang（2006）和 Roca（2007）；(d) 来自 Plint（日期不明）；(e) 来自 Plint（日期不明）和 Tyagi（2009）（镇区 50°带以南）；(f) 来自 Kreitner 和 Plint（2006），以及 Tyagi（2009）（镇区 65°带以南）

图 24-6 中的一系列等厚图表明前陆盆地的沉降模式发生了显著变化，由 NNW 向为主变化为 SE 向为主，逆时针旋转了大约 90°。主要的旋转发生于晚森诺曼期，即上 Dunvegan 组（图 24-6e）至 Doe Creek 地层单元（图 24-6f）。挠曲轴的方向呈现阶段式的突变，推测其部分原因是各种逆冲荷载的相互作用。塞诺曼期无疑是一个过渡阶段，这一时期北部沉积中心开始消亡，而南部沉积中心逐渐出现。对于方向上的变化将在下面进行详细讨论。

24.3.3 "荷载循环"中地层几何学与沉积相的变化

无论是在地层单元尺度还是地层组尺度上，向上变粗、厚度几百米、跨越时间不到几百万年的地层层序在科罗拉多群的岩石中都很常见。地层展布与沉积相之间具有一定的重复性。富含泥质的外滨相与排水较差的海岸平原相被认为可以代表可容空间快速增加，并且在沉积界面上发生了垂直加积。这些"可容空间快速增加"的相可以形成典型的楔状地层段，代表了快速挠曲沉降阶段的沉积（Lumsdon-West 和 Plint，2005；Varban 和 Plint，2008b）。反之，向上变粗的层序的上部主要为强烈进积的临滨砂岩，也有少量的粗粒的河流相砂岩及砾岩，形成了广泛的席状岩石组合。通过海相的 Kaskapau 组及 Cardium 组，以及非海相—边缘海相的 Paddy 组的研究，刻画了沉积相与地层几何学之间的关系。

Kaskapau 组是一个楔形地层组，主要由离岸大于 400km 的异粒泥质陆架相沉积组成（Varban 和 Plint，2005，2008a，2008b；Tyagi，2009；图 24-3 和图 24-7）。泥质陆架沉积向西输运至临滨砂岩中。尽管该地层在垂向上加积了大约 500m，但是其沿盆地边缘向外的进积距离不超过 40km。Kaskapau 组顶部被定义为 GS，此界面为含有细砾岩的海侵界面，代表了相对海平面的下降以及随后的上升（图 24-7）。Cardium 组之上的层段则为席状，并且

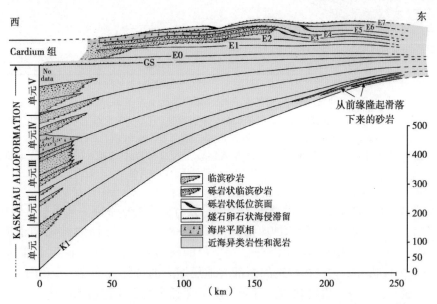

图 24-7 通过 56°N 的 Kaskapau 组中、上部以及 Cardium 组的综合倾向剖面
（据 Hart 和 Plint，1993；Varban 和 Plint，2008b）

Kaskapau 组中、上部的楔体形状单元在 GS 表层突然变为 Cardium 组的层状地层。Kaskapau 组中临滨砂岩只从原有的盆地边缘进积了很短的距离，这种楔体形状反映了较高的可容空间增长速率。与之相反，席状的 Cardium 组代表了一个较小的挠曲阶段，期间存在临滨相向浅水相转变的快速进积过程。沉积速率低的浅水沉积使得海平面变化较快，从而形成了一系列盆地范围内的拆离和 Cardium 组中低位临滨沉积，这些地层逐渐向东尖灭

包含大量的进积临滨砂岩及砾岩,在盆地边缘以东向外延伸达 150km(Hart 和 Plint,1993,2003;Varban 和 Plint,2008b;图 24-7)。

依据 Kaskapau 组及 Cardium 组所反映的几何学与沉积相之间的关系,利用 Jordan 及 Flemings 的数值模型(1991)对挠曲速率的变化进行了解释。高挠曲速率会产生楔形岩体,同时会将粗粒碎屑沉积限制在盆地边缘地区,只有细粒的沉积物才能越过陆架搬运至远处。沉降速率降低后,导致河流坡度变陡,还会导致海水深度下降,从而形成快速进积的临滨面(Varban 和 Plint,2008b 中可见详细等厚图及讨论)。

阿尔布阶的 Paddy 组(Rylanrsdam,2006;Roca 等,2008;图 24-1)时间跨度大约 1Ma,期间形成了 NE—SW 向展布达 300km 的沉积楔形体,其中西南部地层厚达 125m,而向东北部减薄至 10m(图 24-8)。在西南部,Paddy 组主要由互层的泥岩、粉砂岩以及细粒砂岩组成,代表了一种低角度海岸平原沉积环境(Rylaarsdam,2006)。朝东北方向,海岸平原相逐渐变为向上砂质增多的层序,其中含有波痕及咸水生物群,代表了浅海湾和低能三角洲相沉积。富含泥质的潟湖沉积向东逐渐变为富含砂质的近滨潟湖相(Rylaarsdam,2006;图 24-8)。根据潟湖或湖侵界面可以将 Paddy 组划分为 A—I 九个层段(图 24-8;Roca 等,2008)。在西部地区发育有透镜状的粗粒砾石砂岩,中部厚度达 15~25m,宽达几千米。这些透镜体大部分发育在湖侵界面之下,代表了古河谷填充沉积。

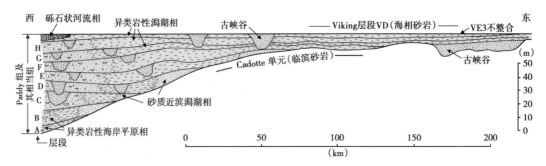

图 24-8　56°N 处 Paddy 组的综合地层倾向剖面(据 Rylaarsdam,2006)

Paddy 组下部的 A—F 段有明显的楔体特征,在东部 Cadotte 组临滨砂岩顶部明显上超于陆上不整合面之上。Paddy 组下部的海岸平原相和潟湖相分布在海平面,向东逐渐超覆在前隆带上。Paddy 组上部的 G、H 和 I 段呈席状覆盖在前隆之上。这些上部层段中的潟湖相沉积逐渐向东西两侧扩展,这意味着该地区有相对海平面上升,研究表明这次海平面上升与 Joli Fou 海的海侵作用有关

Paddy 组 A—F 层段形成了向北逐渐上超于不整合面上的楔形体。此不整合面位于 Cadotte 组近滨砂岩顶部,属于前缘隆起的西南翼(图 24-8)。Paddy 组沉积中心向东延伸的特征在图 24-9(a)—(c)中可以体现出来。与下部的楔形体不同,Paddy 组上部的 G-I 三个层段更多表现为板状沉积,仅在较远的西部有地层增厚的特征(图 24-8 和图 24-9d)。在前陆地区,I 层段发育有砾岩席状体,可能代表了含砾石的辫状河沉积。这个砾岩相带向东突然尖灭。I 层段的顶界面在东部发育有被砂岩填充的下切古河谷,其下切深度约为 25m(Rylaarsdam,2006;Roca 等,2008;图 24-8)。

初始高速挠曲沉降模式可以解释 Paddy 组(A—F 层段)的成因。高速挠曲沉降导致了海岸平原周围快速加积,并使东部或更远处较浅的潟湖形成了洪泛。随着时间的推移,G 段和 H 段的沉降速率逐渐降低,I 段沉积时期,盆地东南边缘发育高梯度、富含砾岩的洪泛体

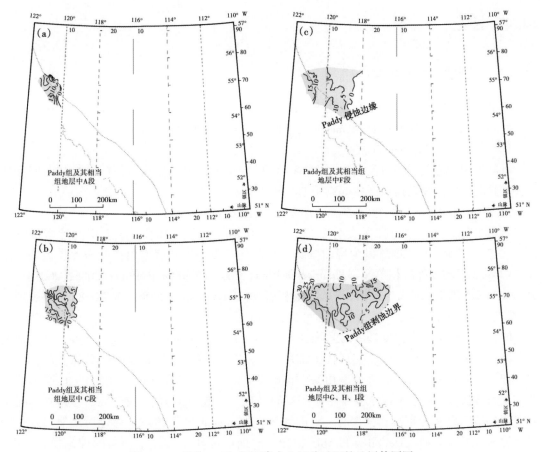

图 24-9 反映 Paddy 组沉降中心迁移过程的地层等厚图

(a) A 段;(b) C 段;(c) F 段;(d) G、H 以及 I 段。东部的空白区是暴露地表并遭受剥蚀的前隆地区,局部被峡谷切割(图 24-8)。Paddy 组沉积中心的南缘为剥蚀边界;对 56°30′N 以北地区的沉积中心没有进行研究。数据来自 Rylaarsdam(2006);(d)包含来自 Leckie 等(1990)的数据

系。Paddy 组 G—I 段中席状可容空间的形成可能受到以下一种或多种因素的控制:由于沉积载荷而导致的区域性均衡沉降、海平面上升、造山带的剥蚀卸载以及随之发生的前缘隆起沉降和动力学沉降。前三个是主要原因。但是长期的可容空间增加很明显是由构造挠曲造成的,Paddy 组各个层段发育的古河道表明了该地区有频繁的海平面升降,从而使得可容空间产生和消失,导致河谷下切和填充交替进行。

Paddy 组顶部(VE3 界面)是一个重要的不整合界面,代表了约 2Ma 的沉积间断。形成这个不整合面之后,在南部出现了向北倾斜的 Joli Fou 组及 Viking 组的 VA 与 VB 地层段(Roca 等,2008;图 24-1 及图 24-8)。Paddy 顶部发育特别宽且深的古河道、I 段为异常粗粒沉积,结合 Paddy 组上部层段内席状体的发育等特征,可以推断在 Paddy 组沉积晚期,挠曲沉降十分有限,并且随后的一个时期内没有发生沉降,并且当河谷下切时甚至还有微小的抬升。这一静止期可能与附近的逆冲带的剥蚀卸载有关。该时期持续到 Viking 组 VD 段初期(图 24-2 和图 24-8)。之后研究区的挠曲沉降集中于西北部地区[Roca 等,2008;图 24-6(a)—(d)]。

577

24.4 讨论

24.4.1 褶皱冲断带的变形机制

本文对科罗拉多群进行了详细的地层划分，发现其具有复杂的沉降模式和构造荷载模式。研究成果可以回顾并回答 Mcmechan 及 Thompson 在 1993 年所提出的一些问题。

1）变形运动的时限能否确定？

沉降沉积中心位置及方向突变（图 24-3 和图 24-6）都可以指示沉降开始的时间和地点，并且能够确定邻近荷载开始起作用的时间。

2）快速造山运动与盆地沉降之间的时间间隔能否确定？

岩石组合在垂向上由楔形变为席状，沉积相也相应地由加积变为进积的砂质近滨相（图 24-7 至图 24-9），都可以为挠曲沉降中心演化过程中沉降速率由快变慢提供证明。组成 Kaskapau 组的多种海相单元（Kreitner 和 Plint, 2006；Varban 和 Plint, 2005, 2008b；图 24-7），以及 Paddy 组从冲积相到边缘海相沉积（Rylaarsdam, 2006；图 24-8），都表现为"由快变慢"的沉降模式。从更长的时间尺度上看，可以将 Kaskapau 组与 Cardium 组作为一个整体，其沉降速率由快变慢的时间跨度大约 5Ma。Puskwaskau 组跨度约 2.5Ma，其内部发育有泥岩为主的加积楔形体，楔形体上部层段 L 为相对薄层的进积型临滨砂岩。这种岩相特征和几何形态也代表了沉降速率由快到慢的变化过程（Hu 和 Plint, 2009）。地层形状由楔形变为席状，且席状地层中发育粗粒碎屑的进积，这成为了 Heller 和 Paola（1992）提出的"反构造"沉积分布模式存在的一个证据。Beaumont 等（1993）通过对犹他州及怀俄明州的 Sevier 构造带（DeCelles 和 Mitra, 1995）的研究建立了一个相关的理论模型，即：可容空间增加率由高到低的变化，可以理解为盆地对超临界状态（向前）逆冲楔变为非临界（向后）楔形体的响应。

3）构造稳定期的持续时间是否能确定？

Viking 组与 Cardium 组（Hart 和 Plint, 1993；Roca, 2007；Varban 和 Plint, 2008b）中的席状岩体表明其形成于浅海沉积环境中的低挠曲沉降率阶段。这些地层组中存在的密集不整合面，可能是由高频快速变化的海平面所导致的。通常区域不整合面的倾角较小，削截 Paddy 组和 Viking 组等下伏的地层单元（Walker, 1995；Roca 等, 2008），并可为区域内细微弯曲及抬升提供证据。这些区域不整合面可能与邻近的科迪勒拉的均衡抬升与剥蚀、外围隆起抬升和较远处的载荷有关。

4）各个构造的活动时间能否确定？

成图结果表明向前推进的逆冲岩席前缘可达 100~300km 宽，并在邻近的前陆盆地中同时产生许多弓形的挠曲沟。沉降中心的突然横向变化（图 24-3、图 24-4 和图 24-6）表明，主要推覆体在停止运动前，向前推进的时间持续了 40 万~80 万年，随后沿走向发育的另一推覆体开始缩短。

从更高分辨率来看，Puskwaskau 组不同层段（平均每层跨越小于 20 万年）的等厚图（图 24-5）表明了不同时期挠曲和逆冲推覆体的位置，它们在不同层段之间沿走向移动了约 100~300km。在不超过几千年的时间内，形成单个海泛面前后时期内，可以存在多个不同沉积中心的形成和消失（对比图 24-5e、f）。一些等厚图中在长条状的区域内地层明显增厚

（图 24-5a、d、e）但是规模很小，这种现象不能用挠曲荷载模式来进行解释，可能是皮斯河拱地区一些断块阶段性活动的产物（Flint 和 Wadsworth，2006）。

对于变形带相对较短部分，其活动沉降轨迹具有局限性，这表明逆冲岩席可以独立活动，并可能被发育高应力的转换带所分开。这些转换带可以调节各个独立逆冲岩体之间的运动特征（Lawton 等，1994）。

对于变形前缘的不连续推进过程本文的研究成果与 Decelles（2004）的结论存在一些分歧。在美国 Sevier 构造带中，他认为"荷载是随着时间连续变化的，而非不连续的"。笔者认为 Decelles（2004）的等厚图地层分辨率不足以解决在加拿大部分盆地所观察到的复杂现象。本文的研究成果表明在跨越几个百万年的岩石组合等厚图中，确实发现了简单的棱状挠曲模式的存在，即存在不连续荷载（图 24-3a 和图 24-4a）。

5）变形前缘的推进速率是否能够确定？

Paddy 组较好地体现了连续层段上超于前缘隆起西翼之上的特征（图 24-8 和图 24-9）。在大约 1Ma 之内，靠近海平面的上超边界向东推进了约 300km（Rylaarsdam，2006；图 24-8）。与之类似的是，Varban 和 Plint（2008b）指出 Kaskapau 组的各个层段上超于前缘隆起的幅度都差不多（大部分在水下），但是强调其中只有一小部分上超作用是由逆冲前缘推进引起的，其速率可能仅为每年几毫米到几厘米（即百万年不超过几千米）。俯冲角度和速率的变化以及动力学板块荷载速率的变化似乎都不能导致 10 万年尺度上能够观测到的侧向尖灭模式的变化。正常情况下，面内应力的变化能够导致前缘隆起细微的抬升和沉降，但是目前还不能确定这种可能性的存在。因此，碎屑岩楔形体向前推进所形成的沉积荷载是前缘隆起方向上超现象发育的一种成因机制。

Varban 和 Plint（2008b）指出，海相泥岩的上超界线可以突然朝造山带方向移动多达 100km，并且会伴随造山带附近的海岸线向西后退。这些侧向尖灭和古地理学上的变化可以反映近源区沉降加速和前隆区的小规模抬升。这些现象可能是由于无序的逆冲作用使荷载和隆起的高部位突然向西移动而引起。

因此，由于受到沉积楔形体推进的均衡效应的影响，使得变形前缘的推进速率和幅度基本上不可能量化，并且它受到了多种因素的影响。至少从我们对小于 10Ma 的地层进行分析的结果来看量化是不可能的。

24.4.2 前人对于构造—沉积相互作用的解释

Cant 和 Stockmal（1989）曾尝试将前陆盆地内主要碎屑岩楔形体的沉积与外来陆块向北美板块的增生联系起来。Underschultz（1991）、Chen 和 Bergman（1999）对阿尔布阶与土伦阶之间挠曲沉积中心的位置与走向的变化进行了研究，他们将其归因于不同的荷载事件以及陆块的局部垂向构造运动。Yang 和 Miall（2008）认为 Fish Scale 组、Kaskapau 组和 Cardium 组是"造山卸载沉积"沉积于沉降前隆之上形成的。Yang 和 Miall（2009）又提出晚塞诺曼期的前缘隆起是对现今落基山脉位置的响应；这一模式与本文中的数据和解释大相径庭（图 24-3）。

24.4.3 沉积模式及区域性构造事件的新解释

1）北美科迪勒拉边缘的形成与中生代陆块和大洋板块的运动有关

以整个白垩纪为研究对象，前陆盆地中充填的沉积物可以被划分为一系列构造地层组合，它们记录了该盆地对科迪勒拉边缘构造事件的响应。在Cant和Stockmal（1989）的研究工作之后，又有许多有关中生代北美板块西部边缘的地块增生和运动的论文发表，这些研究大多以古地磁以及相关资料为基础（Irving等，1996；Butler等，2001；Johnston，2001，2008；Enkin等，2006；Gabrielse等，2006；McCausland等，2006；Colpron等，2007；Evenchick等，2007；Symons等，2009）。

与北美板块边缘有关的地块运动的时间和范围依然存有争议，特别是对于来自白垩纪增生体和周围克拉通地块的古地磁数据的分析，目前有多种解释模式（Buller等，2001；Johnston，2001；Enkin等，2006；McCausland等，2006）。对于这一争论的详细讨论不在本次研究范围之内。现在所考虑的是块体的运动是否能够影响中生代的变形事件和地壳荷载，以及科迪勒拉山和与之毗邻前陆盆地的沉降模式。

本文注意到了晚白垩世早期加拿大科迪勒拉（可能为科迪勒拉带状大陆的一部分）（Johnston，2001，2008）的大规模向南移动（Enkin等，2006），这反映了在阿尔布期至土伦期的过渡阶段，一些地块从北美板块上拆离的过程。北美板块边缘的块体运动变缓的观点（Buller等，2001；McCausland等，2006）更加符合地质约束条件（Gabriolse等，2006；Evonchick等，2007），并且可以为研究科迪勒拉边缘及与之毗邻的前陆盆地内白垩纪事件和块体运动之间的耦合关系提供基础。因此，下面重点讨论McCausland等（2006）提出的块体运动模式。

研究区内主要的增生地体从现今的科迪勒拉前陆及欧明尼卡构造带的所在地区向西移动，形成了现今的山间构造带、海岸及海岛带（图24-10a）。这些地体中最主要的是位于不列颠哥伦比亚省中心的长约1000km的Stikine地体。Stikine地体主要在早侏罗世由年轻火山岛弧以及周边的海相岩石混合而成（Monger，1997；Colpron等，2007）。克拉通边缘的Yukon-Tanana地体沿北部山间带呈南北向展布。与北美大陆边缘邻近地区不同，该地体是一个古生代复合体，其中陆源的变质沉积岩及变质火成岩具有前中生代的变形特征和岩浆特征（Mortensen，1992；Monger，1997；Colpron等，2007）。

Stikine地体与Yukon-Tanana地体是加拿大科迪勒拉增生壳的主要组成部分，其中Stikine地体具有真实的陆壳增厚特征，LITHOPROBE-SNORCLE（岩石圈探针—斯拉夫到北科迪勒拉岩石圈演变）剖面2西南端的反射与折射地震数据证实了这一点（图24-10a；Cook和Erdmer，2005）。地质学证据表明，Stikine地体和Yukon-Tanana地体在早侏罗世便拼接在一起（Monger，1997；Johnston，2001；Colpron等，2007），因此北美大陆边缘自早侏罗世开始具有拼合在一起整体运动和变形的特点（Johnston，2001；McCausland等，2006）。

McCausland等（2006）总结认为，中生代—新生代Stikine地体和Yukon-Tanana地体以及北美板块为整体运动，这一结论主要是根据这些块体上的古地磁数据（图24-10b）。数据表明这些地块在早侏罗世至早—中白垩世相对于北美板块发生了左旋运动，而从中白垩世开始右旋运动，直到始新世这些地块运动到了其现今的位置。这些古地磁记录与相对运动从左旋变为右旋的地质记录一致，同时也与Yukon-Tanana块体在中白垩世最终增生到北美板块上的证据一致（Monger，1997；Evenchick等，2007）。

Gabrielse等（2006）的研究表明，早白垩世晚期（约115—100Ma）左旋偏移开始在BC北部和Yukon地块的主要走滑断层发育区出现，包括Teslin断层、Thibert断层、Kechika断层以及Thudaka断层，同时也在Tintina北部落基山断沟中出现。这些断层在欧明尼卡构

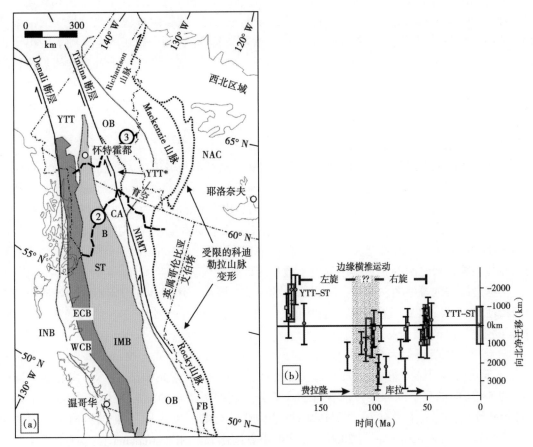

图 24-10 现今科迪勒拉北部的构造特征和北美板块引起的早期地块运动特征

（据 McCausland 等，2006）

（a）科迪勒拉北部的地形地貌—地质带。包含以下构造带：INB—海岛带；WCB—西部海岸构造带；ECB—东部海岸构造带；OB—欧明尼卡构造带；FB—前陆构造带。NAC 代表北美克拉通；NRMT 代表落基山北部的沟槽断层；B 是晚侏罗世 Bowser 盆地。关键的地块有：Yukon-Tanana 地块（YTT）、Stikine（ST）和周缘克拉通 Cassiar 地块（CA）。YTT* 代表 Finlayson 构造带，是中白垩世向北美边缘俯冲的 YTT 的一部分（Murphy 等，2002）。粗虚线是 LITHOPROBE SNORCLE 块体横剖面 2 和横剖面 3 的地表行迹（Cook 和 Erdmer，2005）。（b）根据古地磁数据推测，在早侏罗世—始新世 YTT 和 IMB 地块相对于北美地块向北移动。红色矩形表示 Yukon-Tanana 地块和 Stikine 地块在北美大陆边缘的古地理位置，矩形中 X 分别代表早侏罗世、中白垩世和始新世的地体位置，"0" 处的数值表示与现今位置相比它们向北漂移的距离。对于 YTT-ST 地体的白垩纪演化的其他观点可见 Enkin（2006）或 Johnston（2008）的文献。有点的区域表示科迪勒拉北部地区发育明显的岩浆活动、构造变形或从左旋向右旋运动过渡的关键时期。底部标明了北美板块与法拉隆板块和库拉板块（或重建；Haeussler 等，2003）相互作用的大体时间。McCausland 等（2006）编制的图件中用圆形和方格分别表示 Stikine 地块和 Yukon-Tanana 块体，个别的古地磁结果具有 2-σ 的误差

造带的内部或边缘、北美板块西部、Stikine 地块和 Yukon-Tanana 地块之间发育。中侏罗世—早白垩世欧明尼卡构造带地壳收缩增厚，大量的花岗质岩石伴随这些走滑断层侵入地层当中，这与深部地壳拆离背景下外部陆块向北美板块边缘逆冲运动有关（Evenchick，1991；Evenchick 等，2007）。早白垩世晚期的走滑断层为岩浆的减压熔融和流动提供了前提条件（Gabrielse 等，2006）。

这些地体相对于北美板块发生运动的时间和方向也与基于海底磁异常的板块相对运动模型符合（Engebretson 等，1985）。在从侏罗纪到早白垩世，法拉隆板块与北美板块的汇聚具有一定的左旋性质，但是在大约 115Ma 之后（晚阿普特期），两个板块的会聚则表现出一定的右旋性质（Engebretson 等，1985）。这里存在一个问题，基于海底扩张磁异常的板块相对运动模型，缺乏 125—84Ma 白垩纪超静磁条带的倒转细节（Engebretson 等，1985），这可能会导致插值错误。另外，侏罗纪与白垩纪早期洋壳向北美板块俯冲的消减表明法拉隆板块与北美板块之间曾经存在着微型洋壳板块。

2）区域沉积模式及其构造成因

古地磁证据表明，在早侏罗世到中白垩世之间，相对于北美板块，Stikine 地体与 Yukon-Tanana 地体与洋壳结合更为紧密（McCausland 等，2006；图 24-10b）。前陆盆地钦莫利阶之前的上 Fernie 组、晚侏罗世和早白垩世 Kootenay 组及 Minnes 组（早钦莫利期—晚瓦兰今期：156—134Ma）缺乏西部物源证明了这一点。这几个组构成了一个挠曲沉积中心，延伸范围为 49°N 至 58°N，发育有大约 1000m 的陆源碎屑沉积（Poulton 等，1994）。这些沉积物为地体与北美板块之间的相互作用提供了最早的证据。

根据已发表的等厚图，绘制了一张综合的区域沉降图，用以体现巴雷姆阶到圣通阶的古地理特征（图 24-11）。这些等厚图以加拿大科迪勒拉附近的各构造单元内沉积于 127—85Ma 的主要岩石地层为基本作图单元。

在晚巴雷姆期到早阿尔布期之间（约 127—111Ma，形成了下 Mannville 组与 Bullhead 组），沉降主要集中在 55°N—57°N 的范围内，但发育一个向南延伸至 51°N，向北最终延伸到大约 58°N 向的狭窄前渊。在阿普特期大部分地区的沉降与 Stikine 地体和 Yukon-Tanana 地体沿边缘向南左旋走滑有关，即与欧明尼卡构造带至 Fort St. John 西侧的科迪勒拉山的逆冲荷载相联系。Price 和 Sears（2000）在艾伯塔省南部以及相邻 Montana 省发现了晚侏罗世到早白垩世具有左旋特征的断层。

早—中阿尔布期发育的连续大规模岩石组合（约 111—107Ma；包含了上 Mannville 组与 Fort St. John 组的下部）沿着大陆边缘分布，形成了一个面积较大的前渊（图 24-11b）。如 250m 等厚线所示，此时在 55°N—56°N 范围内的皮斯河拱处于主要沉降期。将这种区域沉降解释为 Stikine 地块与 Yukon-Tanana 块体向科迪勒拉边缘俯冲的结果，是对阿普特期法拉隆板块与北美板块右旋会聚开始的响应。119—115Ma 期间地块从左旋变为右旋，期间的几百万年延迟以及在 111Ma 之后形成的前陆盆地新沉降中心，可能都反映了地体从与板块相结合转变为主要与北美板块相结合（Evenchick 等，2007）的过程。皮斯河拱在 111—107Ma 之间的沉降可能代表了 56°N 处 Stikine 地体北部边缘岬角的瞬间荷载（图 24-11b）。

晚阿尔布期（约 107—99Ma；包括 Viking 组 VD 段、Westgate 组以及 Fish Scales 组的 FA 单元；图 24-1）反映了沉降模式上发生的一个巨大的变化（图 24-11c）。巴雷姆期—中阿尔布期 55°N 以南的前渊地带变得无法识别，取而代之的是发育了一个区域抬升。在 150m 等厚图中 NE—SW 向的皮斯河拱依然可以识别，但是其主要沉降位置更靠北，且沉降中心方向为 N—S 向。在阿尔布期，艾伯塔省北部、不列颠哥伦比亚省东北部以及 Yukon 省南部的海相沉积物主要为厚 2km 左右的泥岩（Stott，1982；Schroder-Adams 和 Pedersen，2003；Jowett 和 Schroder Adams，2005），这为相邻造山带的主要荷载提供了证据。与之相反的是，艾伯塔省南部中—晚阿尔布期的大部分时间都发育不整合（Stott，1982；Leckie 等，2000；Roca 等，2008）。在北部地区，Viking 组 VD 段、Westgate 组以及 Fish Scale 组的等厚图表

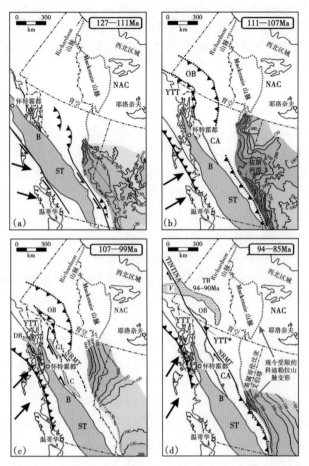

图 24-11 利用主要地层单元的等厚图以及北美板块与增生地体之间相对运动引起的变形事件来推测白垩纪主要沉降中心之间的关系从而重建了研究区的古地理

地体有：CA—Cassiar 地体；ST—Stikine 地体；YTT—Yukon-Tanana 地体，B—中白垩 Bowser 盆地。其他的缩写见图 24-10。海上的箭头表示大洋板块相对于北美板块的运动方向（Engebretson 等，1985）。保持现今的海岸线、国界（灰色）和山脉以及科迪勒拉变形带的边界不变，以便确定北美板块的位置。(a) 巴雷姆期到早阿尔布期（127—111Ma）的等厚图包含了下 Mannville 组与 Bullhead 组（据 Rudkin，1964；Hayes 等，1994）。阿普特期拼接在一起的 YTT-ST 地体和费拉隆板块一起沿着边界向南迁移。欧明尼卡构造带中伴随有左旋运动的逆冲作用导致前陆沉降中心向西北方向迁移。Stikine 地体内的 Bowser 盆地记录了向东南方向的会聚褶皱作用。(b) 早阿尔布期等厚图（111—107Ma）包含了上 Mannville 组与 Fort St. John 组的下部（据 Rudkin，1964；Hayes 等，1994）。地体沿着俯冲面拼接在大陆板块的边缘，从而导致科迪勒拉南部收缩、皮斯河拱的荷载（PRA）以及科迪勒拉北部推覆体广泛发育。(c) 中阿尔布期—早塞诺曼期（107—99Ma）的等厚图包括了 Viking 组 VD 段、Westgate 组以及 Fish Scales 组 FA 单元（据 Rudkin，1964；Leckie 等，1994；图 24-1）。该时期科迪勒拉北部收缩作用继续，并开始发育右旋横推运动。科迪勒拉南部的 ST 经历刚性的顺时针旋转和收缩作用，随后在 109Ma 左右结束。Bowser 盆地东北部会聚褶皱的发育是对右旋压扭作用的响应，这些褶皱垂直叠加在早白垩世的构造之上。Whitehorse 海槽中 YTT—ST 地体以及大不列颠哥伦比亚西北部的主要沉降部分记录了顺时针旋转的信息。这些沉降受到发育在 Cassiar、Glenlyon 和 Dawson 山脉等深成岩体内部的右旋剪切带的调节（分别是 C、DR 和 GL）。(d) 早土伦期到中圣通期（94—85Ma）的等厚图包含了 Kaskapau 组的 II 地层单元到 Puskwaskau 组的 II 地层单元（据 Williams 和 Burk，1964；图 24-1）。科迪勒拉北部 94—90Ma 连续发育的 Fairbanks 和 Tombstone 岩浆构造带（分别为 F 和 TB）表明该地区存在短期的北向俯冲，随后的右旋压扭作用使增生地体向北迁移，沿着 Tintina—落基山北部沟槽断层记录了地体在始新世迁移了 475km。中白垩世—始新世的右旋压扭作用和 Insular Belt 地体的增生作用导致在科迪勒拉南部的幕式收缩，同时在毗邻前陆地区形成了长时间发育的沉降中心。其他的讨论说明见文中明，在晚阿尔布期到早塞诺曼期开始发生较快的沉降，而南部地区则相应地表现为一个前缘隆起（图24-1和图24-12a）。Viking组和Westgate组的等厚图（图24-6（a）至（c））以及Roca等（2008）的电缆测井剖面表明，这一阶段阿尔布期的地层很薄，向南上超在较老地层之上，这表明界面间存在海侵现象。

地体与大陆边缘之间的右旋压扭作用最开始是由阿普特阶走滑断层以及块体内部欧明尼卡带的逆冲作用引起（Gabrielse等，2006）。随后，在阿尔布期右旋的中地壳剪切带、Stikine东部Yukon-Tanana地体及Cassiar地体内钙碱性基底中广泛发育的剪切带也开始影响地体与大陆边缘之间的相互作用（Gabrielse等，2006）。另外，阿尔布期可见大量Yukon-Tanana地体向东北部俯冲至北美板块的Finlayson构造带之上的证据（Murphy等，2002）。同时Yukon-Tanana地体向北美板块逆冲所产生的地壳荷载使得Fort Nelson地区发生沉降（图24-11c）。而与此相反的是，南科迪勒拉地区在109Ma左右（Larson等，2006），逆冲活动中止了约10Ma，然后在土伦期重新复活。

Stikine地体和Yukon-Tanana地体的古地磁数据表明，在阿尔布期Stikine地体至少发生了30°的顺时针旋转（McCausland等，2006），但是还不清楚这种旋转是由地体的整体自转形成的还是局部的垂向旋转。Stikine地体北部以及相邻Yukon-Tanana地体的顺时针旋转情况如图24-11c所示，同时该地区还存在着右旋的横向变形。

在早—中塞诺曼期，西北部发生持续性沉降，形成了由西北部提供物源的Dunvegan三角洲复合体（Stott，1982；Plinl，2000；图24-6d、e）。南部大约53°N的地区在早-中塞诺曼期Fish Scales组发育不整合和凝缩段，这可能与前缘隆起之上沉积有关（Roca等，2008；Plint等，2009）。

在晚塞诺曼期，区域沉降模式开始发生显著的变化，且持续时间较长（图24-11d）。土伦期到圣通期（Kaskapau组Ⅱ地层单元到Puskwaskau组3地层单元；图24-1）沉积中心位于南部，呈NW—SE向展布。土伦阶及康尼亚克阶在56°N以北缺失，仅发育少量圣通阶。晚塞诺曼期在艾伯塔省南部发育的新沉积中心与北部沉降的停止密切相关，同时代表着科迪勒拉附近地区的挤压位置发生了根本性的改变。

Price及Carmichael（1986）指出，55°N北部的Tintina断层—落基山沟槽在白垩纪形成了一个小型的环状构造，随后调节了增生块体的右旋走滑位移。但是在55°N以南，这个断层向东偏移（图24-10a），形成了受限的弧形。这个弧形使断层发生右旋移动并形成了NE向的挤压逆冲和褶皱。构造变形向南一直延伸到53°N，并逐渐增强，使得晚白垩世到始新世期间地层缩短了约200km。在更南部地区的46°30′N处，缩短量下降为20km（Price和Sears，2000）。强烈的NE向逆冲与塞诺曼期之后研究区周缘沉积中心的发育（包含科罗拉多群上部以及更新岩石）之间存在着明确的对应关系（Price，1993；图24-11d）。

3) 主要沉积中心在95Ma左右由NW向转为SW向展布

从地层图中提取出的等厚图，详细地阐明了阿尔布期—土伦期主要沉积中心的迁移（图24-12）。中阿尔布期到中塞诺曼期主要的挠曲沉降大多发育在北部地区，极少发育在南部地区（即53°N以南），因此可以认为南部地区是北部挠曲深沟外部的一个前缘隆起（图24-12a、b）。Viking组及Joli Fou组不仅在这个隆起上受到如上所示的削平剥蚀，并且向南部的科迪勒拉地区逐渐减薄（Roca，2007；Roca等，2008）。以上特征表明至少在晚阿尔布早期，在周缘的变形带中同时发生侵蚀与抬升运动。在51°N以南，只有Joli Fou组和Westgate组再次增厚，从而形成了一个主要由Bow Island组充填的较浅沉积中心（Roca等，2008）。

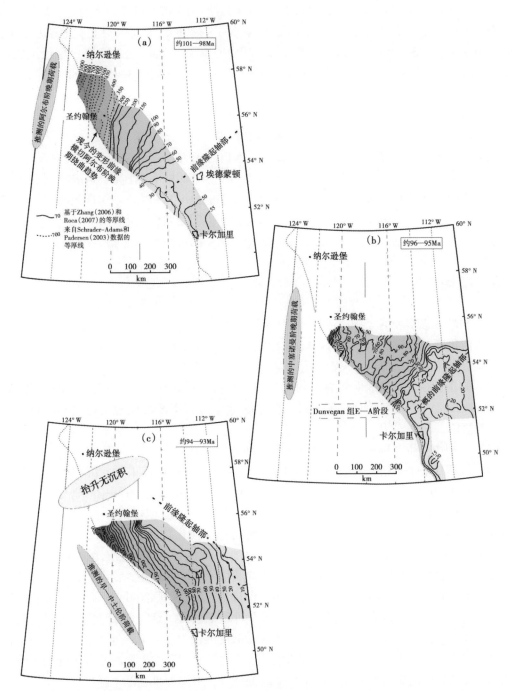

图 24-12 根据研究工作所绘制的等厚图反映了晚阿尔布期到中土伦期的沉降中心迁移情况

(a) Viking 组 VD 段以及 Westgate 组和 Fish Scales 组的总计厚度等厚图。这表明西北部存在厚度大于 1000m、以泥岩为主的沉降中心,而东南部则相应地表现为前隆带。数据来自 Zhang(2006) 和 Roca(2007)。(b) Dunvegan 组上部地层的等厚图表明研究区西北部晚中塞诺曼期开始发育 N—S 向的沉降中心。数据来自 Plint(n.d.)。(c) Kaskapau 组 I—V 地层单元(对应南部的 Blackstine 层)等厚图中的等厚线呈 NW—SE 向,这一特点表明前渊带位于西南部而前隆带位于东北部。Township65°带北部数据据 Varban(2004) 以及 Township65°带南部数据据 Tyagi(2009)。

地层组单元见图 24-1,更多讨论见图 24-1 和文中说明

Dunvegan组上部（层段E—A；图24-12b）的等厚图中依然可见较老的地层呈SW—NE向展布。但是西北部已经形成了一个等厚线呈N—S展布的新沉积中心。地层段的图件（Hay，2006；Hay和Plint，2009）表明新的沉积中心主要由Dunvegan组的A层段组成。上覆Kaskapau组的连续地层单元（图24-3）表明，在晚塞诺曼期到中土伦早期，虽然等厚线在走向上具有短暂的变化，但是整体上发生了逆时针旋转，由N—S向变为NW—SE向，这一模式一直持续到早坎潘期（图24-3，图24-4和图24-12c），这期间上科罗拉多群形成了将近1100m厚的沉积。塞诺曼期之后前缘隆起的位置及旋转参见图24-12c。

主要沉积中心改变较快，这可能与增生地体相对于大洋—北美板块运动的改变有关。Yukon-Tanana地体北部和毗邻北美板块的Tombstone及Fairbanks花岗岩带内花岗质岩石的侵入轨迹表明，新的北东向俯冲运动开始于94—90Ma（图24-11d）。这次俯冲持续的时间较短，可能代表了一个微板块的消亡，随后从晚白垩世到始新世，增生地体沿着大陆边缘向北发生右旋活动，形成了塞诺曼期之后前陆盆地中常见的沉降模式（图24-12c）。

24.5 结论

（1）科罗拉多群发育于中阿尔布晚期到早塞诺曼期，共持续了大约22Ma。其岩石主要为沉积于低角度的海岸平原和浅海斜坡之上的碎屑岩，形成时的水深不超过几十米。由于风暴流的牵引作用使得细粒砂岩、粉砂岩及黏土向远滨扩散。

（2）缺乏斜层理，这表明可容空间形成速率较高，同时被快速填充，从而使得海底保持平坦。Dunvegan组的斜层理可能记录了深水盆地中可容空间形成速率超过了沉积速率的一个反常阶段。

（3）沉积中心也是同期发生沉降的区域，而非沉积高点。

（4）沉积中心的形态与观察时间尺度有关。从百万年的尺度来看，沉积中心表现为似棱柱状。而在40万—80万年，或更小的5万—20万年的尺度上，沉积中心则呈现弧形，并且在侧向上延伸较为有限。从连续沉积中心中可以看出，其在走向和倾向上的延伸范围以及等值线的形态上都具有突变性，并且在较短的时间内可以沿走向移动约300km。

（5）在小于一个或者几个百万年的时间尺度内，海相与陆相的岩石组合中地层段的形成的过程中发育楔状与席状交替的形态。海相楔状体主要为泥质相，而陆相楔状体主要为加积的海岸平原相。相反，席状海相地层段通常含有快速进积的临滨砂岩；而席状陆相层段则在局部存在反常的粗砾碎石冲积相沉积，代表了冲积面较陡。这一现象证实了高可容速率阶段的陆架加积模型的存在。该模型中砂岩和砾石都仅沉积于盆地边界处。粗粒沉积物向盆地内部进积现象仅仅发生于可容速率较低的条件下，而可容速率主要受海平面下降的影响。地层在垂向上由楔状向席状的变化证明单个荷载循环中，沉降速率逐渐下降。

（6）科罗拉多群发育南北两个沉积中心。从中阿尔布期到中塞诺曼期，沉降主要集中于54°N到60°N的南部沉积中心。这个沉积中心在这段时期接受了约1800m厚的细粒海相沉积物。此沉积中心南部的等厚线呈SW—NE向展布，反映了皮斯河拱的沉降差异。本时期艾伯塔省南部的主要特征为沉积间断和不整合。北部沉积中心的沉降主要是由Stikine地体和Yukon-Tanana块体与北美边缘碰撞（约115—95Ma）而形成的地壳增厚及荷载引起。在95Ma左右（晚塞诺曼早期），沉降方向突然逆时针旋转至NW—SE方向，形成了一个平行于现今变形前缘的长轴沉积中心。这个新的沉积中心从晚塞诺曼期（约95Ma）到早坎潘期（及其后期），沉积了科罗拉多群的上部地层。南部地区沉降的开始与北部沉积的突然中断是相对应，北部在晚塞诺曼期到康尼亚克期发育一个不整合界面。本文推测增生地体明显的向北运动始于95Ma。北部的Tintina断裂—落基山沟槽带的运动受到了有效调节带的影响，而55°N以南的断裂带向东挠曲则导致了北东向的挤压逆冲作用以及艾伯塔省南部的沉降。

（7）北美板块西缘的挠曲以及前陆盆地沉积中心的演化模式，可能是由大洋板块与北美板块的会聚作用所控制的。在科迪勒拉山这样大的规模上，主要受到地块增生的影响；而在小于300km的范围内，逆冲推覆体负载、基底薄弱面和其他局部因素也控制了构造变形机制。

（8）根据类比分析，"蹦床构造"这一理论也许能够总结本文的发现：小朋友在蹦床上跳时会产生暂时且分散的"荷载中心"，类似于科罗拉多群中所记录的移动的沉降模式，也能从一定程度上表现出相邻科迪勒拉的荷载移动模式。

致谢（略）

参 考 文 献

Bally, A. W., Gordy, P. L., and Stewart, G. A. (1966) Structure, seismic data, and orogenic evolution of southern Canadian Rocky Mountains. Bulletin of Canadian Petroleum Geology, 14, 337–381.

Barker, I., Moser, D., Kamo, S., and Plint, A. G. (2011) Highprecision ID-TIMS U-Pb Zircon dating of two transcontinental bentonites: Cenomanian Stage, Western Canada Foreland Basin. Canadian Journal of Earth Sciences, 48, 543–556.

Beaumont, C. (1981) Foreland basins. Geophysical Journal of the Royal Astronomical Society, 65, 337–381.

Beaumont, C., Quinlan, G. M. and Stockmal, G. S. (1993) The evolution of the Western Interior Basin: Causes, consequences and unsolved problems. In: Caldwell, W. G. E. and Kauffman, E. G., eds., Evolution of the Western Interior Basin. Geological Association of Canada, Special Paper 39, 97–117.

Bhattacharya, J. P. and Walker, R. G. (1991) Allostratigraphic subdivision of the Upper Cretaceous Dunvegan, Shaftesbury and Kaskapau formations in the northwestern Alberta subsurface. Bulletin of Canadian Petroleum Geology, 39, 145–164.

Bhattacharya, J. P. and MacEachern, J. M. (2009) Hyperpycnal rivers and prodeltaic shelves in the Cretaceous seaway of North America. Journal of Sedimentary Research, 79, 184–209.

Bloch, J., Schroder-Adams, C., Leckie, D. A., McIntyre, D. J., Craig, J. and Staniland, M. (1993) Revised stratigraphy of the lower Colorado Group (Albian to Turonian), Western Canada. Bulletin of Canadian Petroleum Geology, 41, 325–348.

Butler, R. F., Gehrels, G. E. and Kodama, K. P. (2001) A moderate translation alternative to the Baja British Columbia hypothesis. GSA Today, 11, 4–10.

Cant, D. J. and Stockmal, G. S. (1989) The Alberta foreland basin: relationship between stratigraphy and Cordilleran terrane-accretion events. Canadian Journal of Earth Sciences, 26,

1964-1975.

Chen, D. and Bergman, K. M. (1999) Stratal reorientation, depositional processes, and sequence evolution of the Cretaceous in the Peace River Arch region of the Western Canada Sedimentary Basin. Bulletin of Canadian Petroleum Geology, 47, 594-620.

Colpron, M., Nelson, J. L. and Murphy, D. C. (2007) Northern Cordilleran terranes and their interactions through time. GSA Today 17, 4B10.

Cook, F. A. and Erdmer, P. (2005) An 1800kmcross section of the lithosphere throughthe northwestern North American plate: lessons from 4.0 billion years of Earth's history. Canadian Journal of Earth Sciences, 42, 1295-1311.

DeCelles, P. G. (2004) Late Jurassic to Eocene evolution of the Cordilleran thrust belt and foreland basin system, western U. S. A. American Journal of Science, 304, 105-168.

DeCelles, P. G and Mitra, G. (1995) History of the Sevier orogenic wedge in terms of critical taper models, northeast Utah and southeast Wyoming. Geological Society of America Bulletin, 107, 454-462.

Donaldson, W. S., Plint, A. G. and Longstaffe, F. J. (1998) Basement tectonic control on distribution of shallow marine Bad Heart Formation: Peace River Arch area, NW Alberta. Bulletin of Canadian Petroleum Geology, 46, 576-598.

Donaldson, W. S., Plint, A. G. and Longstaffe, F. J. (1999) Tectonic and eustatic control on deposition and preservation of Cretaceous ooidal ironstone and associated facies: Peace River Arch area, NW Alberta, Canada. Sedimentology, 46, 1159-1182.

Engebretson, D. C., Cox, A. and Gordon, R. G. (1985) Relative motions between oceanic and continental plates in the Pacific Basin. Geological Society of America, Special Paper 206, 59 p.

Enkin, R. J. (2006) Paleomagnetism and the case for Baja British Columbia. In: Haggart, J. W., Enkin, R. J. and Monger, J. W. H., eds., Paleogeography of the North American Cordillera: Evidence for and against large-scale displacements. Geological Association of Canada, Special Paper 46, 233-253.

Enkin, R. J., Mahoney, J. B. and Baker, J. (2006) Paleomagnetic signature of the Silverquick/ Powell Creek succession, south-central British Columbia: Reaffirmation of Late Cretaceous large-scale terrane translation. In: Haggart, J. W., Enkin, R. J. and Monger, J. W. H., eds., Paleogeography of the North American Cordillera: Evidence for and against large-scale displacements. Geological Association of Canada, Special Paper 46, 201-220.

Evenchick, C. A. (1991) Geometry, evolution and tectonic framework of the Skeena Fold Belt, north central British Columbia. Tectonics, 10, 527-546.

Evenchick, C. A., McMechan, M. E., McNichol, V. J. and Carr, S. D. (2007) A synthesis of the Jurassic-Cretaceous tectonic evolution of the central and southeastern Canadian Cordillera: Exploring links across the orogen. In: Sears, J. W., Harms, T. A. and Evenchick, C. A., eds., Whence the mountains? Inquiries into the evolution of orogenic systems: A volume in honor of Raymond A. Price. Geological Society of America, Special Paper 433, 117-145.

Gabrielse, H., Murphy, D. C. and Mortensen, J. K. (2006) Cretaceous and Cenozoic dextral orogen – parallel displacements, magmatism, and paleogeography, northcentral Canadian Cordillera. In: Haggart, J. W., Enkin, R. J. and Monger, J. W. H., eds., Paleogeography of the North American Cordillera: Evidence for and against large – scale displacements. Geological Association of Canada, Special Paper 46, 255–276.

Haeussler, P. J., Bradley, D. C., Wells, R. E. and Miller, M. L. (2003) Life and death of the Resurrection plate: Evidence for its existence and subduction in the northeastern Pacific in Paleocene–Eocene time: GSA Bulletin, 115, 867–880.

Hart, B. S. and Plint, A. G. (1993) Tectonic influence on deposition and erosion in a ramp setting: Upper Cretaceous Cardium Formation, Alberta foreland basin. American Association of Petroleum Geologists Bulletin, 77, 2092–2107.

Hart, B. S. and Plint, A. G. (2003) Stratigraphy and sedimentology of shoreface and fluvial conglomerates: insights from the Cardium Formation in NW Alberta and adjacent British Columbia. Bulletin of Canadian Petroleum Geology, 51, 437–464.

Hay, M. J. (2006) Stratigraphy, sedimentology, and paleogeography of the upper Dunvegan Formation, mid–Cenomanian, Alberta, Canada: interactions between deltaic sedimentation, flexural tectonics and eustasy. MSc thesis, University of Western Ontario, London, Ontario, 197 p.

Hay, M. J. and Plint, A. G. (2009) An allostratigraphic framework for a retrogradational delta complex: the uppermost Dunvegan Formation (Cenomanian) in subsurface and outcrop, Alberta and British Columbia. Bulletin of Canadian Petroleum Geology, 57, 323–349.

Hayes, B. J. R., Christopher, J. E., Rosenthal, L., Los, G., McKercher, B., Minken, D., Tremblay, Y. M. and Fennell, J. (1994) Chapter 19, Cretaceous Mannville Group of the Western Canada Sedimentary Basin. In: Mossop, G. and Shetsen, I., eds., Geological Atlas of the Western Canada Sedimentary Basin. Canadian Society of Petroleum Geologists and Alberta Research Council, Calgary, 317–334.

Heller, P. L. and Paola, C. (1992) The large–scale dynamics of grain–size variation in alluvial basins, 2: Application to syntectonic conglomerate. Basin Research, 4, 91–102.

Hu, Y. G. (1997) High – resolution sequence stratigraphic analysis of the Upper Cretaceous Puskwaskau Formation of west–central Alberta and adjacent B. C.: outcrop and subsurface. PhD thesis, University of Western Ontario, 309 p.

Hu, Y. G. and Plint, A. G. (2009) An allostratigraphic correlation of a mudstone–dominated syntectonic wedge: The Puskwaskau Formation (Santonian – Campanian) in outcrop and subsurface, Western Canada foreland basin. Bulletin of Canadian Petroleum Geology, 57, 1–33.

Irving, E., Wynne, P. J., Thorkelson, D. J. and Schiarizza, P. (1996). Large (1000 to 4000 km) northward movements of tectonic domains in the northern Cordillera, 83 to 45 Ma. Journal of Geophysical Research, 101, 17901–17916.

Johnston, S. T. (2001) The Great Alaskan Terrane Wreck: reconciliation of paleomagnetic and geological data in the northern Cordillera. Earth and Planetary Science Letters, 193, 259–272.

Johnston, S. T. (2008) The Cordilleran ribbon continent of North America. Annual Review of Earth and Planetary Sciences, 36, 495-530.

Jordan, T. E. (1981) Thrust loads and foreland basin evolution, Cretaceous, western United States. American Association of Petroleum Geologists Bulletin, 65, 2506-2520.

Jordan, T. E. and Flemings, P. F. (1991) Large-scale stratigraphic architecture, eustatic variation, and unsteady tectonism: A theoretical evaluation. Journal of Geophysical Research, 96B4, 6681-6699.

Jowett, D. M. S. and Schroder-Adams, C. J. (2005) Paleoenvironments and regional stratigraphic framework of the Middle-Upper Albian Lepine Formation in the Liard Basin, northern Canada. Bulletin of Canadian Petroleum Geology, 53, 25-50.

Kauffman, E. G., and Caldwell, W. G. E. (1993) The Western Interior Basin in space and time. In: Caldwell, W. G. E. and Kauffman, E. G., eds., Evolution of the Western Interior Basin. Geological Association of Canada, Special Paper 39, 1-30.

Kreitner, M. A. and Plint, A. G. (2006) Allostratigraphy and paleogeography of the Upper Cenomanian, Lower Kaskapau Formation in subsurface and outcrop, Alberta and British Columbia. Bulletin of Canadian Petroleum Geology, 54, 147-174.

Larson, K. P., Price, R. A. and Archibald, D. A. (2006) Tectonic implications of $^{40}Ar/^{39}Ar$ muscovite dates from the Mt. Haley stock and Lussier River stock, near Fort Steele, British Columbia. Canadian Journal of Earth Sciences, 43, 1673-1684.

Lawton, T. F. (1994) Tectonic setting of Mesozoic sedimentary basins, Rocky Mountain region, United States. In: Caputo, M. V., Peterson, J. A. and Franczyk, K. J., eds., Mesozoic systems of the Rocky Mountain region, USA. Rocky Mountain section of SEPM, Denver, Colorado, 1-25.

Lawton, T. F., Boyer, S. E. and Schmitt, J. G. (1994) Influence of inherited taper on structural variability and conglomerate distribution, Cordilleran fold and thrust belt, western United States. Geology, 22, 339-342.

Leckie, D. A., Staniland, M. R. and Hayes, B. J. (1990) Regional maps of the Albian Peace River and lower Shaftesbury formations on the Peace River Arch, northwestern Alberta and northeastern British Columbia. Bulletin of Canadian Petroleum Geology, 38A, 176-189.

Leckie, D. A., Bhattacharya, J. P., Bloch, J., Gilboy, C. F. and Norris, B. (1994) Chapter 20, Cretaceous Colorado/Alberta Group of the Western Canada Sedimentary Basin. In: Mossop, G. and Shetsen, I., eds., Geological Atlas of the Western Canada Sedimentary Basin. Canadian Society of Petroleum Geologists and Alberta Research Council, Calgary, 335-352.

Leckie, D. A., Schroder-Adams, C. J. and Bloch, J. (2000) The effect of paleotopography on the Late Albian and Cenomanian sea-level record of the Canadian Cretaceous interior seaway. Geological Society of America, Bulletin, 112, 1179-1198.

Liu, S. and Nummedal, D. (2004) Late Cretaceous subsidence in Wyoming: Quantifying the dynamic component. Geology, 32, 397-400.

Lumsdon-West, M. P. and Plint, A. G. (2005) Changing alluvial style in response to changing accommodation rate in a proximal foreland basin setting: Upper Cretaceous Dunvegan Formation, NE British Columbia, Canada. In: Blum, M. D., Marriott, S. B. and Leclair, S., eds., Fluvial Sedimentology VII, International Association of Sedimentologists, Special Publication 35, 493-515.

McCarthy, P. J. and Plint, A. G. (1998) Recognition of interfluve sequence boundaries: integrating paleopedology and sequence stratigraphy. Geology, 26, 387-390.

McCarthy, P. J. and Plint, A. G. (2003) Spatial variability of palaeosols across Cretaceous interfluves in the Dunvegan Formation, NE British Columbia, Canada: palaeohydrologic, palaeogeomorphologic and stratigraphic implications Sedimentology, 50, 1187-1220.

McCarthy, P. J., Faccini, U. F. and Plint, A. G. (1999) Evolution of an ancient floodplain: palaeosols and alluvial architecture in a sequence stratigraphic framework, Cenomanian Dunvegan Formation, NE British Columbia, Canada. Sedimentology, 46, 861-891.

McCausland, P. J. A., Symons, D. T. A., Hart, C. J. R. and Blackburn, W. H. (2006) Assembly of the northern Cordillera: New paleomagnetic evidence for coherent, moderate Jurassic to Eocene motion of the Intermontane belt and Yukon-Tanana terranes. In: Haggart, J. W., Enkin, R. J. and Monger, J. W. H., eds., Paleogeography of the North American Cordillera: Evidence for and against large-scale displacements. Geological Association of Canada, Special Paper 46, 147-170.

McMechan, M. E. and Thompson, R. I. (1993) The Canadian Cordilleran fold and thrust belt south of 66 N and its influence on the Western Interior Basin. In: Caldwell, W. G. E. and Kauffman, E. G., eds., Evolution of the Western Interior Basin. Geological Association of Canada, Special Paper 39, 73-90.

Miall, A. D., Catuneanu, O., Vakarelov, B. K. and Post, R. (2008) Chapter 9, The Western Interior Basin. In: Miall, A. D. (ed.), Sedimentary Basins of the World, Vol. 5, The sedimentary basins of the United States and Canada, Elsevier, Amsterdam, 329-362.

Mitrovica, J. X., Beaumont, C. and Jarvis, G. T. (1989) Tilting of continental interiors by the dynamical effects of subduction. Tectonics, 8, 1079-1094.

Monger, J. W. H. (1993) Cretaceous tectonics of the North American Cordillera. In: Caldwell, W. G. E. and Kauffman, E. G., eds., Evolution of the Western Interior Basin. Geological Association of Canada, Special Paper 39, 31-47.

Monger, J. W. H. (1997) Plate tectonics and northern Cordilleran geology: An unfinished revolution. Geoscience Canada, 24, 189-198.

Monger, J. W. H. and Price, R. A. (2002) The Canadian Cordillera: Geology and tectonic evolution. Canadian Society of Exploration Geophysicists, Recorder, 17, 18-36.

Mortensen, J. K. (1992). Pre-mid-Mesozoic tectonic evolution of the Yukon-Tanana terrane, Yukon and Alaska. Tectonics, 11, 836-853.

Murphy, D. C., Colpron, M., Roots, C. F., Gordey, S. P. and Abbott, J. G., 2002, Finlayson Lake targeted geoscience initiative (southeastern Yukon), part 1: bedrock geology: Yukon Exploration and Geology, 2001, p. 189-207.

Nielsen, K. S., Schroder-Adams, C. J. and Leckie, D. A. (2003) A new stratigraphic framework for the Upper Colorado Group (Cretaceous) in southern Alberta and southwestern Saskatchewan, Canada. Bulletin of Canadian Petroleum Geology, 51, 304-346.

O'Connell, S. (1994) Geological history of the Peace River Arch. In: Mossop, G. and Shetsen, I., eds., Geological Atlas of the Western Canada Sedimentary Basin. Canadian Society of Petroleum Geologists and Alberta Research Council, p. 431-437.

Ogg, J. G., Ogg, G. and Gradstein, F. M. (2009) The concise Geologic Time Scale. Cambridge University Press, 177 p.

Pang, M. and Nummedal, D. (1995) Flexural subsidence and basement tectonics of the Cretaceous Western Interior Basin, United States. Geology, 23, 173-176.

Pattison, S. A. J. and Walker, R. G. (1994) Incision and filling of a lowstand valley: late Albian Viking Formation at Crystal Field, Alberta, Canada. Journal of Sedimentary Research, 64, 365-379.

Plint, A. G. (1990) An allostratigraphic correlation of the Muskiki and Marshybank formations (Coniacian - Santonian) in the Foothills and subsurface of the Alberta Basin. Bulletin of Canadian Petroleum Geology, 38, 288-306.

Plint, A. G. (2000) Sequence stratigraphy and paleogeography of a Cenomanian deltaic complex: The Dunvegan and lower Kaskapau formations in subsurface and outcrop, Alberta and British Columbia, Canada. Bulletin of Canadian Petroleum Geology, 47, 43-79.

Plint, A. G. (2002) Paleo-valley systems in the Upper Cretaceous Dunvegan Formation, Alberta and British Columbia. Bulletin of Canadian Petroleum Geology, 50, 277-296.

Plint, A. G. (2003) Clastic sediment partitioning in a Cretaceous delta system, western Canada: Responses to tectonic and sea-level controls. Geologia Croatica, 56, 39-68.

Plint, A. G. (n. d.) Unpublished data, University of Western Ontario, London. Plint, A. G. and Norris, B. (1991) Anatomy of a ramp margin sequence: Facies successions, paleogeography and sediment dispersal patterns in the Muskiki and Marshybank formations, Alberta Foreland Basin. Bulletin of Canadian Petroleum Geology, 39, 18-42.

Plint, A. G. and Kreitner, M. A. (2007) Extensive, thin sequences spanning Cretaceous foredeep suggest highfrequency eustatic control: Late Cenomanian, Western Canada foreland basin. Geology, 35, 735-738.

Plint, A. G. and Wadsworth, J. A. (2003) Sedimentology and palaeogeomorphology of four large valley systems incising delta plains, Western Canada Foreland Basin: implications for mid-Cretaceous sea-level changes. Sedimentology, 50, 1147-1186.

Plint, A. G. and Wadsworth, J. A. (2006) Delta plain paleodrainage patterns reflect small-scale fault movement and subtle forebulge uplift: Upper Cretaceous Dunvegan Formation, Western Canada Foreland Basin. In: Dalrymple, R. W., Leckie, D. A., and Tillman, R. W., eds., Incised Valley Systems in Time and Space. SEPM Special Publication 85, 219-237.

Plint, A. G., Walker, R. G. and Bergman, K. M. (1986) Cardium Formation 6. Stratigraphic framework of the Cardium in subsurface. Bulletin of Canadian Petroleum Geology, 34, 213-225.

Plint, A. G., Tyagi, A., Hay, M. J., Varban, B. L., Zhang, H. and Roca, X. (2009) Clinoforms, paleobathymetry, and mud dispersal across the Western Canada Cretaceous foreland basin: evidence from the Cenomanian Dunvegan Formation and contiguous strata.

Journal of Sedimentary Research, 79, 144-161.

Poulton, T. P., Christopher, J. E., Hayes, B. J. R., Losert, J., Tittemore, J. and Gilchrist, R. D. (1994) Chapter 18: Jurassic and Lowermost Cretaceous strata of the Western Canada Sedimentary Basin. In: Mossop, G. and Shetsen, I., eds., Geological Atlas of the Western Canada Sedimentary Basin. Canadian Society of Petroleum Geologists and Alberta Research Council, p. 297-316.

Posamentier, H. W. and Chamberlain, C. J. (1993) Sequencestratigraphic analysis of Viking Formation lowstand beach deposits at Joarcam Field, Alberta, Canada. In: Posamentier, H. W., Summerhayes, C. P., Haq, B. U. and Allen, G. P., eds., Sequence Stratigraphy and Facies Associations, International Association of Sedimentologists, Special Publication 18, 469-485.

Price, R. A. (1973) Large-scale gravitational flow of supracrustal rocks, southern Canadian Rockies. In: De Jong, K. E. and Scholten, R., eds., Gravity and tectonics. John Wiley, New York, 491-502.

Price, R. A. (1994) Chapter 2: Cordilleran tectonics and the evolution of the Western Canada Sedimentary Basin. In: Mossop, G. and Shetsen, I., eds., Geological Atlas of the Western Canada Sedimentary Basin. Canadian Society of Petroleum Geologists and Alberta Research Council, p. 13-24.

Price, R. A. and Carmichael, D. M. (1986) Geometric test for Late Cretaceous-Paleogene intracontinental transform faulting in the Canadian Cordillera. Geology, 14, 568-471.

Price, R. A. and Sears, J. W. (2000) Preliminary palinspastic map of the Mesoproterozoic Belt-Purcell Supergroup, Canada and USA: Implications for the plate tectonic setting and structural evolution of the Purcell anticlinorium and the Sullivan Deposit. In: Lydon, J. W., Hoy, T., Slack, J. F. and Knapp, M. E., eds., The geological environment of the Sullivan Deposit, British Columbia. Geological Association of Canada, Mineral Deposits Division, Special Publication 1, 61-81.

Roca, X. (2007) Tectonic and eustatic controls on the allostratigraphy and depositional environments of the Lower Colorado Group (Upper Albian), Central Foothills and adjacent Plains of Alberta, Western Canada foreland basin. PhD thesis, University of Western Ontario, London, Ontario, 323 p.

Roca, X., Rylaarsdam, J. R., Zhang, H., Varban, B. L., Sisulak, C. F., Bastedo, K. and Plint, A. G. (2008) An allostratigraphic correlation of the Lower Colorado Group (Albian) and equivalent strata in Alberta and British Columbia, and Cenomanian rocks of the Upper Colorado Group in southern Alberta. Bulletin of Canadian Petroleum Geology, 56, 259-299.

Rudkin, R. A. (1964) Chapter 11, Lower Cretaceous. In: McCrossan, R. G. and Glaister, R. P., eds., Geological History of Western Canada. Alberta Society of Petroleum Geologists, Calgary, 156-168.

Ryer, T. A. (1993) Speculations on the origins of mid-Cretaceous clastic wedges, central Rocky Mountain region, United States. In: Caldwell, W. G. E. and Kauffman, E. G., eds., Evolution of the Western Interior Basin. Geological Association of Canada, Special Paper 39,

189-198.

Rylaarsdam, J. R. (2006) The stratigraphy and sedimentology of the Upper Albian Paddy Member of the Peace River Formation, and the Walton Creek Member of the Boulder Creek Formation, NW Alberta and adjacent British Columbia. MSc thesis, University of Western Ontario, London, Ontario, 187 p.

Scheiber, J., Southard, J. and Thaisen, K. (2007) Accretion of mudstone beds from migrating floccule ripples. Science, 318, 1760-1762.

Schröder – Adams, C. J. and Pedersen, P. K. 2003, Litho – and biofacies analysis of the Buckinghorse Formation: The Albian Western Interior Sea in northeastern British Columbia (Canada). Bulletin of Canadian Petroleum Geology, 51, 234-252.

Schwartz, R. K. andDeCelles, P. G. (1988) Cordilleran foreland basin evolution in response to interactive Cretaceous thrusting and foreland partitioning, southwestern Montana. In: Schmidt, C. J. and Perry, W. J., eds., Interaction of the Rocky Mountain foreland and the Cordilleran thrust belt. GeologicalSocietyofAmerica, Memoir171, 489-513.

Stott, D. F. (1982) Lower Cretaceous Fort St. John Group and Upper Cretaceous Dunvegan Formation of the Foothills and Plains of Alberta, British Columbia, District of Mackenzie and Yukon Territory. Geological Survey of Canada, Bulletin 328, 124 p.

Symons, D. T. A., Kawasaki, K. and McCausland, P. J. A. (2009) The Yukon – Tanana terrane: Part of North America at 215Ma from paleomagnetism of the Taylor Mountain batholith, Alaska. Tectonophysics, 465, 60-74.

Tyagi, A. (2009) Sedimentology and high-resolution stratigraphy of the Upper Cretaceous (Late Albian to Middle Turonian) Blackstone Formation, Western Interior Basin, Alberta, Canada. PhD thesis, University of Western Ontario, London, Ontario, 699 p.

Tyagi, A., Plint, A. G. and McNeil, D. H. (2007) Correlation of physical surfaces, bentonites, and biozones in the Colorado Group from the Alberta Foothills to south-western Saskatchewan, and a revision of the Belle Fourche/Second White Specks formational boundary. Canadian Journal of Earth Sciences, 44, 871-888.

Underschultz, J. R. (1991) Tectonic loading, sedimentation, and sea – level changes in the foreland basin of north – west Alberta and north – east British Columbia, Canada. Basin Research, 3, 165-174.

Varban, B. L. (2004) Sedimentology and stratigraphy of the Cenomanian – Turonian Kaskapau Formation, northeast British Columbia and northwest Alberta. PhD thesis, University of Western Ontario, London, Ontario, 452 p.

Varban, B. L. and Plint, A. G. (2005) Allostratigraphy of the Kaskapau Formation (Cenomanian-Turonian) in subsurface and outcrop: NE British Columbia and NW Alberta, Western Canada Foreland Basin. Bulletin of Canadian Petroleum Geology, 53, 357-389.

Varban, B. L. and Plint, A. G. (2008a) Palaeoenvironments, palaeogeography, and physiography of a large, shallow, muddyramp: Late Cenomanian-Turonian Kaskapau Formation, Western Canada foreland basin. Sedimentology, 55, 201-233.

Varban, B. L. and Plint, A. G. (2008b) Sequence stacking patterns in the Western Canada foredeep: influence of tectonics, sediment loading and eustasy on deposition of the Upper Cretaceous Kaskapau and Cardium formations. Sedimentology, 55, 395-421.

Walker, R. G. (1995) Sedimentary and tectonic origin of a transgressive surface of erosion: Viking Formation, Alberta, Canada. Journal of Sedimentary Research, 65, 209-221.

Wheeler, J. O. and McFeely, P. (1991) Tectonic assemblage map of the Canadian Cordilleran and adjacent parts of the United States of America: Geological Survey of Canada, Map 1712A, scale 1:2000000.

Williams, G. D and Burk, C. F. (1964) Chapter 12, Upper Cretaceous. In: McCrossan, R. G. and Glaister, R. P., eds., Geological History of Western Canada. Alberta Society of Petroleum Geologists, Calgary, 169-185.

Yang, Y. and Miall, A. D. (2008) Marine transgressions in the mid-Cretaceous of the Cordilleran foreland basin reinterpreted as orogenic unloading deposits. Bulletin of Canadian Petroleum Geology, 56, 179-198.

Yang, Y. and Miall, A. D. (2009) Evolution of the northern Cordilleran foreland basin during the middle Cretaceous. Geological Society of America, Bulletin, 121, 483-501.

Zhang, H. (2006) Allostratigraphy of the Shaftesbury Formation, northern Alberta and northeastern British Columbia: Responses to tectonic and eustatic controls. MSc thesis, University of Western Ontario, London, Ontario, 237 p.

Zonneveld, J-P., Kjarsgaard, B. A., Harvey, S. E., Heaman, L. M., McNeil, D. H. and Marcia, K. Y. (2004) Sedimentologic and stratigraphic constraints on emplacement of the Star Kimberlite, east-central Saskatchewan. Lithos, 76, 115-138.

（崔敏 李政斯译，吴哲 校）

第25章 连续型前陆盆地和破碎型前陆盆地的构造、地貌及沉积特征：以玻利维亚和阿根廷西北部安第斯山脉中部东侧为例

MANFREN R. STRECKER[1], GEORGEE, HILLEY[2],
BODO BOOKHAGEN[3], EDWARD R. SOBEL[4]

（1. Instituteof Geowissenschaften, University at Potsdam, Potsdam；
2. Department of Geological and Environmental Sciences, Stanford University, Stanford, USA；
3. Geography Department, University of California, Santa Barbara, USA；
4. Institute of Geowissenschaften, University at Potsdam, Potsdam, Germany）

摘 要：在本章中，将玻利维亚境内的典型连续性前陆盆地和位于阿根廷西北部的破碎型前陆盆地进行了对比。结果表明这两种前陆盆地在沉积与地貌形态上存在差异，且均受构造条件的制约。一般而言，可容空间是构成前陆盆地的基本单元，这些可容空间是地壳对褶皱—冲断带地形挠曲的响应。可容空间的差异使得前陆盆地内形成了四个独立的沉积带：楔顶带、前渊带、前隆带和隆后带。破碎型前陆盆地主要分布在发育有复活高角度构造的弧后区域。这些构造隆起周缘的地层通常在时间和空间上相互孤立，并非形成一个连续的斜坡，这导致在主要造山带前缘的内部形成一个沿着走向分布的断续型山脉。潜在的高隆升速率受陡立复活逆断层的调节，使得该体系中河源盆地与下游河流水系相隔离，形成了位于隆升山脉后侧的沉积洼地。此外，波长较短且侧向受到地制约的限制性挠曲响应，可能不会形成典型前陆盆地前渊沉积带中的大型可容空间。因此，具有大陆规模的地壳挠曲不一定会形成连续侧向伸展的前陆盆地系统，也可能形成由一系列连接在一起的侧向受限的沉积中心。例如位于阿根廷西北部安第斯山脉周缘的破碎型前陆盆地。地表发生隆升或者个别山脉隆升之后，在山脉周围形成的有限可容空间中发育有多个小型盆地。这种类型的沉积体系缺乏典型前陆盆地应有的地质单元，不发育完整的前渊带、前隆带和隆后带沉积体系。因此，不应该将典型前陆盆地模型应用到破碎型前陆盆地系统中，因为它们具有独特的构造、地形和地球动力学背景。

关键词：连续型前陆盆地 破碎型前陆盆地 褶皱—冲断带 复活构造 安第斯山脉中部

25.1 引言

前陆盆地的结构及其充填地层厚度主要受到构造控制的可容空间的制约，期间伴随横向与垂向上发育的褶皱冲断带（Dickinson, 1974; Jordan, 1981, 1995）。通常根据压缩作用、地壳荷载作用、气候以及沉积物供应和沉积过程之间的相互影响所构成的沉积环境将前陆盆地分为四个构造带/沉积带（Decelles 和 Giles, 1996；图25-1a）。它们分别为：(1) 背驮盆

地,形成于褶皱冲断带外侧,发育盲冲断层,而此褶皱冲断带的外缘就是盆地楔形体的顶部边界,其沉积物沉积于造山带的构造活动地区(Jordan,1995;Decelles 和 Giles,1996);(2) 前渊带,其外侧的挠曲沉降为前渊的沉积物提供了空间,沉积物沿着逆冲楔方向厚度变大、粒度变粗(Dickinson 和 Suczek,1979;Schwab,1986;Decelle 和 Hertel,1989;Decelles 和 Giles,1996);(3) 前缘隆起带,是指前渊外侧末端的翘起部分,由地壳尺度的挠曲作用形成(Jabobi,1981;Karner 和 Watts,1983;Quinlan 和 Beaumont,1984;Crampton 和 Allen,1995;Turcotte 和 Schubert,2002);(4) 隆外凹陷带,是在前缘隆起形成过程中产生的凹陷带,代表了宽阔低能的沉积环境(Decelles 和 Giles,1996;Horton 和 Decelles,1997;Decelles 和 Horton,2003)。当逆冲褶皱带向前陆推进时,不同的构造带和沉积域也会整体向盆地方向迁移,并使得这些区域内的沉积物在垂向上堆积(Decelle 和 Horton,2003)。褶皱冲断带在形态上类似于一个自相似的生长构造楔,可以逐渐通过前缘增生作用向前陆盆地输送沉积物,并沿滑脱面运动,前人在现今安第斯中部玻利维亚的 Subandean 构造带中的研究就揭示了这一点(Sempere 等,1990;Kennan 等,1995;Baby 等,1997;Horton 和 Decelles,1997;McQuarrie,2002;Uba 等,2006)。

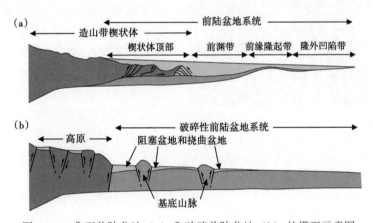

图 25-1　典型前陆盆地(a)和破碎前陆盆地(b)的模型示意图
(a) 前陆盆地系统由楔体顶部发育的同变形沉积、向前推进的褶皱—冲断带前缘发生挠曲而形成的前渊沉积中心、挠曲隆起带上发育的前缘隆起沉积中心,以及挠曲隆起带之后发育的隆外凹陷带组成(改自 DeCelles 和 Giles,1996);
(b) 破碎前陆盆地,由空间上受限的挠曲盆地和隆升地形之后的沉积洼地组成

这个标准模型能够很好地解释现代与古代的周缘前陆盆地(Dickinson,1974)与弧后前陆盆地(Jordan,1981,1995;DeCelles 和 Giles,1996)的主要特征。但该模型仅适用于薄皮褶皱冲断带,不适用于造山带内发育基底隆起的沉积盆地。因为这些以逆断层为边界的基底隆起的形成时期不同,且分布于不同的构造位置(Jordan 和 Allmendinger,1986)。现存的这种前陆盆地包括阿根廷西北部毗邻 Subandes 褶皱冲断带的 Santa Barbara 前陆盆地系统,以及更南部的 Sierras Pampeanas 变质构造岩省(图 25-2;Stelzner,1923;Gonzalez Bonorino,1950;Allmendinger 等,1983;Jordan 和 Allmendinger,1986;Ramos 等,2002)。美国西部的 Laramide 前陆(Steidtmann 等,1983)可以作为这类前陆盆地系统的古代实例(Dickinson 和 Snyder,1978;Jordan 和 Allmendinger,1986)。虽然处于不同的地球动力学背景之下,但是中国和吉尔吉斯斯坦境内的天山及青藏高原北部的祁连山隆起都包含有类似的过程,即前陆遭受挤压剥蚀,基底块体不断隆升(Metivier 等,1998;Sobel 等,2003)。这

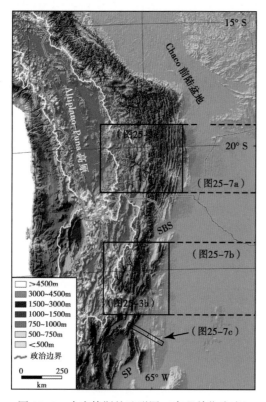

图 25-2 中安第斯的地形图（高程单位为米）
图 25-4c、d 中的经度范围的位置与图 25-3a、b 相同。
图 25-7 中挠曲模型所用的经度范围（横向计算）也有相应说明。SBS 和 SP 分别代表 Santa Barbara 和 Sierras Pampeans 地形构造省。白色实线代表造山带内部 Altilano-Puna 高原的内流水系

种前陆盆地的演化将早期连续沉积的盆地最终肢解，并导致了岩石抬升与沉降，不能利用典型前陆盆地的模型来预测（Jordan 和 Allmendinger，1986）。这种基底范围内的调节变形以及破碎前陆的形成可能在造山带的内侧发育较好（Ramos 等，2002；Strecher 等，2009；Iaffa 等，2011）。值得关注是，这种基底范围内的抬升，会导致与生长造山楔有关的典型前陆盆地内的挠曲响应与预期的结果不同。因此，这类前陆盆地内的沉积体系演化可能从根本上就与毗邻的褶皱—冲断带发育的前陆盆地不同。

在此本文主要研究这种破碎型前陆盆地中的地形地貌和沉积发育情况。通过对比研究 Altiplana-Puna 高原及阿根廷西北部东科迪勒拉附近的沉积体系的地球动力学条件，利用北部玻利维亚 Chaco 前陆盆地中所保存下来的沉积体系来研究其演化过程（图 25-2）。玻利维亚地区与阿根廷西北部在构造演化上的关键性差异是它们的前新生代地史不同（Allmendinger 等，1983；Kley 等，2005），这对两个地区的挠曲变形响应、地貌演化以及周围盆地的沉积特征都具有控制作用。

25.2 中安第斯的弧后地形、构造变形及沉积作用

15°S—35°S 之间中安第斯的弧后地形、构造变形及沉积作用主要受到纳斯卡（Nazca）板块与南美板块的会聚以及俯冲洋壳几何形态的影响（Barazangi 和 Isacks，1976；Bevis 和 Isacks，1984）。北部大约 24°S 地区，南美板块下面的俯冲板块的倾角略陡（约 30°），而在南部 26°S，俯冲板块则变为近水平状（Cahill 和 Isacks，1992）。俯冲板块的几何形态与安第斯山的海拔高度之间存在着一定关系（Gephart，1994）。在 15°S 到 27°S 之间，发育内流水系的 Altiplano-Puna 高原横跨南玻利维亚以及西北阿根廷，其平均海拔为 3.7km（Allmendinger 等，1997）。这个区域发育一个挤压盆地和若干山脉，这些山脉有的孤立存在，有的与沉积盆地相邻。盆地内发育有几千米厚的蒸发岩以及碎屑岩沉积（Jordan 和 Alonso，1987；Alonso 等，1991；Vandervoort 等，1995）。这些盆地以高角度逆断层或较大的火山岩体为边界，而这些断层和岩体要么构成高原的西边界，要么横向上将盆地划分为几部分（Alonso 等，1984；Riller 等，2001）。在东科迪勒拉以东的 Altiplana-Puna 高原和玻利维亚地区，造山带的东部边界为一条楔形的 250km 宽的褶皱冲断带。在阿根廷西北部的南端，

这个褶皱冲断带消失，取而代之的是 Santa Barbara 和 Sierras Pampeanas 山脉范围内出现的区域性厚皮变形（Allmendinger 等，1983；Mon 和 Salfity，1995）。玻利维亚境内褶皱冲断带的空间展布与古生代厚层地层单元有关，这些地层中发育一系列志留系、泥盆系以及白垩系的滑脱面，并控制了造山楔的基底滑脱作用（Sempere 等，1990；McQuarrie，2002；Elger 等，2005）。尽管如此，在 24°S 附近，这些软弱层会逐渐减薄并最终消失，从而结束这种薄皮型的构造变形（Allmendinger 等，1983；Mon 和 Salfity，1995；Cristallini 等，1997；Kley 和 Monaldi，2002）。取而代之的是与白垩纪 Salta 裂谷有关的反向正断层和调节构造，这些构造在新生代安第斯造山运动中起到了调节缩短量的作用（Grier 等，1991；Kley 和 Monaldi，2002；Carrera 等，2006；Hongn 等，2007；Hain 等，2011）。另外，在早古生代（奥陶纪）Ocloyic 造山运动中形成的变质岩体与破碎前陆的西缘拼接在一起，在向 Puna 高原过渡过程中发生隆起，推测其可以影响安第斯山缩短过程中收缩再活动的空间展布特征（Mon 和 Hongn，1991）。这些复活、继承、伸展以及挤压的各向异性形成了高原东部地区不连续的山脉。这些山脉沿着高原的东翼分布，没有进入变形盆地中，从而在高原东部形成多个不连续的山脉。根据早期地壳较弱，构造再活动的力学机制可以推断出相同的结果（Mon 和 Salfity，1995；Hilley 等，2005）。这与哥伦比亚东科迪勒拉的正反转构造几乎一致（Mora 等，2006；Parra 等，2009），而与北美 Laramide 构造省的再活动构造模式不同（Marshak 等，2000）。因此，阿根廷西北部前安第斯的古地理特征为新生代的缩短运动提供了调节带，形成了一个宽阔的变形区。这个区域没有明确边界，不发育活动造山带前缘的变形特征且侧向生长也没有系统性（Allmendinger 等，1983；Grier 等，1991；Mon 和 Hongn，1991；Reynolds 等，2000；Carrera 等，2006；Ramos 等，2006；Hongn 等，2007；Hilley 和 Coutand，2010）。因此，阿根廷境内的破碎前陆构成了一个在空间与时间上具有不同隆升范围并影响盆地形成的地貌单元。在该地貌单元内，位于造山带内部和 Altiplano-Puna 高原周缘的盆地具有相似的构造环境，但是在水文上相互独立（Alonso 等，1984，1991；Jordan 和 Alonso，1987；Kraemer 等，1999；Carrapa 等，2005）。

在中安第斯地区，不同变形样式对河流及沉积系统起到了决定性的作用。在玻利维亚褶皱冲断带内部及东部，沉积物在楔形体顶部发育，然后被搬运到东部未变形的 Chaco 前渊中（Horton 和 DeCelles，1997）。这些前渊的沉积物向东逐渐变薄，据此可以推断前缘隆起的东部地层可能再次变厚（Horton 和 DeCelles，1997），这可以从褶皱冲断带向前推进，从而引起地形荷载挠曲的现象中推断出来（DeCelles 和 Giles，1996）。与之相反的是，阿根廷西北部的沉积体系通常由连接方式不同、局部被分隔的多个沉积中心组成，其边界主要由地壳范围内各向异性而形成的基底隆起构成。与玻利维亚前陆盆地体系以及阿根廷最北部的次安第斯构造省所不同的是，这个区域缺乏较深的侧向前渊，仅发育平缓的前缘隆起和隆外凹陷带。例如在 Santa Barbara 系统中，前陆沉降沉积厚度将近 3000m（Cristalini 等，1997；Bossi 等，2001）。在向次安第斯区域的过渡过程中，沉积厚度逐渐增加（Reynolds 等，2001），在阿根廷以及南玻利维亚地区，次安第斯褶皱冲断带的厚度达到 7000m（Echavarria 等，2003；Uba 等，2006）。这些破碎前陆内部的水文连通性随地质时间的变化而变化。通常干旱—半湿润的盆地都经历了多次内部水系与外部水系的转换（Strecher 等，1989，2007，2009；Mortimer 等，2007），这表明构造运动、水系调节以及气候之间存在着复杂的关系（Starck 和 Anzotegui，2001；Hilley 和 Strecher，2005）。

25.3 玻利维亚与阿根廷西北部东安第斯边缘的地形与地貌特征

玻利维亚与阿根廷西北部的多种变形样式都可以从地形上清晰地反映出来。玻利维亚褶皱冲断带中的变形以发育侧向连续的逆冲断层为特征，其断层面为力学性质较弱的滑脱面。下伏地层在侧向上分布均一，使得部分逆冲断层沿走向连续分布（Allmendinger 等，1983；Sempere 等，1990；Hilley 和 Coutand，2010），这样就形成了与区域缩短方向垂直的连续型山脊（图 25-2 和图 25-3a）。这种地形模式可以通过比较相邻点之间的高程得出。在玻利维亚地区，垂直于缩短方向或沿造山带走向的高程在空间上的连续性较好（大于 50km 尺度的长度上；图 25-4a）。而平行于缩短方向或平行于地形梯度方向上的高程则十分不连续。因此，在南玻利维亚地区，等厚线可以沿造山带走向追溯出很远的距离，其间唯一阻隔地形的就是先成的河谷。另外，褶皱冲断带的变形机制有利于造山楔向褶皱带后部变厚，从而形成向前变薄的几何形态（Dahlen，1984）。因此，这些逆冲断层和褶皱成因的山脊在缩短方向上呈现极大的地形变化，沿褶皱冲断带延伸 300km 左右的平均坡度与高度一致（图 25-4c）。

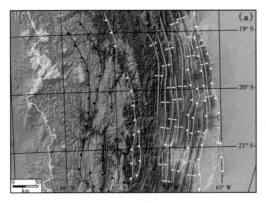 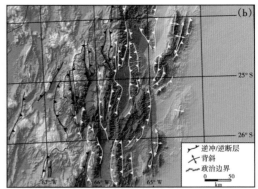

图 25-3　盆地实例

(a) 玻利维亚 Submulunn 褶皱—冲断带的详细地形和构造以及相应的 Chaco 前陆盆地系统。
(b) Puna 高原边缘和 Santa Barbara/Sierras Pampeanas 地貌构造省

相反的是，在阿根廷西北部境内，分散的基底调节了缩短作用，使其在较大尺度内高程十分不连续。在这些地区，垂直于挤压方向，在约 20km 范围内高程与距离有关（图 25-4b）。以逆断层为边界的断块与褶皱冲断带中的构造相比，其连续性更差，这就形成了许多作为排水通道的构造谷地（Trauth 等，2000；Strecker 和 Marrett，1999；Sobel 等，2003）。另外在非厄尔尼诺时期，潮湿的东风可以沿着较大河谷进入干旱的造山带内部（Strecker 等，2007；Bookhagen 和 Strecher，2008，2010）。垂直于缩短方向的平均高程剖面图表现出了更大的高程变化，并且以向前陆延伸的短波长地形为特征，体现了复活构造带的不连续隆升（图 25-4d）。

盆地内变形区地貌的发育特点体现出缩短作用在这两种不同类型前陆盆地中具有不同的调节过程。在玻利维亚地区有大型的、区域性水系穿过褶皱冲断带，且在山脊之间存在网格状的水系（图 25-5a）。在造山带前缘，巨型沉积扇可以容纳来自造山带的沉积物，并通过辫状河平原搬运到东部（Horton 和 DeCelles，2001；Wilkinson 等，2010）。巨型沉积扇发育

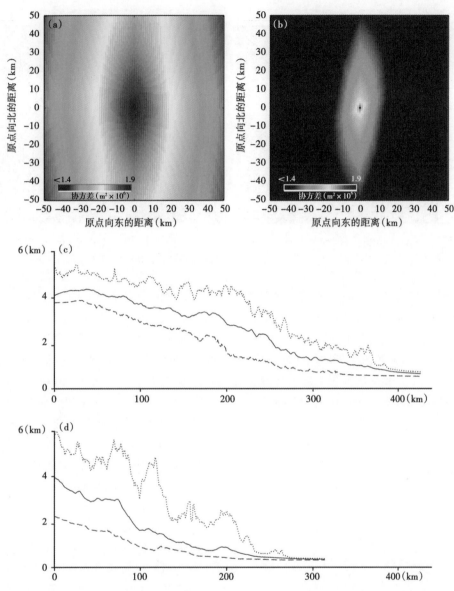

图 25-4 Subandean 褶皱—冲断带与 Altiplano-Puna 边缘的地形方差图和地形高程剖面

阿根廷西北部的（a）Subandean 褶皱—冲断带以及（b）东 Altiplano-Puna 边缘的地形方差图。图中显示出高程与空间位置存在相关性，即高程与南北向的距离存在函数关系。该函数关系可以通过比较数字高程模型（DEM）中各个像素点的高程与对应像素点所在位置上的方向和强度（沿纬度的距离）计算得出。距离较远的两点如果高程相似，则协方差值大。此外，该图显示了对于一个特定点高程如何随方向的变化而变化。例如，在更大距离上 N—S 向高程之间的协方差相比于 E—W 向的协方差，可看作是较大协方差值在 N—S 向的伸长。（a）和（b）中计算方差的地形范围与图 25-3a 和图 25-3b 的范围一致。在 Subandes 地区，N—S 方向上延伸的距离（>50km）与高程相关，而在破碎型前陆中相关地形延伸仅约 20km。这说明了与 Subandean 褶皱—冲断带相比，Puna 高地边缘的地形具有不连续的特征。（c）和（d）分别为横切 Subandean 褶皱—冲断带和 Altiplano-Puna 边缘的纵向地形高程剖面。实线表示平均地形，而虚线为剖面顶部和底部的边界，显示了各个剖面中的最大和最小地形

有从 8Ma 前开始形成的较老沉积岩，这些沉积岩现今成了变形造山楔的组成部分（Uba 等，2006，2007）。在 Santa Barbara 和 Pampeanas 山脉地区，发育有多期短暂的盆地沉积。这些

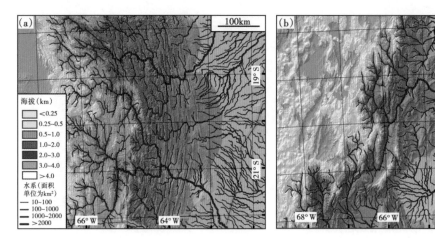

图 25-5 （a）Bolivia 南部 Subandean 褶皱冲断带及毗邻区的水系模式；（b）Argentina 北部破碎前陆盆地及毗邻地区的水系模式。没有展示 Altiplano-Puna 高原内流水系的水系模式

沉积物是从靠近破碎型前陆盆地的水系上游方向的山间盆地中搬运出来的（Hilley 和 Strecher，2005；Mortimer 等，2007；Strecker 等，2007）。这些沉积物通常上超于下游的基底之上，有时也会受到下游基底逆冲作用的影响，这两种现象在本地区都有发育（Bossi 等，2001）。由于湿润的东风受到顺风方向基底山脉的阻隔而不能到达盆地上游区，由此形成了短暂的盆地充填，其中的阻隔基底就形成了有效的地形屏障（Sobel 和 Strecher，2003；Hilley 和 Strecker，2005；Coutand 等，2006；Strecker 等，2007；Bookhagen 和 Strecker，2008）。在以上模型分析基础之上（Sobel 等，2003；Hilley 和 trecker，2005），内部水系和沉积的堆积形成可能是一个幕式的过程，其中下游地形屏障的隆升使穿过这些山区的水道变陡，同时，这些隆升地区的加积作用也与岩石隆升而导致的河床抬升保持同步。岩石抬升速率增快、基岩剥蚀以及沉积作用减慢都有利于内部水系的发育。另外，穿过基岩区的水道如果在其抬升之前就很陡，那么相比于一开始就位于较缓的水道坡环境中，前者更有利于外部水系的发育（图 25-6；Sobel 等，2003）。与之相反的是，如果下游基底区岩石隆升速率缓慢、岩石容易被剥蚀并且降雨明显，那么下切作用、溯源侵蚀以及盆地形成都将更加迅速。这些过程都有利于维持河流与前陆区域的连通（Hilley 和 Strecher，2005）。

玻利维亚褶皱冲断带的变形沿走向具有一致性，而南部基底地区的变形则

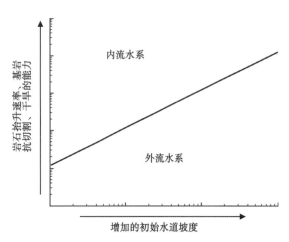

图 25-6 影响内流水系和外流水系因素的示意图
X 轴代表山下地形抬升之前的初始水道坡度，而 Y 轴描述了岩石抬升速率、基岩抗河流切割能力和/或干旱度的影响。实线显示出初始水道坡度与岩石抬升速率/基岩抗河流切割能力/气候共同作用，使得在内流水系边缘产生盆地。在实线之上的环境中，盆地会随着下游的抬升最终形成内流水系，而在实线之下的环境中，盆地会维持与前陆地区的连通。Sobel 等（2003）以及 Hilley 和 Strecker（2005）展示出如何利用每一种影响因素去预测形成内流水系的环境

完全不同，变形发生的位置取决于先期地壳的薄弱位置以及复活断层之上的地形荷载（Bossi 等，2001；Hilley 等，2005；Kley 等，2005；Mortimer 等，2007；Iaffa 等，2011）。另外，在已知缩短量有所增加的情况下，比起玻利维亚低角度逆冲，本区的高角度挤压构造更有利于形成大规模的隆升剥蚀。这可以促进隆起的山脉基底—核部之上的沉积物被快速搬运，也可以使不易被剥蚀的岩石暴露出来（Sobel 和 Strecker，2003）。最终，由于受到安第斯地区的区域性大气环流模式的影响，山脉基底—核部的山区在其迎风面能够聚集降水。例如，这些基底区东缘的降雨量为 1.5~3m/a，而山间盆地通常只有小于 0.3m/a（Bookhagen 和 Strecker，2008）。这导致位于隆起下风向的盆地上游形成了干旱环境。这可能会导致流经这些盆地的河流发生短暂的（Hilley 和 Strecker，2005；Strecker 等，2009）或永久的（Alonso 等，1991；Vandervoort 等，1995；Horton 等，2002）剥蚀能力下降，从而降低了河流与前陆区域的连通性。

  阿根廷西北部境内的不连续变形特征，导致了隆升山脉上游的沉积存在空间差异和穿时性（Dossi 和 Palma，1982；Strecker 等，1989，2009；Hilley 和 Strecker，2005；Carrapa 等，2009）。在 Altiplano-Puna 高原边缘的盆地内，来自 Altiplano-Puna 边缘和山脉的砾岩进入湖相沉积下来，同时有可以用来约束沉积年龄的火山碎屑进入。盆地充填的差异性反映了地理条件，如周围山脉的岩石隆升、侵蚀、周围单独基底山脉的微气候和/或不同的盆地几何形态等因素可能比区域因素起着更重要的作用，其中区域因素是指与时间和空间有关的中尺度的气候变化（Strecker 等，2007）。

  河流为了绕开生长的反转构造，就会沿走向流动更短距离，所以初看起来，可能会认为在破碎前陆地区以断裂为边界的隆起地带的走向长度较短，会促进外流水系的形成。但是相反地，在前陆褶皱—冲断带中，平行于造山带走向的广阔山脊将有利于内流水系的发育。此外，在这两个地貌构造省中，降雨量和排出量，甚至最大蒸发位置都会存在巨大的差异，这就在根本上影响了剥蚀过程。阿根廷西北部破碎前陆地区蒸发量很高，沿着褶皱—冲断带的造山楔在 1500~2000m 海拔范围内蒸发量更加集中（Bookbagen 和 Strecker，2008）。这将影响侵蚀效率———一个区分两个地区特定径流能力的评价指标（Bookbagen 和 Strecker，2010）。北部地区的径流能力一般相对较高，这些地区发育有高侵蚀效率、高降水量、高海拔和陡峭的河道（Safran 等，2005），而破碎前陆地区的径流能力较低。这一趋势将会在非厄尔尼诺时期更加明显，该时期降雨量将会增加三倍（Bookbagen 和 Strecker，2010）。由于厄尔尼诺系统从上新世开始就已存在（Ware 等，2005），所以地貌演化和沉积过程中伴随的蒸发和剥蚀作用的影响将会延长。因此，在气候和侵蚀能力不同，以及褶皱—冲断带与破碎前陆地区变形特征不同的共同影响下，使得南部地区的地形、地貌发育远比北部地区的连续性差。

25.4 中安第斯地区对地形负载的不同挠曲响应

  前陆盆地系统的发育主要由地壳尺度的挠曲作用导致了可容空间的增加而形成（如 Jordan，1981）。地壳尺度的挠曲主要由造山带内部长波长的地形负载形成，并通过弹性应力传递到紧邻的前陆地区。在前陆盆地中，宽阔褶皱—冲断带中的地形负载可以使离造山带前缘很远的地壳发生变形，产生前渊带、前隆带、隆外凹陷带内的可容空间。然而，在本文讨论的破碎前陆盆地中，以薄弱、复活构造为边界的地区，岩层抬升可以沿走向形成陡峭的山脉。因为由挠曲负载形成的可容空间的展布范围和规模与负载的波长有关（Turcotte 和 Schubert，2002），故推断短波长要比长波长形成的可容空间小。

图 25-7 是一个挠曲的可容空间模型，由穿过玻利维亚褶皱—冲断带（图 25-7a）及中安第斯破碎前陆地区（图 25-7b）的典型剖面的现代平均地形制作而成。对于刚性岩石圈，利用虚线标注基准线之上地形形成的挠曲（挠曲刚度，$D=7\times10^{23}\text{Nm}$；Horton 和 DeCelles，1997）；对于软岩石圈用点线标注（$D=2\times10^{22}\text{Nm}$）。该挠曲模型是一维模型，假设平均地形

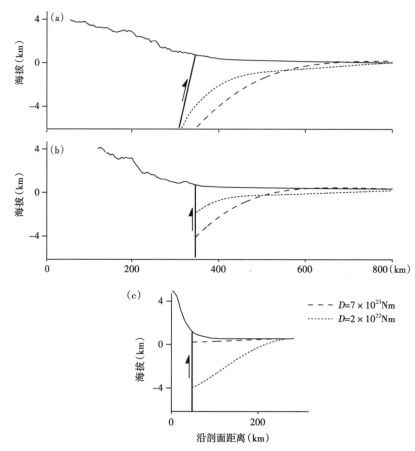

图 25-7 基于在（a）Subandean 褶皱冲断带以及毗邻前陆盆地体系、（b）Puna 边缘以及伴随的破碎前陆盆地体系以及（c）大约 27°S 纬度的 Sierra Aconquija 地区所观测到的平均地形的挠曲调节

在图（a）、(b) 和 (c) 中 x 轴为剖面上的水平距离，y 轴对应图 25-2 所示断面的海拔。利用这两个断面上（显示在图 25-2 中）具有隐式有限差分格式的卫片地形数据，在给定地幔、地壳密度（分别为 3300kg/m³ 和 2700kg/m³）和挠曲强度（$Eh^3/12(1-v^2)$）（其中 E、h 和 v 分别为地壳的杨氏模量、有效弹性厚度和泊松比）的情况下，计算了地壳的挠曲情况（Turcotte 和 Schubert，2002）。假定 $E=50\text{GPa}$，$h=49\text{km}$ 和 15km，$n=0.25$，同时在模型右端有固定的位移和旋转量。该区所示因素中有一个使得左边界伸展，并且在该剖面左边界产生的相应地形荷载。这保证了 Altiplano-Puna 高原荷载的所有均衡未补偿部分都被传送到前陆中。利用基准面之上地形荷载（由平均剖面东端的海拔决定）的隐式有限差分格式可以解决薄板挠曲方程（没有端部荷载）。源自不同假定的有效弹性厚度的挠曲刚度为 $D=7\times10^{23}\text{Nm}$ 和 $D=2\times10^{22}\text{Nm}$。在图（a）—(c) 中用实线表示平均地形，虚线和点线分别表示在硬地壳和软地壳情况下，在沉降盆地中计算出的沉积基底。在 Subandes 地区，由于褶皱冲断带几何形态宽阔，挠曲调节作用能够达到远处的前陆区域，并且比 Santa Barbura 和 Sierras Pampeanas 系统达到更深部的位置

的平面外范围比横切宽度大得多。这个假设对于侧向连续的褶皱—冲断带是合理的，而对破碎前陆地区来说，由于沿着走向的不连续性，这种假设就会导致过量地估计挠曲负载形成的可容纳空间。尽管如此，本文中安第斯地区的两个端元模型清晰地说明了在两种构造背景下形成可容纳空间的主要差异。

在玻利维亚安第斯亚带，宽阔平缓的锥形褶皱—冲断带形成的地形载荷向远处传递进入南美克拉通内部。紧邻褶皱—冲断带的挠曲沉降为前渊沉积带形成了有效的可容空间，同时，在距离造山楔前缘约450km处形成了一个的前隆隆起。在这个区域基岩被剥蚀或者沉积作用受到限制。就像软流圈中地幔向下流动等深部动力学过程可能会增大该区域的沉降一样，隆外凹陷带中的挠曲沉降可能会形成一些额外的沉积可容空间（Mitrovica等，1989；DeCelles和Giles，1996）。随着岩石圈刚度减小，前渊地区的空间范围受到更多限制，这是因为地形前缘远端的盆地基底的挠曲反映的是均衡调整，而不是平缓斜坡地形的挠曲补偿。

在阿根廷西北部的前陆盆地中，由于高原东部边缘的地形波长较短，使得刚性体的挠曲响应受到了抑制（图25-7b）。玻利维亚的锥形负载比陡峭的阿根廷边缘的突变性负载能够通过波长传递更多的应力，这将导致阿根廷地区的前渊的形成受到了更多的限制。此外，挠曲沉降可能发生在单个基底山脉的附近，如Sierra Aconquija附近（Bossi等，2001；Cristallini等，2004；图25-7c）或沿着Cumbres Calchaquies以北的区域（de Urreiztieta，1996）。然而，相对前渊地区来说，周围地区的沉积中心的展布范围都受到一定的制约（图25-7b）。在刚性地壳的情况下，这些地形的载荷将产生少量的沉积可容空间。当地壳刚度比较低时，突变性地形可形成约4km的沉积可容空间。Sierra Aconquija地区的地震反射数据显示紧邻山脉的新生代沉积充填厚度可达约3.5km（Cristallini等，2004），这意味着这些区域的地壳刚度较低。实际上，这一地区的弹性厚度较低也证实了这一推断（Tassra等，2007）。对于这种薄弱地壳，可容空间展布形态与构造变形特征类似。在阿根廷西北部的安第斯地区，在空间上挠曲可容空间可能受限制，表现出不连续的特点，盆地的位置可能最终受控于复活地壳薄弱带的控制（Hilley等，2005；Kley等，2005；Hongn等，2007；Hina等，2011）。此外，由于地壳非均质性和压性断层复活之间存在联系，使得破碎前陆的演化具有无规则和穿时的特点，并可能同时受控于相邻山脉隆升施加的区域岩石圈应力。挠曲作用形成的单个沉积中心可能与隆升基底山脉后缘沉积物注地有关。前隆带和隆后凹陷带的发育受到制约，主要是因为地壳的挠曲作用由来自长波形的均衡补偿而形成，而不是由在前隆带导致抬升而形成。

在中安第斯地区，可容纳空间的形成过程中，地形的起伏和岩石圈的刚性都起到了重要的作用。一般而言，在玻利维亚的安第斯地区，亚安第斯褶皱—冲断带以连续的平缓锥形地形为住。此外，该地区的构造变形沿着盆地内近水平拆离层发生调节变形，表明地壳薄弱现象没有阿根廷南部那么明显。相反地，阿根廷境内陡峭的安第斯前缘与基底抬升相伴生，表明前期构造的复活使得地壳减薄。在安第斯地区，差异的地形负载和可能的低挠曲刚度共同作用于破碎的前陆，从而形成了区域性的可容纳空间；而北部的玻利维亚则表现为连续型的前陆盆地系统。

25.5 结论

前陆盆地系统中的可容空间主要形成于地壳对褶皱—冲断带地形负载的挠曲响应，并最终发展为沉积中心。在破碎前陆背景下，先前的地壳薄弱带的构造复活，使得构造变形延伸进入了前陆地区,在造山带和前陆地区形成了陡峭且波长较短的地形。相对于褶皱—冲断带内的构造来说,构造变形的不连续性以及潜在的快速岩层抬升速率,有利于在较老的复活地质构造之上形成沉积充填。实际上,位于流经阿尔蒂普拉诺—普纳高原边缘的河流系统的源头上的许多沉积盆地,清楚地表明发生了多期的沉积充填,这是由于下游山脉沿着复活伸展构造活动,使岩层抬升并发生调整。然而,由于这些盆地临近的山脉具有丰富的降水,发育较强的溯源侵蚀作用,导致这些盆地常常被改造,并通过河流与前陆其他地区相连通。

沿着阿根廷西北部中安第斯发育了破碎前陆前陆系统,这一前陆系统内早期构造的抬升作用限制了短波长地形沿走向的展布,形成了空间范围比北部玻利维亚褶皱—冲断带更受限的可容空间。取而代之,挠曲盆地直接紧邻单个山脉分布,并且相对于未扰动前陆盆地的前渊沉积带,其空间范围更受限制。此外,对短波长地形负载的有限挠曲响应可能不会形成一个界限清楚的前隆带和隆后凹陷带。因此,将已有前陆盆地模型应用到破碎型前陆盆地系统中应受到一定的制约。研究这些系统时应考虑它们相对于典型前陆盆地环境所具有的不同构造变形样式以及地形特征。

致谢(略)

参 考 文 献

Allmendinger, R. W., Jordan, T. E., Kay, S. M. and Isacks, B. L. (1997) The evolution of the Altiplano-Puna Plateau of the central Andes. Ann. Rev. Earth and Planet. Sci., 25, 139-174.

Allmendinger, R. W., Ramos, V. A., Jordan, T. E., Palma, M. and Isacks, B. L. (1983) Paleogeogrphy and Andean structural geometry, northwest Argentina. Tectonics, 2, 1-16.

Alonso, R. N., Viramonte J, Gutíerrez R. (1984) Puna Austral-Bases para el subprovincialismo geológico de la Puna Argentina. IX Congr. Geol. Arg. S. C. Bariloche. Actas 1, 43-63,

Argentina. Alonso, R. N., Jordan, T. E., Tabbutt, K. T. and Vandervoort, D. S. (1991) Giant evaporite belts of the Neogene central Andes. Geology, 19, 401-404.

Baby, P., Rochat, P., Mascle, G. and Hérail, G. (1997) Neogene shortening contribution to crustal thickening in the back-arc of the Central Andes. Geology, 25, 883-886.

Barazangi, M. and Isacks, B. L. (1976) Spatial distribution of earthquakes and subduction of the Nazca Plate beneath South America. Geology, 4, 686-692.

Bevis, M. and Isacks, B. L. (1984) Hypocentral trend surface analysis: probing the geometry of Benioff Zones. J. Geophys. Res., 89, 6153-6170.

Bookhagen, B. and Strecker, M. R. (2008) High-resolution TRMM rainfall, hillslope angles, and relief along the eastern Andes. Geophys. Res. Lett., 35, L06403, doi: 10.1029/2007GL032011.

Bookhagen, B. and Strecker, M. R. (2010) Modern Andean rainfall variation during ENSO cycles and its impact on the Amazon drainage basin, in Hoorn, C. and Wesselingh, F. P. Amazonia, landscape and species evolution: a look into the past. Oxford, Blackwell Publishing, 223-241.

Bossi, G. E. and Palma, R. M. (1982) Reconsideraci6n de la estratigrafía del Valle de Santa María, Provincia de Catarmarca, Argentina. V. Congr. Lationoamericano Geol., Buenos Aires, Argentina, 155-172.

Bossi, G. E., Georgieff, S. M., Gavriloff, I. J. C., Ibañez, L. M., and Muruaga, C. M. (2001) Cenozoic evolution of the intramontane Santa María basin, Pampean Ranges, northwestern Argentina. J. South Am. Earth Sci., 14, 725-734.

Cahill, T. and Isacks, B. L. (1992) Seismicity and shape of the subducted Nazca plate. J. Geophys. Res., 97, 17503-17529.

Carrapa, B., Adelman, D., Hilley, G. E., Mortimer, E., Sobel, E. R., and Strecker, M. R. (2005) Oligocene uplift and development of plateau morphology in the southern central Andes. Tectonics, 24, 1-19, doi: 10. 1029/2004TC001762.

Carrapa, B. Hauer, J., Schoenbohm, L., Strecker, M. R., Schmitt, A., Villanueva, A., and Sosa Gomez, J. (2009) Dynamics of deformation and sedimentation in the northern Sierras Pampeanas: an integrated study of the Neogene Fiambalá basin, NWArgentina. Bull. Geol. Soc. Am, 120, 1518-1543.

Carrera, N., Muñoz, J. A., Sábat, F., Mon, R., and Roca, E. (2006) The role of inversion tectonics in the structure of the Cordillera Oriental (NW Argentinean Andes) J. Structural Geology, 28, 1921-1932.

Coutand, I., Carrapa, B., Deeken, A., Schmitt, A. K., Sobel. E. R., Strecker, M. R. (2006) Orogenic plateau formation and lateral growth of compressional basins and ranges: insights from sandstone petrography and detrital apatite fission – track thermochronology in the Angastaco Basin, NW-Argentina. Basin Research, 18, 1-26.

Crampton, S. L. and Allen, P. A. (1995) Recognition of forebulge unconformities associated with early stage foreland basin development: example from the North Alpine foreland basin. Am. Assoc. Petrol. Geol. Bull., 79, 1495-1514.

Cristallini, E., Cominguez, A., and Ramos, V. A. (1997) Deep structure of the Metan-Guachipas region: tectonic inversion in Northwestern Argentina. J. South Am. Earth Sci., 10, 403-421.

Cristallini, E., Comínguez, A., Ramos, V. A., and Mercerat, E. D. (2004) Basement double-wedge thrusting in the northern Sierras Pampeanas of Argentina (27S) -constraints from deep seismic reflection, in K. R. McClay (editor) Thrust tectonics and hydrocarbon systems. Am. Assoc. Petrol. Geol. Memoir, 82, 65-90.

Dahlen, F. A. (1984) Non-cohesive Critical Coulomb Wedges: an exact solution. J. Geophys. Res., 89, 10125-10133.

DeCelles, P. G. and Giles, K. A. (1996) Foreland basin systems. Basin Research, 8, 105-123.

DeCelles, P. G. and Hertel, F. (1989) Petrology of fluvial sands from the Amazonian foreland basin, Peru and Bolivia. Bull. Geol. Soc. Am., 101, 1552-1562.

DeCelles, P. G. and Horton, B. K. (2003) Early to middle Tertiary foreland basin development and the history of Andean crustal shortening in Bolivia. Bull. Geol. Soc. Am., 115, 58-77.

de Urreiztieta, M., Gapais, D., LeCorre, C., Cobbold, P. R., and Rosello, E. (1996) Cenozoic dextral transpression and basin development at the southern edge of the Puna Plateau, northwestern Argentina. Tectonophysics, 254, 17-39.

Dickinson, W. R. (1974) Plate tectonics and sedimentation. Spec. Publ. SEPM, 22, 1-27.

Dickinson, W. R. and Snyder, W. S. (1978) Plate tectonics of the Laramide orogeny, in I.

Matthews. V., ed., Laramide folding associated with basement block faulting in the western United States, Geol. Soc. America Memoir, 151.

Denver, CO, Geol. Soc. America, 355–366.

Dickinson, W. R. and Suczek, C. A. (1979) Plate tectonics and sandstone compositions. Am. Assoc. Petrol. Geol. Bull., 63, 2164–2182.

Echavarria, L., Hernandez, R., Allmendinger, R., and Reynolds, J. (2003) Subandean thrust and fold belt of northwestern Argentina: geometry and timing of the Andean evolution. Am. Assoc. Petrol. Geol. Bull., 87, 965–985.

Elger, K., Oncken, O. and Glodny, J. (2005) Plateau–style accumulation of deformation: southern Altiplano. Tectonics, 24, TC4020. doi: 10. 1029/2004TC001675.

Gephart, J. E. (1994) Topography and subduction geometry in the Central Andes: Clues to the mechanics of a noncollisional orogen. J. Geophys. Res., 99, 12279–12288.

González–Bonorino, F. (1950) Algunos problemas geológicos de las Sierras Pampeanas. Rev. Asoc. Geol. Argent., 5, 81–110.

Grier, M. E., Salfity, J. A. and Allmendinger, R. W. (1991) Andean reactivation of the Cretaceous Salta Rift, northwestern Argentina. J. South Am. Earth Sci., 4, 351–372.

Hain, M. P., Strecker, M. R., Bookhagen, B., Alonso, R. N., Pingel, H., and Schmitt, A. K. (2011) Neogene to Quaternary broken foreland formation and sedimentation dynamics in the Andes of NW Argentina (25S). Tectonics, 30, TC2006. doi: 10. 1029/2010TC002703.

Hilley, G. E., Blisniuk, P. M. and Strecker, M. R. (2005) Mechanics and erosion of basement–cored uplift provinces. J. Geophys. Res., 110 B12409, doi: 10. 1029/ 2005JB003704.

Hilley, G. E. and Coutand, I. (2010) Links between topography, erosion, rheological heterogeneity, and deformation in contractional settings: insights from the Central Andes. Tectonophysics, 95, 78–92.

Hilley, G. E. and Strecker, M. R. (2005) Processes of oscillatory basin filling and excavation in a tectonically active orogen: Quebrada del Toro Basin, NW Argentina. Bull. Geol. Soc. Am., 117, 887–901.

Hongn, F., del Papa, C., Powell, J., Petrinovic, I., Mon, R., and Deraco, V. (2007) Middle Eocene deformation and sedimentation in the Puna–Eastern Cordillera transition (23–26 S): control by preexisting heterogeneities on the pattern of initial Andean shortening. Geology, 35, 271–274.

Horton, B. K. and DeCelles, P. G. (1997) The modern foreland basin system adjacent to the Central Andes. Geology, 25, 895–898.

Horton, B. K. and DeCelles, P. G. (2001) Modern and ancient fluvial megafans in the foreland basin system of the central Andes, southern Bolivia: implications for drainage network evolution in fold–thrust belts. Basin Research, 13, 43–63.

Horton, B. K., Hampton, B. A., LaReau, N. and Baldellón, E. (2002) Tertiary provenance history of the northern and central Altiplano (central Andes, Bolivia): a detrital record of plateau–margin tectonics. J. Sediment. Res., 72, 711–726.

Iaffa, D. N., Sàbat, F., Bello, D., Ferrer, O., Mon, R. and Gutierrez, A. A. (2011) Tectonic inversion in a segmented foreland basin from extensional to piggy back settings: the Tucumán basin inNWArgentina. J. South Am. Earth Sci., 31, 457-474.

Jacobi, R. D. (1981) Peripheral bulge: a causal mechanism for the Lower Orodvician unconformity along the western margin of the northern Appalachians. Earth and Planetary Science Letters, 56, 245-251.

Jordan, T. E. (1981) Thrust loads and foreland basin evolution, Cretaceous, western United States. Am. Assoc. Petrol. Geol. Bull., 65, 2506-2520.

Jordan, T. E. (1995) Retroarc foreland and related basins, in C. J. Busby and R. V. Ingersoll, eds., Tectonics of sedimentary basins. Oxford, Blackwell Science, 331-362.

Jordan, T. E. and Allmendinger, R. W. (1986) The Sierras Pampeanas of Argentina: a modern analogue of Rocky Mountain foreland deformation. Am. J. Sci., 286, 737-764.

Jordan T. E. and Alonso, R. N. (1987) Cenozoic stratigraphy and basin tectonics of the Andes mountains, 20-28S lat. Am. Assoc. Petrol. Geol. Bull., 71, 49-64.

Karner, G. D. and Watts, A. B. (1983) Gravity anomalies and flexure of the lithosphere at mountain ranges. J. Geophys. Res., 88, 10449-10477.

Kennan, L., Lamb, S. and Rundle, C. (1995) K-Ar dates from the Altiplano and Cordillera Oriental of Bolivia: implications for Cenozoic stratigraphy and tectonics. J. South Am. Earth Sci., 8, 163-186.

Kley, J. and Monaldi, C. R. (2002) Tectonic inversion of the Santa Barbara System of the central Andean foreland thrust belt, northwestern Argentina. Tectonics, 21, 1061-1072.

Kley, J., Rosello, E. A., Monaldi, C. R., Habighorst, B. (2005) Seismic and field evidence for selective inversion of Cretaceous normal faults, Salta rift, northwest Argentina. Tectonophysics, 399, 155-172.

Kraemer, B., Adelmann, D., Alten, M., Schnurr, W., Erpenstein, K., Kiefer, E., Van den Bogaard, P., and Goerler, K. (1999) Incorporation of the Paleogene foreland into the Neogene Puna Plateau; the Salar de Antofalla area, NW Argentina. J. South Am. Earth Sci., 12, 157-182.

Marshak, S., Karlstrom, K., and Timmons, J. M. (2000) Inversion of Proterozoic extensional faults: an explanation for the pattern of Laramide and ancestral Rockies intracratonic deformation, United States. Geology, 28, 735-738.

McQuarrie, N. (2002) The kinematic history of the central Andean fold-thrust belt, Bolivia: Implications for building a high plateau. Bull. Geol. Soc. Am., 114, 950-963.

M_etivier, F., Y. Gaudemer, P. Tapponnier, and B. Meyer (1998) Northeastward growth of the Tibetan Plateau deduced from balanced reconstructions of two depositional areas: the Qaidam and Hexi corridor basins, China. Tectonics, 17, 823-842.

Mitrovica, J. X., Beaumont, C. and Jarvis, G. T. (1989) Tilting of continental interiors by the dynamical effects of subduction. Tectonics, 8, 1079-1094.

Mon, R., and Hongn, F. (1991) The structure of the Precambrin and lower Paleozoic basement of the central Andes between 22° and 32°S lat. Geologische Rundschau, 80, 745-758.

Mon, R, and Salfity, J. A. (1995) Tectonic evolution of the Andes of northern Argentina, in

Tankard, A. J., Súarez Soruco, R., Welsink, H. J., (editors) Petroleum basins of South America. Am. Assoc. Petrol. Geol. Memoir, 62, 269-83.

Mora, A., Parra, M., Strecker, M. R., Kammer, A., Dimate, C. and Rodriguez, F. (2006) Cenozoic contractional reactivation of Mesozoic extensional structures in the Eastern Cordillera of Colombia. Tectonics, 25, TC2010, doi: 10. 1029/2005TC001854.

Mortimer, E., Carrapa, B., Coutand, I., Schoenbohm, L., Sobel, E. R., Sosa Gomez, J., and Strecker, M. R. (2007) Compartmentalization of a foreland basin in response to plateau growth and diachronous thrusting: El Cajon- Campo Arenal basin, NW Argentina. Bull. Geol. Soc. Am., 119, 637-665.

Parra, M., Mora, A., Jaramillo, C., Strecker, M. R., Sobel, E., Quiroz, L., Rueda, M., and Torres, V. (2009) Orogenic advance in the northern Andes: evidence from the oligomiocene sedimentary record of the Medina Basin, Eastern Cordillera, Colombia. Bull. Geol. Soc. Am., 121, 780-800.

Quinlan, G. M. and Beaumont, C. (1984) Appalachian thrusting, lithospheric flexure, and the Paleozoic stratigraphy of the Eastern Interior of North America. Can. J. Earth Sci., 21, 973-996.

Ramos, V. A., Cristallini, E. O., and Perez, E. J. (2002) The Pampean flat slab of the Central Andes. J. South Am. Earth Sci., 15, 59-78.

Ramos, V. A., R. N. Alonso, and M. R. Strecker (2006) Estructura y Neotectónica de la Lomas de Olmedo, Zona de transiciónentre los sistemas Subandino y de Santa Bárbara Provincia de Salta. Asoc. Geol. Argent. Rev., 4, 579-588.

Reynolds, J. H., Galli, C. I., Hernández, R. M., Idleman, B. D., Kotila, J. M., Hilliard, R. V., and Naeser, C. W. (2000) Middle Miocene tectonic development of the Transition Zone, Salta Province, northwest Argentina: magnetic stratigraphy from the Metán Subgroup, Sierra de González. Bull. Geol. Soc. Am., 112, 1736-1751.

Reynolds, J. H., Hernandez, R. M., Galli, C. I., and Idleman, B. D. (2001) Magnetostratigraphy of the Quebrada La Porcelana section, Sierra de Ramos, Salta Province, Argentina: age limits for the Neogene Orán Group and uplift of the southern Sierras Subandinas. J. South Am. Earth Sci., 14, 681-692.

Riller, U., Petrinovic, I., Ramelow, J., Strecker, M., and Oncken, O. (2001) Late Cenozoic tectonism, caldera and plateau formation in the central Andes. Earth and Planetary Science Letters, 188, 299-311.

Schwab, F. L. (1986) Sedimentary "signatures" of foreland basin assemblages: real or counterfeit, in Allen, P. A., and Homewood, P., eds., Foreland Basins. Spec. Publ. Int. Assoc. Sed. 395-410.

Sempere, T., Hérail, G., Oller, J. and Bonhomme, M. G. (1990) Late Oligocene-Early Miocene major tectonic crisis and related basins in Bolivia. Geology, 18, 946-949.

Sobel, E. R., Hilley, G. E. and Strecker, M. R. (2003) Formation of internally drained contractional basins by ariditylimited bedrock incision. J. Geophys. Res., 108, (B7) 2344, doi: 10. 1029/2002JB001883.

Sobel, E. R. and Strecker, M. R. (2003) Uplift, exhumation and precipitation: tectonic and climatic control of Late Cenozoic landscape evolution in the northern Sierras Pampeanas, Argentina. Basin Research, 15, 431-451.

Starck, D., and Anzotegui, L. M. (2001) The late Miocene climatic change—Persistence of a climatic signal through the orogenic stratigraphic record in northwestern Argentina. J. South Am. Earth Sci., 14, 763-774.

Steidtmann, J. R., McGee, L. C., and Middleton, L. T. (1983) Laramide sedimentation, folding, and faulting in the southern Wind River Range, Wyoming, in Lowell, J. D., ed., Rocky Mountain foreland basins and uplifts. Rocky Mountain Association of Geologists, Denver, 161-167.

Stelzner, A. (1923) Contribución a la geología de la República Argentina, conlaparte límitrofe de los Andes Chilenosentrelos 32° y 33°S. Actas Acad. Nac. Cienc. Cordoba, Argentina 8, 1-28.

Strecker, M. R., Cerveny, P., Bloom, A. L., and Malizia, D. (1989) Late Cenozoic tectonism and landscape development in the foreland of the Andes: Northern Sierras Pampeanas, Argentina, Tectonics, 8, 517-534.

Strecker M. R. and Marrett, R. (1999) Kinematic evolution of fault ramps and its role in development of landslides and lakes in the northwestern Argentine Andes. Geology, 27, 307-310.

Strecker, M. R., Alonso, R. N., Bookhagen, B., Carrapa, B., Hilley, G. E., Sobel, E. R. and Trauth, M. H. (2007) Tectonics and climate of the Southern Central Andes. Ann. Rev. Earth and Planet. Sci., 35, 747-787.

Strecker, M. R., Alonso, R. Bookhagen, B., Carrapa, B. Coutand, I., Hain, M. P., Hilley, G. E., Mortimer, E., Schoenbohm, L., and Sobel. E. R. (2009) Does the topographhic distribution of the central Andean Puna Plateau result from climatic or geodynamic processes? Geology, 37, 643-646.

Tassara, A., Swain, C., Hackney, R., and Kirby, J. (2007) Elastic thickness structure of South America estimated using wavelets and satellite-derived gravity data. Earth and Planetary Science Letters, 253, 17-36.

Trauth, M. H., Alonso, R. N., Haselton, K., Hermanns, R. L. and Strecker, M. R. (2000) Climate change and mass movements in the northwest Argentine Andes: Earth and Planetary Science Letters, 179, 243-256.

Turcotte, D. L. and Schubert, G. (2002) Geodynamics. CambridgeUniversity Press, Cambridge, England, 456 pp.

Uba, C. E., Hulka, C., and Heubeck, C. (2006) Evolution of the late Cenozoic Chaco foreland basin, Southern Bolivia. Basin Research, 18, 145-170.

Uba, C. E., Strecker, M. R. and Schmitt, A. K. (2007) Increased sediment accumulation rates and climatic forcing in the central Andes during the late Miocene. Geology, 35, 979-982.

Vandervoort, D. S., Jordan, T. E., Zeitler, P. K. and Alonso, R. (1995) Chronology of internal drainage development and uplift, southern Puna plateau, Argentine central Andes. Geology, 23, 145-148.

Wara, M. W., Ravelo, A. C., and Delaney, M. L. (2005) Permanent El Niño-like conditions during the Pliocene warm period. Science, 309, 758-761.

Wilkinson, M. J., Marshall, L. G., Lundberg, J. G., and Kreslavsky, M. H. (2010) Megafan environments in northern South America and their impact on Amazon Neogene aquatic ecosystems, in Hoorn, C. and Wesselingh, F. P., eds., Amazonia, landscape and species evolution: a look into the past. Oxford, Blackwell Publishing, 162-184.

(李政斯 译，吴哲 崔敏 校)

第 26 章　逆冲楔/前陆盆地系统

HUGH SINGLAIR

(School of GeoSciences, The University of Edinburgh, UK)

摘　要：逆冲楔/前陆盆地系统的各个部分能够很好地耦合在一起。山脉地形的挠曲负载作用形成了前陆盆地，而侵蚀作用和沉积作用又将物质重新分配。双向会聚形成的小型碰撞山脉可将山脉前侧（前侧前陆盆地）和山脉后侧（后侧前陆盆地）的前陆盆地分隔开来。这两类盆地具有不同的地层特征。在地下荷载缺失的情况下，前陆盆地的沉降主要由俯冲板块的俯冲速率和山脉地形向外增长的程度决定。前侧前陆盆地的发育受到这两种因素的影响，而后侧前陆盆地下部不发育俯冲板块，仅受到逆冲楔增长的影响。逆冲楔生长至稳定的地形状态后，可能出现萎缩。在稳定状态下，前侧前陆盆地由于持续俯冲而持续沉降，而后侧前陆盆地则停止沉降；如果山脉继续遭受挤压，还有可能引发盆地抬升。前陆盆地系统内构造作用的不对称性导致逆冲楔局部发生侵蚀作用，这就影响了前陆盆地中的沉积物通量和物源。逆冲楔内的沉积物聚集使楔体下伏部分保持稳定，从而改变了山脉其他位置的变形。山脉地形对气候和构造扰动作用的响应时间决定了前陆系统中构造作用和地表过程的耦合度：典型的响应时间在 0.5～2.0Ma 之间。

关键词：逆冲断层　楔形体　前陆　盆地　系统

26.1　引言

造山过程中地形变高作用于周围的大陆岩石圈上，使岩石圈向下挠曲（图 26-1）。当边缘坳陷被来自周围山脉的沉积物填充后，就形成了前陆盆地（Beaumout, 1981; Jordan, 1981）。如果位于山脉之下的岩石圈发生挠曲，则在区域挠曲均衡补偿作用下，就会发育高应变区域。此时，在增厚地壳下发育的逆冲断层或者剪切带会在深部发生顺层滑脱。在浅部的脆性地区，摩擦滑动阻力沿基底滑脱面将应力传递给上覆岩石，使得地层加厚并出现山脉地貌。基底滑脱和地形特征结合在一起就形成了变形岩石的"逆冲楔"。陆壳上的逆冲楔与海洋内的增生楔相似，地层的形成可以比作推土机将一堆沙向前推进的过程（Chapple, 1978; Davis 等, 1983）。因此，在最简单的情况下，山脉地形驱动着下伏岩石圈向下挠曲，但是山脉地形自身由俯冲板块顶部滑脱面表层的应力控制，因此，它们能够在力学上耦合在一起。

山脉地形的演化和前陆盆地的形成可以通过山区内的地表物质被侵蚀，并在前陆盆地内以再分配沉积的模式而联系起来（图 26-1）。因此，前陆盆地的沉积充填过程是了解地质历史时期内山脉生长的重要证据。来自山脉的侵蚀流出量与增生流入量相对应，后者在会聚过程中将来自前陆盆地的沉积地层回推至山脉中（图 26-1b）。山脉生长与前陆盆地沉降密切联系，特别是在两种物质流量方式上的完美耦合，这意味着应该把造山过程与盆地形成过程看作一个相互联系的系统。本章内容回顾并综述了过去十五年内，研究逆冲楔和前陆盆地系统内相互作用的大量文献，以求总结一种逆冲楔/前陆盆地系统的模式。

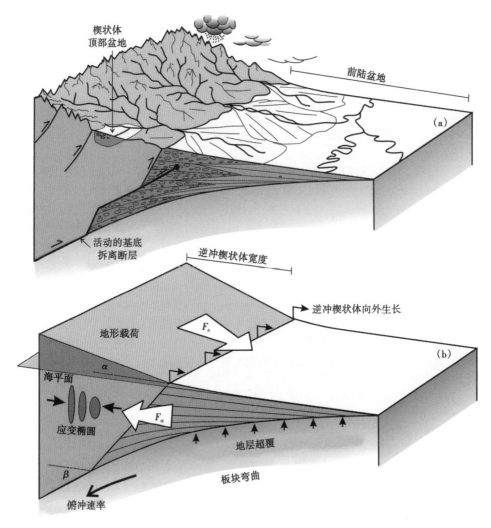

图 26-1 逆冲楔/前陆盆地系统

（a）逆冲楔/前陆盆地系统的地质和地形特征表明逆冲楔的断坡/断坪呈阶梯状，并在会聚过程中增生而形成的前陆盆地下部发生滑脱（Sinclair 和 Naylor，出版中）。在许多实例中，下伏地层层序也在增生（比如阿尔卑斯山脉的滑脱层主要发育于三叠纪继承性被动大陆边缘层序的蒸发岩内部）。图中的河流网络控制了地表的地质作用过程。此外，很多情况下冰川作用也可以控制地表的地质作用过程（例如阿拉斯加的 St Alias 山脉）。这样的前陆盆地沉积充填全部发生在陆相背景下；（b）逆冲楔/前陆盆地系统中重要的可定量化参数：F_α—逆冲楔的剥蚀量；F_e—逆冲楔的增生量；α—逆冲楔的平均地表坡度；β—逆冲楔基底拆离表层的平均坡度

本文首先研究了逆冲楔和前陆盆地相互间的力学耦合关系，之后研究了双向会聚逆冲板块边界环境下，前陆盆地或山脉一侧的沉降和地层演化特征，进而探索了地表过程与逆冲楔和前陆盆地之间的耦合关系，最后讨论了关键构造期的特征和两者的响应。

Beaumont（1981）和 Jordan（1981）最先对前陆盆地进行了定量模拟。Jordan（1981）将每一个新的增生楔单元当作是添加到山脉荷载上的矩形块体，以此，计算出前陆盆地的地层响应，从而模拟分析了爱达荷州和怀俄明州白垩纪塞维尔逆冲带中单一逆冲单元的均衡响

应。这种方法后来被应用于模拟阿巴拉契亚盆地古生代的演化（Quinlan 和 Beaumont, 1984）。Heller 等（1988）在研究变形的逆冲前缘时，考虑到了构造运动期和构造平静期对沉积分散作用的影响。随后，利用一种扩散模型模拟了逆冲楔的地形侵蚀过程和盆地内沉积物的沉积过程。该模型目的是为了研究前陆盆地的演化过程中构造运动与地表地质作用的相互耦合关系（Flemings 和 Jordan，1989；Sinclair 等，1991）。还有一些学者探索了大陆碰撞早期逆冲楔水下生长的意义，以及对前陆盆地的规模和充填程度的影响（Stockmal 等，1986）。所有这些前陆盆地系统的研究工作都是在单一逆冲楔的构造背景下，对单一的地层响应进行研究。Jordan 和 Beaumont（1995）第一次提出了两侧都有前陆盆地的完整山脉模型，两侧盆地分别位于俯冲板块上（前侧前陆盆地）和仰冲板块之上（后侧前陆盆地）。

除了逆冲楔的演化，下伏岩石圈的力学响应是控制前陆盆地的根本因素。通常力学响应由弹性挠曲强度来表示（Karner 和 Watts，1983）。尽管如此，目前仍然存在着一个争论：一种观点认为黏度—弹性和依赖于温度的流变学模型更符合岩石圈的长时间变形特征（＞1Ma），而与之相反的观点则认为岩石圈的变形特征符合可以在短期内恢复的简单弹性模型。想了解该争论的最新进展请参考 Watts（2007）的文章。

26.2 逆冲楔/前陆盆地系统的力学耦合

26.2.1 逆冲楔：首要原则

山脉内部产状平缓的基底滑脱层和与之倾向相反的平均坡度确定了锥形楔的几何形态（图26-1）。这个锥形楔在力学上可作为一个整体进行分析（Chapple，1978）。Davis（1983）和 Dahlen（1984，1990）用最简单的形式定义了锥形逆冲楔的概念，即具有库仑屈服应力的塑性变形岩石在未变形的下边界（也就是下伏岩石圈板块）上滑动就形成了锥形逆冲楔。对于逆冲楔随着时间不断自相似生长，上面提到的物理解释提供了一种理想的楔体形态模型。变形前缘增生逆冲单元的增加控制了逆冲楔的演化，使逆冲楔的整体锥角变小，并且通过反冲断层、杂乱逆冲断层和节理变形使逆冲楔内部地层增厚。由于受到逆冲楔表层的侵蚀作用和沉积搬运作用的影响，锥角可能会变小。因此，逆冲楔/前陆盆地系统可以被认为是一个变形岩石的楔体，该楔体的任何位置都处于失衡的边缘，为了维持一个稳定的楔体状态，楔体使前端增生作用和内部增厚导致的侵蚀作用平衡。逆冲楔和前陆盆地演化之间的基本联系主要有三个方面：（1）稳定的锥角使挠曲支撑作用形成的地表地形长期稳定，也就确定了已知岩石圈强度下所形成前陆盆地的沉降程度；（2）维持锥角的地表地质作用与地下地质作用的耦合特征，指示了前陆盆地内沉积物的来源和搬运过程；（3）逆冲增生过程中板块会聚速率控制了逆冲楔内库仑力的增长，同时也直接影响了前陆盆地的沉降史。

逆冲楔的锥角（图26-1b）可以分成平均地表坡度（α）和基底滑脱面倾角（β）两部分。这两个层面相组成的锥角数值与逆冲楔的内摩擦角（ϕ）（比如强度）和下盘滑脱层的内摩擦角（ϕ_b）有关。另外，在屈服强度内考虑到岩石孔隙流体压力的影响，锥角（$\alpha+\beta$）可以表示为：

$$\alpha + \beta = \frac{(1 - \rho_f/\rho_s)\beta + \tan\phi(1 - \lambda_b)}{(1 - \rho_f/\rho_s) + \dfrac{2(1 - \lambda)\sin\phi}{1 - \sin\phi}}$$

ρ_f 为水或空气的密度，ρ_s 为沉积岩密度，λ 和 λ_b 分别为楔体内部和基底滑脱面上流体压力与静岩压力比（Dahlen，1990）。根据该公式，结合不同孔隙流体压力 α 和 β 的相关图件，可以确定楔形体亚临界状态的各个参数。与内部稳定区相比，楔形体在基底滑脱面上滑动前会发生内部增厚现象。当增厚到能够在基底滑脱面上滑移时，楔形体就位于亚临界状态了（图26-2）。值得注意的是，层面坡度角（α）越小，楔体形成的挠曲变形就越多，因此形成了多个较小的前渊盆地（Sensu DeCelles 和 Giles，1996），但是这种变形特征会使得楔体顶部的沉积作用加强（见后面的讨论部分）。

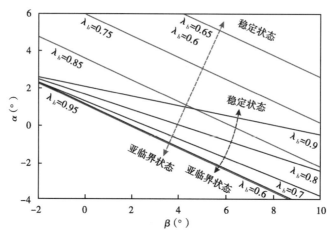

图26-2　根据正文中的公式（26-1），结合平均地表坡度（α）及基底滑脱面倾角（β），分析出的逆冲楔关键锥角的解析解（据Willett 和 Schlunegger，2011；Dahlen 等，1984，修改）
解法是将基底滑脱面上（λ_b）流体与静岩压力比和楔体内部摩擦角上（λ）流体与静岩压力比构成一个方程。红线表示 λ 为常数 0.6、λ_b 为变量时的解析解。蓝线表示 λ_b 为常量 0.95、λ 为变量时的解析解。左下边区域指示楔体位于亚临界状态，右上边区域是楔形体比较稳定的状态，前者楔形体太软弱而不能在基底上滑移，后者楔形体能干性太强而不发生内部变形。层面坡度较低的楔形体有利于楔形体顶部的沉积聚集，增加了基底滑脱面的岩层静应力，因此使楔形体的部分区域进入稳定区。相反，高侵蚀速率使得楔形体进入亚临界状态，有利于发生内部变形

库仑楔形体模型也适用于基底滑脱表现为剪切带和破碎带的脆性变形。在许多实例中，逆冲楔的基底滑脱发育在柔软的盐岩层之上，盐岩的软弱性、黏性和流变性使楔形体的锥角较小（Davis 和 Engelder，1985）。

26.2.2　双向会聚逆冲楔体

楔形体可以简单地描述为推土机在一个层面上推动一堆沙。这个模型中最重要的是将楔形体后部理解为一个刚性挡板。在这种刚性边界条件下，应力的释放伴随着双向会聚逆冲楔形体的形成，这两个逆冲带具有背冲的特点（图26-3），类似于一个小型山脉把水系从中间分开了一样（Malavieille，1984）。在这些模型中，构造边界是由下伏板块来确定，两个板块之间的连接点称为奇点（一个速度边界环境）。这些模型适用于非对称性板块俯冲背景下发育的楔形体。山脉深部构造的层析成像或大地电磁图像上显示的，和（或）两个会聚板块上地壳层序的缩短量是否明显不对称可以反映板块俯冲是否对称。典型实例包括中国台湾、欧洲的阿尔卑斯山和比利牛斯山。在不对称背景下，两个反向楔形体的生长差异导致了下伏板块之上发育"前侧楔形体"，而在上覆板块之上发育"后侧楔形体"（Willett 等，1993）。

前陆盆地形成的重要因素是：俯冲板块之上的盆地与仰冲板块之上的盆地具有不同的盆地演化史。下面将详细说明这些内容。

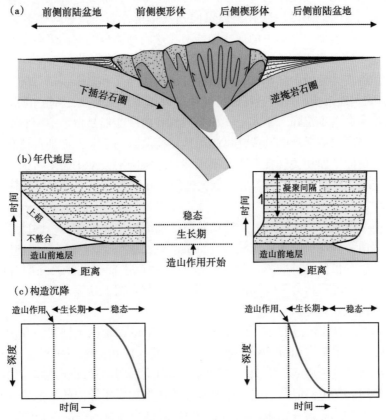

图26-3　前侧前陆盆地和后侧前陆盆地的特征对比
（据Naylor和Sinclair，2008，修改；讨论内容见正文）

26.2.3　逆冲楔的生长期、衰退期和稳定期

逆冲楔的生长是通过前缘增生和（或）基底增生（底侵作用）将下伏板块的岩石转移来实现。随着物质转移到双向会聚的锥形逆冲楔中，为了维持两个相反方向的楔形体锥角不变，逆冲楔整体就必须不断地自我生长。随着逆冲楔的生长，它们的海拔高度迅速增加。但是随着逆冲楔的增大，恒定增生流量所影响的范围也增大，因此，对地表抬升速率的影响会减弱。理论上，地表抬升速率减小的速度与时间的平方根成比例（Dahlen等，1984）。但是当增生逆冲单元厚度增加时，增生流量也会增加。例如：现今已增生到阿尔卑斯—喜马拉雅山链中的中生代特提斯洋边缘就是这种模式。这可能与被动大陆边缘层序增生过程相关。

在没有侵蚀作用的情况下，逆冲楔内岩石颗粒向上运动的速率等于地表的抬升速率，结合逆冲楔生长过程，可以确定随着耦合逆冲楔的生长，地表抬升速率逐渐减小。然而，侵蚀作用会降低地表的抬升速率。因此，为了维持楔形体形态的稳定（作为"稳定状态"参考），就需要地表抬升速率增加，即岩石的抬升速率增加。这一点非常重要，因为地形起伏增大和山地降水增强使得地表相对高差变大，促使侵蚀速率增加。在这种情况下，楔形体内部变形引起的岩石抬升速率将会与增加的侵蚀速率相互竞争，此消彼长。这种侵蚀作用与逆冲楔体内部变形作用的耦合关系，是研究前陆盆地内沉积流量和沉积物源的关键。

如果将该系统简化为增生流入量和侵蚀流出量两部分，那么逆冲楔的地形和总体积在开始时会增加得很明显，但是随后会在两流量间达到平衡，而当侵蚀流出量大于增生流入量时，楔形体就会发生萎缩。变形中的逆冲楔能够在这两种流量的共同影响下达到一个稳定状态，这种看法是一位学者在对台湾地区的研究过程中首先提出的（Suppe，1981），随后的学者研究发现，许多其他的山脉都具有类似的特征（Willett 和 Brandon，2002）。

逆冲楔/前陆盆地系统的最典型特征是逆冲楔内部发生强烈变形时，山脉地形可以不增高，同时前陆盆地会发生沉降。事实上，如果考虑到内部变形是侵蚀速率和范围的响应，那么可以推测逆冲楔内构造活动的阶段性可能与地形荷载减小有关，而地形荷载减小主要与高速侵蚀和沉积流量的增加相关。在沉积物搬运和盆地沉降的模型中需要考虑这种类型的耦合作用（Paola，1988）：这样就可以解决一个关键的未知参数——构造对侵蚀速率变化的响应时间。

随着会聚速率的减小，山脉进入后造山阶段，前缘增生速率降低。因此，持续的侵蚀通量增加将导致双向会聚逆冲楔的体积减小（Stolar 等，2006）。只要楔形体的应力状态失稳，该模型系统将发生自相似性的收缩。这一过程的结束意味着楔形体的外部有块体被分离出来。通常情况下这些沉积物会下沉到山脉内部。有研究表明上新世的阿尔卑斯山就是这种模式（Willett 等，2006）。

在活动山脉宽度减小的过程中，地形荷载减小，前陆盆地向上回弹，在冲积切割作用和侵蚀作用的影响下使得来自山脉的沉积物向四周扩散（Heller 等，1988）。瑞士境内的北阿尔卑斯前陆盆地就是一个典型的实例。尽管阿尔卑斯山顶端的逆冲变形和高侵蚀速率导致了区域均衡回弹作用的发生，但是前陆盆地的侵蚀改造增加了系统内的沉积量，并汇集于远端的沉积中心处，如 Rhone 扇（Cederbom 等，2004）。这个过程的关键点是前陆盆地的均衡回弹不能从构造上表明该山脉的消亡，这仅仅是一个阶段性的收缩和衰减。山脉内变形活动的终止意味着没有机制能够抵消侵蚀作用导致的地壳变薄。因此，在 10~50Ma 之间，整个逆冲楔/前陆盆地系统遭受速率衰减的侵蚀作用（Baldwin 等，2003）。

26.3　前侧前陆盆地与后侧前陆盆地

根据经典的盆地分类方案，可以将前陆盆地分为位于碰撞山脉两侧的"周缘"前陆盆地，以及由于洋壳俯冲作用形成的，位于岩浆岩岛弧之后的"弧后"前陆盆地。这种分类目前被广泛接受（Dickinson，1974）。然而，不同演化模式的"周缘"前陆盆地，其识别方式也不相同，这种分类没有考虑前陆盆地是位于楔形体前侧还是楔形体后侧。Johnson 和 Beaumont（1995）使用"后侧前陆盆地"和"前侧前陆盆地"来区分位于双向会聚逆冲楔模型两侧的两类周缘前陆盆地（图 26-3）。Johnson 和 Beaumont（1995）通过构造/地表地质作用相互耦合的数值模型，探讨了沿山脉出现的地形降水过程中，可能出现的非对称作用。前侧前陆盆地和后侧前陆盆地在地层建造、年代地层学和沉降史上明显不同（Nayor 和 Sinclair，2008）。

位于山脉前侧楔形体之下的岩石圈，在俯冲过程会引起前侧前陆盆地中的沉降加速，但山脉地形无变化。相反，后侧前陆盆地的沉降则仅仅受到楔形体后侧间歇性生长的控制。这种本质上的差异意味着前陆盆地的宽度除以俯冲速率所得数值控制了前侧前陆盆地内所保存地层的时限。前侧前陆盆地中填充的未变形地层通常只是山脉造山史的一部分，大部分前陆盆地的早期沉积保存在逆冲楔中。喜马拉雅山南侧的 Gangetic 前陆盆地就是一个很好的实例（Lyoncaen 和 Molnar，1985）。前陆盆地内保存的最老地层大约为 16Ma 前的地层，近似等于前陆盆地宽度（约 300km）除以缩短速率（约 20km/Ma）所得的数值。同样，北阿尔卑斯前陆盆地的早期演化必须通过研究变形的瑞士阿尔卑斯山来进行重建（Allen 等，1991）。相反，法国南部的阿启坦盆地，代表了比利牛斯山脉后侧前陆盆地，盆地内保存了从早马斯特里赫特期碰撞开始，到中新世造山运动停止这一完整造山运动的沉积记录。而 Ebro 盆地的前缘仅仅保存了始新统（楔形体顶部很好地保存了较老和较新的变形残余地层；例如 Verges 等，1998）。从这方面，对于山脉早期演化的研究，后侧前陆盆地相对于前侧前陆盆地提供了更多的资料。

假定盆地充填为一个常数，那么前侧前陆盆地地层超覆到克拉通盆地边缘的速率等于板块俯冲速率，并且等于逆冲楔向外扩展的速率。因此，这些盆地可以记录下有限时间内的造山运动，期间的主要特征就是快速的地层超覆和盆地转化。后侧前陆盆地的超覆历史仅仅记录了后侧逆冲楔向外扩张的过程（图 26-3）。前侧前陆盆地和后侧前陆盆地最显著的区别在于沉降史，下文将作详细论述。

26.4 前陆盆地沉降过程

正常情况下，地质历史内前陆盆地的沉降表现为逆冲楔的形成过程中的沉降增生（Angevine 等，1990；Miall，1995；DeCeLLes 和 Giles，1996；Allen 和 Allen，2005；Xiangyang 和 Heller，2009）。前侧前陆盆地的一个主要特征是，在该环境中活动俯冲作用引起盆地转换到山脉之下，从而导致了构造沉降速率通常大于 0.05mm/a（图 26-4）。但是后侧前陆盆地和一些后侧弧后前陆盆地内下伏板块的俯冲作用不发育，它们记录了相对较慢的沉降史（通常<0.05mm/a），并缺少在山脉生长过程中增生的证据（图 26-4）。为了解释此类山脉的沉降史，必须将前陆盆地沉降过程中，板块俯冲的控制作用与逆冲楔生长的控制作用（图 26-5）分开来研究。在山脉之下，下伏板块的俯冲作用能够使盆地内岩石圈挠曲面向下移动。如果岩石圈的挠曲强度较高，则盆地内下伏板块呈平面状展布。盆地内所有区域都会沿着挠曲面向下平移，从而形成了一个线性的沉降面。但是，如果岩石圈挠曲强度较弱，盆地下部板块表面的曲率变化明显，并且沿着剖面向下移动的速率将随时间的推移不断增大。因此，在最简单的实例中，正常情况下，沉降速率的增大会使盆地沿着一个由俯冲作用引起的弧形挠曲剖面发生沉降。

除了盆地内的点会沿着挠曲面向下移动之外，山脉的地形荷载也随着时间的推移不断增大。在一个简单的模型实例中，逆冲楔的地形增长以楔形体的自相似增长为特征，楔形体的地表抬升速率会随着时间的推移而减小，当达到稳定状态时，隆升速率可能变为零。随着山脉宽度的不断变大，逆冲楔分散的地形荷载的重心也将在下伏板块之上向外迁移；随着地表的抬升，重心的迁移速率会随着时间的推移逐渐变缓。这种情况下，由生长山脉不断地侵蚀所引起的沉降量通常也会减少，而沉降速率也在变缓。

在大多数前侧前陆盆地的环境中，山脉地形生长所引起的相对缓慢沉降，由山脉下伏板块不断俯冲形成的相对快速沉降所控制；这就为盆地的快速和加速沉降提供了信息。但是在

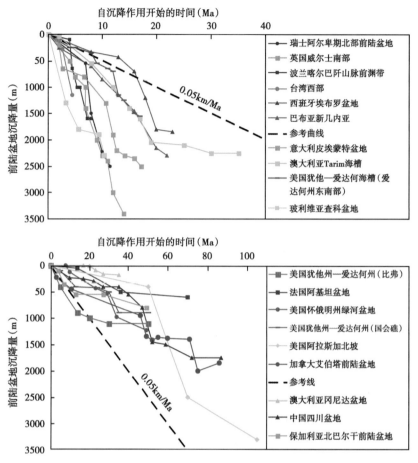

图 26-4 参考文献中对前陆盆地沉降历史的综合分析（据 Sinclair 和 Naylor，修改，出版中）

两个图都能区分出快速和慢速的平均沉降速率。快速沉降速率≥0.05km/Ma，慢速沉降速率为<0.05km/Ma。虚线部分是快速沉降速率和慢速沉降速率的分割线。除了图例中名字旁具有星号的地区外，其他地区的沉降史都基于地层回剥技术来计算构造沉降

没有板块俯冲的后侧前陆盆地和弧后前陆盆地，岩石被搬运到逆冲楔的后部，从而使山脉地形变高。在这种情况下，变形前缘的向外传播过程记录了逆冲楔宽度的绝对增长量，这也是前陆盆地沉降的唯一驱动力。

实际上，虽然位于比利牛斯山脉两侧 Ebro 盆地和 Aquitaine 盆地的沉降史提供了一个前侧前陆盆地和后侧前陆盆地之间的典型对照，但是很难去区别这些信息（Naylor 和 Sinclair，2008）。北美白垩纪 Seaqay 记录的长时间沉降过程表明这些盆地的沉降不是由北美克拉通板块向西的俯冲作用控制，而是由像加拿大滨岸山脉以及塞维尔褶皱—逆冲带一样的弧后褶皱和逆冲带的逐步向外生长控制（Jordan，1981）。相反，中国台湾西部、巴布亚—新几内亚岛、阿尔卑斯山脉北部，以及比利牛斯山脉南部的前侧前陆盆地都记录了沉降速率快速增大的现象，说明了在这样的背景下活动俯冲对盆地沉降的控制作用（图 26-4）。前陆盆地中的其他力学行为也会引起沉降作用，比如弧后背景下与软流圈流动有关的动力学地形以及前侧前陆盆地的"挡板推力"。不管怎样，识别前侧前陆盆地与后侧前陆盆地中普遍存在的差异相当重要。

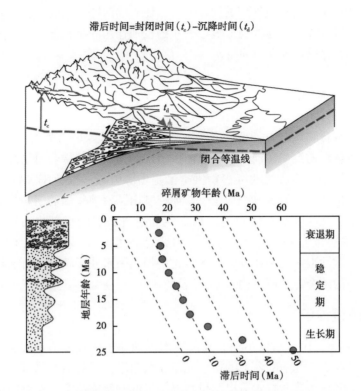

图 26-5　碎屑矿物热年代学和滞后时间（据 Bernet 等，2009，修改）

该图通过锆石裂变径迹等特殊系统的有效封闭温度（时间=T_c）和岩石颗粒到盆地中最后沉降的时间（T_d）揭示了盆地的剥蚀史。滞后时间等于 T_c-T_d，然后将滞后时间与盆地的地层序列结合在一起成图：这可以解释为根据推测的古地温梯度来确定古剥蚀速率随时间的变化（Garver 等，1999）。前陆盆地地层滞后时间的减小、稳定和增加分别代表了山脉的生长、稳定和衰退过程

26.5　逆冲楔/前陆盆地系统中构造作用/地表地质作用的耦合

我们现在认识到地表地质作用可以增强或者减弱变形逆冲楔的构造活动性，甚至可以确定楔形体的整体形态，而不是将侵蚀作用和沉积作用理解为构造挤压的被动响应（Stolar 等，2006）。构造作用/地表地质作用的耦合模型能够从理论上探索这些过程。简单地来说，因为在任何时间，临界楔形体控制了平均地表海拔的分布，所以逆冲楔趋向于稳定形态。楔形体因前缘增生而发生扩展，或者通过内部增厚收缩来维持这种状态。岩石抬升形成了高低不平的地形，而侵蚀作用又使地形变平。侵蚀作用越大，倾向于稳定状态所需要的岩石抬升量也越大。因此，对于增生量已知的高侵蚀速率楔形体而言，可以利用该增生量来维持岩石抬升，以弥补侵蚀作用所带来的损失。与之相反，对于一个已经拥有大增生量低侵蚀速率楔形体来说，它就会向外扩展，并在腹地地区达到海拔稳定的状态。

山脉总体形态和变形历史受到地表地质作用的影响，最典型例子就是安第斯山脉东部的变形史（Strecker 等，2007）。在中安第斯山脉的南部，山脉内部的长期干旱促进了 Altiplano-Puna 高原的形成。当高原的表层抬升达到一个稳定海拔时，随着高原重力的增加，从而使变形向东扩展至前陆盆地。滨岸山脉东部边缘造山作用的增强，导致山脉内部进一步干旱。

这个例子展示了逆冲楔是怎样向外扩展的。此外，还表明前陆盆地的迁移受到系统内气候状况的影响。因此，进一步了解逆冲楔形成过程中的侵蚀作用和沉积作用，以及它们对前陆盆地的沉积物输送和保存的影响非常重要。

26.6 侵蚀和沉积通量

在盆地地层的解释和模型中，来自山脉物源区的沉积通量是约束条件最少的变量之一。来自于山脉的沉积物最初沉积于前陆盆地，所以地层的充填和结构会对物源流域的侵蚀速率变化很敏感。被搬运到前陆盆地的沉积物供给量是山脉流域侵蚀速率和侵蚀范围的一个函数，而这些沉积物主要通过河流进行搬运。长期控制山脉侵蚀作用的因素是最近几年争论较多的一个研究主题，这些因素包括构造过程（例如沿着断裂的局部岩石抬升）、气候过程（例如降水的级别和强度）和岩石类型（岩性和裂缝密度）。研究表明在不同背景下，长期控制侵蚀作用的关键性因素是不相同的。区分这些控制因素并了解它们对前陆盆地沉积地层的影响是盆地分析的两个重要研究内容。

26.6.1 山脉的侵蚀史

在过去的20年中，用于测量不同时间尺度内侵蚀速率的技术方法得到了很大的提高。为了将逆冲楔的长期演化与前陆盆地的地层充填联系起来，需要能够测量1~10Ma尺度内的侵蚀速率技术。低温热年代学技术可以测量百万年尺度的侵蚀速率，如对山脉流域基底样品中的磷灰石和锆石进行裂变径迹和U-Th/He比值分析（Reiners和Brandon，2006）。这些技术能够重建250℃到60℃范围内的盆地冷却历史。结合计算古地温梯度（通常位于20~35℃范围内），可以将这些冷却史转化为地壳上部几千千米内的侵蚀史。通过测量长时期内侵蚀作用的速率和分布范围，使我们对逆冲楔的热演化机制与地表地质作用之间的相互耦合关系的理解有了很大的提高，其中地表地质作用通过将侵蚀作用的产物运移到周围沉积盆地中，而达到重新分配沉积物的目的。

在喜马拉雅山脉中部，利用磷灰石裂变径迹计算出的长期侵蚀速率和季风性降雨之间的空间关系表明，季风性降雨是岩石遭受剥蚀和区域性构造活动的根本驱动力（Thied等，2004）。气候对山脉形成具有明显的影响，阿拉斯加的伊来尔斯山脉就是一个最典型的实例。在中更新世，冰川侵蚀作用的加强导致逆冲楔地表部分的侵蚀作用加速，促使该地区的构造变形增强（Berger等，2008）。相反，台湾地区侵蚀作用的分布范围随着时间的变化发生了迁移，是由局部变形作用发生变化而形成的（Dadson等，2003）。很明显断裂活动能使河流的纵剖面变陡，山坡变得不稳定，并使岩石变得破碎，这些都能够增大侵蚀速率。岩性也是控制河流和山坡地貌形成的关键因素。阿巴拉契亚山脉的山脊和河谷地形等实例表明岩性对长期侵蚀速率具有很大的影响（Hack，1979）。

26.6.2 碎屑矿物年代学

将碎屑矿物年代学技术与一系列物源分析方法相结合，使得沉积学家在研究前陆盆地时，不但能够建立物源区，还能够重建物源区古侵蚀速率随时间变化的细节（Garver等，1999）。这种有效的技术结合将山脉流域基岩侵蚀速率与前陆盆地内沉积通量联系了起来（见Gehrels的文章）。

碎屑矿物年代学的基本原理是：如果磷灰石和锆石系统的矿物冷却年龄在沉积盆地埋藏过程中没有被重置，那么它们就保存了来自源区流域岩石的年龄信息。碎屑矿物颗粒总体年龄和沉积地层年龄之间的差称为"滞后时间"。如果认为颗粒从源区流域到沉积位置的输送过程相对于岩石的形成是瞬时的（比如没有明显的沉积间断），那么滞后时间就代表了研究区埋藏地层的物源区内矿物的冷却年龄。因此，在有效封闭温度和古地温梯度的基础上，这些滞后时间就能被转化为古侵蚀速率（Garver等，1999）。

沉积地层中未被重置的矿物冷却年龄数据会有很多个，其中聚集在一起的数据通常可以反映物源区的形成。每一个组都反映了一个滞后时间，这样就可以得到山脉某些部位的平均古剥蚀速率。如果前陆盆地垂向序列上滞后时间的分布恒定不变，即使存在变形楔体内岩石抬升的自然变化的干扰（Naylor和Sinclair，2007），它仍然可以反映山脉处于剥蚀和热稳定状态（Willett和Brandon，2002）。

理论上，前陆盆地地层层序滞后时间具有初始减小、随后稳定、然后增加的特点，这些特点分别记录了山脉的生长、稳定以及衰退过程（图26-5）。这种方法已被用来研究欧洲阿尔卑斯山脉。在那儿，最近30Ma内盆地地层中最年轻的锆石裂变径迹年龄有大约8Ma的滞后时间，这表明山脉处于一个长期稳定的状态（Bernet等，2009）。但是，这些笔者强调锆石裂变径迹是上部7~10km侵蚀的平均值，因此，相对于磷灰石裂变径迹，锆石裂变径迹对侵蚀速率短期变化的敏感性较低。碎屑矿物裂变径迹要求矿物在埋藏过程中未被重置，这促进了高封闭温度技术的应用，例如，^{39}Ar—^{40}Ar技术被广泛应用于喜马拉雅山脉Siwalik前陆盆地地层的研究中（Najman，2006）。

26.6.3 双向会聚逆冲楔中侵蚀作用的不对称性

双向会聚逆冲楔背景下的构造挤压作用非常不对称。前侧楔形体的增生是由板块的俯冲运动引起的（图26-1和图26-3），而后侧楔形体的生长是岩石沿着山脉中心发生平移（来自前侧）而形成的。此外，仰冲板块之上的前缘增生也对后侧楔形体的生长起一定的作用。所以，后侧楔形体的前缘增生带仅仅记录了山脉向外生长的时间，而前侧楔形体的前部增生记录了生长和板块俯冲作用。这对双向会聚逆冲楔内岩石的运动轨迹产生了影响，从而影响了地表岩石的变质程度和冷却历史（Willett和Brandon，2002）。大部分前侧楔形体内发育有增生的前陆盆地沉积物，这些沉积物来自于前侧楔形体，然后循环回到前侧楔形体中。这样它们从增生到剥蚀的轨迹会相对较短而且埋藏浅，而具体的轨迹则取决于前侧楔形体的厚度。相反，大部分后侧楔形体内的岩石来自于埋藏更深的地层，这些岩石增生到前侧楔形体，随后埋藏于山脉中心区域，然后从后侧楔形体平移并遭受剥蚀，最后裸露于地表。

新西兰的南阿尔卑斯山脉、中国台湾，以及华盛顿州的奥林匹克山脉的研究工作表明变质和冷却过程不对称。在对比利牛斯山脉的研究过程中，考虑前陆盆地演化过程中构造作用不对称影响的情况下，始新世埃布罗盆地的沉积聚集速率（前侧前陆盆地）是Aquitaine盆地（后侧前陆盆地）的8倍（Sinclair等，2005）。将沉积聚集速率的非对称性与磷灰石和锆石裂变径迹技术测得的山脉侵蚀速率的非对称性相对比，结果表明这种非对称性是由于前侧楔形体相对于后侧楔形体来说，前者的挤压作用引起的隆升速率比侵蚀速率更快。

因此，将上述讨论中提到的构造活动和侵蚀作用联系起来，可以确定，前侧前陆盆地和后侧前陆盆地除了在构造边界环境上的不同外，两类逆冲楔在构造挤压上的非对称性也引起两类前陆盆地沉积通量的不对称性。两种楔形体内部源岩差别对不对称性的影响还欠缺考虑：大量山脉的岩石轨迹模型和相关数据（Willett 和 Brandon，2002）表明前侧前陆盆地具有再循环的特征，同时发育相对未变质的矿物组合，而后侧前陆盆地包含来自埋藏更深的蚀变岩石，并且具有很少的循环特征。这些预测还未得到证实。

最后，我们可以推测气候的非对称性对前陆盆地侵蚀作用和沉积通量的非对称性起到明显的作用。山脉与优势降雨方向垂直或斜交将导致山地降水作用增强。数字模型已表明沉积物输送到前陆盆地的潜在重要性（Johnson 和 Beaumout，1995）。新西兰境内南阿尔卑斯山脉降水量曲线的不对称性，反映了侵蚀作用的不对称性。因此，长期侵蚀作用对气候不对称或者山脉构造挤压不对称的响应，控制了进入邻区前陆盆地的沉积通量、颗粒大小和砂岩特征。

26.7 楔体顶部沉积

背驮盆地是生长逆冲楔顶部沉积物的聚集区域（Ori 和 Friend，1984），亚平宁山脉和比利牛斯山脉内就发育有典型的背驮盆地。如果逆冲楔上出现大面积的沉积物聚集，那么对比这些地区与邻区前渊中保存的沉积物演化过程非常重要，因此 DeCelles 和 Giles（1996）提倡使用术语"楔体顶部沉积中心"来代表背驼盆地和披盖在逆冲楔趾部的前陆盆地。

变形逆冲楔顶部的沉积聚集将改变楔体内的应力场（Storti 和 MaClay，1995）。逆冲楔顶部聚集的沉积物增加了岩石圈在基底滑脱面上的压力，使得楔形体进入了稳定状态（图26-2 和图 26-6）。在这种情况下，楔形体将在基底滑脱面上发生滑动，但是不会发生内部变形，直到充足的前缘增生作用将锥形体角度减小至临界状态为止。

在增生作用之前，位于变形逆冲楔外侧的部分前陆盆地地区的偏应力会增加。在接近前缘地区聚集的沉积物增加了下伏天然滑脱面上的岩石圈应力，使楔形体的未变形部分保持稳定和增强（Stockmal 等，2007；Storti 和 McClay，1995）。Stockmal 等（2007）的实验说明前陆盆地的向下挠曲会产生一个楔形的地层充填，从而确定了自身的临界形态为锥形（图26-6）。因此，只要拥有充足的盆地充填，楔形体将不会发生内部变形，但是会在基底滑脱面上滑动，并且使盆地克拉通边缘外侧的背斜发生扩展（图 26-6b）。这个过程可以聚集更多的沉积物，为盆地随后进入稳定状态提供物质基础。

在层面坡度角较小的逆冲楔上，薄弱的基底滑脱面、较高的内部岩石强度和（或）较大的滑脱层坡度会使楔形体顶部沉积的压实作用增强（Ford，2004）。Willett 和 Schlunegger（2010）认为瑞士磨拉石盆地的部分晚中新统是在一个向山脉倾斜的地表坡度面（α）背景下形成的（图26-2），这是由于阿尔卑斯山脉逆冲楔在三叠系的盐岩层中发生滑脱，并向外扩展，然后进入侏罗山而形成的。他们建议使用术语"负α盆地"去描述在倾向山脉的地形上形成的、位于楔形体顶部的盆地。这类盆地的一个重要特征是下伏变形层序稳定，并且可以披覆在周围构造上，而这些构造由于楔形体稳定存在而不再活动。在比利牛斯山脉南部的逆冲带内，始新世和渐新世的地层较厚（图 26-6d、e）。这是因为该时期存在一个下部稳定的逆冲楔体，它的存在与中比利牛斯山内部轴向带被加速侵蚀去顶有关（Sinclair 等，2005）。

综上所述，逆冲楔的平面坡度较小促使楔形体顶部沉积物的聚集，使楔形体的沉积聚集部位稳定，并引起系统内其他部位的收缩。此外，沉积物从逆冲楔前缘地区披覆到前陆盆地近源区域，使该地区处于稳定状态，阻碍了前缘增生作用；沉积断坡会有效地产生必需的临界楔体形态，而不发生内部变形。

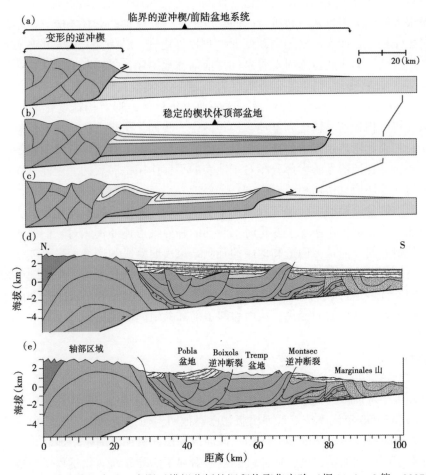

图 26-6 前陆盆地内基于有限元模拟分析的沉积物聚集实验（据 Stockmal 等，2007）

(a) 变形逆冲楔外侧前陆盆地内沉积物的初始聚集，在弯曲的岩石圈上沉积地层变薄。生长逆冲楔的最初楔体形态由沉积面坡度和基底上覆面共同控制。(b) 前缘逆冲带在盆地之下发生滑脱，并向前扩展使盆地卷入到逆冲楔的变形中。这说明盆地的沉积形态提供了一个已经处于临界状态的稳定楔形体（图 26-2）。稳定楔形体顶部盆地持续聚集沉积物，那就是说，只要沉积物聚集能够维持稳定状态所需的层面坡度，就会进行自我调节以保持稳定。(c) 最后，稳定楔形体顶部盆地迁移进入到逆冲楔中的过程使盆地发生了进一步的抬升和变形。(d) 比利牛斯山脉南部楔体顶部的 Pobla 盆地的恢复，渐新世形成的 Montsec 逆断层将它与前缘的埃布罗盆地相分割。与 (b) 和 (c) 相比较，(e) 现今剖面表现出造山运动晚期 Pobla 盆地的持续变形和侵蚀。因此，渐新世的 Pobla 盆地被解释为发育在楔形体顶部的稳定盆地（据 Sinclair，2005）

26.8 系统的时间尺度

随着对逆冲楔与前陆盆地之间的耦合和响应了解得越深入，我们逐渐意识到构造或气候波动的响应时间就越重要。波动的周期与地形的响应时间决定了这个系统是"及时响应"还是"滞后响应"（Allen，2008）。在一个及时响应地形中，波动过程转化为山脉河流输送到前陆盆地内长时期沉积聚集区域的沉积物改变的信号。比利牛斯山脉的 Ebro 前陆盆地研究能够说明逆冲楔变形的沉积响应时间在 0.5~1.0Ma 范围内（Jones 等，2004），同样可以在冰川塑造的地形恢复中发现相同的数值（Hobley等，2010）。这样的时间尺度意味着地层的沉积记录很大程度上来自米兰科维奇时间尺度上的气候波动的滞后响应，但是可以预测到更长期的挤压作用，比如逆冲楔的间断变形（见以下内容）或者更长时期的气候改变。

将沿着逆冲断层的活动位移时期等事件与前陆盆地层序记录相联系时，这些研究内容会变得非常重要。如果变形时间直接用断层岩石定年来测定（例如剪切带中的云母^{39}Ar—^{40}Ar），那么这也是地表位移时间的记录，这些地表位移是地形对构造变化的瞬时响应，然后由河流系统将沉积物信号输送到盆地；换句话说，我们无法将这些与地层信息建立直接的联系。相反，如果利用热年代学的冷却年龄去研究断层，则上盘块体地形的侵蚀响应变化就会被记录下来，并且该信号的大部分相对于地表的平移作用是滞后的。因此，在这样的实例中测定的年龄和地层信息密切相关。

在连续型前陆盆地模型中，尽管我们几乎不了解现存盆地系统的此类信息，但可以假定当地表的侵蚀作用和沉积作用发生变化时，活动的临界锥形逆冲楔体对它们的构造响应，在地质尺度上是瞬时的。但是，构造和地层的观测结果表明褶皱和逆冲变形不连续。数值模型和砂箱实验显示，不连续变形的时间上限是由增生逆冲单元的长度与平均地壳缩短速度之间的比值确定的，许多逆冲楔的比值在 0.1~5.0Ma 之间（Naylor和Sinclair，2007）。这样的不连续性生长导致了基准面的变化，从而引起了前陆盆地内的地层循环（Hoth等，2005）。

26.9 结论

在会聚环境中可以依据变形逆冲楔的物理模型来研究大陆岩石圈均衡响应对山脉地形的支撑作用。20世纪90年代以来一系列出版物已经证明，通过逆冲楔的生长和伴随前陆盆地沉降的大陆板块俯冲之间力学的耦合作用，逆冲楔和前陆盆地之间密切地联系在一起。这种耦合关系存在于两个方面：首先，逆冲楔的地形引起了前陆盆地演化过程中的挠曲压实；其次，前陆盆地地层厚度和盆地迁移进入逆冲楔的速率决定了逆冲楔的生长特点。为了研究这种耦合关系的影响，我们需要综合双向会聚逆冲楔的构造框架来进行解释。同样，逆冲楔地表地质作用和构造挤压也是相互耦合的，这是通过由岩石抬升响应侵蚀过程以及楔体顶部沉积保持变形逆冲楔稳定状态来实现的。因此，基于逆冲楔与前陆盆地在力学上和地表地质作用中的相互耦合关系，应该提倡将这两种地质要素作为一个综合的逆冲楔/前陆盆地系统进行考虑，然后，再进一步探索不同部位的响应。

致谢（略）

参 考 文 献

Allen, P. A. (2008) Timescales of tectonic landscapes and their sediment routing systems, in Gallagher, K., Jones, S. J., and Wainright, J., eds., Landscape evolution: denudation, climate and tectonics over different time and space scales, Geological Society London, Special Publications, 296. Geological Society of London, 7-28.

Allen, P. A., and Allen, J. R. (2005) Basin analysis. Oxford, Blackwell Publishing, 549 p.

Allen, P. A., Crampton, S. L., and Sinclair, H. D. (1991) The inception and early evolution of the North Alpine Foreland Basin. Basin Research, 3, 143-163.

Angevine, C. L., Heller, P. L., and Paola, C. (1990) Quantitative sedimentary basin modeling. AAPG Continuing Education Series, 133.

Baldwin, J. A., Whipple, K. X., and Tucker, G. E. (2003) Implications of the shear stress river incision model for the timescale of postorogenic decay of topography. Journal of Geophysical Research-Solid Earth, 108.

Beaumont, C. (1981) Foreland Basins. Geophysical Journal of the Royal Astronomical Society, 65, 291-329.

Berger, A. L., Gulick, S. P. S., Spotila, J. A., Upton, P., Jaeger, J. M., Chapman, J. B., Worthington, L. A., Pavlis, T. L., Ridgway, K. D., Willems, B. A., and McAleer, R. J. (2008) Quaternary tectonic response to intensified glacial erosion in an orogenic wedge. Nature Geoscience, 1, 793-799.

Bernet, M., Brandon, M., Garver, J., Balestieri, M. L., Ventura, B., and Zattin, M. (2009) Exhuming the Alps through time: clues from detrital zircon fission–track thermochronology. Basin Research, 21, 781-798.

Cederbom, C. E., Sinclair, H. D., Schlunegger, F., and Rahn, M. K. (2004) Climate-induced rebound and exhumation of the European Alps. Geology, 32, 709-712.

Chapple, W. M. (1978) Mechanics of thin-skinned fold and thrust belts. Geological Society of America Bulletin, 89, 1189-1198.

Dadson, S. J., Hovius, N., Chen, H. G., Dade, W. B., Hsieh, M. L., Willett, S. D., Hu, J. C., Horng, M. J., Chen, M. C., Stark, C. P., Lague, D., and Lin, J. C. (2003) Links between erosion, runoff variability and seismicity in the Taiwan orogen. Nature, 426, 648-651.

Dahlen, F. A. (1984) Noncohesive critical Coulomb wedges: an exact solution. Journal of Geophysical Research, 89, 10125-10133.

Dahlen, F. A. (1990) Critical taper model of fold-and-thrust belts and accretionary wedges. Ann. Rev. Earth Planet. Sci., 18, 55-99.

Dahlen, F. A., Suppe, J., and Davis, D. (1984) Mechanics of Fold–and–thrust belts and accretionary wedges: cohesive Coulomb theory. Journal of Geophysical Research, 89, 87-101.

Davis, D., Suppe, J., and Dahlen, F. A. (1983) Mechanics of fold–and–thrust belts and accretionary wedges. Journal of Geophysical Research, 88, 1153-1172.

Davis, D. M., and Engelder, T. (1985) The role of salt in foldand- thrust belts. Tectonophysics, 119, 67-88.

DeCelles, P. G., and Giles, K. A. (1996) Foreland basin systems. Basin Research, 8, 105-123.

Dickinson, W. R. (1974) Plate tectonics and sedimentation, in Dickinson, W. R., ed., Tectonics and sedimentation. Special Publications, 22. Tulsa, OK, Society of Economic Paleontologists and Mineralogists, 1-27.

Flemings, P. B., and Jordan, T. E. (1989) A synthetic stratigraphic model of foreland basin development. Journal of Geophysical Research-Solid Earth and Planets, 94, 3851-3866.

Ford, M. (2004) Depositional wedge tops: interaction between low basal friction external orogenic wedges and flexural foreland basins. Basin Research, 16, 361-375.

Garver, J. I., Brandon, M. T., Roden-Tice, M. K., and Kamp, P. J. J. (1999) Exhumation history of orogenic highlands determined by detrital fission track thermochronology, in Ring, U., Brandon, M. T., Lister, G. S., and Willett, S. D., eds., Exhumation processes: normal faulting, ductile flow and erosion. Special Publications, 154, 283-304.

London, Geological Society. Heller, P. L., Angevine, C. L., Winslow, N. S., and Paola, C. (1988) 2-phase stratigraphic model of foreland-basin sequences. Geology, 16, 501-504.

Hobley, D. J., Sinclair, H. D., and Cowie, P. A. (2010) Fluvial response and paraglacial landscape evolution in the oldest glacial landscape in the Himalayas: processes, rates and timescales. Bulletin of the Geological Society of America, 122, 1569-1584.

Hoth, S., Kukowski, N., and Sinclair, H. D. (2005) Orogenic cycles in foreland basin stratigraphy, in Lacombe, O., Lavé, J., and Roure, F., eds., Thrust belts and foreland basins. Paris, Société Géologique de France, 197-199.

Jones, M. A., Heller, P. L., Roca, E., Garces, M., and Cabrera, L. (2004) Time lag of syntectonic sedimentation across an alluvial basin: theory and example from the Ebro Basin, Spain. Basin Research, 16, 467-488.

Jordan, T. E. (1981) Thrust loads and foreland basin evolution, Cretaceous, Western United-States. Aapg Bulletin- American Association of Petroleum Geologists, 65, 2506-2520.

Karner, G. D., and Watts, A. B. (1983) Gravity-anomalies and flexure of the lithosphere at mountain ranges. Journal of Geophysical Research, 88, 449-477.

Lyon-Caen, H., and Molnar, P. (1985) Gravity-anomalies, flexure of the Indian Plate, and the structure, support and evolution of the Himalaya and Ganga Basin. Tectonics, 4, 513-538.

Malavieille, J. (1984) Experimental-model for imbricated thrusts: comparison with thrust-belts. Bulletin De La Societe Geologique De France, 26, 129-138.

Mellere, D., 1993, Thrust-generated, back-fill stacking of alluvial fan sequences, south-central Pyrenees, Spain (La Pobla de Segur Conglomerates), in Frostick, L. E., and Steel, R. J., eds., Tectonic controls and signatures in sedimentary successions. Special Publication. Ghent, Belgium, International Association of Sedimentologists, 259-276.

Miall, A. D. (1995) Collision-relate foreland basins. Oxford, Blackwell Science, 393 - 424. Molnar, P. (2003) Geomorphology: nature, nurture and landscape. Nature, 426, 612-614.

Najman, Y. (2006) The detrital record of orogenesis: A review of approaches and techniques used in the Himalayan sedimentary basins. Earth-Science Reviews, 74, 1-72.

Naylor, M., and Sinclair, H. D. (2007) Punctuated thrust deformation in the context of doubly vergent thrust wedges: Implications for the localization of uplift and exhumation. Geology, 35, 559-562.

Naylor, M., and Sinclair, H. D. (2008) Pro- vs. retro-foreland basins. Basin Research, 20, 285-303.

Ori, G. G., and Friend, P. F. (1984) Sedimentary Basins Formed and Carried Piggyback on Active Thrust Sheets. Geology, 12, 475-478.

Paola, C. (1988) Subsidence and gravel transport in alluvial basins, in New perspectives in basin analysis (Eds K. L. Kleinspehn and C. Paola). New York, Springer, 231-243.

Quinlan, G. M., and Beaumont, C. (1984) Appalachian thrusting, lithospheric flexure, and the paleozoic stratigraphy of the eastern interior of North-America. Canadian Journal of Earth Sciences, 21, 973-996.

Reiners, P., and Brandon, M. (2006) Using thermochronology to understand orogenic erosion. Annual Review on Earth and Planetary Sciences 34, 419-466.

Sinclair, H. D., Coakley, B. J., Allen, P. A., and Watts, A. B. (1991) Simulation of foreland basin stratigraphy using a diffusion-model of mountain belt uplift and erosion: an example from the Central Alps. Switzerland. Tectonics, 10, 599-620.

Sinclair, H. D., Gibson, M., Naylor, M., and Morris, R. G. (2005) Asymmetric growth of the Pyrenees revealed through measurement and modeling of orogenic fluxes. American Journal of Science, 305, 369-406.

Sinclair, H. D., and Naylor, M. (in press) Foreland Basin subsidence driven by the growth of mountain topography versus plate subduction. Geological Society of America Bulletin. Stockmal, G. S., Beaumont, C., and Boutilier, R. (1986) Geodynamic models of convergent margin tectonics: transition from rifted margin to overthrust belt and consequences for foreland-basin development. AAPG Bulletin-American Association of Petroleum Geologists, 70, 181-190.

Stockmal, G. S., Beaumont, C., Nguyen, M., and Lee, B. (2007) Mechanics of thin-skinned fold-and-thrust belts: insights from numerical models, in Sears, J. W., Harms, T. A., and Evenchick, C. A., eds., Whence the mountains? Inquiries into the evolution of orogenic systems: a volume in honor of Raymond A. Price. Special Paper, 433, 63-98.

Geological Society of America. Stolar, D. B., Willett, S., and Roe, G. H. (2006) Climatic and tectonic forcing of a critical orogen, in Willett, S., Hovius, N., Brandon, M. T., and Fisher, D. M., eds., Tectonics, climate, and landscape evolution. Penrose Conference Series, 241-250.

The Geological Society of America. Storti, F., and McClay, K. (1995) Influence of syntectonic sedimentation on thrust wedges in analog models. Geology, 23, 999-1002.

Strecker, M. R., Alonso, R. N., Bookhagen, B., Carrapa, B., Hilley, G. E., Sobel, E. R., and Trauth, M. H. (2007) Tectonics and climate of the southern central Andes. Annual Review of Earth and Planetary Sciences, 35, 747-787.

Suppe, J. (1981) Mechanics of mountain building and metamorphism in Taiwan Geol. Soc. China Mem., 4, 67-89.

Thiede, R. C., Bookhagen, B., Arrowsmith, J. R., Sobel, E. R., and Strecker, M. R. (2004) Climatic control on rapid exhumation along the Southern Himalayan Front. Earth and Planetary Science Letters, 222, 791-806.

Watts, A. B. (2007) An overview, in Watts, A. B., eds., Crust and lithosphere dynamic. Treatise on geophysics, 6, 1-48.

Amsterdam, Elsevier. Willett, S. D., Beaumont, C., and Fullsack, P. (1993) Mechanical model for the tectonics of doubly vergent compressional orogens. Geology, 21, 371-374.

Willett, S. D., and Brandon, M. T. (2002) On steady states in mountain belts. Geology, 30, 175-178.

Willett, S. D. and Schlunegger, F. (2010) The last phase of deposition in the Swiss Molasse Basin: from foredeep to negative-alpha basin. Basin Research, 22, 623-639.

Willett, S. D., Schlunegger, F., and Picotti, V. (2006) Messinian climate change and erosional destruction of the central European Alps. Geology, 34, 613-616.

Xiangyang, X., and Heller, P. L. (2009) Plate tectonics and basin subsidence history. Geological Society of America Bulletin, 121, 55-64.

(吴哲 译，崔敏 赵梦 校)

第 27 章 挤压背景下生长断层相关褶皱的二维运动学模型

JOSEP POBLET

(Departamento de Geología, Universidad de Oviedo, C/Jesús Arias de Velasco s/n, Spain)

摘 要：本章试图综述目前利用二维运动学模型对挤压环境中伴随有生长地层的断层相关褶皱在构造和沉积作用方面研究的不同认识。研究不同参数在生长地层形成过程中的作用，可以为不同情况下的构造变形提供正确的模型，并且可以揭示在生长地层的分析中能够获得哪些数据。对生长地层特征的分析可以帮助我们更好地理解构造的几何形态和演化过程，并且为油气、矿产和地下水探测，以及评估地震灾害和地下储层提供参考。

关键词：生长地层 同构造沉积 生长逆冲断层相关褶皱 生长三角 递进不整合

27.1 引言

"同构造沉积"是指构造变形过程中形成的沉积地层。通常将生长期存在沉积地层的褶皱和断裂等构造称为"生长构造"，并且将与之相关的同构造期地层称为"生长地层"(Suppe 等，1992)。根据这一定义，褶皱和（或）断裂活动前沉积的地层称为前生长地层，而在变形之后沉积的地层称为"后生长地层"(图 27-1)。在任何构造背景下，与褶皱同时形成的生长地层，在几何学上都具有以下的特点：(1) 厚度上存在变化，通常在构造高部位的地层厚度较薄；而构造低部位处，由于存在较大的可容纳空间，其地层也相对较厚。

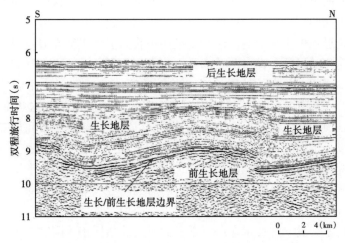

图 27-1 过生长背斜的地震测线（伊比利亚半岛北侧大西洋边缘，比斯开海湾）

(据 Fernández-Viejo 等，2010，细微修改)

展示了前生长地层、后生长地层和生长地层的特征。大多数生长地层超覆在褶皱顶部。生长地层由老到新，其产状逐渐变平，呈现扇状，但褶皱顶部的地层厚度变化不大

(2) 与前生长地层相比，生长地层中褶皱作用较弱，仅部分地层发生了变形，褶皱作用的强度沿地层向上逐渐降低。

局部由断层相关褶皱控制的地区，其生长地层的几何形态受到许多因素的影响，这些因素控制了沉积可容空间，制约了沉积系统并改变了生长地层的原始形态。这些因素包括褶皱扩展机制、构造的类型（根据褶皱和断裂的相互作用进行划分）和动力学演化、褶皱的形态、生长地层的变形机制、沉积速率与褶皱扩展（抬升、翼部变宽以及旋转）速率之间的比值、原始沉积地层产状、生长地层的重力稳定性、特殊的沉积相和沉积样式、侵蚀作用、沉降作用、海平面变化，以及由于压实作用引起的体积缩小等（Suppe 等，1992；Zoetemeijer，1993；Torrente 和 Kligfield，1995；Burbank 等，1996；Hardy 等，1996；Doglion 和 Prosser，1997；Poblet 等，1997；Masaferro 等，2002；Rafini 和 Mercier，2002；Nigro 和 Renda，2004；Shaw 等，2004）。因为变形作用和同构造沉积作用相互影响，所以生长地层的几何形状受构造演化的影响；但是反过来，由于生长地层对位移有阻挡作用，所以生长地层又制约了褶皱的几何形态、运动学演化、活动性以及裂缝的发育。通过与野外类似情况的对比，单个逆冲断层相关褶皱的数值模型和物理模型已经证实了这种影响作用的存在（Barrier 等，2002；Strayer 等，2004）。

在35年前学者们就已经对盆地内的一些沉积层序样式以及沉积与构造变形作用之间的相互影响作了详细的研究，总结了识别标志，并进行了成因解释。最典型的实例就是 Riba（1973，1976）对比利牛斯山脉的 Sant Llorenc de Morunys 构造的研究成果（图27-2）。该构造目前已经成为多家研究机构进行研究的焦点（Ford 等，1997；Suppe 等，1997；Rafini 和 Mercier，2002；Alonso 等，出版中）。此外，还有伊比利亚半岛东北侧（Anadón 等，1986）和加利福尼亚（Medwedeff，1989，1992）的一些构造也成为了研究的热点。但是直到20世纪90年代早期，在 John Suppe 和他的同事提出了生长构造精确的几何学与运动学分析相结合的研究方式之后，由 Oriol Riba 所作的开创性研究才得到了大家的认可。

本章的主要目的是介绍伴随挤压断层相关褶皱的生长地层的特征，但是为了解生长地层的样式，首先需要简单地说明一下基本的褶皱扩展机制。轴面和这些轴面形状之间的生长地层几何形态将在下面两部分内容中进行介绍。这些相互独立并且难以确定的变量，在生长断层相关褶皱的演化过程中相互影响。这些变量使得我们很难明确弄清控制每一个观测特征的参数。为了达到简化的目的，在这两个部分中我们提出了三个假设：(1) 原始生长地层产状水平，褶皱翼部或顶部的生长地层长度和厚度不变；(2) 轴面特征是对尖棱型或膝折型枢纽的响应；(3) 没有压实作用、侵蚀作用和沉降作用的影响。在这两部分内容中笔者描述了生长地层和轴面的几何形态，它们与褶皱扩展机制以及与沉积量和褶皱扩展量比值的关系。但是当发生以下情形时：圆形枢纽取代了膝折型枢纽、生长地层由于垂直/倾斜的剪切作用发生变形、构造的运动学演化发生改变、地层存在原始倾斜，考虑压实和侵蚀作用的影响时，这两部分所描述的关系就会发生改变。随后的内容，笔者简单探讨了一些参数的控制因素。最后一部分展示了生长地层分析技术的重要应用。

有了假三维技术和真三维技术之后，学者们对逆冲断层相关褶皱中生长地层沿走向形态的研究取得了显著的进展（Medwedeff，1992；Poblet 等，1998；Bernal 和 Hardy，2002；Salvini 和 Storti，2002；Fernández 等，2004）。但本章主要介绍逆冲断层相关褶皱的模型和野外实例中，平行于构造运动方向的横剖面上的构造变形样式。这主要是因为二维剖面上，生长地层的形成受到了更大约束，使其便于理解；而对于其三维的样式和控制因素，目前还

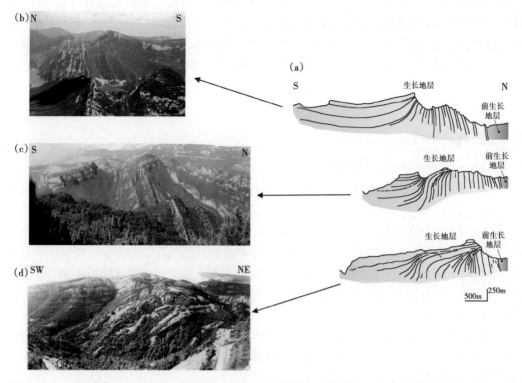

图 27-2 （a）横穿 Sant Llorenc de Morunys（法国比利牛斯山脉南部）生长褶皱南翼生长地层的三个地质剖面（据 Riba，1973，1976，细微修改）和部分横剖面相应的照片（b、c 和 d），显示了生长地层的产状从略微倒转倾斜（向北）到亚平行（向南）的平面展布特征，同时表明可能存在的生长三角带（d）

不是很明确。此外，本章将研究与逆冲断层相关褶皱同时期的"标准"生长地层，但是不包括与活动断层相关褶皱同时代的阶地地貌和地层特征分析，这是因为阶地具有不连续性，需要不同的研究方法（Rockwell 等，1988；Molnar 等，1994；Mueller 和 Suppe，1997；Benedetti 等，2000；Lavé 和 Avouac，2000；以及许多其他实例）。本文的目的是阐述生长逆冲断层相关褶皱中最基本的关系，但文中没有考虑到一些与天然生长地层密切相关的影响因素，比如生长地层的重力稳定性、特殊沉积相和沉积样式、沉降作用、海平面变化以及由于压实作用引起的体积减小等。这些因素对生长地层的形态有显著影响，但超出了本文的研究范畴。

27.2 褶皱扩展机制

挤压背景下断层相关褶皱通常由以下三种褶皱扩展机制形成。

27.2.1 膝折带迁移模式

在这种机制中，地层沿着轴面发生滚动，使得地层倾向"瞬时"发生改变（Suppe，1983；其他文献中）；这个与岩石移动有关的轴面被称为"活动轴面"（Suppe，1992）。大多数情况下，活动轴面向下延伸至断层面的转折处或断层尖灭处。仅仅由膝折带迁移形成的平行断层相关褶皱中，除了在断弯褶皱的位移超过断坡顶部，或者褶皱翼部在整个褶皱演化过程中都保持倾向不变等一些特殊情况外，活动轴面的倾向在褶皱扩展过程中不发生变化。这样，随着更多岩石沿着活动轴面滚动并进入褶皱翼部（自相似行为），褶皱就会随着翼部的变宽而发生扩展增大（Suppe，1983）。典型的断弯褶皱（Suppe，1983）和断层传播褶皱模型（Suppe 和 Medwedeff，1990）、某些双向边界断层传播褶皱模型（Tavani 等，2006）和滑脱褶皱模型（Poblet 和 McClay，1996），以及一些其他模型都是以膝折带迁移作为褶皱的扩展机制。

27.2.2 翼旋转模式

在这种变形机制中，地层倾角逐渐变大（Hardy 和 Poblet，1994，其他文献中），所以轴面会将地层的两个旋转面或者一个旋转面与一个固定面分隔开。两种不同类型的轴面：（1）"固定轴面"是完全固定在地层中，不允许任何地层穿过的轴面；（2）"有限活动轴面"是一种特殊的活动轴面类型（Poblet 和 McClay，1996），这类轴面允许有限的地层穿过轴面，并在轴面固定处逐渐减少为零，这个固定的位置就是轴面的旋转轴。仅仅由翼旋转和有限活动轴面形成的平行断层相关褶皱中，翼部和轴面逐步旋转引起褶皱的扩展，扩展期间翼部的长度近似不变。一些滑脱褶皱模型将翼旋转模式作为上部能干地层单元中的褶皱扩展机制（Homza 和 Wallace，1995；Poblet 和 McClay，1996）。由翼旋转形成的褶皱，会在不同的构造位置发育有限活动轴面和固定轴面。同时，在褶皱的演化过程中一些有限活动轴面可能转变为固定轴面，或者固定轴面转变为有限活动轴面。本文仅仅研究褶皱演化过程中，由有限活动轴面的翼部旋转而形成的褶皱。

27.2.3 膝折带迁移和翼旋转相结合的模式

在断层相关褶皱的形成过程中，岩石的轴面迁移（膝折带迁移机制）和翼部旋转（翼旋转机制）模式，往往同时发生；或者发生在褶皱扩展的不同阶段。这一模式已经被证实，并且作为了断层相关褶皱的理论基础，即剪切断弯褶皱（Suppe 等，2004）、多向剪切断层传播褶皱（Erslev，1991；Hardy 和 Ford，1997；Allmendinger，1998）、位移梯度褶皱（Wickham，1995），以及一些类型的滑脱褶皱（Homza 和 Wallace，1995；Poblet 和 McClay，1996）的理论基础。因为活动轴面、有限活动轴面和固定轴面可能都会出现在这些褶皱中，所以在研究过程中需要具体情况具体分析。

27.3 生长三角带

由活动轴面或者有限活动轴面形成的静止轴面（Suppe 等，2004）或者被动轴面（Poblet 等，1997），可以发育在膝折带迁移、翼部旋转和两种褶皱扩展机制相结合所形成的褶皱中。周围的岩石会沿着活动轴面或有限活动轴面发生滚动迁移。生长地层中的非活动轴面被称为"生长轴面"（Suppe 等，1992），它记录了活动轴面或有限活动轴面处颗粒的沉积轨迹（图27-3）。活动轴面或者有限活动轴面与相应生长轴面之间的距离代表了水平生长地层的长度，反映了水平地层在沉积之后沿着轴面滚动的距离。在横剖面上，由活动或有限活动轴面界定的生长地层面与它们对应的生长轴面共同确定了一个类似三角形状的区域，即"生长三角带"（Suppe 等，1992）（图27-2 至图27-4）。

在存在活动轴面的情况下，生长三角带内生长地层的最大长度，记录了生长地层中活动

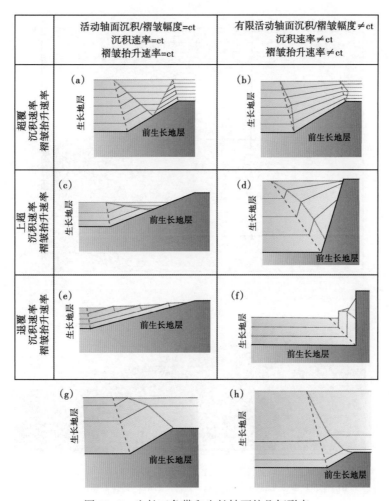

图 27-3　生长三角带和生长轴面的几何形态

图（a）和图（c）据 Suppe 等（1992）修改，图（b）据 Storti 和 Poblet（1997）修改，图（e）据 Shaw 等（2004，2005a）修改。在图（b）中，旋转轴位于前生长地层和生长地层之间，而在图（d）和图（f）中，旋转轴位于生长地层之中。图（g）和图（h）表现了源于活动轴面的生长三角带的特征，其中沉积速率大于褶皱的抬升速率（超覆）。由于沉积速率是变化的，并且抬升速率是一个随时间不发生变化的常量，所以沉积速率与褶皱扩展速率之间的比值是变化的。图（g）表现了沉积速率随着时间变化而减小的情况，图（h）表现了沉积速率随着时间变化而增加的情况。红色虚线代表活动轴面和有限活动轴面；绿色轴线代表生长轴面

轴面活动的时间。如果轴面从生长地层开始沉积时就开始活动了，那么长度最大的地层就是生长地层和前生长地层之间的边界；而如果轴面是在某一特定生长地层的沉积中开始活动的，那么此时就是生长地层最长的时候。上覆于最长地层之上的新地层，通常在活动轴面和生长轴面之间会逐渐变短，形成向上变窄的岩石面和向上抬升的生长三角带（图 27-3a、c、e、g、h 和图 27-4a）。

因为活动轴面没有固定在岩石上，所以可以很容易理解，由活动轴面形成的生长三角带的几何形态。因为有限活动轴面仍然固定在特定地层内部的旋转轴上，所以要定量地研究有限活动轴面形成的生长三角带的几何形态还比较困难。与活动轴面形成的生长三角带不同，有限活动轴面形成的生长三角带内长度最大的地层并不代表有限活动轴面开始活动的时间。

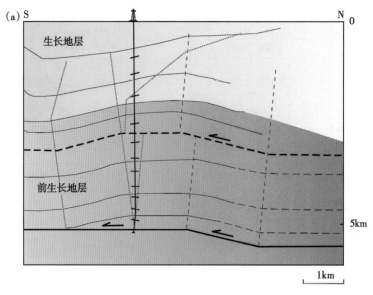

图 27-4 生长三角带实例

（a）Pitas Point 背斜（美国加利福尼亚 Santa Barbarn 海峡东部）中生长断弯褶皱的地震—地质解释剖面，剖面上包含了井位数据（据 Shaw 等作细微修改，2005b）；图中标明了生长三角带、轴面和典型膝折带迁移褶皱的地层样式，其变形特征与图 27-3a 以及将在图 27-5b 模型中描述的情况相似。红色虚线代表活动轴面，绿色点线代表生长轴面。（b）Hortoneda 生长褶皱（法国比利牛斯山脉南部）的南翼（Mellere，1993），图中展示了可能的生长三角带以及典型翼部旋转褶皱的生长地层样式，与图 27-3b、图 27-3d、图 27-5i 和图 27-6g 中描述的模型情况相似。生长地层的产状从陡立向亚平行状逐渐过渡，使年轻生长地层逐步超覆、侵蚀较老地层和前生长地层

在存在活动轴面的生长三角中，上覆于最长地层的新生长地层，通常在有限活动轴面和生长轴面之间会变短，产生一个锥形向上的地层面。最长地层之下的老生长地层也可能变短，从而产生了一个向下变窄的地层面（图 27-3b、d、f 和图 27-4b），并且有限活动轴面和相应的生长轴面会相互接近，直到在旋转轴上合并在一起。

从地层学特征来看，生长三角带上部顶点的出现（活动轴面或有限活动轴面与生长轴面的交点）标志着活动轴面或有限活动轴面停止活动，即生长三角带不再发育（Suppe 等，1992）（图 27-3）。如果在整个褶皱扩展过程中，地层沿着活动轴面或有限活动轴面发生滚动，那么会出现包含生长三角带上部顶点的地层，这就标志着特定构造扩展作用的结束；而如果轴面仅在某些构造扩展阶段发生活动，那么这一地层的出现就意味着，特定的活动轴面或有限活动轴面停止活动，构造会持续扩展。

27.4 生长轴面的几何形态

在仅仅由膝折带迁移或者翼旋转机制形成的褶皱中,活动轴面和有限活动轴面将未褶皱的生长地层与褶皱翼部分隔开。这两种轴面与相应的生长轴面相比,具有特殊的几何形态(图27-3)。活动轴面和有限活动轴面将褶皱两侧生长地层之间的夹角等分,并且控制了生长地层倾角的变化。生长轴面并没有将两侧生长地层之间的夹角等分,这是因为生长地层沿着生长轴面的倾角和厚度发生了变化。沉积厚度的变化是对轴面上可容空间变化的响应,也是影响轴面上沉积速率的一个变量。另外,活动轴面与有限活动轴面在前生长地层和生长地层中的方向相同,而静止轴面在生长地层(生长轴面)中,与前生长地层(实际静止轴面)的方向不同。此外,静止轴面能否对前生长地层中的膝折带产生影响取决于褶皱的运动情况(图27-3)。

引言中提到了一个假设,即由活动轴面演变而来的生长轴面的几何学形态和倾向主要取决于沉积速率和褶皱扩展速率之间的比值以及生长地层表现出的上超—退覆关系。如果沉积速率和褶皱扩展速率之间的比值是一个常数,并且不管沉积速率是大于、等于还是小于褶皱抬升速率,生长地层都表现为上超或超覆的几何形态,那么活动轴面变为面状的生长轴面(图27-3a和图27-3c)。如果沉积速率和褶皱扩展速率之间的比值是一个常数,并且生长地层表现为退覆的几何形态(这种情况下沉积速率通常等于或小于褶皱抬升速率),那么生长轴面表现为多个不连续的陡坡(Shaw等,2004,2005a)(图27-3e)。在最简单的且符合膝折带迁移模式的平行逆冲断层相关褶皱中(包含活动轴面),沿着下滑脱面发生的位移(或褶皱作用下的缩短量),以及翼部倾向、翼部长度等不同的褶皱扩展参数之间,是线性的或者有线性函数关系(Bulnes和Poblet,1999)。因此,使得沉积速率和褶皱扩展速率之间的比值为一个常量相对容易,比如可以通过维持逆冲位移和沉积速率为常数来达到这一目的。当沉积速率与褶皱扩展速率的比值较高时,生长轴面倾角较大,形成顶角较小的生长三角带;而当比值较低时,生长轴面倾角较小,形成顶角较大的生长三角带(Suppe等,1992;Rafini和Mercier,2002)(比较图27-3a、c、e)。当沉积速率与褶皱扩展速率的比值低到一定程度时,生长轴面的包络面会变得水平或者与原始轴面倾向相反(图27-3e)。当沉积速率和褶皱扩展速率之间的比值随时间发生变化时,生长轴面就变成了一个曲面(图27-3g、h)。此时,如果生长轴面向上凸起,则表明沉积速率与褶皱扩展速率之间的比值在不断减小(图27-3g);如果生长轴面向下凹,则表明沉积速率与褶皱扩展速率之间的比值在不断增大(图27-3h)(Suppe等,1992;Rafini和Mercier,2002)。在褶皱扩展过程中,比率的突然变化会形成形态各异的生长轴面。在自然界中,褶皱演化过程中沉积速率和构造速率的变化,使得生长轴面具有复杂多变的几何形态。

由有限活动轴面演变而来的生长轴面与上述由活动轴面演变而来的生长轴面,在倾向和形态上很相似。不论沉积速率与褶皱扩展速率之间的比值如何,生长轴面通常都是一个曲面。这是因为生长轴面与有限活动轴面在两个不同的地层内分别发生会聚,会聚点是有限活动轴面旋转轴固定的位置和有限活动轴面转变为不活动轴面的位置(Storti和Poblet,1997)(图27-3b、d、f)。此外,在最简单且包含翼旋转和有限活动轴面的平行逆冲断层相关褶皱模型中,沿着滑脱层发生的位移量(或褶皱作用下的缩短量)与各个褶皱扩展参数之间呈非线性的函数关系(Bulnes和Poblet,1999)。因此,沉积速率和褶皱扩展速率之间的比值通常不是一个常量。上述所列的影响因素中，沉积速率和褶皱扩展速率之间的比值、生长地层上超—退覆—超覆的几何形态、与有限活动轴面相关的生长轴面形状和倾向等都取决于有限活动轴面的旋转速率以及旋转轴的位置。

27.5 递进不整合

水平沉积的生长地层，其几何形态不受剥蚀作用、压实作用和沉降作用的影响，在变形过程中其地层长度和厚度保持不变。根据褶皱的扩展机制，可以得出，在纯尖棱状或纯膝折状背斜和向斜中，轴面将地层倾斜的翼部与水平的顶部和底部分隔开（Suppe 等，1992；Hardy 和 Poblet，1995；Hardy 等，1996；Poblet 等，1997；Storti 和 Poblet，1997；Rafini 和 Mercier，2002；Poblet，2004；Shaw 等，2004，2005a）（图 27-3 至图 27-6）。

在仅仅由膝折带迁移而形成的生长逆冲断层相关褶皱中，不管两个轴面之间的生长地层是否位于生长三角带之内，它们的厚度和倾向都不发生变化。Suppe 等（1997）提出的断弯和断层传播模型、双边界断层传播模型（Tavani 等，2006），Poblet 等（1997）提出的特殊滑脱褶皱模型等都具有这样的特点。由于生长地层与前生长地层的褶皱变形情况相同，所以在这些褶皱中，生长地层的倾角为零（假设区域地层是水平的）或者生长地层的倾角等于下伏生长地层中的褶皱翼部倾角（图 27-3a、c、e、g、h，图 27-4a，图 27-5a、b、d、e、h 和图 27-6a、b、d、f）。

仅仅由翼旋转而形成的生长逆冲断层相关褶皱中，生长三角带内的生长地层与下伏生长地层一样，具有稳定的厚度和相同的倾角。比如 Poblet 等（1997）提出的滑脱褶皱模型，在这些褶皱中，生长三角带与扇形的生长地层共存。由于生长地层在褶皱变形过程中发生递进式旋转，使得其地层厚度在构造的顶部变薄、倾角变小。因此，生长地层与前生长地层的褶皱变形中显得不谐调。生长楔中老生长地层比新地层倾角大的现象，在本文中称之为"递进不整合"（Riba，1973，1976）（图 27-1，图 27-2，图 27-3b、d、f，图 27-4b，图 27-5i 和图 27-6g）。由膝折带迁移和翼旋转相结合而形成的生长逆冲断层相关褶皱，通常具有平行生长地层和递进不整合共存的特征（Wickham，1995；Hardy 和 Ford，1997；Poblet 等，1997；Storti 和 Poblet，1997；Allmendinger，1998；Suppe 等，2004）（图 27-5c、f、g、j，图 27-6c、e 和图 27-6h）。因此，不管由何种褶皱扩展机制形成的尖棱型或膝折型的生长逆冲断层相关褶皱，其内部都发育有生长三角带，但是仅在发育翼旋转模式的褶皱中表现出递进不整合的特征。因为生长三角带通常集中在有限活动轴面周围，所以递进不整合是识别翼旋转褶皱的最典型的几何学标志（图 27-3、图 27-4b、图 27-5 和图 27-6）。

虽然递进不整合是翼旋转褶皱的一个显著特征，但是当褶皱内不发育薄层轴面，并且不具有尖棱型或膝折型特征时，递进不整合在生长三角带中的演化就会与褶皱的扩展机制无关（图 27-2 和图 27-8）。递进不整合出现在枢纽弯曲的褶皱中，例如由下伏的弯曲断层面而引起的褶皱枢纽弯曲（Suppe 等，1997；Novoa 等，2000；Cristallini 和 Allmendinger，2002；Tavani 等，2005，2007；Hubert-Ferrari 等，2007）（图 27-5）。此外，递进不整合还出现在由物质流变学特征控制的活动轴面宽度有限的区域内（Benesh 等，2007）。因为枢纽弯曲和轴面宽度有限意味着在这些部位发生过地层的旋转，所以递进不整合会在以上两种背景下发育。在翼部较大且倾向固定的褶皱中，如果其膝折带是一个相对较窄的曲面，那么当发生膝折带迁移时，递进不整合就仅在生长三角内部发育（图 27-7a、b）；而发生翼旋转时，递进

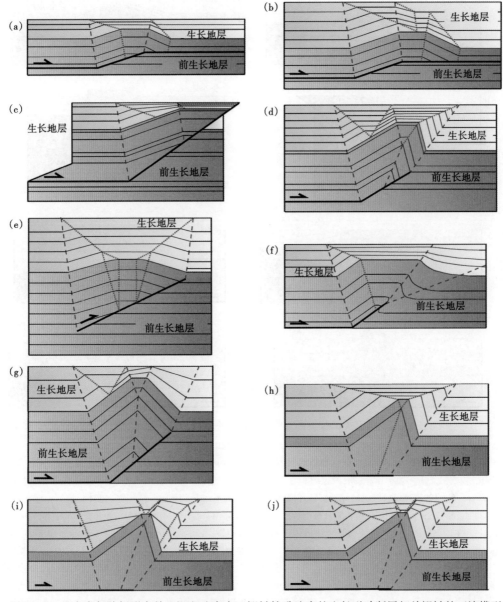

图 27-5 具有类似膝折形态并且沉积速率大于褶皱抬升速率的生长逆冲断层相关褶皱的正演模型
(a) 处于顶部抬升阶段的断弯褶皱（膝折带迁移；Suppe，1983 模型），断距小于断坡长度；(b) 处于顶部扩展阶段的断弯褶皱（膝折带迁移；Suppe，1983 模型），断距大于断坡长度（据 Suppe 等，1992，修改）；(c) 简单剪切的断弯褶皱（膝折带迁移加翼旋转，Suppe 等，2004 模型）（据 Suppe 等，2004，修改）；(d) 断层传播褶皱（膝折带迁移；Suppe 和 Medwadeff，1990 模型）（据 Suppe 等，1992，修改）；(e) 双边界断层传播褶皱（膝折带迁移；Tavani 等，2006 模型）（据 Tavani 等，2007，修改）；(f) 三角剪切断层传播褶皱（膝折带迁移加翼旋转，Erslev，1991 模型）（据 Cristallini 和 Allmendinger，2002，修改）；(g) 断层末端保持稳定的位移梯度褶皱（膝折带迁移加翼旋转；Wickham，1995 模型）（据 Wickham，1995，修改）；(h) 滑脱褶皱（膝折带迁移；Poblet 和 McClay，1996 模型）（据 Poblet 等，1997，修改）；(i) 滑脱褶皱（翼旋转；Poblet 和 McClay，1996 模型）（据 Poblet 等，1997，修改）和 (j) 复合滑脱褶皱（膝折带迁移加翼旋转；Poblet 和 McClay，1996 模型）（据 Poblet 等，1997，修改）。红色虚线表示活动轴面和有限活动轴面；绿色点线表示生长轴面

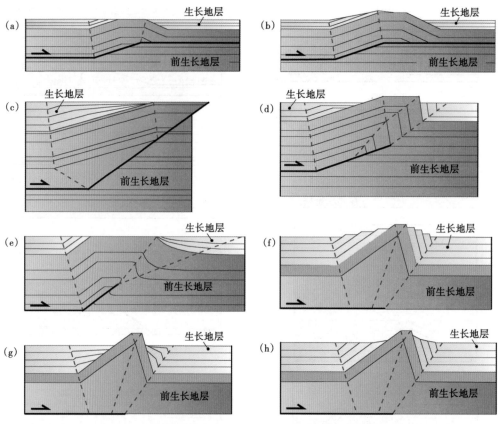

图 27-6　具有类似膝折形态并且沉积速率等于或小于褶皱抬升速率
的生长逆冲断层相关褶皱的正演模型

(a) 处于顶部抬升阶段的断弯褶皱 (膝折带迁移；Suppe, 1983 模型), 断距小于断坡长度 (据 Shaw 等, 2004b, 修改)；(b) 处于顶部扩展阶段的断弯褶皱 (膝折带迁移；Suppe, 1983 模型), 断距大于断坡长度 (据 Suppe 等, 1992, 修改)；(c) 纯剪切断弯褶皱 (膝折带迁移加翼旋转；Suppe 等, 2004 模型) (据 Suppe 等, 2004, 修改)；(d) 断层传播褶皱 (膝折带迁移, Suppe 和 Medwadeff, 1990 模型)) (据 Suppe 等, 1992, 修改)；(e) 三向剪切断层传播褶皱 (膝折带迁移加翼部旋转；Erslev, 1991 模型；据 Cristallini 和 Allmendinger, 2002, 修改)；(f) 滑脱褶皱 (膝折带迁移；Poblet 和 McClay, 1996 模型) (据 Poblet 等, 1997, 修改)；(g) 滑脱褶皱 (翼旋转；Poblet 和 McClay, 1996 模型) (据 Poblet 等, 1997, 修改) 和 (h) 复合滑脱模型 (膝折带迁移加翼旋转；Poblet 和 McClay, 1996 模型) (据 Poblet 等, 1997, 修改)。生长地层中仅标注了活动轴面和有限活动轴面

不整合就会在整个翼部都发育, 很难识别出生长三角带。对于复合褶皱扩展机制形成的生长逆冲断层相关褶皱来说, 各个褶皱扩展机制的重要性、作用时间和作用部位决定了褶皱不同部位的特征 (图 27-7d、e)。在层面连续弯曲的褶皱中 (比如由大量轴面所限定的一小部分倾向固定的层面内, 发生连续弯曲而形成的褶皱), 生长地层在整个褶皱内都发育, 例如某些发育弯曲逆冲断层面的双向边界断层传播褶皱 (Tavani 等, 2007) (图 27-7c)。在一些天然实例中, 膝折带迁移与翼旋转形成的生长褶皱, 在生长地层结构上的区别非常小, 超出了地震剖面和野外数据所能分辨的尺度。在这样的情况下, 生长地层的样式不能作为区分褶皱扩展机制的鉴别标志。

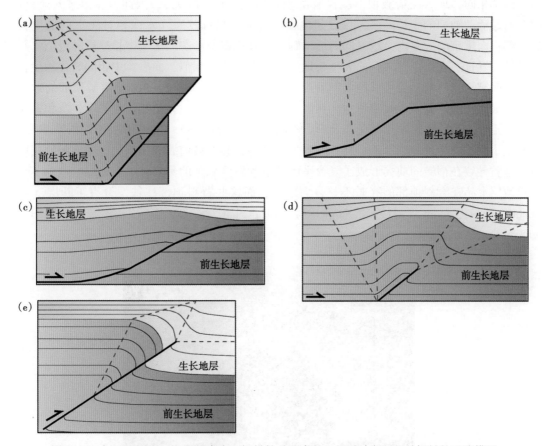

图 27-7 枢纽呈圆形且沉积速率大于褶皱抬升速率的生长逆冲断层相关褶皱的正演模型

(a) Suppe 等 (1997) 和 Novoa 等 (2000) 将突变的逆冲断层断坡上的枢纽进行了圆滑 (膝折带迁移模型; Suppe, 1983) (据 Novoa 等, 2000, 修改); (b) Tavani 等 (2005) 将处于顶部扩展阶段的断弯褶皱的枢纽进行了圆滑 (膝折带迁移模型; Suppe, 1983), 其断距大于断坡长度 (据 Tavani 等, 2005, 修改); (c) 双边界断层扩展褶皱 (膝折带迁移模型; Tavani 等, 2006) (据 Tavani 等, 2007, 修改); (d) 三角剪切断层传播褶皱 (膝折带迁移加上翼旋转模型; Erslev, 1991), 该模型由 Cristallini 和 Allmendinger (2002) 将其枢纽进行了圆滑 (据 Cristallini 和 Allmendinger, 2002, 修改), 和 (e) 前翼发育断层突破的三角剪切断层传播褶皱 (膝折带迁移加上翼旋转模型; Erslev, 1991) (据 Allmendinger, 1998, 修改)

27.6 角度不整合

目前已经证实, 除了上述递进不整合外, 在生长逆冲断层相关褶皱的生长地层内还发育角度不整合 (图 27-1 和图 27-4b)。褶皱运动状态的改变、剥蚀作用或海平面变化等多种因素, 影响着角度不整合的形成。例如, 在停止运动前, 不同类型的褶皱表现出不同的运动学特征: 断弯褶皱在初始阶段表现为顶部抬升, 在其他演化阶段表现为顶部扩展; 而断层传播褶皱和滑脱褶皱会随着逆冲断层的传播, 发生迁移或者断层突破现象。

在膝折型简单生长断弯褶皱中, 如果地层的沉积速率稳定且数值较低, 并且发育水平生长地层, 那么它的演化就能为各种不整合的形成提供典型的实例, 并且能解释在运动学演化过程中, 各种因素对不整合的影响。在顶部抬升阶段, 如果沉积速率与断坡上的褶皱顶部抬升速率相同，那么在褶皱顶部就没有沉积物形成（图 27-9a）。因此，在褶皱的前翼内，没有通过活动轴面的生长地层，仅仅超覆在生长地形相对较高（水深相对浅）的前生长地层顶部；而褶皱的后翼内，越过向斜活动轴面，具有退覆形态的褶皱生长地层和近似水平的生长轴面发育。在这个阶段，前翼生长地层与后生长地层相似，而后翼中相同时代的生长地层发生褶皱。当所有的前生长地层爬升到逆冲断坡顶部时（图 27-9b），褶皱顶部的地形高度达到最大，随后其高程保持恒定，然后进入顶部扩展阶段。在褶皱扩展阶段，沉积速率超过褶皱抬升速率，生长地层超覆在褶皱顶部之上（图 27-9c）。盆地中沉积于构造的后侧，并发生了褶皱的生长地层，从后翼迁移到褶皱顶部，甚至被前翼和顶部之上的生长地层所掩埋。这就导致在褶皱的前翼形成了发育平卧地层和厚度稳定的不对称生长地层。这些不对称生长地层上超在背斜顶部的穿时不整合面之上，而这些背斜顶部往往发育有正断层（Medwedeff，1989；Mount 等，1990；Suppe 等，1992；Shaw 等，2004b）（图 27-9c 和图 27-9d）。

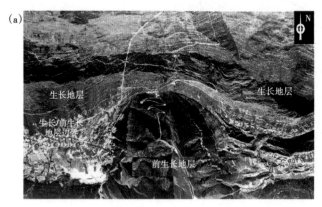

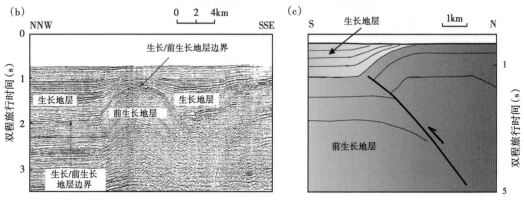

图 27-8 滑脱褶皱实例

(a) Pico del Aguila 背斜内生长滑脱褶皱的航空照片（法国比利牛斯山脉南部）和 (b) 通过 Santaren 背斜内生长滑脱褶皱的地震测线（Bahaman 前陆，Cuban 褶皱和逆冲带）（据 Masaferro 等，1999，细微修改）表明生长地层的产状从陡立—适度倾斜到近水平的扇形形态。最老的生长地层上超在前生长地层上，最新的生长地层超覆在褶皱顶部。此外，最新的生长地层在褶皱顶部存在细微的厚度变化。(c) 过生长三角剪切断层传播褶皱的地震—地质解释剖面模式图（中国塔里木盆地）（地震数据来自 Shaw 等，2005a），与图 27-7d 和图 27-7e 中描述的正演模型在生长地层的样式上具有许多相似的特征

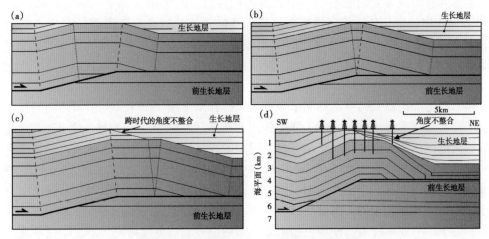

图 27-9 生长断弯褶皱正演模型的演化过程

(a) 初始阶段沉积速率等于褶皱抬升速率的顶部抬升。(b) 初始阶段和下一阶段之间的过渡时期。(c) 最后阶段沉积速率大于褶皱抬升速率的顶部扩展（据 Medwedeff，1989，修改）。红色虚线代表活动轴面；绿色点线代表生长轴面。(d) 通过 Lost 山脉东南部（美国加利福尼亚圣约金河盆地）某个生长断弯褶皱（包含井位信息）的地震—地质解释剖面模式图（据 Medwedeff，1989，细微修改）。该图表明褶皱顶部和东北翼的部分地区的生长地层中，存在不整合。因此，前翼和顶部的生长地层上超在后翼的生长地层上，这与图 27-9 所描述的正演模型相似

27.7 上超、退覆和超覆

不同类型的褶皱—逆冲断层相互作用会对生长地层样式产生很大的影响，包括生长三角带的数目和位置、平行生长地层面、递进不整合面、上超—退覆—超覆以及角度不整合的产状。图 27-3、图 27-5 至图 27-7 反映了不同沉积速率与褶皱抬升速率比值的情况下，几种生长逆冲断层相关褶皱模型的特征。这些模型的几何形态差异较大，因此，对每一种褶皱类型需要进行详细的描述。正常情况下，顶部或顶部附近的构造特征是区别不同类型生长断层相关背斜的关键因素，当沉积供应速率大于褶皱顶部抬升速率时这些特征就表现得很明显；而当侵蚀作用除掉了这些记录后，就很难识别这些特征了（比较图 27-5 和图 27-6）。

不考虑褶皱扩展机制和褶皱/逆冲断层间的相互作用，忽略压实作用、沉降作用和侵蚀作用的影响，同时假设所有沉积颗粒的垂直运动由褶皱的扩展引起，那么超覆在背斜顶部生长地层的存在意味着同构造沉积速率大于褶皱抬升速率。这种情况下，不会发育地形/水深变化（图 27-1，图 27-3a、b，图 27-3g、h，图 27-4a，图 27-5，图 27-7 和图 27-8a、b）。相反，上超在前生长地层之上的生长地层或退覆在褶皱翼部之上生长地层的出现则意味着同构造沉积速率等于或小于背斜抬升速率（即后面实例中的突变构造）（Suppe 等，1992；Hardy 等，1996；Doglioni 和 Prosser，1997；Poblet 等，1997；StortiPoblet，1997；Rafini 和 Mercier，2002；Shaw 等，2004，2005a）（图 27-3c、f，图 27-6，图 27-8c 和图 27-9a、b）。无论是生长地层发生上超、背斜轴面发生倾斜迁移还是生长地层发生退覆都受到许多因素的影响。这些因素包括沉积速率与褶皱扩展速率的比值、侵蚀作用、沉降作用、褶皱扩展机制以及褶皱运动情况等。图 27-6 和图 27-9 说明褶皱的运动学特征影响了上超和退覆的几何学形态。所以，在同一构造的两侧，虽然具有相同褶皱扩展机制和沉积速率与褶皱抬升速率比值，但是由于前翼与后翼的运动学特征不同，使得在断弯褶皱的形成过程中，

同一构造的前翼表现为上超的形态，而后翼表现为退覆的形态（图 27-6a、b 和图 27-9a、b）。图 27-6d 中的生长断层传播褶皱则变现为倾向相反的特征。在图 27-6g 所示的生长滑脱褶皱中，构造两侧的褶皱扩展机制相同，沉积速率与褶皱抬升速率比值相等，前翼变现为退覆的形态，而后翼表现为上超的形态；这种不对称的结果是由褶皱翼部旋转速率不同造成的。

在不同的褶皱扩展阶段内，幕式的速率变化（如沉积和褶皱伸展速率比值）对生长地层的上超、退覆和超覆几何形态起到了至关重要的控制作用。由于受到褶皱运动的控制，生长地层的样式对于不同演化阶段中，强度相同的构造活动的响应有差异。图 27-10a 是一个由翼旋转而形成的生长滑脱褶皱正演模型。模型中，基底滑脱面上两个脉冲式的高速逆冲叠置在一个低速逆冲的背景之上（Hardy 和 Poblet，1994）（图 27-10b）。首次逆冲速率变大，在生长地层内部出现了一个突然性的退覆，这是由于逆冲速率增大而使得抬升速率增加，并且使地层发生旋转而造成的，图 27-10d、e 所描述的野外实例就是这种情况。在持续运动过程中，由于抬升速率降低，使得上超现象在初始退覆之后迅速再现。第二个逆冲速率变大的阶段，在规模上与第一个阶段相同，虽然没有形成任何退覆现象，但是褶皱翼部地区的加积作用却迅速减弱（图 27-10b）。在第二个逆冲速率变大的阶段内，退覆现象不发育，这是因为在这种类型的滑脱褶皱中，其翼部地层的倾角较大，在增加的逆冲速率相同的情况下，相对于翼部平缓的褶皱而言，其隆升速率较小（图 27-10c）。另外，在由翼旋转形成的褶皱中，除非高逆冲速率（包括抬升速率的突然增加）发生在褶皱扩展的最后阶段，否则生长沉积与后生长沉积之间的边界很难识别。因为褶皱收缩速率稳定，所以在构造形成的最

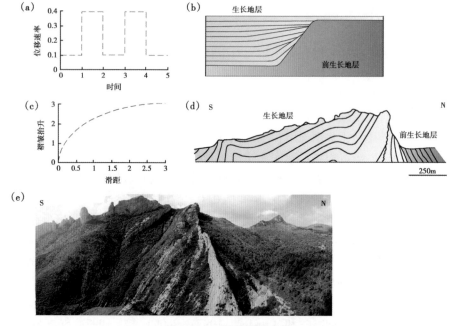

图 27-10　滑脱褶皱实例

(a) 沿着图 27-10b 中所示生长滑脱褶皱正演模型中下部的滑脱面，发生位移速率变化；图中包含了两个强烈构造活动的脉冲变化。(b) 经历了图 27-10a 所示两个构造活动脉冲变化的生长滑脱褶皱正演模型（翼部旋转），和 (b) (c) 中所示生长滑脱褶皱正演模型中顶部收缩构造的地形起伏变化（据 Hardy 和 Poblet，1994，修改）。(d) 横穿 Aguero 生长褶皱（法国比利牛斯山脉南部）南翼生长地层的地质剖面（据 Nichols，1987；Millán 等，1994，细微修改）。(e) Aguero 生长褶皱最北端横剖面的照片，显示了生长地层从倒转到平缓南倾的平面展布特征，以及伴随生长地层中退覆现象出现的上超现象，这些与图 27-10b 所描述的正演模型相似

后阶段，引起的褶皱抬升速率会明显减小（图27-10c）。因此，在褶皱褶皱生长的最后阶段，其周围沉积物的厚度没有明显变化（Poblet 和 Hardy，1995）（图27-1 和图27-8a、b）。

27.8 生长地层的剥蚀作用和原始沉积倾斜

在自然界中，生长逆冲断层相关褶皱中的前生长地层或生长地层的剥蚀作用比较常见（图27-4b 和图27-11）。为了分析剥蚀作用、物质的向下搬运以及原始沉积倾斜对生长地

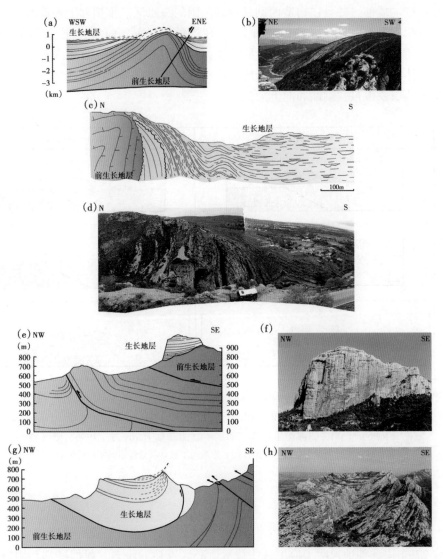

图27-11 比利牛斯山脉和加泰索尼亚沿岸山脉—伊比利亚半岛山脉中，不同构造背景下
生长地层的地质横剖面（a、c、e 和 g）和照片（b、d、f 和 h）

该图所示的生长地层和侵蚀作用的扇形形态与图27-12所描述的正演模型相似。(a) 和 (b) Mediano 背斜内的生长滑脱褶皱（法国比利牛斯山脉南部）（横剖面据 Poblet 等，1998，细微修改）；(c) 和 (d) Almunia del Romeral 生长背斜（法国比利牛斯山脉南部）（横剖面据 McElroy 和 Hirst，1989，细微修改）；(e) 和 (f) Roques de Benet 生长背斜（加泰索尼亚沿岸山脉—伊比利亚半岛山脉）（横剖面据 Lawton 等，1999，细微修改）；以及 (g) 和 (h) Penyagalera 生长向斜（加泰索尼亚沿岸山脉—伊比利亚半岛山脉）（横剖面据 Lawton 等，1999，细微修改）

645

层形成的影响，需要对基准面抬升速率大于背景沉积速率（保证生长地层没有充满所有的容纳空间），同时发育扩散作用的正演模型进行研究（Hardy 和 Poblet，1994，1995；Hardy 等，1996；Hardy 和 Ford，1997；Den-Bezemer 等，1998；Bernal 和 Hardy，2002）（图 27-12）。在这些模型中，在构造低部位生长地层变厚，形成更宽更平滑的褶皱。相对于那些生长地层已经充填至所有可容纳空间的顶部模型，这些模型的生长轴面相对模糊，更加不易识别（比较图 27-12 和图 27-5、图 27-6 以及图 27-9）。模拟水下环境时，大的沉积背景以及局部侵蚀作用引起的派生沉积都可以存在（图 27-12；水下环境）：在这种情况下，生长地层披覆在构造的翼部，并且在构造两侧发育大范围的深水区。模拟陆上环境时，只有局部沉积存在，沉积不发育（图 27-12；地面上）：在这种情况下，前生长地层受到强烈的剥蚀，特别是在倾角较陡的地区，会发生不对称剥蚀；而在褶皱翼部上，发育来自于生长褶皱的生长地层楔形体。当发育上超时，前生长地层内新的生长地层逐渐上超在老地层上；而发育退覆时，生长地层发生自我剥蚀和向上削截。

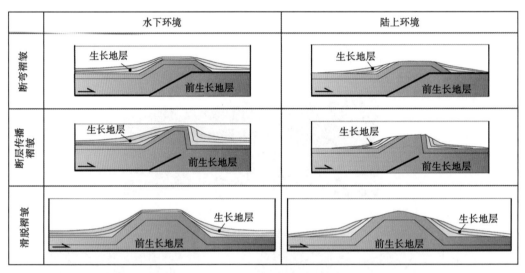

图 27-12　水下和陆上环境中，与断弯褶皱［膝折带迁移模型，Suppe（1983）］断层传播褶皱［膝折带迁移模型，Suppe 和 Medwedff（1990）］以及滑脱褶皱［翼旋转模型，Poblet 和 McClay（1996）］相关的典型生长地层形成示意图（据 Hardy 等，1996，修改）
这些正演模型中包含了侵蚀作用、物质向下迁移和地层原始倾斜的影响

27.9　生长地层的应用

在表 27-1 中详细地说明了应用这些技术所必须遵循的步骤。我们通常利用挤压背景下断层相关褶皱所控制的区域生长地层的年龄来估算褶皱发生扩展的最早时间和最短持续时间，来解释发育有两个或者更多个生长构造的地区内，褶皱和逆冲断层的形成顺序。如果从褶皱扩展开始到扩展结束，同构造沉积都连续发育，那么生长地层就记录了整个的褶皱扩展过程，并且能够估算出构造发育的完整时间。另外，许多研究表明，可以利用现今变形剖面中前生长地层的缩短量、褶皱幅度和翼部倾角除以生长地层发育的时间，来估算传统的平均褶皱缩短量、抬升速率和翼部倾斜速率。

表 27-1 利用各种技术分析生长地层并进行参数估算所需的数据和步骤

目的	所需数据	遵循步骤
确定褶皱开始扩展的时间	最新和最老生长地层的年龄	
估算褶皱扩展的持续时间	最新和最老生长地层的年龄	褶皱扩展的持续时间＝最老生长地层的年龄－最新生长地层的年龄
建立构造演化的时间序列	过天然构造的横剖面和/或天然构造三维视图	确定与本构造相关的生长地层（特别是最新和最老的生长地层）相对于其他构造来说，是前生长地层、后生长地层还是生长地层
确认不同阶段生长构造的几何形态	过天然构造的横剖面和/或天然构造三维视图	根据最符合实际地质情况的算法（地层长度不变、斜剪切等）将横剖面和（或）三维视图上的生长地层按顺序恢复至不同的阶段
	过天然构造的横剖面和/或天然构造三维视图	1. 计算天然构造横剖面和（或）三维视图内大量的参数，比如收缩量、褶皱幅度、翼部倾向等； 2. 利用之前获得的参数，建立生长构造横剖面和（或）三维视图的正演模型，模拟不同阶段天然构造的特征； 3. 将正演模型与过天然构造的横剖面和（或）三维视图进行比较； 4. 不断地重复以上步骤（修正不同输入参数）直到模拟结果与天然构造的几何形态相吻合
估算褶皱的平均扩展速率	1. 过天然构造的剖面 2. 最新和最老生长地层的年龄	1. 总缩短量＝前生长地层中未褶皱地层的长度－构造宽度； 2. 平均缩短速率＝总缩短量／（最老生长地层年龄－最新生长地层年龄）
	1. 过天然构造的剖面 2. 最新和最老生长地层的年龄	1. 褶皱幅度是指前生长地层中褶皱顶部相对于区域来说的相对高度； 2. 平均褶皱抬升速率＝褶皱幅度／（最老生长地层年龄－最新生长地层年龄）
	1. 过天然构造的剖面 2. 最新和最老生长地层的年龄	1. 翼部倾角是指前生长地层中测定的翼部倾角； 2. 平均翼部倾斜速率＝翼部倾角／（最老生长地层年龄－最新生长地层年龄）
估算褶皱扩展速率和（或）断层活动速率	1. 过天然构造的横剖面和（或）天然构造的三维视图 2. 生长地层的年龄	1. 根据最符合实际地质情况的算法（地层长度守恒或斜剪切等算法）将横剖面和（或）三维视图上的生长地层，按顺序恢复至不同阶段； 2. 测量两个不同阶段中生长地层的各种褶皱参数（缩短量、褶皱幅度、翼部倾角）或断层参数（断距）； 3. 褶皱幅度或断层位移速率＝（新生长地层发育阶段中的参数－老生长地层发育阶段中的参数）／（老生长地层年龄－新生长地层年龄）
	过天然构造的横剖面和/或天然构造的三维视图	1. 在天然构造的横剖面上和（或）三维视图内测量大量的参数，比如缩短量、褶皱幅度、翼部倾角等； 2. 利用之前获得的参数，建立生长构造二维和（或）三维的数字和（或）物理正演模型，来模拟不同阶段天然构造的特征； 3. 将正演模型与过天然构造的横剖面和（或）三维视图进行比较； 4. 不断地重复以上步骤（修正不同输入参数）直到模拟结果与天然构造的几何形态相吻合； 5. 测量两个阶段中的各种褶皱参数（缩短量、褶皱幅度、翼部倾角）或断层参数（断距）； 6. 褶皱幅度或断层位移速率＝（新阶段中测定的参数－老阶段中测定的参数）／（老阶段的年龄－新阶段的年龄）

续表

目的	所需数据	遵循步骤
估算断弯褶皱和断层传播褶皱中膝折带迁移（或翼部变宽）的速率	1. 过天然构造的剖面 2. 生长三角带中最新和最老生长地层的年龄	翼部变宽速率=（生长三角带之下的前生长地层中翼部长度/生长三角带中生长地层的最大厚度）/（生长三角带中最老生长地层的年龄-生长三角带中最新生长地层的年龄）
估算断弯褶皱和断层传播褶皱中断层滑移速率	1. 过天然构造的剖面 2. 最新和最老生长地层的年龄	1. 在 Suppe 等（1992）的图解中标注出上盘截切角、不同倾角逆冲断层之间的夹角、翼部宽度和后翼倾角； 2. 滑移量=翼部宽度/图解中 y 轴上获得的数值； 3. 断层滑移速率=滑移量/（最老生长地层年龄-最新生长地层年龄）
利用地层长度估算断层传播褶皱和滑脱褶皱中地层的缩短速率	1. 过天然构造的剖面 2. 生长地层的年龄	1. 利用以下关系确定两个生长地层的缩短量：缩短量=未褶皱地层的长度-构造的宽度； 2. 缩短速率=（较老生长地层的缩短量-新生长地层的缩短量）/（老生长地层年龄-新生长地层年龄）
利用面积估算滑脱褶皱中缩短量与沉积量之间的比值	横切天然构造的剖面	1. 绘制一个在任意高度、平行于区域高程的参照面； 2. 测量不同生长地层相对于参照面的高度（垂直于参照面的生长地层未卷入褶皱变形）； 3. 测量不同生长地层褶皱部位与参照面之间的剩余面积； 4. 绘制生长地层高度（x 轴）与剩余面积（y 轴）之间的关系图； 5. 用线性函数拟合所有数据； 6. 最佳拟合函数的斜率=缩短量/沉积速率
估算褶皱的抬升速率（或顶部构造的地形变化）	1. 过天然构造的剖面 2. 生长地层年龄	1. 分别测量盆地内和抬升区域内生长地层的累计厚度； 2. 将厚度进行去压实校正； 3. 利用以下关系测量两个生长层位形成过程的褶皱抬升量：褶皱抬升量=盆地内某段生长地层去压实后的累计厚度-抬升区域相同生长地层去压实后的累计厚度
估算翼部旋转机制褶皱的翼部倾斜速率	1. 过天然构造的剖面 2. 生长地层年龄	翼部倾斜速率=（老生长地层的倾角-新生长地层倾角）/（老生长地层的年龄-新生长地层的年龄）
估算翼旋转形成褶皱的翼间角的变化速率	1. 过天然构造的剖面 2. 生长地层年龄	翼间角变化速率=[（180°-后翼中新生长地层与前生长地层间的夹角-前翼中新生长地层与前生长地层间的夹角）-（180°-后翼中老生长地层与前生长地层间的夹角-前翼中老生长地层与前生长地层间的夹角）]/（老生长地层年龄-新生长地层年龄）
估算翼部旋转机制平行褶皱的轴面倾角变化速率（褶皱不对称性）	1. 过天然构造的剖面 2. 生长地层年龄	轴面倾角变化速率=[（90°-后翼中新生长地层与前生长地层间的夹角/2-前翼中新生长地层与前生长地层间的夹角/2）-（90°-后翼中老生长地层与前生长地层间的夹角/2-前翼中老生长地层与前生长地层间的夹角/2）]/（老生长地层年龄-新生长地层年龄）
估算褶皱核部面积的变化速率	1. 过天然构造的剖面 2. 生长地层年龄	褶皱核部面积的变化速率=[（新生长地层中未褶皱地层的长度×新生长地层的累计厚度-横剖面上新生长地层与区域基准面之间的面积）-（老生长地层中未褶皱地层的长度×生长地层的累计厚度-横剖面上老生长地层与区域基准面之间的面积）]/（老生长地层年龄-新生长地层年龄）
估算逆冲—断坡褶皱的断层倾角	过天然构造的剖面	断层倾角=$\tan^{-1}$[（老生长地层中未褶皱地层的长度-新生长地层中未褶皱地层的长度）/（盆地内生长层位的累计去压实厚度-顶部区域内生长层位的累计去压实厚度）]

续表

目的	所需数据	遵循步骤
记录构造运动与沉积作用之间的相互作用	1. 过天然构造的剖面 2. 生长地层年龄	1. 利用以下关系估算每一个生长单元的沉积速率：沉积速率＝沉积时期生长单元内累计去压实厚度／（生长单元底部地层的年龄－生长单元顶部地层的年龄）； 2. 估算每一个生长单元内褶皱的抬升速率； 3. 绘制每一个生长单元的沉积速率和褶皱隆升速率柱状图（x 轴代表生长单元由新变老，y 轴代表沉积速率和隆升速率），并将每一时期的结果进行比较； 4. 综合考虑生长地层中上超、退覆和超覆的几何形态
阐明天然褶皱的生长是连续的还是不连续的	过天然构造的剖面	出现一系列横穿构造并具有固定厚度的生长地层，说明褶皱的生长不连续
区分褶皱扩展机制和（或）褶皱—逆冲断层间的相互关系	过天然褶皱的剖面和（或）三维视图	1. 根据最符合实际地质情况的算法（地层长度守恒或斜剪切等算法）将横截面和（或）三维视图上的生长地层，按顺序恢复至各阶段； 2. 测量不同阶段中前生长地层的翼部倾角和翼部长度； 3. 如果在不同的阶段，翼部倾角为一个常数，而翼部长度发生变化，那么它就是由膝折带迁移模式形成的；如果在不同的阶段，翼部倾角发生变化，而翼部长度为一个常数，那么它就是由翼旋转模式形成的；如果在不同的阶段，翼部倾角和翼部长度都发生变化，那么它就是由膝折带迁移和翼旋转模式共同形成的； 4. 在不同的阶段，对褶皱—逆冲断层的相互关系类型进行确认
	过天然构造的横剖面和三维视图	1. 测量天然构造横剖面和（或）三维视图中的各种参数，比如缩短量、褶皱幅度、褶皱翼部倾角等； 2. 利用之前获得的参数，建立生长构造二维和（或）三维的数字和/或物理正演模型，来模拟不同阶段天然构造的特征； 3. 将正演模型与过天然构造的横剖面和（或）三维视图进行比较； 4. 不断地重复以上步骤（修正不同输入参数）直到模拟结果与天然构造的几何形态相吻合； 5. 在正演模型的不同阶段确定褶皱扩展机制和褶皱—逆冲断层的相互作用类型
	过天然构造的剖面	1. 利用以下关系确定不同生长地层的缩短量：缩短量＝生长地层中未褶皱地层的长度－构造宽度； 2. 利用以下关系确定不同生长地层顶部构造的地势起伏情况：顶部构造地势＝盆地内生长地层的累计去压实厚度－顶部区域内生长地层的累计去压实厚度； 3. 利用以下关系测定不同生长地层的翼部倾角：翼部倾角＝生长地层的倾角； 4. 利用以下关系测定不同生长地层的褶皱核部面积：褶皱核部面积＝（生长地层中未褶皱地层的长度×生长地层的累计厚度）－横剖面上生长地层和相邻区域之间的面积； 5. 根据所获得的数据（例如顶部构造地形的起伏变化、翼部倾角和/或褶皱核部面积与缩短量的比值）绘制散点图，将这些数据点进行回归分析，得到适合每个参数的最佳函数关系； 6. 将所得到的函数与天然构造正演模拟推导出的理论函数相比较

续表

目的	所需数据	遵循步骤
确定挤压构造的意义	过天然构造的剖面	1. 研究生长三角带的几何形态； 2. 确定活动轴面和（或）有限活动轴面，以及相应的非活动轴面； 3. 确定穿过活动轴面和（或）有限活动轴面（朝向邻区静止轴面）的生长地层的运动特征； 4. 结合穿过轴面的生长地层的运动特征和地层倾角，给出合理的构造运动学解释
	过天然构造的剖面	研究生长地层中的不整合、上超和退覆样式，并将它们与一系列生长构造的正演模型相比较。这些正演模型与沉积速率/褶皱抬升速率所控制的天然构造类似

最近的研究证实，生长地层可以在不同尺度上对不同时间段内，生长构造进行解释、确认和定量研究，是一种研究生长构造几何形态的有效工具。通过对生长地层的分析可以确定褶皱的运动历史，从而为油气、矿产和地下水探测中生长构造的研究提供证据；同时为地震灾害和地下储层（核废弃物处理，CO_2 俘获）的评价提供借鉴。这些方法主要是单独或者综合利用了以下两种主要策略：

（1）利用地层长度守恒、斜剪切等算法，将自然界中生长断层相关褶皱的二维剖面和（或）三维重建剖面，进行去褶皱和去断距，从而恢复到特定的生长地层发育阶段（按时间序列恢复）（例如 DeCelles 等，1991；Bloch 等，1993；Rowan 等，1993；Buebank 和 Vergés，1994；Vergés 等，1996；Ford 等，1997；Suppe 等，1997；Lawton 等，1999；Novoa 等，2000）。

（2）利用从天然生长构造中获得的参数，建立不同类型的生长断层相关褶皱的二维和（或）三维数字或物理正演模型，并将这些正演模型的结果与天然生长构造的横剖面和（或）三维视图相比较，随后通过修正参数，逐次迭加，然后改进正演模型，最终得到一个与天然构造相吻合的最佳模型（Medwedeff，1989；Mount 等，1990；Poblet 和 Hardy，1995；Ford 等，1997；Suppe 等，1997；Lin 等，2007）。

通过剖面复原、三维块体复原，以及生长断层相关褶皱不同演化阶段的正演模型中一些参数的对比，能够计算出褶皱的扩展速率和/或断层的运动速率。

除了对剖面进行恢复和正演模拟，我们在测量生长断层相关褶皱现今变形剖面中生长地层特征的基础上，就可以利用不同的方法估算褶皱的扩展速率：

（1）断弯褶皱和断层传播褶皱中的膝折带迁移（或翼部变宽）速率（翼部长度与生长地层厚度、时间之间的关系；Suppe 等，1992）；

（2）利用地层长度计算断弯褶皱和断层传播褶皱中断层的滑移速率（上盘截切角度、不同倾角的逆冲断层面之间的夹角、翼部宽度、后翼倾角与生长地层时间之间的关系；Suppe 等，1992；Shaw 和 Suppe，1994；Mueller 和 Suppe，1997；Poblet 等，2004），以及断层传播褶皱和滑脱褶皱中的缩短速率（不同生长地层的长度、构造宽度与生长地层发育时间之间的关系；Schneider 等，1996；Butler 和 Lickorish，1997；Masaferro 等，1999；Poblet 等，2004）；

（3）利用面积来估算滑脱褶皱中缩短量与沉积量之间的比值（生长地层之下面积与参照面之上的高度之间的关系；Daëron 等，2007；Hubert-Ferrari 等，2007；González-Mieres

和 Suppe，出版中）；

（4）褶皱隆升速率（或顶部构造地形或相应沉降速率的变化）（不同生长地层在盆地内去压实厚度以及在褶皱顶部的生长地层的去压实厚度，与生长地层持续时间之间的关系；Suppe 等，1992；Poblet 和 Hardy，1995；Schneider 等，1996；Masaferro 等，1999，2002；Poblet 等，2004）；

（5）由翼旋转形成的平行褶皱中，翼部倾斜速率，翼间角的变化速率和轴面倾角（非对称性）（褶皱两翼中不同生长地层和前生长地层之间的夹角，与生长地层持续时间之间的关系；Holl 和 Anastasio，1993；Espina 等，1996；Butler 和 Lickorish，1997；Masaferro 等，1998）；

（6）褶皱核部面积的变化速率（横剖面上的面积与生长地层持续时间的关系，以及每一生长地层之下的生长地层面积与生长地层持续时间的关系；Poblet 等，2004）；

（7）逆冲断坡褶皱的断层倾角（缩短量和顶部构造地形之间的关系；Schneider 等，1996）。

值得注意的是上部生长地层的翼部宽度不是指示断弯褶皱和断层传播褶皱中膝折带迁移速率和断层滑移速率的稳定指标，因为上述轴部区域仅仅是翼部的一部分（Benesh 等，2007）。此外，应用上述的一些技术需要一定的前提条件，比如变形期间生长地层的长度和（或）厚度不变，但是实际情况是：由于生长地层未固结，其长度和厚度难免有些变化。不能将估算翼部倾斜速率的方法无条件地应用在具有递进不整合特征的褶皱中，因为这些褶皱可能会因翼部旋转而发生扩展，此时，就会得到错误的结果曲线；也不能应用在由膝折带迁移而形成宽翼褶皱中，比如断弯褶皱和断层传播褶皱（Suppe 等，1997；Benesh 等，2007）。

由膝折迁移形成的逆冲断层相关褶皱中，膝折带迁移的速率为 0.5~1.5mm/a（Suppe 等，1992；Poblet 等，1998）。对于各种类型的逆冲断层相关褶皱，前人研究表明其断层滑移速率和缩短速率，通常不会超过 3~5mm/a（Davis 等，1989；Medwedef，1992），但偶尔会达到 20mm/a（例如 Lave 和 Avouac，2000）。褶皱的隆升速率一般都小于 1mm/a，但在特殊情况下可达到 20mm/a。正常的翼部倾斜速率为 0.5°~10°/Ma，但在一些时间较短的快速变形期，翼部倾斜速率可达到 40°/Ma（Holl 和 Anastasio，1993；Butler 和 Lickorish，1997）有的甚至达到 0.32°/ka（Rockwell 等，1988）。这些构造的沉积量与缩短量之间的比值大约在 0.5 到 6.0 之间（González-Mieres 和 Suppe，待刊）。

通过比较生长地层发育期间沉积速率和褶皱抬升速率，结合上超、退覆和超覆的形态特征，不仅可以研究构造运动与耦合沉积作用之间的相互影响（Suppe 等，1992；Burbank 和 Vergés，1994；Torrente 和 Kligfield，1995；Storti 和 Poblet，1997；Hardy 等，1996；Masaferro 等，2002；Shaw 等，2004），还可以通过测量过构造的生长地层的厚度是否发生变化，来衡量天然褶皱的生长是否具有连续性（Masaferro 等，2002；Castelltort 等，2004）。

完全从构造的角度来看，生长地层的几何形态分析已经成为区分不同褶皱扩展机制和不同褶皱—逆冲断层相互作用类型的重要方法，如：（1）按时序对生长构造的横截面和 3D 地表形态进行恢复（参考以上内容）；（2）将生长构造的正演模型与真实实例相比较（Suppe 等，1992；Hardy 和 Poblet，1995；Hardy 等，1996；Torrente 和 Kligfield，1995；Storti 和 Poblet，1997；Poblet 等，1997；Rafini 和 Mercier，2002；Salvini 和 Storti，2002；Shaw 等，2004b）；和（或）（3）分别在翼部倾角、构造地形变化和（或）褶皱核部面积（或这些参数的速率）与缩短量的比值图上，标定出生长地层的实际数据，然后将实际数据与应用正演模拟所获得的理论数据相比较（Butler 和 Lickorish，1997；Poblet 等，2004）。另外，通过分析生长三角带的不对称性，可以确定通过轴面的生长地层的运动特征，并对比不整合/上超/退覆样式与生长构造正演模型中的样式，生长地层就成为确定构造会聚方向的鉴别工具（Suppe 等，1992；Hardy 和 Poblet，1995；Poblet 等，1997；Den-Bezemer 等，1998）。

27.10 结论

实际上生长地层通常伴随着断层相关褶皱的缩短。野外填图和地震研究已经大大提升了我们对这些构造几何形态的认识。数值模拟和物理模拟让我们了解了不同参数对生长地层的影响，这些参数控制了沉积可容空间，影响了沉积系统，并且改变了它们的原始形态。来自野外、地下和模拟研究的数据为定量分析褶皱/断裂运动情况，以及重建该地区的递进式演化过程提供了强有力的工具。虽然现今的生长断层相关褶皱模型的复杂度已经达到了很高的水平，并且是现今许多出版物的研究主题，但是详细的野外和地下观测仍然是检验这些模型适用性的根本方法。因为实践是检验理论和物理模型的最终标准。从纯科学研究和工业应用的观点来看，如果要在生长褶皱的形态和演化的特征上取得进展，未来的研究必须基于详细的、多学科的野外、地下和实验室研究。要想开展新的生长断层相关褶皱的研究，并且在新的概念下去修正旧的解释结果，必须对先前的研究进行再分析。这样将为这种构造类型提供额外的概念模型和数字模型。为了进一步理解生长地层特征与褶皱发育之间的相互作用，需要利用能够对天然构造和理论构造 3D 形态的扩展过程进行可视化的技术，同时需要对正在发育中的天然构造的机制进行模拟的技术。

致谢（略）

参 考 文 献

Allmendinger, R. W. (1998) Inverse and forward numerical modeling of trishear fault-propagation folds. Tectonics, 17, 640-656.

Alonso, J. L., Colombo, F., and Riba, O. (in press) Folding mechanisms in a fault-propagation fold inferred from the analysis of unconformity angles: the Sant Lloren, c growth structure (Pyrenees, Spain), in McClay, K., Shaw, J. H., and Suppe, J., eds., Thrust-related folding. American Association of Petroleum Geologists Memoir.

Anad_on, P., Cabrera, L., Colombo, F., Marzo, M., and Riba, O. (1986) Syntectonic intraformational unconformities in alluvial fan deposits, eastern Ebro Basin margins (NE Spain), in Allen, P., and Homewood, P., eds., Foreland basins. Special publication of the International Association of Sedimentologists, 8, 259-271.

Barrier, L., Nalpas, T., Gapais, D., Proust, J. N., Casas, A., Bourquin, S. (2002) Influence of syntectonic sedimentation on thrust geometry: field examples from the Iberian Chain (Spain) and analogue modelling. Sedimentary Geology, 146, 91-104.

Benedetti, L., Tapponnier, P., King, G. C. P., Meyer, B., and Manighetti, I. (2000) Growth folding and active thrusting in the Montello region, Veneto, northern Italy. Journal of Geophysical Research-Solid Earth, 105, 739-766.

Benesh, N. P., Plesch, A., Shaw, J. H., and Frost, E. K. (2007) Investigation of growth fault bend folding using discrete element modeling: implications for signatures of active folding above blind thrust faults. Journal of Geophysical Research-Solid Earth, 112, B03S04.

Bernal, A., and Hardy, S. (2002) Syn-tectonic sedimentation associated with three-dimensional fault-bend fold structures; a numerical approach. Journal of Structural Geology, 24, 609-635.

Bloch, R. B., Von Huene, R., Hart, P., and Wentworth, C. M. (1993) Style and magnitude of tectonic shortening normal to the San Andreas fault across Pyramid Hills and Kettleman Hills South Dome, California. Geological Society of America Bulletin, 105, 464-478.

Bulnes, M., and Poblet, J. (1999) Estimating the detachment depth in cross sections involving detachment folds. Geological Magazine, 136, 395-412.

Burbank, D. W., and Verges, J. (1994) Reconstruction of topography and related depositional systems during active thrusting. Journal of Geophysical Research- Solid Earth, 99, 20281-20297.

Burbank, D., Meigs, A., and Brozovic, N. (1996) Interactions of growing folds and coeval depositional systems. Basin Research, 8, 199-223.

Butler, R., and Lickorish, W. H. (1997) Using high-resolution stratigraphy to date fold and thrust activity: examples from the Neogene of south-central Sicily. Journal of the Geological Society London, 154, 633-643.

Castelltort, S., Pochat, S., and Van den Driessche, J. (2004) How reliable are growth strata in interpreting short-term (10 s to 100 s ka) growth structures kinematics? Comptes Rendus-Academie des Sciences. Geoscience, 336, 151-158.

Cristallini, E. O., and Allmendinger, R. W. (2002) Backlimb trishear: a kinematic model for curved folds developed over angular fault bends. Journal of Structural Geology, 24, 289-295.

Daeron, M., Avouac, J. P., and Charreau, J. (2007) Modeling the shortening history of a fault tip fold using structural and geomorphic records of deformation. Journal of Geophysical Research-Solid Earth, 112, B03S13.

Davis, T. L., Namson, J., and Yerkes, R. F. (1989) A cross section of the Los Angeles area-seismically active fold and thrust belt, the 1987 Whittier Narrows earthquake and earthquake hazard. Journal of Geophysical Research-Solid Earth and Planets, 94, 9644-9664.

DeCelles, P. G., Gray, M. B., Ridgway, K. D., Cole, R. B., Srivastava, P., Pequera, N., and Pivnik, D. A. (1991) Kinematic history of a foreland uplift from Paleocene synorogenic conglomerate, Beartooth Range, Wyoming and Montana. Geological Society of America Bulletin, 103, 1458-1475.

Den-Bezemer, T., Kooi, H., Podladchikov, Y., and Cloetingh, S. (1998) Numerical modelling of growth strata and grain-size distributions associated with fault-bend folding, in Mascle, A., Puigdefabregas, C., Luterbacher, H. P., and Fernandez, M., eds., Cenozoic foreland basins of Western Europe. Geological Society Special Publications, 134, 381-401.

Doglioni, C., and Prosser, G. (1997) Fold uplift versus regional subsidence and sedimentation rate. Marine and Petroleum Geology, 14, 179-190.

Espina, R. G., Alonso, J. L., and Pulgar, J. A. (1996) Growth and propagation of buckle folds determined from syntectonic sediments (the Ubierna Fold Belt, Cantabrian Mountains, N Spain). Jounal of Structural Geology, 18, 431-441.

Erslev, E. A. (1991) Trishear fault-propagation folding. Geology, 19, 617-620.

Fernández, O., Munoz, J. A., Arbués, P., Falivene, O., and Marzo, M. (2004) Three-dimensional reconstruction of geological surfaces; an example of growth strata and turbidite systems from the Ainsa Basin (Pyrenees, Spain). American Association of Petroleum Geologists Bulletin, 88, 1049-1068.

Fernández-Viejo, G., Gallastegui, J., Pulgar, J. A., Gallart, J., and MARCONI Team (2010) The MARCONI project: a seismic view into the eastern part of the Bay of Biscay. Tectonophysics. doi: 10. 1016/j. tecto. 2010. 06. 020.

Ford, M., Williams, E. A., Artoni, A., Vergés, J., and Hardy, S. (1997) Progressive evolution of a fault-related fold pair from growth strata geometries, Sant Lloren, c de Morunys, SE Pyrenees. Journal of Structural Geology, 19, 413-442.

González-Mieres, R., and Suppe, J. (in press) Shortening histories of active detachment folds based on area of relief methods, in McClay, K., Shaw, J. H., and Suppe, J., eds., Thrust-Related Folding. American Association of Petroleum Geologists Memoir.

Hardy, S., and Ford, M. (1997) Numerical modeling of trishear fault propagation folding. Tectonics, 16, 841-854.

Hardy, S., and Poblet, J. (1994) Geometric and numerical model of progressive limb rotation in detachment folds. Geology, 22, 371-374.

Hardy, S., and Poblet, J. (1995) The velocity description of deformation. Paper 2: sediment geometries associated with fault-bend and fault-propagation folds. Marine and Petroleum Geology, 12, 165-176.

Hardy, S., Poblet, J., McClay, K. R., and Waltham, D. (1996) Mathematical modelling of growth strata associated with fault-related fold structures, in Buchanan, P. G., and Nieuwland, D. A., eds., Modern developments in structural interpretation, validation and modeling. Special Publication Geological Society, 99, 265-282.

Holl, J. E., and Anastasio, D. J. (1993) Palaeomagnetically derived folding rates, southern Pyrenees, Spain. Geology, 21, 271-274.

Homza, T. X., and Wallace, W. K. (1995) Geometric and kinematic models for detachment folds with fixed and variable detachment depths. Journal of Structural Geology, 17, 575-588.

Hubert-Ferrari, A., Suppe, J., González-Mieres, R., and Wang, X. (2007) Mechanisms of active folding of the landscape (southern Tian Shan, China). Journal of Geophysical Research-Solid Earth, 112, B03S09.

Lave, J., and Avouac, J. P. (2000) Active folding of fluvial terraces across the Siwaliks Hills, Himalayas of central Nepal. Journal of Geophysical Research-Solid Earth, 105, 5735-5770.

Lawton, T. F., Roca, E., and Guimerá, J. (1999) Kinematicstratigraphic evolution of a growth syncline and its implications for tectonic development of the proximal foreland basin, southeastern Ebro basin, Catalunya, Spain. Geological Society of America Bulletin, 111, 412-431.

Lin, M. L., Wang, C. P., Chen, W. S., Yang C. N., and Jeng F. S. (2007) Inference of trishear-faulting processes from deformed pregrowth and growth strata. Journal of Structural Geology, 29, 1267-1280.

Masaferro, J. L., Poblet, J., and Bulnes, M. (1998) Cuantificación del crecimiento de pliegues con sedimentos sintectónicos asociados: aplicación al anticlinal de Santaren (orógeno cubano, cuenca de antepais de las Bahamas). Acta Geológica Hispánica, 33, 75-87.

Masaferro, J. L., Poblet, J., Bulnes, M., Eberli, G. P., Dixon, T. H., and McClay, K. (1999) Palaeogene-Neogene/present day (?) growth folding in the Bahamian foreland of the Cuban fold and thrust belt. Journal of the Geological Society, 156, 617-631.

Masaferro, J. L., Bulnes, M., Poblet, J., and Eberli, G. P. (2002) Episodic fold uplift inferred from the geometry of syntectonic carbonate sedimentation: the Santaren anticline, Bahamas foreland. Sedimentary Geology, 146, 11-24.

McElroy, R., and Hirst, J. P. P. (1989) Fourth day. Part D. The Vadiello distributory system, in Friend, P. F., Marzo, M., and Puigdefábregas, C., eds., Pyrenean tectonic control of Oligo-Miocene river systems, Huesca, Aragon, Spain. 4th International Conference on Fluvial Sedimentology, Excursion Guidebook, 4, 127-138.

Medwedeff, D. A. (1989) Growth fault-bend folding at Southeast Lost Hills, San Joaquin Valley, California. American Association of Petroleum Geologists Bulletin, 73, 54-67.

Medwedeff, D. A. (1992) Geometry and kinematics of an active, laterally propagating wedge thrust, Wheeler Ridge, California, in Mitra, S., and Fisher, G. W., eds., Structural geology of fold and thrust belts. The John Hopkins University Press, Baltimore, 3-28.

Millán, H., Martínez-Peña, M. B., Aurell, M., Pocoví, A., Arenas, C., Pardo, G., and Meléndez, A. (1994) Sierras exteriores y sector del Cinca en el Pirineo aragonés: estructura y depositons sintectonicos. Guiade excursiones del congreso del Grupo Espanol del Terciario, Jaca.

Molnar, P., Brown, E. T., Burchfiel, C., Qidong, D., Xianyue, F., JUN, L., Raisbeck, G., Jianbang, S., Zhangming, W., Yiou, F., and Huichuan, Y. (1994) Quaternary climate change and the formation of river terraces across growing anticlines on the north flank of the Tien Shan, China. Journal of Geology, 102, 583-602.

Mount, V. S., Suppe, J., and Hook, S. C. (1990) A forward modeling strategy for balancing cross sections. American Association of Petroleum Geologists Bulletin, 74, 521-531.

Mueller, K., and Suppe, J. (1997) Growth of Wheeler Ridge anticline, California: geomorphic evidence for faultbend folding behaviour during earthquakes. Journal of Structural Geology, 19, 383-396.

Nichols, G. F. (1987) Syntectonic alluvial fan sedimentation, southern Pyrenees. Geological Magazine, 124, 121-133.

Nigro, F., and Renda, P. (2004) Growth pattern of underlithified strata during thrust-related folding. Journal of Structural Geology, 26, 1913-1930.

Novoa, E., Suppe, J., and Shaw, J. (2000) Inclined-shear restoration of growth folds. American Association of Petroleum Geologists Bulletin, 84, 787-804.

Poblet, J. (2004) Geometría y cinemática de pliegues relacionados con cabalgamientos. Trabajos de Geología, 24, 127–146.

Poblet, J., and Hardy, S. (1995) Reverse modelling of detachment folds; application to the Pico del Aguila anticline in the South Central Pyrenees (Spain). Journal of Structural Geology, 17, 1707–1724.

Poblet, J., and McClay, K. R. (1996) Geometry and kinematics of single-layer detachment folds. American Association of Petroleum Geologists Bulletin, 80, 1085–1109.

Poblet, J., Bulnes, M., McClay, K. R., and Hardy, S. (2004) Plots of crestal structural relief and fold area versus shortening: a graphical technique to unravel the kinematics of thrust-related folds, in McClay, K. R., ed., Thrust tectonics and hydrocarbon systems. American Association of Petroleum Geologists Memoir, 82, 372–399.

Poblet, J., McClay, K. R., Storti, F., and Munoz, J. A. (1997) Geometries of syntectonic sediments associated with single-layer detachment folds. Journal of Structural Geology, 19, 369–381.

Poblet, J., Munoz, J. A., Travé, A., and Serra-Kiel, J. (1998) Quantifying the kinematics of detachment folds using the 3D geometry: application to the Mediano anticline (Pyrenees, Spain). Geological Society of America Bulletin, 110, 111–125.

Rafini, S., and Mercier, E. (2002) Forward modelling of foreland basins progressive unconformities. Sedimentary Geology, 146, 75–89.

Riba, O. (1973) Las discordancias sintectónicas del Alto Cardener (Prepirineo Catalán), ensayo de interpretación evolutiva. Acta Geológica Hispánica, 8, 90–99.

Riba, O. (1976) Syntectonic unconformities of the Alto Cardener, Spanish Pyrenees: a genetic interpretation. Sedimentary Geology, 15, 213–233.

Rockwell, T. J., Keller, E. A., and Dembroff, G. R. (1988) Quaternary rate of folding of the Ventura Avenue anticline, western Transverse Ranges, southern California. Geological Society of America Bulletin, 100, 850–858.

Rowan, M. G., Kligfield, R., and Weimer, P. (1993) Processes and rates of deformation: preliminary results from the Mississippi fan foldbelt, deep gulf of Mexico. GCSSEPM Foundation 14th Annual Research Conference, 209–218.

Salvini, F., and Storti, F. (2002) Three-dimensional architecture of growth strata associated to fault-bend, faultpropagation, and decollement anticlines in nonerosional environments. Sedimentary Geology, 146, 57–73.

Schneider, C. L., Hummon, C., Yeats, R., and Hutfile, G. (1996) Structural evolution of the northern Los Angeles basin, California, based on growth strata. Tectonics, 15, 341–355.

Shaw, J. H., and Suppe, J. (1994) Active faulting and growth folding in the eastern Sant Barbara Channel, California. Geological Society of America Bulletin, 106, 607–626.

Shaw, J. H., Novoa, E., and Connors, C. (2004) Structural controls on growth stratigraphy in contractional faultrelated folds, in McClay, K. R., ed., Thrust tectonics and hydrocarbon systems. American Association of Petroleum Geologists Memoir, 82, 400–412.

Shaw, J. H., Connors, C., and Suppe, J. (2005a) Part 1: structural interpretation methods, in

Shaw, J. H., Connors, C., and Suppe, J., eds., Seismic interpretation of contractional fault-related folds. An American Association of Petroleum Geologists seismic atlas. Studies in Geology #53. Tulsa, OK, American Association of Petroleum Geologists, 1-58.

Shaw, J. H., Hook, S. C., and Suppe, J. (2005b) Pitas Point Anticline, California, U. S. A, in Shaw, J. H., Connors, C., and Suppe, J., eds., Seismic interpretation of contractional fault-related folds. An American Association of Petroleum Geologists seismic atlas. Studies in Geology #53. Tulsa, OK, American Association of Petroleum Geologists, 60-62.

Storti, F., and Poblet, J. (1997) Growth stratal architectures associated to décollement folds and fault-propagation folds. Tectonophysics, 282, 353-374.

Strayer, L. M., Erickson, S. G., and Suppe, J. (2004) Influence of growth strata on the evolution of fault-related foldsdistinct-element models, in McClay, K., ed., Thrust tectonics and hydrocarbon systems. American Association of Petroleum Geologists Memoir, 82, 413-437.

Suppe, J. (1983) Geometry and kinematics of fault bend folding. American Journal of Science, 283, 684-721. Suppe, J., and Medwedeff, D. A. (1990) Geometry and kinematics of fault-propagation folding. Eclogae Geologicae Helvetiae, 83, 409-454.

Suppe, J., Chou, G. T., and Hook, S. (1992) Rates of folding and faulting determined from growth strata, in McClay, K. R., ed., Thrust tectonics. London, Chapman and Hall, 105-122.

Suppe, J., Connors, C. D., and Zhang, Y. (2004) Shear faultbend folding, in McClay, K. R., ed., Thrust tectonics and hydrocarbon systems. American Association of Petroleum Geologists Memoir, 82, 303-323.

Suppe, J., Sábat, F., Muñoz, J. A., Poblet, J., Roca, E., and Vergés, J. (1997) Bed-by-bed fold growth by kink band migration: Sant Llorenç, de Morunys, Eastern Pyrenees. Journal of Structural Geology, 19, 443-462.

Tavani, S., Storti, F., and Salvini, F. (2005) Rounding hinges to fault-bend folding: geometric and kinematic implications. Journal of Structural Geology, 27, 3-22.

Tavani, S., Storti, F., and Salvini, F. (2006) Double-edge fault-propagation folding: geometry and kinematics. Journal of Structural Geology, 28, 19-35.

Tavani, S., Storti, F., and Salvini F. (2007) Modelling growth stratal architectures associated with double edge fault-propagation folding. Sedimentary Geology, 196, 119-132.

Torrente, M. M., and Kligfield, R. (1995) Modellizzazione preditiva di pieghe sinsedimentaire. Bolletino de la Societá Geologica Italiana, 114, 293-309.

Vergés, J., Burbank, D. W., and Meigs, A. (1996) Unfolding: an inverse approach to fold kinematics. Geology, 24, 175-179.

Wickham, J. (1995) Fault displacement-gradient folds and the structure at Lost Hills, California (U. S. A.). Journal of Structural Geology, 17, 1293-1302.

Zoetemeijer, R. (1993) Tectonic modelling of foreland basins: thin skinned thrusting, syntectonic sedimentation and lithospheric flexure. PhD thesis, Vrije Universiteit Amsterdam, 148 p.

(吴哲 赵梦 崔敏 译，崔敏 校)

第五部分 板内盆地及其他盆地类型

第28章 板内多阶盆地

CAPIL JOHNSON[1], BRADLEY D. RITTS[2]
(1. Geology and Geophysics, University of Utah, Salt Lake City, USA;
2. Chevron Asia Pacific Exploration and Production, Singapore)

摘 要：很多盆地分类体系强调特定的区域构造背景、沉降驱动机制和局部构造环境。虽然上述基本控盆因素对认识盆地内地层成因非常关键，也是合理的盆地分类框架，但并不能轻易地对很多复合型盆地进行分类。板内多阶盆地是形成于构造活跃板块内部的复合沉积盆地，是板块边界构造条件演变及其与板块内部机制相互作用的结果。板内多阶盆地一般具有分布广、厚度大的沉积盖层，其记录了不同构造运动阶段以及构造环境的转变历史。其中，构造环境的转变包括地体融合（拼贴）导致的逐渐向板块内部的运动和（或）由诸如从聚敛板块边缘向转换边缘的转变，或海沟的迁移引起的板块边界条件的变化。这些构造驱动机制与盆地基底和周围的地壳以及大量沉积物堆积共同作用，通过多个过程产生可容纳空间。板内多阶盆地在时空上具有多变的构造样式，但是往往构造作用是可容纳空间形成的主要因素。大型的板内多阶盆地包括中国的塔里木、鄂尔多斯、四川和准噶尔盆地；亚洲中东部的黑海和里海盆地；较小的盆地群分布在内蒙古、中国西部、台北及中亚的部分地区。欧亚大陆板内多阶盆地是较新的大陆内部强度不均一以及显生宙以来长期构造活动共同作用的产物。

关键词：混合板内盆地　沉积相　沉降机理　砂岩物源区　欧亚大陆

28.1 引言

根据构造和地球动力背景可对现代的板块内部盆地进行大致地分类，如挠曲盆地、伸展盆地和走滑盆地，盆地内的地层实际上也是众所周知的盆地类型的典型记录。然而一些发展成为巨型沉积盆地的板内盆地明显展现出时期长（>100Ma）、阶段多的演化特点，并且具有多个不同沉降阶段的沉积和构造记录，这些记录可能与盆地内或邻近盆地区的多阶段变形有关。因此，这类盆地被称作板内多阶盆地（PIP）。

板内多阶盆地在世界各地均有分布，其中以特提斯构造域北部边缘（中生代—新生代黑海、里海、阿木河、塔里木、准噶尔、鄂尔多斯、四川盆地）及伊朗、埃及和青藏高原最为典型，并以东亚和中亚地区最为有名（图28-1），这归根于东亚与中亚在显生宙多期碰撞引起的大陆拼贴融合过程中形成巨大的、相对年轻的且成分不均匀的基底，这有助于长期演化的多阶段盆地的形成与保存（Molnar 和 Tapponnier，1975，1981）。

板内多阶盆地与多期板内变形联系紧密，一般形成于增生基底，且至少部分受盆地构造和地形地貌控制。板内多阶盆地地层和构造长期演化反映了板块边界数百至数千千米范围外板块的构造类型。板内多阶盆地的沉降控制因素包括板块边界条件（如碰撞、俯冲和板片折返）以及下伏深层的动力学驱动机制（Ziegler 等，1998）。

尽管从定义上看板内多阶盆地为叠合型盆地，但在盆地规模（空间与时间上）、构造控

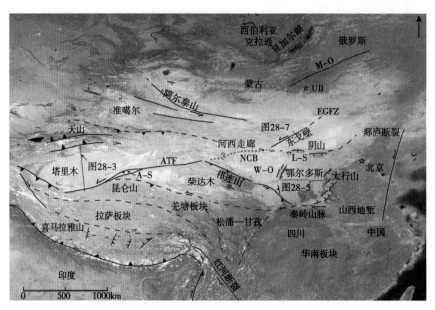

图 28-1 中国和蒙古板内多阶盆地位置简图及主要构造要素（据 Yin 和 Nie, 1996）
断层和显生宙主要缝合线（虚线），逆断层（带三角形的线）、正断层（标记号标记）和平移断层（箭号标记）；
A-S—阿尔金山；ATF—Altyn Tagh 断层；EGFZ—East Gobi 断层线；L-S—Lang Shan；M-O—Mongol-Okhotsk 褶皱带；NCB—华北板块；UB—Ulaan Baatar；W-O—Western Ordos 褶皱和逆冲断层带

制因素、地层地质记录上均具有巨大的多样性。然而，研究揭示了板内多阶盆地普遍存在可以将其划分为一种独特盆地类型的特征，即板内多阶盆地的一系列沉降活化历史与不同时期板块边缘格局和动力学状态紧密关联。此外，这些盆地往往对板块边缘（和板块内部）演化具有详细的记录。因此，更好地了解这类相关盆地的形成与发展是至关重要的。板内多阶盆地已被广泛分为中国式盆地（Bally 和 Snelson, 1980）、碰撞继承盆地（Graham 等, 1993）和板内前陆盆地（Ritts 和 Biffi, 2001）。近期，Carroll 等（2010）提出"围限盆地"的术语来描述中国一些大型的、河流内流的无海相盆地。尽管所有这些盆地类型都属于我们所认为的板内多阶盆地，但是没有一个分类能充分包含这些盆地形成的构造体制和构造样式。板内多阶盆地特征反映了与增生和俯冲带机制相关的构造活动引起的板块边缘收缩、伸展、走滑环境下的复杂拼贴特征。

这里以三个欧亚大陆的板内多阶盆地为例，对该类盆地的多样性和普遍性的特征进行阐述（图28-1）。中国西北部的塔里木盆地和中国北部的鄂尔多斯盆地是典型的巨型完整板内盆地。它们发育在相对稳定的大型陆块之上，并被复杂造山带环绕（经常伴生小的山间盆地）。相比较而言，蒙古东戈壁盆地则是叠加在晚古生代岛弧而非稳定块体之上。在东戈壁断裂带长期发育的离散构造背景影响之下，东戈壁盆地经历多期沉降与构造变形，发展成一个小型的多阶盆地（图28-1）。尽管以上盆地有许多不同之处，但其都具有以下四个共同点：形成于显生宙增生基底、远离同时代生成的板块边缘、盆地雏形分阶段变化和构造背景复杂。

28.2 塔里木盆地

塔里木盆地因其地处中国西北部的地域优势和为各种重要自然资源产地而备受关注。它是一个独特的、巨大的、主体为非海相的原型盆地，由许多板内多阶盆地和前文所述的"中国式盆地"或者碰撞继承盆地组成（Grahan等，1993）。

在区域经历中生代—新生代漫长的盆地整合与生长之后，塔里木盆地最终演化成现在这样巨型、完整、内流、封闭的盆地格局。塔里木盆地中生代基底是形成于相对稳定块体之上的古生代海相至非海相盆地。两者具有不确定的关系，但被认为既不是微大陆（Li等，1996），也不是大洋高原（Sengor等，1996；Guo等，2005）。二叠纪塔里木块体沿着分隔塔里木与准噶尔盆地的天山缝合带向欧亚大陆增生（Carroll等，1995；图28-1）。继这次碰撞之后，欧亚大陆继续向南扩张，三叠纪与羌塘地块碰撞、侏罗纪与拉萨地块碰撞之后，欧亚板块与西藏拼合，其他较小的岛弧和印度板块最终在新生代与欧亚板块拼合（图28-2）。伴随着每一次碰撞，一个新的活动俯冲带在增生的欧亚大陆南边形成。因此，板块边缘会聚作用持续超过200Ma，每一次碰撞塔里木块体离板块边缘越来越远。除这些碰撞事件外，众所周知俯冲带动力是随时间变化的（Kapp等，2000；Yang等，2003），塔里木南部长期发育的构造事件持续驱动中生代—新生代板块内部沉降变形。

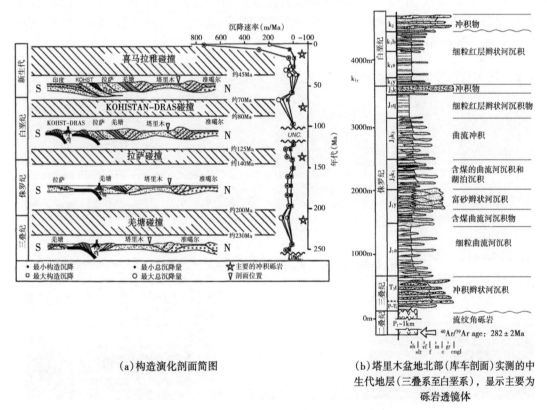

图28-2 塔里木盆地北部构造演化史和地层演化
（据Hendrix，1992；Hendrix等，1992；使用经作者允许）

现今的塔里木盆地是活动的褶皱冲断带和走滑断裂环绕的干旱内陆盆地。尽管距离最近的活动板块边缘在700km以上，但塔里木盆地的应变速率和地震等级都很大。这个构造活动导致快速的沉降，并形成池状可容空间，这也是塔里木盆地大部分地区发育超过1km厚的新近系的主要原因（图28-3；Li等，1996；Yue等，2004；Ritts等，2008）。

663

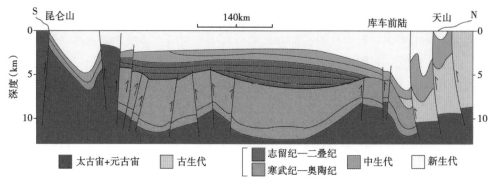

图 28-3 塔里木盆地横向剖面图（据 Li 等修改，1996）
剖面位置见图 28-1，构造上超和交切关系反映了盆地沉积中心的侧向扩大

盆地充填的沉积物主要来自周缘造山带，由辫状冲积扇体系从盆地边缘供给到沉积盆地中。盆地内部远离冲积体系的沉积物在砂质河流体系、浅水蒸发湖泊和风积沙丘地带搬运和沉积（图 28-4）。这些沉积类型从中中新世至今始终如一。尽管如此，在 15Ma 之前，塔里木盆地具有截然不同的沉积样式（Ritts 等，2008）。

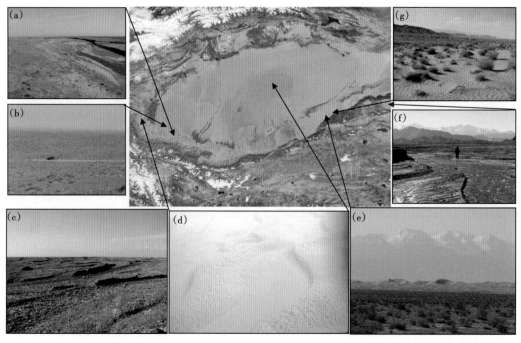

图 28-4 塔里木盆地现代沉积体系是一个封闭干旱的板内多阶盆地的典型实例
（箭头指示每张图片的大概位置）

(a) 砂质辫状河流体系由盆地周边流向盆地内部，向下游流量减小；(b) 内陆位置与四周山系导致雨水盲区使平原干旱植物稀少；(c) 近盆地边缘的具有较大流量的砾石质辫状河沉积体系；(d) 航拍塔里木盆地沙漠，沙丘背风面超过 100m，沙脊延续数千米；(e) 盆地边缘的小规模沙丘（背风面高度几米至几十米）；(f) 盆地边缘辫状河流起源于雪水和冰川融水盆地边缘由连接低海拔盆地中心（800~1400m）和高海拔（1500~2000m 山前带的）大区域的斜坡组成

与现今盆地格局不同，中生代的塔里木盆地不是单一的、池状的叠合盆地（Zhou 和 Chen，1990）。相反，它是由一系列与盆地边缘构造类型有直接联系的次级盆地组成。塔里木盆地北部边缘已被详细研究（Hendrix 等，1992；Charreau 等，2006，2009；Heermance 等，2007）。在中生代期间，塔里木盆地北部地区是板内前陆盆地，它由天山褶皱—冲断带的板内缩短形成。后续的块体碰撞加速了天山缩短和塔北前渊的挠曲再生（图 28-2a）。而块体碰撞期间盆地内重复的幕式砾岩沉积层和近源冲积扇沉积环境是对天山造山带回春和塔里木北部同造山期沉积作用的直接反映（图 28-2b）。相比较而言，由板内巴楚隆起分割的塔里木盆地西部地区是由一系列复杂的与西部盆地边缘缩短和走滑有关的转换挤压盆地和前陆盆地组成，类似地，塔里木盆地南部主要是板内隆起分割的一系列小型构造限定的"破坏前陆盆地"（Dickinson 等，1988）。这些盆地很可能与中生代塔里木盆地东南部相邻的柴达木盆地有紧密联系。

古近纪印度板块的碰撞开启了塔里木盆地格局转变的序幕，直到新近纪完全演化成现今盆地格局。这个转变包括天山活化，如中生代地体碰撞造成天山反复活化一样。虽然塔里木盆地南部边缘的地形演化没有很好地记录，但是古近纪来自喜马拉雅—西藏造山带的形变向北传播，最终在渐新世到达昆仑造山带（Wang 等，2008）。同样在渐新世时期，阿尔金断裂迅速走滑将柴达木盆地和塔里木盆地分隔（Ritts 等，2004；Yue 等，2004b），尽管直到中新世（约 15Ma）阿尔金山才开始形成（Ritts 等，2008）。中中新世的收缩位置在现今的西藏高原北部边界一带，且代表现代自然地理的发育。盆地西部边缘从晚中生代随着帕米尔高原向北推进持续到上新世才最终完全关闭（Coutand 等，2002）。

总之，现在观察到巨型、统一的塔里木板内多阶盆地，实际上是中生代至新生代板块边缘演化的结果。这个演化过程包括盆地的早期分隔沉积逐渐被后续新扩展的盆地沉降整合。类似地，现代的盆地格局同样依靠塔里木盆地南部老造山带的快速活化和新山脉生长（如阿尔金山），最终关闭塔里木盆地南部边缘并使得其向内陆泄水盆地发展。因此，塔里木盆地演化阐明了板内沉积区域扩张如何导致完整的板内多阶盆地的形成。

28.3 鄂尔多斯盆地

中生代的鄂尔多斯盆地与新近纪塔里木盆地一样，是一个四周被活动造山带环绕的巨型板内多阶盆地。不同的是，鄂尔多斯盆地新生代构造变形作用较弱及地层叠加对中生代板内变形和盆地的形成影响较弱，因此，鄂尔多斯盆地是中生代中国板内多阶盆地的最好实例。

鄂尔多斯盆地是被中生代四个造山带所围限的独特盆地，并且各造山带呈直角彼此相连：东西方向分别为北部的阴山造山带和南部的秦岭造山带，南北方向分别为东部的太行山造山带和西鄂尔多斯带（图 28-1）。这些山脉呈复杂的形变样式环绕在鄂尔多斯大陆地块边缘，作为一个前寒武纪陆块，虽然鄂尔多斯盆地均匀沉积了近 7km 厚遍布盆地的中生代未变形地层，但是盆地周边的造山带却表现出复杂的变形样式（图 28-5），这些复杂的变形样式在鄂尔多斯盆地西北部的阴山造山带狼山段和邻接未变形的鄂尔多斯盆地部分贺兰山—桌子山相交的地区具有很好的出露（图 28-6）。

就鄂尔多斯盆地西北部的构造样式而言，前人已经开展了较多的研究工作（图 28-6；Darby 和 Ritts，2002，2007；Ritts 等，2009）。二叠纪主要为碰撞及碰撞后沿 Suolon 缝合线向北会聚的过程（Cope 等，2005），并且一直持续到三叠纪（Darby 等，2001）。在三叠纪，

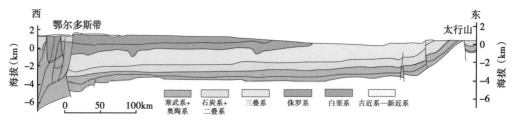

图 28-5　鄂尔多斯横剖面图（据 Hanson 等，2007；Yang 等，2005）
主要地层在整个盆地厚度相对均匀，西鄂尔多斯带具有局部前渊盆地的特征；位置见图 28-1

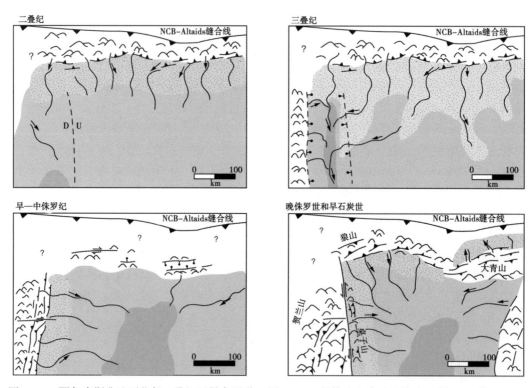

图 28-6　鄂尔多斯盆地西北部二叠纪至早白垩世（图 28-1）的构造和古地理演变（据 Ritts 等，2009）
多阶板内盆地周缘经历多期构造和沉积变迁，同时盆地内部保持持续沉降；在鄂尔多斯盆地西北部，尽管变形样式不同，但是变形作用持续发育在盆地西部和北部边缘；这种情况可能是继承性软弱基底的影响，也可能是盆地内冷的沉积岩厚度不断增加的结果；需要注意的是，造山带内部局部构造控制盆地的形成和鄂尔多斯盆地内部的持续沉降和沉积；此外，鄂尔多斯盆地逐渐孤立和封闭的过程也值得关注；底纹与图案大致代表距源区由近及远的沉积环境下，沉积物颗粒的相对大小

沿鄂尔多斯西部边缘发育伸展型盆地，形成现今贺兰山地区具有较厚的三叠系，并且沉积物主要来源于北部阴山造山带（Ritts 等，2004）。早侏罗世，随着阴山造山带变形向伸展为主的构造背景转变（Darby 等，2001；Ritts 等，2001），鄂尔多斯西部边缘开始向东会聚缩短（Darby 和 Ritts，2002）。最终在晚侏罗世和早白垩世，盆地会聚消减活跃于盆地北、西两个边缘。鄂尔多斯盆地东部边缘也受东西向平缓的收缩影响而关闭（Darby 等，2001；Darby 和 Ritts，2002）。

相比之下，鄂尔多斯盆地北部边缘的阴山造山带在三叠纪为南北方向收缩，早侏罗世为南北向伸展（Darby 等，2001；Ritts 等，2001）。到中侏罗世末期，阴山造山带再次转变为南北向挤压缩短，并持续到早白垩世，同期鄂尔多斯西部褶皱冲断带也为东西向挤压缩短（Darby 等，2001）。与鄂尔多斯西部类似，阴山造山带的挤压缩短伴随着白垩纪的伸展作用（Darby 等，2001；Davis 等，2002；Ritts 等，2010）。

中生代的变形残留保留于稳定鄂尔多斯地块边缘，在这些构造变形强烈控制盆地局部沉积作用和沉积物源的同时，鄂尔多斯盆地内部在该变形时期则表现为垂向相对一致的沉降过程（图28-5）。这种模式可能受盆地继承性的软弱基底导致，也可能是鄂尔多斯盆地本身冷的沉积岩厚度不断增加的结果。每期构造事件都与造山带直接邻的盆地区域的沉积具有紧密联系（图28-6）。例如，三叠纪盆地在西鄂尔多斯带扩张，产生一超过 4km 碎屑充填的扩张盆地，在各方面都符合典型的半地堑盆地模型（Leeder 和 Gawthorpe，1987）。类似地，白垩纪伸展产生半地堑盆地和内克拉通拆离盆地，这取决于构造的类型（Ritts 等，2010）。尽管如此，鄂尔多斯盆地在此期间沉积了贯穿整个鄂尔多斯地块相对均一的厚层地层，并缺少局部的构造驱动。局部构造控制盆地、外围盆地、次盆地及相对均一的广阔区域和完整的盆地沉积的有机结合是构造活动板内多阶盆地的特征。

28.4 蒙古东戈壁盆地

与中国的例子相比，邻近的蒙古东南部的东戈壁盆地则代表多期形变与盆地形成的区域，尽管与塔里木和鄂尔多斯接近，但是东戈壁盆地没有发育成一个巨大的、厚的、整一的沉积盆地。从这方面看，它与中生代的塔里木盆地非常相似。东戈壁盆地发育在古生代阿尔泰构造带岛弧块体拼贴形成的大型增生复合体之上（Sengor 和 Natalin，1996；图28-1）。东戈壁盆地南邻阿尔泰复合体的南部边缘，紧接推测的二叠系缝合带（天山—阴山缝合线；Johnson，2008）的北侧，该缝合带的南侧为中国的诸多微板块（Sengor 等，1993；Badarch 等，2002；Cope 等，2005）。

虽然东戈壁盆地以早白垩世的断陷盆地而被人熟知（Graham 等，2001），但近期的研究资料显示东戈壁盆地至少经历了四个各具特色的板块内部盆地沉降和变形阶段：晚二叠世碰撞后盆地、早中生代左旋剪切和转换挤压前陆盆地、晚中生代断陷盆地及新生代走滑盆地。在地震剖面上可以清楚地看到大部分变形阶段（图28-7），并可与多期可能的构造驱动力联

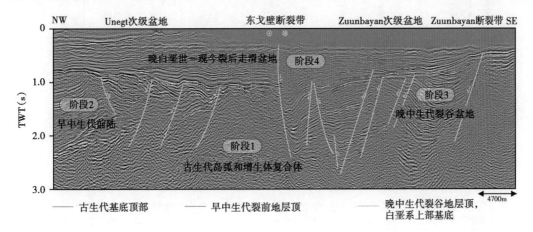

图28-7　东戈壁盆地地震横剖面图（位置见图28-1）及东戈壁盆地主要发育阶段（据 Johnson，2004）

系起来（图 28-8）。需要重点关注的是，实际上沿着东戈壁断裂带主变形期具有清楚的显示（图 28-1 和图 28-7），而东戈壁断裂带是一个始于早中生代剪切带且具有一个复杂多期脆性活化历史的断裂带。

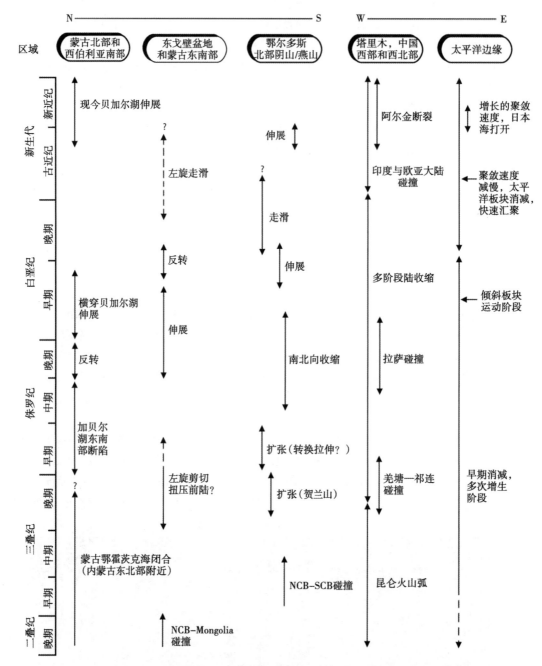

图 28-8　从晚二叠世到新生代主要陆内的碰撞事件和其他变形事件的时间顺序表
（据 Graham 等，2001；Maruyama 等，1997；Northrup 等，1995；RItts 等，2001，2009；
Vincent 和 Allen，1999；Yin 和 Nie，1996；Yue 等，2001；Zonenshain 等，1990）
这些构造事件可能是东戈壁盆地沉降变形的驱动机制；ATF—Altyn tagh 断层；NCB—SCB—华北—华南板块

1）阶段1

随着塔里木板块和华北板块向蒙古南部的石炭—泥盆纪岛弧拼贴块体（阿尔泰山）的移动，古亚洲洋逐渐关闭（Lamb和Badarch，2001），在此期间蒙古南部发育了最后的海相沉积（Dobretsov等，1994）。下—上二叠统强烈变形和变质的浊流沉积地层是该时期残留洋盆的有力证据。浊积岩地层与上覆几乎未变形的晚二叠世—三叠纪冲积、河流和湖相的地层呈不整合接触（Johnson等，2008）。这些沉积岩物源大多来源于晚古生代的火山弧（图28-9），它们构成了东戈壁板内多阶盆地的第一个发育阶段。

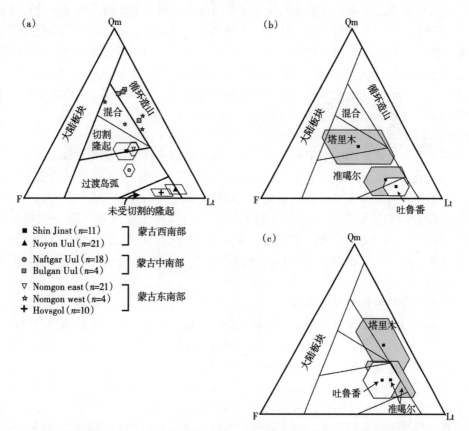

图28-9 砂岩物源的数据显示继承性的物源和局部物源是控制板内多阶盆地砂岩组分的主要因素
(a) 蒙古二叠纪砂岩样本组成模式的三元图解，使用单晶的石英（Qm），钾长石和斜长石（F）和总岩屑（Lt），平均值用标准平均偏差（计算点 $n \geqslant 4$）（原著 Dickinson 等，1983；修订来自 Johnson 等，2008；额外的数据源于 Henmann 等，2008）；(b) 总体颗粒（Qm、F、Lt 不含碳酸盐岩颗粒）显示现代砂岩来自中国西北部盆地（据 Graham 等，1993）；(c) 南部准噶尔盆地、北部塔里木盆地、西部吐鲁番盆地中生代砂岩成分数据，意义和一个标准偏差绘制在 Qm、F、Lt 图上（据 Hendrix，2000）

2）阶段2

主要的左旋剪切带（东戈壁断裂带）分隔了东戈壁盆地（图28-1），剪切构造带中糜棱岩的 $^{40}Ar \backslash ^{39}Ar$ 年龄反映了该期变形发生在晚三叠世（约225~210Ma；Lamb等，1999；Webb 和 Johnson，2006；Webb 等，2007）。在内蒙古南部和西南部，该时间段对应于主要的前陆盆地和山间盆地的形成时期（例如 Noyon Uul；Hendrix 等，1996），并与覆盖在二叠系和三叠系之上（Webb 等，1999）的前寒武纪和早古生代碳酸盐岩飞来峰的推覆剪切时间一

669

致。尽管在东戈壁盆地中三叠纪和早侏罗世的沉积地层没有很好的出露，但推覆构造和巨厚的前白垩纪、二叠纪后地层序列在地震剖面上具有清晰的揭示（图28-7），并且基于露头相互关系和地震数据上解释的裂陷前推覆构造（Heumann等，2008），认为这个序列为早中生代转换挤压前陆盆地。然而，这个前陆盆地的发育时期要晚于东戈壁断裂带的左旋走滑时期。

3）阶段3

该阶段始于晚侏罗世，并在早白垩世达到高峰，东戈壁盆地发育巨厚的断陷期地层（5km），并且遍布华北板块。在东戈壁盆地南北向地震剖面上主要的高角度盆地边界正断层清晰可见（图28-7），并且向着这些典型的地堑和半地堑构造，沉积充填的地层变厚。第二阶段的早中生代剪切带通常被脆性正断层切割，指示了变质组构的构造活化。断陷期的地层序列包括大量的双峰火山岩以及冲积扇、河流、湖泊相地层（Johnson和Graham，2004）。盆地西南部，一个同时代的变质核杂岩体于早白垩世形成（120Ma前），指示了在盆地尺度的伸展格局下，区域应变的分异和变化特征（Johnson等，2001）。

4）阶段4

区域不整合和盆地反转证实了东戈壁盆地的断陷结束于白垩纪中期（图28-7）。超覆其上的为上白垩统—现代沉积，这一直被描述为裂陷期后热沉降阶段伴随轻微变形的沉积特征。然而地下与地表研究表明，上白垩统及更新统的地层被沿着晚三叠世剪切带和活化断层带的脆性左旋走滑断层切割。沉降样式和地层等厚图揭示了强沉降受控于这些脆性构造（Johnson，2004）。由于很少有第四纪或新构造运动的证据，因此主要的走滑很可能发生在早新生代（Heumann等，2008），并可能与渐新世—中新世开始活动的中国阿尔金断裂有关（Darby等，2005）。

尽管对东戈壁盆地每个形成阶段的详尽驱动机制仍有争议，但是欧亚大陆边缘演化的事件对蒙古南部的陆内变形提供了可能的解释（图28-8）。例如，早中生代扭压与蒙古—鄂霍茨克海向北关闭过程中蒙古南部的旋转一致（Zonensnain等，1990），并且该构造事件由于华北与华南板块在南部的碰撞增强，早白垩世裂陷作用在华北和蒙古普遍发育。当有效的驱动机制处于争论的时候，与古太平洋边缘相关的弧后扩张、增厚地壳的重力塌陷及区域的下中地壳流动（Graham等，2001；Meng，2003）都可能是伸展作用的驱动力。总之，东戈壁断裂带在早新生代的活化代表了印度亚洲板块初始碰撞在板内初始大陆挤压阶段，也可能代表初始的阿尔金断裂拉开（Yue和Liou，1999；Yue等，2001，2004b；Webb和Johnson，2006）。这一阶段主要结束于晚中新世，与东亚大陆边缘快速伸展的停止以及青藏高原北部地壳开始变厚的时期吻合（Darby等，2005；Bovet等，2009）。

28.5 讨论

塔里木盆地与鄂尔多斯盆地的对比分析揭示了两盆地都被或曾经被活动的、构造多样的造山带围绕，并演化成巨厚的内陆沉积盆地。东戈壁盆地也由小块体逐渐演化为一个整体盆地，但是它规模较小，沉降阶段主要受盆地内不连续的变形带控制。三个例子都与局部构造密切控制的次盆或周围盆地相关联。尽管这些局部盆地分为断陷、走滑盆地或板内前陆盆地，但是它们也可以与更广阔的均匀沉降区结合，形成整体的板内多阶盆地。

28.6 沉积相和沉积结构

板内多阶盆地没有单一的沉积环境，其沉积环境范围可以从非海相（中国很多中生代—新生代盆地）到深海相（黑海盆地）。然而，由于这些盆地具有典型的池状特征或被周边造山带围限，因此，沉积环境一般从盆地周缘的近源沉积发展到盆地中心的远源沉积（图28-4）。Carroll等（2010）总结了中国现代"围限盆地"干旱气候的显著特点，提出了这些条件和长期发育的内陆泄水盆地之间的可能联系。虽然在地质历史上这些盆地的气候条件具有显著的变化，但是持续的干旱阻止了主要河流越过边界山脉。

盆地边缘大尺度的地层结构或它的次盆地紧密反映了局部的构造背景。例如，塔里木盆地北部边缘挠曲的几何形态类似前渊（图28-3），然而在大区域上板内多阶盆地具有一个相对均匀的地层厚度（图28-5）。这个大规模地层样式暗示垂直沉降机制是盆地边缘相对独立的构造过程。在池状盆地中，盆地边缘的地层超覆可能会局部掩盖构造特征。

28.7 物源与沉积物分布

板内多阶盆地普遍发育在显生宙增生带，特别是在亚洲。因此，在不考虑局部构造和区域构造背景的情况下，盆地充填的碎屑岩组分主要受局部物源的构造继承性控制（一般为不均匀的增生基底）。因此板内多阶盆地中砂岩组分会偏离全球构造环境识别图版中所预测岩性，而这个图版被广泛用来通过砂岩组分判别物源区的构造背景（Dickinson和Suczek，1979；Dickinson等，1983）。在塔里木盆地，新近纪到全新世砂岩指示为火山弧物源区，但是其形成于与再旋回造山带伴生的前陆盆地环境（Graham等，1993；Hendrix，2000）。东戈壁盆地二叠纪砂岩也证明了这一点（图28-9），局部构造对南部蒙古碰撞后次盆地相关的沉积物组成具有强烈的控制作用。这些砂岩主要源自古生代中晚期岛弧和增生地层序列，然而，它们的模式组成则表现为大范围的切割岛弧物源区，甚至为具有高成熟度特征的再旋回造山带的物源特征。

28.8 沉降历史

板内多阶盆地具有多种沉降机制。在这些环境下盆地的部分地区可能经历一些或所有下述的过程：挤压缩短与负载驱动的挠曲沉降、伸展断层驱动相关断层和热沉降、沿走滑断裂带的垂向运动驱动的沉降。该类盆地的另一个沉降历史特征是既不稳定也不统一。例如，塔里木盆地北部沉降分析（Hendrix等，1992；图28-2）揭示沉降作用和粗粒沉积物输入与亚洲板块南部边缘块体的幕式碰撞相关（Heermance等，2007；Charreau等，2009），并指示了天山造山带的收缩复苏。在诸多实例中，小型边缘盆地和与造山带相关的盆地（与贺兰山相邻的鄂尔多斯盆地西部）主要受局部构造机制驱动。

然而，这些沉降机理并不能解释如塔里木盆地和鄂尔多斯盆地这样大区域盆地的垂直沉降的机理。在这些例子中，沉降可能主要与基底非均质性和对负载、热效应及动力沉降的调整相关。巨型盆地的构造封闭特性（如塔里木和鄂尔多斯盆地）都说明闭塞、沉积负载和充填可能导致巨厚地层的积累。相反，东戈壁盆地发育于增生岛而非稳定地块，并且由薄的地壳基底导致其没有完全演化为巨型多阶内陆盆地（Molnar 和 Tapponnier，1981）。因此东戈壁盆地是夭折的巨型板内多阶盆地，它是板内盆地演化的一个可能端元。不管怎样，封闭性在形成这些盆地过程中非常重要，这一点将在下文讨论。

28.9 基底和构造驱动

前文三个研究实例证明，板内多阶盆地要么是继承性岩石圈流变的结果，要么是地壳长期演化过程中由于不同的热力、压力条件的强烈反差发展所致（Carroll 等，2010）。在鄂尔多斯盆地的例子中，板内多阶盆地发展阶段是它南部、北部、西部和东部边缘限制的结果。因此，鄂尔多斯板内多阶盆地来自原先广阔的华北地台的构造分隔。一旦它们的边缘确定，尤其在鄂尔多斯西北部，它们易于在空间上保持相对静止的状态，即使它们经历多种构造样式中的反复变形。作为对比，塔里木地区经过长期复合，形成中生代的板内多阶盆地。然而在塔里木和东戈壁盆地中，最根本的盆地边界如天山和东戈壁断层带在限定盆地边缘中具有关键作用。

鄂尔多斯盆地能特别地说明盆地内部与边缘动力差异如何增强盆地边缘变形，并阻止盆地内部变形（图 28-10）。这些基本的动力差异无论来自继承性基底差异、累积形变、热效应，或来自盆地自身物质与热量再分配（沉积物从周边变形区向盆地输运、增加盆地部分外壳厚度和压制热流体），均允许变形在多阶盆地边缘长时间存留（图 28-10；Braitenberg 等，2003；Carroll 等，2010）。在鄂尔多斯盆地中，不确定的是鄂尔多斯板块与华北板块其他部分之间的继承性的不均一基底是否是板块边缘的局部变形的唯一因素，或者是否在古生代至中生代时期鄂尔多斯盆地地壳的大量增厚促进了盆地局部变形。

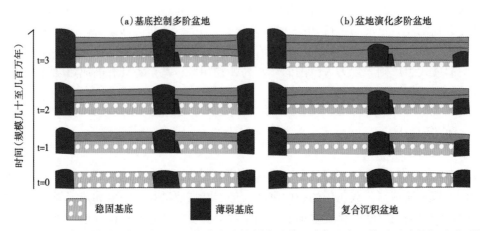

图 28-10 示意图说明基底强度和沉积物分布在控制多阶盆地内部和邻区构造活动的相对重要性
在这两种情况下，盆地因那些仍在活动的构造运动而逐渐孤立或封闭。这些构造开始被基底带限制，盆地基底展示不同力学性质；然而，在某些情况下新增的沉积岩上超或重叠构造，可能会改变盆地不同地区和它边缘的力学性质，结果造成了质量、地壳厚度或热流的再分配；无论盆地主要是由基底强度差异控制或是由沉积所控制，盆地长期演化和多元历程使它们成为多阶盆地

板内多阶盆地边缘（无论是稳定还是扩张边缘）经长时期局部变形的结果是多阶盆地在大部分地质历史上为封闭体系。周围山脉的封闭增强了盆地的可容空间并增强了盆地内的沉积和盆地边缘的变形程度。例如，塔里木盆地和与其相邻的柴达木盆地都被褶皱冲断层带

围限。这些盆地内的沉积物主要来源于周围的褶皱冲断带，并超覆在盆地边缘。物质从周缘造山带大规模转移到多阶盆地前陆必然会降低地面坡度，从而在褶皱冲断层带形成尖灭并推动造山带内部的变形。而陆内的干旱气候可能对板内盆地封闭特征的发育，尤其是在中生代至新生代的中亚地区板内盆地具有一定的影响（Sobel 等，2003；Carroll 等，2010）。

28.10 结论

板内多阶盆地是典型的大而厚的叠合沉积盆地，它远离板块边缘，并经历长期、多阶段构造变形事件影响。它们与板内多期变形密切相关，并且通常形成于不均匀的增生基底，至少部分受盆地构造和地形限制的控制。尽管地层和构造记录复杂多样，但是板内多阶盆地可以根据它们相对板块边缘的位置和长时期记录的不同沉降阶段划分类型。这些沉降阶段在很大程度上是远程的、是对板块边界转化为板块内部增生基底的构造事件的响应。

致谢（略）

参 考 文 献

Badarch, G., Cunningham, W. D., and Windley, B. F. (2002) A new terrane subdivision for Mongolia: implications for the Phanerozoic crustal growth of Central Asia. Journal of Asian Earth Sciences, 21, 87-110.

Bally, A. W., and Snelson, S. (1980) Facts and principles of world petroleum occurrence: realms of subsidence. Canadian Society of Petroleum Geologists, Memoir 6, 9-94.

Bovet, P. M., Ritts, B. D., Gehrels, G. G., Abbink, O. A., Darby, B. J., and Hourigan, J. (2009) Evidence of Miocene Crustal Shortening in the North Qilian Shan From Cenozoic Stratigraphy of the Western Hexi Corridor. American Journal of Science, 309, 290-329. doi: 10.2475/00.4009.02.

Braitenberg, C., Wang, Y., Fang, J., and Hsub, H. T. (2003) Spatial variations of flexure parameters of the Tibet-Qinghai Plateau. Earth and Planetary Science Letters, 205, 211-224.

Carroll, A. R., Graham, S. A., and Smith, M. E. (2010) Walled sedimentary basins of China. Basin Research, 22, 17-32. doi: 10.1111/j.1365-2117.2009.00458.x.

Carroll, A. R., Graham, S. A., Hendrix, M. S., Ying, D., and Zhou, D. (1995) Late Paleozoic tectonic amalgamation of northwestern China: sedimentary record of the northern Tarim, northwestern Turpan and southern Junggar basins. Geological Society of America Bulletin, 107, 571-594.

Charreau, J., Gilder, S., Chen, Y., Dominguez, A., Avouac, J-P., Sen, A., Jolivet, M., Li, Y. A., and Wang, W. M. (2006) Magnetostratigraphy of the Yaha section, Tarim Basin (China): 11 Ma acceleration in erosion and uplift of the Tianshan Mountains. Geology, 34 (3), 181-184. doi: 10.1130/G22106.1.

Charreau, J., Gumiaux, C., Avouac, J-P., Augier, R., Chen, Y., Barrier, L., Gilder, S., Dominguez, S., Charles, N., and Wang, Q. (2009) The Neogene Xiyu Formation, a diachronous prograding gravel wedge at front of the Tianshan: climatic and tectonic implications. Earth and Planetary Science Letters, 287, 298–310. doi: 10. 1016/j. epsl. 2009. 07. 035.

Cope, T., Ritts, B. G., Darby, B. J., Fildani, A., and Graham, S. A. (2005) Late Paleozoic sedimentation on the northern margin of the North China Block; implications for regional tectonics and climate change. International Geology Review, 47 (3), 270–296.

Coutand, I., Strecker, M. R., Arrowsmith, J. R., Hilley, G., Thiede, R. C., Korjenkov, A., and Omuraliev, M. (2002) Late Cenozoic tectonic development of the intramontane Alai Valley (Pamir–Tien Shan region, central Asia): an example of intracontinental deformation due to the Indo– Eurasia collision. Tectonics, 21, 1053. doi: 10. 1029/2002TC001358.

Darby, B. J., and Ritts, B. D. (2002) Mesozoic contractional deformation in the middle of the Asian tectonic collage: the enigmatic Western Ordos fold–thrust belt, China.

Earth and Planetary Science Letters, 205, 13–24.

Darby, B. J. and Ritts, B. D. (2007) Mesozoic structural architecture of the Lang Shan, North–Central China: intraplate contraction, extension, and synorogenic sedimentation. Journal of Structural Geology, 29, 2006–2016.

Darby, B. J., Davis, G. A., and Zheng, Y. (2001) Structural evolution of the southwestern Daqingshan, Yinshan belt, Inner Mongolia, China, in Hendrix, M. S., and Davis, G. A., eds. Paleozoic and Mesozoic tectonic evolution of central and eastern Asia: from continental assembly to intracontinental deformation. Geological Society of America Memoir, 194, p. 199–214.

Darby, B. J., Ritts, B. D., Yue, Y., and Meng, Q. (2005) Did the Altyn Tagh fault extend beyond the Tibetan Plateau? Earth and Planetary Science Letters, 240, 425–435.

Davis, G. A., Darby, B. J., Zheng, Y., and Spell, T. L. (2002) Geometric and temporal evolution of an extensional detachment fault, Hohhot metamorphic core complex, Inner Mongolia, China. Geology, 30, 1003–1006.

Dickinson, W. R., and Suczek, C. A. (1979) Plate tectonics and sandstone compositions. American Association of Petroleum Geologists Bulletin, 63, 2164–2182.

Dickinson, W. R., Beard, L. S., Brakenridge, G. R., Erjavec, J. L., Ferguson, R. C., Inman, K. F., Knepp, R. A., Lindberg, F. A., and Ryberg, P. T. (1983) Provenance of North American Phanerozoic sandstones in relation to tectonic setting. Geological Society of America Bulletin, 94, 222–235.

Dickinson, W. R., Klute, M. A., Hayes, M. J., Janecke, S. U., Lundin, E. R., McKittrick, M. A., and Olivares, M. D. (1988) Paleogeographic and paleotectonic setting of Laramide sedimentary basins in the central Rocky Mountain region. Geological Society of America Bulletin, 100, (7), 1023–1039.

Dobretsov, N. L., Coleman, R. G., and Berzin, N. A., eds. (1994) Geodynamic evolution of the Paleoasian Ocean. Russian Geology and Geophysics, 35 (7–8), 233 p.

Graham, S. A., Hendrix, M. S., Wang, L. B., and Carroll, A. R. (1993) Collisional successor basins of western China; impact of tectonic inheritance on sand composition. Geological Society of America Bulletin, 105, 323–344.

Graham, S. A., Hendrix, M. S., Johnson, C. L., D. Badamgarav, G. Badarch, Amory, J., Porter, M., R. Barsbold, Webb, L. E., and Hacker, B. (2001) Sedimentary record and tectonic implications of Mesozoic rifting in southeast Mongolia. Geological Society of America Bulletin, 113, 1560-1579.

Guo, Z. J., Yin, A., A. Robinson, and Jia, C. Z. (2005) Geochronology and geochemistry of deep-drill-core samples from the basement of the central Tarim basin. Journal of Asian Earth Sciences, 25, 45-56.

Hanson, A. D., Ritts, B. D., and Moldowan, J. M. (2007) Organic geochemistry and thermal maturity of oils and Upper Paleozoic and Mesozoic potential source rocks in the Ordos basin, north-central China. American Association of Petroleum Geologists Bulletin, 91, 1273-1293.

Heermance, R. V., Chen, J., Burbank, D. W., and Wang, C (2007) Chronology and tectonic controls of Late Tertiary deposition in the southwestern Tian Shan foreland, NW China. Basin Research, 19, 599-632. doi: 10. 1111/j. 1365-2117. 2007. 00339. x.

Hendrix, M. S. (1992) Sedimentary basin analysis and petroleum potential of Mesozoic strata, northwest China. PhD dissertation, Stanford University, 565 p.

Hendrix, M. (2000) Evolution of Mesozoic sandstone compositions, southern Junggar, northern Tarim, and western Turpan basins, Northwest China; a detrital record of the ancestral Tian Shan. Journal of Sedimentary Research, 70 (3), 520-532.

Hendrix, M., Graham, S., Carroll, A., Sobel, E., McKnight, C., Schulein, B., and Wang, Z. (1992) Sedimentary record and climatic implications of recurrent deformation in the Tian Shan; evidence from Mesozoic strata of the north Tarim, south Junggar, and Turpan basins, Northwest China. Geological Society of America Bulletin, 104 (1), 53-79.

Hendrix, M. S., Graham, S. A., Amory, J. Y., and Badarch, G. (1996) Noyon Uul Syncline, southern Mongolia; lower Mesozoic sedimentary record of the tectonic amalgamation of Central Asia. Geological Society of America Bulletin, 108, 1256-1274.

Heumann, M. J., Johnson, C. L., Webb, L. E., and Taylor, J. P. (2008) Detrital zircon and sandstone provenance analysis from Permian and Lower Cretaceous sedimentary units to constrain total and incremental left-lateral offset along the East Gobi Fault Zone, southeastern Mongolia. Eos Trans. AGU, 89 (53), Fall Meet. Suppl., Abstract T43E-01.

Johnson, C. L. (2004) Polyphase evolution of the East Gobi basin: sedimentary and structural records of Mesozoic-Cenozoic intraplate deformation in Mongolia. Basin Research, 16, 79-99.

Johnson, C. L., and Graham, S. A. (2004) Sedimentology and reservoir architecture of a synrift lacustrine delta, southeastern Mongolia. Journal of Sedimentary Research, 74 (6), 786-804.

Johnson, C. L., Webb, L. E., Graham, S. A., Hendrix, M. A., and Badarch, G. (2001) Sedimentary and structural records of late Mesozoic high - strain extension and strain partitioning, East Gobi basin, southern Mongolia. Geological Society of America Memoir 194, 413-434.

Johnson, C. L., Amory, J. A., Graham, S. A., Lamb, M. A., G. Badarch, and Affolter, M. (2008) Accretionary tectonics and sedimentation during late Paleozoic arc collision. China-Mongolia border region, in A. Draut, Clift, and D. Scholl, eds., Formation and applications of the sedimentary record in arc collision zones. Geological Society of America Special Paper 436, 363-390.

Kapp, P. P., Yin, A., Manning, C., Murphy, M. A., Harrison, T. M., Ding, L., Deng, X. G., and Wu, C-M. (2000) Blueschist-bearing metamorphic core complexes in the Qiangtang block reveal deep crustal structure of northern Tibet. Geology, 28, 19-22. doi: 10.1130/ 0091-7613 (2000) 28<19: BMCCIT>2. 0. CO; 2.

Kent-Corson, M. L., Ritts, B. D., Zhuang, G., Bovet, P. M., Graham, S. A., and Chamberlain, C. P. (2009) Stable isotopic constraints on the tectonic, topographic, and climatic evolution of the northern margin of the Tibetan Plateau. Earth and Planetary Science Letters, 282, 158-166.

Lamb, M. A., and Badarch, G. (2001) Paleozoic sedimentary basins and volcanic arc systems of southern Mongolia: new geochemical and petrographic constraints. Geological Society of America Memoir 194, 117-147.

Lamb, M. A., Hanson, A. D., Graham, S. A., Badarch, G., and Webb, L. E. (1999) Left-lateral sense offset of upper Proterozoic to Paleozoic features across the Gobi Onon, Tost, and Zuunbayan faults in southern Mongolia and implications for other Central Asian faults. Earth and Planetary Science Letters, 173, 183-194.

Leeder, M. R., and Gawthorpe, R. L. (1987) Sedimentary models for extensional tilt-block/half-graben basins, Geological Society, London, Special Publications, 28, 139-152. doi: 10. 1144/GSL. SP. 1987. 028. 01. 11.

Li, D., Liang, D., Jia, C., Wang, G., Wu, Q., and Dengfa, H. (1996) Hydrocarbon accumulations in the Tarim Basin, China. American Association of Petroleum Geologists Bulletin, 80 (10), 1587-1603.

Maruyama, S., Isozaki, Y., Kimura, G., and Terabayashi, M. (1997) Paleogeographic maps of the Japanese Islands: plate tectonic synthesis from 750 Ma to the present. Island Arc, 6, 121-142.

Meng, Q-R. (2003) What drove late Mesozoic extension of the northern China-Mongolia tract. Tectonophysics, 369 (3-4), 155-174.

Molnar, P., and Tapponnier, P. (1975) Cenozoic tectonics of Asia: effects of a continental collision. Science, 189, 419-426.

Molnar, P., and Tapponnier, P. (1981) A possible dependence of tectonic strength on the age of the crust in Asia. Earth and Planetary Science Letters, 52, 107-114.

Northrup, C. J., Royden, L. H., and Burchfiel, B. C. (1995)

Motion of the Pacific plate relative to Eurasia and its potential relation to Cenozoic extension along the eastern margin of Eurasia. Geology, 23 (8), 719-722.

Ritts, B. D. and Biffi, U. (2001) Mesozoic northeast Qaidam basin: response to contractional reactivation of the Qilian Shan, and implications for the extent of Mesozoic intracontinental deformation in central Asia, in Hendrix, M. S., and Davis, G. A., eds., Paleozoic and Mesozoic tectonic evolution of central Asia: from continental assembly to intracontinental deformation. Geological Society of America Memoir, 194, 293-316.

Ritts, B. D., Darby, B. J., and Cope, T. (2001) Early Jurassic extensional basin formation in the Daqing Shan segment of the Yinshan belt, northern North China Block, Inner Mongolia. Tectonophysics, 339, 235-253.

Ritts, B. D., Yue, Y. J., and Graham, S. A. (2004) Oligocene- Miocene tectonics and sedimentation along the Altyn Tagh Fault, northern Tibetan Plateau: analysis of the Xorkol, Subei, and Aksay basins. Journal of Geology, 112, 207-229.

Ritts B. D., Yue, Y., Graham, S. A., Sobel, E. R., Abbink, O. A., and Stockli, D. (2008) From sea level to high elevation in15 million years: uplift history of the northern Tibetan Plateau margin in the Altun Shan. American Journal of Science, 308, 657-678.

Ritts, B. D., Weislogel, A., Graham, S. A., and Darby, B. J. (2009) Mesozoic tectonics and sedimentation of the giant poly-phase nonmarine intraplate Ordos basin, western North China Block. International Geology Review, 51, 95-115. doi: 10. 1080/00206810802614523.

Ritts, B. D., Berry, A. K., Johnson, C. L., Darby, B. J., and Davis, G. (2010) Early Cretaceous supradetachment basins in the Hohhot metamorphic core complex, Inner Mongolia, China, Basin Research, 22, 45-60.

Sengor, A. M. C., and Natal'in, B. A. (1996) Paleotectonics of Asia: fragments of a synthesis, in Yin, A., and Harrison, M. eds., The tectonic evolution of Asia. Cambridge, Cambridge University Press, 486-640.

Sengor, A. M. C., Natal'in, B. A., and Burtman, V. S. (1993) Evolution of the Altaid tectonic collage and Palaeozoic crustal growth in Eurasia. Nature, 364, 299-307.

Sengor, A. M. C., Graham, S. A., and Biddle, K. T. (1996) Is the Tarim basin underlain by a Neoproterozoic oceanic plateau? Geological Society of America Abstracts with Program, 28, A-67.

Sobel, E. R., Hilley, G. E., and Strecker, M. R. (2003) Formation of internally-drained contractional basins by aridity-limited bedrock incision. Journal of Geophysical Research-Solid Earth, 108 (23), 2344. doi: 10. 1029/2002JB001883.

Vincent, S. J., and Allen, M. B. (1999) Evolution of the Minle and Chaoshui basins, China; implications for Mesozoic strike-slip basin formation in Central Asia. Geological Society of America Bulletin, 111, 725-742.

Wang, C., Zhao, X., Liu, Z., Lippert, P., Graham, S. A., Coe, R. S., Yi, H., Zhu, L., Liu, S., and Li, Y. (2008) Constraints on the early uplift history of the Tibetan Plateau. Proceedings of the National Academy of Sciences, 105 (13), 4987-4992.

Webb, L. E., Graham, S. A., Johnson, C. L., Badarch, G., and Hendrix, M. S. (1999) Occurrence, age, and implications of the Yagan-Onch Hayrhan metamorphic core complex, southern Mongolia. Geology, 27, 143-146.

Webb, L. E., and Johnson, C. L. (2006) Tertiary strike-slip faulting in southeastern Mongolia and implications for Asian tectonics. Earth and Planetary Science Letters, 241, 363-335.

Webb, L. E., Johnson, C. L., and Minjin, C. (2007) Thermochronology of early Mesozoic shear in the East Gobi Fault Zone, Mongolia. EOS trans. AGU Fall Meeting, V23C-1560.

Yang, Y., Ritts, B. D., Zhang, B., Xu, T., and Xi, P. (2003) Upper Triassic-Middle Jurassic strata and sedimentary record in northeast Qaidam basin, Northwest China. Journal of Petroleum Geology, 26, 429–449.

Yang, Y., Li, W., and Ma, L. (2005) Tectonic and stratigraphic controls on hydrocarbon systems in the Ordos Basin: a multicycle cratonic basin in central China. American Association of Petroleum Geologists Bulletin, 89, 255–269.

Yin, A., and Nie, S. (1996) A Phanerozoic palinspastic reconstruction of China and its neighboring regions, in Yin, A., and Harrison, M., eds., The tectonic evolution of Asia. Cambridge, Cambridge University Press, 442–485.

Yue, Y., and Liou, J. G. (1999) Two-stage evolution model for the Altyn Tagh Fault, China. Geology, 27, 227–230.

Yue, Y., Ritts, B. D., and Graham, S. A. (2001) Initiation and long-term slip history of the Altyn Tagh Fault. International Geology Review, 43, 1087–1093.

Yue, Y., Ritts, B. D., Hanson, A. D., and Graham, S. A. (2004a) Sedimentary evidence against large strike-slip translation on the northern Altyn Tagh Fault. Earth and Planetary Science Letters, 228, 311–323.

Yue, Y., Ritts, B. D., Graham, S., Wooden, J., Gehrels, G., and Zhang, Z. (2004b) Slowing extrusion tectonics: lowered estimate of post-Early Miocene long-term slip rate for the Altyn Tagh fault. Earth and Planetary Science Letters, 217, 111–122.

Zhou, Z., and Chen, P. (1990) Biostratigraphy and geologic evolution of Tarim. Beijing, Science Press, 366 p.

Ziegler, P. A., van Wees, J. D., and Cloetingh, S. (1998) Mechanical controls on collision-related compressional intraplate deformation. Tectonophysics, 300, 103–129.

Zonenshain, L., Kuzmin, M. I., and Natapov, L. M. (1990) Geology of the U. S. S. R.: a plate tectonic synthesis. American Geophysical Union, Geodynamics Series, 21, 1–242.

（章志明 译，屈红军 韩银学 蔡露露 校）

第29章 大型格伦维尔沉积事件：罗迪尼亚超大陆形成的记录

ROBERT RAINBIRD[1], PETER CAWOOD[2], GEORGE GEHRELS[3]

(1. Geological Survey of Canada, Ottawa, Canada；
2. School of Earth and Environment, University of Western Australia, Crawley, Australia；
3. Department of Geography and Geosciences, University of St. Andrews, St. Andrews, UK
Department of Geosciences, University of Arizona, Tucson, USA)

摘　要：格伦维尔造山运动是地球上最大的一个造山事件，伴随着大约1200—1000Ma前的中元古代末期罗迪尼亚超大陆的形成而形成。如今的格伦维尔山脉可追溯近12000km，风化和剥蚀作用产生了大量沉积物碎屑，这些沉积物碎屑被巨大的辫状河体系所分散。在化学风化为主的气候条件下，侵蚀、剥蚀和沉积量由于植被缺乏和强烈的大陆风化作用而加强。

随着碎屑锆石物源分析手段的出现，巨大的侵蚀事件和发源于格伦维尔山脉的广阔水系首次被发现。起初，格伦维尔期的锆石颗粒从加拿大西北部的新元古界早期沉积盆地中获得，距最近可能的劳伦大陆东部格伦维尔物源区超过3000km。盆地中厚层河流沉积中的交错层理指示的古水流表明，存在区域上一致的西—北西向沉积搬运过程，这给古地理模型的建立提供了有力证据。与之相关的地层，位于加拿大西北盆地以南数千千米的加拿大和美国的科迪勒拉山脉的相关地层也表现出相似的碎屑锆石年龄分布，这为大规模河流体系的存在进一步提供了证据。这些年代资料也表明河流体系横向分布广泛，并可能来自巨大的格伦维尔山前多个物源。

目前在美国中部地下地层中识别出了沉积体系的近端部分，它们包括几个地层层序，并且这些层序与格伦维尔造山运动构造演变的不同阶段相对应，与北部的中大陆裂谷体系和五大湖区保存的露头也具有很好的相关性。这些同碰撞期裂陷沉积和碰撞后前陆盆地沉积显示出轴向河流的特征，表明它们是平行山前流动的主干河流沉积。类似的地层序列保存在北大西洋的苏格兰、设得兰群岛（位于苏格兰东部的群岛）、东格陵兰岛、斯瓦尔巴特群岛（属于挪威的特罗姆瑟地区）和挪威。这些地层的碎屑锆石颗粒年龄主要是古元古代晚期和中元古代晚期，推测其来源于劳伦大陆东部的格伦维尔区地层。

碎屑锆石地质年代表明，全球许多显生宙地层中的格伦维尔期碎屑为再旋回的产物。一些碎屑直接来源于格伦维尔区岩石的隆起和侵蚀，或是阿巴拉契亚—海西期造山运动和联合古陆形成期间格伦维尔前陆盆地沉积的再旋回碎屑。中元古代晚期的碎屑持续的循环到年轻的地层序列，并且是现代河流沉积物的重要成分之一。

关键词：碎屑锆石地质年代学　新元古代　古地理　罗迪尼亚　河流

29.1 引言

周期性的全球造山运动分隔了地球的演化历史，也是超大陆旋回的一个重要证据（Murphy等，2009；Nance等，1988）。当超大陆形成时，陆块碰撞和地壳沿陆块碰撞边缘的大规模隆升剥蚀，形成了广泛分布的山系。这些山系遭受巨大的风化作用和侵蚀作用，产生大量由大河流系统搬运而形成的沉积物碎屑。在广泛分布的厚层沉积层，这些沉积事件被很好地记录下来（Archer 和 Greb，1955；Cawood 等，2007；Squire 等，2006；Veevers，2004）。全球现代河流沉积物的锆石结晶年龄出现 2.7Ga、1.8Ga、1.1Ga、0.5Ga 和 0.3Ga 的峰值，与推测的超大陆会聚旋回一致（Campbell 和 Allen，2008；Hawkesworth 等，2009；Rino 等，2008）。

格伦维尔造山运动或许是地球历史上最大的造山事件，并以中元古代末期罗迪尼亚超级大陆的形成为标志（Hoffman，1991；Li 等，2008）。现今保存的古格伦维尔山脉链的山根以近于连续的带状延伸了近 12000km，从南墨西哥向东北延伸到北美东部，穿过不列颠岛，至斯堪的纳维亚半岛、俄罗斯东部（图 29-1）。全球许多克拉通上均发现了与它的元素和年龄相似的造山带。随着碎屑锆石地质年代学被用于沉积物源分析，格伦维尔山脉规模巨大的侵蚀事件和广阔的河流体系才被初次发现。碎屑锆石铀铅年龄测定用来证实 Potter（1978）的"大河流"体系的假设，该假设认为加拿大西北元古宙沉积盆地残留巨大河流体系的出现伴随着该盆地东部造山带的隆升（Young，1978）。盆地中厚层河流沉积交错层理的测量，显示古流向为西—北西向（Yong，1979；图 29-2）。起初从阿蒙森盆地获得的新元古代早期砂岩分析显示，近一半的锆石年龄具有劳伦大陆东部格伦维尔省的独特年龄，即这些锆石必须从 3000km 以外的地方搬运来，并最终留存在大陆的另一边（Rainbird 等，1992；图 29-3）。

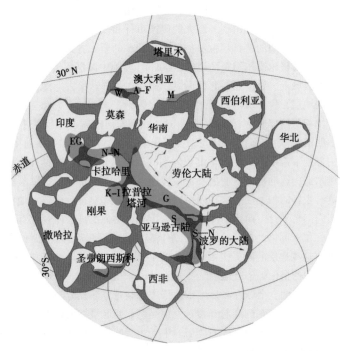

图 29-1 前罗迪尼亚大陆一个可能的组合结构是在 1200Ma 和 1000Ma 前的格伦维尔造山时期出现的大陆碰撞的结果（据 Li 等，2008 简化）

区域中形成的喜马拉雅型山脉用橙色表示（格伦维尔造山带）和沉积物覆盖了超级大陆巨大面积，如红色箭头所示（河流系统或许要比表明的更广泛，但是未获得这些区域的古地理信息）。图 29-2 显示了北美区域的许多详细信息；文中讨论的格伦维尔带可能部分包括 Baltica（S—N）的 Sveco-Norwegian 造山带、Amazonia（S）西缘的 Sunsas 造山带、Mawson 岩体（W）的 Wilkes 区域、澳大利亚的 Albany-Franser（A—F）和 Musgrave（m）造山带、印度（EG）的 Eastern Ghats 带、Zimbabwe 克拉通（N—N）的 Namaqua-Natal 区域和刚果克拉通（K—I）的 Kibaran 和 Irumide 带对加拿大西北盆地 1000km 以外的加拿大北部科迪勒拉山系的相关沉积地层（Rainbird 等，1997）和北美西部边缘古生代地层（Gehrels 等，1995）的追踪研究也发现了格伦维尔期碎屑的广泛分布，这从北加拿大延续了 1000km 的雁列山脉给予初期研究发现以很大的支持。离子探针和激光剥蚀电感耦合等离子体质谱分析技术促进了具有相似年龄岩石、年轻岩石和现代沉积物的碎屑锆石物源分析，并揭示了格伦维尔期沉积事件是广泛分布的。

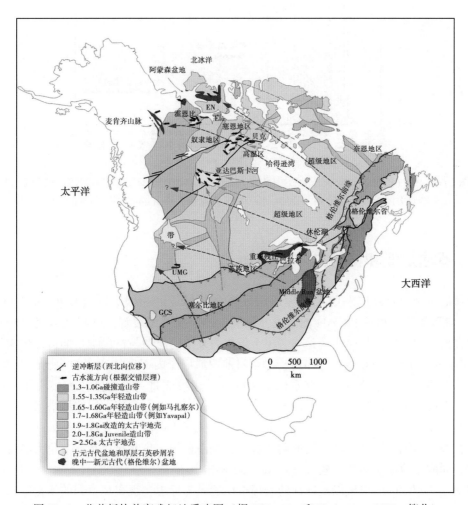

图 29-2　北美板块前寒武纪地质略图（据 Whitmeyer 和 Karlstrom，2007，简化）

展示了元古宙沉积盆地的位置和北美大陆东南部分的造山带；造山带的碎屑搬运到了穿过大陆的大河流体系，向西北搬运并且形成了 Thelon、Athahasca 和其他盆地发现的厚砂岩层序；格伦维尔造山带被解释为深部剥蚀山脉的山根，由从 1200~1000Ma 前形成的多阶段大陆冲积物构成的；来自这些造山带的沉积物古流向为由东向西的（蓝色箭头）且具有大陆另一端数千千米以外的格伦维尔地区碎屑锆石年龄特征（图 29-3）；长的蓝色虚线箭头表明古元古代沉积物的搬运方向，来自于 Yapaval 和 Trans-Hndson 造山带的薄的夹层（地图上分别用绿色和峰表示），虚的红色箭头说明了格伦维尔腹地盆地沉积的古元古代沉积物的搬运方向，盆地沿着劳伦古陆西缘分布［例如，Amundsen 盆地、Mackenzie 山脉、Uinta 山组（UMG）和 GrandGanyon 超群（GCS）］；在北美，格伦维尔沉积体系的远端部分被代表是通过位于 MiddleRun 盆地（地表以下）和 MidcontinentRift-MCR 的沉积物。格伦维尔前缘带指接近冲断层位置的格伦维尔地块内侧变形区；罗迪尼亚超级大陆内的格伦维尔山脉的位置和延伸在图 29-1 中用橙色表示

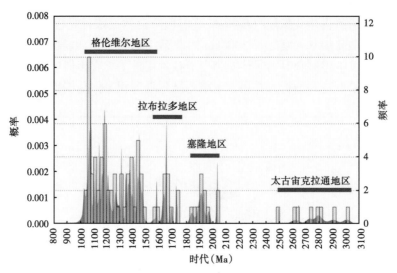

图 29-3 通过北加拿大的 Mackenzie 山脉和 Amundson 盆地的早新元古代砂岩碎屑锆石 U-Pb 年龄直方图与概率密度图的叠加（Id-TIMS 数据来源于 Rainbird 等，1992，1997）

29.2 罗迪尼亚和格伦维尔造山运动

罗迪尼亚是中元古代末期大多数地球板块合并形成的超级大陆（Hoffman，1991；图 29-1）。它可能于 1200—1000Ma 前形成，并于 800Ma 前开始裂解。已有许多罗迪尼亚的重建模式提出，但是大多数模式都是基于单个板块内发生的所谓与格伦维尔期造山带相关联的认识（Dalziel，1991；Moores，1991）。尽管目前罗迪尼亚内部克拉通的结构已经合理构建，但在细节上还有很大的分歧。大多数超级大陆的重建模式将劳伦古陆作为中心或者核心块体，其东南被波罗地古陆、亚马逊古陆、西非克拉通所包围；南部被拉普拉塔河板块和圣弗朗西斯科克拉通所包围；西南被刚果和卡拉哈里克拉通所包围；东北被澳洲大陆和印度、东南极洲板块所包围（图 29-1）。西伯利亚板块、华南板块、华北板块和北劳伦大陆板块的位置则因不同的重建模式而变化较大（Li 等，2008；Pisarevsky 等，2003）。

罗迪尼亚超大陆的形成以全球性的碰撞造山运动为标志，现在观察到的格伦维尔造山带残留部分是一个广泛分布的中—高级变质岩带。在格伦维尔造山带内，现今地球表面的高级变质岩的出现通常表明十多亿年前发生大规模造山作用时地壳发生了几十千米的抬升剥蚀（Jamieson 和 Beaumonl，1989；Jamieson 等，2007）。暴露作用和侵蚀作用的延续时期记录，在泛大陆的水流体系搬运罗迪尼亚超大陆内部沉积盆地大量的碎屑物中，它们将被保存。一些残留的内部盆地在罗迪尼亚和后续泛大陆裂开时发育的大陆边缘得以保存。

格伦维尔造山运动在加拿大东部、纽约州等典型区域均有很好的记录，它反映了中元古代中晚期北西方向为主的上地壳的收缩（Gower 等，2003，2008）。大多数岩石发生了强烈的变形和变质作用，从上角闪岩相到麻粒岩相，并且岩石类型和结晶作用的年龄变化很大。Hoffman（1989）和 Davidson（1995，1998，2008）在格伦维尔省（暴露在加拿大地盾的造山带部分）以往的工作中总结出一些不连续的造山事件的认识。Moore 和 Thompson（1980）提出了由 Elzevirian 和 Ottawan 造山运动组成的格伦维尔造山两阶段旋回理论。Rivers

（1997）提出了紧随 Elzevirian 增生合并事件（1.29~1.19Ga）之后的三个事件：Shawinigan（1.19~1.14Ga）的碰撞和碰撞后岩浆作用、Ottawan（1.08~1.02Ga）和 Rigolet（1.0~0.98Ga）两个年轻的格伦维尔造山期。这些事件的年龄代表了多种造山带增生组分的大陆弧同造山和造山后的花岗岩侵入体，而且这些花岗岩侵入体中锆石丰富（Moecher 和 Samson，2006）。中元古代较早的事件（Tucker 和 Gower，1994）和 1.65Ga 拉布拉多造山运动（Scharer 和 Gower，1988）可能在后来并入格伦维尔造山带的劳伦古陆的东南缘演化中扮演着非常重要的角色。

位于其他克拉通上的中元古代晚期的显著造山带包括波罗的板块的瑞典—挪威造山带（Bingen 等，2008a，2008b），它本来是格伦维尔省向东北延伸，沿着亚马逊古陆西缘分布的 Sunsas 造山带（Santos 等，2008），南极洲东部的 Maud、Rayner 和 Wilkers 省，印度板块的东高止山脉（Mezger 和 Cosca，1999）和沿着津巴布韦克拉通西南缘（Fitzsimons，2000；图 29-1）分布的 Namaqua-Natal 省可能与澳大利亚南部以及中央的 Albany-Fraser 和 Musgrave 造山带有联系（Black 等，1992；Cawood 和 Korsch，2008；Clark 等，2000）。刚果克拉通的 Kibaran 和 Irumide 带是格伦维尔期最显著的造山带，它很可能形成于罗迪尼亚超大陆拼合时期（Kokonyangi 等，2006；De Waele 等，2008）。

29.3 理论：元古宙的大河流体系

29.3.1 厚层成熟的克拉通席状砂岩

最常见的厚层石英砂岩通常沉积于 2500~400Ma 前的岩石中，这个时期代表了近一半的地球演化历史（Soegaard 和 Eriksson，1989）。这些以陆相沉积为主、广泛分布的厚层砂岩，主要沉积在区域不整合面之上，而这些不整合发育在稳定且广泛分布的前寒武克拉通结晶岩上（Eriksson 和 Donaldson，1986）。这些厚层成熟石英砂岩（有时超过 1000m）代表地球早期广泛沉积，并显示出了广泛的辫状河道沉积的证据（北萨斯喀彻温的 Althabasca 群；Ramaekers 和 Catuneanu，2004），德隆河和贝克盆地的 Dubawnt 超群（Rainbird 等，2003）、美国中北部的 Baraboo 地层段（Medaris 等，2003，2007）、南委内瑞拉的 Roraima 超群（Santos 等，2003）、加拿大中南部的 Huron 超群（Young 等，2001）仅代表该砂岩的几个例子（图 29-2 中的北美样品）。这些沉积物中粗粒底载比例相当高，几乎不存在深水河道的证据。这些沉积特征表明一个与现代湿润地区完全不同的水文地质情况，并可能与地表缺少稳定的大陆植被有关（Schumm，1968）。高矿物成熟度砂岩和泥岩夹层的化学组分（富铝、贫碱和碱土元素）表明当时为有利于强烈化学风化气候条件下的大陆风化体制（Nesbitt 等，1996）。

我们与其他研究者的地质年代分析都表明，上述沉积物的年龄介于几次全球规模的造山运动至约 200Ma 前之间（Campbell 和 Allen，2008；Rino 等，2008）。我们认为它们的年龄和交错层理所测的古流向均证明它们是超级大陆合并和巨型山脉形成时期的强烈风化剥蚀产物。许多研究实例中，物源区距离最终的沉积区域很远。这与现代大型河流体系如亚马逊河和密西西比河明显不同（Iizuka 等，2005；Mapes 等，2004），后者的碎屑主要来源于局部物源。物源区特征的差异可能归因于元古宙植被的缺乏导致的更高剥蚀率、沉积碎屑产出量和更有效的沉积物搬运能力。

29.3.2 地层对比和古地理

Fraser 等（1970）首次提出了这样一种观点：晚古元古代到中元古代时期的一些大型陆内沉积盆地或许是残余在加拿大地盾结晶岩不整合面上的席状砂岩沉积。这些盆地由于埋藏和成岩彼此分隔，并在逐渐抬升和侵蚀作用下仅在碟状和槽状凹陷中保存了部分厚层席状砂岩沉积体（图 29-2）。砂岩地层原始分布范围较大的证据包括发现的小型砂岩地区大大超出了现有盆地的边界，许多曾被上覆河流沉积改造的暴露风化层（Hadlari 等，2004）。这些盆地记录了一个早期造山、沉降到沉积作用的综合历史（Aspler 等，2004；Rainbird 等，2003；Ramaekers 等，2007）。这些盆地的层序地层学、沉积学和盆地的沉积填充年代相似，并且数千千米之遥的沉积中心都是可对比的，沉积学研究表明大规模的造山期后砂岩层序是向西流动的辫状河沉积产物。而各盆地中相似的的古流向（Campbell，1979；Rainbird 等，2007；Ramaekers 等，2007；Ross，1983；图 29-2）也支持它们是一系列大型河流沉积的区域席状砂岩的残存部分。

这些观察和较新元古代盆地的分析形成了这样一种观点：保存下来的沉积盆地是东部区域发生造山作用时形成的巨大河流体系的残留（Young，1978，1979；Rainbird 等，1992）。这些情形在北美克拉通演化中至少出现了三次：古元古代的哈德逊造山运动、亚达巴斯卡河沉积岩、塞隆、Hornby 海湾和 Elu 盆地；中元古代的格伦维尔造山作用和同时期的 Middle Run、Torridonian 和 Amundsen 盆地（图 29-2，见下文）；古生代 Appalachian-Hercynian 造山作用、中央 Appalachian（东 Kentucky-northern Tennessee）、东内部盆地（东—伊利诺斯印度）和美国盆地的西南。在每一个实例中，造山作用都能和超级大陆的形成联系起来（分别是哥伦比亚大陆、罗迪尼亚大陆和泛古陆；Rainbird 和 Young，2009）。

29.3.3 检验理论：碎屑锆石地质年代

分析沉积盆地中沉积碎屑的物源区可以为认识有关沉积盆地起源和演化提供重要的信息。物源分析方法随着碎屑锆石地质年代学的引入而取得了革命性的进展，这个技术依靠从碎屑沉积岩中分离出来的锆石颗粒进行 U-Pb 同位素年龄测定（Fedo 等，2004）。与砂岩中其他颗粒组分一样，沉积岩中的锆石颗粒也来自于风化的古老岩浆岩、变质岩或沉积岩，因此可以通过对比碎屑锆石与潜在物源区的年龄特征，推断它们的物源。在过去的 10~15a 中，快速和高精度的取样和源区分析技术使得锆石年代地质学方法取得重大进展。

29.4 格伦维尔沉积事件和罗迪尼亚古地理

罗迪尼亚超级大陆的拼合与格伦维尔造山带宽广的山脉受剥蚀产生的大量碎屑沉积有关。我们认为沉积物被泛大陆河流体系从造山带搬运，近源和远源沉积要素在北美、格陵兰大陆和英国大陆沉积盆地中均有保存（Rainbird 等，1992，1997）。自从大型河流体系的模型发表以后，大量应用锆石地质年代学的研究工作对罗迪尼亚超大陆的古地理谜团不断提供了有价值的证据（Cawood 等，2004；Kirkland 等，2008；Krabbendam 等，2008；Mueller 等，2007；Santos 等，2002；Dehler 等，2010）。

29.4.1 劳伦古陆的格伦维尔前陆盆地

源自格伦维尔山脉的泛大陆河流体系模型验证的最大突破是发现了劳伦大陆水系河流近源部分沉积。在北美，现今地质图的调查以及暴露在加拿大和美国东部的格伦维尔冲断前缘的研究（格伦维尔前缘；图 29-2 和图 29-4）发现格伦维尔省的高级变质岩与 Superior、Southern、Nain 省前陆沉积盆地的改造地壳相邻。COCORP（大陆反射剖面合作项目）OH-1 地震反射剖面和 Ohio 地层钻井测试中首次识别出这些近源沉积物（Shrake 等，1991；Drahovzal 等，1992；图 29-4）。这些数据勾勒了一个不对称的、粗粒的向西楔入的锥体，可能为新元古代时期格伦维尔前缘不成熟碎屑红层——Middle Run 组。这个地震剖面也揭示了一个宽的、界限清楚的东倾反射层，该反射层可能是格伦维尔前缘逆冲构造带（Hauser, 1993）。在 OH-1 井西南地下是一个浅的且与其他地方的 Middle Run 地层相似的东倾沉积地层层序。这些沉积被解释为格伦维尔造山带前陆盆地以前未发现的磨拉石相（Middle Run

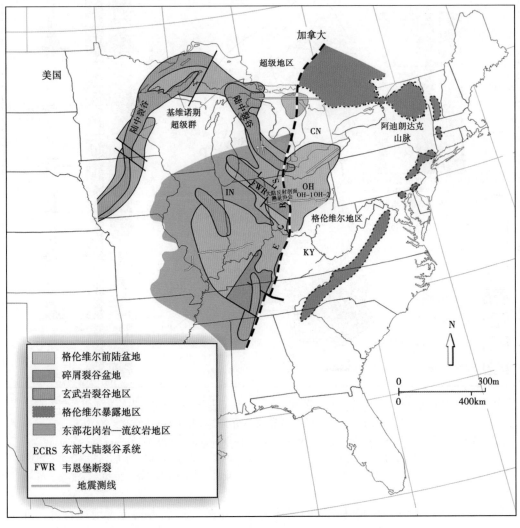

图 29-4 北美格伦维尔新元古代沉积的前陆盆地（据 Baranoski 等修改，2009；图 29-1）
除了一些大湖区的 Keweenawan 超群，大多数的沉积物是根据地表以下的地震剖面和岩心信息推断的

盆地；图 29-2 和图 29-4；Hauser，1993）。从 Middle Run 地层（图 29-5）获得的钻井岩心样品的碎屑锆石地质年代分析证实了上述观点，并揭示了来源于邻近的格伦维尔省剥蚀物对盆地充填比例非常高（Santos 等，2002）。最近基于重新处理的 COCORP 地震反射资料和有限岩心数据，识别出了四个不整合限定的元古宙沉积层序，这些层序代表了格伦维尔造山期的伸展和收缩交替阶段的幕式沉积（Baranoski 等，2009；Rivers，1997）。老的层序记录了一个北西延伸的断陷盆地（如 Fort Wayne 和东大陆裂谷；图 29-4），这个盆地被抬升、剥蚀，随后被一个向西进积的碎屑沉积楔形体埋藏，标志着前陆盆地向格伦维尔造山带的末端发育（Middle Run 盆地；Baranoski 等，2009）。老的裂谷盆地可能由格伦维尔造山开始（Donaldson 和 Irving，1972）的地块碰撞而形成，类似于 Rhine 地堑的撞击裂谷（Sendor 等，1978）或者喜马拉雅造山带中的构造逃逸盆地（Morley 等，2001），也可能是与地幔柱相关的伸展成因（Hauser，1996）。

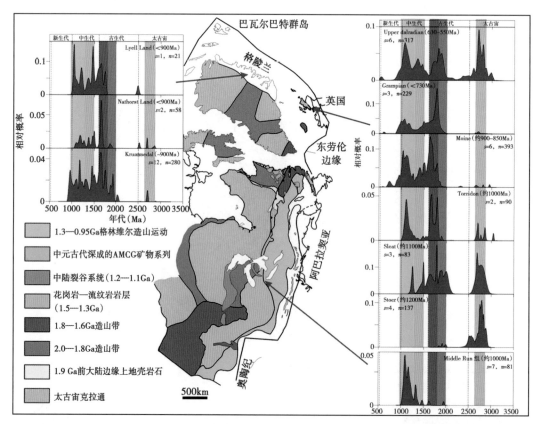

图 29-5 来自于苏格兰岛、东劳伦大陆和格陵兰岛的 Caledonian 造山带内的变质岩、新元古代砂岩的碎屑锆石 U-Pb 年龄的位置和概率密度图（据 Cawood 修改，2007b）

SHRIMP 数据的来源有 Santos 等（2002）的 MiddleRun 组、Kinnaird 等（2007）的 Sleat 群、Rainbird 等（2001）的 Stoer 群和 Torridon 群、Cawood 等（2004）和 Friend 等（2003）的 Moine 地层、Cawood 等（2003）的 Grampian 群、Cawood 等（2003）上 Dalradian 群；AHRIMP 数据来自于 Cawood 等（2007b）的格陵兰岛地层

Baranoski 等（2009）所描述地层的层序特征、年龄阶段和古构造古地理背景与北部中大陆裂谷盆地暴露地表的 Keweenawan 群地层可以对比（图 29-2 和图 29-4）。它由一个下部裂陷伴生的火山岩和火山沉积岩构成，年龄为 1109~1087Ma（见 Hollings 的参考文献，

2007），上覆地层为 Ojakangas（2001）等描述的由 2~3 个不整合面限定的沉积层序。层序下部为典型的裂谷边缘冲积扇粗粒、富岩屑砂岩沉积，并迅速相变为类似于威斯康星和北密西根地区 Oronto 群的河流、湖泊相沉积（Morey 和 Ojakangas，1992）。不整合之上的 Bayfield 群辫状河沉积地层的结构和成分成熟度都较高，类似的岩石包括厚 2100m 的 Fond du Lac 组东流的河流相砾岩和长石砂岩（Morey 和 Ojakangas，1992）。Fond du Lac 组和上覆 Hinckley 砂岩的碎屑锆石地质年代峰值为中元古代，这个峰值与下伏火山岩不同，而来源于邻近的格伦维尔造山带改造的 Grenville 和 Superior 地区，也支持这些砂岩地层可能是上述格伦维尔前陆盆地的一部分的古地理模式。Bayfield 群和类似地层总体的古水流北东流向与格伦维尔造山带前缘主干河流古流向一致。这些砂岩也可以与 Jacobsville 地层对比，为厚度大于 900m 的长石到岩屑类砂岩、砾岩和粉砂岩（Ojakangas 等，2001），它们主要出露在 Michigsn's Keweenaw Peninsula 的北部，向东沿着 Ontario 和 Superior 湖南岸出露（Kalliokoski，1982）。岩石学资料、古水流方向（Kalliokoski，1982；Hedgman，1992）和碎屑锆石年龄（J. Craddock 和 K. Wirth，个人交流，2009）说明南部是较强抬升和物源区。与 Bayfield 群类似，盆地大部分地区河流相的古流向（NE 和 SW）表明轴向搬运的特征，这可能与平行于前陆盆地轴向主干水系的发育有关。

29.4.2　北大西洋地区的相关盆地

古地磁表明波罗的大陆相对于劳伦大陆在中元古代末经历了 95°的顺时针旋转，这个时期为 1265Ma（可能 1120Ma）之后或者 1000Ma 之前的某个时期，与格伦维尔山脉的形成同步（Cawood 等，2010；Pisarevsky 等，2003）。波罗的古陆是从其北界与东格陵兰岛相接的地方北移造成 Scandinavian 边缘与苏格兰岛、Rockall Bank 和南格陵兰岛等相邻（图 29-6）。波罗的古陆相对劳伦古陆的弧形弯曲导致了挪威造山带北部的三角洲洋盆（Aphenochasm）的形成。该残留盆地中元古代末期到新元古代的残余层序，沿北大西洋边缘得以保存，这些地层被分为大约 1000Ma 和 900Ma 构造热事件影响下的构造变形及稳定的两个沉积旋回，伴随有 980—910Ma 的地壳加厚和 840—710Ma 的岩浆活动（Cawood 等，2010）。保存的沉积层序的延伸方向通常平行于劳伦大陆的边缘，说明了大陆边缘的几何形状控制了沉积作用。这些几何形状相同的边缘在盆地充填稳定过程中的收缩变形期重新活动。该地区在新元古代末波罗的古陆分离和古大西洋北开裂所造成的加里东旋回期存在一个岩石圈软弱带（Cawood 和 Nemchin，2001）。

第一个旋回的沉积残余保留在苏格兰岛、设得兰群岛、东格陵兰岛和北挪威（Cawood 等，2007；Gee 和 Tebenkov，2004；Kalsbeek 等，2000；Kirkland 等，2008；Watt 和 Thrane，2001）。碎屑锆石颗粒和叠加的构造热事件限定该套地层沉积于 1030—930Ma 之间。苏格兰变形最弱的地层沉积研究表明，该地层为高能辫状河—平原沉积环境（Friend 等，2003；Krabbendam 等，2008；Nicholson，1993；Stewart，2002；Williams，2001）。由于后期的强烈变形作用中元古代晚期到新元古代早期发育的其他地层单元无法分析其沉积环境。

第二个沉积旋回包括苏格兰岛和北挪威的地层（Cawood 等，2003，2007；Kirkland 等，2007）。这个层序与第一个旋回在空间上是分开的，未看到它们之间的连续地层接触关系。最年轻的碎屑锆石年龄 920Ma 为其上限，构造热事件提供的年龄下限为 840Ma 或更小。推测相似年龄的硅质碎屑岩发育在 Svalbard 地块的西南部，但是缺少广泛的中—新元古代构造

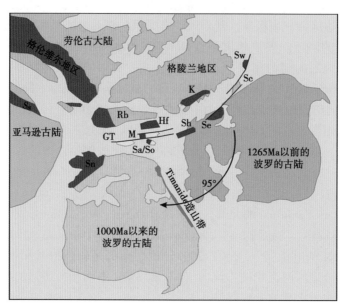

图 29-6　重建的晚中元古代到新元古代的东劳伦古陆、波罗的古陆和北亚马逊古陆的位置图
波罗的古陆 1265Ma 之前（灰色阴影）和 1000Ma 之后（涂色的）相对于劳伦古陆（据 Cawood 等之后，2010）的位置。波罗的古陆的旋转形成了一个三角形的洋盆 Aesir 海。一系列的加里东时期（中元古代）的走滑断层对 Svalbard 群岛的错断以及西、中和东部分隔由 Harland（1997）提供。Gee 和 Tebenkov 的西南苏格兰古陆和 Nordaustlandet 分别位于 Harland（1997）西部和东部。加里东断层把苏格兰地块撕裂成 Grampian 地块、Moine 地块和 Hebridean 前陆。红色部分是第一次旋回有关的沉积层序，黄色则代表第二次。简写：GT—苏格兰古陆 Grampian 地层；M—苏格兰古陆 Moine 地层；Hf—苏格兰 Hebridean 前陆；Rb—东格陵兰古陆的 Krummendal 地层；N—Svalbared 的 Nordaustlandet 地层；Rb—Rockall 河；Sa/So—北挪威的 Svaerholt 和 Sørøy 地层；Sh—Shetland 岛；Sn—Sveconorwegian 造山带；Ss—Sunsas 造山带；Sv—中央 Svalbard；Sw—Svalbard 的西南地层

热事件证据（Gee 和 Tehenkov，2004）。

两套地层的碎屑锆石颗粒年龄均为古元古代晚期和中元古代晚期，推断均来自暴露的东劳伦地块（如 Labradoran，Makkovikian-Ketilidian）和 Grenvulle-Sveconorwegian 造山带（图 29-5），缺乏沉积堆积区附近和下伏地层组成特征的太古宙碎屑（Rainbird 等，2001），表明了该区地层起伏不大（Cawood 等，2007；及其参考文献）。

这些中元古代晚期至新元古代地层的顶面为区域不整合面，其上为新元古代晚期（<750Ma）到早古生代（寒武—奥陶系）的硅质碎屑岩和碳酸盐岩。而碳酸盐岩则是罗迪尼亚超大陆裂解和 Lapetus 洋在阿巴拉契亚—加里东旋回裂开时形成的裂谷和被动大陆边缘沉积（Cawood 和 Nemchin，2001）。从这些单元得到的碎屑锆石表明，格伦维尔山脉碎屑物源的持续输入，但局部有来自太古宙克拉通的碎屑组分（图 29-5；Cawood 和 Nemchin，2001；Cawood 等，2003，2007a，2007b；Dhuime 等，2007）。

新元古代晚期—早古生代地层在阿巴拉契亚造山带沿地层走向可进行对比，它们发育在格伦维尔—瑞典挪威变形带前缘的南段，或者直接上覆在格伦维尔省基岩之上。这些地层的锆石年龄主要为中元古代，很少有更老的年龄，表明格伦维尔造山带为凸出的高地和分水岭，并阻止了劳伦大陆内部太古宙—早中元古代地体的碎屑物质输入（Cawood 等，2007；及其参考文献）。

29.4.3 格伦维尔河流体系的远端沉积

格伦维尔河流体系的远端部分在阿蒙森和麦肯齐盆地均有保存，并覆盖了广阔的加拿大西北地区（图29-2），这表明了格伦维尔沉积事件的潜在宽度和广度。这些沉积体是成熟的石英碎屑砂岩，Mackenzie Mountains 的 Katherine 群的最大厚度可达 1.8km（Long 等，2008）。砂岩主要为辫状河沉积（图29-7），普遍具有北西—西向的古水流特征（图29-2）。河流相砂岩与罗迪尼亚超大陆解体过程中克拉通边缘裂陷盆地之上发育的三角洲沉积、浅海碳酸盐岩沉积以及蒸发岩沉积呈互层发育。现今出露在加利福尼亚科迪勒拉山脉、内华达州、犹他州以及墨西哥北部的劳伦大陆西部边缘沉积盆地新元古代沉积岩中的碎屑锆石显示其具有中元古代晚期的年代学特征（Stewart 等，2001）。虽然一些碎屑锆石年龄具有局部火山岩物源的年龄特征，但是大部分来源于劳伦大陆东部的格伦维尔省（Bickford 等，2000；Barth，2001）。Nankoweap 组是格伦维尔河流体系远端沉积部分的典型研究对象，该层是厚100m 的具交错层理的红色砂岩和少量泥岩，并以三明治结构与美国西南部的 Grand Canyon 超群的 1250—1100Ma 的 Unkar 群（Timmons 等，2005）和 800—742Ma 的 Chuar 群（Dehler 等，2001）之上（图29-2）不整合接触。Nankoweap 组的碎屑锆石年龄特征（Timmons 等，2005；图29-8）与 Amundsen 和 Mackenzie 盆地以及那些近端沉积建立的年龄图版一致（Hinckley 砂岩；Finley-Blasi 等，2006）。另外一个潜在的、与格伦维尔河流体系相关地层是早新元古代 Uinta Mountains 群和东犹他州 Big Cottonwood 组（图29-2；Link 等，1993）。对这些岩石进行的一系列物源分析得到的碎屑锆石 U-Pb 年龄剖面和上述年龄特征相似（图29-8b），同样支持了这个观点：这些沉积是格伦维尔大型河流体系的另一个残留部分，该河流体系向西横穿劳伦大陆，并将碎屑物质搬运到大陆的西部和北部边缘的克拉通边缘盆地或古拉通盆地（图29-1 和图29-2）。

图 29-7　加拿大西北地区 Amundsen 盆地新元古代早期 Shaler 超群 Nelson Head
组格伦维尔辫状河体系的远端沉积（据 Rainbird 等，1996）

厚板状河道砂岩（浅色）和漫滩薄宽透镜状粉细砂岩互层，出露剖面真厚度约35m

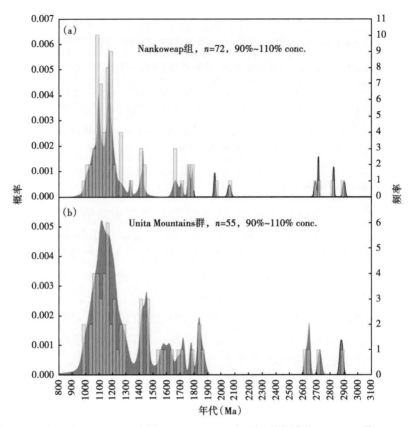

图 29-8　来自于 Nankoweap 地层（GrandCanyon 超群，数据来自 Timmons 等，2005）、BigCottonwood 组和 UintaMoutains 组的新元古代早期砂岩的 LA—ICPMS 碎屑锆石 U-Pb 年龄的叠加直方概率密度图（据 Dehler 等，2010）

29.4.4　劳伦大陆之外的部分

格伦维尔前陆盆地在西伯利亚的一个可能的例子是 Mayamkan 组。该组发育在俄罗斯东部的 Uchur-Maya 坳陷，是一套 1km 厚的新元古代向上变粗的不成熟冲积沉积序列（Rainbird 等，1998）。该套地层的碎屑锆石颗粒以中元古代为主，与 Siberian 克拉通西部的潜在物源区地层的形成时期不同。古地理信息表明盆地接受的碎屑来自于东部物源，推测其物源区可能为罗迪尼亚大陆解体过程中发生裂陷并形成原始的太平洋格伦维尔造山带部分。

南非新元古代盆地大部分形成于 Pan-African 时期，并与冈瓦那超大陆相关。盆地沉积物记录了与罗迪尼亚超级大陆形成相关的中元古代晚期造山带的物源信息。非洲西南部的 Gariep 带以及美国南部同时代的 Dom Feliciano 带新元古代硅质碎屑岩地层的物源研究表明，碎屑锆石年龄以 1200—1000Ma 为主，并且具有 1700Ma 和 2000Ma 的峰值特征（Basei 等，2005）。这些年龄与南非前 Gariep 基底特征可以很好地对比，而该基底具有 Namaqua-Natal 带的中元古代碎屑和 Richtersveld 块体（安第斯型岛弧特征）的古元古代碎屑特征。赞比亚和刚果民主共和国中非铜带的新元古代地层发育 Cu-Co 矿床（Master 等，2005）。碎屑锆石年龄主要是古元古代（2080—1835Ma），可能来源于含有少量 Kibaran 带中元古代信息的 Lufubu 变质杂岩体（图 29-1）。

29.5 年轻沉积物中格伦维尔盆地再旋回沉积的证据

中元古代晚期到新元古代早期发生了大规模的碎屑沉积，格伦维尔时期的碎屑重新沉积到新的地层中。在美国西北部，格伦维尔时期碎屑组成了沉积地层的主要部分，包括美国西南部的新元古代晚期地层（Apache 群；Stowart 等，2001）、科迪勒拉冒地槽地层（Cehrels 等，1995）和寒武—三叠纪的远离大陆架的盆地地层、科罗拉多高原的侏罗纪地层（Dickinson 和 Gehrels，2009）以及白垩纪的前陆盆地地层（Dickinson，2008）。与罗迪尼亚古地理很类似的模型来自于潘基亚超级大陆，在宾夕法尼亚期—三叠纪它由 Appalachian-Hercynian 造山运动拼合而成。潘基亚超大陆的劳亚古陆块体主要为一系列的河流三角洲和轴向辫状河沉积，这些沉积体系充填了 Alleghenian-Appalachian 逆冲前缘的挠曲前陆盆地（Absaroka 层序；Slass，1988）。中阿巴拉契亚盆地的沉积充填被认为代表了像亚马逊规模的排水体系（Archer 和 Greb，1995）。当前陆盆地被填充满时，多余的碎屑通过水流和风力向西搬运穿过了克拉通，最终到达劳伦大陆的西缘（Dickinson 和 Gehrels，2003）。这些沉积物的物源研究显示大多数碎屑锆石具有格伦维尔省的年龄特征（Becker 等，2005；Gray 和 Zeitler，1997；McLennan 等，2001；Thomas 等，2004；Dickinson 和 Gehrels，2003）。这些碎屑来源于阿巴拉契亚造山期岩石的抬升和剥蚀或格伦维尔前陆盆地沉积的碎屑再旋回。

中元古代晚期的碎屑已经持续旋回到年轻的地层层序并且仍是现代河流沉积的重要组分。锆石年龄在 1.2—1.0Ga 的碎屑代表了现代河流碎屑的重要成分，如科罗拉多河（Grove 和 Kimbrough，2008）、密西西比河、几支阿巴拉契亚河支流（Eriksson 等，2003，2004），并且在世界上的大多数其他河流系统中也有发现（Campbell 和 Allen，2008）。Eriksson 注意到一个矛盾是劳伦大陆东部的格伦维尔期岩石的地质年代学定义的不连续事件与来自于阿巴拉契亚河流碎屑锆石年代显示的持续岩浆作用之间存在差异。它们表明现代阿巴拉契亚河流广泛取样的碎屑能更好地反映格伦维尔省和格伦维尔造山带的造山事件。在现代格伦维尔基底的暴露区之外，现代河流的锆石颗粒来自再旋回的沉积岩，而这些沉积岩部分来自已经发生剥蚀搬运或目前被年轻沉积盖层覆盖的格伦维尔期结晶岩石。考虑到格伦维尔时期碎屑的相对丰度，不难理解格伦维尔期的岩石比其他岩浆岩体具有更高的锆石丰度（Dickinson，2008；Moecher 和 Samson，2006）。相应地，格伦维尔期的锆石在沉积体中的代表性需要慎重考虑，特别是在尝试重建沉积物生成量的时候。这些并没有影响我们关于格伦维尔时期碎屑广泛分布的结论，因为这主要是根据 1.2—1.0Ga 的锆石颗粒沉积区距离推测的物源超过 3000km，而不是碎屑锆石的绝对或相对丰度。

29.6 总结

格伦维尔期的构造岩浆事件在 1250—950Ma 到几个百万年，并且形成了广泛的造山带和大量的沉积碎屑，而且这些沉积碎屑在过去的 1000Ma 间多次再旋回沉积。沉积物的产生量和输出量也由于最初有利的化学风化和缺乏固定沉积物的植被而大大增强，直到志留纪陆地植物的出现这种局面才得以改观。大型格伦维尔期沉积事件的证据主要基于北美古陆和邻近劳伦古陆部分的研究，但更为确凿的证据是作为罗迪尼亚超级古陆一部分的其他古陆板块内部的沉积物记录。将来以及正在进行的其他沉积地层的物源研究工作将为格伦维尔碎屑地层的范围和意义提供更为严格的评价。

致谢（略）

参 考 文 献

Archer, A. W., and Greb, S. F. (1995) An Amazon-scale drainage system in the early Pennsylvanian of central North America. Journal of Geology, 103, 611–628.

Aspler, L. B., Chiarenzelli, J. R., and Cousens, B. L. (2004) Fluvial, lacustrine and volcanic sedimentation in the Angikuni sub-basin, and the initiation of 1.84 Ga Baker Lake Basin, western Churchill Province, Nunavut, Canada. Precambrian Research, 129, 225–250.

Baranoski, M. T., Dean, S. L., Wicks, J. L., and Brown, V. M. (2009) Unconformity bounded seismic reflection sequences define Grenville-age rift system and foreland basins beneath the Phanerozoic in Ohio. Geosphere, 5, 140–151.

Barth, A. P., Wooden, J. L., and Coleman, D. S. (2001) SHRIMP-RG U-Pb zircon geochronology of Mesoproterozoic metamorphism and plutonism in the southwesternmost United States. Journal of Geology, 109, 319–327.

Basei, M. A. S., Frimmel, H. E., Nutman, A. P., Preciozzi, F., and Jacob, J. (2005) A connection between the Neoproterozoic Dom Feliciano (Brazil/Uruguay) and Gariep (Namibia/South Africa) orogenic belts: evidence from a reconnaissance provenance study. Precambrian Research, 139, 195–221.

Becker, T. P., Thomas, W. A., Samson, S. D., and Gehrels, G. E. (2005) Detrital zircon evidence of Laurentian crustal dominance in the lower Pennsylvanian deposits of the Alleghanian clastic wedge in eastern North America. Sedimentary Geology, 182, 59–86.

Bickford, M. E., Soegaard, K., Nielsen, K. C., and McLelland, J. M. (2000) Geology and geochronology of Grenville-age rocks in the Van Horn and Franklin Mountains area, West Texas; implications for the tectonic evolution of Laurentia during the Grenville. Geological Society of America Bulletin, 112, 1134–1148.

Bingen, B., Anderson, J., Soderlund, U., and Moller, C. (2008a) The Mesoproterozoic in the Nordic countries. Episodes, 31, 29–34.

Bingen, B., Nordgulen, Ø., and Viola, G. (2008b) A fourphase model for the Sveconorwegian orogeny, SW Scandinavia. Norwegian Journal of Geology, 88, 43–72.

Black, L. P., Harris, L. B., and Delor, C. P. (1992) Reworking of Archean and Early Proterozoic components during a progressive, middle Proterozoic tectonothermal event in the Albany Mobile Belt, Western Australia. Precambrian Research, 59, 95–123.

Campbell, F. H. A. (1979) Stratigraphy and sedimentation in the Helikian Elu Basin and Hiukitak Platform, Bathurst Inlet-Melville Sound, Northwest Territories, Geological Survey of Canada Paper 79, 19.

Campbell, I. H., and Allen, C. M. (2008) Formation of supercontinents linked to increases in atmospheric oxygen. Nature Geoscience, 1, 554–558. Cawood, P. A., and Korsch, R. J. (2008) Assembling Australia: Proterozoic building of a continent. Precambrian Research,

166, 1-35.

Cawood, P. A., and Nemchin, A. A. (2001) Paleogeographic development of the eastern Laurentian margin: constraints from U‒Pb dating of detrital zircons from the northern Appalachians. Geological Society of America Bulletin, 113, 1234-1246.

Cawood, P. A., Nemchin, A. A., Smith, M., and Loewy, S. (2003) Source of the Dalradian Supergroup constrained by U‒Pb dating of detrital zircon and implications for the East Laurentian margin. Journal of the Geological Society, London, 160, 231-246.

Cawood, P. A., Nemchin, A. A., and Strachan, R. (2007a) Provenance record of Laurentian passive‒margin strata in the northern Caledonides: implications for paleodrainageand paleogeography. Geol Soc Am Bull, 119, 993-1003.

Cawood, P. A., Nemchin, A. A., Strachan, R., Prave, T., and Krabbendam, M. (2007b) Sedimentary basin and detrital zircon record along East Laurentia and Baltica during assembly and breakup of Rodinia. Journal of the Geological Society, 164, 257-275.

Cawood, P. A., Nemchin, A. A., Strachan, R. A., Kinny, P. D., and Loewy, S. (2004) Laurentian provenance and an intracratonic tectonic setting for the Moine Supergroup, Scotland, constrained by detrital zircons from the Loch Eil and Glen Urquhart successions. Journal of the Geological Society, London, 161, 861-874.

Cawood, P. A., Strachan, R., Cutts, K., Kinny, P. D., Hand, M., and Pisarevsky, S. (2010) Neoproterozoic orogeny along the margin of Rodinia: Valhalla orogen, North Atlantic. Geology, 38, 99-102.

Clark, D. J., Hensen, B. J., and Kinny, P. D. (2000) Geochronological constraints for a two-stage history of the Albany-Fraser Orogen, Western Australia. Precambrian Research, 102, 155-183.

Condie, K. C., Lee, D., and Farmer, G. L. (2001) Tectonic setting and provenance of the Neoproterozoic Uinta Mountain and Big Cottonwood groups, Northern Utah: constraints from geochemistry, Nd isotopes, and detrital modes. Sedimentary Geology, 141-142, 443-464.

Dalziel, I. W. D. (1991) Pacific margins of Laurentia and East Antarctica‒Australia as a conjugate rift pair: evidence and implications for an Eocambrian supercontinent. Geology, 19, 598-601.

Davidson, A. (1995) A review of the Grenville orogen in its North American type area. AGSO Journal of Australian Geology and Geophysics, 16, 3-24.

Davidson, A. (1998) An overviewof Grenville Province geology, in Lucas, S. B., and St-Onge, M. R., eds., The Geology of North America C-1: geology of the Precambrian Superior and Grenville Provinces and Precambrian fossils in North America. Geological Society of America, 205-270.

Davidson, A. (2008) Late Paleoproterozoic to mid-Neoproterozoic history of northern Laurentia: an overview of central Rodinia. Precambrian Research, 160, 5-22.

Dehler, C. M., Elrick, M., Karlstrom, K. E., Smith, G. A., Crossey, L. J., Timmons, J. M., Eriksson, P. G., Catuneanu, O., Aspler, L. B., Chiarenzelli, J. R., and Martins, N. M. A. (2001) Neoproterozoic Chuar Group (approximately 800-742 Ma) Grand Canyon; a record of cyclic marine deposition during global cooling and supercontinent rifting. Sedimentary Geology, 141-142, 465-499.

Dehler, C. M., Fanning, C. M., Link, P. K., Kingsbury, E. M., and Rybczynski, D., 2010, Maximum depositional age and provenance of the Uinta Mountain Group and Big Cottonwood Formation, northern Utah: paleogeography of rifting western Laurentia. Geological Society of America Bulletin, 122, 1686-1699.

De Waele, B., Johnson, S. P., and Pisarevsky, S. A. (2008) Palaeoproterozoic to Neoproterozoic growth and evolution of the eastern Congo Craton: its role in the Rodinia puzzle. Precambrian Research, 160, 127-141.

Dhuime, B., Bosch, D., Bruguier, O., Caby, R., and Pourtales, S. (2007) Age, provenance and post-deposition metamorphic overprint of detrital zircons from the Nathorst Land group (NE Greenland) -A LA-ICP-MS and SIMS study. Precambrian Research, 155, 24-46.

Dickinson, W. R. (2008) Impact of differential zircon fertility of granitoid basement rocks in North America on age populations of detrital zircons and implications for granite petrogenesis. Earth and Planetary Science Letters, 275, 80-92.

Dickinson, W. R., and Gehrels, G. E. (2003) U-Pb ages of detrital zircons from Permian and Jurassic eolianite sandstones of the Colorado Plateau, USA: paleogeographic implications. Sedimentary Geology, 163, 29-66.

Dickinson, W. R., and Gehrels, G. E. (2008) Sediment delivery to the Cordilleran foreland basin: insights from U-Pb ages of detrital zircons in Upper Jurassic and Cretaceous strata of the Colorado Plateau. American Journal of Science, 308, 1041-1082.

Dickinson, W. R., and Gehrels, G. E. (2009) U-Pb ages of detrital zircons in Jurassic eolian and associated sandstones of the Colorado Plateau: evidence for transcontinental dispersal and intraregional recycling of sediment. Geological Society of America Bulletin, 121, 408-433.

Donaldson, J. A., and Irving, E. (1972) Grenville Front and rifting of the Canadian Shield. Nature, 273, 139-140.

Drahovzal, J. A., Harris, D. C., Wickstrom, L. H., Walker, D., Baranoski, M. T., Keith, B., and Furer, L. C. (1992) The East Continent Rift Basin: a new discovery. Ohio Division of Geological Survey, Information Circular 57, 25 p.

Eriksson, K. A., Campbell, I. H., Palin, J. M., and Allen, C. M. (2003) Predominance of Grenvillian magmatism recorded in detrital zircons from modern Appalachian rivers. Journal of Geology, 111, 707-717.

Eriksson, K. A., Campbell, I. H., Palin, J. M., Allen, C. M., and Bock, B. (2004) Evidence for multiple recycling in Neoproterozoic through Pennsylvanian sedimentary rocks of the central Appalachian basin. Journal of Geology, 112, 261-276.

Eriksson, K. A., and Donaldson, J. A. (1986) Basinal and shelf sedimentation in relation to the Archean-Proterozoic boundary. Precambrian Research, 33, 103-121.

Fedo, C. M., Sircombe, K. N., and Rainbird, R. H. (2004) Detrital zircon analysis of the sedimentary record, Hanchar, J. M., and Hoskin, P. O., eds., Zircon: experiments, isotopes, and trace element investigation. Mineralogical Society of America, Reviews in

Mineralogy and Geochemistry, 53, 277-303.

Finley-Blasi, L., Davidson, C., Wirth, K., Craddock, J., and Vervoort, J. (2006) U-Pb dating of Detrital Zircon from the Neoproterozoic (?) Fond Du Lac and Hinckley Sandstone Formations near Duluth, Minnesota. Abstracts with Programs, 38. Boulder, CO, Geological Society of America, 505. Fitzsimons, I. C. W. (2000) Grenville-age basement provinces in East Antarctica: evidence for three separate collisional orogens. Geology, 28, 879-882.

Fraser, J. A., Donaldson, J. A., Fahrig, W. F., and Tremblay, L. P. (1970) Helikian basins and geosynclines of the northwestern Canadian Shield, in Baer, A. J., ed., Symposium on Basins and Geosynclines of the Canadian Shield, Geological Survey of Canada Paper 70-40, 213-238.

Friend, C. R. L., Strachan, R. A., Kinny, P., and Watt, G. R. (2003) Provenance of the Moine Supergroup of NW Scotland: evidence from geochronology of detrital and inherited zircons from (meta) sedimentary rocks granites and migmatites. Journal of the Geological Society, London, 160, 247-257.

Gee, D. G., and Tebenkov, A. M. (2004) Svalbard: a fragment of the Laurentian margin, in Gee, D. G., and Pease, V. L., eds., The Neoproterozoic Timanide Orogen of Eastern Baltica. Memoir, 30. Geological Society, 191-206.

Gehrels, G. E. (2000) Introduction to detrital zircon studies of Paleozoic and Triassic strata in western Nevada and northern California, in Soreghan, M. J., and Gehrels, G. E., eds., Paleozoic and Triassic paleogeography and tectonics of western Nevada and northern California. Special Paper, 347. Boulder, CO, Geological Society of America, 1-17.

Gehrels, G. E., Dickinson, W. R., Ross, G. M., Stewart, J. H., and Howell, D. G. (1995) Detrital zircon reference for Cambrian to Triassic strata of western North America. Geology, 23, 831-834.

Gower, C. F., Kamo, S., and Krogh, T. E. (2008) Indentor tectonism in the eastern Grenville Province. Precambrian Research, 167, 201-212.

Gower, C. F., and Krogh, T. E. (2003) A U-Pb geochronological review of Pre-Labordian and Labradorian geological history of the eastern Grenville Province, in Brisebois, D., and Clark, T., eds., G_ eologie et ressources min_ erales de la partie est de la Province de Grenville. Qu_ ebec, Minist_ ere des Ressources naturelles, DV 2002-03, 142-172.

Gray, M. B., and Zeitler, P. K. (1997) Comparison of clastic wedge provenances in the Appalachian foreland using U-Pb ages of detrital zircons. Tectonics, 16, 151-160.

Grove, M., and Kimbrough, D. L. (2008) Evolution of the Colorado River: culmination of the major Cenozoic Transformation of SW North America Drainage Patterns, American Geophysical Union, Fall Meeting 2008, abstract #T33D-2101.

Hadlari, T., Rainbird, R. H., and Pehrsson, S. J. (2004) Geology, Schultz Lake, Nunavut, Geological Survey of Canada Open File 1839, 1:250, 000.

Harland, W. B. (1997) The geology of Svalbard. Memoir, 17. London, Geological Society of London, 521 p.

Hauser, E. C. (1993) Grenville foreland thrust belt hidden beneath the U. S. midcontinent.

Geology, 21, 61-64.

Hauser, E. C. (1996) Midcontinent rifting in a Grenville embrace. Geological Society of America Special Paper 308, 67-75.

Hawkesworth, C., Cawood, P., Kemp, T., Storey, C., and Dhuime, B. (2009) Geochemistry: a matter of preservation. Science, 323, 49-50.

Hedgman, C. A. (1992) Petrology and provenance of a conglomerate facies of the Jacobsville sandstone: Ironwood to Bergland Michigan. Geological Society of America, Abstracts with Programs, 24, A329.

Hoffman, P. F. (1989) Precambrian geology and tectonic history of North America, in Bally, A. W., and Palmer, A. R., eds., The geology of North America: an overview, vol. A. Boulder, CO, Geological Society of America, 447-512.

Hoffman, P. F. (1991) Did the breakout of Laurentia turn Gondwanaland inside-out? Science, 252, 1409-1412.

Hollings, P., Fralick, P., and Cousens, B. (2007) Early history of the Midcontinent Rift inferred from geochemistry and sedimentology of the Mesoproterozoic Osler Group, northwestern Ontario. Canadian Journal of Earth Sciences, 44, 389-412.

Iizuka, T., Hirata, T., Komiya, T., Rino, S., Katayama, I., Motoki, A., and Maruyama, S. (2005) U-Pb and Lu-Hf isotope systematics of zircons from the Mississippi River sand: implications for reworking and growth of continental crust. Geology, 33, 485-488.

Jamieson, R. A., and Beaumont, C. (1989) Deformation and metamorphism in convergent orogens: a model for uplift and exhumation of metamorphic terrains. Evolution of metamorphic belts, 117-129.

Jamieson, R. A., Beaumont, C., Nguyen, M. H., and Culshaw, N. G. (2007) Synconvergent ductile flow in variablestrength continental crust: numerical models with application to the western Grenville orogen. Tectonics, 26 TC5005.

Kalliokoski, J. (1982) Jacobsville Sandstone, in Wold, R. J., and Hinze, W. J., eds., Geology and Tectonics of the Lake superior Basin, Geological Society of America, Memoir 156, 147-155.

Kalsbeek, F., Thrane, K., Nutman, A. P., and Jepsen, H. F. (2000) Late Mesoproterozoic to early Neoproterozoic history of the East Greenland Caledonides: evidence for Grenvillian orogenesis. Journal of the Geological Society, London, 157, 1215-1225.

Kirkland, C. L., Daly, J. S., and Whitehouse, M. J. (2007) Provenance and terrane evolution of the Kalak Nappe Complex, Norwegian Caledonides: implications for Neoproterozoic palaeogeography and tectonics. Journal of Geology, 115, 21-41.

Kirkland, C. L., Strachan, R. A., and Prave, A. R. (2008) Detrital zircon signature of the Moine Supergroup, Scotland: contrasts and comparisons with other Neoproterozoic successions within the circum-North Atlantic region. Precambrian Research, 163, 332-350.

Kokonyangi, J. W., Kampunzu, A. B., Armstrong, R., Yoshida, M., Okudaira, T., Arima, M., and Ngulube, D. A. (2006) The Mesoproterozoic Kibaride belt (Katanga, SE D. R. Congo). Journal of African Earth Sciences, 46, 1-35.

Krabbendam, M., Prave, T., and Cheer, D. (2008) A fluvial origin for the Neoproterozoic Morar Group, NW Scotland; implications for Torridon - Morar Group correlation and the Grenville Orogen foreland basin. Journal of the Geological Society, 165, 379-394.

Li, Z. X., Bogdanova, S. V., Collins, A. S., Davidson, A., De Waele, B., Ernst, R. E., Fitzsimons, I. C. W., Fuck, R. A., Gladkochub, D. P., Jacobs, J., Karlstrom, K. E., Lu, S., Natapov, L. M., Pease, V., Pisarevsky, S. A., Thrane, K., and Vernikovsky, V. (2008) Assembly, configuration, and break-up history of Rodinia: a synthesis. Precambrian Research, 160, 179-210.

Link, P. K., Christie-Blick, N., Devlin, W. J., Elston, D. P., Horodyski, R. J., Levy, M., Miller, J. M. G., Pearson, R. C., Prave, A., Stewart, J. H., Winston, D., Wright, L. A., and Wrucke, C. T. (1993) Middle and late Proterozoic stratified rocks of the western U. S. Cordillera, Colorado Plateau, and Basin and Range province, in Reed, J. C. J., Bickford, M. E., Houston, R. S., Link, P. K., Rankin, D. W., Sims, P. K., and Van Schmus, W. R., eds., Precambrian: coterminous U. S. Boulder, Colorado, Geological Society of America, The Geology of North America, C-2, 463-595.

Long, D. G. F., Rainbird, R. H., Turner, E. C., and MacNaughton, R. B. (2008) Early Neoproterozoic strata (Sequence B) of mainland northern Canada and Victoria and Banks islands: a contribution to the Geological Atlas of the Northern Canadian Mainland Sedimentary Basin, Geological Survey of Canada Open File 5700, 22.

Mapes, R. W., Coleman, D. S., Nogueira, A. C. R., and Housh, T. B. (2004) How far do zircons travel? Evaluating the significance of detrital zircon provenance using the modern Amazon River fluvial system, Geological Society of America, Abstracts with Programs, Volume 36, 78.

Master, S., Rainaud, C., Armstrong, R. A., Phillips, D., and Robb, L. J. (2005) Provenance ages of the Neoproterozoic Katanga Supergroup (Central African Copperbelt) with implications for basin evolution. Journal of African Earth Sciences, 42, 41-60.

McLennan, S. M., Bock, B., Compston, W., Hemming, S. R., and McDaniel, D. K. (2001) Detrital zircon geochronology of Taconian and Acadian foreland sedimentary rocks in New England. Journal of Sedimentary Research, 71, 305-317.

Medaris, L. G., Singer, B. S., Dott, R. H., Naymark, A., Johnson, C. M., and Schott, R. C. (2003) Late Paleoproterozoic climate, tectonics, and metamorphism in the Southern Lake Superior region and Proto-North America: evidence from Baraboo interval quartzites. Journal of Geology, 111, 243-257.

Medaris, L. G., Van Schmus, W. R., Loofboro, J., Stonier, P. J., Zhang, X., Holm, D. K., Singer, B. S., and Dott, R. H. (2007) Two Paleoproterozoic (Statherian) siliciclastic metasedimentary sequences in central Wisconsin: the end of the Penokean Orogeny and cratonic stabilization of the southern Lake Superior region. Precambrian Research, 157, 188-202.

Mezger, K., and Cosca, M. A. (1999) The thermal history of the Eastern Ghats Belt (India) as revealed by U-Pb and $^{40}Ar/^{39}Ar$ dating of metamorphic and magmatic minerals: implications for the SWEAT correlation. Precambrian Research, 94, 251-271.

Moecher, D. P., and Samson, S. D. (2006) Differential zircon fertility of source terranes and natural bias in the detrital zircon record: implications for sedimentary provenance analysis. Earth and Planetary Science Letters, 247, 252-266.

Moore, J. M., and Thompson, P. H. (1980) The Flinton Group: a late Precambrian sedimentary sequence in the Grenville Province of eastern Ontario. Canadian Journal of Earth Sciences, 17, 1685-1707. Moores, E. M. (1991) Southwest U. S. -East Antarctic (SWEAT) connection: a hypothesis. Geology, 19, 425-428.

Morey, G. B., and Ojakangas, R. W. (1992) Keweenawan rocks of eastern Minnesota and northwestern Wisconsin, in Wold, R. J., and Hinze, W. J., eds., Geology and Tectonics of the Lake Superior Basin, Geological Society of America Memoir 156, 135-146.

Morley, C. K., Woganan, N., Sankumarn, N., Hoon, T. B., Alief, A., and Simmons, M. (2001) Late Oligocene- Recent stress evolution in rift basins of northern and central Thailand; implications for escape tectonics. Tectonophysics, 334, 115-150.

Mueller, P. A., Foster, D. A., Mogk, D. W., Wooden, J. L., Kamenov, G. D., and Vogl, J. J. (2007) Detrital mineral chronology of the Uinta Mountain Group: implications for the Grenville flood in southwestern Laurentia. Geology, 35, 431-434.

Murphy, J. B., Nance, R. D., and Cawood, P. A. (2009) Contrasting Modes of Supercontinent Formation and the Conundrum of Pangea. Gondwana Research, 15, 408-420. Nance, R. D., Worsley, T. R., and Moody, J. B. (1988) Der Superkontinent-Zyklus. The supercontinent cycle. Spektrum der Wissenschaft, 1988, 80-87.

Nesbitt, H. W., Young, G. M., McLennan, S. M., and Keays, R. R. (1996) Effects of chemical weathering and sorting on the petrogenesis of siliciclastic sediments, with implications for provenance studies. Journal of Geology, 104, 525-542.

Nicholson, P. G. (1993) A basin reappraisal of the Proterozoic Torridon Group, northwest Scotland, in Frostick, L. E., and Steel, R. J., eds., Tectonic controls and signatures in sedimentary successions. Special Publication, 20. Ghent, Belgium, International Association of Sedimentologists, 183-202.

Ojakangas, R. W., Morey, G. B., Southwick, D. L., Eriksson, P. G., Catuneanu, O., Aspler, L. B., Chiarenzelli, J. R., and Martins - Neto, M. A. (2001) Paleoproterozoic basin development and sedimentation in the Lake Superior region, North America. The influence of magmatism, tectonics, sea level change and paleo-climate on Precambrian basin evolution; change over time. Sedimentary Geology, 141-142, 319-341.

Pisarevsky, S. A., Wingate, M. T. D., Powell, C. M., Johnson, S., and Evans, D. A. D. (2003) Models of Rodinia assembly and fragmentation, Geological Society Special Publication, Volume 206, 35-55.

Potter, P. E. (1978) Petrology and chemistry of modern big river sands. Journal of Geology, 86, 423-449.

Rainbird, R. H., Hadlari, T., Aspler, L. B., Donaldson, J. A., LeCheminant, A. N., and Peterson, T. D. (2003) Sequence stratigraphy and evolution of the Paleoproterozoic intracontinental Baker Lake and Thelon basins, western Churchill Province, Nunavut, Canada. Precambrian Research, 125, 21-53.

Rainbird, R. H., Hamiltom, M. A., and Young, G. M. (2001) Detrital zircon geochronology and provenance of the Torridonian, NW Scotland. Journal of the Geological Society [London], 158, 15-27.

Rainbird, R. H., Heaman, L. M., and Young, G. M. (1992) Sampling Laurentia: detrital zircon geochronology offers evidence for an extensive Neoproterozoic river system originating from Grenville orogen. Geology, 20, 351-354.

Rainbird, R. H., Jefferson, C. W., and Young, G. M. (1996) The early Neoproterozoic sedimentary SuccessionBof northwest Laurentia: correlations and paleogeographic significance. Geological Society of America Bulletin, 108, 454-470.

Rainbird, R. H., McNicoll, V. J., Thériault, R. J., Heaman, L. M., Abbott, J. G., Long, D. G. F., and Thorkelson, D. J. (1997) Pan-continental River System Draining Grenville Orogen Recorded by U-Pb and Sm-Nd Geochronology of Neoproterozoic Quartzarenites and Mudrocks, Northwestern Canada. Journal of Geology, 105, 1-18.

Rainbird, R. H., Stern, R. A., Khudoley, A. K., Kropachev, A. P., Heaman, L. M., and Sukhorukov, V. I. (1998) U-Pb geochronology of Riphean sandstone and gabbro from southeast Siberia and its bearing on the Laurentia-Siberia connection. Earth and Planetary Science Letters, 164 (3-4), 409-420.

Rainbird, R. H., Stern, R. A., Rayner, N. M., and Jefferson, C. W. (2007) Age, provenance, and regional correlation of the Athabasca Group, Alberta and Saskatchewan, constrained by igneous and detrital zircon geochronology, in Jefferson, C. W., and Delaney, G. D., eds., EXTECH IV: Geology and Uranium Exploration Technology of the Proterozoic Athabasca Basin, Saskatchewan and Alberta. Geological Survey of Canada Bulletin, 588, 193-209.

Rainbird, R. H., and Young, G. M. (2009) Colossal Rivers, Massive Mountains and Supercontinents. Earth, 54, 52-61.

Ramaekers, P., and Catuneanu, O. (2004) Development and Sequences of the Athabasca Basin, Early Proterozoic, Saskatchewan and Alberta, Canada, in Eriksson, P. G., Altermann, W., Nelson, D. R., Mueller, W. U., and Catuneanu, O., eds., Tempos and events in Precambrian time, Amsterdam, Elsevier, chap. 8. 3.

Ramaekers, P., Jefferson, C. W., Yeo, G. M., Collier, B., Long, D. G. F., Drever, G., McHardy, S., Jiricka, D., Cutts, C., Wheatley, K., Catuneanu, O., Bernier, S., Kupsch, B., and Post, R. (2007) Revised geological map and stratigraphy of the Athabasca Group, Saskatchewan and Alberta; in EXTECH IV: Geology and Uranium Exploration Technology of the Proterozoic Athabasca Basin, Saskatchewan and Alberta, in Jefferson, C. W., and Delaney, G., eds., Geological Survey of Canada, Bulletin 588, 155-192.

Rino, S., Kon, Y., Sato, W., Maruyama, S., Santosh, M., and Zhao, D. (2008) The Grenvillian and Pan-African orogens: world's largest orogenies through geologic time, and their implications on the origin of superplume. Gondwana Research, 14, 51-72.

Rivers, T. (1997) Lithotectonic elements of the Grenville Province: review and tectonic implications. Precambrian Research, 86, 117-154.

Ross, G. M. (1983) Proterozoic aeolian quartz arenites from the Hornby Bay Group, Northwest Territories, Canada: implications for Precambrian aeolian processes. Precambrian Research, 20, 149-160.

Santos, J. O. S., Hartmann, L. A., McNaughton, N. J., Easton, R. M., Rea, R. G., and Potter, P. E. (2002) Sensitive high resolution ion microprobe (SHRIMP) detrital zircon geochronology provides new evidence for a hidden Neoproterozoic foreland basin to the Grenville Orogen in the eastern Midwest U. S. Canadian Journal of Earth Sciences, 39, 1505-1515.

Santos, J. O. S., Potter, P. E., Reis, N. J., Hartmann, L. A., Fletcher, I. R., and McNaughton, N. J. (2003) Age, source, and regional stratigraphy of the Roraima Supergroup and Roraima-like outliers in northern South America based on U-Pb geochronology. Geological Society of America Bulletin, 115, 331-348.

Santos, J. O. S., Rizzotto, G. J., Potter, P. E., McNaughton, N. J., Matos, R. S., Hartmann, L. A., Chemale Jr, F., and Quadros, M. E. S. (2008) Age and autochthonous evolution of the Sunsas Orogen in West Amazon Craton based on mapping and U-Pb geochronology. Precambrian Research, 165, 120-152.

Scharer, U., and Gower, C. F. (1988) Crustal evolution in eastern Labrador; constraints from precise U-Pb ages. Precambrian Research, 38, 405-421.

Schumm, S. A. (1968) Speculations concerning paleohydrologic controls on terrestrial sedimentation. Geological Society of America Bulletin, 79, 1573-1588.

Sengor, A. M. C., Burke, K., and Dewey, J. F. (1978) Rifts at high angles to orogenic belts; tests for their origin and the Upper Rhine Graben as an example. American Journal of Science, 278, 24-40.

Shrake, D. L., Carlton, R. W., Wickstrom, L. H., Potter, P. E., Benjamin, R. H., Wolfe, P. J., and Sitler, G. W. (1991) Pre-Mount Simon basin under the Cincinatti Arch. Geology, 19, 139-142.

Sloss, L. L. (1988) Tectonic evolution of the craton in Phanerozoic time, in Sloss, L. L., ed., Sedimentary cover: North American craton, vol. D-2. Boulder, CO, Geological Society of America.

Soegaard, K., and Eriksson, K. A. (1989) Origin of thick, firstcycle quartz arenite successions: evidence from the 1.7 Ga Ortega Group, northernNewMexico. Precambrian Research, 43, 129-141.

Squire, R. J., Campbell, I. H., Allen, C. M., and Wilson, C. J. L. (2006) Did the Transgondwanan Supermountain trigger the explosive radiation of animals on Earth? Earth and Planetary Science Letters, 250, 116-133.

Stewart, A. D. (2002) The Later Proterozoic Torridonian rocks of Scotland; their sedimentology, geochemistry and origin, Geological Society Memoir No. 24, 136 p.

Stewart, J. H., Gehrels, G. E., Barth, A. P., Link, P. K., Christie, B. N., and Wrucke, C. T. (2001) Detrital zircon provenance of Mesoproterozoic to Cambrian arenites in the Western United States and northwestern Mexico. Geological Society of America Bulletin, 113, 1343-

1356.

Thomas, W. A., Becker, T. P., Samson, S. D., and Hamilton, M. A. (2004) Detrital zircon evidence of a recycled orogenic foreland provenance for Alleghanian clasticwedge sandstones. Journal of Geology, 112, 23-37.

Timmons, J. M., Karlstrom, K. E., Heizler, M. T., Bowring, S. A., Gehrels, G. E., and Crossey, L. J. (2005) Tectonic inferences from the ca. 1255-1100Ma Unkar Group and Nankoweap Formation, Grand Canyon: intracratonic deformation and basin formation during protracted Grenville orogenesis. Geological Society of America Bulletin, 117, 1573-1595.

Tucker, R. D., and Gower, C. F. (1994) A U-Pb geochronological framework for the Pinware terrane, Grenville Province, Labrador, Canada. Journal of Geology, 102, 67-78.

Veevers, J. J. (2004) Gondwanaland from 650-500Ma assembly through 320Ma merger in Pangea to 185-100Ma breakup: supercontinental tectonics via stratigraphy and radiometric dating. Earth-Science Reviews, 68, 1-132.

Watt, G. R., and Thrane, K. (2001) Early Neoproterozoic events in East Greenland. Precambrian Research, 110, 165-184.

Wheeler, J. O., Hoffman, P. F., Card, K. D., Davidson, A., Sanford, B. V., Okulitch, A. V., and Roest, W. R. (1997) Geological Map of Canada, Geological Survey of Canada, Map D1860A.

Whitmeyer, S. J., and Karlstrom, K. E. (2007) Tectonic model for the Proterozoic growth of North America. Geosphere, 3, 220-259.

Williams, G. E. (2001) Neoproterozoic (Torridonian) alluvial fan succession, Northwest Scotland, and its tectonic setting and provenance. Geological Magazine, 138, 471-494.

Wirth, K. R., Vervoort, J., Craddock, J., Davidson, C., Finley-Blasi, L., Kerber, L., Lundquist, B., Vorhies, S., Walker, E. (2006) Source rock ages and patterns of sedimentation in the Lake Superior region: results of preliminary U-Pb detrital zircon studies. Abstracts with Programs, 38. Boulder, CO, Geological Society of America.

Young, G. M. (1978) Proterozoic (<1.7 b. y.) stratigraphy, paleocurrents and orogeny in North America. Egyptian Journal of Geology, 22, 45-64.

Young, G. M. (1979) Correlation of middle and upper Proterozoic strata of the northern rim of the North American craton. Transactions of the Royal Society of Edinburgh, 70, 323-336.

Young, G. M., Long, D. G. F., Fedo, C. M., and Nesbitt, H. W. (2001) Paleoproterozoic Huronian basin: product of a Wilson cycle punctuated by glaciations and a meteorite impact. Sedimentary Geology, 141-142, 233-254.

(白升 译，蒲仁海 韩银学 蔡露露 校)

第 30 章 克拉通盆地

PHILIP A. ALLEN, JOHN J. ARMITAGE

(Department of Earth Science and Engineering, Imperial College,
South Kensington Campus, London, UK)

摘　要：克拉通盆地是大陆岩石圈上广泛分布的、长期的缓慢的沉降区，常被浅水和陆源沉积岩石所充填的一类盆地。在地球动力学方面人们仍然对它了解甚少。目前提出的模式可分为五类：伸展大陆岩石圈的冷却、地幔流的冷却（动态地形）、由相变引起的下伏岩石圈的密度增大、岩浆或地幔柱活动引起的地表响应以及平面应力下的长波变形。

克拉通盆地形成的时期以及空间分布与板块拼贴、超大陆裂解和离散的地球动力阶段有关。许多克拉通盆地形成于新元古代和寒武—奥陶纪。有些克拉通盆地起源于相邻被动大陆边缘倾覆的类斜坡区域，后来被某些次级区域构造活动"个性化"，这些次级构造活动包括陆内造山作用中的构造再活化以及隆起和穹隆的形成。

在漫长的发育过程中，多种不同的机制可能控制着克拉通盆地的地质演化和沉降史。我们提出的低应变速率伸展伴随的下伏岩石圈冷却的模式令人满意地解释了一系列内克拉通盆地长期的沉降史。然而，由大规模地幔对流传送的动态地形在开始或改变大陆内部抬升史过程中所起的准确作用仍然是个进一步研究的有趣焦点。

关键词：大陆岩石圈　伸展　应变速率　沉降　地层

30.1 引言

"克拉通内盆地"、"克拉通盆地"、"内克拉通盆地"以及"陆内凹陷"是位于稳定、相对较厚的大陆岩石圈上的圆—椭圆形地壳凹陷（Sloss 和 Speed，1974；Sloss，1988）。这里的"克拉通盆地"特指那些距离拉伸或聚合大陆边缘一段距离，与明确的大陆扩张形成的裂谷不同的盆地。无论盆地位于结晶地盾、增生体或古褶皱带和裂谷复合体之上，关键的因素是岩石圈必须稳定（Sloss，1988）。

克拉通盆地的特点是长期发育浅水和陆相沉积，地层总体为连续层状（Sloss 和 Speed，1974；Quinlan，1987；Sloss，1990；Leighton 等，1991）。克拉通盆地沉降历史持续很长，偶尔初始阶段有相对快速沉降，随后一段时间沉降速率降低的特点（Nunn 和 Sleep，1984；Stel 等，1993；Xie 和 Heller，2009）（图 30-1），某些程度上类似于大洋盆地（Sleep，1971）。克拉通盆地一般没有明显的以一系列张性断层和相关的地堑、半地堑为特征的初期裂陷阶段，其部分原因可能是保留在陆地上的克拉通盆地基底地震成像较差。克拉通盆地位于远离板块边缘的大陆岩石圈上，但在某些情况下由一个裂谷或夭折的裂谷与海洋连接，如澳大利亚新元古代的中央超级盆地（Walter 等，1995；Lindsay，2002）、美国早古生代的伊利诺伊和俄克拉何马盆地（Braile 等，1986；Kolata 和 Nelson，1990），以及非洲北部中生代的乍得盆地（Burke，1976）。这种几何形态表明许多克拉通盆地位于从伸展板块边缘以高角度延伸到大陆内部的夭折裂谷末端，这可能是先前三联点的位置（Burke 和 Dewey，1973）。

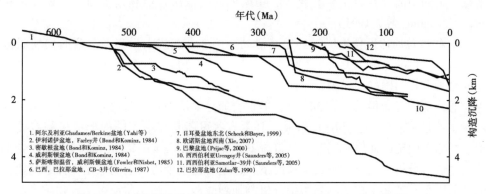

图 30-1　克拉通内盆地沉降曲线（据 Xie 和 Heller，2009，修改）

尽管不同的克拉通盆地有其独特性，但大多数的克拉通盆地也有一些显著的共性，较重要的有以下几点：

（1）盆地地层等厚线和尖灭线通常是圆形或椭圆形的，并且范围广阔。表面积从相对较小的盎格鲁—巴黎盆地（$10^5 km^2$），到大的哈得逊湾（$1.2×10^6 km^2$）和巴拉那盆地（$1.4×10^6 km^2$），再到巨大的中央超级盆地（$2×10^6 km^2$）和西西伯利亚盆地（$3.5×10^6 km^2$）（Leighton 和 Kolata，1990，第 730 页；Sandford，1987；Walter 等，1995，第 173 页；Vyssotski 等，2006）。

（2）在横剖面上，克拉通盆地只是单一的碟形，缺乏主要的同构造断层（扭压作用/张扭作用下的晚期沉积断层更常见），沉积物厚度通常小于 5km，很少达到 6~7km（如西西伯利亚、伊利诺伊州和巴拉那盆地）。在某些情况下，圆形的平面形状是先前更广泛的平台或斜坡被后期分隔的结果，如非洲北部的克拉通盆地，如 AlKhufra、Murzuk 和 Ghadames（Selley，1972，1997；Boote 等，1998）。

（3）沉降持续的时间很长，以亿年来计算（Aleinikov 等，1980）。回剥的构造沉降史一般是亚线性的负指数型（Xie 和 Heller，2009）（图 30-1）。这些长期的沉降通常包括几个被不整合面分隔的盆地阶段，产生的巨型层序反映了控制板块内部构造的转变样式。就最易归为克拉通盆地沉降的巨层序来讲，并忽略长期的巨层序界面的沉积缺失，北美克拉通盆地的沉积速率为 20~30m/Ma（Sloss，1988），这大大低于裂谷、夭折裂谷、年轻的被动大陆边缘、前陆盆地以及走滑盆地的沉积速率（Allen 和 Allen，2005），但比相邻的台地快。横向对应的台地区，如美国横大陆穹隆区，在寒武系和二叠系之间累计约 1km 沉积地层，沉积速率为 3~4m/Ma（Sloss，1988）。

（4）地层主要是陆相—浅海相，这表明沉积速率始终与构造沉降速率保持一致。克拉通盆地巨层序通常开始于大陆广泛区域掀斜作用，如北美东部地区最新的元古宙—早奥陶世"Sauk"序列（Sloss，1963，1988）（见下文）和北非的早古生代（志留纪前）（Selley，1997）。"牛眼"或"水滴"形相带（例如密歇根州盆地的志留系碳酸盐岩和蒸发岩；Nurmi 和 Friedman，1977）表明，一些克拉通盆地的圆形轮廓边界是原始的沉积边界，具同期沉积特征，而不是早先具更广泛的沉积中心的盆地经构造变形和分解而成。低古纬度的克拉通盆地通常受化学沉积控制，这表明碎屑物源供给极为有限，且盆地周边缺乏能提供物源的较高地形。

（5）克拉通盆地的中心一般有规律地相距约1000km。在北美，北美板块早古生代克拉通盆地按一定的距离（几百千米）排列，由南向北，依次是伊利诺伊州、密歇根州和哈得逊湾盆地（Leighton等，1991）。

（6）有些克拉通盆地伴生广泛的岩浆活动，例如玄武岩大量喷发，西西伯利亚盆地和巴拉那盆地（Hompson和Gibson，1991；Saunders等，2007），然而玄武岩火山作用和盆地发展之间是否存在明确的因果关系目前还不清楚。

克拉通盆地的寿命很长（图30-1），但重要的是要认识到：盆地充填通常是由一些不同的巨层序（或层序；Sloss，1963）组成，其中有些可能与完全不同的形成机制有关，如走滑变形、挠褶和明确的伸展。因此，从复合盆地充填（Kingston等，1983）中提取克拉通盆地的巨层序进行分析很重要。在其他情况下，盆地在其整个历史上仍然是克拉通盆地，但存在时间很久以至于被几个构造沉降和抬升机制强烈改造。因此，盆地可能有初始形成机制和后期不同的改造机制。

大量的机制已被用来解释克拉通盆地（Hartley和Allen，1994，表30-1；Klein，1995校阅）。这些模式有部分是交叉重复的，包括以下内容：

①加热之后热收缩（Haxby等，1976；Sleep和Sloss，1980；Kaminski和Jaupart，2000）。

②与地幔柱活动有关的岩浆上涌导致局部扩张（Klemme，1980；Keen，1987；Klein和Hsui，1987；Ziegler和Van Hoorn，1989；Ziegler，1990；Neumann等，1992；Zhao等，1994）。

③深部地壳相变（De Rito等，1983；Fowler和Nesbit，1985；Helwig，1985；Artyushkov和Bear，1990；Artyushkov，1992；Artyushkov等，2008）。

④在平面应力或挠曲负载下先前存在的凹陷活化（Quinlan和Beaumont，1984；Quinlan，1987；Beaumont等，1987）。

⑤非造山岩浆活动中玄武岩基底的侵位，以及非拉张背景下缓慢的热收缩（Stel等，1993）。

⑥负"动力学地形"造成的沉降——这种地形是大规模地幔的下降流或是地幔对流不稳定性产生的应力传送至地球表面的应力表现（Liu，1979；Middleton，1989；Hartley和Allen，1994；Heine等，2008；Farringdon等，2010）或者由冷的大样板块的俯冲造成（Mitrovica等，1989；Burgess等，1997）。

⑦热隆起地表侵蚀伴随着沉积负载（Hsu，1965；Le Pichon等，1973；Sleep和Snell，1976；Sahagian，1993）。

大陆拉伸与伴随的热收缩的负荷挠曲均衡组合在克拉通盆地的许多模式（Sleep和Snell，1976；Kaminski和Jaupart，2000）和分类方案（例如，Klemme，1980；Kingston等，1983）中是明确的或隐含的。在下面的章节中，我们简要地回顾克拉通盆地形成的地质背景，以便支持这一观点：热异常冷却相关的机械伸展在克拉通盆地的形成过程中发挥了基础性的作用。

30.2 克拉通盆地开始的时间和地质背景

要了解克拉通盆地（和一般的沉积盆地），研究它们以前的地质历史是必要的。最根本的是，克拉通盆地开始的时间需要置于相对板块运动与大陆应力状态的框架之内。正如前人认为（Klein 和 Hsui，1987），克拉通盆地的开始在地质时间里并不是均一的，也不是随机分布的；相反，它们开始于超大陆的裂解时期（罗迪尼亚大陆、冈瓦纳大陆，以及过去1000Ma 的泛大陆）（图30-2），特别是在550~500Ma，非洲、北美洲和南美洲的盆地均说明了这一点（见图30-3，表30-1）。

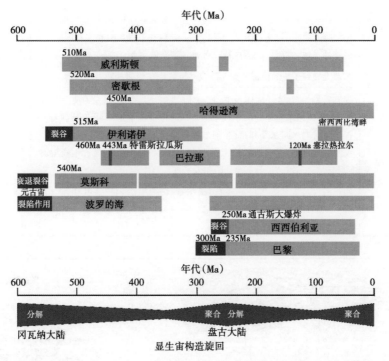

图 30-2　所选的克拉通盆地显示了与两大显生宙构造旋回有关的盆地巨层序充填时间
——一些克拉通盆地巨层序发育在裂谷之前，并伴有一些重要的岩浆事件（蓝色条带）

表 30-1　北美、南美和非洲板块典型盆地及其开始形成时间

盆地类型	盆地名称（位置）	开始年代
南美洲		
与裂谷有关	孔卡沃（巴西）、圣豪尔赫（阿根廷—智利—南大西洋）、帕雷西斯—阿尔托兴古河（巴西）、阿尔托塔帕若斯河（巴西）	侏罗纪
其他克拉通盆地	默伊斯（亚马逊河上游、巴西）、亚马逊州（巴西）、巴拉那（巴西，巴拉圭、阿根廷）	奥陶纪
	巴纳伊巴（巴西）	前寒武纪晚期
北美洲		
克拉通盆地	威利斯顿盆地（加拿大、美国）、俄克拉何马盆地、伊利诺斯盆地（美国）、密歇根州（美国）	寒武纪
	哈得逊湾流域（加拿大）	奥陶纪

续表

盆地类型	盆地名称（位置）	开始年代
非洲		
与裂谷有关	Melrhir 槽（阿尔及利亚、突尼斯）、盖塔拉岭（埃及）、Cyrenaican 平台（利比亚）	寒武纪
	Iullemedden 盆地（尼日尔、马里、尼日利亚、阿尔及利亚、贝宁）	寒武纪
	毕达槽（尼日利亚）、冈戈拉—迈杜古里槽（尼日利亚）、多巴槽（乍得、喀麦隆）、邦戈尔槽（乍得）、Doseo 槽（乍得、中非共和国）、穆格莱德盆地（苏丹）、迈卢特（苏丹、埃塞俄比亚）	白垩纪
	东非大裂谷系统的西部分支、萨拉马特盆地（中非共和国、乍得）	侏罗纪
其他克拉通盆地	提勒盖姆特隆起（阿尔及利亚）、巴沙尔盆地（阿尔及利亚—摩洛哥）、穆维迪尔盆地（阿尔及利亚）、阿赫奈特盆地（阿尔及利亚）、提米蒙盆地（阿尔及利亚）、阿尔库弗腊盆地（乍得、苏丹、利比亚、埃及、尼日尔）、埃及盆地上部（埃及、苏丹、利比亚）、乍得盆地（尼日尔、乍得、尼日利亚、阿尔及利亚、喀麦隆）	寒武纪
	陶丹尼（马里、毛里塔尼亚）、沃尔特（加纳、多哥）、扎伊尔（刚果）、奥万博（纳米比亚、安哥拉）、纳马—喀拉哈里沙漠（博茨瓦纳、纳米比亚、津巴布韦）	前寒武纪

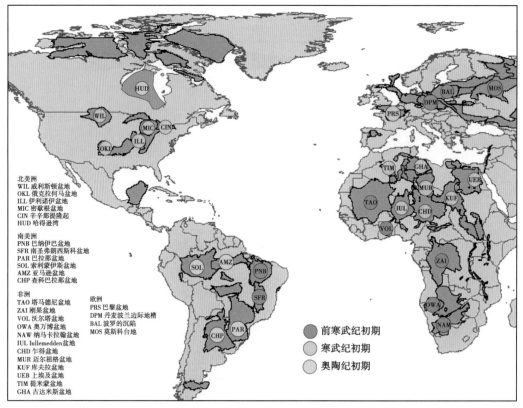

图 30-3 围绕大西洋的大陆陆内盆地分布，主要为典型的克拉通盆地
显示的其他盆地通常与伸展构造有关；盆地是根据其开始形成的时间按颜色编码选择的；
编自 Trond Torsvik 提供的底图和盆地轮廓

在南美,很多盆地可以确定为侏罗纪的裂谷作用,而更典型的克拉通盆地主要在前寒武纪或早古生代开始发育。一个主要的克拉通盆地形成阶段始于奥陶纪,距今 500—450Ma,正好与南美—劳伦古陆从冈瓦纳大陆分离引起的被动大陆边缘的发育一致。应该注意的是,克拉通盆地形成时间滞后于非洲板块联合时间(寒武纪)(见下文)。

在北美,开始于寒武纪的克拉通盆地正好与北美板块的裂解及其周边被动大陆边缘的形成一致。北美克拉通盆地开始时间与非洲许多克拉通盆地相同。

在非洲,许多盆地起因于裂谷或最终的被动大陆边缘,并且主要开始于侏罗纪—白垩纪。这种广泛分布的一套裂谷盆地与大西洋形成时大范围的板块拉张有关。另一套与裂谷相关的盆地开始于前寒武纪晚期—寒武纪,更典型的克拉通盆地开始于寒武纪,与古老的冈瓦纳—劳伦古陆的分裂同期,重要的是它们属于同一时期的同一套裂谷(Guiraud 等,2005)。在寒武纪—奥陶纪,北非地区是一个大范围的缓坡,并在志留纪被进一步分隔。在晚石炭世—二叠纪和晚白垩世—古近纪的收缩构造阶段,盆地进一步特殊化(Boote 等,1998)。因此,克拉通盆地的发生在本质上与裂谷的产生具有相同的应力状态,少部分开始于前寒武纪。

在澳大利亚,也认识到类似的长周期演化盆地。一个大型的(表面积 $2\times10^6 km^2$)克拉通盆地即中央超级盆地是在 Rodinia 大陆裂解时形成(距今 800Ma;Korsch 和 Lindsay,1989;Walter 等,1995),但被 Petermann(570—530Ma)和 Alice Springs(距今 400—300Ma)事件(Hand 和 Sandiford,1999)的陆内造山阶段分隔为各不相同的 Officer、Amadeus 和 Georgina 盆地。中央超级盆地通过 Adelaide 地槽的伸展走廊带与广海相连。

因此,大尺度的地球动力学背景强烈暗示,不管有没有地幔柱参与,克拉通盆地最初形成于广泛的板块拉张应力之下。这被上述盆地详尽的地质历史资料证实。

30.2.1 西西伯利亚盆地

晚二叠世西西伯利亚盆地面积为 $350\times10^4 km^2$,是世界上最大的克拉通盆地。它上覆在受严重侵蚀(在二叠—三叠纪)、褶皱、变质的造山带和晚古生代至联合古陆形成期间多个地体的拼合体之上(Ziegler,1989;Sengor 和 Natal,1996)。莫霍面深度从盆地边缘到盆地中心由 50km 减少至 38km(乌拉尔、Altay-Sayan 褶皱带和东西伯利亚地台)。尽管岩石圈底的区域平面图有很多不确定性(Morozova 等,1999;Artemieva 和 Mooney,2001),但在盆地下方有 200km 深度,并很可能向阿尔泰地区减薄。盆地东北部和喀拉海发现张性裂谷,但向南延伸成 Urengoy 和 Khudosey 两个裂谷带(Nikishin 等,2002;Saunders 等,2005)(图 30-4)。

盆地基底为 250Ma 的通古斯大爆炸玄武岩,其上是全盆地分布的二叠纪陆相沉积,其东北部上覆在裂谷层序之上。因此,最初的拉伸早于岩浆作用,后者又早于克拉通盆地的主要沉降期。裂谷和大量的玄武岩喷发是由于二叠纪地幔柱的侵位作用引起的(Nikishin 等,2002)。玄武岩的分布比地堑更广泛,所以玄武岩的发育与地堑形成并不是密切或直接相关(Vyssotski 等,2006)。火山喷发前裂谷存在争议,但是在 Pur-Taz 地区和喀拉海有据可查:南北向裂谷在地震反射剖面上清晰成像,推测可能是充填了二叠纪的沉积岩。但是目前还没有充分的证据证明南部 Pur-Taz 地区存在前二叠纪裂谷(北纬 64°N)。玄武岩后沉积的三叠系—中侏罗统地层被称为"盆地前期演化",包括低幅度坳陷盆地沉积的大陆硅质碎屑岩和火山岩,裂谷仅限于北部。卡洛夫阶(侏罗系,开始于 159Ma)开始的全盆超覆以及东南

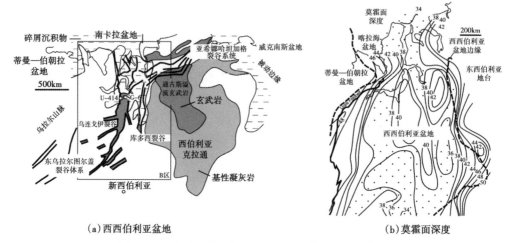

图 30-4　西西伯利亚盆地（Vyssotski 等，2006，修改）

(a) 与拉张板块边缘和西伯利亚地块有关的盆地位置。指示了通古斯大爆炸的玄武岩和基性凝灰岩的目前范围，以及保存在西西伯利亚盆地下面的地堑残留，并标注了裂谷北部沉降分析中使用的钻孔 U-414 和 SG-6。(b) 与盆地侧翼莫霍面深 40km 对比，盆地基底的主裂谷下方莫霍面深 34~36km

主物源的供应标志着西西伯利亚盆地作为一个典型的凹陷型沉积中心开始发育。在启莫里支—巴列姆阶盆地中心水深达到 1200m，表现为未完全充填的盆地，沉积物来自西伯利亚地台（不是乌拉尔）。土伦期的海侵利用了先前拉伸岩石圈的狭长地带重新打开了连接北冰洋与西西伯利亚盆地的通道。盆地在渐新世重新被填充，并且保持在近地表。侏罗纪至古近—新近纪沉降归因于由地幔柱活动导致的岩石圈—软流圈系统热流再平衡（Vyssotski 等，2006）。

30.2.2　巴拉那盆地

巴拉那盆地被认为是"经典"的或是"典型"的克拉通盆地（Zalan 等，1990）（图 30-5）。随着晚侏罗世至早白垩世泛大陆的解体，中生代沉积沿着巴西大陆边缘发育。南美洲板块最南端的裂陷随着巴拉那盆地、坎普斯盆地和桑托斯盆地的溢流玄武岩喷出开始发育（Cainelli 和 Mohriak，1999），推测这些玄武岩可能源于特里斯坦地幔柱（Peate，1997）。与西西伯利亚盆地一致，溢流玄武岩的分布（Serra Ceral 火山岩；137—127Ma；Turner 等，1994）比巴拉那盆地本身更加广阔。这些玄武岩发育在与南美洲板块大陆扩张明显相关的巨厚层序的初始沉降期。在沿海盆地中，裂陷期发育至阿尔布期，从阿尔布期到第四纪为漂移期，所以裂陷—漂移阶段仅仅是巴拉那盆地最新的巨层序。一个重要问题是：巴拉那盆地从什么时候开始成为克拉通盆地？

在西部冈瓦纳区，罗迪尼亚超大陆裂解产生了包括巴拉那残体在内的具有克拉通性质的大陆边缘，形成了一套沉积岩（硅质岩—泥岩—碳酸盐岩系列；Brito Neves，2002）。新元古代时期巴西利亚造山活动造成洋盆关闭，从而形成了冈瓦纳超大陆。造山运动晚期至造山期后形成的盆地被认为是"过渡阶段"（Almeida 等，2000），它处于巴西利亚造山运动和奥陶纪克拉通阶段的广阔稳定地台期（Soares 等，1987）。西部冈瓦纳超大陆被元古宙形成的巨厚沉积物所覆盖，沉积物相继保存在北至索利蒙伊斯河（在亚马逊古陆上）、南至巴拉那地区。Brito Neves 等（1984）指出，与非洲大陆类似，这些盆地是在巴西利亚泛非洲大陆构

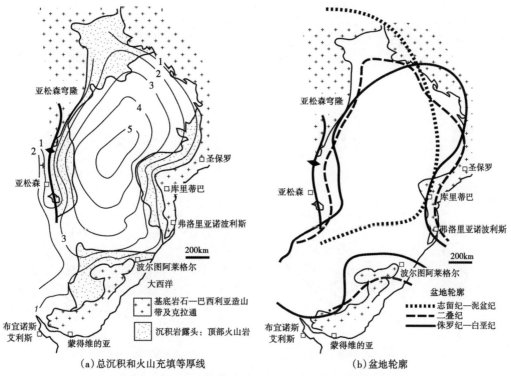

图 30-5 巴拉那盆地（据 Zalan 等，1990，修改）

(a) 总体沉积和火山充填的等厚线，以同心圆模式在盆地中心达到大于 5km；(b) 盆地概括为三个不同的阶段；早古生代盆地向西部 Asuncion 穹隆增厚，而二叠纪和中生代具同一个沉降中心，并与大西洋连通

造旋回后由岩石圈热收缩形成的，例如查科—巴拉那盆地和索利蒙伊斯河盆地，这些克拉通盆地巨厚层序从早—中奥陶世开始，经历了泛大陆之后的热收缩变形（约 230Ma；Veevers，1989）和解体后的拉张变形（225—100Ma），所以巴拉那盆地中的 Serra Geral 溢流玄武岩喷发时间在南美奥陶纪到早三叠世克拉通盆地巨厚层序沉积之后，而最新发现的 Tres Lagoas 玄武岩（443±10Ma）指示了盆地下存裂谷或至少是贯通克拉通内基底断裂的熔融物通道（Julia 等，2008）。

30.2.3 北美克拉通盆地

北美盆地的优势在于其稳定地块的下陷和演化是协调的。从北美克拉通盆地位置图上（图 30-6），我们可以毫无疑问地发现克拉通盆地形成机制与板块边界活动过程密切相关。在北美地区显生宙板块构造演化经历了从罗德里亚分离（约 800—700Ma）到潘基亚超级大陆重组（约 300Ma），再到其后的漂移分离的（丹泽，1991；霍夫曼，1991 年；罗杰斯，1996 年）重要演变历史。这里我们集中关注古生代的构造旋回。

在新元古代到早奥陶世期间（大于 543Ma 至 485Ma），对应 Sloss 的 Sauk 序列（1963），北美克拉通被广泛淹没，留下了 NE—SW 向横贯大陆的凹槽区，并以由凝缩沉积（小于 5m/Ma）或无沉积（图 30-6a）为标志。尽管狭长的线性加速沉降区域以高角度从陆内延伸至板块边缘，如蒙大拿中心地槽将威里斯顿盆地与西部板块边缘连接起来，Reelfoot 裂谷与密歇根州和伊利诺伊州盆地连接，南部的俄克拉荷马州裂谷将 Anadarko 原型盆地与东南部

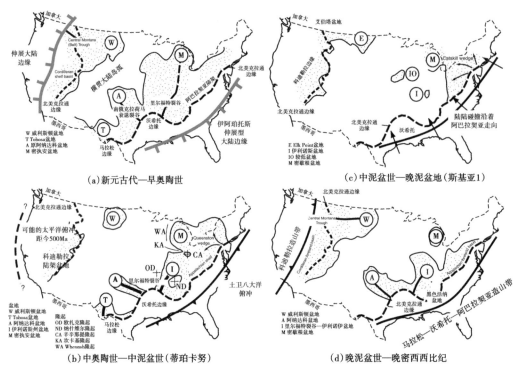

图 30-6　修改自 Sloss（1988）的美国古生代克拉通盆地

(a) 从晚古生代到早奥陶世，对应 Sauk 序列在向两侧边缘有很宽的倾斜的大陆区域，一些克拉通盆地发育在与伸展大陆边缘高角度相交的裂谷之上；(b) 从中奥陶世到早—中泥盆世，对应 Tippecanoe 层序，克拉通盆地开始逐渐地生长成拱形而且沿着东部盆地的边缘俯冲成圆顶状； (c) 从早—中泥盆世到晚泥盆世，对应着 Kaskaskia Ⅰ 层序，大陆和大陆的碰撞引起了北美东部区域隆升，而且大部分粗碎屑岩块脱落到板块东北部；(d) 晚泥盆世到晚密西西比世，对应着 Kaskaskia Ⅱ 层序，美国东部的中心位置的克拉通盆地与 Marathon-Ouachita-Appalachian 造山带和前陆盆地连接，在西北部 Williston 盆地与 Cordilleran 前陆盆地连接

的沃希托边缘连接，但是广泛分布的沉积物堆积表明了克拉通为一个向其边缘掀斜的缓坡。这种大型掀斜发生在劳伦古陆从冈瓦纳大陆分离而漂移的伸展阶段（Sloss，1988；Bally，1989），导致了古大西洋的扩张。因此北美的大部分克拉通盆地是在相邻的大陆边缘伸展时形成，但其典型的圆形平面形态是在后期不同的应力状态下形成的。

北美大陆边缘显生宙的消减可分为古生代 Iapetus 洋壳沿着东缘俯冲与晚古生代到新生代太平洋和法拉隆洋壳沿西缘俯冲两个阶段。在 Sloss 的 Tippecanoe 层序内（1963），古大西洋板块在约 480Ma 开始俯冲，并且延续到约 420Ma（Van der Pluijm 等，1900；McKerrow 等，1911）（图 30-6b）。古大西洋板块向西俯冲受到密歇根盆地东倾的中奥陶统沉积岩的证据支持（Coakley 和 Gurnis，1995），并被 Burgess 等（1997）用于克拉通层系的数值模拟。劳伦古陆边缘山脉的古生代历史尽管被熟知，但比较复杂（Dickinson，2009），劳伦古陆边缘的山脉随罗德里亚裂陷作为大陆边缘开始，但在 Antler 和 Sonnoma 造山期间被地中海式俯冲带反转和弧后扩张强烈改造。太平洋海底俯冲发生在前寒武纪（Scotese 和 Gllonka，1992）和晚古生代（Burchfiel 和 Davis，1972；Miller 等，1992）两个不同时期，Burgess 等（1997；图4，1519）提出了一个开始于 500Ma（寒武—奥陶纪界限附近）的远离现今北美克拉通西部边缘的俯冲模式。

在奥陶纪—密西西比纪（Tippecanoe 和 Kaskaskiad 层序；图 30-6b—d），随着板块边缘的洋壳俯冲和陆陆碰撞作用，克拉通沉积中心的发育各具特色。由于 Iapetus 板块在北美克拉通东南缘的俯冲作用引起上覆板块的远程应力场发生变化，导致了在 Sauk 和 Tippecanoe 层序之间的大规模地表出露和下切不整合的形成（约 480Ma），并且该段的抬升被认为是美国东北部早期造山运动（Taconic）（Hathcher 和 Viele，1982），但 Burgess 等（1997）意识到俯冲驱动的动力学地形导致的抬升作用在克拉通的内部存在变化。随后北美板块东缘经受了洋—陆壳的碰撞和沿着阿巴拉契亚山脉走向的增生作用，从而导致克拉通内的挤压背景。晚志留世，Baltica 板块和 Laurentian 板块之间发生陆—陆碰撞缝合。在这个时期，密歇根盆地独立发展成一个圆形的沉积中心，它的演化与阿巴拉契亚造山带及其前陆盆地系统有联系（Coakley 和 Gurnis，1995）。在早泥盆世，随着大范围的加里东造山运动和冈瓦纳古陆与北美板块发生会聚，横大陆穹隆、Ozark 穹隆、Cincinnati 背斜和 Nashvile 穹隆被抬升。在泥盆纪末到早石炭世，古大西洋板块在南部边缘发生俯冲。普遍抬升与剥蚀使得 sub-Kaskaskia II 层序不整合面被剥蚀（368Ma）。此时，沿着北美克拉通的西缘发育了海沟，它使得威里斯顿盆地从 Alberta 前陆盆地系统中分离出来。随后，密西西比纪末冈瓦纳板块与北美板块的陆—陆碰撞导致 sub-Absaroka 不整合面的广泛抬升剥蚀。因为新的拉伸、俯冲、聚合与碰撞循环集中在克拉通的南侧和西侧，因此，Absaroka I 层序（始于 330Ma）标志着北美克拉通上沉积样式的转变。

北美克拉通盆地明确地显示盆地的沉降开始于向伸展板块边缘呈缓坡状的掀斜大陆，局部的快速沉降凹槽穿过弱沉降的大陆地台区，并以高角度伸至克拉通内部。在北美东部和东南部，洋壳俯冲开始引起了应力场状态的变化，在此期间，形成与之相关的不整合面以及盆地内地层的掀斜。诸如密歇根和伊利诺斯等盆地变成各自独立的近圆形沉降中心，周围则为隆起、穹隆和造山带。克拉通盆地后期的演化被包括大洋板块俯冲、不稳定冷对流、来自会聚板块边缘的平面应力传递、收敛造山带构造载荷挠曲以及硅质碎屑岩从前陆盆地扩散至板块内部的各种动力学地形缓冲。

下面我们讨论低应变速率拉伸机理对较厚的大陆岩石圈之上的克拉通盆地形成和长期演化的影响，上文中提到的一些其他影响作次要考虑。

30.3 克拉通盆地演化的模拟

由于岩石圈减薄，单一的瞬时均匀拉张会导致断层相关沉降，随着岩石圈减薄冷却至其原始状态又会导致裂谷期后的沉降（Mckenzie，1978）。克拉通盆地可能是由拉张系数很小的拉张作用而形成，例如在 Husdon 海湾的拉张系数是 1.05~1.2（Hanne 等，2004），西西伯利亚盆地则小于 1.6（Saunders 等，2005）。假设正常的初始岩石圈厚度是 125km，很小拉张系数下的瞬时拉张产生的沉降太快，以至于不能产生持续的拉力来形成典型的克拉通盆地沉降史（图 30-7）。然而由于受岩石圈时间常数的影响，200km 厚的岩石圈的单一瞬时拉张会持续 250Ma 以上的沉降期，其中，时间常量 $\tau = z_0^2/\pi^2\kappa$ 有关（图 30-7），公式中的 z_0 是初始岩石圈厚度，κ 是扩散系数。

瞬时拉张的概念在拉张应变速率大于 $10^{-15}/s$（约 30nannostrain/a）（图 30-8）的沉积盆地简单模拟中有用。在这个基础上，关于瞬时拉张的假设是合乎情理的（Jarvis 和 McKenzie，1980）。假设一个应变速率常数，我们通过一维均匀拉张模拟来研究，假定瞬时拉张在

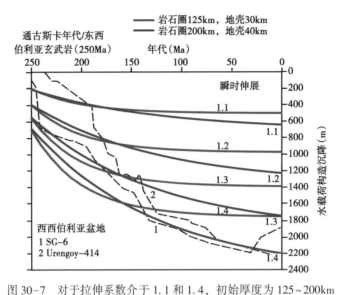

图 30-7　对于拉伸系数介于 1.1 和 1.4，初始厚度为 125~200km
的岩石圈，预测的含水盆地岩石圈瞬时拉张的区域构造沉降曲线

这个沉降预测与 Siberrian 盆地西缘的 SG-6 和 Urengoy-414 钻孔回剥沉降（图 30-3）画在一起
（Saunders 等，2005）；该沉降模式是假设岩石圈底部温度 1330℃，密度随岩石圈内温度变化而变化，
而在岩石圈表面不变（McKenzie，1978）

哪里发生和低应变速率对沉降产生什么影响（Armitage 和 Allen，2010）。利用温度是可以用一维深度剖面计算的。假定岩石圈上部塑性变形并且体积不变，水平的伸展作用会导致垂向上一定比例的减薄。因此，在横向拉张作用和内部放射性衰减加热影响之下，岩石圈热构造受热传导的支配。岩石圈热构造（地热）因不同时间岩石圈的扩张和冷却而异。假设上涌的软流圈成分与它代替的岩石圈地幔物质相同，那么就没有因化学组成和密度变化引起的额外向下负载（Kaminski 和 Jaupart，2000）。

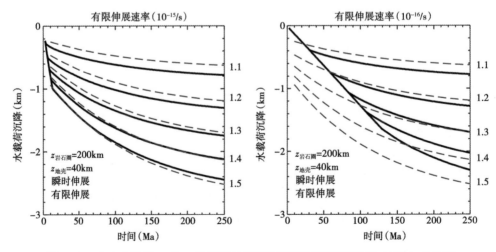

图 30-8　含水盆地以 $10^{-15}/s$ 应变速率有限瞬时扩张和应变速率为 $10^{-16}/s$ 的比较
假设岩石圈的底部 200km，温度 1330℃，地壳厚 40km

计算沉降时，假设岩石圈是局部均衡补偿（Airy），初始结构为海平面。假设密度在地壳内为常数$2900kg/m^3$，并且在地幔岩石圈内随热膨胀呈线性变化。假定岩石圈底部密度为$3400kg/m^3$，温度为$1330℃$，在我们的模型中，盆地沉降对软流圈的温度响应相对不敏感。

在极低速率的扩张下（应变速率为$10^{-16}/s$），物质的热传导和向上的对流具有可比性。当地壳机械拉伸时，较热的岩石圈向上对流。如果这种向上对流速度较快，地壳底部密度较小的地幔上拱就会引起地表隆起；然而，当应变速率足够低时，卜涌的地幔物质冷却，造成伸展点的地热增量小于瞬时拉伸或较高拉伸速率的地热增量（图30-9）。在这种情况下的岩石圈不会作为浮力而抗衡由地壳拉伸产生的沉降。结果当岩石圈在相同拉张系数下以更低的应变速率作用时，拉张沉降持续时间更久更剧烈（图30-8）。由于拉张期

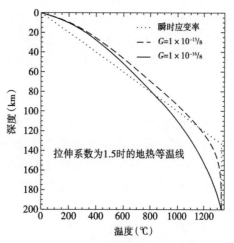

图30-9 地热拉伸系数为1.5时不考虑由于放射性衰变造成的瞬时内部增温图
假设内部增温时伸展模式应变率G为$10^{-15}/s$和$10^{-16}/s$，岩石圈的底部温度假定为$1330℃$，初始岩石圈厚度为200km

后的岩石圈的热扩散比较快，应变速率的拉张沉降量要小，因此，拉伸停止、岩石圈热释放时形成的沉积盆地沉降剖面为一个更稳定的稍有点下弯的碟形（图30-9）。

30.4 讨论

诸多不确定性导致对克拉通盆地的理解存在疑问，如盆地形成时热岩石圈的厚度、矿物相变引起的密度变化对克拉通盆地沉降的影响、盆地开始形成时的大陆板块应力状态。这里简要地对这些问题进行回顾。

我们主要利用过厚层大陆岩石圈的低应变速率拉伸来解释克拉通盆地的沉降史。但是对克拉通盆地形成及其长期演变过程的大陆岩石圈物理状态的评估比较困难。地震横波的速度显示，岩石圈厚度在板块之间和大板块内部均强烈变化（Goes等，2000；Artemieva和Mooney，2001；Goes和Van der Lee，2002；McKenzie等，2005），在主要区域构造线具有很大的梯度变化（Perez-Gussinye和Watts，2005）。岩石圈厚度也被认为受长期地幔冷却的影响（Pollack，1997；Herzberg，2004）。以西西伯利亚盆地为例，俄罗斯地台现今的岩石圈厚度为160~210km，在俄罗斯中央裂谷系统之下的区域较薄（125km），在西伯利亚地盾下的区域较厚（>350km）（Artemieva和Mooney，2001；Artemieva，2003）；相反，薄的岩石圈而非正常地壳被认为可能在长期增生的岛弧带、俯冲带和弧后岩石圈地区（Murphy和Nance1991）。因此，在克拉通盆地开始形成的时间内估算岩石圈的初始厚度是困难的。

沉降源于扩张作用过程中和扩张作用结束后岩石圈密度结构的改变。通常假设密度是温度的函数。由于熔融物保留在地幔，在有熔融发生的裂谷边缘存在额外的浮力（Scott，1992）。固态相变同样可以改变岩石圈的密度结构。岩石圈拉张的简单模式中，石榴石到斜长二辉橄榄岩的相变可能会降低岩石圈密度高达$100kg/m^3$（Podladchikov等，1994；Kaus等，2005）。大陆岩石圈的基本成分具有橄榄岩捕虏体的平均成分的特征。在新太古代岩石

圈中，我们假设岩石圈基本组成成分是尖晶石橄榄岩平均成分（McDonoug，1990）。利用这种基本成分，我们利用热力学的代码 Perple-X 计算出了岩石圈的密度（Connolly，1990，2005；Connolly 和 Kerrick，2002）。对于大陆岩石圈，由石榴石到斜长二辉橄榄岩的相变可以引起岩石轻微的密度变化。因此，矿物形成过程中的相变对岩石圈密度结构的改变不足以导致克拉通盆地内部大规模的伸展沉降。

岩石圈和地壳中更为复杂的相变已被提出来用于解释沉积盆地沉降。例如辉长岩—榴辉岩相变可能导致一些沉降（Joyner，1967；Haxby 等，1976；Stel 等，1993），引起密度增加约 $500kg/m^3$，并且该相变已被用来解释北美威利斯顿和密歇根盆地的形成（Haxby 等，1976；Baird 等，1995）及贝利特盆地的晚期演化（Stelet 等，1993）。尽管如此，产生这种相变必须存在如大量热物质上涌到地壳下部的热源（Baird 等，1995），但是一直没有得到这方面的合理证据。

我们提出，尽管在克拉通盆地长期演化过程中存在次级沉降机制，但其主要机制是大陆岩石圈低应变速率拉伸之后的冷却。远程拉伸应力导致大陆岩石圈处低应变速率状态存在争议，但一个大陆的应力状态与其相对于大地水准面和板块边界应力场的位置有关（Coblentz 和 Sandiford，1994；Coblentz 等，1998；Hillis 和 Reynolds，2000）。如现今的非洲板块和 50Ma 之前的澳大利亚板块被大洋中脊所环绕，所以大陆板块应当经历了弱的拉伸偏应力。

超大陆的裂解及随后的分散过程可能是大陆从大地水准面高点迁移到大地水准面低点造成的（Gurnis，1988）。数值模拟发现，稳定陆块将其下的地幔孤立，从而造成较热地幔发生缓慢横向运动时（Gurnis，1988），遇阻发生地幔物质上涌。这种孤立效应是从稳定的大陆中心推断出的（McKenzie 和 Priestley，2008）。另一种情况是，超级大陆岩石圈底部的热柱上涌也将导致地形隆起（Zhang 等，1996）。这两个作用都将导致大陆岩石圈受到水平拉伸应力。当大陆迁移到另一个较低大地水准面时，张应力得到缓解，并转换到挤压应力状态（Gurnis，1988）。

张应力导致低应变速率解释了克拉通盆地可能是超大陆解体循环中的一部分。假定初始岩石圈厚度为 200km，拉伸应变速率为 $10^{-16}/s$，则西西伯利亚盆地两个钻孔的沉降史对应的拉伸系数为 1.5（图 30-10），而密歇根州和伊利诺伊州盆地的沉降史明显表现更复杂（图

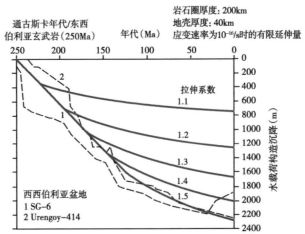

图 30-10　西西伯利亚含水盆地区域构造沉降与岩石圈拉伸的对比图
（图 30-3 中孔洞 SG-6 和 U-414；据 Saunders 等，2005）

岩石圈厚 200km，其中地壳厚 40km，以低应变速率 $10^{-16}/s$ 沉降直到拉伸系数达到 1.1~1.5；
在恒定的应变速率下拉伸的不同阶段可达到不同的拉伸系数

30-11）。模拟了 530~480Ma（Sauk 序列）应变速率分别为 $10^{-16}/s$ 和 $2\times10^{-16}/s$ 的扩张再冷却过程。伊阿珀托斯大洋开始俯冲以及大陆碰撞带（Kaskaskia）（图 30-6）的形成可以被认为是对热沉降背景曲线的偏离。在这些盆地变动的时期，我们预计观察到的沉降史是冷洋壳俯冲引起的冷却、动力学地形、造山带负载有关的挠曲沉降和来自收敛板块东侧的水平应力的传递等多个沉降背景因素的综合体现。

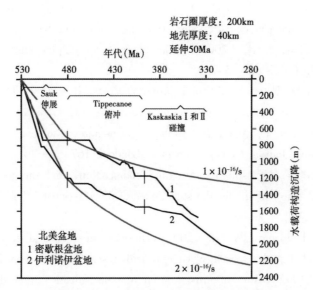

图 30-11 密歇根和伊利诺伊盆地钻孔资料所作的含水区域构造沉降模式，在应变速率分别为 $10^{-16}/s$ 和 $2\times10^{-16}/s$ 下作用 50Ma 后冷却，出现封闭沉降模式；对较低的应变速率来说 50Ma 后拉伸系数为 1.17，对较高的应变速率来说 50Ma 后拉伸系数为 1.37；沉降曲线的各个部分分别对应于 Sauk、Tippecanoe 和 Kaskaskia 层序，两个层序上叠加了次级沉降机制

30.5 结论

　　克拉通盆地是大型碟状沉积盆地，位于相对稳定的大陆岩石圈以及大陆边缘内侧，平面轮廓大致呈圆形。它是以很低的速率在很长的时间内充填。它们的长期性意味着盆地的多阶段充填过程（Kingston 等，1983），许多巨层序代表着盆地不同演化阶段。克拉通盆地的沉降史是不同动力学机制的混合而非一种。

　　典型的克拉通盆地始于水平张应力作用相邻的板块边缘的裂陷和漂移，这也意味着区域翘倾和大陆表面的淹没。在新元古代、早古生代和晚中生代，盆地的形成与超大陆裂解以及大陆碎块的离散有关。随后重要的改造包括圆形沉降中心的独特化可能与俯冲、地幔流动不稳定性，以及后来的大陆碰撞有关。因此，沉降拉伸的主要机制可能与后来多种动力学地形的影响有关，它们包括冷洋块的俯冲或者岩石圈底部的不稳定变形，构造载荷的挠曲沉降以及来自会聚板块边缘的水平压缩。

　　克拉通盆地的沉降史不能被正常厚度的岩石圈瞬时拉张的模式解释。一种可能的情况是克拉通盆地典型的沉降史为一个初始厚度的大陆岩石圈以很低的应变速率持久拉伸的模式。

致谢（略）

参 考 文 献

Aleinikov, A. L., Bellavin, O. V., Bulashevic, Y. P., Tarvin, I. F., Maksimov, E. M., Rudkevich, M. Y., Nalivkin, V. D., Shoblinskoyo, N. V., and Surkev, V. S. (1980) Dynamics of the West Siberian and Russian platforms, in Bally, A. W., Bender, P. L., McGetchin, T. R., and Walcott, R. I., eds., Dynamics of plate interiors. American Geophysical Union and Geological Society America Geodynamic Series, 1, 53-71.

Allen, P. A., and Allen, J. R. (2005) Basin analysis: principles and applications, 2nd ed. Oxford, Blackwell Publishing, 549 p.

Almeida, F. F. M., Brito Neves B. B., and Carneiro, C. D. R. (2000) The origin and evolution of the South American Platform. Earth Science Reviews, 50, 77-111.

Armitage, J. J., and Allen, P. A. (2010) Cratonic basins and the long-term subsidence history of continental interiors. Journal Geological Society London, 167, 61-70.

Artemieva, I. M. (2003) Lithospheric structure, composition, and thermal regime of the East European Craton: implications for the subsidence of the Russian Platform. Earth and Planetary Science Letters, 213, 431-446.

Artemieva, I. M., and Mooney, W. D. (2001) Thermal thickness and evolution of Precambrian lithosphere: a global study. Journal of Geophysical Research, 106, 16387-16414.

Artyushkov, E. V. (1992) Role of crustal stretching on subsidence of the continental crust. Tectonophysics, 215, 187-207.

Artyushkov, E. V., and Bear, M. A. (1990) Formation of hydrocarbon basins: subsidence without stretching in West Siberia, in Pinet, B., and Bois, C., eds., The potential for deep seismic profiling for hydrocarbon exploration. Paris, Technip, 45-61.

Artyushkov, E. V., Tesakov, Y. I., and Chekhovich, P. A. (2008) Ordovician sea-level change and rapid change in crustal subsidence rates in Eastern Siberia and Baltoscandia. Russian Geology and Geophysics, 49, 633-647.

Baird, D. J., Knapp, J. H., Steer, D. N., Brown, L. D., and Nelson, K. D. (1995) Upper-mantle reflectivity beneath the Williston Basin, phase-change Moho, and the origin of intracratonic basins. Geology, 23, 431-434.

Bally, A. W. (1989) Phanerozoic basins of North America, in Bally, A. W., and Palmer, A. R., eds., The geology of North America: an overview, vol. A. Boulder, CO, Geological Society of America, 397-446.

Beaumont, C., Quinlan, G. M., and Hamilton, J. (1987) The Alleghenian orogeny and its relationship to evolution of the Eastern Interior, North America, in Beaumont, C., and Tankard, A. J., eds., Sedimentary basins and basin-forming mechanisms. Memoir, 12. Canadian Society Petroleum Geologists, 425-445.

Boote, D. R. D., Clark-Lowes, D. D., and Traut, M. W. (1998) Paleozoic petroleum systems of North Africa, in Macgregor, D. S., Moody, R. T. J., and Clark-Lowes, D. D., eds., Petroleum geology of North Africa. Special Publication, 132. Geological Society London, 7-68.

Braile, L. W., Hinze, W. J., Keller, G. R., Lidiak, E. G., and Sexton, J. L. (1986) Tectonic development of the New Madrid Rift Complex, Mississippi Embayment, North America. Tectonophysics, 131, 1−21.

Brito Neves, B. B. de (2002) Main stages of the development of the sedimentary basins of South America and their relationship with the tectonics of supercontinents. Gondwana Research, 5, 175−196.

Brito Neves, B. B. De, Fuck, R. A., Cordani, U. G., and Thomaz Filho, A. (1984) Influence of basement structures on the evolution of the major sedimentary basins of Brazil: a case of tectonic heritage. Journal Geodynamics, 1, 495−510.

Burchfiel, B. C., and Davis, G. A. (1972) Structural framework and evolution of the southern part of the Cordilleran orogen, western United States. American Journal of Science, 272, 97−118.

Burgess, P. M., Gurnis, M., and Moresi, L. (1997) Formation of sequences in the cratonic interior of North America by interaction between mantle, eustatic and stratigraphic processes. Bulletin Geological Society America, 108, 1515−1535.

Burke, K. (1976) The Chad Basin: an active intracontinental basin. Tectonophysics, 36, 197−206.

Burke, K., and Dewey, J. F. (1973) Plume-generated triple junctions: key indicators in applying plate tectonics to old rocks. Journal of Geology, 81, 406−433.

Cainelli, C. C., and Mohriak, W. U. (1999) Some remarks on the evolution of sedimentary basins along the Eastern Brazilian Continental Margin. Episodes, 22, 206−216.

Coakley, B., and Gurnis, M. (1995) Far field tilting of Laurentia during the Ordovician and constraints on the evolution of a slab under an ancient continent. Journal Geophysical Research, 100, 6313−6327.

Coblentz, D. D., and Sandiford, M. (1994) Tectonic stresses in the African plate: constraints on the ambient lithospheric stress state. Geology, 22, 831−834.

Coblentz, D. D., Zhou, S., Hillis, R., Richardson, R., and Sandiford, M. (1998) Topography, plate boundary forces and the Indo−Australian Intraplate Stress Field. Journal Geophysical Research, 103, 919−931.

Connolly, J. A. D. (1990) Multivariable Phase diagrams: an algorithm based on generalised thermodynamics. American Journal of Science, 290, 666−718.

Connolly, J. A. D. (2005) Computation of phase equilibria by linear programming: a tool for geodynamic modelling and its application to subduction zone decarbonation. Earth and Planetary Science Letters, 236, 524−541.

Connolly, J. A. D., and Kerrick, D. M. (2002) Metamorphic controls on seismic velocity of subducted oceanic crust at 100−250km depth. Earth and Planetary Science Letters, 204, 61−74.

Dalziel, I. W. D. (1991) Pacific margins of Laurentia and East Antarctica−Australia as a conjugate rift pair: evidence and implications for an Eocambrian supercontinent. Geology, 19, 598−601.

De Rito, R. F., Cozzarelli, F. A., and Hodges, D. S. (1983) Mechanisms of subsidence in ancient cratonic rift basins. Tectonophysics, 94, 141–168.

Dickinson, W. R. (2009) Anatomy and global context of the North American Cordillera, in Kay, S. M., Ramos, V. A., and Dickinson, W. R., eds., Backbone of the Americas: shallow subduction, plateau uplift, and ridge and terrane collision. Memoir, 204. Boulder, CO, Geological Society America, 1–29.

Farringdon, R. J., Stegman, D. R., Moresi, L., and Sandiford, M. (2010) Interactions of 3D mantle flow and continental lithosphere near passive margins. Tectonophysics, 483, 20–28.

Fowler, C. M. R., and Nesbit, E. G., (1985) The subsidence of the Williston Basin. Canadian Journal of Earth Sciences, 22, 408–415.

Goes, S., and Van Der Lee, S. (2002) Thermal structure of the North American uppermost mantle inferred from seismic tomography. Journal Geophysical Research, Solid Earth, 107, B3 2050.

Goes, S., Govers, R., and Vacher, P. (2000) Shallow mantle temperatures under Europe from P and S wave tomography. Journal Geophysical Research–Solid Earth, 105, B5, 11153–11169.

Guiraud, R., Bosworth, W., Thierry, J., and Delplanque, A. (2005) Phanerozoic geological evolution of Northern and Central Africa: An overview. Journal African Earth Sciences, 43, 83–143.

Gurnis, M. (1988) Large - scale mantle convection and the aggregation and dispersal of supercontinents. Nature, 332, 695–699. doi: 10. 1038/332695a0.

Hand, M., and Sandiford, M. (1999) Intraplate deformation in Central Australia, the link between subsidence and fault reactivation. Tectonophysics, 305, 121–140.

Hanne, D., White, N., Butler, A., and Jones, S. (2004) Phanerozoic vertical motions of Hudson Bay. Canadian Journal of Earth Science, 41, 1181–1200.

Hartley, R. W., and Allen, P. A. (1994) Interior cratonic basins of Africa: relation to continental break–up and role of mantle convection. Basin Research, 6, 95–113.

Hatcher, R. D. Jr., and Viele, G. W. (1982) The Appalachian/ Ouachita orogens: United States and Mexico, in perspectives in regional geological synthesis planning for the geology of North America (Ed. By Palmer, A. R.), Geological Society America, D–NAG, Special Publication 1, 67–75.

Haxby, W. F., Turcotte, D. L., and Bird, J. M. (1976) Thermal and mechanical evolution of the Michigan Basin. Tectonophysics, 36, 57–75.

Heine, C., Mueller, R. D., Steinberger, B., and Torsvik, T. H. (2008) Subsidence in intracratonic basins due to dynamic topography. Physics Earth and Planetary Interiors, 171, 252–264.

Helwig, J. A. (1985) Origin and classification of sedimentary basins. Proceedings of the 7th Offshore Technology Conference, 1, 21–32.

Herzberg, C. (2004) Geodynamic information on peridotite petrology. Journal of Petrology, 45, 2507–2530. Hillis, R. R., and Reynolds, S. D. (2000) The Australian stress map. Journal Geological Society London, 157, 915–921.

Hoffman, P. F. (1991) Did the breakout of Laurentia turn Gondwanaland inside–out? Science, 252, 1409–1412.

Hsu, K. J. (1965) Isostasy, crustal thinning, mantle changes, and the disappearance of ancient land masses. American Journal Science, 263, 97-109.

Jarvis, G. T., and Mckenzie, D. P. (1980) Sedimentary basin formation with finite extension rates. Earth and Planetary Science Letters, 48, 42-52.

Joyner, W. B. (1967) Basalt-eclogite transition as a cause for subsidence and uplift. Journal of Geophysics Research, 72, 4977-4998.

Julia, J., Assumpcao, M., and Rocha, M. P. (2008) Deep crustal structure of the Paran_a Basin from receiver functions and Rayleigh-Wave dispersion: evidence for a fragmented cratonic root. Journal Geophysical Research, 113, B08318. doi: 10. 1029/2007jb005374.

Kaminski, E., and Jaupart, C. (2000) Lithosphere structure beneath the Phanerozoic intracratonic basins of North America. Earth and Planetary Science Letters, 178, 139-149.

Kaus, B. J. P., Connelly, J. A. D., Podladchikov, Y. Y., and Schmalholz, S. M. (2005) Effect of mineral phase transitions on sedimentary basin subsidence and uplift. Earth and Planetary Science Letters, 233, 213-228.

Keen, C. E. (1987) Some important consequences of lithospheric extension, in Coward, M. P., Dewey, J. F., and Hancock, P. L., eds., Continental extensional tectonics. Special Publication, 28. Geological Society London, 67-73.

Kingston, D. R., Dishroon, C. P., and Williams, P. A. (1983) Global basin classification system. Bulletin American Association Petroleum Geologists, 67, 2175-2193.

Klein, G. D. (1995) Intracratonic basins, in eds., Tectonics of sedimentary basins. Oxford, Blackwell Science, 459-478.

Klein, G. De V., and Hsui, A. T. (1987) Origins of intracratonic basins. Geology, 17, 1094-1098.

Klemme, H. D. (1980) Petroleum basins: classifications and characteristics. Journal Petroleum Geology, 27, 30-66.

Kolata, D. R., and Nelson, W. J. (1990) Tectonic history of the Illinois Basin, in Leighton, M. W., Kolata, D. R., Oltz, D. F., and Eidel, J. J., eds., Interior cratonic basins. Memoir, 51. American Association of Petroleum Geologists, 263-285.

Korsch, R. J., and Lindsay, J. F. (1989) Relationships between deformation and basin evolution in the Amadeus Basin, Central Australia. Tectonophysics, 158, 5-22.

Le Pichon, X., Francheteau, J., and Bonnin, J. (1973) Plate tectonics: developments. Geodynamics, 6.

Leighton, M. W., Kolata, D. R., Oltz, D. F., and Eidel, J. J. (1991) Cratonic Basins. Memoir, 51. American Association Petroleum Geologists, 819 pp.

Leighton, M. W., and Kolata, D. R. (1990) Selected interior cratonic basins and their place in the scheme of global tectonics: a synthesis. In Cratonic basins. Memoir, 51. American Association Petroleum Geologists, 729-797.

Lindsay, J. F. (2002) Supersequences, superbasins, supercontinents - evidence from the Neoproterozoic-Early Paleozoic basins of Central Australia. Basin Research, 14, 207-223.

Liu, H. S. (1979) Mantle convection and subcrustal stresses under Australia. Modern Geology, 7,

29-36.

Mcdonough, W. F. (1990) Constraints on the composition of the continental lithosphere mantle. Earth and Planetary Science Letters, 101, 1-18.

Mckenzie, D. P. (1978) Some remarks on the development of sedimentary basins. Earth and Planetary Science Letters, 40, 25-32.

Mckenzie, D., and Priestley, K. (2008) The influence of lithospheric thickness variations on continental evolution. Lithos, 102, 1-11. doi: 10. 1016/J. Lithos. 2007. 05. 005.

Mckenzie, D., Jackson, J., and Preistley, K., 2005, Thermal structure of oceanic and continental lithosphere. Earth and Planetary Science Letters, 233, 337-349.

Mckerrow, W. S., Dewey, J. F., and Scotese, C. R. (1991) The Ordovician and Silurian development of the Iapetus Ocean. Special Papers In Paleontology, 44, 165-178.

Middleton, M. F. (1989) A model for the formation of intracratonic sag basins. Geophysical Journal International, 99, 665-676.

Miller, E. L., Miller, M. M., Steevens, C. H., Wright, J. E., and Madrid, R. (1992) Late Paleozoic paleogeographic and tectonic evolution of the Western U. S. Cordillera, in Burchfiel, B. C., Lipman, P. W., and Zoback, M. L., eds., The Cordilleran Orogen: Conterminous U. S., the Geology of North America G-3. Boulder, CO, Geological Society of America, Boulder, CO, 57-106.

Mitrovica, J. X., Beaumont, C., and Jarvis, G. T. (1989) Tilting of continental interiors by the dynamical effects of subduction. Tectonics, 8, 1079-1094.

Murphy, J. B., and Nance, R. D. (1991) Supercontinent model for the contrasting character of Late Proterozoic orogenic belts. Geology, 19, 469-472.

Morozova, E. A., Morozov, I. B., Smithson, S. B., and Solidolov, L. N. (1999) Heterogeneity of the uppermost mantle beneath Russian Eurasia from the ultra-long-range-profile QUARTZ. Journal Geophysical Research, 104 (B9), 20329-20348.

Neumann, E. R., Olsen, K. H., Balbridge, W. S., and Sundvoll, B. (1992) The Oslo Rift: a review, in Geodynamics of rifting, vol. 1, Case history studies of rifts: Europe and Asia (Ed. By Ziegler, P. A.), Tectonophysics, 208, 1-18.

Nikishin, A. M., Ziegler, P. A., Abbott, D., Brunet, M. F., and Cloetingh, S. (2002) Permo-Triassic intraplate magmatism and mantle dynamics. Tectonophysics, 351, 3-39.

Nunn, J. A., and Sleep, N. H. (1984) Thermal contraction and flexure of intracratonic basins: A three dimensional study of the Michigan Basin. Geophysical Journal Royal Astronomical Society, 79, 587-635. Nurmi, R. D., and Friedman, G. M. (1977) Sedimentology and depositional environments of basin-center evaporites, Lower Salina Group (Upper Silurian), Michigan Basin, in Fisher, J. H., eds., Reefs and evaporites: concepts and depositional models. American Association Petroleum Geologists Studies In Geology, 5, 23-52.

Peate, D. W. (1997) The paraná-etendeka province, in large igneous provinces: continental, oceanic and planetary flood volcanism, geophysics monograph series 100 (Ed. By J Mahoney and MCoffin), 217-245, Agu, Washington D. C.

Pérez-Gussinyé, M., and Watts, A. B., 2005, The long-term strength of europe and its implications for plate – forming processes. Nature, 436, 381 – 384. doi: 10. 1038/Nature03854.

Podladchikov, Y. Y., Poliakov, A. N. B., and Yuen, D. A. (1994) The effect of lithospheric phase transitions on subsidence of extending continental lithosphere. Earth and Planetary Science Letters, 124, 95–103.

Pollack, H. N. (1997) Thermal characteristics of the archean, in De Wit, M., and Ashwall, L. D., eds., Greenstone Belts, Oxford Monographs On Geology and Geophysics.

Quinlan, G. M. (1987) Models of subsidence mechanisms in intracratonic basins and their applicability to north American examples, in sedimentary basins and basin-forming mechanisms (Ed. By Beaumont, C., and Tankard, A. J.) 12, 463–481, Memoir Candadian Society Petroleum Geologists.

Quinlan, G. M., and Beaumont, C. (1984) Appalachian thrusting, lithospheric flexure, and the paleozoic stratigraphy of the eastern interior of North America. Canadian Journal of Earth Sciences, 21, 973–996.

Rogers, J. J. (1996) A history of continents in the past three billion years. Journal of Geology, 104, 91–107.

Sahagian, D. (1993) Structural evolution of african basins: stratigraphic synthesis. Basin Research, 5, 41–54.

Sanford, B. V. (1987) Paleozoic geology of the hudson platform, in sedimentary basins and basin-forming mechanisms (Ed. By Beaumont, C. and Tankard, A. J.), 483–505, Canadian Society Petroleum Geologists Memoir, 12.

Saunders, A. D., England, R. W., Reichow, M. K., and White, R. V. (2005) A mantle plume origin for the siberian traps: uplift and extension in the east siberian basin, Russia. Lithos, 79, 407–424.

Saunders, A. D., Jones, S. M., Morgan, L. A., Pierce, K. L., Widdowson, M., and Xu, Y. G. (2007) Regional uplift associated with continental large igneous provinces: the roles of mantle plumes and the lithosphere. Chemical Geology, 241, 282–318.

Scotese, C. R., and Golonka, J. (1992) Paleogeographic Atlas. University Texas At Arlington, Paleomap Project, Arlington, Texas, USA.

Scott, D. R. (1992) Small-scale convection and mantle melting beneath mid-ocean ridges. In Phipps Morgan, J., Blackman, D. K. and Sinton, J. M., eds., Mantle Flow and Melt Generation At Mid-Ocean Ridges. American Geophysical Union.

Selley, R. C. (1972) Diagnosis of marine and non – marine environments from the cambro-ordovician sandstones of Jordan. Journal Geological Society London, 128, 135–150.

Selley, R. C. (1997) The basins of northwest africa: structural evolution, in Selley, R. C. (Ed.) African Basins: Sedimentary Basins of the World, Elsevier, Amsterdam, 17–26.

Sengor, A. M. C., and Natal' in, B. A. (1996) Paleotectonics of Asia: Fragments of a synthesis, in the tectonic evolution of asia (Ed, By Yin, A., Harrison, T. M.), 486–641, Cambridge University Press, New York.

Sleep, N. H. (1971) Thermal effects of the formation of Atlantic continental margins by continental break-up. Geophysical Journal Royal Astronomical Society, 24, 325-350.

Sleep, N. H., and Sloss, L. L. (1980) The Michigan Basin, in dynamics of plate interiors (Eds. By Bally, A. W., Bender, P. L., Mcgetchin, T. R., and Walcott, R. I.), American Geophysical Union/Geological Society of America Geodynamics Series, 1, 93-97.

Sleep, N. H., and Snell, N. S. (1976) Thermal contraction and flexure of mid-continent and Atlantic marginal basins. Geophysical Journal Royal Astronomical Society, 45, 125-154.

Sloss, L. L. (1963) Sequences in the cratonic interior of North America. Bulletin Geological Society of America, 74, 93-114.

Sloss, L. L. (1988) Introduction, in sedimentary cover: North American craton (Ed. By Sloss, L. L.), Geological Society of America, the Geology of North America, D-2, 1-3.

Sloss, L. L. (1990) Epilog, in interior cratonic basins (Eds. By Leighton, M. W., Kolata, D. R., Oltz, D., and Eidel, J. J.), 799-805, American Association Petroleum Geologists Memoir 51.

Sloss, L. L., and Speed, R. C. (1974) Relationship of cratonic and continental amrgin tectonic episodes, in tectonics and sedimentation (Ed. By W. R. Dickinson), 38-55, Special Publication Society Economic Paleontologists and Mineralogists 22.

Soares, P. C., Landim, P. M. B., and Fulfaro, V. J. (1978) Tectonic cycles and sedimentary sequences in the Brasilian intracratonic basins. Bulletin Geological Society America, 89, 181-191.

Stel, H., Cloetingh, S., Heeremans, M., and Van Der Beek, P. (1993) Anorogenic granites, magmatic underplating and the origin of intracratonic basins in a non-extensional setting. Tectonophysics 226, 285-299.

Thompson, R. N., and Gibson, S. A. (1991) Subcontinental mantle plumes, hotspots and preexisting thinspots. Journal of the Geological Society, 148, 973-977.

Turner, S. M., Regelous, S., Kelley, S., Hawkesworth, C. J., and Mantovani, M. S. M. (1994) Magmatism and continental break-up in the South Atlantic: high precision geochronology. Earth Planetary Science Letters, 121, 333-348.

Van Der Pluijm, B. A., Johnson, R. J. E., and Van Der Voo, R. (1990) Early paleozoic paleogeography and accretionary history of the Newfoundland Appalachians. Geology, 18, 898-901.

Veevers, J. J. (1989) Middle/Late Triassic (23.5 Ma) Singularity in stratigraphic and magmatic history of the pangean heat anomaly. Geology, 17, 784-787.

Vyssotski, A. V., Vyssotski, V. N., and Nezhdanov, A. A. (2006) Evolution of the West Siberian basin. Marine and Petroleum Geology, 23, 93-126.

Walter, M. R., Veevers, J. J., Calver, C. R., and Grey, K. (1995) Neoproterozoic stratigraphy of the centralian superbasin, Australia. Precambrian Research, 73, 173-195.

Xie, X., and Heller, P. L. (2009) Plate tectonics and basin subsidence history. Bulletin Geological Society America, 121, 55-64.

Zalan, P. V., Wolff, S., Astolfi, M. A. M., Vieira, 1. S., Concelcao, J. C. J., Appi, V. T., Neto, E. V. S., Cerqueira, J. R., and Marques, A. (1990) The paran_a basin, Brazil, in interior cratonic basins (Eds. By Leighton, M. W., Kolata, D. R., Oltz, D., and Eidel, J. J.), 681-708, American Association Petroleum Geologists Memoir 51.

Zhao, J. X., Mcculloch, M. T., and Korsch, R. J. (1994) Characterization of a plume-related 800ma magmatic event and its implications for basin formation in Central-Southern Australia. Earth and Planetary Science Letters, 121, 349-367.

Zhong, S. J., Gurnis, M., and Moresi, L. (1996) Free-surface formulation of mantle convection. 1: basic theory and application to plumes. Geophysical Journal International, 127, 708-718.

Ziegler, P. A. (1989) Evolution of Laurussia. Kluwer Academic Publishers, London, 102 pp.

Ziegler, P. A. (1990) Geological Atlas of Western and Central Europe. Shell International Petroleum Company/Geological Society of London, 2nd Edition, 239 pp.

Ziegler, P. A., and Van Hoorn, B. (1989) Evolution of the North Sea Rift System. Memoir American Association Petroleum Geologists, 46, 471-500.

(牛宁 译，蒲仁海 韩银学 蔡露露 校)

第31章 内流盆地

GARY NICHOLS

(Department of Earth Sciences, Royal Holloway University of London, Egham, Surrey, UK)

摘　要：内流汇水盆地可以发育在多种构造背景之下。内流盆地的陆相相序特征在很大程度上受没有与海洋直接相通和构造背景控制。气候是决定沉积相总体特征的主控因素，从干旱地区风成主导的环境到潮湿地区的湖盆环境均有发育，主要特征包括具有深的河流下切强烈的加积地层结构，而这种下切结构仅发育在气候控制湖平面波动的湖盆背景下。盆地充填导致向盆地边缘大规模上超，例如供源河谷的回填。在内流盆地内，河流体系广泛分布，可形成由河道和河漫滩组成的盆地规模的扇体。

关键词：内流　内排水　陆相沉积　河流分布体系　湖相盆地

31.1　引言

内流盆地不与开放海域直接相连。河流从周围高地流进盆地内部，地表水没有流出盆地的通道，水通过蒸发进入大气圈或者渗入地下含水层，因此所有这类盆地都是陆相盆地。虽然所有该类盆地可按构造背景分类，但它们又表现出沉积相分布和地层关系密切，后者对盆地的充填序列属性有重要的影响。

由于不与海洋直接相通，因此内流盆地不受全球海平面波动的影响，所以，盆地总体的可容空间取决于盆地边缘地形和区域及盆地内部的构造活动。构造沉降量和沉降速率的小幅度变化对欠补偿内流盆地的影响微弱，因为它相对于盆地的固定沉积基准面没有发生变化。气候控制始终都很重要，它决定了盆地是否受风沙、河流、湖泊条件的影响。此外，沉积期间水量和温度的波动直接会造成相变。

在本文关于内流盆地特征论述当中，现代内流盆地的实例极少，沉积相特征和地层关系的描述仅限于内陆河盆地的演化。毫无疑问，不同时期大量的陆相沉积形成于内流盆地内部，但被描述出来的却寥寥无几。沉积盆地的演化受控于各种因素：构造沉降、海平面升降变化、气候和沉积物供给（图31-1）。海平面和可容空间等控制因素的缺乏很可能形成具有内陆河背景特征的相展布和地层关系模式。

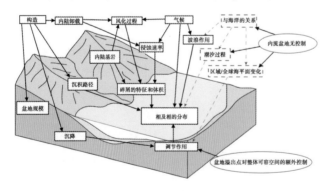

图31-1　构造、气候、沉积物供给等因素对内流盆地的影响（据尼克尔斯，2009，修改）

31.2 现代内流盆地实例

Hartley 等（2010）分析了现代大陆沉积盆地中的河流沉积体系，415 条河流的 45%发育在内流环境。这些以考察河流分布规律为目的而采集的数据资料显示：河流从头到尾大于 30km，这些盆地涵盖所有长或者宽大于 30km 的盆地，但排除了没有大型河流体系的盆地。因此，许多干旱地区的内流盆地不在分析之列，这表明很可能有一半正在沉积的现代陆相盆地具有内汇水环境。

在 Hartley 等的研究中，他们用被认为是内汇水环境下的 175 条河流系统代表各种构造背景：其中大部分形成于前陆盆地环境，其余的分布在克拉通盆地、走滑盆地和张性环境之中，只有 2%发育在弧前盆地。Hartley 等（2010）也分析了每条河流体系的气候环境，发现 84%出现在干旱环境，只有极少部分在极地和大陆气候下发育。值得一提的是，他们也认为内流盆地存在于热带和亚热带地区。虽然分布地域偏向亚洲（大约三分之二），但在非洲、澳大利亚、南美洲和北美洲也有内流盆地的典型代表。盆地的构造和气候背景具有区域性分布的特征，如所有热带气候的内流盆地都发育于非洲的伸展环境，并且主要与东非裂谷系有关。背驮式盆地几乎全部发生在伊朗扎格罗斯褶皱带和中国天山褶皱带。与走滑相关的盆地主要发生在加利福尼亚和中国，澳大利亚的内汇水盆地都发生在克拉通环境，其余的克拉通盆地发生在非洲中西部（乍得）。

最大的现代内流盆地面积超过 $100 \times 10^4 km^2$。例如，澳大利亚中部艾尔湖盆地（图 31-2）面积有 $114 \times 10^4 km^2$，大约为大陆面积的六分之一，并且整个地区的新生代都有缓慢的沉积堆积；而非洲中西部乍得湖盆的面积是艾尔湖盆地规模（$238 \times 10^4 km^2$）的两倍，占非洲面积的 8%，也经历了几千万年的缓慢沉积（Mitchell 和 Reading，1986）。在亚洲，内流盆地的例子包括汇水面积达 $360 \times 10^4 km^2$ 的里海盆地，它形成了当今世界上面积达 $37 \times 10^4 km^2$ 的最大湖泊，并且自晚中新世以来都在进行内部汇水。此外，咸海也是典型的内流盆地，面积达 $155 \times 10^4 km^2$。中型尺寸的盆地包括面积超过 $40 \times 10^4 km^2$ 的塔里木盆地和青藏高原北部

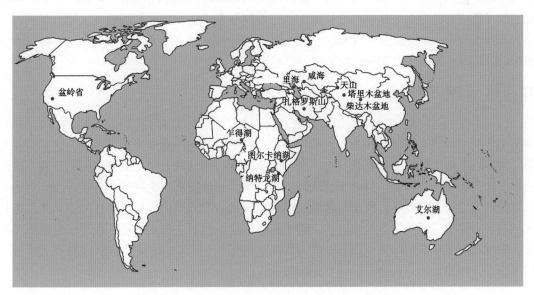

图 31-2 文中所提及的现代主要内流盆地的位置

边缘的柴达木盆地。现代许多更小的内部汇水盆地发育在扩张环境。沿东非裂谷的几个湖泊（例如，纳特龙湖和图卡纳尔湖）均是内流盆地，北美西南部盆地和山脉区域包含许多大陆系统沉积中心的小型伸展型盆地也是内流盆地。

31.3 内流盆地的地层记录

由于很少有学者在他们的分析报告中明确指出盆地是外排水还是内排水系统，因此从地层记录的角度不能计算内流盆地的比例。然而，相对于其他地球历史时期而言，现今普遍发生的内流盆地很可能会是个典型。不过，考虑到目前大陆分布相对分散，因此，当大陆板块拼合成超级大陆时，内流盆地分布可能会更加普遍。例如，随着早古生代 Iapetus 洋的关闭和加里东造山带的形成，内陆盆地在泥盆纪广泛分布（Friend 等，2000）。许多泥盆纪的盆地是张扭构造作用形成的大小各异的沉积—沉降体系，并在挪威、斯瓦尔巴特群岛、东格陵兰、苏格兰、英国南部、爱尔兰和加拿大东部得以保存（Friend 等，2000）。更靠北部的这些盆地，如挪威的霍纳伦盆地、苏格兰奥尼克群岛和东格陵兰的泥盆纪盆地都是内流型盆地且都堆积了厚层的河湖沉积地层。

形成于会聚型板块环境的内流盆地也不乏实例，包括阿根廷西北部的安第斯前陆的冲积扇、冲积平原以及新近纪湖泊相沉积均被认为形成于内排水环境（Mortimer 等，2007）。从晚始新世到晚中新世，在从比斯开湾西部到地中海东部隔绝的盆地环境形成了比利牛斯山脉侧翼的埃布罗河前陆盆地，并堆积了冲积扇相和河流湖泊相的沉积（García-Castellanos 等，2003；Nichols，2004，2007）。

31.4 内流盆地的相分类

大地构造背景决定了盆地的整体形状和大小，但沉积相的特点主要受气候背景以及水的供给和散失之间的平衡影响（García-Castellanos，2006）。由湖泊环境主导的潮湿盆地到风成的干旱环境盆地有一个连续的变化范围。根据气候和水的供排平衡，内流盆地可识别出三种沉积相组合：（1）湖泊主导相；（2）河流主导相；（3）风成主导相（Olsen，1990；Nichols，2004，2005；图31-3）。

31.4.1 湖相主导的盆地

Carroll 和 Bohacs（1999）在其湖泊盆地分类的方案里提出构造控制盆地的可容空间，气候决定湖泊盆地水的供给和蒸发。他们提出三种主要的湖泊盆地类型，分别为水文条件封闭、长期处于内流状态的欠补偿盆地；长期外流的超补偿盆地和介于两者之间的外流、内流交替发育的平衡补偿盆地。欠补偿盆地的湖泊以浅水和蒸发沉积为特征，常常伴生风成和冲积扇沉积，生物分异度低。平衡补给的湖泊经过一定阶段的外流，相组合更加多样化，包括碳酸盐岩、碎屑岩以及富有机质沉积。其中富有机质沉积是在深水湖泊环境下，有机质丰度较高的泥岩层状聚集的结果（Carroll 和 Bohacs，1999）。相反，超补偿湖泊盆地通常以河流相占主导，湖相沉积具有典型的浅水湖泊特征。

苏格兰北部泥盆纪的奥克尼盆地沉积了一套连续的厚层湖相地层（如图31-4），最小的地层厚度为750m（Astin 和 Rogers，1991）。这些湖泊沉积主要是浅水相，表明沉积物供给

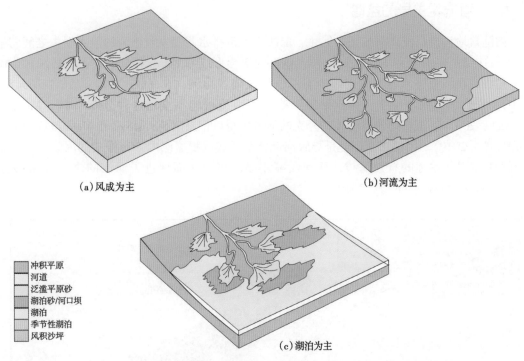

图 31-3 内流盆地的陆地环境范围,从湖泊相、河流相到风成相的相组合
(据尼克尔斯,2005,修改)

与可容空间之间的平衡。而浅水湖相和干旱湖相的交替发育则受控于气候波动(Astin 和 Rogers,1991)。与走滑断层相关的构造沉降构成了盆地的可容空间总和(Mykura,1991),但在内流环境下,大致稳定的湖水深度可以通过盆地范围的沉积物简单替换湖水实现。盆地的沉积相与内流环境的一致性说明加积模式可以持续足够长的时间,沉积几百米的浅湖相沉积。在部分地区的盆地发展史中,有很多现象显示盆地与海相连而不是永久的内陆湖(Marshall 等,1996)。北美东部三叠纪至早侏罗世的纽瓦克超群,为断陷盆地中典型封闭水文系统的湖相沉积(Olsen,1990)。这些沉积表现为盆地中心分选良好的沉积物周期性的沉积作用,包括蒸发岩沉积、盆地边缘的粗粒沉积以及局部物源的沉积都说明其具有封闭盆地的特征。

图 31-4 苏格兰 Orcadian 内流盆地的泥盆系薄层(波浪纹砂岩层;陡崖大约 8m 高)

31.4.2 河流相主导的盆地

内陆盆地河流是流入湖泊还是终止，取决于河流排水与蒸发或渗入的流水损失之间的平衡（图 31-5；Nichols 和 Fisher，2007）。这些河流会形成一个分布格局，在盆地边缘形成湖泊三角洲或在河流进入盆地的连续断裂部位形成扇三角洲（Nichnk，2007；Weissmann 等，2010）。在干旱气候环境下，冲积平原上终止的河流可能不全是这种分布格局（North 和 Warwick，2007），沉积物的扇形体在分支的河流中得以记录（Stanistreet 和 McCarthy，1993；Hartley 等，2010）。内部排水环境是导致河流体系形成沉积盆地内分布样式的因素之一，盆地沉积物提升了盆地的有效海拔，使河流减小了坡度，从而促进了河流的决口（Nichols，2004，2007；Weissrnann 等，2010）。

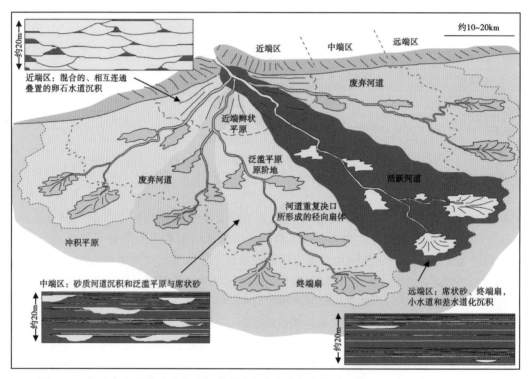

图 31-5 河流相主导的内流盆地河道形成和分布的河流体系（据 Nichols 和 Fisher，2007）

纽瓦克超群主要为河流相沉积，发育在每一个断陷盆地的早期和晚期，中期发育为湖泊相。美国西北部的 Chumstick 城古近纪内流盆地的沉积发育于相对湿润的气候环境（Evans，1991），盆地沉积了厚达 12000m 的河流相沉积物。由于盆地处于走滑构造持续沉降的背景下，在盆地边缘沉积了以粗粒长石砂岩为特征的冲积扇及盆地边缘的细粒河道和河漫滩沉积。洪泛平原相高的有机质含量指示其为潮湿的热带气候（Evans，1991）。

31.4.3 风成相主导的盆地

在干旱环境下，低潜水面和植被的缺乏为细粒度沉积物通过风力重新分配提供了条件。现今世界各地沙漠中的沙子堆积区域通常都是内流水，例如，澳大利亚中部的艾尔湖部分区域和北非、中东、中亚的沙漠地区。与之伴生的沉积相组合则包括短期河流相（河道砾石堆积和冲积扇沉积)、湖相泥质和蒸发沉积。欧洲西北部的早二叠世沉积发育风成沉积占主导的沉积相组合。北海二叠纪盆地的南部是由风成砂占主导且伴随短暂湖泊相和萨勃哈相的内流盆地(Glennie, 1990)。

然而，形成于外排水环境下的风成相的地层序列很可能与形成于内流盆地的风成相相似。与海相沉积交替发育可以证明存在外流水系环境，但是在风成沉积主导下的内流条件很难证明，这是因为造成内流的原因是干旱而非盆地内部限制水体排到海洋，诸如现代的纳米布沙漠。风成地层的堆积与保存部分地受潜水面控制(Kocurek, 1996)，但是要区分内流盆地中潜水面受地下水位的影响还是受海平面的影响非常困难。

31.5 相与沉积结构

31.5.1 加积地层结构

在与海洋直接相连或在海平面波动影响范围之内的盆地中，受海平面影响的基准面是决定垂向地层序列的主要因素。随着基准面的上升，地层呈退积结构，随着基准面的下降，地层呈进积结构，加积模式只形成于沉积物供给量和可容空间的增长量达到平衡的地区(据Van Wagoner等, 1990)。然而，外流盆地的陆相沉积很可能不受海平面的影响，原因可能是地层形成时距离海洋较远，像现代的恒河平原距离孟加拉湾1200km(Gibling等, 2005)，或是因为大陆架坡度比沿海平原的坡度更低(Summerfield, 1985; Miall, 1991; Tailing, 1998; Blum和Tornqvist, 2000)。

在外流水系环境下，基准面是由沉积物供给决定的，因为可容空间在稳定的气候条件下基本不变，同时沉积物在不断累加。一个简单的类比就是将沙子加入半桶水里：沙层之上水的体积基本保持不变直到桶满为止(也就是达到了沉积盆地的溢出点)。总体上，因为水位受气候变化的控制，外流水系盆地将会形成加积地层占主导的结构，局部由于气候引起的水体变化，可能存在相的变化。对于内排水盆地而言，沉积加积模型并不是唯一的，但却可以证明沉积是处于此种环境的特征之一。

发生在湖泊盆地中的相对基准面下降表现为在气候波动的驱动下，水的流入和散失平衡变化导致水位的变化。例如，里海相对短期的湖平面波动便是其中一例(Kroonenberg等, 1997)。这些湖泊水位的变化将会影响水下的可容空间，尤其是湖泊边缘。一次湖泊水位的下降可能会影响河流的级别，并可能存在两种情况：第一种是远离湖岸线的湖底坡度大于流入湖泊的河流坡度，当平衡剖面重新建立时，湖平面的下降会引起河流切割；第二种是湖水很浅且湖底坡度与河流坡度一样低或者更低，在这种情况下，湖面波动不会对河流有任何影响，这种情况就和大陆架坡度和海岸平原具有相似的坡度一样(Miall, 1991; Talling, 1998)。

在风成环境下，地下水位的波动可能造成风成砂的局部变化(Kocurek, 1996)。然而，如果风所携带的沙子没被带出盆地，这将不会导致盆地的收缩，仅仅是风沙沉积物简单地被重新分配到盆地周围。

31.5.2 冲积扇体系

由粗粒度砂岩相组成的邻近盆地边缘起伏地区的冲积扇，会受到所处冲积平原基准面的影响。如果基准面保持不变，冲积扇将会推进并改造早期形成的扇缘。如果相对基准面下降，早期扇体被改造并且普遍遭受河道下切冲刷。在基准面逐渐上升期，冲积扇将会在没有下切的盆地边缘形成垂向堆积序列（除了由构造或气候因素驱动）。例如，位于西班牙埃布罗盆地的中新世地层就是一个很好的以加积为主的地层叠置例子（Nichols 和 Hirst，1998；Nichols，2004），盆地边缘的冲积扇形成了一个厚度可以超过600m、没有重大的相变和内部变形的沉积地层（图31-6）。这归因于内流环境下，河流和湖泊沉积物的积累导致盆地内基准面持续上升。对盆地边缘冲积扇，相对稳定的物源供给导致被动的垂向增生，在活动断裂处便会产生厚层连续的盆地边缘冲积扇沉积，除此之外扇沉积物的变形也可以表明断层活跃，例如在埃布罗盆地边缘（Nichols，1987）和延伸盆地的边缘（Heward，1978）。恒定的盆地边缘相的厚层上超沉积可能成为内排水盆地沉积物的一个有效指示特征。

图 31-6　在边缘或内陆河埃布罗盆地关于中新世冲积扇相的垂直演化
砾岩床几乎无变形地穿过600m厚的层位并且超覆在相邻陡峭的石灰岩床逆冲断层前缘

31.5.3　盆缘沉积

若盆地边缘的断层没有持续限制沉积物的范围，盆地内部沉积物通过长时间的沉积充填作用将逐渐向盆地边缘超覆，盆地范围逐渐扩大（图31-7）。在有关的海平面上升或者盆地沉降的区域，这种地层关系在有海平面波动和盆地沉降的外流盆地中发育。然而，在内流盆地中，地层超覆的程度可以很大，甚至可以达到为盆地提供物源的排水系统内部。在盆地边缘的超覆可以导致提供物源的供水系统回填，导致河流相不整合沉积在远离盆地沉积区域的

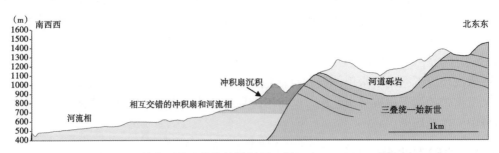

图 31-7　埃布罗盆地北缘的横剖面（据 Nichols，2004，修改）
中新世河流和冲积扇地层超覆和后退到逆冲断层前缘之上；具体区域见图31-8

基底之上。在比利牛斯南部便有一个典型例子，近端河流沉积物以残留物的形式保存于内陆的山区（Vincent，2001），显示出盆地基准面已经高于海平面几百米（Coney 等，1996）。这种地层关系具有非常典型的内流盆地特征，因为它们的沉积基准面远高于海平面。

31.5.4 可容空间

在内流盆地中没有基于海平面来考虑沉积体系的变化，但在湖相盆地中湖平面与海平面以相似的方式影响可容空间。在那些盆地中心有冲积平原或者浅而短暂的湖泊地区，盆地最低点可以作为基准面。然而，陆上盆地基底的可容空间这个概念不容易定义，就某种意义而言，盆地的溢出点为沉积物的堆积提供了总体限制的可容空间。地下水位是影响风沙沉积物保存的重要因素，并充当着基准面的角色（Kocurek，1996）。在风成主导的内流盆地，更高的地下水位可能影响着局部风成沉积的堆积。

31.5.5 沉降

内流盆地可能会沉积数百米厚的地层，甚至在盆地的边缘都没有任何的同沉积变形，更有甚者在无任何构造变形形成可容空间的情况下此类盆地也可以被动地充填。在盆地边缘这个特征尤其明显，并可能会有数百米厚的被动超覆，这与拉分、走滑和前陆盆地的边缘形成鲜明对比。在前陆盆地之中，与沉降变形相关的断层引起盆地边缘的地层变形。沉积载荷沉降会发生在大套地层的加厚地区，但这更可能导致一个广阔的区域挠曲变形，而非盆地边缘的局部变形。

然而内流盆地可能是构造活动的盆地，其沉降可能与地壳形成过程中的伸展和挠曲负载有关，例如新近纪的加州盆地（May 等，1993）和古近纪华盛顿州 Chumstick 盆地（Evans，1991），其地层厚度分别为 11km 和 12km；挪威的 Hornelen 盆地（Nilsen 和 Mclaughlin，1985）发育一套走滑控制的内流盆地厚层连续沉积地层。西班牙埃布罗盆地中逆冲岩片侵位导致内流沉积环境发育过程中的前陆岩石圈负载，并最终导致了盆地边缘局部变形（Nichols，1987）。

31.6 内流水系—外流水系以及它们的相互转换

如果没有构造产生的通道或者沉积物和水体充填盆地到达溢出点并最终与大洋相通，一个盆地内的水系将一直都是内流水系。如果溢出点接近海平面，所有河流体系流入和流出盆地都将受到相对海平面的控制。当河流流过溢出点并开始冲刷盆地内部充填沉积时，处于盆地更高位置的溢出点转变为纯粹的剥蚀退化。西班牙的埃布罗盆地就是一个很明确的例子，因为现代的埃布罗河及从比利牛斯南部流失的支流都切割至渐新世和中新世的河流—湖泊相沉积地层。然而，地层沉积时期，盆地并没有和地中海、大西洋相连（图31-8）。一些学者认为印度河水系曾有一段早期的内排水历史（Sinclair 和 Jaffrey，2001），即使其他学者认为其自始至终都是外排水系统（Searle 和 Owen，1999），不过，在始新世似乎存在一个由内排水到外排水的转化（Clift，2002）。

在大陆裂谷盆地的演化中，随着拉张产生的大洋通道，从内排水到外排水体系的转变是可能存在的（Gawthorpe 和 Leeder，2000）。因此个别裂谷盆地通常显示早期的陆相沉积阶段和随后的同裂陷期海相沉积。设得兰群岛泥盆纪 Lower Chair 群已经发现了内流盆地的沉积作用（Nichols，2005），海相沉积彻底建立之前，该沉积过程发生在晚泥盆世到早石炭世（Allen 和 Mange Rajetzky，1992）。

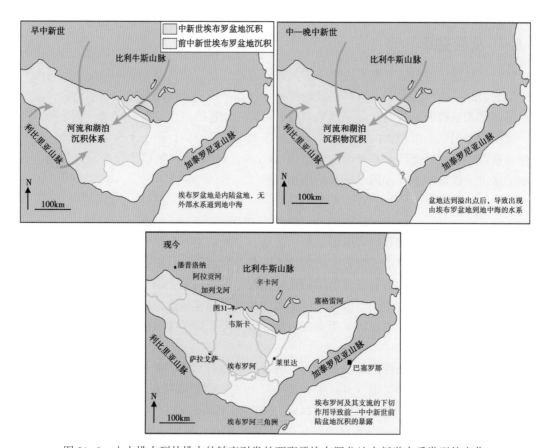

图 31-8　由内排水到外排水的转变引发的西班牙埃布罗盆地中新世水系类型的变化

31.7　结论

能被用来表明内流盆地沉积的特征很多：

（1）一个完全由陆相沉积地层组成的盆地很可能是在内流环境下形成的。对于这种情况的例外是淡水湖相沉积组合的沉积代表着超补偿或平衡补偿的湖泊。盐湖的蒸发相和短暂湖泊相发育在永远不与大洋相通的欠补偿盆地。

（2）虽然河流相沉积的分布格局绝不是内排水盆地所独有，但是内流盆地的河流相可以提供河道分布格局的证据。

（3）盆地地层层序可能会显示一个整体的加积地层结构，特别是在河流相沉积占主导的地区。尽管如此，气候因素会驱动内流盆地河流发生下切作用，同时类似的作用可能会发生在海平面的波动不会使河流下切的外流盆地。

（4）在盆地边缘，发育厚层的垂向加积冲积扇沉积。回填物源的供给通道发育在冲积扇区和规模较大的主河流体系，这些河流体系回填了数千米外的盆地基底。

致谢（略）

参 考 文 献

Allen, P. A., and Mange‑Rajetzky, M. A. (1992) Devonian‑Carboniferous sedimentary evolution of the Clair area, offshore north‑western UK: impact of changing provenance. Marine and Petroleum Geology, 9, 29‑52.

Astin, T. R., and Rogers, D. A. (1991) Subaqueous shrinkage cracks, in the Devonian of Scotland Reinterpreted. Journal of Sedimentary Research, 61, 850‑859.

Blum, M. D., and Tornqvist, T. E. (2000) Fluvial responses to climate and sea‑level change: a review and look forward. Sedimentology, 47, 2‑48.

Carroll, A. R., and Bohacs, K. M. (1999) Stratigraphic classification of ancient lakes: balancing tectonic and climatic controls. Geology, 27, 99‑102.

Clift, P. D. (2002) A brief history of the Indus River, in Clift, P. D., Kroon, D., Gadeicke, C., and Craig, J., eds., The tectonic and climatic evolution of the Arabian Sea Region. Special Publications, 195. Geological Society of London, 237‑258.

Coney, P. J., Munoz, J. A., McClay, K. R., and Evenchick, C. A. (1996) Syntectonic burial and post‑tectonic exhumation of the southern Pyrenees foreland fold‑thrust belt. Journal of the Geological Society, London, 153, 9‑16.

Evans, J. E. (1991) Facies relationships, alluvial architecture, and paleohydrology of a Paleogene, humid‑tropical alluvial‑fan system: the Chumstick Formation, Washington State, USA. Journal of Sedimentary Petrology, 61, 732‑755.

Friend, P. F., Williams, B. P. J., Ford, M., and Williams, E. A. (2000) Kinematics and dynamics of Old Red Sandstone basins, in Friend, P. F. and Williams, B. P. J. eds., New perspectives on the old red sandstone. Special Publications, 180. Geological Society of London, 29‑60.

García‑Castellanos, D. (2006) Long‑term evolution of tectonic lakes: climatic controls on the development of internally‑drained basins. Special Publication, 398. Boulder, CO, Geological Society of America, 283‑294.

García‑Castellanos, D., Verges, J. Gaspar‑Escribano, J., and Cloetingh, S. (2003) Interplay between tectonics, climate, and fluvial transport during the Cenozoic evolution of the Ebro Basin (NE Iberia). Journal of Geophysical Research, 108, 2347.

Gawthorpe, R. L., and Leeder, M. R. (2000) Tectono‑sedimentary evolution of active extensional basins. Basin Research 12, 195‑218.

Gibling, M. R., Tandon, S. K., Sinha, R., and Jain, M. (2005) Discontinuity‑bounded alluvial sequences of the southern Gangetic Plains, India: aggradation and degradation in rtesponse to monsoonal strength. Journal of Sedimentary Research, 75, 369‑385.

Glennie, K. W. (1990) Early Permian‑Rotliegend, in ed., Introduction to the petroleum geology of the North Sea, 3rd ed. Oxford, Blackwell Scientific, 120‑152.

Hartley, A. J., Weissmann, G. S., Nichols, G. J., and Warwick, G. L. (2010) Large distributive fluvial systems: characteristics, distribution and controls on development. Journal of Sedimentary Research, 80, 167‑183.

Heward, A. P. (1978) Alluvial fan sequence and megasequence models: with examples from Westphalian D – Stephanian B coalfields, Northern Spain, in Miall, A. D., et., Fluvial sedimentology. Memoir, 5. Canadian Society of Petroleum Geologists, 669–702.

Kocurek, G. A. (1996) Desert aeolian systems, in Reading, H. G., ed., Sedimentary environments: processes, facies and stratigraphy. Oxford, Blackwell Scientific, 125–153.

Kroonenberg, S. B., Rusakovb, G. V., and Svitochc, A. A. (1997) The wandering of the Volga delta: a response to rapid Caspian sea-level change. Sedimentary Geology, 107, 189–209.

Marshall, J. E. A., Rogers, D. A., and Whiteley M. J. (1996) Devonian marine incursions into the Orcadian Basin, Scotland. Journal of the Geological Society, 153, 451–466.

May, S. R., Ehman, K. D., Gray, G. G., and Crowell, J. C. (1993) Anew angle on the tectonic evolution of the Ridge basin, a "strike-slip" basin in southern California. Geological Society of America Bulletin, 105, 1357–1372.

Miall, A. D. (1991) Stratigraphic sequences and their chronostratigraphic correlation. Journal of Sedimentary Research, 61, 497–505.

Mitchell, A. H. G., and Reading, H. G. (1986) Sedimentation and tectonics, in Reading, H. G., ed., Sedimetrary environments and facies, 2nd ed. Oxford, Blackwell Science, 471–519.

Mortimer, E., Carrapa, B., Coutand, I., Schoenbohm, L., Sobel, E., Sosa Gomez, J., and Strecker, M. (2007) Fragmentation of a foreland basin in response to out-of-sequence basement uplifts and structural reactivation: El Cajón–Campo del Arenal basin, NW Argentina. Geological Society of American Bulletin 119, 637–653.

Mykura, W. (1991) Old red sandstone, in ed., The geology of Scotland. Geological Society of London, 205–251.

Nichols, G. J. (1987) Syntectonic alluvial fan sedimentation, southern Pyrenees. Geological Magazine, 124, 121–133.

Nichols, G. J. (2004) Sedimentation and base level controls in an endorheic basin: the Tertiary of the Ebro Basin, Spain. Boletín Geológico y Minero, 115, 427–438.

Nichols, G. J. (2005) Sedimentary evolution of the Lower Clair Group, Devonian, west of Shetland: climate and sediment supply controls on fluvial, aeolian and lacustrine deposition, in Dore, A. G., and Vining, B. A., eds., Petroleum geology: North West Europe and global perspectives-proceedings of the 6th Petroleum Geology Conference, 957–967.

Nichols, G. J. (2007) Features of fluvial systems in desiccating endorheic basins, in Nichols, G. J., Williams, E. A., and Paola, C., eds., Sedimentary processes, environments and basins-a tribute to Peter Friend. International Association of Sedimentologists Special Publication, 38, 567–587.

Nichols, G. J. (2009) Sedimentology and stratigraphy, 2nd ed. Oxford, Wiley-Blackwell, 419 p. Nichols, G. J., and Fisher, J. A. (2007) Processes, facies and architecture of fluvial distributary system deposits. Sedimentary Geology, 195, 75–90.

Nichols, G. J., and Hirst, J. P. P. (1998) Alluvial fans and fluvial distributary systems, Oligo-Miocene, northern Spain: contrasting processes and products. Journal of Sedimentary Research, 68, 879–889.

Nilsen, T. H., and McLaughlin, R. J. (1985) Comparison of tectonic framework and depositional patterns of Hornelen strike-slip basin in Norway and Ridge and Little Sulphur Creek strike-slip basins of California, in Biddle, K. T., and Christie-Blick, N., eds., Strike-slip deformation and sedimentation. Special Publication, 37. Tulsa, OK, Society of Economic Palaeontologists and Mineralogists, 80-103.

North, C. P., and Warwick, G. L. (2007) Fluvial fans: myths, misconceptions, and the end of the terminal-fan model. Journal of Sedimentary Research, 77, 693-701.

Olsen, P. E. (1990) Tectonic, climatic and biotic modulation of lacustrine ecosystem-examples from Newark Supergroup of Eastern North America, in Katz, B., ed., Lacustrine basin exploration: case studies and modern analogs. Memoir, 50, Tulsa, OK, American Association of Petroleum Geologists, 209-224.

Searle, M. P., and Owen, L. A. (1999) The evolution of the Indus River in relation to topographic uplift, climate and geology of western Tibet, the Trans-Himalayan and High-Himalayan Range, in Meadows, A., and Meadows, P. S., eds., The Indus River: biodiversity, resources, humankind. Oxford, Oxford University Press, 210-230.

Sinclair, H. D., and Jaffrey, N. (2001) Sedimentology of the Indus Group, Ladakh, northern India: implications for the timing of initiation of the palaeo-Indus River. Journal of the Geological Society, London, 158, 151-162.

Smoot, J. P. (1985) The closed-basin hypothesis and its use in facies analysis of the Newark Supergroup, in Robinson, G. R., and Froelich, A. J., eds., Proceedings of the Second US geological Survey Workshop on the Early Mesozoic Basins of the Eastern United States. US Geological Survey Circular, 946, 4-17.

Stanistreet, I. G., and McCarthy, T. S. (1993) The Okavango Fan and the classification of subaerial fan systems. Sedimentary Geology, 85, 115-133.

Summerfield, M. A. (1985) Plate tectonics and landscape development on the African continent. In Morisawa, M., and Hack, J., eds., Tectonic geomorphology. Boston, Allen and Unwin, 27-51.

Talling, P. J. (1998) How and where do incised valleys form if sea level remains above the shelf edge? Geology, 26, 87-90.

Van Wagoner, J. C., Mitchum, R. M., Campion, K. M., and Rahmanian, V. D. (1990) Siliciclastic sequence stratigraphy in well logs, cores and outcrop: concepts for high resolution correlation of time and facies. Methods in Exploration Series, 7. Tulsa, OK, American Association of Petroleum Geologists, 55p.

Vincent, S. J. (2001) The Sis palaeovalley: a record of proximal fluvial sedimentation and drainage basin development in response to Pyrenean mountain building. Sedimentology, 48, 1235-1276.

Weissmann, G. S., Hartley, A. J., Nichols, G. J., Scuderi, L. A., Olson, M., Buehler, H., and Massengill, L. (2010) Fluvial form in modern continental sedimentary basins: distributive fluvial systems (DFS). Geology, 38, 39-42.

(杨志文 译,蒲仁海 韩银学 蔡露露 校)